Forms Involving $\sqrt{u^2 - a^2}$

39 $\displaystyle\int \sqrt{u^2 - a^2}\, du = \frac{u}{2}\sqrt{u^2 - a^2} - \frac{a^2}{2}\ln|u + \sqrt{u^2 - a^2}| + C$

40 $\displaystyle\int u^2 \sqrt{u^2 - a^2}\, du = \frac{u}{8}(2u^2 - a^2)\sqrt{u^2 - a^2} - \frac{a^4}{8}\ln|u + \sqrt{u^2 - a^2}| + C$

41 $\displaystyle\int \frac{\sqrt{u^2 - a^2}}{u}\, du = \sqrt{u^2 - a^2} - a\cos^{-1}\frac{a}{u} + C$

42 $\displaystyle\int \frac{\sqrt{u^2 - a^2}}{u^2}\, du = -\frac{\sqrt{u^2 - a^2}}{u} + \ln|u + \sqrt{u^2 - a^2}| + C$

43 $\displaystyle\int \frac{du}{\sqrt{u^2 - a^2}} = \ln|u + \sqrt{u^2 - a^2}| + C$

44 $\displaystyle\int \frac{u^2\, du}{\sqrt{u^2 - a^2}} = \frac{u}{2}\sqrt{u^2 - a^2} + \frac{a^2}{2}\ln|u + \sqrt{u^2 - a^2}| + C$

45 $\displaystyle\int \frac{du}{u^2\sqrt{u^2 - a^2}} = \frac{\sqrt{u^2 - a^2}}{a^2 u} + C$

46 $\displaystyle\int \frac{du}{(u^2 - a^2)^{3/2}} = -\frac{u}{a^2\sqrt{u^2 - a^2}} + C$

Forms Involving $a + bu$

47 $\displaystyle\int \frac{u\, du}{a + bu} = \frac{1}{b^2}(a + bu - a\ln|a + bu|) + C$

48 $\displaystyle\int \frac{u^2\, du}{a + bu} = \frac{1}{2b^3}[(a + bu)^2 - 4a(a + bu) + 2a^2\ln|a + bu|] + C$

49 $\displaystyle\int \frac{du}{u(a + bu)} = \frac{1}{a}\ln\left|\frac{u}{a + bu}\right| + C$

50 $\displaystyle\int \frac{du}{u^2(a + bu)} = -\frac{1}{au} + \frac{b}{a^2}\ln\left|\frac{a + bu}{u}\right| + C$

51 $\displaystyle\int \frac{u\, du}{(a + bu)^2} = \frac{a}{b^2(a + bu)} + \frac{1}{b^2}\ln|a + bu| + C$

52 $\displaystyle\int \frac{du}{u(a + bu)^2} = \frac{1}{a(a + bu)} - \frac{1}{a^2}\ln\left|\frac{a + bu}{u}\right| + C$

53 $\displaystyle\int \frac{u^2\, du}{(a + bu)^2} = \frac{1}{b^3}\left(a + bu - \frac{a^2}{a + bu} - 2a\ln|a + bu|\right) + C$

54 $\displaystyle\int u\sqrt{a + bu}\, du = \frac{2}{15b^2}(3bu - 2a)(a + bu)^{3/2} + C$

55 $\displaystyle\int \frac{u\, du}{\sqrt{a + bu}} = \frac{2}{3b^2}(bu - 2a)\sqrt{a + bu}$

56 $\displaystyle\int \frac{u^2\, du}{\sqrt{a + bu}} = \frac{2}{15b^3}(8a^2 + 3b^2u^2 - 4abu)\sqrt{a + bu}$

57 $\displaystyle\int \frac{du}{u\sqrt{a + bu}} = \frac{1}{\sqrt{a}}\ln\left|\frac{\sqrt{a + bu} - \sqrt{a}}{\sqrt{a + bu} + \sqrt{a}}\right| + C,\quad \text{if } a > 0$

$\displaystyle\qquad = \frac{2}{\sqrt{-a}}\tan^{-1}\sqrt{\frac{a + bu}{-a}} + C,\quad \text{if } a < 0$

58 $\displaystyle\int \frac{\sqrt{a + bu}}{u}\, du = 2\sqrt{a + bu} + a\int \frac{du}{u\sqrt{a + bu}}$

59 $\displaystyle\int \frac{\sqrt{a + bu}}{u^2}\, du = -\frac{\sqrt{a + bu}}{u} + \frac{b}{2}\int \frac{du}{u\sqrt{a + bu}}$

60 $\displaystyle\int u^n\sqrt{a + bu}\, du = \frac{2}{b(2n + 3)}\left[u^n(a + bu)^{3/2} - na\int u^{n-1}\sqrt{a + bu}\, du\right]$

61 $\displaystyle\int \frac{u^n\, du}{\sqrt{a + bu}} = \frac{2u^n\sqrt{a + bu}}{b(2n + 1)} - \frac{2na}{b(2n + 1)}\int \frac{u^{n-1}\, du}{\sqrt{a + bu}}$

62 $\displaystyle\int \frac{du}{u^n\sqrt{a + bu}} = -\frac{\sqrt{a + bu}}{a(n - 1)u^{n-1}} - \frac{b(2n - 3)}{2a(n - 1)}\int \frac{du}{u^{n-1}\sqrt{a + bu}}$

(Continued inside back cover)

CALCULUS

WITH ANALYTIC GEOMETRY

THIRD EDITION

CALCULUS
WITH ANALYTIC GEOMETRY

EARL W. SWOKOWSKI

MARQUETTE UNIVERSITY

PRINDLE, WEBER & SCHMIDT
Boston, Massachusetts

PWS PUBLISHERS

Prindle, Weber & Schmidt · · Duxbury Press · · PWS Engineering · · Breton Publishers ·
20 Park Plaza · Boston, Massachusetts 02116

PWS Publishers is a division of Wadsworth, Inc.

Portions of this book previously appeared in *Calculus with Analytic Geometry, Alternate Edition* by Earl W. Swokowski. Copyright © 1983 by PWS Publishers.

88 87 86   – 10 9 8 7 6 5 4

ISBN 0-87150-443-X

Library of Congress Cataloging in Publication Data
Swokowski, Earl W.
    Calculus with analytic geometry.

    Includes index.
    1. Calculus.   2. Geometry, Analytic.   I. Title.
QA303.S94 1983b   515′.15   83-23615
ISBN 0-87150-443-X

Production and design: Kathi Townes
Text composition: Composition House Limited
Technical artwork: Vantage Art, Inc.
Cover photo: René Burri/Magnum Photos, Inc.
Cover printer: New England Book Components
Text printer/binder: Kingsport Press

Printed in the United States of America.

PREFACE

Most students study calculus for its use as a tool in areas other than mathematics. They desire information about *why* calculus is important, and *where* and *how* it can be applied. I kept these facts in mind as I worked on this revision. As in previous editions, when introducing concepts I often refer to problems that are familiar to students and that require methods of calculus for solutions. Many new examples and exercises have been designed to further motivate student interest, not only in the mathematical or physical sciences, but in other disciplines as well. Figures are frequently used to bridge the gap between the statement of a problem and its solution.

In addition to achieving a good balance between theory and applications, my primary objective was to write a book that can be read and understood by college freshmen. In each section I have striven for accuracy and clarity of exposition, together with a presentation that makes the transition from precalculus mathematics to calculus as smooth as possible. The comments that follow highlight some of the features of this text and the changes from the second edition.

Chapter 1 contains a review of real numbers, functions, and graphs. Tests for symmetry are discussed in Section 1.2 and, hence, are available for use throughout the text. The material on inverse functions has been moved to Chapter 7, where it is used to define the natural exponential function.

In Chapter 2, figures, examples, and exercises have been added to help students gain a better understanding of limits and continuity. Newton's Method for finding the real zeros of a function now appears in Chapter 3, thus providing an important algebraic application of the derivative early in the text.

The notion of *test value* is introduced in Chapter 4 as a tool for determining intervals in which derivatives are positive or negative. This concept, which is based on the *Intermediate Value Theorem,* is heavily employed to help sketch graphs of functions.

Chapters 5 and 6, on properties and applications of definite integrals, include exercises on numerical integration that require reference to graphs to approximate areas, volumes, work, and force exerted by a liquid.

A major change in this edition is that the topics in analytic geometry (formerly Chapter 7) now constitute Chapter 12. This shift results in an earlier treatment of exponential and logarithmic functions in the new Chapter 7. A discussion of separable and first-order linear differential equations has been added to allow consideration of a greater variety of applications of exponential functions.

Chapters 8 – 10, on transcendental functions, techniques of integration, and improper integrals, contain new exercises that illustrate ideas that did not appear in earlier editions. For example, comparison tests for convergence or divergence of improper integrals are included in the exercises of Chapter 10.

Infinite series are presented in a precise manner in Chapter 11. Chapter 12 consists of a detailed study of conic sections.

Chapters 13 – 15 deal with curves, vectors, and vector-valued functions. These chapters contain many new examples and exercises pertaining to parametric and polar equations, and a strong emphasis is placed on geometric and applied aspects of vectors.

Chapter 16, on functions of several variables, was extensively revised for this edition. The relevance of level curves and surfaces to practical situations is illustrated in numerous examples and exercises. Increments and differentials are motivated by analogous single-variable concepts. The definition of directional derivative does not require the use of direction angles of a line, and considerable stress is given to the gradient of a function. The study of maxima and minima has been expanded to include an examination of boundary extrema. The final section, on Lagrange multipliers, now includes a proof that emphasizes the geometric nature of why the method is valid.

Properties and applications of multiple integrals are considered in Chapter 17.

Vector fields are discussed in Chapter 18, and special attention is given to conservative fields. The physical significance of divergence and curl is brought out by using the theorems of Gauss and Stokes. The last two sections contain results on Jacobians and change of variables in multiple integrals.

The beginning of Chapter 19 briefly reviews the material on separable and first-order linear differential equations, which appears in Chapter 7. A new section contains applications of the theory of second-order differential equations to the analysis of vibrations of a spring.

There is a review section at the end of each chapter consisting of a list of important topics and pertinent exercises. The review exercises are similar to those that appear throughout the text and may be used by students to prepare for examinations. Answers to odd-numbered exercises are given at the end of the text. Instructors may obtain an answer booklet for the even-numbered exercises from the publisher.

I wish to thank the following individuals, who reviewed all, or parts of, the manuscript for this edition or for *Calculus with Analytic Geometry, Alternate Edition,* and offered many helpful suggestions: Alfred Andrew, Georgia Institute of Technology; Jan F. Andrus, University of New Orleans; Jackie Barab, Kansas State University; Robert M. Brooks, University of Utah; Daniel Drucker, Wayne State University; Dennis R. Dunniger, Michigan State University; Joseph M. Egar, Cleveland State University; Ronald D. Ferguson, San Antonio State College; Stuart Goldenberg, California Polytechnic State University; Theodore Guinn, University of New Mexico; Joe A. Guthrie, University of Texas, El Paso; John Higgins, Brigham Young University; Arthur M. Hobbs, Texas A & M; David Hoff, Indiana University; Adam Hulin, University of New Orleans; Michael Iannone, Trenton State College; Walter Jensen, Central New England College; Eleanor Killam, University of Massachusetts, Amherst; W.D. Lichtenstein, University of Georgia; Stanley M. Lukawecki, Clemson University; John Mack, University of Kentucky; Francis Masat, Glassboro State College; Wayne McDaniel, University of Missouri, St. Louis; Judith McKinney, California Polytechnic Institute, Pomona; Donald Miller, University of Nebraska; Louise E. Moser, California State University, Hayward; Norman K. Nystrom, American River College; Richard Patterson, Indiana University/Purdue University; Charles Peele, Marshal University; David A. Petrie, Cypress College; William Robinson, Ventura College; Jean Rubin, Purdue University; John T. Scheick, Ohio State University; Eugene P. Schlereth, University of Tennessee; Jon W. Scott, Montgomery College; Monty J. Strauss, Texas Tech University; Richard G. Vinson, University of South Alabama; Loyd Wilcox, Golden West College; and T.J. Worosz, Metropolitan State College, Denver.

I also wish to express my gratitude to Thomas A. Bronikowski of Marquette University, who authored the student supplement containing detailed solutions for one-third of the exercises; Stephen B. Rodi of Austin Community College, who developed a complete solutions manual; Michael B. Gregory of the University of North Dakota, who supplied a number of challenging exercises; and Christopher L. Morgan, California State University at Hayward, and Howard Pyron, University of Missouri at Rolla, who prepared the computer graphics. Special thanks are due to Stephen J. Merrill of Marquette University for suggesting several interesting examples, including one that indicates how infinite sequences and series may be employed to study the time course of an epidemic, and another that illustrates the use of exponential functions in the field of radiation therapy.

I am grateful for the valuable assistance of the staff of PWS Publishers. In particular, Kathi Townes did a superlative job as production editor, and David Pallai, who supervised the project, was a constant source of information and advice.

In addition to all the persons named here, I express my sincere appreciation to the many unnamed students and teachers who have helped shape my views on how calculus should be presented in the classroom.

Earl W. Swokowski

CONTENTS

18

TOPICS IN VECTOR CALCULUS 825

19

DIFFERENTIAL EQUATIONS 889

APPENDICES A1

ANSWERS TO ODD-NUMBERED EXERCISES A29

INDEX A68

INTRODUCTION: WHAT IS CALCULUS?

Calculus was invented in the seventeenth century to provide a tool for solving problems involving motion. The subject matter of geometry, algebra, and trigonometry is applicable to objects which move at constant speeds; however, methods introduced in calculus are required to study the orbits of planets, to calculate the flight of a rocket, to predict the path of a charged particle through an electromagnetic field and, for that matter, to deal with all aspects of motion.

In order to discuss objects in motion it is essential first to define what is meant by *velocity* and *acceleration*. Roughly speaking, the velocity of an object is a measure of the rate at which the distance traveled changes with respect to time. Acceleration is a measure of the rate at which velocity changes. Velocity may vary considerably, as is evident from the motion of a drag-strip racer or the descent of a space shuttle as it reenters the Earth's atmosphere. In order to give precise meanings to the notions of velocity and acceleration it is necessary to use one of the fundamental concepts of calculus, the *derivative*.

Although calculus was introduced to help solve problems in physics, it has been applied to many different fields. One of the reasons for its versatility is the fact that the derivative is useful in the study of rates of change of many entities other than objects in motion. For example, a chemist may use derivatives to forecast the outcome of various chemical reactions. A biologist may employ it in the investigation of the rate of growth of bacteria in a culture. An electrical engineer uses the derivative to describe the change in current in an electrical circuit. Economists have applied it to problems involving corporate profits and losses.

The derivative is also used to find tangent lines to curves. Although this has some independent geometric interest, the significance of tangent lines is of major importance in physical problems. For example, if a particle moves along a curve, then the tangent line indicates the direction of motion. If we restrict our attention to a sufficiently small portion of the curve, then in a certain sense the tangent line may be used to approximate the position of the particle.

Many problems involving maximum and minimum values may be attacked with the aid of the derivative. Some typical questions that can be answered are: At what angle of elevation should a projectile be fired in order to achieve its maximum range? If a tin can is to hold one gallon of a liquid, what dimensions require the least amount of tin? At what point between two light sources will the illumination be greatest? How can certain corporations maximize their revenue? How can a manufacturer minimize the cost of producing a given article?

Another fundamental concept of calculus is known as the *definite integral*. It, too, has many applications in the sciences. A physicist uses it to find the work required to stretch or compress a spring. An engineer may use it to find the center of mass or moment of inertia of a solid. The definite integral can be used by a biologist to calculate the flow of blood through an arteriole. An economist may employ it to estimate depreciation of equipment in a manufacturing plant. Mathematicians use definite integrals to investigate such concepts as areas of surfaces, volumes of geometric solids, and lengths of curves.

All the examples we have listed, and many more, will be discussed in detail as we progress through this book. There is literally no end to the applications of calculus. Indeed, in the future perhaps *you*, the reader, will discover new uses for this important branch of mathematics.

The derivative and the definite integral are defined in terms of certain limiting processes. The notion of limit is the initial idea which separates calculus from the more elementary branches of mathematics. Sir Isaac Newton (1642–1727) and Gottfried Wilhelm Leibniz (1646–1716) discovered the connection between derivatives and integrals. Because of this, and their other contributions to the subject, they are credited with the invention of calculus. Many other mathematicians have added a great deal to its development.

The preceding discussion has not answered the question "What is calculus?" Actually, there is no simple answer. Calculus could be called the study of limits, derivatives, and integrals; however, this statement is meaningless if definitions of the terms are unknown. Although we have given a few examples to illustrate what can be accomplished with derivatives and integrals, neither of these concepts has been given any meaning. Defining them will be one of the principal objectives of our early work in this text.

1

PREREQUISITES FOR CALCULUS

This chapter contains topics necessary for the study of calculus. After a brief review of real numbers, coordinate systems, and graphs in two dimensions, we turn our attention to one of the most important concepts in mathematics—the notion of *function*.

1.1

REAL NUMBERS

Real numbers are used considerably in precalculus mathematics, and we will assume familiarity with the fundamental properties of addition, subtraction, multiplication, division, exponents, and radicals. Throughout this chapter, unless otherwise specified, lower-case letters a, b, c, ... denote real numbers.

The **positive integers** 1, 2, 3, 4, ... may be obtained by adding the real number 1 successively to itself. The **integers** consist of all positive and negative integers together with the real number 0. A **rational number** is a real number that can be expressed as a quotient a/b, where a and b are integers and $b \neq 0$. Real numbers that are not rational are called **irrational**. The ratio of the circumference of a circle to its diameter is irrational. This real number is denoted by π and the notation $\pi \approx 3.1416$ is used to indicate that π is *approximately equal* to 3.1416. Another example of an irrational number is $\sqrt{2}$.

Real numbers may be represented by nonterminating decimals. For example, the decimal representation for the rational number 7434/2310 is found by division to be 3.2181818 . . . , where the digits 1 and 8 repeat indefinitely. Rational numbers may always be represented by repeating decimals. Decimal representations for irrational numbers may also be obtained; however, they are nonterminating and nonrepeating.

It is possible to associate real numbers with points on a line l in such a way that to each real number a there corresponds one and only one point, and conversely, to each point P there corresponds precisely one real number. Such an association between two sets is referred to as a **one-to-one correspondence**. We first choose an arbitrary point O, called the **origin**, and associate with it the real number 0. Points associated with the integers are then determined by considering successive line segments of equal length on either side of O as illustrated in Figure 1.1. The points corresponding to rational numbers such as $\frac{23}{5}$ and $-\frac{1}{2}$ are obtained by subdividing the equal line segments. Points associated with certain irrational numbers, such as $\sqrt{2}$, can be found by geometric construction. For other irrational numbers such as π, no construction is possible. However, the point corresponding to π can be approximated to any degree of accuracy by locating successively the points corresponding to 3, 3.1, 3.14, 3.141, 3.1415, 3.14159, ... It can be shown that to every irrational number there corresponds a unique point on l and, conversely, every point that is not associated with a rational number corresponds to an irrational number.

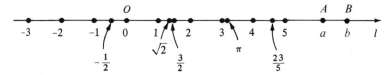

Figure 1.1

The number a that is associated with a point A on l is called the **coordinate** of A. An assignment of coordinates to points on l is called a **coordinate system** for l, and l is called a **coordinate line**, or a **real line**. A direction can be assigned to l by taking the **positive direction** to the right and the **negative direction** to the left. The positive direction is noted by placing an arrowhead on l as shown in Figure 1.1.

The real numbers that correspond to points to the right of O in Figure 1.1 are called **positive real numbers**, whereas those that correspond to points to the left of O are **negative real numbers**. The real number 0 is neither positive nor negative. The collection of positive real numbers is **closed** relative to addition and multiplication; that is, if a and b are positive, then so is the sum $a + b$ and the product ab.

If a and b are real numbers, and $a - b$ is positive, we say that **a is greater than b** and write $a > b$. An equivalent statement is **b is less than a**, written $b < a$. The symbols $>$ or $<$ are called **inequality signs** and expressions such as $a > b$ or $b < a$ are called **inequalities**. From the manner in which we constructed the coordinate line l in Figure 1.1, we see that if A and B are points with coordinates a and b, respectively, then $b > a$ (or $a < b$) *if and only if A lies to the left of B*. Since $a - 0 = a$, it follows that $a > 0$ if and only if a is positive. Similarly, $a < 0$ means that a is negative. The following properties of inequalities can be proved.

(1.1)

If $a > b$ and $b > c$, then $a > c$.

If $a > b$, then $a + c > b + c$.

If $a > b$ and $c > 0$, then $ac > bc$.

If $a > b$ and $c < 0$, then $ac < bc$.

Analogous properties for "less than" can also be established.

The symbol $a \geq b$, which is read ***a* is greater than or equal to *b***, means that either $a > b$ or $a = b$. The symbol $a < b < c$ means that $a < b$ and $b < c$, in which case we say that ***b* is *between a* and *c***. The notations $a \leq b, a < b \leq c$, $a \leq b < c, a \leq b \leq c$, etc., have similar meanings.

Another property, called **completeness**, is needed to characterize the real numbers. This property will be discussed in Chapter 11.

If a is a real number, then it is the coordinate of some point A on a coordinate line l, and the symbol $|a|$ is used to denote the number of units (or distance) between A and the origin, without regard to direction. Referring to Figure 1.2 we see that for the point with coordinate -4 we have $|-4| = 4$. Similarly, $|4| = 4$. In general, if a is negative we change its sign to find $|a|$, whereas if a is nonnegative then $|a| = a$. The nonnegative number $|a|$ is called the **absolute value** of a.

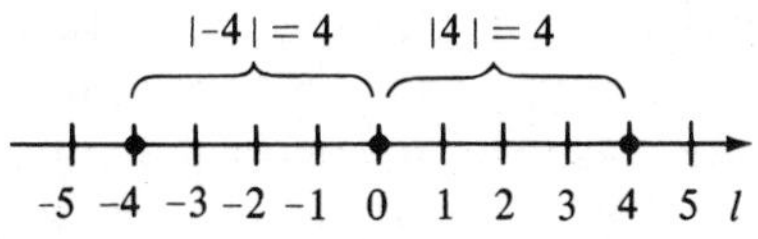

The following definition of absolute value summarizes our remarks.

Figure 1.2

Definition (1.2)

$$|a| = \begin{cases} a & \text{if } a \geq 0 \\ -a & \text{if } a < 0 \end{cases}$$

Example 1   Find $|3|, |-3|, |0|, |\sqrt{2} - 2|$, and $|2 - \sqrt{2}|$.

Solution   Since $3, 2 - \sqrt{2}$, and 0 are nonnegative,

$$|3| = 3, \quad |2 - \sqrt{2}| = 2 - \sqrt{2}, \quad \text{and} \quad |0| = 0.$$

Since -3 and $\sqrt{2} - 2$ are negative, we use the formula $|a| = -a$ of Definition (1.2) to obtain

$$|-3| = -(-3) = 3 \quad \text{and} \quad |\sqrt{2} - 2| = -(\sqrt{2} - 2) = 2 - \sqrt{2}. \quad \blacksquare$$

The following three general properties of absolute values may be established.

(1.3) $$|a| = |-a|, \qquad |ab| = |a||b|, \qquad -|a| \leq a \leq |a|$$

It can also be shown that if b is any positive real number, then

(1.4)
$$\begin{aligned} |a| < b \quad &\text{if and only if} \quad -b < a < b \\ |a| > b \quad &\text{if and only if} \quad a > b \text{ or } a < -b \\ |a| = b \quad &\text{if and only if} \quad a = b \text{ or } a = -b. \end{aligned}$$

It follows from the first and third properties stated in (1.4) that

$$|a| \leq b \quad \text{if and only if} \quad -b \leq a \leq b.$$

The Triangle Inequality (1.5)

$$|a + b| \leq |a| + |b|$$

Proof   From (1.3), $-|a| \le a \le |a|$ and $-|b| \le b \le |b|$. Adding corresponding sides we obtain

$$-(|a| + |b|) \le a + b \le |a| + |b|.$$

Using the remark preceding this theorem gives us the desired conclusion. □

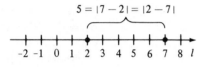

Figure 1.3

We shall use the concept of absolute value to define the distance between any two points on a coordinate line. Let us begin by noting that the distance between the points with coordinates 2 and 7 shown in Figure 1.3 equals 5 units on l. This distance is the difference, $7 - 2$, obtained by subtracting the smaller coordinate from the larger. If we employ absolute values, then, since $|7 - 2| = |2 - 7|$, it is unnecessary to be concerned about the order of subtraction. We shall use this as our motivation for the next definition.

Definition (1.6)

> Let a and b be the coordinates of two points A and B, respectively, on a coordinate line l. The **distance between A and B**, denoted by $d(A,B)$, is
>
> $$d(A, B) = |b - a|.$$

The number $d(A, B)$ is also called the **length of the line segment AB**. Observe that, since $d(B, A) = |a - b|$ and $|b - a| = |a - b|$,

$$d(A, B) = d(B, A).$$

Also note that the distance between the origin O and the point A is

$$d(O, A) = |a - 0| = |a|,$$

which agrees with the geometric interpretation of absolute value illustrated in Figure 1.2.

Example 2   If A, B, C, and D have coordinates $-5, -3, 1$, and 6, respectively, find $d(A, B)$, $d(C, B)$, $d(O, A)$, and $d(C, D)$.

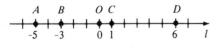

Figure 1.4

Solution   The points are indicated in Figure 1.4. By Definition (1.6),

$$d(A, B) = |-3 - (-5)| = |-3 + 5| = |2| = 2.$$
$$d(C, B) = |-3 - 1| = |-4| = 4.$$
$$d(O, A) = |-5 - 0| = |-5| = 5.$$
$$d(C, D) = |6 - 1| = |5| = 5.$$ ■

The concept of absolute value has uses other than that of finding distances between points. Generally, it is employed whenever one is interested in the *magnitude* or *numerical value* of a real number without regard to its sign.

In order to shorten explanations it is sometimes convenient to use the notation and terminology of sets. A **set** may be thought of as a collection of objects of some type. The objects are called **elements** of the set. Throughout our work $\mathbb{R}$ will denote the set of real numbers. If S is a set, then $a \in S$ means that a is an element of S, whereas $a \notin S$ signifies that a is not an element of S. If every element of a set S is also an element of a set T, then S is called a **subset** of T. Two sets S and T are said to be **equal**, written $S = T$, if S and T contain precisely the same elements. The notation $S \neq T$ means that S and T are not equal. If S and T are sets, their **union** $S \cup T$ consists of the elements that are either in S, in T, or in *both* S and T. The **intersection** $S \cap T$ consists of the elements that the sets have in common.

If the elements of a set S have a certain property, then we write $S = \{x : \quad \}$ where the property describing the arbitrary element x is stated in the space after the colon. For example, $\{x : x > 3\}$ may be used to represent the set of all real numbers greater than 3.

Of major importance in calculus are certain subsets of $\mathbb{R}$ called **intervals**. If $a < b$, the symbol (a, b) is sometimes used for all real numbers between a and b. This set is called an **open interval**. Thus we have:

$$(1.7) \qquad (a, b) = \{x : a < x < b\}.$$

The numbers a and b are called the **endpoints** of the interval.

The **graph** of a set S of real numbers is defined as the points on a coordinate line that correspond to the numbers in S. In particular, the graph of the open interval (a, b) consists of all points between the points corresponding to a and b. In Figure 1.5 we have sketched the graphs of a general open interval (a, b) and the special open intervals $(-1, 3)$ and $(2, 4)$. The parentheses in the figure indicate that the endpoints of the intervals are not to be included. For convenience, we shall use the terms *interval* and *graph of an interval* interchangeably.

If we wish to include an endpoint of an interval, a bracket is used instead of a parenthesis. If $a < b$, then **closed intervals**, denoted by $[a, b]$, and **half-open intervals**, denoted by $[a, b)$ or $(a, b]$, are defined as follows.

$$(1.8) \qquad \begin{aligned} [a, b] &= \{x : a \le x \le b\} \\ [a, b) &= \{x : a \le x < b\} \\ (a, b] &= \{x : a < x \le b\} \end{aligned}$$

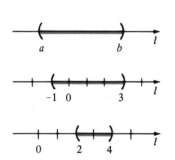

Figure 1.5   Open intervals (a, b), $(-1, 3)$, and $(2, 4)$

Typical graphs are sketched in Figure 1.6, where a bracket indicates that the corresponding endpoint is part of the graph.

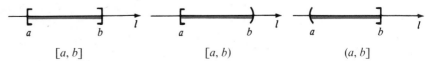

$[a, b]$ $[a, b)$ $(a, b]$

Figure 1.6

In future discussions of intervals, whenever the numbers a and b are not stated explicitly it will always be assumed that $a < b$. If an interval is a subset of another interval I it is called a **subinterval** of I. For example, the closed interval $[2, 3]$ is a subinterval of $[0, 5]$.

We shall sometimes employ the following **infinite intervals**.

$$(1.9) \qquad \begin{aligned} (a, \infty) &= \{x : x > a\} & [a, \infty) &= \{x : x \geq a\} \\ (-\infty, a) &= \{x : x < a\} & (-\infty, a] &= \{x : x \leq a\} \\ (-\infty, \infty) &= \mathbb{R} \end{aligned}$$

For example, $(1, \infty)$ represents all real numbers greater than 1. The symbol ∞ denotes "infinity" and is merely a notational device. It is not to be interpreted as representing a real number. Typical graphs of infinite intervals for an arbitrary real number a are sketched in Figure 1.7, where the absence of a parenthesis or bracket on the left for $(-\infty, a)$ and $(-\infty, a]$ and on the right for (a, ∞) and $[a, \infty)$ indicate that the colored portions extend indefinitely.

As indicated in this section, we frequently make use of letters to denote arbitrary elements of a set. For example, we may use x to denote a real number, although no *particular* real number is specified. A letter that is used to represent any element of a given set is sometimes called a **variable**. Throughout this text, unless otherwise specified, variables will represent real numbers. The **domain of a variable** is the set of real numbers represented by the variable. To illustrate, given the expression $\sqrt{x}$, we note that in order to obtain a real number we must have $x \geq 0$, and hence in this case the domain of x is assumed to be the set of nonnegative real numbers. Similarly, when working with the expression $1/(x - 2)$ we must exclude $x = 2$ (Why?), and consequently we take the domain of x as the set of all real numbers different from 2. The term **constant** is sometimes used for a (fixed) real number.

It is often necessary to consider inequalities that involve variables, such as

$$x^2 - 3 < 2x + 4.$$

If certain numbers such as 4 or 5 are substituted for x, we obtain the false statements $13 < 12$ or $22 < 14$, respectively. Other numbers such as 1 or 2 produce the true statements $-2 < 6$ or $1 < 8$, respectively. In general, if we are given an inequality in x and if a true statement is obtained when x is replaced by a real number a, then a is called a **solution** of the inequality. Thus 1 and 2 are solutions of the inequality $x^2 - 3 < 2x + 4$, whereas 4 and 5 are not solutions. To **solve** an inequality means to find all solutions. We say that two inequalities are **equivalent** if they have exactly the same solutions.

A standard method for solving an inequality is to replace it with a list of equivalent inequalities, terminating in one for which the solutions are obvious. The main tools used in applying this method are properties such as those listed in (1.1), (1.3), and (1.4). For example, if x represents a real number, then adding the same expression in x to both sides leads to an equivalent inequality. We may multiply both sides of an inequality by an expression containing x if we are certain that the expression is positive for all values of x under consideration. If we multiply both sides of an inequality by an expression that is always negative, such as $-7 - x^2$, then the inequality sign is reversed.

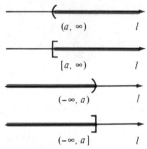

Figure 1.7

The reader should supply reasons for the steps used in finding the solutions of the inequalities in the following examples.

Example 3 Solve the inequality $4x + 3 > 2x - 5$ and represent the solutions graphically.

Solution   The following inequalities are equivalent:

$$4x + 3 > 2x - 5$$
$$4x > 2x - 8$$
$$2x > -8$$
$$x > -4$$

Hence the solutions consist of all real numbers greater than -4, that is, the numbers in the infinite interval $(-4, \infty)$. The graph is sketched in Figure 1.8.   ∎

Figure 1.8

Example 4   Solve the inequality $-5 \le \dfrac{4 - 3x}{2} < 1$ and sketch the graph corresponding to the solutions.

Solution   We may proceed as follows:

$$-5 \le \frac{4 - 3x}{2} < 1$$
$$-10 \le 4 - 3x < 2$$
$$-14 \le -3x < -2$$
$$\frac{14}{3} \ge x > \frac{2}{3}$$
$$\frac{2}{3} < x \le \frac{14}{3}$$

Hence the solutions are the numbers in the half-open interval $(\frac{2}{3}, \frac{14}{3}]$. The graph is sketched in Figure 1.9.   ∎

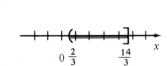

Figure 1.9

Example 5   Solve $x^2 - 7x + 10 > 0$ and represent the solutions graphically.

Solution   Since the inequality may be written

$$(x - 5)(x - 2) > 0,$$

it follows that x is a solution if and only if both factors $x - 5$ and $x - 2$ are positive, or both are negative. The diagram in Figure 1.10 indicates the signs of these factors for various real numbers. Evidently, both factors are positive if x is in the interval $(5, \infty)$ and both are negative if x is in $(-\infty, 2)$. Hence the solutions consist of all real numbers in the union $(-\infty, 2) \cup (5, \infty)$ as illustrated by the sketch in Figure 1.10.   ∎

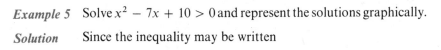
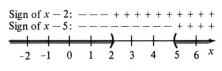

Figure 1.10

Example 6   Solve the inequality $|x - 3| < 0.1$.

Solution   Using (1.4) and (1.1), the inequality is equivalent to each of the following:

$$-0.1 < x - 3 < 0.1$$
$$-0.1 + 3 < (x - 3) + 3 < 0.1 + 3$$
$$2.9 < x < 3.1.$$

Thus the solutions are the real numbers in the open interval $(2.9, 3.1)$.   ∎

Inequalities similar to that in Example 6 will be used to define the concept of *limit* in Section 2.2. Specifically, we shall consider inequalities of the type given in the next example. The Greek letter δ (delta) in Example 7 is used frequently in calculus to denote a small positive real number.

Example 7   Let a and δ denote real numbers, with $\delta > 0$. Solve the inequality

$$0 < |x - a| < \delta$$

and represent the solutions graphically.

Solution   The inequality $0 < |x - a|$ is true if and only if $x \neq a$. The solutions of $|x - a| < \delta$ may be found using (1.4) and (1.1) as follows:

$$|x - a| < \delta$$
$$-\delta < x - a < \delta$$
$$a - \delta < x < a + \delta$$

Thus, the solutions of $0 < |x - a| < \delta$ consist of all real numbers in the open interval $(a - \delta, a + \delta)$ *except* the number a. This is the union

$$(a - \delta, a) \cup (a, a + \delta)$$

of two open intervals. The graph is sketched in Figure 1.11.   ∎

Figure 1.11 $0 < |x - a| < \delta$

Example 8   Solve $|2x - 7| > 3$.

Solution   By (1.4), x is a solution of $|2x - 7| > 3$ if and only if either

$$2x - 7 > 3 \quad \text{or} \quad 2x - 7 < -3.$$

The first of these two inequalities is equivalent to $2x > 10$, or $x > 5$. The second is equivalent to $2x < 4$, or $x < 2$. Hence the solutions of $|2x - 7| > 3$ are the numbers in the union $(-\infty, 2) \cup (5, \infty)$.   ∎

EXERCISES 1.1

In Exercises 1 and 2 replace the comma between each pair of real numbers with the appropriate symbol $<$, $>$, or $=$.

1 (a) $-2, -5$ (b) $-2, 5$
   (c) $6 - 1, 2 + 3$ (d) $\frac{2}{3}, 0.66$
   (e) $2, \sqrt{4}$ (f) $\pi, \frac{22}{7}$

2 (a) $-3, 0$ (b) $-8, -3$
   (c) $8, -3$ (d) $\frac{3}{4} - \frac{2}{3}, \frac{1}{15}$
   (e) $\sqrt{2}, 1.4$ (f) $\frac{4053}{1110}, 3.6513$

Rewrite the expressions in Exercises 3 and 4 without using symbols for absolute values.

3 (a) $|2 - 5|$ (b) $|-5| + |-2|$
   (c) $|5| + |-2|$ (d) $|-5| - |-2|$
   (e) $|\pi - 22/7|$ (f) $(-2)/|-2|$
   (g) $|\frac{1}{2} - 0.5|$ (h) $|(-3)^2|$
   (i) $|5 - x|$ if $x > 5$ (j) $|a - b|$ if $a < b$

4 (a) $|4 - 8|$ (b) $|3 - \pi|$
   (c) $|-4| - |-8|$ (d) $|-4 + 8|$
   (e) $|-3|^2$ (f) $|2 - \sqrt{4}|$
   (g) $|-0.67|$ (h) $-|-3|$
   (i) $|x^2 + 1|$ (j) $|-4 - x^2|$

5 If A, B, and C are points on a coordinate line with coordinates -5, -1, and 7, respectively, find the following distances.
   (a) $d(A, B)$ (b) $d(B, C)$
   (c) $d(C, B)$ (d) $d(A, C)$

6 Rework Exercise 5 if A, B, and C have coordinates 2, -8, and -3, respectively.

Solve the inequalities in Exercises 7–34 and express the solutions in terms of intervals.

7 $5x - 6 > 11$ **8** $3x - 5 < 10$

9 $2 - 7x \leq 16$ **10** $7 - 2x \geq -3$

11 $|2x + 1| > 5$ **12** $|x + 2| < 1$

13 $3x + 2 < 5x - 8$ **14** $2 + 7x < 3x - 10$

15 $12 \geq 5x - 3 > -7$ **16** $5 > 2 - 9x > -4$

17 $-1 < \dfrac{3 - 7x}{4} \leq 6$ **18** $0 \leq 4x - 1 \leq 2$

19 $\dfrac{5}{7 - 2x} > 0$ **20** $\dfrac{4}{x^2 + 9} > 0$

21 $|x - 10| < 0.3$ **22** $\left|\dfrac{2x + 3}{5}\right| < 2$

23 $\left|\dfrac{7 - 3x}{2}\right| \leq 1$ **24** $|3 - 11x| \geq 41$

25 $|25x - 8| > 7$ **26** $|2x + 1| < 0$

27 $3x^2 + 5x - 2 < 0$ **28** $2x^2 - 9x + 7 < 0$

29 $2x^2 + 9x + 4 \geq 0$ **30** $x^2 - 10x \leq 200$

31 $\dfrac{1}{x^2} < 100$ **32** $5 + \sqrt{x} < 1$

33 $\dfrac{3x + 2}{2x - 7} \leq 0$ **34** $\dfrac{3}{x - 9} > \dfrac{2}{x + 2}$

35 The relationship between the Fahrenheit and Celsius temperature scales is given by $C = \frac{5}{9}(F - 32)$. If $60 \leq F \leq 80$, express the corresponding values of C in terms of an inequality.

36 In the study of electricity, Ohm's Law states that if R denotes the resistance of an object (in ohms), E the potential difference across the object (in volts), and I the current that flows through it (in amperes), then $R = E/I$ (see figure). If the voltage is 110, what values of the resistance will result in a current that does not exceed 10 amperes?

Resistance R

Current I

Voltage E

Figure for Exercise 36

37 According to Hooke's Law, the force F (in pounds) required to stretch a certain spring x inches beyond its natural length is given by $F = (4.5)x$ (see figure). If $10 \leq F \leq 18$, what are the corresponding values of x?

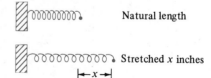

Natural length

Stretched x inches

Figure for Exercise 37

38 Boyle's Law for a certain gas states that $pv = 200$, where p denotes the pressure (lb/in.^{2}) and v denotes the volume (in.^{3}). If $25 \leq v \leq 50$, what are the corresponding values of p?

39 If a baseball is thrown straight upward from level ground with an initial velocity of 72 ft/sec, its altitude s (in feet) after t seconds is given by $s = -16t^2 + 72t$. For what values of t will the ball be at least 32 feet above the ground?

40 The period T (sec) of a simple pendulum of length l (cm) is given by $T = 2\pi\sqrt{l/g}$, where g is a physical constant. If, for the pendulum in a grandfather clock, $g = 980$ and $98 \le l \le 100$, what are the corresponding values of T?

41 Prove that $|a - b| \ge |a| - |b|$.
(*Hint*: Write $|a| = |(a - b) + b|$ and apply (1.5).)

42 If n is any positive integer and $a_1, a_2, \ldots, a_n$ are real numbers, prove that

$$|a_1 + a_2 + \cdots + a_n| \le |a_1| + |a_2| + \cdots + |a_n|.$$

(*Hint*: By (1.5),

$$|a_1 + a_2 + \cdots + a_n| \le |a_1| + |a_2| + \cdots + |a_n|.)$$

43 If $0 < a < b$, or if $a < b < 0$, prove that $(1/a) > (1/b)$.

44 If $0 < a < b$, prove that $a^2 < b^2$. Why is the restriction $0 < a$ necessary?

45 If $a < b$ and $c < d$, prove that $a + c < b + d$.

46 If $a < b$ and $c < d$, is it always true that $ac < bd$? Explain.

47 Prove (1.3).

48 Prove (1.4).

1.2

COORDINATE SYSTEMS IN TWO DIMENSIONS

In Section 1.1 we discussed how coordinates may be assigned to points on a line. Coordinate systems can also be introduced in planes by means of *ordered pairs*. The term **ordered pair** refers to two real numbers, where one is designated as the "first" number and the other as the "second." The symbol (a, b) is used to denote the ordered pair consisting of the real numbers a and b where a is first and b is second. There are many uses for ordered pairs. They were used in Section 1.1 to denote open intervals. In this section they will represent points in a plane. Although ordered pairs are employed in different situations, there is little chance for confusion, since it should always be clear from the discussion whether the symbol (a, b) represents an interval, a point, or some other mathematical object. We consider two ordered pairs (a, b) and (c, d) equal, and write

$$(a, b) = (c, d) \quad \text{if and only if} \quad a = c \text{ and } b = d.$$

This implies, in particular, that $(a, b) \ne (b, a)$ if $a \ne b$. The set of all ordered pairs will be denoted by $\mathbb{R} \times \mathbb{R}$.

A **rectangular**, or **Cartesian,*** **coordinate system** may be introduced in a plane by considering two perpendicular coordinate lines in the plane that intersect in the origin O on each line. Unless specified otherwise, the same unit of length is chosen on each line. Usually one of the lines is horizontal with positive direction to the right, and the other line is vertical with positive direction upward, as indicated by the arrowheads in Figure 1.12. The two lines are called **coordinate axes** and the point O is called the **origin**. The horizontal line is often referred to as the **x-axis** and the vertical line as the **y-axis**, and they are labeled x and y, respectively. The plane is then called a **coordinate plane** or, with the preceding notation for coordinate axes, an **xy-**

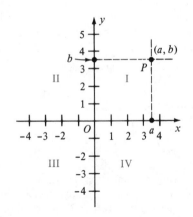

Figure 1.12

---

* The term "Cartesian" is used in honor of the French mathematician and philosopher René Descartes (1596–1650), who was one of the first to employ such coordinate systems.

plane. In certain applications different labels such as d, t, etc., are used for the coordinate lines. The coordinate axes divide the plane into four parts called the **first, second, third,** and **fourth quadrants** and labeled I, II, III, and IV, respectively, as shown in Figure 1.12.

Each point P in an xy-plane may be assigned a unique ordered pair. If vertical and horizontal lines through P intersect the x- and y-axes at points with coordinates a and b, respectively (see Figure 1.12), then P is assigned the ordered pair (a, b). The number a is called the **x-coordinate** (or **abscissa**) of P, and b is called the **y-coordinate** (or **ordinate**) of P. We sometimes say that P *has coordinates* (a, b). Conversely, every ordered pair (a, b) determines a point P in the xy-plane with coordinates a and b. Specifically, P is the point of intersection of lines perpendicular to the x-axis and y-axis at the points having coordinates a and b, respectively. This establishes a one-to-one correspondence between the set of all points in the xy-plane and the set of all ordered pairs. It is sometimes convenient to refer to the *point* (a, b) meaning the point with x-coordinate a and y-coordinate b. The symbol $P(a, b)$ will denote the point P with coordinates (a, b). To **plot a point** $P(a, b)$ means to locate, in a coordinate plane, the point P with coordinates (a, b). This point is represented by a dot in the appropriate position, as illustrated in Figure 1.13.

The next statement provides a formula for finding the distance between two points in a coordinate plane.

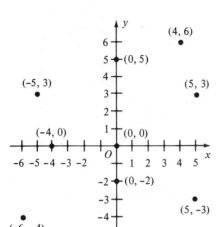

Figure 1.13

Distance Formula **(1.10)**

> The distance $d(P_1, P_2)$ between any two points $P_1(x_1, y_1)$ and $P_2(x_2, y_2)$ in a coordinate plane is
> $$d(P_1, P_2) = \sqrt{(x_2 - x_1)^2 + (y_2 - y_1)^2}.$$

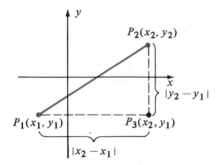

Figure 1.14

Proof   If $x_1 \neq x_2$ and $y_1 \neq y_2$, then as illustrated in Figure 1.14, the points P_1, P_2, and $P_3(x_2, y_1)$ are vertices of a right triangle. By the Pythagorean Theorem,

$$[d(P_1, P_2)]^2 = [d(P_1, P_3)]^2 + [d(P_3, P_2)]^2.$$

Using the fact that $d(P_1, P_3) = |x_2 - x_1|$ and $d(P_3, P_2) = |y_2 - y_1|$ leads to the desired formula.

If $y_1 = y_2$, the points P_1 and P_2 lie on the same horizontal line and

$$d(P_1, P_2) = |x_2 - x_1| = \sqrt{(x_2 - x_1)^2}.$$

Similarly, if $x_1 = x_2$, the points are on the same vertical line and

$$d(P_1, P_2) = |y_2 - y_1| = \sqrt{(y_2 - y_1)^2}.$$

These are special cases of the Distance Formula.

Although we referred to Figure 1.14, the argument used in this proof of the Distance Formula is independent of the positions of the points P_1 and P_2. $\square$

Example 1   Prove that the triangle with vertices $A(-1, -3)$, $B(6, 1)$, and $C(2, -5)$ is a right triangle and find its area.

Solution   By the Distance Formula,

$$d(A, B) = \sqrt{(-1 - 6)^2 + (-3 - 1)^2} = \sqrt{49 + 16} = \sqrt{65}$$
$$d(B, C) = \sqrt{(6 - 2)^2 + (1 + 5)^2} = \sqrt{16 + 36} = \sqrt{52}$$
$$d(A, C) = \sqrt{(-1 - 2)^2 + (-3 + 5)^2} = \sqrt{9 + 4} = \sqrt{13}.$$

Hence $[d(A, B)]^2 = [d(B, C)]^2 + [d(A, C)]^2$; that is, the triangle is a right triangle with hypotenuse AB. The area is $\frac{1}{2}\sqrt{52}\sqrt{13} = 13$ square units. ■

It is easy to obtain a formula for the midpoint of a line segment. Let $P_1(x_1, y_1)$ and $P_2(x_2, y_2)$ be two points in a coordinate plane and let M be the midpoint of the segment P_1P_2. The lines through P_1 and P_2 parallel to the y-axis intersect the x-axis at $A_1(x_1, 0)$ and $A_2(x_2, 0)$ and, from plane geometry, the line through M parallel to the y-axis bisects the segment A_1A_2 (see Figure 1.15). If $x_1 < x_2$, then $x_2 - x_1 > 0$, and hence $d(A_1, A_2) = x_2 - x_1$. Since M_1 is halfway from A_1 to A_2, the x-coordinate of M_1 is

$$x_1 + \tfrac{1}{2}(x_2 - x_1) = x_1 + \tfrac{1}{2}x_2 - \tfrac{1}{2}x_1$$
$$= \tfrac{1}{2}x_1 + \tfrac{1}{2}x_2$$
$$= \frac{x_1 + x_2}{2}.$$

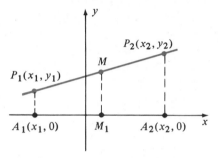

Figure 1.15

It follows that the x-coordinate of M is also $(x_1 + x_2)/2$. It can be shown in similar fashion that the y-coordinate of M is $(y_1 + y_2)/2$. Moreover, these formulas hold for all positions of P_1 and P_2. This gives us the following result.

Midpoint Formula (1.11)

The midpoint of the line segment from $P_1(x_1, y_1)$ to $P_2(x_2, y_2)$ is

$$\left(\frac{x_1 + x_2}{2}, \frac{y_1 + y_2}{2}\right).$$

Example 2   Find the midpoint M of the line segment from $P_1(-2, 3)$ to $P_2(4, -2)$. Plot the points P_1, P_2, M and verify that $d(P_1, M) = d(P_2, M)$.

Solution   Applying the Midpoint Formula (1.11), the coordinates of M are

$$\left(\frac{-2 + 4}{2}, \frac{3 + (-2)}{2}\right) \quad \text{or} \quad \left(1, \frac{1}{2}\right).$$

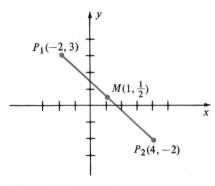

Figure 1.16

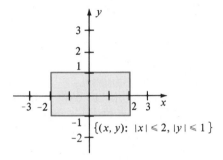

Figure 1.17

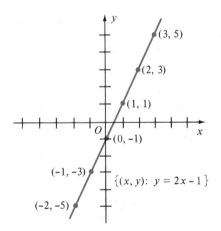

Figure 1.18

The three points P_1, P_2, and M are plotted in Figure 1.16. Using the Distance Formula we obtain

$$d(P_1, M) = \sqrt{(-2 - 1)^2 + (3 - \tfrac{1}{2})^2} = \sqrt{9 + (\tfrac{25}{4})} = \sqrt{61}/2$$

$$d(P_2, M) = \sqrt{(4 - 1)^2 + (-2 - \tfrac{1}{2})^2} = \sqrt{9 + (\tfrac{25}{4})} = \sqrt{61}/2$$

Hence $d(P_1, M) = d(P_2, M)$. ■

If W is a set of ordered pairs, then we may consider the point $P(x, y)$ in a coordinate plane that corresponds to the ordered pair (x, y) in W. The **graph** of W is the set of all points that correspond to the ordered pairs in W. The phrase "sketch the graph W" means to illustrate the significant features of the graph geometrically on a coordinate plane.

Example 3 Sketch the graph of $W = \{(x, y): |x| \leq 2, |y| \leq 1\}$.

Solution The indicated inequalities are equivalent to $-2 \leq x \leq 2$ and $-1 \leq y \leq 1$. Hence the graph of W consists of all points within and on the boundary of the rectangular region shown in Figure 1.17. ■

Example 4 Sketch the graph of $W = \{(x, y): y = 2x - 1\}$.

Solution We begin by finding points with coordinates of the form (x, y) where the ordered pair (x, y) is in W. It is convenient to list these coordinates in the following tabular form, where for each real number x the corresponding value for y is $2x - 1$.

x	-2	-1	0	1	2	3
y	-5	-3	-1	1	3	5

After plotting, it appears that the points with these coordinates lie on a line and we sketch the graph (see Figure 1.18). Ordinarily the few points we have plotted would not be enough to illustrate the graph; however, in this elementary case we can be reasonably sure that the graph is a line. In the next section we will prove that our conjecture is correct. ■

The x-coordinates of points at which a graph intersects the x-axis are called the **x-intercepts** of the graph. Similarly, the y-coordinates of points at which a graph intersects the y-axis are called the **y-intercepts**. In Figure 1.18, there is one x-intercept $1/2$ and one y-intercept -1.

It is impossible to sketch the entire graph in Example 4 since x may be assigned values that are numerically as large as desired. Nevertheless, we often call a drawing of the type given in Figure 1.18 *the graph of W* or *a sketch of the graph* where it is understood that the drawing is only a device for visualizing the actual graph and the line does not terminate as shown in the figure. In general, the sketch of a graph should illustrate enough of the graph so that the remaining parts are evident.

The graph in Example 4 is determined by the equation $y = 2x - 1$ in the sense that for every real number x, the equation can be used to find a number y such that (x, y) is in W. Given an equation in x and y, we say that an ordered pair (a, b) is a **solution** of the equation if equality is obtained when a is substituted for x and b for y. Two equations in x and y are said to be **equivalent** if they have exactly the same solutions. The solutions of an equation in x and y determine a set S of ordered pairs, and we define the **graph of the equation** as the graph of S. Note that the graph of the equation $y = 2x - 1$ is the same as the graph of W (see Figure 1.18).

For some of the equations we shall encounter in this chapter, the technique used for sketching the graph will consist of plotting a sufficient number of points until some pattern emerges, and then sketching the graph accordingly. This is obviously a crude (and often inaccurate) way to arrive at the graph; however, it is a method often employed in elementary courses. As we progress through this text, techniques will be introduced that will enable us to sketch accurate graphs without plotting many points.

Example 5   Sketch the graph of the equation $y = x^2$.

Solution   To obtain the graph, it is necessary to plot more points than in the previous example. Increasing successive x-coordinates by $\frac{1}{2}$, we obtain the following table.

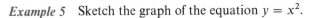

x	-3	$-\frac{5}{2}$	-2	$-\frac{3}{2}$	-1	$-\frac{1}{2}$	0	$\frac{1}{2}$	1	$\frac{3}{2}$	2	$\frac{5}{2}$	3
y	9	$\frac{25}{4}$	4	$\frac{9}{4}$	1	$\frac{1}{4}$	0	$\frac{1}{4}$	1	$\frac{9}{4}$	4	$\frac{25}{4}$	9

Larger numerical values of x produce even larger values of y. For example, the points $(4, 16)$, $(5, 25)$, and $(6, 36)$ are on the graph, as are $(-4, 16)$, $(-5, 25)$, and $(-6, 36)$. Plotting the points given by the table and drawing a smooth curve through these points gives us the sketch in Figure 1.19, where we have labeled several points.   ■

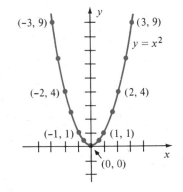

Figure 1.19

The graph in Example 5 is called a **parabola**. The lowest point $(0, 0)$ is called the **vertex** of the parabola and we say that the parabola **opens upward**. If the graph were inverted, as would be the case for $y = -x^2$, then the parabola **opens downward**. The y-axis is called the **axis of the parabola**. Parabolas and their properties will be discussed in detail in Chapter 12, where it will be shown that the graph of every equation of the form $y = ax^2 + bx + c$, with $a \neq 0$, is a parabola whose axis is *parallel* to the y-axis. Parabolas may also open to the right or to the left (cf. Example 6).

If the coordinate plane in Figure 1.19 is folded along the y-axis, then the graph that lies in the left half of the plane coincides with that in the right half. We say that **the graph is symmetric with respect to the y-axis**. As in (i) of Figure 1.20, a graph is symmetric with respect to the y-axis provided that the point $(-x, y)$ is on the graph whenever (x, y) is on the graph. Similarly, as in (ii) of Figure 1.20, **a graph is symmetric with respect to the x-axis** if, whenever a point (x, y) is on the graph, then $(x, -y)$ is also on the graph. In this case if we fold the coordinate plane along the x-axis, the part of the graph that lies above the x-axis will coincide with the part that lies below.

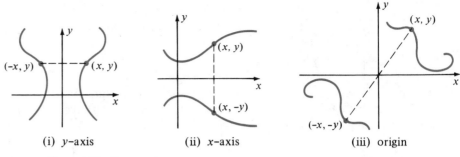

| (i) *y*-axis | (ii) *x*-axis | (iii) origin |

Figure 1.20 Symmetries

Another type of symmetry which certain graphs possess is called **symmetry with respect to the origin**. In this situation, whenever a point (x, y) is on the graph, then $(-x, -y)$ is also on the graph, as illustrated in (iii) of Figure 1.20.

The following tests are useful for investigating these three types of symmetry for graphs of equations in x and y.

Tests for Symmetry (1.12)

> (i)   The graph of an equation is symmetric with respect to the y-axis if substitution of $-x$ for x leads to an equivalent equation.
>
> (ii) The graph of an equation is symmetric with respect to the x-axis if substitution of $-y$ for y leads to an equivalent equation.
>
> (iii) The graph of an equation is symmetric with respect to the origin if the simultaneous substitution of $-x$ for x and $-y$ for y leads to an equivalent equation.

If, in the equation of Example 5, we substitute $-x$ for x, we obtain $y = (-x)^2$, which is equivalent to $y = x^2$. Hence, by Test (i), the graph is symmetric with respect to the y-axis.

If symmetry with respect to an axis exists, then it is sufficient to determine the graph in half of the coordinate plane, since the remainder of the graph is a mirror image, or reflection, of that half.

Example 6 Sketch the graph of the equation $y^2 = x$.

Solution   Since substitution of $-y$ for y does not change the equation, the graph is symmetric with respect to the x-axis. (See Symmetry Test (ii).) It is sufficient, therefore, to plot points with nonnegative y-coordinates and then reflect through the x-axis. Since $y^2 = x$, the y-coordinates of points above the x-axis are given by $y = \sqrt{x}$. Coordinates of some points on the graph are tabulated below. A portion of the graph is sketched in Figure 1.21. The graph is a parabola that opens to the right, with its vertex at the origin. In this case the x-axis is the axis of the parabola.

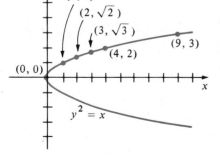

Figure 1.21

x	0	1	2	3	4	9
y	0	1	$\sqrt{2}$	$\sqrt{3}$	2	3

Example 7   Sketch the graph of the equation $4y = x^3$.

Solution   If we substitute $-x$ for x and $-y$ for y, then

$$4(-y) = (-x)^3 \quad \text{or} \quad -4y = -x^3.$$

Multiplying both sides by -1, we see that the last equation has the same solutions as the given equation $4y = x^3$. Hence, from Symmetry Test (iii), the graph is symmetric with respect to the origin. The following table lists some points on the graph.

x	0	$\frac{1}{2}$	1	$\frac{3}{2}$	2	$\frac{5}{2}$
y	0	$\frac{1}{32}$	$\frac{1}{4}$	$\frac{27}{32}$	2	$\frac{125}{32}$

By symmetry (or substitution) we see that the points $(-1, -\frac{1}{4})$, $(-2, -2)$, etc., are on the graph. Plotting points leads to the graph in Figure 1.22. ■

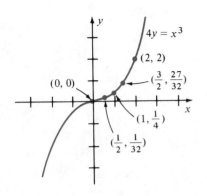

$4y = x^3$
$(2, 2)$
$(0, 0)$
$(\frac{3}{2}, \frac{27}{32})$
$(1, \frac{1}{4})$
$(\frac{1}{2}, \frac{1}{32})$

Figure 1.22

If $C(h, k)$ is a point in a coordinate plane, then a circle with center C and radius $r > 0$ may be defined as the collection of all points in the plane that are r units from C. As indicated in Figure 1.23, a point $P(x, y)$ is on the circle if and only if $d(C, P) = r$ or, by the Distance Formula, if and only if

$$\sqrt{(x - h)^2 + (y - k)^2} = r.$$

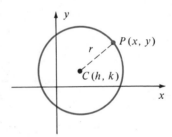

$P(x, y)$
r
$C(h, k)$

Figure 1.23 $(x - h)^2 + (y - k)^2 = r^2$

The following equivalent equation is called the **equation of a circle of radius r and center (h, k)**.

(1.13)

$$(x - h)^2 + (y - k)^2 = r^2$$

If $h = 0$ and $k = 0$, this equation reduces to $x^2 + y^2 = r^2$, which is an equation of a circle of radius r with center at the origin (see Figure 1.24). If $r = 1$, the graph is called a **unit circle**.

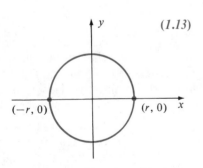

$(-r, 0)$
$(r, 0)$

Figure 1.24 $x^2 + y^2 = r^2$

Example 8   Find an equation of the circle with center $C(-2, 3)$ and containing the point $D(4, 5)$.

Solution   Since D is on the circle, the radius r is $d(C, D)$. By the Distance Formula,

$$r = \sqrt{(-2 - 4)^2 + (3 - 5)^2} = \sqrt{36 + 4} = \sqrt{40}.$$

Using (1.13) with $h = -2$ and $k = 3$, we obtain

$$(x + 2)^2 + (y - 3)^2 = 40,$$

or $\qquad\qquad\qquad x^2 + y^2 + 4x - 6y - 27 = 0.$ ∎

Squaring terms in (1.13) and simplifying, we obtain an equation of the form

$$x^2 + y^2 + ax + by + c = 0$$

where a, b, and c are real numbers. Conversely, if we begin with the last equation, it is always possible, by *completing the squares* in x and y, to obtain an equation of the form

$$(x - h)^2 + (y - k)^2 = d.$$

The method will be illustrated in Example 9. If $d > 0$, the graph is a circle with center (h, k) and radius $r = \sqrt{d}$. If $d = 0$, then, since $(x - h)^2 \geq 0$ and $(y - k)^2 \geq 0$, the only solution of the equation is (h, k), and hence the graph consists of only one point. Finally, if $d < 0$, the equation has no real solutions and there is no graph.

Example 9 Find the center and radius of the circle with equation

$$x^2 + y^2 - 4x + 6y - 3 = 0.$$

Solution We begin by arranging the equation as follows:

$$(x^2 - 4x) + (y^2 + 6y) = 3.$$

Next we complete the squares by adding appropriate numbers within the parentheses. Of course, to obtain equivalent equations we must add the numbers to *both* sides of the equation. In order to complete the square for an expression of the form $x^2 + ax$, we add the square of half the coefficient of x, that is, $(a/2)^2$, to both sides of the equation. Similarly, for $y^2 + by$, we add $(b/2)^2$ to both sides. In this example $a = -4$, $b = 6$, $(a/2)^2 = (-2)^2 = 4$, and $(b/2)^2 = 3^2 = 9$. This leads to

$$(x^2 - 4x + 4) + (y^2 + 6y + 9) = 3 + 4 + 9$$

or $\qquad\qquad\qquad (x - 2)^2 + (y + 3)^2 = 16.$

Hence, by (1.13) the center is $(2, -3)$ and the radius is 4. ∎

EXERCISES 1.2

In Exercises 1–6 find (a) the distance $d(A, B)$ between the points A and B, and (b) the midpoint of the segment AB.

1 $A(6, -2), B(2, 1)$ **2** $A(-4, -1), B(2, 3)$

3 $A(0, -7), B(-1, -2)$ **4** $A(4, 5), B(4, -4)$

5 $A(-3, -2), B(-8, -2)$ **6** $A(11, -7), B(-9, 0)$

In Exercises 7 and 8 prove that the triangle with vertices A, B, and C is a right triangle and find its area.

7 $A(-3, 4), B(2, -1), C(9, 6)$

8 $A(7, 2), B(-4, 0), C(4, 6)$

9 Prove that the following points are vertices of a parallelogram: $A(-4, -1), B(0, -2), C(6, 1), D(2, 2)$.

10 Given $A(-4, -3)$ and $B(6, 1)$, find a formula which expresses the fact that $P(x, y)$ is on the perpendicular bisector of AB.

11 For what values of a is the distance between $(a, 3)$ and $(5, 2a)$ greater than $\sqrt{26}$?

12 Given the points $A(-2, 0)$ and $B(2, 0)$, find a formula not containing radicals that expresses the fact that the sum of the distances from $P(x, y)$ to A and to B, respectively, is 5.

13 Prove that the midpoint of the hypotenuse of any right triangle is equidistant from the vertices. (*Hint*: Label the vertices of the triangle $O(0, 0)$, $A(a, 0)$, and $B(0, b)$.)

14 Prove that the diagonals of any parallelogram bisect each other. (*Hint*: Label three of the vertices of the parallelogram $O(0, 0)$, $A(a, b)$, and $C(c, 0)$.)

In Exercises 15–20 sketch the graph of the set W.

15 $W = \{(x, y): x = 4\}$

16 $W = \{(x, y): y = -3\}$

17 $W = \{(x, y): xy < 0\}$

18 $W = \{(x, y): xy = 0\}$

19 $W = \{(x, y): |x| < 2, |y| > 1\}$

20 $W = \{(x, y): |x| > 1, |y| \le 2\}$

In Exercises 21–38 sketch the graph of the equation and use (1.12) to test for symmetry.

21 $y = 3x + 1$ **22** $y = 4x - 3$

23 $y = -2x + 3$ **24** $y = 2 - 3x$

25 $y = 2x^2 - 1$ **26** $y = -x^2 + 2$

27 $4y = x^2$ **28** $3y + x^2 = 0$

29 $y = -\frac{1}{2}x^3$ **30** $y = \frac{1}{2}x^3$

31 $y = x^3 - 2$ **32** $y = 2 - x^3$

33 $y = \sqrt{x}$ **34** $y = \sqrt{x} - 1$

35 $y = \sqrt{-x}$ **36** $y = \sqrt{x - 1}$

37 $x^2 + y^2 = 16$ **38** $4x^2 + 4y^2 = 25$

In Exercises 39–46 find an equation of a circle satisfying the stated conditions.

39 Center $C(3, -2)$, radius 4

40 Center $C(-5, 2)$, radius 5

41 Center at the origin, passing through $P(-3, 5)$

42 Center $C(-4, 6)$, passing through $P(1, 2)$

43 Center $C(-4, 2)$, tangent to the x-axis

44 Center $C(3, -5)$, tangent to the y-axis

45 Endpoints of a diameter $A(4, -3)$ and $B(-2, 7)$

46 Tangent to both axes, center in the first quadrant, radius 2

In Exercises 47–52 find the center and radius of the circle with the given equation.

47 $x^2 + y^2 + 4x - 6y + 4 = 0$

48 $x^2 + y^2 - 10x + 2y + 22 = 0$

49 $x^2 + y^2 + 6x = 0$

50 $x^2 + y^2 + x + y - 1 = 0$

51 $2x^2 + 2y^2 - x + y - 3 = 0$

52 $9x^2 + 9y^2 - 6x + 12y - 31 = 0$

1.3

LINES

The following concept is fundamental to the study of lines. All lines referred to are considered to be in some fixed coordinate plane.

Definition (1.14)

> Let l be a line that is not parallel to the y-axis, and let $P_1(x_1, y_1)$ and $P_2(x_2, y_2)$ be distinct points on l. The **slope m** of l is
>
> $$m = \frac{y_2 - y_1}{x_2 - x_1}.$$
>
> If l is parallel to the y-axis, then the slope is not defined.

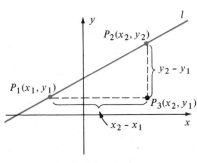

(i) Positive slope

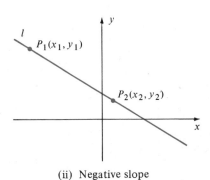

(ii) Negative slope

Figure 1.25

Typical points P_1 and P_2 on a line l are shown in Figure 1.25. The numerator $y_2 - y_1$ in the formula for m measures the vertical change in direction in proceeding from P_1 to P_2 and may be positive, negative, or zero. The denominator $x_2 - x_1$ measures the amount of horizontal change in going from P_1 to P_2, and it may be positive or negative, but never zero, because l is not parallel to the y-axis.

In finding the slope of a line it is immaterial which point is labeled P_1 and which is labeled P_2, since

$$\frac{y_2 - y_1}{x_2 - x_1} = \frac{y_1 - y_2}{x_1 - x_2}.$$

Consequently, we may as well assume that the points are labeled so that $x_1 < x_2$, as in Figure 1.25. In this event $x_2 - x_1 > 0$, and hence the slope is positive, negative, or zero, depending on whether $y_2 > y_1$, $y_2 < y_1$, or $y_2 = y_1$. The slope of the line shown in (i) of Figure 1.25 is positive, whereas the slope of the line shown in (ii) of the figure is negative.

A **horizontal line** is a line that is parallel to the x-axis. Note that a line is horizontal if and only if its slope is 0. A **vertical line** is a line that is parallel to the y-axis. The slope of a vertical line is undefined.

It is important to note that the definition of slope is independent of the two points that are chosen on l, for if other points $P_1'(x_1', y_1')$ and $P_2'(x_2', y_2')$ are used, then as in Figure 1.26, the triangle with vertices P_1', P_2', and $P_3'(x_2', y_1')$ is similar to the triangle with vertices P_1, P_2, and $P_3(x_2, y_1)$. Since the ratios of corresponding sides are equal it follows that

$$\frac{y_2 - y_1}{x_2 - x_1} = \frac{y_2' - y_1'}{x_2' - x_1'}.$$

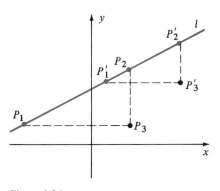

Figure 1.26

Example 1 Sketch the lines through the following pairs of points and find their slopes.

(a) $A(-1, 4)$ and $B(3, 2)$ (b) $A(2, 5)$ and $B(-2, -1)$

(c) $A(4, 3)$ and $B(-2, 3)$ (d) $A(4, -1)$ and $B(4, 4)$.

Solution   The lines are sketched in Figure 1.27. Using Definition (1.14),

(a) $m = \dfrac{2 - 4}{3 - (-1)} = \dfrac{-2}{4} = -\dfrac{1}{2}$

(b) $m = \dfrac{5 - (-1)}{2 - (-2)} = \dfrac{6}{4} = \dfrac{3}{2}$

(c) $m = \dfrac{3 - 3}{-2 - 4} = \dfrac{0}{-6} = 0$

(d)   The slope is undefined since the line is vertical. This is also seen by noting that if the formula for m is used, the denominator is zero.

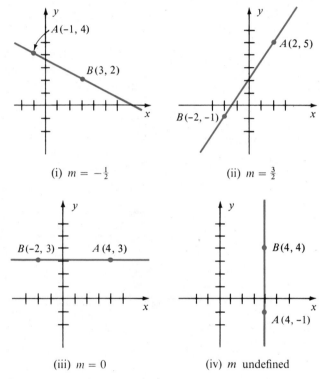

Figure 1.27

Theorem (1.15)

(i)   The graph of the equation $x = a$ is a vertical line with x-intercept a.

(ii)   The graph of the equation $y = b$ is a horizontal line with y-intercept b.

Proof   · The equation $x = a$, where a is a real number, may be considered as an equation in two variables x and y, since we can write it in the form

$$x + (0)y = a.$$

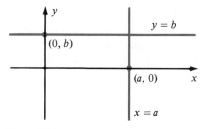

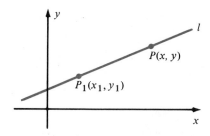

Figure 1.28

Figure 1.29

Some typical solutions of the equation are $(a, -2), (a, 1)$, and $(a, 3)$. Evidently, all solutions of the equation consist of pairs of the form (a, y), where y may have any value and a is fixed. It follows that the graph of $x = a$ is a line parallel to the y-axis with x-intercept a, as illustrated in Figure 1.28. This proves (i). Part (ii) is proved in similar fashion. □

Let us next find an equation of a line l through a point $P_1(x_1, y_1)$ with slope m (only one such line exists). If $P(x, y)$ is any point with $x \neq x_1$ (see Figure 1.29), then P is on l if and only if the slope of the line through P_1 and P is m; that is,

$$\frac{y - y_1}{x - x_1} = m.$$

This equation may be written in the form

$$y - y_1 = m(x - x_1).$$

Note that (x_1, y_1) is also a solution of the last equation and hence the points on l are precisely the points that correspond to the solutions. This equation for l is referred to as the **Point-Slope Form**. Our discussion may be summarized as follows:

Point-Slope Form **(1.16)**

> An equation for the line through the point $P(x_1, y_1)$ with slope m is
>
> $$y - y_1 = m(x - x_1).$$

Example 2 Find an equation of the line through the points $A(1, 7)$ and $B(-3, 2)$.

Solution By Definition (1.14) the slope m of the line is

$$m = \frac{7 - 2}{1 - (-3)} = \frac{5}{4}.$$

We may use the coordinates of either A or B for (x_1, y_1) in the Point-Slope Form (1.16). Using $A(1, 7)$ gives us

$$y - 7 = \tfrac{5}{4}(x - 1)$$

which is equivalent to

$$4y - 28 = 5x - 5 \quad \text{or} \quad 5x - 4y + 23 = 0. \qquad ■$$

The Point-Slope Form may be rewritten as $y = mx - mx_1 + y_1$, which is of the form

$$y = mx + b$$

with $b = -mx_1 + y_1$. The real number b is the y-intercept of the graph, as may be seen by setting $x = 0$. Since the equation $y = mx + b$ displays the slope m and y-intercept b of l, it is called the **Slope-Intercept Form** for the equation of a line. Conversely, if we start with $y = mx + b$, we may write

$$y - b = m(x - 0).$$

Comparing with the Point-Slope Form, we see that the graph is a line with slope m and passing through the point $(0, b)$. This gives us the next result.

Slope-Intercept Form *(1.17)*

> The graph of the equation $y = mx + b$ is a line having slope m and y-intercept b.

We have shown that every line is the graph of an equation of the form

$$ax + by + c = 0$$

where a, b, and c are real numbers, and a and b are not both zero. We call such an equation a **linear equation** in x and y. Let us show, conversely, that the graph of $ax + by + c = 0$ where a and b are not both zero is always a line. On the one hand, if $b \neq 0$, we may solve for y, obtaining

$$y = \left(-\frac{a}{b}\right)x + \left(-\frac{c}{b}\right)$$

which, by the Slope-Intercept Form, is an equation of a line with slope $-a/b$ and y-intercept $-c/b$. On the other hand, if $b = 0$ but $a \neq 0$, then we may solve for x, obtaining $x = -c/a$, which is the equation of a vertical line with x-intercept $-c/a$. This establishes the following important theorem.

Theorem *(1.18)*

> The graph of a linear equation $ax + by + c = 0$ is a line and, conversely, every line is the graph of a linear equation.

For simplicity, we shall use the terminology *the line $ax + by + c = 0$* instead of the more accurate phrase *the line with equation $ax + by + c = 0$*.

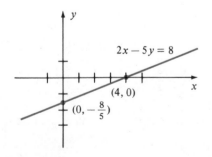

Figure 1.30

Example 3   Sketch the graph of $2x - 5y = 8$.

Solution   From Theorem (1.18) the graph is a line, and hence it is sufficient to find two points on the graph. Let us find the x- and y-intercepts. Substituting $y = 0$ in the given equation, we obtain the x-intercept 4. Substituting $x = 0$, we see that the y-intercept is $-\frac{8}{5}$. This leads to the graph in Figure 1.30.

Another method of solution is to express the given equation in Slope-Intercept Form. To do this we begin by isolating the term involving y on one side of the equals sign, obtaining

$$5y = 2x - 8.$$

Next, dividing both sides by 5 gives us

$$y = \frac{2}{5}x + \left(\frac{-8}{5}\right)$$

which is in the form $y = mx + b$. Hence, the slope is $m = \frac{2}{5}$, and the y-intercept is $b = -\frac{8}{5}$. We may then sketch a line through the point $(0, -\frac{8}{5})$ with slope $\frac{2}{5}$. ∎

The following theorem can be proved geometrically.

Theorem (*1.19*) | Two nonvertical lines are parallel if and only if they have the same slope.

We shall use this fact in the next example.

Example 4 Find an equation of a line through the point $(5, -7)$ that is parallel to the line $6x + 3y - 4 = 0$.

Solution   Let us express the given equation in Slope-Intercept Form. We begin by writing

$$3y = -6x + 4$$

and then divide both sides by 3, obtaining

$$y = -2x + \tfrac{4}{3}.$$

The last equation is in Slope-Intercept Form with $m = -2$, and hence the slope is -2. Since parallel lines have the same slope, the required line also has slope -2. Applying the Point-Slope Form gives us

$$y + 7 = -2(x - 5).$$

This is equivalent to

$$y + 7 = -2x + 10 \quad \text{or} \quad 2x + y - 3 = 0.$$ ∎

The next result specifies conditions for perpendicular lines.

Theorem (*1.20*) | Two lines with slopes m_1 and m_2 are perpendicular if and only if $m_1 m_2 = -1$.

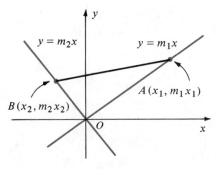

Figure 1.31

Proof   For simplicity, let us consider the special case where the lines intersect at the origin O, as illustrated in Figure 1.31. In this case equations of the lines are $y = m_1 x$ and $y = m_2 x$. If, as in the figure, we choose points $A(x_1, m_1 x_1)$ and $B(x_2, m_2 x_2)$ different from O on the lines, then the lines are perpendicular if and only if angle AOB is a right angle. Applying the Pythagorean Theorem to triangle AOB, this is equivalent to the condition

$$[d(A, B)]^2 = [d(O, B)]^2 + [d(O, A)]^2$$

or, by the Distance Formula,

$$(m_2 x_2 - m_1 x_1)^2 + (x_2 - x_1)^2 = (m_2 x_2)^2 + x_2^2 + (m_1 x_1)^2 + x_1^2.$$

Squaring the indicated terms and simplifying gives us

$$-2m_1 m_2 x_1 x_2 - 2x_1 x_2 = 0.$$

Dividing both sides by $-2x_1 x_2$ we see that the lines are perpendicular if and only if $m_1 m_2 + 1 = 0$, or $m_1 m_2 = -1$.

The same type of proof may be given if the lines intersect at *any* point (a, b).   □

A convenient way to remember the conditions for perpendicularity is to note that m_1 and m_2 must be *negative reciprocals* of one another, that is, $m_1 = -1/m_2$ and $m_2 = -1/m_1$.

Example 5   Find an equation for the perpendicular bisector of the line segment from $A(1, 7)$ to $B(-3, 2)$.

Solution   By the Midpoint Formula (1.11), the midpoint M of the segment AB is $(-1, \frac{9}{2})$. Since the slope of AB is $\frac{5}{4}$ (see Example 2), it follows from Theorem (1.20) that the slope of the perpendicular bisector is $-\frac{4}{5}$. Applying the Point-Slope Form,

$$y - \frac{9}{2} = -\frac{4}{5}(x + 1).$$

Multiplying both sides by 10 and simplifying leads to $8x + 10y - 37 = 0$. ∎

EXERCISES 1.3

In Exercises 1–4 plot the points A and B and find the slope of the line through A and B.

1 $A(-4, 6)$, $B(-1, 18)$

2 $A(6, -2)$, $B(-3, 5)$

3 $A(-1, -3)$, $B(-1, 2)$

4 $A(-3, 4)$, $B(2, 4)$

5   Show that $A(-3, 1)$, $B(5, 3)$, $C(3, 0)$, and $D(-5, -2)$ are vertices of a parallelogram.

6   Show that $A(2, 3)$, $B(5, -1)$, $C(0, -6)$, and $D(-6, 2)$ are vertices of a trapezoid.

7   Prove that the points $A(6, 15)$, $B(11, 12)$, $C(-1, -8)$, and $D(-6, -5)$ are vertices of a rectangle.

8   Prove that the points $A(1, 4)$, $B(6, -4)$, and $C(-15, -6)$ are vertices of a right triangle.

9 If three consecutive vertices of a parallelogram are $A(-1, -3)$, $B(4, 2)$, and $C(-7, 5)$, find the fourth vertex.

10 Let $A(x_1, y_1)$, $B(x_2, y_2)$, $C(x_3, y_3)$, and $D(x_4, y_4)$ denote the vertices of an arbitrary quadrilateral. Prove that the line segments joining midpoints of adjacent sides form a parallelogram.

In Exercises 11–24 find an equation for the line satisfying the given conditions.

11 Through $A(2, -6)$, slope $\frac{1}{2}$

12 Slope -3, y-intercept 5

13 Through $A(-5, -7)$, $B(3, -4)$

14 x-intercept -4, y-intercept 8

15 Through $A(8, -2)$, y-intercept -3

16 Slope 6, x-intercept -2

17 Through $A(10, -6)$, parallel to (a) the y-axis; (b) the x-axis.

18 Through $A(-5, 1)$, perpendicular to (a) the y-axis; (b) the x-axis.

19 Through $A(7, -3)$, perpendicular to the line with equation $2x - 5y = 8$.

20 Through $\left(-\frac{3}{4}, -\frac{1}{2}\right)$, parallel to the line with equation $x + 3y = 1$.

21 Given $A(3, -1)$ and $B(-2, 6)$, find an equation for the perpendicular bisector of the line segment AB.

22 Find an equation for the line that bisects the second and fourth quadrants.

23 Find equations for the altitudes of the triangle with vertices $A(-3, 2)$, $B(5, 4)$, $C(3, -8)$, and find the point at which they intersect.

24 Find equations for the medians of the triangle in Exercise 23, and find their point of intersection.

In Exercises 25–34 use the Slope-Intercept Form (1.17) to find the slope and y-intercept of the line with the given equation and sketch the graph of each line.

25 $3x - 4y + 8 = 0$ **26** $2y - 5x = 1$

27 $x + 2y = 0$

28 $8x = 1 - 4y$

29 $y = 4$

30 $x + 2 = \frac{1}{2}y$

31 $5x + 4y = 20$

32 $y = 0$

33 $x = 3y + 7$

34 $x - y = 0$

35 Find a real number k such that the point $P(-1, 2)$ is on the line $kx + 2y - 7 = 0$.

36 Find a real number k such that the line $5x + ky - 3 = 0$ has y-intercept -5.

37 If a line l has nonzero x- and y-intercepts a and b, respectively, prove that an equation for l is $(x/a) + (y/b) = 1$. (This is called the **intercept form** for the equation of a line.) Express the equation $4x - 2y = 6$ in intercept form.

38 Prove that an equation of the line through $P_1(x_1, y_1)$ and $P_2(x_2, y_2)$ is

$$(y - y_1)(x_2 - x_1) = (y_2 - y_1)(x - x_1).$$

(This is called the **two-point form** for the equation of a line.) Use the two-point form to find an equation of the line through $A(7, -1)$ and $B(4, 6)$.

39 Find all values of r such that the slope of the line through the points $(r, 4)$ and $(1, 3 - 2r)$ is less than 5.

40 Find all values of t such that the slope of the line through $(t, 3t + 1)$ and $(1 - 2t, t)$ is greater than 4.

41 Six years ago a house was purchased for \$59,000. This year it is appraised at \$95,000. Assuming that the value increased by the same amount each year, find an equation that specifies the value at any time after the purchase date. When was the house worth \$73,000?

42 Charles' Law for gases states that if the pressure remains constant, then the relationship between the volume V that the gas occupies and the temperature T (in degrees Celsius), is given by $V = V_0(1 + \frac{1}{273}T)$.

(a) What is the significance of V_0?

(b) What change in temperature is needed to increase the volume from V_0 to $2V_0$?

(c) Sketch the graph of the equation on a TV-plane for the case $V_0 = 100$ and $T \geq -273$.

1.4

FUNCTIONS

The notion of **correspondence** is encountered frequently in everyday life. For example, to each book in a library there corresponds the number of pages in the book. As another example, to each human being there corresponds a birth date. To cite a third example, if the temperature of the air is recorded throughout a day, then at each instant of time there is a corresponding temperature. These examples of correspondences involve two sets, X and Y. In our first example, X denotes the set of books in a library and Y the set of positive integers. To each book x in X there corresponds a positive integer y, namely the number of pages in the book.

Our examples indicate that to each x in X there corresponds *one and only one y* in Y; that is, *y is unique* for a given x. However, the same element of Y may correspond to different elements of X. For example, two different books may have the same number of pages, two different people may have the same birthday, and so on.

In most of our work X and Y will be sets of real numbers. To illustrate, let X and Y both denote the set $\mathbb{R}$ of real numbers, and to each real number x let us assign its square x^2. Thus, to 3 we assign 9, to -5 we assign 25, and to $\sqrt{2}$ the number 2. This gives us a correspondence from $\mathbb{R}$ to $\mathbb{R}$.

All the examples of correspondences we have given are *functions*, as defined below.

Definition (1.21)

> A **function** f from a set X to a set Y is a correspondence that assigns to each element x of X a unique element y of Y. The element y is called the **image** of x under f and is denoted by $f(x)$. The set X is called the **domain** of the function. The **range** of the function consists of all images of elements of X.

The symbol $f(x)$ used for the element associated with x is read "f of x." Sometimes $f(x)$ is called the **value** of f at x.

Functions may be represented pictorially by diagrams of the type shown in Figure 1.32. The curved arrows indicate that the elements $f(x)$, $f(w)$, $f(z)$, and $f(a)$ of Y are associated with the elements $x, w, z,$ and a, respectively, of X. We might imagine a whole family of arrows of this type, where each arrow connects an element of X to some specific element of Y. Although the sets X and Y have been pictured as having no elements in common, this is not required by Definition (1.21). As a matter of fact, we often take $X = Y$. It is important to note that with each x in X there is associated precisely one image $f(x)$; however, different elements such as w and z in Figure 1.32 may have the same image in Y.

Occasionally one of the notations

$$X \xrightarrow{f} Y \quad \text{or} \quad f{:}X \to Y$$

is used to signify that f is a function from X to Y. It is not unusual in this event to say f **maps** X **into** Y. The analogous statement for elements is f **maps** x **into** $f(x)$, denoted by

$$f{:}x \to f(x).$$

Beginning students are sometimes confused by the symbols f and $f(x)$ in Definition (1.21). Remember that f is used to represent the function. It is neither in X nor in Y. However, $f(x)$ is an element of Y, namely the element that f assigns to x.

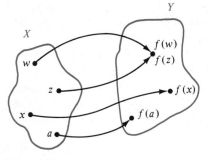

Figure 1.32

If the sets X and Y of Definition (1.21) are intervals, or other sets of real numbers, then instead of using points within regions to represent elements, as in Figure 1.32, we may use two coordinate lines l and l'. This technique is illustrated in Figure 1.33, where two images for a function f are represented graphically.

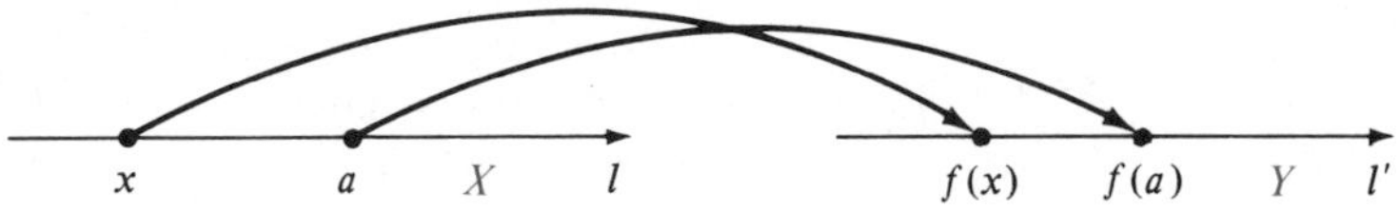

Figure 1.33

Two functions f and g from X to Y are said to be **equal**, and we write

$$f = g \quad \text{provided} \quad f(x) = g(x)$$

for every x in X. For example, if $g(x) = \frac{1}{2}(2x^2 - 6) + 3$ and $f(x) = x^2$ for all x in $\mathbb{R}$, then $g = f$.

Example 1 Let f be the function with domain $\mathbb{R}$ such that $f(x) = x^2$ for every x in $\mathbb{R}$. Find $f(-6)$, $f(\sqrt{3})$, and $f(a + b)$, where a, b are real numbers. What is the range of f?

Solution Values of f (or images under f) may be found by substituting for x in the equation $f(x) = x^2$. Thus

$$f(-6) = (-6)^2 = 36, \qquad f(\sqrt{3}) = (\sqrt{3})^2 = 3,$$

and $$f(a + b) = (a + b)^2 = a^2 + 2ab + b^2.$$

If T denotes the range of f, then by Definition (1.21) T consists of all numbers of the form $f(a)$ where a is in $\mathbb{R}$. Hence T is the set of all squares a^2, where a is a real number. Since the square of any real number is non-negative, T is contained in the set of all nonnegative real numbers. Moreover, every nonnegative real number c is an image under f, since $f(\sqrt{c}) = (\sqrt{c})^2 = c$. Hence the range of f is the set of all nonnegative real numbers. ∎

To describe a function f it is necessary to specify the image $f(x)$ of *each* element x of the domain. A common method for doing this is to use an equation, as in Example 1. In this case, the symbol used for the variable is immaterial. Thus, expressions such as $f(x) = x^2$, $f(s) = s^2$, $g(t) = t^2$, $k(r) = r^2$, etc., define the same function. This is true because if a is any number in the domain, then the image a^2 is obtained no matter which expression is employed.

In the remainder of our work, unless specified otherwise, the phrase *f is a function* will mean that the domain and range are sets of real numbers. If a function is defined by means of some expression as in Example 1 and the domain X is not stated explicitly, then X is considered to be the totality of real numbers for which the given expression is meaningful. To illustrate, if

$f(x) = \sqrt{x}/(x - 1)$, then the domain is assumed to be the set of nonnegative real numbers different from 1. (Why?) If x is in the domain we sometimes say that f **is defined at** x, or that $f(x)$ **exists**. If a set S is contained in the domain we often say that f **is defined on** S. The terminology f **is undefined at** x means that x is not in the domain of f.

Many formulas that occur in mathematics and the sciences determine functions. As an illustration, the formula $A = \pi r^2$ for the area A of a circle of radius r associates with each positive real number r a unique value of A and hence determines a function f where $f(r) = \pi r^2$. The letter r, which represents an arbitrary number from the domain of f, is often called an **independent variable**. The letter A, which represents a number from the range of f, is called a **dependent variable**, since its value depends on the number assigned to r. When two variables r and A are related in this manner, it is customary to use the phrase A *is a function of* r. To cite another example, if an automobile travels at a uniform rate of 50 miles per hour, then the distance d (miles) traveled in time t (hours) is given by $d = 50t$ and we say that d is a function of t.

We have seen that different elements in the domain of a function may have the same image. If images are always different, then the function is called *one-to-one*.

Definition (1.22)

> A function f from X to Y is a **one-to-one function** if, whenever $a \neq b$ in X, then $f(a) \neq f(b)$ in Y.

If f is one-to-one, then each $f(x)$ in the range is the image of *precisely one x* in X. The function illustrated in Figure 1.32 is not one-to-one since two different elements w and z of X have the same image in Y. If the range of f is Y and f is one-to-one, then sets X and Y are said to be in **one-to-one correspondence**. In this case each element of Y is the image of precisely one element of X. The association between real numbers and points on a coordinate line is an example of a one-to-one correspondence.

Example 2

(a) If $f(x) = 3x + 2$, prove that f is one-to-one.

(b) If $g(x) = x^2 + 5$, prove that g is not one-to-one.

Solution

(a) If $a \neq b$, then $3a \neq 3b$ and hence $3a + 2 \neq 3b + 2$, or $f(a) \neq f(b)$. Hence f is one-to-one by Definition (1.22).

(b) The function g is not one-to-one since different numbers in the domain may have the same image. For example, although $-1 \neq 1$, both $g(-1)$ and $g(1)$ are equal to 6. ■

If f is a function from X to X and if $f(x) = x$ for every x, that is, every element x maps into itself, then f is called the **identity function** on X. A function f is a **constant function** if there is some (fixed) element c such that $f(x) = c$ for every x in the domain. If a constant function is represented by a diagram of the type shown in Figure 1.32, then every arrow from X terminates at the same point in Y.

The concept of ordered pair can be used to obtain an alternative approach to functions. We first observe that a function f from X to Y determines the following set W of ordered pairs:

$$W = \{(x, f(x)): x \text{ is in } X\}.$$

Thus W is the totality of ordered pairs for which the first number is in X and the second number is the image of the first. In Example 1, where $f(x) = x^2$, W consists of all pairs of the form (x, x^2) where x is any real number. It is important to note that for each x there is exactly one ordered pair (x, y) in W having x in the first position.

Conversely, if we begin with a set W of ordered pairs such that each x in X appears exactly once in the first position of an ordered pair, and numbers from Y appear in the second position, then W determines a function from X to Y. Specifically, for any x in X there is a unique pair (x, y) in W, and by letting y correspond to x, we obtain a function from X to Y.

We see from the preceding discussion that the following statement could also be used as a definition of function; however, we prefer to think of it as an alternative approach to this concept.

Alternative Definition (1.23)

> A **function** with domain X is a set W of ordered pairs such that, for each x in X, there is exactly one ordered pair (x, y) in W having x in the first position.

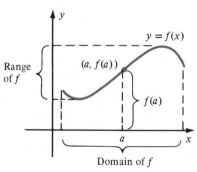

Figure 1.34 Graph of a function f

If f is a function, we may use a graph to exhibit the variation of $f(x)$ as x varies through the domain of f. By definition, the **graph of a function** f is the set of all points $(x, f(x))$ in a coordinate plane, where x is in the domain of f. Thus, the graph of f is the same as the graph of the equation $y = f(x)$, and if $P(a, b)$ is on the graph, then the y-coordinate b is the functional value $f(a)$, as illustrated in Figure 1.34. The figure also exhibits the domain of f (the set of possible values of x) and the range of f (the corresponding values of y). Although we have pictured the domain and range as closed intervals, they may also be infinite intervals or other sets of real numbers. It is important to note that, since there is a unique $f(a)$ for each a in the domain, there is only *one* point on the graph with x-coordinate a. Thus, *every vertical line intersects the graph of a function in at most one point.* Consequently, for graphs of functions it is impossible to obtain a sketch such as that shown in Figure 1.21 where some vertical lines intersect the graph in more than one point.

The x-intercepts of the graph of a function f are the solutions of the equation $f(x) = 0$. These numbers are called the **zeros** of the function. The y-intercept of the graph is $f(0)$, if it exists.

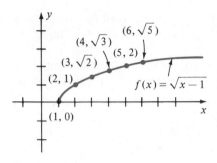

Figure 1.35

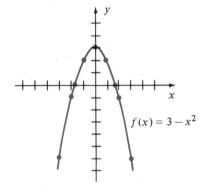

Figure 1.36

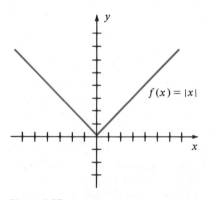

Figure 1.37

Example 3   Sketch the graph of f if $f(x) = \sqrt{x - 1}$. What are the domain and range of f?

Solution   The following table lists some points $(x, f(x))$ on the graph.

x	1	2	3	4	5	6
$f(x)$	0	1	$\sqrt{2}$	$\sqrt{3}$	2	$\sqrt{5}$

Plotting points leads to the sketch shown in Figure 1.35. Note that the x-intercept is 1, and there is no y-intercept.

   The domain of f consists of all real numbers x such that $x \geq 1$, or equivalently, the interval $[1, \infty)$. The range of f is the set of all real numbers y such that $y \geq 0$, or equivalently, $[0, \infty)$.   ∎

Example 4   Sketch the graph of f if $f(x) = 3 - x^2$. What are the domain and range of f?

Solution   We list coordinates $(x, f(x))$ of some points on the graph of f in tabular form, as follows:

x	-3	-2	-1	0	1	2	3
$f(x)$	-6	-1	2	3	2	-1	-6

The x-intercepts are the solutions of the equation $f(x) = 0$, that is, of $3 - x^2 = 0$. These are $\pm\sqrt{3}$. The y-intercept is $f(0) = 3$. Plotting the points given by the table and using the x-intercepts leads to the parabola sketched in Figure 1.36.

   Since x may be assigned any value, the domain of f is $\mathbb{R}$. Referring to the graph we see that the range of f is $(-\infty, 3]$.   ∎

   The solution to Example 4 could have been shortened by observing that since $3 - (-x)^2 = 3 - x^2$, the graph of $y = 3 - x^2$ is symmetric with respect to the y-axis.

Example 5   Sketch the graph of f if $f(x) = |x|$.

Solution   If $x \geq 0$, then $f(x) = x$ and hence the part of the graph to the right of the y-axis is identical to the graph of $y = x$, which is a line through the origin with slope 1. If $x < 0$, then by Definition (1.2), $f(x) = |x| = -x$, and hence the part of the graph to the left of the y-axis is the same as the graph of $y = -x$. The graph is sketched in Figure 1.37.   ∎

Example 6   Sketch the graph of f if $f(x) = \dfrac{1}{x}$.

Solution   The domain of f is the set of all nonzero real numbers. If x is positive, so is $f(x)$, and hence no part of the graph lies in quadrant IV. Quadrant II is also excluded, since if $x < 0$ then $f(x) < 0$. If x is close to zero, the y-coordinate $1/x$ is very large numerically. As x increases through

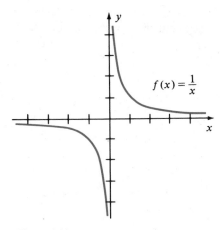

$$f(x) = \frac{1}{x}$$

Figure 1.38

positive values, $1/x$ decreases and is close to zero when x is large. Similarly, if we let x take on numerically large negative values, the y-coordinate $1/x$ is close to zero. Using these remarks and plotting several points gives us the sketch in Figure 1.38.

The graph of f, or equivalently, of the equation $y = 1/x$, is symmetric with respect to the origin. This may be verified by using (iii) of (1.12). ■

Example 7 Describe the graph of a constant function.

Solution If $f(x) = c$, where c is a real number, then the graph of f is the same as the graph of the equation $y = c$ and hence is a horizontal line with y-intercept c. ■

Sometimes functions are described in terms of more than one expression, as in the next examples.

Example 8 Sketch the graph of the function f that is defined as follows:

$$f(x) = \begin{cases} 2x + 3 & \text{if } x < 0 \\ x^2 & \text{if } 0 \le x < 2 \\ 1 & \text{if } x \ge 2. \end{cases}$$

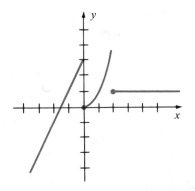

Figure 1.39

Solution If $x < 0$, then $f(x) = 2x + 3$. This means that if x is negative, the expression $2x + 3$ should be used to find functional values. Consequently, if $x < 0$, then the graph of f coincides with the line $y = 2x + 3$ and we sketch that portion of the graph to the left of the y-axis as indicated in Figure 1.39.

If $0 \le x < 2$, we use x^2 to find functional values of f, and therefore this part of the graph of f coincides with the graph of the equation $y = x^2$. We then sketch the part of the graph of f between $x = 0$ and $x = 2$ as indicated in Figure 1.39.

Finally, if $x \ge 2$, the graph of f coincides with the graph of the constant function having values equal to 1. That part of the graph is the horizontal half-line illustrated in Figure 1.39. ■

Example 9 If x is any real number, then there exist consecutive integers n and $n + 1$ such that $n \le x < n + 1$. Let f be the function defined as follows: If $n \le x < n + 1$, then $f(x) = n$. Sketch the graph of f.

Solution The x- and y-coordinates of some points on the graph may be listed as follows:

Values of x	$f(x)$
$\cdots$	$\cdots$
$-2 \le x < -1$	-2
$-1 \le x < 0$	-1
$0 \le x < 1$	0
$1 \le x < 2$	1
$2 \le x < 3$	2
$\cdots$	$\cdots$

Since f is a constant function whenever x is between successive integers, the corresponding part of the graph is a segment of a horizontal line. Part of the graph is sketched in Figure 1.40. The graph continues indefinitely to the right and to the left.

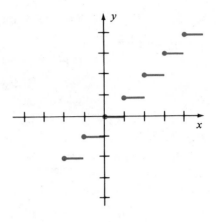

Figure 1.40   ∎

The symbol $[\![x]\!]$ is often used to denote the largest integer n such that $n \le x$. For example $[\![1.6]\!] = 1$, $[\![\sqrt{5}]\!] = 2$, $[\![\pi]\!] = 3$, and $[\![-3.5]\!] = -4$. Using this notation, the function f of Example 9 may be defined by $f(x) = [\![x]\!]$. It is customary to refer to f as the **greatest integer function**.

EXERCISES 1.4

1 If $f(x) = x^3 + 4x - 3$, find $f(1), f(-1), f(0),$ and $f(\sqrt{2})$.

2 If $f(x) = \sqrt{x - 1} + 2x$, find $f(1), f(3), f(5),$ and $f(10)$.

In Exercises 3 and 4 find each of the following, where a and h are real numbers:

(a) $f(a)$

(b) $f(-a)$

(c) $-f(a)$

(d) $f(a + h)$

(e) $f(a) + f(h)$

(f) $\dfrac{f(a + h) - f(a)}{h}$ provided $h \neq 0$

3 $f(x) = 3x^2 - x + 2$ **4** $f(x) = 1/(x^2 + 1)$

In Exercises 5 and 6 find each of the following:

(a) $g(1/a)$ (b) $\dfrac{1}{g(a)}$ (c) $g(a^2)$

(d) $[g(a)]^2$ (e) $g(\sqrt{a})$ (f) $\sqrt{g(a)}$

5 $g(x) = \dfrac{1}{x^2 + 4}$ **6** $g(x) = \dfrac{1}{x}$

In Exercises 7–12 find the largest subset of $\mathbb{R}$ that can serve as the domain of the function f.

7 $f(x) = \sqrt{3x - 5}$ **8** $f(x) = \sqrt{7 - 2x}$

9 $f(x) = \sqrt{4 - x^2}$ **10** $f(x) = \sqrt{x^2 - 9}$

11 $f(x) = \dfrac{x + 1}{x^3 - 9x}$ **12** $f(x) = \dfrac{4x + 7}{6x^2 + 13x - 5}$

In Exercises 13–18 find the number that maps into 4. If $a > 0$, what number maps into a? Find the range of f.

13 $f(x) = 7x - 5$ **14** $f(x) = 3x$

15 $f(x) = \sqrt{x - 3}$ **16** $f(x) = \dfrac{1}{x}$

17 $f(x) = x^3$ **18** $f(x) = \sqrt[3]{x - 4}$

In Exercises 19–26 determine if the function f is one-to-one.

19 $f(x) = 2x + 9$　　　　**20** $f(x) = \dfrac{1}{7x + 9}$

21 $f(x) = 5 - 3x^2$　　　　**22** $f(x) = 2x^2 - x - 3$

23 $f(x) = \sqrt{x}$　　　　**24** $f(x) = x^3$

25 $f(x) = |x|$　　　　**26** $f(x) = 4$

A function f with domain X is termed (i) **even** if $f(-a) = f(a)$ for every a in X, or (ii) **odd** if $f(-a) = -f(a)$ for every a in X. In Exercises 27–36 determine whether f is even, odd, or neither even nor odd.

27 $f(x) = 3x^3 - 4x$　　　　**28** $f(x) = 7x^4 - x^2 + 7$

29 $f(x) = 9 - 5x^2$　　　　**30** $f(x) = 2x^5 - 4x^3$

31 $f(x) = 2$　　　　**32** $f(x) = 2x^3 + x^2$

33 $f(x) = 2x^2 - 3x + 4$　　　　**34** $f(x) = \sqrt{x^2 + 1}$

35 $f(x) = \sqrt[3]{x^3 - 4}$　　　　**36** $f(x) = |x| + 5$

In Exercises 37–60 sketch the graph and determine the domain and range of f.

37 $f(x) = -4x + 3$　　　　**38** $f(x) = 4x - 3$

39 $f(x) = -3$　　　　**40** $f(x) = 3$

41 $f(x) = 4 - x^2$　　　　**42** $f(x) = -(4 + x^2)$

43 $f(x) = \sqrt{4 - x^2}$　　　　**44** $f(x) = \sqrt{x^2 - 4}$

45 $f(x) = \dfrac{1}{x - 4}$　　　　**46** $f(x) = \dfrac{1}{4 - x}$

47 $f(x) = \dfrac{1}{(x - 4)^2}$　　　　**48** $f(x) = -\dfrac{1}{(x - 4)^2}$

49 $f(x) = |x - 4|$　　　　**50** $f(x) = |x| - 4$

51 $f(x) = \dfrac{x}{|x|}$　　　　**52** $f(x) = x + |x|$

53 $f(x) = \sqrt{4 - x}$　　　　**54** $f(x) = 2 - \sqrt{x}$

55 $f(x) = \begin{cases} \dfrac{x^2 - 4}{x - 2} & \text{if } x \neq 2 \\ 3 & \text{if } x = 2 \end{cases}$

56 $f(x) = \begin{cases} \dfrac{x^3 - x}{x} & \text{if } x \neq 0 \\ 1 & \text{if } x = 0 \end{cases}$

57 $f(x) = \begin{cases} -5 & \text{if } x < -5 \\ x & \text{if } -5 \leq x \leq 5 \\ 5 & \text{if } x > 5 \end{cases}$

58 $f(x) = \begin{cases} -x & \text{if } x < 0 \\ 2 & \text{if } 0 \leq x < 1 \\ x^2 & \text{if } x \geq 1 \end{cases}$

59 $f(x) = \begin{cases} x^2 & \text{if } x \leq -1 \\ x^3 & \text{if } |x| < 1 \\ 2x & \text{if } x \geq 1 \end{cases}$

60 $f(x) = \begin{cases} x & \text{if } x \leq 1 \\ -x^2 & \text{if } 1 < x < 2 \\ x & \text{if } x \geq 2 \end{cases}$

If $[\![x]\!]$ denotes values of the greatest integer function, sketch the graph of f in Exercises 61 and 62.

61 (a) $f(x) = [\![2x]\!]$　　　　(b) $f(x) = 2[\![x]\!]$

62 (a) $f(x) = [\![-x]\!]$　　　　(b) $f(x) = -[\![x]\!]$

63 Explain why the graph of the equation $x^2 + y^2 = 1$ is not the graph of a function.

64 (a) Define a function f whose graph is the upper half of a circle with center at the origin and radius 1.

　　(b) Define a function f whose graph is the lower half of a circle with center at the origin and radius 1.

65 Prove that a function f is one-to-one if and only if every horizontal line intersects the graph of f in at most one point.

66 Refer to the remarks preceding Exercises 27–36 and prove that

　　(a) the graph of an even function is symmetric with respect to the y-axis.

　　(b) the graph of an odd function is symmetric with respect to the origin.

67 Find a formula that expresses the radius r of a circle as a function of its circumference C. If the circumference of *any* circle is increased by 12 inches, determine how much the radius increases.

68 Find a formula that expresses the volume of a cube as a function of its surface area. Find the volume if the surface area is 36 square inches.

69 An open box is to be made from a rectangular piece of cardboard having dimensions 20 inches by 30 inches by cutting out identical squares of area x^2 from each corner and turning up the sides (see figure). Express the volume V of the box as a function of x.

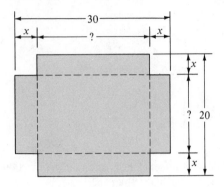

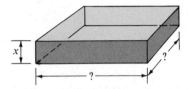

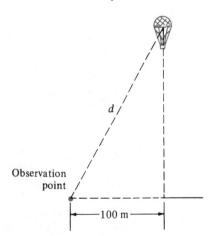

Figure for Exercise 69

70 Find a formula that expresses the area A of an equilateral triangle as a function of the length s of a side.

71 Express the perimeter P of a square as a function of its area A.

72 Express the surface area S of a sphere as a function of its volume V.

73 A hot-air balloon is released at 1:00 P.M. and rises vertically at a rate of 2 meters per second. An observation point is situated 100 meters from a point on the ground directly below the balloon (see figure). If t denotes the time (in seconds) after 1:00 P.M., express the distance d between the balloon and the observation point as a function of t.

74 Two ships leave port at 9:00 A.M., one sailing south at a rate of 16 mph and the other west at a rate of 20 mph (see figure). If t denotes the time (in hours) after 9:00 A.M., express the distance d between the ships as a function of t.

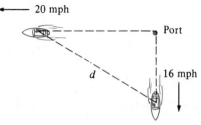

Figure for Exercise 74

75 A company sells running shoes to dealers at a rate of \$20 per pair if less than 50 pairs are ordered. If 50 or more pairs are ordered (up to 600), the price per pair is reduced at a rate of 2 cents times the number ordered. Let A denote the amount of money received when x pairs are ordered. Express A as a function of x.

76 A steel storage tank for propane gas is to be constructed in the shape of a right circular cylinder of altitude 10 ft with a hemisphere attached to each end (see figure). The radius x is yet to be determined.

(a) Express the volume V of the tank as a function of x.

(b) Express the surface area S as a function of x.

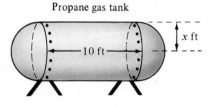

Figure for Exercise 76

Figure for Exercise 73

1.5

COMBINATIONS OF FUNCTIONS

In calculus and its applications it is common to encounter functions that are defined in terms of sums, differences, products, and quotients of various expressions. For example, if $h(x) = x^2 + \sqrt{5x + 1}$, then we may regard $h(x)$ as a sum of values of the simpler functions f and g defined by $f(x) = x^2$ and $g(x) = \sqrt{5x + 1}$. It is natural to refer to the function h as the *sum* of f and g. More generally, if f and g are any functions and D *is the intersection of their domains*, then the **sum** of f and g is the function s defined by

$$s(x) = f(x) + g(x)$$

where x is in D.

It is convenient to denote s by the symbol $f + g$. Since f and g are functions, not numbers, the $+$ used between f and g is not to be considered as addition of real numbers. It is used to indicate that the image of x under $f + g$ is $f(x) + g(x)$, that is,

$$(f + g)(x) = f(x) + g(x).$$

Similarly, the **difference** $f - g$ and the **product** fg of f and g are defined by

$$(f - g)(x) = f(x) - g(x) \quad \text{and} \quad (fg)(x) = f(x)g(x)$$

where x is in D. Finally, the **quotient** f/g of f by g is given by

$$\left(\frac{f}{g}\right)(x) = \frac{f(x)}{g(x)}$$

where x is in D and $g(x) \neq 0$.

Example 1 If $f(x) = \sqrt{4 - x^2}$ and $g(x) = 3x + 1$, find the sum, difference, and product of f and g, and the quotient of f by g.

Solution   The domain of f is the closed interval $[-2, 2]$ and the domain of g is $\mathbb{R}$. Consequently, the intersection of their domains is $[-2, 2]$ and the required functions are given by

$$(f + g)(x) = \sqrt{4 - x^2} + (3x + 1), \quad -2 \leq x \leq 2$$
$$(f - g)(x) = \sqrt{4 - x^2} - (3x + 1), \quad -2 \leq x \leq 2$$
$$(fg)(x) = \sqrt{4 - x^2}(3x + 1), \quad -2 \leq x \leq 2$$
$$(f/g)(x) = \sqrt{4 - x^2}/(3x + 1), \quad -2 \leq x \leq 2, \quad x \neq -\tfrac{1}{3}. \quad \blacksquare$$

If g is a constant function such that $g(x) = c$ for every x, and if f is any function, then cf will denote the product of g and f; that is, $(cf)(x) = cf(x)$ for all x in the domain of f. To illustrate, if f is the function of Example 1, then we have $(cf)(x) = c\sqrt{4 - x^2}$, $-2 \leq x \leq 2$.

Among the most important functions in mathematics are those defined as follows.

Definition (1.24)

> A function f is a **polynomial function** if
>
> $$f(x) = a_n x^n + a_{n-1} x^{n-1} + \cdots + a_1 x + a_0$$
>
> where the coefficients $a_0, a_1, \ldots, a_n$ are real numbers and the exponents are nonnegative integers.

The expression to the right of the equal sign in Definition (1.24) is called a **polynomial in x** (with real coefficients) and each $a_k x^k$ is called a **term** of the polynomial. The number a_0 is called the **constant term**. We often use the phrase *the polynomial $f(x)$* when referring to expressions of this type. If $a_n \neq 0$, then a_n is called the **leading coefficient** of $f(x)$ and we say that f (or $f(x)$) has **degree n**.

If a polynomial function f has degree 0, then $f(x) = c$, where $c \neq 0$, and hence f is a constant function. If a coefficient a_i is zero we often abbreviate (1.24) by deleting the term $a_i x^i$. If *all* the coefficients of a polynomial are zero it is called the **zero polynomial** and is denoted by 0. It is customary not to assign a degree to the zero polynomial.

If some of the coefficients are negative, then for convenience we often use minus signs between appropriate terms. To illustrate, instead of writing $3x^2 + (-5)x + (-7)$, we write $3x^2 - 5x - 7$ for this polynomial of degree 2. Polynomials in other variables may also be considered. For example, $\frac{2}{5}z^2 - 3x^7 + 8 - \sqrt{5}z^4$ is a polynomial in z of degree 7. We ordinarily arrange the terms in order of decreasing powers of the variable and write $-3z^7 - \sqrt{5}z^4 + \frac{2}{5}z^2 + 8$.

According to the definition of degree, if c is a nonzero real number, then c is a polynomial of degree 0. Such polynomials (together with the zero polynomial) are called **constant polynomials**.

If $f(x)$ is a polynomial of degree 1, then $f(x) = ax + b$, where $a \neq 0$. From Section 1.3, the graph of f is a line and, accordingly, f is called a **linear function**.

Any polynomial $f(x)$ of degree 2 may be written

$$f(x) = ax^2 + bx + c,$$

where $a \neq 0$. In this case f is called a **quadratic function**. The graph of f or, equivalently, of the equation $y = ax^2 + bx + c$, is a parabola.

A **rational function** is a quotient of two polynomial functions. Thus q is rational if, for every x in its domain,

$$q(x) = \frac{f(x)}{h(x)}$$

where $f(x)$ and $h(x)$ are polynomials. The domain of a polynomial function is $\mathbb{R}$, whereas the domain of a rational function consists of all real numbers except the zeros of the polynomial in the denominator.

A function f is called **algebraic** if it can be expressed in terms of sums, differences, products, quotients, or roots of polynomial functions. For example, if

$$f(x) = 5x^4 - 2\sqrt[3]{x} + \frac{x(x^2 + 5)}{\sqrt{x^3 + \sqrt{x}}}$$

then f is an algebraic function. Functions that are not algebraic are termed **transcendental**. The trigonometric, exponential, and logarithmic functions considered later in this book are examples of transcendental functions.

We shall conclude this section by describing an important method of using two functions f and g to obtain a third function. Suppose X, Y, and Z are sets of real numbers. Let f be a function from X to Y, and g a function from Y to Z. In terms of the arrow notation we have

$$X \xrightarrow{f} Y \xrightarrow{g} Z,$$

that is, f maps X into Y and g maps Y into Z. A function from X to Z may be defined in a natural way. For every x in X, the number $f(x)$ is in Y. Since the domain of g is Y, we may then find the image of $f(x)$ under g. Of course, this element of Z is written as $g(f(x))$. By associating $g(f(x))$ with x, we obtain a function from X to Z called the **composite function** of g by f. This is illustrated geometrically in Figure 1.41 where we have represented the domain X of f on a coordinate line l, the domain Y of g (and the range of f) on a coordinate line l', and the range of g on a coordinate line l''. The dashes indicate the correspondence defined from X to Z. We sometimes use an operation symbol $\circ$ and denote the latter function $g \circ f$. The next definition summarizes this discussion.

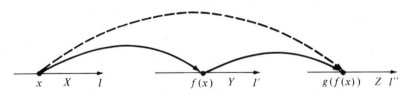

Figure 1.41

Definition (1.25)

> If f is a function from X to Y and g is a function from Y to Z, then the **composite function** $g \circ f$ is the function from X to Z defined by
>
> $$(g \circ f)(x) = g(f(x)),$$
>
> for every x in X.

Actually, it is not essential that the domain of g be all of Y but merely that it *contain* the range of f. In certain cases we may wish to restrict x to some subset of X so that $f(x)$ is in the domain of g. This is illustrated in the next example.

Example 2 If $f(x) = x - 2$ and $g(x) = 5x + \sqrt{x}$, find $(g \circ f)(x)$. What is the domain of $g \circ f$?

Solution Using the definitions of $g \circ f$, f, and g,

$$\begin{aligned}(g \circ f)(x) = g(f(x)) &= g(x - 2) \\ &= 5(x - 2) + \sqrt{x - 2} \\ &= 5x - 10 + \sqrt{x - 2}.\end{aligned}$$

The domain X of f is the set of all real numbers; however, the last equality implies that $(g \circ f)(x)$ is a real number only if $x \geq 2$. Thus, the domain of the composite function $g \circ f$ is the interval $[2, \infty)$. ∎

Given f and g, it may also be possible to find $f(g(x))$. In this case we first obtain the image of x under g and then apply f to $g(x)$. This gives us a composite function from Z to X denoted by $f \circ g$. Thus, by definition,

$$(f \circ g)(x) = f(g(x)),$$

for all x in Z.

Example 3 If $f(x) = x^2 - 1$ and $g(x) = 3x + 5$, find $(f \circ g)(x)$ and $(g \circ f)(x)$.

Solution We may proceed as follows:

$$\begin{aligned}(f \circ g)(x) = f(g(x)) &= f(3x + 5) \\ &= (3x + 5)^2 - 1 \\ &= 9x^2 + 30x + 24.\end{aligned}$$

Similarly,

$$\begin{aligned}(g \circ f)(x) = g(f(x)) &= g(x^2 - 1) \\ &= 3(x^2 - 1) + 5 \\ &= 3x^2 + 2.\end{aligned}$$
∎

Note that in Example 3, $f(g(x))$ and $g(f(x))$ are not the same; that is, $f \circ g \neq g \circ f$.

EXERCISES 1.5

In Exercises 1–6, find the sum, difference, and product of f and g, and the quotient of f by g.

1 $f(x) = 3x^2, g(x) = 1/(2x - 3)$

2 $f(x) = \sqrt{x + 3}, g(x) = \sqrt{x + 3}$

3 $f(x) = x + (1/x), g(x) = x - (1/x)$

4 $f(x) = x^3 + 3x, g(x) = 3x^2 + 1$

5 $f(x) = 2x^3 - x + 5, g(x) = x^2 + x + 2$

6 $f(x) = 7x^4 + x^2 - 1, g(x) = 7x^4 - x^3 + 4x$

In Exercises 7–20 find $(f \circ g)(x)$ and $(g \circ f)(x)$.

7 $f(x) = 2x^2 + 5, g(x) = 4 - 7x$

8 $f(x) = 1/(3x + 1), g(x) = 2/x^2$

9 $f(x) = x^3, g(x) = x + 1$

10 $f(x) = \sqrt{x^2 + 4}, g(x) = 7x^2 + 1$

11 $f(x) = 3x^2 + 2, g(x) = 1/(3x^2 + 2)$

12 $f(x) = 7, g(x) = 4$

13 $f(x) = \sqrt{2x + 1}, g(x) = x^2 + 3$

14 $f(x) = 6x - 12, g(x) = \frac{1}{6}x + 2$

15 $f(x) = |x|, g(x) = -5$

16 $f(x) = \sqrt[3]{x^2 + 1}, g(x) = x^3 + 1$

17 $f(x) = x^2, g(x) = 1/x^2$

18 $f(x) = 1/(x + 1), g(x) = x + 1$

19 $f(x) = 2x - 3, g(x) = (x + 3)/2$

20 $f(x) = x^3 - 1, g(x) = \sqrt[3]{x + 1}$

21 If $f(x)$ and $g(x)$ are polynomials of degree 5 in x, does it follow that $f(x) + g(x)$ has degree 5? Explain.

22 Prove that the degree of the product of two nonzero polynomials equals the sum of the degrees of the polynomials.

23 Using the terminology of Exercise 27 in Section 1.4, prove that (a) the product of two odd functions is even; (b) the product of two even functions is even; and (c) the product of an even function and an odd function is odd.

24 Which parts of Exercise 23 are true if the word "product" is replaced by "sum"?

25 Prove that every function with domain $\mathbb{R}$ can be written as the sum of an even function and an odd function.

26 Show that there exist an infinite number of rational functions f and g such that $f + g = fg$.

27 If $f(x)$ is a polynomial, and if the coefficients of all odd powers of x are 0, show that f is an even function.

28 If $f(x)$ is a polynomial, and if the coefficients of all even powers of x are 0, show that f is an odd function.

29 If f is a linear function and g is a quadratic function, show that $f \circ g$ and $g \circ f$ are quadratic functions.

30 If f and g are polynomial functions of degrees m and n, respectively, show that $f \circ g$ is a polynomial function of degree mn.

1.6

REVIEW

Define or discuss each of the following.

1 Rational and irrational numbers

2 Coordinate line

3 A real number a is greater than a real number b.

4 Inequality

5 Absolute value of a real number

6 Triangle inequality

7 Intervals (open, closed, half-open, infinite)

8 Variable

9 Domain of a variable

10 Ordered pair

11 Rectangular coordinate system in a plane

12 The x- and y-coordinates of a point

13 The Distance Formula

14 The Midpoint Formula

15 The graph of an equation in x and y

16 Tests for symmetry

17 Equation of a circle

18 The slope of a line

19 The Point-Slope Form

20 The Slope-Intercept Form

21 Function

22 The domain of a function

23 The range of a function

24 One-to-one function

25 Constant function

26 The graph of a function

27 The sum, difference, product, and quotient of two functions

28 Polynomial function

29 Rational function

30 The composite function of two functions

EXERCISES 1.6

Solve the inequalities in Exercises 1–8 and express the solutions in terms of intervals.

1 $4 - 3x > 7 + 2x$

2 $\dfrac{7}{2} > \dfrac{1 - 4x}{5} > \dfrac{3}{2}$

3 $|2x - 7| \le 0.01$

4 $|6x - 7| > 1$

5 $2x^2 < 5x - 3$

6 $\dfrac{2x^2 - 3x - 20}{x + 3} < 0$

7 $\dfrac{1}{3x - 1} < \dfrac{2}{x + 5}$

8 $x^2 + 4 \ge 4x$

9 Given the points $A(2, 1)$, $B(-1, 4)$, and $C(-2, -3)$,

    (a) prove that A, B, and C are vertices of a right triangle and find its area.

    (b) find the coordinates of the midpoint of AB.

    (c) find the slope of the line through B and C.

Sketch the graphs of the equations in Exercises 10–13 and discuss symmetries with respect to the x-axis, y-axis, or origin.

10 $3x - 5y = 10$

11 $x^2 + y = 4$

12 $x = y^3$

13 $|x + y| = 1$

In Exercises 14–17 sketch the graph of the set W.

14 $W = \{(x, y) : x > 0\}$

15 $W = \{(x, y) : y > x\}$

16 $W = \{(x, y) : x^2 + y^2 < 1\}$

17 $W = \{(x, y) : |x - 4| < 1, |y + 3| < 2\}$

In Exercises 18–20 find an equation of the circle satisfying the given conditions.

18 Center $C(4, -7)$ and passing through the origin

19 Center $C(-4, -3)$ and tangent to the line with equation $x = 5$

20 Passing through the points $A(-2, 3)$, $B(4, 3)$, and $C(-2, -1)$

21 Find the center and radius of the circle that has equation

$$x^2 + y^2 - 10x + 14y - 7 = 0.$$

Given the points $A(-4, 2)$, $B(3, 6)$, and $C(2, -5)$, solve the problems stated in Exercises 22–26.

22 Find an equation for the line through B that is parallel to the line through A and C.

23 Find an equation for the line through B that is perpendicular to the line through A and C.

24 Find an equation for the line through C and the midpoint of the line segment AB.

25 Find an equation for the line through A that is parallel to the y-axis.

26 Find an equation for the line through C that is perpendicular to the line with equation $3x - 10y + 7 = 0$.

In Exercises 27–30 find the largest subset of $\mathbb{R}$ that can serve as the domain of f.

27 $f(x) = \dfrac{2x - 3}{x^2 - x}$

28 $f(x) = \dfrac{x}{\sqrt{16 - x^2}}$

29 $f(x) = \dfrac{1}{\sqrt{x - 5}\sqrt{7 - x}}$

30 $f(x) = \dfrac{1}{\sqrt{x(x - 2)}}$

31 If $f(x) = 1/\sqrt{x + 1}$ find each of the following.

    (a) $f(1)$ (b) $f(3)$

    (c) $f(0)$ (d) $f(\sqrt{2} - 1)$

    (e) $f(-x)$ (f) $-f(x)$

    (g) $f(x^2)$ (h) $(f(x))^2$

In Exercises 32–35, sketch the graph of f.

32 $f(x) = 1 - 4x^2$

33 $f(x) = 100$

34 $f(x) = -1/(x + 1)$

35 $f(x) = |x + 5|$

In Exercises 36–38 find $(f + g)(x)$, $(f - g)(x)$, $(fg)(x)$, $(f/g)(x)$, $(f \circ g)(x)$, and $(g \circ f)(x)$.

36 $f(x) = x^2 + 3x + 1, g(x) = 2x - 1$

37 $f(x) = x^2 + 4, g(x) = \sqrt{2x + 5}$

38 $f(x) = 5x + 2, g(x) = 1/x^2$

In Exercises 39 and 40, prove that f is one-to-one.

39 $f(x) = 5 - 7x$

40 $f(x) = 4x^2 + 3, x \ge 0$

2

LIMITS AND CONTINUITY
OF FUNCTIONS

The concept of *limit of a function* is one of the fundamental ideas that distinguishes calculus from areas of mathematics such as algebra or geometry. It is not easy to master the formal notion of limit. Indeed, it is usually necessary for the beginner to study the definition many times, looking at it from various points of view, before the meaning becomes clear. In spite of the complexity of the definition, it is easy to develop an intuitive feeling for limits. With this in mind, the discussion in the first section is not rigorous. The mathematically precise description of limit of a function is presented in Section 2.2. The remainder of the chapter contains important theorems and concepts pertaining to limits.

2.1

INTRODUCTION

In calculus and its applications we are often interested in the values $f(x)$ of a function f when x is *very close* to a number a, *but not necessarily equal to a*. As a matter of fact, in many instances the number a is not in the domain of f; that is, $f(a)$ is undefined. Roughly speaking, we ask the following question: As x gets closer and closer to a (but $x \neq a$), does $f(x)$ get closer and closer to some number L? If the answer is *yes*, we say that *the limit of $f(x)$, as x approaches a, equals L*, and we use the following **limit notation**:

$$(2.1) \qquad \lim_{x \to a} f(x) = L.$$

Let us consider a geometric illustration of a limit. The tangent line l at a point P on a circle may be defined as the line that has only the point P in common with the circle, as illustrated in (i) of Figure 2.1. This definition cannot be extended to arbitrary graphs, since a tangent line may intersect a graph several times, as shown in (ii) of Figure 2.1.

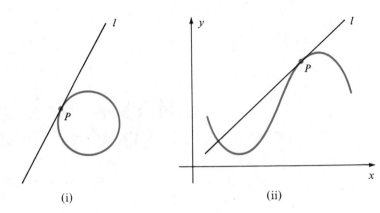

Figure 2.1

To define the tangent line *l* at a point *P* on the graph of an equation it is sufficient to state the slope *m* of *l*, since this completely determines the line. To arrive at *m* we begin by choosing any other point *Q* on the graph and considering the line through *P* and *Q*, as in (i) of Figure 2.2. A line of this type is called a **secant line** for the graph.

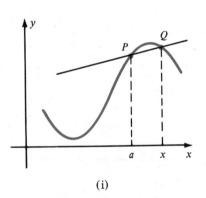

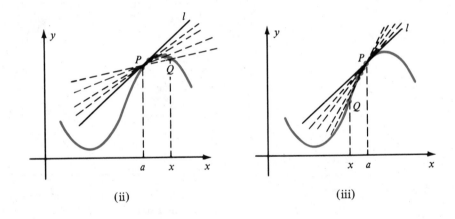

Figure 2.2

Let us consider the variation of this secant line as *Q* gets closer and closer to *P*. If *Q* approaches *P* from the right we have the situation illustrated in (ii) of Figure 2.2, where the dashes indicate several positions of the secant line. It appears that if *Q* is close to *P*, then the slope m_{PQ} of the secant line should be close to the slope *m* of *l*. Part (iii) of the figure illustrates what happens if *Q* approaches *P* from the left. It again appears that m_{PQ} gets closer and closer to the slope *m* of *l*. We could also let *Q* approach *P* by jumping from one side of *P* to the other, in which case the secant line would oscillate back and forth on opposite sides of *l*, with m_{PQ} approaching the slope *m* of *l*. These observations suggest that if the slope m_{PQ} of the secant line has a limiting value as *Q* approaches *P*, then this value should be *defined* as the slope of the tangent line *l* at *P*. If the *x*-coordinate of *P* is *a* and the *x*-coordinate of *Q* is *x* (see (i) of Figure 2.2), then for many graphs the phrase "*Q* approaches *P*" may be replaced by "*x* approaches *a*." This leads to the following definition of the **slope of a tangent line**:

$$m = \lim_{x \to a} m_{PQ}.$$

It is important to observe that $x \neq a$ throughout this limiting process. Indeed, if we let $x = a$, then $P = Q$ and m_{PQ} does not exist.

Example 1 If *a* is any real number, use (2.2) to find the slope of the tangent line to the graph of $y = x^2$ at the point $P(a, a^2)$. Find an equation of the tangent line *l* to the graph at the point $(\frac{3}{2}, \frac{9}{4})$.

Solution   The graph of $y = x^2$ and typical points $P(a, a^2)$ and $Q(x, x^2)$ are illustrated in (i) of Figure 2.3. By Definition (1.14), the slope m_{PQ} of the secant line through *P* and *Q* is

$$m_{PQ} = \frac{x^2 - a^2}{x - a}.$$

Applying (2.2), the slope *m* of the tangent line at *P* is

$$m = \lim_{x \to a} m_{PQ} = \lim_{x \to a} \frac{x^2 - a^2}{x - a}.$$

To find the limit, it is necessary to change the form of the indicated fraction. Since $x \neq a$ in the limiting process it follows that $x - a \neq 0$, and hence we may proceed as follows:

$$m = \lim_{x \to a} \frac{x^2 - a^2}{x - a} = \lim_{x \to a} \frac{(x + a)(x - a)}{(x - a)} = \lim_{x \to a} (x + a).$$

As *x* gets closer and closer to *a*, the expression $x + a$ gets close to $a + a$, or $2a$. Consequently $m = 2a$.

Since the slope of the tangent line *l* at the point $(\frac{3}{2}, \frac{9}{4})$ is the special case in which $a = \frac{3}{2}$, we have $m = 2a = 2(\frac{3}{2}) = 3$, as illustrated in (ii) of Figure 2.2. Using the Point-Slope Form (1.16), an equation of *l* is

$$y - \tfrac{9}{4} = 3(x - \tfrac{3}{2}).$$

This equation simplifies to

$$12x - 4y - 9 = 0. \qquad\blacksquare$$

The preceding discussion and example lack precision because of the haziness of the phrases "very close" and "closer and closer." This will be remedied in the next section, when a formal definition of limit is stated. In the remainder of this section we shall continue in an intuitive manner. As a simple illustration of the limit notation (2.1) suppose $f(x) = \frac{1}{2}(3x - 1)$ and consider $a = 4$. Although 4 is in the domain of the function *f*, we are

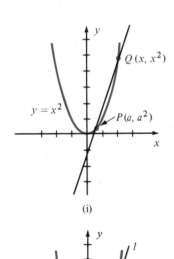

(i)

(ii)

Figure 2.3

primarily interested in values of $f(x)$ when x is *close* to 4, but *not necessarily equal* to 4. The following are some typical values:

$$f(3.9) = 5.35 \qquad\qquad f(4.1) = 5.65$$
$$f(3.99) = 5.485 \qquad\qquad f(4.01) = 5.515$$
$$f(3.999) = 5.4985 \qquad\qquad f(4.001) = 5.5015$$
$$f(3.9999) = 5.49985 \qquad\qquad f(4.0001) = 5.50015$$
$$f(3.99999) = 5.499985 \qquad f(4.00001) = 5.500015$$

It appears that the closer x is to 4, the closer $f(x)$ is to 5.5. This can also be verified by observing that if x is close to 4, then $3x - 1$ is close to 11, and hence $\frac{1}{2}(3x - 1)$ is close to 5.5. Consequently we write

$$\lim_{x \to 4} \tfrac{1}{2}(3x - 1) = 5.5.$$

In this illustration the number 4 could actually have been substituted for x, thereby obtaining 5.5. The next two examples show that it is not always possible to find the limit L in (2.1) by merely substituting a for x.

Example 2 Let $f(x) = \dfrac{x - 9}{\sqrt{x} - 3}$.

(a)   Find $\lim\limits_{x \to 9} f(x)$.

(b)   Sketch the graph of f and verify the limit in part (a) graphically.

Solution
(a)   The number 9 is not in the domain of f since the denominator $\sqrt{x} - 3$ is zero for this value of x. However, if we write

$$f(x) = \frac{x - 9}{\sqrt{x} - 3} = \frac{(\sqrt{x} - 3)(\sqrt{x} + 3)}{\sqrt{x} - 3}$$

it is evident that for all nonnegative values of x, *except* $x = 9$, we have $f(x) = \sqrt{x} + 3$. Thus, the closer x is to 9 (but $x \neq 9$), the closer $f(x)$ is to $\sqrt{9} + 3$, or 6. Using the limit notation and the fact that $x \neq 9$,

$$\lim_{x \to 9} \frac{x - 9}{\sqrt{x} - 3} = \lim_{x \to 9} (\sqrt{x} + 3) = 6.$$

(b)   From the discussion in part (a), the graph of f is the same as the graph of the equation $y = \sqrt{x} + 3$, *except for the point* (9, 6). The fact that (9, 6) is not on the graph of f is illustrated by the hollow circle in Figure 2.4. As x gets closer and closer to 9, the corresponding y-coordinate $f(x)$ on the graph of f gets closer to the number 6. Note that $f(x)$ never actually attains the value 6; however, it can be made as close to 6 as desired by choosing x sufficiently close to 9.   ∎

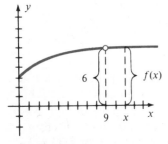

Figure 2.4 $\lim\limits_{x \to 9} f(x) = 6$

Example 3   If $f(x) = \dfrac{2x^2 - 5x + 2}{5x^2 - 7x - 6}$, find $\lim\limits_{x \to 2} f(x)$.

Solution   Note that 2 is not in the domain of f since $0/0$ is obtained when 2 is substituted for x. Factoring the numerator and denominator gives us

$$f(x) = \frac{(x - 2)(2x - 1)}{(x - 2)(5x + 3)}.$$

Thus, if $x \neq 2$, the values of $f(x)$ are the same as those of $(2x - 1)/(5x + 3)$. It follows that if x is close to 2 (but $x \neq 2$), then $f(x)$ is close to $(4 - 1)/(10 + 3)$ or $\frac{3}{13}$. Thus it appears that

$$\lim_{x \to 2} \frac{2x^2 - 5x + 2}{5x^2 - 7x - 6} = \lim_{x \to 2} \frac{(x - 2)(2x - 1)}{(x - 2)(5x + 3)}$$

$$= \lim_{x \to 2} \frac{2x - 1}{5x + 3} = \frac{3}{13}. \qquad \blacksquare$$

The preceding examples demonstrate that algebraic manipulations can sometimes be used to simplify the task of finding limits. In other cases a considerable amount of ingenuity is necessary to determine whether or not a limit exists. This will be especially true when limits of trigonometric, exponential, and logarithmic functions are discussed. For example, it will be shown in Section 8.1 that

$$\lim_{x \to 0} \frac{\sin x}{x} = 1.$$

This important formula cannot be obtained algebraically.

The function f defined by $f(x) = 1/x$ provides an illustration in which no limit exists as x approaches 0. If x is assigned values closer and closer to 0 (but $x \neq 0$), $f(x)$ increases without bound numerically, as illustrated in Figure 1.38. We shall have more to say about this function in Example 2 of the next section.

EXERCISES 2.1

In Exercises 1–16 use algebraic simplifications to help find the limits, if they exist.

1 $\lim\limits_{x \to 2} \dfrac{x^2 - 4}{x - 2}$

2 $\lim\limits_{x \to 3} \dfrac{2x^3 - 6x^2 + x - 3}{x - 3}$

3 $\lim\limits_{x \to 1} \dfrac{x^2 - x}{2x^2 + 5x - 7}$

4 $\lim\limits_{r \to -3} \dfrac{r^2 + 2r - 3}{r^2 + 7r + 12}$

5 $\lim\limits_{x \to 5} \dfrac{3x^2 - 13x - 10}{2x^2 - 7x - 15}$

6 $\lim\limits_{x \to 25} \dfrac{\sqrt{x} - 5}{x - 25}$

7 $\lim\limits_{k \to 4} \dfrac{k^2 - 16}{\sqrt{k} - 2}$

8 $\lim\limits_{h \to 0} \dfrac{(x + h)^3 - x^3}{h}$

9 $\lim\limits_{h \to 0} \dfrac{(x + h)^2 - x^2}{h}$

10 $\lim\limits_{h \to 2} \dfrac{h^3 - 8}{h^2 - 4}$

11 $\lim\limits_{h \to -2} \dfrac{h^3 + 8}{h + 2}$

12 $\lim\limits_{z \to 10} \dfrac{1}{z - 10}$

13 $\lim\limits_{x \to -3/2} \dfrac{2x + 3}{4x^2 + 12x + 9}$

14 $\lim\limits_{s \to -1} \dfrac{1}{s^2 + 2s + 1}$

15 $\lim\limits_{x \to 0} \dfrac{1}{x^2}$

16 $\lim\limits_{t \to 1} \dfrac{(1/t) - 1}{t - 1}$

In Exercises 17–20 (a) find the slope of the tangent line to the graph of f at the point $P(a, f(a))$, and (b) find the equation of the tangent line at the point $P(2, f(2))$.

17 $f(x) = 5x^2 - 4x$

18 $f(x) = 3 - 2x^2$

19 $f(x) = x^3$

20 $f(x) = x^4$

In Exercises 21–24 (a) use (2.2) to find the slope of the tangent line at the point with x-coordinate a on the graph of the given equation; (b) find the equation of the tangent line at the indicated point P; (c) sketch the graph and the tangent line at P.

21 $y = 3x + 2$, $P(1, 5)$

22 $y = \sqrt{x}$, $P(4, 2)$

23 $y = 1/x$, $P(2, \frac{1}{2})$

24 $y = x^{-2}$, $P(2, \frac{1}{4})$

25 Give a geometric argument to show that the graph of $y = |x|$ has no tangent line at the point $(0, 0)$.

26 Give a geometric argument to show that the graph of the greatest integer function has no tangent line at the point $P(1, 1)$.

27 Refer to Exercise 19. Show that the tangent line to the graph of $y = x^3$ at the point $P(0, 0)$ crosses the curve at that point.

28 If $f(x) = ax + b$, prove that the tangent line to the graph of f at any point coincides with the graph of f.

29 Refer to Example 1. Sketch the graph of $y = x^2$ together with tangent lines at the points having x-coordinates -3, -2, -1, 0, 1, 2, and 3. At what point on the graph is the slope of the tangent line equal to 6?

30 Sketch a graph that has three horizontal tangent lines and one vertical tangent line.

The results stated in Exercises 31–36 may be verified by using methods developed later in the text. At this stage of our work, use a calculator to lend support to these results by substituting various real numbers for x. Discuss why this use of a calculator fails to *prove* that the limits exist.

31 $\lim\limits_{x \to 0} (1 + x)^{1/x} \approx 2.72$

32 $\lim\limits_{x \to 0} (1 + 2x)^{3/x} \approx 403.4$

33 $\lim\limits_{x \to 2} \dfrac{3^x - 9}{x - 2} \approx 9.89$

34 $\lim\limits_{x \to 1} \dfrac{2^x - 2}{x - 1} \approx 1.39$

35 $\lim\limits_{x \to 0} \left(\dfrac{4^{|x|} + 9^{|x|}}{2} \right)^{1/|x|} = 6$

36 $\lim\limits_{x \to 0} |x|^x = 1$

2.2

DEFINITION OF LIMIT

Let us return to the illustration $\lim_{x \to 4} \frac{1}{2}(3x - 1) = 5.5$ discussed in Section 2.1 and consider, in more detail, the variation of $f(x) = \frac{1}{2}(3x - 1)$ when x is close to 4. Using the functional values on page 44 we arrive at the following statements.

$$\text{If} \quad 3.9 < x < 4.1 \qquad \text{then} \qquad 5.35 < f(x) < 5.65.$$
$$\text{If} \quad 3.99 < x < 4.01 \qquad \text{then} \qquad 5.485 < f(x) < 5.515.$$
$$\text{If} \quad 3.999 < x < 4.001 \qquad \text{then} \qquad 5.4985 < f(x) < 5.5015.$$
$$\text{If} \quad 3.9999 < x < 4.0001 \qquad \text{then} \quad 5.49985 < f(x) < 5.50015.$$
$$\text{If} \quad 3.99999 < x < 4.00001 \quad \text{then} \quad 5.499985 < f(x) < 5.500015.$$

Each of these statements has the following form, where the Greek letters ε (epsilon) and δ (delta) are used to denote small positive real numbers:

(i) $$\text{If} \quad 4 - \delta < x < 4 + \delta, \quad \text{then} \quad 5.5 - \varepsilon < f(x) < 5.5 + \varepsilon.$$

For example, the first statement follows from (i) by letting $\delta = 0.1$ and $\varepsilon = 0.15$; the second is the case $\delta = 0.01$ and $\varepsilon = 0.015$; for the third let $\delta = 0.001$ and $\varepsilon = 0.0015$; and so on.

We may rewrite (i) in terms of intervals as follows:

(ii) $$\text{If } x \text{ is in the open interval } (4 - \delta, 4 + \delta), \text{ then}$$
$$f(x) \text{ is in the open interval } (5.5 - \varepsilon, 5.5 + \varepsilon).$$

A geometric interpretation of (ii) is given in Figure 2.5, where the curved arrow indicates the correspondence between x and $f(x)$.

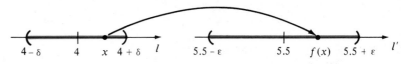

Figure 2.5

Evidently (i) is equivalent to the following statement:

$$\text{If} \quad -\delta < x - 4 < \delta, \quad \text{then} \quad -\varepsilon < f(x) - 5.5 < \varepsilon.$$

Employing absolute values, this may be written:

$$\text{If} \quad |x - 4| < \delta, \quad \text{then} \quad |f(x) - 5.5| < \varepsilon.$$

If we wish to add the condition $x \neq 4$, then it is necessary to demand that $0 < |x - 4|$. This gives us the following extension of (i) and (ii):

(iii) $$\text{If} \quad 0 < |x - 4| < \delta, \quad \text{then} \quad |f(x) - 5.5| < \varepsilon.$$

A statement of type (iii) will appear in the definition of limit; however, it is necessary to change our point of view to some extent. To arrive at (iii), we first considered the domain of f and assigned values to x that were close to 4. We then noted the closeness of $f(x)$ to 5.5. In the definition of limit we shall *reverse* this process by first considering an open interval $(5.5 - \varepsilon, 5.5 + \varepsilon)$ and then, second, determining whether there is an open interval of the form $(4 - \delta, 4 + \delta)$ in the domain of f such that (iii) is true.

The next definition is patterned after the previous remarks.

Definition of Limit (2.3)

> Let f be a function that is defined on an open interval containing a, except possibly at a itself, and let L be a real number. The statement
>
> $$\lim_{x \to a} f(x) = L$$
>
> means that for every $\varepsilon > 0$, there exists a $\delta > 0$, such that
>
> $$\text{if} \quad 0 < |x - a| < \delta, \quad \text{then} \quad |f(x) - L| < \varepsilon.$$

If $\lim_{x \to a} f(x) = L$, we say that **the limit of $f(x)$, as x approaches a, is L**. Since ε can be arbitrarily small, the last two inequalities in this definition are sometimes phrased $f(x)$ *can be made arbitrarily close to L by choosing x sufficiently close to a.*

It is sometimes convenient to use the following form of Definition (2.3), where the two inequalities involving absolute values have been stated in terms of open intervals.

Alternative Definition of Limit (2.4)

> The statement
>
> $$\lim_{x \to a} f(x) = L$$
>
> means that for every $\varepsilon > 0$, there exists a $\delta > 0$, such that if x is in the open interval $(a - \delta, a + \delta)$, and $x \neq a$, then $f(x)$ is in the open interval $(L - \varepsilon, L + \varepsilon)$.

To get a better understanding of the relationship between the positive numbers ε and δ in Definitions (2.3) and (2.4), let us consider a geometric interpretation similar to that in Figure 1.33, where the domain of f is represented by certain points on a coordinate line l, and the range by other points on a coordinate line l'. The limit process may be outlined as follows.

To prove that $\lim_{x \to a} f(x) = L$:

Step 1. For any $\varepsilon > 0$ consider the open interval $(L - \varepsilon, L + \varepsilon)$ (see Figure 2.6).

Figure 2.6

Step 2. Show that there exists an open interval $(a - \delta, a + \delta)$ in the domain of f such that Definition (2.4) is true (see Figure 2.7).

Figure 2.7

It is extremely important to remember that *first* we consider the interval $(L - \varepsilon, L + \varepsilon)$ and then, *second*, we show that an interval $(a - \delta, a + \delta)$ of the required type exists in the domain of f. One scheme for remembering the proper sequence of events is to think of the function f as a cannon that shoots a cannonball from the point on l with coordinate x to the point on l' with coordinate $f(x)$, as illustrated by the curved arrow in Figure 2.7. Step 1 may then be regarded as setting up a target of radius ε with bull's eye at L. To apply Step 2 we must find an open interval containing a in which to place the cannon such that the cannonball hits the target. Incidentally, there is no guarantee that it will hit the bull's eye; however, if $\lim_{x \to a} f(x) = L$ we can make the cannonball land as close as we please to the bull's eye.

It should be clear that the number δ in the limit definition is not unique, for if a specific δ can be found, then any *smaller* positive number δ' will also satisfy the requirements.

Since a function may be described geometrically by means of a graph on a rectangular coordinate system, it is of interest to interpret Definitions (2.3) and (2.4) of limit graphically. Figure 2.8 illustrates the graph of a function f where, for any x in the domain of f, the number $f(x)$ is the y-coordinate of the point on the graph with x-coordinate x. Given any $\varepsilon > 0$, we consider the

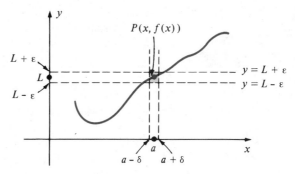

Figure 2.8 $\lim_{x \to a} f(x) = L$

open interval $(L - \varepsilon, L + \varepsilon)$ on the y-axis, and the horizontal lines $y = L \pm \varepsilon$ shown in the figure. If there exists an open interval $(a - \delta, a + \delta)$ such that for every x in $(a - \delta, a + \delta)$, with the possible exception of $x = a$, the point $P(x, f(x))$ lies between the horizontal lines, that is, within the shaded rectangle shown in Figure 2.8, then

$$L - \varepsilon < f(x) < L + \varepsilon$$

and hence $\lim_{x \to a} f(x) = L$.

The next example illustrates how the geometric process pictured in Figure 2.8 may be applied to a specific function.

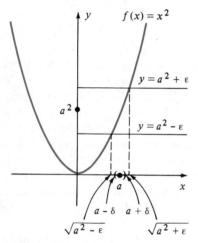

Figure 2.9

Example 1   Prove that $\lim_{x \to a} x^2 = a^2$.

Solution   Let us consider the case $a > 0$. We shall apply the Alternative Definition (2.4) with $f(x) = x^2$ and $L = a^2$. The graph of f is sketched in Figure 2.9, together with typical points on the x- and y-axes corresponding to a and a^2, respectively.

For any positive number ε, consider the horizontal lines $y = a^2 + \varepsilon$ and $y = a^2 - \varepsilon$. These lines intersect the graph of f at points with x-coordinates $\sqrt{a^2 - \varepsilon}$ and $\sqrt{a^2 + \varepsilon}$, as illustrated in the figure. If x is in the open interval $(\sqrt{a^2 - \varepsilon}, \sqrt{a^2 + \varepsilon})$, then

$$\sqrt{a^2 - \varepsilon} < x < \sqrt{a^2 + \varepsilon}.$$

Consequently,

$$a^2 - \varepsilon < x^2 < a^2 + \varepsilon,$$

that is, $f(x) = x^2$ is in the open interval $(a^2 - \varepsilon, a^2 + \varepsilon)$. Geometrically, this means that the point (x, x^2) on the graph of f lies between the horizontal lines. Choose a number δ smaller than both $\sqrt{a^2 + \varepsilon} - a$ and $a - \sqrt{a^2 - \varepsilon}$ as illustrated in Figure 2.9. It follows that if x is in the interval $(a - \delta, a + \delta)$, then x is also in $(\sqrt{a^2 - \varepsilon}, \sqrt{a^2 + \varepsilon})$, and, therefore, $f(x)$ is in the interval $(a^2 - \varepsilon, a^2 + \varepsilon)$. Hence, by Definition (2.4), $\lim_{x \to a} x^2 = a^2$. Although we have considered only $a > 0$, a similar argument applies if $a \le 0$. ∎

To shorten explanations, whenever the notation $\lim_{x \to a} f(x) = L$ is used we shall often assume that all the conditions given in Definition (2.3) are satisfied. Thus, it may not always be pointed out that f is defined on an open interval containing a. Moreover, we shall not always specify L but merely write "$\lim_{x \to a} f(x)$ exists," or "$f(x)$ has a limit as x approaches a." The phrase "find $\lim_{x \to a} f(x)$" means "find a number L such that $\lim_{x \to a} f(x) = L$." If no such L exists we write "$\lim_{x \to a} f(x)$ does not exist."

It can be proved that if $f(x)$ has a limit as x approaches a, then that limit is unique (see Appendix II).

In the following example we return to the function considered at the beginning of this section and *prove* that the limit exists by means of Definition (2.3).

Example 2 Prove that $\lim_{x \to 4} \frac{1}{2}(3x - 1) = \frac{11}{2}$.

Solution Let $f(x) = \frac{1}{2}(3x - 1)$, $a = 4$, and $L = \frac{11}{2}$. According to Definition (2.3) we must show that for every $\varepsilon > 0$, there exists a $\delta > 0$ such that

$$\text{if} \quad 0 < |x - 4| < \delta, \quad \text{then} \quad |\tfrac{1}{2}(3x - 1) - \tfrac{11}{2}| < \varepsilon.$$

A clue to the choice of δ can be found by examining the last inequality involving ε. The following list shows equivalent inequalities.

$$|\tfrac{1}{2}(3x - 1) - \tfrac{11}{2}| < \varepsilon$$
$$\tfrac{1}{2}|(3x - 1) - 11| < \varepsilon$$
$$|3x - 1 - 11| < 2\varepsilon$$
$$|3x - 12| < 2\varepsilon$$
$$3|x - 4| < 2\varepsilon$$
$$|x - 4| < \tfrac{2}{3}\varepsilon$$

The final inequality gives us the needed clue. If we let $\delta = \frac{2}{3}\varepsilon$, then if $0 < |x - 4| < \delta$, the last inequality in the list is true and consequently so is the first. Hence by Definition (2.3), $\lim_{x \to 4} \frac{1}{2}(3x - 1) = \frac{11}{2}$.

It is also possible to give a geometric proof similar to that used in Example 1. ■

It was relatively easy to use the definition of limit in the previous examples because $f(x)$ was a simple expression involving x. Limits of more complicated functions may also be verified by direct applications of the definition; however, the task of showing that for every $\varepsilon > 0$ there exists a suitable $\delta > 0$ often requires a great deal of ingenuity. In Section 2.3 we shall introduce theorems that can be used to find many limits without resorting to a search for the general number δ that appears in Definition (2.3).

The next two examples indicate how the geometric interpretation illustrated in Figure 2.8 may be used to show that certain limits do not exist.

Example 3 Show that $\lim_{x \to 0} \dfrac{1}{x}$ does not exist.

Solution Let us proceed in an indirect manner. Thus, suppose it *were* true that

$$\lim_{x \to 0} \frac{1}{x} = L$$

for some number L. Let us consider any pair of horizontal lines $y = L \pm \varepsilon$ as illustrated in Figure 2.10. Since we are assuming that the limit exists, it

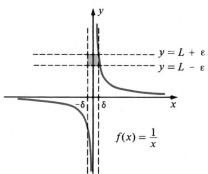

Figure 2.10

should be possible to find an open interval $(0 - \delta, 0 + \delta)$ or equivalently, $(-\delta, \delta)$, containing 0, such that whenever $-\delta < x < \delta$ and $x \neq 0$, the point $(x, 1/x)$ on the graph lies between the horizontal lines. However, since $1/x$ can be made as large as desired by choosing x close to 0, not every point $(x, 1/x)$ with nonzero x-coordinate in $(-\delta, \delta)$ has this property. Consequently our supposition is false; that is, the limit does not exist. ∎

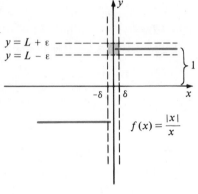

$y = L + \varepsilon$

$y = L - \varepsilon$

$f(x) = \dfrac{|x|}{x}$

Figure 2.11

Example 4   If $f(x) = \dfrac{|x|}{x}$, show that $\lim_{x \to 0} f(x)$ does not exist.

Solution   If $x > 0$, then $|x|/x = x/x = 1$ and hence, to the right of the y-axis, the graph of f coincides with the line $y = 1$. If $x < 0$, then $|x|/x = -x/x = -1$, which means that to the left of the y-axis the graph of f coincides with the line $y = -1$. If it were true that $\lim_{x \to 0} f(x) = L$ for some L, then the preceding remarks imply that $-1 \leq L \leq 1$. As shown in Figure 2.11, if we consider any pair of horizontal lines $y = L \pm \varepsilon$, where $0 < \varepsilon < 1$, then there exist points on the graph that are not between these lines for some nonzero x in *every* interval $(-\delta, \delta)$ containing 0. It follows that the limit does not exist. ∎

EXERCISES 2.2

Establish the limits in Exercises 1–10 by means of Definition (2.3).

1 $\lim_{x \to 2} (5x - 3) = 7$

2 $\lim_{x \to -3} (2x + 1) = -5$

3 $\lim_{x \to -6} (10 - 9x) = 64$

4 $\lim_{x \to 4} (8x - 15) = 17$

5 $\lim_{x \to 5} \left(\dfrac{x}{4} + 2 \right) = \dfrac{13}{4}$

6 $\lim_{x \to 6} \left(9 - \dfrac{x}{6} \right) = 8$

7 $\lim_{x \to 3} 5 = 5$

8 $\lim_{x \to 5} 3 = 3$

9 $\lim_{x \to \pi} x = \pi$

10 $\lim_{x \to a} 2 = 2$

Use the method illustrated in Examples 3 and 4 to show that the limits in Exercises 11–14 do not exist.

11 $\lim_{x \to 3} \dfrac{|x - 3|}{x - 3}$

12 $\lim_{x \to -2} \dfrac{x + 2}{|x + 2|}$

13 $\lim_{x \to -5} \dfrac{1}{x + 5}$

14 $\lim_{x \to 1} \dfrac{1}{(x - 1)^2}$

15 Give an example of a function f that is defined at a and for which $\lim_{x \to a} f(x)$ exists, but $\lim_{x \to a} f(x) \neq f(a)$.

16 If f is the greatest integer function and a is any integer, show that $\lim_{x \to a} f(x)$ does not exist.

17 Let f be defined by the following conditions: $f(x) = 0$ if x is rational and $f(x) = 1$ if x is irrational. Prove that for every real number a, $\lim_{x \to a} f(x)$ does not exist.

18 Why is it impossible to investigate $\lim_{x \to 0} \sqrt{x}$ by means of Definition (2.3)?

Use the graphical technique illustrated in Example 1 to verify the limits in Exercises 19–22.

19 $\lim_{x \to a} x^3 = a^3$

20 $\lim_{x \to a} \sqrt{x} = \sqrt{a}, \quad a > 0$

21 $\lim_{x \to a} \sqrt[3]{x} = \sqrt[3]{a}$

22 $\lim_{x \to a} (x^2 + 1)^2 = (a^2 + 1)^2$

2.3

THEOREMS ON LIMITS

It would be an excruciating task to solve each problem on limits by means of Definition (2.3). The purpose of this section is to introduce theorems that may be used to simplify the process. To prove the theorems it is necessary to employ the definition of limit; however, once they are established it will be possible to determine many limits without referring to an ε or a δ. Several theorems are proved in this section; the remaining proofs may be found in Appendix II.

The simplest limit to consider involves the constant function defined by $f(x) = c$, where c is a real number. If f is represented geometrically by means of coordinate lines l and l', then *every* arrow from l terminates at the same point on l', namely, the point with coordinate c, as indicated in Figure 2.12.

Figure 2.12

It is easy to prove that for every real number a, $\lim_{x \to a} f(x) = c$. Thus, if $\varepsilon > 0$, consider the open interval $(c - \varepsilon, c + \varepsilon)$ on l' as illustrated in the figure. Since $f(x) = c$ is always in this interval, *any* number δ will satisfy the conditions of Definition (2.3); that is, for *every* $\delta > 0$,

$$\text{if } 0 < |x - a| < \delta, \quad \text{then} \quad |f(x) - c| < \varepsilon.$$

The last statement is also evident from the fact that if $f(x) = c$, then

$$|f(x) - c| = |c - c| = 0 < \varepsilon$$

for *every* x. We have proved the following theorem.

Theorem (2.5)

> If a and c are any real numbers, then $\lim_{x \to a} c = c$.

Sometimes Theorem (2.5) is phrased "the limit of a constant is the constant." To illustrate,

$$\lim_{x \to 3} 8 = 8, \quad \lim_{x \to 8} 3 = 3, \quad \lim_{x \to \pi} \sqrt{2} = \sqrt{2}, \quad \text{and} \quad \lim_{x \to a} 0 = 0.$$

The next result tells us that if f is a linear function, then the limit of $f(x)$ as x approaches a may be found by substituting a for x.

Theorem (2.6)

> If a, b, and m are real numbers, then $\lim_{x \to a} (mx + b) = ma + b$.

Proof   If $m = 0$, then $mx + b = b$ and the statement of the theorem reduces to $\lim_{x \to a} b = b$, which was proved in Theorem (2.5).

Next suppose $m \neq 0$. If we let $f(x) = mx + b$ and $L = ma + b$, then according to Definition (2.3) we must show that for every $\varepsilon > 0$ there exists a $\delta > 0$ such that

$$\text{if} \quad 0 < |x - a| < \delta, \quad \text{then} \quad |(mx + b) - (ma + b)| < \varepsilon.$$

As in the solution of Example 2 in the previous section, a clue to the choice of δ can be found by examining the inequality involving ε. All inequalities in the following list are equivalent.

$$|(mx + b) - (ma + b)| < \varepsilon$$
$$|mx - ma| < \varepsilon$$
$$|m||x - a| < \varepsilon$$
$$|x - a| < \frac{\varepsilon}{|m|}$$

The last inequality suggests that we choose $\delta = \varepsilon/|m|$. Thus, given any $\varepsilon > 0$,

$$\text{if} \quad 0 < |x - a| < \delta, \quad \text{where} \quad \delta = \varepsilon/|m|$$

then the last inequality in the list is true, and hence, so is the first inequality, which is what we wished to prove.   □

As special cases of Theorem (2.6), we have

$$\lim_{x \to a} x = a$$
$$\lim_{x \to 4} (3x - 5) = 3 \cdot 4 - 5 = 7$$
$$\lim_{x \to \sqrt{2}} (13x + \sqrt{2}) = 13\sqrt{2} + \sqrt{2} = 14\sqrt{2}.$$

The next theorem states that *if a function f has a positive limit as x approaches a, then f(x) is positive throughout some open interval containing a, with the possible exception of a.*

Theorem (2.7)

> If $\lim_{x \to a} f(x) = L$ and $L > 0$, then there exists an open interval $(a - \delta, a + \delta)$ containing a such that $f(x) > 0$ for all x in $(a - \delta, a + \delta)$, except possibly $x = a$.

Proof   If $\varepsilon = L/2$, then the interval $(L - \varepsilon, L + \varepsilon)$ contains only positive numbers. By Definition (2.4) there exists a $\delta > 0$ such that if x is in the open interval $(a - \delta, a + \delta)$ and $x \neq a$, then $f(x)$ is in $(L - \varepsilon, L + \varepsilon)$, and hence $f(x) > 0$.   □

In like manner, it can be shown that if f has a *negative* limit as x approaches a, then there is an open interval I containing a such that $f(x) < 0$ for every x in I, with the possible exception of $x = a$.

Many functions may be expressed as sums, differences, products, and quotients of other functions. In particular, suppose a function s is a sum of two functions f and g, so that $s(x) = f(x) + g(x)$ for every x in the domain of s. If $f(x)$ and $g(x)$ have limits L and M, respectively, as x approaches a, it is natural to conclude that $s(x)$ has the limit $L + M$ as x approaches a. The fact that this and analogous statements hold for products and quotients are consequences of the next theorem.

Theorem (2.8)

> If $\lim\limits_{x \to a} f(x) = L$ and $\lim\limits_{x \to a} g(x) = M$, then
>
> (i) $\lim\limits_{x \to a} [f(x) + g(x)] = L + M.$
>
> (ii) $\lim\limits_{x \to a} [f(x) \cdot g(x)] = L \cdot M.$
>
> (iii) $\lim\limits_{x \to a} \left[\dfrac{f(x)}{g(x)} \right] = \dfrac{L}{M}$, provided $M \neq 0.$
>
> (iv) $\lim\limits_{x \to a} [cf(x)] = cL.$
>
> (v) $\lim\limits_{x \to a} [f(x) - g(x)] = L - M.$

Although the conclusions of Theorem (2.8) appear to be intuitively evident, the proofs are rather technical and require some ingenuity. Proofs for (i)–(iii) may be found in Appendix II. Part (iv) of the theorem follows readily from part (ii) and Theorem (2.5) as follows:

$$\lim_{x \to a} [cf(x)] = \left[\lim_{x \to a} c \right]\left[\lim_{x \to a} f(x) \right] = cL.$$

Finally, to prove (v) we may write

$$f(x) - g(x) = f(x) + (-1)g(x)$$

and then use (i) and (iv) (with $c = -1$).

The conclusions of Theorem (2.8) are often written as follows:

(i) $\lim\limits_{x \to a} [f(x) + g(x)] = \lim\limits_{x \to a} f(x) + \lim\limits_{x \to a} g(x)$

(ii) $\lim\limits_{x \to a} [f(x) \cdot g(x)] = \lim\limits_{x \to a} f(x) \cdot \lim\limits_{x \to a} g(x)$

(iii) $\lim\limits_{x \to a} \left[\dfrac{f(x)}{g(x)} \right] = \dfrac{\lim\limits_{x \to a} f(x)}{\lim\limits_{x \to a} g(x)}$, if $\lim\limits_{x \to a} g(x) \neq 0$

(iv) $\lim\limits_{x \to a} [cf(x)] = c\left[\lim\limits_{x \to a} f(x) \right]$

(v) $\lim\limits_{x \to a} [f(x) - g(x)] = \lim\limits_{x \to a} f(x) - \lim\limits_{x \to a} g(x).$

Example 1 Find $\lim\limits_{x \to 2} \dfrac{3x + 4}{5x + 7}$.

Solution   The numerator and denominator of the indicated quotient define linear functions whose limits exist by Theorem (2.6). Moreover, the limit of the denominator is not 0. Hence by (iii) of Theorem (2.8) and (2.6),

$$\lim_{x \to 2} \frac{3x + 4}{5x + 7} = \frac{\lim\limits_{x \to 2} (3x + 4)}{\lim\limits_{x \to 2} (5x + 7)} = \frac{3(2) + 4}{5(2) + 7} = \frac{10}{17}.$$ ∎

Notice the simple manner in which the limit in Example 1 was found. It would be a lengthy task to verify the limit by means of Definition (2.3).

Theorem (2.8) may be extended to limits of sums, differences, products, and quotients that involve any number of functions. An application of (ii) to three (equal) functions is given in the next example.

Example 2   Prove that for every real number a, $\lim_{x \to a} x^3 = a^3$.

Solution   Since $\lim_{x \to a} x = a$ we may write

$$\lim_{x \to a} x^3 = \lim_{x \to a} (x \cdot x \cdot x)$$

$$= \left(\lim_{x \to a} x \right) \cdot \left(\lim_{x \to a} x \right) \cdot \left(\lim_{x \to a} x \right)$$

$$= a \cdot a \cdot a = a^3.$$ ∎

The technique used in Example 2 can be extended to x^n, where n is any positive integer. We merely write x^n as a product $x \cdot x \cdots x$ of n factors and then take the limit of each factor. This gives us (i) of the next theorem. Part (ii) may be proved in similar fashion by using (ii) of Theorem (2.8).

Theorem (2.9)

> If n is a positive integer, then
>
> (i) $\lim\limits_{x \to a} x^n = a^n$.
>
> (ii) $\lim\limits_{x \to a} [f(x)]^n = \left[\lim\limits_{x \to a} f(x) \right]^n$,   provided $\lim\limits_{x \to a} f(x)$ exists.

Example 3   Find $\lim_{x \to 2} (3x + 4)^5$.

Solution   Applying (ii) of (2.9) and Theorem (2.6),

$$\lim_{x \to 2} (3x + 4)^5 = \left[\lim_{x \to 2} (3x + 4) \right]^5$$

$$= [3(2) + 4]^5$$

$$= 10^5 = 100,000.$$ ∎

Example 4   Find $\lim_{x \to -2} (5x^3 + 3x^2 - 6)$.

Solution   We may proceed as follows (supply reasons):

$$\lim_{x \to -2} (5x^3 + 3x^2 - 6) = \lim_{x \to -2} (5x^3) + \lim_{x \to -2} (3x^2) - \lim_{x \to -2} (6)$$

$$= 5 \lim_{x \to -2} (x^3) + 3 \lim_{x \to -2} (x^2) - 6$$

$$= 5(-2)^3 + 3(-2)^2 - 6$$

$$= 5(-8) + 3(4) - 6 = -34. \quad \blacksquare$$

Note that the limit in Example 4 is the number obtained by substituting -2 for x in $5x^3 + 3x^2 - 6$. The next theorem states that the same is true for the limit of *every* polynomial.

Theorem (2.10)

> If f is a polynomial function and a is a real number, then
>
> $$\lim_{x \to a} f(x) = f(a).$$

Proof   We may write $f(x)$ in the form

$$f(x) = b_n x^n + b_{n-1} x^{n-1} + \cdots + b_0$$

where the b_i are real numbers. As in Example 4,

$$\lim_{x \to a} f(x) = \lim_{x \to a} (b_n x^n) + \lim_{x \to a} (b_{n-1} x^{n-1}) + \cdots + \lim_{x \to a} b_0$$

$$= b_n \lim_{x \to a} (x^n) + b_{n-1} \lim_{x \to a} (x^{n-1}) + \cdots + \lim_{x \to a} b_0$$

$$= b_n a^n + b_{n-1} a^{n-1} + \cdots + b_0 = f(a). \quad \square$$

Corollary (2.11)

> If q is a rational function and a is in the domain of q, then
>
> $$\lim_{x \to a} q(x) = q(a).$$

Proof   We may write $q(x) = f(x)/h(x)$ where f and h are polynomial functions. If a is in the domain of q, then $h(a) \neq 0$. Using (iii) of Theorem (2.8) and (2.10),

$$\lim_{x \to a} q(x) = \frac{\lim_{x \to a} f(x)}{\lim_{x \to a} h(x)} = \frac{f(a)}{h(a)} = q(a). \quad \square$$

Example 5   Find $\lim\limits_{x \to 3} \dfrac{5x^2 - 2x + 1}{6x - 7}$.

Solution   Applying Corollary (2.11),

$$\lim_{x \to 3} \frac{5x^2 - 2x + 1}{6x - 7} = \frac{5(3)^2 - 2(3) + 1}{6(3) - 7}$$

$$= \frac{45 - 6 + 1}{18 - 7} = \frac{40}{11}.$$ ∎

The following theorem states that for positive integral roots of x, we may determine a limit by substitution. The proof makes use of the definition of limit and may be found in Appendix II.

Theorem (2.12)

If $a > 0$ and n is a positive integer, or if $a \leq 0$ and n is an odd positive integer, then

$$\lim_{x \to a} \sqrt[n]{x} = \sqrt[n]{a}.$$

If m and n are positive integers and $a > 0$, then using (ii) of (2.9) and Theorem (2.12),

$$\lim_{x \to a} (\sqrt[n]{x})^m = \left(\lim_{x \to a} \sqrt[n]{x}\right)^m = (\sqrt[n]{a})^m.$$

In terms of rational exponents,

$$\lim_{x \to a} x^{m/n} = a^{m/n}.$$

This limit formula may be extended to negative exponents by writing $x^{-r} = 1/x^r$ and then using (iii) of (2.8).

Example 6   Find $\lim\limits_{x \to 8} \dfrac{x^{2/3} + 3\sqrt{x}}{4 - (16/x)}$.

Solution   The reader should supply reasons for each of the following steps.

$$\lim_{x \to 8} \frac{x^{2/3} + 3\sqrt{x}}{4 - (16/x)} = \frac{\lim\limits_{x \to 8} (x^{2/3} + 3\sqrt{x})}{\lim\limits_{x \to 8} (4 - (16/x))}$$

$$= \frac{\lim\limits_{x \to 8} x^{2/3} + \lim\limits_{x \to 8} 3\sqrt{x}}{\lim\limits_{x \to 8} 4 - \lim\limits_{x \to 8} (16/x)}$$

$$= \frac{8^{2/3} + 3\sqrt{8}}{4 - (16/8)}$$

$$= \frac{4 + 6\sqrt{2}}{4 - 2} = 2 + 3\sqrt{2}$$ ∎

Theorem (2.13)

> If a function f has a limit as x approaches a, then
>
> $$\lim_{x \to a} \sqrt[n]{f(x)} = \sqrt[n]{\lim_{x \to a} f(x)}$$
>
> provided either n is an odd positive integer or n is an even positive integer and $\lim_{x \to a} f(x) > 0$.

The preceding theorem will be proved in Section 2.5. In the meantime we shall use it whenever applicable to gain experience in finding limits that involve roots of algebraic expressions.

Example 7 Find $\lim_{x \to 5} \sqrt[3]{3x^2 - 4x + 9}$.

Solution Using Theorems (2.13) and (2.10),

$$\lim_{x \to 5} \sqrt[3]{3x^2 - 4x + 9} = \sqrt[3]{\lim_{x \to 5} (3x^2 - 4x + 9)}$$

$$= \sqrt[3]{75 - 20 + 9} = \sqrt[3]{64} = 4. \qquad \blacksquare$$

The beginning student should not be misled by the preceding examples. It is not always possible to find limits merely by substitution. Sometimes other devices must be employed. The next theorem concerns three functions f, h, and g, where $h(x)$ is always "sandwiched" between $f(x)$ and $g(x)$. If f and g have a common limit L as x approaches a, then as stated below, h must have the same limit.

The Sandwich Theorem (2.14)

> If $f(x) \le h(x) \le g(x)$ for all x in an open interval containing a, except possibly at a, and
>
> if $\quad \lim_{x \to a} f(x) = L = \lim_{x \to a} g(x),$
>
> then $\quad \lim_{x \to a} h(x) = L.$

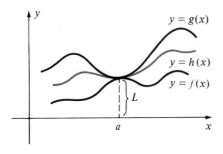

Figure 2.13

A proof of the Sandwich Theorem based on the definition of limit may be found in Appendix II. The result is also clear from geometric considerations. Specifically, if $f(x) \le h(x) \le g(x)$ for all x in an open interval containing x, then the graph of h lies "between" the graphs of f and g in that interval, as illustrated in Figure 2.13. If f and g have the same limit L as x approaches a, then evidently, h also has the limit L.

Example 8   The sine function sin has the property that $-1 \leq \sin t \leq 1$ for every real number t. Use this fact and the Sandwich Theorem (2.14) to prove that

$$\lim_{x \to 0} x \sin \frac{1}{x} = 0.$$

Solution   The limit cannot be found by substituting 0 for x, or by using an algebraic manipulation. However, since all values of the sine function are between -1 and 1 it follows that if $x \neq 0$, $|\sin (1/x)| \leq 1$ and, therefore,

$$\left| x \sin \frac{1}{x} \right| = |x| \left| \sin \frac{1}{x} \right| \leq |x|.$$

Consequently, $$0 \leq \left| x \sin \frac{1}{x} \right| \leq |x|.$$

It is not difficult to show that $\lim_{x \to 0} |x| = 0$. Hence, by the Sandwich Theorem (2.14), with $f(x) = 0$ and $g(x) = |x|$, we see that

$$\lim_{x \to 0} \left| x \sin \frac{1}{x} \right| = 0.$$

It now follows from the Definition of Limit (2.3) that

$$\lim_{x \to 0} x \sin \frac{1}{x} = 0. \qquad \blacksquare$$

EXERCISES 2.3

In Exercises 1–36 find the limits, if they exist.

1 $\lim_{x \to -2} (3x^3 - 2x + 7)$

2 $\lim_{x \to 4} (5x^2 - 9x - 8)$

3 $\lim_{x \to \sqrt{2}} (x^2 + 3)(x - 4)$

4 $\lim_{t \to -3} (3t + 4)(7t - 9)$

5 $\lim_{x \to 4} \sqrt[3]{x^2 - 5x - 4}$

6 $\lim_{x \to -2} \sqrt{x^4 - 4x + 1}$

7 $\lim_{x \to 7} 0$

8 $\lim_{x \to 1/2} \dfrac{4x^2 - 6x + 3}{16x^3 + 8x - 7}$

9 $\lim_{x \to \sqrt{2}} 15$

10 $\lim_{x \to 15} \sqrt{2}$

11 $\lim_{x \to 1/2} \dfrac{2x^2 + 5x - 3}{6x^2 - 7x + 2}$

12 $\lim_{x \to -3} \dfrac{x + 3}{(1/x) + (1/3)}$

13 $\lim_{x \to 2} \dfrac{x - 2}{x^3 - 8}$

14 $\lim_{x \to 2} \dfrac{x^2 - x - 2}{(x - 2)^2}$

15 $\lim_{x \to 16} \dfrac{x - 16}{\sqrt{x} - 4}$

16 $\lim_{x \to -2} \dfrac{x^3 + 8}{x^4 - 16}$

17 $\lim_{s \to 4} \dfrac{6s - 1}{2s - 9}$

18 $\lim_{x \to \pi} (x - 3.1416)$

19 $\lim_{x \to 1} \left(\dfrac{x^2}{x - 1} - \dfrac{1}{x - 1} \right)$

20 $\lim_{x \to 1} \left(\sqrt{x} + \dfrac{1}{\sqrt{x}} \right)^6$

21 $\lim_{x \to 16} \dfrac{2\sqrt{x} + x^{3/2}}{\sqrt[4]{x} + 5}$

22 $\lim_{x \to -8} \dfrac{16x^{2/3}}{4 - x^{4/3}}$

23 $\lim_{x \to 3} \sqrt[3]{\dfrac{2 + 5x - 3x^3}{x^2 - 1}}$

24 $\lim_{x \to \pi} \sqrt[5]{\dfrac{x - \pi}{x + \pi}}$

25 $\lim_{h \to 0} \dfrac{4 - \sqrt{16 + h}}{h}$

26 $\lim_{h \to 0} \left(\dfrac{1}{h} \right) \left(\dfrac{1}{\sqrt{1 + h}} - 1 \right)$

27 $\lim_{x \to 1} \dfrac{(x - 1)^5}{x^5 - 1}$

28 $\lim_{x \to 6} (x + 4)^3 (x - 6)^2$

29 $\lim_{v \to 3} v^2 (3v - 4)(9 - v^3)$

30 $\lim_{k \to 2} \sqrt{3k^2 + 4} \sqrt[3]{3k + 2}$

31 $\lim_{t \to -1} \dfrac{(4t^2 + 5t - 3)^3}{(6t + 5)^4}$

32 $\lim_{t \to 7} \dfrac{\sqrt[5]{3 - 5t}}{(t - 5)^3}$

33 $\lim_{x \to 9} \dfrac{x^2 - 81}{3 - \sqrt{x}}$

34 $\lim_{x \to 8} \dfrac{x - 8}{\sqrt[3]{x} - 2}$

35 $\lim\limits_{h \to 0} \dfrac{(2 + h)^{-2} - 2^{-2}}{h}$ **36** $\lim\limits_{h \to 0} \dfrac{(9 + h)^{-1} - 9^{-1}}{h}$

37 If r is any rational number and $a > 0$, prove that $\lim_{x \to a} x^r = a^r$. Under what conditions will this be true if $a < 0$?

38 If $\lim_{x \to a} f(x) = L \neq 0$ and $\lim_{x \to a} g(x) = 0$, prove that $\lim_{x \to a}[f(x)/g(x)]$ does not exist. (*Hint*: Assume there is a number M such that $\lim_{x \to a}[f(x)/g(x)] = M$ and consider

$$\lim_{x \to a} f(x) = \lim_{x \to a} \left[g(x) \cdot \frac{f(x)}{g(x)} \right].$$

39 Use the Sandwich Theorem and the fact that $\lim_{x \to 0}(|x| + 1) = 1$ to prove that $\lim_{x \to 0}(x^2 + 1) = 1$.

40 Use the Sandwich Theorem with $f(x) = 0$ and $g(x) = |x|$ to prove that

$$\lim_{x \to 0} \frac{|x|}{\sqrt{x^4 + 4x^2 + 7}} = 0.$$

41 If c is a nonnegative real number and $0 \leq f(x) \leq c$ for every x, prove that $\lim_{x \to 0} x^2 f(x) = 0$.

42 Prove that $\lim_{x \to 0} x^4 \sin(1/\sqrt[3]{x}) = 0$. (*Hint*: See Example 8.)

43 Prove that if f has a negative limit as x approaches a, then there is some open interval I containing a such that $f(x)$ is negative for every x in I except possibly $x = a$.

2.4

ONE-SIDED LIMITS

Consider $f(x) = \sqrt{x - 2}$ and the graph of f sketched in Figure 2.14. If $a > 2$, then f is defined throughout an open interval containing a and, by Theorem (2.13),

$$\lim_{x \to a} \sqrt{x - 2} = \sqrt{\lim_{x \to a} (x - 2)} = \sqrt{a - 2}.$$

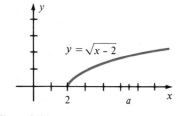

Figure 2.14

The case $a = 2$ is not covered by Definition (2.3) since there is no open interval containing 2 throughout which f is defined (note that $\sqrt{x - 2}$ is not real if $x < 2$). A natural way to extend the definition of limit to include this exceptional case is to restrict x to values *greater* than 2. Thus, we replace the condition $2 - \delta < x < 2 + \delta$, which arises from Definition (2.3), by the condition $2 < x < 2 + \delta$. The corresponding limit is called *the limit of $f(x)$ as x approaches 2 from the right*, or the *right-hand limit* of $\sqrt{x - 2}$ as x approaches 2.

Definition (2.15)

> Let f be a function that is defined on an open interval (a, c), and let L be a real number. The statement
>
> $$\lim_{x \to a^+} f(x) = L$$
>
> means that for every $\varepsilon > 0$, there exists $\delta > 0$, such that
>
> $$\text{if} \quad a < x < a + \delta, \quad \text{then} \quad |f(x) - L| < \varepsilon.$$

If $\lim_{x \to a^+} f(x) = L$, we say that **the limit of $f(x)$, as x approaches a from the right, is L.** We also refer to L as the **right-hand limit** of $f(x)$ as x approaches a. The symbol $x \to a^+$ is used to indicate that values of x are always larger than *a*. Note that the only difference between Definitions (2.15) and (2.3) is that for *right*-hand limits we restrict x to the *right* half $(a, a + \delta)$ of the interval $(a - \delta, a + \delta)$. Definition (2.15) is illustrated geometrically in (i) of Figure 2.15. Intuitively, we think of $f(x)$ getting close to L as x gets close to a, through values *larger* than a.

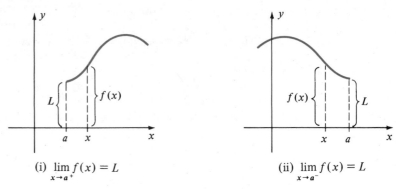

(i) $\lim\limits_{x \to a^+} f(x) = L$ (ii) $\lim\limits_{x \to a^-} f(x) = L$

Figure 2.15

The notion of *left-hand limit* is defined in similar fashion. For example, if $f(x) = \sqrt{2 - x}$, then we restrict x to values *less* than 2. The general definition follows.

Definition (2.16)

> Let f be a function that is defined on an open interval (c, a), and let L be a real number. The statement
>
> $$\lim_{x \to a^-} f(x) = L$$
>
> means that for every $\varepsilon > 0$, there exists a $\delta > 0$, such that
>
> $$\text{if} \quad a - \delta < x < a, \quad \text{then} \quad |f(x) - L| < \varepsilon.$$

If $\lim_{x \to a^-} f(x) = L$, we say that **the limit of $f(x)$ as x approaches a from the left, is L**; or that L is the **left-hand limit** of $f(x)$ as x approaches a. The symbol $x \to a^-$ is used to indicate that x is restricted to values *less* than a. A geometric illustration of Definition (2.16) is given in (ii) of Figure 2.15. Note that for the *left*-hand limit, x is in the *left* half $(a - \delta, a)$ of the interval $(a - \delta, a + \delta)$.

Sometimes Definitions (2.15) and (2.16) are referred to as **one-sided limits** of $f(x)$ as x approaches a. The relation between one-sided limits and limits is stated in the next theorem. The proof is left as an exercise.

Theorem (2.17)

> If f is defined throughout an open interval containing a, except possibly at a itself, then $\lim_{x \to a} f(x) = L$ if and only if both $\lim_{x \to a^-} f(x) = L$ and $\lim_{x \to a^+} f(x) = L$.

The preceding theorem tells us that the limit of $f(x)$ as x approaches a exists if and only if both the right- and left-hand limits exist and are equal.

Theorems similar to the limit theorems of the previous section can be proved for one-sided limits. For example,

$$\lim_{x \to a^+} [f(x) + g(x)] = \lim_{x \to a^+} f(x) + \lim_{x \to a^+} g(x)$$

and

$$\lim_{x \to a^+} \sqrt[n]{f(x)} = \sqrt[n]{\lim_{x \to a^+} f(x)}$$

with the usual restrictions on the existence of limits and nth roots. Analogous results are true for left-hand limits.

Example 1 Find $\lim_{x \to 2^+} (1 + \sqrt{x - 2})$.

Solution Using (one-sided) limit theorems,

$$\lim_{x \to 2^+} (1 + \sqrt{x - 2}) = \lim_{x \to 2^+} 1 + \lim_{x \to 2^+} \sqrt{x - 2}$$
$$= 1 + \sqrt{\lim_{x \to 2^+} (x - 2)}$$
$$= 1 + 0 = 1.$$

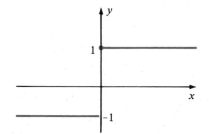

$f(x) = 1 + \sqrt{x - 2}$

Figure 2.16

The graph of $f(x) = 1 + \sqrt{x - 2}$ is sketched in Figure 2.16. Note that there is no left-hand limit, since $\sqrt{x - 2}$ is not a real number if $x < 2$. ∎

Example 2 Suppose $f(x) = |x|/x$ if $x \neq 0$ and $f(0) = 1$. Find $\lim_{x \to 0^+} f(x)$ and $\lim_{x \to 0^-} f(x)$. What is $\lim_{x \to 0} f(x)$?

Solution If $x > 0$, then $|x| = x$ and $f(x) = x/x = 1$. Consequently,

$$\lim_{x \to 0^+} f(x) = \lim_{x \to 0^+} 1 = 1.$$

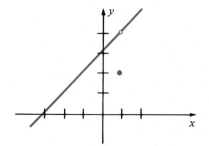

Figure 2.17

If $x < 0$, then $|x| = -x$ and $f(x) = -x/x = -1$. Therefore,

$$\lim_{x \to 0^-} f(x) = \lim_{x \to 0^-} (-1) = -1.$$

Since these right- and left-hand limits are unequal, it follows from Theorem (2.17) that $\lim_{x \to 0} f(x)$ does not exist. The graph of f is sketched in Figure 2.17. ∎

Example 3 Suppose $f(x) = x + 3$ if $x \neq 1$ and $f(1) = 2$. Find $\lim_{x \to 1^-} f(x)$, $\lim_{x \to 1^+} f(x)$, and $\lim_{x \to 1} f(x)$.

Solution The graph of f consists of the point $P(1, 2)$ and all points on the line $y = x + 3$ except the point with coordinates $(1, 4)$ as shown in Figure 2.18. Evidently, $\lim_{x \to 1^+} f(x) = 4 = \lim_{x \to 1^-} f(x)$. Hence by Theorem (2.17), $\lim_{x \to 1} f(x) = 4$. Note that $\lim_{x \to 1} f(x) \neq f(1)$. ∎

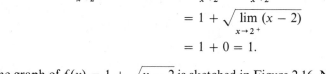

Figure 2.18

EXERCISES 2.4

In Exercises 1–22 find the limits, if they exist

1 (a) $\lim\limits_{x \to 5^-} \sqrt{5 - x}$

(b) $\lim\limits_{x \to 5^+} \sqrt{5 - x}$

(c) $\lim\limits_{x \to 5} \sqrt{5 - x}$

2 (a) $\lim\limits_{x \to 2^-} \sqrt{8 - x^3}$

(b) $\lim\limits_{x \to 2^+} \sqrt{8 - x^3}$

(c) $\lim\limits_{x \to 2} \sqrt{8 - x^3}$

3 (a) $\lim\limits_{x \to 1^-} \sqrt[3]{x^3 - 1}$

(b) $\lim\limits_{x \to 1^+} \sqrt[3]{x^3 - 1}$

(c) $\lim\limits_{x \to 1} \sqrt[3]{x^3 - 1}$

4 (a) $\lim\limits_{x \to -8^+} x^{2/3}$

(b) $\lim\limits_{x \to -8^-} x^{2/3}$

(c) $\lim\limits_{x \to -8} x^{2/3}$

5 (a) $\lim\limits_{x \to 4^-} \dfrac{|x - 4|}{x - 4}$

(b) $\lim\limits_{x \to 4^+} \dfrac{|x - 4|}{x - 4}$

(c) $\lim\limits_{x \to 4} \dfrac{|x - 4|}{x - 4}$

6 (a) $\lim\limits_{x \to -5^+} \dfrac{x + 5}{|x + 5|}$

(b) $\lim\limits_{x \to -5^-} \dfrac{x + 5}{|x + 5|}$

(c) $\lim\limits_{x \to -5} \dfrac{x + 5}{|x + 5|}$

7 $\lim\limits_{x \to 0^+} (4 + \sqrt{x})$

8 $\lim\limits_{x \to 0^+} (4x^{3/2} - \sqrt{x} + 3)$

9 $\lim\limits_{x \to -6^+} (\sqrt{x + 6} + x)$

10 $\lim\limits_{x \to 5/2^-} (\sqrt{5 - 2x} - x^2)$

11 $\lim\limits_{x \to 5^+} (\sqrt{x^2 - 25} + 3)$

12 $\lim\limits_{x \to 3^-} x\sqrt{9 - x^2}$

13 $\lim\limits_{x \to 2^+} \dfrac{4 - x^2}{2 - x}$

14 $\lim\limits_{x \to -4^+} \dfrac{2x^2 + 5x - 12}{x^2 + 3x - 4}$

15 $\lim\limits_{x \to 3^+} \dfrac{\sqrt{(x - 3)^2}}{x - 3}$

16 $\lim\limits_{x \to -10^-} \dfrac{x + 10}{\sqrt{(x + 10)^2}}$

17 $\lim\limits_{x \to 5^+} \dfrac{1 + \sqrt{2x - 10}}{x + 3}$

18 $\lim\limits_{x \to 4^+} \dfrac{\sqrt[4]{x^2 - 16}}{x + 4}$

19 $\lim\limits_{x \to -7^+} \dfrac{x + 7}{|x + 7|}$

20 $\lim\limits_{x \to \pi^-} \dfrac{|\pi - x|}{x - \pi}$

21 $\lim\limits_{x \to 0^+} \dfrac{1}{x}$

22 $\lim\limits_{x \to 8^-} \dfrac{1}{x - 8}$

For each f defined in Exercises 23 and 24, find $\lim\limits_{x \to 2^-} f(x)$, $\lim\limits_{x \to 2^+} f(x)$, and sketch the graph of f.

23 $f(x) = \begin{cases} 3x & \text{if } x \le 2 \\ x^2 & \text{if } x > 2 \end{cases}$

24 $f(x) = \begin{cases} x^3 & \text{if } x \le 2 \\ 4 - 2x & \text{if } x > 2 \end{cases}$

In Exercises 25 and 26 find $\lim\limits_{x \to -3^+} f(x)$, $\lim\limits_{x \to -3^-} f(x)$, and $\lim\limits_{x \to -3} f(x)$, if they exist.

25 $f(x) = \begin{cases} 1/(2 - 3x) & \text{if } x < -3 \\ \sqrt[3]{x + 2} & \text{if } x \ge -3 \end{cases}$

26 $f(x) = \begin{cases} 9/x^2 & \text{if } x \le -3 \\ 4 + x & \text{if } x > -3 \end{cases}$

In Exercises 27–32 sketch the graph of f and for the indicated value of a find (a) $\lim\limits_{x \to a^-} f(x)$; (b) $\lim\limits_{x \to a^+} f(x)$; and (c) $\lim\limits_{x \to a} f(x)$, provided the limits exist.

27 $a = 1$; $f(x) = \begin{cases} x^2 & \text{if } x < 1 \\ 2 & \text{if } x = 1 \\ 4 - x^2 & \text{if } x > 1 \end{cases}$

28 $a = -1$; $f(x) = \begin{cases} 1 - x & \text{if } x < -1 \\ 1 & \text{if } x = -1 \\ x^3 & \text{if } x > -1 \end{cases}$

29 $a = 2$; $f(x) = \begin{cases} \dfrac{x^2 - 4}{x - 2} & \text{if } x \ne 2 \\ 3 & \text{if } x = 2 \end{cases}$

30 $a = 3$; $f(x) = \begin{cases} \dfrac{6 - 2x}{\sqrt{(x - 3)^2}} & \text{if } x \ne 3 \\ 1 & \text{if } x = 3 \end{cases}$

31 $a = 0$; $f(x) = \begin{cases} \dfrac{x^4 + x}{x} & \text{if } x \ne 0 \\ 2 & \text{if } x = 0 \end{cases}$

32 $a = 1$; $f(x) = \begin{cases} \dfrac{4 - 4x - x^2 + x^3}{1 - x} & \text{if } x \ne 1 \\ 4 & \text{if } x = 1 \end{cases}$

In Exercises 33–35, n denotes an arbitrary integer. For each function f, sketch the graph of f and find $\lim\limits_{x \to n^-} f(x)$ and $\lim\limits_{x \to n^+} f(x)$.

33 $f(x) = (-1)^n$ if $n \le x < n + 1$

34 $f(x) = \begin{cases} 0 & \text{if } x = n \\ 1 & \text{if } x \ne n \end{cases}$

35 $f(x) = \begin{cases} x & \text{if } x = n \\ 0 & \text{if } x \ne n \end{cases}$

In Exercises 36 and 37, $[\![\ ]\!]$ denotes the greatest integer function and n is an arbitary integer.

36 Find $\lim\limits_{x \to n^-} [\![x]\!]$ and $\lim\limits_{x \to n^+} [\![x]\!]$.

37 If $f(x) = x - [\![x]\!]$, find $\lim\limits_{x \to n^-} f(x)$ and $\lim\limits_{x \to n^+} f(x)$.

38 If f is a polynomial function, prove that $\lim\limits_{x \to a^+} f(x) = f(a)$ and $\lim\limits_{x \to a^-} f(x) = f(a)$ for every real number a.

39 Prove Theorem (2.17).

40 Sketch geometric interpretations for right-hand limits that are analogous to those in Figures 2.5 and 2.6. Do the same for left-hand limits.

2.5

CONTINUOUS FUNCTIONS

In arriving at the definition of $\lim_{x \to a} f(x)$ we emphasized the restriction $x \neq a$. Several examples in preceding sections have brought out the fact that $\lim_{x \to a} f(x)$ may exist even though f is undefined at a. Let us now turn our attention to the case in which a is in the domain of f. If f is defined at a and $\lim_{x \to a} f(x)$ exists, then this limit may, or may not, equal $f(a)$. If $\lim_{x \to a} f(x) = f(a)$ then f is said to be *continuous* at a according to the next definition.

Definition (2.18)

> A function f is **continuous** at a number a if the following three conditions are satisfied.
>
> (i)　f is defined on an open interval containing a.
>
> (ii)　$\lim_{x \to a} f(x)$ exists.
>
> (iii)　$\lim_{x \to a} f(x) = f(a)$.

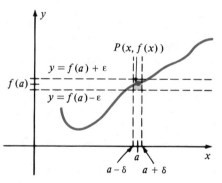

Figure 2.19

If f is not continuous at a, then we say it is **discontinuous** at a, or has a **discontinuity** at a.

If f is continuous at a, then by (i) of Definition (2.18) there is a point $(a, f(a))$ on the graph of f. Moreover, since $\lim_{x \to a} f(x) = f(a)$, the closer x is to a, the closer $f(x)$ is to $f(a)$ or, in geometric terms, the closer the point $(x, f(x))$ on the graph of f is to the point $(a, f(a))$. More precisely, as illustrated in Figure 2.19, for every pair of horizontal lines $y = f(a) \pm \varepsilon$, there exist vertical lines $x = a \pm \delta$, such that if $a - \delta < x < a + \delta$, then the point $(x, f(x))$ on the graph of f lies within the shaded rectangular region.

Functions that are continuous at every number in a given interval are sometimes thought of as functions whose graphs can be sketched without lifting the pencil from the paper; that is, there are no breaks in the graph. Another interpretation of a continuous function f is that a small change in x produces only a small change in the functional value $f(x)$. These are not accurate descriptions, but rather devices to help develop an intuitive feeling for continuous functions.

Example 1

(a)　Prove that a polynomial function is continuous at every real number a.

(b)　Prove that a rational function is continuous at every real number in its domain.

Solution

(a)　A polynomial function f is defined throughout $\mathbb{R}$. Moreover, by Theorem (2.10), $\lim_{x \to a} f(x) = f(a)$ for every real number a. Thus f satisfies conditions (i)–(iii) of Definition (2.18) and hence is continuous at a.

(b)   If q is a rational function, then $q = f/h$, where f and h are polynomial functions. Consequently q is defined for all real numbers *except* the zeros of h. It follows that if $h(a) \neq 0$, then q is defined throughout an open interval containing a. Moreover, by (2.11), $\lim_{x \to a} q(x) = q(a)$. Applying Definition (2.18), q is continuous at a.   ∎

Since the notion of continuity involves the fact that $\lim_{x \to a} f(x) = f(a)$, the following result may be obtained by replacing L in Definition (2.3) by $f(a)$.

Theorem (2.19)

> If a function f is defined on an open interval containing a, then f is continuous at a if for every $\varepsilon > 0$, there exists a $\delta > 0$, such that
>
> $$\text{if} \quad |x - a| < \delta, \quad \text{then} \quad |f(x) - f(a)| < \varepsilon.$$

Note that we do not require $0 < |x - a|$ in Theorem (2.19), since f is defined at a and, moreover, if $x = a$, then

$$|f(x) - f(a)| = |f(a) - f(a)| = 0 < \varepsilon.$$

Graphs of several functions that are *not* continuous at a are sketched in Figure 2.20. The function having the graph illustrated in (i) fails to be continuous since f is undefined at a. The functions whose graphs are sketched in (ii) and (iii) are discontinuous at a since $\lim_{x \to a} f(x)$ does not exist. For the function with a graph as in (iv), both $f(a)$ and $\lim_{x \to a} f(x)$ exist but they are unequal and hence f is discontinuous at a by (iii) of Definition (2.18). The last illustration shows the necessity for checking all three conditions of the definition.

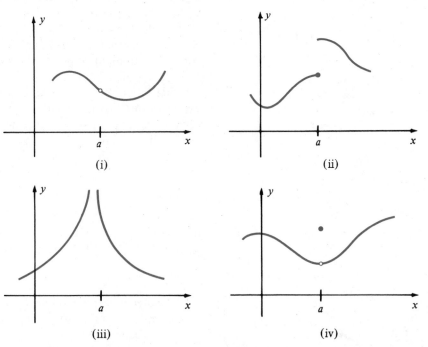

Figure 2.20

As a specific illustration of (iv) in Figure 2.20, consider the function f of Example 3 in Section 2.4, where $f(x) = x + 3$ if $x \neq 1$ and $f(1) = 2$ (see Figure 2.18). Since

$$\lim_{x \to 1} f(x) = 4 \neq f(1),$$

condition (iii) of Definition (2.18) is not satisfied, and hence f has a discontinuity at $x = 1$.

If $f(x) = 1/x$, then f has a discontinuity at $x = 0$. In this case *none* of the conditions of Definition (2.18) is satisfied (see Figure 2.10).

The functions whose graphs are sketched in Figure 2.20 appear to be continuous at numbers other than a. Most functions considered in calculus are of this type; that is, they may be discontinuous at certain numbers of their domains and continuous elsewhere.

If a function f is continuous at every number in an open interval (a, b), we say that **f is continuous on the interval (a, b)**. Similarly, a function is said to be continuous on an infinite interval of the form (a, ∞) or $(-\infty, b)$ if it is continuous at every number in the interval. The next definition covers the case of a closed interval.

Definition (2.20)

> Let a function f be defined on a closed interval $[a, b]$. The **function f is continuous on $[a, b]$** if it is continuous on (a, b) and if, in addition,
>
> $$\lim_{x \to a^+} f(x) = f(a) \quad \text{and} \quad \lim_{x \to b^-} f(x) = f(b).$$

If a function f has either a right-hand or left-hand limit of the type indicated in Definition (2.20) we say that **f is continuous from the right at a** or that **f is continuous from the left at b**, respectively.

Example 2 If $f(x) = \sqrt{9 - x^2}$, sketch the graph of f and prove that f is continuous on the closed interval $[-3, 3]$.

Solution By equation (1.13), the graph of $x^2 + y^2 = 9$ or equivalently $y^2 = 9 - x^2$ is a circle with center at the origin and radius 3. It follows that the graph of $y = \sqrt{9 - x^2}$ and, therefore, the graph of f is the upper half of that circle (see Figure 2.21).

If $-3 < c < 3$ then, using Theorem (2.13),

$$\lim_{x \to c} f(x) = \lim_{x \to c} \sqrt{9 - x^2} = \sqrt{9 - c^2} = f(c).$$

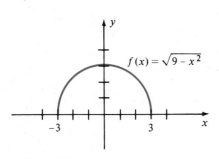

$f(x) = \sqrt{9 - x^2}$

Figure 2.21

Hence, by Definition (2.18), f is continuous at c.

According to Definition (2.20), all that remains is to check the endpoints of the interval using one-sided limits. Since

$$\lim_{x \to -3^+} f(x) = \lim_{x \to -3^+} \sqrt{9 - x^2} = \sqrt{9 - 9} = 0 = f(-3),$$

f is continuous from the right at -3. We also have

$$\lim_{x \to 3^-} f(x) = \lim_{x \to 3^-} \sqrt{9 - x^2} = \sqrt{9 - 9} = 0 = f(3)$$

and hence f is continuous from the left at 3. This completes the proof that f is continuous on $[-3, 3]$. ∎

It should be evident how to define continuity on other types of intervals. For example, a function f is said to be continuous on $[a, b)$ or $[a, \infty)$ if it is continuous at every number greater than a in the interval and if, in addition, f is continuous from the right at a. For intervals of the form $(a, b]$ or $(-\infty, b]$ we require continuity at every number less than b in the interval and also continuity from the left at b.

As illustrated in the next example, when asked to discuss the continuity of a function f we shall list the largest intervals on which f is continuous. Of course f will also be continuous on any subinterval of those intervals.

Example 3 Discuss the continuity of f if $f(x) = \sqrt{x^2 - 9}/(x - 4)$.

Solution   The function is undefined if the denominator $x - 4$ is zero (that is, if $x = 4$) or if the radicand $x^2 - 9$ is negative (that is, if $-3 < x < 3$). Any other real number is in one of the intervals $(-\infty, -3]$, $[3, 4)$, or $(4, \infty)$. The proof that f is continuous on each of these intervals is similar to that given in the solution of Example 2. Thus to prove continuity on $[3, 4)$ it is necessary to show that

$$\lim_{x \to c} f(x) = f(c) \quad \text{if } 3 < c < 4$$

and also that

$$\lim_{x \to 3^+} f(x) = f(3).$$

We shall leave the details of the proof for this, and the other intervals, to the reader. ∎

Limit theorems discussed in Section 2.3 may be used to establish the following important theorem.

Theorem (2.21)

> If the functions f and g are continuous at a, then so are the sum $f + g$, the difference $f - g$, the product fg, and, if $g(a) \neq 0$, the quotient f/g.

Proof   If f and g are continuous at a, then $\lim_{x \to a} f(x) = f(a)$ and $\lim_{x \to a} g(x) = g(a)$. By the definition of sum $(f + g)(x) = f(x) + g(x)$. Consequently,

$$\lim_{x \to a} (f + g)(x) = \lim_{x \to a} [f(x) + g(x)]$$
$$= \lim_{x \to a} f(x) + \lim_{x \to a} g(x)$$
$$= f(a) + g(a)$$
$$= (f + g)(a).$$

This proves that $f + g$ is continuous at a. The remainder of the theorem is proved in similar fashion. □

If f and g are continuous on an interval I it follows that $f + g$, $f - g$, and fg are continuous on I. If, in addition, $g(a) \neq 0$ throughout I, then f/g is continuous on I. These results may be extended to more than two functions; that is, sums, differences, products, or quotients involving any number of continuous functions are continuous (provided zero denominators do not occur).

The next result on limits of composite functions has many applications.

Theorem (2.22)

If f and g are functions such that $\lim_{x \to a} g(x) = b$, and if f is continuous at b, then

$$\lim_{x \to a} f(g(x)) = f(b) = f\left(\lim_{x \to a} g(x)\right).$$

Proof   As was pointed out in Chapter 1 (see Figure 1.41), the composite function $f(g(x))$ may be represented geometrically by means of three real lines l, l', and l'' as shown in Figure 2.22, where to each coordinate x on l there corresponds the coordinate $g(x)$ on l' and then, in turn, $f(g(x))$ on l''. We wish to prove that $f(g(x))$ has the limit $f(b)$ as x approaches a. In terms of Definition (2.3) we must show that for every $\varepsilon > 0$ there exists a $\delta > 0$ such that

(i)   if $0 < |x - a| < \delta$,   then $|f(g(x)) - f(b)| < \varepsilon$.

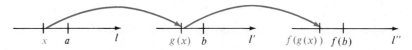

Figure 2.22

Let us begin by considering the interval $(f(b) - \varepsilon, f(b) + \varepsilon)$ on l'' shown in color in Figure 2.23. Since f is continuous at b, $\lim_{z \to b} f(z) = f(b)$ and hence, as illustrated in the figure, there exists a number $\delta_1 > 0$ such that

(ii)   if $|z - b| < \delta_1$,   then $|f(z) - f(b)| < \varepsilon$.

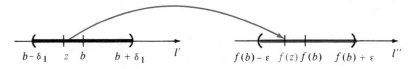

Figure 2.23

In particular, if we let $z = g(x)$ in (ii) it follows that

(iii)   if $|g(x) - b| < \delta_1$,   then $|f(g(x)) - f(b)| < \varepsilon$.

Next, turning our attention to the interval $(b - \delta_1, b + \delta_1)$ on l' and using the definition of $\lim_{x \to a} g(x) = b$, we obtain the fact illustrated in Figure 2.24, that there exists a $\delta > 0$ such that

(iv)   if $0 < |x - a| < \delta$,   then $|g(x) - b| < \delta_1$.

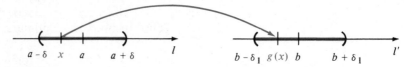

$a - \delta \quad x \quad a \quad a + \delta \qquad l \qquad\qquad b - \delta_1 \quad g(x) \; b \qquad b + \delta_1 \qquad l'$

Figure 2.24

Finally, combining (iv) and (iii) we see that

if $0 < |x - a| < \delta$ then $|f(g(x)) - f(b)| < \varepsilon$

which is the desired conclusion (i).   □

The principal use of Theorem (2.22) is to establish other theorems. To illustrate, if n is a positive integer and $f(x) = \sqrt[n]{x}$, then

$$f(g(x)) = \sqrt[n]{g(x)}$$

and

$$f\left(\lim_{x \to a} g(x)\right) = \sqrt[n]{\lim_{x \to a} g(x)}.$$

If we now use the fact that

$$\lim_{x \to a} f(g(x)) = f\left(\lim_{x \to a} g(x)\right)$$

the result stated in Theorem (2.13) is obtained, that is,

$$\lim_{x \to a} \sqrt[n]{g(x)} = \sqrt[n]{\lim_{x \to a} g(x)},$$

where it is assumed that the indicated nth roots exist.

The next result follows directly from Theorem 2.22.

Theorem (2.23)

> If g is continuous at a and f is continuous at $b = g(a)$, then
>
> $$\lim_{x \to a} f(g(x)) = f\left(\lim_{x \to a} g(x)\right) = f(g(a)).$$

The preceding theorem states that the composite function of f by g is continuous at a. This result may be extended to functions that are continuous on intervals. Sometimes this is expressed by the statement "the composite function of a continuous function by a continuous function is continuous."

Example 4   If $f(x) = |x|$, prove that f is continuous at every real number a.

Solution   Since $|x| = \sqrt{x^2}$ we have, by (2.13) and (2.9),

$$\lim_{x \to a} f(x) = \lim_{x \to a} |x| = \lim_{x \to a} \sqrt{x^2}$$
$$= \sqrt{\lim_{x \to a} x^2} = \sqrt{a^2} = |a| = f(a).$$

Hence, from Definition (2.18), f is continuous at a.   ■

A proof of the following important property of continuous functions may be found in more advanced texts on calculus.

The Intermediate Value Theorem
(2.24)

If a function f is continuous on a closed interval $[a, b]$ and if $f(a) \neq f(b)$, then f takes on every value between $f(a)$ and $f(b)$ in the interval $[a, b]$.

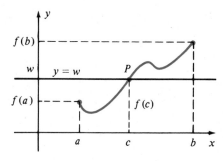

Figure 2.25

Theorem (2.24) states that if w is any number between $f(a)$ and $f(b)$, then there is a number c between a and b such that $f(c) = w$. If the graph of the continuous function f is regarded as extending in an unbroken manner from the point $(a, f(a))$ to the point $(b, f(b))$, as illustrated in Figure 2.25, then for any number w between $f(a)$ and $f(b)$ it appears that a horizontal line with y-intercept w should intersect the graph in at least one point P. The x-coordinate c of P is a number such that $f(c) = w$.

Example 5   Verify the Intermediate Value Theorem (2.24) if $f(x) = \sqrt{x + 1}$ and the interval is $[3, 24]$.

Solution   The function f is continuous on $[3, 24]$ and $f(3) = 2$, $f(24) = 5$. If w is any real number between 2 and 5 we must find a number c in the interval $[3, 24]$ such that $f(c) = w$, that is, $\sqrt{c + 1} = w$. Squaring both sides of the last equation and solving for c we obtain $c = w^2 - 1$. This number c is in the interval $[3, 24]$, since if $2 < w < 5$, then

$$4 < w^2 < 25, \quad \text{or} \quad 3 < w^2 - 1 < 24.$$

To check our work we see that

$$f(c) = f(w^2 - 1) = \sqrt{(w^2 - 1) + 1} = w.$$ ■

A corollary of Theorem (2.24) is that if $f(a)$ and $f(b)$ have opposite signs, then there is a number c between a and b such that $f(c) = 0$; that is, f has a zero at c. Geometrically, this implies that if the point $(a, f(a))$ on the graph of a continuous function lies below the x-axis, and the point $(b, f(b))$ lies above the x-axis, or vice versa, then the graph crosses the x-axis at some point $(c, 0)$, where $a < c < b$.

A useful consequence of the Intermediate Value Theorem is the following, where the interval referred to may be either closed, open, half-open, or infinite.

Theorem (2.25)

> If a function f is continuous and has no zeros on an interval, then either $f(x) > 0$ or $f(x) < 0$ for all x in the interval.

Proof   The conclusion of the theorem states that under the given hypothesis, $f(x)$ has the same sign throughout the interval. If this conclusion were *false*, then there would exist numbers x_1 and x_2 in the interval such that $f(x_1) > 0$ and $f(x_2) < 0$. By our preceding remarks this, in turn, would imply that $f(c) = 0$ for some number c between x_1 and x_2, contrary to hypothesis. Thus, the conclusion must be true.   □

EXERCISES 2.5

In Exercises 1–12 show that the function f is continuous at the number a.

1 $f(x) = \sqrt{2x - 5} + 3x, a = 4$

2 $f(x) = 3x^2 + 7 - 1/\sqrt{-x}, a = -2$

3 $f(x) = x/(x^2 - 4), a = 3$

4 $f(x) = 1/x, a = 10^{-6}$

5 $f(x) = \sqrt[3]{x^2 + 2}, a = -5$

6 $f(x) = \sqrt[3]{x}/(2x + 1), a = 8$

In Exercises 7–10, show that f is continuous on the indicated interval.

7 $f(x) = \sqrt{x - 4}; [4, 8]$

8 $f(x) = \sqrt{16 - x}; (-\infty, 16]$

9 $f(x) = \dfrac{1}{x^2}; (0, \infty)$

10 $f(x) = \dfrac{1}{x - 1}; (1, 3)$

In Exercises 11–22 find all numbers for which the function f is continuous.

11 $f(x) = \dfrac{3x - 5}{2x^2 - x - 3}$

12 $f(x) = \dfrac{x^2 - 9}{x - 3}$

13 $f(x) = \sqrt{2x - 3} + x^2$

14 $f(x) = \dfrac{x}{\sqrt[3]{x - 4}}$

15 $f(x) = \dfrac{x - 1}{\sqrt{x^2 - 1}}$

16 $f(x) = \dfrac{x}{\sqrt{1 - x^2}}$

17 $f(x) = \dfrac{|x + 9|}{x + 9}$

18 $f(x) = \dfrac{x}{x^2 + 1}$

19 $f(x) = \dfrac{5}{x^3 - x^2}$

20 $f(x) = \dfrac{4x - 7}{(x + 3)(x^2 + 2x - 8)}$

21 $f(x) = \dfrac{\sqrt{x^2 - 9}\sqrt{25 - x^2}}{x - 4}$

22 $f(x) = \dfrac{\sqrt{9 - x}}{\sqrt{x - 6}}$

23–28   Find the discontinuities of the functions defined in Exercises 27–32 of Section 2.4.

29 Suppose $f(x) = \begin{cases} cx^2 - 3 & \text{if } x \leq 2 \\ cx + 2 & \text{if } x > 2 \end{cases}$.

Find a value of c such that f is continuous on $\mathbb{R}$.

30 Suppose $f(x) = \begin{cases} c^2 x & \text{if } x < 1 \\ 3cx - 2 & \text{if } x \geq 1 \end{cases}$.

Determine all values of c such that f is continuous on $\mathbb{R}$.

31 Suppose $f(x) = \begin{cases} c & \text{if } x = -3 \\ \dfrac{9 - x^2}{4 - \sqrt{x^2 + 7}} & \text{if } |x| < 3 \\ d & \text{if } x = 3 \end{cases}$.

Find values of c and d such that f is continuous on $[-3, 3]$.

32 Suppose $f(x) = \begin{cases} 4x & \text{if } x \leq -1 \\ cx + d & \text{if } -1 < x \leq 2 \\ -5x & \text{if } x \geq 2 \end{cases}$.

Find values of c and d such that f is continuous on $\mathbb{R}$.

33 Suppose

$$f(x) = x^2 \quad \text{and} \quad g(x) = \begin{cases} -4 & \text{if } x \leq 0 \\ |x - 4| & \text{if } x > 0 \end{cases}.$$

Determine whether the composite functions $f \circ g$ and $g \circ f$ are continuous at 0.

34 Let $f(x) = (x - [\![x]\!])^2$ where $[\![\ ]\!]$ denotes the greatest integer function. If n is any integer, show that (a) f is continuous on the interval $[n, n + 1)$; and (b) f is not continuous on $[n, n + 1]$. Sketch the graph of f.

35 Prove that if $f(x) = 1/x$, then f is continuous on every open interval that does not contain the origin. What is true for open intervals containing the origin?

In Exercises 36 and 37, is f continuous at 3? Explain.

36 $f(x) = \begin{cases} 1 & \text{if } x \neq 3 \\ 0 & \text{if } x = 3 \end{cases}$

37 $f(x) = \begin{cases} \dfrac{|x - 3|}{x - 3} & \text{if } x \neq 3 \\ 1 & \text{if } x = 3 \end{cases}$

38 Suppose $f(x) = \begin{cases} \dfrac{1 - x^2}{1 + x} & \text{if } x \neq -1 \\ 1 & \text{if } x = -1 \end{cases}$.

Is f continuous at -1? Explain.

39 Suppose $f(x) = \begin{cases} \dfrac{2x^2 - x - 6}{x^2 - 3x + 2} & \text{if } x \neq 2 \\ 7 & \text{if } x = 2 \end{cases}$.

Is f continuous at $x = 2$? Explain.

40 Suppose $f(x) = 0$ if x is rational and $f(x) = 1$ if x is irrational. Prove that f is discontinuous at every real number a.

41 Prove that a function f is continuous at a if and only if $\lim_{h \to 0} f(a + h) = f(a)$.

42 If f is continuous on an interval containing c, and if $f(c) > 0$, prove that $f(x)$ is positive throughout an interval containing c. (*Hint*: See Theorem (2.17).)

In Exercises 43–46, verify the Intermediate Value Theorem (2.24) for f on the stated interval $[a, b]$ by showing that if w is any number between $f(a)$ and $f(b)$, then there is a number c in $[a, b]$ such that $f(c) = w$.

43 $f(x) = x^3 + 1; [-1, 2]$

44 $f(x) = -x^3; [0, 2]$

45 $f(x) = x^2 + 4x + 4; [0, 1]$

46 $f(x) = x^2 - x; [-1, 3]$

47 If $f(x) = x^3 - 5x^2 + 7x - 9$, use the Intermediate Value Theorem (2.24) to prove that there is a real number a such that $f(a) = 100$.

48 Prove that the equation $x^5 - 3x^4 - 2x^3 - x + 1 = 0$ has a solution between 0 and 1.

2.6

REVIEW

Define or discuss each of the following.

1 The limit of a function as x approaches a

2 Geometric interpretations of $\lim_{x \to a} f(x) = L$

3 Theorems on limits

4 Limits of polynomial and rational functions

5 The Sandwich Theorem

6 Right- and left-hand limits of functions

7 Continuous function

8 Discontinuities of a function

9 Continuity on an interval

10 The Intermediate Value Theorem

EXERCISES 2.6

In Exercises 1–20 find the limit, if it exists.

1 $\lim_{x \to 3} \dfrac{5x + 11}{\sqrt{x + 1}}$

2 $\lim_{x \to -2} \dfrac{6 - 7x}{(3 + 2x)^4}$

3 $\lim_{x \to -2} (2x - \sqrt{4x^2 + x})$

4 $\lim_{x \to 4^-} (x - \sqrt{16 - x^2})$

5 $\lim_{x \to 3/2} \dfrac{2x^2 + x - 6}{4x^2 - 4x - 3}$

6 $\lim_{x \to 2} \dfrac{3x^2 - x - 10}{x^2 - x - 2}$

7 $\lim_{x \to 2} \dfrac{x^4 - 16}{x^2 - x - 2}$

8 $\lim_{x \to 3^+} \dfrac{1}{x - 3}$

9 $\lim_{x \to 0^+} \dfrac{1}{\sqrt{x}}$

10 $\lim_{x \to 5} \dfrac{(1/x) - (1/5)}{x - 5}$

11 $\lim_{x \to 1/2} \dfrac{8x^3 - 1}{2x - 1}$

12 $\lim_{x \to 2} 5$

13 $\lim_{x \to 3^+} \dfrac{3 - x}{|3 - x|}$

14 $\lim_{x \to 2} \dfrac{\sqrt{x} - \sqrt{2}}{x - 2}$

15 $\lim_{h \to 0} \dfrac{(a + h)^4 - a^4}{h}$

16 $\lim_{x \to -3} \sqrt[3]{\dfrac{x + 3}{x^3 + 27}}$

17 $\lim_{h \to 0} \dfrac{(2 + h)^{-3} - 2^{-3}}{h}$

18 $\lim_{x \to 5} (x^2 + 3)^0$

19 $\lim_{x \to 2^+} \dfrac{\sqrt{(x - 2)^2}}{2 - x}$

20 $\lim_{x \to 1} \dfrac{x - 1}{|x - 1|}$

Find the limits in Exercises 21 and 22, where $[\![\ ]\!]$ denotes the greatest integer function.

21 $\lim_{x \to 3^+} ([\![x]\!] - x^2)$

22 $\lim_{x \to 3^-} ([\![x]\!] - x^2)$

23 Prove, directly from the definition of limit, that $\lim_{x \to 6} (5x - 21) = 9$.

24 Suppose $f(x) = 1$ if x is rational and $f(x) = -1$ if x is irrational. Prove that $\lim_{x \to a} f(x)$ does not exist for any real number a.

In Exercises 25–28, find all numbers for which f is continuous.

25 $f(x) = 2x^4 - \sqrt[3]{x} + 1$

26 $f(x) = \sqrt{(2 + x)(3 - x)}$

27 $f(x) = \dfrac{\sqrt{9 - x^2}}{x^4 - 16}$

28 $f(x) = \dfrac{\sqrt{x}}{x^2 - 1}$

In Exercises 29–32 find the discontinuities of f.

29 $f(x) = \dfrac{|x^2 - 16|}{x^2 - 16}$

30 $f(x) = \dfrac{1}{x^2 - 16}$

31 $f(x) = \dfrac{x^2 - x - 2}{x^2 - 2x}$

32 $f(x) = \dfrac{x + 2}{x^3 - 8}$

33 If $f(x) = 1/x^2$, verify the Intermediate Value Theorem (2.24) for f on the interval $[2, 3]$.

34 Prove that the equation $x^5 + 7x^2 - 3x - 5 = 0$ has a root between -2 and -1.

3

THE DERIVATIVE

The *derivative of a function* is one of the most powerful tools in mathematics. Indeed, it is indispensable for nonelementary investigations in both the natural and human sciences. We shall begin our work by reformulating the notion of tangent line introduced in the preceding chapter and then discussing the velocity of a moving object. These two concepts serve to motivate the definition of derivative given in Section 3.2. The remainder of the chapter is concerned primarily with properties of the derivative.

3.1

INTRODUCTION

Let $P(a, f(a))$ be any point on the graph of a function f. Another point on the graph may be denoted by $Q(a + h, f(a + h))$, where h is the difference between the x-coordinates of Q and P (see (i) of Figure 3.1). By Definition (1.14), the slope m_{PQ} of the secant line through P and Q is

$$m_{PQ} = \frac{f(a + h) - f(a)}{h}.$$

In Chapter 2 the slope m of the tangent line l at P was introduced as the limiting value of m_{PQ} as Q approaches P (see (2.2) and Figure 3.1). If f is continuous, then we can make Q approach P by letting h approach 0. Thus, it is natural to define m as follows.

Definition (3.1)

> Let f be a function that is defined on an open interval containing a. The **slope m of the tangent line** to the graph of f at the point $P(a, f(a))$ is
>
> $$m = \lim_{h \to 0} \frac{f(a + h) - f(a)}{h}$$
>
> provided the limit exists.

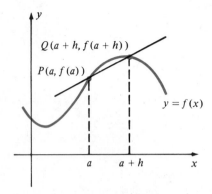
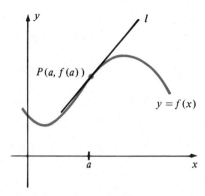

(i) Slope of secant line:

$$m_{PQ} = \frac{f(a + h) - f(a)}{h}$$

(ii) Slope of tangent line l:

$$m = \lim_{h \to 0} \frac{f(a + h) - f(a)}{h}$$

Figure 3.1

If a tangent line is vertical, its slope is undefined and the limit in Definition (3.1) does not exist. Vertical tangent lines will be studied in the next chapter.

Example 1   If $f(x) = x^2$, find the slope of the tangent line to the graph of f at the point $P(a, a^2)$.

Solution   This problem is the same as that stated for equations in Example 1 of Section 2.1 (see Figure 2.3). Using the quotient in Definition (3.1),

$$\begin{aligned}
\frac{f(a + h) - f(a)}{h} &= \frac{(a + h)^2 - a^2}{h} \\
&= \frac{a^2 + 2ah + h^2 - a^2}{h} \\
&= \frac{2ah + h^2}{h} \\
&= 2a + h.
\end{aligned}$$

The slope m of the tangent line is

$$m = \lim_{h \to 0} (2a + h) = 2a. \qquad \blacksquare$$

One of the main reasons for the invention of calculus was the need for a way to study objects in motion. Let us consider the problem of arriving at a satisfactory definition for the velocity, or speed, of an object at a given instant. We shall assume, for simplicity, that the object is moving on a line. Motion on a line is called **rectilinear motion**. It is easy to define the **average velocity** r during an interval of time. We merely use the following formula:

Average velocity (3.2)

$$r = \frac{d}{t}$$

where t denotes the length of the time interval and d is the distance between the initial position of the object and its position after t units of time.

As an elementary illustration, suppose an automobile leaves city A at 1:00 P.M. and travels along a straight highway, arriving at city B, 150 miles from A, at 4:00 P.M. Employing (3.2) we see that its average velocity r during the indicated time interval is 150/3, or 50 miles per hour. This is the velocity that, if maintained for three hours, would enable the automobile to travel the distance from A to B. The average velocity gives no information whatsoever about the velocity at any instant. For example, at 2:30 P.M. the automobile's speedometer may have registered 40, or 30, or the automobile may not even have been moving. If we wish to determine the rate at which the automobile is traveling at 2:30 P.M., information is needed about its motion or position *near* this time. For example, suppose at 2:30 P.M. the automobile is 80 miles from A and at 2:35 P.M. it is 84 miles from A, as illustrated in Figure 3.2. For the interval from 2:30 P.M. to 2:35 P.M. the elapsed time t is 5 minutes, or 1/12 hour, and the distance d is 4 miles. Substituting in (3.2), the average velocity r during this time interval is

$$r = \frac{4}{(1/12)} = 48 \text{ miles per hour.}$$

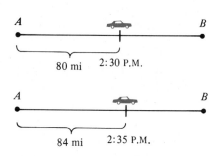

Figure 3.2

However, this is still not an accurate indication of the velocity at 2:30 P.M. since, for example, the automobile may have been traveling very slowly at 2:30 P.M. and then speeded up considerably so as to arrive at the point 84 miles from A at 2:35 P.M. Evidently, a better approximation of the motion would be obtained by considering the average velocity during a smaller time interval, say from 2:30 P.M. to 2:31 P.M. Indeed, it appears that the best procedure would be to take smaller and smaller time intervals near 2:30 P.M. and study the average velocity in each time interval. This leads us into a limiting process similar to that discussed for tangent lines.

To base our discussion on mathematical concepts, let us assume that the position of an object moving rectilinearly may be represented by a point P on a coordinate line l. We shall sometimes refer to the motion of the *point P* on l, or the motion of a *particle* on l whose position is specified by P. We further assume that the position of P is known at every instant in a given interval of time. If $f(t)$ denotes the coordinate of P at time t, then the function f determined in this way is called the **position function** for P. If we keep track of time by means of a clock, then for each t the point P is $f(t)$ units from the origin, as illustrated in Figure 3.3.

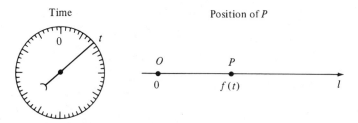

Figure 3.3   Rectilinear motion

To define the velocity of P at time a, we begin by investigating the average velocity in a (small) time interval near a. Thus we consider times a and $a + h$, where h may be positive or negative but not zero. The corresponding positions of P are given by $f(a + h)$ and $f(a)$, as illustrated in Figure 3.4, and hence the amount of change in the position of P is $f(a + h) - f(a)$. This number may be positive, negative, or zero, depending on whether the position of P at time $a + h$ is to the right, to the left, or the same as its position at time a. The number $f(a + h) - f(a)$ is not necessarily the distance traversed by P during the time interval $[a, a + h]$ since, for example, P may have moved beyond the point corresponding to $f(a + h)$ and then returned to that point at time $a + h$.

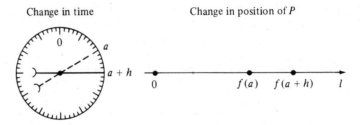

Change in time   Change in position of P

Figure 3.4

By (3.2), the average velocity of P during the time interval $[a, a + h]$ is given by the following formula:

$$\text{Average velocity:} \quad r = \frac{f(a + h) - f(a)}{h}.$$

As in our previous discussion, the smaller h is numerically, the closer this quotient should approximate the velocity of P at time a. Accordingly, we *define* the velocity as the limit, as h approaches zero, of the average velocity, provided the limit exists. This limit is also called the *instantaneous velocity* of P at time a. Summarizing, we have the next definition.

Definition (3.3)

If a point P moves on a coordinate line l such that its coordinate at time t is $f(t)$, then the **velocity** $v(a)$ of P at time a is

$$v(a) = \lim_{h \to 0} \frac{f(a + h) - f(a)}{h}$$

provided the limit exists.

If $f(t)$ is measured in centimeters and t in seconds, then the unit of velocity is centimeters per second, abbreviated cm/sec. If $f(t)$ is in miles and t in hours, then velocity is in miles per hour (mi/hr). Other units of measurement may, of course, be used.

We shall return to the velocity concept in Chapter 4, where it will be shown that if the velocity is positive in a given time interval, then the point is moving in the positive direction on l, whereas if the velocity is negative, the point is moving in the negative direction. Although these facts have not been proved we shall use them in the following example.

Example 2   The position of a point P on a coordinate line l is given by $f(t) = t^2 - 6t$, where $f(t)$ is measured in feet and t in seconds. Find the velocity at time a. What is the velocity at $t = 0$? At $t = 4$? Determine time intervals in which (a) P moves in the positive direction on l and (b) P moves in the negative direction. At what time is the velocity 0?

Solution   From the formula for $f(t)$ we obtain

$$\frac{f(a + h) - f(a)}{h} = \frac{[(a + h)^2 - 6(a + h)] - [a^2 - 6a]}{h}$$

$$= \frac{a^2 + 2ah + h^2 - 6a - 6h - a^2 + 6a}{h}$$

$$= \frac{2ah + h^2 - 6h}{h}$$

$$= 2a + h - 6.$$

Consequently, by Definition (3.3), the velocity $v(a)$ at time a is

$$v(a) = \lim_{h \to 0} \frac{f(a + h) - f(a)}{h} = \lim_{h \to 0} (2a + h - 6) = 2a - 6.$$

In particular, the velocity at $t = 0$ is $v(0) = 2(0) - 6 = -6$ ft/sec. At $t = 4$ it is $v(4) = 2(4) - 6 = 2$ ft/sec.

According to the remarks preceding this example, P moves to the left when the velocity is negative; that is, when $2a - 6 < 0$, or $a < 3$. The particle moves to the right when $2a - 6 > 0$ or $a > 3$. In terms of intervals, the motion is to the left in the time interval $(-\infty, 3)$ and to the right in $(3, \infty)$. The velocity is zero when $2a - 6 = 0$, that is, when $a = 3$ seconds.

■

You have undoubtedly noted the similarity of the limit in Definition (3.3) to that used in the definition of tangent line in (3.1). Indeed, the two expressions are identical! There are many different mathematical and physical applications that lead to precisely this same limit. Several of these will be discussed in the next chapter.

EXERCISES 3.1

In Exercises 1–4 find the slope of the tangent line at the point $P(a, f(a))$ on the graph of f. Sketch the graph and show the tangent lines at various points.

1   $f(x) = 2 - x^3$ 2 $f(x) = 3x - 5$

3   $f(x) = \sqrt{x} + 1$ 4 $f(x) = (1/x) - 1$

In Exercises 5 and 6, the position of a point P moving on a coordinate line l is given by $f(t)$, where t is measured in seconds and $f(t)$ in centimeters.

(a) Find the average velocity of P in the following time intervals: $[1, 1.2]$; $[1, 1.1]$; $[1, 1.01]$; $[1, 1.001]$.

(b) Find the velocity of P at $t = 1$.

(c) Determine the time intervals in which P moves in the positive direction.

(d) Determine the time intervals in which P moves in the negative direction.

5 $f(t) = 4t^2 + 3t$ **6** $f(t) = t^3$

7 A projectile is fired directly upward from the ground with an initial velocity of 112 ft/sec, and its distance above the ground after t seconds is $112t - 16t^2$ ft. What is the velocity of the projectile at $t = 2$, $t = 3$, and $t = 4$? At what time does it reach its maximum height? When does it strike the ground? What is its velocity at the moment of impact?

8 If a balloonist drops ballast (a sand bag) from a balloon 500 ft above the ground, then its distance above ground after t seconds is $500 - 16t^2$ ft. Find the velocity at $t = 1$, $t = 2$, and $t = 3$. With what velocity does the ballast strike the ground?

9 If the position of an object moving on a coordinate line is given by a polynomial function of degree 1, prove that the velocity is constant.

10 If the position function of a rectilinearly moving object is a constant function, prove that the velocity is 0 at all times. Describe the motion of the particle.

3.2

THE DERIVATIVE OF A FUNCTION

In the preceding section the same limiting process occurred in two different situations. The limit in Definitions (3.1) and (3.3) is one of the fundamental concepts of calculus. In the remainder of the chapter some rules pertaining to this concept will be developed. Let us begin by introducing the terminology and notation associated with this limit.

Definition (3.4)

> Let f be a function that is defined on an open interval containing a. The **derivative of f at a**, written $f'(a)$, is
> $$f'(a) = \lim_{h \to 0} \frac{f(a + h) - f(a)}{h}$$
> provided the limit exists.

The symbol $f'(a)$ is read "f prime of a." The terminology "$f'(a)$ exists" will mean that the limit in Definition (3.4) exists. If $f'(a)$ exists we say that the function f **is differentiable at a** or f **has a derivative at a**.

It is important to observe that if f is differentiable at a, then by Definition (3.1), $f'(a)$ *is the slope of the tangent line to the graph of f at the point* $(a, f(a))$.

We say that a function f **is differentiable on an open interval (a, b)** if it is differentiable at every number c in (a, b). In like manner, we refer to functions that are differentiable on intervals of the form (a, ∞), $(-\infty, a)$, or $(-\infty, \infty)$. For closed intervals we use the following convention, which is analogous to the definition of continuity on a closed interval given in (2.20).

Definition (3.5)

A function f is **differentiable on a closed interval** $[a, b]$ if it is differentiable on the open interval (a, b) and if the following limits exist:

$$\lim_{h \to 0^+} \frac{f(a + h) - f(a)}{h} \quad \text{and} \quad \lim_{h \to 0^-} \frac{f(b + h) - f(b)}{h}.$$

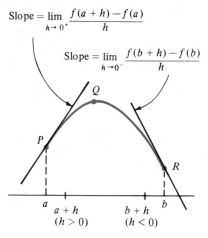

Slope $= \lim_{h \to 0^+} \dfrac{f(a + h) - f(a)}{h}$

Slope $= \lim_{h \to 0^-} \dfrac{f(b + h) - f(b)}{h}$

a $a + h$ $(h > 0)$ $b + h$ $(h < 0)$ b

Figure 3.5

The one-sided limits in Definition (3.5) are sometimes referred to as the **right-hand** and **left-hand derivatives** of f at a and b, respectively. Note that for the right-hand derivative we have $h \to 0^+$. Consequently $h > 0$ and $a + h > a$ (see Figure 3.5). Thus we may think of $a + h$ as approaching *a from the right*. For the left-hand derivative, $h \to 0^-$ and in this case, $h < 0$ and $b + h < b$. Hence for the left-hand derivative we may regard $b + h$ as approaching *b from the left*.

If f is defined on a closed interval $[a, b]$ and is undefined elsewhere, then the right-hand and left-hand derivatives allow us to define the slopes of the tangent lines at the points $P(a, f(a))$ and $R(b, f(b))$, respectively, as illustrated in Figure 3.5. Thus, for the slope of the tangent line at P we take the limiting value of the slope m_{PQ} of the secant line through P and Q as Q approaches P from the right. For the tangent line at R, the point Q approaches R from the left.

Differentiability on an interval of the form $[a, b), [a, \infty), (a, b]$, or $(-\infty, b]$ is defined in the obvious way, using a one-sided limit at an endpoint.

It should be clear that if f is defined on an open interval containing a, then $f'(a)$ *exists if and only if both the right-hand and left-hand derivatives exist at a, and are equal*. The functions whose graphs are sketched in Figure 3.6 have right-hand and left-hand derivatives at a that give the slopes of the indicated lines l_1 and l_2, respectively. However, since the slopes of l_1 and l_2 are unequal, $f'(a)$ does not exist. Generally, if the graph of f has a *corner* at $(a, f(a))$, then f is not differentiable at a.

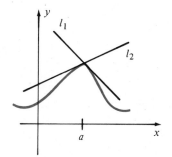
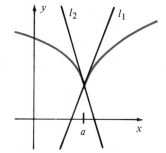
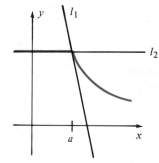

Figure 3.6

Given a function f, let $S = \{a: f'(a) \text{ exists}\}$. If f is defined on an interval, then the terminology "$f'(a)$ exists" may refer to the existence of certain right-hand or left-hand limits of the types specified in Definition (3.5). If, with each x in S, we associate the number $f'(x)$, we obtain a function f' with domain

S called the **derivative of f**. The value of f' at x is given by the following (or an appropriate one-sided) limit:

(3.6)
$$f'(x) = \lim_{h \to 0} \frac{f(x + h) - f(x)}{h}.$$

It is important to note that in determining $f'(x)$, the number x is fixed, but arbitrary, and the limit is taken as h approaches zero. The terminology "differentiate $f(x)$" or "find the derivative of $f(x)$" means to find $f'(x)$. In statements of definitions and theorems, the expression "f is differentiable" will mean that $f'(x)$ exists for all numbers x in some set of real numbers. The domain of f' will not always be stated explicitly; however, for specific problems, it may often be found by inspection.

Example 1 If $f(x) = 3x^2 - 5x + 4$, find $f'(x)$. What is the domain of f'? Find $f'(2)$, $f'(-\sqrt{2})$, and $f'(a)$.

Solution   Employing (3.6),

$$f'(x) = \lim_{h \to 0} \frac{f(x + h) - f(x)}{h}$$

$$= \lim_{h \to 0} \frac{[3(x + h)^2 - 5(x + h) + 4] - (3x^2 - 5x + 4)}{h}$$

$$= \lim_{h \to 0} \frac{(3x^2 + 6xh + 3h^2 - 5x - 5h + 4) - (3x^2 - 5x + 4)}{h}$$

$$= \lim_{h \to 0} \frac{6xh + 3h^2 - 5h}{h} = \lim_{h \to 0} (6x + 3h - 5)$$

$$= 6x - 5.$$

Since $f'(x) = 6x - 5$ for all x, the domain of f' is $\mathbb{R}$. Substituting for x, we have

$$f'(2) = 6(2) - 5 = 7$$
$$f'(-\sqrt{2}) = 6(-\sqrt{2}) - 5 = -(6\sqrt{2} + 5)$$
$$f'(a) = 6a - 5. \qquad \blacksquare$$

Example 2 Find $f'(x)$ if $f(x) = \sqrt{x}$. What is the domain of f'?

Solution   The domain of f consists of all nonnegative real numbers. We shall examine the cases $x > 0$ and $x = 0$ separately. If $x > 0$, then by (3.6),

$$f'(x) = \lim_{h \to 0} \frac{\sqrt{x + h} - \sqrt{x}}{h}.$$

To find the limit we begin by multiplying numerator and denominator of the indicated quotient by $\sqrt{x + h} + \sqrt{x}$. Thus

$$f'(x) = \lim_{h \to 0} \frac{\sqrt{x + h} - \sqrt{x}}{h} \cdot \frac{\sqrt{x + h} + \sqrt{x}}{\sqrt{x + h} + \sqrt{x}}$$

$$= \lim_{h \to 0} \frac{(x + h) - x}{h(\sqrt{x + h} + \sqrt{x})}$$

$$= \lim_{h \to 0} \frac{1}{\sqrt{x + h} + \sqrt{x}}$$

$$= \frac{1}{\sqrt{x} + \sqrt{x}} = \frac{1}{2\sqrt{x}}.$$

Since $x = 0$ is an endpoint of the domain of f, we must use a one-sided limit to determine whether $f'(0)$ exists. Specifically, if f is differentiable at 0, then using Definition (3.5) with $x = 0$,

$$f'(0) = \lim_{h \to 0^+} \frac{\sqrt{0 + h} - \sqrt{0}}{h}$$

$$= \lim_{h \to 0^+} \frac{\sqrt{h}}{h} = \lim_{h \to 0^+} \frac{1}{\sqrt{h}}.$$

Since the last limit does not exist (see Exercise 38 of Section 2.3), $f'(0)$ does not exist. Hence the domain of f' is the set of positive real numbers. ∎

Example 3 Show that the function f defined by $f(x) = |x|$ is not differentiable at 0.

Solution The graph of f was discussed in Example 5 of Section 1.4 and is resketched in Figure 3.7. It is geometrically evident that f has no derivative at 0 since the graph has a corner at the origin. We can prove that $f'(0)$ does not exist by showing that the right-hand and left-hand derivatives of f at 0 are not equal. Using the limits in Definition (3.5) with $a = 0$ and $b = 0$,

$$\lim_{h \to 0^+} \frac{f(0 + h) - f(0)}{h} = \lim_{h \to 0^+} \frac{|0 + h| - |0|}{h} = \lim_{h \to 0^+} \frac{|h|}{h} = 1$$

$$\lim_{h \to 0^-} \frac{f(0 + h) - f(0)}{h} = \lim_{h \to 0^-} \frac{|0 + h| - |0|}{h} = \lim_{h \to 0^-} \frac{|h|}{h} = -1.$$

Thus $f'(0)$ does not exist. ∎

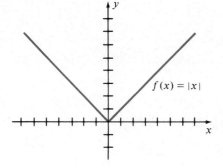

$f(x) = |x|$

Figure 3.7

It follows from Example 3 that the graph of $y = |x|$ does not have a tangent line at the point $P(0, 0)$.

In Definition (3.4), the derivative was defined as a certain limit. There is another important limit formula for $f'(a)$. To see how it arises geometrically, let us begin by labeling the graph of f as shown in Figure 3.8.

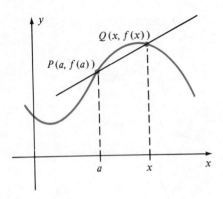

Figure 3.8

By Definition (1.14), the slope m_{PQ} of the secant line through P and Q is

$$m_{PQ} = \frac{f(x) - f(a)}{x - a}.$$

If f is continuous at a, we can make Q approach P by letting x approach a. Thus it appears that the slope of the tangent line at P is

$$m = \lim_{x \to a} \frac{f(x) - f(a)}{x - a}$$

provided the limit exists. Since $m = f'(a)$ this leads to the alternative formula for the derivative of f stated in the next theorem. A nongeometric proof of the formula may be found in Appendix II.

Theorem (3.7)

> If f is defined on an open interval containing a, then
>
> $$f'(a) = \lim_{x \to a} \frac{f(x) - f(a)}{x - a}$$
>
> provided the limit exists.

Example 4 If $f(x) = x^{1/3}$ and $a \neq 0$, find $f'(a)$.

Solution   By Theorem (3.7),

$$f'(a) = \lim_{x \to a} \frac{x^{1/3} - a^{1/3}}{x - a}$$

provided the limit exists. The indicated quotient may be expressed as follows:

$$\lim_{x \to a} \frac{x^{1/3} - a^{1/3}}{x - a} = \lim_{x \to a} \frac{x^{1/3} - a^{1/3}}{(x^{1/3})^3 - (a^{1/3})^3}.$$

We may factor the denominator by using the formula

$$p^3 - q^3 = (p - q)(p^2 + pq + q^2)$$

with $p = x^{1/3}$ and $q = a^{1/3}$. This gives us

$$\lim_{x \to a} \frac{x^{1/3} - a^{1/3}}{(x^{1/3} - a^{1/3})(x^{2/3} + x^{1/3}a^{1/3} + a^{2/3})}.$$

Dividing numerator and denominator of the last quotient by $x^{1/3} - a^{1/3}$ and taking the limit, we obtain

$$f'(a) = \frac{1}{a^{2/3} + a^{2/3} + a^{2/3}} = \frac{1}{3a^{2/3}}. \qquad \blacksquare$$

Theorem (3.8)

> If a function f is differentiable at a, then f is continuous at a.

Proof If x is in the domain of f and $x \neq a$, then $f(x)$ may be written as follows:

$$f(x) = f(a) + \frac{f(x) - f(a)}{x - a}(x - a).$$

Employing limit theorems and Theorem (3.7),

$$\lim_{x \to a} f(x) = \lim_{x \to a} f(a) + \lim_{x \to a} \frac{f(x) - f(a)}{x - a} \cdot \lim_{x \to a} (x - a)$$
$$= f(a) + f'(a) \cdot 0 = f(a).$$

Thus, by Definition (2.18), f is continuous at a. $\qquad\square$

By using one-sided limits, Theorem (3.8) can be extended to functions that are differentiable on a closed interval.

The converse of Theorem (3.8) is false; that is, *there exist continuous functions that are not differentiable*. To illustrate, if $f(x) = |x|$, then f is continuous at 0; however, as was proved in Example 3, f is not differentiable at 0 (see Figure 3.7).

We shall conclude this section by introducing the following notations for derivatives, where $y = f(x)$:

(3.9)
$$f'(x) = D_x[f(x)] = D_x y = y'.$$

In (3.9) the subscript x on D is employed to designate the independent variable. For example, if the independent variable is t we shall write $f'(t) = D_t[f(t)]$. The symbols D_x, D_t, etc., are referred to as **differential operators**. Standing alone, D_x has no practical significance; however, if an expression in x is placed to the right of it, then the derivative is obtained. To illustrate, using Example 1,

$$D_x(3x^2 - 5x + 4) = 6x - 5.$$

We say that D_x *operates* on the expression $3x^2 - 5x + 4$. Sometimes $D_x y$ is referred to as **the derivative of y with respect to x**. As indicated in (3.9), y' may be used as an abbreviation for this derivative.

In Section 3.4 we shall introduce still another notation for derivatives. All of these representations are used in mathematics and applications, and it is advisable for students to become familiar with the various forms.

EXERCISES 3.2

In Exercises 1–10 use (3.6) to find $f'(x)$.

1 $f(x) = 37$

2 $f(x) = 17 - 6x$

3 $f(x) = 9x - 2$

4 $f(x) = 7x^2 - 5$

5 $f(x) = 2 + 8x - 5x^2$

6 $f(x) = x^3 + x$

7 $f(x) = 1/(x - 2)$

8 $f(x) = (1 + \sqrt{3})^2$

9 $f(x) = \sqrt{3x + 1}$

10 $f(x) = 1/(2x)$

In Exercises 11–14 find $D_x y$.

11 $y = 7/\sqrt{x}$ **12** $y = (2x + 3)^2$

13 $y = 2x^3 - 4x + 1$ **14** $y = x/(3x + 4)$

In Exercises 15–20 find $f'(a)$ by means of Theorem (3.7).

15 $f(x) = x^2$ **16** $f(x) = \sqrt{2}x$

17 $f(x) = 6/x^2$ **18** $f(x) = 8 - x^3$

19 $f(x) = 1/(x + 5)$ **20** $f(x) = \sqrt{x}$

In Exercises 21 and 22 use right-hand and left-hand derivatives to prove that f is not differentiable at $x = 5$.

21 $f(x) = |x - 5|$

22 $f(x) = [\![x]\!]$ (the greatest integer function)

23 Prove that if n is any integer, then the greatest integer function f is not differentiable at n.

24 Suppose $f(x) = 0$ if x is rational and $f(x) = 1$ if x is irrational. Prove that f is not differentiable at any real number a.

25 If $f(x)$ is a polynomial of degree 1, prove that $f'(x)$ is a polynomial of degree 0. What if $f(x)$ is a polynomial of degree 2 or 3?

26 Prove that $D_x c = 0$ for every real number c.

In Exercises 27–30, sketch the graph of f and find the domain of f'.

27 $f(x) = \begin{cases} 2x & \text{if } x \leq 0 \\ x^2 & \text{if } x > 0 \end{cases}$

28 $f(x) = \begin{cases} 2x - 1 & \text{if } x \leq 1 \\ x^2 & \text{if } x > 1 \end{cases}$

29 $f(x) = \begin{cases} |x| & \text{if } |x| \leq 1 \\ 2 - |x| & \text{if } |x| > 1 \end{cases}$

30 $f(x) = \begin{cases} x - [\![x]\!] & \text{if } n \leq x < n + 1 \text{ and } n \text{ is an even integer} \\ 1 - x + [\![x]\!] & \text{if } n \leq x < n + 1 \text{ and } n \text{ is an odd integer} \end{cases}$

where $[\![\ ]\!]$ denotes the greatest integer function.

31 Let $f(x) = |x|$. Prove that $f'(x) = 1$ if $x > 0$ and that $f'(x) = -1$ if $x < 0$.

32 Let $f(x) = |x|/x$. Find (a) the domain of f'; (b) $f'(x)$ for every x in the domain of f'.

3.3

RULES FOR FINDING DERIVATIVES

This section contains some general rules that simplify the task of finding derivatives. In statements of theorems we shall use the differential operator symbol D_x to denote derivatives (see (3.9)). The first result is sometimes phrased: *the derivative of a constant is zero*.

Theorem (3.10)

$$\boxed{D_x c = 0}$$

Proof   Let f be the constant function such that $f(x) = c$ for every x. We wish to show that $f'(x) = 0$. Since every value of f is c it follows that $f(x + h) = c$ for all h. Hence

$$f'(x) = \lim_{h \to 0} \frac{f(x + h) - f(x)}{h} = \lim_{h \to 0} \frac{c - c}{h} = \lim_{h \to 0} 0 = 0$$

where the last step follows from Theorem (2.5).   □

Theorem (3.11)

$$\boxed{D_x(x) = 1}$$

Proof   If we let $f(x) = x$, then using (3.6),

$$f'(x) = \lim_{h \to 0} \frac{f(x + h) - f(x)}{h}$$

$$= \lim_{h \to 0} \frac{(x + h) - x}{h}$$

$$= \lim_{h \to 0} \frac{h}{h} = \lim_{h \to 0} 1 = 1. \qquad \square$$

In the proof of the next rule we shall make use of the *Binomial Theorem*, which states that the following formula is true for all real numbers a and b and every positive integer n.

$$(3.12) \qquad (a + b)^n = a^n + na^{n-1}b + \frac{n(n-1)}{2!} a^{n-2}b^2$$

$$+ \cdots + \binom{n}{r} a^{n-r}b^r + \cdots + nab^{n-1} + b^n$$

where the symbol $\binom{n}{r}$ used for the coefficient of the term involving b^r is defined by

$$\binom{n}{r} = \frac{n(n-1)(n-2)\cdots(n-r+1)}{r(r-1)(r-2)\cdots 1} = \frac{n!}{(n-r)!r!}.$$

The validity of (3.12) may be established by mathematical induction. The special cases $n = 2$, $n = 3$, and $n = 4$ are

$$(a + b)^2 = a^2 + 2ab + b^2$$
$$(a + b)^3 = a^3 + 3a^2b + 3ab^2 + b^3$$
$$(a + b)^4 = a^4 + 4a^3b + 6a^2b^2 + 4ab^3 + b^4.$$

The Power Rule (3.13)

$$\boxed{\text{If } n \text{ is a positive integer, then } D_x(x^n) = nx^{n-1}.}$$

Proof   Let $f(x) = x^n$. We wish to show that $f'(x) = nx^{n-1}$. By (3.6),

$$f'(x) = \lim_{h \to 0} \frac{(x + h)^n - x^n}{h}$$

provided the limit exists. Employing the Binomial Theorem (3.12) with $a = x$ and $b = h$ gives us

$$(x + h)^n = x^n + nx^{n-1}h + \frac{n(n-1)}{2!} x^{n-2}h^2 + \cdots + nxh^{n-1} + h^n.$$

It is important to observe that every term after the first contains h to some positive integral power. If we subtract x^n and divide by h we obtain

$$f'(x) = \lim_{h \to 0} \left[nx^{n-1} + \frac{n(n-2)}{2!} x^{n-2}h + \cdots + nxh^{n-2} + h^{n-1} \right].$$

Since each term within the brackets except the first contains a power of h, we see that $f'(x) = nx^{n-1}$.

If $x \neq 0$, then (3.13) is also true if $n = 0$, for in this case $f(x) = x^0 = 1$ and by Theorem (3.10), $f'(x) = 0 = 0 \cdot x^{0-1}$. $\quad\square$

Example 1
(a) Find $D_x(x^3)$ and $D_x(x^8)$.
(b) Find y' if $y = x^{100}$.

Solution   Applying the Power Rule (3.13) gives us

(a) $D_x(x^3) = 3x^2$ and $D_x(x^8) = 8x^7$.

(b) $y' = D_x y = D_x(x^{100}) = 100x^{99}$. $\quad\blacksquare$

If symbols other than x are used for the independent variable, then the Power Rule may be written as

$$D_t(t^n) = nt^{n-1}, \quad D_z(z^n) = nz^{n-1}, \quad D_v(v^n) = nv^{n-1},$$

etc. In Section 3.6 we shall prove that the Power Rule is valid for every rational number n.

In the statements of Theorems (3.14)–(3.17) it is assumed that the functions f and g are differentiable.

Theorem (3.14)

$$\boxed{D_x[cf(x)] = cD_x[f(x)]}$$

Proof   If we let $g(x) = cf(x)$, then

$$D_x[cf(x)] = D_x[g(x)]$$

$$= \lim_{h \to 0} \frac{g(x + h) - g(x)}{h}$$

$$= \lim_{h \to 0} \frac{cf(x + h) - cf(x)}{h}$$

$$= c \lim_{h \to 0} \frac{f(x + h) - f(x)}{h}$$

$$= cf'(x) = cD_x[f(x)]. \quad\square$$

For the special case $f(x) = x^n$, Theorems (3.14) and (3.13) lead to the next formula, which is true for every real number c and every positive integer n:

$$D_x(cx^n) = cnx^{n-1}$$

Thus, to differentiate cx^n we multiply the coefficient c by the exponent n, and then reduce the exponent by 1.

Example 2

(a) Find $D_x(7x^4)$.

(b) Find $F'(z)$ if $F(z) = -3z^{15}$.

Solution   By the remarks preceding this example we have

(a) $D_x(7x^4) = (7)4x^3 = 28x^3$.

(b) $F'(z) = (-3)(15)z^{14} = -45z^{14}$.   ■

The theorems that follow involve several functions. As usual we assume that every number x is chosen from the intersection of the domains of the functions.

Theorem (3.15)

$$D_x[f(x) + g(x)] = D_x[f(x)] + D_x[g(x)]$$

$$D_x[f(x) - g(x)] = D_x[f(x)] - D_x[g(x)]$$

Proof   To prove the first formula, let $k(x) = f(x) + g(x)$. We wish to show that $k'(x) = f'(x) + g'(x)$. This may be done as follows.

$$k'(x) = \lim_{h \to 0} \frac{k(x + h) - k(x)}{h}$$

$$= \lim_{h \to 0} \frac{[f(x + h) + g(x + h)] - [f(x) + g(x)]}{h}$$

$$= \lim_{h \to 0} \left[\frac{f(x + h) - f(x)}{h} + \frac{g(x + h) - g(x)}{h} \right]$$

$$= \lim_{h \to 0} \frac{f(x + h) - f(x)}{h} + \lim_{h \to 0} \frac{g(x + h) - g(x)}{h}$$

$$= f'(x) + g'(x)$$

The formula for $D_x[f(x) - g(x)]$ is proved in similar fashion.   □

The preceding theorem may be extended to sums or differences of any number of functions.

Since a polynomial is a sum of terms of the form cx^n where n is a non-negative integer, we may use results on sums and differences to obtain the derivative, as illustrated in the next example.

Example 3   Find $f'(x)$ if $f(x) = 2x^4 - 5x^3 + x^2 - 4x + 1$.

Solution $f'(x) = D_x(2x^4 - 5x^3 + x^2 - 4x + 1)$
$$= D_x(2x^4) - D_x(5x^3) + D_x(x^2) - D_x(4x) + D_x(1)$$
$$= 8x^3 - 15x^2 + 2x - 4 \qquad\blacksquare$$

The Product Rule (3.16)

$$D_x[f(x)g(x)] = f(x)D_x[g(x)] + g(x)D_x[f(x)]$$

Proof   Let $k(x) = f(x)g(x)$. We wish to show that
$$k'(x) = f(x)g'(x) + g(x)f'(x).$$

If $k'(x)$ exists, then
$$k'(x) = \lim_{h \to 0} \frac{k(x + h) - k(x)}{h}$$
$$= \lim_{h \to 0} \frac{f(x + h)g(x + h) - f(x)g(x)}{h}.$$

To change the form of the quotient so that the limit may be evaluated, we subtract and add the expression $f(x + h)g(x)$ in the numerator. Thus
$$k'(x) = \lim_{h \to 0} \frac{f(x + h)g(x + h) - f(x + h)g(x) + f(x + h)g(x) - f(x)g(x)}{h}$$

which may be written
$$k'(x) = \lim_{h \to 0} \left[f(x + h) \cdot \frac{g(x + h) - g(x)}{h} + g(x) \cdot \frac{f(x + h) - f(x)}{h} \right]$$
$$= \lim_{h \to 0} f(x + h) \cdot \lim_{h \to 0} \frac{g(x + h) - g(x)}{h}$$
$$+ \lim_{h \to 0} g(x) \cdot \lim_{h \to 0} \frac{f(x + h) - f(x)}{h}.$$

Since f is differentiable at x, it is continuous at x (see Theorem (3.8)). Hence $\lim_{h \to 0} f(x + h) = f(x)$. Also, $\lim_{h \to 0} g(x) = g(x)$ since x is fixed in this limiting process. Finally, applying the definition of derivative to $f(x)$ and $g(x)$ we obtain
$$k'(x) = f(x)g'(x) + g(x)f'(x). \qquad\square$$

The Product Rule may be phrased as follows: *The derivative of a product equals the first factor times the derivative of the second factor, plus the second times the derivative of the first.*

Example 4 Find $f'(x)$ if $f(x) = (x^3 + 1)(2x^2 + 8x - 5)$.

Solution Using the Product Rule (3.16),

$$
\begin{aligned}
f'(x) &= (x^3 + 1)D_x(2x^2 + 8x - 5) + (2x^2 + 8x - 5)D_x(x^3 + 1) \\
&= (x^3 + 1)(4x + 8) + (2x^2 + 8x - 5)(3x^2) \\
&= 4x^4 + 8x^3 + 4x + 8 + 6x^4 + 24x^3 - 15x^2 \\
&= 10x^4 + 32x^3 - 15x^2 + 4x + 8.
\end{aligned}
$$

We could also find $f'(x)$ by first multiplying the two factors $x^3 + 1$ and $2x^2 + 8x - 5$ and then differentiating the resulting polynomial. ■

The Quotient Rule (3.17)

$$
D_x\left[\frac{f(x)}{g(x)}\right] = \frac{g(x)D_x[f(x)] - f(x)D_x[g(x)]}{[g(x)]^2},
$$

provided $g(x) \neq 0$.

Proof Let $k(x) = f(x)/g(x)$. We wish to show that

$$
k'(x) = \frac{g(x)f'(x) - f(x)g'(x)}{[g(x)]^2}.
$$

Using the definitions of $k'(x)$ and $k(x)$,

$$
\begin{aligned}
k'(x) &= \lim_{h \to 0} \frac{k(x + h) - k(x)}{h} \\[2mm]
&= \lim_{h \to 0} \frac{\dfrac{f(x + h)}{g(x + h)} - \dfrac{f(x)}{g(x)}}{h} \\[2mm]
&= \lim_{h \to 0} \frac{g(x)f(x + h) - f(x)g(x + h)}{hg(x + h)g(x)}.
\end{aligned}
$$

Subtracting and adding $g(x)f(x)$ in the numerator of the last quotient,

$$
k'(x) = \lim_{h \to 0} \frac{g(x)f(x + h) - g(x)f(x) + g(x)f(x) - f(x)g(x + h)}{hg(x + h)g(x)}
$$

or equivalently,

$$
k'(x) = \lim_{h \to 0} \frac{g(x)\left[\dfrac{f(x + h) - f(x)}{h}\right] - f(x)\left[\dfrac{g(x + h) - g(x)}{h}\right]}{g(x + h)g(x)}.
$$

Taking the limit of the numerator and denominator gives us the desired formula. □

The Quotient Rule may be stated as follows: *The derivative of a quotient equals the denominator times the derivative of the numerator minus the numerator times the derivative of the denominator, divided by the square of the denominator.*

Example 5   Find y' if $y = \dfrac{3x^2 - x + 2}{4x^2 + 5}$.

Solution   By the Quotient Rule (3.17),

$$
\begin{aligned}
y' &= \frac{(4x^2 + 5)D_x(3x^2 - x + 2) - (3x^2 - x + 2)D_x(4x^2 + 5)}{(4x^2 + 5)^2} \\[2mm]
&= \frac{(4x^2 + 5)(6x - 1) - (3x^2 - x + 2)(8x)}{(4x^2 + 5)^2} \\[2mm]
&= \frac{(24x^3 - 4x^2 + 30x - 5) - (24x^3 - 8x^2 + 16x)}{(4x^2 + 5)^2} \\[2mm]
&= \frac{4x^2 + 14x - 5}{(4x^2 + 5)^2}.
\end{aligned}
$$

■

It is now a simple matter to extend the Power Rule (3.13) to the case in which the exponent is a negative integer.

Theorem (3.18)

> If n is a positive integer, then $D_x(x^{-n}) = -nx^{-n-1}$.

Proof   Using the definition of x^{-n} and the Quotient Rule (3.17),

$$
\begin{aligned}
D_x(x^{-n}) = D_x\!\left(\frac{1}{x^n}\right) &= \frac{x^n D_x(1) - 1 D_x(x^n)}{(x^n)^2} \\[2mm]
&= \frac{x^n(0) - 1(nx^{n-1})}{(x^n)^2} = \frac{-nx^{n-1}}{x^{2n}} \\[2mm]
&= (-n)x^{(n-1)-2n} = (-n)x^{-n-1}.
\end{aligned}
$$

□

Example 6   Differentiate the following: (a) $g(w) = 1/w^4$.   (b) $H(s) = 3/s$.

Solution

(a)   Writing $g(w) = w^{-4}$ and using Theorem (3.18) (with w as the independent variable),

$$
g'(w) = D_w(w^{-4}) = -4w^{-5} = -\frac{4}{w^5}.
$$

(b)   Since $H(s) = 3s^{-1}$,

$$
H'(s) = D_s(3s^{-1}) = 3(-1)s^{-2} = -\frac{3}{s^2}.
$$

■

The differentiation formulas (3.15), (3.16), and (3.17) are stated in terms of arbitrary values $f(x)$ and $g(x)$ of the functions f and g. If we wish to state these rules without referring to the variable x, we may use the following forms, where it is assumed that f and g are differentiable functions.

(3.19)
$$(f + g)' = f' + g' \qquad (f - g)' = f' - g'$$
$$(fg)' = fg' + gf' \qquad \left(\frac{f}{g}\right)' = \frac{gf' - fg'}{g^2}$$

EXERCISES 3.3

Differentiate the functions defined in Exercises 1–32.

1 $f(x) = 10x^2 + 9x - 4$

2 $f(x) = 6x^3 - 5x^2 + x + 9$

3 $f(s) = 15 - s + 4s^2 - 5s^4$

4 $f(t) = 12 - 3t^4 + 4t^6$

5 $g(x) = (x^3 - 7)(2x^2 + 3)$

6 $k(x) = (2x^2 - 4x + 1)(6x - 5)$

7 $h(r) = r^2(3r^4 - 7r + 2)$

8 $g(s) = (s^3 - 5s + 9)(2s + 1)$

9 $f(x) = (4x - 5)/(3x + 2)$

10 $h(x) = (8x^2 - 6x + 11)/(x - 1)$

11 $h(z) = (8 - z + 3z^2)/(2 - 9z)$

12 $f(w) = 2w/(w^3 - 7)$

13 $f(x) = 3x^3 - 2x^2 + 4x - 7$

14 $g(z) = 5z^4 - 8z^2 + z$

15 $F(t) = t^2 + (1/t^2)$

16 $s(x) = 2x + (2x)^{-1}$

17 $g(x) = (8x^2 - 5x)(13x^2 + 4)$

18 $H(y) = (y^5 - 2y^3)(7y^2 + y - 8)$

19 $G(v) = (v^3 - 1)/(v^3 + 1)$

20 $f(t) = (8t + 15)/(t^2 - 2t + 3)$

21 $f(x) = 1/(1 + x + x^2 + x^3)$

22 $p(x) = 1 + (1/x) + (1/x^2) + (1/x^3)$

23 $g(z) = z(2z^3 - 5z - 1)(6z^2 + 7)$

24 $N(v) = 4v(v - 1)(2v - 3)$

25 $K(s) = (3s)^{-4}$

26 $W(s) = (3s)^4$

27 $h(x) = (5x - 4)^2$

28 $g(r) = (5r - 4)^{-2}$

29 $f(t) = \dfrac{(3/5t) - 1}{(2/t^2) + 7}$

30 $S(w) = (2w + 1)^3$

31 $M(x) = (2x^3 - 7x^2 + 4x + 3)/x^2$

32 $f(x) = (3x^2 - 5x + 8)/7$

In Exercises 33 and 34 find y' by means of (a) the Quotient Rule (3.17), (b) the Product Rule (3.16), and (c) simplifying algebraically and using the Power Rule (3.13).

33 $y = (3x - 1)/x^2$

34 $y = (x^2 + 1)/x^4$

In Exercises 35 and 36 find y' by (a) using the Product Rule and (b) first multiplying the two factors.

35 $y = (12x - 17)(5x + 3)$

36 $y = 8x^2(5x - 9)$

37 If f, g, and h are differentiable, use the Product Rule to prove that

$$D_x[f(x)g(x)h(x)] = f(x)g(x)h'(x) + f(x)h(x)g'(x)$$
$$+ h(x)g(x)f'(x).$$

As a corollary, let $f = g = h$ to prove that

$$D_x[f(x)]^3 = 3[f(x)]^2 f'(x).$$

38 Extend Exercise 37 to the derivative of a product of four functions and then find a formula for $D_x[f(x)]^4$.

In Exercises 39 and 40, use Exercise 37 to find y'.

39 $y = (8x - 1)(x^2 + 4x + 7)(x^3 - 5)$

40 $y = (3x^4 - 10x^2 + 8)(2x^2 - 10)(6x + 7)$

41 Find an equation of the tangent line to the graph of $y = 5/(1 + x^2)$ at each point.

   (a) $P(0, 5)$ (b) $P(1, \frac{5}{2})$ (c) $P(-2, 1)$

42 Find an equation of the tangent line to the graph of $y = 2x^3 + 4x^2 - 5x - 3$ at each point.

   (a) $P(0, -3)$ (b) $P(-1, 4)$ (c) $P(1, -2)$

43 Find the x-coordinates of all points on the graph of $y = x^3 + 2x^2 - 4x + 5$ at which the tangent line is (a) horizontal; (b) parallel to the line $2y + 8x - 5 = 0$.

44 Find point P on the graph of $y = x^3$ such that the tangent line at P has x-intercept 4.

In Exercises 45 and 46 the position of a point P moving on a coordinate line l is given by $f(t)$. Determine the time intervals in which P moves in (a) the positive direction; (b) the negative direction. When is the velocity 0?

45 $f(t) = 2t^3 + 9t^2 - 60t + 1$

46 $f(t) = 3t^5 - 5t^3$

If f and g are differentiable functions such that $f(2) = 3$, $f'(2) = -1$, $g(2) = -5$, and $g'(2) = 2$, find the numbers in Exercises 47 and 48.

47 (a) $(f + g)'(2)$ (b) $(f - g)'(2)$
   (c) $(4f)'(2)$ (d) $(fg)'(2)$
   (e) $(f/g)'(2)$

48 (a) $(g - f)'(2)$ (b) $(g/f)'(2)$
   (c) $(4g)'(2)$ (d) $(ff)'(2)$

In Exercises 49 and 50 find the points at which the graphs of f and f' intersect.

49 $f(x) = x^3 - x^2 + x + 1$ **50** $f(x) = x^2 + 2x + 1$

51 Find an equation of the tangent line to the graph of $y = 3x^2 + 4x - 6$ that is parallel to the line $5x - 2y - 1 = 0$.

52 Find equations of the tangent lines to the graph of $y = x^3$ that are parallel to the line $16x - 3y + 17 = 0$.

53 Find an equation of the line through the point $P(5, 9)$ that is tangent to the graph of $y = x^2$.

54 Find equations of the lines through $P(3, 1)$ that are tangent to the graph of $xy = 4$.

55 A weather balloon is released and rises vertically such that its distance $s(t)$ above the ground during the first 10 seconds of flight is given by $s(t) = 6 + 2t + t^2$, where $s(t)$ is in feet and t is in seconds.

   (a) Find the velocity of the balloon at $t = 1$, $t = 4$, and $t = 8$.

   (b) Find the velocity of the balloon at the instant that it is 50 feet above the ground.

56 A ball rolls down an inclined plane such that the distance (in cm) it rolls in t seconds is given by $s(t) = 2t^3 + 3t^2 + 4$, where $0 \le t \le 3$ (see figure).

   (a) What is the velocity of the ball at $t = 2$?

   (b) At what time is the velocity 30 cm/sec?

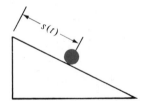

Figure for Exercise 56

3.4

INCREMENTS AND DIFFERENTIALS

Let us consider the equation $y = f(x)$ where f is a function. In many applications the independent variable x is subjected to a small change and it is necessary to find the corresponding change in the dependent variable y. If x changes from x_1 to x_2, then the amount of change is often denoted by the symbol Δx (read "delta x"); that is,

$$\Delta x = x_2 - x_1.$$

The number Δx is called an **increment** of x. Note that $x_2 = x_1 + \Delta x$; that is, the new value of x equals the initial value plus the change, Δx, in x. Similarly, Δy will denote the change in the dependent variable y that corresponds to the change Δx. Thus

$$\Delta y = f(x_2) - f(x_1) = f(x_1 + \Delta x) - f(x_1).$$

The geometric representations for these increments in terms of the graph of f are shown in Figure 3.9.

Sometimes it is convenient to let x represent the initial value of the independent variable. In this case we use the phrase x *is given an increment* Δx to describe a (small) change in this variable. Our definition of Δy then has the following form.

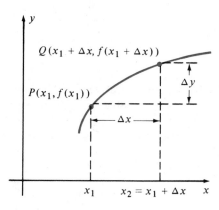

Figure 3.9

Definition (3.20)

> Let $y = f(x)$, where f is a function. If x is given an increment Δx, then the **increment Δy of y** is
>
> $$\Delta y = f(x + \Delta x) - f(x).$$

Example 1 Suppose $y = 3x^2 - 5$.

(a) If x is given an increment Δx, find Δy.

(b) Use Δy to calculate the numerical change in y if x changes from 2 to 2.1.

Solution

(a) Let $f(x) = 3x^2 - 5$. Applying Definition (3.20) gives us

$$\begin{aligned}
\Delta y &= f(x + \Delta x) - f(x) \\
&= [3(x + \Delta x)^2 - 5] - [3x^2 - 5] \\
&= 3[x^2 + 2x(\Delta x) + (\Delta x)^2] - 5 - [3x^2 - 5] \\
&= 3x^2 + 6x(\Delta x) + 3(\Delta x)^2 - 5 - 3x^2 + 5 \\
&= 6x(\Delta x) + 3(\Delta x)^2
\end{aligned}$$

(b) We wish to find Δy if $x = 2$ and $\Delta x = 0.1$. Substituting in the formula for Δy gives us

$$\begin{aligned}
\Delta y &= 6(2)(0.1) + 3(0.1)^2 \\
&= 12(0.1) + 3(0.01) = 1.2 + 0.03 = 1.23.
\end{aligned}$$

Thus, y changes by an amount 1.23 if x changes from 2 to 2.1. We could also find Δy directly as follows:

$$\Delta y = f(x + \Delta x) - f(x) = f(2.1) - f(2)$$
$$= [3(2.1)^2 - 5] - [3(2)^2 - 5] = 1.23. \qquad \blacksquare$$

The increment notation may be used in the definition of the derivative of a function. All that is necessary is to substitute Δx for h in (3.6) as follows:

$$(3.21) \qquad f'(x) = \lim_{\Delta x \to 0} \frac{f(x + \Delta x) - f(x)}{\Delta x} = \lim_{\Delta x \to 0} \frac{\Delta y}{\Delta x}.$$

In words (3.21) may be phrased as follows: "The derivative of f is the limit of the ratio of the increment Δy of the dependent variable to the increment Δx of the independent variable as Δx approaches zero." In Figure 3.9, note that $\Delta y/\Delta x$ is the slope of the secant line through P and Q. It follows from (3.21) that if $f'(x)$ exists, then

$$\frac{\Delta y}{\Delta x} \approx f'(x) \quad \text{if } \Delta x \approx 0.$$

Geometrically this implies that if Δx is close to 0, then the slope $\Delta y/\Delta x$ of the secant line through P and Q is close to the slope $f'(x)$ of the tangent line at P. We may also write

$$\Delta y \approx f'(x)\,\Delta x \quad \text{if } \Delta x \approx 0.$$

We give the expression $f'(x)\,\Delta x$ a special name in the next definition.

Definition (3.22)

Let $y = f(x)$, where f is differentiable, and let Δx be an increment of x.

(i)   The **differential dx** of the independent variable x is $dx = \Delta x$.

(ii) The **differential dy** of the dependent variable y is

$$dy = f'(x)\,\Delta x = f'(x)\,dx.$$

Observe that dy depends, for its value, on *both* x and Δx. From (i) we see that as far as the independent variable x is concerned, there is no difference between the increment Δx and the differential dx.

It follows from our previous discussion and (ii) of Definition (3.22) that if Δx is small, then

$$(3.23) \qquad \Delta y \approx dy = f'(x)\,dx = (D_x y)\,dx.$$

Consequently, if $y = f(x)$, then dy can be used as an approximation to the exact change Δy of the dependent variable corresponding to a small change Δx in x. This observation is useful in applications where only a rough estimate of the change in y is desired.

Example 2 If $y = 3x^2 - 5$, use dy to approximate Δy if x changes from 2 to 2.1.

Solution   If $f(x) = 3x^2 - 5$, then by Example 1, $\Delta y = 1.23$. Using Definition (3.22),

$$dy = f'(x)\, dx = 6x\, dx.$$

In the present example, $x = 2$, $\Delta x = dx = 0.1$, and

$$dy = (6)(2)(0.1) = 1.2.$$

Hence our approximation is correct to the nearest tenth.   ∎

Example 3 If $y = x^3$ and Δx is an increment of x, find (a) Δy; (b) dy; (c) $\Delta y - dy$; (d) the value of $\Delta y - dy$ if $x = 1$ and $\Delta x = 0.02$.

Solution

(a)   Using (3.20) with $f(x) = x^3$ we obtain

$$\Delta y = f(x + \Delta x) - f(x) = (x + \Delta x)^3 - x^3.$$

Applying the Binomial Theorem (3.12) with $n = 3$ gives us

$$\Delta y = x^3 + 3x^2(\Delta x) + 3x(\Delta x)^2 + (\Delta x)^3 - x^3$$
$$= 3x^2(\Delta x) + 3x(\Delta x)^2 + (\Delta x)^3.$$

(b)   By (ii) of Definition (3.22),

$$dy = f'(x)\, dx = 3x^2\, dx = 3x^2(\Delta x).$$

(c)   From parts (a) and (b),

$$\Delta y - dy = 3x^2(\Delta x) + 3x(\Delta x)^2 + (\Delta x)^3 - 3x^2(\Delta x)$$
$$= 3x(\Delta x)^2 + (\Delta x)^3.$$

(d)   Substituting $x = 1$ and $\Delta x = 0.02$ in part (c),

$$\Delta y - dy = 3(1)(0.02)^2 + (0.02)^3$$
$$= 3(0.0004) + (0.000008) = 0.001208.$$

This shows that if dy is used to approximate Δy when x changes from 1 to 1.02, then the error involved is approximately 0.001.   ∎

It is instructive to study the geometric interpretations of dx and dy. Consider the graphs in Figure 3.10, where l is the tangent line at the point P.

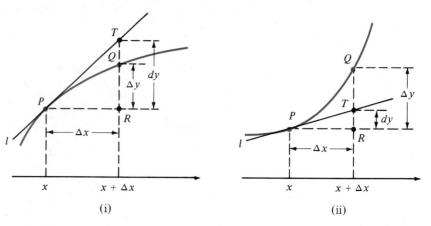

(i)   (ii)

Figure 3.10

We have pictured both Δx and Δy as positive; however, they may also take on negative values. If T is the point of intersection of l and the vertical line through Q, and if R is the point with coordinates $(x + \Delta x, f(x))$, then it can be seen from triangle PRT that the slope $f'(x)$ of the tangent line is the ratio $r/\Delta x$, where r denotes the length of the segment RT. Hence, $r/\Delta x = f'(x)$ or $r = f'(x)\,\Delta x = dy$. A similar result holds if increments are negative. In general, if x is given an increment Δx, then dy is the amount that the *tangent line* rises (or falls) when the independent variable changes from x to $x + \Delta x$. This is in contrast to the amount Δy that the *graph* rises (or falls) between P and Q. This geometric interpretation also emphasizes the fact that $dy \approx \Delta y$ if Δx is small.

The next example illustrates how differentials may be used to find certain approximation formulas.

Example 4

(a)   Use differentials to obtain a formula for approximating the volume of a thin cylindrical shell of altitude h, inner radius r, and thickness t.

(b)   What error is involved in using the formula?

Solution

(a)   A typical cylindrical shell is illustrated in Figure 3.11, where the thickness t has been denoted by Δr. We wish to find the volume contained between the "inner" cylinder of radius r and the "outer" cylinder of radius $r + \Delta r$. The volume V of the inner cylinder is

$$V = \pi r^2 h.$$

If we increase r by an amount Δr, then the volume of the shell is the change ΔV in V. Using differentials,

$$\Delta V \approx dV = (D_r V)\,\Delta r = (2\pi rh)\,\Delta r$$

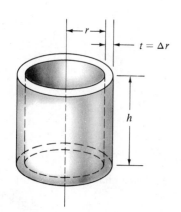

Figure 3.11

and hence a formula for approximating the volume of the shell is

$$\Delta V \approx (2\pi rh)\,\Delta r = (2\pi rh)t.$$

In words,

Volume $\approx$ (Area of inner cylinder wall) $\times$ (Thickness).

(b) The exact volume of the shell is

$$\Delta V = \pi(r + \Delta r)^2 h - \pi r^2 h$$

which simplifies to $\Delta V = (2\pi rh)\,\Delta r + \pi h(\Delta r)^2.$

The error involved in using dV to approximate ΔV is

$$\Delta V - dV = \pi h(\Delta r)^2.$$

This indicates that the approximation dV is quite accurate if Δr is small in comparison to h. ■

Differentials may sometimes be used to estimate errors, as illustrated by the next example.

Example 5   The radius of a spherical balloon is estimated to be 12 inches with a maximum error in measurement of 0.05 inches. Estimate the maximum error in the calculated volume of the sphere.

Solution   If x denotes the radius of the balloon and V the volume, then $V = \frac{4}{3}\pi x^3$. If the maximum error in the measured value of x is denoted by dx (or equivalently, by Δx), then the exact value of the radius is between $x - dx$ and $x + dx$. If we let ΔV denote the error in V caused by the error dx in x, then we may approximate ΔV by means of the differential

$$dV = (D_x V)\,dx = 4\pi x^2\,dx.$$

Letting $x = 12$ and $dx = \pm 0.05$, we obtain

$$dV = 4\pi(12)^2(\pm 0.05) = \pm(28.8)\pi \approx \pm 90.$$

Thus the maximum error in volume due to the error in measurement (± 0.05 inches) of the radius is approximately 90 cubic inches. ■

The terminology introduced in the next definition may be used to describe an error in the measurement of a quantity.

Definition (3.24)

> (a) **Average error** $= \dfrac{\textbf{error in measurement}}{\textbf{measured value}}$
>
> (b) **Percentage error** $=$ **(average error)** $\times$ **100**

The average error is also called the **relative error.** For example, if the measured length of an object is 20 inches with a possible error of 0.1 inches, then from Definition (3.24), the average error is 0.1/20 or 0.005. The significance of this number is that the error involved is, *on the average*, 0.005 inches per inch. The percentage error is defined as the relative error multiplied by 100. In the illustration just given, the percentage error is (0.005)(100) or 0.5%.

In terms of differentials, if w represents a measurement with a possible error of dw, then from our previous remarks, the average error is dw/w. Of course, if dw is an *approximation* to the error in w, then dw/w is an *approximation* to the average error. These remarks are illustrated in the next example.

Example 6   The radius of a spherical balloon is estimated to be 12 inches with a maximum error in measurement of 0.05 inches. Approximate the average error and the percentage error for

(a)   the measured value of the radius.

(b)   the calculated value of the volume.

Solution   In Example 5 we estimated the maximum error in the calculated volume. As in that example, we take the measured value of the radius as $x = 12$ inches, with an error in measurement of approximately $dx = \pm 0.05$ inches. Using (3.24) and the error approximation dx gives us the following.

(a)   For the measured value of the radius,

$$\text{Average error} \approx \frac{dx}{x} = \frac{\pm 0.05}{12} \approx \pm 0.004 \text{ inches per inch}$$

$$\text{Percentage error} \approx (\pm 0.004)(100) = \pm 0.4\%$$

(b)   The calculated volume is $V = \frac{4}{3}\pi(12)^3 = 2304\pi$. It was found, in Example 5, that $dV = \pm(28.8)\pi$. Hence

$$\text{Average error} \approx \frac{dV}{V} = \frac{\pm(28.8)\pi}{2304\pi} \approx \pm 0.01$$

$$\text{Percentage error} \approx (\pm 0.01)(100) = 1\%. \qquad \blacksquare$$

By (ii) of Definition (3.22), $dy = f'(x)dx$. If both sides of this equation are divided by dx (assuming that $dx \neq 0$) we obtain the following notation for the derivative of $y = f(x)$.

(3.25)

$$\frac{dy}{dx} = f'(x) = \lim_{\Delta x \to 0} \frac{\Delta y}{\Delta x}$$

Thus, the derivative $f'(x)$ may be expressed as a quotient of two differentials. Indeed, this is one of the reasons for defining dx and dy as in (3.22). The notation dy/dx was employed by Leibniz in his pioneering work on calculus in the late 1600s and is still widely used in mathematics and the applied sciences. It is not always essential to regard dy/dx as a quotient of two differentials. As a matter of fact dy/dx *is often thought of formally as just another symbol for the derivative* $f'(x)$. For example,

$$\text{if} \quad y = 7x^3 - 4x^2 + 2, \quad \text{then} \quad \frac{dy}{dx} = 21x^2 - 8x.$$

The symbol d/dx is used in the same manner as the operator D_x as follows.

$$(3.26) \qquad f'(x) = D_x[f(x)] = \frac{d}{dx}[f(x)] = \frac{d[f(x)]}{dx}$$

In particular, rules for differentiating sums, differences, products, and quotients may be stated as follows, where $u = f(x)$ and $v = g(x)$.

$$\frac{d}{dx}(u + v) = \frac{du}{dx} + \frac{dv}{dx} \qquad \frac{d}{dx}(u - v) = \frac{du}{dx} - \frac{dv}{dx}$$

(3.27)

$$\frac{d}{dx}(uv) = u\frac{dv}{dx} + v\frac{du}{dx} \qquad \frac{d}{dx}\left(\frac{u}{v}\right) = \frac{v\dfrac{du}{dx} - u\dfrac{dv}{dx}}{v^2}$$

EXERCISES 3.4

In Exercises 1–4 (a) use Definition (3.20) to find a general formula for Δy; (b) use Δy to calculate the change in y corresponding to the stated values of x and Δx.

1   $y = 2x^2 - 4x + 5, x = 2, \Delta x = -0.2$

2   $y = x^3 - 4, x = -1, \Delta x = 0.1$

3   $y = 1/x^2, x = 3, \Delta x = 0.3$

4   $y = 1/(2 + x), x = 0, \Delta x = -0.03$

In Exercises 5–12 find (a) Δy; (b) dy; and (c) $dy - \Delta y$.

5   $y = 3x^2 + 5x - 2$ 6 $y = 4 - 7x - 2x^2$

7   $y = 1/x$ 8 $y = 7x + 12$

9   $y = 4 - 9x$ 10 $y = 8$

11   $y = x^4$ 12 $y = x^{-2}$

In Exercises 13 and 14 use differentials to approximate the change in $f(x)$ if x changes from a to b.

13   $f(x) = 4x^5 - 6x^4 + 3x^2 - 5; a = 1, b = 1.03$

14   $f(x) = -3x^3 + 8x - 7; a = 4, b = 3.96$

15   If $f(z) = z^3 - 3z^2 + 2z - 7$ and $w = f(z)$, find dw. Use dw to approximate the change in w if z changes from 4 to 3.95.

16   If $F(t) = 1/(2 - t^2)$ and $s = F(t)$, find ds. Use ds to approximate the change in s if t changes from 1 to 1.02.

17   The radius of a circular manhole cover is estimated to be 16 in. with a maximum error in measurement of 0.06 in. Use differentials to estimate the maximum error in the calculated area of one side of the cover. What is the approximate average error and the approximate percentage error?

18   The length of a side of a square floor tile is estimated as 1 ft with a maximum error in measurement of $\frac{1}{16}$ in. Use differentials to estimate the maximum error in the calculated area. What is the approximate average error and the approximate percentage error?

19   Use differentials to approximate the increase in volume of a cube if the length of each edge changes from 10 in. to 10.1 in. What is the exact change in volume?

20   A spherical balloon is being inflated with gas. Use differentials to approximate the increase in surface area of the balloon if the diameter changes from 2 ft to 2.02 ft.

21   One side of a house has the shape of a square surmounted by an equilateral triangle. If the length of the base is measured as 48 ft, with a maximum error in measurement of 1 in., calculate the area of the side and use differentials to estimate the maximum error in the calculation. What is the approximate average error and the approximate percentage error?

22 Small errors in measurements of dimensions of large containers can have a marked effect on calculated volumes. A silo has the shape of a right circular cylinder surmounted by a hemisphere (see figure). The altitude of the cylinder is exactly 50 ft. The circumference of the base is measured as 30 ft, with a maximum error in measurement of 6 in. Calculate the volume of the silo from these measurements and use differentials to estimate the maximum error in the calculation. What is the approximate average error and the approximate percentage error?

Figure for Exercise 22

23 As sand leaks out of a container, it forms a conical pile whose altitude is always the same as the radius. If, at a certain instant, the radius is 10 cm, use differentials to approximate the change in radius that will increase the volume of the pile by 2 cm^3.

24 Use the technique illustrated in Example 4 of this section to find formulas that can be used to approximate the volume of a thin shell-shaped solid whose surface is (a) spherical; (b) cubical.

25 Newton's Law of Gravitation states that the force F of attraction between two particles having masses m_1 and m_2 is given by $F = gm_1m_2/s^2$, where g is a constant and s is the distance between the particles. If $s = 20$ cm, use differentials to approximate the change in s that will increase F by 10%.

26 Boyle's Law states that the pressure p and volume v of a confined gas are related by the formula $pv = c$, where c is a constant, or equivalently, by $p = c/v$, where $v \neq 0$. Show that dp and dv are related by means of the formula $p\,dv + v\,dp = 0$.

27 Use differentials to approximate $(0.98)^4$. (*Hint*: Let $y = f(x) = x^4$ and consider $f(x + \Delta x) = f(x) + \Delta y$, where $x = 1$ and $\Delta x = -0.02$.) What is the exact value of $(0.98)^4$?

28 Use differentials to approximate

$$N = (2.01)^4 - 3(2.01)^3 + 4(2.01)^2 - 5.$$

What is the exact value of N?

29 The area A of a square of side s is given by $A = s^2$. If s increases by an amount Δs, give geometric illustrations of dA and $\Delta A - dA$.

30 The volume V of a cube of edge x is given by $V = x^3$. If x increases by an amount Δx, give geometric illustrations of dV and $\Delta V - dV$.

31 Constriction of arterioles is a cause of high blood pressure. It has been verified experimentally that as blood flows through an arteriole of fixed length, the pressure difference between the two ends of the arteriole is inversely proportional to the fourth power of the radius. If the radius of an arteriole decreases by 10%, use differentials to find the percentage change in the pressure difference.

32 The electrical resistance R of a wire is directly proportional to its length and inversely proportional to the square of its diameter. If the length is fixed, how accurately must the diameter be measured (in terms of percentage error) to keep the percentage error in R between -3% and 3%?

3.5

THE CHAIN RULE

The rules for derivatives obtained in previous sections are limited in scope because they can only be used for sums, differences, products, and quotients that involve x^n, where n is an integer. At this stage of our work there is no rule that can be applied *directly* to an expression such as $(x^2 + 1)^3$. Clearly

$$D_x(x^2 + 1)^3 \neq 3(x^2 + 1)^2$$

for if we change the form of the expression and write

$$y = (x^2 + 1)^3 = x^6 + 3x^4 + 3x^2 + 1$$

then

$$\begin{aligned} D_x y &= 6x^5 + 12x^3 + 6x \\ &= 6x(x^4 + 2x^2 + 1) \\ &= 6x(x^2 + 1)^2. \end{aligned}$$

Hence

$$D_x(x^2 + 1)^3 = 6x(x^2 + 1)^2.$$

Since similar manipulations are rather cumbersome for higher powers such as $(x^2 + 1)^{10}$, let us find a simpler method of finding the derivative. The method we shall employ is based on expressing y as a *composite* function of x. Recall that if f and g are functions such that

$$y = f(u) \quad \text{and} \quad u = g(x),$$

and if $g(x)$ is in the domain of f, then we may write

$$y = f(u) = f(g(x)),$$

that is, y is a function of x. Indeed, the function is the same as the composite function $f \circ g$ defined in Section 1.5. Note that $(x^2 + 1)^3$ may be expressed in this way by letting

$$y = u^3 \quad \text{and} \quad u = x^2 + 1.$$

Thus, if we can obtain a general rule for differentiating $f(g(x))$, then it may be applied to $y = (x^2 + 1)^3$ and, in fact, to any expression of the form $y = [f(x)]^n$, where n is an integer.

To get some idea of the type of rule to expect, let us return to the equations

$$y = f(u) \quad \text{and} \quad u = g(x).$$

Our goal is to find a formula for the derivative dy/dx of the composite function given by $y = f(g(x))$. If f and g are differentiable, then using the differential notation (3.25),

$$\frac{dy}{du} = f'(u) \quad \text{and} \quad \frac{du}{dx} = g'(x).$$

If we consider the product

$$\frac{dy}{du}\frac{du}{dx}$$

and treat the derivatives as quotients of differentials, then the product *suggests* the following rule:

$$\frac{dy}{dx} = \frac{dy}{du}\frac{du}{dx} = f'(u)g'(x).$$

Note that this rule leads to the correct derivative of $y = (x^2 + 1)^3$, for if we write

$$y = u^3 \quad \text{and} \quad u = x^2 + 1,$$

then using the rule we obtain

$$\frac{dy}{dx} = \frac{dy}{du}\frac{du}{dx} = (3u^2)(2x) = 6xu^2 = 6x(x^2 + 1)^2,$$

which agrees with our previous result.

Although we have not *proved* that this technique is valid, it makes the next theorem plausible. In the statement of the theorem it is assumed that variables are chosen such that the composite function $f \circ g$ is defined, and that if g has a derivative at x, then f has a derivative at $g(x)$.

The Chain Rule (3.28)

If $y = f(u), u = g(x)$, and the derivatives $\dfrac{dy}{du}$ and $\dfrac{du}{dx}$ both exist, then the composite function defined by $y = f(g(x))$ has a derivative given by

$$\frac{dy}{dx} = \frac{dy}{du}\frac{du}{dx} = f'(u)g'(x) = f'(g(x))g'(x).$$

Partial Proof   Let Δx be an increment such that both x and $x + \Delta x$ are in the domain of the composite function. Since $y = f(g(x))$, the corresponding increment of y is

$$\Delta y = f(g(x + \Delta x)) - f(g(x)).$$

If the composite function has a derivative at x, then by (3.25),

$$\frac{dy}{dx} = \lim_{\Delta x \to 0} \frac{\Delta y}{\Delta x}.$$

Next consider $u = g(x)$ and let Δu be the increment of u that corresponds to Δx, that is,

$$\Delta u = g(x + \Delta x) - g(x).$$

Since

$$g(x + \Delta x) = g(x) + \Delta u = u + \Delta u$$

we may express the formula for Δy as

$$\Delta y = f(u + \Delta u) - f(u).$$

If $y = f(u)$ is differentiable at u, then

$$\frac{dy}{du} = f'(u) = \lim_{\Delta u \to 0} \frac{\Delta y}{\Delta u}.$$

Similarly, if $u = g(x)$ is differentiable at x, then

$$\frac{du}{dx} = g'(x) = \lim_{\Delta x \to 0} \frac{\Delta u}{\Delta x}.$$

Let us assume that there exists an open interval I containing x such that whenever $x + \Delta x$ is in I and $\Delta x \neq 0$, then $\Delta u \neq 0$. In this case we may write

$$\frac{dy}{dx} = \lim_{\Delta x \to 0} \frac{\Delta y}{\Delta x} = \lim_{\Delta x \to 0} \left(\frac{\Delta y}{\Delta u} \frac{\Delta u}{\Delta x} \right) = \left(\lim_{\Delta x \to 0} \frac{\Delta y}{\Delta u} \right) \left(\lim_{\Delta x \to 0} \frac{\Delta u}{\Delta x} \right)$$

provided the limits exist. Since g is differentiable at x, it is continuous at x. Hence if $\Delta x \to 0$, then $g(x + \Delta x)$ approaches $g(x)$ and, therefore, $\Delta u \to 0$. It follows that the last limit formula displayed above may be written

$$\frac{dy}{dx} = \left(\lim_{\Delta u \to 0} \frac{\Delta y}{\Delta u} \right) \left(\lim_{\Delta x \to 0} \frac{\Delta u}{\Delta x} \right)$$

$$= \left(\frac{dy}{du} \right) \left(\frac{du}{dx} \right) = f'(u)g'(x) = f'(g(x))g'(x)$$

which is what we wished to prove.

In most applications of the Chain Rule, $u = g(x)$ has the property assumed in the preceding paragraph. If g does not satisfy this property, then every open interval containing x contains a number $x + \Delta x$, with $\Delta x \neq 0$, such that $\Delta u = 0$. In this case our proof is invalid, since Δu occurs in a denominator. To construct a proof that takes functions of this type into account, it is necessary to introduce techniques that are more sophisticated than those we have used. A complete proof of the Chain Rule is given in Appendix II. □

Example 1   If $y = (3x^2 - 7x + 1)^5$, use the Chain Rule to find dy/dx.

Solution   We may express y as a composite function of x by letting

$$y = u^5 \quad \text{and} \quad u = 3x^2 - 7x + 1.$$

Applying the Chain Rule (3.28),

$$\frac{dy}{dx} = \frac{dy}{du}\frac{du}{dx} = 5u^4(6x - 7)$$
$$= 5(3x^2 - 7x + 1)^4(6x - 7). \qquad \blacksquare$$

One of the main uses for the Chain Rule is to establish other differentiation formulas. As a first illustration we shall obtain a formula for the derivative of a power of a function.

The Power Rule for Functions (3.29)

If g is a differentiable function and n is an integer, then
$$D_x[g(x)]^n = n[g(x)]^{n-1}D_x[g(x)].$$

Proof   If we let $y = u^n$ and $u = g(x)$, then $y = [g(x)]^n$ and, by the Chain Rule (3.28),

$$\frac{dy}{dx} = \frac{dy}{du}\frac{du}{dx} = nu^{n-1}D_x u = n[g(x)]^{n-1}D_x[g(x)].$$

This completes the proof.   □

The following form of the Power Rule for Functions is easier to remember than (3.29):

(3.30)
$$D_x(u^n) = nu^{n-1}D_x u$$

where $u = g(x)$ and n is an integer. Note that if $u = x$, then $D_x u = 1$ and (3.30) reduces to $D_x(x^n) = nx^{n-1}$.

Example 2   Find $f'(x)$ if $f(x) = (x^5 - 4x + 8)^7$.

Solution   Using (3.30) with $u = x^5 - 4x + 8$ and $n = 7$,

$$f'(x) = D_x(x^5 - 4x + 8)^7 = 7(x^5 - 4x + 8)^6 D_x(x^5 - 4x + 8)$$
$$= 7(x^5 - 4x + 8)^6(5x^4 - 4). \qquad \blacksquare$$

Example 3 Find y' if $y = \dfrac{1}{(4x^2 + 6x - 7)^3}$.

Solution Writing $y = (4x^2 + 6x - 7)^{-3}$ and applying the Power Rule (3.30) with $u = 4x^2 + 6x - 7$ and $n = -3$,

$$
\begin{aligned}
y' &= D_x(4x^2 + 6x - 7)^{-3} \\
&= -3(4x^2 + 6x - 7)^{-4}D_x(4x^2 + 6x - 7) \\
&= -3(4x^2 + 6x - 7)^{-4}(8x + 6) \\
&= \frac{-6(4x + 3)}{(4x^2 + 6x - 7)^4}.
\end{aligned}
$$

∎

Example 4 If $F(z) = (2z + 5)^3(3z - 1)^4$, find $F'(z)$.

Solution Using first the Product Rule, second the Power Rule, and then factoring the result, we have

$$
\begin{aligned}
F'(z) &= (2z + 5)^3 D_z(3z - 1)^4 + (3z - 1)^4 D_z(2z + 5)^3 \\
&= (2z + 5)^3 \cdot 4(3z - 1)^3(3) + (3z - 1)^4 \cdot 3(2z + 5)^2(2) \\
&= 6(2z + 5)^2(3z - 1)^3[2(2z + 5) + (3z - 1)] \\
&= 6(2z + 5)^2(3z - 1)^3(7z + 9).
\end{aligned}
$$

∎

The Chain Rule will have far-reaching consequences in our later work with differentiation and integration. For example, when we discuss the trigonometric functions in Chapter 8 it will be shown that for every real number x,

$$D_x \sin x = \cos x.$$

As a corollary, if $y = \sin u$, and $u = g(x)$, then,

$$\frac{dy}{dx} = \frac{dy}{du}\frac{du}{dx} = (\cos u)\frac{du}{dx}$$

that is,

$$D_x[\sin g(x)] = [\cos g(x)]g'(x).$$

In like manner, if we have *any* rule for the derivative of a function of x, then we can immediately use the Chain Rule (3.28) to extend it to a function of u, where $u = g(x)$.

If g is a differentiable function and $y = g(x)$, then (3.30) may be written in any of the forms

(3.31) $$D_x(y^n) = ny^{n-1}D_x y = ny^{n-1}y' = ny^{n-1}\frac{dy}{dx}.$$

The dependent variable y represents the expression $g(x)$ and consequently, when differentiating y^n it is *essential* to multiply ny^{n-1} by the derivative y'. A common error is to write $D_x y^r = ry^{r-1}$. Formula (3.31) will be extremely important in our work with implicit functions in the next section.

EXERCISES 3.5

Differentiate the functions defined in Exercises 1–26.

1 $f(x) = (x^2 - 3x + 8)^3$

2 $f(x) = (4x^3 + 2x^2 - x - 3)^2$

3 $g(x) = (8x - 7)^{-5}$

4 $k(x) = (5x^2 - 2x + 1)^{-3}$

5 $f(x) = x/(x^2 - 1)^4$

6 $g(x) = (x^4 - 3x^2 + 1)/(2x + 3)^4$

7 $f(x) = (8x^3 - 2x^2 + x - 7)^5$

8 $g(w) = (w^4 - 8w^2 + 15)^4$

9 $F(v) = (17v - 5)^{1000}$

10 $K(x) = (3x^2 - 5x + 7)^{-1}$

11 $s(t) = (4t^5 - 3t^3 + 2t)^{-2}$

12 $p(s) = 1/(8 - 5s + 7s^2)^{10}$

13 $N(x) = (6x - 7)^3(8x^2 + 9)^2$

14 $f(w) = (2w^2 - 3w + 1)(3w + 2)^4$

15 $g(z) = \left(z^2 - \dfrac{1}{z^2}\right)^6$

16 $S(t) = \left(\dfrac{3t + 4}{6t - 7}\right)^3$

17 $k(u) = (u^2 + 1)^3/(4u - 5)^5$

18 $g(x) = (3x - 8)^{-2}(7x^2 + 4)^{-3}$

19 $f(x) = \left(\dfrac{3x^2 - 5}{2x^2 + 7}\right)^2$

20 $M(z) = \dfrac{9z^3 + 2z}{(6z + 1)^3}$

21 $G(s) = (s^{-4} + 3s^{-2} + 2)^{-6}$

22 $F(v) = (v^{-1} - 2v^{-2})^{-3}$

23 $h(x) = [(2x + 1)^{10} + 1]^{10}$

24 $f(t) = \left[\left(1 + \dfrac{1}{t}\right)^{-1} + 1\right]^{-1}$

25 $F(t) = 2t(2t + 1)^2(2t + 3)^3$

26 $N(x) = \dfrac{7x(x^2 + 1)^2}{(3x + 10)^4}$

For Exercises 27–30 find (a) an equation of the tangent line to the graph of the given equation at the indicated point; and (b) the points on the graph at which the tangent line is horizontal.

27 $y = (4x^2 - 8x + 3)^4$, $P(2, 81)$

28 $y = \left(x + \dfrac{1}{x}\right)^5$, $P(1, 32)$

29 $y = (2x - 1)^{10}$, $P(1, 1)$

30 $y = (x^2 - 1)^7$, $P(0, -1)$

31 If $y = (x^4 - 3x^2 + 1)^{10}$, find dy and use it to approximate the change in y if x changes from 1 to 1.01.

32 If $w = z^3(z - 1)^5$, find dw and use it to approximate the change in w if z changes from 2 to 1.98.

33 If $w = f(z)$ and $z = g(s)$, (a) express the Chain Rule formula for dw/ds in terms of the differential notation; and (b) find dw/ds if $w = z^3 - (2/z)$ and $z = (s^2 + 1)^5$.

34 If $v = F(u)$ and $u = G(t)$, (a) express the Chain Rule formula for dv/dt in terms of the differential notation; and (b) find dv/dt if $v = (u^4 + 2u^2 + 1)^3$ and $u = 4t^2$.

35 If a body of mass m moves rectilinearly with a velocity v, then its *kinetic energy* K is given by $K = \frac{1}{2}mv^2$. If v is a function of time t, use the Chain Rule to find a formula for dK/dt.

36 As a spherical weather balloon is being inflated, its radius r is a function of time t. If V is the volume of the balloon, use the Chain Rule to find a formula for dV/dt.

37 If $f(x) = x^4 - 3x^3 + 3x + 2$ and $g(x) = x^3 - 3x^2 + 2x + 1$, use differentials to approximate the change in $f(g(x))$ if x changes from 1 to 0.99.

38 If $f(x) = x^5 + x^4 + x^3 + x^2 + x + 1$, approximate the change in $f(f(x))$ if x changes from 0 to 0.01.

39 If $k(x) = f(g(x))$ with $f(2) = -4$, $g(2) = 2$, $f'(2) = 3$, and $g'(2) = 5$, find $k'(2)$.

40 If r, s, and t are functions such that $r(x) = s(t(x))$, and if $s(0) = -1$, $t(0) = 0$, $s'(0) = -3$, and $r'(0) = 2$, find $t'(0)$.

41 If $z = k(y)$, $y = f(u)$, and $u = g(x)$, show that under suitable restrictions

$$\frac{dz}{dx} = \frac{dz}{dy}\frac{dy}{du}\frac{du}{dx}.$$

Extend this result to any number of functions.

42 A function f is *even* if $f(-x) = f(x)$ for all x in its domain D, whereas f is *odd* if $f(-x) = -f(x)$ for all x in D. Suppose f is differentiable. Use the Chain Rule to prove that (a) if f is even, then f' is odd, and, (b) if f is odd, then f' is even. Use polynomial functions to give examples of (a) and (b).

43 The **normal line** at a point $P(x_1, y_1)$ on the graph of a differentiable function f is defined as the line through P that is perpendicular to the tangent line. If $f'(x_1) \neq 0$, prove that an equation of the normal line at $P(x_1, y_1)$ is

$$y - y_1 = -\frac{1}{f'(x_1)}(x - x_1).$$

44 Refer to Exercise 43. What is an equation of the normal line if $f'(x_1) = 0$?

In Exercises 45–48 find an equation of the normal line to the graph of f at the indicated point P. Sketch the graph, showing the normal line at P.

45 $f(x) = x^2 + 1$, $P(1, 2)$

46 $f(x) = 8 - x^3$, $P(1, 7)$

47 $f(x) = 1/(x^2 + 1)$, $P(1, \frac{1}{2})$

48 $f(x) = (x - 1)^4$, $P(2, 1)$

49 Find an equation of the line with slope 4 that is normal to the graph of $y = 1/(8x + 3)^2$.

50 Find the points on the graph of $y = (2x - 1)^3$ at which the normal line is parallel to the line $x + 12y = 36$.

51 In Chapter 7 a function denoted by ln is defined that has the property $D_x \ln x = 1/x$ for all $x > 0$. Use the Chain Rule to find $D_x \ln (x^4 + 2x^2 + 5)$.

52 In Chapter 7 a function f is defined that has the property $D_x[f(x)] = f(x)$ for all x. Find a formula for $D_x[f(g(x))]$, where g is any differentiable function.

3.6

IMPLICIT DIFFERENTIATION

Given an equation of the form

$$y = 2x^2 - 3$$

it is customary to say that y *is a function of* x, since we may write

$$y = f(x), \quad \text{where} \quad f(x) = 2x^2 - 3.$$

The equation

$$4x^2 - 2y = 6$$

defines the same function f, since solving for y gives us

$$-2y = -4x^2 + 6, \quad \text{or} \quad y = 2x^2 - 3.$$

In the case $4x^2 - 2y = 6$ we say that y (or f) is an **implicit function** of x, or that f is determined *implicitly* by the equation. Substituting $f(x)$ for y in $4x^2 - 2y = 6$ we obtain

$$4x^2 - 2f(x) = 6$$
$$4x^2 - 2(2x^2 - 3) = 6$$
$$4x^2 - 4x^2 + 6 = 6$$

which is an identity since it is true for every x in the domain of f. This fact is characteristic of every function f determined implicitly by an equation in x and y; that is, f is implicit if and only if substitution of $f(x)$ for y leads to an identity. Since $(x, f(x))$ is a point on the graph of f, the last statement implies that *the graph of the implicit function f coincides with a portion (or all) of the graph of the equation*. This is illustrated in the next example, where it is also shown that an equation in x and y may determine more than one implicit function.

Example 1 How many different functions are determined implicitly by the equation $x^2 + y^2 = 1$?

Solution   The graph of $x^2 + y^2 = 1$ is the unit circle with center at the origin. Solving the equation for y in terms of x we obtain

$$y = \pm\sqrt{1 - x^2}.$$

Two obvious implicit functions are given by

$$f(x) = \sqrt{1 - x^2} \quad \text{and} \quad g(x) = -\sqrt{1 - x^2}.$$

The graphs of f and g are the upper and lower halves, respectively, of the unit circle (see (i) and (ii) of Figure 3.12). To find other implicit functions we may let a be any number between -1 and 1 and then define the function k by

$$k(x) = \begin{cases} \sqrt{1 - x^2} & \text{if } -1 \le x \le a \\ -\sqrt{1 - x^2} & \text{if } a < x \le 1. \end{cases}$$

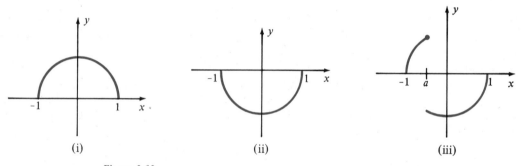

(i)   (ii)   (iii)

Figure 3.12

The graph of k is sketched in (iii) of Figure 3.12. Note that there is a discontinuity at $x = a$. The function k is determined implicitly by the equation $x^2 + y^2 = 1$, since

$$x^2 + [k(x)]^2 = 1$$

for every x in the domain of k. By letting a take on different values, it is possible to obtain as many implicit functions as desired. There are also many other types of functions that are determined implicitly by $x^2 + y^2 = 1$. As mentioned previously, the graph of each implicit function is a portion of the graph of the equation.   ■

It is not obvious that the equation

$$y^4 + 3y - 4x^3 = 5x + 1$$

determines an implicit function f in the sense that if $f(x)$ is substituted for y, then

$$[f(x)]^4 + 3[f(x)] - 4x^3 = 5x + 1$$

is an identity for all x in the domain of f. It is possible to state conditions under which an implicit function exists and is differentiable at numbers in its domain; however, the proof requires advanced methods and hence is omitted here. In the examples that follow it will be assumed that a given equation in x and y determines a differentiable function f such that if $f(x)$ is substituted for y, the equation is an identity for all x in the domain of f. The derivative of f may then be found by the method of **implicit differentiation** as illustrated in the following example.

Example 2 Assuming that the equation $y^4 + 3y - 4x^3 = 5x + 1$ determines, implicitly, a differentiable function f such that $y = f(x)$, find its derivative.

Solution   We think of y as a symbol that denotes $f(x)$ and regard the equation as an identity for all x in the domain of f. It follows that the derivatives of both sides are equal, that is,

$$D_x(y^4 + 3y - 4x^3) = D_x(5x + 1), \quad \text{or}$$

(∗) $$D_x(y^4) + D_x(3y) - D_x(4x^3) = D_x(5x) + D_x(1).$$

It is important to remember that, in general, $D_x y^4 \neq 4y^3$. Instead, since $y = f(x)$, we use (3.31), obtaining $D_x(y^4) = 4y^3 y'$. In like manner, $D_x(3y) = 3D_x y = 3y'$. Substituting in (∗),

$$4y^3 y' + 3y' - 12x^2 = 5 + 0.$$

We now solve for y'. Thus

$$(4y^3 + 3)y' = 12x^2 + 5$$

and $$y' = \frac{12x^2 + 5}{4y^3 + 3}$$

provided the denominator is not zero. In terms of f, this becomes

$$f'(x) = \frac{12x^2 + 5}{4(f(x))^3 + 3}. \qquad \blacksquare$$

The last two equations in the solution of Example 2 bring out a disadvantage in using the method of implicit differentiation, namely that the formula for y' (or $f'(x)$) may contain the expression y (or $f(x)$). However, these formulas can still be extremely useful in analyzing f and its graph.

In the next example, implicit differentiation is used to find the slope of the tangent line at a point $P(a, b)$ on the graph of an equation. In problems of this type we shall assume that the equation determines an implicit function f whose graph coincides with the graph of the equation for all x in some open interval containing a. It is also important to note that, since $P(a, b)$ is a point on the graph, the ordered pair (a, b) must be a solution of the equation.

Example 3 Find the slope of the tangent line to the graph of

$$y^4 + 3y - 4x^3 = 5x + 1$$

at the point $P(1, -2)$.

Solution   Note that $P(1, -2)$ is on the graph since

$$(-2)^4 + 3(-2) - 4(1)^3 = 5(1) + 1.$$

By Definition (3.1) the slope m of the tangent line at $P(1, -2)$ is the value of the derivative y' when $x = 1$ and $y = -2$. The given equation is the same as that in Example 2, where it was found that $y' = (12x^2 + 5)/(4y^3 + 3)$. Substituting 1 for x and -2 for y gives us

$$m = \frac{12(1)^2 + 5}{4(-2)^3 + 3} = -\frac{17}{29}. \qquad \blacksquare$$

Example 4 Assume that $x^2 + y^2 = 1$ determines a function f such that $y = f(x)$. Find y'.

Solution   As in Example 2 we differentiate both sides with respect to x, obtaining

$$D_x(x^2) + D_x(y^2) = D_x(1).$$

Consequently,

$$2x + 2yy' = 0, \quad \text{or} \quad yy' = -x$$

and
$$y' = -\frac{x}{y}, \quad \text{if } y \neq 0. \qquad \blacksquare$$

The equation $x^2 + y^2 = 1$ determines many implicit functions (see Example 1). It can be shown that the method of implicit differentiation provides the derivative of *any* differentiable function determined by an equation in two variables. To illustrate, the slope of the tangent line at the point (x, y) on any of the graphs in Figure 3.12 is given by $y' = -x/y$, provided the derivative exists.

Example 5 Find y' if $4xy^3 - x^2y + x^3 - 5x + 6 = 0$.

Solution   Differentiating both sides of the equation with respect to x,

$$D_x(4xy^3) - D_x(x^2y) + D_x(x^3) - D_x(5x) + D_x(6) = D_x(0).$$

Since y denotes $f(x)$, where f is a function, the Product Rule must be applied to $D_x(4xy^3)$ and $D_x(x^2y)$. Thus

$$
\begin{aligned}
D_x(4xy^3) &= 4xD_x(y^3) + y^3D_x(4x) \\
&= 4x(3y^2y') + y^3(4) \\
&= 12xy^2y' + 4y^3
\end{aligned}
$$

and
$$D_x(x^2y) = x^2 D_x y + y D_x(x^2)$$
$$= x^2 y' + y(2x).$$

Substituting these in the first equation of the solution and differentiating the other terms leads to

$$(12xy^2 y' + 4y^3) - (x^2 y' + 2xy) + 3x^2 - 5 = 0.$$

Collecting the terms containing y' and transposing the remaining terms to the right side of the equation gives us

$$(12xy^2 - x^2)y' = 5 - 3x^2 + 2xy - 4y^3.$$

Consequently,
$$y' = \frac{5 - 3x^2 + 2xy - 4y^3}{12xy^2 - x^2}. \qquad \blacksquare$$

EXERCISES 3.6

Find at least one function f determined implicitly by each of the equations in Exercises 1–8 and state the domain of f.

1 $3x - 2y + 4 = 2x^2 + 3y - 7x$

2 $x^3 - xy + 4y = 1$

3 $x^2 + y^2 - 16 = 0$

4 $3x^2 - 4y^2 = 12$

5 $x^2 - 2xy + y^2 = x$

6 $3x^2 - 4xy + y^2 = 0$

7 $\sqrt{x} + \sqrt{y} = 1$

8 $|x - y| = 2$

Assuming that each of the equations in Exercises 9–20 determines a function f such that $y = f(x)$, find y'.

9 $8x^2 + y^2 = 10$

10 $4x^3 - 2y^3 = x$

11 $2x^3 + x^2y + y^3 = 1$

12 $5x^2 + 2x^2y + y^2 = 8$

13 $5x^2 - xy - 4y^2 = 0$

14 $x^4 + 4x^2y^2 - 3xy^3 + 2x = 0$

15 $(1/x^2) + (1/y^2) = 1$

16 $x = y + 2y^2 + 3y^3$

17 $x^2y^3 + 4xy + x - 6y = 2$

18 $4 - 7xy = (y^2 + 4)^5$

19 $(y^2 - 9)^4 = (4x^2 + 3x - 1)^2$

20 $(1 + xy)^3 = 2x^2 - 9$

In Exercises 21–24 find an equation of the tangent line to the graph of the given equation at the indicated point P.

21 $xy + 16 = 0$, $P(-2, 8)$

22 $y^2 - 4x^2 = 5$, $P(-1, 3)$

23 $2x^3 - x^2y + y^3 - 1 = 0$, $P(2, -3)$

24 $(1/x) + (3/y) = 1$, $P(2, 6)$

25 Show that the equation $x^2 + y^2 + 1 = 0$ does not determine a function f such that $y = f(x)$.

26 Show that the equation $y^2 = x$ determines an infinite number of implicit functions. Sketch the graphs of four such functions.

27 How many implicit functions are determined by the following equations?
(a) $x^4 + y^4 - 1 = 0$ (b) $x^4 + y^4 = 0$
(c) $y^2 + \sqrt{x} + 4 = 0$

28 Use implicit differentiation to prove that if P is any point on the circle $x^2 + y^2 = a^2$, then the tangent line at P is perpendicular to OP.

29 Suppose that $3x^2 - x^2y^3 + 4y = 12$ determines a differentiable function f such that $y = f(x)$. If $f(2) = 0$, use differentials to approximate the change in $f(x)$ if x changes from 2 to 1.97.

30 Suppose that $x^3 + xy + y^4 = 19$ determines a differentiable function f such that $y = f(x)$. If $P(1, 2)$ is a point on the graph of f, use differentials to approximate the y-coordinate b of the point $Q(1.10, b)$ on the graph.

3.7

DERIVATIVES INVOLVING POWERS OF FUNCTIONS

The Power Rule (3.13) can be extended to the case where the exponent is a rational number. Thus, if

$$y = x^{m/n}$$

where m and n are integers and $n \neq 0$, then

$$y^n = x^m.$$

Assuming that y' exists and no zero denominators occur we have, by implicit differentiation,

$$ny^{n-1} \frac{dy}{dx} = mx^{m-1},$$

or

$$\frac{dy}{dx} = \frac{m}{n} x^{m-1} y^{1-n} = \frac{m}{n} x^{m-1} (x^{m/n})^{1-n}.$$

Simplifying the right side by means of laws of exponents gives us the following **Power Rule for Rational Exponents:** *If m and n are integers and n ≠ 0, then*

$$D_x(x^{m/n}) = \frac{m}{n} x^{(m/n)-1}.$$

Our proof of this formula is incomplete in the sense that we have assumed the existence of y'. To give a complete proof the definition of derivative (3.6) could be used.

Example 1   Find y' if $y = 6\sqrt[3]{x^4} + 4/\sqrt{x}$.

Solution   Introducing rational exponents, $y = 6x^{4/3} + 4x^{-1/2}$. Using the Power Rule,

$$y' = 6(4/3)x^{(4/3)-1} + 4(-1/2)x^{-(1/2)-1}$$
$$= 8x^{1/3} - 2x^{-3/2}$$
$$= 8x^{1/3} - 2/x^{3/2}$$

or, in terms of radicals,

$$y' = 8\sqrt[3]{x} - 2/\sqrt{x^3}.\qquad\blacksquare$$

It can also be shown that the Power Rule for Functions (3.29) is true for every *rational* exponent. The following examples illustrate uses of this rule.

Example 2   Find $f'(x)$ if $f(x) = \sqrt[3]{5x^2 - x + 4}$.

Solution   Writing $f(x) = (5x^2 - x + 4)^{1/3}$ and using the Power Rule for Functions with $n = 1/3$,

$$f'(x) = \tfrac{1}{3}(5x^2 - x + 4)^{-2/3} D_x(5x^2 - x + 4)$$
$$= \left(\frac{1}{3}\right) \frac{1}{(5x^2 - x + 4)^{2/3}} (10x - 1)$$
$$= \frac{10x - 1}{3\sqrt[3]{(5x^2 - x + 4)^2}}.\qquad\blacksquare$$

Example 3 Find dy/dx if $y = (3x + 1)^6\sqrt{2x - 5}$.

Solution   Since $y = (3x + 1)^6(2x - 5)^{1/2}$, we have, by the Product and Power Rules,

$$\frac{dy}{dx} = (3x + 1)^6 \tfrac{1}{2}(2x - 5)^{-1/2}(2) + (2x - 5)^{1/2}6(3x + 1)^5(3)$$

$$= \frac{(3x + 1)^6}{\sqrt{2x - 5}} + 18(3x + 1)^5\sqrt{2x - 5}$$

$$= \frac{(3x + 1)^6 + 18(3x + 1)^5(2x - 5)}{\sqrt{2x - 5}}$$

$$= \frac{(3x + 1)^5(39x - 89)}{\sqrt{2x - 5}}.$$ ∎

The next example is of interest because it illustrates the fact that after the Power Rule is applied to $[g(x)]^r$, it may be necessary to apply it again to find $g'(x)$.

Example 4 Find $f'(x)$ if $f(x) = (7x + \sqrt{x^2 + 6})^4$.

Solution   Applying the Power Rule,

$$f'(x) = 4(7x + \sqrt{x^2 + 6})^3 D_x(7x + \sqrt{x^2 + 6})$$
$$= 4(7x + \sqrt{x^2 + 6})^3[D_x(7x) + D_x\sqrt{x^2 + 6}].$$

Again applying the Power Rule,

$$D_x\sqrt{x^2 + 6} = D_x(x^2 + 6)^{1/2} = \tfrac{1}{2}(x^2 + 6)^{-1/2}D_x(x^2 + 6)$$

$$= \frac{1}{2\sqrt{x^2 + 6}}(2x) = \frac{x}{\sqrt{x^2 + 6}}.$$

Therefore,

$$f'(x) = 4(7x + \sqrt{x^2 + 6})^3\left[7 + \frac{x}{\sqrt{x^2 + 6}}\right].$$ ∎

EXERCISES 3.7

Differentiate the functions defined in Exercises 1–26.

1 $f(x) = \sqrt[3]{x^2} + 4\sqrt{x^3}$

2 $f(x) = 10\sqrt[5]{x^3} + \sqrt[3]{x^5}$

3 $k(r) = \sqrt[3]{8r^3 + 27}$

4 $h(z) = (2z^2 - 9z + 8)^{-2/3}$

5 $F(v) = 5/\sqrt[5]{v^5 - 32}$

6 $k(s) = 1/\sqrt{3s - 4}$

7 $f(x) = \sqrt{2x}$

8 $g(x) = \sqrt[5]{1/x}$

9 $f(z) = 10\sqrt{z^3} + 3/\sqrt[3]{z}$

10 $f(t) = \sqrt[3]{t^2} - (1/\sqrt{t^3})$

11 $g(w) = (w^2 - 4w + 3)/w^{3/2}$

12 $K(x) = 8x^2\sqrt{x} + 3x\sqrt[3]{x}$

13 $M(x) = \sqrt{4x^2 - 7x + 4}$

14 $F(s) = \sqrt[3]{5s - 8}$

15 $f(t) = 4/(9t^2 + 16)^{2/3}$

16 $k(v) = 1/\sqrt{(v^4 + 7v^2)^3}$

17 $H(u) = \sqrt{\dfrac{3u + 8}{2u + 5}}$

18 $G(x) = \left(\dfrac{x}{x^2 + 1}\right)^{5/2}$

19 $k(s) = \sqrt[4]{s^2 + 9}\,(4s + 5)^4$

20 $g(y) = (15y + 2)(y^2 - 2)^{3/4}$

21 $h(x) = (x^2 + 4)^{5/3}(x^3 + 1)^{3/5}$

22 $f(w) = \sqrt{w^3(9w + 1)^5}$

23 $g(z) = \sqrt[3]{2z + 3}/\sqrt{3z + 2}$

24 $H(x) = (2x + 3)/\sqrt{4x^2 + 9}$

25 $f(x) = (7x + \sqrt{x^2 + 3})^6$

26 $p(z) = \sqrt{1 + \sqrt{1 + 2z}}$

In Exercises 27–30 find an equation of the tangent line to the graph of the equation at the point P.

27 $y = \sqrt{2x^2 + 1}$, $P(-1, \sqrt{3})$

28 $y = (5x - 8)^{1/3}$, $P(7, 3)$

29 $y = 4x^{2/3} + 2x^{-1/3} - 10$, $P(-8, 5)$

30 $y = 4x/\sqrt{x + 1}$, $P(3, 6)$

31   Find the point P on the graph of $y = \sqrt{2x - 4}$ such that the tangent line at P passes through the origin.

32   Find the points on the graph of $y = x^{5/3} + x^{1/3}$ at which the tangent line is perpendicular to the line $2y + x = 7$.

Assuming that each of the equations in Exercises 33–38 determines a function f such that $y = f(x)$, find y'.

33 $\sqrt{x} + \sqrt{y} = 100$

34 $x^{2/3} + y^{2/3} = 4$

35 $6x + \sqrt{xy} - 3y = 4$

36 $xy^2 + \sqrt[3]{xy} + x^4 = 7$

37 $3x^2 + \sqrt[3]{xy} = 2y^2 + 20$

38 $2x - \sqrt{xy} + y^3 = 16$

In Exercises 39 and 40 find dy.

39 $y = \sqrt[3]{6x + 11}$

40 $y = 1/\sqrt{x^2 + 1}$

41   Use differentials to approximate $\sqrt[3]{65}$.

42   Use differentials to approximate $\sqrt{99}$.

43   Use differentials to approximate the change in $f(x) = (4x^2 + 9)^{3/2}$ if x changes from 2 to 1.998.

44   Use differentials to approximate

$$N = 5(1.01)^{3/5} - 3(1.01)^{1/5} + 7.$$

45   The curved surface area S of a right circular cone having altitude h and base radius r is given by $S = \pi r\sqrt{r^2 + h^2}$. For a certain cone, $r = 6$ cm and the altitude is measured as 8 cm, with a maximum error in measurement of 0.1 cm. Calculate S from these measurements and use differentials to estimate the maximum error in the calculation. What is the approximate percentage error?

46   The period T of a simple pendulum of length l may be calculated by means of the formula $T = 2\pi\sqrt{l/g}$, where g is a constant. Use differentials to approximate the change in l that will increase T by 1 %.

In Exercises 47–50, use the fact that $|a| = \sqrt{a^2}$ to find $f'(x)$. In addition, find the domain of f' and sketch the graph of f.

47 $f(x) = |1 - x|$ **48** $f(x) = |x^3 - x|$

49 $f(x) = |1 - x^2|$ **50** $f(x) = x/|x|$

3.8

HIGHER ORDER DERIVATIVES

The derivative of a function f leads to another function f'. If f' has a derivative, it is denoted by f'' and is called the **second derivative** of f. Thus

$$f''(x) = D_x(f'(x)) = D_x(D_x(f(x)))$$

for every x such that the indicated derivatives exist. The expression on the right in this equation is abbreviated $D_x^2 f(x)$. In like manner the **third derivative** f''' of f is the derivative of the second derivative. Specifically,

$$f'''(x) = D_x(f''(x)) = D_x(D_x^2 f(x)) = D_x^3 f(x).$$

In general, if n is a positive integer, then $f^{(n)}$ denotes the nth derivative of f and is found by starting with f and differentiating, successively, n times. Using operator notation, $f^{(n)}(x) = D_x^n f(x)$. The integer n is called the **order** of the derivative $f^{(n)}(x)$.

If $y = f(x)$, then the first n derivatives are denoted by

$$D_x y, \quad D_x^2 y, \quad D_x^3 y, \quad \ldots, \quad D_x^n y$$

or

$$y', \quad y'', \quad y''', \quad \ldots, \quad y^{(n)}.$$

If the differential notation is used we write

$$\frac{dy}{dx}, \quad \frac{d^2 y}{dx^2}, \quad \frac{d^3 y}{dx^3}, \quad \ldots, \quad \frac{d^n y}{dx^n}.$$

In this case, the symbols used for higher derivatives should *not* be interpreted as quotients.

Example 1   If $f(x) = 4x^2 - 5x + 8 - (3/x)$, find the first four derivatives of $f(x)$.

Solution   Since $f(x) = 4x^2 - 5x + 8 - 3x^{-1}$,

$$f'(x) = 8x - 5 + 3x^{-2} = 8x - 5 + \frac{3}{x^2}$$

$$f''(x) = 8 - 6x^{-3} = 8 - \frac{6}{x^3}$$

$$f'''(x) = 18x^{-4} = \frac{18}{x^4}$$

$$f^{(4)}(x) = -72x^{-5} = -\frac{72}{x^5}. \qquad \blacksquare$$

The next example illustrates a technique for finding higher derivatives of implicit functions.

Example 2   Find y'' if $y^4 + 3y - 4x^3 = 5x + 1$.

Solution   The equation was investigated in Example 3 of Section 3.6, where we found that

$$y' = (12x^2 + 5)/(4y^3 + 3).$$

Hence, $$y'' = D_x(y') = D_x\left(\frac{12x^2 + 5}{4y^3 + 3}\right).$$

We now use the quotient rule, differentiating implicitly as follows:

$$y'' = \frac{(4y^3 + 3)D_x(12x^2 + 5) - (12x^2 + 5)D_x(4y^3 + 3)}{(4y^3 + 3)^2}$$

$$= \frac{(4y^3 + 3)(24x) - (12x^2 + 5)(12y^2 y')}{(4y^3 + 3)^2}.$$

Substituting for y' yields

$$y'' = \frac{(4y^3 + 3)(24x) - (12x^2 + 5) \cdot 12y^2\left(\dfrac{12x^2 + 5}{4y^3 + 3}\right)}{(4y^3 + 3)^2}$$

$$= \frac{(4y^3 + 3)^2(24x) - 12y^2(12x^2 + 5)^2}{(4y^3 + 3)^3}.$$ ∎

There are numerous mathematical and physical applications of higher derivatives. Some of them will be discussed in the next chapter.

EXERCISES 3.8

Find the first and second derivatives of the functions defined in Exercises 1–10.

1 $f(x) = 3x^4 - 4x^2 + x - 2$

2 $g(x) = 3x^8 - 2x^5$

3 $H(s) = \sqrt[3]{s} + (2/s^2)$

4 $F(t) = t^{3/2} - 2t^{1/2} + 4t^{-1/2}$

5 $g(z) = \sqrt{3z + 1}$

6 $k(s) = (s^2 + 4)^{2/3}$

7 $k(r) = (4r + 7)^5$

8 $f(x) = \sqrt[5]{10x + 7}$

9 $f(x) = \sqrt{x^2 + 4}$

10 $h(x) = 1$

In Exercises 11–16 find $D_x^3 y$.

11 $y = 2x^5 + 3x^3 - 4x + 1$

12 $y = \sqrt{2 - 5x}$

13 $y = (2x - 3)/(3x + 1)$

14 $y = 1/(x^2 + 4)$

15 $y = \sqrt[3]{2 - 9x}$

16 $y = (3x + 1)^4$

Assuming that each of the equations in Exercises 17–20 determines a function f, such that $y = f(x)$, find y''.

17 $x^3 - y^3 = 1$

18 $x^2 y^3 = 1$

19 $x^2 - 3xy + y^2 = 4$

20 $\sqrt{xy} - y + x = 0$

21 Find all the nonzero derivatives of

$$f(x) = x^6 - 2x^4 + 3x^3 - x + 2.$$

22 Find all the nonzero derivatives of $f(x) = (x^2 - 1)^3$.

23 If $f(x) = 1/x$, find a formula for $f^{(n)}(x)$ where n is any positive integer. What is $f^{(n)}(1)$?

24 If $f(x) = \sqrt{x}$, find a formula for $f^{(n)}(x)$ where n is any positive integer.

25 If $f(x)$ is a polynomial of degree n, show that $f^{(k)}(x) = 0$ if $k > n$.

26 Find a polynomial $f(x)$ of degree 2 such that $f(1) = 5$, $f'(1) = 3$, and $f''(1) = -4$.

27 If $f(x) = 2x^3 + 3x^2 - 12x + 7$, find the value of f'' at each zero of f'.

28 If $f(x) = x^4 - x^3 - 6x^2 + 7x$, find an equation of the tangent line to the graph of f' at the point $P(2, 3)$.

29 If $f(x) = x^4 - 10x^2 + x + 2$, use differentials to approximate the change in $f'(x)$ if x changes from 2 to 2.005.

30 Suppose $u = f(x)$ and $v = g(x)$ where f and g have derivatives of all orders. If $y = uv$, prove that

$$y'' = u''v + 2u'v' + uv''$$

$$y''' = u'''v + 3u''v' + 3u'v'' + uv'''$$

$$y^{(4)} = u^{(4)}v + 4u'''v' + 6u''v'' + 4u'v''' + uv^{(4)}.$$

Formulate a conjecture for $y^{(n)}$.

31 If $y = f(g(x))$ and f'' and g'' exist, use the Chain Rule to express $D_x^2 y$ in terms of the first and second derivatives of f and g.

32 Suppose f has a second derivative. Prove that (a) if f is an even function, then f'' is even; and (b) if f is an odd function, then f'' is odd. Illustrate these facts by using polynomial functions. (*Hint*: See Exercise 42 of Section 3.5.)

3.9

NEWTON'S METHOD

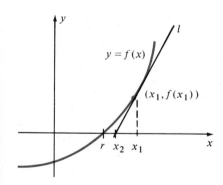

Figure 3.13

In this section we shall describe a method for approximating a real zero of a differentiable function f, that is, a real number r such that $f(r) = 0$. To use the method we begin by making a first approximation x_1 to the zero r. Since r is an x-intercept of the graph of f, a suitable number x_1 can usually be found by referring to a rough sketch of the graph. If we consider the tangent line l to the graph of f at the point $(x_1, f(x_1))$, and if x_1 is sufficiently close to r, then as illustrated in Figure 3.13, the x-intercept x_2 of l should be a better approximation to r.

Since the slope of l is $f'(x_1)$, an equation of the tangent line is

$$y - f(x_1) = f'(x_1)(x - x_1).$$

The x-intercept x_2 is given by

$$0 - f(x_1) = f'(x_1)(x_2 - x_1).$$

If $f'(x_1) \neq 0$, the preceding equation is equivalent to

$$x_2 = x_1 - \frac{f(x_1)}{f'(x_1)}.$$

If we take x_2 as a second approximation to r, then the process may be repeated by using the tangent line at $(x_2, f(x_2))$. If $f'(x_2) \neq 0$, this leads to a third approximation x_3, given by

$$x_3 = x_2 - \frac{f(x_2)}{f'(x_2)}.$$

The process is continued until the desired degree of accuracy is obtained. This technique of successive approximations of real zeros is referred to as **Newton's Method**, which we state as follows.

Newton's Method (3.32)

Let f be a differentiable function and suppose r is a real zero of f. If x_n is an approximation to r, then the next approximation is

$$x_{n+1} = x_n - \frac{f(x_n)}{f'(x_n)}$$

provided $f'(x_n) \neq 0$.

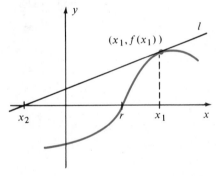

Figure 3.14

Newton's Method does not guarantee that x_{n+1} is a better approximation to r than x_n for every n. In particular, some care must be exercised in choosing the first approximation x_1. Indeed, if x_1 is not sufficiently close to r, it is possible for the second approximation x_2 to be worse than x_1, as illustrated in Figure 3.14. It is evident that we should not choose a number x_n such that $f'(x_n)$ is close to 0, for then the tangent line l is almost horizontal.

We shall use the following rule when applying Newton's Method. If an approximation to k decimal places is required, we shall approximate each of the numbers $x_2, x_3, \ldots$ to k decimal places, continuing the process until two consecutive approximations are the same. This is illustrated in the following examples.

Example 1 Use Newton's Method to approximate $\sqrt{7}$ to five decimal places.

Solution The stated problem is equivalent to that of approximating the positive real zero r of $f(x) = x^2 - 7$. Since $f(2) = -3$ and $f(3) = 2$, it follows from the continuity of f that $2 < r < 3$. Moreover, since f is increasing there can be only one zero in the interval $(2, 3)$.

If x_n is any approximation to r, then by (3.32) the next approximation x_{n+1} is given by

$$x_{n+1} = x_n - \frac{f(x_n)}{f'(x_n)} = x_n - \frac{x_n^2 - 7}{2x_n}.$$

Let us choose $x_1 = 2.5$ as a first approximation. Using the formula for x_{n+1} with $n = 1$ gives us

$$x_2 = 2.5 - \frac{(2.5)^2 - 7}{2(2.5)} = 2.65000.$$

Again using the formula (with $n = 2$) we obtain the next approximation

$$x_3 = 2.65000 - \frac{(2.65000)^2 - 7}{2(2.65000)} \approx 2.64575.$$

Repeating the procedure (with $n = 3$),

$$x_4 = 2.64575 - \frac{(2.64575)^2 - 7}{2(2.64575)} \approx 2.64575.$$

Since two consecutive values of x_n are the same (to the desired degree of accuracy) we have $\sqrt{7} \approx 2.64575$. It can be shown that to nine decimal places, $\sqrt{7} \approx 2.645751311$. ∎

Example 2 Find, to four decimal places, the largest positive real root of the equation $x^3 - 3x + 1 = 0$.

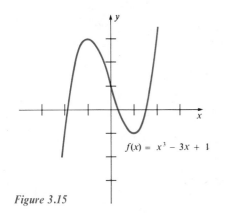

$f(x) = x^3 - 3x + 1$

Figure 3.15

Solution If we let $f(x) = x^3 - 3x + 1$, then the problem is equivalent to finding the largest positive real *zero* of f. The graph of f is sketched in Figure 3.15. Note that f has three real zeros. We wish to find the zero that lies between 1 and 2. Since $f'(x) = 3x^2 - 3$, the formula for x_{n+1} in Newton's Method is

$$x_{n+1} = x_n - \frac{x_n^3 - 3x_n + 1}{3x_n^2 - 3}.$$

Referring to the graph we take, as a first approximation, $x_1 = 1.5$ and proceed as follows:

$$x_2 = 1.5 - \frac{(1.5)^3 - 3(1.5) + 1}{3(1.5)^2 - 3} \approx 1.5333$$

$$x_3 = 1.5333 - \frac{(1.5333)^3 - 3(1.5333) + 1}{3(1.5333)^2 - 3} \approx 1.5321$$

$$x_4 = 1.5321 - \frac{(1.5321)^3 - 3(1.5321) + 1}{3(1.5321)^2 - 3} \approx 1.5321.$$

Thus the desired approximation is 1.5321. The remaining two real roots can be approximated in similar fashion (see Exercise 15). ∎

EXERCISES 3.9

In Exercises 1 and 2 use Newton's Method to approximate the given number to four decimal places.

1 $\sqrt[3]{2}$

2 $\sqrt[5]{3}$

In Exercises 3–8 use Newton's Method to approximate the indicated real root to four decimal places.

3 The positive root of $x^3 + 5x - 3 = 0$

4 The largest root of $2x^3 - 4x^2 - 3x + 1 = 0$

5 The root of $x^4 + 2x^3 - 5x^2 + 1 = 0$ that is between 1 and 2

6 The root of $x^4 - 5x^2 + 2x - 5 = 0$ that is between 2 and 3

7 The root of $x^5 + x^2 - 9x - 3 = 0$ that is between -2 and -1.

8 The root of $x^5 + x^3 + 2x - 5 = 0$ that is between 1 and 2.

In Exercises 9 and 10 approximate the largest zero of $f(x)$ to five decimal places.

9 $f(x) = x^4 - 11x^2 - 44x - 24$

10 $f(x) = x^3 - 36x - 84$

In Exercises 11–18 use Newton's Method to find all real roots of the equation to two decimal places.

11 $x^4 = 125$

12 $10x^2 - 1 = 0$

13 $x^4 - x - 2 = 0$

14 $x^5 - 2x^2 + 4 = 0$

15 $x^3 - 3x + 1 = 0$

16 $x^3 + 2x^2 - 8x - 3 = 0$

17 $x^4 - x^3 - 10x^2 - x + 1 = 0$

18 $4x^3 - 4x + 1 = 0$

3.10

REVIEW

Define or explain each of the following.

1 The slope of the tangent line to the graph of a function

2 The velocity of a point moving rectilinearly

3 The derivative of a function

4 Differentiable function

5 Increment notation

6 Differentials

7 Right-hand derivative of a function

8 Left-hand derivative of a function

9 Differentiability on an interval

10 The relationship between continuity and differentiability of a function

11 The Product Rule

12 The Quotient Rule

13 The Chain Rule

14 The Power Rule for Functions

15 Implicit differentiation

16 Higher order derivatives of functions

17 Newton's Method

EXERCISES 3.10

In Exercises 1 and 2 find $f'(x)$ directly from the definition of derivative.

1 $f(x) = 4/(3x^2 + 2)$

2 $f(x) = \sqrt{5 - 7x}$

Find the first derivatives in Exercises 3–30.

3 $f(x) = 2x^3 - 7x + 2$

4 $k(x) = 1/(x^4 - x^2 + 1)$

5 $g(t) = \sqrt{6t + 5}$

6 $h(t) = 1/\sqrt{6t + 5}$

7 $F(z) = \sqrt[3]{7z^2 - 4z + 3}$

8 $f(w) = \sqrt[5]{3w^2}$

9 $G(x) = 6/(3x^2 - 1)^4$

10 $H(x) = (3x^2 - 1)^4/6$

11 $F(y) = (y^2 - y^{-2})^{-2}$

12 $h(z) = [(z^2 - 1)^5 - 1]^5$

13 $g(x) = \sqrt[5]{(3x + 2)^4}$

14 $P(x) = (x + x^{-1})^2$

15 $r(s) = \left(\dfrac{8s^2 - 4}{1 - 9s^3}\right)^4$

16 $g(w) = \dfrac{(w - 1)(w - 3)}{(w + 1)(w + 3)}$

17 $F(x) = (x^6 + 1)^5(3x + 2)^3$

18 $k(z) = (z^2 + (z^2 + 9)^{1/2})^{1/2}$

19 $g(y) = (7y - 2)^{-2}(2y + 1)^{2/3}$

20 $p(x) = (2x^4 + 3x^2 - 1)/x^2$

21 $f(x) = (x^2 + 1)(x^2 + 2)(x^2 + 3)$

22 $H(t) = (t^6 + t^{-6})^6$

23 $h(x) = \sqrt{x + \sqrt{x + \sqrt{x}}}$

24 $K(r) = \sqrt{r}\sqrt{r + 1}\sqrt{r + 2}$

25 $f(x) = \sqrt[3]{2x + 3}/\sqrt{3x + 2}$

26 $f(x) = 6x^2 - (5/x) + (2/\sqrt[3]{x^2})$

27 $g(z) = (9z^{5/3} - 5z^{3/5})^3$

28 $F(t) = (5t^2 - 7)/(t^2 + 2)$

29 $k(s) = (2s^2 - 3s + 1)(9s - 1)^4$

30 $f(w) = \sqrt{(2w + 5)/(7w - 9)}$

Assuming that each of the equations in Exercises 31–34 determines a function f such that $y = f(x)$, find y'.

31 $5x^3 - 2x^2y^2 + 4y^3 - 7 = 0$

32 $3x^2 - xy^2 + y^{-1} = 1$

33 $(\sqrt{x} + 1)/(\sqrt{y} + 1) = y$

34 $y^2 - \sqrt{xy} + 3x = 2$

In Exercises 35–38, find an equation of the tangent line to the graph of the equation at the indicated point P.

35 $y = 2x - (4/\sqrt{x})$, $P(4, 6)$ **36** $y = (x^3 + 2)^5$, $P(-1, 1)$

37 $x^2y - y^3 = 8$, $P(-3, 1)$ **38** $x = y^{2/3}$, $P(0, 0)$

39 Find the x-coordinates of all points on the graph of $y = x^3 - 2x^2 + x - 2$ at which the tangent line is perpendicular to the line $4x + 2y = 5$.

40 If $y = 2x^3 - x^2 - 3x$, (a) find the x-coordinates of all points on the graph at which the tangent line is horizontal; (b) find the slope of the tangent line at the points where the graph crosses the x-axis.

In Exercises 41 and 42, find y', y'', and y'''.

41 $y = 5x^3 + 4\sqrt{x}$

42 $y = (1/x^2) + (1/x)$

43 If $x^2 + 4xy - y^2 = 8$, find y'' by implicit differentiation.

44 If $f(x) = x^3 - x^2 - 5x + 2$,

(a) find the x-coordinates of all points on the graph of f at which the tangent line is parallel to the line through $A(-3, 2)$ and $B(1, 14)$.

(b) find the values of f'' at each zero of f'.

45 If $f(x) = 1/(1 - x)$, find a formula for $f^{(n)}(x)$, where n is any positive integer.

46 If $y = 5x/(x^2 + 1)$, find dy and use it to approximate the change in y if x changes from 2 to 1.98. What is the exact change in y?

47 The side of an equilateral triangle is estimated to be 4 in. with a maximum error of 0.03 in. Use differentials to estimate the maximum error in the calculated area of the triangle. What is the approximate percentage error?

48 If $s = 3r^2 - 2\sqrt{r + 1}$ and $r = t^3 + t^2 + 1$, use the Chain Rule to find the value of ds/dt at $t = 1$.

49 If $f(x) = 2x^3 + x^2 - x + 1$ and $g(x) = x^5 + 4x^3 + 2x$, use differentials to approximate the change in $g(f(x))$ if x changes from -1 to -1.01.

50 Use differentials to approximate $\sqrt[3]{64.2}$.

51 Suppose f and g are functions such that $f(2) = -1$, $f'(2) = 4, f''(2) = -2, g(2) = -3, g'(2) = 2$ and $g''(2) = 1$. Find the values of each of the following at $x = 2$.

(a) $(2f - 3g)'$ (b) $(2f - 3g)''$ (c) $(fg)'$

(d) $(fg)''$ (e) $\left(\dfrac{f}{g}\right)'$ (f) $\left(\dfrac{f}{g}\right)''$

52 Let V and S denote the volume and surface area, respectively, of a spherical balloon. If the diameter is 8 cm and the volume increases by 12 cm^3, use differentials to approximate the change in S.

53 According to *Stefan's Law*, the radiant energy emitted from the surface of a body is given by $R = kT^4$, where R denotes the rate of emission per unit area, T is the temperature (in degrees Kelvin) and k is a constant. If there is a 0.5% error in the measurement of T, find the resulting percentage error in the calculated value of R.

54 The intensity of illumination from a source of light is inversely proportional to the square of the distance from the source. If a student works at a desk that is a certain distance from a lamp, use differentials to find the percentage change in distance that will increase the intensity by 10%.

55 Use Newton's Method to approximate, to four decimal places, the root of $7x^4 + 2x^3 - 3x^2 - 7x - 5 = 0$ that is between 0 and -1.

56 Use Newton's Method to approximate $\sqrt[4]{5}$ to three decimal places.

APPLICATIONS OF THE DERIVATIVE

In this chapter the derivative is used as a tool to investigate some mathematical and physical problems. Included are methods for determining maximum and minimum values of functions, applications of extrema to various situations, and the graphical concepts of *concavity* and *points of inflection*. We shall also examine how the derivative may be used to find rates at which quantities change, not only with respect to time, but also with respect to other variables. The concept of *antiderivative* is introduced near the end of the chapter and applied to problems on rectilinear motion. The final section contains a discussion of how derivatives may be used in economics.

4.1

LOCAL EXTREMA OF FUNCTIONS

Suppose that the graph illustrated in Figure 4.1 was made by a recording instrument used to measure the variation of a physical quantity with respect to time. In this case the x-axis represents time and the y-coordinates of points on the graph denote magnitudes of the quantity. For example, y-values might represent measurements such as temperature, resistance in an

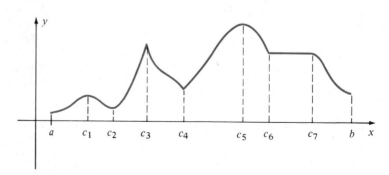

Figure 4.1

electrical circuit, blood pressure of an individual, the amount of chemical in a solution, or the bacteria count in a culture.

Referring to the graph we see that the quantity increased in the time interval $[a, c_1]$, decreased in $[c_1, c_2]$, increased in $[c_2, c_3]$, etc. If we restrict our attention to the interval $[c_1, c_4]$, the quantity had its largest (or maximum) value at c_3 and its smallest (or minimum) value at c_2. In other intervals there were different largest or smallest values. For example, over the entire interval $[a, b]$, the maximum value occurred at c_5 and the minimum value at a.

If Figure 4.1 is a sketch of the graph of a function f, it is convenient to use similar terminology to describe the behavior of $f(x)$ as x varies. The mathematical terms that are employed are included in the next two definitions.

Definition (4.1)

Let a function f be defined on an interval I and let x_1, x_2 denote numbers in I.

(i) f is **increasing** on I if $f(x_1) < f(x_2)$ whenever $x_1 < x_2$.

(ii) f is **decreasing** on I if $f(x_1) > f(x_2)$ whenever $x_1 < x_2$.

(iii) f is **constant** on I if $f(x_1) = f(x_2)$ for every x_1 and x_2.

Geometric interpretations of the preceding definition are given in Figure 4.2, where the interval I is not indicated. Of course, if f is constant on I, then the graph is part of a horizontal line.

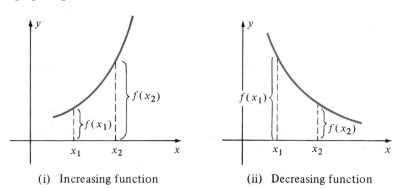

(i) Increasing function   (ii) Decreasing function

Figure 4.2

We shall use the phrases "f is increasing" and "$f(x)$ is increasing" interchangeably. This will also be done for the term "decreasing." If a function is increasing, then the graph rises as x increases, as illustrated in (i) of Figure 4.2. If a function is decreasing, then the graph falls as x increases, as in (ii) of the figure. If the sketch in Figure 4.1 represents the graph of a function f, then f is increasing on the intervals $[a, c_1]$, $[c_2, c_3]$, and $[c_4, c_5]$. It is decreasing on $[c_1, c_2]$, $[c_3, c_4]$, $[c_5, c_6]$, and $[c_7, b]$. The function is constant on the interval $[c_6, c_7]$.

The next definition introduces terminology we shall use for largest and smallest values of functions on an interval.

Definition (4.2)

> Let a function f be defined on an interval I and let c be a number in I.
>
> (i)　$f(c)$ is the **maximum value** of f on I if $f(x) \le f(c)$ for every x in I.
>
> (ii)　$f(c)$ is the **minimum value** of f on I if $f(x) \ge f(c)$ for every x in I.

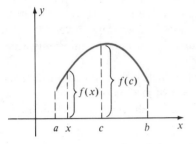

(i)　Maximum value $f(c)$

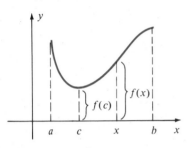

(ii)　Minimum value $f(c)$

Figure 4.3

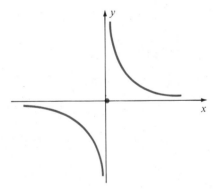

Figure 4.4

Maximum and minimum values are illustrated graphically in Figure 4.3. We have pictured I as a closed interval $[a, b]$; however, Definition (4.2) may be applied to any interval. Although the graphs in the figure have horizontal tangent lines at the point $(c, f(c))$, maximum and minimum values can also occur at points where graphs have corners or at endpoints of domains.

If $f(c)$ is the maximum value of f on I, we say that f *takes on* its maximum value at c. If we restrict our attention to points with x-coordinates in I, then the point $(c, f(c))$ is a highest point on the graph, as illustrated in (i) of Figure 4.3. Similarly, if $f(c)$ is the minimum value of f on I we say that f *takes on* its minimum value at c, and, as in (ii) of the figure, $(c, f(c))$ is a lowest point on the graph. Maximum and minimum values are sometimes called **extreme values** or **extrema** of f. A function can take on a maximum or minimum value more than once. Indeed, if f is a constant function, then $f(c)$ is both a maximum and a minimum value of f for *every* real number c.

Certain functions may have a maximum value but no minimum value on an interval, or vice versa. Other functions may have neither a maximum nor minimum value. To illustrate, suppose $f(x) = 1/x$ if $x \ne 0$ and $f(0) = 0$. The graph of f is sketched in Figure 4.4.

Note that f is not continuous at 0. On the closed interval $[-1, 1]$, the function has neither a maximum value nor a minimum value. (Why?) The same is true on the open interval $(0, 1)$, for if $0 < a < 1$, then there always exist numbers x_1 and x_2 in $(0, 1)$ such that $f(x_1) > f(a)$ and $f(x_2) < f(a)$. On the half-open interval $(0, 1]$ there is a minimum value $f(1)$, but no maximum value. On $[-1, 0)$ there is a maximum value $f(-1)$, but no minimum value. On the closed interval $[1, 2]$ f has a maximum value $f(1)$ and a minimum value $f(2)$.

From the preceding illustration it appears that the existence of maximum or minimum values may depend on both the continuity of the function and whether the interval under consideration is open, closed, or half-open. The next theorem provides conditions under which a function takes on a maximum or minimum value on an interval. The interested reader may consult more advanced texts on calculus for the proof.

Theorem (4.3)

> If a function f is continuous on a closed interval $[a, b]$, then f takes on a minimum value and a maximum value at least once in $[a, b]$.

The extrema are also called the **absolute minimum** and **absolute maximum values** for f on an interval. We shall also be interested in *local extrema* of f, defined as follows.

Definition (4.4)

Let c be a number in the domain of a function f.

(i) $f(c)$ is a **local maximum** of f if there exists an open interval (a, b) containing c such that $f(x) \leq f(c)$ for all x in (a, b).

(ii) $f(c)$ is a **local minimum** of f if there exists an open interval (a, b) containing c such that $f(x) \geq f(c)$ for all x in (a, b).

The term "*local*" is used because we *localize* our attention to a sufficiently small open interval containing c such that f takes on a largest (or smallest) value at c. Outside of that interval f may take on larger (or smaller) values. Sometimes the word "relative" is used in place of "local." Each local maximum or minimum will be called a **local extremum** of f and the totality of such numbers are the **local extrema** of f. Several examples of local extrema are illustrated in Figure 4.5. As indicated in the figure, it is possible for a local minimum to be *larger* than a local maximum.

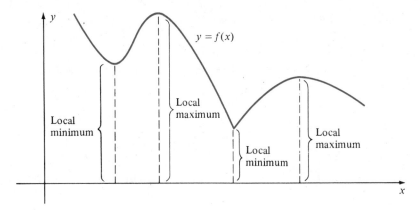

Figure 4.5

For the function whose graph is sketched in Figure 4.1, local maxima occur at c_1, c_3, and c_5 whereas local minima occur at c_2 and c_4. The functional values that correspond to numbers in the open interval (c_6, c_7) are *both* local maxima and local minima. (Why?) The local extrema may not include the absolute minimum or maximum values of f. For example, with reference to Figure 4.1, $f(a)$ is the minimum value of f on $[a, b]$, although it is not a local minimum since there is no *open* interval I contained in $[a, b]$ such that $f(a)$ is the least value of f on I. The number $f(c_5)$ is both a local maximum and the absolute maximum for f on $[a, b]$.

At a point corresponding to a local extremum of the function graphed in Figure 4.5, the tangent line is either horizontal or there is a corner. The x-coordinates of these points are numbers at which the derivative is zero or does not exist. The next theorem brings out the fact that this is generally true.

Theorem (4.5)

If a function f has a local extremum at a number c in an open interval, then either $f'(c) = 0$ or $f'(c)$ does not exist.

Proof   Suppose f has a local extremum at c. If $f'(c)$ does not exist there is nothing more to prove. If $f'(c)$ exists, then precisely one of the following occurs: (i) $f'(c) > 0$, (ii) $f'(c) < 0$, or (iii) $f'(c) = 0$. We shall arrive at (iii) by proving that neither (i) nor (ii) can occur. Thus, suppose $f'(c) > 0$. Employing Theorem (3.7),

$$f'(c) = \lim_{x \to c} \frac{f(x) - f(c)}{x - c} > 0$$

and hence by Theorem (2.7), there exists an open interval (a, b) containing c such that

$$\frac{f(x) - f(c)}{x - c} > 0$$

for all x in (a, b) different from c. The last inequality implies that if $a < x < b$ and $x \neq c$, then $f(x) - f(c)$ and $x - c$ are either both positive or both negative; that is,

$$\begin{cases} f(x) - f(c) < 0 & \text{whenever } x - c < 0, \text{ and} \\ f(x) - f(c) > 0 & \text{whenever } x - c > 0. \end{cases}$$

Another way of stating these facts is that if x is in (a, b) and $x \neq c$, then

$$\begin{cases} f(x) < f(c) & \text{whenever } x < c, \text{ and} \\ f(x) > f(c) & \text{whenever } x > c. \end{cases}$$

It follows that $f(c)$ is neither a local maximum nor a local minimum for f, contrary to hypothesis. Consequently (i) cannot occur. In like manner, the assumption that $f'(c) < 0$ leads to a contradiction. Hence (iii) must hold and the theorem is proved.   □

Corollary (4.6)

> If $f'(c)$ exists and $f'(c) \neq 0$, then $f(c)$ is not a local extremum of the function f.

A result similar to Theorem (4.5) is true for the *absolute* maximum and minimum values of a function that is continuous on a closed interval $[a, b]$, provided the extrema occur on the *open* interval (a, b). The theorem may be stated as follows.

Theorem (4.7)

> If a function f is continuous on a closed interval $[a, b]$ and has its maximum or minimum value at a number c in the open interval (a, b), then either $f'(c) = 0$ or $f'(c)$ does not exist.

The proof is exactly the same as that of Theorem (4.5) with the word "local" deleted.

It follows from Theorems (4.5) and (4.7) that the numbers at which the derivative either is zero or does not exist play a crucial role in the search for extrema of a function. Because of this, we give these numbers a special name in the next definition.

Definition (4.8)

> A number c in the domain of a function f is a **critical number** of f if either $f'(c) = 0$ or $f'(c)$ does not exist.

Referring to Theorem (4.7) we see that if f is continuous on a closed interval $[a, b]$, then the absolute maximum and minimum values occur either at a critical number of f, or at the endpoints a or b of the interval. If either $f(a)$ or $f(b)$ is an absolute extremum of f on $[a, b]$ it is called an **endpoint extremum**. The sketches in Figure 4.6 illustrate this concept.

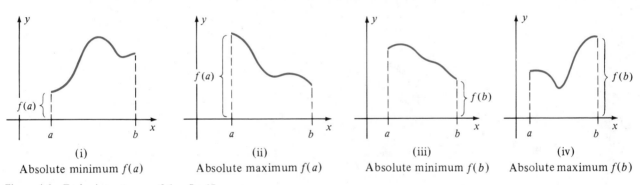

(i)	(ii)	(iii)	(iv)
Absolute minimum $f(a)$	Absolute maximum $f(a)$	Absolute minimum $f(b)$	Absolute maximum $f(b)$

Figure 4.6 Endpoint extrema of f on $[a, b]$

The preceding discussion enables us to state the following.

(4.9) *Guidelines for Finding the Absolute Extrema of a Continuous Function f on a Closed Interval $[a, b]$*

1 Find all the critical numbers of f.
2 Calculate $f(c)$ for each critical number c.
3 Calculate $f(a)$ and $f(b)$.
4 The absolute maximum and minimum of f on $[a, b]$ are the largest and smallest of the functional values calculated in 2 and 3.

Example 1 If $f(x) = x^3 - 12x$, find the absolute maximum and minimum values of f on the closed interval $[-3, 5]$. Sketch the graph of f.

Solution Let us follow the Guidelines (4.9). Thus we begin by finding the critical numbers of f. Differentiating,

$$f'(x) = 3x^2 - 12 = 3(x^2 - 4) = 3(x + 2)(x - 2).$$

Since the derivative exists everywhere, the only critical numbers are those for which the derivative is zero, that is, -2 and 2. Since f is continuous on $[-3, 5]$, it follows from our discussion that the absolute maximum and minimum are among the numbers $f(-2), f(2), f(-3)$, and $f(5)$. Calculating these values (see Guidelines 2 and 3), we obtain

$$f(-2) = (-2)^3 - 12(-2) = -8 + 24 = 16$$
$$f(2) = 2^3 - 12(2) = 8 - 24 = -16$$
$$f(-3) = (-3)^3 - 12(-3) = -27 + 36 = 9$$
$$f(5) = 5^3 - 12(5) = 125 - 60 = 65.$$

Thus, by Guideline 4, the minimum value of f on $[-3, 5]$ is $f(2) = -16$ and the maximum value is the endpoint extremum $f(5) = 65$.

Using the functional values we have calculated and plotting several more points leads to the sketch in Figure 4.7, where for clarity, different scales have been used on the x- and y-axes. The tangent line is horizontal at the points on the graph corresponding to the critical numbers -2 and 2. It will follow from our work in Section 4.4 that $f(-2) = 16$ is a *local* maximum for f, as indicated by the graph. ∎

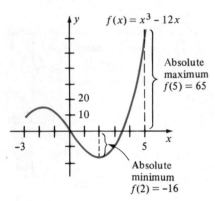

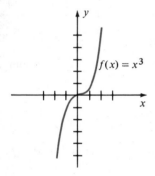

Figure 4.7

We see from Theorem (4.5) that if a function has a *local* extremum, then it *must* occur at a critical number; however, not every critical number leads to a local extremum, as illustrated by the following example.

Example 2   If $f(x) = x^3$, prove that f has no local extremum.

Solution   The graph of f is sketched in Figure 4.8. The derivative of f is $f'(x) = 3x^2$, which exists for all x and is zero only if $x = 0$. Consequently 0 is the only critical number. However, if $x < 0$, then $f(x)$ is negative whereas if $x > 0$, then $f(x)$ is positive. Here $f(0)$ is neither a local maximum nor a local minimum. Since a local extremum *must* occur at a critical number (see Theorem (4.5)) it follows that f has no local extrema. Note that the tangent line is horizontal at the point $P(0, 0)$, which has the critical number 0 as x-coordinate. ∎

Figure 4.8

In later sections methods are developed for finding local extrema of functions. At that time it may be necessary to obtain critical numbers of functions that are defined by fairly complicated expressions, as illustrated in the next example.

Example 3   Find the critical numbers of f if $f(x) = (x + 5)^2 \sqrt[3]{x - 4}$.

Solution   Differentiating $f(x) = (x + 5)^2(x - 4)^{1/3}$, we obtain

$$f'(x) = (x + 5)^2 \tfrac{1}{3}(x - 4)^{-2/3} + 2(x + 5)(x - 4)^{1/3}.$$

As an aid to finding the critical numbers we simplify $f'(x)$ as follows:

$$f'(x) = \frac{(x + 5)^2}{3(x - 4)^{2/3}} + 2(x + 5)(x - 4)^{1/3}$$

$$= \frac{(x + 5)^2 + 6(x + 5)(x - 4)}{3(x - 4)^{2/3}}$$

$$= \frac{(x + 5)[(x + 5) + 6(x - 4)]}{3(x - 4)^{2/3}}$$

$$= \frac{(x + 5)(7x - 19)}{3(x - 4)^{2/3}}.$$

Consequently, $f'(x) = 0$ if $x = -5$ or $x = \frac{19}{7}$. The derivative $f'(x)$ does not exist at $x = 4$. Thus f has three critical numbers, namely -5, $\frac{19}{7}$, and 4. ∎

EXERCISES 4.1

In Exercises 1–4, find the absolute maximum and minimum of f on the indicated closed interval.

1 $f(x) = 5 - 6x^2 - 2x^3; [-3, 1]$

2 $f(x) = 3x^2 - 10x + 7; [-1, 3]$

3 $f(x) = 1 - x^{2/3}; [-1, 8]$

4 $f(x) = x^4 - 5x^2 + 4; [0, 2]$

5 (a) If $f(x) = x^{1/3}$, prove that 0 is the only critical number of f, and that $f(0)$ is not a local extremum.

(b) If $f(x) = x^{2/3}$, prove that 0 is the only critical number of f, and that $f(0)$ is a local minimum of f.

6 If $f(x) = |x|$, prove that 0 is the only critical number of f; that $f(0)$ is a local minimum of f; and that the graph of f has no tangent line at the point $(0, 0)$.

In Exercises 7 and 8, prove that f has no local extrema. Sketch the graph of f. Prove that f is continuous on the interval $(0, 1)$, but f has neither a maximum nor minimum value on $(0, 1)$. Why doesn't this contradict Theorem (4.3)?

7 $f(x) = x^3 + 1$ **8** $f(x) = 1/x^2$

Find the critical numbers of the functions defined in Exercises 9–26.

9 $f(x) = 4x^2 - 3x + 2$

10 $g(x) = 2x + 5$

11 $s(t) = 2t^3 + t^2 - 20t + 4$

12 $K(z) = 4z^3 + 5z^2 - 42z + 7$

13 $F(w) = w^4 - 32w$

14 $k(r) = r^5 - 2r^3 + r - 12$

15 $f(z) = \sqrt{z^2 - 16}$

16 $M(x) = \sqrt[3]{x^2 - x - 2}$

17 $g(t) = t^2 \sqrt[3]{2t - 5}$

18 $T(v) = (4v + 1)\sqrt{v^2 - 16}$

19 $G(x) = (2x - 3)/(x^2 - 9)$

20 $f(s) = s^2/(5s + 4)$

21 $f(t) = (t^3 - 9t)^3$

22 $g(x) = x^3 + (6/x)$

23 $F(x) = x^{2/3}(x^2 - 9)$

24 $H(u) = u^{1/3}(5u - 2)$

25 $f(x) = (x + 5)^4(2x - 3)^3$

26 $G(r) = (r - 1)^2(1 - 6r)^{2/3}$

27 Prove that a polynomial function of degree 1 has no local or absolute extrema on the interval $(-\infty, \infty)$. What is true on a closed interval $[a, b]$?

28 If f is a constant function and (a, b) is any open interval, prove that $f(c)$ is both a local and an absolute extremum of f for every number c in (a, b).

29 If f is the greatest integer function, prove that every number is a critical number of f.

30 Let f be defined by the following conditions: $f(x) = 0$ if x is rational and $f(x) = 1$ if x is irrational. Prove that every number is a critical number of f.

31 Prove that a quadratic function has exactly one critical number on $(-\infty, \infty)$.

32 Prove that a polynomial function of degree 3 has either two, one, or no critical numbers on $(-\infty, \infty)$. Sketch graphs that illustrate how each of these possibilities can occur.

33 If $f(x) = x^n$, where n is a positive integer, prove that f has either one or no local extrema on $(-\infty, \infty)$ according as n is even or odd, respectively. Sketch typical graphs illustrating each case.

34 Prove that a polynomial function of degree n can have at most $n - 1$ local extrema on $(-\infty, \infty)$.

4.2

ROLLE'S THEOREM AND THE MEAN VALUE THEOREM

Sometimes it is extremely difficult to find the critical numbers of a function. As a matter of fact, there is no guarantee that critical numbers even exist. The next theorem, credited to the French mathematician Michel Rolle (1652–1719), provides sufficient conditions for the existence of a critical number. The theorem is stated for a function f that is continuous on an interval $[a, b]$, differentiable on (a, b), and such that $f(a) = f(b)$. Some typical graphs of functions of this type are sketched in Figure 4.9.

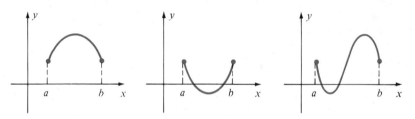

Figure 4.9

Referring to the sketches in Figure 4.9, it seems reasonable to expect that there is at least one number c between a and b such that the tangent line at the point $(c, f(c))$ is horizontal, or equivalently, such that $f'(c) = 0$. This is precisely the conclusion of the following theorem.

Rolle's Theorem (4.10)

> If a function f is continuous on a closed interval $[a, b]$, differentiable on the open interval (a, b), and if $f(a) = f(b)$, then $f'(c) = 0$ for at least one number c in (a, b).

Proof   The function f must fall into at least one of the following three categories.

(i) $f(x) = f(a)$ *for all x in (a, b).* In this case f is a constant function and hence $f'(x) = 0$ for all x. Consequently *every* number c in (a, b) is a critical number.

(ii) $f(x) > f(a)$ *for some x in (a, b).* In this case the maximum value of f in $[a, b]$ is greater than $f(a)$ or $f(b)$ and, therefore, must occur at some number c in the *open* interval (a, b). Since the derivative exists throughout (a, b), we conclude from Theorem (4.7) that $f'(c) = 0$.

(iii) $f(x) < f(a)$ *for some x in* (a, b). In this case the minimum value of f in $[a, b]$ is less than $f(a)$ or $f(b)$ and must occur at some number c in (a, b). As in (ii), $f'(c) = 0$. □

Corollary (*4.11*)

If f is continuous on a closed interval $[a, b]$ and if $f(a) = f(b)$, then f has at least one critical number in the open interval (a, b).

Proof   On the one hand, if f' does not exist at some number c in (a, b), then by Definition (4.8), c is a critical number. On the other hand, if f' exists throughout (a, b), then a critical number exists by Rolle's Theorem. □

Before discussing the next theorem, which may be considered a generalization of Rolle's Theorem to the case in which $f(a) \neq f(b)$, let us consider the points $P(a, f(a))$ and $R(b, f(b))$ on the graph of f as illustrated by any of the sketches in Figure 4.10.

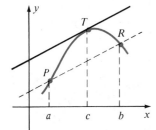 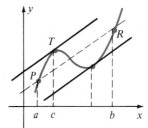

Figure 4.10

If f' exists throughout the open interval (a, b), then it appears geometrically evident that there is at least one point $T(c, f(c))$ on the graph at which the tangent line is parallel to the secant line through P and R. This fact may be expressed in terms of slopes as follows:

$$f'(c) = \frac{f(b) - f(a)}{b - a}$$

where the quotient on the right is obtained by using the formula for the slope of the line through P and R. If we multiply both sides of this equation by $b - a$, we obtain the formula stated in the next theorem.

The Mean Value Theorem (*4.12*)

If a function f is continuous on a closed interval $[a, b]$ and is differentiable on the open interval (a, b), then there exists a number c in (a, b) such that

$$f(b) - f(a) = f'(c)(b - a).$$

Proof   Let us define a function g as follows:

$$g(x) = f(x) - f(a) - \left[\frac{f(b) - f(a)}{b - a}\right](x - a)$$

for all x in $[a, b]$. Although we have seemingly "pulled g out of the air," there is an interesting geometric interpretation for $g(x)$ (see Exercise 22). Since f is continuous on $[a, b]$ and differentiable on (a, b), the same is true for g. Differentiating, we obtain

$$g'(x) = f'(x) - \frac{f(b) - f(a)}{b - a}.$$

Moreover, by direct substitution we see that $g(a) = g(b) = 0$, and hence the function g satisfies the hypotheses of Rolle's Theorem. Consequently, there exists a number c in (a, b) such that $g'(c) = 0$, or equivalently,

$$f'(c) - \frac{f(b) - f(a)}{b - a} = 0.$$

The last equation may be written in the form stated in the conclusion of the theorem.   □

The Mean Value Theorem is also called **The Theorem of the Mean**. It will be employed later to help establish several very important results. The following example provides a numerical illustration of the Mean Value Theorem.

Example   Prove that the function f defined by $f(x) = x^3 - 8x - 5$ satisfies the hypotheses of the Mean Value Theorem on the interval $[1, 4]$, and find a number c in the interval $(1, 4)$ that satisfies the conclusion of the theorem.

Solution   Since f is a polynomial function it is continuous and differentiable for all real numbers. In particular, it is continuous on $[1, 4]$ and differentiable on the open interval $(1, 4)$. According to the Mean Value Theorem, there exists a number c in $(1, 4)$ such that

$$f(4) - f(1) = f'(c)(4 - 1).$$

Since $f(4) = 27$, $f(1) = -12$, and $f'(x) = 3x^2 - 8$, this is equivalent to

$$27 - (-12) = (3c^2 - 8)(3), \quad \text{or} \quad 39 = 3(3c^2 - 8).$$

We leave it to the reader to show that the last equation implies that $c = \pm\sqrt{7}$. Hence the desired number in the interval $(1, 4)$ is $c = \sqrt{7}$.   ■

EXERCISES 4.2

1 If $f(x) = |x|$, show that $f(-1) = f(1)$ but $f'(c) \neq 0$ for every number c in the open interval $(-1, 1)$. Why doesn't this contradict Rolle's Theorem?

2 If $f(x) = 5 + 3(x - 1)^{2/3}$, show that $f(0) = f(2)$, but $f'(c) \neq 0$ for every number c in the open interval $(0, 2)$. Why doesn't this contradict Rolle's Theorem?

3 If $f(x) = 4/x$, prove that there is no number c such that $f(4) - f(-1) = f'(c)[4 - (-1)]$. Why doesn't this contradict the Mean Value Theorem applied to the interval $[-1, 4]$?

4 If f is the greatest integer function and if a and b are real numbers such that $b - a \geq 1$, prove that there is no number c such that $f(b) - f(a) = f'(c)(b - a)$. Why doesn't this contradict the Mean Value Theorem?

In Exercises 5–8 show that f satisfies the hypotheses of Rolle's Theorem on the indicated interval $[a, b]$ and find all numbers c in (a, b) such that $f'(c) = 0$.

5 $f(x) = 3x^2 - 12x + 11$, $[0, 4]$

6 $f(x) = 5 - 12x - 2x^2$, $[-7, 1]$

7 $f(x) = x^4 + 4x^2 + 1$, $[-3, 3]$

8 $f(x) = x^3 - x$, $[-1, 1]$

In Exercises 9–18 determine whether the function f satisfies the hypotheses of the Mean Value Theorem on the indicated interval $[a, b]$ and if so, find all numbers c in (a, b) such that $f(b) - f(a) = f'(c)(b - a)$.

9 $f(x) = x^3 + 1$, $[-2, 4]$

10 $f(x) = 5x^2 - 3x + 1$, $[1, 3]$

11 $f(x) = x + (4/x)$, $[1, 4]$

12 $f(x) = 3x^5 + 5x^3 + 15x$, $[-1, 1]$

13 $f(x) = x^{2/3}$, $[-8, 8]$

14 $f(x) = 1/(x - 1)^2$, $[0, 2]$

15 $f(x) = 4 + \sqrt{x - 1}$, $[1, 5]$

16 $f(x) = 1 - 3x^{1/3}$, $[-8, -1]$

17 $f(x) = x^3 - 2x^2 + x + 3$, $[-1, 1]$

18 $f(x) = |x - 3|$, $[-1, 4]$

19 Prove that if f is a linear function, then f satisfies the hypotheses of the Mean Value Theorem on every closed interval $[a, b]$, and that *every* number c satisfies the conclusion of the theorem.

20 If f is a quadratic function and $[a, b]$ is any closed interval, prove that there is precisely one number c in the interval (a, b) that satisfies the conclusion of the Mean Value Theorem.

21 If f is a polynomial function of degree 3 and $[a, b]$ is any closed interval, prove that there are at most two numbers in (a, b) that satisfy the conclusion of the Mean Value Theorem. Sketch graphs that illustrate the various possibilities. What can be said of a polynomial function of degree 4? Illustrate with sketches. Generalize to polynomial functions of degree n, where n is any positive integer.

22 Prove that if $g(x)$ is the function defined in the proof of the Mean Value Theorem (4.12), then $|g(x)|$ is the distance (measured along a vertical line with x-intercept x) between the graph of f and the line through $P(a, f(a))$ and $R(b, f(b))$.

23 If f is continuous on $[a, b]$ and if $f'(x) = c$ for every x in (a, b), use the Mean Value Theorem to prove that $f(x) = cx + d$ for some real number d.

24 If $f(x)$ is a polynomial of degree 3, use Rolle's Theorem to prove that f has at most three real zeros. Extend this result to polynomials of degree n.

25 A straight highway 50 miles long connects two cities A and B. Use the Mean Value Theorem to prove that it is impossible to travel from A to B by automobile in one hour without the speedometer reading 50 mi/hr at least once.

4.3

THE FIRST DERIVATIVE TEST

The following theorem indicates how the derivative may be used to determine intervals on which a function is increasing or decreasing.

Theorem (4.13)

> Let f be a function that is continuous on a closed interval $[a, b]$ and differentiable on the open interval (a, b).
>
> (i)   If $f'(x) > 0$ for all x in (a, b), then f is increasing on $[a, b]$.
>
> (ii) If $f'(x) < 0$ for all x in (a, b), then f is decreasing on $[a, b]$.

Proof   (i) Suppose $f'(x) > 0$ for all x in (a, b) and consider any numbers x_1, x_2 in $[a, b]$ such that $x_1 < x_2$. We wish to show that $f(x_1) < f(x_2)$. Applying the Mean Value Theorem (4.12) to the interval $[x_1, x_2]$,

$$f(x_2) - f(x_1) = f'(w)(x_2 - x_1)$$

where w is in the open interval (x_1, x_2). Since $x_2 - x_1 > 0$ and since, by hypothesis, $f'(w) > 0$, the right-hand side of the previous equation is positive; that is, $f(x_2) - f(x_1) > 0$. Hence $f(x_2) > f(x_1)$, which is what we wished to show. The proof of (ii) is similar and is left as an exercise.   □

Figures 4.11 and 4.12 give geometric illustrations of the preceding theorem. A typical tangent line l to the graph is shown at a point whose x-coordinate x is in the interval (a, b). As illustrated in Figure 4.11, if $f'(x) > 0$ the tangent line rises and, as we have proved, so does the graph of f. If $f'(x) < 0$ both the tangent line and the graph of f fall, as illustrated in Figure 4.12.

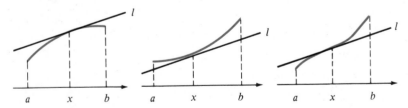

Figure 4.11 $f'(x) > 0$; f increasing on $[a, b]$

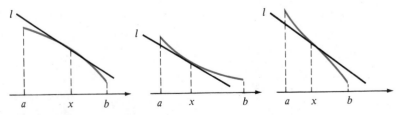

Figure 4.12 $f'(x) < 0$; f decreasing on $[a, b]$

The proof given for Theorem (4.13) may also be used to show that if $f'(x) > 0$ throughout an infinite interval of the form $(-\infty, a)$ or (b, ∞), then f is increasing on $(-\infty, a]$ or $[b, \infty)$, respectively, provided f is continuous on those intervals. An analogous result holds for decreasing functions if $f'(x) < 0$.

To apply Theorem (4.13) it is necessary to determine intervals in which the derivative $f'(x)$ is either always positive, or always negative. Theorem (2.25) is useful in this respect. Specifically, if the *derivative* f' is continuous and has no zeros on an interval, then either $f'(x) > 0$ or $f'(x) < 0$ for all x in the interval. Thus, if we choose *any* number k in the interval, and if $f'(k) > 0$, then $f'(x) > 0$ for *all* x in the interval. Similarly, if $f'(k) < 0$, then $f'(x) < 0$ throughout the interval. We shall call $f'(k)$ a **test value** of $f'(x)$ for the interval. The use of test values is demonstrated in the following examples.

Example 1 If $f(x) = x^3 + x^2 - 5x - 5$, find the intervals on which f is increasing and the intervals on which f is decreasing. Sketch the graph of f.

Solution   Differentiating, we obtain

$$f'(x) = 3x^2 + 2x - 5 = (3x + 5)(x - 1).$$

By Theorem (4.13) it is sufficient to find the intervals in which $f'(x) > 0$ and those in which $f'(x) < 0$. The factored form of $f'(x)$ and the critical numbers $-\frac{5}{3}$ and 1 suggest that we consider the open intervals $(-\infty, -\frac{5}{3})$, $(-\frac{5}{3}, 1)$, and $(1, \infty)$. Note that on any of these intervals, f' is continuous and has no zeros and, therefore, $f'(x)$ has the same sign throughout the interval. This sign can be determined by choosing a suitable test value for the interval.

If we choose -2 in the interval $(-\infty, -\frac{5}{3})$, we obtain the test value

$$f'(-2) = 3(-2)^2 + 2(-2) - 5 = 12 - 4 - 5 = 3.$$

Since 3 is positive, $f'(x)$ is positive throughout $(-\infty, -\frac{5}{3})$.

If we choose 0 in $(-\frac{5}{3}, 1)$, the test value is

$$f'(0) = 3(0)^2 + 2(0) - 5 = -5.$$

Since -5 is negative, $f'(x) < 0$ throughout $(-\frac{5}{3}, 1)$.

Finally, choosing 2 in $(1, \infty)$ we obtain

$$f'(2) = 3(2)^2 + 2(2) - 5 = 12 + 4 - 5 = 11.$$

Since $11 > 0$, it follows that $f'(x) > 0$ throughout $(1, \infty)$.

In future examples we will display our work in tabular form as follows, where the last row is a consequence of Theorem (4.13).

Interval	$(-\infty, -\frac{5}{3})$	$(-\frac{5}{3}, 1)$	$(1, \infty)$
k	-2	0	2
Test value $f'(k)$	3	-5	11
Sign of $f'(x)$	$+$	$-$	$+$
Variation of f	increasing on $(-\infty, -\frac{5}{3}]$	decreasing on $[-\frac{5}{3}, 1]$	increasing on $[1, \infty)$

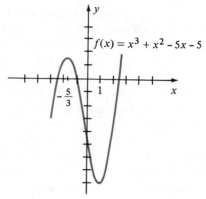

Figure 4.13

Note that $f(x)$ may be written as $f(x) = x^2(x + 1) - 5(x + 1) = (x^2 - 5)(x + 1)$ and hence the x-intercepts of the graph are $\sqrt{5}$, $-\sqrt{5}$, and -1. The y-intercept is $f(0) = -5$. The points corresponding to the critical numbers are $(-\frac{5}{3}, \frac{40}{27})$ and $(1, -8)$. Plotting these six points and using the information on where f is increasing or decreasing gives us the sketch shown in Figure 4.13.   ∎

As we saw in Section 4.1, if a function has a local extremum, then it must occur at a critical number; however, not every critical number leads to a local extremum (see Example 2 of Section 4.1). To find the local extrema, we begin by locating all the critical numbers of the function. Next, each critical number is tested to determine whether or not a local extremum occurs. There are several methods for conducting this test. The following theorem is based on the sign of the first derivative of f. Roughly speaking, it states that if $f'(x)$ changes sign as x increases through a critical number c, then f has a local maximum or minimum at c. If $f'(x)$ does not change sign, then there is no extremum at c.

The First Derivative Test (4.14)

Suppose c is a critical number of a function f and (a, b) is an open interval containing c. Suppose further that f is continuous on $[a, b]$ and differentiable on (a, b), except possibly at c.

(i)   If $f'(x) > 0$ for $a < x < c$ and $f'(x) < 0$ for $c < x < b$, then $f(c)$ is a local maximum of f.

(ii) If $f'(x) < 0$ for $a < x < c$ and $f'(x) > 0$ for $c < x < b$, then $f(c)$ is a local minimum of f.

(iii) If $f'(x) > 0$ or if $f'(x) < 0$ for all x in (a, b) except $x = c$, then $f(c)$ is not a local extremum of f.

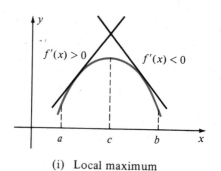

(i) Local maximum

(ii) Local minimum

Figure 4.14

Proof   If the sign of $f'(x)$ varies as indicated in (i), then by Theorem (4.13) f is increasing on $[a, c]$ and decreasing on $[c, b]$, and thus $f(x) < f(c)$ for all x in (a, b) different from c. Hence, by Definition (4.4), $f(c)$ is a local maximum for f. Parts (ii) and (iii) may be proved in like manner.   □

A device that may be used to remember the first derivative test is to think of graphs of the type sketched in Figure 4.14. In the case of a local maximum, as illustrated in (i) of Figure 4.14, the slope of the tangent line at $P(x, f(x))$ is positive if $x < c$ and negative if $x > c$. In the case of a local minimum, as illustrated in (ii) of Figure 4.14, the opposite situation occurs. Similar drawings can be sketched if the graph has a corner at the point $(c, f(c))$.

Example 2   Find the local extrema of f if $f(x) = x^3 + x^2 - 5x - 5$.

Solution   This function is the same as that considered in Example 1. The critical numbers of f are $-\frac{5}{3}$ and 1. We see from the table in Example 1 that the sign of $f'(x)$ changes from positive to negative as x increases through $-\frac{5}{3}$.

Hence, by the First Derivative Test, f has a local maximum at $-\frac{5}{3}$. This maximum value is $f(-\frac{5}{3}) = \frac{40}{27}$. Similarly, a local minimum occurs at 1 since the sign of $f'(x)$ changes from negative to positive as x increases through 1. This minimum value is $f(1) = -8$. The reader should refer to Figure 4.13 for the geometric significance of these local extrema. ∎

Example 3 Find the local maxima and minima of f if $f(x) = x^{1/3}(8 - x)$. Sketch the graph of f.

Solution   By the Product Rule,

$$f'(x) = x^{1/3}(-1) + (8 - x)\tfrac{1}{3}x^{-2/3}$$
$$= \frac{-3x + (8 - x)}{3x^{2/3}} = \frac{4(2 - x)}{3x^{2/3}}$$

and hence the critical numbers of f are 0 and 2. As in Example 1, these suggest that we consider the signs of $f'(x)$ corresponding to the intervals $(-\infty, 0)$, $(0, 2)$, and $(2, \infty)$. Since f' is continuous and has no zeros on each interval, we may determine the sign of $f'(x)$ by using a suitable test value $f'(k)$. The following table summarizes this work (check each entry).

Interval	$(-\infty, 0)$	$(0, 2)$	$(2, \infty)$
k	-1	1	8
Test value $f'(k)$	4	$\frac{4}{3}$	-2
Sign of $f'(x)$	$+$	$+$	$-$
Variation of f	increasing on $(-\infty, 0]$	increasing on $[0, 2]$	decreasing on $[2, \infty)$

The number 8 was chosen in $(2, \infty)$ since $8^{2/3}$ is an integer. One may, of course, choose *any* number k in $(2, \infty)$ and use a calculator to find $f'(k)$.

By the First Derivative Test, f has a local maximum at 2 since the sign of $f'(x)$ changes from $+$ to $-$ as x increases through 2. This local maximum is $f(2) = 2^{1/3}(8 - 2) = 6\sqrt[3]{2} \approx 7.6$. The function does not have an extremum at 0 since the sign of $f'(x)$ does not change as x increases through 0. To sketch the graph we first plot points corresponding to the critical numbers. From the formula for $f(x)$ it is evident that the x-intercepts of the graph are 0 and 8. The graph is sketched in Figure 4.15. ∎

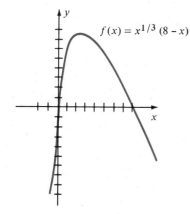

$f(x) = x^{1/3}(8 - x)$

Figure 4.15

Example 4 Find the local extrema of f if $f(x) = x^{2/3}(x^2 - 8)$. Sketch the graph of f.

Solution   By the Product Rule,

$$f'(x) = x^{2/3}(2x) + (x^2 - 8)(\tfrac{2}{3}x^{-1/3})$$
$$= \frac{6x^2 + 2(x^2 - 8)}{3x^{1/3}} = \frac{8(x^2 - 2)}{3x^{1/3}}.$$

The critical numbers are $-\sqrt{2}, 0,$ and $\sqrt{2}$. These suggest an examination of the sign of $f'(x)$ in the intervals $(-\infty, -\sqrt{2}), (-\sqrt{2}, 0), (0, \sqrt{2}),$ and $(\sqrt{2}, \infty)$. Arranging our work in tabular form as in previous examples we obtain the following.

Interval	$(-\infty, -\sqrt{2})$	$(-\sqrt{2}, 0)$	$(0, \sqrt{2})$	$(\sqrt{2}, \infty)$
k	-8	-1	1	8
Test value $f'(k)$	$-\frac{248}{3}$	$\frac{8}{3}$	$-\frac{8}{3}$	$\frac{248}{3}$
Sign of $f'(x)$	$-$	$+$	$-$	$+$
Variation of f	decreasing on $(-\infty, -\sqrt{2}]$	increasing on $[-\sqrt{2}, 0]$	decreasing on $[0, \sqrt{2}]$	increasing on $[\sqrt{2}, \infty)$

By the First Derivative Test, f has local minima at $-\sqrt{2}$ and $\sqrt{2}$ and a local maximum at 0. The corresponding functional values are $f(0) = 0$ and

$$f(\sqrt{2}) = -6\sqrt[3]{2} = f(-\sqrt{2}).$$

Note that the derivative does not exist at 0. The graph is sketched in Figure 4.16. ∎

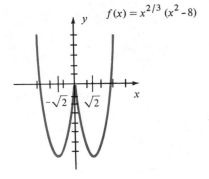

$f(x) = x^{2/3}(x^2 - 8)$

Figure 4.16

It was pointed out in Section 4.1 that an absolute maximum or minimum of a function may not be included among the local extrema. Recall from (4.9) that if a function f is continuous on a closed interval $[a, b]$ and we wish to find the absolute maximum and minimum values, then all the local extrema should be found and *in addition* the values $f(a)$ and $f(b)$ of f at the endpoints a and b of the interval $[a, b]$ should be calculated. The largest number among the local extrema and the values $f(a)$ and $f(b)$ is the absolute maximum of f on $[a, b]$, whereas the smallest of these numbers is the absolute minimum of f on $[a, b]$. To illustrate these remarks, let us consider the function discussed in the previous example, but restrict our attention to certain intervals.

Example 5 If $f(x) = x^{2/3}(x^2 - 8)$, find the absolute maximum and minimum values of f in each of the following intervals: (a) $[-1, \frac{1}{2}]$, (b) $[-1, 3]$, (c) $[-3, -2]$.

Solution The graph in Figure 4.16 indicates the local extrema and the intervals in which f is increasing or decreasing. Figure 4.17 illustrates the part of the graph of f that corresponds to each interval in (a)–(c). Referring to these sketches we obtain the following table (check the entries).

Interval	Absolute minimum	Absolute maximum
$[-1, \frac{1}{2}]$	$f(-1) = -7$	$f(0) = 0$
$[-1, 3]$	$f(\sqrt{2}) = -6\sqrt[3]{2}$	$f(3) = \sqrt[3]{9}$
$[-3, -2]$	$f(-2) = -4\sqrt[3]{4}$	$f(-3) = \sqrt[3]{9}$

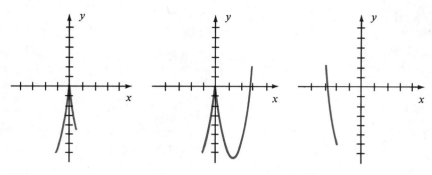

Figure 4.17

Note that on some intervals the maximum or minimum value of f is also a local extremum, whereas on other intervals this is not the case. ■

EXERCISES 4.3

In Exercises 1–16, find the local extrema of f. Describe the intervals in which f is increasing or decreasing, and sketch the graph of f.

1 $f(x) = 5 - 7x - 4x^2$

2 $f(x) = 6x^2 - 9x + 5$

3 $f(x) = 2x^3 + x^2 - 20x + 1$

4 $f(x) = x^3 - x^2 - 40x + 8$

5 $f(x) = x^4 - 8x^2 + 1$

6 $f(x) = x^3 - 3x^2 + 3x + 7$

7 $f(x) = x^{4/3} + 4x^{1/3}$

8 $f(x) = x^{2/3}(8 - x)$

9 $f(x) = x^2 \sqrt[3]{x^2 - 4}$

10 $f(x) = x\sqrt{4 - x^2}$

11 $f(x) = x^{2/3}(x - 7)^2 + 2$

12 $f(x) = 4x^3 - 3x^4$

13 $f(x) = x^3 + (3/x)$

14 $f(x) = 8 - \sqrt[3]{x^2 - 2x + 1}$

15 $f(x) = 10x^3(x - 1)^2$

16 $f(x) = (x^2 - 10x)^4$

In Exercises 17–22, find the local extrema of f.

17 $f(x) = \sqrt[3]{x^3 - 9x}$

18 $f(x) = x^2/\sqrt{x + 7}$

19 $f(x) = (x - 2)^3(x + 1)^4$

20 $f(x) = x^2(x - 5)^4$

21 $f(x) = (2x - 5)/(x + 3)$

22 $f(x) = (x^2 + 3)/(x - 1)$

23–26 For the functions defined in Exercises 1–4, find the absolute maximum and minimum values on each of the following intervals:

(a) $[-1, 1]$ (b) $[-4, 2]$ (c) $[0, 5]$

In Exercises 27 and 28 sketch the graph of a differentiable function f that satisfies the given conditions.

27 $f'(-5) = 0$; $f'(0) = 0$; $f'(5) = 0$; $f'(x) > 0$ if $|x| > 5$; $f'(x) < 0$ if $0 < |x| < 5$.

28 $f'(a) = 0$ for $a = 1, 2, 3, 4, 5$, and $f'(x) > 0$ for all other values of x.

In Exercises 29 and 30 find the local extrema of f'. Describe the intervals in which f' is increasing or decreasing. Sketch the graph of f and study the variation of the slope of the tangent line as x increases through the domain of f.

29 $f(x) = x^4 - 6x^2$ **30** $f(x) = 4x^3 - 3x^4$

31 If $f(x) = ax^3 + bx^2 + cx + d$, determine values for a, b, c, and d such that f has a local maximum 2 at $x = -1$ and a local minimum -1 at $x = 1$.

32 If $f(x) = ax^4 + bx^3 + cx^2 + dx + e$, determine values of a, b, c, d, and e such that f has a local maximum 2 at $x = 0$, and a local minimum -14 at $x = -2$ and $x = 2$, respectively.

4.4

CONCAVITY AND THE SECOND DERIVATIVE TEST

The concept of *concavity* is useful for describing the graph of a differentiable function f. If $f'(c)$ exists, then f has a tangent line with slope $f'(c)$ at the point $P(c, f(c))$. Figure 4.18 illustrates three possible situations that may occur if $f'(c) > 0$. Similar situations occur if $f'(c) < 0$ or $f'(c) = 0$. Note that in (i) of Figure 4.18 there is an open interval (a, b) containing c such that for all x in (a, b), except $x = c$, the point $Q(x, f(x))$ on the graph of f lies above the point on the tangent line having x-coordinate x. In this case we say that on the interval (a, b) the graph of f is *above* the tangent line through P. In (ii) of Figure 4.18 we say that the graph of f is *below* the tangent line through P. In (iii), for every open interval (a, b) containing c, the tangent line is neither above nor below the graph but instead *crosses* the graph at that point.

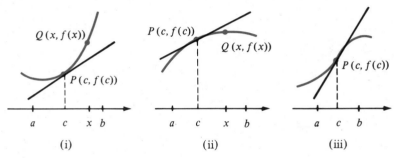

Figure 4.18

The terminology introduced in the next definition is used to describe the graphs illustrated in (i) and (ii) of Figure 4.18.

Definition (4.15)

> Let f be a function that is differentiable at a number c.
>
> (i) The graph of f is **concave upward** at the point $P(c, f(c))$ if there exists an open interval (a, b) containing c such that on (a, b) the graph of f is above the tangent line through P.
>
> (ii) The graph of f is **concave downward** at $P(c, f(c))$ if there exists an open interval (a, b) containing c such that on (a, b) the graph of f is below the tangent line through P.

Consider a function f whose graph is concave upward at the point $P(c, f(c))$. Let (a, b) be an open interval containing c such that on (a, b) the graph of f is above the tangent line through P. If f is differentiable on (a, b), then the graph of f has a tangent line at every point $Q(x, f(x))$ with x in (a, b). If we consider a number of these tangent lines, as indicated in Figure 4.19, it appears that as the x-coordinate x of the point of tangency increases, the slope $f'(x)$ of the tangent line also increases. Thus in (i) of Figure 4.19, a larger positive slope is obtained as P moves to the right. In (ii) of the figure, the slope of the tangent line also increases, becoming less negative as P moves to the right.

In like manner, if the graph of f is concave downward at $P(c, f(c))$, then the situations illustrated in Figure 4.20 may occur. Here the slope $f'(x)$ of the tangent line *decreases* as P moves to the right.

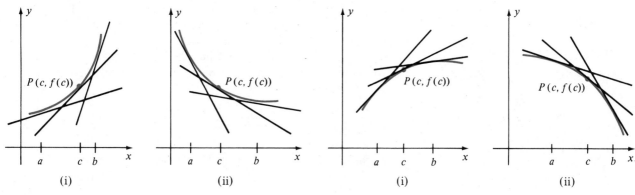

Figure 4.19 Upward concavity   *Figure 4.20* Downward concavity

Conversely, we might expect that if the derivative $f'(x)$ increases as x increases through c, then the graph is concave upward at $P(c, f(c))$, whereas if $f'(x)$ decreases, then the graph is concave downward. According to Theorem (4.13), if the derivative of a function is positive on an interval, then the function is increasing. Consequently, if the values of the *second* derivative f'' are positive on an interval, then the *first* derivative f' is increasing. Similarly, if f'' is negative, then f' is decreasing. This suggests that the sign of f'' can be used to test concavity as stated in the next theorem.

Test for Concavity (4.16)

Suppose a function f is differentiable on an open interval containing c, and $f''(c)$ exists.

(i)   If $f''(c) > 0$, the graph is concave upward at $P(c, f(c))$.

(ii) If $f''(c) < 0$, the graph is concave downward at $P(c, f(c))$.

Proof   Applying the formula in Theorem (3.7) to the function f',

$$f''(c) = \lim_{x \to c} \frac{f'(x) - f'(c)}{x - c}.$$

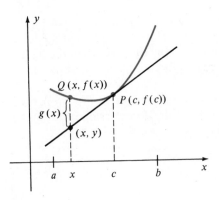

Figure 4.21

If $f''(c) > 0$, then by Theorem (2.7) there is an open interval (a, b) containing c such that

$$\frac{f'(x) - f'(c)}{x - c} > 0$$

for all x in (a, b) different from c. Hence $f'(x) - f'(c)$ and $x - c$ both have the same sign for all x in (a, b) different from c. We next show that this implies upward concavity.

For any x in the interval (a, b), consider the point (x, y) on the tangent line through $P(c, f(c))$ and the point $Q(x, f(x))$ on the graph of f (see Figure 4.21). If we let $g(x) = f(x) - y$, then we can establish upward concavity by showing that $g(x)$ is positive for all x such that $x \ne c$.

Since an equation of the tangent line at P is $y - f(c) = f'(c)(x - c)$, the y-coordinate of the point (x, y) on the tangent line is given by $y = f(c) + f'(c)(x - c)$. Consequently,

$$g(x) = f(x) - y = f(x) - f(c) - f'(c)(x - c).$$

Applying the Mean Value Theorem (4.12) to f and the interval $[x, c]$, we see that there exists a number w in the open interval (x, c) such that

$$f(x) - f(c) = f'(w)(x - c).$$

Substituting for $f(x) - f(c)$ in the preceding formula for $g(x)$ leads to

$$g(x) = f'(w)(x - c) - f'(c)(x - c)$$

which can be factored as follows:

$$g(x) = [f'(w) - f'(c)](x - c).$$

Since w is in the interval (a, b) we know, from the first paragraph of the proof, that $f'(w) - f'(c)$ and $w - c$ have the same sign. Moreover, since w is between x and c, $w - c$ and $x - c$ have the same sign. Thus $f'(w) - f'(c)$ and $x - c$ have the same sign for all x in (a, b) such that $x \ne c$. It follows from the factored form of $g(x)$ that $g(x)$ is positive if $x \ne c$, which is what we wished to prove. Part (ii) may be established in similar fashion.   □

If the second derivative $f''(x)$ changes sign as x increases through a number k, then by (4.16) the concavity changes from upward to downward, or from downward to upward. The point $(k, f(k))$ is called a *point of inflection* according to the next definition.

Definition (4.17)

> A point $P(k, f(k))$ on the graph of a function f is a **point of inflection** if there exists an open interval (a, b) containing k such that one of the following statements holds.
>
> (i) $f''(x) > 0$ if $a < x < k$ and $f''(x) < 0$ if $k < x < b$; or
>
> (ii) $f''(x) < 0$ if $a < x < k$ and $f''(x) > 0$ if $k < x < b$.

The sketch in Figure 4.22 displays typical points of inflection on a graph. A graph is said to be **concave upward** (or **downward**) **on an interval** if it is concave upward (or downward) at every number in the interval. Intervals on which the graph in Figure 4.22 is concave upward or concave downward are abbreviated CU or CD, respectively. Observe that a corner may, or may not, be a point of inflection.

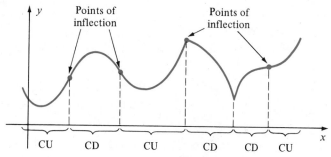

Figure 4.22

Example 1   If $f(x) = x^3 + x^2 - 5x - 5$, determine intervals on which the graph of f is concave upward and intervals on which the graph is concave downward.

Solution   The function f was considered in Examples 1 and 2 of the preceding section. Since $f'(x) = 3x^2 + 2x - 5$, we have

$$f''(x) = 6x + 2 = 2(3x + 1).$$

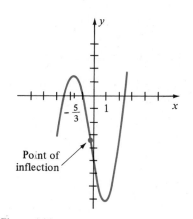

Point of inflection

Figure 4.23

Hence $f''(x) < 0$ if $3x + 1 < 0$, that is, if $x < -\frac{1}{3}$. It follows from the Test for Concavity (4.16) that the graph is concave downward on the infinite interval $(-\infty, -\frac{1}{3})$. Similarly $f''(x) > 0$ if $x > -\frac{1}{3}$ and, therefore, the graph is concave upward on $(-\frac{1}{3}, \infty)$. By Definition (4.17) the point $P(-\frac{1}{3}, -\frac{88}{27})$ at which $f''(x)$ changes sign (and the concavity changes from downward to upward) is a point of inflection. The graph of f obtained previously (see Figure 4.13) is sketched again in Figure 4.23 to show the point of inflection. ∎

If $P(c, f(c))$ is a point of inflection on the graph of f and if f'' is continuous on an open interval containing c, then necessarily $f''(c) = 0$. This is a consequence of the Intermediate Value Theorem (2.24) applied to the function f''. Thus, to locate points of inflection for a function whose second derivative is continuous, we begin by finding all numbers x such that $f''(x) = 0$. Each of these numbers is then tested to determine whether it is an x-coordinate of a point of inflection. Before giving an example of this technique, we state the following useful test for local maxima and minima.

The Second Derivative Test (4.18)

Suppose a function f is differentiable on an open interval containing c and $f'(c) = 0$.

(i) If $f''(c) < 0$, then f has a local maximum at c.

(ii) If $f''(c) > 0$, then f has a local minimum at c.

Proof   If $f'(c) = 0$, then the tangent line to the graph at $P(c, f(c))$ is horizontal. If, in addition, $f''(c) < 0$, then the graph is concave downward at c and hence there is an interval (a, b) containing c such that the graph lies below the tangent line at P. It follows that $f(c)$ is a local maximum for f as illustrated in (i) of Figure 4.24. A similar proof may be given for part (ii), which is illustrated graphically in (ii) of Figure 4.24.   ☐

If $f''(c) = 0$, the Second Derivative Test is not applicable. In such cases the First Derivative Test should be employed (see Example 3).

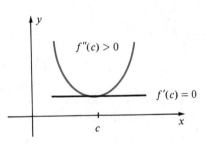

(i) Local maximum

(ii) Local minimum

Figure 4.24

Example 2   If $f(x) = 12 + 2x^2 - x^4$, use the Second Derivative Test to find the local maxima and minima of f. Discuss concavity, find the points of inflection, and sketch the graph of f.

Solution   We begin by finding the first and second derivatives and factoring them as follows:

$$f'(x) = 4x - 4x^3 = 4x(1 - x^2)$$
$$f''(x) = 4 - 12x^2 = 4(1 - 3x^2).$$

The expression for $f'(x)$ is used to find the critical numbers $0, 1$, and -1. The values of f'' at these numbers are

$$f''(0) = 4 > 0, \quad f''(1) = -8 < 0, \quad \text{and} \quad f''(-1) = -8 < 0.$$

Hence, by the Second Derivative Test there is a local minimum at 0 and local maxima at 1 and -1. The corresponding functional values are $f(0) = 12$ and $f(1) = 13 = f(-1)$. The following table summarizes the preceding discussion.

Critical number c	$f''(c)$	Sign of $f''(c)$	Conclusion
-1	-8	$-$	Local max: $f(-1) = 13$
0	4	$+$	Local min: $f(0) = 12$
1	-8	$-$	Local max: $f(1) = 13$

To locate the possible points of inflection we solve the equation $f''(x) = 0$, that is, $4(1 - 3x^2) = 0$. Evidently, the solutions are $-\sqrt{3}/3$ and $\sqrt{3}/3$. This suggests an examination of the sign of $f''(x)$ in each of the intervals $(-\infty, -\sqrt{3}/3)$, $(-\sqrt{3}/3, \sqrt{3}/3)$, and $(\sqrt{3}/3, \infty)$. Since f'' is continuous and has no zeros on each interval, we may use test values to determine the sign of $f''(x)$. Let us arrange our work in tabular form as follows, where the last row is a consequence of (4.18).

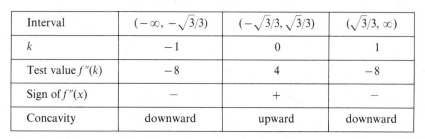

Interval	$(-\infty, -\sqrt{3}/3)$	$(-\sqrt{3}/3, \sqrt{3}/3)$	$(\sqrt{3}/3, \infty)$
k	-1	0	1
Test value $f''(k)$	-8	4	-8
Sign of $f''(x)$	$-$	$+$	$-$
Concavity	downward	upward	downward

Since the sign of $f''(x)$ changes as x increases through $-\sqrt{3}/3$ and $\sqrt{3}/3$, the corresponding points $(\pm\sqrt{3}/3, 113/9)$ on the graph are points of inflection. These are the points at which the sense of concavity changes. Indeed, as shown in the table, the graph is concave upward on the open interval $(-\sqrt{3}/3, \sqrt{3}/3)$ and concave downward outside of $[-\sqrt{3}/3, \sqrt{3}/3]$. The graph is sketched in Figure 4.25. ∎

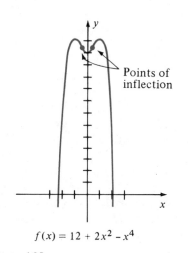

Points of inflection

$f(x) = 12 + 2x^2 - x^4$

Figure 4.25

Example 3 If $f(x) = x^5 - 5x^3$, use the Second Derivative Test to find the local extrema of f. Discuss concavity, find the points of inflection, and sketch the graph of f.

Solution   Differentiating, we obtain

$$f'(x) = 5x^4 - 15x^2 = 5x^2(x^2 - 3)$$
$$f''(x) = 20x^3 - 30x = 10x(2x^2 - 3).$$

Solving the equation $f'(x) = 0$ gives us the critical numbers 0, $-\sqrt{3}$, and $\sqrt{3}$. As in Example 2 we construct the following table (check the entries).

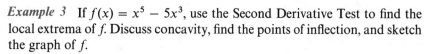

Critical number c	$f''(c)$	Sign of $f''(c)$	Conclusion
$-\sqrt{3}$	$-30\sqrt{3}$	$-$	Local max: $f(-\sqrt{3}) = 6\sqrt{3}$
0	0	none	No conclusion
$\sqrt{3}$	$30\sqrt{3}$	$+$	Local min: $f(\sqrt{3}) = -6\sqrt{3}$

Since $f''(0) = 0$, the Second Derivative Test is not applicable at 0 and hence we turn to the First Derivative Test. The reader should check, using test values, that if $-\sqrt{3} < x < 0$, then $f'(x) < 0$, and if $0 < x < \sqrt{3}$, then $f'(x) < 0$. Since $f'(x)$ does not change sign, there is neither a maximum nor a minimum at $x = 0$.

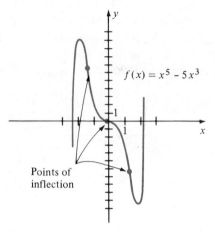

$f(x) = x^5 - 5x^3$

Points of inflection

Figure 4.26

To find possible points of inflection we consider the equation $f''(x) = 0$, that is, $10x(2x^2 - 3) = 0$. The solutions of this equation, in order of magnitude, are $-\sqrt{6}/2, 0,$ and $\sqrt{6}/2$. As in Example 2 we construct the following table.

Interval	$(-\infty, -\sqrt{6}/2)$	$(-\sqrt{6}/2, 0)$	$(0, \sqrt{6}/2)$	$(\sqrt{6}/2, \infty)$
k	-2	-1	1	2
Test value $f''(k)$	-100	10	-10	100
Sign of $f''(x)$	$-$	$+$	$-$	$+$
Concavity	downward	upward	downward	upward

Since the sign of $f''(x)$ changes as x increases through $-\sqrt{6}/2, 0,$ and $\sqrt{6}/2$, the points $(0, 0)$, $(-\sqrt{6}/2, 21\sqrt{6}/8)$, and $(\sqrt{6}/2, -21\sqrt{6}/8)$ are points of inflection. The graph is sketched in Figure 4.26 where, for clarity, we have used different scales on the x- and y-axes.   ■

In all of the preceding examples f'' was continuous. It is also possible for $(c, f(c))$ to be a point of inflection if either $f'(c)$ or $f''(c)$ does not exist, as illustrated in the next example.

Example 4   If $f(x) = 1 - x^{1/3}$, find the local extrema, discuss concavity, find the points of inflection, and sketch the graph of f.

Solution   Differentiating, we obtain

$$f'(x) = -\frac{1}{3}x^{-2/3} = -\frac{1}{3x^{2/3}}$$

$$f''(x) = \frac{2}{9}x^{-5/3} = \frac{2}{9x^{5/3}}.$$

The first derivative does not exist at $x = 0$, and 0 is the only critical number for f. Since $f''(0)$ is undefined, the Second Derivative Test is not applicable. However, if $x \neq 0$, then $x^{2/3} > 0$ and $f'(x) = -1/3x^{2/3} < 0$, which means that f is decreasing throughout its domain. Consequently $f(0)$ is not a local extremum.

The fact that f'' is undefined at $x = 0$ leads us to examine whether or not the point $(0, 1)$ on the graph of f is a point of inflection. Let us apply Definition (4.17) with $c = 0$. If $x < 0$, then $x^{5/3} < 0$. Hence

$$f''(x) = \frac{2}{9x^{5/3}} < 0 \quad \text{if } x < 0$$

which implies that the graph of f is concave downward on the interval $(-\infty, 0)$. If $x > 0$, then $x^{5/3} > 0$. Thus

$$f''(x) = \frac{2}{9x^{5/3}} > 0 \quad \text{if } x > 0$$

which means that the graph is concave upward on the interval $(0, \infty)$. By (4.17), the point $(0, 1)$ is a point of inflection. Using this information and plotting several points gives us the sketch in Figure 4.27.

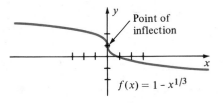

Figure 4.27

■

EXERCISES 4.4

In Exercises 1–18, use the Second Derivative Test (whenever applicable) to find the local extrema of f. Discuss concavity, find x-coordinates of points of inflection, and sketch the graph of f.

1 $f(x) = x^3 - 2x^2 + x + 1$

2 $f(x) = x^3 + 10x^2 + 25x - 50$

3 $f(x) = 3x^4 - 4x^3 + 6$

4 $f(x) = 8x^2 - 2x^4$

5 $f(x) = 2x^6 - 6x^4$

6 $f(x) = 3x^5 - 5x^3$

7 $f(x) = (x^2 - 1)^2$

8 $f(x) = x - (16/x)$

9 $f(x) = \sqrt[5]{x} - 1$

10 $f(x) = (x + 4)/\sqrt{x}$

11 $f(x) = x^2 - (27/x^2)$

12 $f(x) = x^{2/3}(1 - x)$

13 $f(x) = x/(x^2 + 1)$

14 $f(x) = x^2/(x^2 + 1)$

15 $f(x) = \sqrt[3]{x^2}(3x + 10)$

16 $f(x) = x^4 - 4x^3 + 10$

17 $f(x) = 8x^{1/3} + x^{4/3}$

18 $f(x) = x\sqrt{4 - x^2}$

In Exercises 19–26 sketch the graph of a continuous function f that satisfies all of the stated conditions.

19 $f(0) = 1; f(2) = 3; f'(0) = f'(2) = 0;$
$f'(x) < 0$ if $|x - 1| > 1; f'(x) > 0$ if $|x - 1| < 1;$
$f''(x) > 0$ if $x < 1; f''(x) < 0$ if $x > 1.$

20 $f(0) = 4; f(2) = 2; f(5) = 6; f'(0) = f'(2) = 0;$
$f'(x) > 0$ if $|x - 1| > 1; f'(x) < 0$ if $|x - 1| < 1;$
$f''(x) < 0$ if $x < 1$ or if $|x - 4| < 1;$
$f''(x) > 0$ if $|x - 2| < 1$ or if $x > 5.$

21 $f(0) = 2; f(2) = f(-2) = 1; f'(0) = 0;$
$f'(x) > 0$ if $x < 0; f'(x) < 0$ if $x > 0;$
$f''(x) < 0$ if $|x| < 2; f''(x) > 0$ if $|x| > 2.$

22 $f(1) = 4; f'(x) > 0$ if $x < 1; f'(x) < 0$ if $x > 1;$
$f''(x) > 0$ for all $x \neq 1.$

23 $f(-2) = f(6) = -2; f(0) = f(4) = 0; f(2) = f(8) = 3;$
f' is undefined at 2 and 6; $f'(0) = 1; f'(x) > 0$ throughout $(-\infty, 2)$ and $(6, \infty); f'(x) < 0$ if $|x - 4| < 2; f''(x) < 0$ throughout $(-\infty, 0), (4, 6),$ and $(6, \infty); f''(x) > 0$ throughout $(0, 2)$ and $(2, 4).$

24 $f(0) = 2; f(2) = 1; f(4) = f(10) = 0; f(6) = -4; f'(2) = f'(6) = 0;$ $f'(x) < 0$ throughout $(-\infty, 2), (2, 4), (4, 6),$ and $(10, \infty);$ $f'(x) > 0$ throughout $(6, 10);$ $f'(4)$ and $f'(10)$ do not exist; $f''(x) > 0$ throughout $(-\infty, 2),$ $(4, 10),$ and $(10, \infty); f''(x) < 0$ throughout $(2, 4).$

25 If n is an odd integer, then $f(n) = 1$ and $f'(n) = 0$; if n is an even integer, then $f(n) = 0$ and $f'(n)$ does not exist; if n is any integer then
(a) $f'(x) > 0$ whenever $2n < x < 2n + 1;$
(b) $f'(x) < 0$ whenever $2n - 1 < x < 2n;$
(c) $f''(x) < 0$ whenever $2n < x < 2n + 2.$

26 $f(x) = x$ if $x = -1, 2, 4,$ or $8; f'(x) = 0$ if $x = -1, 4, 6,$ or $8; f'(x) < 0$ throughout $(-\infty, -1), (4, 6),$ and $(8, \infty);$ $f'(x) > 0$ throughout $(-1, 4)$ and $(6, 8); f''(x) > 0$ throughout $(-\infty, 0), (2, 3),$ and $(5, 7); f''(x) < 0$ throughout $(0, 2), (3, 5),$ and $(7, \infty).$

27 Prove that the graph of a quadratic function has no points of inflection. State conditions under which the graph is always (a) concave upward; (b) concave downward. Illustrate with sketches.

28 Prove that the graph of a polynomial function of degree 3 has exactly one point of inflection. Illustrate this fact with sketches.

29 Prove that the graph of a polynomial function of degree $n > 2$ has at most $n - 2$ points of inflection.

30 If $f(x) = x^n$, where $n > 1$, prove that the graph of f has either one or no points of inflection, depending on whether n is odd or even. Illustrate with sketches.

4.5

HORIZONTAL AND VERTICAL ASYMPTOTES

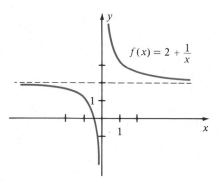

Figure 4.28

Thus far we have used the derivative primarily as a tool for sketching graphs. In particular, we have developed techniques for determining where a graph rises or falls, the high and low points on a graph, concavity, and points of inflection. Before turning to physical applications of derivatives, we shall discuss several other interesting characteristics of some graphs. When applied to the derivative f', our discussion will provide information about vertical tangent lines to the graph of f'.

Let us begin by considering the graph of $f(x) = 2 + (1/x)$, which is sketched in Figure 4.28, and concentrate on values of $f(x)$ if x is large and positive. For example, $f(100) = 2.01$, $f(1000) = 2.001$, $f(10,000) = 2.0001$, and $f(100,000) = 2.00001$. It is evident that we can make $f(x)$ as close to 2 as desired or, equivalently, we can make $|f(x) - 2|$ arbitrarily small by choosing x sufficiently large. This behavior of $f(x)$ is expressed by writing

$$\lim_{x \to \infty} \left(2 + \frac{1}{x}\right) = 2$$

which may be read "the limit of $2 + (1/x)$ as x becomes infinite (or as x increases without bound) is 2."

The general definition that describes the behavior of $f(x)$ in the preceding illustration may be phrased as follows.

Definition (4.19)

> Let f be defined on an interval (c, ∞). The statement
>
> $$\lim_{x \to \infty} f(x) = L$$
>
> means that for every $\varepsilon > 0$, there corresponds a positive number N, such that
>
> if $x > N$, then $|f(x) - L| < \varepsilon$.

If $\lim_{x \to \infty} f(x) = L$ we say that **the limit of $f(x)$ is L as x becomes infinite** (or **as x increases without bound**). To emphasize the fact that x is large and positive, we often say that the limit of $f(x)$ is L as x becomes *positively* infinite.

It is important to remember that the symbol ∞ does not represent a real number. It is used here to denote that x increases without bound. Later, in Definition (4.22), we shall employ it in a different context.

It is of interest to study the graphical significance of the preceding definition. Suppose $\lim_{x \to \infty} f(x) = L$, and consider any horizontal lines $y = L \pm \varepsilon$ (see Figure 4.29). According to Definition (4.19), if x is larger than some number N, then all points $P(x, f(x))$ lie between these horizontal lines. Roughly speaking, the graph of f gets closer and closer to the line $y = L$ as x gets larger and larger. The line $y = L$ is called a **horizontal asymptote** for the graph of f. For example, the line $y = 2$ in Figure 4.28 is a horizontal asymptote for the graph of $f(x) = 2 + (1/x)$.

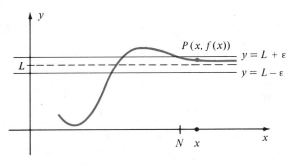

Figure 4.29 $\lim_{x \to \infty} f(x) = L$

In Figure 4.29, the graph of f approaches the asymptote $y = L$ from above, that is, with $f(x) > L$. However, we could also have $f(x) < L$ or other variations.

The next definition covers the situation in which $|x|$ is large and x is *negative*.

Definition (4.20)

> Let f be defined on an interval $(-\infty, c)$. The statement
>
> $$\lim_{x \to -\infty} f(x) = L$$
>
> means that for every $\varepsilon > 0$, there corresponds a negative number N, such that
>
> $$\text{if} \quad x < N, \quad \text{then} \quad |f(x) - L| < \varepsilon.$$

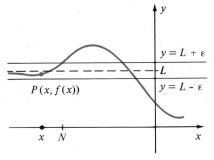

Figure 4.30 $\lim_{x \to -\infty} f(x) = L$

If $\lim_{x \to -\infty} f(x) = L$ we say that **the limit of $f(x)$ is L as x becomes negatively infinite** (or **as x decreases without bound**).

A geometric illustration of (4.20) is given in Figure 4.30. If we consider any lines $y = L \pm \varepsilon$, then all points $P(x, f(x))$ on the graph lie between these lines if x is *less* than some number N. As before, we refer to the line $y = L$ as a horizontal asymptote for the graph of f.

Limit theorems that are analogous to those in Chapter 2 may be established. In particular, Theorem (2.8) concerning limits of sums, products, and quotients is true for the cases $x \to \infty$ or $x \to -\infty$. Similarly, Theorem (2.13) on the limit of $\sqrt[n]{f(x)}$ holds if $x \to \infty$ or $x \to -\infty$. Finally, it is trivial to show that

$$\lim_{x \to \infty} c = c \quad \text{and} \quad \lim_{x \to -\infty} c = c.$$

The next theorem provides an important tool for calculating specific limits.

Theorem (4.21)

If k is a positive rational number and c is any real number, then

$$\lim_{x \to \infty} \frac{c}{x^k} = 0 \quad \text{and} \quad \lim_{x \to -\infty} \frac{c}{x^k} = 0$$

provided x^k is always defined.

Proof   To prove that $\lim_{x \to \infty} (c/x^k) = 0$ by means of (4.19) we must show that for every $\varepsilon > 0$, there corresponds a number N such that

$$\text{if} \quad x > N, \quad \text{then} \quad \left| \frac{c}{x^k} - 0 \right| < \varepsilon.$$

If $c = 0$, any $N > 0$ will suffice. If $c \neq 0$, the following four inequalities are equivalent.

$$\left| \frac{c}{x^k} - 0 \right| < \varepsilon; \qquad \frac{|x|^k}{|c|} > \frac{1}{\varepsilon}; \qquad |x|^k > \frac{|c|}{\varepsilon}; \qquad x > \left(\frac{|c|}{\varepsilon} \right)^{1/k}.$$

If we let $N = (|c|/\varepsilon)^{1/k}$, then whenever $x > N$ the fourth, and hence the first, inequality is true, which is what we wished to show. The second part of the theorem may be proved in similar fashion.

If f is a rational function, then limits as $x \to \infty$ or $x \to -\infty$ may be found by first dividing numerator and denominator of $f(x)$ by a suitable power of x and then applying Theorem (4.21). Specifically, suppose $g(x)$ and $h(x)$ are polynomials and we wish to investigate $f(x) = g(x)/h(x)$. If the degree of $g(x)$ is not greater than the degree of $h(x)$, then we divide numerator and denominator by x^k, where k is the degree of $h(x)$. If the degree of $g(x)$ is greater than the degree of $h(x)$, it can be shown that $f(x)$ has no limit as $x \to \infty$ or as $x \to -\infty$.

Example 1   Investigate $\lim_{x \to -\infty} (2x^2 - 5)/(3x^2 + x + 2)$.

Solution   Since we are interested only in negative values of x, we may assume that $x \neq 0$. Using the rule stated in the preceding paragraph we first divide numerator and denominator of the given expression by x^2 and then employ the appropriate limit theorems. Thus

$$\lim_{x \to -\infty} \frac{2x^2 - 5}{3x^2 + x + 2} = \lim_{x \to -\infty} \frac{2 - (5/x^2)}{3 + (1/x) + (2/x^2)}$$

$$= \frac{\lim\limits_{x \to -\infty} [2 - (5/x^2)]}{\lim\limits_{x \to -\infty} [3 + (1/x) + (2/x^2)]}$$

$$= \frac{\lim\limits_{x \to -\infty} 2 - \lim\limits_{x \to -\infty} (5/x^2)}{\lim\limits_{x \to -\infty} 3 + \lim\limits_{x \to -\infty} (1/x) + \lim\limits_{x \to -\infty} (2/x^2)} = \frac{2 - 0}{3 + 0 + 0} = \frac{2}{3}.$$

It follows that the line $y = \frac{2}{3}$ is a horizontal asymptote for the graph of f. ∎

Example 2   If $f(x) = 4x/(x^2 + 9)$, determine the horizontal asymptotes, find the relative maxima and minima and the points of inflection, discuss concavity, and sketch the graph of f.

Solution   We divide numerator and denominator of $f(x)$ by x^2 and use limit theorems as follows:

$$\lim_{x \to \infty} f(x) = \lim_{x \to \infty} \frac{4x}{x^2 + 9} = \lim_{x \to \infty} \frac{(4/x)}{1 + (9/x^2)}$$

$$= \frac{\lim\limits_{x \to \infty} (4/x)}{\lim\limits_{x \to \infty} 1 + \lim\limits_{x \to \infty} (9/x^2)} = \frac{0}{1 + 0} = \frac{0}{1} = 0.$$

Hence $y = 0$ (the x-axis) is a horizontal asymptote for the graph of f. Similarly, $\lim_{x \to -\infty} f(x) = 0$.

It is left to the reader to show that the first and second derivatives of f are given by

$$f'(x) = \frac{4(9 - x^2)}{(x^2 + 9)^2}; \qquad f''(x) = \frac{8x(x^2 - 27)}{(x^2 + 9)^3}.$$

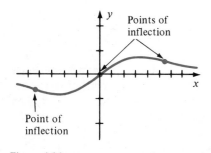

Points of inflection

Point of inflection

Figure 4.31

The critical numbers of f are ± 3. (Why?) Using either the First or Second Derivative Test, we can show that $f(-3)$ is a relative minimum and $f(3)$ is a relative maximum. Points of inflection have x-coordinates 0 and $\pm\sqrt{27} = \pm 3\sqrt{3}$. The reader may verify that the graph is concave downward in the intervals $(-\infty, -3\sqrt{3})$ and $(0, 3\sqrt{3})$ and concave upward in $(-3\sqrt{3}, 0)$ and $(3\sqrt{3}, \infty)$. The graph is sketched in Figure 4.31, where for clarity we have used different scales on the axes. ∎

Example 3   Investigate $\lim_{x \to \infty} \frac{\sqrt{9x^2 + 2}}{3 - 4x}$.

Solution   If x is large and positive, the numerator $\sqrt{9x^2 + 2}$ may be approximated by $\sqrt{9x^2}$, or $3x$. This suggests that we divide numerator and denominator by $x = \sqrt{x^2}$. Since $\sqrt{9x^2 + 2}/\sqrt{x^2} = \sqrt{(9x^2 + 2)/x^2} = \sqrt{9 + (2/x^2)}$ we obtain

$$\lim_{x \to \infty} \frac{\sqrt{9x^2 + 2}}{3 - 4x} = \lim_{x \to \infty} \frac{\sqrt{9 + (2/x^2)}}{(3/x) - 4} = \frac{\lim\limits_{x \to \infty} \sqrt{9 + (2/x^2)}}{\lim\limits_{x \to \infty} [(3/x) - 4]}$$

$$= \frac{\sqrt{\lim\limits_{x \to \infty} (9 + (2/x^2))}}{\lim\limits_{x \to \infty} (3/x) - \lim\limits_{x \to \infty} 4} = \frac{\sqrt{\lim\limits_{x \to \infty} 9 + \lim\limits_{x \to \infty} (2/x^2)}}{0 - 4}$$

$$= \frac{\sqrt{9 + 0}}{-4} = -\frac{3}{4}. \qquad \blacksquare$$

Let us next consider the function f defined by $f(x) = 1/(x - 3)^2$. If x is close to 3 (but $x \neq 3$) the denominator $(x - 3)^2$ is close to zero, and hence $f(x)$ is very large. Indeed, $f(x)$ can be made as large as desired by choosing x sufficiently close to 3. This behavior of $f(x)$ is symbolized by writing

$$\lim_{x \to 3} \frac{1}{(x - 3)^2} = \infty.$$

In general, we have the following definition.

Definition (4.22)

> Let f be defined on an open interval containing a (except possibly at $x = a$). The statement **$f(x)$ becomes infinite** (or **increases without bound**) as x approaches a, written
>
> $$\lim_{x \to a} f(x) = \infty$$
>
> means that for every positive number M, there corresponds a $\delta > 0$ such that
>
> $$\text{if} \quad 0 < |x - a| < \delta, \quad \text{then} \quad f(x) > M.$$

In Definition (4.22) we sometimes state that $f(x)$ becomes *positively* infinite as x approaches a. A geometric interpretation in terms of the graph of f is shown in Figure 4.32. If we consider any horizontal line $y = M$, then when x is in a suitable interval $(a - \delta, a + \delta)$, but $x \neq a$, the points on the graph of f lie *above* the horizontal line.

As we mentioned earlier, the symbol ∞ is used as a device for describing certain types of functional behavior. It is *not* a real number. In particular, $\lim_{x \to a} f(x) = \infty$ does not mean that the limit *exists* as x approaches a. The limit notation is merely used to denote the variation of $f(x)$ that we have described.

The notion of $f(x)$ increasing without bound as x approaches a from the right or left can also be introduced. This is indicated by writing

$$\lim_{x \to a^+} f(x) = \infty \quad \text{or} \quad \lim_{x \to a^-} f(x) = \infty,$$

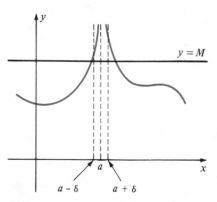

Figure 4.32 $\lim\limits_{x \to a} f(x) = \infty$

respectively. Only minor modifications of Definition (4.22) are needed to define these concepts. Thus, for $x \to a^+$, it is assumed that $f(x)$ exists on some open interval (a, c), and the inequality $0 < |x - a| < \delta$ in Definition (4.22) is changed to $a < x < a + \delta$; that is, x may only take on values *larger* than a. Similar statements may be made for $x \to a^-$. Graphs illustrating these ideas are sketched in Figure 4.33.

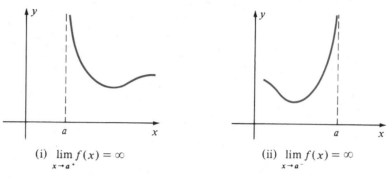

Figure 4.33

<table>
<tr><td>Definition (4.23)</td><td>Let f be defined on an open interval containing a (except possibly at $x = a$). The statement $f(x)$ **becomes negatively infinite** (or **decreases without bound**) as x approaches a, written

$$\lim_{x \to a} f(x) = -\infty$$

means that for every negative number M there corresponds a $\delta > 0$ such that

if $\quad 0 < |x - a| < \delta, \quad$ then $\quad f(x) < M$.</td></tr>
</table>

Graphical illustrations of Definition (4.23) together with the cases $x \to a^+$ and $x \to a^-$ are sketched in Figure 4.34. If either of Definitions (4.22) or (4.23) occurs (including the cases of one-sided limits), then the line $x = a$ is called a **vertical asymptote** for the graph of f. We also say that f has an **infinite discontinuity** at $x = a$. The line $x = a$ in Figures (4.32)–(4.34) is a vertical asymptote for the graph.

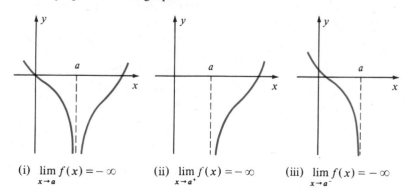

Figure 4.34

The next theorem, stated without proof, is useful in the investigation of certain limits.

Theorem (4.24)

(i) If n is an even positive integer, then
$$\lim_{x \to a} \frac{1}{(x - a)^n} = \infty.$$

(ii) If n is an odd positive integer, then
$$\lim_{x \to a^+} \frac{1}{(x - a)^n} = \infty \quad \text{and} \quad \lim_{x \to a^-} \frac{1}{(x - a)^n} = -\infty.$$

It follows that if n is a positive integer and $f(x) = 1/(x - a)^n$, then the graph of f has the line $x = a$ as a vertical asymptote. Note also that since $\lim_{x \to \infty} 1/(x - a)^n = 0$, the x-axis is a horizontal asymptote.

It is unnecessary to memorize Theorem (4.24), since for any specific problem it is usually possible to determine the answer intuitively. As an illustration, consider $\lim_{x \to -5^-} 1/(x + 5)^3$. If x is close to -5 and $x < -5$, then $x + 5$ is close to 0 and is *negative*. Consequently $(x + 5)^3$ is close to 0 and negative and, therefore, $1/(x + 5)^3$ is a numerically large negative number. This suggests that
$$\lim_{x \to -5^-} \frac{1}{(x + 5)^3} = -\infty.$$

To investigate $x \to -5^+$, we first note that if $x > -5$, then $x + 5 > 0$, and reason that
$$\lim_{x \to -5^+} \frac{1}{(x + 5)^3} = \infty.$$

Example 4 If $f(x) = 1/(x - 2)^3$, discuss $\lim_{x \to 2^+} f(x)$, $\lim_{x \to 2^-} f(x)$, and sketch the graph of f.

Solution   If x is very close to 2, but larger than 2, then $x - 2$ is a small positive number and hence $1/(x - 2)^3$ is a large positive number. Thus it is evident that
$$\lim_{x \to 2^+} \frac{1}{(x - 2)^3} = \infty.$$

This fact also follows from (ii) of Theorem (4.24) with $a = 2$ and $n = 3$.

If x is close to 2 but *less* than 2, then $x - 2$ is close to 0 and negative. Hence
$$\lim_{x \to 2^-} \frac{1}{(x - 2)^3} = -\infty.$$

This is also a consequence of (ii) of (4.24). The graph of f is sketched in Figure 4.35.

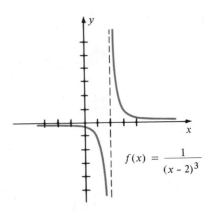

$$f(x) = \frac{1}{(x - 2)^3}$$

Figure 4.35

The line $x = 2$ is a vertical asymptote. The reader should verify that

$$\lim_{x \to \infty} \frac{1}{(x - 2)^3} = 0 \quad \text{and} \quad \lim_{x \to -\infty} \frac{1}{(x - 2)^3} = 0.$$

Hence $y = 0$ (the x-axis) is a horizontal asymptote. ∎

Several properties of sums, products, and quotients of functions that become infinite are stated below. Analogous results hold for the cases $x \to a^+$ and $x \to a^-$.

Theorem (4.25)

> If $\lim_{x \to a} f(x) = \infty$ and $\lim_{x \to a} g(x) = c$ then:
>
> (i) $\displaystyle\lim_{x \to a} [g(x) + f(x)] = \infty.$
>
> (ii) If $c > 0$, then $\displaystyle\lim_{x \to a} [g(x)f(x)] = \infty$ and $\displaystyle\lim_{x \to a} \frac{f(x)}{g(x)} = \infty.$
>
> (iii) If $c < 0$, then $\displaystyle\lim_{x \to a} [g(x)f(x)] = -\infty$ and $\displaystyle\lim_{x \to a} \frac{f(x)}{g(x)} = -\infty.$
>
> (iv) $\displaystyle\lim_{x \to a} \frac{g(x)}{f(x)} = 0.$

The conclusions of Theorem (4.25) appear intuitively evident. In (i), for example, if $g(x)$ is close to c and if $f(x)$ increases without bound as x approaches a, then it is reasonable to expect that $g(x) + f(x)$ can be made arbitrarily large by choosing x sufficiently close to a. In (iii), if $\lim_{x \to a} g(x) = c < 0$, then $g(x)$ is negative when x is close to a. Consequently, if $f(x)$ is large and positive, then when x is close to a the product $g(x)f(x)$ is numerically large and *negative*. This suggests that $\lim_{x \to a} g(x)f(x) = -\infty$. The reader should supply similar arguments for the remaining parts of Theorem (4.25). We shall not give formal proofs for these results. A similar theorem can be stated for $\lim_{x \to a} f(x) = -\infty$. Theorems involving more than two functions can also be proved.

Graphs of rational functions frequently have vertical and horizontal asymptotes. Illustrations of this are shown in Figures 4.28, 4.31, and 4.35. Another is given in the following example.

Example 5   If $f(x) = 2x^2/(9 - x^2)$, find the vertical and horizontal asymptotes for the graph of f. Sketch the graph of f.

Solution   We begin by writing

$$f(x) = \frac{2x^2}{(3 - x)(3 + x)}.$$

The denominator is zero at $x = 3$ and $x = -3$, and hence the corresponding lines are candidates for vertical asymptotes. (Why?) Since

$$\lim_{x \to 3^-} \frac{1}{3 - x} = \infty \quad \text{and} \quad \lim_{x \to 3^-} \frac{2x^2}{3 + x} = 3$$

we see from (ii) of Theorem (4.25) that

$$\lim_{x \to 3^-} f(x) = \lim_{x \to 3^-} \left(\frac{1}{3 - x} \right) \left(\frac{2x^2}{3 + x} \right) = \infty ;$$

that is, $f(x)$ becomes positively infinite as x approaches 3 from the left. Moreover, since

$$\lim_{x \to 3^+} \frac{1}{3 - x} = -\infty \quad \text{and} \quad \lim_{x \to 3^+} \frac{2x^2}{3 + x} = 3 > 0$$

it follows that

$$\lim_{x \to 3^+} f(x) = \lim_{x \to 3^+} \left(\frac{1}{3 - x} \right) \left(\frac{2x^2}{3 + x} \right) = -\infty .$$

Thus $f(x)$ becomes negatively infinite as x approaches 3 from the right. This variation of $f(x)$ near the vertical asymptote $x = 3$ is illustrated in Figure 4.36.

In like manner, since

$$\lim_{x \to -3^-} \frac{1}{3 + x} = -\infty \quad \text{and} \quad \lim_{x \to -3^-} \frac{2x^2}{3 - x} = 3 > 0$$

we conclude that

$$\lim_{x \to -3^-} f(x) = \lim_{x \to -3^-} \left(\frac{1}{3 + x} \right) \left(\frac{2x^2}{3 - x} \right) = -\infty .$$

Similarly,

$$\lim_{x \to -3^+} \frac{1}{3 + x} = \infty \quad \text{implies that} \quad \lim_{x \to -3^+} f(x) = \infty .$$

This behavior of $f(x)$ near $x = -3$ is illustrated in Figure 4.36.

To find the horizontal asymptotes we consider

$$\lim_{x \to \infty} f(x) = \lim_{x \to \infty} \frac{2x^2}{9 - x^2} = \lim_{x \to \infty} \frac{2}{(9/x^2) - 1} = -2.$$

The same limit is obtained if $x \to -\infty$. Consequently $y = -2$ is a horizontal asymptote. Using this information and plotting several points gives us the sketch in Figure 4.36.

If more detailed information about the graph is desired, we could use the first derivative to show that f is decreasing on the intervals $(-\infty, -3)$ and $(-3, 0]$, and increasing on $[0, 3)$ and $(3, \infty)$, with a minimum occurring at $x = 0$ as shown in the figure. The second derivative could be used to investigate concavity. ∎

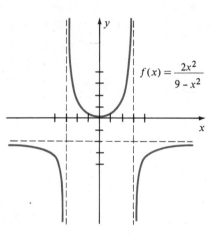

$f(x) = \dfrac{2x^2}{9 - x^2}$

Figure 4.36

If $f(x) = g(x)/h(x)$ where $g(x)$ and $h(x)$ are polynomials, and *if the degree of $g(x)$ is* 1 *greater than the degree of $h(x)$*, then the graph of f has an **oblique asymptote** $y = ax + b$, in the sense that the graph approaches this line as x increases or decreases without bound. To establish this fact we may use long division to express $f(x)$ in the form

$$f(x) = \frac{g(x)}{h(x)} = (ax + b) + \frac{r(x)}{h(x)}$$

where either $r(x) = 0$, or the degree of $r(x)$ is less than the degree of $h(x)$. It follows that $\lim_{x\to\infty} r(x)/h(x) = 0$ and $\lim_{x\to-\infty} r(x)/h(x) = 0$ (see Exercise 15). Consequently, $f(x)$ gets closer and closer to $ax + b$ as x either increases or decreases without bound. The next example illustrates a special case of this procedure.

Example 6   Find all the asymptotes and sketch the graph of f if

$$f(x) = \frac{x^2 - 9}{2x - 4}.$$

Solution   As in previous examples, a vertical asymptote occurs if $2x - 4 = 0$, that is, if $x = 2$. Moreover, since the degree of the numerator $x^2 - 9$ is 1 greater than the degree of the denominator $2x - 4$, the graph has an oblique asymptote. It may be verified by division that

$$\frac{x^2 - 9}{2x - 4} = (\tfrac{1}{2}x + 1) - \frac{5}{2x - 4}.$$

As in the discussion preceding this example, the line $y = \tfrac{1}{2}x + 1$ is an oblique asymptote. This line and the vertical asymptote $x = 2$ are sketched (with dashes) in Figure 4.37.

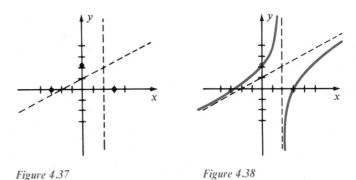

Figure 4.37   Figure 4.38

The x-intercepts of the graph are the solutions of the equation $x^2 - 9 = 0$, and hence are 3 and -3. The y-intercept is $f(0) = \tfrac{9}{4}$. The corresponding points are plotted in Figure 4.37. It is now easy to show that the graph has the shape indicated in Figure 4.38.   ■

If we regard a tangent line l as a limiting position of a secant line (see Figure 2.2), then it is possible for l to be vertical. In such cases, as the secant line approaches l its slope increases or decreases without bound. With this in mind we formulate the next definition.

Definition (4.26)

The graph of a function f is said to have a **vertical tangent line** at the point $P(a, f(a))$ if f is continuous at a and

$$\lim_{x \to a} |f'(x)| = \infty.$$

Example 7   If $f(x) = (x - 8)^{1/3} + 1$ and $g(x) = (x - 8)^{2/3} + 1$, show that the graphs of f and g have vertical tangent lines at $P(8, 1)$. Sketch the graphs, showing the tangent lines at P.

Solution   Both f and g are continuous at 8. Differentiating we obtain

$$f'(x) = \frac{1}{3}(x - 8)^{-2/3} = \frac{1}{3(x - 8)^{2/3}}$$

$$g'(x) = \frac{2}{3}(x - 8)^{-1/3} = \frac{2}{3(x - 8)^{1/3}}.$$

Evidently,

$$\lim_{x \to 8} |f'(x)| = \infty \quad \text{and} \quad \lim_{x \to 8} |g'(x)| = \infty.$$

The graphs are sketched in Figure 4.39. Note that for f the slope of the tangent line becomes positively infinite whether x approaches 8 from the left or from the right. However, for g the slope becomes negatively infinite if $x \to 8^-$ and positively infinite if $x \to 8^+$. ∎

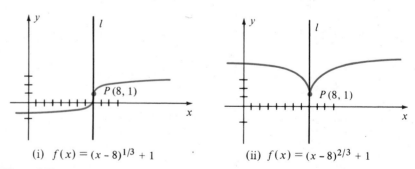

(i) $f(x) = (x - 8)^{1/3} + 1$ (ii) $f(x) = (x - 8)^{2/3} + 1$

Figure 4.39

Definition (4.26) may be modified to include vertical tangent lines at an endpoint of the domain of a function. Thus, if f is continuous on $[a, b]$, but is undefined outside of this interval, then the graph is said to have a vertical tangent line at $P(a, f(a))$ or $Q(b, f(b))$ if

$$\lim_{x \to a^+} |f'(x)| = \infty \quad \text{or} \quad \lim_{x \to b^-} |f'(x)| = \infty,$$

respectively. For example, if $f(x) = \sqrt{1 - x^2}$, then the graph of f is the upper half of the unit circle with center at the origin, and there are vertical tangent lines at the points $(-1, 0)$ and $(1, 0)$. (Check this fact.)

EXERCISES 4.5

Find the limits in Exercises 1–12.

1 $\lim\limits_{x \to \infty} \dfrac{5x^2 - 3x + 1}{2x^2 + 4x - 7}$

2 $\lim\limits_{x \to \infty} \dfrac{3x^3 - x + 1}{6x^3 + 2x^2 - 7}$

3 $\lim\limits_{x \to -\infty} \dfrac{4 - 7x}{2 + 3x}$

4 $\lim\limits_{x \to -\infty} \dfrac{(3x + 4)(x - 1)}{(2x + 7)(x + 2)}$

5 $\lim\limits_{x \to \infty} \sqrt[3]{\dfrac{8 + x^2}{x(x + 1)}}$

6 $\lim\limits_{x \to -\infty} \dfrac{4x - 3}{\sqrt{x^2 + 1}}$

7 $\lim\limits_{x \to \infty} \dfrac{\sqrt{4x + 1}}{10 - 3x}$

8 $\lim\limits_{x \to \infty} \dfrac{2x^2 - x + 3}{x^3 + 1}$

9 $\lim\limits_{x \to \infty} \sqrt{x^2 + 5x} - x$

10 $\lim\limits_{x \to \infty} (x - \sqrt{x^2 - 3x})$

11 $\lim\limits_{x \to -\infty} \dfrac{1 + \sqrt[5]{x}}{1 - \sqrt[5]{x}}$

12 $\lim\limits_{x \to \infty} \dfrac{\sqrt[3]{x}}{x^3 + 1}$

13 Determine the horizontal asymptotes and sketch the graph of $y = 4x^2/(1 + x^2)$.

14 Determine the horizontal asymptotes and sketch the graph of $y = 4/(9 + x^2)$.

15 If f and g are polynomial functions of degree n, find $\lim_{x \to \infty} f(x)/g(x)$ and $\lim_{x \to -\infty} f(x)/g(x)$. What is true if the degree of f is less than the degree of g?

16 If f is any polynomial function of degree greater than 0, prove that $\lim_{x \to \infty} 1/f(x) = 0$.

17 Prove, by a direct application of Definition (4.19), that $\lim_{x \to \infty} [2 + (1/x)] = 2$.

18 Define $\lim_{x \to a^-} f(x) = \infty$ in a manner similar to (4.22).

19 If f and g are functions such that $0 < f(x) < g(x)$ for all x, and if $\lim_{x \to \infty} g(x) = 0$, prove that $\lim_{x \to \infty} f(x) = 0$.

20 Prove that if $f(x)$ has a limit as x becomes infinite, then the limit is unique.

In Exercises 21–30, find $\lim_{x \to a^+} f(x)$ and $\lim_{x \to a^-} f(x)$. Identify the vertical and horizontal asymptotes, find the local maxima and minima, and sketch the graph of f.

21 $f(x) = \dfrac{5}{x - 4}, a = 4$

22 $f(x) = \dfrac{5}{4 - x}, a = 4$

23 $f(x) = \dfrac{8}{(2x + 5)^3}, a = -\frac{5}{2}$

24 $f(x) = \dfrac{-4}{7x + 3}, a = -\frac{3}{7}$

25 $f(x) = \dfrac{3x}{(x + 8)^2}, a = -8$

26 $f(x) = \dfrac{3x^2}{(2x - 9)^2}, a = \frac{9}{2}$

27 $f(x) = \dfrac{2x^2}{x^2 - x - 2}, a = -1, a = 2$

28 $f(x) = \dfrac{4x}{x^2 - 4x + 3}, a = 1, a = 3$

29 $f(x) = \dfrac{1}{x(x - 3)^2}, a = 0, a = 3$

30 $f(x) = \dfrac{x^2}{x + 1}, a = -1$

In Exercises 31–40 determine the vertical and horizontal asymptotes and sketch the graph of f.

31 $f(x) = \dfrac{1}{x^2 - 4}$

32 $f(x) = \dfrac{5x}{4 - x^2}$

33 $f(x) = \dfrac{2x^4}{x^4 + 1}$

34 $f(x) = \dfrac{3x}{x^2 + 1}$

35 $f(x) = \dfrac{1}{x^3 + x^2 - 6x}$

36 $f(x) = \dfrac{x^2 - x}{16 - x^2}$

37 $f(x) = \dfrac{x^2 + 3x + 2}{x^2 + 2x - 3}$

38 $f(x) = \dfrac{x^2 - 5x}{x^2 - 25}$

39 $f(x) = \dfrac{x + 4}{x^2 - 16}$

40 $f(x) = \dfrac{\sqrt[3]{16 - x^2}}{4 - x}$

In Exercises 41–46 find the vertical and oblique asymptotes and sketch the graph of f.

41 $f(x) = \dfrac{x^2 - x - 6}{x + 1}$ **42** $f(x) = \dfrac{2x^2 - x - 3}{x - 2}$

43 $f(x) = \dfrac{8 - x^3}{2x^2}$ **44** $f(x) = \dfrac{x^3 + 1}{x^2 - 9}$

45 $f(x) = \dfrac{x^4 - 4}{x^3 - 1}$ **46** $f(x) = \dfrac{1 - x^4}{2x^3 - 8x}$

47 If f and g are polynomial functions and the degree of $f(x)$ is greater than the degree of $g(x)$, what is $\lim_{x \to \infty} f(x)/g(x)$?

48 Prove Theorem (4.24).

In Exercises 49–52, find the points where the graph of f has a vertical tangent line.

49 $f(x) = x(x + 2)^{3/5}$ **50** $f(x) = \sqrt{x + 2}$

51 $f(x) = \sqrt{16 - 9x^2} + 3$ **52** $f(x) = \sqrt[3]{x} - 5$

53 Prove that the graph of a rational function has no vertical tangent lines.

54 Sketch the graph of $y = |x^3 - x|$ and determine where the graph has (a) a horizontal tangent line; (b) a vertical tangent line; (c) no tangent line.

55 Prove that if $q(x) = f(x)/g(x)$ where $f(x)$ and $g(x)$ are polynomials, and if the degree of $g(x)$ is n, then the graph of q can have at most n vertical asymptotes. Give examples illustrating that the graph may have anywhere from 0 to n vertical asymptotes.

56 Prove that the graph of a function can have at most two different horizontal asymptotes.

57 Basic processes governing bacterial growth in a culture may be investigated by utilizing a laboratory device called a *chemostat* (see figure). The rate of bacterial growth is controlled by supplying all nutrients that are essential for excess growth except one, the *rate limiting nutrient*.

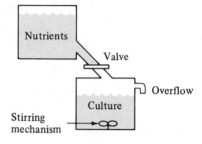

Figure for Exercise 57

Let x denote the concentration of the rate limiting nutrient. If N is the bacterial population and $r(x)$ is the rate of growth corresponding to x, then Monod* suggested that

$$r(x) = \frac{axN}{b + x}$$

where a and b are constants. The *relative growth rate* (rate of growth per unit population) $R(x)$ is defined by

$$R(x) = \frac{r(x)}{N} = \frac{ax}{b + x}.$$

The relationship between the constants a and b may be determined by analyzing the graph of the rational function R.

Show that R has no relative extrema, but that the relative growth rate approaches $a = 2R(b)$ as the concentration of the rate limiting nutrient increases without bound. Sketch the graph of R.

---

* See *Ann. Inst. Pasteur*, Vol. 79 (1950), p. 390.

4.6

APPLICATIONS OF EXTREMA

The theory we have developed for finding extrema of functions can be applied to certain practical problems. These problems may either be described orally, or stated in written words, as is the case in textbooks. To solve them, it is necessary to convert verbal statements into the language of mathematics by introducing formulas, functions, or equations. Since the types of applications are unlimited, it is difficult to state specific rules for finding solutions. However, it is possible to develop a general strategy for attacking such problems. The following guidelines are often helpful.

Guidelines for Solving Applied Problems Involving Extrema

1 Read the problem carefully several times and think about the given facts, together with the unknown quantities that are to be found.

2 If possible, sketch a picture or diagram and label it appropriately, introducing variables for unknown quantities. Words such as "what," "find," "how much," "how far," or "when" should alert you to the unknown quantities.

3 Write down the known facts together with any relationships involving the variables. A relationship may often be described by means of an equation.

4 Determine which variable is to be maximized or minimized and express this variable as a function of *one* of the other variables.

5 Find the critical numbers of the function obtained in Guideline 4 and test each of them for maxima or minima.

6 Check to see whether extrema occur at the endpoints of the domain of the function obtained in Guideline 4.

7 Don't become discouraged if you are unable to solve a given problem. It takes a great deal of effort and practice to become proficient in solving applied problems. Keep trying!

The use of these guidelines is illustrated in the following examples.

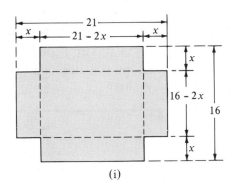

(i)

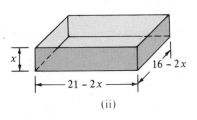

(ii)

Figure 4.40

Example 1 An open box with a rectangular base is to be constructed from a rectangular piece of cardboard 16 in. wide and 21 in. long by cutting out a square from each corner and then bending up the sides. Find the size of the corner square that will produce a box having the largest possible volume.

Solution   Applying Guideline 2, let us begin by drawing a picture of the cardboard as in (i) of Figure 4.40, where we have introduced a variable x to denote the length of the side of the square to be cut out from each corner. Using Guideline 3, we specify the known facts (the size of the rectangle) by labeling the figure appropriately.

Next (see Guideline 4), we determine that the quantity to be maximized is the volume V of the box to be constructed by folding along the dashed lines (see (ii) of Figure 4.40). Continuing with Guideline 4, let us express V as a function of the variable x. Referring to (ii) of the figure,

$$V = x(16 - 2x)(21 - 2x) = 2(168x - 37x^2 + 2x^3).$$

This equation expresses V as a function of x, and we proceed to find the critical numbers and test them for maxima or minima (see Guideline 5). Differentiating with respect to x,

$$D_x V = 2(168 - 74x + 6x^2)$$
$$= 4(3x^2 - 37x + 84)$$
$$= 4(3x - 28)(x - 3).$$

Thus the possible critical numbers are $\frac{28}{3}$ and 3. Since $\frac{28}{3}$ is outside the domain of x (Why?) the only critical number is 3.

Let us employ the Second Derivative Test for extrema. The second derivative of V is

$$D_x^2 V = 2(-74 + 12x) = 4(6x - 37).$$

Substituting the critical number 3 for x we obtain

$$D_x^2 V = 4(18 - 37) = -76 < 0$$

and hence V has a relative maximum at $x = 3$.

Finally, we check for endpoint extrema (see Guideline 6). Since $0 \leq x \leq 8$ (Why?) and since $V = 0$ if either $x = 0$ or $x = 8$, the maximum value of V does not occur at an endpoint of the domain of x. Consequently, a three-inch square should be cut from each corner of the cardboard in order to maximize the volume of the resulting box. ∎

In the remaining examples we shall not always point out the guidelines that are employed. The reader should be able to determine specific guidelines by studying the solutions.

Example 2   A circular cylindrical container, open at the top and having a capacity of 24π in.^{3}, is to be manufactured. If the cost of the material used for the bottom of the container is three times that used for the curved part and if there is no waste of material, find the dimensions that will minimize the cost.

Solution   Let us begin by drawing a picture as in Figure 4.41, letting r denote the length (in inches) of the radius of the base of the container and h the altitude (in inches). Since the volume is 24π in.^{3}, we have

$$\pi r^2 h = 24\pi.$$

This gives us the relationship

$$h = \frac{24}{r^2}$$

between r and h.

Our goal is to minimize the cost C of the material used to manufacture the container.

If a denotes the cost per square inch of the material to be used for the curved part, then the material to be used for the bottom costs $3a$ per square inch. The cost of the material for the curved part is, therefore, $a(2\pi rh)$ and the cost for the base is $3a(\pi r^2)$. The total cost C for material is

$$C = 3a(\pi r^2) + a(2\pi rh) = a\pi(3r^2 + 2rh)$$

or, since $h = 24/r^2$,

$$C = a\pi\left(3r^2 + \frac{48}{r}\right).$$

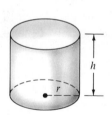

Figure 4.41

Since a is fixed, this expresses C as a function of one variable r as recommended in Guideline 4. To find the value of r that will lead to the smallest value for C, we find the critical numbers. Differentiating the last equation with respect to r gives us

$$D_r C = a\pi\left(6r - \frac{48}{r^2}\right) = 6a\pi\left(\frac{r^3 - 8}{r^2}\right).$$

Since $D_r C = 0$ if $r = 2$, we see that 2 is the only critical number. (The number 0 is not a critical number since C is undefined if $r = 0$.) Since $D_r C < 0$ if $r < 2$ and $D_r C > 0$ if $r > 2$, it follows from the First Derivative Test that C has its minimum value if the radius of the cylinder is 2 inches. The corresponding value for the altitude (obtained from $h = 24/r^2$) is $\frac{24}{4}$, or 6 in.

Since the domain of the variable r is the infinite interval $(0, \infty)$, there can be no endpoint extrema. ■

Example 3   Find the maximum volume of a right circular cylinder that can be inscribed in a cone of altitude 12 cm and base radius 4 cm, if the axes of the cylinder and cone coincide.

Solution   The problem is illustrated in Figure 4.42 where (ii) represents a cross-section through the axis of the cone and cylinder. The volume V of the cylinder is $V = \pi r^2 h$. To express V in terms of one variable (see Guideline 4) we must find a relationship between r and h. Referring to (ii) of Figure 4.42 and using similar triangles we see that

$$\frac{h}{4 - r} = \frac{12}{4} = 3 \quad \text{or} \quad h = 3(4 - r).$$

Consequently,

$$V = \pi r^2 h = \pi r^2[3(4 - r)] = 3\pi(4r^2 - r^3).$$

If either $r = 0$ or $r = 4$ we see that $V = 0$, and hence the maximum volume is not an endpoint extremum. It is sufficient, therefore, to search for relative maximum values. Since

$$D_r V = 3\pi(8r - 3r^2) = 3\pi r(8 - 3r)$$

the critical numbers for V are $r = 0$ and $r = \frac{8}{3}$. Let us apply the Second Derivative Test for $r = \frac{8}{3}$. Differentiating $D_r V$ gives us

$$D_r^2 V = 3\pi(8 - 6r).$$

Substituting $\frac{8}{3}$ for r, we obtain

$$3\pi[8 - 6(\tfrac{8}{3})] = 3\pi(8 - 16) = -24\pi < 0$$

which means that a maximum occurs. The corresponding value for $h = 3(4 - r)$ is

$$h = 3(4 - \tfrac{8}{3}) = 3(\tfrac{4}{3}) = 4.$$

Hence the maximum volume of the inscribed cylinder is

$$V = \pi\left(\frac{8}{3}\right)^2(4) = \pi\left(\frac{64}{9}\right)(4) = \frac{256\pi}{9}\,\text{cm}^3.$$ ■

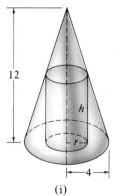

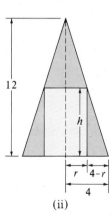

Figure 4.42

Example 4. A North–South highway intersects an East–West highway at a point P. An automobile crosses P at 10:00 A.M., traveling east at a constant speed of 20 mph. At that same instant another automobile is 2 mi north of P, traveling south at 50 mph. Find the time at which they are closest to each other and approximate the minimum distance between the automobiles.

Solution Typical positions of the automobiles are illustrated in Figure 4.43. If t denotes the number of hours after 10:00 A.M., then the slower automobile is $20t$ mi east of P, as indicated in the figure. The faster automobile is $50t$ mi south of its position at 10:00 A.M. and hence its distance from P is $2 - 50t$. By the Pythagorean Theorem, the distance d between the automobiles is

$$d = \sqrt{(2 - 50t)^2 + (20t)^2}$$
$$= \sqrt{4 - 200t + 2500t^2 + 400t^2}$$
$$= \sqrt{4 - 200t + 2900t^2}.$$

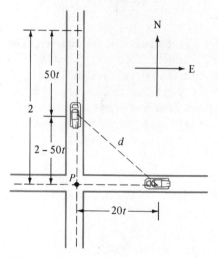

Figure 4.43

Evidently d has its smallest value when the expression under the radical is minimal. Thus we may simplify our work by letting

$$f(t) = 4 - 200t + 2900t^2$$

and finding the value of t for which f has a minimum. Since

$$f'(t) = -200 + 5800t$$

the only critical number for f is

$$t = \frac{200}{5800} = \frac{1}{29}.$$

Moreover, since $f''(t) = 5800$, the second derivative is always positive and, therefore, f has a relative minimum at $t = \frac{1}{29}$, and $f(\frac{1}{29}) \approx 0.55$. Since the domain of t is $[0, \infty)$ and since $f(0) = 4$, there is no endpoint extremum. Consequently the automobiles will be closest at $\frac{1}{29}$ hours (or approximately 2.07 minutes) after 10:00 A.M. The minimal distance is

$$\sqrt{f(\tfrac{1}{29})} \approx \sqrt{0.55} \approx 0.74 \text{ mi.} \qquad \blacksquare$$

Example 5 A man in a rowboat 2 mi from the nearest point on a straight shoreline wishes to reach his house 6 mi further down the shore. If he can row at a rate of 3 mph and walk at a rate of 5 mph; how should he proceed in order to arrive at his house in the shortest amount of time?

Solution A diagram of the problem is shown in Figure 4.44 where A denotes the position of the boat, B the nearest point on shore, and C the man's house. As shown in the figure, D is the point at which the boat will reach shore and x denotes the distance between B and D. Thus x is restricted to the interval $[0, 6]$.

By the Pythagorean Theorem, the distance between A and D is $\sqrt{x^2 + 4}$. Using the formula *time* = *distance*/*rate*, the time it takes the man to row from

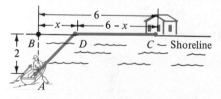

Figure 4.44

A to D is $\sqrt{x^2 + 4}/3$, whereas the time it takes the man to walk from D to C is $(6 - x)/5$. Hence the total time T for the trip is given by

$$T = \frac{\sqrt{x^2 + 4}}{3} + \frac{6 - x}{5}$$

or equivalently,

$$T = \tfrac{1}{3}(x^2 + 4)^{1/2} + \tfrac{6}{5} - \tfrac{1}{5}x.$$

We wish to find the minimum value for T. Note that the case $x = 0$ corresponds to the extreme situation in which the man rows directly to B and then walks the entire distance from B to his house. If $x = 6$, then the man rows directly from A to his house. These numbers may be considered as endpoints of the domain of T. If $x = 0$, then from the formula for T,

$$T = \frac{\sqrt{4}}{3} + \frac{6}{5} - 0 = \frac{28}{15} \approx 1.87$$

which is approximately 1 hour and 52 minutes. If $x = 6$, then

$$T = \frac{\sqrt{40}}{3} + \frac{6}{5} - \frac{6}{5} = \frac{2\sqrt{10}}{3} \approx 2.11$$

or approximately 2 hours and 7 minutes.

Differentiating the general formula for T we see that

$$D_x T = \frac{1}{3} \cdot \frac{1}{2}(x^2 + 4)^{-1/2}(2x) - \frac{1}{5}$$

$$= \frac{x}{3(x^2 + 4)^{1/2}} - \frac{1}{5}.$$

To find the critical numbers we let $D_x T = 0$. This leads to the following chain of equations:

$$5x = 3(x^2 + 4)^{1/2}$$
$$25x^2 = 9(x^2 + 4)$$
$$16x^2 = 36$$
$$x^2 = \tfrac{36}{16}$$
$$x = \tfrac{6}{4} = \tfrac{3}{2}.$$

Thus $\tfrac{3}{2}$ is the only critical number. We next employ the Second Derivative Test. Applying the quotient rule to the formula for $D_x T$,

$$D_x^2 T = \frac{3(x^2 + 4)^{1/2} - x \cdot 3(\tfrac{1}{2})(x^2 + 4)^{-1/2}(2x)}{9(x^2 + 4)}$$

$$= \frac{3(x^2 + 4) - 3x^2}{9(x^2 + 4)^{3/2}}$$

$$= \frac{4}{3(x^2 + 4)^{3/2}}$$

which is positive if $x = \frac{3}{2}$. Consequently T has a local minimum at $x = \frac{3}{2}$. The time T that corresponds to $x = \frac{3}{2}$ is

$$T = \tfrac{1}{3}(\tfrac{9}{4} + 4)^{1/2} + \tfrac{6}{5} - \tfrac{3}{10} = \tfrac{26}{15}$$

or, equivalently, 1 hour and 44 minutes.

We have already examined the values of T at the endpoints of the domain, obtaining approximately 1 hour 52 minutes and over 2 hours, respectively. Hence the minimum value of T occurs at $x = \frac{3}{2}$ and, therefore, the boat should land between B and C, $1\frac{1}{2}$ mi from B. For a similar problem, but one in which the endpoints of the domain lead to minimum time, see Exercise 28. ∎

Example 6   A wire 60 in. long is to be cut into two pieces. One of the pieces will be bent into the shape of a circle and the other into the shape of an equilateral triangle. Where should the wire be cut so that the sum of the areas of the circle and triangle is (a) a maximum; (b) minimum?

Solution   If the wire is cut and x denotes the length of one of the pieces of wire, then the length of the other piece is $60 - x$. Let the piece of length x be bent to form a circle of radius r so that $2\pi r = x$. If the remaining piece is bent into an equilateral triangle of side s, then $3s = 60 - x$ (see Figure 4.45). The area of the circle is

$$\pi r^2 = \pi\left(\frac{x}{2\pi}\right)^2 = \left(\frac{1}{4\pi}\right)x^2$$

and the area of the triangle is

$$\frac{\sqrt{3}}{4}s^2 = \frac{\sqrt{3}}{4}\left(\frac{60 - x}{3}\right)^2.$$

Consequently, the sum A of the areas can be expressed in terms of one variable x as follows:

$$A = \left(\frac{1}{4\pi}\right)x^2 + \left(\frac{\sqrt{3}}{36}\right)(60 - x)^2.$$

We now find the critical numbers. Differentiating,

$$D_x A = \left(\frac{1}{2\pi}\right)x - \left(\frac{\sqrt{3}}{18}\right)(60 - x)$$

$$= \left(\frac{1}{2\pi} + \frac{\sqrt{3}}{18}\right)x - \frac{10\sqrt{3}}{3}$$

and, therefore, $D_x A = 0$ if and only if

$$x = \frac{10\sqrt{3}/3}{(1/2\pi) + (\sqrt{3}/18)} \approx 22.61.$$

The second derivative

$$D_x^2 A = \frac{1}{2\pi} + \frac{\sqrt{3}}{18}$$

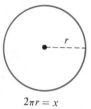
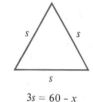

$2\pi r = x$ $3s = 60 - x$

Figure 4.45

is always positive and hence the indicated critical number will yield a minimum value for A. This minimum value is approximated by

$$A \approx \frac{1}{4\pi}(22.61)^2 + \frac{\sqrt{3}}{36}(60 - 22.61)^2 \approx 107.94.$$

Since there are no other critical numbers, the maximum value of A must be an endpoint extremum. If $x = 0$, then all the wire is used to form a triangle and

$$A = \frac{\sqrt{3}}{36}(60)^2 \approx 173.21.$$

If $x = 60$, then all the wire is used for the circle, and

$$A = \frac{1}{4\pi}(60)^2 \approx 268.48.$$

Thus the maximum value of A occurs if the wire is not cut and the entire length of wire is bent into the shape of a circle.   ■

EXERCISES 4.6

1 Find two real numbers whose difference is 40 and whose product is a minimum.

2 Find two positive real numbers whose sum is 40 and whose product is a maximum.

3 If a box with a square base and open top is to have a volume of $4\,\text{ft}^3$, find the dimensions that require the least material (neglect the thickness of the material and waste in construction).

4 Work Exercise 3 if the box has a closed top.

5 A fence 8 ft tall stands on level ground and runs parallel to a tall building (see figure). If the fence is 1 ft from the building, find the length of the shortest ladder that will extend from the ground over the fence to the wall of the building.

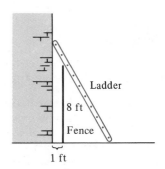

Figure for Exercise 5

6 A page of a book is to have an area of 90 in.^{2}, with 1-inch margins at the bottom and sides and a $\frac{1}{2}$-inch margin at the top. Find the dimensions of the page that will allow the largest printed area.

7 Find the dimensions of the rectangle of maximum area that can be inscribed in a semicircle of radius a, if two vertices lie on the diameter (see figure).

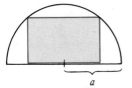

Figure for Exercise 7

8 Find the dimensions of the rectangle of maximum area that can be inscribed in an equilateral triangle of side a, if two vertices of the rectangle lie on one of the sides of the triangle.

9 Of all possible right circular cones that can be inscribed in a sphere of radius a, find the volume of the one that has maximum volume.

10 Find the dimensions of the right circular cylinder of maximum volume that can be inscribed in a sphere of radius a.

11 A metal cylindrical container with an open top is to hold one cubic foot. If there is no waste in construction, find the dimensions that require the least amount of material.

12 If the circular base of the container in Exercise 11 is cut from a square sheet and the remaining metal is discarded, find the dimensions that require the least amount of material.

13 A builder intends to construct a storage shed having a volume of 900 cubic feet, a flat roof, and a rectangular base whose width is three-fourths the length. The cost per square foot of the materials is $4 for the floor, $6 for the sides, and $3 for the roof. What dimensions will minimize the cost?

14 A water cup in the shape of a right circular cone is to be constructed by removing a circular sector from a circular sheet of paper of radius a and then joining the two straight edges of the remaining paper (see figure). Find the volume of the largest cup that can be constructed.

Figure for Exercise 14

15 A farmer has k feet of fencing and wishes to enclose a rectangular field by using a barn as part of one side of the field, as illustrated in the figure. Prove that the area of the field is greatest when the rectangle is a square.

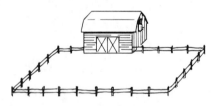

Figure for Exercise 15

16 Suppose the farmer in Exercise 15 wants the area of the rectangular field to be A square feet. Prove that the least amount of fencing is required when the rectangle is a square.

17 A hotel that charges $80 per day for a room gives special rates to organizations that reserve between 30 and 60 rooms. If more than 30 rooms are reserved, the charge per room is decreased by $1 times the number of rooms over 30. Under these conditions, how many rooms must be rented if the hotel is to receive the maximum income per day?

18 Refer to Exercise 17. Suppose that for each room rented it costs the hotel $6 per day for cleaning and maintenance. In this case how many rooms must be rented to obtain the greatest net income?

19 A long rectangular sheet of metal, 12 in. wide, is to be made into a rain gutter by turning up two sides at right angles to the sheet. How many inches should be turned up to give the gutter its greatest capacity?

20 Work Exercise 19 if the sides of the gutter make an angle of 120° with the base.

21 Prove that the rectangle of largest area having a given perimeter p is a square.

22 A right circular cylinder is generated by rotating a rectangle of perimeter p about one of its sides. What dimensions of the rectangle will generate the cylinder of maximum volume?

23 The strength of a rectangular beam varies jointly as the product of its width and the square of the depth of a cross section. Find the dimensions of the strongest beam that can be cut from a cylindrical log of radius a (see figure).

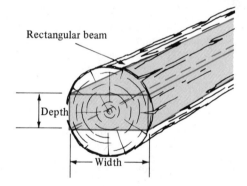

Figure for Exercise 23

24 A window has the shape of a rectangle surmounted by a semicircle. If the perimeter of the window is 15 ft, find the dimensions that will allow the maximum amount of light to enter.

25 Find the point on the graph of $y = x^2 + 1$ that is closest to the point (3, 1).

26 Find the x-coordinate of the point on the graph of $y = x^3$ that is closest to the point (4, 0).

27 A company sells running shoes to dealers at a rate of $20 per pair if less than 50 pairs are ordered. If 50 or more pairs are ordered (up to 600), the price per pair is reduced at a rate of 2 cents times the number ordered. What size order will produce the maximum amount of money for the company?

28 Refer to Example 5 of this section. If the man is in a motor-boat that can travel at an average rate of 15 mph, how should he proceed in order to arrive at his house in the least time?

29 The illumination from a light source is directly proportional to the strength of the source and inversely proportional to the square of the distance from the source. If two light sources of strengths S_1 and S_2 are d units apart, at what point on the line segment joining the two sources is the illumination minimal?

30 At 1:00 P.M. ship A is 30 miles due south of ship B and is sailing north at a rate of 15 mph. If ship B is sailing west at a rate of 10 mph, find the time at which the distance between the ships is minimal (see figure).

Figure for Exercise 30

31 A veterinarian has 100 ft of fencing and wishes to construct six dog kennels by first building a fence around a rectangular region, and then subdividing that region into six smaller rectangles by placing five fences parallel to one of the sides. What dimensions of the region will maximize the total area?

32 A paper cup having the shape of a right circular cone is to be constructed. If the volume desired is 36π in.^{3}, find the dimensions that require the least amount of paper (neglect any waste that may occur in the construction).

33 A steel storage tank for propane gas is to be constructed in the shape of a right circular cylinder with a hemisphere at each end (see figure). The cost per square foot of constructing the end pieces is twice that of the cylindrical piece. If the desired capacity is 10π ft^3, what dimensions will minimize the cost of construction?

Propane gas tank

Figure for Exercise 33

34 A pipeline for transporting oil will connect two points A and B that are 3 miles apart and on opposite banks of a straight river one mile wide (see figure). Part of the pipeline will run under water from A to a point C on the opposite bank, and then above ground from C to B. If the cost per mile of running the pipeline under water is four times the cost per mile of running it above ground, find the location of C that will minimize the cost (ignore the slope of the river bed).

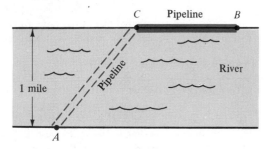

Figure for Exercise 34

35 A wire 36 cm long is to be cut into two pieces. One of the pieces will be bent into the shape of an equilateral triangle and the other into the shape of a rectangle whose length is twice its width. Where should the wire be cut if the combined area of the triangle and rectangle is (a) a minimum; (b) a maximum?

36 An isosceles triangle has base b and equal sides of length a. Find the dimensions of the rectangle of maximum area that can be inscribed in the triangle if one side of the rectangle lies on the base of the triangle.

37 Two vertical poles of lengths 6 ft and 8 ft stand on level ground, with their bases 10 ft apart. Approximate the minimal length of cable that can reach from the top of one pole to some point on the ground between the poles, and then to the top of the other pole.

38 Find the dimensions of the rectangle of maximum area having two vertices on the x-axis and two vertices above the x-axis, on the graph of $y = 4 - x^2$.

39 A window has the shape of a rectangle surmounted by an equilateral triangle. If the perimeter of the window is 12 ft, find the dimensions of the rectangle that will produce the largest area for the window.

40 If a trapezoid has three nonparallel sides of length 8, find the length of the fourth side that will maximize the area.

41 The owner of an apple orchard estimates that if 24 trees are planted per acre, then each mature tree will yield 600 apples per year. For each additional tree planted per acre the number of apples produced by each tree decreases by 12 per year. How many trees should be planted per acre to obtain the most apples?

42 A real estate company owns 180 apartments that are fully occupied when the rent is $300 per month. The company estimates that for each $10 increase in rent, five apartments will become unoccupied. What rent should be charged in order to obtain the largest gross income?

43 A package can be sent by parcel post only if the sum of its length and girth (the perimeter of the base) is not more than 96 in. Find the dimensions of the box of maximum volume that can be sent if the base of the box is a square.

44 A North–South highway A and an East–West highway B intersect at a point P. At 10:00 A.M. an automobile crosses P traveling north on highway A at a speed of 50 mph. At that same instant, an airplane flying east at a speed of 200 mph and an altitude of 26,400 ft is directly above the point on highway B that is 100 miles west of P. If the automobile and airplane maintain the same speed and direction, at what time will they be closest to one another?

45 Two factories A and B that are 4 miles apart emit particles in smoke that pollute the area that lies between the factories. Suppose that the number of particles emitted from each factory is directly proportional to the amount of smoke and inversely proportional to the cube of the distance from the factory. If factory A emits twice as much smoke as factory B, at what point between A and B is the pollution minimal?

46 An oil field contains 8 wells that produce a total of 1600 barrels of oil per day. For each additional well that is drilled, the average production per well decreases by 10 barrels per day. How many additional wells should be drilled to obtain the maximum amount of oil per day?

47 The accompanying figure illustrates a battery having voltage E and fixed internal resistance r, connected to a circuit that has resistance R. By Ohm's Law, the current I in the circuit is $I = E/(R + r)$. If the power output P is given by $P = I^2R$, show that the maximum power occurs if $R = r$.

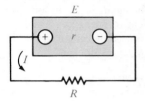

Figure for Exercise 47

48 The power output P of an automobile battery is given by $P = EI - I^2r$, where E is the voltage, I is the current, and r is the internal resistance of the battery. What current corresponds to the maximum power?

49 Use the First Derivative Test to prove that the shortest distance from a point (x_1, y_1) to the line $ax + by + c = 0$ is given by

$$d = \frac{|ax_1 + by_1 + c|}{\sqrt{a^2 + b^2}}.$$

50 Prove that the shortest distance from a point (x_1, y_1) to the graph of a differentiable function f is measured along a normal line to the graph, that is, a line perpendicular to the tangent line.

4.7

THE DERIVATIVE AS A RATE OF CHANGE

Most quantities encountered in everyday life change with time. This is especially evident in scientific investigations. For example, a chemist may be interested in the rate at which a certain substance dissolves in water. An electrical engineer may wish to know the rate of change of current in part of an electrical circuit. A biologist may be concerned with the rate at which the bacteria in a culture increase or decrease. Numerous other examples could be cited, including many from fields other than the natural sciences. Let us consider the following general situation, which can be applied to all of the preceding examples.

Suppose a variable w is a function of time such that at time t, w is given by $w = g(t)$, where g is a differentiable function. The difference between the initial and final values of w in the time interval $[t, t + h]$ is $g(t + h) - g(t)$. As in our development of the velocity concept, we formulate the following definition.

Definition (4.27)

> Let $w = g(t)$, where g is a differentiable function and t represents time.
>
> (i)   The **average rate of change** of $w = g(t)$ in the interval $[t, t + h]$ is
> $$\frac{g(t + h) - g(t)}{h}.$$
>
> (ii)   The **rate of change** of $w = g(t)$ with respect to t is
> $$\frac{dw}{dt} = g'(t) = \lim_{h \to 0} \frac{g(t + h) - g(t)}{h}.$$

The units to be used in Definition (4.27) depend on the nature of the quantity represented by w. Sometimes dw/dt is referred to as the **instantaneous rate of change** of w with respect to t.

Example 1   A scientist finds that if a certain substance is heated, the Celsius temperature after t minutes, where $0 \le t \le 5$, is given by the formula $g(t) = 30t + 6\sqrt{t} + 8$.

(a)   Find the average rate of change of $g(t)$ during the time interval $[4, 4.41]$.

(b)   Find the rate of change of $g(t)$ at $t = 4$.

Solution

(a)   Letting $t = 4$ and $h = 0.41$ in (i) of Definition (4.27), the average rate of change of g in $[4, 4.41]$ is

$$\frac{g(4.41) - g(4)}{0.41} = \frac{[30(4.41) + 6\sqrt{4.41} + 8] - [120 + 6\sqrt{4} + 8]}{0.41}$$

$$= \frac{12.9}{0.41} \approx 31.46°C/min.$$

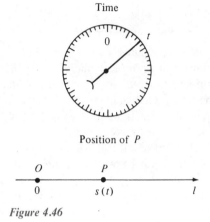

Time

Position of P

O P

0 $s(t)$ l

Figure 4.46

(b)   From (ii) of Definition (4.27) the rate of change of g at time t is given by $g'(t) = 30 + 3/\sqrt{t}$. In particular,

$$g'(4) = 30 + \frac{3}{\sqrt{4}} = 31.5°C/min. \qquad \blacksquare$$

The velocity of a point P that moves on a coordinate line l was defined in Chapter 3. In the future we shall often use the letter s to denote functions whose values are distances. If the coordinate of P on l is $s(t)$, as illustrated in Figure 4.46, then s is called the **position function** of P. By Definition (3.3), the velocity $v(t)$ of P at time t is $s'(t)$. Note that in terms of Definition (4.27), the velocity is the rate of change of $s(t)$ with respect to time. The next definition summarizes this discussion and introduces some new terminology.

Definition (4.28)

> Let the position of a point P on a coordinate line l at time t be given by $s(t)$, where s is a differentiable function.
>
> (i)   The **velocity** $v(t)$ of P is $v(t) = s'(t)$.
>
> (ii)   The **speed** of P at time t is $|v(t)|$.
>
> (iii)   The **acceleration** $a(t)$ of $P(t)$ is $a(t) = v'(t) = s''(t)$.

We shall call v the **velocity function** of P and a the **acceleration function** of P. Using another notation,

$$v = \frac{ds}{dt} \quad \text{and} \quad a = \frac{dv}{dt} = \frac{d^2s}{dt^2}.$$

If t is measured in seconds and $s(t)$ in centimeters, then the unit for $v(t)$ is cm/sec (read "centimeters per second"). Similarly, if t is measured in hours and $s(t)$ in miles, then $v(t)$ is in mi/hr (miles per hour), etc. By Definition (4.27), the acceleration $a(t)$ is the rate of change of velocity of P at time t. The units for $a(t)$ are cm/sec^2 (centimeters per second per second), mi/hr^2 (miles per hour per hour), etc.

The results of Section 4.3 may be used to determine the direction of motion of P. If the velocity $v(t)$ is positive in a time interval I, then $s'(t) > 0$ and hence, by Theorem (4.13), $s(t)$ is increasing; that is, P is moving in the positive direction on l. Similarly, if $v(t)$ is negative, then P is moving in the negative direction. The speed $|v(t)|$ indicates the rate at which P is moving without specifying the direction of motion.

Since $v(t) = s'(t)$, the times at which the velocity is zero are critical numbers for the position function s and hence lead to possible local maxima or minima for s. If a local maximum or minimum for s occurs at t_1, then t_1 is usually a time at which P reverses direction.

Applying the theory of increasing and decreasing functions developed in Section 4.3, it follows that if $a(t) > 0$, then the velocity is increasing at time t. If $a(t) < 0$, the velocity is decreasing. Note that if *both* $a(t) < 0$ and $v(t) < 0$, then the velocity is decreasing in the sense of becoming more negative. In this event, even though the velocity is decreasing, the speed $|v(t)|$ is increasing. An illustration of this occurs in Example 2.

Since the acceleration function a is the second derivative of the position function s, it can sometimes be used in conjunction with the Second Derivative Test to find the local extrema of s. For example, if $v(t_1) = 0$ and $a(t_1) < 0$, then $s(t_1)$ is a local maximum for s and hence P changes direction from right to left at time t_1. Similarly, if $v(t_1) = 0$ and $a(t_1) > 0$, then $s(t_1)$ is a local minimum for s and P changes direction from left to right at time t_1.

Example 2   The position function s of a point P on a coordinate line is given by

$$s(t) = t^3 - 12t^2 + 36t - 20,$$

where t is measured in seconds and $s(t)$ in centimeters. Describe the motion of P during the time interval $[-1, 9]$.

Solution   Differentiating, we obtain

$$v(t) = s'(t) = 3t^2 - 24t + 36 = 3(t - 2)(t - 6),$$

$$a(t) = v'(t) = 6t - 24 = 6(t - 4).$$

Consequently, the velocity is 0 at $t = 2$ and $t = 6$. These critical numbers for s lead us to examine the time intervals $(-1, 2)$, $(2, 6)$, and $(6, 9)$. The following table describes the direction of motion of P. The signs of $v(t)$ may be determined by using test values.

Time Interval	Sign of $v(t)$	Direction of Motion
$(-1, 2)$	$+$	right
$(2, 6)$	$-$	left
$(6, 9)$	$+$	right

The table shows that the function s has a local maximum at time $t = 2$ and a local minimum at $t = 6$. This can also be verified by noting that $a(2) = -12 < 0$ and $a(6) = 12 > 0$. The next table displays the values of the position, velocity, and acceleration functions at important times, namely the endpoints of the time interval $[-1, 9]$, and the times at which the velocity or acceleration is zero.

t	-1	2	4	6	9
$s(t)$	-69	12	-4	-20	61
$v(t)$	63	0	-12	0	63
$a(t)$	-30	-12	0	12	30

It is convenient to represent the motion of P schematically as in Figure 4.47. The curve above the coordinate line is not the path of the point, but is intended to show the manner in which P moves on l. As indicated by the

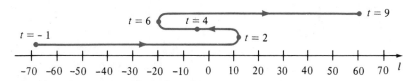

Figure 4.47

table and Figure 4.47, at $t = -1$ the point is 69 cm to the left of the origin and is moving to the right with a velocity of 63 cm/sec. The negative acceleration -30 cm/sec^2 indicates that the velocity is decreasing at a rate of 30 cm per second each second. The point continues to move to the right and to slow down until it has zero velocity at $t = 2$, 12 cm to the right of the origin. The point P then reverses direction and moves until, at $t = 6$, it is 20 cm to the left of the origin. It then again reverses direction and moves to the right for the remainder of the time interval, with increasing velocity. The direction of motion is indicated by the arrows on the curve in Figure 4.47.

The time $t = 4$, at which the acceleration is zero, is a critical number for the velocity function. (Why?) The significance of this time is clearer if we consider the speed $|v(t)| = |3(t - 2)(t - 6)|$. In the time interval $[2, 6]$, the speed increases from 0 at $t = 2$ to a local maximum $|-12| = 12$ at $t = 4$ and then decreases to 0 at $t = 6$. ∎

Example 3   A projectile is fired straight upward with a velocity of 400 ft/sec. Its distance above the ground t seconds after being fired is given by $s(t) = -16t^2 + 400t$. Find the time and the velocity at which the projectile hits the ground. What is the maximum altitude achieved by the projectile? What is the acceleration at any time t?

Solution   The path of the projectile is on a vertical coordinate line with origin at ground level and positive direction upward. The projectile will be on the ground when $-16t^2 + 400t = 0$, that is, when $-16t(t - 25) = 0$. Hence the projectile hits the ground after 25 sec. The velocity at time t is

$$v(t) = s'(t) = -32t + 400.$$

In particular, at $t = 25$,

$$v(25) = -32(25) + 400 = -400 \text{ ft/sec.}$$

The maximum altitude occurs when $s'(t) = 0$, that is, when $-32t + 400 = 0$. Solving for t gives us $t = \frac{400}{32} = \frac{25}{2}$ and hence its maximum altitude is

$$s(\tfrac{25}{2}) = -16(\tfrac{25}{2})^2 + 400(\tfrac{25}{2}) = 2500 \text{ feet.}$$

Finally, the acceleration at any time t is $a(t) = v'(t) = -32 \text{ ft/sec}^2$. ∎

We may investigate rates of change with respect to variables other than time, as indicated by the next definition.

Definition (4.29)

Let $y = f(x)$, where x is any variable.

(i)   The **average rate of change of y with respect to x in the interval** $[x, x + h]$ is the quotient

$$\frac{f(x + h) - f(x)}{h}.$$

(ii)   The **rate of change of y with respect to x** is the limit of the average rate of change as $h \to 0$, that is, dy/dx.

By (ii) of (4.29), if the variable x changes, then y changes at a rate of dy/dx units per unit change in x. To illustrate, suppose a quantity of gas is enclosed in a spherical balloon. If the gas is heated or cooled and if the pressure remains constant, then the balloon expands or contracts and its volume V is a function of the temperature T. The derivative dV/dT gives us the rate of change of volume with respect to temperature.

Example 4 The current I (in amperes) in a certain electrical circuit is given by $I = 100/R$, where R denotes resistance (in ohms). Find the rate of change of I with respect to R when the resistance is 20 ohms.

Solution Since $dI/dR = -100/R^2$, we see that if $R = 20$ then $dI/dR = -\frac{100}{400} = -\frac{1}{4}$. Thus, if R is increasing, then when $R = 20$, the current I is decreasing at a rate of $\frac{1}{4}$ amperes per ohm. ∎

EXERCISES 4.7

1 As a spherical balloon is being inflated, its radius (in centimeters) after t minutes is given by $r(t) = 3\sqrt[3]{t + 8}$, where $0 \leq t \leq 10$. What is the rate of change with respect to t of each of the following at $t = 8$?

(a) $r(t)$

(b) The volume of the balloon

(c) The surface area

2 The volume of a spherical hot air balloon (in ft^3) t hours after 1:00 P.M. is given by $V(t) = \frac{4}{3}\pi(9 - 2t)^3$, where $0 \leq t \leq 4$. What is the rate of change with respect to t of each of the following at 4:00 P.M.?

(a) $V(t)$

(b) The radius of the balloon

(c) The surface area

3 Suppose that t seconds after starting to run, an individual's pulse rate (in beats/min) is given by $P(t) = 56 + 2t^2 - t$, where $0 \leq t \leq 7$. Find the rate of change of $P(t)$ with respect to t at (a) $t = 2$, (b) $t = 4$, and (c) $t = 6$.

4 The temperature T (degrees Celsius) of a solution at time t (minutes) is given by $T(t) = 10 + 4t + 3/(t + 1)$, where $1 \leq t \leq 10$. Find the rate of change of $T(t)$ with respect to t at (a) $t = 2$, (b) $t = 5$, and (c) $t = 9$.

5 A stone is dropped into a pond, causing water waves that form concentric circles. If, after t seconds, the radius of one of the waves is $40t$ cm, find the rate of change with respect to t of the area of the circle caused by the wave at (a) $t = 1$, (b) $t = 2$, and (c) $t = 3$.

6 Boyle's Law for gases states that $pv = c$, where p denotes the pressure, v the volume, and c is a constant. Suppose that at time t (minutes) the pressure is $20 + 2t$ gm/cm^2,

where $0 \leq t \leq 10$. If the volume is 60 cm^3 at $t = 0$, find the rate at which the volume is changing with respect to t at $t = 5$.

Position functions of points moving rectilinearly are defined in Exercises 7–16. Find the velocity and acceleration at time t and describe the motion of the point during the indicated time interval. Illustrate the motion by means of a diagram of the type shown in Figure 4.47.

7 $s(t) = 3t^2 - 12t + 1, [0, 5]$

8 $s(t) = t^2 + 3t - 6, [-2, 2]$

9 $s(t) = t^3 - 9t + 1, [-3, 3]$

10 $s(t) = 24 + 6t - t^3, [-2, 3]$

11 $s(t) = t + (4/t), [1, 4]$

12 $s(t) = 2\sqrt{t} + (1/\sqrt{t}), [1, 4]$

13 $s(t) = 2t^4 - 6t^2, [-2, 2]$

14 $s(t) = 2t^3 - 6t^5, [-1, 1]$

15 $s(t) = t^3 + 1, [0, 4]$

16 $s(t) = \sqrt[3]{t}, [-8, 0]$

17 A projectile is fired directly upward with a velocity of 144 ft/sec. Its height $s(t)$ in feet above the ground after t sec is given by $s(t) = 144t - 16t^2$. What are the velocity and acceleration after t sec? After 3 sec? What is the maximum height? When does it strike the ground?

18 A ball rolls down an inclined plane such that the distance $s(t)$ (in inches) it rolls in t seconds is given by $s(t) = 5t^2 + 2$. What is the velocity after 1 sec? 2 sec? When will the velocity be 28 in./sec?

19 The position function s of a point moving rectilinearly is given by $s(t) = 2t^3 - 15t^2 + 48t - 10$ where t is in seconds and $s(t)$ is in meters. Find the acceleration when the velocity is 12 m/sec. Find the velocity when the acceleration is 10 m/sec^2.

20 Work Exercise 19 if $s(t) = t^5 - (5/3)t^3 - 48t$.

21 The illumination from a light source is directly proportional to the strength of the source and inversely proportional to the square of the distance s from the source. At a distance of 2 ft from a glowing fire, a photographer's light meter registers 120 units. If the photographer backs away from the fire, find the rate of change of the meter reading with respect to s when the photographer is 20 ft from the fire.

22 Show that the rate of change of the volume of a sphere with respect to its radius is numerically equal to the surface area of the sphere.

23 Show that the rate of change of the radius of a circle with respect to its circumference is independent of the size of the circle. Illustrate this fact by using great circles on two spheres—one the size of a basketball and the other the size of the Earth.

24 The formula for the adiabatic expansion of air is $pv^{1.4} = c$, where p is the pressure, v the volume, and c is a constant. Find a formula for the rate of change of pressure with respect to volume.

25 The relationship between the temperature F on the Fahrenheit scale and the temperature C on the Celsius scale is given by $C = \frac{5}{9}(F - 32)$. What is the rate of change of F with respect to C?

26 If a sum of money P_0 is invested at an interest rate of $100r$ per cent per year, compounded monthly, then the principal P at the end of one year is $P = P_0(1 + r/12)^{12}$. If r is regarded as a real number, find the rate of change of P with respect to r when $P_0 = \$1,000$ and $r = 0.12$.

27 The period T of a simple pendulum, that is, the time required for one complete oscillation, varies directly as the square root of its length l. What can be said about the rate of change of T with respect to l?

28 The electrical resistance R of a copper wire of fixed length is inversely proportional to the square of its diameter d. What is the rate of change of R with respect to d?

29 An open box is to be constructed from a rectangular piece of cardboard 40 cm wide and 60 cm long by cutting out a square of side s cm from each corner and then bending up the cardboard. Express the volume V of the box as a function of s and find the rate of change of V with respect to s.

30 The formula $1/f = (1/p) + (1/q)$ is used in optics, where f is the focal length of a convex lens, and p and q are the distances from the lens to the object and image, respectively. If f is fixed, find a general formula for the rate of change of q with respect to p.

4.8

RELATED RATES

In applications it is not unusual to encounter two variables x and y that are differentiable functions of time t, say $x = f(t)$ and $y = g(t)$. In addition, x and y may be related by means of some equation such as

$$x^2 - y^3 - 2x + 7y^2 - 2 = 0.$$

Differentiating with respect to t and using the Chain Rule produces an equation involving the rates of change dx/dt and dy/dt. As an illustration, the previous equation leads to

$$2x\frac{dx}{dt} - 3y^2\frac{dy}{dt} - 2\frac{dx}{dt} + 14y\frac{dy}{dt} = 0.$$

The derivatives dx/dt and dy/dt in this equation are called **related rates**, since they are related by means of the equation. The last equation can be used to find one of the rates when the other is known. This observation has many practical applications. The following examples give several illustrations.

Example 1 A ladder 20 ft long leans against a vertical building. If the bottom of the ladder slides away from the building horizontally at a rate of 2 ft/sec, how fast is the ladder sliding down the building when the top of the ladder is 12 ft above the ground?

Solution   We begin by representing a general position of the ladder schematically, as in Figure 4.48, introducing a variable x to denote the distance from the base of the building to the bottom of the ladder, and a variable y to denote the distance from the ground to the top of the ladder.

Since x is changing at a rate of 2 ft/sec, we may write

$$\frac{dx}{dt} = 2 \text{ ft/sec.}$$

Our objective is to find dy/dt, the rate at which the top of the ladder is sliding down the building, at the instant that $y = 12$ ft.

The relationship between x and y may be determined by applying the Pythagorean Theorem to the triangle in the figure. This gives us

$$x^2 + y^2 = 400.$$

Differentiating with respect to t leads to

$$2x\frac{dx}{dt} + 2y\frac{dy}{dt} = 0$$

from which it follows that

$$\frac{dy}{dt} = -\frac{x}{y}\frac{dx}{dt}.$$

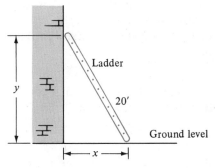

Ladder

20′

Ground level

y

x

Figure 4.48

The last equation is a *general* formula relating the two rates of change under consideration. Let us now consider the special case $y = 12$. The corresponding value of x may be determined from

$$x^2 + 144 = 400, \quad \text{or} \quad x^2 = 400 - 144 = 256.$$

Thus $x = \sqrt{256} = 16$ when $y = 12$. Substituting in the general formula for dy/dt we obtain

$$\frac{dy}{dt} = -\frac{16}{12}(2) = -\frac{8}{3} \text{ ft/sec.} \qquad \blacksquare$$

Some of the Guidelines listed in Section 4.6 for applications of extrema are useful when attempting to solve a related rate problem. In particular, the first three Guidelines—(1) read the problem carefully, (2) sketch and label a suitable diagram, and (3) write down known facts and relationships— are strongly recommended. After following these steps, it is essential to formulate a *general* equation that relates the variables involved in the problem, as we did in Example 1. *A common error is to introduce specific values for the rates and variable quantities too early in the solution.* Always remember to obtain a *general* formula that involves the rates of change at *any* time t. Specific values should not be substituted for variables until the final steps of the solution.

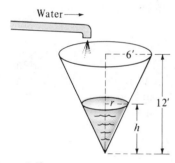

Figure 4.49

Example 2   A water tank has the shape of an inverted right circular cone of altitude 12 ft and base radius 6 ft. If water is being pumped into the tank at a rate of 10 gal/min, approximate the rate at which the water level is rising when it is 3 ft deep (1 gal ≈ 0.1337 ft^3).

Solution   We begin by sketching the tank as in Figure 4.49, letting r denote the radius of the surface of the water when the depth is h. Note that r and h are functions of time t. The volume V of water in the tank corre- sponding to depth h is

$$V = \tfrac{1}{3}\pi r^2 h.$$

Let us express V in terms of one variable. Referring to similar triangles in Figure 4.49 we see that r and h are related by the equations

$$\frac{r}{h} = \frac{6}{12}, \quad \text{or} \quad r = \frac{h}{2}.$$

Consequently, at depth h,

$$V = \frac{1}{3}\pi\left(\frac{h}{2}\right)^2 h = \frac{1}{12}\pi h^3.$$

Differentiating with respect to t gives us the following general relationship between the rates of change of V and h at any time:

$$\frac{dV}{dt} = \frac{1}{4}\pi h^2 \frac{dh}{dt}.$$

An equivalent formula is

$$\frac{dh}{dt} = \frac{4}{\pi h^2}\frac{dV}{dt}.$$

In particular, if $h = 3$, and $dV/dt = 10$ gal/min ≈ 1.337 ft^3/min, we see that

$$\frac{dh}{dt} \approx \frac{4}{\pi(9)}(1.337) \approx 0.189 \text{ ft/min}. \qquad \blacksquare$$

Example 3 At 1:00 P.M., ship A is 25 mi due south of ship B. If ship A is sailing west at a rate of 16 mi/hr and ship B is sailing south at a rate of 20 mi/hr, find the rate at which the distance between the ships is changing at 1:30 P.M.

Solution   Let x and y denote the miles covered by ships A and B, respectively, in t hours after 1:00 P.M. We then have the situation sketched in Figure 4.50, where P and Q are their respective positions at 1:00 P.M. and z is the distance between the ships at time t. By the Pythagorean Theorem,

$$z^2 = x^2 + (25 - y)^2.$$

Differentiating with respect to t, we obtain

$$2z\frac{dz}{dt} = 2x\frac{dx}{dt} + 2(25 - y)\left(-\frac{dy}{dt}\right)$$

or

$$z\frac{dz}{dt} = x\frac{dx}{dt} + (y - 25)\frac{dy}{dt}.$$

It is given that

$$\frac{dx}{dt} = 16 \text{ mi/hr} \quad \text{and} \quad \frac{dy}{dt} = 20 \text{ mi/hr}.$$

Our objective is to find dz/dt.

At 1:30 P.M. the ships have traveled for half an hour and we have $x = 8$, $y = 10$, $25 - y = 15$, and, therefore, $z^2 = 64 + 225 = 289$, or $z = 17$. Substituting in the equation involving dz/dt gives us

$$17\frac{dz}{dt} = 8(16) + (-15)(20)$$

or

$$\frac{dz}{dt} = -\frac{172}{17} \approx -10.12 \text{ mi/hr}.$$

The negative sign indicates that the distance between the ships is decreasing at 1:30 P.M.

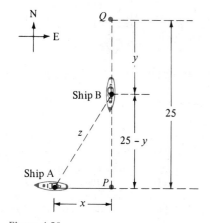

Figure 4.50

Another method of solution is to write $x = 16t$, $y = 20t$, and

$$z = [x^2 + (25 - y)^2]^{1/2} = [256t^2 + (25 - 20t)^2]^{1/2}.$$

The derivative dz/dt may then be found, and substitution of $\frac{1}{2}$ for t produces the desired rate of change.   ■

EXERCISES 4.8

1 A ladder 20 ft long leans against a vertical building. If the bottom of the ladder slides away from the building horizontally at a rate of 3 ft/sec, how fast is the ladder sliding down the building when the top of the ladder is 8 ft from the ground?

2 As a circular metal griddle is being heated, its diameter changes at a rate of 0.01 cm/min. When the diameter is 30 cm at what rate is the area of one side changing?

3 Gas is being pumped into a spherical balloon at a rate of 5 ft³/min. If the pressure is constant, find the rate at which the radius is changing when the diameter is 18 in.

4 A girl starts at a point A and runs east at a rate of 10 ft/sec. One minute later, another girl starts at A and runs north at a rate of 8 ft/sec. At what rate is the distance between them changing 1 minute after the second girl starts?

5 A light is at the top of a 16-ft pole. A boy 5 ft tall walks away from the pole at a rate of 4 ft/sec (see figure). At what rate is the tip of his shadow moving when he is 18 ft from the pole? At what rate is the length of his shadow increasing?

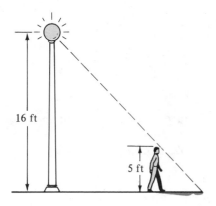

16 ft

5 ft

Figure for Exercise 5

6 A man on a dock is pulling in a boat by means of a rope attached to the bow of the boat 1 ft above water level and passing through a simple pulley located on the dock 8 ft above water level (see figure). If he pulls in the rope at a rate of 2 ft/sec, how fast is the boat approaching the dock

when the bow of the boat is 25 ft from a point that is 7 ft directly below the pulley?

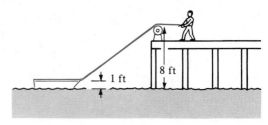

1 ft   8 ft

Figure for Exercise 6

7 The top of a silo has the shape of a hemisphere of diameter 20 ft. If it is coated uniformly with a layer of ice, and if the thickness is decreasing at a rate of $\frac{1}{4}$ in./hr, how fast is the volume of ice changing when the ice is 2 in. thick?

8 As sand leaks out of a hole in a container, it forms a conical pile whose altitude is always the same as its radius. If the height of the pile is increasing at a rate of 6 in./min, find the rate at which the sand is leaking out when the altitude is 10 in.

9 A boy flying a kite holds the string 5 ft above ground level and string is payed at a rate of 2 ft/sec as the kite moves horizontally at an altitude of 105 ft (see figure). Assuming there is no sag in the string, find the rate at which the kite is moving when 125 ft of string has been payed out.

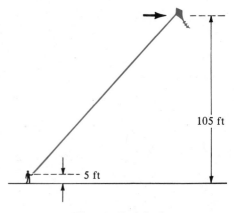

105 ft

5 ft

Figure for Exercise 9

10 A weather balloon is rising vertically at a rate of 2 ft/sec. An observer is situated 100 yd from a point on the ground directly below the balloon. At what rate is the distance between the balloon and the observer changing when the altitude of the balloon is 500 ft?

11 Boyle's Law for gases states that $pv = c$, where p denotes the pressure, v the volume, and c is a constant. At a certain instant the volume is 75 in.^{3}, the pressure is 30 lb/in.^{2}, and the pressure is decreasing at a rate of 2 lb/in.^{2} every minute. At what rate is the volume changing at this instant?

12 Suppose a spherical snowball is melting and the radius is decreasing at a constant rate, changing from 12 in. to 8 in. in 45 minutes. How fast was the volume changing when the radius was 10 in.?

13 The ends of a water trough 8 ft long are equilateral triangles whose sides are 2 ft long (see figure). If water is being pumped into the trough at a rate of 5 ft^3/min, find the rate at which the water level is rising when the depth is 8 in.

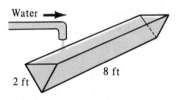

Figure for Exercise 13

14 Work Exercise 13 if the ends of the trough have the shape of the graph of $y = 2|x|$ between the points $(-1, 2)$ and $(1, 2)$.

15 A point $P(x, y)$ moves on the graph of the equation $y = x^3 + x^2 + 1$, the x-coordinate changing at a rate of 2 units/sec. How fast is the y-coordinate changing at the point $(1, 3)$?

16 A point $P(x, y)$ moves on the graph of $y^2 = x^2 - 9$ such that $dx/dt = 1/x$. Find dy/dt at the point $(5, 4)$.

17 The area of an equilateral triangle is decreasing at a rate of 4 cm^2/min. Find the rate at which the length of a side is changing when the area of the triangle is 200 cm^2.

18 Gas is escaping from a spherical balloon at a rate of 10 ft^3/hr. At what rate is the radius changing when the volume is 400 ft^3?

19 A stone is dropped into a lake, causing circular waves whose radii increase at a constant rate of 0.5 m/sec. At what rate is the circumference of a wave changing when its radius is 4 m?

20 A softball diamond has the shape of a square with sides 60 ft long. If a player is running from second base to third at a speed of 24 ft/sec, at what rate is her distance from home plate changing when she is 20 ft from third?

21 When two electrical resistances R_1 and R_2 are connected in parallel (see figure), the total resistance R is given by $1/R = (1/R_1) + (1/R_2)$. If R_1 and R_2 are increasing at rates of 0.01 ohm/sec and 0.02 ohm/sec, respectively, at what rate is R changing at the instant that $R_1 = 30$ ohms and $R_2 = 90$ ohms?

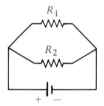

Figure for Exercise 21

22 The formula for the adiabatic expansion of air is $pv^{1.4} = c$, where p denotes pressure, v denotes volume, and c is a constant. At a certain instant the pressure is 40 dynes/cm^2 and is increasing at a rate of 3 dynes/cm^2 per second. If, at that same instant, the volume is 60 cm^3, find the rate at which the volume is changing.

23 If a spherical tank of radius a contains water that has a maximum depth h, then the volume V of water in the tank is given by $V = \frac{1}{3}\pi h^2(3a - h)$. Suppose a spherical tank of radius 16 ft is being filled at a rate of 100 gal/min. Approximate the rate at which the water level is rising when $h = 4$ ft (1 gal ≈ 0.1337 ft^3).

24 A spherical water storage tank for a small community is coated uniformly with a 2-inch layer of ice. If the volume of ice is melting at a rate that is directly proportional to its surface area, show that the outside diameter is decreasing at a constant rate.

25 From the edge of a cliff that overlooks a lake 200 ft below, a boy drops a stone and then, two seconds later, drops another stone from exactly the same position. Discuss the rate at which the distance between the two stones is changing during the next second. (Assume that the distance an object falls in t seconds is $16t^2$ ft.)

26 A metal rod has the shape of a right circular cylinder. As it is being heated, its length is increasing at a rate of 0.005 cm/min and its diameter is increasing at 0.002 cm/min. At what rate is the volume changing when the rod has length 40 cm and diameter 3 cm?

27 An airplane, flying at a constant speed of 360 mi/hr and climbing at an angle of 45°, passes over an air traffic control tower on the ground at an altitude of 10,560 ft. Find the rate at which its distance from the tower is changing one minute later (neglect the height of the tower).

28 Refer to Exercise 44 of Section 4.6. At what rate is the distance between the airplane and the automobile changing at 10:15 A.M.?

29 A paper cup containing water has the shape of a frustum of a right circular cone of altitude 6 in. and lower and upper base radii 1 in. and 2 in., respectively. If water is leaking out of the cup at a rate of 3 in.³/hr, at what rate is the water level decreasing when its depth is 4 in.? (*Note:* The volume V of a frustum of a right circular cone of altitude h and base radii a and b is given by $V = \frac{1}{3}\pi h(a^2 + b^2 + ab)$.)

30 The top part of a swimming pool is a rectangle of length 60 ft and width 30 ft. The depth of the pool varies uniformly from 4 ft to 9 ft through a horizontal distance of 40 ft and then is level for the remaining 20 ft, as illustrated by the cross-sectional view in the accompanying figure. If the pool is being filled with water at a rate of 500 gal/min, approximate the rate at which the water level is rising when the depth at the deep end is 4 ft (1 gal $\approx$ 0.1337 ft³).

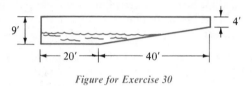

Figure for Exercise 30

4.9

ANTIDERIVATIVES

In Chapter 3 certain problems were stated in the form "*Given a function g, find the derivative g'.*" We shall now consider the converse problem, namely, "*Given the derivative g', find the function g.*" An equivalent way of stating this converse problem is, "Given a function f, find a function F such that $F' = f$." As a simple illustration, suppose $f(x) = 8x^3$. In this case it is easy to find a function F that has f as its derivative. We know that differentiating a power of x *reduces* the exponent by 1 and therefore to obtain F we must *increase* the given exponent by 1. Thus $F(x) = ax^4$ for some number a. Differentiating, we obtain $F'(x) = 4ax^3$ and, in order for this to equal $f(x)$, a must equal 2. Consequently, the function F defined by $F(x) = 2x^4$ has the desired property. According to the next definition, F is called an *antiderivative* of f.

Definition (4.30)

> A function F is an **antiderivative** of a function f if $F' = f$.

For convenience, we shall use the phrase "$F(x)$ is an antiderivative of $f(x)$" synonymously with "F is an antiderivative of f." The domain of an antiderivative will not usually be specified. It follows from the Fundamental Theorem of Calculus (5.21) in the next chapter that a suitable domain is any closed interval $[a, b]$ on which f is continuous.

Antiderivatives are never unique. Indeed, since the derivative of a constant is zero, it follows that if F is an antiderivative of f, so is the function G defined by $G(x) = F(x) + C$, for every number C. For example, if $f(x) = 8x^3$, then functions defined by expressions such as $2x^4 + 7$, $2x^4 - \sqrt{3}$, and $2x^4 + \frac{2}{5}$ are antiderivatives of f. The next theorem brings out the fact that functions of this type are the only possible antiderivatives of f.

Theorem (4.31)

> If F_1 and F_2 are differentiable functions such that $F_1'(x) = F_2'(x)$ for all x in a closed interval $[a, b]$, then $F_2(x) = F_1(x) + C$ for some number C and all x in $[a, b]$.

Proof　　　If we define the function g by

$$g(x) = F_2(x) - F_1(x)$$

then

$$g'(x) = F_2'(x) - F_1'(x) = 0$$

for all x in $[a, b]$. If x is any number such that $a < x \le b$, then applying the Mean Value Theorem (4.12) to the function g and the closed interval $[a, x]$, there exists a number z in the open interval (a, x) such that

$$g(x) - g(a) = g'(z)(x - a) = 0 \cdot (x - a) = 0.$$

Hence $g(x) = g(a)$ for all x in $[a, b]$. Substitution in the first equation stated in the proof gives us

$$g(a) = F_2(x) - F_1(x).$$

Adding $F_1(x)$ to both sides we obtain the desired conclusion, with $C = g(a)$.

The graphical significance of the theorem is illustrated in Figure 4.51. The figure indicates that if the slopes m_1 and m_2 of the tangent lines to the graphs of F_1 and F_2 are the same for every x in $[a, b]$, then one graph can be obtained from the other by a vertical shift through a distance $|C|$. □

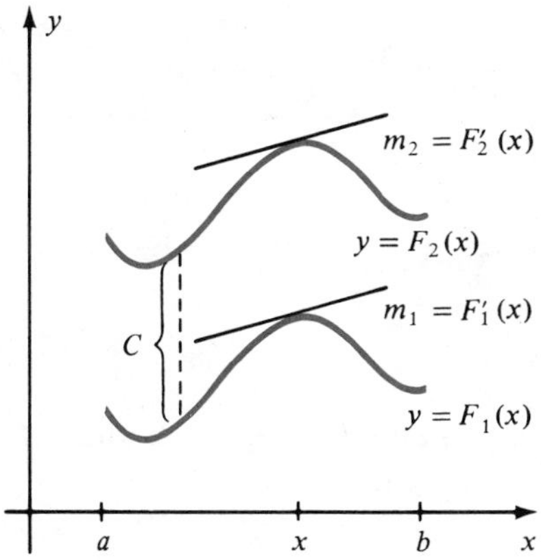

Figure 4.51

If F_1 and F_2 are antiderivatives of the same function f, then $F_1' = f = F_2'$ and hence, by the theorem just proved, $F_2(x) = F_1(x) + C$ for some C. In other words, if $F(x)$ is an antiderivative of $f(x)$, then every other antiderivative has the form $F(x) + C$ where C is an arbitrary constant (that is, an unspecified real number). We shall refer to $F(x) + C$ as the **most general antiderivative** of $f(x)$.

Theorem (4.32)

> If $f'(x) = 0$ for all x in $[a, b]$, then f is a constant function on $[a, b]$.

Proof　　　Denote f by F_2 and let the function F_1 be defined by $F_1(x) = 0$ for all x. Since $F_1'(x) = 0$ and $F_2'(x) = f'(x) = 0$, we see that $F_1'(x) = F_2'(x)$ for all x in $[a, b]$. Applying Theorem (4.31), there is a number C such that $F_2(x) = F_1(x) + C$; that is, $f(x) = C$ for all x in $[a, b]$. This completes the proof. □

Rules for derivatives may be used to obtain formulas for antiderivatives, as in the proof of the following important result.

Power Rule for Antidifferentiation
(4.33)

> Let a be any real number, r any rational number different from -1, and C an arbitrary constant.
>
> $$\text{If}\quad f(x) = ax^r, \quad\text{then}\quad F(x) = \left(\frac{a}{r+1}\right)x^{r+1} + C$$
>
> is the most general antiderivative of $f(x)$.

Proof   It is sufficient to show that $F'(x) = f(x)$. This fact follows readily from the Power Rule for Derivatives (3.30), since

$$F'(x) = \left(\frac{a}{r+1}\right)(r+1)x^r = ax^r = f(x). \qquad \square$$

Example 1 Find the most general antiderivative of
(a) $4x^5$ (b) $7/x^3$ (c) $\sqrt[3]{x^2}$.

Solution

(a) Using the Power Rule (4.33) with $a = 4$ and $r = 5$ gives us the antiderivative

$$\tfrac{4}{6}x^6 + C, \quad\text{or}\quad \tfrac{2}{3}x^6 + C.$$

(b) Writing $7/x^3$ as $7x^{-3}$ and using Rule (4.33) with $a = 7$ and $r = -3$ leads to

$$\frac{7}{-2}x^{-2} + C, \quad\text{or}\quad -\frac{7}{2x^2} + C.$$

(c) Since $\sqrt[3]{x^2} = x^{2/3}$ we may apply the Power Rule with $a = 1$ and $r = \tfrac{2}{3}$, obtaining

$$\frac{1}{\tfrac{5}{3}}x^{5/3} + C, \quad\text{or}\quad \tfrac{3}{5}x^{5/3} + C. \qquad \blacksquare$$

To avoid algebraic errors, *it is highly recommended to check solutions of problems involving antidifferentiation by differentiating the final antiderivative.* In each case the given expression should be obtained.

Theorem (4.34)

> If $F(x)$ and $G(x)$ are antiderivatives of $f(x)$ and $g(x)$, respectively, then
> (i) $F(x) + G(x)$ is an antiderivative of $f(x) + g(x)$.
> (ii) $cF(x)$ is an antiderivative of $cf(x)$, where c is any real number.

Proof   By hypothesis, $D_x[F(x)] = f(x)$ and $D_x[G(x)] = g(x)$. Hence

$$D_x[F(x) + G(x)] = D_x[F(x)] + D_x[G(x)] = f(x) + g(x).$$

This proves (i). To prove (ii) we merely note that

$$D_x[cF(x)] = cD_x[F(x)] = cf(x). \qquad \square$$

Part (i) of the preceding theorem can be extended to any finite sum of functions. This fact may be stated as follows: "*An antiderivative of a sum is the sum of the antiderivatives.*" As usual, when working with several functions we assume that the domain is restricted to the intersection of the domains of the individual functions. A similar result is true for differences.

Example 2   Find the most general antiderivative $F(x)$ of

$$f(x) = 3x^4 - x + 4 + (5/x^3).$$

Solution   Writing the last term of $f(x)$ as $5x^{-3}$ and applying the Power Rule for Antidifferentiation to each term gives us

$$F(x) = \tfrac{3}{5}x^5 - \tfrac{1}{2}x^2 + 4x - \tfrac{5}{2}x^{-2} + C.$$

It is unnecessary to introduce an arbitrary constant for each of the four antiderivatives, since they could be added together to produce the one constant C. ∎

Equations that involve derivatives of an unknown function f are very common in mathematical applications. Such equations are called **differential equations**. The function f is called a **solution** of the differential equation. To **solve** a differential equation means to find all solutions. Sometimes, in addition to the differential equation we may know certain values of f, called **boundary values**, as illustrated in the next example.

Example 3   Solve the differential equation $f'(x) = 6x^2 + x - 5$ with boundary value $f(0) = 2$.

Solution   From our discussion of antiderivatives,

$$f(x) = 2x^3 + \tfrac{1}{2}x^2 - 5x + C$$

for some number C. Letting $x = 0$ and using the given boundary value, we obtain

$$f(0) = 0 + 0 - 0 + C = 2$$

and hence $C = 2$. Consequently, the solution f of the differential equation with the given boundary value is

$$f(x) = 2x^3 + \tfrac{1}{2}x^2 - 5x + 2. \qquad ∎$$

If a point P is moving rectilinearly, then its position function s is an antiderivative of its velocity function, that is, $s'(t) = v(t)$. Similarly, since $v'(t) = a(t)$, the velocity function is an antiderivative of the acceleration function. If the velocity or acceleration function is known, then given sufficient boundary conditions it is possible to determine the position function. The particular boundary conditions corresponding to $t = 0$ are sometimes called the **initial conditions**.

Example 4 A motorboat moves away from a dock along a straight line, with an acceleration at time t given by $a(t) = 12t - 4$ ft/sec^2. If, at time $t = 0$, the boat had a velocity of 8 ft/sec and was 15 ft from the dock, find its distance $s(t)$ from the dock at the end of t seconds.

Solution   From $v'(t) = 12t - 4$ we obtain, by antidifferentiation,

$$v(t) = 6t^2 - 4t + C$$

for some number C. Substitution of 0 for t and use of the fact that $v(0) = 8$ gives us $8 = 0 - 0 + C = C$. Thus

$$v(t) = 6t^2 - 4t + 8$$

or, equivalently, $s'(t) = 6t^2 - 4t + 8.$

The most general antiderivative of $s'(t)$ is

$$s(t) = 2t^3 - 2t^2 + 8t + D$$

where D is some number. Substitution of 0 for t and use of the fact that $s(0) = 15$ leads to $15 = 0 - 0 + 0 + D = D$ and, consequently, the desired position function is given by

$$s(t) = 2t^3 - 2t^2 + 8t + 15.$$ ∎

An object on or near the surface of the Earth is acted upon by a force called **gravity**, which produces a constant acceleration denoted by g. The approximation to g that is employed for most problems is 32 ft/sec^2 or 980 cm/sec^2. The use of this important physical constant is illustrated in the following example.

Example 5 A stone is thrown vertically upward from a position 144 ft above the ground with a velocity of 96 ft/sec. If air resistance is neglected, find its distance above the ground after t sec. For what length of time does the stone rise? When, and with what velocity, does it strike the ground?

Solution   The motion of the stone may be represented by a point moving rectilinearly on a vertical coordinate line l with origin at ground level and positive direction upward (see Figure 4.52). The distance above the ground at time t is $s(t)$ and the initial conditions are $s(0) = 144$ and $v(0) = 96$.

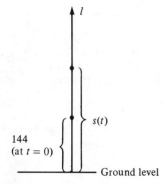

Figure 4.52

Since the velocity is decreasing, $v'(t) < 0$; that is, the acceleration is negative. Hence, by the remarks preceding this example,

$$a(t) = -32.$$

Since v is an antiderivative of a,

$$v(t) = -32t + C,$$

for some number C. Substituting 0 for t and using the fact that $v(0) = 96$ gives us $96 = 0 + C = C$ and, consequently,

$$v(t) = -32t + 96.$$

Since $s'(t) = v(t)$ we obtain, by antidifferentiation,

$$s(t) = -16t^2 + 96t + D$$

for some number D. Letting $t = 0$ and using the fact that $s(0) = 144$ leads to $144 = 0 + 0 + D = D$. It follows that the distance from the ground to the stone at time t is given by

$$s(t) = -16t^2 + 96t + 144.$$

The stone will rise until $v(t) = 0$, that is, until $-32t + 96 = 0$, or $t = 3$. The stone will strike the ground when $s(t) = 0$, that is when $-16t^2 + 96t + 144 = 0$ or, equivalently, $t^2 - 6t - 9 = 0$. Applying the quadratic formula, $t = 3 \pm 3\sqrt{2}$. The solution $3 - 3\sqrt{2}$ is extraneous (Why?) and hence the stone strikes the ground after $3 + 3\sqrt{2}$ sec. The velocity at that time is

$$\begin{aligned} v(3 + 3\sqrt{2}) &= -32(3 + 3\sqrt{2}) + 96 \\ &= -96\sqrt{2} \approx -135.8 \text{ ft/sec.} \quad \blacksquare \end{aligned}$$

EXERCISES 4.9

Find the most general antiderivatives of the functions defined in Exercises 1–22.

1 $f(x) = 9x^2 - 4x + 3$

2 $f(x) = 4x^2 - 8x + 1$

3 $f(x) = 2x^3 - x^2 + 3x - 7$

4 $f(x) = 10x^4 - 6x^3 + 5$

5 $f(x) = \dfrac{1}{x^3} - \dfrac{3}{x^2}$

6 $f(x) = \dfrac{4}{x^7} - \dfrac{7}{x^4} + x$

7 $f(x) = 3\sqrt{x} + (1/\sqrt{x})$

8 $f(x) = \sqrt{x^3} - \tfrac{1}{2}x^{-2} + 5$

9 $f(x) = (6/\sqrt[3]{x}) - (\sqrt[3]{x}/6) + 7$

10 $f(x) = 3x^5 - \sqrt[3]{x^5}$

11 $f(x) = 2x^{5/4} + 6x^{1/4} + 3x^{-4}$

12 $f(x) = \left(x - \dfrac{1}{x}\right)^2$

13 $f(x) = (3x + 4)^2/x^4$

14 $f(x) = (2x + 1)^3$

15 $f(x) = (3x - 1)^2$

16 $f(x) = (2x - 5)(3x + 1)$

17 $f(x) = (8x - 5)/\sqrt[3]{x}$

18 $f(x) = (2x^2 - x + 3)/\sqrt{x}$

19 $f(x) = \sqrt[5]{32x^4}$

20 $f(x) = \sqrt[3]{64x^5}$

21 $f(x) = (x^3 - 1)/(x - 1)$

22 $f(x) = \dfrac{x^3 + 3x^2 - 9x - 2}{x - 2}$

Solve the differential equations in Exercises 23–28 subject to the given boundary conditions.

23 $f'(x) = 12x^2 - 6x + 1, f(1) = 5$

24 $f'(x) = 9x^2 + x - 8, f(-1) = 1$

25 $f''(x) = 4x - 1, f'(2) = -2, f(1) = 3$

26 $f'''(x) = 6x, f''(0) = 2, f'(0) = -1, f(0) = 4$

27 $f'''(t) = 3/\sqrt{t}, f(1) = 4, f'(1) = 2, f''(1) = 8$

28 $f''(t) = \sqrt[3]{t^2}, f'(1) = 2, f(1) = 3$

In Exercises 29–32 a point is moving rectilinearly subject to the given conditions. Find $s(t)$.

29 $a(t) = 2 - 6t, v(0) = -5, s(0) = 4$

30 $a(t) = 3t^2, v(0) = 20, s(0) = 5$

31 $a(t) = -32, v(0) = 80, s(0) = 240$

32 $a(t) = -980, v(0) = -100, s(0) = 400$

33 A projectile is fired vertically upward from ground level with a velocity of 1600 ft/sec. If air resistance is neglected, find its distance $s(t)$ above ground at time t. What is its maximum height?

34 An object is dropped from a height of 1000 ft. Neglecting air resistance, find the distance it falls in t seconds. What is its velocity at the end of 3 sec? When will it strike the ground?

35 A stone is thrown directly downward from a height of 96 ft with a velocity of 16 ft/sec. Find (a) its distance above the ground after t seconds; (b) when it will strike the ground; and (c) the velocity at which it strikes the ground.

36 The gravitational constant for objects near the surface of the moon is approximately 5.3 ft/sec². If an astronaut throws a stone directly upward with an initial velocity of 60 ft/sec, find its maximum altitude. If, after returning to earth, the astronaut throws the same stone directly upward with the same initial velocity, find the maximum altitude.

37 If a projectile is fired vertically upward from a height of s_0 ft above the ground with a velocity of v_0 ft/sec, prove that if air resistance is neglected, its distance $s(t)$ above the ground after t sec is given by $s(t) = -\frac{1}{2}gt^2 + v_0 t + s_0$, where g is the gravitational constant.

38 A ball rolls down an inclined plane with an acceleration of 2 ft/sec². If the ball is given no initial velocity, how far will it roll in t sec? What initial velocity must be given for the ball to roll 100 feet in 5 sec?

39 If an automobile starts from rest, what constant acceleration will enable it to travel 500 ft in 10 sec?

40 If a car is traveling at a speed of 60 mi/hr, what constant (negative) acceleration will enable it to stop in 9 sec?

41 If C and F denote Celsius and Fahrenheit temperature readings, then the rate of change of F with respect to C is given by $dF/dC = \frac{9}{5}$. If $F = 32$ when $C = 0$, use antidifferentiation to obtain a general formula for F in terms of C.

42 The rate of change of the temperature T of a solution is given by $dT/dt = \frac{1}{4}t + 10$, where t is the time in minutes and T is measured in degrees Celsius. If the temperature is 5°C at $t = 0$, find a formula that gives T at time t.

43 The volume V of a balloon is changing with respect to time t at a rate given by $dV/dt = 3\sqrt{t} + (t/4)$ ft³/sec. If, at $t = 4$, the volume is 20 ft³, express V as a function of t.

44 Suppose the slope of the tangent line at any point P on the graph of an equation equals the square of the x-coordinate of P. Find the equation if the graph contains (a) the origin; (b) the point $(3, 6)$; (c) the point $(-1, 1)$. Sketch the graph in each case.

45 Suppose F and G are antiderivatives of f and g, respectively. Prove or disprove that (a) FG is an antiderivative of fg, and (b) F/G is an antiderivative of f/g.

46 Suppose F is an antiderivative of f. Prove that (a) if F is an even function, then f is odd; (b) if F is an odd function, then f is even. Use polynomial functions to give examples of (a) and (b).

4.10

APPLICATIONS TO ECONOMICS

Calculus has become an important tool for problems that occur in economics. As will be seen in this section, if a function f is used to describe some economic entity, then the adjective *marginal* is employed to specify the derivative f'.

If the cost of producing x units of a certain commodity is denoted by $C(x)$, then C is called a **cost function**. The **average cost** $c(x)$ of one unit is defined by $c(x) = C(x)/x$. In order to use the techniques of calculus, x is regarded as a real number, even though this variable may take on only integer values. We always assume that $x \geq 0$, since the production of a negative number of units has no practical significance. The derivative C' of the cost function is called the **marginal cost function**. If we interpret the derivative as a rate of change (see Section 4.7), then $C'(x)$ is the rate at which the cost changes with respect to the number of units produced. The number $C'(x)$ is referred to as the **marginal cost** associated with the production of x units. Evidently $C'(x) > 0$, since the cost should increase as more units are produced.

If C is a cost function and n is a positive integer, then by Theorem (3.7)

$$C'(n) = \lim_{h \to 0} \frac{C(n + h) - C(n)}{h}.$$

Hence, if h is small, then

$$C'(n) \approx \frac{C(n + h) - C(n)}{h}.$$

If the number n of units produced is large, economists often let $h = 1$ in the last formula to approximate the marginal cost, obtaining

$$C'(n) \approx C(n + 1) - C(n).$$

In this context, the marginal cost associated with the production of n units is (approximately) the cost of producing one more unit.

Some companies find that the cost $C(x)$ of producing x units of a certain commodity is given by a formula such as

$$C(x) = a + bx + dx^2 + kx^3.$$

The constant a represents a fixed overhead charge for items such as rent, heat, light, etc., that are independent of the number of units produced. If the cost of producing one unit were b dollars and no other factors were involved, then the second term bx in the formula would represent the cost of producing x units. If x becomes very large, then the terms dx^2 and kx^3 may significantly affect production costs.

Example 1 An electronics company estimates that the cost (in dollars) of producing x components used in electronic toys is given by

$$C(x) = 200 + 0.05x + 0.0001x^2.$$

Find the cost, the average cost, and the marginal cost of producing
(a) 500 units   (b) 1000 units   (c) 5000 units.

Solution   The average cost of producing x components is given by

$$c(x) = \frac{C(x)}{x} = \frac{200}{x} + 0.05 + 0.0001x.$$

The marginal cost is

$$C'(x) = 0.05 + 0.0002x.$$

We leave it to the reader to verify the entries in the following table, where numbers in the last three columns represent dollars, rounded off to the nearest cent.

x	$C(x)$	$c(x) = \dfrac{C(x)}{x}$	$C'(x)$
500	250.00	0.50	0.15
1000	350.00	0.35	0.25
5000	2950.00	0.59	1.05

The derivative $c'(x)$ of the average cost $c(x)$ is called the **marginal average cost**. Applying the quotient rule to $c(x) = C(x)/x$ we obtain

$$c'(x) = \frac{xC'(x) - C(x)}{x^2}$$

and consequently the average cost will have an extremum if

$$xC'(x) - C(x) = 0$$

that is, if

$$C'(x) = \frac{C(x)}{x} = c(x).$$

Thus, a minimum average cost can occur only when the marginal and average costs are equal.

Example 2   In Example 1, find (a) the number of components that will minimize the average cost, and (b) the minimum average cost.

Solution

(a)   From the preceding discussion we must have $C'(x) = c(x)$, that is,

$$0.05 + 0.0002x = \frac{200}{x} + 0.05 + 0.0001x.$$

This equation reduces to

$$0.0001x = \frac{200}{x}$$

or
$$x^2 = \frac{200}{0.0001} = 2,000,000.$$

Consequently,

$$x = \sqrt{2,000,000} \approx 1414.$$

The First or Second Derivative Test can be used to verify that the average cost is minimal for this number of units.

(b)   Using the result obtained in part (a), the minimum average cost is

$$c(1414) = \frac{200}{1414} + 0.05 + 0.0001(1414) \approx 0.33. \qquad \blacksquare$$

Example 3   A company determines that the cost $C(x)$ of manufacturing x units of a commodity may be approximated by

$$C(x) = 100 + \frac{10}{x} + \frac{x^2}{200}.$$

How many units should be produced in order to minimize the cost?

Solution   Since the marginal cost is

$$C'(x) = -\frac{10}{x^2} + \frac{x}{100}$$

the cost function will have an extremum if

$$-\frac{10}{x^2} + \frac{x}{100} = 0$$

or, equivalently,

$$\frac{-1000 + x^3}{100x^2} = 0.$$

Solving for x gives us $x = 10$. It is left to the reader to show that C has an absolute minimum at $x = 10$. $\qquad \blacksquare$

A company must consider many factors in order to determine a selling price. In addition to the cost of production and the profit desired, the seller should be aware of the manner in which consumer demand will vary if the price increases. For some commodities there is a constant demand, and changes in price have little effect on sales. For items that are not necessities of life, a price increase will probably lead to a decrease in the number of units sold. Suppose a company knows from past experience that it can sell x units when the price per unit is given by $p(x)$, where p is some type of function. We sometimes say that $p(x)$ is the price per unit when there is a **demand** for x units, and we refer to p as the **demand function** for the commodity. The total income, or total revenue, is the number of units sold times the price per unit, that is, $x \cdot p(x)$. For this reason the function R, defined by

$$R(x) = xp(x)$$

is called the **total revenue function**. The derivatives p' and R' are called the **marginal demand** and **marginal revenue** functions, respectively. They are used to find the rates of change of the demand and total revenue functions with respect to the number of units sold.

If we let $S = p(x)$, then S is the selling price per unit associated with a demand of x units. Since a decrease in S would ordinarily be associated with an increase in x, a demand function p is usually decreasing, that is $p'(x) < 0$ for all x. Demand functions are sometimes defined implicitly by an equation involving S and x, as in the next example.

Example 4 The demand for x units of a certain commodity is related to a selling price of S dollars per unit by means of the equation $2x + S^2 - 12,000 = 0$. Find the demand function, the marginal demand function, the total revenue function, and the marginal revenue function. Find the number of units and the price per unit that yield the maximum revenue. What is the maximum revenue?

Solution   Since $S^2 = 12,000 - 2x$ and S is positive, we see that the demand function p is given by

$$S = p(x) = \sqrt{12,000 - 2x}.$$

The domain of p consists of all x such that $12,000 - 2x > 0$, or equivalently, $2x < 12,000$. Thus $0 \le x < 6000$. The graph of p is sketched in Figure 4.53. In theory, there are no sales if the selling price is $\sqrt{12,000}$, or approximately $109.54, and when the selling price is close to $0 the demand is close to 6000.

The marginal demand function p' is given by

$$p'(x) = \frac{-1}{\sqrt{12,000 - 2x}}.$$

The negative sign indicates that a decrease in price is associated with an increase in demand.

The total revenue function R is given by

$$R(x) = xp(x) = x\sqrt{12,000 - 2x}.$$

Differentiating and simplifying gives us the marginal revenue function R', where

$$R'(x) = \frac{12,000 - 3x}{\sqrt{12,000 - 2x}}.$$

Thus, $x = 12,000/3 = 4000$ is a critical number for the total revenue function R. Since $R'(x)$ is positive if $0 \le x < 4000$ and negative if $4000 < x < 6000$,

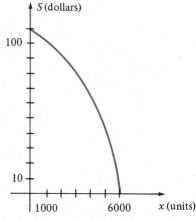

Figure 4.53 Graph of demand function

there is a maximum total revenue when 4000 units are produced. This corresponds to a selling price of

$$p(4000) = \sqrt{12,000 - 2(4000)} \approx \$63.25.$$

The maximum revenue, obtained from selling 4000 units at this price is

$$4000(63.25) = \$253,000. \qquad \blacksquare$$

If x units of a commodity are sold at a price of $p(x)$ per unit, then the **profit** $P(x)$ is given by

$$P(x) = R(x) - C(x)$$

where R and C are the total revenue and cost functions, respectively. We call P the **profit function** and P' the **marginal profit function**. The critical numbers of P are solutions of the equation

$$P'(x) = R'(x) - C'(x) = 0.$$

This shows that there may be a maximum (or minimum) profit when $R'(x) = C'(x)$. Thus if P has an extremum then the marginal revenue and the marginal cost are equal.

Example 5 A furniture company estimates that the weekly cost (in dollars) of manufacturing x hand-finished reproductions of a Colonial period desk is given by

$$C(x) = x^3 - 3x^2 - 80x + 500.$$

Each desk produced is sold for \$2,800. What weekly production rate will maximize the profit? What is the largest possible profit per week?

Solution Since the income obtained from selling x desks is $2800x$, the total revenue function R is given by $R(x) = 2800x$. From the discussion preceding this example, the maximum profit occurs if $C'(x) = R'(x)$, that is, if

$$3x^2 - 6x - 80 = 2800.$$

This equation reduces to

$$x^2 - 2x - 960 = 0, \quad \text{or} \quad (x - 32)(x + 30) = 0$$

and hence either $x = 32$ or $x = -30$. Since the negative solution is extraneous, it suffices to check $x = 32$.

The second derivative of the profit function $P = R - C$ is given by

$$P''(x) = R''(x) - C''(x) = 0 - (6x - 6).$$

Consequently,

$$P''(32) = -6(32) + 6 < 0$$

which means that a maximum profit occurs if 32 desks per week are manufactured. The maximum weekly profit is

$$P(32) = R(32) - C(32)$$
$$= 2800(32) - [(32)^3 - 3(32)^2 - 80(32) + 500]$$
$$= \$61,964.$$

∎

As a final remark on economic applications, if a marginal function is
known, then it is sometimes possible to use antidifferentiation to find the
function, as illustrated in the next example.

Example 6 A company that supplies a manufacturer of office equipment
finds that the marginal cost (in dollars) associated with the production of x
units of a photo-copier component is given by $30 - 0.02x$. If the cost of
producing one unit is \$35, find the cost function and the cost of producing
100 units.

Solution   If C is the cost function, then the marginal cost is the rate of
change of C with respect to x, that is

$$C'(x) = 30 - 0.02x.$$

Antidifferentiation gives us

$$C(x) = 30x - 0.01x^2 + K$$

for some real number K. Letting $x = 1$ and using $C(1) = 35$ we obtain

$$35 = 30 - 0.01 + K$$

and hence $K = 5.01$. Consequently

$$C(x) = 30x - 0.01x^2 + 5.01.$$

In particular, the cost of producing 100 units is

$$C(100) = 3000 - 100 + 5.01 = \$2,905.01.$$

∎

EXERCISES 4.10

In Exercises 1–4, C is the cost function for a certain commodity.
Find (a) the cost of producing 100 units; (b) the average and
marginal cost functions, and their values at $x = 100$; (c) the
minimum average cost; (d) if possible, the value of x that will
minimize the cost.

1 $C(x) = 800 + 0.04x + 0.0002x^2$

2 $C(x) = 6400 + 6.5x + 0.003x^2$

3 $C(x) = 250 + 100x + 0.001x^3$

4 $C(x) = 200 + (100/\sqrt{x}) + (\sqrt{x}/1000)$

5 A manufacturer of small motors estimates that the cost
(in dollars) of producing x motors per day is given by
$C(x) = 25 + 2x + (256/\sqrt{x})$. What daily production will
minimize the average cost?

6 A company is pilot-testing the production of a new
industrial solvent. It finds that the cost of producing x
liters of each pilot run is given by $C(x) = 3 + x + (10/x)$.
Compare the marginal cost of producing ten liters with
the cost of producing the eleventh liter.

In each of Exercises 7–10 the given equation relates the demand for x units of a certain commodity to a selling price of S dollars per unit. Find the demand function, the marginal demand function, the total revenue function, and the marginal revenue function. If possible, find the number of units and the price per unit that will yield the maximum revenue.

7 $3S + 4x - 800 = 0$

8 $x^2 + S + xS - 400 = 0$

9 $S^2 - 8S - x + 16 = 0$

10 $Sx^2 - 1000x + 144S - 5000 = 0$

In Exercises 11–14, R denotes the total revenue function when there is a demand for x units of a certain commodity. Find (a) the marginal revenue function; (b) the maximum total revenue; and (c) the demand function.

11 $R(x) = 2x\sqrt{400 - x}$ **12** $R(x) = 70x - x^2$

13 $R(x) = x(300 - x^2)$ **14** $R(x) = 300x - 2x^{3/2}$

In Exercises 15 and 16 the demand and cost functions of a commodity are denoted by p and C, respectively. Find (a) the marginal demand function; (b) the total revenue function; (c) the profit function; (d) the marginal profit function; (e) the maximum profit; and (f) the marginal cost when the demand is 10 units.

15 $p(x) = 50 - (x/10); C(x) = 10 + 2x$

16 $p(x) = 80 - \sqrt{x - 1}; C(x) = 75x + 2\sqrt{x - 1}$

17 A travel agency estimates that to sell x "package deal" vacations, the price per vacation should be $1800 - 2x$ dollars, where $1 \le x \le 100$. If the cost to the agency for x vacations is $1000 + x + 0.01x^2$ dollars, find (a) the total revenue function; (b) the profit function; (c) the number of vacations that will maximize the profit; and (d) the maximum profit.

18 A manufacturer determines that x units of a product will be sold if the selling price is $400 - 0.05x$ dollars for each unit. If the production cost for x units is $500 + 10x$, find (a) the total revenue function; (b) the profit function; (c) the number of units that maximize the profit; and (d) the price per unit when the marginal revenue is 300.

19 A kitchen specialty company determines that the cost of manufacturing and packaging x pepper grinders per day is $500 + 0.02x + 0.001x^2$ dollars. If each item is sold for $8.00, what rate of production will maximize the profit? What is the maximum daily profit?

20 A company that conducts bus tours found that when the price was $9.00 per person, it averaged 1000 customers per week. When it reduced the price to $7.00 per person, the average number of customers increased to 1500 per week. Assuming that the demand function is linear, what price should be charged to obtain the greatest weekly revenue?

21 When a company sold a certain commodity at $50 per unit, there was a demand for 1000 units per week. After the price rose to $70, the demand dropped to 800 units per week. Assuming that the demand function p is linear, find p and the total revenue function.

22 If a demand function p is defined by $p(x) = ax^2 + b$ where $a < 0$ and $b > 0$, what value of x will maximize the total revenue?

23 Prove that if the demand function of a commodity is linear and decreasing for all $x > 0$, then so is the marginal revenue function.

24 Show that the marginal revenue of a commodity can be found by adding to the selling price the product of the number of units sold and the marginal demand.

25 A sportswear manufacturer determines that the marginal cost of producing x sweatsuits is given in dollars by $20 - 0.015x$. If the cost of producing one suit is $25, find the cost function and the cost of producing 50 suits.

26 If the marginal cost function is given by $2/(x + 6)^{1/3}$, and if the cost of producing two units is $20, find the cost function and the cost of producing 120 units.

27 If, in a certain company, the marginal revenue is given by $x^2 - 6x + 15$, find the total revenue function, and the marginal demand function.

28 Work Exercise 27 if the marginal revenue is given by $4/(x + 2)^{3/2}$.

4.11

REVIEW

Define or discuss each of the following.

1 Local maximum or local minimum of a function

2 Absolute maximum or absolute minimum of a function

3 Critical numbers of a function

4 Rolle's Theorem

5 The Mean Value Theorem

6 The First Derivative Test

7 Endpoint extrema of a function

8 Upward or downward concavity

9 Tests for concavity

10 Point of inflection

11 The Second Derivative Test

12 $\lim_{x \to \infty} f(x) = L$; $\lim_{x \to -\infty} f(x) = L$

13 $\lim_{x \to a} f(x) = \infty$; $\lim_{x \to a} f(x) = -\infty$

14 Vertical and horizontal asymptotes

15 The derivative as a rate of change

16 Velocity and acceleration in rectilinear motion

17 Related rates

18 Antiderivative of a function

19 The power rule for antidifferentiation

20 Differential equation

EXERCISES 4.11

1 Find equations of the tangent and normal lines to the graph of the equation $6x^2 - 2xy + y^3 = 9$ at the point $P(2, -3)$.

2 Find equations of the tangent lines to the graph of the equation $x^2 + y^2 = 1$ that contain the point $P(2, 1)$.

In Exercises 3 and 4, find the x-coordinates of points on the graph of the given equation where the tangent line is either horizontal or vertical.

3 $y = (x^2 + 3x + 2)^{1/3}$

4 $x^2 - 2xy + y^2 - 4 = 0$

5 If $f(x) = x^3 + x^2 + x + 1$, find a number c that satisfies the conclusion of the Mean Value Theorem on the interval $[0, 4]$.

In Exercises 6–8 find the local extrema of f by means of the First Derivative Test. Describe the intervals in which f is increasing or decreasing and sketch the graph of f.

6 $f(x) = -x^3 + 4x^2 - 3x$

7 $f(x) = 1/(1 + x^2)$

8 $f(x) = (4 - x^2)x^{1/3}$

In Exercises 9–12 find the local extrema of f by means of the Second Derivative Test. Discuss concavity, find x-coordinates of points of inflection, and sketch the graph of f.

9 $f(x) = -x^3 + 4x^2 - 3x$

10 $f(x) = 1/(1 + x^2)$

11 $f(x) = 40x^3 - x^6$

12 $f(x) = x^4 + 4x^3 - 2x^2 - 12x + 6$

13 Find the local extrema, discuss concavity, and sketch the graph of f if $f(x) = |x^3 - 6x|$.

14 The position function of a point moving rectilinearly is given by $s(t) = (t^2 + 3t + 1)/(t^2 + 1)$. Find the velocity and acceleration at time t and describe the motion of the point during the time interval $[-2, 2]$.

15 The position of a moving point on a coordinate line is given by $s(t) = 3t^4 - 4t^3 - 12t^2 + 5$. Find the velocity and acceleration at time t and describe the motion of the point for $-3 \le t \le 3$.

16 A ladder 12 ft long leans against a house as shown in the figure. If the lower end slides to the right at a rate of 3 ft/sec, how fast is the ratio y/x changing when the lower end is 6 ft from the house?

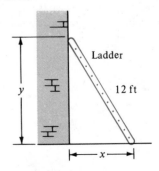

Figure for Exercise 16

17 A stone is thrown directly downward from a height of 900 ft with a velocity of 30 ft/sec. Find its distance above ground after t seconds. What is its velocity after 5 sec? When will it strike the ground?

In Exercises 18–23 find the most general antiderivative of f.

18 $f(x) = \dfrac{8x^2 - 4x + 5}{x^4}$

19 $f(x) = 3x^5 + 2x^3 - x$

20 $f(x) = 100$

21 $f(x) = x^{3/5}(2x - \sqrt{x})$

22 $f(x) = (2x + 1)^3$

23 $f(x) = 0$

24 Solve the differential equation $f''(x) = x^{1/3} - 5$ if $f'(1) = 2$ and $f(1) = -8$.

25 A man wishes to put a fence around a rectangular field and then subdivide this field into three smaller rectangular plots by placing two fences parallel to one of the sides. If he can afford only 1000 yd of fencing, what dimensions will give him the maximum area?

26 A V-shaped water gutter is to be constructed from two rectangular sheets of metal 10 in. wide. Prove that the carrying capacity of the gutter is greatest when the sheets are perpendicular to each other.

27 Find the altitude of the right circular cylinder of maximum curved surface area that can be inscribed in a sphere of radius a.

28 A wire 5 ft long is to be cut into two pieces. One of the pieces is to be bent into the shape of a circle and the other into the shape of a square. Where should the wire be cut in order that the sum of the areas of the circle and square is (a) a maximum; (b) a minimum?

29 The interior of a half-mile race track consists of a rectangle with semicircles at two opposite ends. Find the dimensions that will maximize the area of the rectangle.

30 A rectangle has its vertices on the x-axis, the y-axis, the origin, and the graph of $y = 4 - x^2$. Of all such rectangles, find the dimensions of the one with maximum area.

31 A water tank has the shape of a right circular cone of altitude 12 ft and base radius 4 ft, with vertex at the bottom of the tank. If water is being taken out of the tank at a rate of 10 ft^3/min, how fast is the water level falling when the depth is 5 ft?

32 The ends of a horizontal water trough 10 ft long are isosceles trapezoids with lower base 3 ft, upper base 5 ft, and altitude 2 ft. If the water level is rising at a rate of $\frac{1}{4}$ in./min when the depth is 1 ft, how fast is water entering the trough?

33 Two cars are approaching the same intersection along roads that run at right angles to each other. Car A is traveling at 20 mi/hr and car B is traveling at 40 mi/hr. If, at a certain instant, A is $\frac{1}{4}$ mile from the intersection and B is $\frac{1}{2}$ mile from the intersection, find the rate at which they are approaching one another at that instant.

34 A point $P(x, y)$ moves on the graph of $y^2 = 2x^3$ such that $dy/dt = x$, where t is time. Find dx/dt at the point $(2, 4)$.

35 Boyle's Law states that $pv = c$, where p is pressure, v is volume, and c is a constant. Find a formula for the rate of change of p with respect to v.

36 A railroad bridge is 20 ft above a river and crosses the river at right angles. A man in a train traveling 60 mi/hr passes over the center of the bridge at the same instant that a man in a motor boat traveling 20 mi/hr passes under the center of the bridge (see figure). How fast are the two men separating 10 seconds later?

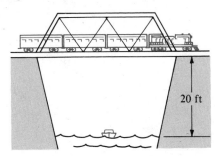

Figure for Exercise 36

Find each of the limits in Exercises 37–44, if it exists.

37 $\displaystyle\lim_{x \to -\infty} \frac{(2x - 5)(3x + 1)}{(x + 7)(4x - 9)}$

38 $\displaystyle\lim_{x \to \infty} \frac{2x + 11}{\sqrt{x + 1}}$

39 $\displaystyle\lim_{x \to -\infty} \frac{6 - 7x}{(3 + 2x)^4}$

40 $\displaystyle\lim_{x \to -3} \sqrt[3]{\frac{x + 3}{x^3 + 27}}$

41 $\displaystyle\lim_{x \to 2/3^+} \frac{x^2}{4 - 9x^2}$

42 $\displaystyle\lim_{x \to 3/5^-} \frac{1}{5x - 3}$

43 $\displaystyle\lim_{x \to 0^+} \left(\sqrt{x} - \frac{1}{\sqrt{x}} \right)$

44 $\displaystyle\lim_{x \to 1} \frac{x - 1}{\sqrt{(x - 1)^2}}$

Find horizontal, vertical, and oblique asymptotes, and sketch the graph of f in Exercises 45–48.

45 $f(x) = \dfrac{3x^2}{9x^2 - 25}$

46 $f(x) = \dfrac{x^2}{(x - 1)^2}$

47 $f(x) = \dfrac{x^2 + 2x - 8}{x + 3}$

48 $f(x) = \dfrac{x^4 - 16}{x^3}$

5
THE DEFINITE INTEGRAL

Calculus consists of two main parts, *differential calculus* and *integral calculus*. Differential calculus is based upon the derivative. In this chapter we define the concept that is the basis for integral calculus: the *definite integral*. One of the most important results we shall discuss is the *Fundamental Theorem of Calculus*. This theorem demonstrates that differential and integral calculus are very closely related.

5.1

AREA

In our development of the definite integral we shall employ sums of many numbers. To express such sums compactly, it is convenient to use **summation notation**. To illustrate, given a collection of numbers $\{a_1, a_2, \ldots, a_n\}$, the symbol $\sum_{i=1}^{n} a_i$ represents their sum, that is,

$$(5.1) \qquad \sum_{i=1}^{n} a_i = a_1 + a_2 + a_3 + \cdots + a_n.$$

In (5.1), the Greek capital letter Σ (sigma) indicates a sum, and the symbol a_i represents the ith term. The letter i is called the **index of summation** or the **summation variable**, and the numbers 1 and n indicate the extreme values of the summation variable.

Example 1   Find $\sum_{i=1}^{4} i^2(i - 3)$.

Solution   In this case, $a_i = i^2(i - 3)$. To find the indicated sum we merely substitute, in succession, the integers 1, 2, 3, and 4 for i and add the resulting terms. Thus,

$$\sum_{i=1}^{4} i^2(i - 3) = 1^2(1 - 3) + 2^2(2 - 3) + 3^2(3 - 3) + 4^2(4 - 3)$$

$$= (-2) + (-4) + 0 + 16 = 10.$$ ∎

The letter used for the summation variable is arbitrary. The sum in Example 1 may be written

$$\sum_{i=1}^{4} i^2(i - 3) = \sum_{k=1}^{4} k^2(k - 3) = \sum_{j=1}^{4} j^2(j - 3)$$

or in many other ways.

If $a_i = c$ for each i then, for example,

$$\sum_{i=1}^{2} a_i = a_1 + a_2 = c + c = 2c = \sum_{i=1}^{2} c$$

$$\sum_{i=1}^{3} a_i = a_1 + a_2 + a_3 = c + c + c = 3c = \sum_{i=1}^{3} c.$$

In general, the following result is true for every positive integer n.

(5.2)
$$\sum_{i=1}^{n} c = nc$$

The domain of the summation variable does not have to begin at 1. For example, the following is self-explanatory:

$$\sum_{i=4}^{8} a_i = a_4 + a_5 + a_6 + a_7 + a_8.$$

Example 2   Find $\sum_{i=0}^{3} \frac{2^i}{(i + 1)}$.

Solution $\sum_{i=0}^{3} \frac{2^i}{(i + 1)} = \frac{2^0}{(0 + 1)} + \frac{2^1}{(1 + 1)} + \frac{2^2}{(2 + 1)} + \frac{2^3}{(3 + 1)}$

$$= 1 + 1 + \frac{4}{3} + 2 = \frac{16}{3}.$$ ∎

Theorem (5.3)

> If n is any positive integer and $\{a_1, a_2, \ldots, a_n\}$ and $\{b_1, b_2, \ldots, b_n\}$ are sets of real numbers, then
>
> (i) $\displaystyle\sum_{i=1}^{n} (a_i + b_i) = \sum_{i=1}^{n} a_i + \sum_{i=1}^{n} b_i$;
>
> (ii) $\displaystyle\sum_{i=1}^{n} ca_i = c\left(\sum_{i=1}^{n} a_i\right)$, for any real number c;
>
> (iii) $\displaystyle\sum_{i=1}^{n} (a_i - b_i) = \sum_{i=1}^{n} a_i - \sum_{i=1}^{n} b_i$.

Proof   To prove formula (i) we begin by writing

$$\sum_{i=1}^{n} (a_i + b_i) = (a_1 + b_1) + (a_2 + b_2) + (a_3 + b_3) + \cdots + (a_n + b_n).$$

The terms on the right may be rearranged to produce

$$\sum_{i=1}^{n} (a_i + b_i) = (a_1 + a_2 + a_3 + \cdots + a_n) + (b_1 + b_2 + b_3 + \cdots + b_n).$$

Expressing the right side in summation notation gives us formula (i).

For formula (ii) we have

$$\sum_{i=1}^{n} (ca_i) = ca_1 + ca_2 + ca_3 + \cdots + ca_n$$

$$= c(a_1 + a_2 + a_3 + \cdots + a_n) = c\left(\sum_{i=1}^{n} a_i\right).$$

The proof of (iii) is left as an exercise.   □

Before we state the definition of the definite integral, it will be instructive to consider the area of a certain region in a plane. Physical examples could be used; however, we prefer to postpone them until the next chapter. It is important to remember that the discussion of area in this section is *not* to be considered as the definition of the definite integral. It is included only to help motivate the work in Section 5.2 in the same way that slopes of tangent lines and velocities were used to motivate the definition of derivative.

It is easy to calculate the area of a plane region bounded by straight lines. For example, the area of a rectangle is the product of its length and width. The area of a triangle is one-half the product of the altitude and base. The area of any polygon can be found by subdividing it into triangles. To find areas of more complicated regions, whose boundaries involve graphs of functions, it is necessary to introduce a limiting process and then use methods of calculus. In particular, let us consider a region S in a coordinate plane, bounded by vertical lines with x-intercepts a and b, by the x-axis, and by the graph of a function f that is continuous and nonnegative on the closed interval $[a, b]$. A region of this type is illustrated in Figure 5.1. Since $f(x) \geq 0$ for every x in $[a, b]$, no part of the graph lies below the x-axis. For convenience we shall refer to S as **the region under the graph of f from a to b.** Our objective is to define the area of S.

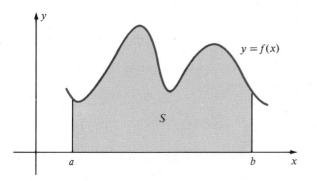

Figure 5.1

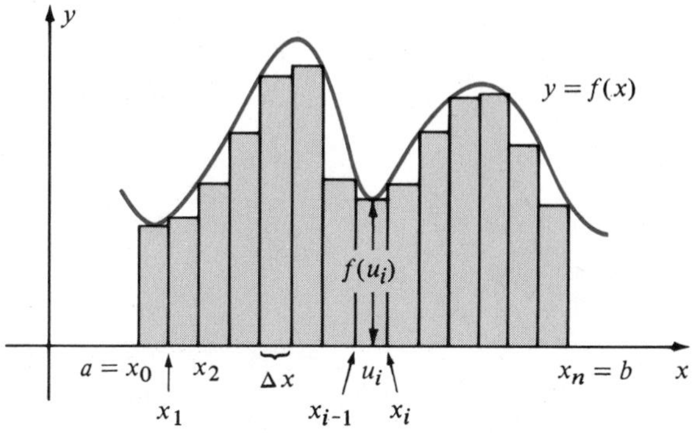

Figure 5.2

If n is any positive integer, let us begin by dividing the interval $[a, b]$ into n subintervals, all having the same length $(b - a)/n$. This can be accomplished by choosing numbers $x_0, x_1, x_2, \ldots, x_n$ where $a = x_0, b = x_n$, and

$$x_i - x_{i-1} = \frac{b - a}{n}$$

for $i = 1, 2, \ldots, n$. If the length $(b - a)/n$ of each subinterval is denoted by Δx, then for each i we have

$$\Delta x = x_i - x_{i-1}, \quad \text{and} \quad x_i = x_{i-1} + \Delta x$$

as illustrated in Figure 5.2.

Note that

$$x_0 = a, \quad x_1 = a + \Delta x, \quad x_2 = a + 2\,\Delta x, \quad \ldots,$$

$$x_i = a + i\,\Delta x, \quad \ldots, \quad x_n = a + n\,\Delta x = b.$$

Since f is continuous on each subinterval $[x_{i-1}, x_i]$, it follows from Theorem (4.3) that f takes on a minimum value at some number u_i in $[x_{i-1}, x_i]$. For each i, let us construct a rectangle of width $\Delta x = x_i - x_{i-1}$ and length equal to the minimum distance $f(u_i)$ from the x-axis to the graph of f, as shown in Figure 5.2. The area of the ith rectangle is $f(u_i)\,\Delta x$. The boundary of the region formed by the totality of these rectangles is called the **inscribed rectangular polygon** associated with the subdivision of $[a, b]$ into n subintervals. The area of this inscribed polygon is the sum of the areas of the n rectangles, that is,

$$f(u_1)\,\Delta x + f(u_2)\,\Delta x + \cdots + f(u_n)\,\Delta x.$$

Using summation notation we may write:

$$\text{Area of inscribed rectangular polygon} = \sum_{i=1}^{n} f(u_i)\,\Delta x$$

where $f(u_i)$ is the minimum value of f on $[x_{i-1}, x_i]$.

Referring to Figure 5.2, we see that if n is very large or, equivalently, if Δx is very small, then the sum of the rectangular areas appears to be close to what we wish to consider as the area of the region S. Indeed, reasoning intuitively, if there exists a number A that has the property that the sum $\sum_{i=1}^{n} f(u_i)\,\Delta x$ gets closer and closer to A as Δx gets closer and closer to 0 (but $\Delta x \neq 0$), then we shall call A the **area** of S and write

$$A = \lim_{\Delta x \to 0} \sum_{i=1}^{n} f(u_i)\,\Delta x.$$

The meaning of this "limit of a sum" is not the same as that for limit of a function, introduced in Chapter 2. To eliminate the hazy phrase "closer and closer" and arrive at a satisfactory definition of A, let us take a slightly different point of view. If A denotes the area of S, then the difference

$$A - \sum_{i=1}^{n} f(u_i)\,\Delta x$$

is the area of the unshaded portion in Figure 5.2 that lies under the graph of f and over the inscribed rectangular polygon. This number may be thought of as the error involved in using the area of the inscribed rectangular polygon to approximate A. If we have the proper notion of area, then we should be able to make this error arbitrarily small by choosing the width Δx of the rectangles sufficiently small. This is the motivation for the following definition of the area A of S, where the notation is the same as that used in the preceding discussion.

Definition (5.4)

> Let f be continuous and nonnegative on $[a, b]$. Let A be a real number and let $f(u_i)$ be the minimum value of f on $[x_{i-1}, x_i]$. The statement
>
> $$A = \lim_{\Delta x \to 0} \sum_{i=1}^{n} f(u_i)\,\Delta x$$
>
> means that for every $\varepsilon > 0$ there corresponds a $\delta > 0$ such that if $0 < \Delta x < \delta$, then
>
> $$A - \sum_{i=1}^{n} f(u_i)\,\Delta x < \varepsilon.$$

If A is the indicated limit and we let $\varepsilon = 10^{-9}$, then Definition (5.4) states that by using sufficiently thin rectangles, the difference between A and the area of the inscribed polygon is less than one-billionth of a square unit! Similarly, if $\varepsilon = 10^{-12}$ we can make this difference less than one-trillionth of a square unit. In general, the difference can be made less than *any* preassigned ε.

If f is continuous on $[a, b]$, then, as it is shown in more advanced texts, a number A satisfying Definition (5.4) actually exists. We shall call A **the area under the graph of f from a to b.**

The area A may also be obtained by means of **circumscribed rectangular polygons** of the type illustrated in Figure 5.3. In this case we select the number v_i in each interval $[x_{i-1}, x_i]$ such that $f(v_i)$ is the maximum value of f on $[x_{i-1}, x_i]$.

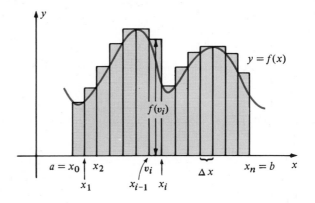

Figure 5.3

We may then write:

$$\text{Area of circumscribed rectangular polygon} = \sum_{i=1}^{n} f(v_i)\,\Delta x.$$

The limit of this sum as $\Delta x \to 0$ is defined as in (5.4), where the only change is that we use

$$\sum_{i=1}^{n} f(v_i)\,\Delta x - A < \varepsilon$$

in the definition, since we want this difference to be nonnegative. It can be proved that the same number A is obtained using either inscribed or circumscribed rectangles.

The following formulas will be useful in some illustrations of Definition (5.4).

(5.5)

(i) $\displaystyle \sum_{i=1}^{n} i = 1 + 2 + \cdots + n = \frac{n(n+1)}{2}$

(ii) $\displaystyle \sum_{i=1}^{n} i^2 = 1^2 + 2^2 + \cdots + n^2 = \frac{n(n+1)(2n+1)}{6}$

(iii) $\displaystyle \sum_{i=1}^{n} i^3 = 1^3 + 2^3 + \cdots + n^3 = \left[\frac{n(n+1)}{2}\right]^2$

These may be proved by means of mathematical induction (see Appendix I).

The next two examples provide specific illustrations of how summation properties may be used in conjunction with Definition (5.4) to find the areas of certain regions in a coordinate plane.

Example 3   If $f(x) = 16 - x^2$, find the area of the region under the graph of f from 0 to 3.

Solution   The region is illustrated in Figure 5.4 where for clarity we have used different scales on the x- and y-axes. If the interval $[0, 3]$ is divided into n equal subintervals, then the length Δx of a typical subinterval is $3/n$. Employing the notation used in Figure 5.2, with $a = 0$ and $b = 3$,

$$x_0 = 0, \; x_1 = \Delta x, \; x_2 = 2(\Delta x), \; \ldots, \; x_i = i(\Delta x), \; \ldots, \; x_n = n(\Delta x) = 3.$$

Using the fact that $\Delta x = 3/n$ we may write

$$x_i = i(\Delta x) = i\left(\frac{3}{n}\right) = \frac{3i}{n}.$$

Since f is decreasing on $[0, 3]$, the number u_i in $[x_{i-1}, x_i]$ at which f takes on its minimum value is always the right-hand endpoint x_i of the subinterval, that is, $u_i = x_i = 3i/n$. Since

$$f(u_i) = f\left(\frac{3i}{n}\right) = 16 - \left(\frac{3i}{n}\right)^2 = 16 - \frac{9i^2}{n^2},$$

the summation in Definition (5.4) may be written

$$\sum_{i=1}^{n} f(u_i)\,\Delta x = \sum_{i=1}^{n} \left(16 - \frac{9i^2}{n^2}\right)\left(\frac{3}{n}\right)$$

$$= \sum_{i=1}^{n} \left(\frac{48}{n} - \frac{27i^2}{n^3}\right).$$

Using Theorem (5.3) and (5.2), the last sum may be simplified as follows:

$$\sum_{i=1}^{n} \frac{48}{n} - \sum_{i=1}^{n} \frac{27i^2}{n^3} = \left(\frac{48}{n}\right)n - \frac{27}{n^3} \sum_{i=1}^{n} i^2.$$

Next, applying (ii) of (5.5), we obtain

$$\sum_{i=1}^{n} f(u_i)\,\Delta x = 48 - \frac{27}{n^3}\left[\frac{n(n+1)(2n+1)}{6}\right]$$

$$= 48 - \frac{9}{2n^3}[2n^3 + 3n^2 + n].$$

In order to find the area, we must now let Δx approach 0. Since $\Delta x = (b - a)/n$, this can be accomplished by letting n increase without bound. Although our discussion of limits involving infinity in Section 4.5 was concerned with a real variable x, a similar discussion can be given for the

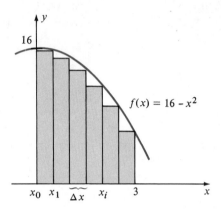

Figure 5.4

variable n, where n is an integer. Assuming that this is true, and that we can replace $\Delta x \to 0$ by $n \to \infty$, we have

$$\lim_{\Delta x \to 0} \sum_{i=1}^{n} f(u_i)\,\Delta x = \lim_{n \to \infty} \left\{ 48 - \frac{9}{2n^3}\,[2n^3 + 3n^2 + n] \right\}$$

$$= \lim_{n \to \infty} 48 - \frac{9}{2} \lim_{n \to \infty} \left[\frac{2n^3 + 3n^2 + n}{n^3} \right]$$

$$= 48 - \frac{9}{2} \lim_{n \to \infty} \left[2 + \frac{3}{n} + \frac{1}{n^2} \right]$$

$$= 48 - \tfrac{9}{2}[2 + 0 + 0] = 48 - 9 = 39. \qquad \blacksquare$$

Because of the assumptions we made, the preceding solution is not completely rigorous. Indeed, one reason for introducing the definite integral is to enable us to solve problems of this type in a simple and precise manner.

The area in the preceding example may also be found by using circumscribed rectangular polygons. In this case we select, in each subinterval $[x_{i-1}, x_i]$, the number $v_i = (i-1)(3/n)$ at which f takes on its maximum value.

The next example illustrates the use of circumscribed rectangles in finding an area.

Example 4 If $f(x) = x^3$, find the area under the graph of f from 0 to b, where $b > 0$.

Solution If we subdivide the interval $[0, b]$ into n equal parts, then as Figure 5.5 illustrates, we obtain a typical circumscribed rectangular polygon where, as in Example 3,

$$\Delta x = \frac{b}{n} \quad \text{and} \quad x_i = i(\Delta x).$$

For clarity different scales have been used on the x- and y-axes.

Since f is an increasing function, the maximum value $f(v_i)$ in the interval $[x_{i-1}, x_i]$ occurs at the right-hand endpoint, that is,

$$v_i = x_i = i(\Delta x) = i\left(\frac{b}{n}\right) = \frac{bi}{n}.$$

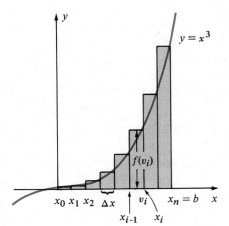

Figure 5.5

The sum of the areas of the circumscribed rectangles is

$$\sum_{i=1}^{n} f(v_i)\,\Delta x = \sum_{i=1}^{n} \left(\frac{bi}{n}\right)^3 \left(\frac{b}{n}\right) = \sum_{i=1}^{n} \frac{b^4}{n^4}\,i^3$$

$$= \frac{b^4}{n^4} \sum_{i=1}^{n} i^3 = \frac{b^4}{n^4} \left[\frac{n(n+1)}{2} \right]^2$$

where the final step follows from (iii) of (5.5). The reader may now verify that

$$\sum_{i=1}^{n} f(v_i)\,\Delta x = \frac{b^4}{4}\left(\frac{n^4 + 2n^3 + n^2}{n^4}\right) = \frac{b^4}{4}\left(1 + \frac{2}{n} + \frac{1}{n^2}\right).$$

If we let Δx approach 0, then n increases without bound and the expression in parentheses approaches 1. It follows that the area under the graph is

$$\lim_{\Delta x \to 0} \sum_{i=1}^{n} f(v_i)\,\Delta x = \frac{b^4}{4}. \qquad \blacksquare$$

EXERCISES 5.1

Determine the sums in Exercises 1–10.

1 $\displaystyle\sum_{i=1}^{5} (3i - 10)$

2 $\displaystyle\sum_{i=1}^{6} (9 - 2i)$

3 $\displaystyle\sum_{j=1}^{4} (j^2 + 1)$

4 $\displaystyle\sum_{n=1}^{10} [1 + (-1)^n]$

5 $\displaystyle\sum_{k=0}^{5} k(k-1)$

6 $\displaystyle\sum_{k=0}^{4} (k-2)(k-3)$

7 $\displaystyle\sum_{i=1}^{8} 2^i$

8 $\displaystyle\sum_{s=1}^{6} (5/s)$

9 $\displaystyle\sum_{i=1}^{50} 10$

10 $\displaystyle\sum_{k=1}^{1000} 2$

11 Prove (iii) of Theorem (5.3).

12 Extend (i) of Theorem (5.3) to $\sum_{i=1}^{n} (a_i + b_i + c_i)$.

13 Find $\sum_{i=1}^{n} (i^2 + 3i + 5)$. (*Hint*: Write the sum as $\sum_{i=1}^{n} i^2 + 3\sum_{i=1}^{n} i + \sum_{i=1}^{n} 5$ and employ (5.5) and (5.2).)

14 Find $\displaystyle\sum_{i=1}^{n} (3i^2 - 2i + 1)$.

15 Find $\displaystyle\sum_{k=1}^{n} (2k - 3)^2$.

16 Find $\displaystyle\sum_{k=1}^{n} (k^3 + 2k^2 - k + 4)$.

In Exercises 17–26 find the area under the graph of f from a to b using (a) inscribed rectangles; (b) circumscribed rectangles.

In each case sketch the graph and typical rectangles, labeling the drawing as in Figures 5.4 and 5.5.

17 $f(x) = 2x + 3$; $a = 0, b = 4$

18 $f(x) = 8 - 3x$; $a = 0, b = 2$

19 $f(x) = x^2$; $a = 0, b = 5$

20 $f(x) = x^2 + 2$; $a = 1, b = 3$

21 $f(x) = 3x^2 + 5$; $a = 1, b = 4$ (*Hint*: $x_i = 1 + (3i/n)$.)

22 $f(x) = 7$; $a = -2, b = 6$

23 $f(x) = 9 - x^2$; $a = 0, b = 3$

24 $f(x) = 4x^2 + 3x + 2$; $a = 1, b = 5$

25 $f(x) = x^3 + 1$; $a = 1, b = 2$

26 $f(x) = 4x + x^3$; $a = 0, b = 2$

27 Use Definition (5.4) to prove that the area of a right triangle of altitude h and base b is $\frac{1}{2}bh$. (*Hint*: Consider the area under the graph of $f(x) = (h/b)x$ from 0 to b.)

28 If $f(x) = px^2 + qx + r$ and $f(x) \geq 0$ for all x, prove that the area under the graph of f from 0 to b is

$$p(b^3/3) + q(b^2/2) + rb.$$

29 Use mathematical induction to prove (5.5).

30 Prove (i) of (5.5) by writing

$$S = 1 + \quad 2 \quad + \cdots + n$$
$$S = n + (n-1) + \cdots + 1$$

and then adding corresponding sides of these equations.

5.2

DEFINITION OF DEFINITE INTEGRAL

Limiting processes similar to the one for areas used in the preceding section arise frequently in mathematics and its applications. The situations that occur often lead to limits of the form

$$\lim_{\Delta x \to 0} \sum_{i=1}^{n} f(w_i) \, \Delta x.$$

A special instance of this limit appeared in Definition (5.4); however, in the general case, there are several major differences from our work with areas. In our discussion of area in the previous section we made the following assumptions:

1 The function f is continuous on a closed interval $[a, b]$.

2 $f(x)$ is nonnegative for all x in $[a, b]$.

3 All the subintervals $[x_{i-1}, x_i]$ determined by the subdivision of $[a, b]$ have the same length Δx.

4 The numbers w_i are chosen such that $f(w_i)$ is always the minimum (or maximum) value of f on $[x_{i-1}, x_i]$.

These four conditions are not always present in applied problems. For this reason it is necessary to allow the following changes in 1–4.

1' The function f may be discontinuous at some numbers in $[a, b]$.

2' $f(x)$ may be negative for some x in $[a, b]$.

3' The lengths of the subintervals $[x_{i-1}, x_i]$ may be different.

4' The number w_i is *any* number in $[x_{i-1}, x_i]$.

We shall begin by introducing some new terminology and notation. A **partition** P of a closed interval $[a, b]$ is any decomposition of $[a, b]$ into subintervals of the form

$$[x_0, x_1], [x_1, x_2], [x_2, x_3], \ldots, [x_{n-1}, x_n]$$

where n is a positive integer and the x_i are numbers such that

$$a = x_0 < x_1 < x_2 < x_3 < \cdots < x_{n-1} < x_n = b.$$

The length of the ith subinterval $[x_{i-1}, x_i]$ will be denoted by Δx_i, that is,

$$\Delta x_i = x_i - x_{i-1}.$$

A typical partition of $[a, b]$ is illustrated in Figure 5.6. The largest of the numbers $\Delta x_1, \Delta x_2, \ldots, \Delta x_n$ is called the **norm** of the partition P and is denoted by $\|P\|$.

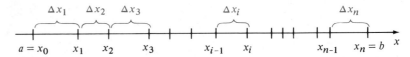

Figure 5.6 A partition of $[a, b]$

The following concept, named after the mathematician G. F. B. Riemann (1826–1866), is fundamental for the definition of the definite integral.

Definition (5.6)

> Let f be a function that is defined on a closed interval $[a, b]$ and let P be a partition of $[a, b]$. A **Riemann sum** of f for P is any expression R_P of the form
>
> $$R_P = \sum_{i=1}^{n} f(w_i)\, \Delta x_i$$
>
> where w_i is some number in $[x_{i-1}, x_i]$ for $i = 1, 2, \ldots, n$.

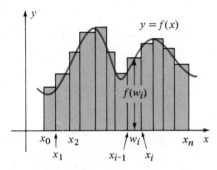

Figure 5.7

The sum in Definition (5.4), which represents a sum of areas of certain inscribed rectangles, is a special type of Riemann sum. In Definition (5.6), $f(w_i)$ is not necessarily a maximum or minimum value of f on $[x_{i-1}, x_i]$. Thus, if we construct a rectangle of length $f(w_i)$ and width Δx_i as illustrated in Figure 5.7, the rectangle may be neither inscribed nor circumscribed. Since $f(x)$ may be negative for some x in $[a, b]$, some terms of R_P in Definition (5.6) may be negative. Consequently, a Riemann sum does not always represent a sum of areas of rectangles. It is possible to interpret a Riemann sum geometrically as follows. If R_P is defined as in (5.6), then for each subinterval $[x_{i-1}, x_i]$ let us construct a horizontal line segment through the point $(w_i, f(w_i))$, thereby obtaining a collection of rectangles. If $f(w_i)$ is positive, the rectangle lies above the x-axis as illustrated by the shaded rectangles in Figure 5.8, and the product $f(w_i)\, \Delta x_i$ is the area of this rectangle. If $f(w_i)$ is negative, then the rectangle lies below the x-axis as illustrated by the unshaded rectangles in Figure 5.8. In this case the product $f(w_i)\, \Delta x_i$ is the *negative* of the area of a rectangle. It follows that R_P is the sum of the areas of the rectangles that lie above the x-axis and the *negatives* of the areas of the rectangles that lie below the x-axis.

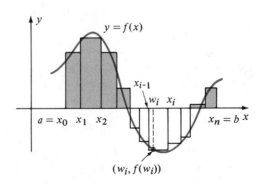

Figure 5.8

Example   Suppose $f(x) = 8 - (x^2/2)$ and P is the partition of $[0, 6]$ into the five subintervals determined by $x_0 = 0$, $x_1 = 1.5$, $x_2 = 2.5$, $x_3 = 4.5$, $x_4 = 5$, and $x_5 = 6$. Find (a) the norm of the partition and (b) the Riemann sum R_P if $w_1 = 1$, $w_2 = 2$, $w_3 = 3.5$, $w_4 = 5$, and $w_5 = 5.5$.

Solution   The graph of f is sketched in Figure 5.9. Also shown in the figure are the points on the x-axis that correspond to x_i and the rectangles of lengths $|f(w_i)|$ for $i = 1, 2, 3, 4$, and 5.

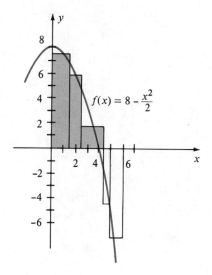

Figure 5.9

Thus, $\Delta x_1 = 1.5, \Delta x_2 = 1, \Delta x_3 = 2, \Delta x_4 = 0.5, \Delta x_5 = 1$

and hence the norm $\|P\|$ of the partition is Δx_3, or 2. By Definition (5.6),

$$
\begin{aligned}
R_P &= f(w_1)\,\Delta x_1 + f(w_2)\,\Delta x_2 + f(w_3)\,\Delta x_3 + f(w_4)\,\Delta x_4 + f(w_5)\,\Delta x_5 \\
&= f(1)(1.5) + f(2)(1) + f(3.5)(2) + f(5)(0.5) + f(5.5)(1) \\
&= (7.5)(1.5) + (6)(1) + (1.875)(2) + (-4.5)(0.5) + (-7.125)(1)
\end{aligned}
$$

which reduces to $R_P = 11.625$.   ■

In the future we shall not always specify the number n of subintervals in a partition P of $[a, b]$. In this event a Riemann sum (5.6) will be written

$$R_P = \sum_i f(w_i)\,\Delta x_i$$

where it is understood that terms of the form $f(w_i)\,\Delta x_i$ are to be summed over all subintervals $[x_{i-1}, x_i]$ of the partition P.

In a manner similar to that used in formulating Definition (5.4), we may define what is meant by

$$\lim_{\|P\| \to 0} \sum_i f(w_i)\,\Delta x_i = I$$

where I is a real number. Intuitively, it will mean that if the norm $\|P\|$ of the partition P is close to 0, then every Riemann sum for P is close to I.

Definition (5.7)

Let f be a function that is defined on a closed interval $[a, b]$, and let I be a real number. The statement

$$\lim_{\|P\| \to 0} \sum_i f(w_i) \Delta x_i = I$$

means that for every $\varepsilon > 0$ there exists a $\delta > 0$, such that if P is a partition of $[a, b]$ with $\|P\| < \delta$, then

$$\left| \sum_i f(w_i) \Delta x_i - I \right| < \varepsilon$$

for any choice of numbers w_i in the subintervals $[x_{i-1}, x_i]$ of P. The number I is called a **limit of a sum**.

Note that for every $\delta > 0$ there are infinitely many partitions P such that $\|P\| < \delta$. Moreover, for each such partition P there are infinitely many ways of choosing the numbers w_i in $[x_{i-1}, x_i]$. Consequently, there may be an infinite number of different Riemann sums associated with *each* partition P. However, if the limit I exists, then for any ε, each of these Riemann sums is within ε units of I, provided a small enough norm is chosen. Although Definition (5.7) differs from the definition of limit of a function, a proof similar to that given in Appendix II may be used to show that if the limit I exists, then it is unique.

We next define the definite integral as a limit of a sum.

Definition (5.8)

Let f be a function that is defined on a closed interval $[a, b]$. The **definite integral of f from a to b**, denoted by $\int_a^b f(x)\, dx$, is

$$\int_a^b f(x)\, dx = \lim_{\|P\| \to 0} \sum_i f(w_i) \Delta x_i$$

provided the limit exists.

If the definite integral of f from a to b exists, then f is said to be **integrable** on the closed interval $[a, b]$, or we say that the integral $\int_a^b f(x)\, dx$ **exists**. The process of finding the number represented by the limit is called **evaluating the integral**.

The symbol $\int$ in Definition (5.8) is called an **integral sign**. It may be thought of as an elongated letter S (the first letter of the word *sum*) and is used to indicate the connection between definite integrals and Riemann sums. The numbers a and b are referred to as the **limits of integration**, a being called the

lower limit and b the **upper limit**. Note that the word *limit* in this terminology is used in conjunction with the smallest and largest numbers in the interval $[a, b]$ and has no connection with any definitions of limits given earlier. The expression $f(x)$ that appears to the right of the integral sign (we sometimes say "*behind* the integral sign") is called the **integrand**. Finally, the symbol dx that follows $f(x)$ should not be confused with the differential of x defined in Section 3.4. At this stage of our work it is merely used to indicate the variable. Later in the text the use of the differential symbol will have certain practical advantages.

Letters other than x may be used in the notation for the definite integral. This follows from the fact that when we describe a function, the symbol used for the independent variable is immaterial. Thus, if f is integrable on $[a, b]$, then

$$\int_a^b f(x)\, dx = \int_a^b f(s)\, ds = \int_a^b f(t)\, dt$$

etc. For this reason the letter x in Definition (5.8) is sometimes referred to as a **dummy variable**.

Whenever an interval $[a, b]$ is employed it is assumed that $a < b$. Consequently, Definition (5.8) does not take into account the cases in which the lower limit of integration is greater than or equal to the upper limit. The definition may be extended to include the case where the lower limit is greater than the upper limit, as follows.

Definition (5.9)

$$\text{If } c > d, \text{ then } \int_c^d f(x)\, dx = -\int_d^c f(x)\, dx.$$

In words, Definition (5.9) may be phrased as "interchanging the limits of integration changes the sign of the integral." One reason for the form of Definition (5.9) will become apparent later, after we have considered the Fundamental Theorem of Calculus.

The case in which the lower and upper limits of integration are equal is covered by the next definition.

Definition (5.10)

$$\text{If } f(a) \text{ exists, then } \int_a^a f(x)\, dx = 0.$$

Not every function f is integrable. For example, if $f(x)$ becomes positively or negatively infinite at some number in $[a, b]$, then the definite integral does not exist. To illustrate, let f be defined on $[a, b]$ and $\lim_{x \to a+} f(x) = \infty$. In the first subinterval $[x_0, x_1]$ of any partition P of $[a, b]$, a number w_1 can be found such that $f(w_1) \Delta x_1$ is larger than any given number M. It follows that for any partition P, we can form a Riemann sum $\sum_i f(w_i) \Delta x_i$ that is arbitrarily large. Hence if I is any real number and P any partition of $[a, b]$, then there exist Riemann sums R_P such that $R_P - I$ is arbitrarily large. This implies that f is not integrable. A similar argument can be given if f becomes infinite at any other number in $[a, b]$. Consequently, *if a function f is integrable on $[a, b]$, then it is bounded on $[a, b]$*; that is, there is a real number M such that $|f(x)| \leq M$ for all x in $[a, b]$.

The reader may be tempted to conjecture that if a function is discontinuous somewhere in $[a, b]$, then it is not integrable. This conjecture is false. Definite integrals of discontinuous functions may or may not exist, depending on the nature of the discontinuities. However, according to the next theorem, continuous functions are *always* integrable.

Theorem (5.11)

> If f is continuous on $[a, b]$, then f is integrable on $[a, b]$.

A proof of Theorem 5.11 may be found in texts on advanced calculus.

If f is integrable, then the limit in Definition (5.8) exists for all choices of w_i in $[x_{i-1}, x_i]$. This fact allows us to specialize w_i if we wish to do so. For example, we could always choose w_i as the smallest number x_{i-1} in the subinterval, or as the largest number x_i, or as the midpoint of the subinterval, or as the number that always produces the minimum or maximum value in $[x_{i-1}, x_i]$, etc. In addition, since the limit is independent of the partitions P of $[a, b]$ (provided that $\|P\|$ is sufficiently small) we may specialize the partitions to the case in which all the subintervals $[x_{i-1}, x_i]$ have the same length Δx. A partition of this type is called a **regular partition**. We will see in the next chapter that the specializations we have described are often used in applications. As an immediate illustration, we have the following important result.

Theorem (5.12)

> If f is continuous and $f(x) \geq 0$ for all x in $[a, b]$, then the area A of the region under the graph of f from a to b is
>
> $$A = \int_a^b f(x)\, dx.$$

Proof   The area A was defined, using (5.4), as the limit of the sum $\sum_i f(u_i) \Delta x$, where $f(u_i)$ is the minimum value of f in $[x_{i-1}, x_i]$. Since this is a special type of Riemann sum, the conclusion follows from Definition (5.8). $\square$

EXERCISES 5.2

In Exercises 1–4, the given numbers $\{x_0, x_1, x_2, \ldots, x_n\}$ determine a partition P of the indicated interval $[a, b]$. Find $\Delta x_1, \Delta x_2, \ldots, \Delta x_n$ and the norm $\|P\|$ of the partition.

1 $[0, 5]$; $\{0, 1.1, 2.6, 3.7, 4.1, 5\}$

2 $[2, 6]$; $\{2, 3, 3.7, 4, 5.2, 6\}$

3 $[-3, 1]$; $\{-3, -2.7, -1, 0.4, 0.9, 1\}$

4 $[1, 4]$; $\{1, 1.6, 2, 3.5, 4\}$

In Exercises 5 and 6 find the Riemann sum R_P of f, where P is the regular partition of $[1, 5]$ into the four equal subintervals determined by $x_0 = 1, x_1 = 2, x_2 = 3, x_3 = 4, x_4 = 5$, and

(a) w_i is the right-hand endpoint x_i of $[x_{i-1}, x_i]$.
(b) w_i is the left-hand endpoint x_{i-1} of $[x_{i-1}, x_i]$.
(c) w_i is the midpoint of $[x_{i-1}, x_i]$.

5 $f(x) = 2x + 3$

6 $f(x) = 3 - 4x$

7 If $f(x) = 8 - (x^2/2)$, find the Riemann sum R_P of f where P is the regular partition of $[0, 6]$ into the six equal subintervals determined by $x_0 = 0, x_1 = 1, x_2 = 2, x_3 = 3, x_4 = 4, x_5 = 5, x_6 = 6$, and w_i is the midpoint of the interval $[x_{i-1}, x_i]$.

8 If $f(x) = 8 - (x^2/2)$, find the Riemann sum R_P of f where P is the partition of $[0, 6]$ into the four subintervals determined by $x_0 = 0, x_1 = 1.5, x_2 = 3, x_3 = 4.5, x_4 = 6$, and $w_1 = 1, w_2 = 2, w_3 = 4$, and $w_4 = 5$.

9 Suppose $f(x) = x^3$ and P is the partition of $[-2, 4]$ into the four subintervals determined by $x_0 = -2$, $x_1 = 0, x_2 = 1, x_3 = 3$, and $x_4 = 4$. Find the Riemann sum R_P if $w_1 = -1, w_2 = 1, w_3 = 2$, and $w_4 = 4$.

10 Suppose $f(x) = \sqrt{x}$ and P is the partition of $[1, 16]$ into the five subintervals determined by $x_0 = 1, x_1 = 3, x_2 = 5, x_3 = 7, x_4 = 9$, and $x_5 = 16$. Find the Riemann sum R_P if $w_1 = 1, w_2 = 4, w_3 = 5, w_4 = 9$, and $w_5 = 9$.

In Exercises 11–14, use Definition (5.8) to express each limit as a definite integral on the indicated closed interval $[a, b]$.

11 $\lim\limits_{\|P\| \to 0} \sum\limits_{i=1}^{n} (3w_i^2 - 2w_i + 5) \Delta x_i$; $[-1, 2]$

12 $\lim\limits_{\|P\| \to 0} \sum\limits_{i=1}^{n} \pi(w_i^2 - 4) \Delta x_i$; $[2, 3]$

13 $\lim\limits_{\|P\| \to 0} \sum\limits_{i=1}^{n} 2\pi w_i(1 + w_i^3) \Delta x_i$; $[0, 4]$

14 $\lim\limits_{\|P\| \to 0} \sum\limits_{i=1}^{n} (\sqrt[3]{w_i} + 4w_i) \Delta x_i$; $[-4, -3]$

15 If $\int_1^4 \sqrt{x}\, dx = \frac{14}{3}$, find $\int_4^1 \sqrt{x}\, dx$.

16 Find $\int_3^3 x^2\, dx$.

17 If $\int_1^2 (5x^4 - 1)\, dx = 30$, find $\int_1^2 (5r^4 - 1)\, dr$.

18 If $\int_{-1}^8 \sqrt[3]{s}\, ds = \frac{45}{4}$, find $\int_8^{-1} \sqrt[3]{t}\, dt$.

In Exercises 19–22 find the value of the definite integral by interpreting it as the area under the graph of a function f.

19 $\int_{-3}^2 (2x + 6)\, dx$

20 $\int_{-1}^2 (7 - 3x)\, dx$

21 $\int_0^3 \sqrt{9 - x^2}\, dx$

22 $\int_{-a}^a \sqrt{a^2 - x^2}\, dx, a > 0$

23 Find $\int_0^5 x^3\, dx$. (*Hint*: See Example 4 of Section 5.1.)

24 Let c be an arbitrary real number and suppose $f(x) = c$ for all x. If P is any partition of $[a, b]$, show that every Riemann sum R_P of f equals $c(b - a)$. Use this fact to prove that $\int_a^b c\, dx = c(b - a)$. Interpret this geometrically if $c > 0$.

25 Give an example of a function that is continuous on the interval $(0, 1)$ such that $\int_0^1 f(x)\, dx$ does not exist. Why doesn't this contradict Theorem (5.11)?

26 Give an example of a function that is not continuous on $[0, 1]$ such that $\int_0^1 f(x)\, dx$ exists.

5.3

PROPERTIES OF THE DEFINITE INTEGRAL

This section contains some fundamental properties of the definite integral. Most of the proofs are rather technical and have been placed in Appendix II, where the reader may study them whenever time permits.

Theorem (5.13)

$$\int_a^b k\,dx = k(b-a)$$

Proof   If f is the constant function defined by $f(x) = k$ for all x in $[a, b]$ and P is a partition of $[a, b]$, then for every Riemann sum of f,

$$\sum_i f(w_i)\,\Delta x_i = \sum_i k\,\Delta x_i = k\sum_i \Delta x_i = k(b-a),$$

since the sum $\sum_i \Delta x_i$ is the length of the interval $[a, b]$. Consequently,

$$\left|\sum_i f(w_i)\,\Delta x_i - k(b-a)\right| = |k(b-a) - k(b-a)| = 0,$$

which is less than any positive number ε *regardless* of the size of $\|P\|$. Therefore, by Definition (5.7),

$$\lim_{\|P\|\to 0}\sum_i f(w_i)\,\Delta x_i = \lim_{\|P\|\to 0}\sum_i k\,\Delta x_i = k(b-a).$$

By Definition (5.8), this means that

$$\int_a^b f(x)\,dx = \int_a^b k\,dx = k(b-a). \qquad \square$$

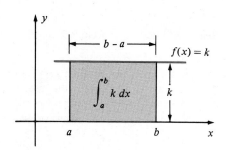

Figure 5.10

Theorem (5.13) is in agreement with the discussion of area in Section 5.1, for if $k > 0$, then the graph of f is a horizontal line k units above the x-axis, and the region under the graph from a to b is a rectangle with sides of length k and $b - a$ as illustrated in Figure 5.10. Hence the area $\int_a^b f(x)\,dx$ of the rectangle is $k(b - a)$.

Example 1 Evaluate $\displaystyle\int_{-2}^3 7\,dx.$

Solution   Using (5.13),

$$\int_{-2}^3 7\,dx = 7[3-(-2)] = 7(5) = 35. \qquad \blacksquare$$

For the special case of (5.13) with $k = 1$ we shall abbreviate the integrand by writing

$$\int_a^b dx = b - a.$$

If a function f is integrable on $[a, b]$ and k is a real number, then by (ii) of Theorem (5.3) a Riemann sum of the function kf may be written

$$\sum_i kf(w_i)\, \Delta x_i = k \sum_i f(w_i)\, \Delta x_i.$$

It is proved in Appendix II that the limit of the sum on the left is equal to k times the limit of the sum on the right. Restating this fact in terms of definite integrals gives us the next theorem.

Theorem (5.14)

> If f is integrable on $[a, b]$ and k is any real number, then kf is integrable on $[a, b]$ and
>
> $$\int_a^b kf(x)\, dx = k \int_a^b f(x)\, dx.$$

The conclusion of Theorem (5.14) is sometimes stated "*a constant factor in the integrand may be taken outside the integral sign.*" It is *not* permissible to take expressions involving variables outside the integral sign in this manner.

If two functions f and g are defined on $[a, b]$, then by (i) of Theorem (5.3), a Riemann sum of $f + g$ may be written

$$\sum_i [f(w_i) + g(w_i)]\, \Delta x_i = \sum_i f(w_i)\, \Delta x_i + \sum_i g(w_i)\, \Delta x_i.$$

It can be shown that if f and g are integrable, then the limit of the sum on the left may be found by adding the limits of the two sums on the right. This fact is stated in integral form in (i) of the next theorem. A proof may be found in Appendix II. The analogous result for differences is stated in (ii).

Theorem (5.15)

> If f and g are integrable on $[a, b]$, then $f + g$ and $f - g$ are integrable on $[a, b]$ and
>
> (i) $\displaystyle \int_a^b [f(x) + g(x)]\, dx = \int_a^b f(x)\, dx + \int_a^b g(x)\, dx.$
>
> (ii) $\displaystyle \int_a^b [f(x) - g(x)]\, dx = \int_a^b f(x)\, dx - \int_a^b g(x)\, dx.$

Theorem (5.15) may be extended to any finite number of functions. Thus, if $f_1, f_2, \ldots, f_n$ are integrable on $[a, b]$, then so is their sum and

$$\int_a^b [f_1(x) + f_2(x) + \cdots + f_n(x)] \, dx$$

$$= \int_a^b f_1(x) \, dx + \int_a^b f_2(x) \, dx + \cdots + \int_a^b f_n(x) \, dx.$$

Example 2 Given $\int_0^2 x^3 \, dx = 4$ and $\int_0^2 x \, dx = 2$, evaluate $\int_0^2 (5x^3 - 3x + 6) \, dx$.

Solution   We may proceed as follows.

$$\int_0^2 (5x^3 - 3x + 6) \, dx = \int_0^2 5x^3 \, dx - \int_0^2 3x \, dx + \int_0^2 6 \, dx$$

$$= 5 \int_0^2 x^3 \, dx - 3 \int_0^2 x \, dx + 6(2 - 0)$$

$$= 5(4) - 3(2) + 12 = 26 \qquad \blacksquare$$

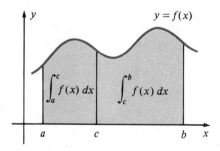

Figure 5.11

If f is continuous on $[a, b]$ and $f(x) \geq 0$ for all x in $[a, b]$, then by Theorem (5.12), the integral $\int_a^b f(x) \, dx$ is the area under the graph of f from a to b. In like manner, if $a < c < b$, then the integrals $\int_a^c f(x) \, dx$ and $\int_c^b f(x) \, dx$ are the areas under the graph of f from a to c and from c to b, respectively, as illustrated in Figure 5.11. Since the area from a to b is the sum of the two smaller areas, we have

$$\int_a^b f(x) \, dx = \int_a^c f(x) \, dx + \int_c^b f(x) \, dx.$$

The next theorem shows that the last equality is true under a more general hypothesis. The proof is given in Appendix II.

Theorem (5.16)

> If $a < c < b$, and if f is integrable on both $[a, c]$ and $[c, b]$, then f is integrable on $[a, b]$ and
>
> $$\int_a^b f(x) \, dx = \int_a^c f(x) \, dx + \int_c^b f(x) \, dx.$$

The following result shows that Theorem (5.16) can be generalized to the case where c is not necessarily between a and b.

Theorem (5.17)

> If f is integrable on a closed interval and if a, b, and c are any three numbers in the interval, then
>
> $$\int_a^b f(x)\,dx = \int_a^c f(x)\,dx + \int_c^b f(x)\,dx.$$

Proof   If a, b, and c are all different, then there are six possible ways of ordering these three numbers. The theorem should be verified for each of these cases as well as for the cases in which two, or all three, of the numbers are equal. We shall verify one case and leave the remaining parts as exercises. Thus, suppose the numbers are ordered such that $c < a < b$. Using Theorem (5.16),

$$\int_c^b f(x)\,dx = \int_c^a f(x)\,dx + \int_a^b f(x)\,dx$$

which, in turn, may be written

$$\int_a^b f(x)\,dx = -\int_c^a f(x)\,dx + \int_c^b f(x)\,dx.$$

The desired conclusion now follows, since interchanging the limits of integration changes the sign of the integral (see Definition (5.9)).

If f and g are continuous on $[a, b]$ and $f(x) \geq g(x) \geq 0$ for all x in $[a, b]$, then the area under the graph of f from a to b is greater than or equal to the area under the graph of g from a to b. The corollary to the next theorem is a generalization of this fact to arbitrary integrable functions. The proof of the theorem is given in Appendix II.

Theorem (5.18)

> If f is integrable on $[a, b]$ and if $f(x) \geq 0$ for all x in $[a, b]$, then
>
> $$\int_a^b f(x)\,dx \geq 0.$$

Corollary (5.19)

> If f and g are integrable on $[a, b]$ and $f(x) \geq g(x)$ for all x in $[a, b]$, then
>
> $$\int_a^b f(x)\,dx \geq \int_a^b g(x)\,dx.$$

Proof   Since $f - g$ is integrable and $f(x) - g(x) \geq 0$ for all x in $[a, b]$, then by Theorem (5.18),

$$\int_a^b [f(x) - g(x)] \, dx \geq 0.$$

Applying (ii) of Theorem (5.15) leads to the desired conclusion.

It can be shown that Theorem (5.18) and Corollary (5.19) are true if $\geq$ is replaced by $>$ (see Exercise 28).

EXERCISES 5.3

Evaluate the definite integrals in Exercises 1–6.

1 $\int_{-2}^4 5 \, dx$ **2** $\int_1^{10} \sqrt{2} \, dx$

3 $\int_6^2 3 \, dx$ **4** $\int_4^{-3} dx$

5 $\int_{-1}^1 dx$ **6** $\int_2^2 100 \, dx$

It will follow from our work in Section 5.5 that

$$\int_1^4 x^2 \, dx = 21, \quad \int_1^4 x \, dx = \tfrac{15}{2}, \quad \text{and} \quad \int_1^4 \sqrt{x} \, dx = \tfrac{14}{3}.$$

Use these facts to evaluate the integrals in Exercises 7–14.

7 $\int_1^4 (3x^2 + 5) \, dx$ **8** $\int_1^4 (6x - 1) \, dx$

9 $\int_1^4 (2 - 9x - 4x^2) \, dx$ **10** $\int_1^4 (3x + 2)^2 \, dx$

11 $\int_4^1 \sqrt{5x} \, dx$ **12** $\int_1^4 2x(x + 1) \, dx$

13 $\int_1^4 (\sqrt{x} - 5)^2 \, dx$ **14** $\int_4^1 (3\sqrt{x} + 1)(\sqrt{x} - 2) \, dx$

Verify the inequalities in Exercises 15 and 16 without evaluating the integrals.

15 $\int_1^2 (3x^2 + 4) \, dx \geq \int_1^2 (2x^2 + 5) \, dx$

16 $\int_2^4 (5x^2 - 4\sqrt{x} + 2) \, dx > 0$

In Exercises 17 and 18 assume that f is integrable on $[a, b]$.

17 If $f(x) \leq M$ for all x in $[a, b]$, prove that

$$\int_a^b f(x) \, dx \leq M(b - a).$$

Illustrate this result graphically.

18 If $m \leq f(x)$ for all x in $[a, b]$, prove that

$$m(b - a) \leq \int_a^b f(x) \, dx.$$

Illustrate this result graphically.

In Exercises 19–22 express each sum or difference as a single integral of the form $\int_a^b f(x) \, dx$.

19 $\int_5^1 f(x) \, dx + \int_{-3}^5 f(x) \, dx$

20 $\int_4^1 f(x) \, dx + \int_6^4 f(x) \, dx$

21 $\int_c^{c+h} f(x) \, dx - \int_c^h f(x) \, dx$

22 $\int_{-2}^6 f(x) \, dx - \int_{-2}^2 f(x) \, dx$

23 Prove (ii) of Theorem (5.15). (*Hint*: Let $f(x) - g(x) = f(x) + [-g(x)]$ and use (i) of (5.15) and Theorem (5.14).)

24 If f and g are integrable functions on $[a, b]$ and if p and q are any real numbers, prove that

$$\int_a^b [pf(x) + qg(x)] \, dx = p \int_a^b f(x) \, dx + q \int_a^b g(x) \, dx.$$

(*Hint*: Use Theorems (5.15) and (5.14).)

25 Complete the proof of Theorem (5.17) by considering all other orderings of the numbers a, b, and c.

26 Use Theorem (5.17) to prove that if f is integrable on a closed interval and if $c_1, c_2, \ldots, c_n$ are any numbers in the interval, then

$$\int_{c_1}^{c_n} f(x)\, dx = \sum_{i=1}^{n-1} \int_{c_i}^{c_{i+1}} f(x)\, dx.$$

27 If f is integrable on $[a, b]$ and if $f(x) \le 0$ for all x in $[a, b]$, prove that $\int_a^b f(x)\, dx \le 0$. (*Hint*: If $f(x) \le 0$, then $-f(x) \ge 0$.)

28 If f is continuous on $[a, b]$ and $f(x) > 0$ for all x in $[a, b]$, prove that $\int_a^b f(x)\, dx > 0$. (*Hint*: Let $f(u)$ be the minimum

value of f on $[a, b]$ and prove that $R_P \ge f(u)(b - a)$, where P is any partition of $[a, b]$ and R_P is any Riemann sum of f for P.)

29 If f is continuous on $[a, b]$, prove that

$$\left| \int_a^b f(x)\, dx \right| \le \int_a^b |f(x)|\, dx.$$

30 Refer to Exercise 29. Suppose $f_1, f_2, \ldots, f_n$ are continuous functions on $[a, b]$, where n is any positive integer. Use mathematical induction to prove that

$$\left| \sum_{i=1}^{n} \int_a^b f_i(x)\, dx \right| \le \sum_{i=1}^{n} \int_a^b |f_i(x)|\, dx.$$

5.4

THE MEAN VALUE THEOREM FOR DEFINITE INTEGRALS

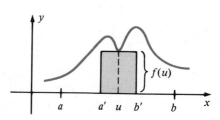

Figure 5.12

Suppose f is continuous and $f(x) \ge 0$ for all x in a closed interval $[a, b]$. If $f(c) > 0$ for some c in $[a, b]$, then $\lim_{x \to c} f(x) > 0$ and, by an argument similar to that used in the proof of Theorem (2.7), there is an interval $[a', b']$ contained in $[a, b]$ throughout which $f(x)$ is positive. Let $f(u)$ be the minimum value of f on $[a', b']$, as illustrated in Figure 5.12. It follows that the area under the graph of f from a to b is at least as large as the area $f(u)(b' - a')$ of the pictured rectangle. Consequently, by Theorem (5.12), $\int_a^b f(x)\, dx > 0$. This result can also be proved directly from the definition of the definite integral.

Suppose the functions f and g are continuous on $[a, b]$. If $f(x) \ge g(x)$ for all x in $[a, b]$, but $f \ne g$, then $f(x) - g(x) > 0$ for some x and, by the previous discussion, $\int_a^b [f(x) - g(x)]\, dx > 0$. Consequently, $\int_a^b f(x)\, dx > \int_a^b g(x)\, dx$. This fact will be used in the proof of the next theorem.

The Mean Value Theorem for Definite Integrals
(5.20)

> If f is continuous on a closed interval $[a, b]$, then there is a number z in the open interval (a, b) such that
>
> $$\int_a^b f(x)\, dx = f(z)(b - a).$$

Proof   If f is a constant function, then the result follows trivially from Theorem (5.13) where z is *any* number in (a, b). Next assume that f is not a constant function and suppose that m and M are the minimum and maximum values of f, respectively, on $[a, b]$. Let $f(u) = m$ and $f(v) = M$ where u

and v are in $[a, b]$. Since f is not a constant function, $m < f(x) < M$ for some x in $[a, b]$ and hence by the remark immediately preceding this theorem,

$$\int_a^b m \, dx < \int_a^b f(x) \, dx < \int_a^b M \, dx.$$

Employing Theorem (5.13),

$$m(b - a) < \int_a^b f(x) \, dx < M(b - a).$$

Dividing by $b - a$ and replacing m and M by $f(u)$ and $f(v)$, respectively, we obtain

$$f(u) < \frac{1}{b - a} \int_a^b f(x) \, dx < f(v).$$

Since $[1/(b - a)] \int_a^b f(x) \, dx$ is a number between $f(u)$ and $f(v)$, it follows from the Intermediate Value Theorem (2.24) that there is a number z, strictly between u and v, such that

$$f(z) = \frac{1}{b - a} \int_a^b f(x) \, dx.$$

Multiplying both sides by $b - a$ gives us the conclusion of the theorem.　□

The number z of Theorem (5.20) is not necessarily unique. Indeed, as pointed out in the proof, if f is a constant function then *any* number z can be used. The theorem guarantees that at *least* one number z will produce the desired result.

The Mean Value Theorem has an interesting geometric interpretation if $f(x) \geq 0$ on $[a, b]$. In this case $\int_a^b f(x) \, dx$ is the area under the graph of f from a to b, and the number $f(z)$ in Theorem (5.20) is the y-coordinate of the point P on the graph of f having x-coordinate z (see Figure 5.13). If a horizontal line is drawn through P, then the area of the rectangular region bounded by this line, the x-axis, and the lines $x = a$ and $x = b$ is $f(z)(b - a)$ which, according to Theorem (5.20), is the same as the area under the graph of f from a to b.

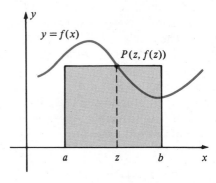

Figure 5.13

Example　It can be proved that $\int_0^3 [4 - (x^2/4)] \, dx = \frac{39}{4}$. Find a number that satisfies the conclusion of the Mean Value Theorem for this integral.

Solution　According to the Mean Value Theorem for Definite Integrals, there is a number z between 0 and 3 such that

$$\int_0^3 \left(4 - \frac{x^2}{4}\right) dx = \left(4 - \frac{z^2}{4}\right)(3 - 0)$$

or, equivalently,

$$\frac{39}{4} = \left(\frac{16 - z^2}{4}\right)(3).$$

Multiplying both sides of the last equation by $\frac{4}{3}$ leads to $13 = 16 - z^2$ and, therefore, $z^2 = 3$. Consequently, $\sqrt{3}$ satisfies the conclusion of Theorem (5.20). ∎

The Mean Value Theorem for Definite Integrals can be used to help prove a number of important theorems. One of the most important is the *Fundamental Theorem of Calculus*, which appears in the next section.

EXERCISES 5.4

The definite integrals given in Exercises 1–10 may be verified by methods to be introduced later. In each exercise find numbers that satisfy the conclusion of the Mean Value Theorem for Definite Integrals (5.20).

1 $\displaystyle\int_0^3 3x^2 \, dx = 27$

2 $\displaystyle\int_{-1}^3 (3x^2 - 2x + 3) \, dx = 32$

3 $\displaystyle\int_{-1}^8 3\sqrt{x + 1} \, dx = 54$

4 $\displaystyle\int_{-2}^{-1} 8x^{-3} \, dx = -3$

5 $\displaystyle\int_1^2 (4x^3 - 1) \, dx = 14$

6 $\displaystyle\int_1^4 (2 + 3\sqrt{x}) \, dx = 20$

7 $\displaystyle\int_2^7 (x + 3)^{-2} \, dx = 1/10$

8 $\displaystyle\int_1^8 4\sqrt[3]{x} \, dx = 45$

9 $\displaystyle\int_0^a \sqrt{a^2 - x^2} \, dx = \frac{\pi}{4} a^2, \, a > 0$

10 $\displaystyle\int_1^3 (x^2 + x^{-2}) \, dx = 28/3$

11 If $f(x) = k$ for all x in $[a, b]$, prove that every number z in $[a, b]$ satisfies the conclusion of Theorem (5.20). Interpret this fact geometrically.

12 If $f(x) = x$ and $0 < a < b$, find (without integrating) a number z in (a, b) such that $\int_a^b f(x) \, dx = f(z)(b - a)$.

5.5

THE FUNDAMENTAL THEOREM OF CALCULUS

The task of evaluating a definite integral by means of Definition (5.8) is quite difficult even in the simplest cases. This section contains a theorem that can be used to find the definite integral without using limits of sums. Due to its importance in evaluating definite integrals, and because it exhibits the connection between differentiation and integration, the theorem is aptly called *The Fundamental Theorem of Calculus*. This theorem was discovered independently by Sir Isaac Newton (1642–1727) in England and by Gottfried Wilhelm Leibniz (1646–1716) in Germany. It is primarily because of this discovery that these outstanding mathematicians are credited with the invention of calculus.

To avoid confusion, in the following discussion we shall use the variable t and denote the definite integral of f from a to b by $\int_a^b f(t) \, dt$. If f is continuous on $[a, b]$ and if x is in $[a, b]$, then f is continuous on $[a, x]$ and hence by Theorem (5.11) f is integrable on $[a, x]$. Consequently, the equation

$$G(x) = \int_a^x f(t)\, dt$$

defines a function G with domain $[a, b]$, since for each x in $[a, b]$ there corresponds a unique number $G(x)$. The next theorem brings out the remarkable fact that G is an antiderivative of f. In addition, it shows how any antiderivative may be used to find a definite integral of f.

The Fundamental Theorem of Calculus (5.21)

Suppose f is continuous on a closed interval $[a, b]$.

Part I If the function G is defined by

$$G(x) = \int_a^x f(t)\, dt$$

for all x in $[a, b]$, then G is an antiderivative of f on $[a, b]$.

Part II If F is any antiderivative of f, then

$$\int_a^b f(x)\, dx = F(b) - F(a).$$

Proof   To establish Part I we must show that if x is in $[a, b]$, then $G'(x) = f(x)$, that is,

$$\lim_{h \to 0} \frac{G(x + h) - G(x)}{h} = f(x).$$

Before giving a formal proof, it is instructive to consider some geometric aspects of this formula. If $f(x) \geq 0$ throughout $[a, b]$, then $G(x)$ is the area under the graph of f from a to x, as illustrated in Figure 5.14. If $h > 0$, then the difference $G(x + h) - G(x)$ is the area under the graph of f from x to $x + h$, the number h is the length of the interval $[x, x + h]$, and $f(x)$ is the y-coordinate of the point with x-coordinate x on the graph of f. We will show that $[G(x + h) - G(x)]/h = f(z)$, where z is between x and $x + h$. Reasoning intuitively, it appears that if $h \to 0$, then $z \to x$ and $f(z) \to f(x)$, which is what we wish to prove.

Let us now give a rigorous proof that $G'(x) = f(x)$. If x and $x + h$ are in $[a, b]$, then using the definition of G, together with Definition (5.9) and Theorem (5.16),

$$G(x + h) - G(x) = \int_a^{x+h} f(t)\, dt - \int_a^x f(t)\, dt$$

$$= \int_a^{x+h} f(t)\, dt + \int_x^a f(t)\, dt$$

$$= \int_x^{x+h} f(t)\, dt.$$

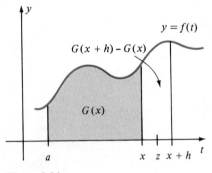

Figure 5.14

Consequently, if $h \neq 0$,

$$\frac{G(x+h) - G(x)}{h} = \frac{1}{h} \int_x^{x+h} f(t)\, dt.$$

If $h > 0$, then by the Mean Value Theorem for Integrals (5.20), there is a number z (depending on h) in the open interval $(x, x + h)$ such that

$$\int_x^{x+h} f(t)\, dt = f(z)h$$

and, therefore,

$$\frac{G(x+h) - G(x)}{h} = f(z).$$

Since $x < z < x + h$ it follows from the continuity of f that

$$\lim_{h \to 0^+} f(z) = \lim_{z \to x^+} f(z) = f(x)$$

and hence

$$\lim_{h \to 0^+} \frac{G(x+h) - G(x)}{h} = \lim_{h \to 0^+} f(z) = f(x).$$

If $h < 0$, then we may prove in similar fashion that

$$\lim_{h \to 0^-} \frac{G(x+h) - G(x)}{h} = f(x).$$

The last two one-sided limits imply that

$$G'(x) = \lim_{h \to 0} \frac{G(x+h) - G(x)}{h} = f(x)$$

which is what we wished to prove.

To prove Part II, let F be any antiderivative of f and let G be the special antiderivative defined in Part I. It follows from Theorem (4.31) that F and G differ by a constant; that is, there is a number C such that $G(x) - F(x) = C$ for all x in $[a, b]$. Hence, from the definition of G,

$$\int_a^x f(t)\, dt - F(x) = C$$

for all x in $[a, b]$. If we let $x = a$ and use the fact that $\int_a^a f(t)\, dt = 0$, we obtain $0 - F(a) = C$. Consequently,

$$\int_a^x f(t)\, dt - F(x) = -F(a).$$

Since this is an identity for all x in $[a, b]$ we may substitute b for x, obtaining

$$\int_a^b f(t)\, dt - F(b) = -F(a).$$

Adding $F(b)$ to both sides of this equation and replacing the variable t by x gives us the desired conclusion.

It is customary to denote the difference $F(b) - F(a)$ either by the symbol $F(x)]_a^b$ or by $[F(x)]_a^b$. Part II of the Fundamental Theorem may then be expressed as follows.

Theorem (5.22)

> If f is continuous on $[a, b]$ and F is any antiderivative of f, then
> $$\int_a^b f(x)\, dx = F(x)\Big]_a^b = F(b) - F(a).$$

The formula in Theorem (5.22) is also valid if $a \geq b$. Thus, if $a > b$, then by Definition (5.9),

$$\int_a^b f(x)\, dx = -\int_b^a f(x)\, dx$$
$$= -[F(a) - F(b)]$$
$$= F(b) - F(a).$$

If $a = b$, then by Definition (5.10),

$$\int_a^a f(x)\, dx = 0 = F(a) - F(a).$$

Example 1 Evaluate $\displaystyle\int_{-2}^3 (6x^2 - 5)\, dx$.

Solution An antiderivative of $6x^2 - 5$ is given by $F(x) = 2x^3 - 5x$. Applying Theorem (5.22),

$$\int_{-2}^3 (6x^2 - 5)\, dx = \left[2x^3 - 5x\right]_{-2}^3$$
$$= [2(3)^3 - 5(3)] - [2(-2)^3 - 5(-2)]$$
$$= [54 - 15] - [-16 + 10] = 45. \qquad\blacksquare$$

Note that if $F(x) + C$ is used in place of $F(x)$ in Theorem (5.22), the same result is obtained, since

$$\left[F(x) + C\right]_a^b = \{F(b) + C\} - \{F(a) + C\}$$

$$= F(b) - F(a) = \left[F(x)\right]_a^b.$$

This is in keeping with the statement that *any* antiderivative F may be employed in the Fundamental Theorem. Also note that for any number k,

$$\left[kF(x)\right]_a^b = kF(b) - kF(a) = k[F(b) - F(a)] = k\left[F(x)\right]_a^b;$$

that is, a constant factor can be "taken out" of the bracket when this notation is used. This result is analogous to Theorem (5.14) on definite integrals.

Theorem (5.23)

> If k is a real number, and r is a rational number such that $r \neq -1$, then
>
> $$\int_a^b kx^r \, dx = \left[\left(\frac{k}{r+1}\right)x^{r+1}\right]_a^b = \left(\frac{k}{r+1}\right)(b^{r+1} - a^{r+1}).$$

Proof   If $f(x) = kx^r$, then the function F defined by $F(x) = [k/(r+1)]x^{r+1}$ is an antiderivative of f, whenever f is defined. Applying Theorem (5.22) gives us the conclusion.   □

If an integrand is a sum of terms of the form kx^r where $r \neq -1$, then (5.23) may be applied to each term, as illustrated in the next example.

Example 2   Evaluate $\displaystyle\int_{-1}^{2} (x^3 + 1)^2 \, dx.$

Solution   Squaring the integrand and then applying Theorem (5.23) to each term gives us

$$\int_{-1}^{2} (x^3 + 1)^2 \, dx = \int_{-1}^{2} (x^6 + 2x^3 + 1) \, dx$$

$$= \left[\frac{1}{7}x^7 + \frac{2}{4}x^4 + x\right]_{-1}^{2}$$

$$= \left[\frac{1}{7}(2)^7 + \frac{1}{2}(2)^4 + 2\right] - \left[\frac{1}{7}(-1)^7 + \frac{1}{2}(-1)^4 + (-1)\right]$$

$$= \frac{405}{14}. \qquad\blacksquare$$

In the preceding example it is very important to note that

$$\int_{-1}^{2} (x^3 + 1)^2 \, dx \neq \frac{(x^3 + 1)^3}{3}\Bigg]_{-1}^{2}.$$

Example 3   Evaluate $\int_1^4 \left(5x - 2\sqrt{x} + \dfrac{32}{x^3}\right) dx$.

Solution   We begin by changing the form of the integrand so that Theorem (5.23) may be applied to each term. Thus

$$\int_1^4 (5x - 2x^{1/2} + 32x^{-3}) \, dx = \left[\frac{5}{2}x^2 - \frac{2}{(3/2)}x^{3/2} + \frac{32}{-2}x^{-2}\right]_1^4$$

$$= \left[\frac{5}{2}x^2 - \frac{4}{3}x^{3/2} - \frac{16}{x^2}\right]_1^4$$

$$= \left[\frac{5}{2}(4)^2 - \frac{4}{3}(4)^{3/2} - \frac{16}{4^2}\right] - \left[\frac{5}{2} - \frac{4}{3} - 16\right]$$

$$= 259/6. \qquad \blacksquare$$

Example 4   Evaluate $\int_{-2}^3 |x| \, dx$.

Solution   By Definition (1.2), $|x| = -x$ if $x < 0$ and $|x| = x$ if $x \geq 0$. This suggests that we use Theorem (5.16) to express the given integral as a sum of two definite integrals as follows:

$$\int_{-2}^3 |x| \, dx = \int_{-2}^0 |x| \, dx + \int_0^3 |x| \, dx$$

$$= \int_{-2}^0 (-x) \, dx + \int_0^3 x \, dx$$

$$= -\left[\frac{x^2}{2}\right]_{-2}^0 + \left[\frac{x^2}{2}\right]_0^3$$

$$= -\left[0 - \frac{4}{2}\right] + \left[\frac{9}{2} - 0\right]$$

$$= 2 + \frac{9}{2} = \frac{13}{2}. \qquad \blacksquare$$

The technique of defining a function by means of a definite integral, as in Part I of the Fundamental Theorem of Calculus (5.21), will have a very important application in Chapter 7 (see Definition (7.3)). Recall, from (5.21), that if f is continuous on $[a, b]$ and the function G is defined by $G(x) = \int_a^x f(t) \, dt$, where $a \leq x \leq b$, then G is an antiderivative of f; that is, $D_x G(x) = f(x)$. This may be stated in integral form as follows:

$$D_x \int_a^x f(t) \, dt = f(x).$$

The preceding formula is generalized in the next theorem.

Theorem (5.24)

Let f be continuous on $[a, b]$. If $a \leq c \leq b$, then for all x in $[a, b]$,

$$D_x \int_c^x f(t) \, dt = f(x).$$

Proof   If F is an antiderivative of f, then

$$D_x \int_c^x f(t) \, dt = D_x[F(x) - F(c)]$$
$$= D_x F(x) - D_x F(c)$$
$$= f(x) - 0 = f(x). \qquad \square$$

Example 5   If $G(x) = \int_1^x \dfrac{1}{t} \, dt$, where $x > 0$, find $G'(x)$.

Solution   Let $f(x) = 1/x$ and consider any interval $[a, b]$ where $a > 0$ and $b \geq 1$. If $a \leq x \leq b$, then by Theorem (5.24),

$$G'(x) = D_x \int_1^x \frac{1}{t} \, dt = \frac{1}{x}. \qquad \blacksquare$$

EXERCISES 5.5

Evaluate the definite integrals in Exercises 1–32.

1 $\displaystyle\int_1^4 (x^2 - 4x - 3) \, dx$

2 $\displaystyle\int_{-2}^3 (5 + x - 6x^2) \, dx$

3 $\displaystyle\int_{-2}^3 (8z^3 + 3z - 1) \, dz$

4 $\displaystyle\int_0^2 (w^4 - 2w^3) \, dw$

5 $\displaystyle\int_7^{12} dx$

6 $\displaystyle\int_{-6}^{-1} 8 \, dx$

7 $\displaystyle\int_1^2 [5/(8x^6)] \, dx$

8 $\displaystyle\int_1^4 \sqrt{16x^5} \, dx$

9 $\displaystyle\int_4^9 \frac{t - 3}{\sqrt{t}} \, dt$

10 $\displaystyle\int_{-1}^{-2} \frac{2s - 7}{s^3} \, ds$

11 $\displaystyle\int_{-8}^8 (\sqrt[3]{s^2} + 2) \, ds$

12 $\displaystyle\int_1^0 t^2(\sqrt[3]{t} - \sqrt{t}) \, dt$

13 $\displaystyle\int_0^1 (2x - 3)(5x + 1) \, dx$

14 $\displaystyle\int_{-1}^1 (x^2 + 1)^2 \, dx$

15 $\displaystyle\int_{-1}^0 (2w + 3)^2 \, dw$

16 $\displaystyle\int_5^5 \sqrt[3]{x^2} + \sqrt{x^5 + 1} \, dx$

17 $\displaystyle\int_3^2 \frac{x^2 - 1}{x - 1} \, dx$

18 $\displaystyle\int_0^{-1} \frac{x^3 + 8}{x + 2} \, dx$

19 $\int_1^1 (4x^2 - 5)^{100}\, dx$

20 $\int_1^2 (4u^{-5} + 6u^{-4})\, du$

21 $\int_1^3 \dfrac{2x^3 - 4x^2 + 5}{x^2}\, dx$

22 $\int_{-2}^{-1} \left(r - \dfrac{1}{r}\right)^2 dr$

23 $\int_{-3}^6 |x - 4|\, dx$

24 $\int_{-1}^1 (x + 1)(x + 2)(x + 3)\, dx$

25 $\int_0^4 \sqrt{3t}(\sqrt{t} + \sqrt{3})\, dt$

26 $\int_{-1}^5 |2x - 3|\, dx$

Verify the identities in Exercises 27 and 28 by first using the Fundamental Theorem of Calculus (5.21) and then differentiating.

27 $D_x \int_0^x (t^3 - 4\sqrt{t} + 5)\, dt = x^3 - 4\sqrt{x} + 5$ if $x \ge 0$

28 $D_x \int_0^x (5t + 3)^2\, dt = (5x + 3)^2$

29 Find $D_x \int_0^x \dfrac{1}{t + 1}\, dt$

30 Find $D_x \int_0^x \dfrac{1}{\sqrt{1 - t^2}}\, dt$, if $|x| < 1$.

31 If $f(x) = x^2 + 1$, find the area of the region under the graph of f from -1 to 2.

32 If $f(x) = x^3$, find the area of the region under the graph of f from 1 to 3.

Verify the formulas in Exercises 33–36.

33 $\int_a^b \dfrac{x}{\sqrt{(x^2 + 1)^3}}\, dx = \dfrac{-1}{\sqrt{x^2 + 1}}\Big]_a^b$

34 $\int_a^b \dfrac{x^3}{\sqrt{x^2 + 1}}\, dx = \left[\dfrac{1}{3}\sqrt{(x^2 + 1)^3} - \sqrt{x^2 + 1}\right]_a^b$

35 $\int_a^b \dfrac{x}{(x^2 + c^2)^{n+1}}\, dx = \dfrac{-1}{2n(x^2 + c^2)^n}\Big]_a^b$, $n \neq 0$

36 $\int_a^b \dfrac{x^2}{(x^3 + c^3)^{n+1}}\, dx = \dfrac{-1}{3n(x^3 + c^3)^n}\Big]_a^b$, $n \neq 0$

Find numbers that satisfy the conclusion of the Mean Value Theorem for Integrals (5.20) for the definite integrals in Exercises 37–40.

37 $\int_0^4 (\sqrt{x} + 1)\, dx$ **38** $\int_{-1}^1 (2x + 1)^2\, dx$

39 $\int_{-1}^2 (3x^3 + 2)\, dx$ **40** $\int_1^9 (3/x^2)\, dx$

41 If f is continuous on $[a, b]$, then the **average value of f on $[a, b]$** is defined as the number

$$\frac{1}{b - a} \int_a^b f(x)\, dx.$$

Suppose a point P moves on a coordinate line with a continuous velocity function v. Show that the average velocity during the time interval $[a, b]$ equals the average value of v on $[a, b]$.

42 Refer to Exercise 41. If a function f has a continuous derivative on $[a, b]$, show that the average rate of change of $f(x)$ with respect to x on $[a, b]$ (see Section 4.7) equals the average value of f' on $[a, b]$.

In Exercises 43 and 44, find the average value of the function f on the given interval.

43 $f(x) = x^2 + 3x - 1$, $[-1, 2]$

44 $f(x) = x^3 + 1$, $[2, 4]$

45 If g is differentiable and f is continuous for all x, prove that $D_x \int_a^{g(x)} f(t)\, dt = f(g(x))g'(x)$. (*Hint*: Use Part I of Theorem (5.21) and the Chain Rule.)

46 Extend the formula in Exercise 45 to

$$D_x \int_{k(x)}^{g(x)} f(t)\, dt = f(g(x))g'(x) - f(k(x))k'(x).$$

Use Exercises 45 and 46 to find the derivatives in the following exercises.

47 $D_x \int_2^{x^4} \dfrac{t}{\sqrt{t^3 + 2}}\, dt$

48 $D_x \int_0^{x^2} \sqrt[3]{t^4 + 1}\, dt$

49 $D_x \int_{3x}^{x^3} (t^3 + 1)^{10}\, dt$

50 $D_x \int_{1/x}^{\sqrt{x}} \sqrt{t^4 + t^2 + 4}\, dt$

5.6

INDEFINITE INTEGRALS AND CHANGE OF VARIABLES

The connection between antiderivatives and definite integrals provided by the Fundamental Theorem of Calculus has made it customary to use integral signs to denote antiderivatives. In order to distinguish antiderivatives from definite integrals, no limits of integration are attached to the integral sign, as in the following definition.

Definition (5.25)

The **indefinite integral** $\int f(x)\,dx$ of f (or of $f(x)$) is defined by

$$\int f(x)\,dx = F(x) + C$$

where $F'(x) = f(x)$ and C is an arbitrary constant.

Note that the indefinite integral is merely another way of specifying the most general antiderivative of f. Instead of using the terminology *antidifferentiation* for the process of finding F when f is given, we now use the phrase **indefinite integration**. The arbitrary constant C is called the **constant of integration**, $f(x)$ is called the **integrand**, and x is called the **variable of integration**. We often refer to the process of finding $F(x) + C$ in Definition (5.25) as **evaluating the indefinite integral**. The domain of F will not usually be stated explicitly. It is always assumed that a suitable interval over which f is integrable has been chosen. In particular, a closed interval on which f is continuous could be used. As with definite integrals, the symbol employed for the variable of integration is insignificant since, for example, $\int f(t)\,dt$, $\int f(u)\,du$, etc., give rise to the same function F as $\int f(x)\,dx$.

Since the indefinite integral of f is an antiderivative, the Fundamental Theorem of Calculus gives us the following relationship between definite and indefinite integrals.

Theorem (5.26)

$$\int_a^b f(x)\,dx = \left[\int f(x)\,dx \right]_a^b$$

Thus, if the indefinite integral of a function f is known, then definite integrals of f can be evaluated. In future chapters, methods for finding indefinite integrals of various functions will be developed. At the present time it is possible to state several useful rules. The following **power rule for (indefinite) integration** may be proved by differentiating the expression on the right-hand side of the equation and showing that the integrand is obtained.

Theorem (5.27)

$$\int x^r \, dx = \frac{1}{r+1} x^{r+1} + C$$

where the exponent r is a rational number and $r \neq -1$.

Example 1 Evaluate the following:

(a) $\displaystyle\int x^2 \, dx$ (b) $\displaystyle\int \left(8t^3 - 5 + \frac{1}{t^3} \right) dt$

Solution   We only need to find an antiderivative for each integrand and then add the arbitrary constant C. Thus

(a) $\displaystyle\int x^2 \, dx = \tfrac{1}{3}x^3 + C.$

(b) $\displaystyle\int \left(8t^3 - 5 + \frac{1}{t^2} \right) dt = 2t^4 - 5t - \frac{1}{t} + C.$ ■

The next theorem indicates what happens if an indefinite integration is followed by a differentiation, or vice versa.

Theorem (5.28)

(i) $D_x \displaystyle\int f(x) \, dx = f(x)$

(ii) $\displaystyle\int D_x[f(x)] \, dx = f(x) + C$

Proof   To prove (i) we may use Theorem (5.25) as follows:

$$D_x \int f(x) \, dx = D_x[F(x) + C] = F'(x) = f(x).$$

The formula in (ii) follows from the fact that $f(x) + C$ is an antiderivative of $D_x[f(x)]$.

The following theorem is analogous to Theorems (5.14) and (5.15) for definite integrals.

Theorem (5.29)

(i) $\displaystyle\int kf(x) \, dx = k \int f(x) \, dx$

(ii) $\displaystyle\int [f(x) + g(x)] \, dx = \int f(x) \, dx + \int g(x) \, dx$

Proof   To prove (i) we merely differentiate the right-hand side and show that the integrand on the left is obtained. Thus,

$$D_x\left[k\int f(x)\,dx\right] = k \cdot D_x \int f(x)\,dx = kf(x)$$

where we have used (i) of (5.28). Part (ii) may be proved in similar fashion.

Every rule for differentiation can be transformed into a corresponding rule for indefinite integration. For example,

$$D_x\sqrt{x^2+5} = \frac{x}{\sqrt{x^2+5}} \quad \text{implies that} \quad \int \frac{x}{\sqrt{x^2+5}}\,dx = \sqrt{x^2+5} + C;$$

$$D_x(x^3+2x)^{10} = 10(x^3+2x)^9(3x^2+2) \quad \text{implies that}$$

$$\int 10(x^3+2x)^9(3x^2+2)\,dx = (x^3+2x)^{10} + C.$$

A similar technique may be used in conjunction with the Chain Rule to obtain a useful formula for indefinite integration. Thus, suppose F is an antiderivative of a function f, and g is a differentiable function such that $g(x)$ is in the domain of F for every x in some closed interval $[a, b]$. We may then consider the composite function defined by $F(g(x))$ for all x in $[a, b]$. Applying the Chain Rule (3.28) and the fact that $F' = f$, we obtain

$$D_x F(g(x)) = F'(g(x))g'(x) = f(g(x))g'(x).$$

This, in turn, gives us the integration formula

$$\int f(g(x))g'(x)\,dx = F(g(x)) + C, \qquad \text{where } F' = f.$$

There is a simple way to remember this formula. If we let $u = g(x)$ and formally replace $g'(x)\,dx$ by the differential du, we obtain

$$\int f(u)\,du = F(u) + C, \qquad \text{where } F' = f$$

which has the same form as Theorem (5.25). This memorization device indicates that $g'(x)\,dx$ may be regarded as the product of $g'(x)$ and dx, and is one of the main reasons for using the symbol dx in the integral notation. Finding indefinite integrals in this way will be referred to as a **change of variable** or as **the method of substitution**. We may summarize our discussion as follows, where f and g are assumed to have the properties described previously.

Theorem **(5.30)**

Given $\int f(g(x))g'(x)\,dx$, let $u = g(x)$ and $du = g'(x)\,dx$. If F is an antiderivative of f, then

$$\int f(g(x))g'(x)\,dx = \int f(u)\,du = F(u) + C = F(g(x)) + C.$$

As a first application, we may extend Theorem (5.27) to powers of functions. Thus, if $f(x) = x^r$ in (5.30), then $f(g(x)) = [g(x)]^r$. If we let

$$F(x) = \frac{1}{r+1}x^{r+1}, \quad \text{then} \quad F(g(x)) = \frac{1}{r+1}[g(x)]^{r+1}$$

and the conclusion of Theorem (5.30) takes on the form

$$\int [g(x)]^r g'(x)\, dx = \frac{1}{r+1}[g(x)]^{r+1} + C.$$

This may be checked by differentiating the expression on the right-hand side of the equation. A convenient way of expressing the last formula is stated in the following theorem.

Theorem (5.31)

> Suppose $u = g(x)$, where g is a differentiable function. If r is a rational number such that $r \neq -1$, then
> $$\int u^r\, du = \frac{1}{r+1}u^{r+1} + C.$$

Example 2   Find $\int (2x^3 + 1)^7 x^2\, dx$.

Solution   If we let $u = 2x^3 + 1$, then $du = 6x^2\, dx$. To obtain the form (5.31), it is necessary to introduce the factor 6 in the integrand. Doing this, and compensating by multiplying the integral by 1/6 (this is legitimate by (i) of Theorem (5.29)), we obtain

$$\int (2x^3 + 1)^7 x^2\, dx = \frac{1}{6}\int (2x^3 + 1)^7 6x^2\, dx.$$

Making the indicated substitution and integrating,

$$\int (2x^3 + 1)^7 x^2\, dx = \frac{1}{6}\int u^7\, du = \frac{1}{6}\left[\frac{1}{8}u^8 + K\right]$$

where K is a constant. It is now necessary to return to the original variable x. Since $u = 2x^3 + 1$, the last equality gives us

$$\int (2x^3 + 1)^7 x^2\, dx = \frac{1}{6}\left[\frac{1}{8}(2x^3 + 1)^8 + K\right]$$
$$= \frac{1}{48}(2x^3 + 1)^8 + \frac{1}{6}K.$$

Instead of employing constants of integration such as $\frac{1}{6}K$, it is customary to write this result as

$$\int (2x^3 + 1)^7 x^2\, dx = \frac{1}{48}(2x^3 + 1)^8 + C.$$

The relationship between K and C is $C = \frac{1}{6}K$; however, there is no practical advantage in remembering this fact. In the future we shall manipulate constants of integration in various ways without making explicit mention of relationships that exist. Moreover, instead of proceeding as above, it should be clear that it is permissible to integrate as follows:

$$\frac{1}{6}\int u^7 \, du = \frac{1}{6}\left[\frac{1}{8}u^8\right] + C.$$ ∎

Formal substitutions in indefinite integrals can be made in different ways. To illustrate, another method for solving Example 2 is to let

$$u = 2x^3 + 1, \qquad du = 6x^2 \, dx, \qquad \frac{1}{6} \, du = x^2 \, dx.$$

We then substitute directly for $x^2 \, dx$ instead of introducing the number 6 in the integrand as follows:

$$\int (2x^3 + 1)^7 x^2 \, dx = \int u^7 \frac{1}{6} \, du = \frac{1}{6}\int u^7 \, du$$

$$= \frac{1}{48}u^8 + C = \frac{1}{48}(2x^3 + 1)^8 + C.$$

The change of variable technique for indefinite integrals is a powerful tool, provided the student can recognize that the integrand is of the form $f(g(x))g'(x)$, or $f(g(x))kg'(x)$ where k is a real number. The ability to recognize this form is directly proportional to the number of exercises worked!

Example 3 Find $\int x\sqrt[3]{7 - 6x^2} \, dx$.

Solution   Note that the integrand contains the term $x \, dx$. If the factor x were missing, the change of variable technique could not be applied. Letting

$$u = 7 - 6x^2, \qquad du = -12x \, dx,$$

introducing the factor -12 in the integrand, and then compensating by multiplying the integral by $-\frac{1}{12}$, gives us

$$\int x\sqrt[3]{7 - 6x^2} \, dx = -\frac{1}{12}\int \sqrt[3]{7 - 6x^2}(-12)x \, dx$$

$$= -\frac{1}{12}\int \sqrt[3]{u} \, du = -\frac{1}{12}\int u^{1/3} \, du$$

$$= -\frac{1}{12}\frac{u^{4/3}}{(4/3)} + C = -\frac{1}{16}(7 - 6x^2)^{4/3} + C.$$

As in the second solution to Example 2, we could begin with $du = -12x \, dx$ and write $x \, dx = (-\frac{1}{12}) \, du$. The change of variables then takes on the form

$$\int \sqrt[3]{7 - 6x^2}\, x \, dx = \int \sqrt[3]{u}\left(-\frac{1}{12}\right) du = -\frac{1}{12}\int \sqrt[3]{u} \, du.$$

The remainder of the solution now proceeds exactly as before. ∎

The method of substitution may also be used to evaluate definite integrals. We could use Theorem (5.30) to find an indefinite integral (that is, an anti-derivative) and then apply the Fundamental Theorem of Calculus. Another method, which is sometimes shorter, is to change the limits of integration. Using (5.30) together with the Fundamental Theorem gives us the following formula, where $F' = f$:

$$\int_a^b f(g(x))g'(x)\,dx = F(g(x))\Big]_a^b.$$

The number on the right of this equality may be written

$$F(g(b)) - F(g(a)) = F(u)\Big]_{g(a)}^{g(b)} = \int_{g(a)}^{g(b)} f(u)\,du.$$

This gives us the following important result, where for brevity we have not restated the restrictions on f and g.

Theorem (5.32)

$$\int_a^b f(g(x))g'(x)\,dx = \int_{g(a)}^{g(b)} f(u)\,du, \qquad \text{where } u = g(x).$$

Theorem (5.32) states that after making the substitutions $u = g(x)$ and $du = g'(x)\,dx$, we may use the values of g that correspond to $x = a$ and $x = b$, respectively, as the limits of the integral involving u. It is then unnecessary to return to the variable x after integrating. This technique is illustrated in the next example.

Example 4 Evaluate $\displaystyle\int_2^{10} \frac{3}{\sqrt{5x - 1}}\,dx$.

Solution Let us begin by writing the integral as

$$3 \int_2^{10} \frac{1}{\sqrt{5x - 1}}\,dx.$$

The form of the integrand suggests the substitution $u = 5x - 1$. Of course,

$$\text{if } u = 5x - 1, \quad \text{then} \quad du = 5\,dx.$$

To change the limits of integration we note that

$$\text{if } x = 2, \quad \text{then} \quad u = 9,$$

and $\qquad\qquad\qquad\qquad \text{if } x = 10, \quad \text{then} \quad u = 49.$

Applying Theorem (5.32) leads to the following evaluation:

$$3 \int_2^{10} \frac{1}{\sqrt{5x-1}}\, dx = \frac{3}{5} \int_2^{10} \frac{1}{\sqrt{5x-1}}\, 5\, dx$$

$$= \frac{3}{5} \int_9^{49} \frac{1}{\sqrt{u}}\, du = \frac{3}{5} \int_9^{49} u^{-1/2}\, du$$

$$= \left(\frac{3}{5}\right) 2u^{1/2} \Big]_9^{49} = \frac{6}{5}\left[49^{1/2} - 9^{1/2}\right]$$

$$= 24/5. \qquad \blacksquare$$

EXERCISES 5.6

Evaluate the integrals in Exercises 1–22.

1 $\int (3x+1)^4\, dx$

2 $\int (2x^2 - 3)^5 x\, dx$

3 $\int \sqrt{t^3 - 1}\, t^2\, dt$

4 $\int \sqrt{9 - z^2}\, z\, dz$

5 $\int \frac{x-2}{(x^2 - 4x + 3)^3}\, dx$

6 $\int \frac{x^2 + x}{(4 - 3x^2 - 2x^3)^4}\, dx$

7 $\int \frac{s}{\sqrt[3]{1 - 2s^2}}\, ds$

8 $\int \sqrt[5]{t^4 - t^2}\, (10t^3 - 5t)\, dt$

9 $\int \frac{(\sqrt{u} + 3)^4}{\sqrt{u}}\, du$

10 $\int \left(1 + \frac{1}{u}\right)^{-3} \left(\frac{1}{u^2}\right) du$

11 $\int_1^4 \sqrt{5 - x}\, dx$

12 $\int_1^5 \sqrt[3]{2x - 1}\, dx$

13 $\int_{-1}^1 (t^2 - 1)^3 t\, dt$

14 $\int_{-2}^0 \frac{v^2}{(v^3 - 2)^2}\, dv$

15 $\int_0^1 \frac{1}{(3 - 2v)^2}\, dv$

16 $\int_0^4 \frac{x}{\sqrt{x^2 + 9}}\, dx$

17 $\int (x^2 + 1)^3\, dx$

18 $\int (3 - x^3)^2 x\, dx$

19 $\int 5\sqrt{8x + 5}\, dx$

20 $\int \frac{6}{\sqrt{4 - 5t}}\, dt$

21 $\int_1^4 \frac{1}{\sqrt{x}(\sqrt{x} + 1)^3}\, dx$

22 $\int (3 - x^4)^3 x^3\, dx$

Evaluate the integrals in Exercises 23–26 by (a) the method of substitution and (b) expanding the integrand. In what way do the constants of integration differ?

23 $\int (x + 4)^2\, dx$

24 $\int (x^2 + 4)^2 x\, dx$

25 $\int \frac{(\sqrt{x} + 3)^2}{\sqrt{x}}\, dx$

26 $\int \left(1 + \frac{1}{x}\right)^2 \frac{1}{x^2}\, dx$

Verify the identities in Exercises 27 and 28 by first integrating and then differentiating.

27 $D_x \int x^3 \sqrt{x^4 + 5}\, dx = x^3 \sqrt{x^4 + 5}$

28 $D_x \int (3x + 2)^7\, dx = (3x + 2)^7$

29 Find $D_x \int \dfrac{1}{\sqrt{x^3 + x + 5}}\, dx$.

30 Find $\int D_x \dfrac{x}{\sqrt[3]{x + 1}}\, dx$.

31 Find $\int_0^3 D_x \sqrt{x^2 + 16}\, dx$.

32 Find $D_x \int_0^1 \sqrt{x^2 + 4x}\, dx$.

33 If $f(x) = \sqrt{x + 1}$, find the area of the region under the graph of f from 0 to 3.

34 If $f(x) = x/(x^2 + 1)^2$, find the area of the region under the graph of f from 1 to 2.

In Exercises 35–38 find numbers that satisfy the conclusion of the Mean Value Theorem for Integrals.

35 $\displaystyle\int_0^4 \dfrac{x}{\sqrt{x^2 + 9}}\, dx$

36 $\displaystyle\int_{-2}^0 \sqrt[3]{x + 1}\, dx$

37 $\displaystyle\int_0^5 \sqrt{x + 4}\, dx$

38 $\displaystyle\int_{-3}^2 \sqrt{6 - x}\, dx$

Verify the formulas in Exercises 39 and 40.

39 $\displaystyle\int \dfrac{\sqrt{x^2 - a^2}}{x^4}\, dx = \dfrac{\sqrt{(x^2 - a^2)^3}}{3a^2 x^3} + C$

40 $\displaystyle\int \dfrac{1}{x^2 \sqrt{x^2 - a^2}}\, dx = \dfrac{\sqrt{x^2 - a^2}}{a^2 x} + C$

41 Let f be continuous on $[-a, a]$. If f is an even function, show that $\int_{-a}^a f(x)\, dx = 2 \int_0^a f(x)\, dx$ and interpret this result geometrically. Verify the result for the special case $f(x) = x^4 - 3x^2 + 1$.

42 Let f be continuous on $[-a, a]$. If f is an odd function show that $\int_{-a}^a f(x)\, dx = 0$ and interpret this result geometrically. Verify the result for the special case $f(x) = x^3 - 4x$.

5.7

NUMERICAL INTEGRATION

To evaluate a definite integral $\int_a^b f(x)\, dx$ by means of the Fundamental Theorem of Calculus, it is necessary to find an antiderivative of f. If an antiderivative cannot be found, then numerical methods may be used to approximate the integral to any degree of accuracy. For example, if the norm of a partition of $[a, b]$ is small, then by Definition (5.8) the definite integral can be approximated by any Riemann sum of f. In particular, if we use a regular partition with $\Delta x = (b - a)/n$, then

$$\int_a^b f(x)\, dx \approx \sum_{i=1}^n f(w_i)\, \Delta x$$

where w_i is any number in the ith subinterval $[x_{i-1}, x_i]$ of the partition. Of course, the accuracy of the approximation depends upon the nature of f and the magnitude of Δx. It may be necessary to make Δx very small in order to obtain the desired degree of accuracy. This, in turn, means that n is large, and hence the preceding sum contains many terms. Figure 5.2 illustrates the case in which $f(w_i)$ is the minimum value of f in $[x_{i-1}, x_i]$. In this case the error involved in the approximation is numerically the same as the area of the unshaded region that lies under the graph of f and over the inscribed rectangles.

If we take $w_i = x_{i-1}$, that is, if f is evaluated at the left-hand endpoint of each subinterval $[x_{i-1}, x_i]$, we may write

$$\int_a^b f(x)\, dx \approx \sum_{i=1}^n f(x_{i-1})\, \Delta x.$$

If we let $w_i = x_i$, that is, if f is evaluated at the right-hand endpoint of $[x_{i-1}, x_i]$, then

$$\int_a^b f(x)\, dx \approx \sum_{i=1}^n f(x_i)\, \Delta x.$$

Another, and usually more accurate approximation, can be obtained by using the average of the last two approximations, that is,

$$\frac{1}{2}\left[\sum_{i=1}^n f(x_{i-1})\, \Delta x + \sum_{i=1}^n f(x_i)\, \Delta x\right].$$

With the exception of $f(x_0)$ and $f(x_n)$, each $f(x_i)$ appears twice, and hence we may write this expression as

$$\frac{\Delta x}{2}\left[f(x_0) + \sum_{i=1}^{n-1} 2f(x_i) + f(x_n)\right].$$

Since $\Delta x = (b - a)/n$, this gives us the following rule.

Trapezoidal Rule **(5.33)**

> If f is continuous on $[a, b]$ and if a regular partition of $[a, b]$ is determined by $a = x_0, x_1, \ldots, x_n = b$, then
>
> $$\int_a^b f(x)\, dx \approx \frac{b - a}{2n}\left[f(x_0) + 2f(x_1) + 2f(x_2) + \cdots + 2f(x_{n-1}) + f(x_n)\right].$$

The term *trapezoidal* arises from the case in which $f(x)$ is nonnegative on $[a, b]$. As illustrated in Figure 5.15, if P_k is the point with x-coordinate x_k on the graph of $y = f(x)$, then for each $i = 1, 2, \ldots, n$, the points on the

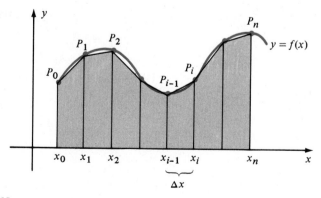

Figure 5.15

x-axis with x-coordinates x_{i-1} and x_i, together with P_{i-1} and P_i, are vertices of a trapezoid having area

$$\frac{\Delta x}{2}[f(x_{i-1}) + f(x_i)].$$

The sum of the areas of these trapezoids is the same as the sum in Rule (5.33). Hence, in geometric terms, the Trapezoidal Rule gives us an approximation to the area under the graph of f from a to b by means of trapezoids instead of the rectangles associated with Riemann sums.

The next result provides information about the maximum error that can occur if the Trapezoidal Rule is used to approximate a definite integral. The proof requires advanced methods and is therefore omitted.

Error Estimate for the Trapezoidal Rule (5.34)

If M is a positive real number such that $|f''(x)| \leq M$ for all x in $[a, b]$, then the error involved in using the Trapezoidal Rule (5.33) is not greater than $M(b - a)^3/12n^2$.

Example 1   Approximate $\int_1^2 (1/x)\, dx$ by using the Trapezoidal Rule with $n = 10$. Estimate a maximum error in the approximation.

Solution   Note that at this stage of our work it is impossible to find a function F such that $F'(x) = 1/x$ and hence we cannot use the Fundamental Theorem of Calculus to evaluate the integral. It is convenient to arrange our work as follows, where each $f(x_i)$ was obtained with a calculator and is accurate to four decimal places.

i	x_i	$f(x_i)$	m	$mf(x_i)$
0	1.0	1.0000	1	1.0000
1	1.1	0.9091	2	1.8182
2	1.2	0.8333	2	1.6666
3	1.3	0.7692	2	1.5384
4	1.4	0.7143	2	1.4286
5	1.5	0.6667	2	1.3334
6	1.6	0.6250	2	1.2500
7	1.7	0.5882	2	1.1764
8	1.8	0.5556	2	1.1112
9	1.9	0.5263	2	1.0526
10	2.0	0.5000	1	0.5000

The sum of the numbers in the last column is 13.8754. Since

$$(b - a)/2n = (2 - 1)/20 = 0.05,$$

it follows from (5.33) that

$$\int_1^2 \frac{1}{x}\, dx \approx (0.05)(13.8754) \approx 0.6938.$$

The error in the approximation may be estimated by means of (5.34). Since $f(x) = 1/x$ we have $f'(x) = -1/x^2$ and $f''(x) = 2/x^3$. The maximum value of $f''(x)$ on the interval $[1, 2]$ occurs at $x = 1$ and hence

$$|f''(x)| \le \frac{2}{(1)^3} = 2.$$

Applying (5.34) with $M = 2$ we see that the maximum error is not greater than

$$\frac{2(2 - 1)^3}{12(10)^2} = \frac{1}{600} < 2 \times 10^{-3}. \qquad \blacksquare$$

It will be shown in Chapter 7 that the integral in Example 1 equals the natural logarithm of 2, denoted by $\ln 2$. Using a calculator it is possible to verify that to five decimal places, $\ln 2$ is approximated by 0.69315. To obtain this approximation by means of the Trapezoidal Rule, it is necessary to use a very large value of n.

The following rule is usually more accurate than the Trapezoidal Rule.

Simpson's Rule *(5.35)*

> Suppose f is continuous on $[a, b]$ and n is an even integer. If a regular partition is determined by $a = x_0, x_1, \ldots, x_n = b$, then
>
> $$\int_a^b f(x)\, dx \approx \frac{b - a}{3n} [f(x_0) + 4f(x_1) + 2f(x_2) + 4f(x_3) + \cdots$$
> $$+ 2f(x_{n-2}) + 4f(x_{n-1}) + f(x_n)].$$

The idea behind the proof of Simpson's Rule is that instead of using trapezoids to approximate the graph of f, we use portions of graphs of equations of the form $y = cx^2 + dx + e$. If $c \ne 0$, it can be shown that the graph of such an equation is a parabola. If $P_0(x_0, y_0)$, $P_1(x_1, y_1)$, and $P_2(x_2, y_2)$ are points on the parabola, where $x_0 < x_1 < x_2$, then substituting the coordinates of P_0, P_1, and P_2, respectively, into this equation produces three equations that may be solved for c, d, and e. As a special case, suppose h, y_0, y_1, and y_2 are positive, and consider the points $P_0(-h, y_0)$, $P_1(0, y_1)$, and $P_2(h, y_2)$, as illustrated in (i) of Figure 5.16.

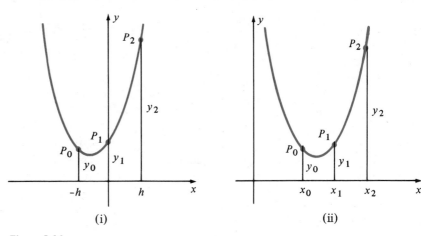

(i)   (ii)

Figure 5.16

The area A under the graph of the equation from $-h$ to h is

$$A = \int_{-h}^{h} (cx^2 + dx + e)\, dx = \frac{cx^3}{3} + \frac{dx^2}{2} + ex \Big]_{-h}^{h} = \frac{h}{3}(2ch^2 + 6e).$$

Since the coordinates of P_0, P_1, and P_2 are solutions of $y = cx^2 + dx + e$, we have

$$y_0 = ch^2 - dh + e$$
$$y_1 = e$$
$$y_2 = ch^2 + dh + e$$

from which it follows that

$$y_0 + 4y_1 + y_2 = 2ch^2 + 6e.$$

Consequently, $$A = \frac{h}{3}(y_0 + 4y_1 + y_2).$$

If the points P_0, P_1, and P_2 are translated horizontally, as illustrated in (ii) of Figure 5.16, then the area under the graph remains the same. Consequently, the preceding formula for A is true for *any* points P_0, P_1, and P_2, provided $x_1 - x_0 = x_2 - x_1$.

If $f(x) \geq 0$ on $[a, b]$, then Simpson's Rule is obtained by regarding the definite integral as the area under the graph of f from a to b. Thus, suppose n is an even integer and $h = (b - a)/n$. We divide $[a, b]$ into n subintervals, each of length h, by choosing numbers $a = x_0, x_1, \ldots, x_n = b$. Let $P_k(x_k, y_k)$ be the point on the graph of f with x-coordinate x_k, as illustrated in Figure 5.17.

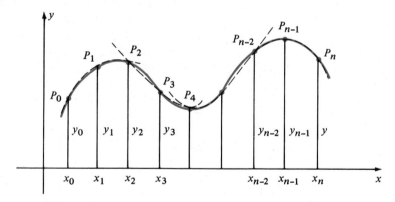

Figure 5.17

If the arc through P_0, P_1, and P_2 is approximated by the graph of an equation $y = cx^2 + dx + e$, then, as we have seen, the area under the graph of f from x_0 to x_2 is approximated by

$$\frac{h}{3}(y_0 + 4y_1 + y_2).$$

If the arc through P_2, P_3, and P_4 is approximated in similar fashion, then the area under the graph of f from x_2 to x_4 is approximately

$$\frac{h}{3}(y_2 + 4y_3 + y_4).$$

We continue in this manner until we reach the last triple of points P_{n-2}, P_{n-1}, P_n and the corresponding approximation to the area under the graph, namely,

$$\frac{h}{3}(y_{n-2} + 4y_{n-1} + y_n).$$

Summing these approximations gives us

$$\int_a^b f(x)\,dx \approx \frac{h}{3}(y_0 + 4y_1 + 2y_2 + 4y_3 + \cdots + 2y_{n-2} + 4y_{n-1} + y_n)$$

which is the same as the sum in (5.35). If f is negative in $[a, b]$, then negatives of areas may be used to establish Simpson's Rule.

The following result may be established using advanced methods.

Error Estimate for
Simpson's Rule
(5.36)

> If M is a positive real number such that $|f^{(4)}(x)| \le M$ for all x in $[a, b]$, then the error involved in using Simpson's Rule (5.35) is not greater than $M(b - a)^5/180n^4$.

Example 2 Approximate $\int_1^2 (1/x)\,dx$ by using Simpson's Rule with $n = 10$. Estimate the error in the approximation.

Solution   This is the same integral considered in Example 1. We arrange our work as follows:

i	x_i	$f(x_i)$	m	$mf(x_i)$
0	1.0	1.0000	1	1.0000
1	1.1	0.9091	4	3.6364
2	1.2	0.8333	2	1.6666
3	1.3	0.7692	4	3.0768
4	1.4	0.7143	2	1.4286
5	1.5	0.6667	4	2.6668
6	1.6	0.6250	2	1.2500
7	1.7	0.5882	4	2.3528
8	1.8	0.5556	2	1.1112
9	1.9	0.5263	4	2.1052
10	2.0	0.5000	1	0.5000

The sum of the numbers in the last column is 20.7944. Since

$$\frac{b - a}{3n} = \frac{2 - 1}{30}$$

it follows from (5.35) that

$$\int_1^2 \frac{1}{x}\, dx \approx \left(\frac{1}{30}\right)(20.7944) \approx 0.6932.$$

We shall use (5.36) to estimate the error in the approximation. If $f(x) = 1/x$, the reader may verify that $f^{(4)}(x) = 24/x^5$. The maximum value of $f^{(4)}(x)$ on the interval $[1, 2]$ occurs at $x = 1$ and hence

$$|f^{(4)}(x)| \le \frac{24}{(1)^5} = 24.$$

Applying (5.36) with $M = 24$, we see that the maximum error in the approximation is not greater than

$$\frac{24(2 - 1)^5}{180(10)^4} = \frac{2}{150000} < 1.4 \times 10^{-5}.$$

Note that this estimated error is much less than that obtained using the Trapezoidal Rule in Example 1. ∎

At the beginning of this section we noted that a definite integral may be approximated by a Riemann sum $\sum_i f(w_i)\, \Delta x$. We then considered the special case where w_i was always a left-hand (or right-hand) endpoint of the ith subinterval $[x_{i-1}, x_i]$ of a regular partition of $[a, b]$. Another technique is to choose w_i as the *midpoint* of $[x_{i-1}, x_i]$. This leads to the next rule.

The Midpoint Rule (5.37)

> If f is continuous on $[a, b]$ and if a regular partition of $[a, b]$ is determined by $a = x_0, x_1, \ldots, x_n = b$, then
>
> $$\int_a^b f(x)\, dx \approx \frac{b - a}{n}[f(\bar{x}_1) + f(\bar{x}_2) + \cdots + f(\bar{x}_n)]$$
>
> where $\bar{x}_i = (x_{i-1} + x_i)/2$ is the midpoint of $[x_{i-1}, x_i]$.

The proof of (5.37) merely consists of noting that $\Delta x = (b - a)/n$ for each i. It can be shown that if M is a positive real number and $|f''(x)| \le M$ for all x in $[a, b]$, then the error involved in using the Midpoint Rule is not greater than $M(b - a)^3/24n^2$.

Example 3 Approximate $\int_1^2 (1/x)\, dx$ by using the Midpoint Rule with $n = 10$.

Solution   The integral is the same as that considered in the preceding two examples. Since the partition is determined by 1, 1.1, 1.2, …, 1.9, 2, the midpoints of the subintervals are 1.05, 1.15, ..., 1.95, and the Midpoint Rule gives us

$$\int_1^2 \frac{1}{x}\, dx \approx 0.1\left[\frac{1}{1.05} + \frac{1}{1.15} + \frac{1}{1.25} + \frac{1}{1.35} + \frac{1}{1.45} + \frac{1}{1.55} + \frac{1}{1.65}\right.$$
$$\left. + \frac{1}{1.75} + \frac{1}{1.85} + \frac{1}{1.95}\right]$$

It is left to the reader to verify that this gives us the approximation $(0.1)(6.928) = 0.6928$. This number compares favorably with the values 0.6938 and 0.6932 obtained from the Trapezoidal Rule and Simpson's Rule. ∎

An important aspect of numerical integration is that it can be used to approximate the integral of a function that is described in tabular form. To illustrate, suppose it is found experimentally that two physical variables x and y are related as shown in the following table.

x	1.0	1.5	2.0	2.5	3.0	3.5	4.0
y	3.1	4.0	4.2	3.8	2.9	2.8	2.7

If we regard y as a function of x, say $y = f(x)$, where f is continuous, then the definite integral $\int_1^4 f(x)\, dx$ may have an important physical meaning. In the present illustration, *the integral may be approximated without knowing the explicit formula for $f(x)$*. In particular, if the Trapezoidal Rule (5.33) is used, with $n = 6$ and $(b - a)/2n = (4 - 1)/12 = 0.25$, then

$$\int_1^4 f(x)\, dx \approx 0.25[3.1 + 2(4.0) + 2(4.2) + 2(3.8) + 2(2.9) + 2(2.8) + 2.7]$$

$$\approx 10.3.$$

Since the number of subdivisions is even, we could also approximate the integral by means of Simpson's Rule.

EXERCISES 5.7

In Exercises 1–8 use (a) the Trapezoidal Rule (5.33) and (b) Simpson's Rule (5.35) to approximate the definite integral for the stated value of n. Use approximations to four decimal places for $f(x_i)$ and round off answers to two decimal places. (It is advisable to use a calculator for these exercises.)

1 $\displaystyle\int_1^4 \frac{1}{x}\, dx, n = 6$

2 $\displaystyle\int_0^3 \frac{1}{1 + x}\, dx, n = 8$

3 $\displaystyle\int_0^1 \frac{1}{\sqrt{1 + x^2}}\, dx, n = 4$

4 $\displaystyle\int_2^3 \sqrt{1 + x^3}\, dx, n = 4$

5 $\displaystyle\int_0^2 \frac{1}{4 + x^2}\, dx, n = 10$

6 $\displaystyle\int_0^{0.6} \frac{1}{\sqrt{4 - x^2}}\, dx, n = 6$

7 $\displaystyle\int_1^{5/2} \sqrt[3]{x^2 + 8}\, dx, n = 6$

8 $\displaystyle\int_1^3 \frac{x}{x^4 + 1}\, dx, n = 4$

9–12 Rework Exercises 1–4 using the Midpoint Rule (5.37).

13 Use the Trapezoidal Rule (5.33) with $(b - a)/n = 0.1$ to show that

$$\int_1^{2.7} \frac{1}{x}\, dx < 1 < \int_1^{2.8} \frac{1}{x}\, dx.$$

14 Find upper bounds for the errors in parts (a) and (b) of Exercise 1.

Suppose the tables in Exercises 15 and 16 were obtained experimentally, where x and y are physical variables. Assuming that $y = f(x)$ where f is continuous, approximate $\int_2^4 f(x)\,dx$ by means of (a) the Trapezoidal Rule; (b) Simpson's Rule.

15

x	y
2.00	4.12
2.25	3.76
2.50	3.21
2.75	3.58
3.00	3.94
3.25	4.15
3.50	4.69
3.75	5.44
4.00	7.52

16

x	y
2.0	12.1
2.2	11.4
2.4	9.7
2.6	8.4
2.8	6.3
3.0	6.2
3.2	5.8
3.4	5.4
3.6	5.1
3.8	5.9
4.0	5.6

17 The graph in the accompanying figure was recorded by an instrument used to measure a physical quantity. Estimate y-coordinates of points on the graph and approximate the area of the shaded region by using (with $n = 6$) (a) the Trapezoidal Rule; (b) Simpson's Rule; (c) the Midpoint Rule.

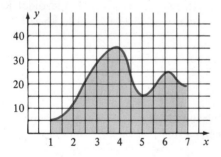

Figure for Exercise 17

18 A man-made lake has the shape illustrated in the accompanying figure, where adjacent measurements are 20 ft apart. Use the Trapezoidal Rule to estimate the surface area of the lake.

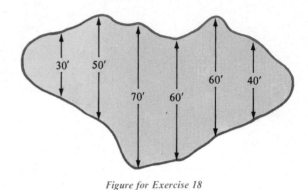

Figure for Exercise 18

19 Find the least integer n such that the error involved in approximating $\int_{1/2}^1 (1/x)\,dx$ is less that 0.0001 if (a) the Trapezoidal Rule is used; (b) Simpson's Rule is used.

20 Find the least integer n such that the error involved in approximating $\int_0^3 (1 + x)^{-1}\,dx$ is less than 0.001 if (a) the Trapezoidal Rule is used; (b) Simpson's Rule is used.

21 If $f(x)$ is a polynomial of degree less than 4, prove that Simpson's Rule gives the exact value of $\int_a^b f(x)\,dx$.

22 If f is continuous and has nonnegative values, and if $f''(x) > 0$ throughout $[a, b]$, prove that $\int_a^b f(x)\,dx$ is less than the number given by the Trapezoidal Rule.

5.8

REVIEW

Define or discuss each of the following.

1 The area under the graph of a nonnegative continuous function f from a to b

2 Partition of $[a, b]$

3 Norm of a partition

4 Riemann sum

5 $\displaystyle\lim_{\|P\| \to 0} \sum_i f(w_i)\,\Delta x_i = I$

6 The definite integral of f from a to b

7 Upper and lower limits of integration

8 Integrand

9 Properties of the definite integral

10 The Mean Value Theorem for Definite Integrals

11 The Fundamental Theorem of Calculus

12 Indefinite integral

13 Constant of integration

14 Power rule for indefinite integration

15 The method of substitution

16 Numerical integration

EXERCISES 5.8

Determine the sums in Exercises 1–3.

1 $\displaystyle\sum_{k=1}^{5}(k^2+3)$ **2** $\displaystyle\sum_{k=1}^{50}8$ **3** $\displaystyle\sum_{k=0}^{4}(-1/2)^{k-2}$

4 Suppose f is defined on the interval $[-2,3]$ by $f(x) = 1 - x^2$, and let P be the partition of $[-2,3]$ into five equal subintervals. Find the Riemann sum R_P if f is evaluated at the midpoint of each subinterval.

5 Work Exercise 4 if f is evaluated at the right-hand endpoint of each subinterval.

6 Given $\int_1^4 (x^2 + 2x - 5)\,dx$, find numbers that satisfy the conclusion of the Mean Value Theorem for Integrals.

Evaluate the integrals in Exercises 7–30.

7 $\displaystyle\int_0^1 \sqrt[3]{8x^7}\,dx$ **8** $\displaystyle\int \sqrt[3]{5t+1}\,dt$

9 $\displaystyle\int_0^1 \frac{z^2}{(1+z^3)^2}\,dz$ **10** $\displaystyle\int (x^2+4)^2\,dx$

11 $\displaystyle\int (1-2x^2)^3 x\,dx$ **12** $\displaystyle\int \frac{(1+\sqrt{x})^2}{\sqrt[3]{x}}\,dx$

13 $\displaystyle\int_1^2 \frac{w+1}{\sqrt{w^2+2w}}\,dw$ **14** $\displaystyle\int_1^2 \frac{s^2+2}{s^2}\,ds$

15 $\displaystyle\int \frac{1}{\sqrt{x}(1+\sqrt{x})^2}\,dx$ **16** $\displaystyle\int_1^2 \frac{x^2-x-6}{x+2}\,dx$

17 $\displaystyle\int (3-2x-5x^3)\,dx$ **18** $\displaystyle\int (y+y^{-1})^2\,dy$

19 $\displaystyle\int_0^2 x^2\sqrt{x^3+1}\,dx$ **20** $\displaystyle\int_1^1 3x^2\sqrt{x^3+x}\,dx$

21 $\displaystyle\int (4t+1)(4t^2+2t-7)^2\,dt$

22 $\displaystyle\int \frac{\sqrt[4]{1-v^{-1}}}{v^2}\,dv$

23 $\displaystyle\int (2x^{-3}-3x^{-2})\,dx$

24 $\displaystyle\int_1^9 \sqrt{2r+7}\,dr$

25 $\displaystyle\int_{-1}^2 D_x\sqrt{(x+1)/(x+2)}\,dx$

26 $\displaystyle\int D_x(2x^3+3x^2-4x-5)\,dx$

27 $\displaystyle\int D_y\sqrt[5]{y^4+2y^2+1}\,dy$

28 $\displaystyle\int_0^7 D_x(x^2/\sqrt{3x+4})\,dx$

29 $\displaystyle D_x \int_0^1 (x^3+x^2-7)^5\,dx$

30 $\displaystyle D_z \int_0^z (x^2+1)^{10}\,dx$

31 Evaluate $\int_0^{10}\sqrt{1+x^4}\,dx$ by using (a) the Trapezoidal Rule with $n=5$ and (b) Simpson's Rule with $n=8$. Use approximations to four decimal places for $f(x_i)$ and round off answers to two decimal places.

32 If $f(x) = x^4\sqrt{x^5+4}$, find the area of the region under the graph of f from 0 to 2.

33 Find the area of the region under the graph of $y = x^3/(x^4+1)^2$ from $x=1$ to $x=2$.

34 Use a definite integral to prove that the area of a right triangle of altitude a and base b is $ab/2$. (*Hint*: Take the vertices at the points $(0,0)$, $(b,0)$, and (b,a).)

6

APPLICATIONS OF THE DEFINITE INTEGRAL

The definite integral is useful for solving a large variety of applied problems. In this chapter we shall discuss area, volume, work, liquid force, lengths of curves, and some problems from economics, biology, physics, and engineering. Other applications will be considered later in the text.

6.1

AREA

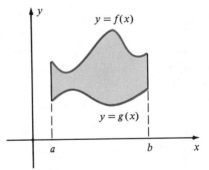

Figure 6.1

If a function f is continuous on a closed interval $[a, b]$, and $f(x) \geq 0$ for all x in $[a, b]$, then by Theorem (5.12) the area under the graph of f from a to b equals $\int_a^b f(x)\, dx$. If g is another continuous nonnegative valued function on $[a, b]$, and if $f(x) \geq g(x)$ for all x in $[a, b]$, then the area A of the region bounded by the graphs of f, g, $x = a$, and $x = b$ (see Figure 6.1) can be found by subtracting the area under the graph of g from the area under the graph of f, that is,

$$A = \int_a^b f(x)\, dx - \int_a^b g(x)\, dx = \int_a^b [f(x) - g(x)]\, dx.$$

This formula for A can be extended to the case in which f or g is negative for some x in $[a, b]$, as illustrated in (i) of Figure 6.2. To find the area of the indicated region, let us choose a negative number d less than the minimum value of g on $[a, b]$ and consider the two functions f_1 and g_1 defined by

$$f_1(x) = f(x) - d, \qquad g_1(x) = g(x) - d$$

for all x in $[a, b]$. Since d is negative, values of f_1 and g_1 are found by adding the positive number $|d|$ to corresponding values of f and g, respectively. Geometrically, this amounts to raising the graphs of f and g a distance $|d|$,

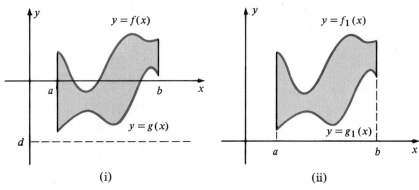

Figure 6.2

giving us a region having the same shape as the original region, but lying entirely above the x-axis.

If A is the area of the region in (ii) of Figure 6.2, then

$$A = \int_a^b [f_1(x) - g_1(x)] \, dx$$

$$= \int_a^b \{[f(x) - d] - [g(x) - d]\} \, dx$$

$$= \int_a^b [f(x) - g(x)] \, dx.$$

The preceding discussion may be summarized as follows.

Theorem (6.1)

> If f and g are continuous and $f(x) \geq g(x)$ for all x in $[a, b]$, then the area A of the region bounded by the graphs of f, g, $x = a$, and $x = b$ is
>
> $$A = \int_a^b [f(x) - g(x)] \, dx.$$

The formula for A in Theorem (6.1) may be interpreted as a limit of a sum. If we define the function h by $h(x) = f(x) - g(x)$, and if w is in $[a, b]$, then $h(w)$ is the distance from the point on the graph of g with x-coordinate w to the point on the graph of f with x-coordinate w. As in the discussion of Riemann sums in Chapter 5, let P denote a partition of $[a, b]$ determined by the numbers $a = x_0, x_1, \ldots, x_n = b$. For each i, let $\Delta x_i = x_i - x_{i-1}$, and let w_i be an arbitrary number in the ith subinterval $[x_{i-1}, x_i]$ of P. By the definition of h,

$$h(w_i) \, \Delta x_i = [f(w_i) - g(w_i)] \, \Delta x_i$$

which, in geometric terms, is the area of a rectangle of length $f(w_i) - g(w_i)$ and width Δx_i (see Figure 6.3).

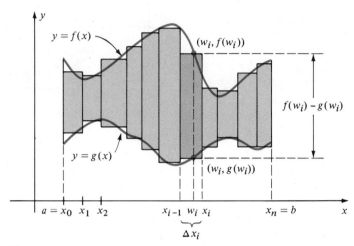

Figure 6.3

The Riemann sum

$$\sum_i h(w_i)\,\Delta x_i = \sum_i [f(w_i) - g(w_i)]\,\Delta x_i$$

is the sum of the areas of all rectangles pictured in Figure 6.3, and may be thought of as an approximation to the area of the region bounded by the graphs of f and g from $x = a$ to $x = b$. By the definition of the definite integral,

$$\lim_{\|P\| \to 0} \sum_i h(w_i)\,\Delta x_i = \int_a^b h(x)\,dx.$$

Using the definition of h gives us the following.

Theorem **(6.2)**

> If f and g are continuous and $f(x) \geq g(x)$ for all x in $[a, b]$, then the area A of the region bounded by the graphs of $f, g, x = a$, and $x = b$ is
>
> $$A = \lim_{\|P\| \to 0} \sum_i [f(w_i) - g(w_i)]\,\Delta x_i = \int_a^b [f(x) - g(x)]\,dx.$$

When we use Theorem (6.2) we initially think of approximating the region by means of rectangles of the type shown in Figure 6.3. After writing a formula for the area of a typical rectangle, we sum all such rectangles and, as in (6.2), take the limit of this sum to obtain the area of the region. This technique is illustrated in the following examples.

Example 1 Find the area of the region bounded by the graphs of the equations $y = x^2$ and $y = \sqrt{x}$.

Solution   We shall employ the Riemann sum approach. The region and a typical rectangle are sketched in Figure 6.4.

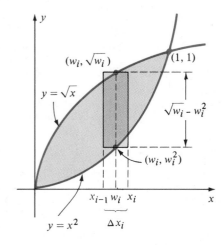

Figure 6.4

As indicated in the figure, the length of a typical rectangle is $\sqrt{w_i} - w_i^2$ and its area is $(\sqrt{w_i} - w_i^2)\,\Delta x_i$. Using Theorem (6.2) with $a = 0$ and $b = 1$, we obtain

$$A = \lim_{\|P\| \to 0} \sum_i (\sqrt{w_i} - w_i^2)\,\Delta x_i$$

$$= \int_0^1 (\sqrt{x} - x^2)\,dx$$

$$= \left[\frac{2}{3}x^{3/2} - \frac{1}{3}x^3 \right]_0^1 = \frac{2}{3} - \frac{1}{3} = \frac{1}{3}.$$

The area can also be found by direct substitution in Theorem (6.1), with $f(x) = \sqrt{x}$ and $g(x) = x^2$. ■

As illustrated in Example 1, to find areas by using limits of sums, we always begin by sketching the region with at least one typical rectangle, labeling the drawing appropriately. The reason for stressing the summation technique is that similar limiting processes will be employed later for calculating many other mathematical and physical quantities. Treating areas as limits of sums will make it easier to understand those future applications. At the same time it will help solidify the meaning of the definite integral.

Example 2 Find the area of the region bounded by the graphs of $y + x^2 = 6$ and $y + 2x - 3 = 0$.

Solution   The region and a typical rectangle are sketched in Figure 6.5. The points of intersection $(-1, 5)$ and $(3, -3)$ of the two graphs may be

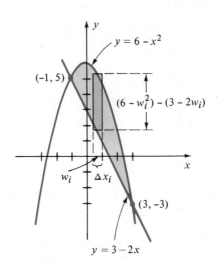

Figure 6.5

found by solving the two given equations simultaneously. To use Theorem (6.1) or (6.2) it is necessary to solve each equation for y in terms of x, obtaining

$$y = 6 - x^2 \quad \text{and} \quad y = 3 - 2x.$$

The functions f and g are then given by

$$f(x) = 6 - x^2 \quad \text{and} \quad g(x) = 3 - 2x.$$

As shown in Figure 6.5, the length of a typical rectangle is

$$(6 - w_i^2) - (3 - 2w_i)$$

where w_i is some number in the ith subinterval of a partition P of $[-1, 3]$. The area of this rectangle is

$$[(6 - w_i^2) - (3 - 2w_i)]\, \Delta x_i.$$

Applying Theorem (6.2),

$$
\begin{aligned}
A &= \lim_{\|P\| \to 0} \sum_i [(6 - w_i^2) - (3 - 2w_i)]\, \Delta x_i \\
&= \int_{-1}^{3} [(6 - x^2) - (3 - 2x)]\, dx \\
&= \int_{-1}^{3} (3 - x^2 + 2x)\, dx \\
&= \left[3x - \frac{x^3}{3} + x^2 \right]_{-1}^{3} \\
&= \left[9 - \frac{27}{3} + 9 \right] - \left[-3 - \left(-\frac{1}{3} \right) + 1 \right] = \frac{32}{3}.
\end{aligned}
$$

∎

After students thoroughly understand the limit of a sum technique illustrated in Examples 1 and 2, it is customary to bypass the subscript part of the solution and proceed immediately to "setting up" the integral. To illustrate, in Example 2 we could regard dx as the width of a typical rectangle. The length of the rectangle is represented by the distance

$$(6 - x^2) - (3 - 2x)$$

between the upper and lower boundaries of the region, where x is an arbitrary number in $[-1, 3]$. Thus the area of a rectangle may be represented, in nonsubscript notation, by

$$[(6 - x^2) - (3 - 2x)]\, dx.$$

Placing the integral sign $\int_{-1}^{3}$ in front of this expression may be regarded as summing all such terms and simultaneously letting the widths of the rectangles approach 0.

Since one of our objectives is to emphasize the definition of the definite integral (5.8), we will continue to use limits of sums in many of the remaining examples in this chapter. The symbol w_i employed in solutions always

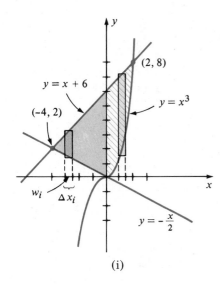

(i)

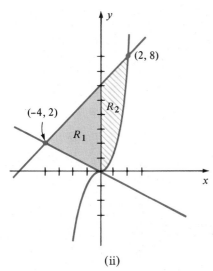

(ii)

Figure 6.6

denotes a number in the ith subinterval of a partition. Those who wish to use the nonsubscript approach may (at the discretion of the instructor) proceed immediately to the definite integral that follows a stated limit of a sum.

The following example illustrates that it is sometimes necessary to subdivide a region and use Theorem (6.1) or (6.2) more than once in order to find the area.

Example 3   Find the area of the region R that is bounded by the graphs of $y - x = 6$, $y - x^3 = 0$, and $2y + x = 0$.

Solution   The graphs and region are sketched in (i) of Figure 6.6 where, as indicated, each equation has been solved for y in terms of x so that appropriate functions may be introduced. Typical rectangles are shown extending from the lower boundary to the upper boundary of R. Since the lower boundary consists of portions of two different graphs, the area cannot be found by using only one definite integral. However, if R is divided into two subregions R_1 and R_2, as shown in (ii) of Figure 6.6, then we can determine the area of each and add them together.

For region R_1 the upper and lower boundaries are the lines $y = x + 6$ and $y = -x/2$, respectively, and hence the length of a typical rectangle is

$$(w_i + 6) - (-w_i/2).$$

Applying Theorem (6.2), the area A_1 of R_1 is

$$
\begin{aligned}
A_1 &= \lim_{\|P\| \to 0} \sum_i \left[(w_i + 6) - \left(-\frac{w_i}{2} \right) \right] \Delta x_i \\
&= \int_{-4}^{0} \left[(x + 6) - \left(-\frac{x}{2} \right) \right] dx \\
&= \int_{-4}^{0} \left[\frac{3}{2} x + 6 \right] dx = \left[\frac{3}{4} x^2 + 6x \right]_{-4}^{0} \\
&= 0 - (12 - 24) = 12.
\end{aligned}
$$

Region R_2 has the same upper boundary $y = x + 6$ as R_1; however, the lower boundary is given by $y = x^3$. In this case the length of a typical rectangle is

$$(w_i + 6) - w_i^3$$

and the area A_2 of R_2 is

$$
\begin{aligned}
A_2 &= \lim_{\|P\| \to 0} \sum_i \left[(w_i + 6) - w_i^3 \right] \Delta x_i \\
&= \int_{0}^{2} \left[(x + 6) - x^3 \right] dx \\
&= \left[\frac{x^2}{2} + 6x - \frac{x^4}{4} \right]_{0}^{2} \\
&= (2 + 12 - 4) - 0 = 10.
\end{aligned}
$$

The area of the entire region R is, therefore, $A_1 + A_2 = 12 + 10 = 22.$ ∎

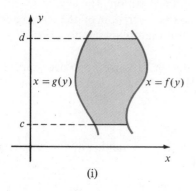

(i)

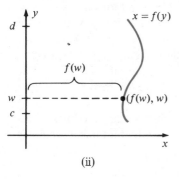

(ii)

Figure 6.7

Sometimes it is necessary to find the area A of the region bounded by the graphs of $y = c$ and $y = d$, and of two equations of the form $x = f(y)$ and $x = g(y)$, where f and g are continuous functions and $f(y) \geq g(y)$ for all y in $[c, d]$ (see (i) of Figure 6.7). In a manner similar to our earlier discussion, but with the roles of x and y interchanged, we obtain the formula

$$A = \int_c^d [f(y) - g(y)] \, dy$$

where we now regard y as the independent variable. We shall refer to this technique as **integration with respect to** y, whereas the method used in Theorem (6.1) is called **integration with respect to** x. As indicated in (ii) of Figure 6.7, if an equation is solved for x in terms of y such that $x = f(y)$, and if a value w is assigned to y, then $f(w)$ is the x-coordinate of the corresponding point on the graph.

Summation techniques can also be applied to regions of the type shown in Figure 6.7. In this case we select points on the y-axis with y-coordinates $y_0, y_1, \ldots, y_n$, where $y_0 = c$ and $y_n = d$, thereby obtaining a partition of the interval $[c, d]$ into subintervals of width $\Delta y_i = y_i - y_{i-1}$. For each i we choose a number w_i in $[y_{i-1}, y_i]$ and consider rectangles that have areas $[f(w_i) - g(w_i)] \Delta y_i$, as illustrated in Figure 6.8. This leads to

$$A = \lim_{\|P\| \to 0} \sum_i [f(w_i) - g(w_i)] \Delta y_i = \int_c^d [f(y) - g(y)] \, dy,$$

where the last equality follows from the definition of the definite integral.

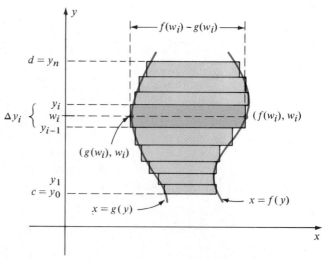

Figure 6.8

Example 4 Find the area of the region bounded by the graphs of the equations $2y^2 = x + 4$ and $x = y^2$.

Solution   Two sketches of the region are shown in Figure 6.9, where (i) illustrates the situation that occurs if we use vertical rectangles (integration with respect to x), and (ii) is the case if we use horizontal rectangles (integration with respect to y). Referring to (i) of the figure we see that several definite

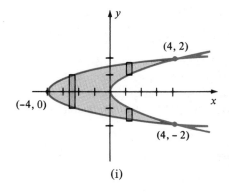

(i)

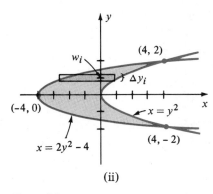

(ii)

Figure 6.9

integrals are required to find the area. (Why?) However, in (ii) we can use integration with respect to y to find the area with only one integration. Letting $f(y) = y^2$, $g(y) = 2y^2 - 4$, and referring to (ii) of Figure 6.9, the length $f(w_i) - g(w_i)$ of a *horizontal* rectangle is

$$w_i^2 - (2w_i^2 - 4).$$

Since the width is Δy_i, the area of the rectangle is

$$[w_i^2 - (2w_i^2 - 4)] \, \Delta y_i.$$

Hence, the area of R is

$$
\begin{aligned}
A &= \lim_{\|P\| \to 0} \sum_i [w_i^2 - (2w_i^2 - 4)] \, \Delta y_i \\
&= \int_{-2}^{2} [y^2 - (2y^2 - 4)] \, dy \\
&= \int_{-2}^{2} (4 - y^2) \, dy \\
&= \left[4y - \frac{y^3}{3} \right]_{-2}^{2} = \left[8 - \frac{8}{3} \right] - \left[-8 - \left(-\frac{8}{3} \right) \right] = \frac{32}{3}.
\end{aligned}
$$

Actually, the integration could have been simplified further. For example, since the x-axis bisects the region, it is sufficient to find the area of that part of the region that lies above the x-axis and double it, obtaining

$$A = 2 \int_{0}^{2} [y^2 - (2y^2 - 4)] \, dy. \qquad \blacksquare$$

Throughout this section we have assumed that the graphs of the functions (or equations) do not cross one another in the interval under discussion. For example, in Theorem (6.1) we demanded that $f(x) \geq g(x)$ for all x in $[a, b]$. If the graphs of f and g cross at one point $P(c, d)$, where $a < c < b$, and we wish to find the area bounded by the graphs from $x = a$ to $x = b$, then the theory developed in this section may still be used; however, *two* integrations are required, one corresponding to the interval $[a, c]$ and the other to $[c, b]$. If the graphs cross *several* times, then several integrals may be necessary. Problems whose graphs cross one or more times appear in Exercises 33–36.

In scientific investigations, physical interpretations are often attached to areas. One illustration of this occurs in the *theory of elasticity* where, to test the strength of a material, an investigator records values of strain that correspond to different loads (stresses). The sketch in Figure 6.10 is a typical stress-strain diagram for a sample of an elastic material such as vulcanized rubber. Note that it is customary to assign stress values in the vertical direction. Referring to the figure, we see that as the load applied to the material (the stress) was increased, the strain (indicated by the arrows on the solid part of the graph) increased until the material stretched to five times its original length. As the load was decreased, the elastic material returned to its original length; however, the same graph was not retraced. Instead, the

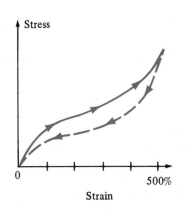

Figure 6.10 Stress-strain diagram for an elastic material

graph indicated by the dashes was obtained. This phenomenon is called *elastic hysteresis*. (A similar occurrence takes place in the study of magnetic materials, where it is called *magnetic hysteresis*.) The two curves in the figure make up a *hysteresis loop* for the material. It can be shown that the area of the region enclosed by this loop is directly proportional to the energy dissipated within the elastic (or magnetic) material during the test. In the case of vulcanized rubber, the larger the area, the better the material is for absorbing vibrations.

EXERCISES 6.1

In Exercises 1–20 sketch the region bounded by the graphs of the given equations, show a typical vertical or horizontal rectangle, and find the area of the region.

1   $y = 1/x^2, y = -x^2, x = 1, x = 2$

2   $y = \sqrt{x}, y = -x, x = 1, x = 4$

3   $y^2 = -x, x - y = 4, y = -1, y = 2$

4   $x = y^2, y - x = 2, y = -2, y = 3$

5   $y = x^2 + 1, y = 5$

6   $y = 4 - x^2, y = -4$

7   $y = x^2, y = 4x$

8   $y = x^3, y = x^2$

9   $y = 1 - x^2, y = x - 1$

10   $x + y = 3, y + x^2 = 3$

11   $y^2 = 4 + x, y^2 + x = 2$

12   $x = y^2, x - y - 2 = 0$

13   $y = x, y = 3x, x + y = 4$

14   $x - y + 1 = 0, 7x - y - 17 = 0, 2x + y + 2 = 0$

15   $y = x^3 - x, y = 0$

16   $y = x^3 - x^2 - 6x, y = 0$

17   $x = 4y - y^3, x = 0$

18   $x = y^{2/3}, x = y^2$

19   $y = x\sqrt{4 - x^2}, y = 0$

20   $y = x\sqrt{x^2 - 9}, y = 0, x = 5$

21   Let R be the region bounded by the graphs of $x - 2y = 0$, $x - 2y - 4 = 0, y = 3$, and $y = 0$. Sketch R and express its area A as a limit of a sum. Find A by using (a) integration; (b) a formula from geometry.

22   Let R be the region bounded by the graphs of $x = \sqrt{1 - y^2}$ and $x = 0$. Sketch R and express its area A as a limit of a sum. Find A without integrating.

23   If R is the region bounded by the graph of the equation $(x - 4)^2 + y^2 = 9$, express the area A of R as a limit of a sum. Find A without integrating.

24   If R is the region bounded by the graphs of $2x + 3y = 6$, $x = 0$, and $y = 0$, express the area A of R as a limit of a sum. Find A without integrating.

Each of Exercises 25–32 represents a limit of a sum for a function f on the indicated interval. Interpret each limit as an area of a region R and find its value. Sketch the region R.

25   $\lim\limits_{\|P\| \to 0} \sum\limits_i (4w_i + 1) \Delta x_i; [0, 1]$

26   $\lim\limits_{\|P\| \to 0} \sum\limits_i (w_i - w_i^3) \Delta x_i; [0, 1]$

27   $\lim\limits_{\|P\| \to 0} \sum\limits_i (4 - w_i^2) \Delta y_i; [0, 1]$

28   $\lim\limits_{\|P\| \to 0} \sum\limits_i \sqrt{3w_i + 1} \Delta y_i; [0, 1]$

29   $\lim\limits_{\|P\| \to 0} \sum\limits_i [w_i/(w_i^2 + 1)^2] \Delta x_i; [2, 5]$

30   $\lim\limits_{\|P\| \to 0} \sum\limits_i (5w_i/\sqrt{9 + w_i^2}) \Delta x_i; [1, 4]$

31   $\lim\limits_{\|P\| \to 0} \sum\limits_i [(5 + \sqrt{w_i})/\sqrt{w_i}] \Delta y_i; [1, 4]$

32   $\lim\limits_{\|P\| \to 0} \sum\limits_i w_i^{-2} \Delta y_i; [-5, -2]$

In Exercises 33–36 find the area of the region that lies between the graphs of f and g when x is restricted to the given interval $[a, b]$. (*Hint*: $|f(x) - g(x)|$ is the distance between the graphs when x is in $[a, b]$.)

33   $f(x) = 6 - 3x^2, g(x) = 3x; [0, 2]$

34   $f(x) = x^2 - 4, g(x) = x + 2; [1, 4]$

35   $f(x) = x^3 - 4x + 2, g(x) = 2; [-1, 3]$

36   $f(x) = x^2, g(x) = x^3; [-1, 2]$

37   The shape of a particular stress-strain diagram is shown in the accompanying figure (refer to the last paragraph of

this section). Estimate *y*-coordinates and approximate the area of the region enclosed by the hysteresis loop using (with $n = 6$) (a) the Trapezoidal Rule; (b) Simpson's Rule.

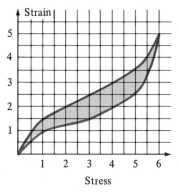

Figure for Exercise 37

38 Suppose the functional values of f and g in the following table were obtained empirically. Assuming that f and g are continuous, approximate the area between their graphs from $x = 1$ to $x = 5$ using (with $n = 8$) (a) the Trapezoidal Rule; (b) Simpson's Rule.

x	1	1.5	2	2.5	3	3.5	4	4.5	5
$f(x)$	3.5	2.5	3	4	3.5	2.5	2	2	3
$g(x)$	1.5	2	2	1.5	1	0.5	1	1.5	1

6.2

SOLIDS OF REVOLUTION

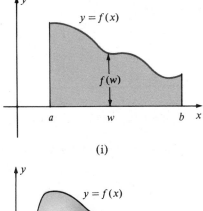

(i)

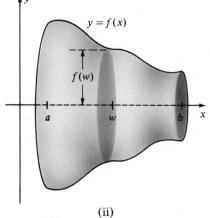

(ii)

Figure 6.11

The volume of an object plays an important role in many problems that arise in the physical sciences. For example, it is essential to know the volume in order to find the center of gravity or moment of inertia of a homogeneous solid. (These concepts will be discussed later in the text.) The task of determining the volume of an irregularly shaped object is often difficult, if not impossible. For this reason, we shall begin with objects that have simple shapes. Included in this category are the solids of revolution discussed here and in the next section.

If a region in a plane is revolved about a line in the plane, the resulting solid is called a **solid of revolution**, and the solid is said to be **generated** by the region. The line about which the revolution takes place is called an **axis of revolution**. If the region bounded by the graph of a continuous nonnegative valued function f, the x-axis, and the graphs of $x = a$ and $x = b$ (see (i) of Figure 6.11) is revolved about the x-axis, a solid of the type shown in (ii) of Figure 6.11 is generated. For example, if f is a constant function, then the region is rectangular and the solid generated is a right circular cylinder. If the graph of f is a semicircle with endpoints of a diameter at the points $(a, 0)$ and $(b, 0)$ where $b > a$, then the solid of revolution is a sphere with diameter $b - a$. If the given region is a right triangle with base on the x-axis and two vertices at the points $(a, 0)$ and $(b, 0)$ with the right angle at one of these points, then a right circular cone is generated.

If a plane perpendicular to the x-axis intersects the solid shown in (ii) of Figure 6.11, a circular cross section is obtained. If, as indicated in the figure, the plane passes through the point on the x-axis with x-coordinate w, then the radius of the circle is $f(w)$ and hence its area is $\pi[f(w)]^2$. We shall arrive at a definition for the volume of such a solid of revolution by using Riemann sums in a manner similar to that used for areas in the previous section.

Suppose f is continuous and $f(x) \geq 0$ for all x in $[a, b]$. Consider a Riemann sum $\sum_i f(w_i) \Delta x_i$, where w_i is any number in the ith subinterval $[x_{i-1}, x_i]$ of a partition P of $[a, b]$. Geometrically, this gives us a sum of areas of rectangles of the types shown in (i) of Figure 6.12. The solid generated by the polygon formed by these rectangles has the appearance shown in (ii) of the figure. Observe that the ith rectangle generates a circular disc (that is, a "flat" right circular cylinder) of base radius $f(w_i)$ and altitude, or "thickness," $\Delta x_i = x_i - x_{i-1}$. The volume of this disc is the area of the base times the altitude, that is, $\pi[f(w_i)]^2 \Delta x_i$. The sum of the volumes of all such discs is the volume of the solid shown in (ii) of Figure 6.12 and is given by

$$\sum_i \pi[f(w_i)]^2 \Delta x_i.$$

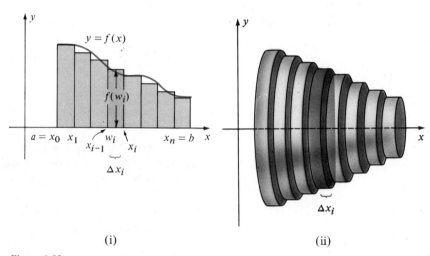

(i)   (ii)

Figure 6.12

The sum may be regarded as a Riemann sum for the function h defined by $h(x) = \pi[f(x)]^2$. Intuitively, it appears that if $\|P\|$ is close to zero, then the sum is close to the volume of the solid. It is natural, therefore, to define the volume of revolution as the limit of this sum as follows.

Definition (6.3)

Let f be continuous on $[a, b]$. The **volume** V of the solid of revolution generated by revolving the region bounded by the graphs of f, $x = a$, $x = b$, and the x-axis about the x-axis is

$$V = \lim_{\|P\| \to 0} \sum_i \pi[f(w_i)]^2 \Delta x_i = \int_a^b \pi[f(x)]^2 \, dx.$$

The fact that the limit of the sum in Definition (6.3) equals $\int_a^b \pi[f(x)]^2 \, dx$ follows from the definition of the definite integral. Hereafter, when considering limits of Riemann sums, the meanings of all symbols will not be explicitly pointed out. Instead, it will be assumed that the reader is aware of the significance of symbols such as $\|P\|$, w_i, and Δx_i.

The requirement that $f(x) \geq 0$ for all x in $[a, b]$ was omitted in Definition (6.3). If f is negative for some x, as illustrated in (i) of Figure 6.13, and if the region bounded by the graphs of f, $x = a$, $x = b$, and the x-axis is revolved about the x-axis, a solid of the type shown in (ii) of the figure is obtained. This solid is the same as that generated by revolving the region under the graph of $y = |f(x)|$ from a to b about the x-axis. Since $|f(x)|^2 = [f(x)]^2$, the limit in Definition (6.3) is the desired volume.

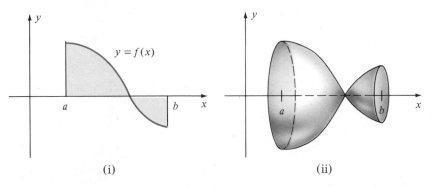

(i)   (ii)

Figure 6.13

Example 1 If $f(x) = x^2 + 1$, find the volume of the solid generated by revolving the region under the graph of f from -1 to 1 about the x-axis.

Solution   The solid is illustrated in Figure 6.14. Included in the sketch is a typical rectangle and the disc that it generates. Since the radius of the disc is $w_i^2 + 1$, its volume is

$$\pi(w_i^2 + 1)^2 \, \Delta x_i$$

and, as in (6.3),

$$V = \lim_{\|P\| \to 0} \sum_i \pi(w_i^2 + 1)^2 \, \Delta x_i$$

$$= \int_{-1}^{1} \pi(x^2 + 1)^2 \, dx = \pi \int_{-1}^{1} (x^4 + 2x^2 + 1) \, dx$$

$$= \pi \left[\frac{1}{5} x^5 + \frac{2}{3} x^3 + x \right]_{-1}^{1}$$

$$= \pi \left[\left(\frac{1}{5} + \frac{2}{3} + 1 \right) - \left(-\frac{1}{5} - \frac{2}{3} - 1 \right) \right] = \frac{56}{15} \pi. \qquad \blacksquare$$

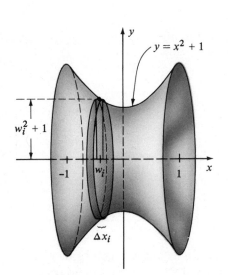

Figure 6.14

If we had used the symmetry of the graph in Example 1, then the volume could have been calculated by integrating from 0 to 1 and doubling the result. Another solution consists of substitution of $x^2 + 1$ for $f(x)$ in the formula $V = \int_a^b \pi [f(x)]^2\, dx$. We shall not ordinarily specify the units of measure. If the unit of linear measurement is inches, the volume is expressed in cubic inches. If x is in cm, then V is in cm^3, etc.

Sometimes it is convenient to find volumes by integrating with respect to y. For example, consider a region bounded by horizontal lines with y-intercepts c and d, by the y-axis, and by the graph of $x = g(y)$, where the function g is continuous for all y in $[c, d]$. If this region is revolved about the y-axis, the volume V of the resulting solid may be found by interchanging the roles of x and y in Definition (6.3). Specifically, let P be the partition of the interval $[c, d]$ determined by the numbers $c = y_0, y_1, \ldots, y_n = d$. Let $\Delta y_i = y_i - y_{i-1}$, let w_i be any number in the ith subinterval, and consider the rectangles of length $g(w_i)$ and width Δy_i as illustrated in (i) of Figure 6.15. The solid generated by revolving these rectangles about the y-axis is illustrated in (ii) of the figure.

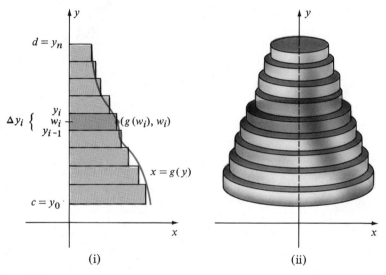

(i)   (ii)

Figure 6.15

The volume of the disc generated by the ith rectangle is $\pi [g(w_i)]^2\, \Delta y_i$. Summing and taking the limit gives us the following analogue of Definition (6.3).

Definition (6.4)

> Let g be continuous on $[c, d]$. The **volume** V of the solid of revolution generated by revolving the region bounded by the graphs of $x = g(y)$, $y = c$, $y = d$, and the y-axis about the y-axis is
>
> $$V = \lim_{\|P\| \to 0} \sum_i \pi [g(w_i)]^2\, \Delta y_i = \int_c^d \pi [g(y)]^2\, dy.$$

Example 2 The region bounded by the y-axis and the graphs of $y = x^3$, $y = 1$, and $y = 8$ is revolved about the y-axis. Find the volume of the resulting solid.

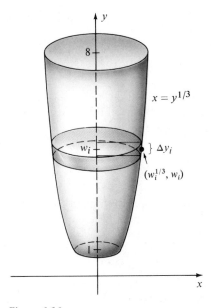

Figure 6.16

Solution   The solid is sketched in Figure 6.16 together with a disc generated by a typical horizontal rectangle. Since we plan to integrate with respect to y, we solve the equation $y = x^3$ for x in terms of y, obtaining $x = y^{1/3}$. If we let $x = g(y) = y^{1/3}$, then as shown in Figure 6.16, the radius of a typical disc is $g(w_i) = w_i^{1/3}$ and its volume is $\pi(w_i^{1/3})^2 \, \Delta y_i$. Applying Definition (6.4) with $g(y) = y^{1/3}$ gives us

$$V = \lim_{\|P\| \to 0} \sum_i \pi(w_i^{1/3})^2 \, \Delta y_i$$

$$= \int_1^8 \pi(y^{1/3})^2 \, dy = \pi \int_1^8 y^{2/3} \, dy$$

$$= \pi \left(\frac{3}{5}\right) \left[y^{5/3} \right]_1^8 = \frac{3}{5}\pi[8^{5/3} - 1] = \frac{93}{5}\pi. \qquad \blacksquare$$

Let us next consider a region bounded by the graphs of $x = a$, $x = b$, and of two continuous functions f and g where $f(x) \geq g(x) \geq 0$ for all x in $[a, b]$, as illustrated in (i) of Figure 6.17. If this region is revolved about the x-axis, a solid of the type illustrated in (ii) of the figure is obtained. Note that if $g(x) > 0$ for all x in $[a, b]$, then there is a hole through the solid.

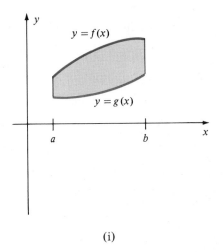

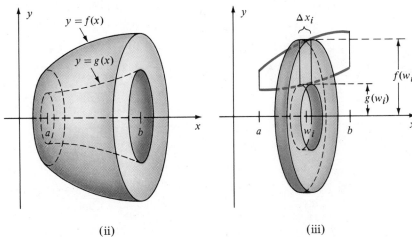

(i)   (ii)   (iii)

Figure 6.17

The volume V may be found by subtracting the volume of the solid generated by the smaller region from the volume of the solid generated by the larger region. Using Definition (6.3) gives us

$$V = \int_a^b \pi[f(x)]^2 \, dx - \int_a^b \pi[g(x)]^2 \, dx = \int_a^b \pi\{[f(x)]^2 - [g(x)]^2\} \, dx.$$

The last integral has an interesting interpretation as a limit of a sum. As illustrated in (iii) of Figure 6.17, a rectangle extending from the graph of g

to the graph of f, through the points with x-coordinate w_i, generates a washer-shaped solid whose volume is

$$\pi[f(w_i)]^2\,\Delta x_i - \pi[g(w_i)]^2\,\Delta x_i = \pi\{[f(w_i)]^2 - [g(w_i)]^2\}\,\Delta x_i.$$

Summing the volumes of all such washers and taking the limit gives us the desired integral formula. When working problems of this type it is often convenient to use the following general formula:

(6.5)   Volume of a washer $= \pi[(\text{outer radius})^2 - (\text{inner radius})^2]\cdot(\text{thickness}).$

In integration problems the thickness will be given by either Δx_i or Δy_i. To find the volume we take a limit of a sum of volumes of these washers.

Example 3   The region bounded by the graphs of the equations $x^2 = y - 2$, $2y - x - 2 = 0$, $x = 0$, and $x = 1$ is revolved about the x-axis. Find the volume of the resulting solid.

Solution   The region and a typical rectangle are sketched in (i) of Figure 6.18. Since we wish to integrate with respect to x we solve the first two equations for y in terms of x, obtaining $y = x^2 + 2$ and $y = \frac{1}{2}x + 1$. The washer generated by the rectangle in (i) is illustrated in (ii) of Figure 6.18.

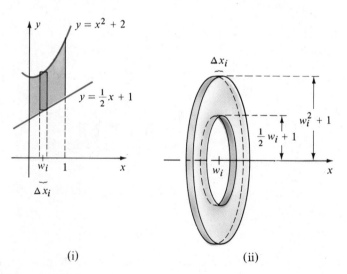

(i)   (ii)

Figure 6.18

Since the outer radius of the washer is $w_i^2 + 2$ and the inner radius is $\frac{1}{2}w_i + 1$, its volume (see (6.5)) is

$$\pi\left[(w_i^2 + 2)^2 - \left(\frac{1}{2}w_i + 1\right)^2\right]\Delta x_i.$$

Taking the limit of a sum of such volumes gives us

$$V = \int_0^1 \pi \left[(x^2 + 2)^2 - \left(\frac{1}{2}x + 1 \right)^2 \right] dx$$

$$= \pi \int_0^1 \left(x^4 + \frac{15}{4}x^2 - x + 3 \right) dx$$

$$= \pi \left[\frac{1}{5}x^5 + \frac{5}{4}x^3 - \frac{1}{2}x^2 + 3x \right]_0^1 = \frac{79\pi}{20}. \quad \blacksquare$$

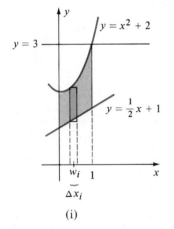

(i)

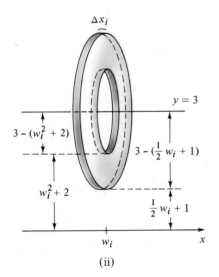

(ii)

Figure 6.19

Example 4 Find the volume of the solid generated by revolving the region described in Example 3 about the line $y = 3$.

Solution   The region and a typical rectangle are resketched in (i) of Figure 6.19, together with the axis of revolution $y = 3$. The washer generated by the rectangle is illustrated in (ii) of Figure 6.19. Note that the radii of this washer are as follows:

$$\text{Inner radius} = 3 - (w_i^2 + 2) = 1 - w_i^2$$

$$\text{Outer radius} = 3 - \left(\frac{1}{2}w_i + 1 \right) = 2 - \frac{1}{2}w_i.$$

Using (6.5), the volume of the washer is

$$\pi \left[\left(2 - \frac{1}{2}w_i \right)^2 - (1 - w_i^2)^2 \right] \Delta x_i.$$

Taking a limit of a sum of such terms gives us

$$V = \int_0^1 \pi \left[\left(2 - \frac{1}{2}x \right)^2 - (1 - x^2)^2 \right] dx$$

$$= \pi \int_0^1 \left[\left(4 - 2x + \frac{1}{4}x^2 \right) - (1 - 2x^2 + x^4) \right] dx$$

$$= \pi \int_0^1 \left(3 - 2x + \frac{9}{4}x^2 - x^4 \right) dx$$

$$= \pi \left[3x - x^2 + \frac{3}{4}x^3 - \frac{1}{5}x^5 \right]_0^1$$

$$= \pi \left[3 - 1 + \frac{3}{4} - \frac{1}{5} \right] = \frac{51}{20}\pi \approx 8.01. \quad \blacksquare$$

By interchanging the roles of x and y we can apply the techniques discussed in this section to solids generated by revolving regions about the y-axis, as in the next example.

Example 5 The region in the first quadrant bounded by the graphs of $y = \frac{1}{8}x^3$ and $y = 2x$ is revolved about the y-axis. Find the volume of the resulting solid.

Solution   The region and a typical rectangle are shown in (i) of Figure 6.20. Since we wish to integrate with respect to y, we solve the given equations for x in terms of y, obtaining

$$x = \frac{1}{2}y \quad \text{and} \quad x = 2y^{1/3}.$$

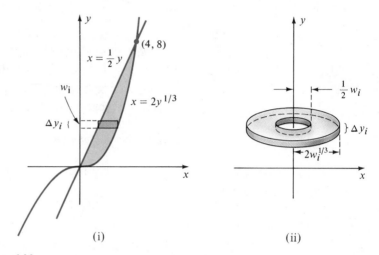

Figure 6.20

As shown in (ii) of Figure 6.20, the inner and outer radii of the washer generated by the rectangle are $\frac{1}{2}w_i$ and $2w_i^{1/3}$, respectively. Since the thickness is Δy_i it follows from (6.5) that the volume of the washer is

$$\pi\left[(2w_i^{1/3})^2 - \left(\frac{1}{2}w_i\right)^2\right]\Delta y_i = \pi\left[4w_i^{2/3} - \frac{1}{4}w_i^2\right]\Delta y_i.$$

Taking a limit of a sum of such terms gives us

$$V = \int_0^8 \pi\left[4y^{2/3} - \frac{1}{4}y^2\right]dy = \pi\left[\frac{12}{5}y^{5/3} - \frac{1}{12}y^3\right]_0^8$$

$$= \pi\left[\frac{12}{5}(8^{5/3}) - \frac{1}{12}(8^3)\right] = \frac{512}{15}\pi \approx 107.2. \quad \blacksquare$$

EXERCISES 6.2

In Exercises 1–12 sketch the region R bounded by the graphs of the given equations and find the volume of the solid generated by revolving R about the indicated axis. In each case show a typical rectangle, together with the disc or washer it generates. (*Remark*: Although the solids described in the following exercises are mathematical entities, it should be kept in mind that many objects encountered in everyday life have similar shapes.)

1 $y = 1/x, x = 1, x = 3, y = 0$; about the x-axis.

2 $y = \sqrt{x}, y = 0, x = 4$; about the x-axis.

3 $y = x^2$, $y = 2$; about the y-axis.

4 $y = 1/x$, $x = 0$, $y = 1$, $y = 3$; about the y-axis.

5 $y = x^2 - 4x$, $y = 0$; about the x-axis.

6 $y = x^3$, $x = -2$, $y = 0$; about the x-axis.

7 $y^2 = x$, $2y = x$; about the y-axis.

8 $y = 2x$, $y = 4x^2$; about the y-axis.

9 $y = x^2$, $y = 4 - x^2$; about the x-axis.

10 $x = y^3$, $x^2 + y = 0$; about the x-axis.

11 $x = y^2$, $y - x + 2 = 0$; about the y-axis.

12 $x + y = 1$, $y = x + 1$, $x = 2$; about the y-axis.

13 The region bounded by the graphs of $x = y^3$, $x = 8$, and $y = 0$ is revolved about the line $x = 8$. Find the volume of the resulting solid.

14 The region bounded by the graphs of $y = x^4$ and $y = 1$ is revolved about the line $y = 1$. Find the volume of the resulting solid.

15 If the region described in Exercise 13 is revolved about the line $y = 3$, find the volume of the resulting solid.

16 If the region described in Exercise 14 is revolved about the line $x = 2$, find the volume of the resulting solid.

17 Find the volume of the solid generated by revolving the region bounded by the graphs of $y = x^2$ and $y = 4$ (a) about the line $y = 4$; (b) about the line $y = 5$; (c) about the line $x = 2$.

18 Find the volume of the solid generated by revolving the region bounded by the graphs of $y = \sqrt{x}$, $y = 0$, and $x = 4$ (a) about the line $x = 4$; (b) about the line $x = 6$; (c) about the line $y = 2$.

In Exercises 19–24 sketch the region R bounded by the graphs of the equations given in (a) and then set up (but do not evaluate) integrals needed to find the volume of the solid obtained by revolving R about the line given in (b). As usual, show typical rectangles and the corresponding discs or washers.

19 (a) $y = x^3$, $y = 4x$
    (b) $y = 8$

20 (a) $y = x^3$, $y = 4x$
    (b) $x = 4$

21 (a) $x + y = 3$, $y + x^2 = 3$
    (b) $x = 2$

22 (a) $y = 1 - x^2$, $x - y = 1$
    (b) $y = 3$

23 (a) $x^2 + y^2 = 1$
    (b) $x = 5$

24 (a) $y = x^{2/3}$, $y = x^2$
    (b) $y = -1$

In Exercises 25–28 use a definite integral to derive a formula for the volume of the indicated solid.

25 A right circular cone of altitude h and radius of base r.

26 A sphere of radius r.

27 A frustum of a right circular cone of altitude h, lower base radius r_1, and upper base radius r_2.

28 The volume of a spherical segment of height h and radius of sphere r.

Exercises 29 and 30 each represent a limit of a sum for a function f on the interval $[0, 1]$. Interpret each limit as a volume and find its value.

29 $\lim\limits_{\|P\| \to 0} \sum\limits_i \pi(w_i^4 - w_i^6)\, \Delta x_i$

30 $\lim\limits_{\|P\| \to 0} \sum\limits_i \pi(w_i - w_i^8)\, \Delta y_i$

6.3

VOLUMES USING CYLINDRICAL SHELLS

There is another method for finding volumes of solids of revolution that, in certain cases, is simpler to apply than those discussed in Section 6.2. The method employs hollow circular cylinders, that is, thin cylindrical shells of the type illustrated in Figure 6.21.

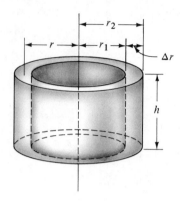

Figure 6.21

The volume of a shell having outer radius r_2, inner radius r_1, and altitude h is $\pi r_2^2 h - \pi r_1^2 h$. This expression may also be written

$$\pi(r_2^2 - r_1^2)h = \pi(r_2 + r_1)(r_2 - r_1)h$$
$$= 2\pi\left(\frac{r_2 + r_1}{2}\right)h(r_2 - r_1).$$

If we let $r = (r_2 + r_1)/2$ (the **average radius** of the shell) and $\Delta r = r_2 - r_1$ (the **thickness** of the shell), then the volume of the shell is given by $2\pi rh\,\Delta r$, that is,

(6.6)   Volume of a shell $= 2\pi$(average radius)(altitude)(thickness).

Let the function f be continuous and $f(x) \geq 0$ for all x in $[a, b]$, where $0 \leq a < b$. Let R be the region bounded by the graph of f, the x-axis, and by the graphs of $x = a$ and $x = b$, as illustrated in (i) of Figure 6.22. The solid generated by revolving R about the y-axis is illustrated in (ii) of the figure. Note that if $a > 0$, there is a hole through the solid. Let P be a partition of $[a, b]$ and consider a rectangle with base corresponding to the interval $[x_{i-1}, x_i]$ and altitude $f(w_i)$, where w_i is the midpoint of $[x_{i-1}, x_i]$. If this rectangle is revolved about the y-axis, then as illustrated in (iii) of Figure 6.22 there results a cylindrical shell with average radius w_i, altitude $f(w_i)$, and thickness $\Delta x_i = x_i - x_{i-1}$. By (6.6) we may express its volume as

$$2\pi w_i f(w_i)\,\Delta x_i.$$

Doing this for each subinterval in the partition and adding gives us

$$\sum_i 2\pi w_i f(w_i)\,\Delta x_i.$$

Geometrically, this sum represents the volume of a solid of the type illustrated in (iv) of Figure 6.22. Evidently, the smaller the norm $\|P\|$ of the partition, the better the sum approximates the volume V of the solid generated by R.

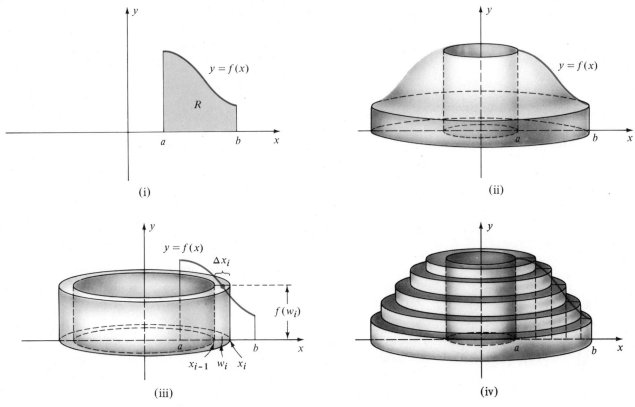

Figure 6.22

Indeed, it appears that the volume of the solid illustrated in (ii) of the figure is the limit of the sum. This gives us the following.

Definition (6.7)

Let f be continuous on $[a, b]$ where $0 \le a < b$. The **volume** V of the solid of revolution generated by revolving the region bounded by the graphs of f, $x = a$, $x = b$, and the x-axis about the y-axis is

$$V = \lim_{\|P\| \to 0} \sum_i 2\pi w_i f(w_i) \, \Delta x_i = \int_a^b 2\pi x f(x) \, dx.$$

The last equality follows from the definition of the definite integral. It can be proved that if the methods of Section 6.2 are also applicable, then both methods lead to the same answer.

Example 1 The region bounded by the graph of $y = 2x - x^2$ and the x-axis is revolved about the y-axis. Find the volume of the resulting solid.

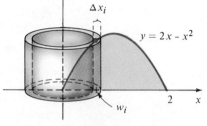

Figure 6.23

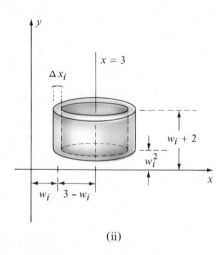

(i)

(ii)

Figure 6.24

Solution   The region and a shell generated by a typical rectangle are sketched in Figure 6.23, where w_i is the midpoint of $[x_{i-1}, x_i]$. Since the average radius of the shell is w_i, the altitude is $2w_i - w_i^2$, and the thickness is Δx_i, it follows from (6.6) that the volume of the shell is

$$2\pi w_i(2w_i - w_i^2)\, \Delta x_i.$$

Consequently, as in Definition (6.7), the volume V of the solid is

$$V = \lim_{\|P\| \to 0} \sum_i 2\pi w_i(2w_i - w_i^2)\, \Delta x_i$$

$$= \int_0^2 2\pi x(2x - x^2)\, dx = 2\pi \int_0^2 (2x^2 - x^3)\, dx$$

$$= 2\pi \left[\frac{2}{3}x^3 - \frac{1}{4}x^4 \right]_0^2 = \frac{8\pi}{3}.$$

The volume V can also be found using washers; however, the calculations would be more involved since the given equation would have to be solved for x in terms of y.   ■

Example 2   The region bounded by the graphs of $y = x^2$ and $y = x + 2$ is revolved about the line $x = 3$. Express the volume of the resulting solid as a definite integral.

Solution   The region and a typical rectangle are shown in (i) of Figure 6.24, where w_i represents the midpoint of the ith subinterval $[x_{i-1}, x_i]$. The cylindrical shell generated by the rectangle is illustrated in (ii) of Figure 6.24. This shell has the following dimensions:

$$\text{altitude} = (w_i + 2) - w_i^2$$
$$\text{average radius} = 3 - w_i$$
$$\text{thickness} = \Delta x_i.$$

Hence, by (6.6) its volume is

$$2\pi(3 - w_i)[(w_i + 2) - w_i^2]\, \Delta x_i.$$

Taking the limit of a sum of such terms gives us

$$V = \int_{-1}^2 2\pi(3 - x)(x + 2 - x^2)\, dx.$$ ■

The definite integral in the preceding example could be evaluated by multiplying the factors in the integrand and then integrating each term. Since we have worked a sufficient number of problems of this type, it would not be very instructive to carry out all of these details. For convenience we shall refer to the process of expressing V in terms of an integral as *setting up the integral for V*.

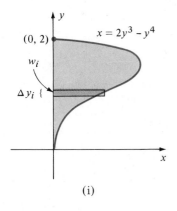

(i)

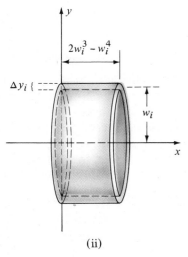

(ii)

Figure 6.25

By interchanging the roles of x and y we can find certain volumes by using shells and integrating with respect to y, as illustrated in the next example.

Example 3 The region in the first quadrant bounded by the graph of $x = 2y^3 - y^4$ and the y-axis is revolved about the x-axis. Set up the integral for the volume of the resulting solid.

Solution The region is sketched in (i) of Figure 6.25 together with a typical (horizontal) rectangle. The cylindrical shell generated by the rectangle is illustrated in (ii) of Figure 6.25. The shell in the figure has the following dimensions:

$$\text{altitude} = 2w_i^3 - w_i^4$$
$$\text{average radius} = w_i$$
$$\text{thickness} = \Delta y_i.$$

Hence, by (6.6) its volume is

$$2\pi w_i(2w_i^3 - w_i^4)\,\Delta y_i.$$

Taking a limit of a sum of such terms and remembering that y is the independent variable gives us

$$V = \int_0^2 2\pi y(2y^3 - y^4)\,dy. \qquad \blacksquare$$

It is worth noting that in the preceding example we were forced to use shells and to integrate with respect to y, since use of washers and integration with respect to x would require that the equation $x = 2y^3 - y^4$ be solved for y in terms of x, a rather formidable task.

EXERCISES 6.3

In Exercises 1–10 sketch the region R bounded by the graphs of the given equations and use the methods of this section to find the volume of the solid generated by revolving R about the indicated axis. In each case show a typical rectangle together with the cylindrical shell it generates.

1   $y = \sqrt{x}, x = 4, y = 0$; about the y-axis.

2   $y = 1/x, x = 1, x = 2, y = 0$; about the y-axis.

3   $y = x^2, y^2 = 8x$; about the y-axis.

4   $y = x^2 - 5x, y = 0$; about the y-axis.

5   $2x - y - 12 = 0, x - 2y - 3 = 0, x = 4$; about the y-axis.

6   $y = x^3 + 1, x + 2y = 2, x = 1$; about the y-axis.

7   $x^2 = 4y, y = 4$; about the x-axis.

8   $y^3 = x, y = 3, x = 0$; about the x-axis.

9   $y = 2x, y = 6, x = 0$; about the x-axis.

10   $2y = x, y = 4, x = 1$; about the x-axis.

11   The region bounded by the graphs of $yx^2 = 1, y = 1$, and $y = 4$ is revolved about the line $y = 5$. Find the volume of the resulting solid.

12 The region bounded by the graphs of $y^2 = x$, $x = 0$, $y = -1$, and $y = 1$ is revolved about the line $y = 2$. Find the volume of the resulting solid.

13 Find the volume of the solid generated by revolving the region bounded by the graphs of $y = x^2 + 1$, $x = 0$, $x = 2$, and $y = 0$ (a) about the line $x = 3$; (b) about the line $x = -1$.

14 Find the volume of the solid generated by revolving the region bounded by the graphs of $y = 4 - x^2$ and $y = 0$ (a) about the line $x = 2$; (b) about the line $x = -3$.

15–26 Use the methods of this section to solve Exercises 17–28 of Section 6.2.

The expressions in Exercises 27 and 28 each represent a limit of a sum for a function f on the interval $[0, 1]$. Interpret each limit as a volume and find its value.

27 $\lim\limits_{\|P\| \to 0} \sum\limits_{i} 2\pi(w_i^2 - w_i^3) \Delta x_i$

28 $\lim\limits_{\|P\| \to 0} \sum\limits_{i} 2\pi(w_i^5 + w_i^{3/2}) \Delta x_i$

6.4

VOLUMES BY SLICING

If a plane intersects a solid, then the region common to the plane and the solid is called a **cross-section** of the solid. In Section 6.2 we encountered circular and washer-shaped cross sections. We shall now consider solids that have the property that for every x in a closed interval $[a, b]$ on a coordinate line l, the plane perpendicular to l at the point with coordinate x intersects the solid in a cross section whose area is given by $A(x)$, where A is a continuous function on $[a, b]$. Figures 6.26 and 6.27 illustrate solids of the type we wish to discuss.

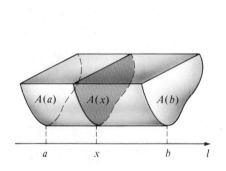

Figure 6.26

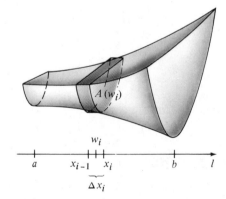

Figure 6.27

The solid is called a **cylinder** if, as illustrated in Figure 6.26, all cross sections are the same. If we are only interested in that part of the graph bounded by planes through the points with coordinates a and b, then the cross sections determined by these planes are called the **bases** of the cylinder and the distance between the bases is called the **altitude**. By definition, the volume of such a cylinder is the area of a base multiplied by the altitude. As a special case, a right circular cylinder of base radius r and altitude h has volume $\pi r^2 h$.

The solid illustrated in Figure 6.27 is not a cylinder since cross sections by planes perpendicular to l are not all the same. To find the volume we begin with a partition P of $[a, b]$ by choosing $a = x_0, x_1, x_2, \ldots, x_n = b$. The planes perpendicular to l at the points with these coordinates slice the solid into smaller pieces. The ith such slice is shown in Figure 6.27. As usual, let $\Delta x_i = x_i - x_{i-1}$ and choose any number w_i in $[x_{i-1}, x_i]$. It appears that if Δx_i is small, then the volume of the slice can be approximated by the volume of the cylinder of base area $A(w_i)$ and altitude Δx_i, that is, by $A(w_i)\,\Delta x_i$. Consequently, the total volume of the solid is approximated by the Riemann sum $\sum_{i=1}^{n} A(w_i)\,\Delta x_i$. Since the approximation improves as $\|P\|$ gets smaller, we define the volume V of the solid as the limit of this sum. Our discussion may be summarized as follows, where it is assumed that the conditions on $A(x)$ are those we have imposed previously.

(6.8)
$$V = \lim_{\|P\| \to 0} \sum_i A(w_i)\,\Delta x_i = \int_a^b A(x)\,dx$$

where $A(x)$ is the area of a cross section corresponding to the number x in $[a, b]$.

Example 1 Find the volume of a right pyramid that has altitude h and square base of side a.

Solution If, as shown in Figure 6.28, we introduce a coordinate line l along the axis of the pyramid, with origin O at the vertex, then cross sections by planes perpendicular to l are squares. If $A(x)$ is the cross-sectional area determined by the plane that intersects the axis x units from O, then

$$A(x) = (2y)^2 = 4y^2$$

where y is the distance indicated in the figure. By similar triangles

$$\frac{y}{x} = \frac{a/2}{h}, \quad \text{or} \quad y = \frac{ax}{2h}$$

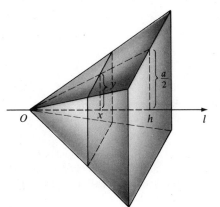

Figure 6.28

and hence
$$A(x) = 4y^2 = \frac{4a^2x^2}{4h^2} = \frac{a^2}{h^2}x^2.$$

Applying (6.8),

$$V = \int_0^h \left(\frac{a^2}{h^2}\right)x^2\,dx = \left(\frac{a^2}{h^2}\right)\left[\frac{x^3}{3}\right]_0^h = \frac{a^2h}{3}. \qquad \blacksquare$$

Example 2 A solid has, as its base, the circular region in the xy-plane bounded by the graph of $x^2 + y^2 = a^2$, where $a > 0$. Find the volume of the solid if every cross section by a plane perpendicular to the x-axis is an equilateral triangle with one side in the base.

Solution   A typical cross section by a plane x units from the origin is illustrated in Figure 6.29.

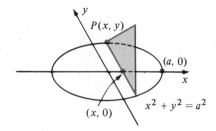

Figure 6.29

If the point $P(x, y)$ is on the circle, then the length of a side of the triangle is $2y$ and the altitude is $\sqrt{3}y$. Hence the area $A(x)$ of the pictured triangle is

$$A(x) = \tfrac{1}{2}(2y)(\sqrt{3}y) = \sqrt{3}y^2 = \sqrt{3}(a^2 - x^2).$$

Applying (6.8) gives us

$$V = \int_{-a}^{a} \sqrt{3}(a^2 - x^2)\, dx = \sqrt{3}\left[a^2 x - \frac{x^3}{3}\right]_{-a}^{a} = \frac{4\sqrt{3}a^3}{3}. \quad \blacksquare$$

EXERCISES 6.4

1 The base of a solid is the circular region in the xy-plane bounded by the graph of $x^2 + y^2 = a^2$, where $a > 0$. Find the volume of the solid if every cross section by a plane perpendicular to the x-axis is a square.

2 Work Exercise 1 if every cross section is an isosceles triangle with base on the xy-plane and altitude equal to the length of the base.

3 A solid has as its base the region in the xy-plane bounded by the graphs of $y = 4$ and $y = x^2$. Find the volume of the solid if every cross section by a plane perpendicular to the x-axis is an isosceles right triangle with hypotenuse on the xy-plane.

4 Work Exercise 3 if each cross section is a square.

5 Find the volume of a pyramid of the type illustrated in Figure 6.28 if the altitude is h and the base is a rectangle of dimensions a and $2a$.

6 A solid has as its base the region in the xy-plane bounded by the graphs of $y = x$ and $y^2 = x$. Find the volume of the solid if every cross section by a plane perpendicular to the x-axis is a semicircle with diameter in the xy-plane.

7 A solid has as its base the region in the xy-plane bounded by the graphs of $y^2 = 4x$ and $x = 4$. If every cross section by a plane perpendicular to the y-axis is a semicircle, find the volume of the solid.

8 A solid has as its base the region in the xy-plane bounded by the graphs of $x^2 = 16y$ and $y = 2$. Every cross section by a plane perpendicular to the y-axis is a rectangle whose height is twice that of the side in the xy-plane. Find the volume of the solid.

9 A log having the shape of a right circular cylinder of radius a is lying on its side. A wedge is removed from the log by making a vertical cut and another cut at an angle of 45°, both cuts intersecting at the center of the log (see figure). Find the volume of the wedge.

Figure for Exercise 9

10 The axes of two right circular cylinders of radius a intersect at right angles. Find the volume of the solid bounded by the cylinders. (*Hint*: Use square cross sections.)

11 The base of a solid is the circular region in the xy-plane bounded by the graph of $x^2 + y^2 = a^2$, where $a > 0$. Find the volume of the solid if every cross section by a plane perpendicular to the x-axis is an isosceles triangle of constant altitude b. (*Hint*: Interpret $\int_{-a}^{a} \sqrt{a^2 - x^2}\, dx$ as an area.)

12 Cross sections of a horn-shaped solid by planes perpendicular to its axis are circles. If a cross section that is s inches from the smaller end of the solid has diameter $6 + (s^2/36)$ in., and if the length of the solid is 2 ft, find its volume.

13 A tetrahedron has three mutually perpendicular faces and three mutually perpendicular edges of lengths 2, 3, and 4 cm, respectively. Find its volume.

14 A hole of diameter d is bored through a spherical solid of radius r such that the axis of the hole coincides with a diameter of the sphere. Find the volume of the solid that remains.

15 The base of a solid is an isosceles right triangle whose equal sides have length a. Find the volume if cross sections that are perpendicular to the base and to one of the equal sides are semi-circular.

16 Work Exercise 15 if the cross sections are regular hexagons with one side in the base.

17 A **prismatoid** is a solid whose cross-sectional areas by planes parallel to, and a distance x from, a fixed plane can be expressed as $ax^2 + bx + c$ where a, b, and c are real numbers. Prove that the volume V of a prismatoid is

$$V = \frac{h}{6}(B_1 + B_2 + 4B)$$

where B_1 and B_2 are the areas of the bases, B is the cross-sectional area parallel to and halfway between the bases, and h is the distance between the bases. Show that the frustum of a right circular cone of base radii r_1 and r_2 and altitude h is a prismatoid and find its volume.

18 *Cavalieri's Theorem* states that if two solids have equal altitudes and if all cross sections by planes parallel to their bases and at the same distances from their bases have equal areas, then the solids have the same volume. Prove this theorem.

19 Show that the disc and washer methods discussed in Section 6.2 are special cases of (6.8).

20 A circular swimming pool has diameter 28 ft. The depth of the water changes slowly from 3 ft at a point A on one side of the pool to 9 ft at a point B diametrically opposite A (see figure). Depth readings (in ft) taken along the diameter AB are given in the following table, where x is the distance (in ft) from A.

x	0	4	8	12	16	20	24	28
Water depth	3	3.5	4	5	6.5	8	8.5	9

Use the Trapezoidal Rule (with $n = 7$) to estimate the volume of water in the pool. Approximate the number of gallons of water contained in the pool (1 gal ≈ 0.134 ft^3).

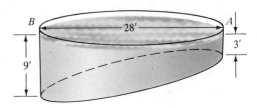

Figure for Exercise 20

6.5

WORK

The concept of **force** may be thought of intuitively as the physical entity that is used to describe a push or pull on an object. For example, a force is needed to push or pull an object along a horizontal plane, to lift an object off the ground, or to move a charged particle through an electromagnetic field.

Forces are often measured in pounds. If an object weighs ten pounds, then by definition the force required to lift it (or hold it off the ground) is ten pounds. A force of this type is a **constant force**, since its magnitude does not change while it is applied to the object.

The concept of *work* is used when a force moves an object from one place to another. The following definition covers the simplest case, where the object moves along a line in the same direction as the applied force.

Definition (6.9)

> If a constant force F is applied to an object, moving it a distance d in the direction of the force, then the **work** W done on the object is
>
> $$W = Fd.$$

In the British system, if F is measured in pounds and d in feet, the unit for W is the foot-pound (ft-lb). If F is in pounds and d in inches, then the unit of work is the inch-pound (in.-lb). In the metric system, two different units of force are used, depending on whether the cgs (centimeter-gram-second) system or the mks (meter-kilogram-second) system is employed. In the cgs system, a **dyne** is defined as the force that, when applied to a mass of 1 gram, induces an acceleration of 1 cm/sec^2. If F is expressed in dynes and d in centimeters, then the unit of work is the dyne-centimeter (dyn-cm), or **erg**. In the mks system, 1 **Newton** is the magnitude of the force required to impart an acceleration of 1 m/sec^2 to a mass of 1 kg. If F is given in Newtons and d in meters, then the unit of work is the Newton-meter (N-m), or **joule** (J). It can be shown that 1 joule $= 10^7$ ergs ≈ 0.74 ft-lb.

Example 1 Find the work done in pushing an automobile along a level road from a point A to another point B, 20 feet from A, while exerting a constant force of 90 pounds.

Solution The problem is illustrated in Figure 6.30, where we have pictured the road as part of a line l. Since the constant force is $F = 90$ lb and the distance the automobile moves is $d = 20$ ft, it follows from Definition (6.9) that the work done is

$$W = (90)(20) = 1,800 \text{ ft-lb.}$$

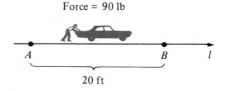

Force = 90 lb

Figure 6.30

As illustrated in the next example, it is sometimes necessary to determine F before applying Definition (6.9). In this event the formula $F = ma$ from physics may be used, where m is the mass of the object and a is the acceleration.

Example 2   Approximate the work done if an object of mass 15 kg is lifted vertically, a distance of 4 meters.

Solution   The required force F is given by $F = ma$, where if the mass m is measured in kg, and the acceleration a in m/sec², then F is in Newtons. In this example, a is the acceleration due to gravity, which is, in the mks system, approximately 9.81 m/sec². Hence

$$F = ma \approx (15)(9.81) = 147.15 \, \text{N}.$$

Applying Definition (6.9) and the mks system,

$$W = Fd \approx (147.15)(4) = 588.6 \, \text{N-m} = 588.6 \, \text{joules}. \qquad \blacksquare$$

Anyone who has pushed an automobile (or some other object) is aware of the fact that the force applied usually varies from one point to another. Thus, if an automobile is stalled it may require a larger force to get it moving than that which is needed after it is in motion. In addition, the force could vary considerably because of friction since, for example, part of the road may be smooth concrete and another part rough gravel. Similarly, someone inside the automobile could change the force required to move it by applying the brakes from time to time. Forces that are not constant are sometimes referred to as *variable forces*.

If a variable force is applied to an object, moving it a certain distance in the direction of the force, then methods of calculus are needed to find the work done. For the present we shall assume that the object moves along a line l. The case of motion along a nonlinear path will be considered in Chapter 18. Let us introduce a coordinate system on l and assume that the object moves from the point A with coordinate a to the point B with coordinate b where $b > a$. To solve the problem, it is essential to know the force at the point on l with coordinate x, for every x in the interval $[a, b]$. This force will be denoted by $f(x)$. For simplicity, it will be assumed that the function f obtained in this manner is continuous on $[a, b]$.

Let P denote the partition of the interval $[a, b]$ determined by the numbers $a = x_0, x_1, \ldots, x_n = b$ and let $\Delta x_i = x_i - x_{i-1}$ (see Figure 6.31). If w_i is a number in $[x_{i-1}, x_i]$, then the force at the point Q with coordinate w_i is $f(w_i)$. If Δx_i is small, then since f is continuous, the values of f change very little in the interval $[x_{i-1}, x_i]$. Roughly speaking, the function f is almost constant on $[x_{i-1}, x_i]$. It appears, therefore, that the work W_i done as the object moves through the ith subinterval may be approximated by means of Definition (6.9); that is,

$$W_i \approx f(w_i) \, \Delta x_i.$$

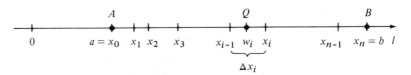

Figure 6.31

It seems evident that the smaller we choose Δx_i, the better $f(w_i)\,\Delta x_i$ approximates the work done in the interval $[x_{i-1}, x_i]$. If it is also assumed that work is additive, in the sense that the work W done as the object moves from A to B can be found by adding the work done over each subinterval, then

$$W \approx \sum_{i=1}^{n} f(w_i)\,\Delta x_i.$$

Since we expect this approximation to improve as the norm $\|P\|$ of the partition becomes smaller, it is natural to define W as the limit of the preceding sum. This limit leads to a definite integral.

Definition (6.10)

> Let the force at the point with coordinate x on a coordinate line l be $f(x)$, where f is continuous on $[a, b]$. The **work** W done in moving an object from the point with coordinate a to the point with coordinate b is
>
> $$W = \lim_{\|P\|\to 0} \sum_{i} f(w_i)\,\Delta x_i = \int_{a}^{b} f(x)\,dx.$$

The formula in Definition (6.10) can be used to find the work done in stretching or compressing a spring. To solve problems of this type it is necessary to use the following law from physics.

Hooke's Law (6.11)

> The force $f(x)$ required to stretch a spring x units beyond its natural length is
>
> $$f(x) = kx$$
>
> where k is a constant called the **spring constant**.

The formula in (6.11) is also used to find the work done in compressing a spring x units from its natural length.

Example 3 A force of 9 lb is required to stretch a spring from its natural length of 6 in. to a length of 8 in. Find the work done in stretching the spring (a) from its natural length to a length of 10 in.; (b) from a length of 7 in. to a length of 9 in.

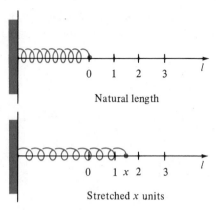

Natural length

Stretched x units

Figure 6.32

Solution

(a)   Let us introduce a coordinate line *l* as shown in Figure 6.32, where one end of the spring is attached to some point to the left of the origin and the end to be pulled is located at the origin. According to Hooke's Law (6.11), the force $f(x)$ required to stretch a spring x units beyond its natural length is

$$f(x) = kx$$

for some constant k. Using the given data, $f(2) = 9$. Substituting in $f(x) = kx$ we obtain $9 = k \cdot 2$, and hence the spring constant is $k = \frac{9}{2}$. Consequently, for this spring, Hooke's Law has the form

$$f(x) = \frac{9}{2}x.$$

By Definition (6.10) the work done in stretching the spring 4 in. is

$$W = \int_0^4 \frac{9}{2}x\,dx = \frac{9}{4}x^2\bigg]_0^4 = 36 \text{ in.-lb.}$$

(b)   We use the same function f but change the interval to $[1, 3]$, obtaining

$$W = \int_1^3 \frac{9}{2}x\,dx = \frac{9}{4}x^2\bigg]_1^3 = \frac{81}{4} - \frac{9}{4} = 18 \text{ in.-lb.} \quad \blacksquare$$

Example 4   A right circular conical tank of altitude 20 ft and radius of base 5 ft has its vertex at ground level and axis vertical. If the tank is full of water, find the work done in pumping the water over the top of the tank.

Solution   We begin by introducing a coordinate system as shown in Figure 6.33. The cone intersects the xy-plane along the line of slope 4 through the origin. An equation for this line is $y = 4x$.

Let P denote the partition of the interval $[0, 20]$ determined by $0 = y_0$, $y_1, y_2, \ldots, y_n = 20$, let $\Delta y_i = y_i - y_{i-1}$, and let x_i be the x-coordinate of the point on $y = 4x$ with y-coordinate y_i. If the cone is subdivided by means of planes perpendicular to the y-axis at each y_i, then we may think of the water as being sliced into n parts. As illustrated in Figure 6.33, the volume of the ith slice may be approximated by the volume $\pi x_i^2\,\Delta y_i$ of a circular disc or, since $x_i = y_i/4$, by $\pi(y_i/4)^2\,\Delta y_i$. This leads to the approximation

$$\text{Volume of } i\text{th slice} \approx \pi\left(\frac{y_i^2}{16}\right)\Delta y_i.$$

Assuming that water weighs 62.5 lb/ft³, the weight of the disc in Figure 6.33 is approximately $62.5\pi(y_i^2/16)\,\Delta y_i$. By (6.9), the work done in lifting the disc to the top of the tank is the product of the distance $20 - y_i$ and the weight, that is,

$$\text{Work done in lifting } i\text{th slice} \approx (20 - y_i)62.5\pi\left(\frac{y_i^2}{16}\right)\Delta y_i.$$

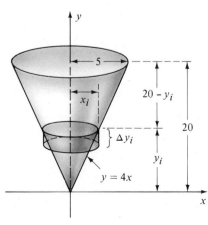

Figure 6.33

Since the last number is an approximation to the work done in lifting the ith slice to the top, the work done in emptying the entire tank is approximately

$$\sum_{i=1}^{n} (20 - y_i)62.5\pi\left(\frac{y_i^2}{16}\right)\Delta y_i.$$

The actual work W is obtained by taking the limit of this sum as the norm $\|P\|$ approaches zero. This gives us

$$W = \int_0^{20} (20 - y)62.5\pi\left(\frac{y^2}{16}\right) dy$$

$$= \frac{62.5\pi}{16} \int_0^{20} (20y^2 - y^3) \, dy$$

$$= \frac{62.5\pi}{16} \left[\frac{20y^3}{3} - \frac{y^4}{4}\right]_0^{20}$$

$$= \frac{62.5\pi}{16} \left(\frac{40,000}{3}\right) \approx 163,625 \text{ ft-lb.} \quad \blacksquare$$

Example 5 The pressure p (lb/in.^{2}) and volume v (in.^{3}) of an enclosed expanding gas are related by the formula $pv^k = c$, where k and c are constants. If the gas expands from $v = a$ to $v = b$, show that the work done (in.-lb) is given by

$$W = \int_a^b p \, dv.$$

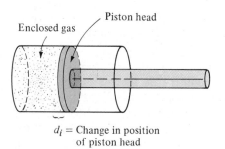

Enclosed gas

Piston head

d_i = Change in position of piston head

Figure 6.34

Solution Since the work done is independent of the shape of the container, we may assume that the gas is enclosed in a right circular cylinder of radius r and that the expansion takes place against a piston head as illustrated in Figure 6.34. Let P denote a partition of $[a, b]$ determined by $a = v_0, v_1, \ldots, v_n = b$, and let $\Delta v_i = v_i - v_{i-1}$. We shall regard the expansion as taking place in volume increments $\Delta v_1, \Delta v_2, \ldots, \Delta v_n$, and let $d_1, d_2, \ldots, d_n$ be the corresponding distances that the piston head moves (see Figure 6.34). It follows that for each i,

$$\Delta v_i = \pi r^2 d_i \quad \text{or} \quad d_i = \left(\frac{1}{\pi r^2}\right)\Delta v_i.$$

If p_i represents a value of the pressure p corresponding to the ith increment, then the force against the piston head is the product of p_i and the area of the piston head, that is, $p_i \pi r^2$. Hence the work done in this increment is

$$(p_i \pi r^2)d_i = (p_i \pi r^2)\left(\frac{1}{\pi r^2}\right)\Delta v_i = p_i \Delta v_i$$

and, therefore, $$W \approx \sum_i p_i \, \Delta v_i.$$

Since this approximation improves as the Δv_i approach zero, we conclude that

$$W = \lim_{\|P\| \to 0} \sum_i p_i \, \Delta v_i = \int_a^b p \, dv. \qquad \blacksquare$$

EXERCISES 6.5

1   A spring of natural length 10 in. stretches 1.5 in. under a weight of 8 lb. Find the work done in stretching the spring (a) from its natural length to a length of 14 in.; (b) from a length of 11 in. to a length of 13 in.

2   A force of 25 N is required to compress a spring of natural length 0.80 m to a length of 0.75 m. Find the work done in compressing the spring from its natural length to a length of 0.70 m.

3   If a spring is 12 cm long, compare the work done in stretching it from 12 cm to 13 cm with that done in stretching it from 13 cm to 14 cm.

4   It requires 60 ergs of work to stretch a certain spring from a length of 6 cm to 7 cm, and another 120 ergs of work to stretch it from 7 cm to 8 cm. Find the spring constant and the natural length of the spring.

5   A fishtank has a rectangular base of width 2 ft and length 4 ft, and rectangular sides of height 3 ft. If the tank is filled with water, what work is required to pump the water out over the top of the tank?

6   Generalize Example 4 of this section to the case of a conical tank of altitude h and radius of base a which is filled with a liquid of density (weight per unit volume) ρ.

7   A freight elevator of mass 1500 kg is supported by a cable 4 m long whose mass is 7 kg per linear meter. Approximate the work required to lift the elevator 3 m by winding the cable onto a winch.

8   A 170-pound man climbs a vertical telephone pole 15 ft high. What work is done if he reaches the top in (a) 10 sec? (b) 5 sec?

9   A vertical cylindrical tank of diameter 3 ft and height 6 ft is full of water. Find the work required to pump the water (a) out over the top of the tank; (b) out through a pipe which rises to a height 4 ft above the top of the tank.

10 Answer parts (a) and (b) of Exercise 9 if the tank is only half full of water.

11 The ends of a water trough 8 ft long are equilateral triangles having sides of width 2 ft. If the trough is full of water, find the work required to pump it out over the top.

12 A cistern has the shape of a hemisphere of radius 5 ft, with the circular part at the top. If the cistern is full of water, find the work required to pump all the water to a point 4 ft above the top of the cistern.

13 The force (in dyn) with which two electrons repel one another is inversely proportional to the square of the distance (in cm) between them.
(a) If one electron is held fixed at the point $(5, 0)$, find the work done in moving a second electron along the x-axis from the origin to the point $(3, 0)$.
(b) If two electrons are held fixed at the points $(5, 0)$ and $(-5, 0)$, respectively, find the work done in moving a third electron from the origin to $(3, 0)$.

14 A uniform cable 12 m long and having mass 30 kg hangs vertically from a pulley system at the top of a building (see figure). If a steel beam of mass 250 kg is attached to the end of the cable, what work is required to pull it to the top?

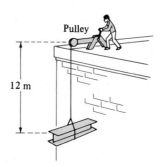

Figure for Exercise 14

15 A bucket containing water is lifted vertically at a constant rate of 1.5 ft/sec by means of a rope of negligible weight. As it rises, water leaks out at the rate of 0.25 lb/sec. If the bucket weighs 4 lb when empty, and if it contained 20 lb of water at the instant that the lifting began, determine the work done in raising the bucket 12 ft.

16 In Exercise 15, find the work required to raise the bucket until half the water has leaked out.

17 The volume and pressure of a certain gas vary in accordance with the law $pv^{1.2} = 115$, where the units of measurement are inches and pounds. Find the work done if the gas expands from 32 to 40 in.^{3}. (*Hint*: See Example 5.)

18 The pressure and volume of a quantity of enclosed steam are related by the formula $pv^{1.14} = c$, where c is a constant. If the initial pressure and volume are p_0 and v_0, respectively, find a formula for the work done if the steam expands to twice its volume. (*Hint*: See Example 5.)

19 Newton's Law of Gravitation states that the force F of attraction between two particles having masses m_1 and m_2 is given by $F = gm_1m_2/s^2$, where g is a constant and s is the distance between the particles. If the mass m_1 of the Earth is regarded as concentrated at the center of the Earth, and a rocket of mass m_2 is on the surface (a distance of 4,000 mi from the center), find a general formula for the work done in firing the rocket vertically upward to an altitude h (see figure).

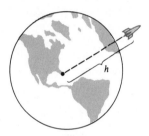

Figure for Exercise 19

20 In the study of electricity the formula $F = kq/r^2$, where k is a constant, is used to find the force (in dyn) with which a positive charge Q of strength q units repels a unit positive charge located r cm from Q. Find the work done in moving a unit charge from a point d cm from Q to a point $d/2$ cm from Q.

Suppose the tables in Exercises 21 and 22 were obtained experimentally, where $f(x)$ denotes the force acting at the point with coordinate x on a coordinate line l. Use the Trapezoidal Rule (5.35) to approximate the work done on the interval $[a, b]$, where a and b are the smallest and largest values of x, respectively.

21	x ft	$f(x)$ lb
	0	7.4
	0.5	8.1
	1.0	8.4
	1.5	7.8
	2.0	6.3
	2.5	7.1
	3.0	5.9
	3.5	6.8
	4.0	7.0
	4.5	8.0
	5.0	9.2

22	x m	$f(x)$ N
	1	125
	2	120
	3	130
	4	146
	5	165
	6	157
	7	150
	8	143
	9	140

23 If the force function is constant, show that Definition (6.10) reduces to Definition (6.9).

6.6

FORCE EXERTED BY A LIQUID

In physics the *pressure* p at a depth h in a liquid is defined as the weight of the liquid contained in a column having a cross-sectional area of one square unit and altitude h. Pressure may also be regarded as the force per unit area exerted by the liquid, as in the following definition.

Definition (6.12)

> If the **density** (the weight per unit volume) of a liquid is denoted by ρ, then the **pressure** p at a depth h is
>
> $$p = \rho h.$$

It can be verified experimentally that the pressure at any depth is exerted equally in all directions.

Example 1   If the density of the water in a lake is 62.5 lb/ft^3, find the pressure at a depth of (a) 4 ft; (b) 10 ft.

Solution   Using Definition (6.12) we obtain

(a) $p = \rho h = (62.5)(4) = 250 \text{ lb/ft}^2.$

(b) $p = \rho h = (62.5)(10) = 625 \text{ lb/ft}^2.$ ∎

If a rectangular tank, such as an ordinary fish aquarium, is filled with water, then the total force exerted by the water on the (horizontal) base can be found by multiplying the pressure at the bottom of the tank by the area of the base. For example, if the depth of water is 4 ft and the area of the base is 10 ft^2, then from (a) of Example 1, the pressure at the bottom is 250 lb/ft^2 and the total force acting on the base is $(250)(10) = 2500$ lb. This corresponds to 10 columns of water, each having cross-sectional area 1 ft^2 and each weighing 250 lb.

It is more complicated to find the force exerted on one of the sides of the tank since the pressure is not constant there, but increases as the depth increases. Instead of investigating this particular problem, we shall consider a more general situation.

Consider a flat plate that is submerged in a liquid of density ρ such that the face of the plate is perpendicular to the surface of the liquid. Suppose that the shape of the plate is the same as that of the region in an xy-plane bounded by the graphs of the equations $y = c$, $y = d$, $x = f(y)$, and $x = g(y)$, where f and g are continuous functions on $[c, d]$ and $f(y) \geq g(y)$ for all y in $[c, d]$. Furthermore, suppose that the surface of the water contains the line $y = k$. A region of this type is illustrated in Figure 6.35, although in general it is not required that the region lie entirely above the x-axis as shown in the figure.

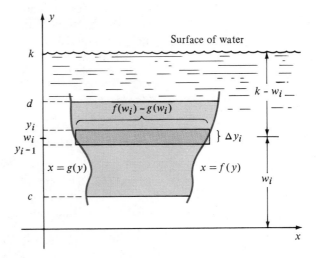

Figure 6.35

Let P denote the partition of $[c, d]$ determined by $c = y_0, y_1, \ldots, y_n = d$ and let $\Delta y_i = y_i - y_{i-1}$. Select a number w_i in each subinterval $[y_{i-1}, y_i]$ and consider the rectangle of area $[f(w_i) - g(w_i)] \Delta y_i$. If Δy_i is small, then all points in the rectangular region are roughly the same distance, $k - w_i$, from the surface of the liquid and hence by Definition (6.12) the pressure at any point within the rectangle may be approximated by $\rho(k - w_i)$. It appears, therefore, that the force on the ith rectangle is approximately equal to this pressure times the area of the rectangle, that is,

$$\rho(k - w_i)[f(w_i) - g(w_i)] \Delta y_i.$$

The total force on the region shown in Figure 6.35 may be approximated by the Riemann sum

$$\sum_i \rho(k - w_i)[f(w_i) - g(w_i)] \Delta y_i.$$

Since we expect this approximation to improve as the norm of the partition decreases we arrive at the following definition.

Definition (6.13)

> The **force F exerted by a liquid** of constant density ρ on a region of the type illustrated in Figure 6.35, where the functions f and g are continuous on $[c, d]$, is
>
> $$F = \lim_{\|P\| \to 0} \sum_i \rho(k - w_i)[f(w_i) - g(w_i)] \Delta y_i$$
>
> $$= \int_c^d \rho(k - y)[f(y) - g(y)] \, dy.$$

If a more complicated region is divided into subregions of the type used in (6.13), we apply the definition to each subregion and add the resulting numbers. The coordinate system may be introduced in various ways. In Example 3 we shall choose the x-axis along the surface of the liquid and the positive direction of the y-axis downward. In this case, Definition (6.13) must be changed accordingly.

It is often convenient to use the following formula to approximate the **force on a thin horizontal rectangle**:

(6.14)   Force $\approx$ (density) $\cdot$ (depth) $\cdot$ (area of rectangle)

After using (6.14), we take a limit of a sum to obtain the total force on the region, as illustrated in the following examples.

Example 2   The ends of a water trough 8 ft long have the shape of isosceles trapezoids of lower base 4 ft, upper base 6 ft, and altitude 4 ft. Find the total force on one end if the trough is full of water.

Solution   Figure 6.36 illustrates one of the ends of the trough superimposed on a rectangular coordinate system and appropriately labeled. As indicated in the figure, the equation of the line through the points $(2, 0)$ and $(3, 4)$ is $y = 4x - 8$ or, equivalently, $x = \frac{1}{4}(y + 8)$. It follows that the rectangle has area

$$2x_i \, \Delta y_i = 2\left(\frac{1}{4}\right)(w_i + 8) \, \Delta y_i = \left(\frac{1}{2}\right)(w_i + 8) \, \Delta y_i.$$

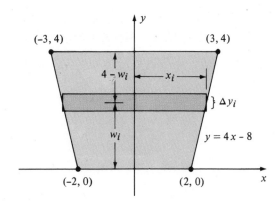

Figure 6.36

Using (6.14), we find that the force exerted by the liquid on this rectangle is approximately

$$(62.5)(4 - w_i)\left(\frac{1}{2}\right)(w_i + 8)\,\Delta y_i.$$

Summing and taking the limit gives us the total force. Thus,

$$F = \int_0^4 (62.5)(4 - y)\left(\frac{1}{2}\right)(y + 8)\,dy$$

$$= 31.25\int_0^4 (32 - 4y - y^2)\,dy$$

$$= 31.25\left[32y - 2y^2 - \frac{y^3}{3}\right]_0^4$$

$$= 31.25\left[128 - 32 - \frac{64}{3}\right] = \frac{7{,}000}{3}\,\text{lb.} \qquad \blacksquare$$

Example 3 A cylindrical oil storage tank 6 ft in diameter and 10 ft long is lying on its side. If the tank is half full of oil weighing 58 lb/ft^3, find the force exerted by the oil on one side of the tank.

Solution Let us introduce a coordinate system so that the end of the tank is a circle of radius 3 ft with center at the origin. The equation of the circle is $x^2 + y^2 = 9$. If we choose the positive direction of the *y*-axis *downward*, then as shown in Figure 6.37, w_i represents the depth at a point in a typical horizontal rectangle. From the equation of the circle, the length of the rectangle is $2\sqrt{9 - w_i^2}$. Using (6.14), the force acting on the pictured rectangle is

$$58w_i(2\sqrt{9 - w_i^2})\,\Delta y_i.$$

As usual, we take a limit of a sum of such terms to obtain the total force F on the end of the tank. This gives us

$$F = \int_0^3 58y(2\sqrt{9 - y^2})\,dy.$$

The integral may be evaluated by letting

$$u = 9 - y^2 \quad \text{and} \quad du = -2y\,dy.$$

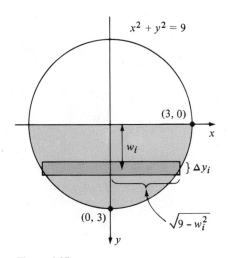

Figure 6.37

Note that if $y = 0$, then $u = 9$, whereas if $y = 3$, then $u = 0$. We may, therefore, transform the integral as follows:

$$F = -58 \int_0^3 \sqrt{9 - y^2}(-2y \, dy)$$

$$= -58 \int_9^0 u^{1/2} \, du = -58 \frac{u^{3/2}}{(3/2)} \Big]_9^0$$

$$= -\frac{116}{3} \left[u^{3/2} \right]_9^0 = -\frac{116}{3} [0 - 9^{3/2}] = 1044 \text{ lb.} \quad \blacksquare$$

EXERCISES 6.6

1 A glass tank to be used as an aquarium is 3 ft long and has square ends of width 1 ft. If the tank is filled with water, find the force exerted by the water on (a) one end and (b) one side.

2 If one of the square ends of the tank in Exercise 1 is divided into two parts by means of a diagonal, find the force exerted on each part.

3 The ends of a water trough 6 ft long have the shape of isosceles triangles with equal sides of length 2 ft and the third side of length $2\sqrt{3}$ ft at the top of the trough. Find the force exerted by the water on one end of the trough if the trough is (a) full of water; (b) half full of water.

4 Answer (a) and (b) of Exercise 3 if the (vertical) altitude of the triangle is h feet, where $0 < h < 2$.

5 A cylindrical oil storage tank 4 ft in diameter and 5 ft long is lying on its side. If the tank is half full of oil weighing 60 lb/ft³, find the force exerted by the oil on one end of the tank.

6 A rectangular gate in a dam is 5 ft long and 3 ft high. If the gate is vertical, with the top of the gate parallel to the surface of the water and 6 ft below it, find the force of the water against the gate.

7 A rectangular swimming pool is 20 ft wide and 40 ft long. The depth of the water in the pool varies uniformly from 3 ft at one end to 9 ft at the other end. Find the total force exerted by the water on the bottom of the pool.

8 Find the force exerted by the water on the side of the swimming pool described in Exercise 7.

9 A plate having the shape of an isosceles trapezoid with upper base 4 ft long and lower base 8 ft long is submerged vertically in water such that the bases are parallel to the surface. If the distances from the surface of the water to the lower and upper bases are 10 ft and 6 ft, respectively, find the force exerted by the water on one side of the plate.

10 A circular plate of radius 2 ft is submerged vertically in water. If the distance from the surface of the water to the center of the plate is 6 ft, find the force exerted by the water on one side of the plate.

11 The ends of a water trough have the shape of the region bounded by the graphs of $y = x^2$ and $y = 4$ where x and y are measured in feet. If the trough is full of water, find the force on one end.

12 A flat plate has the shape of the region bounded by the graphs of $y = x^4$ and $y = 1$ where x and y are measured in feet. The plate is submerged vertically in water with the straight part of its boundary parallel to (and closest to) the surface. If the distance from the surface of the water to the straight part of the boundary is 4 ft, find the force exerted by the water on one side of the plate.

13 A rectangular plate 3 ft wide and 6 ft long is submerged vertically in oil with its short side parallel to, and 2 ft below, the surface. If the oil weighs 50 lb/ft³, find the total force exerted on one side of the plate.

14 If the plate in Exercise 13 is divided into two parts by means of a diagonal, find the force exerted on each part.

15 A flat, irregularly-shaped plate is submerged vertically in water (see figure). Measurements of its width, taken at

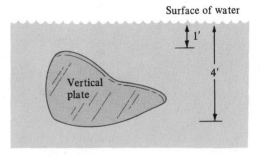

Figure for Exercise 15

successive depths at intervals of 0.5 ft, are compiled in the following table.

Water depth (ft)	1	1.5	2	2.5	3	3.5	4
Width of plate (ft)	0	2	3	5.5	4.5	3.5	0

Estimate the force of the water on one side of the plate by using (with $n = 6$) (a) the Trapezoidal Rule; (b) Simpson's Rule.

6.7

ARC LENGTH

To solve certain problems in the sciences it is essential to consider the *length* of the graph of a function. For example, if a projectile moves along a parabolic course, we may wish to determine the distance it travels during a specified interval of time. Similarly, it may be necessary to find the length of a twisted piece of wire. Of course, if the wire is flexible, we could simply straighten it and find the linear length with a ruler (or by means of the Distance Formula). However, if the wire is not flexible, then other methods must be employed. As we shall see, the key to defining the length of a graph is to divide the graph into many small pieces and then approximate each piece by means of a line segment. Next, we take a limit of a sum of the lengths of all such line segments. This leads to a definite integral. To guarantee that the integral exists, it is necessary to place restrictions on the function, as indicated in the following discussion.

A function f is said to be **smooth** on an interval if it has a derivative f' that is continuous throughout the interval. Roughly speaking, this means that a small change in x produces a small change in the slope $f'(x)$ of the tangent line to the graph of f. Thus, there are no sharp corners on the graph. We intend to define what is meant by the **length of arc** between the two points A and B on the graph of a smooth function.

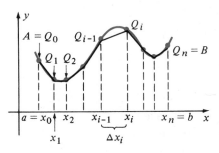

Figure 6.38

If f is smooth on a closed interval $[a, b]$, the points $A(a, f(a))$ and $B(b, f(b))$ will be called the **endpoints** of the graph of f. Let us consider the partition P of $[a, b]$ determined by $a = x_0, x_1, x_2, \ldots, x_n = b$, and let Q_i denote the point with coordinates $(x_i, f(x_i))$. This gives us $n + 1$ points $Q_0, Q_1, Q_2, \ldots, Q_n$ on the graph of f. If, as illustrated in Figure 6.38, we connect each Q_{i-1} to Q_i by a line segment of length $d(Q_{i-1}, Q_i)$, then the length L_P of the resulting broken line is

$$L_P = \sum_{i=1}^{n} d(Q_{i-1}, Q_i).$$

If the norm $\|P\|$ of the partition is small, then Q_{i-1} is close to Q_i for each i, and we expect L_P to be an approximation to the length of arc between A and B. This gives us a clue to a suitable definition of arc length. Specifically, we shall consider the limit of the sum L_P as $\|P\| \to 0$. To formulate this concept precisely, and at the same time arrive at a formula for calculating arc length, let us proceed as follows. By the Distance Formula,

$$d(Q_{i-1}, Q_i) = \sqrt{(x_i - x_{i-1})^2 + [f(x_i) - f(x_{i-1})]^2}.$$

Applying the Mean Value Theorem (4.12),

$$f(x_i) - f(x_{i-1}) = f'(w_i)(x_i - x_{i-1})$$

where w_i is in the open interval (x_{i-1}, x_i). Substituting this into the preceding formula and letting $\Delta x_i = x_i - x_{i-1}$, we obtain

$$d(Q_{i-1}, Q_i) = \sqrt{(\Delta x_i)^2 + [f'(w_i)\,\Delta x_i]^2}$$
$$= \sqrt{1 + [f'(w_i)]^2}\,\Delta x_i.$$

Consequently,

$$L_P = \sum_{i=1}^{n} \sqrt{1 + [f'(w_i)]^2}\,\Delta x_i.$$

Observe that L_P is a Riemann sum for the function g defined by $g(x) = \sqrt{1 + [f'(x)]^2}$. In addition, g is continuous on $[a, b]$ since f' is continuous. As we have mentioned, if the norm $\|P\|$ is small, then the length L_P of the broken line should approximate the length of the graph of f from A to B. Moreover, this approximation should improve as $\|P\|$ decreases. It is natural, therefore, to define the *length* (also called the *arc length*) of the graph of f from A to B as the limit of the sum L_P. Since $g = \sqrt{1 + (f')^2}$ is a continuous function, the limit exists and equals the definite integral $\int_a^b \sqrt{1 + [f'(x)]^2}\,dx$. This arc length will be denoted by the symbol L_a^b.

Definition (6.15)

> Let the function f be smooth on a closed interval $[a, b]$. The **arc length of the graph** of f from $A(a, f(a))$ to $B(b, f(b))$ is
>
> $$L_a^b = \int_a^b \sqrt{1 + [f'(x)]^2}\,dx.$$

Definition (6.15) will be extended to more general graphs in Chapter 13. A graph that has arc length is said to be **rectifiable**. If a function f is defined implicitly by an equation in x and y, then we shall also refer to the *arc length of the graph of the equation*.

Example 1 If $f(x) = 3x^{2/3} - 10$, find the arc length of the graph of f from the point $A(8, 2)$ to $B(27, 17)$.

Solution The graph of f is sketched in Figure 6.39. Since $f'(x) = 2x^{-1/3} = 2/(x^{1/3})$, we have

$$L_8^{27} = \int_8^{27} \sqrt{1 + \left(\frac{2}{x^{1/3}}\right)^2}\,dx = \int_8^{27} \sqrt{1 + \frac{4}{x^{2/3}}}\,dx$$
$$= \int_8^{27} \frac{\sqrt{x^{2/3} + 4}}{x^{1/3}}\,dx.$$

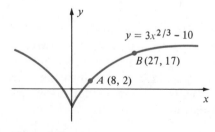

Figure 6.39

To evaluate this integral, let

$$u = x^{2/3} + 4 \quad \text{and} \quad du = \frac{2}{3} x^{-1/3} \, dx = \frac{2}{3} \frac{1}{x^{1/3}} \, dx.$$

The integral can be expressed in a suitable form for integration by introducing the factor 2/3 in the integrand and compensating for this by multiplying the integral by 3/2. Thus

$$L_8^{27} = \frac{3}{2} \int_8^{27} \sqrt{x^{2/3} + 4} \left(\frac{2}{3} \frac{1}{x^{1/3}} \right) dx.$$

If $x = 8$, then $u = (8)^{2/3} + 4 = 8$, whereas if $x = 27$, then $u = (27)^{2/3} + 4 = 13$. Making the substitution and changing the limits of integration,

$$L_8^{27} = \frac{3}{2} \int_8^{13} \sqrt{u} \, du = u^{3/2} \Big]_8^{13} = 13^{3/2} - 8^{3/2} \approx 24.2. \qquad \blacksquare$$

The next result is the analogue of Definition (6.15) for integration with respect to y.

Definition (6.16)

Let the function g be defined by $x = g(y)$, where g is smooth on the interval $[c,d]$. The **arc length of the graph** of g from the point $(g(c), c)$ to the point $(g(d), d)$ is

$$L_c^d = \int_c^d \sqrt{1 + [g'(y)]^2} \, dy.$$

Example 2 Set up an integral for finding the arc length of the graph of $y = y^3 - x$ from $A(0, -1)$ to $B(6, 2)$. Approximate the integral by using Simpson's Rule (5.37) with $n = 6$.

Solution The equation is not of the form $y = f(x)$ and hence Definition (6.15) cannot be applied. However, if we write $x = y^3 - y$, then we can employ Definition (6.16) with $g(y) = y^3 - y$. The graph of the equation is sketched in Figure 6.40. Using (6.16) with $c = -1$ and $d = 2$ gives us

$$L_{-1}^2 = \int_{-1}^2 \sqrt{1 + (3y^2 - 1)^2} \, dy$$

$$= \int_{-1}^2 \sqrt{9y^4 - 6y^2 + 2} \, dy.$$

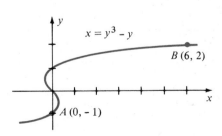

Figure 6.40

To use Simpson's Rule, we let $f(y) = \sqrt{9y^4 - 6y^2 + 2}$ and arrange our work as we did in Section 5.7. The reader may verify the entries in the following table.

i	y_i	$f(y_i)$	m	$mf(y_i)$
0	-1.0	2.2361	1	2.2361
1	-0.5	1.0308	4	4.1232
2	0.0	1.4142	2	2.8284
3	0.5	1.0308	4	4.1232
4	1.0	2.2361	2	4.4722
5	1.5	5.8363	4	23.3452
6	2.0	11.0454	1	11.0454

The sum of the numbers in the last column is 52.1737. Hence by Simpson's Rule (5.35) with $a = -1$, $b = 2$, and $n = 6$,

$$\int_{-1}^{2} \sqrt{9y^4 - 6y^2 + 2}\, dy \approx \frac{1}{6}[52.1737] \approx 8.7. \qquad \blacksquare$$

If a graph can be decomposed into a finite number of parts, each of which is the graph of a smooth function, then the arc length of the graph is defined as the sum of the arc lengths of the individual graphs. A function of this type is said to be **piecewise smooth** on its domain.

To avoid any misunderstanding in the following discussion, the variable of integration will be denoted by t. In this case the arc length formula given in Definition (6.15) becomes

$$L_a^b = \int_a^b \sqrt{1 + [f'(t)]^2}\, dt.$$

If f is smooth on $[a, b]$, then f is smooth on $[a, x]$ for every x in $[a, b]$, and the length of the graph from the point $A(a, f(a))$ to the point $Q(x, f(x))$ is

$$L_a^x = \int_a^x \sqrt{1 + [f'(t)]^2}\, dt.$$

If we change the notation and use the symbol $s(x)$ in place of L_a^x, then s may be regarded as a function with domain $[a, b]$, since to each x in $[a, b]$ there corresponds a unique number $s(x)$. We shall call s the *arc length function* for the graph of f, as in the next definition.

Definition (6.17)

Let the function f be smooth on $[a, b]$. The **arc length function** s for the graph of f on $[a, b]$ is defined by

$$s(x) = \int_a^x \sqrt{1 + [f'(t)]^2}\, dt$$

where $a \leq x \leq b$.

As shown in Figure 6.41, the values $s(x)$ of s may be represented geometrically as lengths of arc of the graph of f from $A(a, f(a))$ to $Q(x, f(x))$.

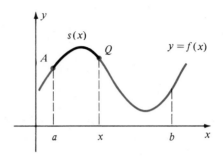

Figure 6.41

The following theorem is important for problems involving the arc length function.

Theorem (6.18)

Let f be smooth on $[a, b]$, and let s be the arc length function for the graph of $y = f(x)$ on $[a, b]$. If dx and dy are differentials of x and y, then

(i) $ds = \sqrt{1 + [f'(x)]^2}\, dx$

(ii) $(ds)^2 = (dx)^2 + (dy)^2.$

Proof   By Definition (6.17) and Theorem (5.24),

$$D_x[s(x)] = D_x\left[\int_a^x \sqrt{1 + [f'(t)]^2}\, dt\right] = \sqrt{1 + [f'(x)]^2}.$$

Consequently, by Definition (3.22),

$$ds = s'(x)\, dx = \sqrt{1 + [f'(x)]^2}\, dx$$

This proves (i). Squaring both sides of this equation, we obtain

$$(ds)^2 = \{1 + [f'(x)]^2\}(dx)^2$$
$$= (dx)^2 + [f'(x)\, dx]^2.$$

Since, by Definition (3.22), $dy = f'(x)\, dx$, this gives us (ii). □

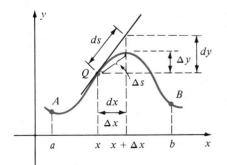

Figure 6.42 $(ds)^2 = (dx)^2 + (dy)^2$

There is an interesting geometric interpretation for formula (ii) of the preceding theorem. Consider $y = f(x)$ and give x an increment Δx. Let Δy denote the change in y and Δs the change in arc length corresponding to Δx. These increments are illustrated in Figure 6.42 where dy is the amount that the tangent line rises or falls if the independent variable changes from x to $x + \Delta x$ (see also Figure 3.10). Since $(ds)^2 = (dx)^2 + (dy)^2$, ds may be regarded as the length of the hypotenuse of a right triangle with sides $|dx|$

and $|dy|$ as illustrated in Figure 6.42. This figure also indicates that if Δx is small, then ds may be used to approximate the increment Δs of arc length.

Example 3   Use differentials to approximate the arc length of the graph of $y = x^3 + 2x$ from $A(1, 3)$ to $B(1.2, 4.128)$.

Solution   If we let $f(x) = x^3 + 2x$, then by (i) of Theorem (6.18),

$$ds = \sqrt{1 + (3x^2 + 2)^2}\ dx.$$

An approximation may be obtained by letting $x = 1$ and $dx = 0.2$. Thus

$$ds = \sqrt{1 + 5^2}\,(0.2) = \sqrt{26}\,(0.2) \approx 1.02. \qquad \blacksquare$$

EXERCISES 6.7

In Exercises 1–4 use Definition (6.15) to find the arc length of the graph of the given equation from A to B.

1    $8x^2 = 27y^3$; $A(1, \frac{2}{3})$, $B(8, \frac{8}{3})$

2    $(y + 1)^2 = (x - 4)^3$; $A(5, 0)$, $B(8, 7)$

3    $y = 5 - \sqrt{x^3}$; $A(1, 4)$, $B(4, -3)$

4    $y = 6\sqrt[3]{x^2} + 1$; $A(-1, 7)$, $B(-8, 25)$

5–6   Solve Exercises 1 and 2 by means of Definition (6.16).

In Exercises 7–10 find the arc length from A to B of the graph of the given equation.

7    $y = \dfrac{x^3}{12} + \dfrac{1}{x}$; $A(1, \frac{13}{12})$, $B(2, \frac{7}{6})$

8    $y + \dfrac{1}{4x} + \dfrac{x^3}{3} = 0$; $A(2, \frac{67}{24})$, $B(3, \frac{109}{12})$

9    $30xy^3 - y^8 = 15$; $A(\frac{8}{15}, 1)$, $B(\frac{271}{240}, 2)$

10    $x = \dfrac{y^4}{16} + \dfrac{1}{2y^2}$; $A(\frac{9}{8}, -2)$, $B(\frac{9}{16}, -1)$

In Exercises 11 and 12 set up (but do not evaluate) the integral for finding the arc length of the graph of the given equation from A to B.

11    $2y^3 - 7y + 2x - 8 = 0$; $A(3, 2)$, $B(4, 0)$

12    $11x - 4x^3 - 7y + 7 = 0$; $A(1, 2)$, $B(0, 1)$

13   Find the arc length of the graph of $x^{2/3} + y^{2/3} = 1$. (*Hint*: Consider symmetry with respect to the line $y = x$.)

14   Find the arc length of the graph of $y = (3x^8 + 5)/30x^3$ from the point $(1, 4/5)$ to the point $(2, 773/240)$.

15   If f is defined by $f(x) = \sqrt[3]{x^2}$, what is the arc length function s? If x increases from 1 to 1.1, find Δs and ds.

16   Work Exercise 15 if $f(x) = \sqrt{x^3}$.

17   Use differentials to approximate the arc length of the graph of $y = x^2$ from $A(2, 4)$ to $B(2.1, 4.41)$. Illustrate this approximation graphically and compare it with $d(A, B)$.

18   Use differentials to approximate the arc length of the graph of $y + x^3 = 0$ from $A(1, -1)$ to $B(1.1, -1.331)$. Illustrate this approximation graphically and compare it with $d(A, B)$.

In Exercises 19–22 use Simpson's Rule (5.35) with $n = 4$ to approximate the arc length of the graph of the given equation from A to B.

19    $xy = 2$; $A(1, 2)$, $B(2, 1)$

20    $y = x^2 + x + 3$; $A(-2, 5)$, $B(2, 9)$

21    $y = x^3$; $A(0, 0)$, $B(2, 8)$

22    $4y = x^4$; $A(0, 0)$, $B(2, 4)$

6.8

OTHER APPLICATIONS

In this section we shall consider several miscellaneous applications to give the reader some idea of the versatility of the definite integral. The first three have to do with the study of economics; the remaining applications are from the physical sciences.

It is important for people in business to plan for depreciation of equipment. The most elementary technique employed is the **straight-line method**, in which the rate of depreciation is considered constant. For example, if the depreciation each year is $200, then the total depreciation $f(t)$ at the end of t years is given by $f(t) = 200t$. Note that the rate of depreciation is $f'(t)$. In many instances the rate of depreciation is not constant. To illustrate, an automobile depreciates very rapidly during the first few years after it is purchased and then more slowly as it gets older. For other items the rate of depreciation may increase from year to year. In general, suppose that the rate of depreciation of a certain piece of equipment over the time interval $[0, t]$ may be approximated by $g(t)$, where g is a continuous function. If the total depreciation in $[0, t]$ is $f(t)$, then $f'(t) = g(t)$ and we may write

$$f(t) = \int_0^t f'(x)\,dx = \int_0^t g(x)\,dx.$$

Suppose further that an additional fixed cost C is required to overhaul the equipment after time t. Thus, in the time interval $[0, t]$, the expense $h(t)$ connected with the equipment is given by

$$h(t) = C + \int_0^t g(x)\,dx.$$

The average expense, that is, the expense per year, is

$$k(t) = \frac{h(t)}{t}.$$

If no other factors are involved, then the best time to overhaul the equipment corresponds to the value of t for which the function k has a relative minimum. This occurs when

$$k'(t) = \frac{th'(t) - h(t)}{t^2} = 0$$

or equivalently, when

$$h'(t) = \frac{h(t)}{t} = k(t).$$

Since $h'(t) = g(t)$, the best time to overhaul occurs when $g(t) = k(t)$, that is, when the rate of depreciation is the same as the average expense.

Example 1   Using the notation of the preceding discussion, suppose that for a given piece of equipment $g(t) = 300\sqrt{t}$ and $C = 500$, where t is in years. When should the equipment be overhauled in order to minimize the average expense?

Solution   From the preceding discussion, the average expense is given by

$$k(t) = h(t) \div t = \left[C + \int_0^t g(x)\,dx \right] \div t$$
$$= \left[500 + \int_0^t 300\sqrt{x}\,dx \right] \div t$$
$$= (500 + 200t^{3/2})/t.$$

From the previous discussion, k will have a minimum value if

$$(500 + 200t^{3/2})/t = 300t^{1/2}.$$

Solving for t we obtain $t = 5^{2/3} \approx 2.924$ or, to the nearest month, 2 years and 11 months.   ∎

In economics, the process that a corporation uses to increase its accumulated wealth is called **capital formation**. If the amount K of capital at time t can be approximated by $K = f(t)$, where f is a differentiable function, then the rate of change of K with respect to t is called the **net investment flow**. Hence, if I denotes the investment flow, then

$$I = \frac{dK}{dt} = f'(t).$$

Conversely, if I is given by $g(t)$, where the function g is continuous on an interval $[a, b]$, then the increase in capital over this time interval is

$$\int_a^b g(t)\,dt = f(b) - f(a).$$

Example 2   Suppose a corporation wishes to have its net investment flow approximated by $g(t) = t^{1/3}$, where t is in years and $g(t)$ is in millions of dollars per year. If $t = 0$ corresponds to the present time, estimate the amount of capital formation over the next eight years.

Solution   From our discussion, the increase in capital over the next eight years is

$$\int_0^8 g(t)\,dt = \int_0^8 t^{1/3}\,dt = \frac{3}{4} t^{4/3} \Big]_0^8 = 12.$$

Consequently, the amount of capital formation is $12,000,000.   ∎

In many types of employment, a worker must perform the same assignment repeatedly. For example, a boy hired by a bicycle shop may be asked to assemble new bicycles. As he assembles one bicycle after another, the boy should become more proficient and, up to a certain point, should assemble each one in less time than the preceding one. Another example of this process of learning by repetition is that of a key-punch operator who must translate information from written forms to punched cards. The time required to punch each card should decrease as the number of cards increases. As a final illustration, the time required for a person to trace a path through a maze should improve with practice.

Let us consider a general situation in which a certain task is to be repeated many times. Suppose experience has shown that the time required to perform the task for the ith time can be approximated by $f(i)$, where f is a continuous decreasing function on a suitable interval. The total time required to perform the task k times is given by the sum

$$\sum_{i=1}^{k} f(i) = f(1) + f(2) + \cdots + f(k).$$

If we consider the graph of f, then as illustrated in Figure 6.43 the preceding sum equals the area of the pictured inscribed rectangular polygon and, therefore, may be approximated by the definite integral $\int_0^k f(x)\,dx$. Evidently, the approximation will be close to the actual sum if f decreases slowly on $[0, k]$. If f changes rapidly per unit change in x, then an integral should not be used as an approximation.

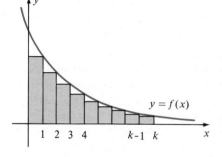

Figure 6.43

Example 3 A company that conducts polls by means of telephone interviews finds that the time required by an employee to complete one interview depends on the number of interviews that the employee has completed previously. Suppose it is estimated that for a certain survey, the number of minutes required to complete the ith interview is given by $f(i) = 6(1 + i)^{-1/5}$. Use a definite integral to approximate the time required for an employee to complete (a) 100 interviews; (b) 200 interviews. If an interviewer receives \$4.80 per hour, estimate how much more expensive it is to have two employees each conduct 100 interviews than it is to have one employee conduct 200 interviews.

Solution As in the preceding discussion, the time required for 100 interviews is approximately

$$\int_0^{100} 6(1 + x)^{-1/5}\,dx = 6 \cdot \frac{5}{4}(1 + x)^{4/5}\bigg]_0^{100}$$

$$= \frac{15}{2}[(101)^{4/5} - 1]$$

$$\approx 293.5 \text{ min.}$$

The time required for 200 interviews is approximately

$$\int_0^{200} 6(1 + x)^{-1/5}\,dx = \frac{15}{2}[(201)^{4/5} - 1]$$

$$\approx 514.4 \text{ min.}$$

Since an interviewer receives \$0.08 per minute, the cost for one employee to conduct 200 interviews is roughly (\$0.08)(514.4) or \$41.15. If two employees each conduct 100 interviews the cost is about 2(\$0.08)(293.5) or \$46.96, which is \$5.81 more than the cost of one employee. Note, however, that the time saved in using two people is approximately 220 minutes.

A computer may be used to show that

$$\sum_{i=1}^{100} 6(1 + i)^{-1/5} \approx 291.75$$

and

$$\sum_{i=1}^{200} 6(1 + i)^{-1/5} \approx 512.57.$$

Hence the results obtained by integration (the area under the graph of f) are roughly 2 minutes more than the value of the corresponding sum (the area of the inscribed rectangular polygon). ■

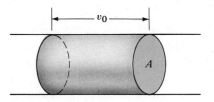

Figure 6.44

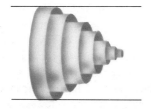

Figure 6.45

Our next application of the definite integral is taken from the field of biology. If a liquid flows through a cylindrical tube and if the velocity is a constant v_0, then the volume of liquid passing a fixed point per unit time is given by $v_0 A$, where A is the area of a cross section of the tube (see Figure 6.44). A more complicated formula is required to study the flow of blood in an arteriole. In this case the flow is in layers, as illustrated in Figure 6.45. In the layer closest to the wall of the arteriole, the blood tends to stick to the wall, and its velocity may be considered zero. The velocity increases as the layers approach the center of the arteriole.

For computational purposes, we may regard the blood flow as consisting of thin cylindrical shells that slide along, with the outer shell fixed and the velocity of the shells increasing as the radii of the shells decrease (see Figure 6.45). If the velocity in each shell is considered constant, then from the theory of liquids in motion, the velocity $v(r)$ in a shell having average radius r is

$$v(r) = \frac{P}{4vl}(R^2 - r^2)$$

where R is the radius of the arteriole (in cm), l is the length of the arteriole (in cm), P is the pressure difference between the two ends of the arteriole (in dyn/cm^2) and v is the viscosity of the blood (in dyn-sec/cm^2). Note that the formula gives zero velocity if $r = R$ and maximum velocity $PR^2/4vl$ as r approaches 0. If the radius of the kth shell is r_k and the thickness of the shell is Δr_k, then by (6.6) the volume of blood in this shell is

$$2\pi r_k v(r_k) \, \Delta r_k = \frac{2\pi r_k P}{4vl}(R^2 - r_k^2) \, \Delta r_k.$$

If there are n shells, then the total flow in the arteriole per unit time may be approximated by

$$\sum_{k=1}^{n} \frac{2\pi r_k P}{4vl}(R^2 - r_k^2) \, \Delta r_k.$$

To estimate the total flow F, that is, the volume of blood per unit time, we take the limit of this sum as n increases without bound. This leads to the following definite integral:

$$F = \int_0^R \frac{2\pi r P}{4vl} (R^2 - r^2)\, dr$$

$$= \frac{2\pi P}{4vl} \int_0^R (R^2 r - r^3)\, dr$$

$$= \frac{\pi P}{2vl} \left[\frac{1}{2} R^2 r^2 - \frac{1}{4} r^4 \right]_0^R.$$

Substituting the limits of integration gives us

$$F = \frac{\pi P R^4}{8vl} \text{ cm}^3.$$

This formula for F is not exact since the thickness of the shells cannot be made arbitrarily small. Indeed, the lower limit is the width of a red blood cell, or approximately 2×10^{-4} cm. However, we may assume that the formula gives a reasonable estimate. It is interesting to observe that a small change in the radius of an arteriole produces a large change in the flow, since F is directly proportional to the fourth power of R. A small change in the pressure difference has a lesser effect, since P appears to only the first power.

It should be evident from our work in this chapter that if a quantity can be approximated by a sum of many terms, then it is a candidate for representation as a definite integral. The main requirement is that as the number of terms increases, the sum approaches a limit. Similarly, any quantity that can be interpreted as an area of a region in a plane may be investigated by means of a definite integral. (See, for example, the discussion of hysteresis at the end of Section 6.1.) Conversely, definite integrals allow us to represent physical quantities as areas. Let us conclude this section by reconsidering several earlier concepts from this point of view.

Suppose $v(t)$ is the velocity, at time t, of an object that is moving on a coordinate line. If s is the position function, then $s'(t) = v(t)$, and

$$\int_a^b v(t)\, dt = \int_a^b s'(t)\, dt = s(t) \Big]_a^b = s(b) - s(a).$$

If $v(t) > 0$ throughout the time interval $[a, b]$, these equalities tell us that the area under the graph of the function v from a to b is the distance the object travels, as illustrated in Figure 6.46. This observation is useful to an engineer or physicist, who may not have an explicit form for $v(t)$, but merely a graph (or table) indicating the velocity at various times. The distance traveled may then be estimated by approximating the area under the graph.

Incidentally, if $v(t) < 0$ at certain times in $[a, b]$, the graph of v may have the appearance illustrated in Figure 6.47. The figure indicates that the object moved in the negative direction from $t = c$ to $t = d$. The distance it traveled during that time is $\int_c^d |v(t)|\, dt$. (Why?) It follows that $\int_a^b |v(t)|\, dt$ is the *total* distance traveled in $[a, b]$, whether $v(t)$ is positive or negative.

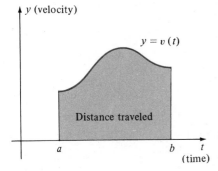

Figure 6.46

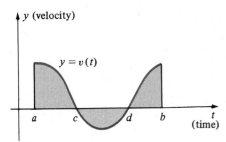

Figure 6.47

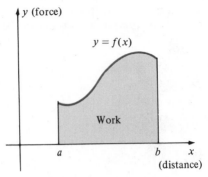

Figure 6.48

In Section 6.5 we considered the work W done by a variable force $f(x)$ as its point of application moves along a coordinate line from $x = a$ to $x = b$. The value of W is given by $W = \int_a^b f(x)\, dx$ (see Definition (6.10)). Suppose $f(x) \geq 0$ for all x on $[a, b]$. If we sketch the graph of f, then, as illustrated in Figure 6.48, the work W is the area under the graph from a to b.

As a final illustration, suppose that the amount of a physical entity such as oil, water, electric power, money supply, bacteria count, blood flow, etc., is increasing or decreasing in some manner, and that $R(t)$ is the rate at which it is changing at time t. If $Q(t)$ is the amount of the entity present at time t, and if Q is differentiable, then we know from Section 4.7 that $Q'(t) = R(t)$. If $R(t) > 0$ (or $R(t) < 0$) in a time interval $[a, b]$, then the amount that the entity increases (or decreases) between $t = a$ and $t = b$ is

$$Q(b) - Q(a) = \int_a^b Q'(t)\, dt = \int_a^b R(t)\, dt.$$

As before, this number may be represented as the area of the region in a ty-plane bounded by the graphs of R, $t = a$, $t = b$, and $y = 0$.

Example 4 Starting at 9:00 A.M., oil is pumped into a storage tank at a rate of $150t^{1/2} + 25$ gal/hr, where t is the time (in hours) after 9:00 A.M. How many gallons will have been pumped into the tank at 1:00 P.M.?

Solution   Letting $R(t) = 150t^{1/2} + 25$ in the preceding discussion, we obtain

$$\int_0^4 (150t^{1/2} + 25)\, dt = 100t^{3/2} + 25t \Big]_0^4$$

$$= 100(4)^{3/2} + 25(4)$$

$$= 100(8) + 100 = 900 \text{ gal.} \quad \blacksquare$$

We have given only a few illustrations of the use of definite integrals. The interested reader may find many more in books on physics and engineering.

EXERCISES 6.8

1 The rate of depreciation of a certain piece of equipment over the time interval $[0, 3]$ may be approximated by $g(t) = 1 - (t^2/9)$ where t is in years and $g(t)$ is in hundreds of dollars. Find the total depreciation at the end of (a) 6 months; (b) 1 year; (c) 18 months; (d) 2 years.

2 The rate of depreciation of a certain piece of equipment over the time interval $[0, 6]$ may be approximated by

$g(t) = \sqrt{10 - t}$, where t is in years and $g(t)$ is in thousands of dollars. If, after six years, an additional cost of \$420 is needed to overhaul the equipment, what is the average expense for the six years?

3 If, in Example 1, $g(t) = 100t^{2/3}$ and $C = 400$, find the best time to overhaul the equipment.

4 If, in Example 2, the rate of investment is approximated by $g(t) = 2t(3t + 1)$, where $g(t)$ is in thousands of dollars, find the amount of capital formation over the intervals $[0, 5]$ and $[5, 10]$.

5 A key-punch operator is required to transfer registration data of college students from written sheets to punched cards. It is estimated that the number of minutes required to punch the ith card is approximately $f(i) = 6(1 + i)^{-1/3}$. Use a definite integral to estimate the time required for (a) one operator to punch 600 cards; (b) two operators to punch 300 cards each.

6 It is estimated that the number of minutes needed for a person to trace a path through a certain maze without error is given by $f(i) = 5i^{-1/2}$, where i is the number of trials previously completed. Approximately how much time is required to complete 10 trials?

7 A small parts manufacturer estimates that the time required for a worker to assemble a certain item depends on the number of items the worker has previously assembled. If the time (in minutes) required to assemble the nth item is given by $f(n) = 20(n + 1)^{-.04} + 3$, approximate the time required to assemble (a) 1 item; (b) 4 items; (c) 8 items; (d) 16 items.

8 Use a definite integral to approximate the sum

$$\sum_{k=1}^{100} k(k^2 + 1)^{-1/4}.$$

9 A motorboat uses gasoline at the rate of $t\sqrt{9 - t^2}$ gal/hr. If the motor is started at $t = 0$, how much gasoline is used in the next two hours?

10 The population of a city has increased since 1980 at a rate of $1.5 + 0.3\sqrt{t} + 0.006t^2$ thousand people per year, where t is the number of years after 1980. Assuming that this rate continues and that the population was 50,000 in 1980, estimate the population in 1989.

11 The accompanying figure indicates the velocity of an automobile (in mi/hr) as it traveled along a freeway over a twelve minute interval. Approximate the distance traveled.

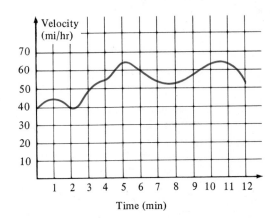

Figure for Exercise 11

12 The acceleration (in ft/sec^2) of an automobile over a period of 8 seconds is indicated in the accompanying figure. Approximate the net change in velocity in this time period.

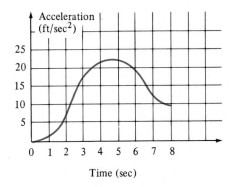

Figure for Exercise 12

6.9

REVIEW

Define or discuss each of the following.

1 The area between the graphs of two continuous functions

2 Solid of revolution

3 Methods of finding volumes of solids of revolution

4 Volumes by slicing

5 Work

6 Force exerted by a liquid

7 Smooth function

8 Arc length of a graph

9 Piecewise smooth function

10 The arc length function

EXERCISES 6.9

In Exercises 1 and 2, sketch the region bounded by the graphs of the equations and find the area in two ways, first by integrating with respect to x, and second by integrating with respect to y.

1 $y = -x^2, y = x^2 - 8$

2 $y^2 = 4 - x, x + 2y - 1 = 0$

In Exercises 3–6 find the area of the region bounded by the graphs of the equations.

3 $x = y^2, x + y = 1$

4 $y + x^3 = 0, y = \sqrt{x}, 3y + 7x - 10 = 0$

5 $y = x^3 - x^2 - 6x + 3, y = 3$

6 $y^3 - 2y^2 - y = x - 2, x = 0$

In Exercises 7–10 sketch the region R bounded by the graphs of the given equations and find the volume of the solid generated by revolving R about the indicated axis.

7 $y = \sqrt{4x + 1}, y = 0, x = 0, x = 2$; about the x-axis

8 $y = x^4, y = 0, x = 1$; about the y-axis

9 $y = x^3 + 1, x = 0, y = 2$; about the y-axis

10 $y = \sqrt[3]{x}, y = \sqrt{x}$; about the x-axis

11 Find the volume of the solid generated by revolving the region bounded by the graphs of the equations $y = 4x^2$ and $4x + y - 8 = 0$ (a) about the x-axis; (b) about the line $x = 1$; (c) about the line $y = 16$.

12 Find the volume of the solid generated by revolving the region bounded by the graphs of $y = x^3, x = 2$, and $y = 0$ (a) about the x-axis; (b) about the y-axis; (c) about the line $x = 2$; (d) about the line $x = 3$; (e) about the line $y = 8$; (f) about the line $y = -1$.

13 Find the arc length of the graph of $(x + 3)^2 = 8(y - 1)^3$ from the point $A(-2, \frac{3}{2})$ to the point $B(5, 3)$.

14 A solid has, for its base, the region in the xy-plane bounded by the graphs of $y^2 = 4x$ and $x = 4$. Find the volume of the solid if every cross section by a plane perpendicular to the x-axis is an isosceles right triangle with one of the equal sides on the base of the solid.

15 An above-ground swimming pool has the shape of a right circular cylinder of diameter 12 ft and height 5 ft. If the depth of the water in the pool is 4 ft, find the work required to empty the pool by pumping the water out over the top.

16 As a bucket is raised a distance of 30 ft from the bottom of a well, water leaks out at a uniform rate. Find the work done if the bucket originally contains 24 lb of water and one-third leaks out. Assume that the weight of the empty bucket is 4 lb and neglect the weight of the rope.

17 A square plate of side 4 ft is submerged vertically in water such that one of the diagonals is parallel to the surface. If the distance from the surface to the center of the plate is 6 ft, find the force exerted by the water on one side of the plate.

18 Use differentials to approximate the arc length of the graph of $y = 2 \sin (x/3)$ between the points with x-coordinates π and $91\pi/90$.

If $\lim_{\|P\| \to 0} \sum_i \pi w_i^4 \, \Delta x_i$ represents the limit of a sum for a function f on the interval $[0, 1]$, solve Exercises 19–22.

19 Find the value of the limit.

20 Interpret the limit as the area of a region in the xy-plane.

21 Interpret the limit as the volume of a solid of revolution.

22 Interpret the limit as the work done by a force.

7

EXPONENTIAL AND LOGARITHMIC FUNCTIONS

Most of the functions considered in preceding chapters are algebraic functions. Functions that are not algebraic are called *transcendental*. In this chapter we introduce two important transcendental functions and investigate some of their properties. Since the two functions to be defined are inverse functions of one another, we shall begin the chapter by discussing this fundamental concept.

7.1

INVERSE FUNCTIONS

In Section 1.5 we defined the composite function $f \circ g$ of two functions f and g by means of the formula $(f \circ g)(x) = f(g(x))$. Similarly, $(g \circ f)(x) = g(f(x))$. Usually $f(g(x))$ and $g(f(x))$ are not the same, that is, $f \circ g \neq g \circ f$. In certain cases it may happen that equality *does* occur. Of major importance is the case in which $f(g(x))$ and $g(f(x))$ are not only identical, but both are equal to x. Needless to say, f and g must be very special functions for this to happen. In the following discussion we indicate the manner in which they will be restricted.

Suppose f is a *one-to-one function* with domain X and range Y (see Definition (1.22)). This implies that each element y of Y is the image of precisely one element x of X. We can also say that *each element y of Y can be written in one and only one way in the form $f(x)$, where x is in X.* We may then define a function g from Y to X by demanding that

$$g(y) = g(f(x)) = x \quad \text{for every } x \text{ in } X.$$

This amounts to *reversing* the correspondence given by f. If f is represented geometrically by drawing arrows as in (i) of Figure 7.1, then g can be represented by simply *reversing* these arrows as illustrated in (ii) of the figure.

(i) $y = f(x)$

(ii) $x = g(y)$

Figure 7.1

As illustrated in Figure 7.1, if $f(x) = y$, then $x = g(y)$. This implies that

$$f(g(y)) = y \quad \text{for every } y \text{ in } Y.$$

Since the notation used for the variable is immaterial, we may write

$$f(g(x)) = x \quad \text{for every } x \text{ in } Y.$$

The functions f and g are called *inverse functions* of one another, according to the following definition.

Definition (7.1)

> Let f be a one-to-one function with domain X and range Y. A function g with domain Y and range X is called the **inverse function of f** if
>
> $$g(f(x)) = x \quad \text{for every } x \text{ in } X$$
>
> and $\qquad\qquad f(g(x)) = x \quad \text{for every } x \text{ in } Y.$

If f is one-to-one there is only one inverse function g of f and, moreover, g is one-to-one (see Exercises 16 and 18). It follows from Definition (7.1) that f is the inverse function of g. We often say that f and g are *inverse functions of one another*. The following relationship (illustrated in Figure 7.1) always holds between a one-to-one function f and its inverse function g:

(7.2) $$y = f(x) \quad \text{if and only if} \quad x = g(y).$$

The symbol f^{-1} is often used to denote the inverse function of f. Employing this notation,

$$f^{-1}(f(x)) = x \quad \text{for every } x \text{ in } X$$

$$f(f^{-1}(x)) = x \quad \text{for every } x \text{ in } Y.$$

The symbol -1 used here should not be mistaken for an exponent; that is, $f^{-1}(x)$ does not mean $1/f(x)$. The reciprocal $1/f(x)$ may be denoted by $[f(x)]^{-1}$.

Inverse functions are very important in the study of trigonometry. In this chapter we shall discuss several other important inverse functions. It is important to remember that to define the inverse of a function f, *it is absolutely essential that f be one-to-one*. The most common examples of one-to-one functions are those that are increasing or decreasing on their domains, for in this case, if $a \neq b$ in the domain, then $f(a) \neq f(b)$ in the range.

The diagrams in Figure 7.1 provide a hint for finding the inverse of a one-to-one function f in certain cases. If possible, *we solve the equation $y = f(x)$ for x in terms of y*, obtaining an equation of the form $x = g(y)$. If the two conditions $f(g(x)) = x$ and $g(f(x)) = x$ are true for all x in the domains of f and g, then g is the required inverse function f^{-1}. The success of this method depends on the nature of the equation $y = f(x)$. The next two examples illustrate the method if f is either a linear or quadratic function.

Example 1 If $f(x) = 3x - 5$, find the inverse function of f.

Solution The graph of the linear function f is a line of slope 3, and hence f is increasing for all x. It follows that f is a one-to-one function with domain and range $\mathbb{R}$, and hence the inverse function g exists. If we let

$$y = 3x - 5$$

and then solve for x in terms of y, we obtain

$$x = \frac{y + 5}{3}.$$

Letting

$$g(y) = \frac{y + 5}{3}$$

gives us a function g that reverses the correspondence determined by f. Since the symbol used for the independent variable is immaterial, we may replace y by x in the expression for g, obtaining

$$g(x) = \frac{x + 5}{3}.$$

To verify that g is actually the inverse function of f, we must verify that the two conditions of Definition (7.1), $f(g(x)) = x$ and $g(f(x)) = x$, are fulfilled. Thus,

$$f(g(x)) = f\left(\frac{x + 5}{3}\right) = 3\left(\frac{x + 5}{3}\right) - 5 = x.$$

Similarly,

$$g(f(x)) = g(3x - 5) = \frac{(3x - 5) + 5}{3} = x.$$

This proves that g is the inverse function of f. Using the f^{-1} notation,

$$f^{-1}(x) = \frac{x + 5}{3}.$$ ∎

Example 2 Find the inverse function of f if the domain X is the interval $[0, \infty)$ and $f(x) = x^2 - 3$ for all x in X.

Solution The domain has been restricted so that f is increasing and hence is one-to-one. The range of f is the interval $[-3, \infty)$. As in Example 1 we begin by considering the equation

$$y = x^2 - 3.$$

Solving for x gives us

$$x = \pm\sqrt{y + 3}.$$

Since x is nonnegative, we reject $x = -\sqrt{y + 3}$ and, as in the preceding example, we let

$$g(y) = \sqrt{y + 3} \quad \text{or equivalently,} \quad g(x) = \sqrt{x + 3}.$$

We now check the two conditions in Definition (7.1), obtaining

$$f(g(x)) = f(\sqrt{x + 3}) = (\sqrt{x + 3})^2 - 3 = (x + 3) - 3 = x$$

and

$$g(f(x)) = g(x^2 - 3) = \sqrt{(x^2 - 3) + 3} = x.$$

This proves that

$$f^{-1}(x) = \sqrt{x + 3}, \quad \text{where } x \geq -3. \qquad \blacksquare$$

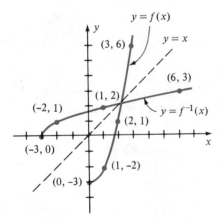

Figure 7.2

There is an interesting relationship between the graphs of a function f and its inverse function f^{-1}. We first note that f maps a into b if and only if f^{-1} maps b into a; that is, $b = f(a)$ means the same thing as $a = f^{-1}(b)$. These equations imply that the point (a, b) is on the graph of f if and only if the point (b, a) is on the graph of f^{-1}. As an illustration, in Example 2 we found that the functions f and f^{-1} given by

$$f(x) = x^2 - 3 \quad \text{and} \quad f^{-1}(x) = \sqrt{x + 3}$$

are inverse functions of one another, provided x is suitably restricted. Some points on the graph of f are $(0, -3), (1, -2), (2, 1)$, and $(3, 6)$. Corresponding points on the graph of f^{-1} are $(-3, 0), (-2, 1), (1, 2)$, and $(6, 3)$. The graphs of f and f^{-1} are sketched on the same coordinate axes in Figure 7.2. If the page is folded along the line $y = x$, which bisects quadrants I and III (as indicated by the dashes in the figure), then the graphs of f and f^{-1} coincide. The two graphs are said to be *reflections* of one another through the line l (or *symmetric* with respect to l). This is typical of the graphs of all functions f that have inverse functions f^{-1}.

EXERCISES 7.1

In Exercises 1–4 prove that f and g are inverse functions of one another.

1 $f(x) = 9x + 2, g(x) = \frac{1}{9}x - \frac{2}{9}$

2 $f(x) = x^3 + 1, g(x) = \sqrt[3]{x - 1}$

3 $f(x) = \sqrt{2x + 1}, x \geq -\frac{1}{2}; g(x) = \frac{1}{2}x^2 - \frac{1}{2}, x \geq 0$

4 $f(x) = 1/(x - 1), x > 1; g(x) = (1 + x)/x, x > 0$

In Exercises 5–14 find the inverse function of f.

5 $f(x) = 8 + 11x$

6 $f(x) = 1/(8 + 11x), x > -\frac{8}{11}$

7 $f(x) = 6 - x^2, 0 \leq x \leq \sqrt{6}$

8 $f(x) = 2x^3 - 5$

9 $f(x) = \sqrt{7x - 2}, x \geq \frac{2}{7}$

10 $f(x) = \sqrt{1 - 4x^2}, 0 \le x \le \frac{1}{2}$

11 $f(x) = 7 - 3x^3$

12 $f(x) = x$

13 $f(x) = (x^3 + 8)^5$

14 $f(x) = x^{1/3} + 2$

15 Sketch the graphs of f and g given in Exercise 1 on the same coordinate plane. Do the same for the functions defined in Exercise 3.

16 If f is a one-to-one function with domain X and range Y, and if g is the inverse function of f, prove that g is a one-to-one function with domain Y and range X.

17 Prove that the linear function f defined by $f(x) = ax + b$ has an inverse function if $a \ne 0$. Does a constant function have an inverse? Does an identity function have an inverse?

18 Prove that a function f can have at most one inverse function.

19 Prove that not every polynomial function has an inverse function.

20 If f has an inverse function f^{-1}, prove that f^{-1} has an inverse function and that $(f^{-1})^{-1} = f$.

7.2

THE NATURAL LOGARITHMIC FUNCTION

If a is a positive real number, it is easy to define a^n for every integral or rational exponent n. However, the extension to irrational exponents requires more complex concepts than those discussed in elementary mathematics courses. To give meaning to a symbol such as a^π, we could express π as a nonterminating decimal $3.14159\ldots$ and consider the numbers a^3, $a^{3.1}$, $a^{3.14}$, $a^{3.141}$, $a^{3.1415}$, $a^{3.14159}$, etc. Since we expect successive powers to get closer to a^π, it would seem natural to define a^π as $\lim_{r \to \pi} a^r$, where r is restricted to *rational* numbers. Unfortunately, there are several disadvantages to this technique. In addition to showing that the limit exists and is unique, we would also have to prove that laws of exponents such as $a^u a^v = a^{u+v}$ are true for all real numbers u and v, a formidable task, to say the least.

Irrational exponents are sometimes defined by employing the Completeness Property for real numbers discussed in Chapter 11 (see (11.9)). This method is rigorous, but it requires a great deal of painstaking work to establish properties of exponents.

Due to the difficulties we have mentioned, it is customary, in precalculus courses, to *assume* that if $a > 0$, then a^u exists for every real number u, and that the usual laws of exponents are true for real exponents. Logarithms with base a are introduced by demanding that

$$u = \log_a v \quad \text{if and only if} \quad a^u = v.$$

Properties of logarithms may then be obtained using the (assumed) laws of exponents. Although this development is suitable in elementary algebra, it is inappropriate in advanced courses, where the standards of mathematical rigor are higher.

In this chapter we shall use concepts developed in calculus to state definitions of $\log_a x$ for every positive real number x and of a^x for every real number x. Our approach to these definitions will be first to use a definite integral to introduce the *natural logarithmic function*, denoted by ln. Later, the natural logarithmic function will be used to define the *natural exponential function.* Finally, we will give meaning to the expressions a^x and $\log_a x$. The reason for proceeding in this fashion is that it provides a precise method of defining general logarithmic and exponential functions without invoking specialized limiting processes. Our technique will also allow us to establish results on continuity, derivatives, and integrals in a very simple manner. As will be seen, the end result will be the laws involving exponents and logarithms that students encounter before taking calculus.

If a function f is continuous on a closed interval $[a, b]$, then as in the proof of Theorem (5.21) we can define a function F by

$$F(x) = \int_a^x f(t)\, dt$$

where x is any number in $[a, b]$. If $f(t) \geq 0$ throughout $[a, b]$, then $F(x)$ equals the area under the graph of f from a to x. Let us consider the special case $f(t) = t^n$, where n is any integer different from -1. By the Fundamental Theorem of Calculus,

$$F(x) = \int_a^x t^n\, dt = \frac{1}{n+1} t^{n+1} \Big]_a^x$$

$$= \frac{1}{n+1}(x^{n+1} - a^{n+1})$$

provided t^n is defined throughout $[a, x]$. It is necessary to exclude $f(t) = 1/t$ in the last formula since $1/(n + 1)$ is undefined if $n = -1$. Indeed, up to this stage of our work it has been impossible to find an antiderivative for $f(x) = 1/x$. We shall now remedy this situation by introducing a function whose derivative is $1/x$.

Definition (7.3)

> The **natural logarithmic function**, denoted by **ln**, is defined by
>
> $$\ln x = \int_1^x \frac{1}{t}\, dt$$
>
> for all $x > 0$.

The expression $\ln x$ is called the **natural logarithm of x**. The restriction $x > 0$ is necessary since $\int_1^x (1/t)\, dt$ does not exist if $x \leq 0$. (Why?) The reason that the term *logarithmic* is used in Definition (7.3) will become clear after we have proved that ln satisfies the laws of logarithms studied in precalculus courses (see Theorem (7.6)).

It follows from Theorem (5.24) that

$$D_x \int_1^x \frac{1}{t}\, dt = \frac{1}{x}$$

for all $x > 0$ (see Example 5 of Section 5.5). This gives us the next theorem.

Theorem (7.4)

$$D_x(\ln x) = \frac{1}{x}$$

Thus, ln x is an *antiderivative of* $1/x$. Since ln x is differentiable, and its derivative $1/x$ is positive for all $x > 0$, it follows from Theorems (3.8) and (4.13) that *the natural logarithmic function is continuous and increasing throughout its domain.*

Let us sketch the graph of $y = \ln x$. The y-coordinates of points on the graph are given by the integral in Definition (7.3). In particular, if $x = 1$, then by Definition (5.10),

$$\ln 1 = \int_1^1 \frac{1}{t}\, dt = 0.$$

Thus the graph of $y = \ln x$ has x-intercept 1. Since ln is an increasing function, it follows that if $x > 1$, then the point (x, y) on the graph lies above the x-axis, whereas if $0 < x < 1$, then (x, y) lies below the x-axis. To estimate y if $x \neq 1$ we may apply either the Trapezoidal Rule or Simpson's Rule. If $x = 2$, then by Example 2 in Section 5.7,

$$\ln 2 = \int_1^2 \frac{1}{t}\, dt \approx 0.6932 \approx 0.69.$$

It will be shown in Theorem (7.6) that if $a > 0$, then $\ln a^r = r \ln a$ for every rational number r. Using this result, we obtain

$$\ln 4 = \ln 2^2 = 2 \ln 2 \approx 2(0.69) = 1.38$$
$$\ln 8 = \ln 2^3 = 3 \ln 2 \approx 2.07$$
$$\ln \tfrac{1}{2} = \ln 2^{-1} = -\ln 2 \approx -0.69$$
$$\ln \tfrac{1}{4} = \ln 2^{-2} = -2 \ln 2 \approx -1.38$$
$$\ln \tfrac{1}{8} = \ln 2^{-3} = -3 \ln 2 \approx -2.07.$$

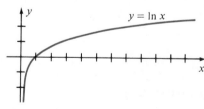

Table C provides a list of natural logarithms of many other numbers correct to three decimal places. Calculators may also be used to estimate values of ln.

Plotting the points that correspond to the coordinates we have found and using the fact that ln is continuous and increasing gives us the sketch in Figure 7.3.

At the end of this section it will be proved that

$$\lim_{x \to \infty} \ln x = \infty$$

that is, the values of ln can be made arbitrarily large by choosing x sufficiently large. It will also be shown that

$$\lim_{x \to 0^+} \ln x = -\infty$$

which means that the y-axis is a vertical asymptote for the graph.

Figure 7.3

Finally, to investigate concavity we note that

$$D_x^2(\ln x) = D_x\left(\frac{1}{x}\right) = -\frac{1}{x^2}$$

which is negative for every $x > 0$. Hence by (4.16), the graph of the natural logarithmic function is concave downward at every point $P(c, \ln c)$ on the graph.

The next result generalizes Theorem (7.4).

Theorem (7.5)

> If $y = \ln u$ and $u = g(x)$, where g is differentiable and $g(x) > 0$, then
>
> $$D_x \ln u = \frac{1}{u} D_x u.$$

Proof   By the Chain Rule,

$$\frac{dy}{dx} = \frac{dy}{du}\frac{du}{dx} = \frac{1}{u} D_x u. \qquad \square$$

Note that if $u = x$, then Theorem (7.5) reduces to (7.4).

Example 1

(a)   Find $f'(x)$ if $f(x) = \ln(x^2 + 6)$.

(b)   Find y' if $y = \ln\sqrt{x + 1}$, where $x > -1$.

Solution

(a)   Letting $u = x^2 + 6$ in Theorem (7.5), we obtain

$$f'(x) = D_x \ln(x^2 + 6) = \frac{1}{x^2 + 6} D_x(x^2 + 6) = \frac{2x}{x^2 + 6}.$$

(b)   Letting $u = \sqrt{x + 1}$ in Theorem (7.5) gives us

$$y' = \frac{1}{\sqrt{x + 1}} D_x\sqrt{x + 1} = \frac{1}{\sqrt{x + 1}} \cdot \frac{1}{2}(x + 1)^{-1/2}$$

$$= \frac{1}{\sqrt{x + 1}} \cdot \frac{1}{2}\frac{1}{\sqrt{x + 1}} = \frac{1}{2(x + 1)}. \qquad \blacksquare$$

The next theorem shows that natural logarithms obey the same laws of logarithms studied in precalculus mathematics courses.

Theorem (7.6)

> If $p > 0$ and $q > 0$, then
>
> (i) $\ln pq = \ln p + \ln q.$
>
> (ii) $\ln \left(\dfrac{p}{q}\right) = \ln p - \ln q.$
>
> (iii) $\ln p^r = r \ln p$, where r is any rational number.

Proof

(i)   If $p > 0$, then using Theorem (7.5) with $u = px$ gives us

$$D_x \ln (px) = \frac{1}{px} D_x(px) = \frac{1}{px} p = \frac{1}{x}.$$

Since $\ln px$ has the same derivative as $\ln x$ for all $x > 0$, it follows from Theorem (4.31) that these expressions differ by a constant; that is,

$$\ln px = \ln x + C$$

for some real number C. Substituting 1 for x we obtain

$$\ln p = \ln 1 + C.$$

Since $\ln 1 = 0$ we see that $C = \ln p$, and hence

$$\ln px = \ln x + \ln p.$$

Substituting q for x in the last equation gives us

$$\ln pq = \ln q + \ln p$$

which is what we wished to prove.

(ii) Using the formula established in part (i), with $p = 1/q$, we see that

$$\ln \frac{1}{q} + \ln q = \ln \left(\frac{1}{q} \cdot q\right) = \ln 1 = 0$$

and hence $\ln \dfrac{1}{q} = -\ln q.$

Consequently,

$$\ln \frac{p}{q} = \ln \left(p \cdot \frac{1}{q}\right) = \ln p + \ln \frac{1}{q} = \ln p - \ln q.$$

(iii) If r is any rational number and $x > 0$, then by Theorem (7.5) with $u = x^r$,

$$D_x(\ln x^r) = \frac{1}{x^r} D_x(x^r) = \frac{1}{x^r} rx^{r-1} = r\left(\frac{1}{x}\right).$$

However, by Theorems (3.14) and (7.4), we may also write

$$D_x(r \ln x) = r D_x(\ln x) = r\left(\frac{1}{x}\right).$$

Consequently, $D_x(\ln x^r) = D_x(r \ln x).$

Since $\ln x^r$ and $r \ln x$ have the same derivative, it follows from Theorem (4.31) that

$$\ln x^r = r \ln x + C$$

for some constant C. If we let $x = 1$ in the last formula we obtain

$$\ln 1 = r \ln 1 + C.$$

Since $\ln 1 = 0$, this implies that $C = 0$ and, therefore,

$$\ln x^r = r \ln x.$$

In Section 7.5 we shall extend this law to irrational exponents.   □

The next example illustrates the fact that it is sometimes convenient to use Theorem (7.6) before differentiating.

Example 2   Find $f'(x)$ if $f(x) = \ln [\sqrt{6x - 1}\,(4x + 5)^3]$, where $x > \frac{1}{6}$.

Solution   We first write $\sqrt{6x - 1} = (6x - 1)^{1/2}$ and then use (i) and (iii) of Theorem (7.6), obtaining

$$f(x) = \ln [(6x - 1)^{1/2}(4x + 5)^3]$$
$$= \ln (6x - 1)^{1/2} + \ln (4x + 5)^3$$
$$= \frac{1}{2} \ln (6x - 1) + 3 \ln (4x + 5).$$

We may now differentiate with ease, using Theorem (7.5). Thus,

$$f'(x) = \frac{1}{2} \cdot \frac{1}{6x - 1} (6) + 3 \cdot \frac{1}{4x + 5} (4)$$
$$= \frac{3}{6x - 1} + \frac{12}{4x + 5}$$
$$= \frac{84x + 3}{(6x - 1)(4x + 5)}.$$ ■

In future examples and exercises, if a function is defined in terms of the natural logarithmic function, its domain will not usually be stated explicitly. Instead it will be assumed that x is restricted to values for which the given expression has meaning.

Example 3   Find y' if $y = \ln \sqrt[3]{\dfrac{x^2 - 1}{x^2 + 1}}$.

Solution   We first use Theorem (7.6) to change the form of y as follows:

$$y = \ln \left(\frac{x^2 - 1}{x^2 + 1}\right)^{1/3} = \frac{1}{3} \ln \left(\frac{x^2 - 1}{x^2 + 1}\right)$$

$$= \frac{1}{3} [\ln (x^2 - 1) - \ln (x^2 + 1)].$$

Next we apply Theorem (7.5), obtaining

$$y' = \frac{1}{3} \left[\frac{1}{x^2 - 1}(2x) - \frac{1}{x^2 + 1}(2x)\right]$$

$$= \frac{2x}{3} \left[\frac{1}{x^2 - 1} - \frac{1}{x^2 + 1}\right]$$

$$= \frac{2x}{3} \left[\frac{2}{(x^2 - 1)(x^2 + 1)}\right] = \frac{4x}{3(x^4 - 1)}. \qquad \blacksquare$$

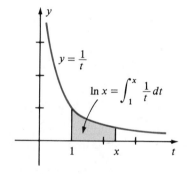

$$\ln x = \int_1^x \frac{1}{t} \, dt$$

Figure 7.4

We shall conclude this section by investigating the variation of $\ln x$ as $x \to \infty$ and as $x \to 0^+$.

If $x > 1$ and t is in the interval $[1, x]$, then $1/t > 0$, and hence we may interpret the integral $\int_1^x (1/t) \, dt = \ln x$ as the area of the region in a ty-coordinate system bounded by the graphs of $y = 1/t$, $t = 1$, $t = x$, and the t-axis. A region of this type is illustrated in Figure 7.4.

Next, observe that the sum of the areas of the three rectangles shown in Figure 7.5 is

$$\frac{1}{2} + \frac{1}{3} + \frac{1}{4} = \frac{13}{12}.$$

Since the area under the graph of $y = 1/t$ from $t = 1$ to $t = 4$ is $\ln 4$ we see that

$$\ln 4 > \frac{13}{12} > 1.$$

It follows that if M is any positive rational number, then

$$M \ln 4 > M, \quad \text{or} \quad \ln 4^M > M.$$

If $x > 4^M$, then since $\ln$ is an increasing function,

$$\ln x > \ln 4^M > M.$$

This proves that $\ln x$ can be made as large as desired by choosing x sufficiently large, that is,

$$\lim_{x \to \infty} \ln x = \infty.$$

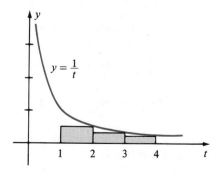

Figure 7.5

Next we note that

$$-\ln \frac{1}{x} = -(\ln 1 - \ln x) = \ln x$$

and hence

$$\lim_{x \to 0^+} \ln x = \lim_{x \to 0^+} \left(-\ln \frac{1}{x} \right).$$

As x approaches zero through positive values, $1/x$ becomes positively infinite and, therefore, so does $\ln (1/x)$. Consequently, $-\ln (1/x)$ becomes negatively infinite, that is,

$$\lim_{x \to 0^+} \ln x = -\infty.$$

EXERCISES 7.2

In Exercises 1–28 find $f'(x)$ if $f(x)$ equals the given expression.

1 $\ln (9x + 4)$

2 $\ln (x^4 + 1)$

3 $\ln (2 - 3x)^5$

4 $\ln (5x^2 + 1)^3$

5 $\ln \sqrt{7 - 2x^3}$

6 $\ln \sqrt[3]{6x + 7}$

7 $\ln (3x^2 - 2x + 1)$

8 $\ln (4x^3 - x^2 + 2)$

9 $\ln \sqrt[3]{4x^2 + 7x}$

10 $(\ln x) \ln (x + 5)$

11 $x \ln x$

12 $x^2/(\ln x)$

13 $\ln \sqrt{x} + \sqrt{\ln x}$

14 $\ln x^3 + (\ln x)^3$

15 $\dfrac{1}{\ln x} + \ln \left(\dfrac{1}{x} \right)$

16 $\ln \sqrt{\dfrac{4 + x^2}{4 - x^2}}$

17 $\ln (5x - 7)^4 (2x + 3)^3$

18 $\ln \sqrt[3]{4x - 5} (3x + 8)^2$

19 $\ln \dfrac{\sqrt{x^2 + 1}}{(9x - 4)^2}$

20 $\ln \dfrac{x^2(2x - 1)^3}{(x + 5)^2}$

21 $\ln \sqrt[3]{\dfrac{x^2 - 1}{x^2 + 1}}$

22 $\ln (\ln x)$

23 $\ln (x + \sqrt{x^2 - 1})$

24 $\ln (x + \sqrt{x^2 + 1})$

25 $(\ln \sqrt{x^2 + 1})^2$

26 $(\ln \sqrt{x}) / \sqrt{\ln x}$

27 $\sqrt{\ln (x^2 + 1)}$

28 $\ln |x|$

In Exercises 29–34 use implicit differentiation to find y'.

29 $3y - x^2 + \ln xy = 2$

30 $y^2 + \ln (x/y) - 4x + 3 = 0$

31 $x \ln y - y \ln x = 1$

32 $y^3 + x^2 \ln y = 5x + 3$

33 $y = \ln (x + y)$

34 $(\ln x)(\ln y) = xy - 1$

35 Find an equation of the tangent line to the graph of $y = x^2 + \ln (2x - 5)$ at the point on the graph with x-coordinate 3.

36 Find an equation of the tangent line to the graph of $y = x + \ln x$ that is perpendicular to the line $2x + 6y = 5$.

37 What is the difference between the graphs of $y = \ln (x^2)$ and $y = 2 \ln x$?

38 If $0 < a < b$, show that the natural logarithmic function satisfies the hypotheses of the Mean Value Theorem (4.12) on $[a, b]$, and find a general formula for the number c in the conclusion of (4.12).

39 Find the points on the graph of $y = x^2 + 4 \ln x$ at which the tangent line is parallel to the line $y - 6x + 3 = 0$.

40 Find an equation of the tangent line to the graph of $x^3 - x \ln y + y^3 = 2x + 5$ at the point $(2, 1)$.

41 If $\ln 2.00 \approx 0.6932$ use differentials to approximate $\ln 2.01$.

42 If $f(x) = \ln x$, find a formula for the nth derivative $f^{(n)}(x)$, where n is any positive integer.

43 The position function of a point moving on a coordinate line is given by $s(t) = t^2 - 4 \ln (t + 1)$, where $0 \le t \le 4$. Find the velocity and acceleration at time t and describe the motion of the point during the time interval $[0, 4]$.

44 The equation

$$t = \frac{1}{c(a - b)} \ln \frac{b(a - x)}{a(b - x)}$$

occurs in the study of certain chemical reactions, where x is the concentration of a substance at time t and a, b, c are constants. Prove that $dx/dt = c(a - x)(b - x)$.

45 Use Table C or a calculator to help sketch the graph of $y = \ln x$. Find the slope of the tangent line to the graph at the points with x-coordinates 1, 5, 10, 100, and 1000. What is true as the x-coordinate a of the point of tangency increases without bound? What is true if a approaches 0?

46 Sketch the graphs of (a) $y = \ln |x|$; (b) $y = |\ln x|$.

47 Prove that for every $x > 0$, $(x - 1)/x \le \ln x \le x - 1$.

48 Use Exercise 47 to prove that $\lim\limits_{x \to 0} \dfrac{\ln (x + 1)}{x} = 1$.

Verify the formulas in Exercises 49–52.

49 $\displaystyle\int \ln x \, dx = x \ln x - x + C$

50 $\displaystyle\int (\ln x)^2 \, dx = x(\ln x)^2 - 2x \ln x + 2x + C$

51 $\displaystyle\int \frac{1}{\sqrt{x^2 + a^2}} \, dx = \ln (x + \sqrt{x^2 + a^2}) + C$

52 $\displaystyle\int \frac{1}{a^2 - x^2} \, dx = \frac{1}{2a} \ln \frac{a + x}{a - x} + C, a^2 > x^2$

7.3

THE NATURAL EXPONENTIAL FUNCTION

In Section 7.2 we proved that

$$\lim_{x \to \infty} \ln x = \infty \quad \text{and} \quad \lim_{x \to 0^+} \ln x = -\infty.$$

These facts are used in the proof of the following important result.

Theorem (7.7)

> To every real number x there corresponds a unique positive real number y such that $\ln y = x$.

Proof   First note that if $x = 0$, then $\ln 1 = 0$. Moreover, since $\ln$ is an increasing function, 1 is the only value of y such that $\ln y = 0$.

If x is positive, then we may choose a number b such that

$$\ln 1 < x < \ln b.$$

Since $\ln$ is continuous, the Intermediate Value Theorem (2.24) guarantees the existence of a number y between 1 and b such that $\ln y = x$. Again, since $\ln$ is an increasing function, there is only one such number.

Finally, if x is negative, then there is a $b > 0$ such that

$$\ln b < x < \ln 1$$

and as before, there is precisely one number y between b and 1 such that $\ln y = x$. $\square$

It follows from Theorem (7.7) that the range of ln is $\mathbb{R}$. Since ln is an increasing function, it is one-to-one and therefore has an inverse function, to which we give the following special name.

Definition (7.8)

> The **natural exponential function**, denoted by **exp**, is the inverse of the natural logarithmic function.

The reason for the term *exponential* in this definition will become clear shortly. Since exp is the inverse of ln, its domain is $\mathbb{R}$ and its range is the set of nonnegative real numbers. Moreover, as in (7.2),

$$y = \exp x \quad \text{if and only if} \quad x = \ln y$$

where x is any real number and $y > 0$. By Definition (7.1) we may also write

$$\ln (\exp x) = x \quad \text{and} \quad \exp (\ln y) = y.$$

As pointed out in Section 7.1, if two functions are inverses of one another, then their graphs are reflections through the line $y = x$. Hence the graph of $y = \exp x$ can be obtained by reflecting the graph of $y = \ln x$ through this line, as illustrated in Figure 7.6. Evidently,

$$\lim_{x \to \infty} \exp x = \infty \quad \text{and} \quad \lim_{x \to -\infty} \exp x = 0.$$

By Theorem (7.7) there is a unique positive real number whose natural logarithm is 1. This number is denoted by e. The great Swiss mathematician Leonhard Euler (1707–1783) was among the first to study its properties extensively.

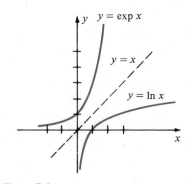

Figure 7.6

Definition of e (7.9)

> The letter e denotes the positive real number such that $\ln e = 1$.

Several values of ln were calculated in Section 7.2. It can be shown, by means of the Trapezoidal Rule, that

$$\int_1^{2.7} \frac{1}{t}\, dt < 1 < \int_1^{2.8} \frac{1}{t}\, dt$$

(see Exercise 13 in Section 5.7). Consequently, by Definition (7.3),

$$\ln 2.7 < \ln e < \ln 2.8$$

which implies that

$$2.7 < e < 2.8.$$

Later, in Theorem (7.24) it will be shown that e may be expressed as the following limit:

$$e = \lim_{h \to 0} (1 + h)^{1/h}.$$

This formula can be used to approximate e to any degree of accuracy. In particular, to five decimal places (see page 331),

$$e \approx 2.71828.$$

It can be shown that e is an irrational number.

If r is any rational number, then using (iii) of Theorem (7.6),

$$\ln e^r = r \ln e = r(1) = r.$$

This equation may be used to motivate a definition of e^x, where x is any real number. Specifically, we shall *define e^x* as the real number y such that $\ln y = x$. The following equivalent statement is a convenient way to remember this fact.

Definition of e^x **(7.10)**

> If x is any real number, then
>
> $$e^x = y \quad \text{if and only if} \quad \ln y = x.$$

Recall, from Definition (7.8), that

$$\exp x = y \quad \text{if and only if} \quad \ln y = x.$$

Comparing this relationship with Definition (7.10) we see that

$$e^x = \exp x \qquad \text{for every } x.$$

This is the reason for calling exp an *exponential* function. Indeed, it is often called the **exponential function with base e**. The fact that $\ln (\exp x) = x$ for every x, and $\exp (\ln x) = x$ for every $x > 0$ may now be written as follows.

Theorem **(7.11)**

> (i) $\ln e^x = x$ for every x.
>
> (ii) $e^{\ln x} = x$ for every $x > 0$.

Since e^x was not defined as a power of e, it is not obvious that the usual laws of exponents hold. However, these laws may be established easily, as indicated in the proof of the next theorem.

Theorem (7.12)

> If p and q are real numbers and r is a rational number, then
>
> (i) $e^p e^q = e^{p+q}$
>
> (ii) $\dfrac{e^p}{e^q} = e^{p-q}$
>
> (iii) $(e^p)^r = e^{pr}$.

Proof   Using Theorems (7.6) and (7.11),

$$\ln e^p e^q = \ln e^p + \ln e^q = p + q = \ln e^{p+q}.$$

Since the natural logarithmic function is one-to-one it follows that

$$e^p e^q = e^{p+q}.$$

This proves (i). The proofs for (ii) and (iii) are similar and are left as an exercise. It will be shown in Section 7.5 that (iii) is also true if r is irrational.   □

It is not difficult to show that if x is rational, then e^x has the same meaning as e raised to the power x. For example, we may use Theorem (7.11) as follows:

$$e^0 = e^{\ln 1} = 1$$

$$e^1 = e^{\ln e} = e.$$

Next, from Theorem (7.12),

$$e^2 = e^{1+1} = e^1 e^1 = ee$$

$$e^3 = e^{2+1} = e^2 e^1 = (ee)e$$

and in general, if n is a positive integer, e^n is a product of n factors, all equal to e. Negative exponents also have the usual properties, that is,

$$e^{-1} = e^{0-1} = \frac{e^0}{e^1} = \frac{1}{e}.$$

Similarly, $e^{-n} = \dfrac{1}{e^n}$

if n is any positive integer. Rational powers of e may also be interpreted as they are in elementary algebra.

The graph of $y = e^x$ is the same as that of $y = \exp x$ illustrated in Figure 7.6. Hereafter *we shall use e^x instead of* $\exp x$ *to denote values of the natural exponential function.*

In precalculus mathematics, graphs of equations such as $y = 2^x$ and $y = 3^x$ are sketched by *assuming* that these exponential expressions are defined for all real x and increase as x increases. Using this intuitive point of view, a rough sketch of the graph of $y = e^x$ can be obtained by sketching $y = (2.7)^x$.

It will be shown in Section 7.8 that the inverse function of a differentiable function is differentiable. Anticipating this result, let us find the derivative of the natural exponential function implicitly. Thus, if

$$y = e^x, \quad \text{then} \quad \ln y = x.$$

Differentiating implicitly gives us

$$D_x(\ln y) = D_x(x), \quad \text{or} \quad \frac{1}{y} D_x y = 1.$$

Multiplying both sides of the last equation by y we obtain

$$D_x y = y.$$

This gives us the following important formula, which is true for every real number x.

Theorem (7.13)

$$\boxed{D_x e^x = e^x}$$

Note that the preceding theorem states that the natural exponential function is its own derivative.

Example 1 Find $f'(x)$ if $f(x) = x^2 e^x$.

Solution By the Product Rule,

$$f'(x) = x^2 D_x e^x + e^x D_x x^2$$
$$= x^2 e^x + e^x(2x) = (x + 2)xe^x. \quad \blacksquare$$

The next result is a generalization of Theorem (7.13).

Theorem (7.14)

> If $u = g(x)$ and g is differentiable, then
>
> $$D_x e^u = e^u D_x u.$$

Proof Letting $y = e^u$, $u = g(x)$, and using the Chain Rule,

$$\frac{dy}{dx} = \frac{dy}{du}\frac{du}{dx} = e^u D_x u.$$

If $u = x$ then Theorem (7.14) reduces to (7.13).

Example 2 Find y' if $y = e^{\sqrt{x^2+1}}$.

Solution   By Theorem (7.14),

$$D_x e^{\sqrt{x^2+1}} = e^{\sqrt{x^2+1}} D_x \sqrt{x^2+1} = e^{\sqrt{x^2+1}} D_x(x^2+1)^{1/2}$$

$$= e^{\sqrt{x^2+1}}(\tfrac{1}{2})(x^2+1)^{-1/2}(2x) = e^{\sqrt{x^2+1}} \cdot \frac{x}{\sqrt{x^2+1}}$$

$$= \frac{xe^{\sqrt{x^2+1}}}{\sqrt{x^2+1}}.$$ ∎

Example 3 The function f defined by $f(x) = e^{-x^2/2}$ arises in the branch of mathematics called *probability*. Find the local extrema of f, discuss concavity, find the points of inflection, and sketch the graph of f.

Solution   By Theorem (7.14),

$$f'(x) = e^{-x^2/2} D_x\left(-\frac{x^2}{2}\right) = e^{-x^2/2}\left(-\frac{2x}{2}\right) = -xe^{-x^2/2}.$$

Since $e^{-x^2/2}$ is always positive, the only critical number of f is 0. If $x < 0$, then $f'(x) > 0$ whereas if $x > 0$, then $f'(x) < 0$. It follows from the First Derivative Test that f has a local maximum at 0. The maximum value is $f(0) = e^{-0} = 1$.

Applying the Product Rule to $f'(x)$,

$$f''(x) = -xe^{-x^2/2}(-2x/2) - e^{-x^2/2}$$
$$= e^{-x^2/2}(x^2 - 1)$$

and hence the second derivative is zero at -1 and 1. If $-1 < x < 1$, then $f''(x) < 0$ and, by (4.16), the graph of f is concave downward in the open interval $(-1, 1)$. If $x < -1$ or $x > 1$, then $f''(x) > 0$ and, therefore, the graph is concave upward throughout the infinite intervals $(-\infty, -1)$ and $(1, \infty)$. Consequently, $P(-1, e^{-1/2})$ and $Q(1, e^{-1/2})$ are points of inflection. Writing

$$f(x) = \frac{1}{e^{x^2/2}}$$

it is evident that as x increases numerically, $f(x)$ approaches 0. It is left as an exercise to prove that $\lim_{x \to \infty} f(x) = 0$ and $\lim_{x \to -\infty} f(x) = 0$; that is, the x-axis is a horizontal asymptote. The graph of f is sketched in Figure 7.7.   ∎

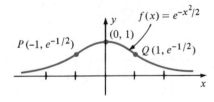

Figure 7.7

Exponential functions play an important role in the field of *radiotherapy*, where tumors are treated by means of radiation. Of major interest is the fraction of a tumor population that survives a treatment. This *surviving fraction* depends not only on the energy and nature of the radiation, but also on the depth, size, and characteristics of the tumor itself. The exposure to radiation may be thought of as a number of potentially damaging events, where only one "hit" is required to kill a tumor cell. Suppose that each cell has exactly one "target" that must be hit. If k denotes the average target size of a tumor cell, and if x is the number of damaging events (the *dose*), then the surviving fraction $f(x)$ is given by

$$f(x) = e^{-kx}.$$

This is called the *one target–one hit surviving fraction.*

Suppose next that each cell has n targets and that hitting any one of the targets results in the death of a cell. In this case, the *n target–one hit surviving fraction* is given by

$$f(x) = 1 - (1 - e^{-kx})^n.$$

The graph of f may be analyzed to determine what effect increasing the dosage x will have on decreasing the surviving fraction of tumor cells. Note that $f(0) = 1$; that is, if there is no dose, then all cells survive. By differentiation, we obtain

$$f'(x) = 0 - n(1 - e^{-kx})^{n-1}D_x(-e^{-kx}) = -nke^{-kx}(1 - e^{-kx})^{n-1}.$$

Let us consider the special case $n = 2$. In this event

$$f'(x) = -2ke^{-kx}(1 - e^{-kx}).$$

Since $f'(x) < 0$ and $f'(0) = 0$, the function f is decreasing and the graph has a horizontal tangent line at the point $(0, 1)$. It may be verified that the second derivative is

$$f''(x) = 2k^2e^{-kx}(1 - 2e^{-kx}).$$

Evidently, $f''(x) = 0$ if $1 - 2e^{-kx} = 0$; that is, if $e^{-kx} = \frac{1}{2}$, or equivalently, $-kx = \ln \frac{1}{2} = -\ln 2$. This gives us

$$x = \frac{1}{k} \ln 2.$$

It can be shown that if $0 \le x < (1/k) \ln 2$, then $f''(x) < 0$, and hence the graph is concave downward. If $x > (1/k) \ln 2$ we have $f''(x) > 0$, and the graph is concave upward. This implies that there is a point of inflection with x-coordinate $(1/k) \ln 2$. The y-coordinate of this point is

$$f\left(\frac{1}{k} \ln 2\right) = 1 - (1 - e^{-\ln 2})^2$$

$$= 1 - \left(1 - \frac{1}{2}\right)^2 = \frac{3}{4}.$$

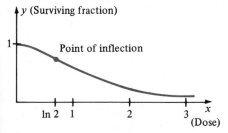

Figure 7.8 Surviving fraction of tumor cells after a radiation treatment

The graph is sketched in Figure 7.8 for the case $k = 1$. The "shoulder" on the curve near the point $(0, 1)$ represents the threshold nature of the treatment; that is, a small dose results in very little tumor elimination. Note that for large

x, an increase in dosage has little effect on the surviving fraction. To determine the ideal dose that should be administered to a given patient, specialists in radiation therapy must also take into account the number of healthy cells that are killed during a treatment.

EXERCISES 7.3

In Exercises 1–24 find $f'(x)$ if $f(x)$ equals the given expression.

1 e^{-5x}

2 e^{3x}

3 e^{3x^2}

4 e^{1-x^3}

5 $\sqrt{1 + e^{2x}}$

6 $1/(e^x + 1)$

7 $e^{\sqrt{x+1}}$

8 xe^{-x}

9 $x^2 e^{-2x}$

10 $\sqrt{e^{2x} + 2x}$

11 $e^x/(x^2 + 1)$

12 x/e^{x^2}

13 $(e^{4x} - 5)^3$

14 $(e^{3x} - e^{-3x})^4$

15 $e^{1/x} + 1/e^x$

16 $e^{\sqrt{x}} + \sqrt{e^x}$

17 $\dfrac{e^x - e^{-x}}{e^x + e^{-x}}$

18 $e^{(ex)}$

19 $e^{-2x} \ln x$

20 $e^{x \ln x}$

21 $\ln \dfrac{e^x + 1}{e^x - 1}$

22 $\dfrac{\ln (e^x + 1)}{\ln (e^x - 1)}$

23 $\sqrt{\ln (e^{2x} + e^{-2x})}$

24 $\ln \sqrt{e^{2x} + e^{-2x}}$

In Exercises 25–28 use implicit differentiation to find y'.

25 $e^{xy} - x^3 + 3y^2 = 11$

26 $xe^y + 2x - \ln (y + 1) = 3$

27 $y^3 + xe^y = 3x^2 - 10$

28 $xe^y - ye^x = 2$

29 Find an equation of the tangent line to the graph of $y = (x - 1)e^x + 3 \ln x + 2$ at the point $P(1, 2)$.

30 Find an equation of the tangent line to the graph of $y = x - e^{-x}$ that is parallel to the line $6x - 2y = 7$.

31 Show that the natural exponential function satisfies the hypotheses of the Mean Value Theorem (4.12) on every closed interval $[a, b]$, and find a general formula for the number c in the conclusion of (4.12).

32 Compare the graphs of $y = \sqrt{e^x}$ and $y = e^{\sqrt{x}}$.

33 Find a point on the graph of $y = e^{2x}$ at which the tangent line passes through the origin.

34 Find an equation of the normal line to the graph of $y = 4xe^{x^2 - 1}$ at the point $P(1, 4)$.

In Exercises 35–40 find the local extrema of f. Determine where f is increasing or decreasing, discuss concavity, find the points of inflection, and sketch the graph of f.

35 $f(x) = xe^x$

36 $f(x) = x^2 e^{-2x}$

37 $f(x) = e^{-2x}$

38 $f(x) = xe^{-x}$

39 $f(x) = x \ln x$

40 $f(x) = e^x + e^{-x}$

41 A radioactive substance decays according to the law $q(t) = q_0 e^{-ct}$, where q_0 is the initial amount of the substance, c is a positive constant, and $q(t)$ is the amount remaining after t units of time. Show that the rate at which the substance decays is proportional to $q(t)$.

42 The current $I(t)$ at time t in a certain electrical circuit is given by $I(t) = I_0 e^{-Rt/L}$, where R and L denote the resistance and inductance, respectively, and I_0 is the current at time $t = 0$. Show that the rate of change of the current at any time t is proportional to $I(t)$.

43 If a drug is injected into the blood stream, then its concentration t minutes later is given by

$$C(t) = \frac{k}{a - b} (e^{-bt} - e^{-at})$$

where a, b, and k are positive constants.

(a) At what time does the maximum concentration occur?

(b) What can be said about the concentration after a long period of time? (Justify your answer.)

44 If a beam of light that has intensity k is projected vertically downward into water, then its intensity $I(x)$ at a depth of x meters is $I(x) = ke^{-1.4x}$.

    (a) At what rate is the intensity changing with respect to depth at 1 meter? 5 meters? 10 meters?

    (b) At what depth is the intensity one-half its value at the surface? One-tenth its value?

45 Use differentials to approximate the change in $f(x) = xe^{x^2}$ if x changes from 1.00 to 1.01. What is the approximate value of $f(1.01)$?

46 If $f(x) = e^{2x}$, find a formula for the nth derivative $f^{(n)}(x)$.

In Exercises 47–50 find values of c such that $y = e^{cx}$ is a solution of the differential equation.

47 $y'' - 3y' + 2y = 0$

48 $y'' - 9y = 0$

49 $y''' - y'' - 4y' + 4y = 0$

50 $y''' - y'' - 6y' = 0$

51 Prove the following.

    (a) $\lim_{x \to \infty} e^{-x^2/2} = 0$ (b) $\lim_{x \to -\infty} e^{-x^2/2} = 0$

52 Prove the following.

    (a) $\lim_{x \to -\infty} e^x = 0$ (b) $\lim_{x \to \infty} e^x = \infty$

53 In statistics the *normal distribution function f* is defined by

$$f(x) = \frac{1}{\sigma\sqrt{2\pi}} e^{(-1/2)[(x-\mu)/\sigma]^2}$$

where μ and σ are constants, and $\sigma > 0$. (μ is called the *mean* and σ the *variance* of the distribution.) Find the local extrema of f and determine where f is increasing

or decreasing. Discuss concavity, find points of inflection, find $\lim_{x \to \infty} f(x)$ and $\lim_{x \to -\infty} f(x)$, and sketch the graph of f.

54 Prove that $e^x \geq 1 + x$ for all x. (*Hint*: Use the Mean Value Theorem.)

55 Prove (ii) and (iii) of Theorem (7.12).

56 The integral $\int_a^b e^{-x^2} dx$ has important applications. Use the Trapezoidal Rule (with $n = 10$) and a calculator to approximate this integral if $a = 0$ and $b = 1$.

57 Nerve impulses in the human body travel along nerve fibers that consist of an *axon*, which transports the impulse, and an insulating coating surrounding the axon called the *myelin sheath* (see figure). *Nodes of Ranvier* serve as relay stations along the fibers; however, between nodes the nerve fiber is similar to an insulated cylindrical cable. It is known that for such cables, the velocity v of an impulse is given by $v = -k(r/R)^2 \ln(r/R)$ where r is the radius of the cable, R is the insulation radius, and k is a constant. Find the value of r/R that maximizes v. (In most nerve fibers $r/R \approx 0.6$.)

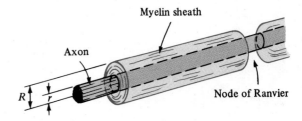

Figure for Exercise 63

58 Sketch the graph of the *three target–one hit surviving fraction* (with $k = 1$) given by $f(x) = 1 - (1 - e^{-x})^3$. Compare the graph with Figure 7.8.

7.4

DIFFERENTIATION AND INTEGRATION

The formula for the derivative of $\ln u$ established in Section 7.2 can be extended as indicated in the next theorem.

Theorem (7.15)

If $u = g(x)$, where g is differentiable and $g(x) \neq 0$, then

$$D_x \ln |u| = \frac{1}{u} D_x u.$$

Proof   If $x < 0$, then $\ln |x| = \ln (-x)$ and by Theorem (7.5),

$$D_x \ln (-x) = \frac{1}{(-x)} D_x(-x) = \frac{1}{(-x)} (-1) = \frac{1}{x}.$$

Consequently,

$$D_x \ln |x| = \frac{1}{x} \quad \text{for all } x \neq 0.$$

The Chain Rule may now be used to complete the proof. $\square$

Observe that if $u = g(x) > 0$, then Theorem (7.15) is the same as (7.5).

Example 1   Find $f'(x)$ if $f(x) = \ln |4 + 5x - 2x^3|$.

Solution   Using Theorem (7.15) with $u = 4 + 5x - 2x^3$,

$$f'(x) = \frac{1}{4 + 5x - 2x^3} (5 - 6x^2) = \frac{5 - 6x^2}{4 + 5x - 2x^3}.$$ ∎

We may use differentiation formulas for ln to obtain rules for integration. In particular, since

$$D_x \ln |g(x)| = \frac{1}{g(x)} g'(x)$$

the following integration formula holds.

Theorem (7.16)

> If $u = g(x) \neq 0$, and g is differentiable,
>
> $$\int \frac{1}{u} du = \ln |u| + C.$$

Of course, if $u > 0$, then the absolute value sign may be deleted. A special case of Theorem (7.16) is

$$\int \frac{1}{x} dx = \ln |x| + C.$$

Example 2   Find $\displaystyle\int \frac{x}{3x^2 - 5} dx$.

Solution   If we let $u = 3x^2 - 5$, then $du = 6x\, dx$. We may arrive at the form (7.16) by introducing a factor 6 in the integrand as follows:

$$\int \frac{x}{3x^2 - 5}\, dx = \frac{1}{6} \int \frac{1}{3x^2 - 5}\, 6x\, dx = \frac{1}{6} \int \frac{1}{u}\, du$$

$$= \frac{1}{6} \ln |u| + C = \frac{1}{6} \ln |3x^2 - 5| + C.$$

Another technique is to replace the expression $x\, dx$ in the integral by $\frac{1}{6}\, du$ and then integrate.   ∎

Example 3   Evaluate $\displaystyle\int_2^4 \frac{1}{9 - 2x}\, dx$.

Solution   Since $1/(9 - 2x)$ is continuous on $[2, 4]$, the definite integral exists. One method of evaluation consists of using an indefinite integral to find an antiderivative of $1/(9 - 2x)$. If we let $u = 9 - 2x$, then $du = -2\, dx$ and we may proceed as follows:

$$\int \frac{1}{9 - 2x}\, dx = -\frac{1}{2} \int \frac{1}{9 - 2x}\, (-2)\, dx$$

$$= -\frac{1}{2} \int \frac{1}{u}\, du = -\frac{1}{2} \ln |u| + C$$

$$= -\frac{1}{2} \ln |9 - 2x| + C.$$

Applying the Fundamental Theorem of Calculus,

$$\int_2^4 \frac{1}{9 - 2x}\, dx = -\frac{1}{2} \left[\ln |9 - 2x| \right]_2^4$$

$$= -\frac{1}{2} [\ln 1 - \ln 5] = \frac{1}{2} \ln 5.$$

Another method is to use the same substitution in the *definite* integral and change the limits of integration. Referring to $u = 9 - 2x$ we see that if $x = 2$ then $u = 5$, and if $x = 4$ then $u = 1$. Consequently,

$$\int_2^4 \frac{1}{9 - 2x}\, dx = -\frac{1}{2} \int_2^4 \frac{1}{9 - 2x}\, (-2)\, dx$$

$$= -\frac{1}{2} \int_5^1 \frac{1}{u}\, du = -\frac{1}{2} \left[\ln |u| \right]_5^1$$

$$= -\frac{1}{2} [\ln 1 - \ln 5] = \frac{1}{2} \ln 5.$$ ∎

Example 4   Find $\int \dfrac{\sqrt{\ln x}}{x}\,dx$.

Solution   If we let $u = \ln x$, then $du = (1/x)\,dx$ and

$$\int \frac{\sqrt{\ln x}}{x}\,dx = \int \sqrt{\ln x} \cdot \frac{1}{x}\,dx = \int u^{1/2}\,du = \frac{2}{3}u^{3/2} + C$$

$$= \frac{2}{3}(\ln x)^{3/2} + C. \qquad \blacksquare$$

It is easy to find integration formulas for the natural exponential function. Since $D_x e^{g(x)} = e^{g(x)}g'(x)$ we obtain the following.

Theorem (7.17)

> If $u = g(x)$, and g is differentiable,
>
> $$\int e^u\,du = e^u + C.$$

As a special case of Theorem (7.17),

$$\int e^x\,dx = e^x + C.$$

Example 5   Find $\int \dfrac{e^{3/x}}{x^2}\,dx$.

Solution   If we let $u = 3/x$, then $du = (-3/x^2)\,dx$ and the integrand may be written in the form (7.17) by introducing the factor -3. Doing this and compensating by multiplying the integral by $-\frac{1}{3}$,

$$\int \frac{e^{3/x}}{x^2}\,dx = -\frac{1}{3}\int e^{3/x}\left(-\frac{3}{x^2}\right)dx$$

$$= -\frac{1}{3}\int e^u\,du$$

$$= -\frac{1}{3}e^u + C$$

$$= -\frac{1}{3}e^{3/x} + C. \qquad \blacksquare$$

Example 6 Find $\int_1^2 \dfrac{e^{3/x}}{x^2}\,dx$.

Solution An antiderivative was found in Example 5. Applying the Fundamental Theorem of Calculus,

$$\int_1^2 \frac{e^{3/x}}{x^2}\,dx = -\frac{1}{3}\left[e^{3/x}\right]_1^2$$

$$= -\frac{1}{3}(e^{3/2} - e^3).$$

This example can also be solved by using the method of substitution. If, as in Example 5 we let $u = 3/x$, then $du = (-3/x^2)\,dx$. Next we note that if $x = 1$, then $u = 3$, and if $x = 2$, then $u = \frac{3}{2}$. Consequently,

$$\int_1^2 \frac{e^{3/x}}{x^2}\,dx = -\frac{1}{3}\int_3^{3/2} e^u\,du$$

$$= -\frac{1}{3}\left[e^u\right]_3^{3/2} = \left(-\frac{1}{3}\right)(e^{3/2} - e^3). \qquad \blacksquare$$

Example 7 Find the area of the region bounded by the graphs of the equations $y = e^x$, $y = \sqrt{x}$, $x = 0$, and $x = 1$.

Solution The region and a typical rectangle of the type considered in Chapter 6 are shown in Figure 7.9. The area A is

$$A = \lim_{\|P\|\to 0} \sum_i (e^{w_i} - \sqrt{w_i})\,\Delta x_i$$

$$= \int_0^1 (e^x - \sqrt{x})\,dx$$

$$= \left[e^x - \frac{2}{3}x^{3/2}\right]_0^1$$

$$= \left(e - \frac{2}{3}\right) - (e^0 - 0) = e - \frac{5}{3}.$$

If an approximation is desired, then to the nearest hundredth,

$$A \approx 2.72 - 1.67 = 1.05. \qquad \blacksquare$$

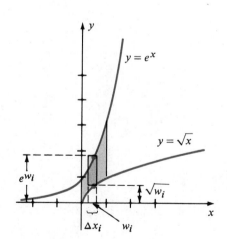

Figure 7.9

Given $y = f(x)$ it is sometimes convenient to find $D_x y$ by a process called **logarithmic differentiation**. This method is especially useful if $f(x)$ involves complicated products, quotients, or powers. Later in the text we shall also apply it to expressions such as $y = (x^2 + 1)^x$. The process may be outlined as follows where, at the outset, it is assumed that $f(x) > 0$.

Steps in Logarithmic Differentiation

1. $y = f(x)$ (Given)

2. $\ln y = \ln f(x)$ (Take natural logarithms and simplify)

3. $D_x[\ln y] = D_x[\ln f(x)]$ (Differentiate implicitly)

4. $\dfrac{1}{y} D_x y = D_x[\ln f(x)]$ (By Theorem (7.5))

5. $D_x y = f(x) D_x[\ln f(x)]$ (Multiply by $y = f(x)$)

Of course, to complete the solution it is necessary to differentiate $\ln f(x)$ at some stage after Step 3. If $f(x) < 0$ for some x, then Step 2 is invalid, since $\ln f(x)$ is undefined. However, in this event we could replace Step 1 by $|y| = |f(x)|$ and take natural logarithms, obtaining $\ln|y| = \ln|f(x)|$. If we now differentiate implicitly and use Theorem (7.15) we again arrive at Step 4. Thus negative values of $f(x)$ do not change the outcome and it is unnecessary to be concerned whether $f(x)$ is positive or negative. The method should not be used to find $f'(a)$ if $f(a) = 0$, since $\ln 0$ is undefined.

Example 8   Use logarithmic differentiation to find $D_x y$ if $y = \dfrac{(5x-4)^3}{\sqrt{2x+1}}$.

Solution   As in the outline we begin by taking the natural logarithm of each side and simplifying, obtaining

$$\ln y = 3 \ln (5x - 4) - \left(\frac{1}{2}\right) \ln (2x + 1).$$

Differentiating both sides with respect to x, we obtain

$$\frac{1}{y} D_x y = 3 \frac{1}{5x - 4} (5) - \left(\frac{1}{2}\right) \frac{1}{2x + 1} (2)$$

$$= \frac{25x + 19}{(5x - 4)(2x + 1)}.$$

Finally, multiplying both sides of the last equation by y (that is, by $(5x - 4)^3/\sqrt{2x + 1}$),

$$D_x y = \frac{25x + 19}{(5x - 4)(2x + 1)} \cdot \frac{(5x - 4)^3}{\sqrt{2x + 1}}$$

$$= \frac{(25x + 19)(5x - 4)^2}{(2x + 1)^{3/2}}.$$

This result may be checked by applying the quotient rule to y. ∎

EXERCISES 7.4

Evaluate the integrals in Exercises 1–28.

1 $\displaystyle\int \frac{x}{x^2 + 1}\,dx$

2 $\displaystyle\int \frac{1}{8x + 3}\,dx$

3 $\displaystyle\int \frac{1}{7 - 5x}\,dx$

4 $\displaystyle\int \frac{x^3}{x^4 - 5}\,dx$

5 $\displaystyle\int \frac{x - 2}{x^2 - 4x + 9}\,dx$

6 $\displaystyle\int \frac{(2 + \ln x)^3}{x}\,dx$

7 $\displaystyle\int \frac{x^2}{x^3 + 1}\,dx$

8 $\displaystyle\int_1^2 \frac{3x}{x^2 + 4}\,dx$

9 $\displaystyle\int_{-2}^1 \frac{1}{2x + 7}\,dx$

10 $\displaystyle\int_{-1}^0 \frac{1}{4 - 5x}\,dx$

11 $\displaystyle\int_1^4 \frac{2}{\sqrt{x}(\sqrt{x} + 4)}\,dx$

12 $\displaystyle\int \frac{x - 1}{3x^2 - 6x + 2}\,dx$

13 $\displaystyle\int (x + e^{5x})\,dx$

14 $\displaystyle\int (1 + e^{-3x})\,dx$

15 $\displaystyle\int \frac{\ln x}{x}\,dx$

16 $\displaystyle\int \frac{1}{x(\ln x)^2}\,dx$

17 $\displaystyle\int_1^3 e^{-4x}\,dx$

18 $\displaystyle\int_0^1 e^{2x+3}\,dx$

19 $\displaystyle\int \frac{e^{\sqrt{x}}}{\sqrt{x}}\,dx$

20 $\displaystyle\int x e^{x^2}\,dx$

21 $\displaystyle\int \frac{(e^x + 1)^2}{e^x}\,dx$

22 $\displaystyle\int \frac{e^x}{(e^x + 1)^2}\,dx$

23 $\displaystyle\int \frac{e^x - e^{-x}}{e^x + e^{-x}}\,dx$

24 $\displaystyle\int \frac{e^x}{e^x + 1}\,dx$

25 $\displaystyle\int \frac{1}{x^2 + 2x + 1}\,dx$

26 $\displaystyle\int \frac{(x^2 - 4)^2}{2x}\,dx$

27 $\displaystyle\int \frac{2x^2 - 5x - 7}{x - 3}\,dx$

28 $\displaystyle\int \frac{x^2 + 3x + 1}{x}\,dx$

In Exercises 29–32 find the area of the region bounded by the graphs of the given equations.

29 $y = e^{2x}$, $y = 0$, $x = 0$, $x = \ln 3$

30 $xy = 1$, $y = 0$, $x = 1$, $x = e$

31 $y = e^{-x}$, $xy = 1$, $x = 1$, $x = 2$

32 $y = e^{-2x}$, $y = -e^x$, $x = 0$, $x = 2$

33 The region bounded by the graphs of $y = e^{-x^2}$, $y = 0$, $x = 0$, and $x = 1$ is revolved about the y-axis. Find the volume of the resulting solid.

34 The region bounded by the graphs of $y = 1/\sqrt{x}$, $y = 0$, $x = 1$, and $x = 4$ is revolved about the x-axis. Find the volume of the resulting solid.

35 A particle moves on a coordinate line such that its velocity at time t is e^{-3t}. If the particle is at the origin at $t = 0$, how far does it travel during the time interval $[0, 2]$?

36 A solid has, for its base, the region in the xy-plane bounded by the graphs of $y = e^{2x}$, $y = e^x$, and $x = 1$. Find the volume of the solid if every cross section by a plane perpendicular to the x-axis is a square.

In Exercises 37 and 38 (a) set up an integral for finding the arc length L of the graph of the given equation from A to B; (b) write out the formula for approximating L by means of the Trapezoidal Rule with $n = 5$; and (c) use a calculator to find the value of the approximation given by the formula in part (b).

37 $y = e^x$; $A(0, 1)$, $B(1, e)$

38 $y = e^{-x^2}$; $A(0, 1)$, $B(1, e^{-1})$

In Exercises 39–46 find $D_x y$ by logarithmic differentiation.

39 $y = (5x + 2)^3 (6x + 1)^2$

40 $y = \sqrt{4x + 7}\,(x - 5)^3$

41 $y = (x^2 + 3)^5 / \sqrt{x + 1}$

42 $y = (x + 1)^2 (x + 2)^3 (x + 3)^4$

43 $y = \sqrt[3]{2x + 1}\,(4x - 1)^2 (3x + 5)^4$

44 $y = (2x - 3)^2 / \sqrt{x + 1}\,(7x + 2)^3$

45 $y = \sqrt{(3x^2 + 2)\sqrt{6x - 7}}$

46 $y = (x^2 + 3)^{2/3} (3x - 4)^4 / \sqrt{x}$

In Exercises 47–50 find $f'(x)$.

47 $f(x) = \ln |3 - 2x|$

48 $f(x) = \ln |4 - 5x^3|^2$

49 $f(x) = \ln |1 - e^{-2x}|^3$

50 $f(x) = 1/\ln |2x - 15|$

51 The pressure p and volume v of an expanding gas are related by the equation $pv = k$, where k is a constant. Show that the work done if the gas expands from v_0 to v_1 is $k \ln (v_1/v_0)$. (*Hint*: See Example 4 in Section 6.5.)

7.5

GENERAL EXPONENTIAL AND LOGARITHMIC FUNCTIONS

We shall next define a^x, where $a > 0$ and x is any real number. If the exponent is a *rational* number r, then applying (ii) of Theorem (7.11) and (iii) of Theorem (7.6) we see that

$$a^r = e^{\ln a^r} = e^{r \ln a}.$$

This formula is the motivation for the following definition of a^x.

Definition of a^x (7.18)

$$a^x = e^{x \ln a}$$

for every $a > 0$ and every real number x.

The function f defined by $f(x) = a^x$ is called the **exponential function with base a**. Since e^x is positive for all x, so is a^x. To approximate values of a^x, we may refer to tables of logarithmic and exponential functions, or use a calculator.

It is now possible to prove that the law of logarithms stated in (iii) of Theorem (7.6) is also true for irrational exponents. Thus, if u is any *real* number, then by Definition (7.18) and (i) of Theorem (7.11),

$$\ln a^u = \ln e^{u \ln a} = u \ln a.$$

The next theorem states that properties of rational exponents studied in elementary algebra are also true for real exponents.

Laws of Exponents (7.19)

Suppose $a > 0$ and $b > 0$. If u and v are any real numbers, then

$$a^u a^v = a^{u+v} \qquad (a^u)^v = a^{uv} \qquad (ab)^u = a^u b^u$$

$$\frac{a^u}{a^v} = a^{u-v} \qquad \left(\frac{a}{b}\right)^u = \frac{a^u}{b^u}$$

Proof   To show that $a^u a^v = a^{u+v}$ we use Definition (7.18) and (i) of Theorem (7.12) as follows:

$$a^u a^v = e^{u \ln a} e^{v \ln a}$$
$$= e^{u \ln a + v \ln a}$$
$$= e^{(u+v) \ln a}$$
$$= a^{u+v}.$$

To prove that $(a^u)^v = a^{uv}$ we first use Definition (7.18) with a^u in place of a and $v = x$ to write

$$(a^u)^v = e^{v \ln a^u}.$$

Using the fact that $\ln a^u = u \ln a$, and then applying Definition (7.18), we obtain

$$(a^u)^v = e^{vu \ln a} = a^{vu} = a^{uv}.$$

The proofs of the remaining laws are left as exercises. $\quad\square$

Theorem (7.20)

> (i) $\quad D_x(a^x) = a^x \ln a$
>
> (ii) $\quad D_x(a^u) = (a^u \ln a)D_x u, \quad$ where $u = g(x)$.

Proof Applying Definition (7.18) and Theorem (7.14),

$$D_x(a^x) = D_x(e^{x \ln a}) = e^{x \ln a}D_x(x \ln a) = e^{x \ln a}(\ln a).$$

Again using (7.18), we obtain

$$D_x(a^x) = a^x \ln a.$$

This gives us (i). Formula (ii) may be obtained by using the Chain Rule. $\quad\square$

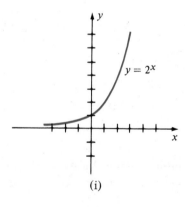

(i)

It should be noted that if $a = e$, then (i) of Theorem (7.20) reduces to (7.13). If $a > 1$, then $\ln a > 0$ and, therefore, $D_x(a^x) > 0$. Hence a^x is increasing on the interval $(-\infty, \infty)$ if $a > 1$. If $0 < a < 1$, then $\ln a < 0$ and $D_x(a^x) < 0$; that is, a^x is decreasing for all x.

The graphs of $y = 2^x$ and $y = (\tfrac{1}{2})^x = 2^{-x}$ illustrated in Figure 7.10 may be sketched by plotting some representative points. The graph of the equation $y = a^x$ has the general shape illustrated in (i) or (ii) of Figure 7.10, depending on whether $a > 1$ or $0 < a < 1$, respectively.

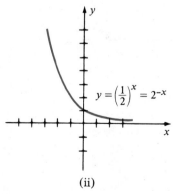

(ii)

Example 1 Find y' if $y = 3^{\sqrt{x}}$.

Solution Using (ii) of Theorem (7.20) with $a = 3$ and $u = \sqrt{x}$,

$$y' = (3^{\sqrt{x}} \ln 3)\left(\frac{1}{2}x^{-1/2}\right) = \frac{3^{\sqrt{x}} \ln 3}{2\sqrt{x}}. \qquad \blacksquare$$

If $u = g(x)$ it is very important to distinguish between expressions of the form a^u and u^a where a is a real number. To differentiate a^u we use (7.20), whereas for u^a the Power Rule must be employed, as illustrated in the next example.

Figure 7.10

Example 2   Find y' if $y = (x^2 + 1)^{10} + 10^{x^2+1}$.

Solution   Using the Power Rule for Functions and Theorem (7.20), we
obtain

$$y' = 10(x^2 + 1)^9(2x) + 10^{x^2+1} \ln 10(2x)$$
$$= 20x[(x^2 + 1)^9 + 10^{x^2} \ln 10]. \qquad \blacksquare$$

The integration formula in (i) of the next theorem may be verified by
showing that the integrand is the derivative of the expression on the right side
of the equation. Formula (ii) follows from (ii) of Theorem (7.20).

Theorem (7.21)

> (i) $\displaystyle \int a^x \, dx = \left(\frac{1}{\ln a}\right)a^x + C$
>
> (ii) $\displaystyle \int a^u \, du = \left(\frac{1}{\ln a}\right)a^u + C, \quad \text{where } u = g(x).$

Example 3   Evaluate $\displaystyle \int 3^{x^2}x \, dx$.

Solution   If we let $u = x^2$, then $du = 2x \, dx$ and we may proceed as follows:

$$\int 3^{x^2}x \, dx = \frac{1}{2} \int 3^{x^2}(2x) \, dx = \frac{1}{2} \int 3^u \, du$$

$$= \frac{1}{2} \left(\frac{1}{\ln 3}\right)3^u + C = \left(\frac{1}{2 \ln 3}\right)3^{x^2} + C. \qquad \blacksquare$$

If $a \neq 1$ and $f(x) = a^x$, then f is a one-to-one function. Its inverse function
is denoted by $\log_a$ and is called the *logarithmic function with base a*. Another
way of stating this relationship is as follows.

Definition of $\log_a x$ (7.22)

> $$y = \log_a x \quad \text{if and only if} \quad x = a^y.$$

The expression $\log_a x$ is called **the logarithm of x with base a**. Using this
terminology, natural logarithms are logarithms with base e, that is,

$$\ln x = \log_e x.$$

The fact that laws of logarithms similar to Theorem (7.6) are true for logarithms
with base a is left as an exercise.

In order to obtain the relationship between $\log_a$ and $\ln$, consider $y = \log_a x$ or, equivalently, $x = a^y$. Taking the natural logarithm of both sides of the last equation gives us $\ln x = y \ln a$, or $y = \ln x / \ln a$. This proves that

$$\log_a x = \frac{\ln x}{\ln a}.$$

Differentiating both sides of the last equation leads to (i) of the next theorem. Using the Chain Rule and generalizing to absolute values as in Theorem (7.15), gives us (ii).

Theorem (7.23)

(i) $D_x \log_a x = \dfrac{1}{x \ln a}$.

(ii) $D_x \log_a |u| = \dfrac{1}{u \ln a} D_x u,$ where $u = g(x) \neq 0.$

Logarithms with base 10 are useful for certain applications (see Exercises 43–45). It is customary to refer to such logarithms as **common logarithms** and to use the symbol **log x** as an abbreviation for $\log_{10} x$. This notation is used in the next example.

Example 4   Find $f'(x)$ if $f(x) = \log \sqrt[3]{(2x + 5)^2}$.

Solution   We begin by using the analogue of (iii) of Theorem (7.6) for logarithms with any base together with the fact that $(2x + 5)^2 = |2x + 5|^2$. Thus

$$f(x) = \frac{1}{3} \log |2x + 5|^2 = \frac{2}{3} \log |2x + 5|.$$

Next, employing (ii) of Theorem (7.23) with $a = 10$,

$$f'(x) = \frac{2}{3} \frac{1}{(2x + 5) \ln 10} (2) = \frac{4}{(6x + 15) \ln 10}. \qquad \blacksquare$$

Since irrational exponents have been given meaning, we may now consider the **general power function** f defined by $f(x) = x^c$ where c is any real number. If c is irrational, then by definition, the domain of f is the set of positive real numbers. Using Definition (7.18), and Theorems (7.14) and (7.4),

$$D_x x^c = D_x(e^{c \ln x}) = e^{c \ln x} D_x(c \ln x)$$

$$= e^{c \ln x}\left(\frac{c}{x}\right) = x^c\left(\frac{c}{x}\right) = cx^{c-1}.$$

This proves that the Power Rule is true for irrational as well as rational exponents. The Power Rule for Functions may also be extended to irrational exponents.

Example 5 Find y' if (a) $y = x^{\sqrt{2}}$; (b) $y = (1 + e^{2x})^{\pi}$.

Solution

$$\text{(a)} \quad y' = \sqrt{2}x^{\sqrt{2}-1}$$

$$\text{(b)} \quad y' = \pi(1 + e^{2x})^{\pi-1}D_x(1 + e^{2x})$$

$$= \pi(1 + e^{2x})^{\pi-1}(2e^{2x})$$

$$= 2\pi e^{2x}(1 + e^{2x})^{\pi-1} \qquad \blacksquare$$

Example 6 Find $D_x y$ if $y = x^x$, where $x > 0$.

Solution   Since the exponent in x^x is a variable, the Power Rule may not be used. Similarly, Theorem (7.20) is not applicable since the base a is not a fixed real number. However, by Definition (7.18), $x^x = e^{x \ln x}$ for all $x > 0$, and hence

$$D_x(x^x) = D_x(e^{x \ln x})$$

$$= e^{x \ln x}D_x(x \ln x)$$

$$= e^{x \ln x}\left[x\left(\frac{1}{x}\right) + (1) \ln x\right]$$

$$= x^x(1 + \ln x).$$

Another way of solving this problem is to use the method of logarithmic differentiation introduced in the previous section. In this case we take the natural logarithm of both sides of the equation $y = x^x$, and then differentiate implicitly as follows:

$$\ln y = \ln x^x = x \ln x$$

$$D_x(\ln y) = D_x(x \ln x)$$

$$\frac{1}{y}D_x y = 1 + \ln x$$

$$D_x y = y(1 + \ln x) = x^x(1 + \ln x). \qquad \blacksquare$$

We shall conclude this section by expressing the number e as a limit.

Theorem (7.24)

$$\text{(i)} \quad \lim_{h \to 0}(1 + h)^{1/h} = e \qquad\qquad \text{(ii)} \quad \lim_{n \to \infty}\left(1 + \frac{1}{n}\right)^n = e$$

Proof   Applying the derivative formula (3.6) to $f(x) = \ln x$ and using laws of logarithms gives us

$$f'(x) = \lim_{h \to 0} \frac{\ln(x + h) - \ln x}{h}$$

$$= \lim_{h \to 0} \frac{1}{h} \ln \frac{x + h}{x}$$

$$= \lim_{h \to 0} \frac{1}{h} \ln \left(1 + \frac{h}{x}\right)$$

$$= \lim_{h \to 0} \ln \left(1 + \frac{h}{x}\right)^{1/h}.$$

Since $f'(x) = 1/x$ we have, for $x = 1$,

$$1 = \lim_{h \to 0} \ln (1 + h)^{1/h}.$$

We next observe, from Theorem (7.11), that

$$(1 + h)^{1/h} = e^{\ln(1 + h)^{1/h}}.$$

Since the natural exponential function is continuous at 1, it follows from Theorem (2.27) that

$$\lim_{h \to 0} (1 + h)^{1/h} = \lim_{h \to 0} \left[e^{\ln(1 + h)^{1/h}}\right]$$

$$= e^{\left[\lim_{h \to 0} \ln(1 + h)^{1/h}\right]}$$

$$= e^1 = e.$$

This establishes part (i) of the theorem. The limit in part (ii) may be obtained by introducing the change of variable $n = 1/h$ (see Exercise 46).   □

The formulas in Theorem (7.24) are very important in mathematics. Indeed, they are often used to *define* the number e. If a calculator is available the reader will find it instructive to calculate $(1 + h)^{1/h}$ for numerically small values of h. Some approximate values are given in the following table.

h	$(1 + h)^{1/h}$	h	$(1 + h)^{1/h}$
0.01	2.704814	−0.01	2.731999
0.001	2.716924	−0.001	2.719642
0.0001	2.718146	−0.0001	2.718418
0.00001	2.718268	−0.00001	2.718295
0.000001	2.718280	−0.000001	2.718283

To five decimal places, $e \approx 2.71828$.

EXERCISES 7.5

In Exercises 1–22 find $f'(x)$ if $f(x)$ is defined by the given expression.

1 7^x

2 5^{-x}

3 8^{x^2+1}

4 $9^{\sqrt{x}}$

5 $\log(x^4 + 3x^2 + 1)$

6 $\log_3 |6x - 7|$

7 5^{3x-4}

8 3^{2-x^2}

9 $(x^2 + 1)10^{1/x}$

10 $(10^x + 10^{-x})^{10}$

11 $7^{\sqrt{x^4+9}}$

12 $x/(6^x + x^6)$

13 $\log(3x^2 + 2)^5$

14 $\log\sqrt{x^2 + 1}$

15 $\log_5 \left| \dfrac{6x + 4}{2x - 3} \right|$

16 $\log \left| \dfrac{1 - x^2}{2 - 5x^3} \right|$

17 $\log \ln x$

18 $\ln \log x$

19 $x^e + e^x$

20 $x^\pi \pi^x$

21 $(x + 1)^x$

22 x^{x^2+4}

Evaluate the integrals in Exercises 23–36.

23 $\displaystyle\int 10^{3x}\,dx$

24 $\displaystyle\int 5^{-2x}\,dx$

25 $\displaystyle\int x(3^{-x^2})\,dx$

26 $\displaystyle\int \frac{(2^x + 1)^2}{2^x}\,dx$

27 $\displaystyle\int \frac{2^x}{2^x + 1}\,dx$

28 $\displaystyle\int \frac{3^x}{\sqrt{3^x + 4}}\,dx$

29 $\displaystyle\int_1^2 5^{-2x}\,dx$

30 $\displaystyle\int_{-1}^1 2^{3x-1}\,dx$

31 $\displaystyle\int x^2 2^{x^3}\,dx$

32 $\displaystyle\int \frac{10^{\sqrt{x}}}{\sqrt{x}}\,dx$

33 $\displaystyle\int (3^x + 3^{-x})^2\,dx$

34 $\displaystyle\int 4^x(4^x + 1)^3\,dx$

35 $\displaystyle\int \frac{1}{x \log x}\,dx$

36 $\displaystyle\int \frac{10^x + 10^{-x}}{10^x - 10^{-x}}\,dx$

37 Find the area of the region bounded by the graphs of $y = 2^x$, $x + y = 1$, and $x = 1$.

38 The region under the graph of $y = 3^{-x}$ from $x = 1$ to $x = 2$ is revolved about the x-axis. Find the volume of the resulting solid.

39 Find equations of the tangent and normal lines to the graph of the equation $y = x3^x$ at the point $P(1, 3)$.

40 Find the x-coordinate of the point on the graph of $y = 2^x$ at which the tangent line is parallel to the line with equation $2y - 8x + 3 = 0$.

41 The number of bacteria in a certain culture at time t is given by $Q(t) = 2(3^t)$, where t is measured in hours and $Q(t)$ in thousands. Show that the rate of change of $Q(t)$ at time t is proportional to $Q(t)$.

42 If 10 g of salt is added to a certain quantity of water, then the amount $q(t)$ that remains undissolved after t minutes is given by $q(t) = 10(4/5)^t$. Show that the rate of change of $q(t)$ at time t is proportional to $q(t)$.

43 Chemists use a number denoted by pH to describe quantitatively the acidity or basicity of solutions. By definition, $\text{pH} = -\log[H^+]$, where $[H^+]$ is the hydrogen ion concentration in moles per liter. For a certain brand of vinegar it is estimated (with a maximum percentage error of 0.5%) that $[H^+] \approx 6.3 \times 10^{-3}$. Calculate the pH and use differentials to estimate the maximum error in the calculation.

44 Using the Richter scale, the magnitude R of an earthquake of intensity I may be found by means of the formula $R = \log(I/I_0)$, where I_0 is a certain minimum intensity. Suppose it is estimated that an earthquake has an intensity that is 100 times that of I_0. If there is a maximum percentage error of 1% in the estimate, use differentials to approximate the maximum percentage error in the calculated value of R.

45 The loudness of sound, as experienced by the human ear, is based upon intensity levels. A formula used for finding the intensity level α that corresponds to a sound intensity I is $\alpha = 10 \log(I/I_0)$ decibels where I_0 is a special value of I agreed to be the weakest sound that can be detected by the ear under certain conditions. Find the rate of change of α with respect to I if

(a) I is 10 times as great as I_0.

(b) I is 1,000 times as great as I_0.

(c) I is 10,000 times as great as I_0. (This is the intensity level of the average voice.)

46 Establish (ii) of Theorem (7.24) by using the limit in part (i) and the change of variable $n = 1/h$.

In Exercises 47 and 48 sketch the graphs of the two given equations on the same coordinate plane and discuss the relationship between the graphs.

47 $y = 10^x$, $y = \log x$ **48** $y = 2^x$, $y = \log_2 x$

49 Prove that for all real numbers u and v, and any $a > 0$, $b > 0$: $(ab)^u = a^u b^u$; $a^u/a^v = a^{u-v}$.

50 Prove the Laws of Logarithms:

(i) $\log_a (uv) = \log_a u + \log_a v$

(ii) $\log_a \dfrac{u}{v} = \log_a u - \log_a v$

(iii) $\log_a u^r = r \log_a u$

7.6

SEPARABLE DIFFERENTIAL EQUATIONS AND LAWS OF GROWTH

Differential equations are equations that involve derivatives or differentials of unknown functions (cf. page 187). An equation that involves x, y, y', y'', ..., and $y^{(n)}$, where y is a function of x and $y^{(n)}$ is the nth derivative of y with respect to x, is called an **ordinary differential equation of order n**. The following are examples of ordinary differential equations of orders 1, 2, 3, and 4, respectively.

$$y' = 2x$$

$$\frac{d^2 y}{dx^2} + x^2 \left(\frac{dy}{dx}\right)^3 - 15y = 0$$

$$(y''')^4 - x^2(y'')^5 + 4xy = xe^x$$

$$\left(\frac{d^4 y}{dx^4}\right)^2 - 1 = x^3 \frac{dy}{dx}$$

If a function f has the property that substitution of $f(x)$ for y in a differential equation results in an identity for all x in some interval, then $f(x)$ (or f) is called a **solution** of the differential equation. For example, if C is any real number, then a solution of the differential equation $y' = 2x$ is

$$f(x) = x^2 + C$$

because substitution of $f(x)$ for y leads to the identity $2x = 2x$. We call $x^2 + C$ the **general solution** of $y' = 2x$, since every solution has this form.

Example 1 If C_1 and C_2 are any real numbers, prove that

$$f(x) = C_1 e^{5x} + C_2 e^{-5x}$$

is a solution of $y'' - 25y = 0$.

Solution   Since

$$f'(x) = 5C_1 e^{5x} - 5C_2 e^{-5x}$$

and

$$f''(x) = 25C_1 e^{5x} + 25C_2 e^{-5x}$$

we have

$$y'' - 25y = (25C_1 e^{5x} + 25C_2 e^{-5x}) - 25(C_1 e^{5x} + C_2 e^{-5x}) = 0$$

for every x. This shows that $f(x)$ is a solution of $y'' - 25y = 0$. ∎

It can be shown that the solution $C_1 e^{5x} + C_2 e^{-5x}$ in Example 1 is the general solution of $y'' - 25y = 0$. Observe that the differential equation is of order 2 and the general solution contains two arbitrary constants (called **parameters**) C_1 and C_2. The precise definition of general solution involves the concept of **independent parameters** and is left for more advanced courses. It can be shown that general solutions of nth-order differential equations contain n independent parameters $C_1, C_2, \ldots, C_n$. A **particular solution** is obtained by assigning specific values to the parameters. Some differential equations have solutions that are not special cases of the general solution. These so-called **singular solutions** will not be discussed in this text.

Example 2 Find the particular solution of $y' = 2x$ that satisfies the condition that $y = 5$ when $x = 2$.

Solution The general solution of $y' = 2x$ may be written $y = x^2 + C$. If $y = 5$ when $x = 2$, then $5 = 4 + C$, or $C = 1$. Hence the desired particular solution is $y = x^2 + 1$. ∎

Conditions of the type stated in Example 2 are often called **boundary conditions** for the differential equation. If the general solution contains one parameter, then one boundary condition is sufficient to find a particular solution. If, as in Example 1, the general solution contains two parameters, then two boundary conditions are needed for particular solutions. Similar statements hold if more than two parameters are involved. Values of $y, y', \ldots$, that correspond to "first" values of the independent variable x are sometimes referred to as **initial conditions**. For example, if x represents time, we often take $x = 0$ and the values of $y, y', \ldots$ at $x = 0$ as initial conditions.

The solutions we have considered express y explicitly in terms of x. Solutions of certain differential equations are stated implicitly. In this case, implicit differentiation is used to check the solution, as illustrated in the following example.

Example 3 Show that $x^3 + x^2 y - 2y^3 = C$ is an implicit solution of

$$(x^2 - 6y^2)y' + 3x^2 + 2xy = 0.$$

Solution If the first equation is satisfied by $y = f(x)$, then differentiating implicitly,

$$3x^2 + 2xy + x^2 y' - 6y^2 y' = 0$$

or

$$(x^2 - 6y^2)y' + 3x^2 + 2xy = 0.$$

Thus $f(x)$ is a solution of the differential equation. ∎

One of the simplest types of differential equations is

$$M(x) + N(y)y' = 0, \quad \text{or} \quad M(x) + N(y)\frac{dy}{dx} = 0$$

where M and N are continuous functions. If $y = f(x)$ is a solution, then

$$M(x) + N(f(x))f'(x) = 0.$$

If $f'(x)$ is continuous, then indefinite integration leads to

$$\int M(x)\, dx + \int N(f(x))f'(x)\, dx = C$$

or

$$\int M(x)\, dx + \int N(y)\, dy = C.$$

The last equation is an (implicit) solution of the differential equation. A device that is useful for remembering this method of solution is to change the equation $M(x) + N(y)(dy/dx) = 0$ to the following *differential form*:

$$M(x)\, dx + N(y)\, dy = 0.$$

The solution is then found by formally integrating each term. The differential equation $M(x) + N(y)y' = 0$ is said to be **separable**, since the variables x and y may be separated as we have indicated.

Example 4   Solve the differential equation $y^4 e^{2x} + \dfrac{dy}{dx} = 0$.

Solution   Writing the equation in differential form and then separating the variables gives us

$$y^4 e^{2x}\, dx + dy = 0$$

and

$$e^{2x}\, dx + \frac{1}{y^4}\, dy = 0.$$

Integrating each term, we obtain the (implicit) solution

$$\frac{1}{2} e^{2x} - \frac{1}{3y^3} = C.$$

If we multiply both sides by 6, then another form of the solution is

$$3e^{2x} - \frac{2}{y^3} = K$$

where $K = 6C$. If an explicit form is desired we may solve for y, obtaining

$$y = \left(\frac{2}{3e^{2x} - K}\right)^{1/3}. \qquad \blacksquare$$

Example 5   Solve the differential equation $2y + (xy + 3x)\dfrac{dy}{dx} = 0$.

Solution   We may express the equation in differential form as

$$2y\, dx + (xy + 3x)\, dy = 0$$

or

$$2y\, dx + x(y + 3)\, dy = 0.$$

The variables may be separated by dividing both sides of the last equation by xy. This gives us

$$\frac{2}{x}\,dx + \left(\frac{y + 3}{y}\right)dy = 0$$

or $$\frac{2}{x}\,dx + \left(1 + \frac{3}{y}\right)dy = 0.$$

Integrating each term leads to the following equivalent equations.

$$2 \ln |x| + y + 3 \ln |y| = c,$$
$$\ln |x|^2 + \ln |y|^3 = c - y,$$
$$\ln |x|^2|y|^3 = c - y.$$

Changing to exponential form we obtain

$$|x|^2|y|^3 = e^{c-y} = e^c e^{-y} = ke^{-y}$$

where $k = e^c$. It can be shown, by differentiation, that

$$x^2 y^3 = ke^{-y}, \quad \text{or} \quad x^2 y^3 e^y = k$$

is an implicit solution of the differential equation.   ■

Separable differential equations are useful in the investigation of growth processes that occur in the sciences. In particular, suppose that a physical quantity varies with time and that the magnitude of the quantity at time t may be approximated by $q(t)$, where q is a differentiable function and $q(t) > 0$ for all t. The derivative $q'(t)$ may be used to measure the rate of change of $q(t)$ with respect to time. In many applications, the rate of change at a given instant is directly proportional to the magnitude of the quantity at that instant. This fact can be expressed by means of the differential equation

$$q'(t) = cq(t)$$

where c is a constant. The number of bacteria in certain cultures behaves in this way. If the number of bacteria is small, then the rate of increase is small; however, as the number of bacteria increases, the rate of increase becomes larger. In many cases, this change in the rate of change is very great and we may regard the result as a microscopic view of a population explosion. The decay of a radioactive substance obeys a similar law, since as the amount of matter decreases, the rate of decay—that is, the amount of radiation—also decreases. As a final illustration, suppose an electrical condenser is allowed to discharge. If there is a large charge on the condenser at the outset, the rate of discharge is also large, but as the charge becomes weaker, the condenser discharges less rapidly. There are many other examples of this phenomenon.

If we let $y = q(t)$, then the differential equation $q'(t) = cq(t)$ may be written

$$\frac{dy}{dt} = cy, \quad \text{or} \quad dy = (cy)\, dt.$$

Dividing both sides of the last equation by y we obtain

$$\frac{1}{y}\, dy = c\, dt.$$

Integrating both sides,

$$\ln y = ct + k.$$

It follows that

$$y = e^{ct+k} = e^k e^{ct}.$$

If y_0 denotes the initial value of y (that is, the value corresponding to $t = 0$), then letting $t = 0$ in the last equation gives us

$$y_0 = e^k e^0 = e^k$$

and hence the solution $y = e^k e^{ct}$ may be written

$$y = y_0 e^{ct}.$$

We have proved the following theorem, where the symbols have the same meanings as in the preceding discussion.

Theorem (7.25)

$$\boxed{\text{If} \quad \frac{dy}{dt} = cy, \quad \text{then} \quad y = y_0 e^{ct}.}$$

The preceding theorem states that *if the rate of change of $y = q(t)$ is directly proportional to y, then y may be expressed in terms of an exponential function.* If y increases with t, the formula $y = y_0 e^{ct}$ is called a **law of growth**, whereas if y decreases, it is referred to as a **law of decay**.

Example 6 The number of bacteria in a certain culture increases from 600 to 1800 in 2 hours. Assuming that the rate of increase is directly proportional to the number of bacteria present, find a formula for the number of bacteria in the culture at any time t. What is the number of bacteria at the end of 4 hours?

Solution Let $y = q(t)$ denote the number of bacteria after t hours. Thus $y_0 = q(0) = 600$ and, at time $t = 2$, the value of y is $q(2) = 1800$. By hypothesis,

$$\frac{dy}{dt} = cy.$$

Following exactly the same procedure used in the proof of Theorem (7.25) gives us

$$y = y_0 e^{ct} = 600e^{ct}.$$

Letting $t = 2$ we obtain

$$1800 = 600e^{2c}, \quad \text{or} \quad e^{2c} = 3.$$

Consequently,

$$2c = \ln 3, \quad \text{or} \quad c = \tfrac{1}{2} \ln 3.$$

Thus the formula for y is

$$y = 600e^{[(1/2)\ln 3]t}.$$

If we use the fact that $e^{\ln 3} = 3$ (see (Definition (7.10)), then this law of growth can be expressed in terms of an exponential function with base 3 as follows:

$$y = 600(e^{\ln 3})^{t/2} = 600(3)^{t/2}.$$

In particular, at the end of 4 hours the number of bacteria is

$$600(3)^{4/2} = 600(9) = 5400. \qquad \blacksquare$$

Example 7 Radium decays exponentially and has a half-life of approximately 1600 years; that is, given any quantity, one-half of it will disintegrate in 1600 years. Find a formula for the amount y remaining from 50 mg of pure radium after t years. When will there be 20 mg left?

Solution   If we let $y = q(t)$, then $y_0 = q(0) = 50$ and $q(1600) = 25$. Since $dy/dt = cy$ for some c, it follows from Theorem (7.25) that

$$y = 50e^{ct}.$$

Letting $t = 1600$ we obtain

$$25 = 50e^{1600c}, \quad \text{or} \quad e^{1600c} = \frac{1}{2}.$$

Hence

$$1600c = \ln \frac{1}{2} = \ln 1 - \ln 2 = -\ln 2$$

and

$$c = -\frac{\ln 2}{1600}.$$

Consequently,

$$y = 50e^{-(\ln 2/1600)t}.$$

As in Example 1 we may write this formula in terms of a different base as follows:

$$y = 50(e^{\ln 2})^{-t/1600}$$

or

$$y = 50(2)^{-t/1600}.$$

To find the value of t at which $y = 20$ it is necessary to solve the equation

$$20 = 50(2)^{-t/1600} \quad \text{or} \quad 2^{t/1600} = \tfrac{5}{2}.$$

Taking the natural logarithm of both sides,

$$\frac{t}{1600} \ln 2 = \ln \frac{5}{2}, \quad \text{or} \quad t = \frac{1600 \ln \frac{5}{2}}{\ln 2}.$$

If an approximation is desired, then using a calculator or Table C,

$$t \approx 1600(0.916)/(0.693) \approx 2{,}115 \text{ years.} \qquad \blacksquare$$

Example 8 According to Newton's Law of Cooling, the rate at which an object cools is directly proportional to the difference in temperature between the object and the surrounding medium. If a certain object cools from 125° to 100° in half an hour when surrounded by air at a temperature of 75°, find its temperature at the end of another half hour.

Solution   If y denotes the temperature of the object after t hours of cooling, then by Newton's Law,

$$\frac{dy}{dt} = c(y - 75)$$

where c is a constant. Separating variables we obtain

$$\frac{1}{y - 75} \, dy = c \, dt.$$

Integrating both sides leads to

$$\ln (y - 75) = ct + b$$

where b is a constant or, equivalently,

$$y - 75 = e^{ct+b} = e^b e^{ct}.$$

If we let $k = e^b$, then the last formula may be written

$$y = ke^{ct} + 75.$$

Since $y = 125$ when $t = 0$ we see that

$$125 = ke^0 + 75 = k + 75 \quad \text{or} \quad k = 50.$$

Hence

$$y = 50e^{ct} + 75.$$

Since $y = 100$ when $t = \tfrac{1}{2}$ hour,

$$100 = 50e^{c/2} + 75 \quad \text{or} \quad e^{c/2} = \tfrac{25}{50} = \tfrac{1}{2}.$$

This implies that

$$c = 2 \ln \tfrac{1}{2} = \ln \tfrac{1}{4}.$$

Consequently, the temperature y after t hours is given by

$$y = 50e^{t \ln(1/4)} + 75.$$

In particular, if $t = 1$,

$$y = 50e^{\ln(1/4)} + 75$$
$$= 50(\tfrac{1}{4}) + 75 = 87.5°.$$ ∎

In many cases, growth circumstances are more stable than that described in Theorem (7.25). Typical situations that occur involve populations, sales of products, and values of assets. In biology, a function G is sometimes used as follows to estimate the size of a quantity at time t:

$$G(t) = ke^{(-Ae^{-Bt})}$$

where k, A, and B are positive constants. The function G is called a **Gompertz growth function**. It is always positive and increasing, but has a limit as t increases without bound. The graph of G is called a **Gompertz growth curve**.

Example 9 Discuss and sketch the graph of the Gompertz growth function G.

Solution   We first observe that the y-intercept is $G(0) = ke^{-A}$ and that $G(t) > 0$ for all t. Differentiating, we obtain

$$G'(t) = ke^{(-Ae^{-Bt})}D_t(-Ae^{-Bt})$$
$$= ABke^{(-Bt-Ae^{-Bt})}$$

and

$$G''(t) = ABke^{(-Bt-Ae^{-Bt})}D_t(-Bt - Ae^{-Bt})$$
$$= ABk(-B + ABe^{-Bt})e^{-Bt-Ae^{-Bt}}.$$

Since $G'(t) > 0$ for all t, the function G is increasing on $[0, \infty)$. The second derivative $G''(t)$ is zero if

$$-B + ABe^{-Bt} = 0, \quad \text{or} \quad e^{Bt} = A.$$

Solving the last equation for t gives us $t = (1/B)\ln A$, which is a critical number for the function G'. It is left as an exercise to show that at this time the rate of growth G' has a maximum value Bk/e. It is also left to the reader to show that

$$\lim_{t \to \infty} G'(t) = 0 \quad \text{and} \quad \lim_{t \to \infty} G(t) = k.$$

Hence, as t increases without bound, the rate of growth approaches 0 and the graph of G has a horizontal asymptote $y = k$. A typical graph is sketched in Figure 7.11.   ∎

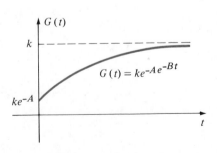

Figure 7.11

EXERCISES 7.6

In Exercises 1–6 prove that y is a solution of the indicated differential equation.

1 $y = C_1 e^x + C_2 e^{2x}$; $y'' - 3y' + 2y = 0$

2 $y = Ce^{-3x}$; $y' + 3y = 0$

3 $y = Cx^{-2/3}$; $2xy^3 + 3x^2 y^2 \dfrac{dy}{dx} = 0$

4 $y = Cx^3$; $x^3 y''' + x^2 y'' - 3xy' - 3y = 0$

5 $y^2 - x^2 - xy = C$; $(x - 2y)\dfrac{dy}{dx} + 2x + y = 0$

6 $x^2 - y^2 = C$; $y\dfrac{dy}{dx} = x$

Solve the differential equations in Exercises 7–14.

7 $x\dfrac{dy}{dx} - y = 0$

8 $(xy - 4x) + (x^2 y + y)\dfrac{dy}{dx} = 0$

9 $y' = x - 1 + xy - y$

10 $(y + yx^2)y' + (x + xy^2) = 0$

11 $e^{x+2y} - e^{2x-y}\dfrac{dy}{dx} = 0$

12 $y(1 + x^3)\dfrac{dy}{dx} + x^2(1 + y^2) = 0$

13 $x^2 y' - yx^2 = y$

14 $xy + y'e^{x^2}\ln y = 0$

In Exercises 15–18 find the particular solution of the differential equation that satisfies the given boundary conditions.

15 $2y^2 y' = 3y - y'$; $y = 1$ when $x = 3$

16 $\sqrt{x}y' - \sqrt{y} = x\sqrt{y}$; $y = 4$ when $x = 9$

17 $x\,dy - (2x + 1)e^{-y}\,dx = 0$; $y = 2$ when $x = 1$

18 $(xy + x)\,dx + \sqrt{4 + x^2}\,dy = 0$; $y = 1$ when $x = 0$

19 The number of bacteria in a certain culture increases from 5,000 to 15,000 in 10 hours. Assuming that the rate of increase is proportional to the number of bacteria present, find a formula for the number of bacteria in the culture at any time t. Estimate the number at the end of 20 hours. When will the number be 50,000?

20 The polonium isotope ^{210}Po has a half-life of approximately 140 days. If a sample weighs 20 mg initially, how much remains after t days? Approximately how much will be left after two weeks?

21 If the temperature is constant, then the rate of change of barometric pressure p with respect to altitude h is proportional to p. If $p = 30$ in. at sea level and $p = 29$ in. when $h = 1000$ ft, find the pressure at an altitude of 5000 ft.

22 The population of a certain city is increasing at the rate of 5 percent per year. If the present population is 500,000, what will the population be in 10 years?

23 It is usually assumed that $\frac{1}{4}$ acre of land is required to provide food for one person. It is also estimated that there are 10 billion acres of tillable land in the world, and hence a maximum population of 40 billion people can be sustained if no other food source is available. The world population at the beginning of 1980 was approximately 4.5 billion. Assuming that the population increases at a rate of 2 percent per year, when will the maximum population be reached?

24 A metal plate that has been heated cools from 180° to 150° in 20 minutes when surrounded by air at a temperature of 60°. Use Newton's Law of Cooling (see Example 8) to approximate its temperature at the end of one hour of cooling. When will the temperature be 100°?

25 An outdoor thermometer registers a temperature of 40°. Five minutes after it is brought into a room where the temperature is 70°, the thermometer registers 60°. When will it register 65°? (See Example 8.)

26 The rate at which salt dissolves in water is directly proportional to the amount that remains undissolved. If 10 lb of salt is placed in a container of water and 4 lb dissolves in 20 minutes, how long will it take 2 more pounds to dissolve?

27 According to one of Kirchoff's rules for electrical circuits, $E = Ri + L(di/dt)$ where the constants E, R, and L denote the electromotive force, the resistance, and the inductance, respectively, and i denotes the current at time t (see figure). If the electromotive force is terminated at time $t = 0$ and if the current is I at the instant of removal, prove that $i = Ie^{-Rt/L}$.

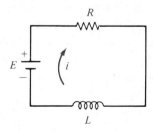

Figure for Exercise 27

28 The rate at which sugar decomposes in solution is directly proportional to the amount of sugar remaining. If a solution contains k pounds of sugar initially and if y is the amount that decomposes in time t, express y in terms of t.

29 If a sum of money P is invested at an interest rate of $100r$ percent per year, compounded m times per year, then the principal at the end of t years is given by $P(1 + rm^{-1})^{mt}$. If we regard m as a real number and let m increase without bound, then the interest is said to be *compounded continuously*. Use Theorem (7.24) to show that in this case the principal after t years is Pe^{rt}.

30 Use the results of Exercise 29 to find the principal after one year if $1000 is invested at a rate of 10 percent, compounded (a) monthly; (b) daily; (c) hourly; (d) continuously.

31 In Example 9, verify that (a) Bk/e is a maximum value for G'; (b) $\lim_{t \to \infty} G'(t) = 0$; (c) $\lim_{t \to \infty} G(t) = k$.

32 Sketch the graph of G in Example 9 if $k = 10$, $A = 1/2$, and $B = 1$.

33 In the **Law of Logistic Growth**, it is assumed that at time t, the rate of growth $f'(t)$ of a quantity $f(t)$ is given by $f'(t) = Af(t)[B - f(t)]$ where A and B are constants. If $f(0) = C$, show that

$$f(t) = \frac{BC}{C + (B - C)e^{-ABt}}.$$

34 The graph of $f(x) = a + b(1 - e^{-cx})$ where a, b, and c are positive constants may be used to describe certain learning processes. To illustrate, suppose a manufacturer estimates that a new employee can produce five items the first day on the job. As the employee becomes more proficient, more items per day can be produced until a certain maximum production is reached. Suppose that on the nth day on the job, the number $f(n)$ of items produced is approximated by $f(n) = 3 + 20(1 - e^{-0.1n})$.

   (a) Estimate the number of items produced on the fifth day; the ninth day; the twenty-fourth day; the thirtieth day.

   (b) Sketch the graph of f from $n = 0$ to $n = 30$. (Graphs of this type are called **learning curves**, and are used frequently in education and psychology.)

   (c) What happens as n increases without bound?

35 Suppose that a radioactive substance decays according to the law $q(t) = q_0 e^{-ct}$, where $q_0 = q(0)$. Find its half-life if (a) $c = 100$; (b) $c = 1000$; (c) c is any positive constant.

36 A physicist finds that an unknown radioactive substance registers 2000 counts per minute on a Geiger counter.

Ten days later it registers 1500 counts per minute. Approximate the half-life of the substance.

37 A machine purchased for $20,000 had a value of $16,000 after two years of use. Assuming that the value decreases exponentially, approximate the value of the machine at the end of another year.

38 The cost of real estate in a certain city has appreciated at a rate of 10 percent per year since 1975. If this rate continues, approximately what will a house that was purchased for $60,000 in 1980 be worth in 1990?

39 Veterinarians use sodium pentobarbital to anesthetize animals. Suppose that to anesthetize a dog, 30 mg are required for each kilogram of body weight. If sodium pentobarbital is eliminated exponentially from the bloodstream, and half is eliminated in four hours, approximate the single dose that will anesthetize a dog that weighs 20 kg for 45 minutes.

40 In the study of lung physiology the following differential equation is used to describe the transport of a substance across a capillary wall:

$$\frac{dh}{dt} = -\frac{V}{Q}\left(\frac{h}{k + h}\right)$$

where h is the hormone concentration in the bloodstream, t is time, V is the maximum transport rate, Q is the volume of the capillary, and k is a constant that measures the affinity between the hormones and enzymes that assist with the transport process. Find the general solution of the differential equation.

41 The technique of **carbon dating** is used to determine the age of archeological or geological specimens. This method is based on the fact that the unstable carbon isotope C^{14} is present in the CO_2 in the atmosphere. Plants take in carbon from the atmosphere; when they die the C^{14} that has accumulated begins to decay, with a half-life of approximately 5700 years. By measuring the amount of C^{14} that remains in a specimen, it is possible to approximate when the organism died. Suppose that a bone which is discovered contains 20 percent as much C^{14} as an equal amount of carbon in present day bone. Approximate the age of the bone.

42 Refer to Exercise 41. The hydrogen isotope $^{3}_{1}H$, which has a half life of 12.3 years, is produced in the atmosphere by cosmic rays and is brought to earth by rain. If the wood siding of an old house contains 10 percent as much $^{3}_{1}H$ as the siding on a similar new house, approximate the age of the old house.

7.7

FIRST-ORDER LINEAR DIFFERENTIAL EQUATIONS

Definition (7.26)

The following type of differential equation arises frequently in the study of physical phenomena.

> A **first-order linear differential equation** is an equation of the form
>
> $$y' + P(x)y = Q(x)$$
>
> where P and Q are continuous functions.

If, in Definition (7.26), $Q(x) = 0$ for all x, we obtain $y' + P(x)y = 0$, which is separable. Specifically, we may write

$$\frac{1}{y}\frac{dy}{dx} = -P(x), \quad \text{or} \quad \frac{1}{y}\,dy = -P(x)\,dx$$

provided $y \neq 0$. Integration gives us

$$\ln|y| = -\int P(x)\,dx + \ln|C|.$$

We have expressed the constant of integration as $\ln|C|$ in order to change the form of the last equation as follows:

$$\ln|y| - \ln|C| = -\int P(x)\,dx$$

$$\ln\left|\frac{y}{C}\right| = -\int P(x)\,dx$$

$$\frac{y}{C} = e^{-\int P(x)\,dx}$$

$$ye^{\int P(x)\,dx} = C.$$

We next observe that

$$D_x[ye^{\int P(x)\,dx}] = y'e^{\int P(x)\,dx} + P(x)ye^{\int P(x)\,dx}$$
$$= e^{\int P(x)\,dx}[y' + P(x)y].$$

Consequently, if we multiply both sides of $y' + P(x)y = Q(x)$ by $e^{\int P(x)\,dx}$, then the resulting equation may be written

$$D_x[ye^{\int P(x)\,dx}] = Q(x)e^{\int P(x)\,dx}.$$

Integrating both sides gives us the following implicit solution of the first-order differential equation in Definition (7.26):

$$ye^{\int P(x)\,dx} = \int Q(x)e^{\int P(x)\,dx}\,dx + K.$$

Solving this equation for y leads to an explicit solution. The expression $e^{\int P(x)\,dx}$ is called an **integrating factor** of the differential equation. We have proved the following result.

Theorem (7.27)

> The first-order linear differential equation $y' + P(x)y = Q(x)$ may be transformed into a separable differential equation by multiplying both sides by the integrating factor $e^{\int P(x)\,dx}$.

Example 1   Solve the differential equation $\dfrac{dy}{dx} - 3x^2 y = x^2$.

Solution   The equation has the form (7.26) with $P(x) = -3x^2$ and $Q(x) = x^2$. By Theorem (7.27) an integrating factor is

$$e^{\int -3x^2\,dx} = e^{-x^3}.$$

It is unnecessary to introduce a constant of integration since $e^{-x^3+c} = e^c e^{-x^3}$, which differs from e^{-x^3} by a constant factor e^c. Multiplying both sides of the differential equation by the integrating factor e^{-x^3} gives us

$$e^{-x^3}\frac{dy}{dx} - 3x^2 e^{-x^3}y = x^2 e^{-x^3}$$

or $$D_x(e^{-x^3}y) = x^2 e^{-x^3}.$$

Integrating both sides of the last equation, we obtain

$$e^{-x^3}y = \int x^2 e^{-x^3}\,dx = -\tfrac{1}{3}e^{-x^3} + C.$$

Finally, multiplying by e^{x^3} yields the following explicit solution:

$$y = -\tfrac{1}{3} + Ce^{x^3}. \qquad \blacksquare$$

Example 2   Solve the differential equation $x^2 y' + 5xy + 3x^5 = 0$ where $x \neq 0$.

Solution   To find an integrating factor we begin by expressing the differential equation in the "standardized" form (7.26), where the coefficient of y' is 1. Thus, dividing both sides by x^2, we obtain

$$y' + \frac{5}{x}y = -3x^3$$

which has the form given in (7.26) with $P(x) = 5/x$. By Theorem (7.27) an integrating factor is

$$e^{\int P(x)\,dx} = e^{5\ln|x|} = e^{\ln|x|^5} = |x|^5.$$

If $x > 0$ then $|x|^5 = x^5$, whereas if $x < 0$ then $|x|^5 = -x^5$. In either case, multiplying both sides of the standardized form by $|x|^5$ yields

$$x^5 y' + 5x^4 y = -3x^8,$$

or
$$D_x(x^5 y) = -3x^8.$$

Integrating both sides of the last equation gives us the solution

$$x^5 y = \int -3x^8\,dx = -\frac{x^9}{3} + C$$

or
$$y = -\frac{x^4}{3} + \frac{C}{x^5}. \qquad \blacksquare$$

We have previously used antidifferentiation to derive laws of motion for falling bodies, assuming that air resistance could be neglected (see Example 5 in Section 4.9). This is a valid assumption for small, slowly moving objects; however, in many cases air resistance must be taken into account. Indeed, this frictional force often increases as the speed of the object increases. In the following example we shall derive the law of motion for a falling body under the assumption that the resistance due to the air is directly proportional to the speed of the body.

Example 3   An object of mass m is released from a hot-air balloon. Find the distance it falls in t seconds, if the force of resistance due to the air is directly proportional to the speed of the object.

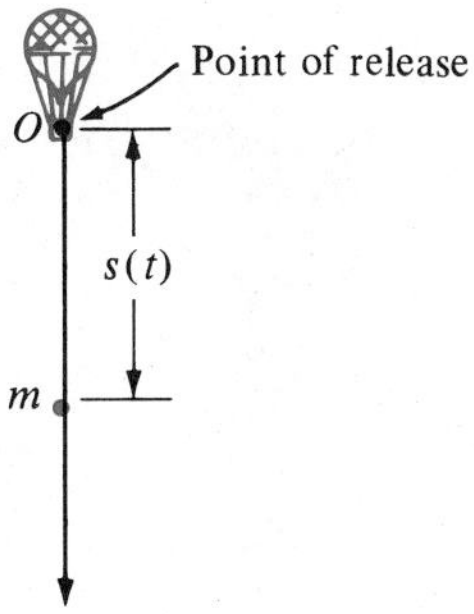

Solution   Let us introduce a vertical axis with positive direction downward and origin at the point of release as illustrated in Figure 7.12. We wish to find the distance $s(t)$ from the origin to the object at time t. The speed of the object is $v = s'(t)$ and the magnitude of the acceleration is $a = dv/dt = s''(t)$. If g is the gravitational constant, then the object is attracted toward the earth with a force of magnitude mg. By hypothesis, the force of resistance due to the air is kv for some constant k, and this force is directed opposite to the motion. It follows that the downward force F on the object is $mg - kv$. Since Newton's Second Law of Motion states that $F = ma = m(dv/dt)$, we arrive at the following differential equation:

Point of release

O

$s(t)$

m

Figure 7.12

$$m\frac{dv}{dt} = mg - kv$$

or equivalently
$$\frac{dv}{dt} + \frac{k}{m}v = g.$$

If we denote the constant k/m by c, this equation may be written

$$\frac{dv}{dt} + cv = g,$$

which is a first-order differential equation with t as the independent variable. By (7.27), an integrating factor is

$$e^{\int c\,dt} = e^{ct}.$$

Multiplying both sides of the last differential equation by e^{ct} gives us

$$e^{ct}\frac{dv}{dt} + ce^{ct}v = ge^{ct}$$

or $\qquad\qquad\qquad\qquad D_t(ve^{ct}) = ge^{ct}.$

Integrating both sides we obtain

$$ve^{ct} = \frac{g}{c}e^{ct} + K$$

or $\qquad\qquad\qquad\qquad v = \frac{g}{c} + Ke^{-ct}$

where K is a constant.

If we let $t = 0$, then $v = 0$ and hence

$$0 = \frac{g}{c} + K, \quad \text{or} \quad K = -\frac{g}{c}.$$

Consequently,

$$v = \frac{g}{c} - \frac{g}{c}e^{-ct}.$$

Integrating both sides of this equation with respect to t and using the fact that $v = s'(t)$ we see that

$$s(t) = \frac{g}{c}t + \frac{g}{c^2}e^{-ct} + E.$$

The constant E may be found by letting $t = 0$. Since $s(0) = 0$,

$$0 = 0 + \frac{g}{c^2} + E, \quad \text{or} \quad E = -\frac{g}{c^2}.$$

Thus, the distance the object falls in t seconds is

$$s(t) = \frac{g}{c}t + \frac{g}{c^2}e^{-ct} - \frac{g}{c^2}.$$

It is interesting to compare this formula for $s(t)$ with that obtained when the air resistance is neglected. In the latter case the differential equation $m(dv/dt) = mg - kv$ reduces to $dv/dt = g$ and it follows that $s'(t) = v = gt$. Integrating both sides leads to the much simpler formula $s(t) = \frac{1}{2}gt^2$. ■

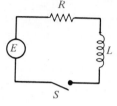

Figure 7.13

Example 4   A simple electrical circuit consists of a resistance R and an inductance L connected in series, as illustrated schematically in Figure 7.13, where E is a constant electromotive force. If the switch S is closed at $t = 0$, then it follows from one of Kirchoff's rules for electrical circuits that if $t > 0$, the current i satisfies the differential equation

$$L\frac{di}{dt} + Ri = E.$$

Express i as a function of t.

Solution   The differential equation may be written

$$\frac{di}{dt} + \frac{R}{L}i = \frac{E}{L}$$

which is a first-order linear differential equation. Applying Theorem (7.27) we multiply both sides by the integrating factor

$$e^{\int (R/L)\, dt} = e^{(R/L)t}$$

obtaining

$$e^{(R/L)t}\frac{di}{dt} + \frac{R}{L}e^{(R/L)t}i = \frac{E}{L}e^{(R/L)t}$$

or

$$D_t(ie^{(R/L)t}) = \frac{E}{L}e^{(R/L)t}.$$

Integration with respect to t yields

$$ie^{(R/L)t} = \int \frac{E}{L}e^{(R/L)t}\, dt = \frac{E}{R}e^{(R/L)t} + C.$$

Since $i = 0$ when $t = 0$, it follows that $C = -E/R$. Substituting for C leads to

$$ie^{(R/L)t} = \frac{E}{R}e^{(R/L)t} - \frac{E}{R}.$$

Finally, multiplying both sides by $e^{-(R/L)t}$ gives us

$$i = \frac{E}{R}[1 - e^{-(R/L)t}].$$

Observe that as t increases without bound i approaches E/R, which is the current when no inductance is present. ■

EXERCISES 7.7

Solve the differential equations in Exercises 1–12.

1 $y' + 2y = e^{2x}$ **2** $y' - 3y = 2$

3 $x \dfrac{dy}{dx} - 3y = x^5$ **4** $\dfrac{dy}{dx} - 6xy = x$

5 $xy' + y + x = e^x$ **6** $xy' + (1 + x)y = 5$

7 $x^2 \dfrac{dy}{dx} + (2xy - e^x) = 0$

8 $x^2 \dfrac{dy}{dx} + (x - 3xy + 1) = 0$

9 $xy' + (2 + 3x)y = xe^{-3x}$

10 $(x + 4)y' + 5y = x^2 + 8x + 16$

11 $\dfrac{dy}{dx} - 5y = e^{5x}$

12 $\dfrac{dy}{dx} + 3x^2 y = x^2 + e^{-x^3}$

In Exercises 13–16 find the particular solution of the differential equation that satisfies the given boundary conditions.

13 $x \dfrac{dy}{dx} - y = x^2 + x;\ y = 2$ when $x = 1$

14 $\dfrac{dy}{dx} + 2y = e^{-3x};\ y = 2$ when $x = 0$

15 $xy' + y + xy = e^{-x};\ y = 0$ when $x = 1$

16 $y' + 2xy - e^{-x^2} = x;\ y = 1$ when $x = 0$

Solve each of the differential equations in Exercises 17 and 18 by (a) using an integrating factor; (b) separating the variables.

17   The equation

$$R\frac{dQ}{dt} + \frac{Q}{C} = E$$

describes the charge Q on a condenser of capacity C during a charging process involving a resistance R and electromotive force E. If the charge is 0 when $t = 0$, express Q as a function of t.

18   The equation

$$R\frac{di}{dt} + \frac{i}{C} = \frac{dE}{dt}$$

describes an electrical circuit consisting of an electromotive force E with a resistance R and capacity C connected in series. Express i as a function of t if $i = i_0$ when $t = 0$ and E is a constant.

19   At time $t = 0$ a tank contains K lb of salt dissolved in 80 gal of water. Suppose that water containing $\frac{1}{3}$ lb of salt per gallon is being added to the tank at a rate of 6 gal/min, and that the well-stirred solution is being drained from the tank at the same rate. Find a formula for the amount $f(t)$ of salt in the tank at time t.

20   An object of mass m is moving rectilinearly subject to a force given by $F(t) = e^{-t}$, where t is time. The motion is resisted by a frictional force that is numerically equal to twice the speed of the object. If $v = 0$ at $t = 0$, find a formula for v at any time $t > 0$.

21   *Learning curves* are used to describe how certain skills are acquired. To illustrate, suppose a manufacturer estimates that a new employee will produce A items the first day on the job, and that as the proficiency of the employee increases, items will be produced more rapidly, until a maximum of M items per day is reached. Let $f(t)$ denote the number produced on day t, where $t \geq 1$. Suppose that the rate of production $f'(t)$ is proportional to $M - f(t)$.

(a) Find a formula for $f(t)$.

(b) If $M = 30$, $f(1) = 5$, and $f(2) = 8$, estimate the number of items produced on day 20.

22   A room having dimensions 10 ft $\times$ 15 ft $\times$ 8 ft originally contains 0.001 percent carbon monoxide. Suppose that at time $t = 0$, fumes containing 5 percent carbon monoxide begin entering the room at a rate of 0.12 ft^3/min, and that the well-circulated mixture is eliminated from the room at the same rate.

(a) Find a formula for the volume $f(t)$ of carbon monoxide in the room at time t.

(b) After approximately how many minutes will the room contain the "dangerous-to-your-health" level of 0.015 percent carbon monoxide?

7.8

DERIVATIVES OF INVERSE FUNCTIONS

When inverse functions were defined in Section 7.1, we introduced techniques for finding $f^{-1}(x)$ if $f(x)$ is stated in terms of a simple algebraic expression. It was also pointed out that the graphs of f and f^{-1} are reflections of one another through the line $y = x$. According to Definition (7.8), the natural logarithmic and exponential functions are inverse functions of one another. Inverses of trigonometric functions will be discussed in Chapter 8. In this section we shall prove several general theorems that hold for all inverse functions, and obtain a formula for finding their derivatives.

Theorem (7.28)

> If a function f is continuous and increasing on an interval $[a, b]$, then f has an inverse function f^{-1} that is continuous and increasing on the interval $[f(a), f(b)]$.

Proof The theorem is intuitively evident if we regard the graph of f^{-1} as a reflection, through the line $y = x$, of the graph of f. (See, for example, Figure 7.2.) However, since a graphical illustration does not constitute a proof, we shall proceed as follows.

To establish the existence of f^{-1}, it is sufficient to show that f is a one-to-one function with range $[f(a), f(b)]$. Consider numbers x_1 and x_2 in $[a, b]$. If $x_1 < x_2$, then, since f is increasing, $f(x_1) < f(x_2)$. Similarly if $x_2 < x_1$, then $f(x_2) < f(x_1)$. Consequently, if $x_1 \neq x_2$, then $f(x_1) \neq f(x_2)$, that is, f is one-to-one. By the Intermediate Value Theorem (2.24), f takes on every value between $f(a)$ and $f(b)$ and, therefore, the range of f is $[f(a), f(b)]$. Thus the inverse function f^{-1} exists, with domain $[f(a), f(b)]$ and range $[a, b]$.

To prove that f^{-1} is increasing, it must be shown that if $w_1 < w_2$ in $[f(a), f(b)]$, then $f^{-1}(w_1) < f^{-1}(w_2)$ in $[a, b]$. We shall give an indirect proof of this fact. Thus, *suppose* $f^{-1}(w_2) \leq f^{-1}(w_1)$. Since f is increasing, $f(f^{-1}(w_2)) \leq f(f^{-1}(w_1))$ and hence $w_2 \leq w_1$, which is a contradiction. Consequently $f^{-1}(w_1) < f^{-1}(w_2)$.

It remains to prove that f^{-1} is continuous on $[f(a), f(b)]$. Since f^{-1} is the inverse of f, it follows that $y = f(x)$ if and only if $x = f^{-1}(y)$. In particular, if y_0 is in the open interval $(f(a), f(b))$, let x_0 denote the number in the interval (a, b) such that $y_0 = f(x_0)$ or, equivalently, $x_0 = f^{-1}(y_0)$. We wish to show that

(∗)
$$\lim_{y \to y_0} f^{-1}(y) = f^{-1}(y_0) = x_0.$$

It is enlightening to use a geometric representation of a function and its inverse as described in Section 7.1. In this event, the domain $[a, b]$ of f is represented by points on an x-axis and the domain $[f(a), f(b)]$ by points on a y-axis. Arrows are drawn from one axis to the other to represent functional values. To prove (∗), consider any interval $(x_0 - \varepsilon, x_0 + \varepsilon)$ where $\varepsilon > 0$. It is sufficient to find an interval $(y_0 - \delta, y_0 + \delta)$, of the type sketched in (i) of

Figure 7.14, such that whenever y is in $(y_0 - \delta, y_0 + \delta)$, then $f^{-1}(y)$ is in $(x_0 - \varepsilon, x_0 + \varepsilon)$. We may assume, without loss of generality, that $x_0 - \varepsilon$ and $x_0 + \varepsilon$ are in $[a, b]$. As illustrated in (ii) of Figure 7.14, let $\delta_1 = y_0 - f(x_0 - \varepsilon)$ and $\delta_2 = f(x_0 + \varepsilon) - y_0$. Since the function f determines a one-to-one correspondence between the numbers in the intervals $(x_0 - \varepsilon, x_0 + \varepsilon)$ and $(y_0 - \delta_1, y_0 + \delta_2)$, f^{-1} maps the numbers in $(y_0 - \delta_1, y_0 + \delta_2)$ onto the numbers in $(x_0 - \varepsilon, x_0 + \varepsilon)$. Consequently, if δ denotes the smaller of δ_1 and δ_2, then whenever y is in $(y_0 - \delta, y_0 + \delta)$, $f^{-1}(y)$ is in $(x_0 - \varepsilon, x_0 + \varepsilon)$. This proves that f^{-1} is continuous on the open interval $(f(a), f(b))$. The continuity at the endpoints $f(a)$ and $f(b)$ is proved in a similar manner using one-sided limits.

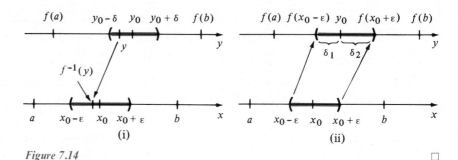

Figure 7.14

The proof of the next result is omitted since it is similar to that of Theorem (7.28).

Theorem (7.29)

> If a function f is continuous and decreasing on an interval $[a, b]$, then f has an inverse function that is continuous and decreasing on the interval $[f(b), f(a)]$.

Example 1 Verify Theorem (7.28) if $f(x) = x^2 - 3$ on the interval $[0, 6]$.

Solution Since f is a polynomial function, it is continuous on $[0, 6]$. Moreover, since $f'(x) = 2x$, $f'(x) > 0$ for all x in $(0, 6]$ and, therefore, f is increasing on $[0, 6]$. It was shown in Example 2 of Section 7.1 that f^{-1} is given by

$$f^{-1}(x) = \sqrt{x + 3}, \quad \text{where } x \geq -3.$$

In the present example the domain of f^{-1} is $[f(0), f(6)]$, that is, $[-3, 33]$. Since $\lim_{x \to a} \sqrt{x + 3} = \sqrt{a + 3}$ for every a in $(-3, 33]$, f^{-1} is continuous on this interval. It is also continuous at -3. Finally, since the derivative

$$D_x f^{-1}(x) = \frac{1}{2\sqrt{x + 3}}$$

is positive on $(-3, 33]$, it follows that f^{-1} is increasing on $[-3, 33]$. ■

The next theorem provides a method for finding the derivative of an inverse function.

Theorem (7.30)

> If a differentiable function f has an inverse function g, and if $f'(g(c)) \neq 0$, then g is differentiable at c and
>
> $$g'(c) = \frac{1}{f'(g(c))}.$$

Proof   Using the formula for the derivative in Theorem (3.7),

$$g'(c) = \lim_{x \to c} \frac{g(x) - g(c)}{x - c}.$$

Let us introduce a new variable z such that $z = g(x)$ and let $a = g(c)$. Since f and g are inverse functions of one another, it follows that

$$g(x) = z \quad \text{if and only if} \quad f(z) = x,$$
$$g(c) = a \quad \text{if and only if} \quad f(a) = c.$$

The statement $x \to c$ in the limit formula for $g'(c)$ may be replaced by $z \to a$, since if $x \to c$, then $g(x) \to g(c)$, that is, $z \to a$. However, if $z \to a$, then $f(z) \to f(a)$. Thus we may write

$$g'(c) = \lim_{z \to a} \frac{z - a}{f(z) - f(a)}$$

$$= \lim_{z \to a} \frac{1}{\left[\dfrac{f(z) - f(a)}{z - a}\right]}$$

$$= \frac{1}{f'(a)} = \frac{1}{f'(g(c))}. \qquad \square$$

It is convenient to restate Theorem (7.30) as follows.

Theorem (7.31)

> If g is the inverse function of a differentiable function f, and if $f'(g(x)) \neq 0$, then
>
> $$g'(x) = \frac{1}{f'(g(x))}.$$

Example 2   Use Theorem (7.31) and the fact that the natural exponential function is the inverse of the natural logarithmic function to prove that $D_x e^x = e^x$.

Solution   If $f(x) = \ln x$ and $g(x) = e^x$, then $f'(x) = 1/x$ and $f'(g(x)) = f'(e^x) = 1/e^x$. Hence, from Theorem (7.31),

$$g'(x) = \frac{1}{(1/e^x)} = e^x.$$ ∎

Example 3   If $f(x) = x^3 + 2x - 1$, prove that f has an inverse function g and find the slope of the tangent line to the graph of g at the point $P(2, 1)$.

Solution   Since f is continuous and $f'(x) = 3x^2 + 2 > 0$ for all x, we may conclude from Theorem (7.30) that f has an inverse function g. Since $f(1) = 2$ it follows that $g(2) = 1$ and consequently the point $P(2, 1)$ is on the graph of g. Using Theorem (7.31), the slope of the desired tangent line is

$$g'(2) = \frac{1}{f'(g(2))} = \frac{1}{f'(1)} = \frac{1}{5}.$$ ∎

An easy way to remember Theorem (7.31) is to let $y = f(x)$. If g is the inverse function of f, then $g(y) = g(f(x)) = x$. From (7.31),

$$g'(y) = \frac{1}{f'(g(y))} = \frac{1}{f'(x)}$$

or, using differential notation,

$$\frac{dx}{dy} = \frac{1}{\left(\dfrac{dy}{dx}\right)}.$$

This shows that in a sense, the derivative of the inverse function g is the reciprocal of the derivative of f. A disadvantage of using the last two formulas is that neither is stated in terms of the independent variable for the inverse function. To illustrate, suppose in Example 3 we let $y = x^3 + 2x - 1$ and $x = g(y)$. Then

$$\frac{dx}{dy} = \frac{1}{(dy/dx)} = \frac{1}{3x^2 + 2},$$

that is, $$g'(y) = \frac{1}{3x^2 + 2} = \frac{1}{3(g(y))^2 + 2}.$$

This may also be written in the form

$$g'(x) = \frac{1}{3(g(x))^2 + 2}.$$

Consequently, to find $g'(x)$ it is necessary to know $g(x)$, just as in Theorem (7.31).

EXERCISES 7.8

In Exercises 1–10 prove that the function f, defined on the given interval, has an inverse function f^{-1} and state its domain. Find $f^{-1}(x)$. Sketch the graphs of f and f^{-1} on the same coordinate plane. Find $D_x f^{-1}(x)$ directly and also by means of Theorem (7.31).

1 $f(x) = \sqrt{2x + 3}, [1, 11]$

2 $f(x) = \sqrt[3]{5x + 2}, [0, 5]$

3 $f(x) = 4 - x^2, [0, 7]$

4 $f(x) = x^2 - 4x + 5, [-1, 1]$

5 $f(x) = 1/x, (0, \infty)$

6 $f(x) = \sqrt{9 - x^2}, [0, 3]$

7 $f(x) = e^{-x^2}, [0, \infty)$

8 $f(x) = \ln(3 - 2x), (-\infty, \frac{3}{2})$

9 $f(x) = e^x - e^{-x}, (-\infty, \infty)$

10 $f(x) = e^x + e^{-x}, [0, \infty)$

In Exercises 11–14 prove that f has an inverse function f^{-1} on every closed interval $[a, b]$ and find the slope of the tangent line to the graph of f^{-1} at the indicated point P.

11 $f(x) = x^5 + 3x^3 + 2x - 1, P(5, 1)$

12 $f(x) = 2 - x - x^3, P(-8, 2)$

13 $f(x) = (e^{2x} - 1)/(e^{2x} + 1), P(0, 0)$

14 $f(x) = e^{2 - 3x}, P(1/e, 1)$

In Exercises 15–18, prove that if the domain of f is $\mathbb{R}$, then f has no inverse function. Also prove that if the domain is suitably restricted, then f^{-1} exists.

15 $f(x) = x^4 + 3x^2 + 7$

16 $f(x) = |x - 2|$

17 $f(x) = 10^{-x^2}$

18 $f(x) = \ln(x^2 + 1)$

19 Complete the proof of Theorem (7.28) by showing that f^{-1} is continuous at $f(a)$ and $f(b)$.

20 Prove Theorem (7.29).

7.9

REVIEW

Define or discuss each of the following.

1 Inverse functions

2 The natural logarithmic function

3 The natural exponential function

4 Laws of logarithms

5 Derivative formulas for $\ln u$ and e^u

6 Logarithmic differentiation

7 a^x, where $a > 0$

8 The number e

9 $\log_a$

10 Derivative formulas for $\log_a u$ and a^u

11 General power function

12 Separable differential equation

13 Laws of growth and decay

14 First-order linear differential equation

15 Derivatives of inverse functions.

EXERCISES 7.9

Find $f'(x)$ if $f(x)$ is defined as in Exercises 1–24.

1 $(1 - 2x)\ln|1 - 2x|$

2 $\sqrt{\ln\sqrt{x}}$

3 $\ln\dfrac{(3x + 2)^4\sqrt{6x - 5}}{8x - 7}$

4 $\log\left|\dfrac{2 - 9x}{1 - x^2}\right|$

5 $\dfrac{1}{\ln(2x^2 + 3)}$

6 $\dfrac{\ln x}{e^{2x} + 1}$

7 $e^{\ln(x^2 + 1)}$

8 $\ln(e^{4x} + 9)$

9 $10^x \log x$

10 $\ln\sqrt[4]{x/(3x + 5)}$

11 $x^{\ln x}$

12 $4^{\sqrt{2x+3}}$

13 $x^2 e^{1-x^2}$

14 $\sqrt{e^{3x} + e^{-3x}}$

15 $2^{-1/x}/(x^3 + 4)$

16 $5^{3x} + (3x)^5$

17 $(1 + \sqrt{x})^e$

18 $\ln e^{\sqrt{x}}$

19 $10^{\ln x}$

20 $(x^2 + 1)^{2x}$

21 $\ln |x^2 - 5x - 3|$

22 $7^{\ln |x|}$

23 $(\ln x)^{\ln x}$

24 $\ln [(\ln x)^x]$

Find y' in Exercises 25 and 26.

25 $1 + xy = e^{xy}$

26 $\ln (x + y) + x^2 - 2y^3 = 1$

27 Show, by substitution, that $y = c_1 e^{ax} + c_2 x e^{ax}$ is a solution of the differential equation $y'' - 2ay' + a^2 y = 0$, where c_1, c_2, and a are any constants.

28 The position of a point on a coordinate line during the time interval $[-5, 5]$ is given by $f(t) = (t^2 + 2t)e^{-t}$.

(a) When is the velocity zero?

(b) When is the acceleration zero?

(c) When does the point move in the positive direction?

(d) When does the point move in the negative direction?

Evaluate the integrals in Exercises 29–46.

29 $\displaystyle\int \frac{(1 + e^x)^2}{e^{2x}} dx$

30 $\displaystyle\int_0^1 e^{-3x+2} dx$

31 $\displaystyle\int_1^4 \frac{1}{\sqrt{x}e^{\sqrt{x}}} dx$

32 $\displaystyle\int \frac{1}{x \ln x} dx$

33 $\displaystyle\int \frac{1}{x - x \ln x} dx$

34 $\displaystyle\int_1^2 \frac{x^2 + 1}{x^3 + 3x} dx$

35 $\displaystyle\int \frac{x^2}{3x + 2} dx$

36 $\displaystyle\int \frac{e^{1/x}}{x^2} dx$

37 $\displaystyle\int_0^1 x 4^{x^2} dx$

38 $\displaystyle\int \frac{(e^{2x} + e^{3x})^2}{e^{5x}} dx$

39 $\displaystyle\int \frac{1}{x\sqrt{\log x}} dx$

40 $\displaystyle\int \frac{x}{x^4 + 2x^2 + 1} dx$

41 $\displaystyle\int \frac{x^2 + 1}{x + 1} dx$

42 $\displaystyle\int \frac{2e^x}{1 + e^x} dx$

43 $\displaystyle\int 5^x e^x dx$

44 $\displaystyle\int x 10^{x^2} dx$

45 $\displaystyle\int x^e dx$

46 $\displaystyle\int 5^x \sqrt{1 + 5^x} dx$

Solve the differential equations in Exercises 47–50.

47 $(x^3 y^2 + y^2)y' + x^2 = 0$

48 $xyy' - \sqrt{1 - y^2}(1 + x^2) = 0$

49 $x^2 y' + 3y = 1$

50 $xy' + 4y = x^{-3} e^{2x}$

In Exercises 51 and 52 find the particular solution of the differential equation that satisfies the given boundary conditions.

51 $6y^2(1 - x^2)y' + x(y^3 + 1) = 0$; $y = 2$ when $x = 0$

52 $\dfrac{dy}{dx} = 8x^3 - 4x^3 y$; $y = 5$ when $x = 0$

53 A particle moves on a coordinate line with an acceleration at time t of $e^{t/2}$ cm/sec^2. At $t = 0$ the particle is at the origin and its velocity is 6 cm/sec. How far does it travel during the time interval $[0, 4]$?

54 Find the local extrema for $f(x) = x^2 \ln x$, $x > 0$. Discuss concavity, find the points of inflection, and sketch the graph of f.

55 Find an equation of the tangent line to the graph of $y = xe^{1/x^3} + \ln |2 - x^2|$ at the point $P(1, e)$.

56 Find the area of the region bounded by the graphs of the equations $y = e^{2x}$, $y = x/(x^2 + 1)$, $x = 0$, and $x = 1$.

57 The region bounded by the graphs of $y = e^{4x}$, $x = -2$, $x = -3$, and $y = 0$ is revolved about the x-axis. Find the volume of the resulting solid.

58 If $f(x) = 2x^3 - 8x + 5$ on $[-1, 1]$, prove that f has an inverse function g, and find $g'(5)$.

59 Suppose $f(x) = e^{2x} + 2e^x + 1$, where $x \geq 0$.

(a) Prove that f has an inverse function f^{-1} and state its domain.

(b) Find $f^{-1}(x)$ and $D_x f^{-1}(x)$.

(c) Find the slope of the tangent line to the graph of f at the point $(0, 4)$ and the slope of the tangent line to the graph of f^{-1} at $(4, 0)$.

60 A certain radioactive substance has a half-life of five days. How long will it take for an amount A to disintegrate to the time at which only 1 percent of A remains?

61 The rate at which sugar decomposes in water is proportional to the amount that remains undecomposed.

Suppose that 10 lb of sugar is placed in a container of water at 1:00 P.M., and one-half is dissolved at 4:00 P.M.

(a) How long will it take two more pounds to dissolve?

(b) How much of the 10 lb will be dissolved at 8:00 P.M.?

62 According to Newton's Law of Cooling, the rate at which an object cools is directly proportional to the difference in temperature between the object and its surrounding medium. If $f(t)$ denotes the temperature at time t, show that

$f(t) = T + [f(0) - T]e^{-kt}$ where T is the temperature of the surrounding medium and k is a positive constant.

63 The bacterium *E. coli* undergoes cell division approximately every 20 minutes. Starting with 100,000 cells, determine the number of cells after two hours.

64 The differential equation $p\,dv + cv\,dp = 0$ describes the adiabatic change of state of air, where p and v are the pressure and volume, respectively, and c is a constant. Solve for p as a function of v.

8

OTHER TRANSCENDENTAL FUNCTIONS

In this chapter we shall develop formulas for limits, derivatives, and integrals of trigonometric and inverse trigonometric functions. The last two sections contain a discussion of the hyperbolic and inverse hyperbolic functions. A review of the trigonometric functions and trigonometric identities is provided in Appendix III. Students who need a refresher on these topics should read this appendix carefully before proceeding to Section 8.1.

8.1

LIMITS OF TRIGONOMETRIC FUNCTIONS

Whenever we discuss limits of trigonometric expressions involving $\sin t$, $\cos x$, $\tan \theta$, etc., *we shall assume that each variable represents a real number or the radian measure of an angle.* The following result is important for future developments.

Theorem (8.1)

$$\lim_{t \to 0} \sin t = 0$$

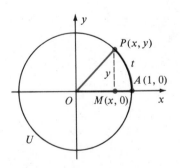

Figure 8.1

Proof Let us first prove that $\lim_{t \to 0^+} \sin t = 0$. Since we are only interested in positive values of t near zero, there is no loss of generality in assuming that $0 < t < \pi/2$. Let U be the circle of radius 1 with center at the origin of a rectangular coordinate system, and let A be the point $(1, 0)$. If, as illustrated in Figure 8.1, $P(x, y)$ is the point on U such that $\overset{\frown}{AP} = t$, then the radian measure of angle AOP is t. Referring to the figure we see that

$$0 < y < t$$

or, since $y = \sin t$,

$$0 < \sin t < t.$$

Since $\lim_{t \to 0^+} t = 0$, it follows from the Sandwich Theorem (2.14) that $\lim_{t \to 0^+} \sin t = 0$.

To complete the proof it is sufficient to show that $\lim_{t \to 0^-} \sin t = 0$. If $-\pi/2 < t < 0$, then $0 < -t < \pi/2$ and hence, from the first part of the proof,

$$0 < \sin(-t) < -t.$$

Multiplying the last inequality by -1 and using the trigonometric identity $\sin(-t) = -\sin t$ gives us

$$t < \sin t < 0.$$

Since $\lim_{t \to 0^-} t = 0$, it follows from the Sandwich Theorem that $\lim_{t \to 0^-} \sin t = 0$.   □

Corollary (8.2)

$$\lim_{t \to 0} \cos t = 1$$

Proof   Since $\sin^2 t + \cos^2 t = 1$, it follows that $\cos t = \pm\sqrt{1 - \sin^2 t}$. If $-\pi/2 < t < \pi/2$, then $\cos t$ is positive, and hence $\cos t = \sqrt{1 - \sin^2 t}$. Consequently,

$$\lim_{t \to 0} \cos t = \lim_{t \to 0} \sqrt{1 - \sin^2 t} = \sqrt{\lim_{t \to 0}(1 - \sin^2 t)}$$
$$= \sqrt{1 - 0} = 1.$$ □

For our work in Section 8.2 it will be essential to know the limits of $(\sin t)/t$ and $(1 - \cos t)/t$ as t approaches 0. These are established in Theorems (8.4) and (8.5). In the proof of (8.4) we shall make use of the following result.

Theorem (8.3)

If θ is the radian measure of a central angle of a circle of radius r, then the area A of the sector determined by θ is

$$A = \tfrac{1}{2}r^2\theta.$$

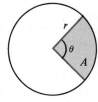

Figure 8.2

Proof   A typical central angle θ and the sector it determines are shown in Figure 8.2. The area of the sector is directly proportional to θ, that is,

$$A = k\theta$$

for some real number k. For example, the area determined by an angle of 2 radians is twice the area determined by an angle of 1 radian. In particular, if $\theta = 2\pi$, then the sector is the entire circle, and $A = \pi r^2$. Thus

$$\pi r^2 = k(2\pi), \quad \text{or} \quad k = \tfrac{1}{2}r^2$$

and therefore

$$A = \tfrac{1}{2}r^2\theta.$$ □

Theorem (8.4)

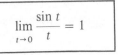

$$\lim_{t \to 0} \frac{\sin t}{t} = 1$$

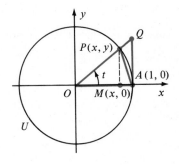

Figure 8.3

Proof   If $0 < t < \pi/2$ we have the situation illustrated in Figure 8.3 where U is a unit circle. If A_1 is the area of triangle AOP, A_2 the area of circular sector AOP, and A_3 the area of triangle AOQ, then

$$A_1 < A_2 < A_3.$$

Using the formula for the area of a triangle and Theorem (8.3) we obtain

$$A_1 = \tfrac{1}{2}(1)d(M, P) = \tfrac{1}{2}y = \tfrac{1}{2}\sin t$$
$$A_2 = \tfrac{1}{2}(1)^2 t = \tfrac{1}{2}t$$
$$A_3 = \tfrac{1}{2}(1)d(A, Q) = \tfrac{1}{2}\tan t.$$

Thus $\tfrac{1}{2}\sin t < \tfrac{1}{2}t < \tfrac{1}{2}\tan t.$

Dividing by $\tfrac{1}{2}\sin t$ and using the fact that $\tan t = \sin t/\cos t$ gives us

$$1 < \frac{t}{\sin t} < \frac{1}{\cos t}$$

or, equivalently,

(∗) $$1 > \frac{\sin t}{t} > \cos t.$$

If $-\pi/2 < t < 0$, then $0 < -t < \pi/2$, and from the result just established,

$$1 > \frac{\sin(-t)}{-t} > \cos(-t).$$

Since $\sin(-t) = -\sin t$ and $\cos(-t) = \cos t$, this inequality reduces to (∗). This shows that (∗) is also true if $-\pi/2 < t < 0$, and hence is true for every t in the open interval $(-\pi/2, \pi/2)$ except $t = 0$. Since $\lim_{t \to 0} \cos t = 1$, and $(\sin t)/t$ is always between $\cos t$ and 1, it follows from the Sandwich Theorem that

$$\lim_{t \to 0} \frac{\sin t}{t} = 1. \qquad \square$$

Roughly speaking, Theorem (8.4) implies that if t is close to 0, then $(\sin t)/t$ is close to 1. Another way of stating this is to write $\sin t \approx t$ for small values of t. It is important to remember that if t denotes an angle, then *radian measure must be used in Theorem (8.4)* and in the approximation formula $\sin t \approx t$. To illustrate, trigonometric tables or a calculator show that to five decimal places,

$$\sin(0.06) \approx 0.05996$$
$$\sin(0.05) \approx 0.04998$$
$$\sin(0.04) \approx 0.03999$$
$$\sin(0.03) \approx 0.03000.$$

Theorem (8.5)

$$\lim_{t \to 0} \frac{1 - \cos t}{t} = 0$$

Proof   We may change the form of $(1 - \cos t)/t$ as follows:

$$\frac{1 - \cos t}{t} = \frac{1 - \cos t}{t} \cdot \frac{1 + \cos t}{1 + \cos t}$$

$$= \frac{1 - \cos^2 t}{t(1 + \cos t)}$$

$$= \frac{\sin^2 t}{t(1 + \cos t)}$$

$$= \frac{\sin t}{t} \cdot \frac{\sin t}{1 + \cos t}.$$

Consequently,

$$\lim_{t \to 0} \frac{1 - \cos t}{t} = \lim_{t \to 0} \left(\frac{\sin t}{t} \cdot \frac{\sin t}{1 + \cos t} \right)$$

$$= \left(\lim_{t \to 0} \frac{\sin t}{t} \right) \left(\lim_{t \to 0} \frac{\sin t}{1 + \cos t} \right)$$

$$= 1 \cdot \left(\frac{0}{1 + 1} \right) = 1 \cdot 0 = 0. \qquad \square$$

Example 1   Find $\lim_{x \to 0} (\sin 5x)/2x$.

Solution   We cannot apply Theorem (8.4) directly, since the given expression is not in the form $(\sin t)/t$. However, we may introduce this form (with $t = 5x$) by using the following algebraic manipulation:

$$\lim_{x \to 0} \frac{\sin 5x}{2x} = \lim_{x \to 0} \frac{1}{2} \frac{\sin 5x}{x}$$

$$= \lim_{x \to 0} \frac{5}{2} \frac{\sin 5x}{5x}$$

$$= \frac{5}{2} \lim_{x \to 0} \frac{\sin 5x}{5x}.$$

It follows from the definition of limit that $x \to 0$ may be replaced by $5x \to 0$. Hence, by Theorem (8.4), with $t = 5x$, we see that

$$\lim_{x \to 0} \frac{\sin 5x}{2x} = \frac{5}{2} (1) = \frac{5}{2}.$$

Warning: When working similar problems, note that $\sin 5x \neq 5 \sin x$.   ∎

Example 2　Find $\lim_{t \to 0} (\tan t)/2t$.

Solution　Using the fact that $\tan t = \sin t/\cos t$,

$$\lim_{t \to 0} \frac{\tan t}{2t} = \lim_{t \to 0} \left(\frac{1}{2} \cdot \frac{\sin t}{t} \cdot \frac{1}{\cos t} \right)$$
$$= \tfrac{1}{2} \cdot 1 \cdot 1 = \tfrac{1}{2}.$$

■

Example 3　Find $\lim_{x \to 0} (2x + 1 - \cos x)/3x$.

Solution　We plan to use Theorem (8.5). With this in mind we begin by isolating the part of the quotient that involves $(1 - \cos x)/x$ and then proceed as follows.

$$\lim_{x \to 0} \frac{2x + 1 - \cos x}{3x} = \lim_{x \to 0} \left(\frac{2x}{3x} + \frac{1 - \cos x}{3x} \right)$$
$$= \lim_{x \to 0} \left(\frac{2x}{3x} \right) + \lim_{x \to 0} \frac{1}{3} \left(\frac{1 - \cos x}{x} \right)$$
$$= \lim_{x \to 0} \frac{2}{3} + \frac{1}{3} \lim_{x \to 0} \frac{1 - \cos x}{x}$$
$$= \tfrac{2}{3} + \tfrac{1}{3} \cdot 0 = \tfrac{2}{3}$$

■

EXERCISES 8.1

Find the limits in Exercises 1–26.

1 $\lim_{x \to 0} \dfrac{x}{\sin x}$

2 $\lim_{x \to 0} \dfrac{\sin x}{\sqrt[3]{x}}$

3 $\lim_{t \to 0} \dfrac{\sin^3 t}{(2t)^3}$

4 $\lim_{\theta \to 0} \dfrac{3\theta + \sin \theta}{\theta}$

5 $\lim_{x \to 0} \dfrac{2 + \sin x}{3 + x}$

6 $\lim_{t \to 0} \dfrac{1 - \cos 3t}{t}$

7 $\lim_{\theta \to 0} \dfrac{2 \cos \theta - 2}{3\theta}$

8 $\lim_{x \to 0} \dfrac{x^2 + 1}{x + \cos x}$

9 $\lim_{x \to 0} \dfrac{\sin (-3x)}{4x}$

10 $\lim_{x \to 0} \dfrac{x \sin x}{x^2 + 1}$

11 $\lim_{x \to 0} \dfrac{1 - \cos x}{x^{2/3}}$

12 $\lim_{x \to 0} \dfrac{1 - 2x^2 - 2 \cos x + \cos^2 x}{x^2}$

13 $\lim_{t \to 0} \dfrac{4t^2 + 3t \sin t}{t^2}$

14 $\lim_{x \to 0} \dfrac{x \cos x - x^2}{2x}$

15 $\lim_{t \to 0} \dfrac{\cos t}{1 - \sin t}$

16 $\lim_{t \to 0} \dfrac{\sin t}{1 + \cos t}$

17 $\lim_{t \to 0} \dfrac{1 - \cos t}{\sin t}$

18 $\lim_{x \to 0} \dfrac{\sin (x/2)}{x}$

19 $\lim_{x \to 0} \dfrac{x + \tan x}{\sin x}$

20 $\lim_{t \to 0} \dfrac{\sin^2 2t}{t^2}$

21 $\lim_{x \to 0} x \cot x$

22 $\lim_{x \to 0} \dfrac{\csc 2x}{\cot x}$

23 $\lim_{\alpha \to 0} \alpha^2 \csc^2 \alpha$

24 $\lim_{x \to 0} \dfrac{\sin 3x}{\sin 5x}$

25 $\lim_{v \to 0} \dfrac{\cos (v + \pi/2)}{v}$

26 $\lim_{x \to 0} \dfrac{\sin^2 (x/2)}{\sin x}$

Establish the limits in Exercises 27–30, where a and b are any nonzero real numbers.

27 $\lim_{x \to 0} \dfrac{\sin ax}{bx} = \dfrac{a}{b}$

28 $\lim_{x \to 0} \dfrac{1 - \cos ax}{bx} = 0$

29 $\lim_{x \to 0} \dfrac{\sin ax}{\sin bx} = \dfrac{a}{b}$

30 $\lim_{x \to 0} \dfrac{\cos ax}{\cos bx} = 1$

8.2

DERIVATIVES OF TRIGONOMETRIC FUNCTIONS

Let us begin by listing formulas for the derivatives of all six trigonometric functions. In the statement of the theorem it is assumed that $u = g(x)$, where g is differentiable, and that x is restricted to values for which the indicated functions are defined.

Theorem (8.6)

$$D_x \sin u = \cos u \, D_x u \qquad\qquad D_x \csc u = -\csc u \cot u \, D_x u$$

$$D_x \cos u = -\sin u \, D_x u \qquad\qquad D_x \sec u = \sec u \tan u \, D_x u$$

$$D_x \tan u = \sec^2 u \, D_x u \qquad\qquad D_x \cot u = -\csc^2 u \, D_x u$$

Proof   To obtain the first formula it is sufficient to show that $D_x \sin x = \cos x$, since $D_x \sin u$ may then be found by using the Chain Rule. Thus, let $f(x) = \sin x$. If $f'(x)$ exists,

$$f'(x) = \lim_{h \to 0} \frac{f(x + h) - f(x)}{h}$$

$$= \lim_{h \to 0} \frac{\sin (x + h) - \sin x}{h}.$$

Using the addition formula for the sine function,

$$f'(x) = \lim_{h \to 0} \frac{\sin x \cos h + \cos x \sin h - \sin x}{h}$$

$$= \lim_{h \to 0} \frac{\sin x (\cos h - 1) + \cos x \sin h}{h}$$

$$= \lim_{h \to 0} \left[\sin x \left(\frac{\cos h - 1}{h} \right) + \cos x \left(\frac{\sin h}{h} \right) \right].$$

By Theorems (8.5) and (8.4),

$$\lim_{h \to 0} \left(\frac{\cos h - 1}{h} \right) = 0 \quad \text{and} \quad \lim_{h \to 0} \frac{\sin h}{h} = 1.$$

Hence,

$$f'(x) = (\sin x)(0) + (\cos x)(1) = \cos x,$$

that is,

$$D_x \sin x = \cos x.$$

If $y = \sin u$ and $u = g(x)$, where g is differentiable, then by the Chain Rule,

$$\frac{dy}{dx} = \frac{dy}{du} \frac{du}{dx} = \cos u \frac{du}{dx}.$$

Using the D_x notation, this may be written

$$D_x \sin u = \cos u \, D_x u.$$

The remaining five formulas may be obtained from that of the sine function with the aid of trigonometric identities. If we use the addition formulas for sines and cosines we see that $\cos x = \sin(\pi/2 - x)$ and $\sin x = \cos(\pi/2 - x)$. Using these facts together with the formula for $D_x \sin u$,

$$\begin{aligned}
D_x \cos x &= D_x \sin(\pi/2 - x) \\
&= \cos(\pi/2 - x) \, D_x(\pi/2 - x) \\
&= (\sin x)(-1) = -\sin x.
\end{aligned}$$

As before, the Chain Rule gives us the formula for $D_x \cos u$.

Applying the Quotient Rule to $\tan x = \sin x/\cos x$,

$$\begin{aligned}
D_x \tan x &= \frac{\cos x \, D_x \sin x - \sin x \, D_x \cos x}{\cos^2 x} \\
&= \frac{\cos x(\cos x) - \sin x(-\sin x)}{\cos^2 x} \\
&= \frac{\cos^2 x + \sin^2 x}{\cos^2 x} = \frac{1}{\cos^2 x} = \sec^2 x.
\end{aligned}$$

Similarly, for the secant function,

$$\begin{aligned}
D_x \sec x &= D_x \frac{1}{\cos x} = \frac{(\cos x) \, D_x 1 - (1) \, D_x \cos x}{\cos^2 x} \\
&= \frac{0 + \sin x}{\cos^2 x} = \frac{1}{\cos x} \frac{\sin x}{\cos x} \\
&= \sec x \tan x.
\end{aligned}$$

Proofs for $\cot x$ and $\csc x$ are left as exercises.   □

Example 1   Find y' if $y = \cos(5x^3)$.

Solution   Using the formula for $D_x \cos u$ with $u = 5x^3$,

$$\begin{aligned}
D_x \cos(5x^3) &= [-\sin(5x^3)]D_x(5x^3) \\
&= [-\sin(5x^3)](15x^2) \\
&= -15x^2 \sin(5x^3).
\end{aligned}$$ ■

Example 2   Find $f'(x)$ if $f(x) = \sin^3(4x)$.

Solution   First note that $f(x) = \sin^3(4x) = [\sin(4x)]^3$. Applying the Power Rule for Functions with $u = \sin(4x)$,

$$f'(x) = 3[\sin(4x)]^2 D_x(\sin 4x) = (3 \sin^2 4x)D_x \sin 4x.$$

Next, by Theorem (8.6),

$$D_x \sin 4x = (\cos 4x)D_x(4x) = (\cos 4x)(4) = 4 \cos 4x.$$

Consequently,

$$f'(x) = (3 \sin^2 4x)(4 \cos 4x) = 12 \sin^2 4x \cos 4x.$$

Example 3 Find $f'(x)$ if (a) $f(x) = e^{-2x} \tan 4x$;   (b) $f(x) = \sec^2 (3x - 1)$.

Solution

(a) First using the Product Rule and then Theorems (8.6) and (7.14),

$$f'(x) = e^{-2x} D_x \tan 4x + (\tan 4x) D_x e^{-2x}$$
$$= e^{-2x}(\sec^2 4x)(4) + (\tan 4x)(e^{-2x})(-2)$$
$$= 2e^{-2x}(2 \sec^2 4x - \tan 4x).$$

(b) By the Power Rule and Theorem (8.6),

$$f'(x) = 2 \sec (3x - 1) D_x \sec (3x - 1)$$
$$= 2 \sec (3x - 1) \sec (3x - 1) \tan (3x - 1)(3)$$
$$= 6 \sec^2 (3x - 1) \tan (3x - 1). \qquad \blacksquare$$

Since the sine and cosine functions are differentiable at every real number x, they are continuous throughout $\mathbb{R}$ (see Theorem (3.8)). Similarly, each of the remaining four trigonometric functions is continuous at every number in its domain.

Portions of the graphs of the trigonometric functions are sketched in Figure 8.4. The dashed lines indicate vertical asymptotes. The reader should

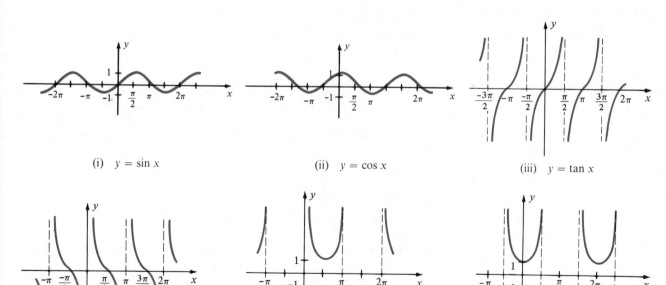

(i) $y = \sin x$

(ii) $y = \cos x$

(iii) $y = \tan x$

(iv) $y = \cot x$

(v) $y = \csc x$

(vi) $y = \sec x$

Figure 8.4

verify each graph by plotting several points and investigating symmetry, periodicity, maxima, minima, and concavity.

Example 4 If $f(x) = 2 \sin x + \cos 2x$, find the local extrema and sketch the graph of f on the interval $[0, 2\pi]$.

Solution   Differentiating, we obtain

$$f'(x) = 2 \cos x - 2 \sin 2x.$$

Since $\sin 2x = 2 \sin x \cos x$, this may be rewritten

$$f'(x) = 2 \cos x - 4 \sin x \cos x$$
$$= 2 \cos x(1 - 2 \sin x).$$

The derivative exists for all x, and $f'(x) = 0$ if either $\sin x = \frac{1}{2}$ or $\cos x = 0$. Hence the critical numbers of f in the interval $[0, 2\pi]$ are $\pi/6$, $5\pi/6$, $\pi/2$, and $3\pi/2$.

The second derivative of f is

$$f''(x) = -2 \sin x - 4 \cos 2x.$$

Substituting the critical numbers for x we obtain

$$f''(\pi/6) = -3, \quad f''(5\pi/6) = -3, \quad f''(\pi/2) = 2, \quad f''(3\pi/2) = 6.$$

Applying the Second Derivative Test we see that there are local maxima at $\pi/6$ and $5\pi/6$, and local minima at $\pi/2$ and $3\pi/2$. This information, together with the following table, leads to the sketch in Figure 8.5.

x	0	$\pi/6$	$\pi/2$	$5\pi/6$	π	$3\pi/2$	2π
$f(x)$	1	$\frac{3}{2}$	1	$\frac{3}{2}$	1	-3	1

■

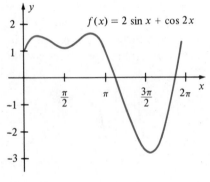

$f(x) = 2 \sin x + \cos 2x$

Figure 8.5

Example 5 Suppose $\quad f(x) = \begin{cases} x \sin \dfrac{1}{x} & \text{if } x \neq 0 \\[2mm] 0 & \text{if } x = 0. \end{cases}$

Prove that f is continuous at every real number and sketch the graph of f.

Solution   Let us show that $\lim_{x \to a} f(x) = f(a)$ for every real number a. If $a \neq 0$, then

$$\lim_{x \to a} \left(x \sin \frac{1}{x} \right) = \left(\lim_{x \to a} x \right)\left(\lim_{x \to a} \sin \frac{1}{x} \right)$$

$$= a \sin \frac{1}{a} = f(a).$$

From Example 8 in Section 2.3,

$$\lim_{x \to 0} x \sin \frac{1}{x} = 0 = f(0).$$

Hence by Definition (2.18), f is continuous at every real number.

The graph of $y = f(x)$ is symmetric with respect to the y-axis, since substitution of $-x$ for x gives us

$$y = -x \sin \frac{1}{-x} = -x\left(-\sin \frac{1}{x}\right) = x \sin \frac{1}{x}.$$

Thus we may restrict our discussion to points (x, y) with $x \geq 0$, since the remainder of the graph can be determined by a reflection through the y-axis.

To find the x-intercepts we solve the equation $x \sin (1/x) = 0$, obtaining $x = 0$ and $1/x = n\pi$, where n is any integer. In particular, in the interval $[0, 1]$ the graph has an infinite number of x-intercepts of the form $1/n\pi$. Several of these are

$$\frac{1}{\pi}, \quad \frac{1}{2\pi}, \quad \frac{1}{3\pi}, \quad \frac{1}{4\pi}, \quad \text{and} \quad \frac{1}{5\pi}.$$

It is of interest to note that the function f is continuous on the interval $[0, 1]$, but its graph cannot be sketched without lifting the pencil from the paper since there is no "first" x-intercept greater than 0.

It is not difficult to show that

$$f(x) < 0 \quad \text{if} \quad \frac{1}{2\pi} < x < \frac{1}{\pi}$$

and

$$f(x) > 0 \quad \text{if} \quad \frac{1}{3\pi} < x < \frac{1}{2\pi}.$$

Similar inequalities can be obtained if $1/((n + 1)\pi) < x < 1/n\pi$, where n is any positive integer.

If $x \neq 0$, then $|\sin (1/x)| \leq 1$ and hence

$$|f(x)| = \left| x \sin \frac{1}{x} \right| = |x| \left| \sin \frac{1}{x} \right| \leq |x|.$$

It follows that the graph of f lies between (and on) the lines $y = x$ and $y = -x$. The graph intersects these lines if $|f(x)| = |x|$, that is, if $|\sin(1/x)| = 1$. Observe that $|\sin (1/x)| = 1$ if

$$\frac{1}{x} = \frac{\pi}{2} + n\pi = \frac{\pi(1 + 2n)}{2}, \quad \text{or} \quad x = \frac{2}{\pi(1 + 2n)}$$

where n is any integer. The x-coordinates of several such points of intersection are

$$\frac{2}{\pi}, \quad \frac{2}{3\pi}, \quad \frac{2}{5\pi}, \quad \frac{2}{7\pi}, \quad \text{and} \quad \frac{2}{9\pi}.$$

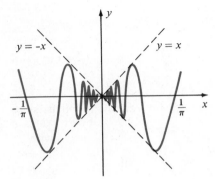

Figure 8.6

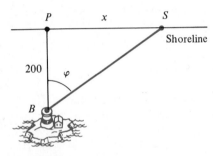

Figure 8.7

In Exercise 70 you are asked to show that the graph of f is tangent to $y = x$ or $y = -x$ at each of these points, and that the graph has a horizontal tangent line at $(x, f(x))$ if $\tan (1/x) = 1/x$.

Using the information we have obtained leads to the graph in Figure 8.6. ∎

Example 6 A revolving beacon in a lighthouse that is 200 ft from the nearest point P on a straight shoreline makes one revolution every 15 seconds. Find the rate at which a ray from the light moves along the shore at a point 400 ft from P.

Solution   The problem is diagrammed in Figure 8.7, where B denotes the position of the beacon and φ is the angle between BP and a ray to a point S on the shore x units from P. Since the light revolves four times per minute, the angle φ changes at a rate of 8π radians per minute, that is, $d\varphi/dt = 8\pi$. Using triangle PBS we see that $\tan \varphi = x/200$, or $x = 200 \tan \varphi$. Consequently,

$$\frac{dx}{dt} = 200 \sec^2 \varphi \, \frac{d\varphi}{dt} = (200 \sec^2 \varphi)(8\pi) = 1600\pi \sec^2 \varphi.$$

If $x = 400$, then

$$\tan \varphi = 400/200 = 2, \quad \text{and} \quad \sec \varphi = \sqrt{1 + \tan^2 \varphi} = \sqrt{1 + 4} = \sqrt{5}.$$

Hence,

$$\frac{dx}{dt} = 1600\pi(\sqrt{5})^2 = 8000\pi \text{ ft/min}. \qquad ∎$$

The sine and cosine functions play major roles in the study of physical entities that vibrate or move in a periodic manner. In particular, one of the most important types of rectilinear motion is described in the following definition.

Definition (8.7)

> A point P that moves on a coordinate line l such that its distance $s(t)$ from the origin at time t is given by either
>
> $$s(t) = a \cos (\omega t + b) \quad \text{or} \quad s(t) = a \sin (\omega t + b)$$
>
> where a, b, and $\omega > 0$ are real numbers, is said to be in **simple harmonic motion**.

Another way of defining simple harmonic motion is to demand that the point P move on l such that the acceleration $s''(t)$ satisfies the condition

$$s''(t) = -\omega^2 s(t)$$

for all t. It can be shown (see Section 19.6) that this condition is equivalent to Definition (8.7).

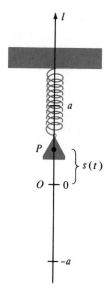

Figure 8.8

In simple harmonic motion the point P oscillates between the points on l with coordinates $-a$ and a. The **amplitude** of the motion is the maximum displacement $|a|$ of the point from the origin. The **period** is the time $2\pi/\omega$ required for one complete oscillation. The **frequency** $\omega/2\pi$ is the number of oscillations per unit of time.

Simple harmonic motion takes place in many different types of wave motion, such as water waves, sound waves, radio waves, light waves, and distortional waves that are present in vibrating bodies. Another example of simple harmonic motion can be obtained by considering a spring with an attached weight, which is oscillating vertically relative to a coordinate line, as illustrated in Figure 8.8. The number $s(t)$ represents the coordinate of a fixed point P in the weight, and we assume that the amplitude a of the motion is constant. In this case, there is no frictional force retarding the motion. If friction is present, then the amplitude decreases with time, and the motion is said to be *damped*.

Example 7 Suppose the weight shown in Figure 8.8 is oscillating according to the law

$$s(t) = 10 \cos\left(\frac{\pi}{6}t\right)$$

where t is measured in seconds and $s(t)$ in centimeters. Discuss the motion of the weight.

Solution By Definition (8.7), the motion is simple harmonic. The amplitude a is 10 cm. Since $\omega = \pi/6$, the period is $2\pi/\omega = 2\pi/(\pi/6) = 12$. Thus, it requires 12 seconds for one complete oscillation. The frequency is $\omega/2\pi = (\pi/6)/2\pi = \frac{1}{12}$, that is, $\frac{1}{12}$ of an oscillation takes place each second.

Let us examine the motion during the time interval $[0, 12]$. The velocity and acceleration functions are given by

$$v(t) = s'(t) = -10 \sin\left(\frac{\pi}{6}t\right) \cdot \left(\frac{\pi}{6}\right) = -\frac{5\pi}{3}\sin\left(\frac{\pi}{6}t\right),$$

$$a(t) = v'(t) = -\frac{5\pi}{3}\cos\left(\frac{\pi}{6}t\right) \cdot \left(\frac{\pi}{6}\right) = -\frac{5\pi^2}{18}\cos\left(\frac{\pi}{6}t\right).$$

The velocity is 0 at $t = 0$, $t = 6$ and $t = 12$, since $\sin((\pi/6)t) = 0$ for these values of t. The acceleration is 0 at $t = 3$ and $t = 9$, since in these cases $\cos((\pi/6)t) = 0$. The times at which the velocity and acceleration are zero lead us to examine the time intervals $(0, 3)$, $(3, 6)$, $(6, 9)$, and $(9, 12)$. The following table displays the main characteristics of the motion.

Time Interval	Sign of $v(t)$	Direction of Motion	Sign of $a(t)$	Variation of $v(t)$	Speed $|v(t)|$
$(0, 3)$	$-$	Downward	$-$	Decreasing	Increasing
$(3, 6)$	$-$	Downward	$+$	Increasing	Decreasing
$(6, 9)$	$+$	Upward	$+$	Increasing	Increasing
$(9, 12)$	$+$	Upward	$-$	Decreasing	Decreasing

The reader should verify all entries in the table. Note that if $0 < t < 3$, the velocity $v(t)$ is negative and decreasing. This implies that the speed $|v(t)|$ is *increasing*. If $3 < t < 6$, the velocity is negative and increasing. In this case the speed of P is *decreasing* in the time interval $(3, 6)$. Similar remarks can be made for the intervals $(6, 9)$ and $(9, 12)$.

We may summarize the motion of P as follows. At $t = 0$, $s(0) = 10$ and the point P is 10 cm above the origin O. It then moves downward, gaining speed until it reaches O at $t = 3$. It then slows down until it reaches a point 10 cm below O at the end of 6 sec. The direction of motion is then reversed, and the weight moves upward, gaining speed until it reaches O at $t = 9$, after which it slows down until it returns to its original position at the end of 12 sec. The direction of motion is then reversed again, and the same pattern is repeated indefinitely. ∎

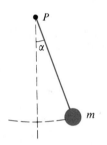

Figure 8.9

The definition of simple harmonic motion is usually extended to include situations where $s(t)$ is *any* mathematical or physical quantity (not necessarily a distance). In the next illustration $s(t)$ is an angle.

Historically, the first type of harmonic motion to be scientifically investigated involved pendulums. Figure 8.9 illustrates a *simple pendulum* consisting of a bob of mass m attached to one end of a string, with the other end of the string attached to a fixed point P. In the ideal case it is assumed that the string is weightless, there is no air resistance, and the only force acting on the bob is gravity. If the bob is displaced sideways and released, the pendulum oscillates back and forth in a vertical plane. Let α denote the *angular displacement* at time t (see Figure 8.9). If the bob moves through a small arc (say $|\alpha| < 5°$), then it can be shown, by using physical laws, that

$$\alpha = \beta \cos(\omega t + \alpha_0)$$

where α_0 is the initial displacement, ω is the frequency of oscillation of the angle α, and β is the (angular) amplitude of oscillation. This means that the *angle* α is in simple harmonic motion and hence we refer to the motion as *angular* simple harmonic motion.

As a final illustration, in electrical circuits an alternating electromotive force (emf) and the current may vary harmonically. For example, the emf e (measured in volts) at time t may be given by

$$e = E \sin \omega t$$

where E is the maximum value of e. If an emf of this type is impressed on a circuit containing only a resistance R, then by Ohm's Law, the current i at time t is

$$i = \frac{e}{R} = \frac{E}{R} \sin \omega t = I \sin \omega t$$

where $I = E/R$. A schematic drawing of an electrical circuit of this type is illustrated in Figure 8.10.

In this case, the maximum value I of i occurs at the same time as the maximum value E of e. In other situations, these maximum values may occur

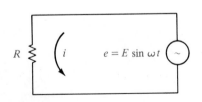

Figure 8.10

at different times, in which case we say there is a **phase difference** between e and i. For example, if $e = E \sin \omega t$, we could have either

$$\text{(a)} \quad i = I \sin(\omega t - \varphi) \quad \text{or} \quad \text{(b)} \quad i = I \sin(\omega t + \varphi)$$

where $\varphi > 0$. In case (a), the current is said to **lag** the emf by an amount φ/ω, and the graph of i can be obtained by shifting the graph of $i = I \sin \omega t$ an amount φ/ω to the *right*. In case (b), we shift the graph to the *left*, and the current is said to **lead** the emf by an amount φ/ω.

EXERCISES 8.2

1 Verify the formulas for $D_x \cot u$ and $D_x \csc u$.

2 Verify the graphs of the trigonometric functions. In each case discuss continuity, determine where the function increases or decreases, and find the local extrema. Discuss concavity, find the points of inflection, and locate the vertical asymptotes.

In Exercises 3–40 find $f'(x)$ if $f(x)$ equals the given expression.

3 $\sin(x^2 + 2)$

4 $\cos(4 - 3x)$

5 $\cos^5 3x$

6 $\sin^4(x^3)$

7 $x^3 \cos(1/x)$

8 $\sin x/(1 + \cos x)$

9 $\tan(8x + 3)$

10 $\cot(x/2)$

11 $\sec \sqrt{x - 1}$

12 $\csc(x^2 + 4)$

13 $\cot(x^3 - 2x)$

14 $\tan \sqrt[3]{5 - 6x}$

15 $\cos 3x^2 + \cos^2 3x$

16 $\tan^3 6x$

17 $\csc^2 2x$

18 $\sec e^{-2x}$

19 $x^2 \csc 5x$

20 $x \csc(1/x)$

21 $\tan^2 x \sec^3 x$

22 $x^2 \sec^3 4x$

23 $(\sin 5x - \cos 5x)^5$

24 $\sin \sqrt{x} + \sqrt{\sin x}$

25 $\cot^3(3x + 1)$

26 $e^{\cos 2x}$

27 $(\cos 4x)/(1 - \sin 4x)$

28 $(\sec 2x)/(\tan 2x + 1)$

29 $\ln|\csc x + \cot x|$

30 $\sin(2x + 3)^4$

31 $e^{-3x} \tan \sqrt{x}$

32 $\ln \csc^2 3x$

33 $\ln \ln \sec 2x$

34 $\csc(\cot 4x)$

35 $\tan^3 2x - \sec^3 2x$

36 $(\tan 2x - \sec 2x)^3$

37 $\dfrac{\csc 3x}{x^3 + 1}$

38 $4x^3 - x^2 \cot^3(1/x)$

39 $x^{\cot x}$

40 $(\tan x)^{3x}$

In Exercises 41–46 find dy/dx and d^2y/dx^2.

41 $y = \sec^2 3x$

42 $y = \cot^3 5x$

43 $y = \sin x - x \cos x$

44 $y = \ln \tan x$

45 $y = \sqrt{\tan x}$

46 $y = \dfrac{\cos x - 1}{\cos x + 1}$

In Exercises 47–50 use implicit differentiation to find y'.

47 $y^2 = x \cos y$

48 $xy = \tan xy$

49 $e^x \cot y = xe^{2y}$

50 $y^2 + 1 = x^2 \sec y$

In Exercises 51–56, find the local extrema of f on the interval $[0, 2\pi]$, and determine where f is increasing or decreasing on $[0, 2\pi]$. Sketch the graph of f corresponding to $0 \le x \le 2\pi$.

51 $f(x) = \cos x + \sin x$

52 $f(x) = \cos x - \sin x$

53 $f(x) = \frac{1}{2}x - \sin x$

54 $f(x) = x + 2 \cos x$

55 $f(x) = 2 \cos x + \sin 2x$

56 $f(x) = 2 \cos x + \cos 2x$

57 Find the local extrema and sketch the graph of $f(x) = e^{-x} \sin x$ for $0 \le x \le 4\pi$.

58 Find the local extrema of $f(x) = e^{-x} \sec x$ for $0 \le x \le 2\pi$.

In Exercises 59 and 60 find the absolute maximum and minimum values of f on the given interval.

59 $f(x) = \tan x - \sqrt{2} \sec x$, $[0, \pi/3]$

60 $f(x) = 3 \cot x + 4x$, $[\pi/6, 5\pi/6]$

In Exercises 61 and 62 find equations of the tangent and normal lines to the graph of the equation at the point P.

61 $y = (1/8) \csc^3 x$, $P(\pi/6, 1)$

62 $y = \tan x - \sqrt{2} \sin x$; $P(3\pi/4, -2)$

63   At a point 20 ft from the base of a flagpole, the angle of elevation of the top of the pole is measured as 60°, with a possible error of 15'. Use differentials to approximate the error in the calculated height of the pole.

64   Use differentials to approximate the change in $\cot x$ if x changes from 45° to 46°.

65   An airplane at an altitude of 20,000 ft is flying at a constant speed on a line that will take it directly over an observer on the ground. If, at a given instant, the observer notes that the angle of elevation of the airplane is 45° and is increasing at a rate of 1° per second, find the speed of the airplane.

66   A missile is fired vertically from a point that is 5 mi from a tracking station and at the same elevation (see figure). For the first 20 seconds of flight its angle of elevation θ changes at a constant rate of 2° per second. Find the velocity of the missile when the angle of elevation is 30°.

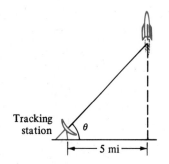

Figure for Exercise 66

67   A billboard 20 ft high is located on top of a building, with its lower edge 60 ft above the level of a viewer's eye (see figure). How far from a point directly below the sign should a viewer stand to maximize the angle θ between the lines of sight of the top and bottom of the billboard? (This angle should result in the best view of the billboard).

68   A man on a small island I, which is k mi from the closest point A on a straight shoreline, wishes to reach a camp that is d mi downshore from A (see figure). He plans to swim to some point P on shore and then walk the rest of the way. Suppose that c_1 calories per mile are expended while swimming and c_2 calories per mile while walking, where $c_1 > c_2$.

(a) Find a formula for the total number c of calories expended in completing the trip.

(b) Show that the minimum value of c depends only on angle AIP, and not on the distance k or d.

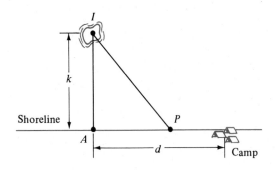

Figure for Exercise 68

69   A stuntman falls from a hot-air balloon that is hovering at a constant altitude, 100 ft above a lake. A television camera on shore, 200 ft from a point directly below the balloon, follows his descent (see figure). At what rate is the angle of elevation θ of the camera changing 2 seconds after the stuntman falls? (Neglect the height of the camera.)

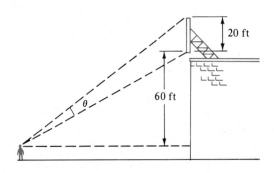

Figure for Exercise 67

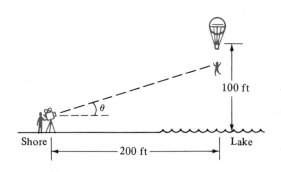

Figure for Exercise 69

70 Let f be the function defined in Example 5.

(a) Prove that the graph of f has a horizontal tangent line at $(x, f(x))$ if $\tan(1/x) = 1/x$. Show that there are an infinite number of such points by sketching the graphs of $y = \tan t$ and $y = t$ on the same coordinate axes.

(b) Show that the lines $y = x$ and $y = -x$ are tangent to the graph of f at an infinite number of points.

71 If an object of weight W lb is pulled along a horizontal plane by a force applied to a rope that is attached to the object, and if the rope makes an angle θ with the horizontal, then it can be shown that the magnitude of the force is given by

$$F = \frac{\mu W}{\mu \sin \theta + \cos \theta}$$

where μ is a constant called the *coefficient of friction*. Suppose a man is pulling a 100-pound box along a floor, and that $\mu = 0.2$ (see figure). If θ is changed from 45° to 46°, use differentials to approximate the change in the force that must be applied.

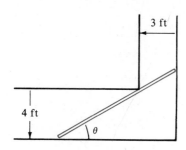

Figure for Exercise 71

72 If $\mu = \sqrt{3}/3$ in Exercise 71, find the angle that will minimize F.

73 Two corridors 3 ft and 4 ft wide, respectively, meet at right angles. Find the length of the longest thin straight rod that will pass horizontally around the corner (see figure).

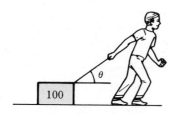

Figure for Exercise 73

74 When light travels from one point to another it takes the path that requires the least amount of time. Suppose that light has velocity v_1 in air and v_2 in water, where $v_1 > v_2$. If light travels from a point P in air to a point Q in water (see figure), show that the path requiring the least amount of time occurs if

$$\frac{\sin \theta_1}{\sin \theta_2} = \frac{v_1}{v_2}.$$

(This is an example of **Snell's Law of Refraction**.)

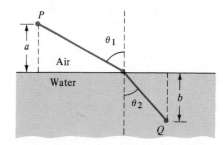

Figure for Exercise 74

75 A water cup in the shape of a right circular cone is to be constructed by removing a circular sector from a circular sheet of paper and then joining the two straight edges of the remaining paper. Find the central angle of the sector that will maximize the volume of the cup.

76 An isosceles triangle has equal sides of length 12 in. If the angle θ between these sides is increased from 30° to 33°, use differentials to approximate the change in the area of the triangle.

77 Use differentials to approximate the change in $\cos \theta$ if θ changes from 30° to 28°.

78 A cork bobs up and down in a pail of water such that the distance from the bottom of the pail to the center of the cork at time $t \geq 0$ is given by $s(t) = 12 + \cos(\pi t)$, where $s(t)$ is in inches and t is in seconds (see figure).

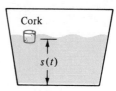

Figure for Exercise 78

(a) Find the velocity of the cork at $t = 0$, $t = \frac{1}{2}$, $t = 1$, $t = \frac{3}{2}$, and $t = 2$.

(b) During what time intervals is the cork rising? When is it falling?

In Exercises 79 and 80, s is the position function of a particle that is moving rectilinearly. Find (a) the velocity and acceleration at the indicated time t, and (b) the subintervals of $[0, 2\pi]$ in which the particle moves in the positive direction.

79 $s(t) = \sqrt{3}t + \cos 2t, \; t = \pi/4$

80 $s(t) = 2 \sin^2 t - \sin 2t, \; t = \pi/6$

In Exercises 81–84, s is the position function of a particle moving in simple harmonic motion, and t is the time in seconds. Find the amplitude, period, and frequency, and describe the motion of the particle during one complete oscillation.

81 $s(t) = 5 \cos (\pi/4)t$ **82** $s(t) = 4 \sin \pi t$

83 $s(t) = 6 \sin (2\pi/3)t$ **84** $s(t) = 3 \cos 2t$

85 A point $P(x, y)$ is moving at a constant rate around a circle that has the equation $x^2 + y^2 = a^2$. Prove that the projection $Q(x, 0)$ of P on the x-axis is in simple harmonic motion.

86 If a point P moves on a coordinate line such that

$$s(t) = a \cos \omega t + b \sin \omega t,$$

show that P is in simple harmonic motion by

(a) using the remark following Definition (8.7).

(b) using only trigonometric methods. (*Hint*: Show that $s(t) = a \cos (\omega t - c)$ for some c.)

In Exercises 87 and 88 use Newton's Method to approximate the indicated real root to four decimal places.

87 The positive root of $2x - 3 \sin x = 0$

88 The root of $\cos x + x = 2$

In Exercises 89 and 90 use Newton's Method to find all real roots of the equation to two decimal places.

89 $2x - 5 - \sin x = 0$ **90** $x^2 - \cos 2x = 0$

In Exercises 91 and 92, use mathematical induction to prove that the given statement is true for every positive integer n.

91 If $f(x) = \sin x$, then $f^{(n)}(x) = \sin [x + (n\pi)/2]$.

92 If $f(x) = \cos x$, then $f^{(n)}(x) = \cos [x + (n\pi)/2]$.

8.3

INTEGRALS OF TRIGONOMETRIC FUNCTIONS

Formulas for indefinite integrals of important trigonometric expressions are listed in the next theorem, where $u = g(x)$ and g is differentiable.

Theorem (8.8)

$$\int \sin u \, du = -\cos u + C \qquad \int \cos u \, du = \sin u + C$$

$$\int \tan u \, du = \ln |\sec u| + C \qquad \int \cot u \, du = \ln |\sin u| + C$$

$$\int \sec u \, du = \ln |\sec u + \tan u| + C \qquad \int \csc u \, du = \ln |\csc u - \cot u| + C$$

$$\int \sec^2 u \, du = \tan u + C \qquad \int \csc^2 u \, du = -\cot u + C$$

$$\int \sec u \tan u \, du = \sec u + C \qquad \int \csc u \cot u \, du = -\csc u + C$$

Proof   It suffices to consider the case $u = x$, since the more general formulas then follow from Theorem (5.30).

The integration formula $\int \sin x \, dx = -\cos x + C$ follows from the fact that $D_x(-\cos x) = \sin x$. Similarly, $\int \cos x \, dx = \sin x + C$ is true since $\sin x$ is an antiderivative of $\cos x$.

The integral of $\tan x$ may be obtained by first writing

$$\int \tan x \, dx = \int \frac{\sin x}{\cos x} \, dx.$$

If we now let $v = \cos x$, then $dv = -\sin x \, dx$ and hence

$$\int \tan x \, dx = -\int \frac{1}{\cos x}(-\sin x) \, dx = -\int \frac{1}{v} \, dv$$

$$= -\ln|v| + C.$$

Consequently, if $\cos x \neq 0$, then

$$\int \tan x \, dx = -\ln|\cos x| + C.$$

Since $-\ln|\cos x| = \ln(1/|\cos x|) = \ln|\sec x|$, we may also write

$$\int \tan x \, dx = \ln|\sec x| + C.$$

Similarly, since $\cot x = \cos x/\sin x$,

$$\int \cot x \, dx = \int \frac{1}{\sin x} \cos x \, dx.$$

Letting $v = \sin x$ and $dv = \cos x \, dx$ we obtain

$$\int \cot x \, dx = \int \frac{1}{v} \, dv = \ln|v| + C = \ln|\sin x| + C.$$

The integral formula for $\sec x$ may be found by using the following technique:

$$\int \sec x \, dx = \int \sec x \, \frac{\sec x + \tan x}{\sec x + \tan x} \, dx$$

$$= \int \frac{\sec^2 x + \sec x \tan x}{\sec x + \tan x} \, dx.$$

If $v = \tan x + \sec x$, then $dv = (\sec^2 x + \sec x \tan x) \, dx$ and formal substitution gives us

$$\int \sec x \, dx = \int \frac{1}{v} \, dv$$

$$= \ln|v| + C$$

$$= \ln|\sec x + \tan x| + C.$$

The proofs of the remaining formulas are left as exercises. $\square$

Example 1   Evaluate $\int \sin 5x \, dx$.

Solution   If we let $u = 5x$, then $du = 5 \, dx$. Introducing the factor 5 in the integrand and compensating for this by multiplying the integral by $\frac{1}{5}$ we obtain

$$\int \sin 5x \, dx = \frac{1}{5} \int (\sin 5x)5 \, dx$$

$$= \frac{1}{5} \int \sin u \, du$$

$$= -\frac{1}{5} \cos u + C$$

$$= -\frac{1}{5} \cos 5x + C. \qquad \blacksquare$$

Find $\int x \cos x^2 \, dx$.

If $u = x^2$, then $du = 2x \, dx$ and the form $\int \cos u \, du$ may be obtained by introducing the factor 2 in the integrand. This leads to

$$\int x \cos x^2 \, dx = \frac{1}{2} \int \cos x^2 (2x) \, dx$$

$$= \frac{1}{2} \int \cos u \, du = \frac{1}{2} \sin u + C$$

$$= \frac{1}{2} \sin x^2 + C. \qquad \blacksquare$$

Example 3   Find $\int \sec x(\sec x + \tan x) \, dx$.

Solution   We may proceed as follows:

$$\int \sec x(\sec x + \tan x) \, dx = \int (\sec^2 x + \sec x \tan x) \, dx$$

$$= \int \sec^2 dx + \int \sec x \tan x \, dx$$

$$= \tan x + \sec x + C. \qquad \blacksquare$$

Example 4   Find $\int \left(\frac{1}{\sqrt{x}}\right) \csc^2 \sqrt{x} \, dx$.

Solution   If

$$u = \sqrt{x}, \quad \text{then} \quad du = \left(\frac{1}{2\sqrt{x}}\right) dx$$

and the integrand can be written in the form $\int \csc^2 u \, du$ by introducing the factor $\frac{1}{2}$. Thus,

$$\int \left(\frac{1}{\sqrt{x}}\right) \csc^2 \sqrt{x} \, dx = 2 \int \csc^2 \sqrt{x} \left(\frac{1}{2\sqrt{x}}\right) dx$$

$$= 2 \int \csc^2 u \, du = -2 \cot u + C$$

$$= -2 \cot \sqrt{x} + C. \qquad \blacksquare$$

Example 5 Evaluate $\displaystyle\int_0^{\pi/2} \tan \frac{x}{2} \, dx$.

Solution   If $u = x/2$, then $du = \frac{1}{2} \, dx$. Making this change of variable and noting that $u = 0$ if $x = 0$, and $u = \pi/4$ if $x = \pi/2$, we obtain

$$\int_0^{\pi/2} \tan \frac{x}{2} \, dx = 2 \int_0^{\pi/2} \tan \frac{x}{2} \cdot \frac{1}{2} \, dx$$

$$= 2 \int_0^{\pi/4} \tan u \, du = 2 \ln \sec u \Big]_0^{\pi/4}.$$

It is unnecessary to employ the absolute value sign as in Theorem (8.8) since $\sec u$ is positive if u is between 0 and $\pi/4$. Since $\ln \sec \pi/4 = \ln \sqrt{2} = (\frac{1}{2}) \ln 2$, and $\ln \sec 0 = \ln 1 = 0$, it follows that

$$\int_0^{\pi/2} \tan \frac{x}{2} \, dx = 2 \cdot \frac{1}{2} \ln 2 = \ln 2. \qquad \blacksquare$$

Example 6 Evaluate $\displaystyle\int e^{3x} \sec e^{3x} \, dx$.

Solution   If we let $u = e^{3x}$, then $du = 3e^{3x} \, dx$, and

$$\int e^{3x} \sec e^{3x} \, dx = \frac{1}{3} \int \sec e^{3x}(3e^{3x}) \, dx$$

$$= \frac{1}{3} \int \sec u \, du$$

$$= \frac{1}{3} \ln |\sec u + \tan u| + C$$

$$= \frac{1}{3} \ln |\sec e^{3x} + \tan e^{3x}| + C. \qquad \blacksquare$$

Example 7 Evaluate $\int (\csc x - 1)^2 \, dx$.

Solution $\int (\csc x - 1)^2 \, dx = \int (\csc^2 x - 2 \csc x + 1) \, dx$

$$= \int \csc^2 x \, dx - 2 \int \csc x \, dx + \int dx$$

$$= -\cot x - 2 \ln |\csc x - \cot x| + x + C. \quad \blacksquare$$

Find $\int \sqrt{\cos 5x} \, \sin 5x \, dx$.

If we let $u = \cos 5x$, then $du = -5 \sin 5x \, dx$ and

$$\int \sqrt{\cos 5x} \, \sin 5x \, dx = -\frac{1}{5} \int \sqrt{\cos 5x} (-5 \sin 5x) \, dx$$

$$= -\frac{1}{5} \int \sqrt{u} \, du = -\frac{1}{5} \int u^{1/2} \, du$$

$$= -\frac{1}{5} \left(\frac{2}{3} u^{3/2} \right) + C$$

$$= -\frac{2}{15} (\cos 5x)^{3/2} + C. \quad \blacksquare$$

Additional methods for integrating trigonometric expressions will be discussed in Chapter 9.

EXERCISES 8.3

Evaluate the integrals in Exercises 1–34.

1 $\int \cos 4x \, dx$

2 $\int \sec^2 5x \, dx$

3 $\int \tan 3x \sec 3x \, dx$

4 $\int x^2 \cot x^3 \csc x^3 \, dx$

5 $\int (\tan 3x + \sec 3x) \, dx$

6 $\int \dfrac{1}{\sec 2x} \, dx$

7 $\int \dfrac{1}{\cos 2x} \, dx$

8 $\int \dfrac{\cot \sqrt[3]{x}}{\sqrt[3]{x^2}} \, dx$

9 $\int x \csc^2 (x^2 + 1) \, dx$

10 $\int (x + \csc 8x) \, dx$

11 $\int \cot 6x \sin 6x \, dx$

12 $\int \sin 2x \tan 2x \, dx$

13 $\int \cos 3x \sqrt[3]{\sin 3x} \, dx$

14 $\int \dfrac{\sin 2x}{\sqrt{1 - \cos 2x}} \, dx$

15 $\int \sin x (1 + \sqrt{\cos x})^2 \, dx$

16 $\int \sin^3 x \cos x \, dx$

17 $\int \dfrac{\sin x}{\cos^2 x} \, dx$

18 $\int \dfrac{\cos x}{(1 - \sin x)^2} \, dx$

19 $\int_0^{\pi/4} \tan x \sec^2 x \, dx$

20 $\int \csc^2 x \cot x \, dx$

21 $\int \dfrac{\tan^2 2x}{\sec 2x} \, dx$

22 $\int_{\pi/6}^{\pi/2} \dfrac{\cos^2 x}{\sin x} \, dx$

23 $\int_{\pi/6}^{\pi} \sin x \cos x \, dx$

24 $\int \dfrac{\cos^2 x}{\csc x} \, dx$

25 $\int \dfrac{1 - \sin x}{x + \cos x}\,dx$

26 $\int \dfrac{e^x}{\cos e^x}\,dx$

27 $\int_{\pi/4}^{\pi/3} \dfrac{1 + \sin x}{\cos^2 x}\,dx$

28 $\int_0^{\pi/4} (1 + \sec x)^2\,dx$

29 $\int e^x (1 + \tan e^x)\,dx$

30 $\int (\csc^2 x) 2^{\cot x}\,dx$

31 $\int \dfrac{e^{\cos x}}{\csc x}\,dx$

32 $\int \dfrac{\tan e^{-3x}}{e^{3x}}\,dx$

33 $\int \dfrac{\sec^2 x}{2 \tan x + 1}\,dx$

34 $\int \dfrac{\sec x \tan x}{1 + 3 \sec x}\,dx$

35 Find the area of the region under the graph of $y = \sin(x/2)$ from $x = 0$ to $x = \pi$.

36 Find the area of the region under the graph of $y = 2 \tan x$ from $x = 0$ to $x = \pi/4$.

37 Find the area of the region bounded by the graphs of $y = \sec x$, $y = x$, $x = -\pi/4$, and $x = \pi/4$.

38 Find the area of the region bounded by the graphs of $y = \sin x$, $y = \cos x$, $x = -\pi/2$, and $x = \pi/6$.

39 The region bounded by the graphs of the equations $y = \sec x$, $x = -\pi/3$, $x = \pi/3$, and $y = 0$ is revolved about the x-axis. Find the volume of the resulting solid.

40 The region between the graphs of $y = \tan(x^2)$ and the x-axis, from $x = 0$ to $x = \sqrt{\pi/2}$, is revolved about the y-axis. Find the volume of the resulting solid.

41 Verify the formula for $\int \csc u\,du$.

42 Verify the formula for $\int \sec u \tan u\,du$

Verify the formulas in Exercises 43 and 44 by evaluating the integral in different ways. How can the answers be reconciled?

43 $\int \tan x \sec^2 x\,dx = \frac{1}{2} \tan^2 x + C = \frac{1}{2} \sec^2 x + D$

44 $\int \sin x \cos x\,dx = \frac{1}{2} \sin^2 x + C$

$= -\frac{1}{2} \cos^2 x + D = -\frac{1}{4} \cos 2x + E$

In Exercises 45 and 46, (a) set up an integral for finding the arc length L, from A to B, of the graph of the given equation; (b) write out the formula for approximating L by means of Simpson's Rule (5.35) with $n = 4$; (c) use a calculator to find the approximation given by the formula in part (b).

45 $y = \sec(x/2)$; $A(0, 1)$, $B(\pi/2, \sqrt{2})$

46 $y = \tan x$; $A(0, 0)$, $B(\pi/4, 1)$

47 The ends of a water trough 6 ft long have the shape of the graph of $y = \sec x$ from $x = -\pi/3$ to $x = \pi/3$, where x is in feet. Suppose the trough is full of water.

(a) Set up a definite integral with respect to y that equals the work required to pump the water out over the top.

(b) Change the integral in (a) to an integral with respect to x and use Simpson's Rule with $n = 4$ on the interval $[-\pi/3, \pi/3]$ to approximate the work required.

48 Given the full water trough described in Exercise 47, (a) set up a definite integral with respect to y that equals the total force on one of the ends and (b) change the integral in (a) to an integral with respect to x and use the Trapezoidal Rule with $n = 8$ on the interval $[-\pi/3, \pi/3]$ to approximate the total force.

In Exercises 49 and 50 use (a) the Trapezoidal Rule and (b) Simpson's Rule to approximate the definite integral for the stated value of n. Use approximations to four decimal places for $f(x_i)$ and round off answers to two decimal places.

49 $\int_0^\pi \sqrt{\sin x}\,dx$, $n = 6$

50 $\int_0^\pi \sin \sqrt{x}\,dx$, $n = 4$

51 In **seasonal population growth** the population $q(t)$ at time t increases during spring and summer, but decreases during the fall and winter. A differential equation that is sometimes used to describe this type of growth is $q'(t)/q(t) = k \sin 2\pi t$, where t is in years and $t = 0$ corresponds to the first day of spring. (a) Show that the population $q(t)$ is seasonal. (b) If $q_0 = q(0)$, find an explicit formula for $q(t)$.

8.4

THE INVERSE TRIGONOMETRIC FUNCTIONS

If f is a one-to-one function with domain Y and range X, then its inverse function f^{-1} has domain X and range Y. Moreover,

$$y = f^{-1}(x) \quad \text{if and only if} \quad f(y) = x$$

for every x in X and every y in Y. Since the trigonometric functions are not one-to-one, they do not have inverses. However, by restricting the domains it is possible to obtain functions that vary in the same way as the trigonometric functions (over the smaller domains), but do possess inverse functions.

Let us first consider the sine function whose domain is $\mathbb{R}$ and range is the closed interval $[-1, 1]$. The sine function is not one-to-one since, for example, numbers such as $\pi/6$, $5\pi/6$, and $-7\pi/6$ lead to the same functional value, $\frac{1}{2}$. It is easy to find a subset S of $\mathbb{R}$ with the property that if x ranges through S, then $\sin x$ takes on each value between -1 and 1 once and only once. It is convenient to choose the interval $[-\pi/2, \pi/2]$ for S. The new function obtained by restricting the domain of the sine function to $[-\pi/2, \pi/2]$ is continuous and increasing and hence, by Theorem (7.28), has an inverse function that is continuous and increasing on $[-1, 1]$. This leads to the following definition.

Definition (8.9)

> The **inverse sine function**, denoted $\sin^{-1}$, is defined by
>
> $$y = \sin^{-1} x \quad \text{if and only if} \quad \sin y = x$$
>
> where $-1 \leq x \leq 1$ and $-\pi/2 \leq y \leq \pi/2$.

It is also customary to refer to this function as the **arcsine function** and use $\arcsin x$ in place of $\sin^{-1} x$. We shall employ both notations in our work. The -1 in $\sin^{-1}$ is not to be regarded as an exponent, but rather as a means of denoting this inverse function. Observe that by Definition (8.9),

$$-\frac{\pi}{2} \leq \sin^{-1} x \leq \frac{\pi}{2} \quad \text{or} \quad -\frac{\pi}{2} \leq \arcsin x \leq \frac{\pi}{2}.$$

It follows from (8.9) that the graph of $y = \sin^{-1} x$ (or $y = \arcsin x$) may be found by sketching the graph of $x = \sin y$, where $-\pi/2 \leq y \leq \pi/2$ (see Figure 8.11).

Since $\sin$ and $\sin^{-1}$ are inverse functions of one another,

$$\sin^{-1} (\sin x) = x \quad \text{if} \quad -\frac{\pi}{2} \leq x \leq \frac{\pi}{2}$$

and

$$\sin (\sin^{-1} x) = x \quad \text{if} \quad -1 \leq x \leq 1.$$

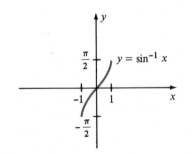

Figure 8.11

These formulas may also be written

$$\arcsin (\sin x) = x \quad \text{and} \quad \sin (\arcsin x) = x.$$

Example 1 Find $\sin^{-1}(\sqrt{2}/2)$ and $\arcsin (-\frac{1}{2})$.

Solution   If $y = \sin^{-1} (\sqrt{2}/2)$, then $\sin y = \sqrt{2}/2$ and consequently $y = \pi/4$. Note that it is essential to choose y in the interval $[-\pi/2, \pi/2]$. A number such as $3\pi/4$ is incorrect, even though $\sin (3\pi/4) = \sqrt{2}/2$.
In like manner, if $y = \arcsin (-\frac{1}{2})$, then $\sin y = -\frac{1}{2}$ and hence $y = -\pi/6$.
∎

Similar discussions may be given for the other trigonometric functions. If the domain of the cosine function is restricted to the interval $[0, \pi]$ we obtain a continuous, decreasing function that has a continuous decreasing inverse function by Theorem (7.29). This leads to the next definition.

Definition (8.10)

> The **inverse cosine function**, denoted by $\cos^{-1}$, is defined by
>
> $$y = \cos^{-1} x \quad \text{if and only if} \quad \cos y = x$$
>
> where $-1 \leq x \leq 1$ and $0 \leq y \leq \pi$.

The inverse cosine function is also referred to as the **arccosine function** and the notation $\arccos x$ is used interchangeably with $\cos^{-1} x$. Note that

$$\cos^{-1} (\cos x) = \arccos (\cos x) = x \quad \text{if} \quad 0 \leq x \leq \pi$$

$$\cos (\cos^{-1} x) = \cos (\arccos x) = x \quad \text{if} \quad -1 \leq x \leq 1.$$

If we restrict the domain of the tangent function to the open interval $(-\pi/2, \pi/2)$, then a one-to-one function is obtained. We may, therefore, adopt the following definition.

Definition (8.11)

> The **inverse tangent function** or **arctangent function**, denoted by $\tan^{-1}$ or arctan, is defined by
>
> $$y = \tan^{-1} x = \arctan x \quad \text{if and only if} \quad \tan y = x$$
>
> where x is any real number and $-\pi/2 < y < \pi/2$.

The domain of the arctangent function is $\mathbb{R}$ and the range is the open interval $(-\pi/2, \pi/2)$.

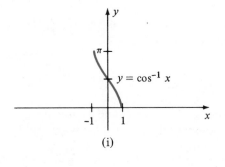

(i)

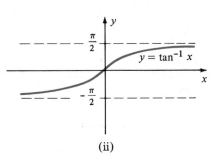

(ii)

Figure 8.12

The graphs of the inverse cosine and inverse tangent functions are sketched in Figure 8.12.

Example 2   Find sec (arctan $\frac{2}{3}$) without using tables or calculators.

Solution   If $y = \arctan \frac{2}{3}$, then $\tan y = \frac{2}{3}$. Since $\sec^2 y = 1 + \tan^2 y$ and $0 < y < \pi/2$,

$$\sec y = \sqrt{1 + \tan^2 y} = \sqrt{1 + \left(\frac{2}{3}\right)^2}$$

$$= \sqrt{1 + \frac{4}{9}} = \sqrt{\frac{13}{9}} = \frac{\sqrt{13}}{3}.$$

Hence, sec (arctan $\frac{2}{3}$) = sec $y = \sqrt{13}/3$.   ∎

If y varies through the two intervals $[0, \pi/2)$ and $[\pi, 3\pi/2)$, then sec y takes on each of its values once and only once and an inverse function may be defined.

Definition (8.12)

> The **inverse secant** or **arcsecant function**, denoted by $\sec^{-1}$ or arcsec, is defined by
>
> $$y = \sec^{-1} x = \text{arcsec } x \quad \text{if and only if} \quad \sec y = x$$
>
> where $|x| \geq 1$ and y is in $[0, \pi/2)$ or in $[\pi, 3\pi/2)$.

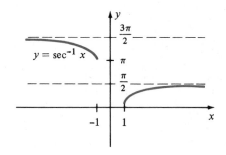

Figure 8.13

The graph of $y = \sec^{-1} x$ is sketched in Figure 8.13. The main reason for choosing y as in Definition (8.12) instead of the more natural intervals $[0, \pi/2)$ and $(\pi/2, \pi]$ is that the differentiation formula for the inverse secant is simpler.

The following examples illustrate some of the manipulations that may be carried out with inverse trigonometric functions.

Example 3   If $-1 \leq x \leq 1$, rewrite cos (sin^{-1} x) as an algebraic expression in x.

Solution   Let $y = \sin^{-1} x$, so that $\sin y = x$. We wish to express cos y in terms of x. Since $-\pi/2 \leq y \leq \pi/2$, it follows that cos $y \geq 0$, and hence

$$\cos y = \sqrt{1 - \sin^2 y} = \sqrt{1 - x^2}.$$

Consequently, $\cos (\sin^{-1} x) = \sqrt{1 - x^2}$.   ∎

Example 4 Verify the identity

$$\frac{1}{2}\cos^{-1} x = \tan^{-1}\sqrt{\frac{1 - x}{1 + x}}$$

where $|x| < 1$.

Solution　Let $y = \cos^{-1} x$. We wish to show that

$$\frac{y}{2} = \tan^{-1}\sqrt{\frac{1 - x}{1 + x}}.$$

By a half-angle formula,

$$\left|\tan \frac{y}{2}\right| = \sqrt{\frac{1 - \cos y}{1 + \cos y}}.$$

Since $y = \cos^{-1} x$ and $|x| < 1$, it follows that $0 < y < \pi$ and hence that $0 < (y/2) < \pi/2$. Consequently, $\tan (y/2) > 0$, and we may drop the absolute value sign, obtaining

$$\tan \frac{y}{2} = \sqrt{\frac{1 - \cos y}{1 + \cos y}}.$$

Using the fact that $\cos y = x$ gives us

$$\tan \frac{y}{2} = \sqrt{\frac{1 - x}{1 + x}}$$

or, equivalently,　　　$$\frac{y}{2} = \tan^{-1}\sqrt{\frac{1 - x}{1 + x}}$$

which is what we wished to show.　　　　　　　■

EXERCISES 8.4

Find the exact values in Exercises 1–18 without the use of tables or calculators.

1 (a) $\sin^{-1} \sqrt{3}/2$
　(b) $\sin^{-1} (-\sqrt{3}/2)$

2 (a) $\sin^{-1} 0$
　(b) $\arccos 0$

3 (a) $\cos^{-1} \sqrt{2}/2$
　(b) $\cos^{-1} (-\sqrt{2}/2)$

4 (a) $\arcsin (-1)$
　(b) $\cos^{-1} (-1)$

5 (a) $\tan^{-1} \sqrt{3}$
　(b) $\arctan (-\sqrt{3})$

6 (a) $\tan^{-1} (-1)$
　(b) $\arccos (1/2)$

7 $\sin [\cos^{-1} \sqrt{3}/2]$

8 $\cos [\sin^{-1} 0]$

9 $\sin (\arccos \frac{3}{5})$

10 $\tan [\tan^{-1} 10]$

11 $\arcsin (\sin \sqrt{5})$

12 $\tan^{-1} (\cos 0)$

13 $\cos [\sin^{-1} \frac{3}{5} + \tan^{-1} \frac{4}{3}]$

14 $\sin [\arcsin \frac{1}{2} + \arccos 0]$

15 $\tan [\arctan \frac{3}{4} + \arccos \frac{3}{5}]$

16 $\cos [2 \sin^{-1} \frac{8}{17}]$

17 $\sin\left[2\arccos\left(-\tfrac{4}{5}\right)\right]$

18 $\sin\left(\arctan\tfrac{1}{2} - \arccos\tfrac{4}{5}\right)$

Rewrite the expressions in Exercises 19–22 as algebraic expressions in x.

19 $\sin\left(\tan^{-1} x\right)$ **20** $\tan\left(\arccos x\right)$

21 $\cos\left(\tfrac{1}{2}\arccos x\right)$ **22** $\cos\left(2\tan^{-1} x\right)$

Verify the identities in Exercises 23–30.

23 $\sin^{-1} x + \cos^{-1} x = \pi/2$ (*Hint*: Let $\alpha = \sin^{-1} x$, $\beta = \cos^{-1} x$, and consider $\sin(\alpha + \beta)$.)

24 $\arctan x + \arctan(1/x) = \pi/2,\ x > 0$

25 $\arcsin\dfrac{2x}{1 + x^2} = 2\arctan x,\ |x| \le 1$

26 $2\cos^{-1} x = \cos^{-1}(2x^2 - 1),\ 0 \le x \le 1$

27 $\sin^{-1}(-x) = -\sin^{-1} x$

28 $\arccos(-x) = \pi - \arccos x$

29 $\sin^{-1} x = \tan^{-1}\dfrac{x}{\sqrt{1 - x^2}}$

30 $\tan^{-1} x + \tan^{-1} y = \tan^{-1}\dfrac{x + y}{1 - xy}$, where $|x| < 1$ and $|y| < 1$.

31 Define $\cot^{-1}$ by restricting the domain of cot to the interval $(0, \pi)$.

32 Define $\csc^{-1}$ by restricting the domain of csc to $(0, \pi/2]$ and $(\pi, 3\pi/2]$.

Sketch the graphs of the equations in Exercises 33–42.

33 $y = \sin^{-1} 2x$ **34** $y = \cos^{-1}(x/2)$

35 $y = \tfrac{1}{2}\sin^{-1} x$ **36** $y = 2\cos^{-1} x$

37 $y = 2\tan^{-1} x$ **38** $y = \tan^{-1} 2x$

39 $y = \sin(\sin^{-1} x)$ **40** $y = \sin(\arccos x)$

41 $y = \sin^{-1}(\sin x)$ **42** $y = \arccos(\cos x)$

Show that the equations in Exercises 43 and 44 are *not* identities.

43 $\tan^{-1} x = 1/\tan x$

44 $(\arcsin x)^2 + (\arccos x)^2 = 1$.

8.5

DERIVATIVES AND INTEGRALS INVOLVING INVERSE TRIGONOMETRIC FUNCTIONS

In this section we shall concentrate on the inverse sine, cosine, tangent, and secant functions. Formulas for their derivatives and for integrals that result in inverse trigonometric functions are listed in the next three theorems, where $u = g(x)$ and x is restricted to values for which the indicated expressions are defined.

Theorem (8.13)

$$D_x \sin^{-1} u = \frac{1}{\sqrt{1 - u^2}}\, D_x u \qquad D_x \cos^{-1} u = -\frac{1}{\sqrt{1 - u^2}}\, D_x u$$

$$D_x \tan^{-1} u = \frac{1}{1 + u^2}\, D_x u \qquad D_x \sec^{-1} u = \frac{1}{u\sqrt{u^2 - 1}}\, D_x u$$

Proof   As usual, we shall consider the special case $u = x$, since the general formula may then be obtained by applying the Chain Rule.

If we let $f(x) = \sin x$ and $g(x) = \sin^{-1} x$ in Theorem (7.30), then it follows that the inverse sine function g is differentiable if $|x| < 1$. We shall use implicit differentiation to find $g'(x)$. First note that by Definition (8.9),

$$y = \sin^{-1} x \quad \text{if and only if} \quad \sin y = x$$

where $-1 < x < 1$, and $-\pi/2 < y < \pi/2$. Differentiating the last equation implicitly,

$$\cos y \, D_x y = 1$$

and hence

$$D_x y = D_x \sin^{-1} x = \frac{1}{\cos y}.$$

Since $-\pi/2 < y < \pi/2$, $\cos y$ is positive and, therefore,

$$\cos y = \sqrt{1 - \sin^2 y} = \sqrt{1 - x^2}.$$

Substitution in the previous formula for $D_x y$ gives us

$$D_x \sin^{-1} x = \frac{1}{\sqrt{1 - x^2}}$$

provided $|x| < 1$. The inverse sine function is not differentiable at ± 1. This fact is geometrically evident from Figure 8.11, since vertical tangent lines occur at the endpoints of the graph.

In similar fashion, to find the derivative of the inverse cosine function we begin with the equivalent equations

$$y = \cos^{-1} x \quad \text{and} \quad \cos y = x$$

where $|x| < 1$ and $0 < y < \pi$. Differentiating the second equation implicitly gives us

$$-\sin y \, D_x y = 1$$

and, therefore,

$$D_x y = D_x \cos^{-1} x = -\frac{1}{\sin y}.$$

Since $0 < y < \pi$, $\sin y$ is positive, and $\sin y = \sqrt{1 - \cos^2 y} = \sqrt{1 - x^2}$. Consequently, if $|x| < 1$, then

$$D_x \cos^{-1} x = -\frac{1}{\sqrt{1 - x^2}}.$$

It follows from Theorem (7.30) that the inverse tangent function is differentiable at every real number. If we consider the equivalent equations

$$y = \tan^{-1} x \quad \text{and} \quad \tan y = x$$

where $-\pi/2 < y < \pi/2$, and differentiate the second equation implicitly, the result is

$$\sec^2 y \, D_x y = 1.$$

Consequently,

$$D_x y = D_x \tan^{-1} x = \frac{1}{\sec^2 y}.$$

Using the fact that $\sec^2 y = 1 + \tan^2 y = 1 + x^2$ gives us

$$D_x \tan^{-1} x = \frac{1}{1 + x^2}.$$

Finally, consider the equivalent equations

$$y = \sec^{-1} x \quad \text{and} \quad \sec y = x$$

where y is either in $[0, \pi/2)$ or $[\pi, 3\pi/2)$. Differentiating the second equation implicitly gives us

$$\sec y \tan y \, D_x y = 1$$

or, if $\tan y \neq 0$,

$$D_x y = D_x \sec^{-1} x = \frac{1}{\sec y \tan y}.$$

Using the fact that $\tan y = \sqrt{\sec^2 y - 1} = \sqrt{x^2 - 1}$, we have

$$D_x \sec^{-1} x = \frac{1}{x\sqrt{x^2 - 1}}$$

provided $|x| > 1$. The inverse secant function is not differentiable at $x = \pm 1$. Indeed, the graph has vertical tangent lines at the points with these x-coordinates (see Figure 8.13).   □

Example 1   Find dy/dx if $y = \sin^{-1} 3x - \cos^{-1} 3x$.

Solution   Applying Theorem (8.13) with $u = 3x$,

$$\frac{dy}{dx} = \frac{3}{\sqrt{1 - 9x^2}} + \frac{3}{\sqrt{1 - 9x^2}} = \frac{6}{\sqrt{1 - 9x^2}}. \qquad \blacksquare$$

Example 2   Find $f'(x)$ if $f(x) = \tan^{-1} e^{2x}$.

Solution   Using (8.13) with $u = e^{2x}$,

$$f'(x) = \frac{1}{1 + (e^{2x})^2} D_x e^{2x} = \frac{2e^{2x}}{1 + e^{4x}}. \qquad \blacksquare$$

Example 3   Find y' if $y = \sec^{-1}(x^2)$.

Solution   Applying Theorem (8.13) with $u = x^2$,

$$y' = \frac{1}{x^2\sqrt{x^4 - 1}} (2x) = \frac{2}{x\sqrt{x^4 - 1}}. \qquad \blacksquare$$

Following the same pattern employed many times before, we may use the differentiation formulas for the inverse trigonometric functions to obtain integration formulas. The proof of the next theorem is left to the reader.

Theorem (8.14)

$$\int \frac{1}{\sqrt{1 - u^2}} \, du = \sin^{-1} u + C$$

$$\int \frac{1}{1 + u^2} \, du = \tan^{-1} u + C$$

$$\int \frac{1}{u\sqrt{u^2 - 1}} \, du = \sec^{-1} u + C$$

Example 4 Evaluate $\int \frac{e^{2x}}{\sqrt{1 - e^{4x}}} \, dx$.

Solution If we let $u = e^{2x}$, then $du = 2e^{2x} \, dx$ and the integral may be written as in the first formula of Theorem (8.14). Thus,

$$\int \frac{e^{2x}}{\sqrt{1 - e^{4x}}} \, dx = \frac{1}{2} \int \frac{2e^{2x}}{\sqrt{1 - (e^{2x})^2}} \, dx$$

$$= \frac{1}{2} \int \frac{1}{\sqrt{1 - u^2}} \, du$$

$$= \frac{1}{2} \sin^{-1} u + C$$

$$= \frac{1}{2} \sin^{-1} e^{2x} + C.$$ ∎

The formulas in Theorem (8.14) can be extended as follows, where a is a nonzero real number.

Theorem (8.15)

$$\int \frac{1}{\sqrt{a^2 - u^2}} \, du = \sin^{-1} \frac{u}{a} + C$$

$$\int \frac{1}{a^2 + u^2} \, du = \frac{1}{a} \tan^{-1} \frac{u}{a} + C$$

$$\int \frac{1}{u\sqrt{u^2 - a^2}} \, du = \frac{1}{a} \sec^{-1} \frac{u}{a} + C$$

Proof   Let us prove the second formula. If $u = g(x)$ and g is differentiable, then by Theorem (8.13),

$$D_x\left[\frac{1}{a}\tan^{-1}\frac{g(x)}{a}\right] = \frac{1}{a}\frac{1}{1 + [g(x)/a]^2}D_x\left[\frac{1}{a}g(x)\right]$$

$$= \frac{1}{a}\frac{a^2}{a^2 + [g(x)]^2}\frac{1}{a}g'(x)$$

$$= \frac{1}{a^2 + [g(x)]^2}g'(x).$$

Consequently,

$$\int\frac{1}{a^2 + [g(x)]^2}g'(x)\,dx = \frac{1}{a}\tan^{-1}\frac{g(x)}{a} + C = \frac{1}{a}\tan^{-1}\frac{u}{a} + C.$$

The verifications of the remaining formulas are left as exercises.   □

Example 5   Evaluate $\displaystyle\int\frac{x^2}{5 + x^6}\,dx$.

Solution   If we let $u = x^3$, then $du = 3x^2\,dx$ and the integral may be written as in the second formula of Theorem (8.15). Thus,

$$\int\frac{x^2}{5 + x^6}\,dx = \frac{1}{3}\int\frac{3x^2}{5 + (x^3)^2}\,dx$$

$$= \frac{1}{3}\int\frac{1}{(\sqrt{5})^2 + u^2}\,du$$

$$= \frac{1}{3}\cdot\frac{1}{\sqrt{5}}\tan^{-1}\frac{u}{\sqrt{5}} + C$$

$$= \frac{\sqrt{5}}{15}\tan^{-1}\frac{x^3}{\sqrt{5}} + C. \qquad\blacksquare$$

Example 6   Evaluate $\displaystyle\int\frac{1}{x\sqrt{x^4 - 9}}\,dx$.

Solution   If we let $u = x^2$, then $du = 2x\,dx$ and the integral may be transformed into the proper form as follows:

$$\int\frac{1}{x\sqrt{x^4 - 9}}\,dx = \frac{1}{2}\int\frac{1}{x^2\sqrt{(x^2)^2 - 9}}2x\,dx$$

$$= \frac{1}{2}\int\frac{1}{u\sqrt{u^2 - 9}}\,du$$

$$= \frac{1}{2}\cdot\frac{1}{3}\sec^{-1}\frac{u}{3} + C$$

$$= \frac{1}{6}\sec^{-1}\frac{x^2}{3} + C. \qquad\blacksquare$$

EXERCISES 8.5

In Exercises 1–24 find $f'(x)$ if $f(x)$ equals the given expression.

1 $\tan^{-1}(3x - 5)$

2 $\sin^{-1}(x/3)$

3 $\sin^{-1}\sqrt{x}$

4 $\tan^{-1}x^2$

5 $e^{-x}\operatorname{arcsec} e^{-x}$

6 $\sqrt{\operatorname{arcsec} 3x}$

7 $x^2\arctan x^2$

8 $\tan^{-1}\sin 2x$

9 $(1 + \cos^{-1}3x)^3$

10 $x^2\sec^{-1}5x$

11 $\ln\arctan x^2$

12 $\arcsin\ln x$

13 $1/\sin^{-1}x$

14 $\arctan\dfrac{x + 1}{x - 1}$

15 $\sec^{-1}\sqrt{x^2 - 1}$

16 $\left(\dfrac{1}{x} - \arcsin\dfrac{1}{x}\right)^4$

17 $(\arctan x)/(x^2 + 1)$

18 $\cos^{-1}\cos e^x$

19 $\sqrt{x}\sec^{-1}\sqrt{x}$

20 $e^{2x}/(\sin^{-1}5x)$

21 $3^{\arcsin x^3}$

22 $x\arccos\sqrt{4x + 1}$

23 $(\tan x)^{\arctan x}$

24 $(\tan^{-1}4x)e^{\tan^{-1}4x}$

Find y' in Exercises 25 and 26.

25 $x^2 + x\sin^{-1}y = ye^x$

26 $\ln(x + y) = \tan^{-1}xy$

Evaluate the integrals in Exercises 27–42.

27 $\displaystyle\int_0^4 \frac{1}{x^2 + 16}\,dx$

28 $\displaystyle\int_0^1 \frac{e^x}{1 + e^{2x}}\,dx$

29 $\displaystyle\int_0^{\sqrt{2}/2} \frac{x}{\sqrt{1 - x^4}}\,dx$

30 $\displaystyle\int_{2/\sqrt{3}}^2 \frac{1}{x\sqrt{x^2 - 1}}\,dx$

31 $\displaystyle\int \frac{\sin x}{\cos^2 x + 1}\,dx$

32 $\displaystyle\int \frac{\cos x}{\sqrt{9 - \sin^2 x}}\,dx$

33 $\displaystyle\int \frac{1}{\sqrt{x}(1 + x)}\,dx$

34 $\displaystyle\int \frac{1}{e^x\sqrt{1 - e^{-2x}}}\,dx$

35 $\displaystyle\int \frac{e^x}{\sqrt{16 - e^{2x}}}\,dx$

36 $\displaystyle\int \frac{\sec x\tan x}{1 + \sec^2 x}\,dx$

37 $\displaystyle\int \frac{1}{x\sqrt{x^6 - 4}}\,dx$

38 $\displaystyle\int \frac{x}{\sqrt{36 - x^2}}\,dx$

39 $\displaystyle\int \frac{x}{x^2 + 9}\,dx$

40 $\displaystyle\int \frac{1}{x\sqrt{x - 1}}\,dx$

41 $\displaystyle\int \frac{1}{\sqrt{e^{2x} - 25}}\,dx$

42 $\displaystyle\int \frac{e^x}{\sqrt{4 - e^x}}\,dx$

43 Find the area of the region bounded by the graphs of the equations $y = 4/\sqrt{16 - x^2}$, $x = -2$, $x = 2$, and $y = 0$.

44 If $f(x) = x^2/(1 + x^6)$, find the area of the region under the graph of f from $x = 0$ to $x = 1$.

45 The floor of a storage shed has the shape of a right triangle. The sides opposite and adjacent to an acute angle θ of the triangle are measured as 10 ft and 7 ft, respectively, with a possible error of $\frac{1}{2}$ inch in the 10-foot measurement. Use the differential of an inverse trigonometric function to approximate the error in the calculated value of θ.

46 Use differentials to approximate the change in $\arcsin x$ if x changes from 0.25 to 0.26.

47 An airplane at an altitude of 5 mi and a speed of 500 mi/hr is flying in a direction away from an observer on the ground. Use inverse trigonometric functions to find the rate at which the angle of elevation is changing when the airplane is over a point 2 mi from the observer.

48 A searchlight located $\frac{1}{8}$ mi from the nearest point P on a straight road is trained on an automobile traveling on the road at a rate of 50 mi/hr. Use inverse trigonometric functions to find the rate at which the searchlight is rotating when the car is $\frac{1}{4}$ mi from P.

49 A billboard 20 ft high is located on top of a building, with its lower edge 60 ft above the level of a viewer's eye. Use inverse trigonometric functions to find how far from a point directly below the sign a viewer should stand to maximize the angle between the lines of sight of the top and bottom of the billboard (cf. Exercise 67 of Section 8.2).

50 Given points $A(3, 1)$ and $B(6, 4)$ on a rectangular coordinate system, find the x-coordinate of the point P on the x-axis such that angle APB has its largest value.

51 If $f(x) = \sin^{-1}(\sin x)$ find $f'(x)$, $f''(x)$, and the local extrema of f. Discuss concavity and sketch the graph of f.

52 Work Exercise 51 if $f(x) = \arccos(\cos x)$.

53 Find equations for the tangent and normal lines to the graph of $y = \sin^{-1}(x - 1)$ at the point $(\frac{3}{2}, \pi/6)$.

54 Find the points on the graph of $y = \tan^{-1}2x$ at which the tangent line is parallel to the line $13y - 2x + 5 = 0$.

55 Find the intervals in which the graph of $y = \tan^{-1}x$ is (a) concave upward; (b) concave downward.

56 The velocity, at time t, of a point moving on a coordinate line is $(1 + t^2)^{-1}$ ft/sec. If the point is at the origin at $t = 0$, find its position at the instant that the acceleration and velocity have the same absolute value.

57 The region bounded by the graphs of the equations $y = e^x$, $y = 1/\sqrt{x^2 + 1}$, and $x = 1$ is revolved about the x-axis. Find the volume of the resulting solid.

58 Use differentials to approximate the arc length of the graph of $y = \tan^{-1} x$ from the point $(0, 0)$ to the point $(0.1, \tan^{-1} 0.1)$.

59 A missile is fired vertically from a point that is 5 mi from a tracking station and at the same elevation. For the first 20 seconds of flight its angle of elevation changes at a constant rate of 2° per second. Use inverse trigonometric functions to find the velocity of the missile when the angle of elevation is 30° (cf. Exercise 66, Section 8.2).

60 Prove that $\int_0^1 [4/(1 + x^2)]\, dx = \pi$ and then apply Simpson's Rule with $n = 10$ to approximate π.

61 Verify the formula for $\displaystyle\int \frac{1}{\sqrt{a^2 - u^2}}\, du$.

62 Verify the formula for $\displaystyle\int \frac{1}{u\sqrt{u^2 - a^2}}\, du$.

63 As blood flows through a blood vessel there is a loss of energy due to friction. According to **Poiseuille's Law**, this energy loss E is given by $E = kl/r^4$, where k is a constant, and where r and l are the radius and length,

respectively, of the blood vessel. Suppose a blood vessel of radius r_2 and length l_2 branches off, at an angle θ, from a blood vessel of radius r_1 and length l_1, as illustrated in the accompanying figure, where the colored arrows indicate the direction of blood flow. The energy loss is then the sum of the individual energy losses; that is,

$$E = (kl_1/r_1^4) + (kl_2/r_2^4).$$

Express l_1 and l_2 in terms of a, b, and θ, and find the angle that minimizes the energy loss.

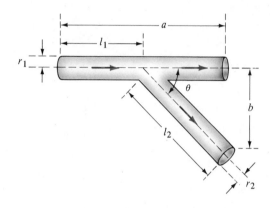

Figure for Exercise 63

8.6

THE HYPERBOLIC FUNCTIONS

The exponential expressions

$$\frac{e^x - e^{-x}}{2} \quad \text{and} \quad \frac{e^x + e^{-x}}{2}$$

occur frequently in applied mathematics and engineering. Their behavior is, in many ways, similar to that of $\sin x$ and $\cos x$. For reasons to be discussed later in this section, the expressions are called the *hyperbolic sine* and the *hyperbolic cosine* of x and are used to define the following **hyperbolic functions**.

Definition (8.16)

> The **hyperbolic sine function**, denoted by **sinh**, and the **hyperbolic cosine function**, denoted **cosh**, are defined by
>
> $$\sinh x = \frac{e^x - e^{-x}}{2} \quad \text{and} \quad \cosh x = \frac{e^x + e^{-x}}{2}$$
>
> where x is any real number.

The graph of $y = \cosh x$ may be found by the method called **addition of ordinates**. To use this technique, the graphs of $y = \frac{1}{2}e^x$ and $y = \frac{1}{2}e^{-x}$ are sketched as indicated by the dashes in Figure 8.14. The y-coordinates of points on the graph of $y = \cosh x$ may then be obtained by adding y-coordinates of points on the other two graphs. This results in the sketch shown in the figure. It is evident from the graph that the range of cosh is $[1, \infty)$. This can also be proved directly.

The graph of $y = \sinh x$ may be obtained by adding y-coordinates of the graphs of $y = \frac{1}{2}e^x$ and $y = -\frac{1}{2}e^{-x}$ as illustrated in Figure 8.15.

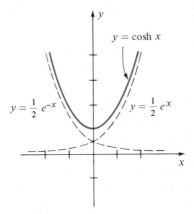

Figure 8.14

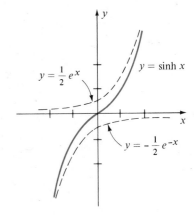

Figure 8.15

The hyperbolic cosine function can be used to describe the shape of a uniform flexible cable, or chain, whose ends are supported from the same height. This is often the case for telephone or power lines, as illustrated in Figure 8.16. At first glance the shape of the cable has the general appearance of a parabola, but this is not the case. If we introduce a coordinate system as indicated in the figure, then it can be shown that an equation that corresponds to the shape of the cable is $y = a \cosh (x/a)$, where a is a real number. The graph is called a **catenary**, after the Latin word for *chain*.

Another application of the hyperbolic cosine function occurs in the analysis of motion in a resisting medium. If an object is dropped from a given height, and if air resistance is neglected, then the distance y that it falls in t seconds is $y = \frac{1}{2}gt^2$, where g is a gravitational constant. However, air resistance cannot always be neglected. Indeed, as the velocity of the object increases, air resistance may significantly affect its motion. For example, if the air resistance is directly proportional to the square of the velocity, then it can be shown that the distance y that the object falls in t seconds is

$$y = A \ln (\cosh Bt)$$

where A and B are constants.

Many identities similar to those for trigonometric functions hold for the hyperbolic sine and cosine functions. For example, if $\cosh^2 x$ and $\sinh^2 x$ denote $(\cosh x)^2$ and $(\sinh x)^2$, respectively, we have the following.

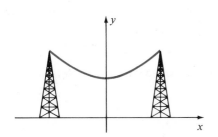

Figure 8.16 Catenary; $y = a \cosh \dfrac{x}{a}$

Theorem **(8.17)**

$$\boxed{\cosh^2 x - \sinh^2 x = 1}$$

Proof   By Definition (8.16),

$$\cosh^2 x - \sinh^2 x = \left(\frac{e^x + e^{-x}}{2}\right)^2 - \left(\frac{e^x - e^{-x}}{2}\right)^2$$

$$= \frac{e^{2x} + 2 + e^{-2x}}{4} - \frac{e^{2x} - 2 + e^{-2x}}{4}$$

$$= \frac{e^{2x} + 2 + e^{-2x} - e^{2x} + 2 - e^{-2x}}{4}$$

$$= \frac{4}{4} = 1. \qquad \square$$

Theorem (8.17) is analogous to the identity $\cos^2 x + \sin^2 x = 1$. Many other hyperbolic identities may be established. Some will be found in the exercises. In each case it is sufficient to express the hyperbolic functions in terms of exponential functions and show that one side of the equation can be transformed into the other. The hyperbolic identities are similar to (but not always identical to) certain trigonometric identities. Any differences that occur usually involve signs of terms.

An interesting geometric relationship exists between the points $(\cos t, \sin t)$ and $(\cosh t, \sinh t)$, where t is any real number. Let us consider the graphs of $x^2 + y^2 = 1$ and $x^2 - y^2 = 1$ sketched in Figure 8.17. The graph in (i) of the figure is a unit circle with center at the origin. The graph in (ii) is called a *hyperbola*. (Hyperbolas and their properties will be discussed in detail in Chapter 12.) Note first that since $\cos^2 t + \sin^2 t = 1$, the point $P(\cos t, \sin t)$ is on the circle $x^2 + y^2 = 1$. Next, by Theorem (8.17), $\cosh^2 t - \sinh^2 t = 1$, and hence the point $P(\cosh t, \sinh t)$ is on the hyperbola $x^2 - y^2 = 1$. These are the reasons for referring to cos and sin as *circular* functions (see Appendix III) and to cosh and sinh as *hyperbolic* functions.

There is another analogy between the graphs in Figure 8.17. If $0 < t < \pi/2$, then t is the radian measure of angle *POB* shown in (i) of the figure. By Theorem (8.3), the area A of the shaded circular sector is $A = (\frac{1}{2})(1)^2 t = t/2$

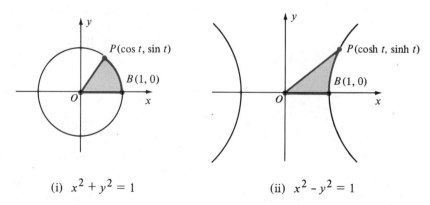

(i) $x^2 + y^2 = 1$ (ii) $x^2 - y^2 = 1$

Figure 8.17

and hence $t = 2A$. It can be shown that if $P(\cosh t, \sinh t)$ is the point in (ii) of Figure 8.17, then $t = 2A$, where A is the area of the shaded *hyperbolic* sector (see Exercise 53).

The striking analogies between the trigonometric and hyperbolic sine and cosine make it natural to introduce functions that correspond to the four remaining trigonometric functions. Specifically, the hyperbolic tangent, cotangent, secant, and cosecant functions, denoted by tanh, coth, sech, and csch, respectively, are defined as follows.

Definition (8.18)

$$\tanh x = \frac{\sinh x}{\cosh x} = \frac{e^x - e^{-x}}{e^x + e^{-x}}$$

$$\coth x = \frac{\cosh x}{\sinh x} = \frac{e^x + e^{-x}}{e^x - e^{-x}}, \quad x \neq 0$$

$$\operatorname{sech} x = \frac{1}{\cosh x} = \frac{2}{e^x + e^{-x}}$$

$$\operatorname{csch} x = \frac{1}{\sinh x} = \frac{2}{e^x - e^{-x}}, \quad x \neq 0$$

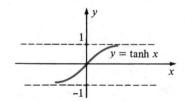

$y = \tanh x$

Figure 8.18

The graph of $y = \tanh x$ is sketched in Figure 8.18. The lines $y = 1$ and $y = -1$ are horizontal asymptotes. We leave the verification of this, and the sketching of the graphs of the remaining hyperbolic functions, as exercises.

If we divide both sides of (8.17) by $\cosh^2 x$ and use the definitions of tanh x and sech x, we obtain the first formula of the next theorem. The second formula may be obtained by dividing both sides of (8.17) by $\sinh^2 x$.

Theorem (8.19)

$$1 - \tanh^2 x = \operatorname{sech}^2 x$$
$$\coth^2 x - 1 = \operatorname{csch}^2 x$$

Note the differences between these and the analogous identities studied in trigonometry.

Derivative formulas for the hyperbolic functions are listed in the next theorem, where $u = g(x)$, and g is differentiable.

Theorem (8.20)

$$D_x \sinh u = \cosh u \, D_x u \qquad\qquad D_x \cosh u = \sinh u \, D_x u$$

$$D_x \tanh u = \operatorname{sech}^2 u \, D_x u \qquad\qquad D_x \coth u = -\operatorname{csch}^2 u \, D_x u$$

$$D_x \operatorname{sech} u = -\operatorname{sech} u \tanh u \, D_x u \qquad\qquad D_x \operatorname{csch} u = -\operatorname{csch} u \coth u \, D_x u$$

Proof   As usual, we consider only the case $u = x$. Since $D_x e^x = e^x$ and $D_x e^{-x} = -e^{-x}$,

$$D_x \sinh x = D_x\left(\frac{e^x - e^{-x}}{2}\right) = \frac{e^x + e^{-x}}{2} = \cosh x$$

$$D_x \cosh x = D_x\left(\frac{e^x + e^{-x}}{2}\right) = \frac{e^x - e^{-x}}{2} = \sinh x.$$

To differentiate $\tanh x$ we apply the Quotient Rule as follows.

$$D_x \tanh x = D_x \frac{\sinh x}{\cosh x}$$

$$= \frac{\cosh x\, D_x \sinh x - \sinh x\, D_x \cosh x}{\cosh^2 x}$$

$$= \frac{\cosh^2 x - \sinh^2 x}{\cosh^2 x}$$

$$= \frac{1}{\cosh^2 x} = \operatorname{sech}^2 x$$

The proofs of the remaining formulas are left as exercises.   □

Example 1   Find $f'(x)$ if $f(x) = \cosh(x^2 + 1)$.

Solution   Applying Theorem (8.20), with $u = x^2 + 1$,

$$f'(x) = \sinh(x^2 + 1) \cdot D_x(x^2 + 1)$$

$$= 2x \sinh(x^2 + 1).$$ ∎

The integration formulas corresponding to the derivative formulas in Theorem (8.20) are as follows.

Theorem (8.21)

$$\int \sinh u\, du = \cosh u + C \qquad \int \cosh u\, du = \sinh u + C$$

$$\int \operatorname{sech}^2 u\, du = \tanh u + C \qquad \int \operatorname{csch}^2 u\, du = -\coth u + C$$

$$\int \operatorname{sech} u \tanh u\, du = -\operatorname{sech} u + C \qquad \int \operatorname{csch} u \coth u\, du = -\operatorname{csch} u + C$$

Example 2   Evaluate $\int x^2 \sinh (x^3)\, dx$.

Solution   If we let $u = x^3$, then $du = 3x^2\, dx$ and

$$\int x^2 \sinh (x^3)\, dx = \frac{1}{3} \int \sinh (x^3)(3x^2)\, dx$$

$$= \frac{1}{3} \int \sinh u \, du = \frac{1}{3} \cosh u + C$$

$$= \frac{1}{3} \cosh x^3 + C. \qquad \blacksquare$$

EXERCISES 8.6

Verify the identities in Exercises 1–14.

1 $\cosh x + \sinh x = e^x$

2 $\cosh x - \sinh x = e^{-x}$

3 $\sinh (-x) = -\sinh x$

4 $\cosh (-x) = \cosh x$

5 $\sinh (x + y) = \sinh x \cosh y + \cosh x \sinh y$

6 $\cosh (x + y) = \cosh x \cosh y + \sinh x \sinh y$

7 $\sinh 2x = 2 \sinh x \cosh x$

8 $\cosh 2x = \cosh^2 x + \sinh^2 x$

9 $\tanh (x + y) = \dfrac{\tanh x + \tanh y}{1 + \tanh x \tanh y}$

10 $\tanh 2x = \dfrac{2 \tanh x}{1 + \tanh^2 x}$

11 $\cosh \dfrac{x}{2} = \sqrt{\dfrac{1 + \cosh x}{2}}$

12 $\tanh \dfrac{x}{2} = \dfrac{\sinh x}{1 + \cosh x}$

13 $\sinh x + \sinh y = 2 \sinh \frac{1}{2}(x + y) \cosh \frac{1}{2}(x - y)$

14 $(\cosh x + \sinh x)^n = \cosh nx + \sinh nx$ for every positive integer n (*Hint:* Use Exercise 1.)

In Exercises 15–30 find $f'(x)$ if $f(x)$ equals the given expression.

15 $\sinh 5x$

16 $\cosh \sqrt{4x^2 + 3}$

17 $\sqrt{x} \tanh \sqrt{x}$

18 $x \operatorname{csch} e^{4x}$

19 $(\operatorname{sech} x^2)/(x^2 + 1)$

20 $(\coth x)/(\cot x)$

21 $\cosh x^3$

22 $\sinh (x^2 + 1)$

23 $\cosh^3 x$

24 $\sinh^2 3x$

25 $\ln \sinh 2x$

26 $\arctan \tanh x$

27 $e^{3x} \operatorname{sech} x$

28 $\sqrt{\operatorname{sech} 5x}$

29 $1/(\tanh x + 1)$

30 $(1 + \cosh x)/(1 - \cosh x)$

31   Find y' if $\sinh xy = ye^x$.

32   Find y' if $x^2 \tanh y = \ln y$.

Evaluate the integrals in Exercises 33–44.

33 $\displaystyle\int \frac{\sinh \sqrt{x}}{\sqrt{x}}\,dx$

34 $\displaystyle\int \frac{\cosh \ln x}{x}\,dx$

35 $\displaystyle\int \coth x\,dx$

36 $\displaystyle\int \frac{1}{\cosh^2 3x}\,dx$

37 $\displaystyle\int \sinh x \cosh x\,dx$

38 $\displaystyle\int \operatorname{sech}^2 x \tanh x\,dx$

39 $\displaystyle\int \tanh 3x \operatorname{sech} 3x\,dx$

40 $\displaystyle\int \sinh x\sqrt{\cosh x}\,dx$

41 $\displaystyle\int \tanh^2 3x \operatorname{sech}^2 3x\,dx$

42 $\displaystyle\int \tanh x\,dx$

43 $\displaystyle\int \frac{\operatorname{sech}^2 x}{1 - 2\tanh x}\,dx$

44 $\displaystyle\int \frac{e^{\sinh x}}{\operatorname{sech} x}\,dx$

45 Find the area of the region bounded by the graphs of $y = \sinh 3x$, $y = 0$, and $x = 1$.

46 Find the arc length of the graph of $y = \cosh x$ from $x = 0$ to $x = 1$.

47 Find the points on the graph of $y = \sinh x$ at which the tangent line has slope 2.

48 The region bounded by the graphs of $y = \cosh x$, $x = -1$, $x = 1$, and $y = 0$ is revolved about the x-axis. Find the volume of the resulting solid.

49 Verify the graph of $y = \tanh x$ in Figure 8.18.

Sketch the graphs of the equations in Exercises 50–52.

50 $y = \coth x$

51 $y = \operatorname{sech} x$

52 $y = \operatorname{csch} x$

53 If A is the shaded region in (ii) of Figure 8.17, show that $t = 2A$.

54 Sketch the graph of $x^2 - y^2 = 1$ and show that as t varies, the point $P(\cosh t, \sinh t)$ traces the part of the graph in quadrants I and IV.

[handwritten notes:]
$u = 1 - 2\tanh x$
$du = -2\operatorname{sech}^2 x\,dx$
$43)$
$-\frac{1}{2}du =$
$-\frac{1}{2}\int \frac{du}{u} -\frac{1}{2}\ln|1-2\tanh x| + C$

8.7

THE INVERSE HYPERBOLIC FUNCTIONS

The hyperbolic sine function is continuous and increasing for all x and hence, by Theorem (7.28), has a continuous, increasing inverse function, denoted by $\sinh^{-1}$. Since $\sinh x$ is defined in terms of e^x, we might expect that $\sinh^{-1}$ can be expressed in terms of the inverse, $\ln$, of the natural exponential function. The first formula of the next theorem shows that this is, indeed, the case.

Theorem (8.22)

$$\sinh^{-1} x = \ln\left(x + \sqrt{x^2 + 1}\right)$$

$$\cosh^{-1} x = \ln\left(x + \sqrt{x^2 - 1}\right), \quad x \ge 1$$

$$\tanh^{-1} x = \frac{1}{2}\ln\frac{1 + x}{1 - x}, \quad |x| < 1$$

$$\operatorname{sech}^{-1} x = \ln\left(\frac{1 + \sqrt{1 - x^2}}{x}\right), \quad 0 < x \le 1$$

Proof   To prove the formula for $\sinh^{-1} x$ we begin by noting that

$$y = \sinh^{-1} x \quad \text{if and only if} \quad x = \sinh y.$$

The last equation can be used to find an explicit form for $\sinh^{-1} x$. Thus, if

$$x = \sinh y = \frac{e^y - e^{-y}}{2}$$

then
$$e^y - 2x - e^{-y} = 0.$$

Multiplying both sides by e^y yields

$$e^{2y} - 2xe^y - 1 = 0$$

and, applying the Quadratic Formula,

$$e^y = \frac{2x \pm \sqrt{4x^2 + 4}}{2} \quad \text{or} \quad e^y = x \pm \sqrt{x^2 + 1}.$$

Since e^y is never negative, the minus sign must be discarded. Doing this and then taking the natural logarithm of both sides of the equation, we obtain

$$y = \ln (x + \sqrt{x^2 + 1}),$$

that is,
$$\sinh^{-1} x = \ln (x + \sqrt{x^2 + 1}).$$

The formulas for the remaining inverse hyperbolic functions are obtained in similar fashion. As was the case with trigonometric functions, it is sometimes necessary to restrict the domain for an inverse function to exist. For example, if the domain of cosh is restricted to the set of nonnegative real numbers, then the resulting function is continuous and increasing and its inverse function $\cosh^{-1}$ is defined by

$$y = \cosh^{-1} x \quad \text{if and only if} \quad \cosh y = x, \quad y \geq 0.$$

Employing techniques similar to those used for $\sinh^{-1} x$ leads to the logarithmic formula for $\cosh^{-1} x$.

In like manner, we write

$$y = \tanh^{-1} x \quad \text{if and only if} \quad \tanh y = x.$$

Using Definition (8.18), the last equation may be written

$$\frac{e^y - e^{-y}}{e^y + e^{-y}} = x.$$

Solving for y gives us the logarithmic form for $\tanh^{-1} x$.

Finally, if the domain of sech is restricted to nonnegative numbers, there results a one-to-one function and we define

$$y = \text{sech}^{-1} x \quad \text{if and only if} \quad \text{sech } y = x, \quad y \geq 0.$$

Again introducing the exponential form leads to the fourth formula in the statement of the theorem.   □

In the next theorem $u = g(x)$, where g is differentiable and x is suitably restricted.

Theorem (8.23)

$$D_x \sinh^{-1} u = \frac{1}{\sqrt{u^2 + 1}} D_x u$$

$$D_x \cosh^{-1} u = \frac{1}{\sqrt{u^2 - 1}} D_x u, \quad u > 1$$

$$D_x \tanh^{-1} u = \frac{1}{1 - u^2} D_x u, \quad |u| < 1$$

$$D_x \operatorname{sech}^{-1} u = \frac{-1}{u\sqrt{1 - u^2}} D_x u, \quad 0 < u < 1$$

Proof   It follows from Theorem (8.22) that

$$D_x \sinh^{-1} x = \frac{1}{x + \sqrt{x^2 + 1}} \left(1 + \frac{x}{\sqrt{x^2 + 1}}\right)$$

$$= \frac{\sqrt{x^2 + 1} + x}{(x + \sqrt{x^2 + 1})\sqrt{x^2 + 1}}$$

$$= \frac{1}{\sqrt{x^2 + 1}}.$$

This formula can be extended to $D_x \sinh^{-1} u$ by applying the Chain Rule in the usual way. The proofs of the remaining formulas are left as exercises.   □

Example 1   Find y' if $y = \sinh^{-1} (\tan x)$.

Solution   Using Theorem (8.23) with $u = \tan x$,

$$y' = \frac{1}{\sqrt{\tan^2 x + 1}} D_x \tan x = \frac{1}{\sqrt{\sec^2 x}} \sec^2 x$$

$$= \frac{1}{|\sec x|} |\sec x|^2 = |\sec x|.$$ ■

The following theorem may be verified by differentiating the right-hand side of each formula, where $u = g(x)$.

Theorem (8.24)

$$\int \frac{1}{\sqrt{u^2 + a^2}} \, du = \sinh^{-1} \frac{u}{a} + C, \quad a > 0$$

$$\int \frac{1}{\sqrt{u^2 - a^2}} \, du = \cosh^{-1} \frac{u}{a} + C, \quad u > a > 0$$

$$\int \frac{1}{a^2 - u^2} \, du = \frac{1}{a} \tanh^{-1} \frac{u}{a} + C, \quad a > 0, \quad |u| < a$$

$$\int \frac{1}{u\sqrt{a^2 - u^2}} \, du = -\frac{1}{a} \operatorname{sech}^{-1} \frac{|u|}{a} + C, \quad a > 0, \quad 0 < |u| < a$$

Example 2 Evaluate $\displaystyle\int \frac{1}{\sqrt{9x^2 + 25}} \, dx$.

Solution The integral may be expressed as in the first formula of Theorem (8.24) by letting $u = 3x$ and $du = 3 \, dx$. Thus,

$$\int \frac{1}{\sqrt{9x^2 + 25}} \, dx = \frac{1}{3} \int \frac{1}{\sqrt{(3x)^2 + (5)^2}} \, 3 \, dx$$

$$= \frac{1}{3} \sinh^{-1} \frac{3x}{5} + C. \qquad \blacksquare$$

Example 3 Evaluate $\displaystyle\int \frac{e^x}{16 - e^{2x}} \, dx$.

Solution Letting $u = e^x$, $du = e^x \, dx$, and applying Theorem (8.24), we obtain

$$\int \frac{e^x}{16 - e^{2x}} \, dx = \int \frac{1}{4^2 - (e^x)^2} \, e^x \, dx$$

$$= \int \frac{1}{4^2 - u^2} \, du$$

$$= \frac{1}{4} \tanh^{-1} \frac{u}{4} + C$$

$$= \frac{1}{4} \tanh^{-1} \frac{e^x}{4} + C$$

provided $e^x < 4$. $\qquad \blacksquare$

EXERCISES 8.7

1 Derive the formula for $\cosh^{-1} x$ in Theorem (8.22).

2 Derive the formula for $\tanh^{-1} x$ in Theorem (8.22).

3 Verify the formula for $D_x \cosh^{-1} u$.

4 Verify the formula for $D_x \tanh^{-1} u$.

5 Verify the formula for $\int \dfrac{1}{\sqrt{u^2 - a^2}} \, du$.

6 Verify the formula for $\int \dfrac{1}{a^2 - u^2} \, du$.

Sketch the graphs of the equations in Exercises 7–10.

7 $y = \sinh^{-1} x$ **8** $y = \cosh^{-1} x$

9 $y = \tanh^{-1} x$ **10** $y = \operatorname{sech}^{-1} x$

In Exercises 11–20 find $f'(x)$ if $f(x)$ equals the given expression.

11 $\sinh^{-1} 5x$ **12** $\sinh^{-1} e^x$

13 $\cosh^{-1} \sqrt{x}$ **14** $\sqrt{\cosh^{-1} x}$

15 $\tanh^{-1}(x^2 - 1)$ **16** $\tanh^{-1} \sin 3x$

17 $x \sinh^{-1}(1/x)$ **18** $1/\sinh^{-1} x^2$

19 $\ln \cosh^{-1} 4x$ **20** $\cosh^{-1} \ln 4x$

Evaluate the integrals in Exercises 21–28.

21 $\int \dfrac{1}{\sqrt{9x^2 + 25}} \, dx$ **22** $\int \dfrac{1}{\sqrt{16x^2 - 9}} \, dx$

23 $\int \dfrac{1}{49 - 4x^2} \, dx$ **24** $\int \dfrac{\sin x}{\sqrt{1 + \cos^2 x}} \, dx$

25 $\int \dfrac{e^x}{\sqrt{e^{2x} - 16}} \, dx$ **26** $\int \dfrac{2}{5 - 3x^2} \, dx$

27 $\int \dfrac{1}{x\sqrt{9 - x^4}} \, dx$ **28** $\int \dfrac{1}{\sqrt{5 - e^{2x}}} \, dx$

29 Prove that if $a > 0$, then
$$\int \frac{1}{\sqrt{u^2 + a^2}} \, du = \ln(u + \sqrt{u^2 + a^2}) + C.$$

30 Prove that if $u > a > 0$, then
$$\int \frac{1}{\sqrt{u^2 - a^2}} \, du = \ln(u + \sqrt{u^2 - a^2}) + C.$$

8.8

REVIEW

Define or discuss each of the following.

1 Limits of trigonometric functions

2 Differentiation formulas for trigonometric functions

3 Simple harmonic motion

4 Integration formulas for trigonometric functions

5 The inverse trigonometric functions

6 Differentiation and integration formulas involving inverse trigonometric functions

7 Hyperbolic functions

8 Inverse hyperbolic functions

EXERCISES 8.8

In Exercises 1–6 find the limit, if it exists.

1 $\lim\limits_{x \to 0} \dfrac{x^2}{\sin x}$ **2** $\lim\limits_{x \to 0} \dfrac{x^2 + \sin^2 x}{4x^2}$

3 $\lim\limits_{x \to 0} \dfrac{\sin^2 x + \sin 2x}{3x}$ **4** $\lim\limits_{x \to 0} \dfrac{2 - \cos x}{1 + \sin x}$

5 $\lim\limits_{x \to 0} \dfrac{2\cos x + 3x - 2}{5x}$ **6** $\lim\limits_{x \to 0} \dfrac{3x + 1 - \cos^2 x}{\sin x}$

Find $f'(x)$ if $f(x)$ is defined as in Exercises 7–50.

7 $\cos \sqrt{3x^2 + x}$ **8** $x^2 \cot 2x$

9 $(\sec x + \tan x)^5$ **10** $\sqrt[3]{x^3 + \csc 6x}$

11 $x^2 \operatorname{arcsec} x^2$ **12** $\tan^{-1}(\ln 3x)$

13 $\dfrac{(3x + 7)^4}{\sin^{-1} 5x}$ **14** $\dfrac{\sin 8x}{4x^2 - x}$

15 $(\cos x)^{x+1}$

16 $5^{\tan 2x}$

17 $\sqrt{2x^2 + \text{sech } 4x}$

18 $x \sec^{-1} 4x$

19 $\ln (\csc^3 2x)$

20 $\log (\sec 2x)$

21 $\dfrac{1}{2x + \sec^2 x}$

22 $\cot \dfrac{1}{x} + \dfrac{1}{\cot x}$

23 $e^{\cos x} + (\cos x)^e$

24 $\dfrac{\ln \sinh x}{x}$

25 $\cosh e^{-5x}$

26 $(\cos x)^{\cot x}$

27 $\tanh^{-1} (\tanh \sqrt[3]{x})$

28 $\sec (\sec x)$

29 $2^{\arctan 2x}$

30 $(1 + \text{arcsec } 2x)^{\sqrt{2}}$

31 $\sin^3 e^{-2x}$

32 $\ln \sin (\pi/3)$

33 $e^{-x^2} \cot x^2$

34 $\sec 5x \tan 5x$

35 $\dfrac{\csc x + 1}{\cot x + 1}$

36 $\dfrac{1 - x^2}{\arccos x}$

37 $\sin^{-1} \sqrt{1 - x^2}$

38 $\sqrt{\sin^{-1} (1 - x^2)}$

39 $\tan (\sin 3x)$

40 $\sqrt{\sin \sqrt{x}}$

41 $(\tan x + \tan^{-1} x)^4$

42 $e^{4x} \sec^{-1} e^{4x}$

43 $\tan^{-1} (\tan^{-1} x)$

44 $e^{x \cosh x}$

45 $e^{-x} \sinh e^{-x}$

46 $\ln \tanh (5x + 1)$

47 $\dfrac{\sinh x}{\cosh x - \sinh x}$

48 $\dfrac{1}{x} \tanh \dfrac{1}{x}$

49 $\sinh^{-1} x^2$

50 $\cosh^{-1} \tan x$

Evaluate the integrals in Exercises 51–88.

51 $\displaystyle\int \cos (5 - 3x)\, dx$

52 $\displaystyle\int \csc \dfrac{x}{2} \cot \dfrac{x}{2}\, dx$

53 $\displaystyle\int \dfrac{\sec^2 \sqrt{x}}{\sqrt{x}}\, dx$

54 $\displaystyle\int \dfrac{\sec (1/x)}{x^2}\, dx$

55 $\displaystyle\int (\cot 9x + \csc 9x)\, dx$

56 $\displaystyle\int x \csc^2 (3x^2 + 4)\, dx$

57 $\displaystyle\int e^x \tan e^x\, dx$

58 $\displaystyle\int \cot 2x \csc 2x\, dx$

59 $\displaystyle\int (\csc 3x + 1)^2\, dx$

60 $\displaystyle\int \dfrac{\sin x + 1}{\cos x}\, dx$

61 $\displaystyle\int \dfrac{\sin 4x}{\tan 4x}\, dx$

62 $\displaystyle\int x \cot x^2\, dx$

63 $\displaystyle\int x^2 \cos (2x^3)\, dx$

64 $\displaystyle\int_0^{\pi/4} \sin 2x \cos^2 2x\, dx$

65 $\displaystyle\int_0^{\pi/2} \cos x \sqrt{3 + 5 \sin x}\, dx$

66 $\displaystyle\int_0^{\pi/4} D_x(x \sin^2 x)\, dx$

67 $\displaystyle\int \dfrac{\cos 3x}{\sin^3 3x}\, dx$

68 $\displaystyle\int \dfrac{\cos 2x}{1 - 2 \sin 2x}\, dx$

69 $\displaystyle\int \dfrac{x}{4 + 9x^2}\, dx$

70 $\displaystyle\int \dfrac{1}{4 + 9x^2}\, dx$

71 $\displaystyle\int \dfrac{e^{2x}}{\sqrt{1 - e^{2x}}}\, dx$

72 $\displaystyle\int \dfrac{e^x}{\sqrt{1 - e^{2x}}}\, dx$

73 $\displaystyle\int \dfrac{x}{\text{sech } x^2}\, dx$

74 $\displaystyle\int \dfrac{1}{x\sqrt{x^4 - 1}}\, dx$

75 $\displaystyle\int_{-1/2}^{1/2} \dfrac{1}{\sqrt{1 - x^2}}\, dx$

76 $\displaystyle\int_0^{\pi/2} \dfrac{\cos x}{1 + \sin^2 x}\, dx$

77 $\displaystyle\int \sec^2 x(1 + \tan x)^2\, dx$

78 $\displaystyle\int (1 + \cos^2 x) \sin x\, dx$

79 $\displaystyle\int \dfrac{\csc^2 x}{2 + \cot x}\, dx$

80 $\displaystyle\int (\sin x)e^{1 + \cos x}\, dx$

81 $\displaystyle\int \dfrac{\sinh (\ln x)}{x}\, dx$

82 $\displaystyle\int \text{sech}^2 (1 - 2x)\, dx$

83 $\displaystyle\int \dfrac{1}{\sqrt{9 - 4x^2}}\, dx$

84 $\displaystyle\int \dfrac{x}{\sqrt{9 - 4x^2}}\, dx$

85 $\displaystyle\int \dfrac{1}{x\sqrt{9 - 4x^2}}\, dx$

86 $\displaystyle\int \dfrac{1}{x\sqrt{4x^2 - 9}}\, dx$

87 $\displaystyle\int \dfrac{x}{\sqrt{25x^2 + 36}}\, dx$

88 $\displaystyle\int \dfrac{1}{\sqrt{25x^2 + 36}}\, dx$

89 Find the points on the graph of $y = \sin^{-1} 3x$ at which the tangent line is parallel to the line through $A(2, -3)$ and $B(4, 7)$.

90 Find an equation of the tangent line to the graph of $y = \pi \tan (y/x) + x^2 - 16$ at the point $(4, \pi)$.

91 Find the local extrema of $f(x) = 8 \sec x + \csc x$ on the interval $(0, \pi/2)$ and describe where $f(x)$ is increasing or decreasing on that interval.

92 Find the points of inflection and discuss the concavity of the graph of $y = x \sin^{-1} x$.

93 The region between the graphs of $y = \tan x$ and the x-axis, from $x = 0$ to $x = \pi/4$, is revolved about the x-axis. Find the volume of the resulting solid.
(*Hint*: $\tan^2 x = \sec^2 x - 1$.)

94 Find the area of the region bounded by the graphs of $y = x/(1 + x^4)$, $x = 1$, and $y = 0$.

95 Use differentials to approximate $(0.98)^2 \tan^{-1}(0.98)$.

96 Find the arclength of the graph of $y = \ln \tanh(x/2)$ from $x = 1$ to $x = 2$.

97 A person on level ground, $\frac{1}{2}$ km from a point at which a balloon was released, observes its vertical ascent. If the balloon is rising at a constant rate of 2 m/sec, find the rate at which the angle of elevation of the observer's line of sight is changing at the instant the balloon is 100 m above the level of the observer's eyes.

98 A square picture having sides 2 ft long is hung on a wall such that the base is 6 ft above the floor. A person whose eye level is 5 ft above the floor approaches the picture at a rate of 2 ft/sec. If θ is the angle between the line of sight and the top and bottom of the picture, find (a) the rate at which θ is changing when the person is 8 ft from the wall, and (b) the distance from the wall at which θ has its maximum value.

99 The position of a moving point on a coordinate line is given by

$$s(t) = a \sin(kt + m) + b \cos(kt + m)$$

where $a, b, k,$ and m are constants. Prove that the magnitude of the acceleration is directly proportional to the distance from the origin.

100 Use the Mean Value Theorem to prove that if u and v are any real numbers, then $|\sin u - \sin v| \le |u - v|$.

9

ADDITIONAL TECHNIQUES AND APPLICATIONS OF INTEGRATION

In this chapter we shall discuss techniques that can be used to evaluate many types of integrals. We also consider applications of the definite integral to the problems of finding moments and centers of mass of certain solids. The formulas for integrals listed below were obtained previously, where if $u = f(x)$, then $du = f'(x)\, dx$.

$$\int u^n\, du = \frac{1}{n+1}\, u^{n+1} + C, \ n \neq -1 \qquad \int \frac{1}{u}\, du = \ln|u| + C$$

$$\int e^u\, du = e^u + C \qquad \int a^u\, du = \frac{1}{\ln a}\, a^u + C$$

$$\int \sin u\, du = -\cos u + C \qquad \int \cos u\, du = \sin u + C$$

$$\int \tan u\, du = \ln|\sec u| + C \qquad \int \cot u\, du = \ln|\sin u| + C$$

$$\int \sec u\, du = \ln|\sec u + \tan u| + C \qquad \int \csc u\, du = \ln|\csc u - \cot u| + C$$

$$\int \sec u \tan u\, du = \sec u + C \qquad \int \csc u \cot u\, du = -\csc u + C$$

$$\int \sec^2 u\, du = \tan u + C \qquad \int \csc^2 u\, du = -\cot u + C$$

$$\int \frac{1}{\sqrt{a^2 - u^2}}\, du = \sin^{-1}\frac{u}{a} + C \qquad \int \frac{1}{u\sqrt{u^2 - a^2}}\, du = \frac{1}{a}\sec^{-1}\frac{u}{a} + C$$

$$\int \frac{1}{a^2 + u^2}\, du = \frac{1}{a}\tan^{-1}\frac{u}{a} + C$$

9.1

INTEGRATION BY PARTS

The following result is useful for simplifying certain types of integrals.

Integration by Parts Formula (9.1)

If $u = f(x)$ and $v = g(x)$, where f' and g' are continuous, then

$$\int u\, dv = uv - \int v\, du.$$

Proof   By the Product Rule,

$$D_x[f(x)g(x)] = f(x)g'(x) + g(x)f'(x)$$

or, equivalently,

$$f(x)g'(x) = D_x[f(x)g(x)] - g(x)f'(x).$$

Integrating both sides of the previous equation gives us

$$\int f(x)g'(x)\, dx = \int D_x[f(x)g(x)]\, dx - \int g(x)f'(x)\, dx.$$

By Theorem (5.28) the first integral on the right side equals $f(x)g(x) + C$. Since we obtain another constant of integration from the second integral, it is unnecessary to include C in the formula; that is,

$$\int f(x)g'(x)\, dx = f(x)g(x) - \int g(x)f'(x)\, dx.$$

Since $du = f'(x)\, dx$ and $dv = g'(x)\, dx$, the preceding formula may be written as in (9.1).

When applying Formula (9.1) to a given integral, we begin by letting one part of the integrand correspond to dv. The expression chosen for dv must include the differential dx. After selecting dv, the remaining part of the integrand is designated by u. Since this process involves splitting the integrand into two parts, the use of (9.1) is referred to as **integrating by parts**. A proper choice for dv is crucial. A good rule of thumb is to choose the most complicated part of the integrand that can be readily integrated. The following examples illustrate this important method of integration.

Example 1   Find $\int xe^{2x}\, dx$.

Solution   There are four possible choices for dv, namely dx, $x\, dx$, $e^{2x}\, dx$, or $xe^{2x}\, dx$. If we let $dv = e^{2x}\, dx$, then the remaining part of the integrand is u; that is, $u = x$. To find v we integrate dv, obtaining $v = \frac{1}{2}e^{2x}$. Note that a constant of integration is not added at this stage of the solution. (In Exercise 51 you are asked to prove that if a constant *is* added to v, the same result is obtained.). Since $u = x$ we see that $du = dx$. For ease of reference it is convenient to display these expressions as follows:

$$dv = e^{2x}\, dx \qquad u = x$$

$$v = \tfrac{1}{2}e^{2x} \qquad du = dx.$$

Substituting these expressions in Formula (9.1), that is, *integrating by parts*, we obtain

$$\int xe^{2x}\, dx = x(\tfrac{1}{2}e^{2x}) - \int \tfrac{1}{2}e^{2x}\, dx.$$

The integral on the right side may be found by means of Theorem (7.21). This gives us

$$\int xe^{2x}\, dx = \tfrac{1}{2}xe^{2x} - \tfrac{1}{4}e^{2x} + C. \qquad \blacksquare$$

It takes considerable practice to become proficient in making a suitable choice for dv. To illustrate, if we had chosen $dv = x\, dx$ in Example 1, then it would have been necessary to let $u = e^{2x}$, giving us

$$dv = x\, dx \qquad u = e^{2x}$$

$$v = \tfrac{1}{2}x^2 \qquad du = 2e^{2x}\, dx.$$

Integrating by parts, we obtain

$$\int xe^{2x}\, dx = \tfrac{1}{2}x^2 e^{2x} - \int x^2 e^{2x}\, dx.$$

Since the exponent associated with x has increased, the integral on the right is more complicated than the given integral. This indicates an incorrect choice for dv.

Example 2 Evaluate $\displaystyle\int x \sec^2 x\, dx$.

Solution   Since $\sec^2 x$ can be integrated readily, we let $dv = \sec^2 x\, dx$. The remaining part of the integrand is x and hence we must let $u = x$. Thus

$$dv = \sec^2 x\, dx \qquad u = x$$

$$v = \tan x \qquad du = dx$$

and integration by parts gives us

$$\int x \sec^2 x\, dx = x \tan x - \int \tan x\, dx$$

$$= x \tan x - \ln |\sec x| + C. \qquad \blacksquare$$

We did not state a formula for $\int \ln x \, dx$ in Chapter 7. The reason for not doing so is that integration by parts is needed to find an antiderivative of the natural logarithmic function, as shown in the next example.

Example 3   Find $\displaystyle\int \ln x \, dx$.

Solution   Let

$$dv = dx \qquad u = \ln x$$

$$v = x \qquad du = \frac{1}{x} dx.$$

Integrating by parts, we obtain

$$\int \ln x \, dx = x \ln x - \int x\left(\frac{1}{x}\right) dx$$

$$= x \ln x - \int dx$$

$$= x \ln x - x + C. \qquad\qquad \blacksquare$$

Sometimes it is necessary to use integration by parts more than once in the same problem. This is illustrated in the next example.

Example 4   Find $\displaystyle\int x^2 e^{2x} \, dx$.

Solution   Let

$$dv = e^{2x} \, dx \qquad u = x^2$$

$$v = \tfrac{1}{2} e^{2x} \qquad du = 2x \, dx.$$

Integrating by parts, we obtain

$$\int x^2 e^{2x} \, dx = x^2(\tfrac{1}{2} e^{2x}) - \int (\tfrac{1}{2} e^{2x}) 2x \, dx$$

$$= \tfrac{1}{2} x^2 e^{2x} - \int x e^{2x} \, dx.$$

To evaluate the integral on the right side of the last equation we must again integrate by parts. Proceeding exactly as in Example 1 leads to

$$\int x^2 e^{2x} \, dx = \tfrac{1}{2} x^2 e^{2x} - \tfrac{1}{2} x e^{2x} + \tfrac{1}{4} e^{2x} + C. \qquad\qquad \blacksquare$$

The following example illustrates another device for finding certain integrals by means of two applications of the integration by parts formula.

Example 5 Find $\int e^x \cos x \, dx$.

Solution Let

$$dv = \cos x \, dx \qquad u = e^x$$
$$v = \sin x \qquad du = e^x \, dx.$$

Integrating by parts,

(a)
$$\int e^x \cos x \, dx = e^x \sin x - \int e^x \sin x \, dx.$$

We next apply integration by parts to the integral on the right side of equation (a). Letting

$$dv = \sin x \, dx \qquad u = e^x$$
$$v = -\cos x \qquad du = e^x \, dx$$

and integrating by parts leads to

(b)
$$\int e^x \sin x \, dx = -e^x \cos x + \int e^x \cos x \, dx.$$

If we now use equation (b) to substitute on the right side of equation (a) we obtain

$$\int e^x \cos x \, dx = e^x \sin x - \left[-e^x \cos x + \int e^x \cos x \, dx \right]$$

or
$$\int e^x \cos x \, dx = e^x \sin x + e^x \cos x - \int e^x \cos x \, dx.$$

Adding $\int e^x \cos x \, dx$ to both sides gives us

$$2 \int e^x \cos x \, dx = e^x (\sin x + \cos x).$$

Finally, dividing both sides by 2 and adding the constant of integration, we have

$$\int e^x \cos x \, dx = \tfrac{1}{2} e^x (\sin x + \cos x) + C. \qquad \blacksquare$$

Some care must be taken when evaluating integrals of the type given in Example 5. To illustrate, suppose in the evaluation of the integral on the right in equation (a) of the solution we had used

$$dv = e^x \, dx \qquad u = \sin x$$
$$v = e^x \qquad du = \cos x \, dx.$$

In this event integration by parts leads to

$$\int e^x \sin x \, dx = e^x \sin x - \int e^x \cos x \, dx.$$

If we now substitute in (a), we obtain

$$\int e^x \cos x \, dx = e^x \sin x - \left[e^x \sin x - \int e^x \cos x \, dx \right]$$

which reduces to

$$\int e^x \cos x \, dx = \int e^x \cos x \, dx.$$

Although this is a true statement, it is not a solution to the problem! Incidentally, the integral in Example 5 *can* be evaluated by using $dv = e^x \, dx$ for *both* the first and second applications of the integration by parts formula.

Integration by parts may sometimes be employed to obtain **reduction formulas** for integrals. Such formulas can be used to write an integral involving powers of an expression in terms of integrals that involve lower powers of the expression.

Example 6 Find a reduction formula for $\int \sin^n x \, dx$.

Solution   Let

$$dv = \sin x \, dx \qquad u = \sin^{n-1} x$$

$$v = -\cos x \qquad du = (n-1)\sin^{n-2} x \cos x \, dx.$$

Integrating by parts,

$$\int \sin^n x \, dx = -\cos x \sin^{n-1} x + (n-1) \int \sin^{n-2} x \cos^2 x \, dx.$$

Since $\cos^2 x = 1 - \sin^2 x$, we may write

$$\int \sin^n x \, dx = -\cos x \sin^{n-1} x + (n-1) \int \sin^{n-2} x \, dx - (n-1) \int \sin^n x \, dx.$$

Consequently,

$$\int \sin^n x \, dx + (n-1) \int \sin^n x \, dx = -\cos x \sin^{n-1} x + (n-1) \int \sin^{n-2} x \, dx.$$

The left side of the last equation reduces to $n \int \sin^n x \, dx$. Dividing both sides by n, we obtain

$$\int \sin^n x \, dx = -\frac{1}{n} \cos x \sin^{n-1} x + \frac{(n-1)}{n} \int \sin^{n-2} x \, dx.$$ ∎

Example 7 Use the reduction formula in Example 6 to evaluate $\int \sin^4 x \, dx$.

Solution Using the formula with $n = 4$ gives us

$$\int \sin^4 x \, dx = -\tfrac{1}{4} \cos x \sin^3 x + \tfrac{3}{4} \int \sin^2 x \, dx.$$

Applying the reduction formula, with $n = 2$, to the integral on the right,

$$\int \sin^2 x \, dx = -\tfrac{1}{2} \cos x \sin x + \tfrac{1}{2} \int dx$$

$$= -\tfrac{1}{2} \cos x \sin x + \tfrac{1}{2}x + C_1.$$

Consequently,

$$\int \sin^4 x \, dx = -\tfrac{1}{4} \cos x \sin^3 x - \tfrac{3}{8} \cos x \sin x + \tfrac{3}{8}x + C. \qquad \blacksquare$$

It should be evident that by repeated applications of the formula in Example 6 we can find $\int \sin^n x \, dx$ for any positive integer n, because these reductions end with either $\int \sin x \, dx$ or $\int dx$, and each of these can be evaluated easily.

EXERCISES 9.1

Evaluate the integrals in Exercises 1–38.

1 $\int xe^{-x} \, dx$

2 $\int x \sin x \, dx$

3 $\int x^2 e^{3x} \, dx$

4 $\int x^2 \sin 4x \, dx$

5 $\int x \cos 5x \, dx$

6 $\int xe^{-2x} \, dx$

7 $\int x \sec x \tan x \, dx$

8 $\int x \csc^2 3x \, dx$

9 $\int x^2 \cos x \, dx$

10 $\int x^3 e^{-x} \, dx$

11 $\int \tan^{-1} x \, dx$

12 $\int \sin^{-1} x \, dx$

13 $\int \sqrt{x} \ln x \, dx$

14 $\int x^2 \ln x \, dx$

15 $\int x \csc^2 x \, dx$

16 $\int x \tan^{-1} x \, dx$

17 $\int e^{-x} \sin x \, dx$

18 $\int e^{3x} \cos 2x \, dx$

19 $\int \sin x \ln \cos x \, dx$

20 $\int_0^1 x^3 e^{-x^2} \, dx$

21 $\int \sec^3 x \, dx$

22 $\int \csc^5 x \, dx$

23 $\int_0^1 \frac{x^3}{\sqrt{x^2 + 1}} \, dx$

24 $\int \sin \ln x \, dx$

25 $\int_0^{\pi/2} x \sin 2x \, dx$

26 $\int_{\pi/6}^{\pi/4} x \sec^2 x \, dx$

27 $\int x(2x + 3)^{99} \, dx$

28 $\int \frac{x^5}{\sqrt{1 - x^3}} \, dx$

29 $\int e^{4x} \sin 5x \, dx$

30 $\int x^3 \cos (x^2) \, dx$

31 $\int (\ln x)^2 \, dx$

32 $\int x 2^x \, dx$

33 $\int x^3 \sinh x \, dx$

34 $\int (x + 4) \cosh 4x \, dx$

35 $\int \cos \sqrt{x}\, dx$

36 $\int \cot^{-1} 3x\, dx$

37 $\int \cos^{-1} x\, dx$

38 $\int (x + 1)^{10}(x + 2)\, dx$

Use integration by parts to derive the reduction formulas in Exercises 39–42.

39 $\int x^m e^x\, dx = x^m e^x - m \int x^{m-1} e^x\, dx$

40 $\int x^m \sin x\, dx = -x^m \cos x + m \int x^{m-1} \cos x\, dx$

41 $\int (\ln x)^m\, dx = x(\ln x)^m - m \int (\ln x)^{m-1}\, dx$

42 $\int \sec^m x\, dx = \dfrac{\sec^{m-2} x \tan x}{m - 1} + \dfrac{m - 2}{m - 1} \int \sec^{m-2} x\, dx,$

provided that $m \neq 1$.

43 Use Exercise 39 to evaluate $\int x^5 e^x\, dx$.

44 Use Exercise 41 to evaluate $\int (\ln x)^4\, dx$.

45 If $f(x) = \sin \sqrt{x}$, find the area of the region under the graph of f from 0 to π^2.

46 The region between the graph of $y = x\sqrt{\sin x}$ and the x-axis from $x = 0$ to $x = \pi/2$ is revolved about the x-axis. Find the volume of the resulting solid.

47 The region bounded by the graphs of $y = \ln x$, $y = 0$, and $x = e$ is revolved about the y-axis. Find the volume of the resulting solid.

48 Suppose the force $f(x)$ acting at the point with coordinate x on a coordinate line l is given by $f(x) = x^5\sqrt{x^3 + 1}$. Find the work done in moving an object from $x = 0$ to $x = 1$.

49 The ends of a water trough have the shape of the region between the graph of $y = \sin x$ and the x-axis from $x = \pi$ to $x = 2\pi$. If the trough is full of water, find the total force on one end.

50 The velocity (at time t) of a point moving along a coordinate line is t/e^{2t} ft/sec. If the point is at the origin at $t = 0$, find its position at time t.

51 When applying the Integration by Parts Formula (9.1), show that if, after choosing dv, we use $v + C$ in place of v, the same result is obtained.

52 In Chapter 6 the argument given for finding volumes by means of cylindrical shells (see Definition (6.7)) was incomplete because it was not shown that the same result is obtained if the disc method is also applicable. Use integration by parts to prove that if f is differentiable and either $f'(x) > 0$ on $[a, b]$ or $f'(x) < 0$ on $[a, b]$, and if V is the volume of the solid obtained by revolving the region bounded by the graphs of f, $x = a$, and $x = b$ about the x-axis, then the same value of V is obtained using either the disc method or the shell method. (*Hint:* Let g be the inverse function of f and use integration by parts on $\int_a^b \pi[f(x)]^2\, dx$.)

53 Criticize the following use of Formula (9.1). Given $\int (1/x)\, dx$, let $dv = dx$ and $u = 1/x$, so that $v = x$ and $du = (-1/x^2)\, dx$. Hence

$$\int \frac{1}{x}\, dx = \left(\frac{1}{x}\right)x - \int x\left(-\frac{1}{x^2}\right) dx$$

or

$$\int \frac{1}{x}\, dx = 1 + \int \frac{1}{x}\, dx.$$

Consequently, $0 = 1$.

54 Prove that the analogue of Formula (9.1) for definite integrals is

$$\int_a^b u\, dv = uv \Big]_a^b - \int_a^b v\, du.$$

9.2

TRIGONOMETRIC INTEGRALS

In Example 6 of Section 9.1 a reduction formula was obtained for $\int \sin^n x\, dx$. Integrals of this type may also be found without resorting to integration by parts. If n is an odd positive integer, we begin by writing

$$\int \sin^n x\, dx = \int \sin^{n-1} x \sin x\, dx.$$

Since the integer $n - 1$ is even, we may then use the trigonometric identity $\sin^2 x = 1 - \cos^2 x$ to obtain a form that is easy to integrate.

Example 1 Evaluate $\int \sin^5 x \, dx$.

Solution   As in the preceding discussion we have

$$\int \sin^5 x \, dx = \int \sin^4 x \sin x \, dx$$

$$= \int (\sin^2 x)^2 \sin x \, dx$$

$$= \int (1 - \cos^2 x)^2 \sin x \, dx$$

$$= \int (1 - 2\cos^2 x + \cos^4 x) \sin x \, dx.$$

We next employ the method of substitution, letting

$$u = \cos x \quad \text{and} \quad du = -\sin x \, dx.$$

Thus

$$\int \sin^5 x \, dx = -\int (1 - 2\cos^2 x + \cos^4 x)(-\sin x) \, dx$$

$$= -\int (1 - 2u^2 + u^4) \, du$$

$$= -u + \tfrac{2}{3}u^3 - \tfrac{1}{5}u^5 + C$$

$$= -\cos x + \tfrac{2}{3}\cos^3 x - \tfrac{1}{5}\cos^5 x + C. \qquad \blacksquare$$

A similar technique may be employed for odd powers of $\cos x$. Specifically, we write

$$\int \cos^n x \, dx = \int \cos^{n-1} x \cos x \, dx$$

and use the fact that $\cos^2 x = 1 - \sin^2 x$ to obtain an integrable form.

If the integrand is $\sin^n x$ or $\cos^n x$ and n is *even*, then the half-angle formulas

$$\sin^2 x = \frac{1 - \cos 2x}{2} \quad \text{or} \quad \cos^2 x = \frac{1 + \cos 2x}{2}$$

may be used to simplify the integrand.

Example 2 Evaluate $\int \cos^2 x \, dx$.

Solution   Using a half-angle formula,

$$\int \cos^2 x \, dx = \tfrac{1}{2} \int (1 + \cos 2x) \, dx$$

$$= \tfrac{1}{2}x + \tfrac{1}{4}\sin 2x + C. \qquad \blacksquare$$

Example 3   Evaluate $\int \sin^4 x \, dx$.

Solution $\int \sin^4 x \, dx = \int (\sin^2 x)^2 \, dx$

$$= \int \left(\frac{1 - \cos 2x}{2} \right)^2 dx$$

$$= \tfrac{1}{4} \int (1 - 2 \cos 2x + \cos^2 2x) \, dx.$$

We apply a half-angle formula again and write

$$\cos^2 2x = \tfrac{1}{2}(1 + \cos 4x) = \tfrac{1}{2} + \tfrac{1}{2} \cos 4x.$$

Substituting in the last integral and simplifying gives us

$$\int \sin^4 x \, dx = \tfrac{1}{4} \int (\tfrac{3}{2} - 2 \cos 2x + \tfrac{1}{2}\cos 4x) \, dx$$

$$= \tfrac{3}{8}x - \tfrac{1}{4} \sin 2x + \tfrac{1}{32} \sin 4x + C. \qquad \blacksquare$$

Integrals involving only products of $\sin x$ and $\cos x$ may be evaluated using the following guidelines.

(9.2)   *Guidelines for evaluating integrals of the form $\int \sin^m x \cos^n x \, dx$*

1   If both m and n are even integers, use half-angle formulas for $\sin^2 x$ and $\cos^2 x$ to reduce the exponents by one-half.

2   If n is an odd integer, write the integral as

$$\int \sin^m x \cos^n x \, dx = \int \sin^m x \cos^{n-1} x \cos x \, dx$$

and express $\cos^{n-1} x$ in terms of $\sin x$ by using the trigonometric identity $\cos^2 x = 1 - \sin^2 x$. The substitution $u = \sin x$ then leads to an integrand that can be handled easily.

3   If m is an odd integer, write the integral as

$$\int \sin^m x \cos^n x \, dx = \int \sin^{m-1} x \cos^n x \sin x \, dx$$

and express $\sin^{m-1} x$ in terms of $\cos x$ by using the trigonometric identity $\sin^2 x = 1 - \cos^2 x$. Use the substitution $u = \cos x$ to evaluate the resulting integral.

Example 4 Evaluate $\int \cos^3 x \sin^4 x \, dx$.

Solution Using Guideline 2 of (9.2),

$$\int \cos^3 x \sin^4 x \, dx = \int \cos^2 x \sin^4 x \cos x \, dx$$

$$= \int (1 - \sin^2 x)\sin^4 x \cos x \, dx.$$

If we let $u = \sin x$, then $du = \cos x \, dx$ and the integral may be written

$$\int \cos^3 x \sin^4 x \, dx = \int (1 - u^2)u^4 \, du = \int (u^4 - u^6) \, du$$

$$= \tfrac{1}{5}u^5 - \tfrac{1}{7}u^7 + C$$

$$= \tfrac{1}{5} \sin^5 x - \tfrac{1}{7} \sin^7 x + C. \qquad \blacksquare$$

(9.3) *Guidelines for evaluating integrals of the form* $\int \tan^m x \sec^n x \, dx$

1 If n is an even integer, write the integral as

$$\int \tan^m x \sec^{n-2} x \sec^2 x \, dx$$

and express $\sec^{n-2} x$ in terms of $\tan x$ by using the trigonometric identity $\sec^2 x = 1 + \tan^2 x$. The substitution $u = \tan x$ leads to a simple integral.

2 If m is an odd integer, write the integral as

$$\int \tan^{m-1} x \sec^{n-1} x \sec x \tan x \, dx.$$

Since $m - 1$ is even, $\tan^{m-1} x$ may be expressed in terms of $\sec x$ by means of the identity $\tan^2 x = \sec^2 x - 1$. The substitution $u = \sec x$ then leads to a form that is readily integrable.

3 If n is odd and m is even, then another method such as integration by parts should be used.

Example 5 Evaluate $\int \tan^2 x \sec^4 x \, dx$.

Solution Using Guideline 1 of (9.3),

$$\int \tan^2 x \sec^4 x \, dx = \int \tan^2 x \sec^2 x \sec^2 x \, dx$$

$$= \int \tan^2 x (\tan^2 x + 1) \sec^2 x \, dx.$$

If we let $u = \tan x$, then $du = \sec^2 x\, dx$ and

$$\int \tan^2 x \sec^4 x\, dx = \int u^2(u^2 + 1)\, du$$

$$= \int (u^4 + u^2)\, du$$

$$= \tfrac{1}{5}u^5 + \tfrac{1}{3}u^3 + C$$

$$= \tfrac{1}{5}\tan^5 x + \tfrac{1}{3}\tan^3 x + C. \qquad\blacksquare$$

Example 6 Evaluate $\displaystyle\int \tan^3 x \sec^5 x\, dx$.

Solution   Using Guideline 2 of (9.3),

$$\int \tan^3 x \sec^5 x\, dx = \int \tan^2 x \sec^4 x\,(\sec x \tan x)\, dx$$

$$= \int (\sec^2 x - 1)\sec^4 x\,(\sec x \tan x)\, dx.$$

Substituting $u = \sec x$ and $du = \sec x \tan x\, dx$, we obtain

$$\int \tan^3 x \sec^5 x\, dx = \int (u^2 - 1)u^4\, du$$

$$= \int (u^6 - u^4)\, du$$

$$= \tfrac{1}{7}u^7 - \tfrac{1}{5}u^5 + C$$

$$= \tfrac{1}{7}\sec^7 x - \tfrac{1}{5}\sec^5 x + C. \qquad\blacksquare$$

Integrals of the form $\int \cot^m x \csc^n x\, dx$ may be evaluated in similar fashion. Finally, integrals of the form $\int \sin mx \cos nx\, dx$ may be evaluated by means of the product formulas (see Appendix III), as illustrated in the next example.

Example 7 Evaluate $\displaystyle\int \cos 5x \cos 3x\, dx$.

Solution   We may write

$$\cos 5x \cos 3x = \tfrac{1}{2}(\cos 8x + \cos 2x).$$

Consequently,

$$\int \cos 5x \cos 3x \, dx = \tfrac{1}{2} \int (\cos 8x + \cos 2x) \, dx$$

$$= \tfrac{1}{16} \sin 8x + \tfrac{1}{4} \sin 2x + C. \qquad \blacksquare$$

EXERCISES 9.2

Evaluate the integrals in Exercise 1–30.

1 $\displaystyle\int \cos^3 x \, dx$

2 $\displaystyle\int \sin^2 2x \, dx$

3 $\displaystyle\int \sin^2 x \cos^2 x \, dx$

4 $\displaystyle\int \cos^7 x \, dx$

5 $\displaystyle\int \sin^3 x \cos^2 x \, dx$

6 $\displaystyle\int \sin^5 x \cos^3 x \, dx$

7 $\displaystyle\int \sin^6 x \, dx$

8 $\displaystyle\int \sin^4 x \cos^2 x \, dx$

9 $\displaystyle\int \tan^3 x \sec^4 x \, dx$

10 $\displaystyle\int \sec^6 x \, dx$

11 $\displaystyle\int \tan^3 x \sec^3 x \, dx$

12 $\displaystyle\int \tan^5 x \sec x \, dx$

13 $\displaystyle\int \tan^6 x \, dx$

14 $\displaystyle\int \cot^4 x \, dx$

15 $\displaystyle\int \sqrt{\sin x} \cos^3 x \, dx$

16 $\displaystyle\int \frac{\cos^3 x}{\sqrt{\sin x}} \, dx$

17 $\displaystyle\int (\tan x + \cot x)^2 \, dx$

18 $\displaystyle\int \cot^3 x \csc^3 x \, dx$

19 $\displaystyle\int_0^{\pi/4} \sin^3 x \, dx$

20 $\displaystyle\int_0^1 \tan^2 (\pi x/4) \, dx$

21 $\displaystyle\int \sin 5x \sin 3x \, dx$

22 $\displaystyle\int_0^{\pi/4} \cos x \cos 5x \, dx$

23 $\displaystyle\int_0^{\pi/2} \sin 3x \cos 2x \, dx$

24 $\displaystyle\int \sin 4x \cos 3x \, dx$

25 $\displaystyle\int \csc^4 x \cot^4 x \, dx$

26 $\displaystyle\int (1 + \sqrt{\cos x})^2 \sin x \, dx$

27 $\displaystyle\int \frac{\cos x}{2 - \sin x} \, dx$

28 $\displaystyle\int \frac{\tan^2 x - 1}{\sec^2 x} \, dx$

29 $\displaystyle\int \frac{\sec^2 x}{(1 + \tan x)^2} \, dx$

30 $\displaystyle\int \frac{\sec x}{\cot^5 x} \, dx$

31 The region bounded by the x-axis and the arc of $y = \cos^2 x$ from $x = 0$ to $x = 2\pi$ is revolved about the x-axis. Find the volume of the resulting solid.

32 The region between the graphs of $y = \tan^2 x$ and $y = 0$ from $x = 0$ to $x = \pi/4$ is revolved about the x-axis. Find the volume of the resulting solid.

33 Suppose the velocity (at time t) of a point moving on a coordinate line is $\cos^2 \pi t$ ft/sec. How far does the point travel in 5 seconds?

34 The acceleration (at time t) of a point moving along a coordinate line is $\sin^2 t \cos t$ ft/sec^2. At $t = 0$ the point is at the origin and its velocity is 10 ft/sec. Find its position at time t.

35 (a) Prove that if m and n are positive integers,

$$\int \sin mx \sin nx \, dx$$

$$= \begin{cases} \dfrac{\sin(m-n)x}{2(m-n)} - \dfrac{\sin(m+n)x}{2(m+n)} + C & \text{if } m \neq n \\[2ex] \dfrac{x}{2} - \dfrac{\sin 2mx}{4m} + C & \text{if } m = n \end{cases}$$

(b) Obtain formulas similar to that in part (a) for

$$\int \sin mx \cos nx \, dx \quad \text{and} \quad \int \cos mx \cos nx \, dx.$$

36 (a) Use (a) of Exercise 35 to prove that

$$\int_{-\pi}^{\pi} \sin mx \sin nx \, dx = \begin{cases} 0 & \text{if } m \neq n \\ \pi & \text{if } m = n \end{cases}$$

(b) Find

(i) $\displaystyle\int_{-\pi}^{\pi} \sin mx \cos nx \, dx$

(ii) $\displaystyle\int_{-\pi}^{\pi} \cos mx \cos nx \, dx$

9.3

TRIGONOMETRIC SUBSTITUTIONS

If an integrand contains one of the expressions $\sqrt{a^2 - x^2}$, $\sqrt{a^2 + x^2}$, or $\sqrt{x^2 - a^2}$, where $a > 0$, the radical sign can be eliminated by using the trigonometric substitution listed in the following table.

Expression	Trigonometric substitution
$\sqrt{a^2 - x^2}$	$x = a \sin \theta$
$\sqrt{a^2 + x^2}$	$x = a \tan \theta$
$\sqrt{x^2 - a^2}$	$x = a \sec \theta$

When making a trigonometric substitution we shall assume that θ is in the range of the corresponding inverse trigonometric function. Thus, for the substitution $x = a \sin \theta$ we have $-\pi/2 \le \theta \le \pi/2$. In this event, $\cos \theta \ge 0$ and

$$
\begin{aligned}
\sqrt{a^2 - x^2} &= \sqrt{a^2 - a^2 \sin^2 \theta} \\
&= \sqrt{a^2(1 - \sin^2 \theta)} \\
&= \sqrt{a^2 \cos^2 \theta} \\
&= a \cos \theta.
\end{aligned}
$$

If $\sqrt{a^2 - x^2}$ occurs in a denominator, we add the restriction $|x| \ne a$, or equivalently, $-\pi/2 < \theta < \pi/2$.

Example 1   Evaluate $\displaystyle\int \frac{1}{x^2\sqrt{16 - x^2}} \, dx$.

Solution   Let $x = 4 \sin \theta$, where $-\pi/2 < \theta < \pi/2$. It follows that

$$\sqrt{16 - x^2} = \sqrt{16 - 16 \sin^2 \theta} = 4\sqrt{1 - \sin^2 \theta} = 4 \cos \theta.$$

Since $x = 4 \sin \theta$, we have $dx = 4 \cos \theta \, d\theta$. Substituting in the given integral,

$$
\begin{aligned}
\int \frac{1}{x^2\sqrt{16 - x^2}} \, dx &= \int \frac{1}{(16 \sin^2 \theta)4 \cos \theta} 4 \cos \theta \, d\theta \\
&= \frac{1}{16} \int \frac{1}{\sin^2 \theta} \, d\theta \\
&= \frac{1}{16} \int \csc^2 \theta \, d\theta \\
&= -\frac{1}{16} \cot \theta + C.
\end{aligned}
$$

It is now necessary to return to the original variable of integration, x. Since $\theta = \arcsin (x/4)$, we could write $-\frac{1}{16} \cot \theta$ as $-\frac{1}{16} \cot \arcsin (x/4)$.

However, since the given integral involves $\sqrt{16 - x^2}$, it is desirable to have the evaluated form also contain this radical. A method for accomplishing this is to use the following geometric device. If $0 < \theta < \pi/2$, then since $\sin \theta = x/4$, we may interpret θ as an acute angle of a right triangle having opposite side and hypotenuse of lengths x and 4, respectively (see Figure 9.1). The length $\sqrt{16 - x^2}$ of the adjacent side is calculated by means of the Pythagorean Theorem. Referring to the triangle we see that

$$\cot \theta = \frac{\sqrt{16 - x^2}}{x}.$$

It can be shown that the last formula is also true if $-\pi/2 < \theta < 0$. Thus, Figure 9.1 may be used whether θ is positive or negative.

Substituting $\sqrt{16 - x^2}/x$ for $\cot \theta$ in our integral evaluation gives us

$$\int \frac{1}{x^2\sqrt{16 - x^2}} \, dx = -\frac{1}{16} \cdot \frac{\sqrt{16 - x^2}}{x} + C$$

$$= -\frac{\sqrt{16 - x^2}}{16x} + C. \quad \blacksquare$$

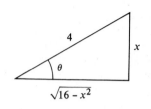

Figure 9.1 $\dfrac{x}{4} = \sin \theta$

If an integrand contains $\sqrt{a^2 + x^2}$, where $a > 0$, then the substitution $x = a \tan \theta$ will eliminate the radical sign. When using this substitution we will assume that θ is in the range of the inverse tangent function; that is, $-\pi/2 < \theta < \pi/2$. In this event $\sec \theta > 0$ and

$$\sqrt{a^2 + x^2} = \sqrt{a^2 + a^2 \tan^2 \theta}$$
$$= \sqrt{a^2(1 + \tan^2 \theta)}$$
$$= \sqrt{a^2 \sec^2 \theta}$$
$$= a \sec \theta.$$

After making this substitution and evaluating the resulting trigonometric integral, it is necessary to return to the variable x. The preceding formulas show that

$$\tan \theta = \frac{x}{a} \quad \text{and} \quad \sec \theta = \frac{\sqrt{a^2 + x^2}}{a}.$$

As in the solution of Example 1, the trigonometric functions of θ can be found by referring to the triangle in Figure 9.2, whether θ is positive or negative.

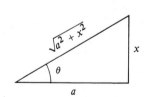

Figure 9.2 $\dfrac{x}{a} = \tan \theta$

Example 2 Evaluate $\displaystyle\int \frac{1}{\sqrt{4 + x^2}} \, dx$.

Solution Let us substitute as follows:

$$x = 2 \tan \theta, \qquad dx = 2 \sec^2 \theta \, d\theta.$$

Consequently,

$$\sqrt{4 + x^2} = \sqrt{4 + 4\tan^2\theta} = 2\sqrt{1 + \tan^2\theta} = 2\sqrt{\sec^2\theta} = 2\sec\theta$$

and

$$\int \frac{1}{\sqrt{4 + x^2}}\,dx = \int \frac{1}{2\sec\theta}\,2\sec^2\theta\,d\theta$$

$$= \int \sec\theta\,d\theta$$

$$= \ln|\sec\theta + \tan\theta| + C.$$

Since $\tan\theta = x/2$ we see from the triangle in Figure 9.3 that

$$\sec\theta = \frac{\sqrt{4 + x^2}}{2}$$

and hence

$$\int \frac{1}{\sqrt{4 + x^2}}\,dx = \ln\left|\frac{\sqrt{4 + x^2}}{2} + \frac{x}{2}\right| + C.$$

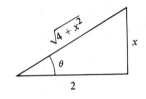

Figure 9.3 $\dfrac{x}{2} = \tan\theta$

The expression on the right may be written

$$\ln\left|\frac{\sqrt{4 + x^2} + x}{2}\right| + C = \ln|\sqrt{4 + x^2} + x| - \ln 2 + C.$$

Since $\sqrt{4 + x^2} + x > 0$ for all x, the absolute value sign is unnecessary. If we also let $D = -\ln 2 + C$, then

$$\int \frac{1}{\sqrt{4 + x^2}}\,dx = \ln(\sqrt{4 + x^2} + x) + D. \qquad \blacksquare$$

For integrands containing $\sqrt{x^2 - a^2}$ we substitute $x = a\sec\theta$, where θ is chosen in the range of the inverse secant function; that is, either $0 \le \theta < \pi/2$ or $\pi \le \theta < 3\pi/2$. In this case $\tan\theta \ge 0$ and

$$\sqrt{x^2 - a^2} = \sqrt{a^2\sec^2\theta - a^2}$$
$$= \sqrt{a^2(\sec^2\theta - 1)}$$
$$= \sqrt{a^2\tan^2\theta}$$
$$= a\tan\theta.$$

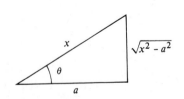

Figure 9.4 $\dfrac{x}{a} = \sec\theta$

Since $\sec\theta = \dfrac{x}{a}$ and $\tan\theta = \dfrac{\sqrt{x^2 - a^2}}{a}$

it follows that we may refer to the triangle in Figure 9.4 when changing from the variable θ to the variable x.

Example 3 Evaluate $\int \dfrac{\sqrt{x^2 - 9}}{x}\, dx$.

Solution Let us substitute as follows:

$$x = 3 \sec \theta, \qquad dx = 3 \sec \theta \tan \theta\, d\theta.$$

Consequently,

$$\sqrt{x^2 - 9} = \sqrt{9 \sec^2 \theta - 9} = 3\sqrt{\sec^2 \theta - 1} = 3\sqrt{\tan^2 \theta} = 3 \tan \theta$$

and, therefore,

$$\int \frac{\sqrt{x^2 - 9}}{x}\, dx = \int \frac{3 \tan \theta}{3 \sec \theta}\, 3 \sec \theta \tan \theta\, d\theta$$

$$= 3 \int \tan^2 \theta\, d\theta$$

$$= 3 \int (\sec^2 \theta - 1)\, d\theta$$

$$= 3(\tan \theta - \theta) + C.$$

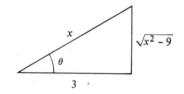

Figure 9.5 $\dfrac{x}{3} = \sec \theta$

Since $\sec \theta = x/3$ we may refer to the triangle in Figure 9.5 and write

$$\int \frac{\sqrt{x^2 - 9}}{x}\, dx = 3\left[\frac{\sqrt{x^2 - 9}}{3} - \sec^{-1}\left(\frac{x}{3}\right) \right] + C$$

$$= \sqrt{x^2 - 9} - 3 \sec^{-1}\left(\frac{x}{3}\right) + C. \qquad \blacksquare$$

Hyperbolic functions may also be used to simplify certain integrations. For example, $\cosh^2 u = 1 + \sinh^2 u$ and hence, if an integrand contains the expression $\sqrt{a^2 + x^2}$, the substitution $x = a \sinh u$ leads to

$$\sqrt{a^2 + x^2} = \sqrt{a^2 + a^2 \sinh^2 u}$$

$$= \sqrt{a^2(1 + \sinh^2 u)}$$

$$= \sqrt{a^2 \cosh^2 u}$$

$$= a \cosh u.$$

Example 4 Evaluate $\int \dfrac{1}{\sqrt{4 + x^2}}\, dx$ by using hyperbolic functions.

Solution The given integral is the same as that considered in Example 2. Let us substitute as follows:

$$x = 2 \sinh u, \qquad dx = 2 \cosh u\, du.$$

It follows that

$$\sqrt{4 + x^2} = \sqrt{4 + 4\sinh^2 u} = \sqrt{4\cosh^2 u} = 2\cosh u$$

and hence

$$\int \frac{1}{\sqrt{4 + x^2}}\,dx = \int \frac{1}{2\cosh u}\,2\cosh u\,du$$

$$= \int du = u + C$$

$$= \sinh^{-1}\left(\frac{x}{2}\right) + C.$$

The fact that this is equivalent to the solution in Example 2 follows from Theorem (8.22). ∎

Although additional integration techniques are now available, the reader should also keep earlier methods in mind. For example, the integral $\int (x/\sqrt{9 + x^2})\,dx$ could be evaluated by means of the trigonometric substitution $x = 3\tan\theta$. However, it is simpler to use the algebraic substitution $u = 9 + x^2$ and $du = 2x\,dx$, for in this event the integral takes on the form $\frac{1}{2}\int u^{-1/2}\,du$, which is readily integrated by means of the Power Rule. In the following exercises we shall include integrals that can be evaluated using simpler techniques than trigonometric substitutions.

EXERCISES 9.3

Evaluate the integrals in Exercises 1–22.

1 $\int \dfrac{x^2}{\sqrt{4 - x^2}}\,dx$

2 $\int \dfrac{\sqrt{4 - x^2}}{x^2}\,dx$

3 $\int \dfrac{1}{x\sqrt{9 + x^2}}\,dx$

4 $\int \dfrac{1}{x^2\sqrt{x^2 + 9}}\,dx$

5 $\int \dfrac{1}{x^2\sqrt{x^2 - 25}}\,dx$

6 $\int \dfrac{1}{x^3\sqrt{x^2 - 25}}\,dx$

7 $\int \dfrac{x}{\sqrt{4 - x^2}}\,dx$

8 $\int \dfrac{x}{x^2 + 9}\,dx$

9 $\int \dfrac{1}{(x^2 - 1)^{3/2}}\,dx$

10 $\int \dfrac{1}{\sqrt{4x^2 - 25}}\,dx$

11 $\int \dfrac{1}{(36 + x^2)^2}\,dx$

12 $\int \dfrac{1}{(16 - x^2)^{5/2}}\,dx$

13 $\int \sqrt{9 - 4x^2}\,dx$

14 $\int \dfrac{\sqrt{x^2 + 1}}{x}\,dx$

15 $\int \dfrac{x}{(16 - x^2)^2}\,dx$

16 $\int x\sqrt{x^2 - 9}\,dx$

17 $\int \dfrac{x^3}{\sqrt{9x^2 + 49}}\,dx$

18 $\int \dfrac{1}{x\sqrt{25x^2 + 16}}\,dx$

19 $\int \dfrac{1}{x^4\sqrt{x^2 - 3}}\,dx$

20 $\int \dfrac{x^2}{(1 - 9x^2)^{3/2}}\,dx$

21 $\int \dfrac{(4 + x^2)^2}{x^3}\,dx$

22 $\int \dfrac{3x - 5}{\sqrt{1 - x^2}}\,dx$

23–28 Use trigonometric substitutions to obtain the formulas in Theorems (8.14) and (8.15).

29 The region bounded by the graphs of $y = x(x^2 + 25)^{-1/2}$, $y = 0$, and $x = 5$ is revolved about the y-axis. Find the volume of the resulting solid.

30 Find the area of the region bounded by the graph of $y = x^3(10 - x^2)^{-1/2}$, the x-axis, and the line $x = 1$.

31 Find the arc length of the graph of $y = x^2/2$ from $A(0, 0)$ to $B(2, 2)$.

32 The region bounded by the graphs of $y = 1/(x^2 + 1)$, $y = 0$, $x = 0$, and $x = 2$ is revolved about the x-axis. Find the volume of the resulting solid.

33 Find the area of the region bounded by the graphs of $y = \sqrt{x^2 + 4}$, $y = 0$, $x = 0$, and $x = 3$.

34 Find the area of the region enclosed by the graph of $4x^2 + y^2 = 16$.

35 Suppose that $y = f(x)$ and $x\, dy - \sqrt{x^2 - 16}\, dx = 0$. If $f(4) = 0$, find $f(x)$.

36 Suppose two variables x and y are related such that $\sqrt{1 - x^2}\, dy = x^3\, dx$. If $y = 0$ when $x = 0$, express y as a function of x.

Evaluate the integrals in Exercises 37–40 by means of a hyperbolic substitution.

37 $\displaystyle \int \frac{1}{x^2\sqrt{25 + x^2}}\, dx$

38 $\displaystyle \int \frac{x^2}{(x^2 + 9)^{3/2}}\, dx$

39 $\displaystyle \int \frac{1}{x^2\sqrt{1 - x^2}}\, dx$ (*Hint*: Let $x = \tanh u$ and use (8.19).

40 $\displaystyle \int \frac{1}{16 - x^2}\, dx$

In Exercises 41–46 use trigonometric substitutions to establish the indicated formulas in the Table of Integrals (see inside front cover).

41 Formula 21   **42** Formula 27

43 Formula 31   **44** Formula 36

45 Formula 41   **46** Formula 44

9.4

PARTIAL FRACTIONS

It is easy to verify that

$$\frac{2}{x^2 - 1} = \frac{1}{x - 1} + \frac{-1}{x + 1}.$$

The expression on the right side of the equation is called the *partial fraction decomposition* of $2/(x^2 - 1)$. This decomposition may be used to find the indefinite integral of $2/(x^2 - 1)$. We merely integrate each of the fractions that make up the decomposition, obtaining

$$\int \frac{2}{x^2 - 1}\, dx = \int \frac{1}{x - 1}\, dx + \int \frac{-1}{x + 1}\, dx$$

$$= \ln|x - 1| - \ln|x + 1| + C$$

$$= \ln\left|\frac{x - 1}{x + 1}\right| + C.$$

It is theoretically possible to write *any* rational expression $f(x)/g(x)$ as a sum of rational expressions whose denominators involve powers of polynomials of degree not greater than two. Specifically, if $f(x)$ and $g(x)$ are polynomials *and the degree of $f(x)$ is less than the degree of $g(x)$*, then it follows from a theorem in algebra that

$$\frac{f(x)}{g(x)} = F_1 + F_2 + \cdots + F_k$$

where each F_i has one of the forms

$$\frac{A}{(px + q)^m} \quad \text{or} \quad \frac{Cx + D}{(ax^2 + bx + c)^n}$$

for some nonnegative integers m and n, and where $ax^2 + bx + c$ is **irreducible**, in the sense that this quadratic expression has no real zeros, that is, $b^2 - 4ac < 0$. In this event $ax^2 + bx + c$ cannot be expressed as a product of two first-degree polynomials. The sum $F_1 + F_2 + \cdots + F_k$ is called the **partial fraction decomposition** of $f(x)/g(x)$ and each F_i is called a **partial fraction**. We shall not prove this algebraic result but will, instead, give guidelines for obtaining the decomposition.

To find the partial fraction decomposition of a rational expression $f(x)/g(x)$ *it is essential* that $f(x)$ have lower degree than $g(x)$. If this is not the case, then long division should be employed to arrive at such an expression. For example, given

$$\frac{x^3 - 6x^2 + 5x - 3}{x^2 - 1}$$

we obtain, by long division,

$$\frac{x^3 - 6x^2 + 5x - 3}{x^2 - 1} = x - 6 + \frac{6x - 9}{x^2 - 1}.$$

The partial fraction decomposition is then found for $(6x - 9)/(x^2 - 1)$.

(9.4) *Guidelines for finding partial fraction decompositions of $f(x)/g(x)$*

A   If the degree of $f(x)$ is not lower than the degree of $g(x)$, use long division to obtain the proper form.

B   Express $g(x)$ as a product of linear factors $px + q$ or irreducible quadratic factors $ax^2 + bx + c$, and collect repeated factors so that $g(x)$ is a product of *different* factors of the form $(px + q)^m$ or $(ax^2 + bx + c)^n$, where m and n are nonnegative integers.

C   Apply the following rules.

Rule 1.   For each factor of the form $(px + q)^m$ where $m \geq 1$, the partial fraction decomposition contains a sum of m partial fractions of the form

$$\frac{A_1}{px + q} + \frac{A_2}{(px + q)^2} + \cdots + \frac{A_m}{(px + q)^m}$$

where each A_i is a real number.

Rule 2.   For each factor of the form $(ax^2 + bx + c)^n$ where $n \geq 1$ and $ax^2 + bx + c$ is irreducible, the partial fraction decomposition contains a sum of n partial fractions of the form

$$\frac{A_1 x + B_1}{ax^2 + bx + c} + \frac{A_2 x + B_2}{(ax^2 + bx + c)^2} + \cdots + \frac{A_n x + B_n}{(ax^2 + bx + c)^n}$$

where each A_i and B_i is a real number.

Example 1　Evaluate $\displaystyle\int \frac{4x^2 + 13x - 9}{x^3 + 2x^2 - 3x}\,dx.$

Solution　The denominator of the integrand has the factored form $x(x + 3)(x - 1)$. Each of the linear factors is handled under Rule 1 of (9.4), with $m = 1$. Thus, for the factor x there corresponds a partial fraction of the form A/x. Similarly, for the factors $x + 3$ and $x - 1$ there correspond partial fractions $B/(x + 3)$ and $C/(x - 1)$, respectively. Thus the partial fraction decomposition has the form

$$\frac{4x^2 + 13x - 9}{x(x + 3)(x - 1)} = \frac{A}{x} + \frac{B}{x + 3} + \frac{C}{x - 1}.$$

Multiplying by the lowest common denominator gives us

(∗)　　　$4x^2 + 13x - 9 = A(x + 3)(x - 1) + Bx(x - 1) + Cx(x + 3),$

where we have used the symbol (∗) for later reference. In a case such as this, in which the factors are all linear and nonrepeated, the values for A, B, and C can be found by substituting values for x that make the various factors zero. If we let $x = 0$ in (∗), then

$$-9 = -3A \quad \text{or} \quad A = 3.$$

Letting $x = 1$ in (∗) gives us

$$8 = 4C \quad \text{or} \quad C = 2.$$

Finally, if $x = -3$, then

$$-12 = 12B \quad \text{or} \quad B = -1.$$

The partial fraction decomposition is, therefore,

$$\frac{4x^2 + 13x - 9}{x(x + 3)(x - 1)} = \frac{3}{x} + \frac{-1}{x + 3} + \frac{2}{x - 1}.$$

Integrating,

$$\int \frac{4x^2 + 13x - 9}{x(x + 3)(x - 1)}\,dx = \int \frac{3}{x}\,dx + \int \frac{-1}{x + 3}\,dx + \int \frac{2}{x - 1}\,dx$$

$$= 3\ln|x| - \ln|x + 3| + 2\ln|x - 1| + D$$

$$= \ln|x^3| - \ln|x + 3| + \ln|x - 1|^2 + D$$

$$= \ln\left|\frac{x^3(x - 1)^2}{x + 3}\right| + D.$$

Another technique for finding A, B, and C is to compare coefficients of x. If the right-hand side of (∗) is expanded and like powers of x are collected, then

$$4x^2 + 13x - 9 = (A + B + C)x^2 + (2A - B + 3C)x - 3A.$$

We now use the fact that if two polynomials are equal, then coefficients of like powers are the same. Thus

$$A + B + C = 4$$
$$2A - B + 3C = 13$$
$$-3A = -9$$

It is left to the reader to show that the solution of this system of equations is $A = 3$, $B = -1$, and $C = 2$.   ∎

Example 2   Evaluate $\displaystyle\int \frac{3x^3 - 18x^2 + 29x - 4}{(x + 1)(x - 2)^3}\,dx.$

Solution   By Rule 1 of (9.4), there is a partial fraction of the form $A/(x + 1)$ corresponding to the factor $x + 1$ in the denominator of the integrand. For the factor $(x - 2)^3$ we apply Rule 1 (with $m = 3$), obtaining a sum of three partial fractions $B/(x - 2)$, $C/(x - 2)^2$, and $D/(x - 2)^3$. Consequently, the partial fraction decomposition has the form

$$\frac{3x^3 - 18x^2 + 29x - 4}{(x + 1)(x - 2)^3} = \frac{A}{x + 1} + \frac{B}{x - 2} + \frac{C}{(x - 2)^2} + \frac{D}{(x - 2)^3}.$$

Multiplying both sides by $(x + 1)(x - 2)^3$ gives us

$$(*) \qquad 3x^3 - 18x^2 + 29x - 4$$
$$= A(x - 2)^3 + B(x + 1)(x - 2)^2 + C(x + 1)(x - 2) + D(x + 1).$$

Two of the unknown constants may be determined easily. If we let $x = 2$ in $(*)$, then

$$24 - 72 + 58 - 4 = 3D, \quad 6 = 3D, \quad \text{and} \quad D = 2.$$

Similarly, letting $x = -1$ in $(*)$,

$$-3 - 18 - 29 - 4 = -27A, \quad -54 = -27A, \quad \text{and} \quad A = 2.$$

The remaining constants may be found by comparing coefficients. If the right side of $(*)$ is expanded and like powers of x collected, we see that the coefficient of x^3 is $A + B$. This must equal the coefficient of x^3 on the left, that is,

$$A + B = 3.$$

Since $A = 2$, it follows that $B = 3 - A = 3 - 2 = 1$. Finally, we compare the constant terms in $(*)$ by letting $x = 0$. This gives us

$$-4 = -8A + 4B - 2C + D.$$

Substituting the values we have found for A, B, and D leads to

$$-4 = -16 + 4 - 2C + 2$$

which has the solution $C = -3$. The partial fraction decomposition is, therefore,

$$\frac{3x^3 - 18x^2 + 29x - 4}{(x + 1)(x - 2)^3} = \frac{2}{x + 1} + \frac{1}{x - 2} + \frac{-3}{(x - 2)^2} + \frac{2}{(x - 2)^3}.$$

To find the given integral we integrate each of the partial fractions on the right side of the last equation. This gives us

$$2 \ln |x + 1| + \ln |x - 2| + \frac{3}{x - 2} - \frac{1}{(x - 2)^2} + E$$

which may be written in the more compact form

$$\ln (x + 1)^2 |x - 2| + \frac{3x - 7}{(x - 2)^2} + E. \qquad \blacksquare$$

Example 3 Evaluate $\displaystyle\int \frac{x^2 - x - 21}{2x^3 - x^2 + 8x - 4} \, dx.$

Solution   The denominator may be factored by grouping as follows:

$$2x^3 - x^2 + 8x - 4 = x^2(2x - 1) + 4(2x - 1) = (x^2 + 4)(2x - 1).$$

Applying Rule 2 of (9.4) to the irreducible quadratic factor $x^2 + 4$ we see that one of the partial fractions has the form $(Ax + B)/(x^2 + 4)$. By Rule 1, there is also a partial fraction $C/(2x - 1)$ corresponding to the factor $2x - 1$. Consequently,

$$\frac{x^2 - x - 21}{2x^3 - x^2 + 8x - 4} = \frac{Ax + B}{x^2 + 4} + \frac{C}{2x - 1}.$$

As in previous examples, this leads to

(∗) $$x^2 - x - 21 = (Ax + B)(2x - 1) + C(x^2 + 4).$$

Substituting $x = \frac{1}{2}$ we obtain $\frac{1}{4} - \frac{1}{2} - 21 = \frac{17}{4}C$, which has the solution $C = -5$. The remaining constants may be found by comparing coefficients. Rearranging the right side of (∗) gives us

$$x^2 - x - 21 = (2A + C)x^2 + (-A + 2B)x - B + 4C.$$

Comparing the coefficients of x^2 we see that $2A + C = 1$. Since $C = -5$ it follows that $2A = 6$ or $A = 3$. Similarly, comparing the constant terms, $-B + 4C = -21$ and hence $-B - 20 = -21$ or $B = 1$. Thus the partial fraction decomposition of the integrand is

$$\frac{x^2 - x - 21}{2x^3 - x^2 + 8x - 4} = \frac{3x + 1}{x^2 + 4} + \frac{-5}{2x - 1}$$

$$= \frac{3x}{x^2 + 4} + \frac{1}{x^2 + 4} - \frac{5}{2x - 1}.$$

The given integral may now be found by integrating the right side of the last equation. This gives us

$$\frac{3}{2} \ln (x^2 + 4) + \frac{1}{2} \tan^{-1} \frac{x}{2} - \frac{5}{2} \ln |2x - 1| + D. \qquad \blacksquare$$

Example 4 Evaluate $\int \dfrac{5x^3 - 3x^2 + 7x - 3}{(x^2 + 1)^2} \, dx.$

Solution   Applying Rule 2 of (9.4), with $n = 2$,

$$\frac{5x^3 - 3x^2 + 7x - 3}{(x^2 + 1)^2} = \frac{Ax + B}{x^2 + 1} + \frac{Cx + D}{(x^2 + 1)^2}$$

and, therefore,

$$5x^3 - 3x^2 + 7x - 3 = (Ax + B)(x^2 + 1) + Cx + D$$

or $5x^3 - 3x^2 + 7x - 3 = Ax^3 + Bx^2 + (A + C)x + (B + D).$

Comparing the coefficients of x^3 and x^2 we obtain $A = 5$ and $B = -3$. From the coefficients of x we see that $A + C = 7$ or $C = 7 - A = 7 - 5 = 2$. Finally, the constant terms give us $B + D = -3$ or $D = -3 - B = -3 - (-3) = 0$. Therefore,

$$\frac{5x^3 - 3x^2 + 7x - 3}{(x^2 + 1)^2} = \frac{5x - 3}{x^2 + 1} + \frac{2x}{(x^2 + 1)^2}$$

$$= \frac{5x}{x^2 + 1} - \frac{3}{x^2 + 1} + \frac{2x}{(x^2 + 1)^2}.$$

Integrating, we obtain

$$\int \frac{5x^3 - 3x^2 + 7x - 3}{(x^2 + 1)^2} \, dx = \tfrac{5}{2} \ln (x^2 + 1) - 3 \tan^{-1} x - \frac{1}{x^2 + 1} + E. \qquad \blacksquare$$

EXERCISES 9.4

Evaluate the integrals in Exercises 1–32.

1 $\displaystyle\int \frac{5x - 12}{x(x - 4)} \, dx$

2 $\displaystyle\int \frac{x + 34}{(x - 6)(x + 2)} \, dx$

3 $\displaystyle\int \frac{37 - 11x}{(x + 1)(x - 2)(x - 3)} \, dx$

4 $\displaystyle\int \frac{4x^2 + 54x + 134}{(x - 1)(x + 5)(x + 3)} \, dx$

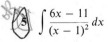
5 $\displaystyle\int \frac{6x - 11}{(x - 1)^2} \, dx$

6 $\displaystyle\int \frac{-19x^2 + 50x - 25}{x^2(3x - 5)} \, dx$

7 $\int \dfrac{x + 16}{x^2 + 2x - 8} \, dx$

8 $\int \dfrac{11x + 2}{2x^2 - 5x - 3} \, dx$

9 $\int \dfrac{5x^2 - 10x - 8}{x^3 - 4x} \, dx$

10 $\int \dfrac{4x^2 - 5x - 15}{x^3 - 4x^2 - 5x} \, dx$

11 $\int \dfrac{2x^2 - 25x - 33}{(x + 1)^2(x - 5)} \, dx$

12 $\int \dfrac{2x^2 - 12x + 4}{x^3 - 4x^2} \, dx$

13 $\int \dfrac{9x^4 + 17x^3 + 3x^2 - 8x + 3}{x^5 + 3x^4} \, dx$

14 $\int \dfrac{5x^2 + 30x + 43}{(x + 3)^3} \, dx$

15 $\int \dfrac{x^3 + 3x^2 + 3x + 63}{(x^2 - 9)^2} \, dx$

16 $\int \dfrac{1}{(x - 7)^5} \, dx$

17 $\int \dfrac{5x^2 + 11x + 17}{x^3 + 5x^2 + 4x + 20} \, dx$

18 $\int \dfrac{4x^3 - 3x^2 + 6x - 27}{x^4 + 9x^2} \, dx$

19 $\int \dfrac{x^2 + 3x + 1}{x^4 + 5x^2 + 4} \, dx$

20 $\int \dfrac{4x}{(x^2 + 1)^3} \, dx$

21 $\int \dfrac{2x^3 + 10x}{(x^2 + 1)^2} \, dx$

22 $\int \dfrac{x^4 + 2x^2 + 4x + 1}{(x^2 + 1)^3} \, dx$

23 $\int \dfrac{x^3 + 3x - 2}{x^2 - x} \, dx$

24 $\int \dfrac{x^4 + 2x^2 + 3}{x^3 - 4x} \, dx$

25 $\int \dfrac{x^6 - x^3 + 1}{x^4 + 9x^2} \, dx$

26 $\int \dfrac{x^5}{(x^2 + 4)^2} \, dx$

27 $\int \dfrac{2x^3 - 5x^2 + 46x + 98}{(x^2 + x - 12)^2} \, dx$

28 $\int \dfrac{-2x^4 - 3x^3 - 3x^2 + 3x + 1}{x^2(x + 1)^3} \, dx$

29 $\int \dfrac{4x^3 + 2x^2 - 5x - 18}{(x - 4)(x + 1)^3} \, dx$

30 $\int \dfrac{10x^2 + 9x + 1}{2x^3 + 3x^2 + x} \, dx$

31 $\int \dfrac{2x^4 - 2x^3 + 6x^2 - 5x + 1}{x^3 - x^2 + x - 1} \, dx$

32 $\int \dfrac{x^5 - x^4 - 2x^3 + 4x^2 - 15x + 5}{(x^2 + 1)^2(x^2 + 4)} \, dx$

Use partial fractions to evaluate the integrals in Exercises 33–36 (see formulas 19, 49, 50, and 52 of the Table of Integrals that appears on the inside front cover).

33 $\int \dfrac{1}{a^2 - u^2} \, du$

34 $\int \dfrac{1}{u(a + bu)} \, du$

35 $\int \dfrac{1}{u^2(a + bu)} \, du$

36 $\int \dfrac{1}{u(a + bu)^2} \, du$

37 If $f(x) = x/(x^2 - 2x - 3)$, find the area of the region under the graph of f from $x = 0$ to $x = 2$.

38 The region bounded by the graphs of $y = 1/(x - 1)(4 - x)$, $y = 0$, $x = 2$, and $x = 3$ is revolved about the y-axis. Find the volume of the resulting solid.

39 If the region described in Exercise 38 is revolved about the x-axis, find the volume of the resulting solid.

40 Suppose the velocity of a point moving along a coordinate line is $(t + 3)/(t^3 + t)$ ft/sec, where t is the time in seconds. How far does the point travel during the time interval $[1, 2]$?

41 As an alternative to partial fractions, show that an integral of the form

$$\int \dfrac{1}{ax^2 + bx} \, dx$$

may be evaluated by writing it as

$$\int \dfrac{(1/x^2)}{a + (b/x)} \, dx$$

and using the substitution $u = a + (b/x)$.

42 Generalize Exercise 41 to integrals of the form

$$\int \dfrac{1}{ax^n + bx} \, dx.$$

43 Suppose $g(x) = (x - c_1)(x - c_2) \cdots (x - c_n)$, where n is a positive integer and the real numbers $c_1, c_2, \ldots, c_n$ are all different. If $f(x)$ is a polynomial of degree less than n, show that

$$\frac{f(x)}{g(x)} = \frac{A_1}{x - c_1} + \frac{A_2}{x - c_2} + \cdots + \frac{A_n}{x - c_n}$$

where $A_i = f(c_i)/g'(c_i)$ for $i = 1, 2, \ldots, n$. (This is a method for finding the partial fraction decomposition if the denominator can be factored into distinct linear factors.)

44 Use Exercise 43 to find the partial fraction decomposition of

$$\frac{2x^4 - x^3 - 3x^2 + 5x + 7}{x^5 - 5x^3 + 4x}.$$

9.5

QUADRATIC EXPRESSIONS

Partial fraction decompositions may lead to integrands containing an irreducible quadratic expression $ax^2 + bx + c$. If $b \neq 0$ it is often necessary to complete the square as follows:

$$ax^2 + bx + c = a\left(x^2 + \frac{b}{a}x\right) + c$$

$$= a\left(x + \frac{b}{2a}\right)^2 + c - \frac{b^2}{4a}.$$

The substitution $u = x + (b/2a)$ may then lead to an integrable form.

Example 1 Evaluate $\displaystyle\int \frac{2x - 1}{x^2 - 6x + 13}\, dx$.

Solution Note that the quadratic expression $x^2 - 6x + 13$ is irreducible, since $b^2 - 4ac = 36 - 52 = -16 < 0$. We complete the square as follows:

$$x^2 - 6x + 13 = (x^2 - 6x \qquad) + 13$$
$$= (x^2 - 6x + 9) + 13 - 9 = (x - 3)^2 + 4.$$

If we let $u = x - 3$, then $x = u + 3$, $dx = du$, and hence

$$\int \frac{2x - 1}{(x - 3)^2 + 4}\, dx = \int \frac{2(u + 3) - 1}{u^2 + 4}\, du$$

$$= \int \frac{2u + 5}{u^2 + 4}\, du$$

$$= \int \frac{2u}{u^2 + 4}\, du + 5 \int \frac{1}{u^2 + 4}\, du$$

$$= \ln (u^2 + 4) + \frac{5}{2} \tan^{-1} \frac{u}{2} + C$$

$$= \ln (x^2 - 6x + 13) + \frac{5}{2} \tan^{-1} \frac{x - 3}{2} + C. \quad \blacksquare$$

The technique of completing the square may also be employed if quadratic expressions appear under a radical sign.

Example 2   Evaluate $\int \dfrac{1}{\sqrt{8 + 2x - x^2}}\, dx$.

Solution   We may complete the square for the quadratic expression $8 + 2x - x^2$ as follows:

$$8 + 2x - x^2 = 8 - (x^2 - 2x) = 8 + 1 - (x^2 - 2x + 1)$$
$$= 9 - (x - 1)^2.$$

Next, letting $u = x - 1$ we have $du = dx$, and hence

$$\int \frac{1}{\sqrt{8 + 2x - x^2}}\, dx = \int \frac{1}{\sqrt{9 - u^2}}\, du$$

$$= \sin^{-1} \frac{u}{3} + C$$

$$= \sin^{-1} \frac{x - 1}{3} + C. \qquad \blacksquare$$

In the next example it is necessary to make a trigonometric substitution after completing the square.

Example 3   Evaluate $\int \dfrac{1}{\sqrt{x^2 + 8x + 25}}\, dx$.

Solution   We complete the square for the quadratic expression as follows:

$$x^2 + 8x + 25 = (x^2 + 8x \qquad) + 25$$
$$= (x^2 + 8x + 16) + 25 - 16$$
$$= (x + 4)^2 + 9.$$

Hence $\displaystyle \int \frac{1}{\sqrt{x^2 + 8x + 25}}\, dx = \int \frac{1}{\sqrt{(x + 4)^2 + 9}}\, dx.$

If we next make the trigonometric substitution

$$x + 4 = 3 \tan \theta$$

then $dx = 3 \sec^2 \theta\, d\theta$

and $\sqrt{(x + 4)^2 + 9} = \sqrt{9 \tan^2 \theta + 9} = 3\sqrt{\tan^2 \theta + 1} = 3 \sec \theta.$

Hence

$$\int \frac{1}{\sqrt{x^2 + 8x + 25}}\, dx = \int \frac{1}{3 \sec \theta}\, 3 \sec^2 \theta\, d\theta$$

$$= \int \sec \theta\, d\theta$$

$$= \ln|\sec \theta + \tan \theta| + C.$$

To return to the variable x, we use the triangle in Figure 9.6. This gives us

$$\int \frac{1}{\sqrt{x^2 + 8x + 25}}\, dx = \ln\left|\frac{\sqrt{x^2 + 8x + 25}}{3} + \frac{x + 4}{3}\right| + C$$

$$= \ln|\sqrt{x^2 + 8x + 25} + x + 4| + D$$

where $D = C - \ln 3$. ■

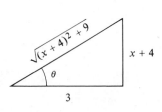

Figure 9.6 $\dfrac{x + 4}{3} = \tan \theta$

EXERCISES 9.5

Evaluate the integrals in Exercises 1–18.

1 $\displaystyle\int \frac{1}{x^2 - 4x + 8}\, dx$

2 $\displaystyle\int \frac{1}{\sqrt{7 + 6x - x^2}}\, dx$

3 $\displaystyle\int \frac{1}{\sqrt{4x - x^2}}\, dx$

4 $\displaystyle\int \frac{1}{x^2 - 2x + 2}\, dx$

5 $\displaystyle\int \frac{2x + 3}{\sqrt{9 - 8x - x^2}}\, dx$

6 $\displaystyle\int \frac{x + 5}{9x^2 + 6x + 17}\, dx$

7 $\displaystyle\int \frac{1}{x^3 - 1}\, dx$

8 $\displaystyle\int \frac{x^3}{x^3 - 1}\, dx$

9 $\displaystyle\int \frac{1}{(x^2 + 4x + 5)^2}\, dx$

10 $\displaystyle\int \frac{1}{x^4 - 4x^3 + 13x^2}\, dx$

11 $\displaystyle\int \frac{1}{(x^2 + 6x + 13)^{3/2}}\, dx$

12 $\displaystyle\int \sqrt{x(6 - x)}\, dx$

13 $\displaystyle\int \frac{1}{2x^2 - 3x + 9}\, dx$

14 $\displaystyle\int \frac{2x}{(x^2 + 2x + 5)^2}\, dx$

15 $\displaystyle\int_2^3 \frac{x^2 - 4x + 6}{x^2 - 4x + 5}\, dx$

16 $\displaystyle\int \frac{x}{2x^2 + 3x - 4}\, dx$

17 $\displaystyle\int \frac{e^x}{e^{2x} + 3e^x + 2}\, dx$

18 $\displaystyle\int_0^1 \frac{x - 1}{x^2 + x + 1}\, dx$

19 Find the area of the region bounded by the graphs of $y = (x^3 + 1)^{-1}$, $y = 0$, $x = 0$, and $x = 1$.

20 The region bounded by the graph of $y = 1/(x^2 + 2x + 10)$, the coordinate axes, and the line $x = 2$ is revolved about the x-axis. Find the volume of the resulting solid.

21 If the region described in Exercise 20 is revolved about the y-axis, find the volume of the resulting solid.

22 The velocity (at time t) of a point moving along a coordinate line is $(75 + 10t - t^2)^{-1/2}$ ft/sec. How far does the point travel during the time interval $[0, 5]$?

9.6

MISCELLANEOUS SUBSTITUTIONS

We have often used a change of variables to aid in the evaluation of a definite or indefinite integral. In this section we shall consider additional substitutions that are sometimes useful. The first example indicates that if an integral contains an expression of the form $\sqrt[n]{f(x)}$, then one of the substitutions $u = \sqrt[n]{f(x)}$ or $u = f(x)$ may simplify the evaluation.

Example 1   Evaluate $\displaystyle\int \frac{x^3}{\sqrt[3]{x^2 + 4}}\, dx$.

Solution 1   The substitution $u = \sqrt[3]{x^2 + 4}$ leads to the following equivalent equations:

$$u = \sqrt[3]{x^2 + 4}, \qquad u^3 = x^2 + 4, \qquad x^2 = u^3 - 4.$$

Taking the differential of each side of the last equation, we obtain

$$2x\, dx = 3u^2\, du, \quad \text{or} \quad x\, dx = \tfrac{3}{2}u^2\, du.$$

We now substitute in the given integral as follows:

$$\int \frac{x^3}{\sqrt[3]{x^2 + 4}}\, dx = \int \frac{x^2}{\sqrt[3]{x^2 + 4}} \cdot x\, dx$$

$$= \int \frac{u^3 - 4}{u} \cdot \frac{3}{2} u^2\, du = \frac{3}{2} \int (u^4 - 4u)\, du$$

$$= \frac{3}{2}\left(\frac{1}{5} u^5 - 2u^2\right) + C = \frac{3}{10} u^2(u^3 - 10) + C$$

$$= \frac{3}{10}(x^2 + 4)^{2/3}(x^2 - 6) + C.$$

Solution 2   If we substitute u for the expression *underneath* the radical, then

$$u = x^2 + 4 \qquad x^2 = u - 4$$

$$2x\, dx = du \qquad x\, dx = \tfrac{1}{2}\, du.$$

In this case we may write

$$\int \frac{x^3}{\sqrt[3]{x^2 + 4}}\, dx = \int \frac{x^2}{\sqrt[3]{x^2 + 4}} \cdot x\, dx$$

$$= \int \frac{u - 4}{u^{1/3}} \cdot \frac{1}{2}\, du = \frac{1}{2} \int (u^{2/3} - 4u^{-1/3})\, du$$

$$= \frac{1}{2}\left[\frac{3}{5} u^{5/3} - 6u^{2/3}\right] + C = \frac{3}{10} u^{2/3}[u - 10] + C$$

$$= \frac{3}{10}(x^2 + 4)^{2/3}(x^2 - 6) + C. \qquad\blacksquare$$

Example 2 Evaluate $\int \dfrac{1}{\sqrt{x} + \sqrt[3]{x}}\, dx$.

Solution   If we let $z = \sqrt[6]{x}$, then

$$x = z^6, \qquad \sqrt{x} = z^3, \qquad \sqrt[3]{x} = z^2, \qquad dx = 6z^5\, dz$$

and

$$\int \frac{1}{\sqrt{x} + \sqrt[3]{x}}\, dx = \int \frac{1}{z^3 + z^2}\, 6z^5\, dz = 6 \int \frac{z^3}{z + 1}\, dz.$$

By long division,

$$\frac{z^3}{z + 1} = z^2 - z + 1 - \frac{1}{z + 1}.$$

Consequently,

$$\int \frac{1}{\sqrt{x} + \sqrt[3]{x}}\, dx = 6 \int \left[z^2 - z + 1 - \frac{1}{z + 1} \right] dz$$

$$= 6(\tfrac{1}{3}z^3 - \tfrac{1}{2}z^2 + z - \ln|z + 1|) + C$$

$$= 2\sqrt{x} - 3\sqrt[3]{x} + 6\sqrt[6]{x} - 6 \ln (\sqrt[6]{x} + 1) + C. \qquad \blacksquare$$

If an integrand is a rational expression in $\sin x$ and $\cos x$, then the substitution $z = \tan (x/2)$ where $-\pi < x < \pi$ will transform it into a rational (algebraic) expression in z. To prove this, first note that

$$\cos \frac{x}{2} = \frac{1}{\sec (x/2)} = \frac{1}{\sqrt{1 + \tan^2 (x/2)}} = \frac{1}{\sqrt{1 + z^2}},$$

$$\sin \frac{x}{2} = \tan \frac{x}{2} \cos \frac{x}{2} = z\, \frac{1}{\sqrt{1 + z^2}}.$$

Consequently,

$$\sin x = 2 \sin \frac{x}{2} \cos \frac{x}{2} = \frac{2z}{1 + z^2},$$

$$\cos x = 1 - 2 \sin^2 \frac{x}{2} = 1 - \frac{2z^2}{1 + z^2} = \frac{1 - z^2}{1 + z^2}.$$

Moreover, since $x/2 = \tan^{-1} z$, we have $x = 2 \tan^{-1} z$ and, therefore,

$$dx = \frac{2}{1 + z^2}\, dz.$$

The following theorem summarizes this discussion.

Theorem (9.5)

If an integrand is a rational expression in sin x and cos x, the following substitutions will produce a rational expression in z:

$$\sin x = \frac{2z}{1 + z^2}, \qquad \cos x = \frac{1 - z^2}{1 + z^2}, \qquad dx = \frac{2}{1 + z^2} \, dz,$$

where $z = \tan (x/2)$.

Example 3 Evaluate $\displaystyle\int \frac{1}{4 \sin x - 3 \cos x} \, dx$.

Solution Applying Theorem (9.5) and simplifying the integrand,

$$\int \frac{1}{4 \sin x - 3 \cos x} \, dx = \int \frac{1}{4\left(\dfrac{2z}{1 + z^2}\right) - 3\left(\dfrac{1 - z^2}{1 + z^2}\right)} \cdot \frac{2}{1 + z^2} \, dz$$

$$= \int \frac{2}{8z - 3(1 - z^2)} \, dz$$

$$= 2 \int \frac{1}{3z^2 + 8z - 3} \, dz.$$

Using partial fractions,

$$\frac{1}{3z^2 + 8z - 3} = \frac{1}{10} \left(\frac{3}{3z - 1} - \frac{1}{z + 3}\right)$$

and hence

$$\int \frac{1}{4 \sin x - 3 \cos x} \, dx = \frac{1}{5} \int \left(\frac{3}{3z - 1} - \frac{1}{z + 3}\right) dz$$

$$= \frac{1}{5} (\ln |3z - 1| - \ln |z + 3|) + C$$

$$= \frac{1}{5} \ln \left|\frac{3z - 1}{z + 3}\right| + C$$

$$= \frac{1}{5} \ln \left|\frac{3 \tan (x/2) - 1}{\tan (x/2) + 3}\right| + C. \qquad \blacksquare$$

Other substitutions are sometimes useful; however, it is impossible to state rules that apply to all situations. Whether or not one can express an integrand in a suitable form is often a matter of individual ingenuity.

EXERCISES 9.6

Evaluate the integrals in Exercises 1–28.

1 $\int x\sqrt[3]{x+9}\,dx$

2 $\int x^2\sqrt{2x+1}\,dx$

3 $\int \dfrac{x}{\sqrt[5]{3x+2}}\,dx$

4 $\int \dfrac{5x}{(x+3)^{2/3}}\,dx$

5 $\int_4^9 \dfrac{1}{\sqrt{x}+4}\,dx$

6 $\int_0^{25} \dfrac{1}{\sqrt{4+\sqrt{x}}}\,dx$

7 $\int \dfrac{\sqrt{x}}{1+\sqrt[3]{x}}\,dx$

8 $\int \dfrac{1}{\sqrt[4]{x}+\sqrt[3]{x}}\,dx$

9 $\int \dfrac{1}{(x+1)\sqrt{x-2}}\,dx$

10 $\int_0^4 \dfrac{2x+3}{\sqrt{1+2x}}\,dx$

11 $\int \dfrac{x+1}{(x+4)^{1/3}}\,dx$

12 $\int \dfrac{x^{1/3}+1}{x^{1/3}-1}\,dx$

13 $\int e^{3x}\sqrt{1+e^x}\,dx$

14 $\int \dfrac{e^{2x}}{\sqrt[3]{1+e^x}}\,dx$

15 $\int \dfrac{e^{2x}}{e^x+4}\,dx$

16 $\int \dfrac{\sin 2x}{\sqrt{1+\sin x}}\,dx$

17 $\int \sin\sqrt{x+4}\,dx$

18 $\int \sqrt{x}\,e^{\sqrt{x}}\,dx$

19 $\int_2^3 \dfrac{x}{(x-1)^6}\,dx$ (*Hint:* Let $u=x-1$.)

20 $\int \dfrac{x^2}{(3x+4)^{10}}\,dx$ (*Hint:* Let $u=3x+4$.)

21 $\int \dfrac{1}{2+\sin x}\,dx$

22 $\int \dfrac{1}{3+2\cos x}\,dx$

23 $\int \dfrac{1}{1+\sin x+\cos x}\,dx$

24 $\int \dfrac{1}{\tan x+\sin x}\,dx$

25 $\int \dfrac{\sec x}{4-3\tan x}\,dx$

26 $\int \dfrac{1}{\sin x-\sqrt{3}\cos x}\,dx$

27 $\int \dfrac{\sin 2x}{\sin^2 x-2\sin x-8}\,dx$ (*Hint:* Let $u=\sin x$.)

28 $\int \dfrac{\sin x}{5\cos x+\cos^2 x}\,dx$ (*Hint:* Let $u=\cos x$.)

Use an appropriate substitution to establish the identities in Exercises 29 and 30 for any real numbers n and m.

29 $\int_0^{\pi/2} \sin^n x\,dx = \int_0^{\pi/2} \cos^n x\,dx$

30 $\int_0^1 x^m(1-x)^n\,dx = \int_0^1 x^n(1-x)^m\,dx$

31 Use an appropriate substitution to show that

$$\int_x^1 \frac{1}{1 + t^2} \, dt = \int_1^{1/x} \frac{1}{1 + t^2} \, dt \quad \text{for } x > 0$$

and then use this result to establish the trigonometric identity

$$\arctan x + \arctan \frac{1}{x} = \frac{\pi}{2}$$

(cf. Exercise 24 of Section 8.4.).

32 Show that

$$\int \frac{1}{(x + a)^{3/2} + (x - a)^{3/2}} \, dx$$

may be transformed into the integral of a rational function by means of the substitution $x = (a/2)(t^2 + t^{-2})$.

Use Theorem (9.5) to verify the formulas in Exercises 33 and 34.

33 $\displaystyle\int \sec x \, dx = \ln \left| \frac{1 + \tan (x/2)}{1 - \tan (x/2)} \right| + C$

34 $\displaystyle\int \csc x \, dx = \frac{1}{2} \ln \left(\frac{1 - \cos x}{1 + \cos x} \right) + C$

9.7

TABLES OF INTEGRALS

Mathematicians and scientists who use integrals in their work often refer to tables of integrals. Many of the formulas contained in these tables may be obtained by methods we have studied. In general, tables of integrals should not be used until the student has had sufficient experience with the standard methods of integration. Indeed, for complicated integrals it is often necessary to make substitutions, or to use partial fractions, integration by parts, or other techniques to obtain integrands for which the table is applicable.

The following examples illustrate the use of several formulas stated in the brief Table of Integrals printed on the inside covers of this text. To guard against errors when using the table, answers should always be checked by differentiation.

Evaluate $\displaystyle\int x^3 \cos x \, dx.$

We first use reduction Formula 85 in the Table of Integrals with $n = 3$ and $u = x$, obtaining

$$\int x^3 \cos x \, dx = x^3 \sin x - 3 \int x^2 \sin x \, dx.$$

Next we apply Formula 84 with $n = 2$, and then Formula 83, obtaining

$$\int x^2 \sin x \, dx = -x^2 \cos x + 2 \int x \cos x \, dx$$

$$= -x^2 \cos x + 2[\cos x + x \sin x] + C.$$

Substitution in the first expression gives us

$$\int x^3 \cos x \, dx = x^3 \sin x + 3x^2 \cos x - 6 \cos x - 6x \sin x + C. \quad \blacksquare$$

Example 2   Evaluate $\int \dfrac{1}{x^2\sqrt{3 + 5x^2}}\, dx$, where $x > 0$.

Solution   The integrand suggests that we use that part of the table dealing with the form $\sqrt{a^2 + u^2}$. Specifically, Formula 28 states that

$$\int \frac{du}{u^2\sqrt{a^2 + u^2}} = -\frac{\sqrt{a^2 + u^2}}{a^2 u} + C$$

where for compactness the differential *du* is placed in the numerator instead of to the right of the integrand. To use this formula we must adjust the given integral so that it matches *exactly* with the formula. If we let

$$a^2 = 3 \quad \text{and} \quad u^2 = 5x^2$$

then the expression underneath the radical is taken care of; however, we also need

(i) u^2 to the left of the radical.

(ii) *du* in the numerator.

We can achieve (i) by writing the integral as

$$5 \int \frac{1}{5x^2\sqrt{3 + 5x^2}}\, dx.$$

To accomplish (ii) we note that

$$u = \sqrt{5}\,x \quad \text{and} \quad du = \sqrt{5}\, dx$$

and write the preceding integral as

$$5 \cdot \frac{1}{\sqrt{5}} \int \frac{1}{5x^2\sqrt{3 + 5x^2}} \sqrt{5}\, dx.$$

The integral now matches exactly with that in Formula 28 and hence

$$\int \frac{1}{x^2\sqrt{3 + 5x^2}}\, dx = \sqrt{5}\left[-\frac{\sqrt{3 + 5x^2}}{3(\sqrt{5}\,x)} \right] + C$$

$$= -\frac{\sqrt{3 + 5x^2}}{3x} + C. \qquad \blacksquare$$

As illustrated in the next example, it may be necessary to make a substitution of some type before a table can be used to help evaluate an integral.

Example 3   Evaluate $\int \dfrac{\sin 2x}{\sqrt{3 - 5\cos x}}\, dx$.

Solution   Let us begin by writing the integral as

$$\int \frac{2 \sin x \cos x}{\sqrt{3 - 5 \cos x}}\, dx.$$

Since there are no formulas in the table that have this form, we consider making the substitution $u = \cos x$. In this case $du = -\sin x\, dx$ and the integral may be written

$$2 \int \frac{\sin x \cos x}{\sqrt{3 - 5 \cos x}}\, dx = -2 \int \frac{\cos x}{\sqrt{3 - 5 \cos x}}(-\sin x)\, dx$$

$$= -2 \int \frac{u}{\sqrt{3 - 5u}}\, du.$$

Referring to the Table of Integrals we see that Formula 55 is

$$\int \frac{u\, du}{\sqrt{a + bu}} = \frac{2}{3b^2}(bu - 2a)\sqrt{a + bu}.$$

Using this result with $a = 3$ and $b = -5$ gives us

$$-2 \int \frac{u}{\sqrt{3 - 5u}}\, du = -2\left(\frac{2}{75}\right)(-5u - 6)\sqrt{3 - 5u} + C.$$

Finally, since $u = \cos x$, we obtain

$$\int \frac{\sin 2x}{\sqrt{3 - 5 \cos x}}\, dx = \frac{4}{75}(5 \cos x + 6)\sqrt{3 - 5 \cos x} + C. \qquad ■$$

EXERCISES 9.7

Use the Table of Integrals printed on the inside covers of this text to evaluate the following integrals.

1 $\displaystyle\int \frac{\sqrt{4 + 9x^2}}{x}\, dx$

2 $\displaystyle\int \frac{1}{x\sqrt{2 + 3x^2}}\, dx$

3 $\displaystyle\int (16 - x^2)^{3/2}\, dx$

4 $\displaystyle\int x^2\sqrt{4x^2 - 16}\, dx$

5 $\displaystyle\int x\sqrt{2 - 3x}\, dx$

6 $\displaystyle\int x^2\sqrt{5 + 2x}\, dx$

7 $\displaystyle\int \sin^6 3x\, dx$

8 $\displaystyle\int x \cos^5 (x^2)\, dx$

9 $\displaystyle\int \csc^4 x\, dx$

10 $\displaystyle\int \sin 5x \cos 3x\, dx$

11 $\displaystyle\int x \sin^{-1} x\, dx$

12 $\displaystyle\int x^2 \tan^{-1} x\, dx$

13 $\displaystyle\int e^{-3x} \sin 2x\, dx$

14 $\displaystyle\int x^5 \ln x\, dx$

15 $\displaystyle\int \frac{\sqrt{5x - 9x^2}}{x}\, dx$

16 $\displaystyle\int \frac{1}{x\sqrt{3x - 2x^2}}\, dx$

17 $\displaystyle\int \frac{x}{5x^4 - 3}\, dx$

18 $\displaystyle\int \cos x\sqrt{\sin^2 x - 4}\, dx$

19 $\displaystyle\int e^{2x} \cos^{-1} e^x\, dx$

20 $\displaystyle\int \sin^2 x \cos^3 x\, dx$

21 $\displaystyle\int x^3\sqrt{2 + x}\, dx$

22 $\displaystyle\int \frac{7x^3}{\sqrt{2 - x}}\, dx$

23 $\displaystyle\int \frac{\sin 2x}{4 + 9 \sin x}\, dx$ **24** $\displaystyle\int \frac{\tan x}{\sqrt{4 + 3 \sec x}}\, dx$ **27** $\displaystyle\int \frac{1}{x(4 + \sqrt[3]{x})}\, dx$ **28** $\displaystyle\int \frac{1}{2x^{3/2} + 5x^2}\, dx$

25 $\displaystyle\int \frac{\sqrt{9 + 2x}}{x}\, dx$ **26** $\displaystyle\int \sqrt{8x^3 - 3x^2}\, dx$ **29** $\displaystyle\int \sqrt{16 - \sec^2 x}\, \tan x\, dx$ **30** $\displaystyle\int \frac{\cot x}{\sqrt{4 - \csc^2 x}}\, dx$

9.8

MOMENTS AND CENTROIDS OF PLANE REGIONS

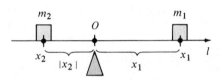

Figure 9.7

Some applications of the definite integral were considered in Chapter 6. In this section and the next, several other applications are discussed.

Let l be a coordinate line and P a point on l with coordinate x. If a particle of mass m is located at P, then the **moment** (more precisely, the **first moment**) of the particle with respect to the origin O is defined as the product mx. Suppose l is horizontal with positive direction to the right and let us imagine that l is free to rotate about O as if a fulcrum were positioned as shown in Figure 9.7. If an object has its mass m_1 concentrated at a point with positive coordinate x_1, then the moment $m_1 x_1$ is positive, and l would rotate in a clockwise direction. If an object of mass m_2 is at a point with negative coordinate x_2, then its moment $m_2 x_2$ is negative, and l would rotate in a counterclockwise direction. The system consisting of both objects is said to be in **equilibrium** if $m_1 x_1 = m_2 |x_2|$. Since $x_2 < 0$, this is equivalent to $m_1 x_1 = -m_2 x_2$, or

$$m_1 x_1 + m_2 x_2 = 0;$$

that is, *the sum of the moments with respect to the origin is zero*. This situation is similar to a seesaw that balances at the point O if two persons having masses of m_1 and m_2, respectively, are located as indicated in Figure 9.7.

If the system is not in equilibrium, then as illustrated in Figure 9.8 there is a "balance" point P with coordinate $\bar{x}$, in the sense that

$$m_1(x_1 - \bar{x}) = m_2 |x_2 - \bar{x}| = -m_2(x_2 - \bar{x})$$

where $|x_2 - \bar{x}| = -(x_2 - \bar{x})$ since $x_2 < \bar{x}$.

To locate P we may solve for $\bar{x}$ as follows:

$$m_1(x_1 - \bar{x}) + m_2(x_2 - \bar{x}) = 0$$
$$m_1 x_1 + m_2 x_2 - (m_1 + m_2)\bar{x} = 0$$
$$(m_1 + m_2)\bar{x} = m_1 x_1 + m_2 x_2$$
$$\bar{x} = \frac{m_1 x_1 + m_2 x_2}{m_1 + m_2}.$$

Thus, to find $\bar{x}$ we divide the sum of the moments with respect to the origin by the total mass $m = m_1 + m_2$. The point with coordinate $\bar{x}$ is called the

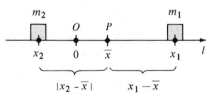

Figure 9.8

center of mass (or **center of gravity**) of the system. The next definition is an extension of this concept to more than two objects.

Definition (9.6)

Let l be a coordinate line and let a system of particles of masses m_1, m_2, ..., m_n be located at points with coordinates x_1, x_2, ..., x_n, respectively.

(i) The **moment of the system with respect to the origin** is the sum

$$\sum_{i=1}^{n} m_i x_i .$$

(ii) Let $m = \sum_{i=1}^{n} m_i$ be the total mass of the system. The **center of mass** (or **center of gravity**) of the system is the point with coordinate $\bar{x}$, such that

$$\bar{x} = \frac{\sum_{i=1}^{n} m_i x_i}{m} \quad \text{or} \quad m\bar{x} = \sum_{i=1}^{n} m_i x_i .$$

The number $m\bar{x}$ in Definition (9.6) may be regarded as the moment with respect to the origin of a particle of mass m located at the point with coordinate $\bar{x}$. The formula in (ii) of (9.6) then states that $\bar{x}$ gives the position at which the total mass m could be concentrated without changing the moment of the system with respect to the origin. The point with coordinate $\bar{x}$ is the balance point of the system in the sense of our seesaw illustration. If $\bar{x} = 0$, then by (9.6) $\sum_{i=1}^{n} m_i x_i = 0$ and the system is said to be in **equilibrium**. In this event the origin is the center of mass.

Example 1 Three particles of masses 40, 60, and 100 grams are located at points with coordinates -2, 3, and 7, respectively, on a coordinate line l. Find the center of mass of the system.

Solution If we denote the three masses by m_1, m_2, and m_3, then we have the situation illustrated in Figure 9.9, where $x_1 = -2$, $x_2 = 3$, and $x_3 = 7$.

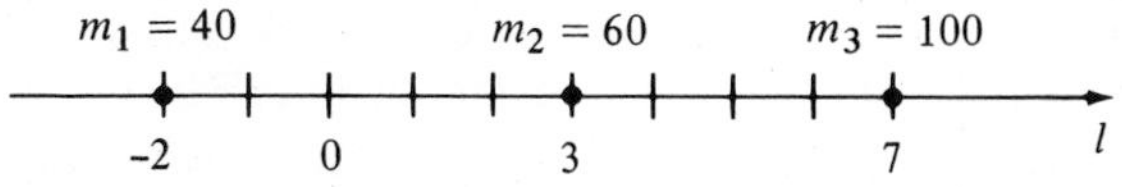

Figure 9.9

Applying Definition (9.6), the coordinate $\bar{x}$ of the center of mass is given by

$$\bar{x} = \frac{40(-2) + 60(3) + 100(7)}{40 + 60 + 100}$$

$$= \frac{-80 + 180 + 700}{200} = \frac{800}{200} = 4. \qquad \blacksquare$$

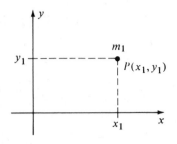

Figure 9.10

The concepts defined in (9.6) may be extended to two dimensions. Specifically, if a particle of mass m_1 is located at a point $P(x_1, y_1)$ in a coordinate plane, as illustrated in Figure 9.10, then the moments M_x and M_y of the particle with respect to the x-axis and y-axis, respectively, are defined by

$$M_x = m_1 y_1 \quad \text{and} \quad M_y = m_1 x_1.$$

Note that if x_1 and y_1 are both positive, then to find M_x we multiply the mass of the particle by its distance y_1 from the x-axis, whereas M_y is found by multiplying the mass by the distance x_1 from the y-axis. If either x_1 or y_1 is negative, then M_x or M_y is also negative.

As indicated in the next definition, to find M_x and M_y for a collection of particles, we add the moments of the individual particles.

Definition (9.7)

Let n particles of masses $m_1, m_2, \ldots, m_n$ be located at points $P_1(x_1, y_1)$, $P_2(x_2, y_2), \ldots, P_n(x_n, y_n)$, respectively, in a coordinate plane.

(i)   The **moments M_x and M_y of the system** with respect to the x-axis and y-axis, respectively, are

$$M_x = \sum_{i=1}^{n} m_i y_i \quad \text{and} \quad M_y = \sum_{i=1}^{n} m_i x_i.$$

(ii)   Let $m = \sum_{i=1}^{n} m_i$ be the total mass of the system. The **center of mass** (or **center of gravity**) of the system is the point $P(\bar{x}, \bar{y})$ such that

$$m\bar{x} = M_y \quad \text{and} \quad m\bar{y} = M_x.$$

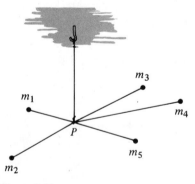

Figure 9.11

The number $m\bar{x}$ may be regarded as the moment with respect to the y-axis of a particle of mass m located at the point $P(\bar{x}, \bar{y})$, and $m\bar{y}$ may be thought of as its moment with respect to the x-axis. Consequently, the equations in (ii) of Definition (9.7) imply that the center of mass of the system of particles is the point at which the total mass could be concentrated without changing the moments of the system with respect to the coordinate axes. To get an intuitive picture of this situation, we may think of the n particles as fastened to the center of mass P by weightless rods, in a manner similar to the way spokes of a wheel are attached to the center of the wheel. The system would then balance when supported from the ceiling of a room by a cord attached to P, as illustrated in Figure 9.11. The appearance would be similar to a mobile; in our case the mobile has all its objects in the same horizontal plane.

Example 2 Particles of masses 4, 8, 3, and 2 kg are located at the points $P_1(-2, 3), P_2(2, -6), P_3(7, -3), P_4(5, 1)$, respectively. Find the moments M_x and M_y and the coordinates of the center of mass of the system.

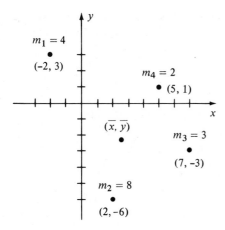

Figure 9.12

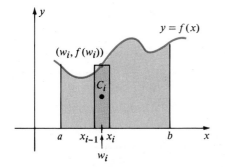

Figure 9.13

Solution   The system is illustrated in Figure 9.12, where we have also anticipated the position of $(\bar{x}, \bar{y})$. Applying Definition (9.7),

$$M_x = (4)(3) + (8)(-6) + (3)(-3) + (2)(1) = -43$$
$$M_y = (4)(-2) + (8)(2) + (3)(7) + (2)(5) = 39.$$

Since $m = 4 + 8 + 3 + 2 = 17$, it follows from (ii) of (9.7) that

$$\bar{x} = \frac{M_y}{m} = \frac{39}{17} \approx 2.3, \qquad \bar{y} = \frac{M_x}{m} = -\frac{43}{17} \approx -2.5. \qquad \blacksquare$$

Let us next consider a thin sheet of material (called a **lamina**) that is **homogeneous**, that is, has a constant density. We wish to define the center of mass P so that it is the balance point, in a sense analogous to that for systems of particles. If a lamina has the shape of a rectangle, then it is evident that the center of mass is the intersection of the diagonals. If the rectangle lies in an xy-plane, then we assume that the mass of the rectangle (more precisely, of the rectangular lamina) can be concentrated at the center of mass without changing its moments with respect to the x- or y-axes.

For simplicity let us begin by considering a lamina that has the shape of a region of the type illustrated in Figure 9.13, where f is continuous on the closed interval $[a, b]$. We partition $[a, b]$ by choosing numbers $a = x_0 < x_1 < \cdots < x_n = b$ and, for each i, choose w_i as the midpoint of the subinterval $[x_{i-1}, x_i]$, that is, $w_i = (x_{i-1} + x_i)/2$. The Riemann sum $\sum_{i=1}^{n} f(w_i) \Delta x_i$ may be thought of as the sum of areas of rectangles of the type shown in Figure 9.13.

If the **area density** (the mass per unit area) is denoted by ρ, then the mass corresponding to the ith rectangle is $\rho f(w_i) \Delta x_i$, and consequently the mass of the rectangular polygon associated with the partition is the sum $\sum_i \rho f(w_i) \Delta x_i$. As the norm $\|P\|$ of the partition approaches zero, the area of the rectangular polygon approaches the area of the face of the lamina, and the sum should approach the mass of the lamina. It is natural, therefore, to define the mass m of the lamina by

$$m = \lim_{\|P\| \to 0} \sum_i \rho f(w_i) \Delta x_i = \rho \int_a^b f(x) \, dx.$$

The center of mass of the ith rectangular lamina illustrated in Figure 9.13 is located at the point $C_i(w_i, \frac{1}{2}f(w_i))$. If we assume that the mass is concentrated at C_i, then its moment with respect to the x-axis can be found by multiplying the distance $\frac{1}{2}f(w_i)$ from the x-axis to C_i by the mass $\rho f(w_i) \Delta x_i$. Using the additive property of moments, the moment of the rectangular polygon associated with the partition is $\sum_i \frac{1}{2}f(w_i) \cdot \rho f(w_i) \Delta x_i$. The moment M_x of the lamina is defined as the limit of this sum, that is,

$$M_x = \lim_{\|P\| \to 0} \sum_i \frac{1}{2}f(w_i) \cdot \rho f(w_i) \Delta x_i = \rho \int_a^b \frac{1}{2}f(x) \cdot f(x) \, dx.$$

In like manner, using the distance w_i from the y-axis to the center of mass of the ith rectangle, we arrive at the definition of the moment M_y of the lamina with respect to the y-axis. Specifically,

$$M_y = \lim_{\|P\| \to 0} \sum_i w_i \cdot \rho f(w_i) \, \Delta x_i = \rho \int_a^b x \cdot f(x) \, dx.$$

As with particles (see Definition (9.7)), the coordinates $\bar{x}$ and $\bar{y}$ of the center of mass of the lamina are defined by $m\bar{x} = M_y$ and $m\bar{y} = M_x$.

The following definition summarizes this discussion.

Definition (9.8)

Let the function f be continuous and nonnegative on $[a, b]$. If a homogeneous lamina of area density ρ has the shape of the region under the graph of f from a to b, then

(i) the **mass** of the lamina is $m = \rho \int_a^b f(x) \, dx.$

(ii) the **moments M_x and M_y** of the lamina are

$$M_x = \rho \int_a^b \tfrac{1}{2} f(x) \cdot f(x) \, dx \quad \text{and} \quad M_y = \rho \int_a^b x \cdot f(x) \, dx.$$

(iii) the **center of mass** (or **center of gravity**) of the lamina is the point $P(\bar{x}, \bar{y})$ such that

$$m\bar{x} = M_y \quad \text{and} \quad m\bar{y} = M_x.$$

Substituting the integral forms into (iii) of Definition (9.8) and solving for $\bar{x}$ and $\bar{y}$ gives us

$$\bar{x} = \frac{M_y}{m} = \frac{\rho \int_a^b x \cdot f(x) \, dx}{\rho \int_a^b f(x) \, dx}, \qquad \bar{y} = \frac{M_x}{m} = \frac{\rho \int_a^b \tfrac{1}{2} f(x) \cdot f(x) \, dx}{\rho \int_a^b f(x) \, dx}.$$

Since the constant ρ in these formulas may be canceled, we see that the coordinates of the center of mass of a homogeneous lamina are independent of the density ρ; that is, they depend only on the shape of the lamina and not on the density. For this reason the point $(\bar{x}, \bar{y})$ is sometimes referred to as the center of mass of a *region* in the plane, or as the **centroid** of the region. We can obtain formulas for centroids by letting $\rho = 1$ in the preceding formulas.

Example 3 Find the coordinates of the centroid of the region bounded by the graphs of $y = e^x$, $x = 0$, $x = 1$, and $y = 0$.

Solution   The region is sketched in Figure 9.14. Using Definition (9.8) with $\rho = 1$,

$$m = \int_0^1 e^x \, dx = e^x \Big]_0^1 = e - 1$$

$$M_x = \int_0^1 \tfrac{1}{2} e^x \cdot e^x \, dx = \int_0^1 \tfrac{1}{2} e^{2x} \, dx = \tfrac{1}{4} e^{2x} \Big]_0^1 = \tfrac{1}{4}(e^2 - 1).$$

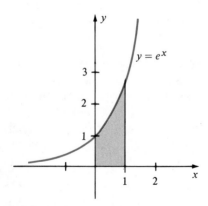

Figure 9.14

Consequently,

$$\bar{y} = \frac{M_x}{m} = \frac{\frac{1}{4}(e^2 - 1)}{e - 1} = \frac{e + 1}{4} \approx 0.93.$$

Next, by (ii) of (9.8),

$$M_y = \int_0^1 xe^x \, dx.$$

Integrating by parts (with $u = x$ and $dv = e^x \, dx$), we obtain

$$M_y = \left[xe^x - e^x \right]_0^t = (e - e) - (0 - 1) = 1$$

and, therefore,

$$\bar{x} = \frac{M_y}{m} = \frac{1}{e - 1} \approx 0.58. \qquad \blacksquare$$

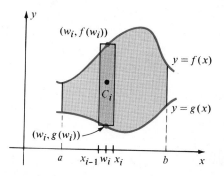

Figure 9.15

Formulas similar to those in Definition (9.8) may be obtained for more complicated regions. Thus, consider a lamina of constant density ρ which has the shape illustrated in Figure 9.15, where f and g are continuous functions and $f(x) \geq g(x)$ for all x in $[a, b]$. Partitioning $[a, b]$, choosing w_i as before, and then applying the Midpoint Formula, we see that the center of mass of the ith rectangular lamina pictured in Figure 9.15 is the point

$$C_i(w_i, \tfrac{1}{2}[f(w_i) + g(w_i)]).$$

Using an argument similar to that given previously, the moment of the ith rectangle with respect to the x-axis is the distance from the x-axis to C_i multiplied by the mass; that is,

$$\tfrac{1}{2}[f(w_i) + g(w_i)] \cdot \rho[f(w_i) - g(w_i)] \, \Delta x_i.$$

Summing and taking the limit as the norm of the partition approaches zero gives us

$$M_x = \rho \int_a^b \tfrac{1}{2}[f(x) + g(x)] \cdot [f(x) - g(x)] \, dx$$

$$= \rho \int_a^b \tfrac{1}{2}\{[f(x)]^2 - [g(x)]^2\} \, dx.$$

In like manner,

$$M_y = \rho \int_a^b x[f(x) - g(x)] \, dx.$$

The formulas in (iii) of (9.8) may then be used to find $\bar{x}$ and $\bar{y}$.

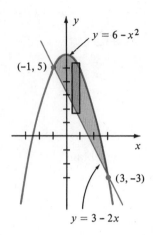

Figure 9.16

Example 4 Find the coordinates of the centroid of the region bounded by the graphs of $y + x^2 = 6$ and $y + 2x - 3 = 0$.

Solution   The region is the same as that considered in Example 2 of Section 6.1 where it was found that the area equals 32/3. The region is resketched in Figure 9.16. If we let $f(x) = 6 - x^2$ and $g(x) = 3 - 2x$, then as in the preceding discussion, with $\rho = 1$,

$$M_x = \int_{-1}^{3} \tfrac{1}{2}[(6 - x^2) + (3 - 2x)][(6 - x^2) - (3 - 2x)]\, dx$$

$$= \tfrac{1}{2} \int_{-1}^{3} [(6 - x^2)^2 - (3 - 2x)^2]\, dx$$

$$= \tfrac{1}{2} \int_{-1}^{3} (x^4 - 16x^2 + 12x + 27)\, dx.$$

It is left to the reader to show that $M_x = 416/15$. Hence,

$$\bar{y} = \frac{M_x}{m} = \frac{416/15}{32/3} = \frac{13}{5}.$$

Similarly, $$M_y = \int_{-1}^{3} x[(6 - x^2) - (3 - 2x)]\, dx$$

$$= \int_{-1}^{3} x(3 - x^2 + 2x)\, dx$$

$$= \int_{-1}^{3} (3x - x^3 + 2x^2)\, dx$$

from which it follows that $M_y = 32/2$. Consequently,

$$\bar{x} = \frac{M_y}{m} = \frac{32/3}{32/3} = 1.$$ ■

Formulas for moments may also be obtained for other types of regions; however, it is advisable to remember the *technique* of finding moments of rectangular laminas (by multiplying a distance from an axis by a mass), instead of memorizing formulas that cover all possible cases.

EXERCISES 9.8

1 Particles of masses 2, 7, and 5 kg are located at the points $A(4, -1)$, $B(-2, 0)$, and $C(-8, -5)$, respectively. Find the moments M_x and M_y and the coordinates of the center of mass of this system.

2 Particles of masses 10, 3, 4, 1, and 8 g are located at the points $A(-5, -2)$, $B(3, 7)$, $C(0, -3)$, $D(-8, -3)$, and $O(0, 0)$. Find the moments M_x and M_y and the coordinates of the center of mass of this system.

In Exercises 3–12, sketch the region bounded by the graphs of the given equations and find the centroid of the region.

3 $y = x^3, y = 0, x = 1$

4 $y = \sqrt{x}, y = 0, x = 9$

5 $y = \sin x, y = 0, x = 0, x = \pi$

6 $y = \sec^2 x, y = 0, x = -\pi/4, x = \pi/4$

7 $y = 1 - x^2, y = x - 1$

8 $x = y^2, x - y - 2 = 0$

9 $y = 1/\sqrt{16 + x^2}, y = 0, x = 0, x = 3$

10 $xy = 1, y = 0, x = 1, x = e$

11 $y = e^{2x}, y = 0, x = -1, x = 0$

12 $y = \cosh x, y = 0, x = -1, x = 1$

13 Prove that the centroid of a triangle coincides with the intersection of the medians. (*Hint:* Take the vertices at the points $(0, 0)$, (a, b), and $(0, c)$, where a, b, and c are positive.)

14 Find the centroid of the region in the first quadrant bounded by the circle $x^2 + y^2 = a^2$ and the coordinate axes.

15 Find the centroid of a semicircular region of radius a.

16 Let $0 < a < b$ and let the points P, Q, R, and S have coordinates $(-b, 0)$ $(-a, 0)$, $(a, 0)$ and $(b, 0)$, respectively. Find the centroid of the region bounded by the graphs of $y = \sqrt{b^2 - x^2}$, $y = \sqrt{a^2 - x^2}$, and the line segments PQ and RS. (*Hint:* Use Exercise 15).

17 A plane region has the shape of a square of side $2a$ surmounted by a semicircle of radius a. Find the centroid. (*Hint:* Use Exercise 15 and the fact that moments are additive.)

18 A region has the shape of a square of side a surmounted by an equilateral triangle of side a. Find the centroid. (*Hint:* Use Exercise 13 and the fact that moments are additive.)

19 Find the centroid of the region in the first quadrant that is bounded by the coordinate axes and the graph of $x^{2/3} + y^{2/3} = a^{2/3}$.

20 Let R be the plane region in the first quadrant bounded by part of the parabola $y^2 = cx$ where $c > 0$, the x-axis, and the vertical line through the point (a, b) on the parabola, as shown in the figure. Prove that the centroid of R is $(\frac{3}{5}a, \frac{3}{8}b)$.

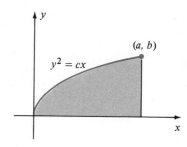

Figure for Exercise 20

9.9

CENTROIDS OF SOLIDS OF REVOLUTION

The center of mass of a homogeneous solid of revolution is a certain point on the axis of revolution. In the special case of a right circular cylinder of finite altitude, the center of mass is the point on the axis that lies halfway between the two bases. To develop formulas for locating the center of mass of a solid we begin with an xy-plane and introduce a third coordinate axis, called the **z-axis**, perpendicular to both the x- and y-axes at the origin, as illustrated in Figure 9.17. The y- and z-axes determine a coordinate plane called the

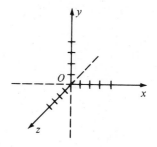

Figure 9.17

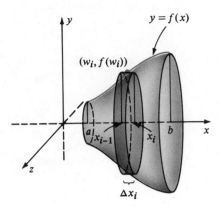

Figure 9.18

yz-plane. Similarly, the plane determined by the *x*- and *z*-axes is referred to as the **xz-plane**.

Let us consider a solid of revolution of the type shown in Figure 9.18, that is, a solid obtained by revolving, about the *x*-axis, the region bounded by the graph of a continuous function *f* and the lines $x = a$, $x = b$, and $y = 0$. It will be assumed that the resulting solid has constant **density** (mass per unit volume) ρ and that its mass *m* is the product of ρ and the volume of the solid. Thus, as in Definition (6.3),

$$m = \rho \int_a^b \pi [f(x)]^2 \, dx.$$

Since the center of mass *C* of the solid in Figure 9.18 is on the *x*-axis, it is sufficient to define the *x*-coordinate $\bar{x}$ of *C*. Let us partition $[a, b]$ in the usual way, choose w_i as the midpoint of the *i*th subinterval $[x_{i-1}, x_i]$, and consider the rectangle with its base on $[x_{i-1}, x_i]$ and with altitude $f(w_i)$. The mass of the disc generated by revolving this rectangle about the *x*-axis is $\rho \pi [f(w_i)]^2 \, \Delta x_i$. The center of mass of this disc is the point on the *x*-axis with coordinate w_i. The **moment with respect to the yz-plane of the ith disc** is defined as the product of the distance w_i from the *yz*-plane to the center of mass of the *i*th disc and the mass of the disc, that is,

$$w_i \cdot \rho \pi [f(w_i)]^2 \, \Delta x_i.$$

Assuming that moments are additive, the moment with respect to the *yz*-plane of the solid that consists of the discs determined by all such rectangles is

$$\sum_i w_i \cdot \rho \pi [f(w_i)]^2 \, \Delta x_i.$$

As in the next definition, the moment of the solid is the limit

$$\int_a^b x \cdot \rho \pi [f(x)]^2 \, dx,$$

of this sum.

Definition (9.9)

Let *f* be continuous on $[a, b]$ and let *R* be the region bounded by the graphs of $y = f(x)$, $x = a$, $x = b$, and $y = 0$. Let *Q* be the solid of revolution obtained by revolving *R* about the *x*-axis. If *Q* has constant density ρ, then

(i)   the **mass** of *Q* is $m = \rho \int_a^b \pi [f(x)]^2 \, dx$.

(ii) the **moment** M_{yz} of *Q* with respect to the *yz*-plane is

$$M_{yz} = \int_a^b x \cdot \rho \pi [f(x)]^2 \, dx.$$

(iii) the *x*-coordinate $\bar{x}$ of the **center of mass** of *Q* is defined by

$$m\bar{x} = M_{yz} \quad \text{or} \quad \bar{x} = \frac{M_{yz}}{m}.$$

If the integral forms in (i) and (ii) of Definition (9.9) are substituted in (iii), the density factor ρ cancels. Consequently, the center of mass of a homogeneous solid of revolution is independent of the density; that is, it depends only on the shape of the solid. For this reason we shall often refer to the **centroid of a geometric solid** instead of to center of mass. For convenience we may let $\rho = 1$ in (9.9) to find centroids of geometric solids of revolution.

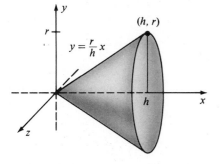

Figure 9.19

Example 1   Find the center of mass of a homogeneous right circular cone of altitude h and radius of base r.

Solution   If the triangular region with vertices at the points $(0, 0)$, $(h, 0)$, and (h, r) is revolved about the x-axis, a cone of the desired type is generated (see Figure 9.19). An equation of the line through $(0,0)$ and (h, r) is $y = (r/h)x$, and hence $f(x) = (r/h)x$ defines a function f of the type used in the preceding discussion. Applying Definition (9.9),

$$M_{yz} = \rho\pi \int_0^h x\left(\frac{r}{h}x\right)^2 dx = \rho\pi\left(\frac{r^2}{h^2}\right)\int_0^h x^3\,dx$$

$$= \rho\pi\left(\frac{r^2}{h^2}\right)\left[\frac{x^4}{4}\right]_0^h = \rho\cdot\frac{1}{4}\pi r^2 h^2.$$

The mass of the cone could be found from Definition (9.9); however, since the volume of a right circular cone of altitude h and base radius r is $\frac{1}{3}\pi r^2 h$, the mass m is $\rho\cdot\frac{1}{3}\pi r^2 h$. Finally, using (iii) of (9.9),

$$\bar{x} = \frac{M_{yz}}{m} = \frac{\rho\cdot\frac{1}{4}\pi r^2 h^2}{\rho\cdot\frac{1}{3}\pi r^2 h} = \frac{3}{4}h.$$

Thus, the center of mass is on the axis of the cone, three-fourths of the way from the vertex to the base.   ■

Formulas analogous to those in (9.9) can be obtained for solids that are generated by revolving regions about the y-axis (see Figure 9.20). All that is

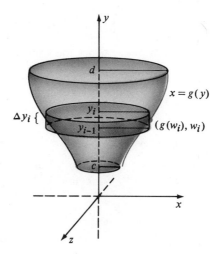

Figure 9.20

necessary is to interchange the roles of x and y in our previous work, as in the next definition.

Definition (9.10)

Let g be continuous on $[c, d]$ and let R be the region bounded by the graphs of $x = g(y)$, $y = c$, $y = d$, and $x = 0$. Let Q be the solid of revolution obtained by revolving R about the y-axis. If Q has constant density ρ, then

(i)   the **mass** of Q is $m = \displaystyle\int_c^d \rho\pi[g(y)]^2 \, dy$.

(ii)   the **moment** M_{xz} of Q with respect to the xz-plane is

$$M_{xz} = \int_c^d y \cdot \rho\pi[g(y)]^2 \, dy.$$

(iii)   the y-coordinate $\bar{y}$ of the **center of mass** of Q is defined by

$$m\bar{y} = M_{xz} \quad \text{or} \quad \bar{y} = \frac{M_{xz}}{m}.$$

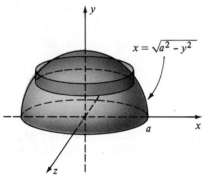

$x = \sqrt{a^2 - y^2}$

Figure 9.21

Example 2   Find the centroid of a hemisphere of radius a.

Solution   If the region in the first quadrant under the graph of the equation $x^2 + y^2 = a^2$ is revolved about the y-axis, there results a hemisphere of radius a (see Figure 9.21). Solving the equation of the circle for x in terms of y gives us $x = \sqrt{a^2 - y^2}$, and hence $g(y) = \sqrt{a^2 - y^2}$ defines a function g of the type described in Definition (9.10). Using (9.10) with $\rho = 1$,

$$M_{xz} = \pi \int_0^a y(\sqrt{a^2 - y^2})^2 \, dy = \pi \int_0^a (a^2 y - y^3) \, dy$$

$$= \pi\left[\frac{1}{2} a^2 y^2 - \frac{1}{4} y^4\right]_0^a = \pi\left[\frac{1}{2} a^4 - \frac{1}{4} a^4\right] = \frac{1}{4} \pi a^4.$$

Since the volume of a sphere of radius a is $\frac{4}{3}\pi a^3$, the hemisphere has volume $\frac{2}{3}\pi a^3$. Hence

$$\bar{y} = \frac{M_{xz}}{V} = \frac{\frac{1}{4}\pi a^4}{\frac{2}{3}\pi a^3} = \frac{3a}{8}. \qquad \blacksquare$$

Centroids may also be found by using washer-shaped elements of volume or cylindrical shells. The next example illustrates the shell method.

Example 3   The region bounded by the graph of $y = 2x - x^2$ and the x-axis is revolved about the y-axis. Find the centroid of the resulting solid.

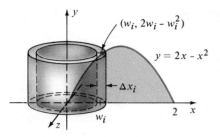

Figure 9.22

Solution   The region is the same as that considered in Example 1 of Section 6.3, and is resketched in Figure 9.22 along with the shell generated by a typical rectangle. As in the solution of that example, the volume of the shell is $2\pi w_i(2w_i - w_i^2)\,\Delta x_i$. The centroid of the shell is on the y-axis and its distance from the x-axis is one-half the altitude of the shell, or $\frac{1}{2}(2w_i - w_i^2)$. Consequently, the moment of the shell with respect to the xz-plane is

$$\tfrac{1}{2}(2w_i - w_i^2)\cdot 2\pi w_i(2w_i - w_i^2)\,\Delta x_i.$$

Summing and taking the limit as $\|P\|$ approaches zero gives us the moment M_{xz} for the solid, namely

$$M_{xz} = \int_0^2 \frac{1}{2}(2x - x^2)\cdot 2\pi x(2x - x^2)\,dx$$

$$= \pi \int_0^2 (4x^3 - 4x^4 + x^5)\,dx$$

$$= \pi\left[x^4 - \frac{4}{5}x^5 + \frac{1}{6}x^6\right]_0^2 = \frac{16\pi}{15}.$$

From Example 1 in Section 6.3, $m = 1V = 8\pi/3$ and therefore

$$\bar{y} = \frac{M_{xz}}{m} = \frac{16\pi/15}{8\pi/3} = \frac{2}{5}. \qquad \blacksquare$$

We shall conclude this chapter by stating a useful theorem about solids of revolution. To illustrate a special case of the theorem, let f and g be continuous functions such that $f(x) \geq g(x)$ for all x in an interval $[a, b]$, where $a \geq 0$. Let R denote the region bounded by the graphs of f, g, and the vertical lines $x = a$ and $x = b$. A typical region of this type is sketched in Figure 9.15. If R is revolved about the y-axis, then the volume of the resulting solid is

$$V = \int_a^b 2\pi x[f(x) - g(x)]\,dx.$$

Suppose the region R has centroid $(\bar{x}, \bar{y})$ and area A. If we use (iii) of Definition (9.8) with $\rho = 1$ (in which case $m = \rho A = A$), we obtain

$$\bar{x} = \frac{M_y}{A} = \frac{\int_a^b x[f(x) - g(x)]\,dx}{A} = \frac{V/2\pi}{A} = \frac{V}{2\pi A}$$

and hence $V = 2\pi\bar{x}A$. Since $\bar{x}$ is the distance from the y-axis to the centroid of R, the last formula states that the volume V of the solid of revolution may be found by multiplying the area A of R by the distance $2\pi\bar{x}$ which the centroid travels when R is revolved once about the y-axis. A similar statement is true if $g(x) \geq 0$ and R is revolved about the x-axis. In general, it is possible to prove the following theorem, which is named after the mathematician Pappus of Alexandria (*ca.* 300 A.D.).

Theorem of Pappus (9.11)

> Let R be a region in a plane that lies entirely on one side of a line l in the plane. If R is revolved once about l, then the volume of the resulting solid is the product of the area of R and the distance traveled by the centroid of R.

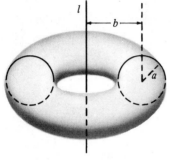

Figure 9.23

Example 4 The region enclosed by a circle of radius a is revolved about a line l in the plane of the circle that is a distance b from the center of the circle, where $b > a$ (see Figure 9.23). Find the volume V of the resulting solid. (The surface of this doughnut-shaped solid is called a **torus**.)

Solution   The region bounded by the circle has area πa^2 and the distance traveled by the centroid is $2\pi b$. Hence by the Theorem of Pappus,

$$V = (2\pi b)(\pi a^2) = 2\pi^2 a^2 b.$$ ∎

EXERCISES 9.9

In Exercises 1–6, find the centroid of the solid generated by revolving the region bounded by the graphs of the given equations about the x-axis.

1 $y = 1/x, y = 0, x = 1, x = 2$

2 $y = (\sin x)^{1/2}, y = 0, x = 0, x = \pi$

3 $y = e^x, y = 0, x = 0, x = 1$

4 $y = 1/\sqrt{16 + x^2}, y = 0, x = 0, x = 4$

5 $y = x^2, x = y^3$

6 $x^2 = y, y = 2x$

In Exercises 7–12, find the centroid of the solid generated by revolving the region bounded by the graphs of the given equations about the y-axis.

7 $y^2 - x^2 = 4, y = 4$

8 $x = \sqrt{y - 1}, x = 0, y = 1, y = 5$

9 $y = e^{2x}, y = 0, x = 0, x = 2$

10 $y = e^{-x}, y = 0, x = 0, x = -1$

11 $y = 1/(x^2 + 25), y = 0, x = 0, x = 5$

12 $y = 1/(9 - x^2), y = 0, x = 0, x = 2$

13 The region between the graph of $y = \cos x$ and the x-axis, from $x = 0$ to $x = \pi/2$, is revolved about the y-axis. Find the centroid of the resulting solid.

14 If the region described in Exercise 13 is revolved about the x-axis, find the centroid of the resulting solid.

15 A homogeneous solid has the shape of a right circular cylinder of altitude h and base radius a, surmounted by a hemisphere of radius a (see figure). Find the center of mass of the solid. (*Hint:* Use Example 2.)

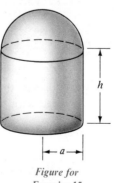

Figure for Exercise 15

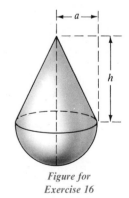

Figure for Exercise 16

16 A homogeneous solid has the shape of a hemisphere of radius a surmounted by a right circular cone of base radius a and altitude h (see figure). Find the center of mass of the solid. (*Hint:* Use Examples 1 and 2).

In Exercises 17 and 18 use the Theorem of Pappus to find the volume generated by revolving the quadrilateral with vertices A, B, C, D about (a) the y-axis; (b) the x-axis.

17 $A(1, 0)$; $B(3, 6)$; $C(11, 6)$; $D(9, 0)$

18 $A(2, 2)$; $B(1, 3)$; $C(4, 6)$; $D(5, 5)$

Solve Exercises 19–23 by using the Theorem of Pappus.

19   Find the centroid of the region in the first quadrant bounded by the graph of $y = \sqrt{a^2 - x^2}$ and the coordinate axes.

20   Find the centroid of the triangle with vertices $O(0, 0)$, $A(0, a)$, $B(b, 0)$.

21   Find the volume of the solid generated by revolving the region in the xy-plane bounded by $y = x^2$ and $y = 8 - x^2$ about the x-axis.

22   Find the volume of the right circular cylinder of altitude h and base radius a.

23   Find the volume of the right circular cone of altitude h and base radius a.

9.10

REVIEW

Discuss each of the following.

1   Integration by parts

2   Trigonometric substitutions

3   Partial fractions

4   Moments and centroids of plane regions

5   Moments and centroids of solids of revolution

EXERCISES 9.10

Evaluate the integrals in Exercises 1–100.

1 $\displaystyle\int x \sin^{-1} x \, dx$

2 $\displaystyle\int \csc^3 x \, dx$

3 $\displaystyle\int_0^1 \ln(1 + x) \, dx$

4 $\displaystyle\int_0^1 e^{\sqrt{x}} \, dx$

5 $\displaystyle\int \cos^3 2x \sin^2 2x \, dx$

6 $\displaystyle\int \cos^4 x \, dx$

7 $\displaystyle\int \tan x \sec^5 x \, dx$

8 $\displaystyle\int \tan x \sec^6 x \, dx$

9 $\displaystyle\int \frac{1}{(x^2 + 25)^{3/2}} \, dx$

10 $\displaystyle\int \frac{1}{x^2\sqrt{16 - x^2}} \, dx$

11 $\displaystyle\int \frac{\sqrt{4 - x^2}}{x} \, dx$

12 $\displaystyle\int \frac{x}{(x^2 + 1)^2} \, dx$

13 $\displaystyle\int \frac{x^3 + 1}{x(x - 1)^3} \, dx$

14 $\displaystyle\int \frac{1}{x + x^3} \, dx$

15 $\displaystyle\int \frac{x^3 - 20x^2 - 63x - 198}{x^4 - 81} \, dx$

16 $\displaystyle\int \frac{x - 1}{(x + 2)^5} \, dx$

17 $\displaystyle\int \frac{x}{\sqrt{4 + 4x - x^2}} \, dx$

18 $\displaystyle\int \frac{x}{x^2 + 6x + 13} \, dx$

19 $\displaystyle\int \frac{\sqrt[3]{x+8}}{x}\, dx$

20 $\displaystyle\int \frac{\sin x}{2\cos x + 3}\, dx$

21 $\displaystyle\int e^{2x} \sin 3x\, dx$

22 $\displaystyle\int \cos(\ln x)\, dx$

23 $\displaystyle\int \sin^3 x \cos^3 x\, dx$

24 $\displaystyle\int \cot^2 3x\, dx$

25 $\displaystyle\int \frac{x}{\sqrt{4-x^2}}\, dx$

26 $\displaystyle\int \frac{1}{x\sqrt{9x^2+4}}\, dx$

27 $\displaystyle\int \frac{x^5 - x^3 + 1}{x^3 + 2x^2}\, dx$

28 $\displaystyle\int \frac{x^3}{x^3 - 3x^2 + 9x - 27}\, dx$

29 $\displaystyle\int \frac{1}{x^{3/2} + x^{1/2}}\, dx$

30 $\displaystyle\int \frac{2x+1}{(x+5)^{100}}\, dx$

31 $\displaystyle\int e^x \sec e^x\, dx$

32 $\displaystyle\int x \tan x^2\, dx$

33 $\displaystyle\int x^2 \sin 5x\, dx$

34 $\displaystyle\int \sin 2x \cos x\, dx$

35 $\displaystyle\int \sin^3 x \cos^{1/2} x\, dx$

36 $\displaystyle\int \sin 3x \cot 3x\, dx$

37 $\displaystyle\int e^x \sqrt{1 + e^x}\, dx$

38 $\displaystyle\int x(4x^2 + 25)^{-1/2}\, dx$

39 $\displaystyle\int \frac{x^2}{\sqrt{4x^2 + 25}}\, dx$

40 $\displaystyle\int \frac{3x+2}{x^2 + 8x + 25}\, dx$

41 $\displaystyle\int \sec^2 x \tan^2 x\, dx$

42 $\displaystyle\int \sin^2 x \cos^5 x\, dx$

43 $\displaystyle\int x \cot x \csc x\, dx$

44 $\displaystyle\int (1 + \csc 2x)^2\, dx$

45 $\displaystyle\int x^2 (8 - x^3)^{1/3}\, dx$

46 $\displaystyle\int x(\ln x)^2\, dx$

47 $\displaystyle\int \sqrt{x} \sin \sqrt{x}\, dx$

48 $\displaystyle\int x\sqrt{5 - 3x}\, dx$

49 $\displaystyle\int \frac{e^{3x}}{1 + e^x}\, dx$

50 $\displaystyle\int \frac{e^{2x}}{4 + e^{4x}}\, dx$

51 $\displaystyle\int \frac{x^2 - 4x + 3}{\sqrt{x}}\, dx$

52 $\displaystyle\int \frac{\cos^3 x}{\sqrt{1 + \sin x}}\, dx$

53 $\displaystyle\int \frac{x^3}{\sqrt{16 - x^2}}\, dx$

54 $\displaystyle\int \frac{x}{25 - 9x^2}\, dx$

55 $\displaystyle\int \frac{1 - 2x}{x^2 + 12x + 35}\, dx$

56 $\displaystyle\int \frac{7}{x^2 - 6x + 18}\, dx$

57 $\displaystyle\int \tan^{-1} 5x\, dx$

58 $\displaystyle\int \sin^4 3x\, dx$

59 $\displaystyle\int \frac{e^{\tan x}}{\cos^2 x}\, dx$

60 $\displaystyle\int \frac{x}{\csc 5x^2}\, dx$

61 $\displaystyle\int \frac{1}{\sqrt{7 + 5x^2}}\, dx$

62 $\displaystyle\int \frac{2x + 3}{x^2 + 4}\, dx$

63 $\displaystyle\int \cot^6 x\, dx$

64 $\displaystyle\int \cot^5 x \csc x\, dx$

65 $\displaystyle\int x^3\sqrt{x^2 - 25}\, dx$

66 $\displaystyle\int (\sin x)10^{\cos x}\, dx$

67 $\displaystyle\int (x^2 - \operatorname{sech}^2 4x)\, dx$

68 $\displaystyle\int x \cosh x\, dx$

69 $\displaystyle\int x^2 e^{-4x}\, dx$

70 $\displaystyle\int x^5\sqrt{x^3 + 1}\, dx$

71 $\displaystyle\int \frac{3}{\sqrt{11 - 10x - x^2}}\, dx$

72 $\displaystyle\int \frac{12x^3 + 7x}{x^4}\, dx$

73 $\displaystyle\int \tan 7x \cos 7x\, dx$

74 $\displaystyle\int e^{1 + \ln 5x}\, dx$

75 $\displaystyle\int \frac{4x^2 - 12x - 10}{(x - 2)(x^2 - 4x + 3)}\, dx$

76 $\displaystyle\int \frac{1}{x^4\sqrt{16 - x^2}}\, dx$

77 $\displaystyle\int (x^3 + 1) \cos x\, dx$

78 $\displaystyle\int (x - 3)^2(x + 1)\, dx$

79 $\displaystyle\int \frac{\sqrt{9 - 4x^2}}{x^2}\, dx$

80 $\displaystyle\int \frac{4x^3 - 15x^2 - 6x + 81}{x^4 - 18x^2 + 81}\, dx$

81 $\displaystyle\int (5 - \cot 3x)^2\, dx$

82 $\displaystyle\int x(x^2 + 5)^{3/4}\, dx$

83 $\displaystyle\int \frac{1}{x(\sqrt{x} + \sqrt[4]{x})}\, dx$

84 $\displaystyle\int \frac{x}{\cos^2 4x}\, dx$

85 $\displaystyle\int \frac{\sin x}{\sqrt{1 + \cos x}}\, dx$

86 $\displaystyle\int \frac{4x^2 - 6x + 4}{(x^2 + 4)(x - 2)}\, dx$

87 $\displaystyle\int \frac{x^2}{(25 + x^2)^2}\, dx$

88 $\displaystyle\int \sin^4 x \cos^3 x\, dx$

89 $\displaystyle\int \tan^3 x \sec x\, dx$

90 $\displaystyle\int \frac{x}{\sqrt{4 + 9x^2}}\, dx$

91 $\displaystyle\int \frac{2x^3 + 4x^2 + 10x + 13}{x^4 + 9x^2 + 20}\, dx$

92 $\displaystyle\int \frac{\sin x}{(1 + \cos x)^3}\, dx$

93 $\displaystyle\int \frac{(x^2 - 2)^2}{x}\, dx$

94 $\displaystyle\int \cot^2 x \csc x\, dx$

95 $\displaystyle\int x^{3/2} \ln x\, dx$

96 $\displaystyle\int \frac{x}{\sqrt[3]{x} - 1}\, dx$

97 $\displaystyle\int \frac{x^2}{\sqrt[3]{2x + 3}}\, dx$

98 $\displaystyle\int \frac{1 - \sin x}{\cot x}\,dx$

99 $\displaystyle\int x^3 e^{x^2}\,dx$

100 $\displaystyle\int (x + 2)^2 (x + 1)^{10}\,dx$

101 The region between the graph of $y = \sin x$ and the x-axis from $x = 0$ to $x = \pi$ is revolved about the y-axis. Find the volume of the resulting solid.

102 The region bounded by the graphs of $y = \tan x$, $y = 0$, $x = \pi/6$, and $x = \pi/4$ is revolved about the x-axis. Find the volume of the resulting solid.

103 Find the arc length of the graph of $y = \ln \sec x$ from $A(0, 0)$ to $B(\pi/3, \ln 2)$.

104 Find the area of the region bounded by the coordinate axes and the graphs of $y = (9 + 4x^2)^{-1/2}$ and $x = 2$.

In Exercises 105 and 106 sketch the region bounded by the graphs of the given equations and find the centroid.

105 $y = x^3, y = x^2$

106 $y = \cos x, y = 0, x = 0, x = \pi/2$

In Exercises 107 and 108 find the centroid of the solid generated by revolving the region bounded by the graphs of the given equations about the x-axis.

107 $y = \sqrt{x}, y = 0, x = 4$

108 $y = \sec x, y = 0, x = 0, x = \pi/4$

109 The region bounded by the graphs of $y = e^{-3x}$, $y = 0$, $x = 0$, and $x = 1$ is revolved about the y-axis. Find the centroid of the resulting solid.

110 The region bounded by the graphs of $y = \sin (x^2)$, $y = 0$, $x = 0$, and $x = \sqrt{\pi}$ is revolved about the y-axis. Find the centroid of the resulting solid.

10

INDETERMINATE FORMS, IMPROPER INTEGRALS, AND TAYLOR'S FORMULA

In this chapter we introduce techniques that are useful in the investigation of certain limits. We shall also study definite integrals that have discontinuous integrands or infinite limits of integration. The final section contains a method for approximating functions by means of polynomials. These topics have many mathematical and physical applications. Our most important use for them will occur in the next chapter, when *infinite series* are studied.

10.1

THE INDETERMINATE FORMS 0/0 AND ∞/∞

In our early work with limits we encountered expressions of the form $\lim_{x \to c} f(x)/g(x)$, where both f and g have the limit 0 as x approaches c. In this event, $f(x)/g(x)$ is said to have the **indeterminate form 0/0** at $x = c$. The word *indeterminate* is used because a further analysis is necessary to conclude whether or not the limit exists. Perhaps the most important examples of the indeterminate form 0/0 occur in the use of the derivative formula

$$f'(c) = \lim_{x \to c} \frac{f(x) - f(c)}{x - c}.$$

If f and g become positively or negatively infinite as x approaches c, we say that $f(x)/g(x)$ has the **indeterminate form ∞/∞** at $x = c$.

Indeterminate forms can sometimes be investigated by employing algebraic manipulations. To illustrate, in Chapter 2 we considered

$$\lim_{x \to 2} \frac{2x^2 - 5x + 2}{5x^2 - 7x - 6}.$$

The indicated quotient has the indeterminate form $0/0$; however, the limit may be found as follows:

$$\lim_{x \to 2} \frac{(x-2)(2x-1)}{(x-2)(5x+3)} = \lim_{x \to 2} \frac{2x-1}{5x+3} = \frac{3}{13}.$$

Other indeterminate forms require more complicated techniques. For example, in Chapter 8 a geometric argument was used to show that

$$\lim_{x \to 0} \frac{\sin x}{x} = 1.$$

In this section we shall establish **L'Hôpital's Rule** and illustrate how it can be used to investigate many indeterminate forms. The proof makes use of the following formula, which bears the name of the famous French mathematician A. Cauchy (1789–1857).

Cauchy's Formula (10.1)

> If the functions f and g are continuous on a closed interval $[a, b]$, differentiable on the open interval (a, b), and if $g'(x) \neq 0$ for all x in (a, b), then there is a number w in (a, b) such that
>
> $$\frac{f(b) - f(a)}{g(b) - g(a)} = \frac{f'(w)}{g'(w)}.$$

Proof   We first note that $g(b) - g(a) \neq 0$, for otherwise $g(a) = g(b)$, and by Rolle's Theorem (4.10) there is a number c in (a, b) such that $g'(c) = 0$, contrary to hypothesis.

It is convenient to introduce a new function h as follows:

$$h(x) = [f(b) - f(a)]g(x) - [g(b) - g(a)]f(x)$$

for all x in $[a, b]$. It follows that h is continuous on $[a, b]$, differentiable on (a, b), and that $h(a) = h(b)$. By Rolle's Theorem, there is a number w in (a, b) such that $h'(w) = 0$, that is,

$$[f(b) - f(a)]g'(w) - [g(b) - g(a)]f'(w) = 0.$$

The preceding equation may be written in the form stated in the conclusion of the theorem.   □

As a special case, if we let $g(x) = x$ in Formula (10.1), then the conclusion has the form

$$\frac{f(b) - f(a)}{b - a} = \frac{f'(w)}{1}$$

which is equivalent to

$$f(b) - f(a) = f'(w)(b - a).$$

This shows that Cauchy's Formula is a generalization of the Mean Value Theorem (4.12).

The next result is the main theorem on indeterminate forms.

L'Hôpital's Rule* **(10.2)**

> Suppose the functions f and g are differentiable on an open interval (a, b) containing c, except possibly at c itself. If $g'(x) \neq 0$ for $x \neq c$, and if $f(x)/g(x)$ has the indeterminate form $0/0$ or ∞/∞ at $x = c$, then
>
> $$\lim_{x \to c} \frac{f(x)}{g(x)} = \lim_{x \to c} \frac{f'(x)}{g'(x)}$$
>
> provided $f'(x)/g'(x)$ has a limit or becomes infinite as x approaches c.

Proof   Suppose $f(x)/g(x)$ has the indeterminate form $0/0$ at $x = c$ and $\lim_{x \to c} f'(x)/g'(x) = L$ for some number L. We wish to prove that $\lim_{x \to c} f(x)/g(x) = L$. Let us introduce two new functions F and G where

$$F(x) = f(x) \quad \text{if } x \neq c \text{ and } F(c) = 0,$$

$$G(x) = g(x) \quad \text{if } x \neq c \text{ and } G(c) = 0.$$

Since

$$\lim_{x \to c} F(x) = \lim_{x \to c} f(x) = 0 = F(c),$$

the function F is continuous at c and hence is continuous *throughout* the interval (a, b). Similarly, G is continuous on (a, b). Moreover, at every $x \neq c$ we have $F'(x) = f'(x)$ and $G'(x) = g'(x)$. It follows from Cauchy's Formula, applied either to the interval $[c, x]$ or to $[x, c]$, that there is a number w between c and x such that

$$\frac{F(x) - F(c)}{G(x) - G(c)} = \frac{F'(w)}{G'(w)} = \frac{f'(w)}{g'(w)}.$$

Using the fact that $F(x) = f(x)$, $G(x) = g(x)$, and $F(c) = G(c) = 0$ gives us

$$\frac{f(x)}{g(x)} = \frac{f'(w)}{g'(w)}.$$

---

* G. L'Hôpital (1661–1704) was a French nobleman who published the first calculus book. The rule appeared in that book; however, it was actually discovered by his teacher, the Swiss mathematician Johann Bernoulli (1667–1748), who communicated the result to L'Hôpital in 1694.

Since w is always between c and x it follows that

$$\lim_{x \to c} \frac{f(x)}{g(x)} = \lim_{x \to c} \frac{f'(w)}{g'(w)} = \lim_{w \to c} \frac{f'(w)}{g'(w)} = L$$

which is what we wished to prove. A similar argument may be given if $f'(x)/g'(x)$ becomes infinite as x approaches c. The proof for the indeterminate form ∞/∞ is more difficult and may be found in texts on advanced calculus.

Beginning students sometimes use L'Hôpital's Rule incorrectly by applying the quotient rule to $f(x)/g(x)$. Note that (10.2) states that the derivatives of $f(x)$ and $g(x)$ are taken *separately*, after which the limit of the quotient $f'(x)/g'(x)$ is investigated.

Example 1   Find $\lim\limits_{x \to 0} \dfrac{\cos x + 2x - 1}{3x}$.

Solution   The quotient has the indeterminate form $0/0$ at $x = 0$. By L'Hôpital's Rule (10.2),

$$\lim_{x \to 0} \frac{\cos x + 2x - 1}{3x} = \lim_{x \to 0} \frac{-\sin x + 2}{3} = \frac{2}{3}. \qquad \blacksquare$$

To be completely rigorous in Example 1 we should have determined whether or not $\lim_{x \to 0} (-\sin x + 2)/3$ existed *before* equating it to the given expression; however, to simplify solutions it is customary to proceed as indicated.

Sometimes it is necessary to employ L'Hôpital's Rule several times in the same problem, as illustrated in the next example.

Example 2   Find $\lim\limits_{x \to 0} \dfrac{e^x + e^{-x} - 2}{1 - \cos 2x}$.

Solution   The quotient has the indeterminate form $0/0$. By L'Hôpital's Rule,

$$\lim_{x \to 0} \frac{e^x + e^{-x} - 2}{1 - \cos 2x} = \lim_{x \to 0} \frac{e^x - e^{-x}}{2 \sin 2x}$$

provided the second limit exists. Since the last quotient has the indeterminate form $0/0$, we apply L'Hôpital's Rule a second time, obtaining

$$\lim_{x \to 0} \frac{e^x - e^{-x}}{2 \sin 2x} = \lim_{x \to 0} \frac{e^x + e^{-x}}{4 \cos 2x} = \frac{1}{2}.$$

It follows that the given limit exists and equals $\frac{1}{2}$. $\qquad \blacksquare$

L'Hôpital's Rule is also valid for one-sided limits, as illustrated in the following example.

Example 3   Find $\displaystyle\lim_{x\to(\pi/2)^-}\frac{4\tan x}{1+\sec x}$.

Solution   The indeterminate form is ∞/∞. By L'Hôpital's Rule,

$$\lim_{x\to(\pi/2)^-}\frac{4\tan x}{1+\sec x}=\lim_{x\to(\pi/2)^-}\frac{4\sec^2 x}{\sec x\tan x}=\lim_{x\to(\pi/2)^-}\frac{4\sec x}{\tan x}.$$

The last quotient again has the indeterminate form ∞/∞ at $x=\pi/2$; however, any additional applications of L'Hôpital's Rule always produce the form ∞/∞. (Check this fact.) In this case the limit may be found by using trigonometric identities to change the quotient as follows:

$$\frac{4\sec x}{\tan x}=\frac{4/\cos x}{\sin x/\cos x}=\frac{4}{\sin x}.$$

Consequently,

$$\lim_{x\to(\pi/2)^-}\frac{4\tan x}{1+\sec x}=\lim_{x\to(\pi/2)^-}\frac{4}{\sin x}=4.\qquad\blacksquare$$

It can be shown that the statement of L'Hôpital's Rule remains true if the symbol $x\to c$ is replaced by $x\to\infty$ or $x\to-\infty$. Let us give a partial proof of this fact. Suppose

$$\lim_{x\to\infty}f(x)=\lim_{x\to\infty}g(x)=0.$$

If we let $u=1/x$ and apply L'Hôpital's Rule,

$$\lim_{x\to\infty}\frac{f(x)}{g(x)}=\lim_{u\to0^+}\frac{f(1/u)}{g(1/u)}=\lim_{u\to0^+}\frac{D_u f(1/u)}{D_u g(1/u)}.$$

By the Chain Rule,

$$D_u f(1/u)=f'(1/u)(-1/u^2)\quad\text{and}\quad D_u g(1/u)=g'(1/u)(-1/u^2).$$

Substituting in the last limit and simplifying, we obtain

$$\lim_{x\to\infty}\frac{f(x)}{g(x)}=\lim_{u\to0^+}\frac{f'(1/u)}{g'(1/u)}=\lim_{x\to\infty}\frac{f'(x)}{g'(x)}.$$

We shall also refer to this as L'Hôpital's Rule. The next two examples illustrate the application of the rule to the form ∞/∞.

Example 4 Find $\lim\limits_{x \to \infty} \dfrac{\ln x}{\sqrt{x}}$.

Solution   The indeterminate form is ∞/∞. By L'Hôpital's Rule,

$$\lim_{x \to \infty} \frac{\ln x}{\sqrt{x}} = \lim_{x \to \infty} \frac{1/x}{1/(2\sqrt{x})} = \lim_{x \to \infty} \frac{2\sqrt{x}}{x} = \lim_{x \to \infty} \frac{2}{\sqrt{x}} = 0. \qquad \blacksquare$$

Example 5 Find $\lim\limits_{x \to \infty} \dfrac{e^{3x}}{x^2}$, if it exists.

Solution   The indeterminate form is ∞/∞. In this case we must apply L'Hôpital's Rule twice, as follows.

$$\lim_{x \to \infty} \frac{e^{3x}}{x^2} = \lim_{x \to \infty} \frac{3e^{3x}}{2x} = \lim_{x \to \infty} \frac{9e^{3x}}{2} = \infty$$

Thus the given quotient increases without bound as x becomes infinite. $\blacksquare$

It is extremely important to verify that a given quotient has the indeterminate form 0/0 or ∞/∞ before using L'Hôpital's Rule. Indeed, if the rule is applied to a nonindeterminate form, an incorrect conclusion may be obtained, as illustrated in the next example.

Example 6 Find $\lim\limits_{x \to 0} \dfrac{e^x + e^{-x}}{x^2}$, if it exists.

Solution   Suppose we overlook the fact that the quotient does *not* have either of the indeterminate forms 0/0 or ∞/∞ at $x = 0$. If we (incorrectly) apply L'Hôpital's Rule we obtain

$$\lim_{x \to 0} \frac{e^x + e^{-x}}{x^2} = \lim_{x \to 0} \frac{e^x - e^{-x}}{2x}.$$

Since the last quotient has the indeterminate form 0/0 we may apply L'Hôpital's Rule, obtaining

$$\lim_{x \to 0} \frac{e^x - e^{-x}}{2x} = \lim_{x \to 0} \frac{e^x + e^{-x}}{2} = \frac{1 + 1}{2} = 1.$$

This would lead us to the (wrong) conclusion that the given limit exists and equals 1.

A correct method for investigating the limit is to observe that

$$\lim_{x \to 0} \frac{e^x + e^{-x}}{x^2} = \lim_{x \to 0} (e^x + e^{-x})\left(\frac{1}{x^2}\right).$$

Since

$$\lim_{x \to 0} (e^x + e^{-x}) = 2 \quad \text{and} \quad \lim_{x \to 0} \frac{1}{x^2} = \infty,$$

it follows from (ii) of Theorem (4.25) that

$$\lim_{x \to 0} \frac{e^x + e^{-x}}{x^2} = \infty.$$

■

EXERCISES 10.1

Find the limits in Exercises 1–54, if they exist.

1 $\displaystyle\lim_{x \to 0} \frac{\sin x}{2x}$

2 $\displaystyle\lim_{x \to 0} \frac{5x}{\tan x}$

3 $\displaystyle\lim_{x \to 5} \frac{\sqrt{x-1}-2}{x^2-25}$

4 $\displaystyle\lim_{x \to 4} \frac{x-4}{\sqrt[3]{x+4}-2}$

5 $\displaystyle\lim_{x \to 2} \frac{2x^2-5x+2}{5x^2-7x-6}$

6 $\displaystyle\lim_{x \to -3} \frac{x^2+2x-3}{2x^2+3x-9}$

7 $\displaystyle\lim_{x \to 1} \frac{x^3-3x+2}{x^2-2x-1}$

8 $\displaystyle\lim_{x \to 2} \frac{x^2-5x+6}{2x^2-x-7}$

9 $\displaystyle\lim_{x \to 0} \frac{\sin x - x}{\tan x - x}$

10 $\displaystyle\lim_{x \to 0} \frac{\sin x}{x - \tan x}$

11 $\displaystyle\lim_{x \to 0} \frac{x+1-e^x}{x^2}$

12 $\displaystyle\lim_{x \to 0^+} \frac{x+1-e^x}{x^3}$

13 $\displaystyle\lim_{x \to 0} \frac{x - \sin x}{x^3}$

14 $\displaystyle\lim_{x \to \pi/2} \frac{1 - \sin x}{\cos x}$

15 $\displaystyle\lim_{x \to \pi/2} \frac{1 + \sin x}{\cos^2 x}$

16 $\displaystyle\lim_{x \to 0^+} \frac{\cos x}{x}$

17 $\displaystyle\lim_{x \to (\pi/2)^-} \frac{2 + \sec x}{3 \tan x}$

18 $\displaystyle\lim_{x \to 0^+} \frac{\ln x}{\cot x}$

19 $\displaystyle\lim_{x \to \infty} \frac{x^2}{\ln x}$

20 $\displaystyle\lim_{x \to \infty} \frac{\ln x}{x^2}$

21 $\displaystyle\lim_{x \to 0^+} \frac{\ln \sin x}{\ln \sin 2x}$

22 $\displaystyle\lim_{x \to 0} \frac{2x}{\tan^{-1} x}$

23 $\displaystyle\lim_{x \to 0} \frac{e^x - e^{-x} - 2 \sin x}{x \sin x}$

24 $\displaystyle\lim_{x \to 2} \frac{\ln (x-1)}{x-2}$

25 $\displaystyle\lim_{x \to 0} \frac{x \cos x + e^{-x}}{x^2}$

26 $\displaystyle\lim_{x \to 0} \frac{2e^x - 3x - e^{-x}}{x^2}$

27 $\displaystyle\lim_{x \to \infty} \frac{2x^2+3x+1}{5x^2+x+4}$

28 $\displaystyle\lim_{x \to \infty} \frac{x^3+x+1}{3x^3+4}$

29 $\displaystyle\lim_{x \to \infty} \frac{x \ln x}{x + \ln x}$

30 $\displaystyle\lim_{x \to \infty} \frac{e^{3x}}{\ln x}$

31 $\displaystyle\lim_{x \to \infty} \frac{x^n}{e^x}, n > 0$

32 $\displaystyle\lim_{x \to \infty} \frac{e^x}{x^n}, n > 0$

33 $\lim\limits_{x\to 2^+} \dfrac{\ln(x-1)}{(x-2)^2}$

34 $\lim\limits_{x\to 0} \dfrac{\sin^2 x + 2\cos x - 2}{\cos^2 x - x\sin x - 1}$

35 $\lim\limits_{x\to 0} \dfrac{\sin^{-1} 2x}{\sin^{-1} x}$

36 $\lim\limits_{x\to\infty} \dfrac{\ln(\ln x)}{\ln x}$

37 $\lim\limits_{x\to 0} \dfrac{\tan x - \sin x}{x^3 \tan x}$

38 $\lim\limits_{x\to 1} \dfrac{2x^3 - 5x^2 + 6x - 3}{x^3 - 2x^2 + x - 1}$

39 $\lim\limits_{x\to -\infty} \dfrac{3 - 3^x}{5 - 5^x}$

40 $\lim\limits_{x\to 0} \dfrac{2 - e^x - e^{-x}}{1 - \cos^2 x}$

41 $\lim\limits_{x\to 1} \dfrac{x^4 - x^3 - 3x^2 + 5x - 2}{x^4 - 5x^3 + 9x^2 - 7x + 2}$

42 $\lim\limits_{x\to 1} \dfrac{x^4 + x^3 - 3x^2 - x + 2}{x^4 - 5x^3 + 9x^2 - 7x + 2}$

43 $\lim\limits_{x\to 0} \dfrac{x - \tan^{-1} x}{x\sin x}$

44 $\lim\limits_{x\to\infty} \dfrac{e^{-x}}{1 + e^{-x}}$

45 $\lim\limits_{x\to\infty} \dfrac{x^{3/2} + 5x - 4}{x\ln x}$

46 $\lim\limits_{x\to 0} \dfrac{x\sin^{-1} x}{x - \sin x}$

47 $\lim\limits_{x\to\infty} \dfrac{\sqrt{x^2 + 1}}{\tan^{-1} x}$

48 $\lim\limits_{x\to\infty} \dfrac{3^x + 2x}{x^3 + 1}$

49 $\lim\limits_{x\to\infty} \dfrac{2e^{3x} + \ln x}{e^{3x} + x^2}$

50 $\lim\limits_{x\to\pi/2} \dfrac{\tan x}{\cot 2x}$

51 $\lim\limits_{x\to\infty} \dfrac{\sqrt{x^2 + 1}}{x}$

52 $\lim\limits_{x\to 0} \dfrac{e^{-1/x}}{x}$

53 $\lim\limits_{x\to\infty} \dfrac{x - \cos x}{x}$

54 $\lim\limits_{x\to\infty} \dfrac{x + \cosh x}{x^2 + 1}$

55 The current I at time t in an electrical circuit is given by $I = (E/R)(1 - e^{-Rt/L})$ where E, R and L are the electromotive force, resistance, and inductance, respectively. Find the following limits, where all variables except those indicated in the limit notation are positive constants.

(a) $\lim\limits_{R\to 0^+} I$ (b) $\lim\limits_{L\to 0^+} I$ (c) $\lim\limits_{t\to\infty} I$

56 Let $x > 0$. If $n \neq -1$ we know that $\int_1^x t^n\, dt = [t^{n+1}/(n+1)]_1^x$. Show that

$$\lim_{n\to -1} \int_1^x t^n\, dt = \int_1^x t^{-1}\, dt.$$

In Exercises 57 and 58 find $\lim_{x\to\infty} f(x)/g(x)$.

57 $f(x) = \displaystyle\int_0^x e^{t^2}\, dt, \quad g(x) = e^{x^2}$

58 $f(x) = \displaystyle\int_1^x (\sin t)^{2/3}\, dt, \quad g(x) = x^2$

10.2

OTHER INDETERMINATE FORMS

If $\lim_{x\to c} f(x) = 0$ and $\lim_{x\to c} g(x) = \infty$ or $\lim_{x\to c} g(x) = -\infty$, then $f(x)g(x)$ is said to have the **indeterminate form $0 \cdot \infty$** at $x = c$. The same terminology is used for one-sided limits or if x becomes positively or negatively infinite. This form may be changed to one of the indeterminate forms $0/0$ or ∞/∞ by writing

$$f(x)g(x) = \frac{f(x)}{1/g(x)} \quad \text{or} \quad f(x)g(x) = \frac{g(x)}{1/f(x)}.$$

Example 1 Find $\lim\limits_{x \to 0^+} x^2 \ln x$.

Solution The indeterminate form is $0 \cdot \infty$. We first write

$$x^2 \ln x = \frac{\ln x}{(1/x^2)}$$

and then apply L'Hôpital's Rule to the resulting indeterminate form ∞/∞. Thus

$$\lim_{x \to 0^+} x^2 \ln x = \lim_{x \to 0^+} \frac{\ln x}{(1/x^2)} = \lim_{x \to 0^+} \frac{(1/x)}{(-2/x^3)}.$$

The last quotient has the indeterminate form ∞/∞; however, further applications of L'Hôpital's Rule would again lead to ∞/∞. In this case we simplify the quotient algebraically and find the limit as follows:

$$\lim_{x \to 0^+} \frac{(1/x)}{(-2/x^3)} = \lim_{x \to 0^+} \frac{x^3}{-2x} = \lim_{x \to 0^+} \frac{x^2}{-2} = 0. \qquad \blacksquare$$

Example 2 Find $\lim\limits_{x \to \pi/2} (2x - \pi) \sec x$.

Solution The indeterminate form is $0 \cdot \infty$. Hence we begin by writing

$$(2x - \pi) \sec x = \frac{2x - \pi}{1/\sec x} = \frac{2x - \pi}{\cos x}.$$

Since the last expression has the indeterminate form $0/0$ at $x = \pi/2$, L'Hôpital's Rule may be applied as follows:

$$\lim_{x \to \pi/2} \frac{2x - \pi}{\cos x} = \lim_{x \to \pi/2} \frac{2}{-\sin x} = -2. \qquad \blacksquare$$

Indeterminate forms denoted by $\mathbf{0^0}$, $\mathbf{\infty^0}$, and $\mathbf{1^\infty}$ arise from expressions such as $f(x)^{g(x)}$. One method for dealing with these forms is to write

$$y = f(x)^{g(x)}$$

and take the natural logarithm of both sides, obtaining

$$\ln y = \ln f(x)^{g(x)} = g(x) \ln f(x).$$

Note that if the indeterminate form for y is 0^0 or ∞^0, then the indeterminate form for $\ln y$ is $0 \cdot \infty$, which may be handled using previous methods. Similarly, if the form for y is 1^∞, then the indeterminate form for $\ln y$ is $\infty \cdot 0$. It follows that

$$\text{if} \quad \lim_{x \to c} \ln y = L, \quad \text{then} \quad \lim_{x \to c} y = \lim_{x \to c} e^{\ln y} = e^L,$$

that is,

$$\lim_{x \to c} f(x)^{g(x)} = e^L.$$

This procedure may be summarized as follows.

Guidelines for investigating $\lim\limits_{x \to c} f(x)^{g(x)}$ *if the indeterminate form is* $0^0, 1^\infty, or \, \infty^0$

1.   Let $y = f(x)^{g(x)}$.
2.   Take logarithms: $\ln y = \ln f(x)^{g(x)} = g(x) \ln f(x)$.
3.   Find $\lim_{x \to c} \ln y$, if it exists.
4.   If $\lim_{x \to c} \ln y = L$, then $\lim_{x \to c} y = e^L$.

A common error is to stop after showing $\lim_{x \to c} \ln y = L$ and conclude that the given expression has the limit L. Remember that *we wish to find the limit of y*, and if $\ln y$ has the limit L, then y has the limit e^L. The guidelines may also be used if $x \to \infty$, or $x \to -\infty$, or for one-sided limits.

Example 3   Find $\lim\limits_{x \to 0} (1 + 3x)^{1/2x}$.

Solution   The indeterminate form is 1^∞. Following the guidelines we begin by writing

1. $$y = (1 + 3x)^{1/2x}.$$

2. $$\ln y = \frac{1}{2x} \ln (1 + 3x) = \frac{\ln (1 + 3x)}{2x}.$$

The last expression has the indeterminate form $0/0$ at $x = 0$. By L'Hôpital's Rule,

3. $$\lim_{x \to 0} \ln y = \lim_{x \to 0} \frac{\ln (1 + 3x)}{2x} = \lim_{x \to 0} \frac{3/(1 + 3x)}{2} = \frac{3}{2}.$$

Consequently,

4. $$\lim_{x \to 0} (1 + 3x)^{1/2x} = \lim_{x \to 0} y = e^{3/2}. \qquad \blacksquare$$

If $\lim_{x \to c} f(x) = \lim_{x \to c} g(x) = \infty$, then $f(x) - g(x)$ has the indeterminate form $\infty - \infty$ at $x = c$. In this case the expression should be changed so that one of the forms we have discussed is obtained.

Example 4   Find $\lim\limits_{x \to 0} \left(\dfrac{1}{e^x - 1} - \dfrac{1}{x} \right)$.

Solution   The form is $\infty - \infty$; however, if the difference is written as a single fraction, then

$$\lim_{x \to 0} \left(\frac{1}{e^x - 1} - \frac{1}{x} \right) = \lim_{x \to 0} \frac{x - e^x + 1}{xe^x - x}.$$

This gives us the indeterminate form 0/0. It is necessary to apply L'Hôpital's Rule twice, since the first application leads to the indeterminate form 0/0. Thus

$$\lim_{x \to 0} \frac{x - e^x + 1}{xe^x - x} = \lim_{x \to 0} \frac{1 - e^x}{xe^x + e^x - 1}$$

$$= \lim_{x \to 0} \frac{-e^x}{xe^x + 2e^x} = -\frac{1}{2}. \qquad \blacksquare$$

EXERCISES 10.2

Find the limits in Exercises 1–42, if they exist.

1 $\displaystyle\lim_{x \to 0^+} x \ln x$

2 $\displaystyle\lim_{x \to (\pi/2)^-} \tan x \ln \sin x$

3 $\displaystyle\lim_{x \to \infty} (x^2 - 1)e^{-x^2}$

4 $\displaystyle\lim_{x \to \infty} x(e^{1/x} - 1)$

5 $\displaystyle\lim_{x \to 0} e^{-x} \sin x$

6 $\displaystyle\lim_{x \to -\infty} x \tan^{-1} x$

7 $\displaystyle\lim_{x \to 0^+} \sin x \ln \sin x$

8 $\displaystyle\lim_{x \to \infty} x\left(\frac{\pi}{2} - \tan^{-1} x\right)$

9 $\displaystyle\lim_{x \to \infty} x \sin \frac{1}{x}$

10 $\displaystyle\lim_{x \to \infty} e^{-x} \ln x$

11 $\displaystyle\lim_{x \to 0} x \sec^2 x$

12 $\displaystyle\lim_{x \to 0} (\cos x)^{x+1}$

13 $\displaystyle\lim_{x \to \infty} \left(1 + \frac{1}{x}\right)^{5x}$

14 $\displaystyle\lim_{x \to 0^+} (e^x + 3x)^{1/x}$

15 $\displaystyle\lim_{x \to 0^+} (e^x - 1)^x$

16 $\displaystyle\lim_{x \to 0^+} x^x$

17 $\displaystyle\lim_{x \to \infty} x^{1/x}$

18 $\displaystyle\lim_{x \to (\pi/2)^-} (\tan x)^{\cos x}$

19 $\displaystyle\lim_{x \to (\pi/2)^-} (\tan x)^x$

20 $\displaystyle\lim_{x \to 2^+} (x - 2)^x$

21 $\displaystyle\lim_{x \to 0^+} (2x + 1)^{\cot x}$

22 $\displaystyle\lim_{x \to 0^+} (1 + 3x)^{\csc x}$

23 $\displaystyle\lim_{x \to \infty} \left(\frac{x^2}{x - 1} - \frac{x^2}{x + 1}\right)$

24 $\displaystyle\lim_{x \to 1} \left(\frac{1}{x - 1} - \frac{1}{\ln x}\right)$

25 $\displaystyle\lim_{x \to 0} \left(\frac{1}{x} - \frac{1}{\sin x}\right)$

26 $\displaystyle\lim_{x \to (\pi/2)^-} (\sec x - \tan x)$

27 $\displaystyle\lim_{x \to 1^-} (1 - x)^{\ln x}$

28 $\displaystyle\lim_{x \to \infty} (1 + e^x)^{e^{-x}}$

29 $\displaystyle\lim_{x \to 0} \left(\frac{1}{\sqrt{x^2 + 1}} - \frac{1}{x}\right)$

30 $\displaystyle\lim_{x \to 0} (\cot^2 x - \csc^2 x)$

31 $\displaystyle\lim_{x \to 0} \cot 2x \tan^{-1} x$

32 $\displaystyle\lim_{x \to \infty} x^3 \, 2^{-x}$

33 $\displaystyle\lim_{x \to 0} (\cot^2 x - e^{-x})$

34 $\lim_{x \to \infty} (\sqrt{x^2 + 4} - \tan^{-1} x)$

35 $\lim_{x \to (\pi/2)^-} (1 + \cos x)^{\tan x}$

36 $\lim_{x \to 0} (1 + ax)^{b/x}$

37 $\lim_{x \to -3} \left(\dfrac{x}{x^2 + 2x - 3} - \dfrac{4}{x + 3} \right)$

38 $\lim_{x \to \infty} (\sqrt{x^4 + 5x^2 + 3} - x^2)$

39 $\lim_{x \to 0} (x + \cos 2x)^{\csc 3x}$

40 $\lim_{x \to \pi/2} \sec x \cos 3x$

41 $\lim_{x \to \infty} (\sinh x - x)$

42 $\lim_{x \to \infty} [\ln (4x + 3) - \ln (3x + 4)]$

Sketch the graphs of the functions defined in Exercises 43 and 44. Find the local extrema and discuss the behavior of $f(x)$ near $x = 0$. Find horizontal asymptotes, if they exist.

43 $f(x) = x^{1/x}, \ x > 0$

44 $f(x) = x^x, \ x > 0$

45 The *geometric mean* of two positive real numbers a and b is defined as $\sqrt{ab}$. Use L'Hôpital's Rule to prove that

$$\sqrt{ab} = \lim_{x \to \infty} \left(\frac{a^{1/x} + b^{1/x}}{2} \right)^x.$$

46 If a sum of money P is invested at an interest rate of $100r$ percent per year, compounded m times per year, then the principal at the end of t years is given by $P(1 + rm^{-1})^{mt}$. If we regard m as a real number and let m increase without bound, then the interest is said to be *compounded continuously*. Use L'Hôpital's Rule to show that in this case the principal after t years is Pe^{rt} (cf. Exercise 29 in Section 7.6).

10.3

INTEGRALS WITH INFINITE LIMITS OF INTEGRATION

Suppose a function f is continuous and nonnegative on an infinite interval $[a, \infty)$ and $\lim_{x \to \infty} f(x) = 0$. If $t > a$, then the area $A(t)$ under the graph of f from a to t, as illustrated in (i) of Figure 10.1, is

$$A(t) = \int_a^t f(x) \, dx.$$

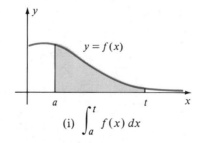

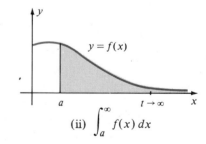

Figure 10.1

If $\lim_{t \to \infty} A(t)$ exists, then the limit may be interpreted as the area of the *unbounded* region that lies under the graph of f, over the x-axis, and to the right of $x = a$, as illustrated in (ii) of the figure. The symbol $\int_a^\infty f(x) \, dx$ is used to denote this number.

Part (i) of the next definition generalizes the preceding remarks to the case where $f(x)$ may be negative in $[a, \infty)$.

Definition (10.3)

> (i)   If f is continuous on $[a, \infty)$, then
>
> $$\int_a^\infty f(x)\, dx = \lim_{t \to \infty} \int_a^t f(x)\, dx.$$
>
> (ii)   If f is continuous on $(-\infty, a]$, then
>
> $$\int_{-\infty}^a f(x)\, dx = \lim_{t \to -\infty} \int_t^a f(x)\, dx.$$

If $f(x) \geq 0$ for all x, then the limit in (ii) may be regarded as the area under the graph of f, over the x-axis, and to the *left* of $x = a$.

The expressions in Definition (10.3) are called **improper integrals**. They differ from definite integrals because one of the limits of integration is not a real number. These integrals are said to **converge** if, as $|t|$ increases without bound, the limits on the right side of the equations exist. The limits are called the **values** of the improper integrals. If the limits do not exist, the improper integrals are said to **diverge**.

Improper integrals may have *two* infinite limits of integration, as in the following definition.

Definition (10.4)

> Let f be continuous for all x. If a is any real number, then
>
> $$\int_{-\infty}^\infty f(x)\, dx = \int_{-\infty}^a f(x)\, dx + \int_a^\infty f(x)\, dx.$$

The integral on the left in Definition (10.4) is said to **converge** if and only if *both* of the integrals on the right converge. If one of the integrals diverges, then $\int_{-\infty}^\infty f(x)\, dx$ is said to **diverge**. It can be shown that (10.4) is independent of the real number a (see Exercise 40). We may also show that $\int_{-\infty}^\infty f(x)\, dx$ is not necessarily the same as $\lim_{t \to \infty} \int_{-t}^t f(x)\, dx$ (see Exercise 39).

Example 1   Determine whether the following integrals converge or diverge.

(a) $\displaystyle \int_2^\infty \frac{1}{(x-1)^2}\, dx$ (b) $\displaystyle \int_2^\infty \frac{1}{x-1}\, dx$

Solution

(a)   By (i) of Definition (10.3),

$$\int_2^\infty \frac{1}{(x-1)^2}\, dx = \lim_{t \to \infty} \int_2^t \frac{1}{(x-1)^2}\, dx = \lim_{t \to \infty} \frac{-1}{x-1}\Big]_2^t$$

$$= \lim_{t \to \infty} \left(\frac{-1}{t-1} + \frac{1}{2-1} \right) = 0 + 1 = 1.$$

Thus the integral converges and has the value 1.

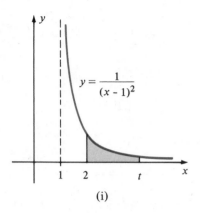

(i)

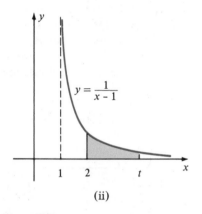

(ii)

Figure 10.2

(b)   By (i) of (10.3),

$$\int_2^\infty \frac{1}{x-1}\,dx = \lim_{t \to \infty} \int_2^t \frac{1}{x-1}\,dx$$

$$= \lim_{t \to \infty} \ln(x-1)\Big]_2^t$$

$$= \lim_{t \to \infty} [\ln(t-1) - \ln(2-1)]$$

$$= \lim_{t \to \infty} \ln(t-1) = \infty.$$

Consequently, this improper integral diverges.   ■

The graphs of the two functions defined by the integrands in parts (a) and (b) of Example 1 are sketched in Figure 10.2. Note that although the graphs have the same general shape for $x \geq 2$, we may assign an area to the region under the graph shown in (i) of the figure, whereas this is not true for the graph in (ii).

There is an interesting sidelight to the graph in (ii) of Figure 10.2. If the region under the graph of $y = 1/(x-1)$ from 2 to t is revolved about the x-axis, then the volume of the resulting solid is

$$\pi \int_2^t \frac{1}{(x-1)^2}\,dx.$$

The improper integral

$$\pi \int_2^\infty \frac{1}{(x-1)^2}\,dx$$

may be regarded as the volume of the *unbounded* solid obtained by revolving, about the x-axis, the region under the graph of $y = 1/(x-1)$ for $x \geq 2$. By (a) of Example 1, the value of this improper integral is $\pi \cdot 1$ or π. This gives us the rather curious fact that although the area of the region is infinite, the volume of the solid of revolution it generates is finite.

Example 2   Assign an area to the region that lies under the graph of $y = e^x$, over the x-axis, and to the left of $x = 1$.

Solution   The region bounded by the graphs of $y = e^x$, $y = 0$, $x = 1$, and $x = t$, where $t < 1$, is sketched in Figure 10.3.

By our previous remarks, the desired area is

$$\int_{-\infty}^1 e^x\,dx = \lim_{t \to -\infty} \int_t^1 e^x\,dx = \lim_{t \to -\infty} e^x\Big]_t^1$$

$$= \lim_{t \to -\infty} (e - e^t) = e - 0 = e.$$ ■

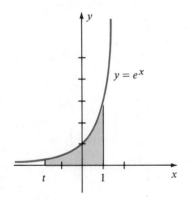

Figure 10.3

Example 3 Evaluate $\int_{-\infty}^{\infty} \dfrac{1}{1 + x^2}\, dx$. Sketch the graph of $f(x) = \dfrac{1}{1 + x^2}$ and interpret the integral as an area.

Solution   Using Definition (10.4) with $a = 0$,

$$\int_{-\infty}^{\infty} \frac{1}{1 + x^2}\, dx = \int_{-\infty}^{0} \frac{1}{1 + x^2}\, dx + \int_{0}^{\infty} \frac{1}{1 + x^2}\, dx.$$

Next, applying (i) of Definition (10.3),

$$\int_{0}^{\infty} \frac{1}{1 + x^2}\, dx = \lim_{t \to \infty} \int_{0}^{t} \frac{1}{1 + x^2}\, dx = \lim_{t \to \infty} \arctan x \Big]_{0}^{t}$$

$$= \lim_{t \to \infty} (\arctan t - \arctan 0) = \pi/2 - 0 = \pi/2.$$

Similarly, we may show, by using (ii) of (10.3), that

$$\int_{-\infty}^{0} \frac{1}{1 + x^2}\, dx = \frac{\pi}{2}.$$

Consequently, the given improper integral converges and has the value $\pi/2 + \pi/2 = \pi$.

The graph of $y = 1/(1 + x^2)$ is sketched in Figure 10.4. As in our previous discussion, the unbounded region that lies under the graph and above the x-axis may be assigned an area of π square units. ∎

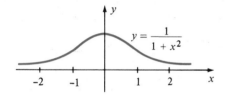

$y = \dfrac{1}{1 + x^2}$

Figure 10.4

Figure 10.5

Improper integrals with infinite limits of integration have many applications in the physical world. To illustrate, suppose a and b are the coordinates of two points A and B on a coordinate line l, as shown in Figure 10.5. If $F(x)$ is the force acting at the point P with coordinate x, then by Definition (6.10), the work done as P moves from A to B is given by

$$W = \int_{a}^{b} F(x)\, dx.$$

In similar fashion, the improper integral $\int_{a}^{\infty} F(x)\, dx$ is used to define the work done as P moves indefinitely to the right (in applications, the terminology "to infinity" is sometimes used). For example, if $F(x)$ is the force of attraction between a particle fixed at point A and a (movable) particle at P, and if $c > a$, then $\int_{c}^{\infty} F(x)\, dx$ represents the work required to move P from the point with coordinate c to infinity (see Exercises 33 and 34).

EXERCISES 10.3

Determine whether the integrals in Exercises 1–24 converge or diverge, and evaluate those that converge.

1 $\displaystyle\int_{1}^{\infty} \frac{1}{x^{4/3}}\, dx$

2 $\displaystyle\int_{-\infty}^{0} \frac{1}{(x - 1)^3}\, dx$

3 $\displaystyle\int_{1}^{\infty} \frac{1}{x^{3/4}}\, dx$

4 $\displaystyle\int_{0}^{\infty} \frac{x}{1 + x^2}\, dx$

5 $\displaystyle\int_{-\infty}^{2} \frac{1}{5 - 2x}\, dx$

6 $\displaystyle\int_{-\infty}^{\infty} \frac{x}{x^4 + 9}\, dx$

7 $\displaystyle\int_0^\infty e^{-2x}\,dx$

8 $\displaystyle\int_{-\infty}^0 e^x\,dx$

9 $\displaystyle\int_{-\infty}^{-1} \frac{1}{x^3}\,dx$

10 $\displaystyle\int_0^\infty \frac{1}{\sqrt[3]{x+1}}\,dx$

11 $\displaystyle\int_{-\infty}^0 \frac{1}{(x-8)^{2/3}}\,dx$

12 $\displaystyle\int_1^\infty \frac{x}{(1+x^2)^2}\,dx$

13 $\displaystyle\int_0^\infty \frac{\cos x}{1+\sin^2 x}\,dx$

14 $\displaystyle\int_{-\infty}^2 \frac{1}{x^2+4}\,dx$

15 $\displaystyle\int_{-\infty}^\infty xe^{-x^2}\,dx$

16 $\displaystyle\int_{-\infty}^\infty \cos^2 x\,dx$

17 $\displaystyle\int_1^\infty \frac{\ln x}{x}\,dx$

18 $\displaystyle\int_3^\infty \frac{1}{x^2-1}\,dx$

19 $\displaystyle\int_0^\infty \cos x\,dx$

20 $\displaystyle\int_{-\infty}^{\pi/2} \sin 2x\,dx$

21 $\displaystyle\int_{-\infty}^\infty \operatorname{sech} x\,dx$

22 $\displaystyle\int_0^\infty xe^{-x}\,dx$

23 $\displaystyle\int_{-\infty}^0 \frac{1}{x^2-3x+2}\,dx$

24 $\displaystyle\int_4^\infty \frac{x+18}{x^2+x-12}\,dx$

In Exercises 25–28 assign, if possible, a value to (a) the area of the region R, and (b) the volume of the solid obtained by revolving R about the x-axis.

 25 $R = \{(x, y): x \geq 1, 0 \leq y \leq 1/x\}$

26 $R = \{(x, y): x \geq 1, 0 \leq y \leq 1/\sqrt{x}\}$

27 $R = \{(x, y): x \geq 4, 0 \leq y \leq x^{-3/2}\}$

28 $R = \{(x, y): x \geq 8, 0 \leq y \leq x^{-2/3}\}$

29 The infinite region to the right of the y-axis and between the graphs of $y = e^{-x^2}$ and $y = 0$ is revolved about the y-axis. Show that a volume can be assigned to the resulting infinite solid. What is the volume?

30 Find the arc length of the curve $x^{2/3} + y^{2/3} = 1$.

31 Find all values of n for which the integral $\int_1^\infty x^n\,dx$ (a) converges; (b) diverges.

32 Find all integral values of n for which the integral $\int_{-\infty}^{-1} x^n\,dx$ (a) converges; (b) diverges.

33 In Figure 10.5 take $a = 0$ and let A represent the center of the earth. If P represents a point above the surface, then the force exerted by gravity at P is given by $F(x) = k/x^2$, where k is a constant. If the radius of the earth is

4000 mi, find the work required to project an object weighing 100 lb along l, from the surface to infinity. (*Hint*: $F(4000) = 100$.)

34 The force (in dynes) with which two electrons repel one another is inversely proportional to the square of the distance (in cm) between them. If, in Figure 10.5, one electron is fixed at A, find the work done if another electron is repelled along l from a point B, which is one cm from A, to infinity.

It can be shown that if f and g are continuous functions and if $0 \leq f(x) \leq g(x)$ for all x in $[a, \infty)$, then the following **comparison tests for improper integrals** are true:

(i) If $\int_a^\infty g(x)\,dx$ converges, then $\int_a^\infty f(x)\,dx$ converges.

(ii) If $\int_a^\infty f(x)\,dx$ diverges, then $\int_a^\infty g(x)\,dx$ diverges.

Use these tests to determine the convergence or divergence of the improper integrals in Exercises 35–38.

35 $\displaystyle\int_0^\infty \frac{1}{1+x^4}\,dx$

36 $\displaystyle\int_2^\infty \frac{1}{\sqrt[3]{x^2-1}}\,dx$

37 $\displaystyle\int_2^\infty \frac{1}{\ln x}\,dx$

38 $\displaystyle\int_1^\infty e^{-x^2}\,dx$

39 Find a function f such that $\lim_{t\to\infty} \int_{-t}^t f(x)\,dx$ exists and $\int_{-\infty}^\infty f(x)\,dx$ diverges.

40 Prove that if $\int_{-\infty}^a f(x)\,dx = L$ and $\int_a^\infty f(x)\,dx = K$, then

$$\int_{-\infty}^b f(x)\,dx + \int_b^\infty f(x)\,dx = L + K$$

for every real number b.

41 Prove that if $\int_a^\infty f(x)\,dx$ and $\int_a^\infty g(x)\,dx$ both converge, then $\int_a^\infty [f(x) + g(x)]\,dx$ converges. Show that the converse of this result is false.

42 If $f(x)$ is continuous on $[a, b]$, prove that

$$\lim_{t\to b^-} \int_a^t f(x)\,dx = \int_a^b f(x)\,dx$$

and

$$\lim_{t\to a^+} \int_t^b f(x)\,dx = \int_a^b f(x)\,dx.$$

In the theory of differential equations, if f is a function, then the **Laplace Transform** L of $f(x)$ is defined by

$$L[f(x)] = \int_0^\infty e^{-sx} f(x)\,dx$$

for every real number s for which the improper integral converges. In Exercises 43–48 find $L[f(x)]$ if $f(x)$ is the indicated expression.

43 1

44 x

45 $\cos x$

46 $\sin x$

47 e^{ax}

48 $\sin ax$

49 The **gamma function** Γ is defined by $\Gamma(n) = \int_0^\infty x^{n-1} e^{-x}\, dx$ where n is any positive real number.

(a) Find $\Gamma(1)$, $\Gamma(2)$, and $\Gamma(3)$.

(b) Prove that $\Gamma(n+1) = n\Gamma(n)$. (*Hint*: Integrate by parts.)

(c) Use mathematical induction to prove that if n is any positive integer, then $\Gamma(n+1) = n!$. (This shows that factorials are special values of the gamma function.)

10.4

INTEGRALS WITH DISCONTINUOUS INTEGRANDS

If a function f is continuous on a closed interval $[a, b]$, then by Theorem (5.11) the definite integral $\int_a^b f(x)\, dx$ exists. If f has an infinite discontinuity at some number in the interval it may still be possible to assign a value to the integral. Suppose, for example, that f is continuous and nonnegative on the half-open interval $[a, b)$ and $\lim_{x \to b^-} f(x) = \infty$. If $a < t < b$, then the area $A(t)$ under the graph of f from a to t (see (i) of Figure 10.6) is

$$A(t) = \int_a^t f(x)\, dx.$$

If $\lim_{t \to b^-} A(t)$ exists, then the limit may be interpreted as the area of the unbounded region that lies under the graph of f, over the x-axis, and between $x = a$ and $x = b$. It is natural to denote this number by $\int_a^b f(x)\, dx$.

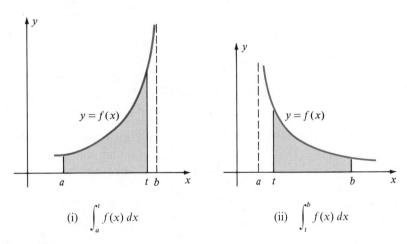

(i) $\displaystyle\int_a^t f(x)\, dx$ (ii) $\displaystyle\int_t^b f(x)\, dx$

Figure 10.6

In like manner, for the situation illustrated in (ii) of the figure, where $\lim_{x \to a^+} f(x) = \infty$, we define $\int_a^b f(x)\, dx$ as the limit of $\int_t^b f(x)\, dx$ as $t \to a^+$.

These remarks serve as motivation for the following definition.

Definition (10.5)

(i)   If f is continuous on $[a, b)$ and discontinuous at b, then

$$\int_a^b f(x)\,dx = \lim_{t \to b^-} \int_a^t f(x)\,dx.$$

(ii)   If f is continuous on $(a, b]$ and discontinuous at a, then

$$\int_a^b f(x)\,dx = \lim_{t \to a^+} \int_t^b f(x)\,dx.$$

As in the preceding section, the integrals defined in (10.5) are referred to as **improper integrals** and they are said to **converge** if the indicated limits exist. The limits are called the **values** of the improper integrals. If the limits do not exist, the improper integrals are said to **diverge**.

Another type of improper integral is defined as follows.

Definition (10.6)

If f has a discontinuity at a number c in the open interval (a, b) but is continuous elsewhere in $[a, b]$, then

$$\int_a^b f(x)\,dx = \int_a^c f(x)\,dx + \int_c^b f(x)\,dx$$

provided *both* of the integrals on the right converge. If both converge, then the value of the integral $\int_a^b f(x)\,dx$ is the sum of the two values.

The graph of a function satisfying the conditions of Definition (10.6) is sketched in Figure 10.7.

A definition similar to (10.6) is used if f has any finite number of discontinuities in (a, b). For example, if f has discontinuities at c_1 and c_2, where $c_1 < c_2$, but is continuous elsewhere in $[a, b]$, then we choose a number d between c_1 and c_2 and express $\int_a^b f(x)\,dx$ as a sum of four improper integrals over the intervals $[a, c_1]$, $[c_1, d]$, $[d, c_2]$, and $[c_2, b]$, respectively. The given integral converges if and only if each of the improper integrals in the sum converges. Finally, if f is continuous on (a, b) but becomes infinite at *both* a and b, then we again define $\int_a^b f(x)\,dx$ by means of (10.6).

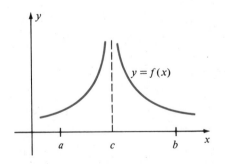

Figure 10.7

Example 1   Evaluate $\displaystyle\int_0^3 \frac{1}{\sqrt{3 - x}}\,dx$.

Solution   Since the integrand has an infinite discontinuity at $x = 3$, we apply (i) of Definition (10.5) as follows:

$$\int_0^3 \frac{1}{\sqrt{3-x}}\, dx = \lim_{t \to 3^-} \int_0^t \frac{1}{\sqrt{3-x}}\, dx$$

$$= \lim_{t \to 3^-} \left[-2\sqrt{3-x} \right]_0^t$$

$$= \lim_{t \to 3^-} \left[-2\sqrt{3-t} + 2\sqrt{3} \right]$$

$$= 0 + 2\sqrt{3} = 2\sqrt{3}. \qquad \blacksquare$$

Example 2   Determine whether $\int_0^1 \frac{1}{x}\, dx$ converges or diverges.

Solution   The integrand is undefined at $x = 0$. Applying (ii) of (10.5),

$$\int_0^1 \frac{1}{x}\, dx = \lim_{t \to 0^+} \int_t^1 \frac{1}{x}\, dx = \lim_{t \to 0^+} \left[\ln x \right]_t^1$$

$$= \lim_{t \to 0^+} \left[0 - \ln t \right] = \infty.$$

Since the limit does not exist, the improper integral diverges. $\blacksquare$

Example 3   Determine whether $\int_0^4 \frac{1}{(x-3)^2}\, dx$ converges or diverges.

Solution   The integrand is undefined at $x = 3$. Since this number is in the interior of the interval $[0, 4]$, we use Definition (10.6) with $c = 3$, obtaining

$$\int_0^4 \frac{1}{(x-3)^2}\, dx = \int_0^3 \frac{1}{(x-3)^2}\, dx + \int_3^4 \frac{1}{(x-3)^2}\, dx.$$

For the integral on the left to converge, *both* integrals on the right must converge. Applying (i) of Definition (10.5) to the first integral,

$$\int_0^3 \frac{1}{(x-3)^2}\, dx = \lim_{t \to 3^-} \int_0^t \frac{1}{(x-3)^2}\, dx$$

$$= \lim_{t \to 3^-} \frac{-1}{x-3} \Big]_0^t$$

$$= \lim_{t \to 3^-} \left(\frac{-1}{t-3} - \frac{1}{3} \right) = \infty.$$

It follows that the given improper integral diverges. $\blacksquare$

It is important to note that the Fundamental Theorem of Calculus cannot be applied to the integral in Example 3 since the function given by the integrand is not continuous on $[0, 4]$. Indeed, if we had applied the Fundamental Theorem we would have obtained

$$\int_0^4 \frac{1}{(x-3)^2} \, dx = \left. \frac{-1}{(x-3)} \right]_0^4$$

$$= -1 - \frac{1}{3} = -\frac{4}{3}.$$

This result is obviously incorrect since the integrand is never negative.

Example 4 Evaluate $\displaystyle\int_{-2}^7 \frac{1}{(x+1)^{2/3}} \, dx$.

Solution   The integrand is undefined at $x = -1$, a number between -2 and 7. Consequently, we apply Definition (10.6) with $c = -1$, as follows:

$$\int_{-2}^7 \frac{1}{(x+1)^{2/3}} \, dx = \int_{-2}^{-1} \frac{1}{(x+1)^{2/3}} \, dx + \int_{-1}^7 \frac{1}{(x+1)^{2/3}} \, dx.$$

We next investigate each integral on the right. Using (i) of (10.5) with $b = -1$ gives us

$$\int_{-2}^{-1} \frac{1}{(x+1)^{2/3}} \, dx = \lim_{t \to -1^-} \int_{-2}^t \frac{1}{(x+1)^{2/3}} \, dx$$

$$= \lim_{t \to -1^-} \left[3(x+1)^{1/3} \right]_{-2}^t$$

$$= \lim_{t \to -1^-} \left[3(t+1)^{1/3} - 3(-1)^{1/3} \right]$$

$$= 0 + 3 = 3.$$

In similar fashion, using (ii) of (10.5) with $a = -1$,

$$\int_{-1}^7 \frac{1}{(x+1)^{2/3}} \, dx = \lim_{t \to -1^+} \int_t^7 \frac{1}{(x+1)^{2/3}} \, dx$$

$$= \lim_{t \to -1^+} \left[3(x+1)^{1/3} \right]_t^7$$

$$= \lim_{t \to -1^+} \left[3(8)^{1/3} - 3(t+1)^{1/3} \right]$$

$$= 6 - 0 = 6.$$

Since both integrals converge, the given integral converges and has the value $3 + 6 = 9$.   ∎

An improper integral may have both a discontinuity in the integrand and an infinite limit of integration. Integrals of this type may be investigated by expressing them as sums of improper integrals, each of which has one of the forms previously defined. As an illustration, since the integrand of

$\int_0^\infty (1/\sqrt{x})\, dx$ is discontinuous at $x = 0$, we choose any number (for example, 1) greater than 0 and write

$$\int_0^\infty \frac{1}{\sqrt{x}}\, dx = \int_0^1 \frac{1}{\sqrt{x}}\, dx + \int_1^\infty \frac{1}{\sqrt{x}}\, dx.$$

It is easy to show that the first integral on the right side of the equation converges and the second diverges. (Verify these facts.) Hence (by definition) the given integral diverges.

   Improper integrals of the types considered in this section arise in certain physical applications. Figure 10.8 is a schematic drawing of a spring with an attached weight that is oscillating between points with coordinates $-d$ and d on a coordinate line y (the y-axis has been positioned at the right for clarity).

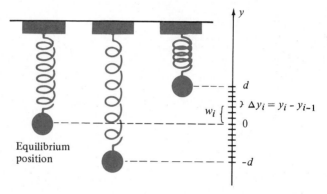

Figure 10.8

   The **period** T is the time required for one complete oscillation, that is, *twice* the time required for the weight to cover the interval $[-d, d]$. Let $v(y)$ denote the velocity of the weight when it is at the point with coordinate y in $[-d, d]$. We partition $[-d, d]$ in the usual way and let $\Delta y_i = y_i - y_{i-1}$ denote the distance the weight travels during the time interval Δt_i. If w_i is any number in $[y_{i-1}, y_i]$, then $v(w_i)$ is the velocity of the weight when it is at the point with coordinate w_i. If the norm of the partition is small and we assume v is a continuous function, then the distance Δy_i may be approximated by the product $v(w_i)\, \Delta t_i$, that is,

$$\Delta y_i \approx v(w_i)\, \Delta t_i.$$

Hence the time required for the weight to cover the distance Δy_i may be approximated by

$$\Delta t_i \approx \frac{1}{v(w_i)}\, \Delta y_i$$

and, therefore,

$$T = 2 \sum_i \Delta t_i \approx 2 \sum_i \frac{1}{v(w_i)}\, \Delta y_i.$$

Taking the limit of the sum on the right, and using the definition of the definite integral,

$$T = 2 \int_{-d}^{d} \frac{1}{v(y)} \, dy.$$

Since $v(d) = 0$ and $v(-d) = 0$, the integral is improper (but necessarily convergent).

EXERCISES 10.4

Determine whether the integrals in Exercises 1–30 converge or diverge and evaluate those that converge.

1 $\int_{0}^{8} \frac{1}{\sqrt[3]{x}} \, dx$

2 $\int_{0}^{9} \frac{1}{\sqrt{x}} \, dx$

3 $\int_{-3}^{1} \frac{1}{x^2} \, dx$

4 $\int_{-2}^{-1} \frac{1}{(x+2)^{5/4}} \, dx$

5 $\int_{0}^{\pi/2} \sec^2 x \, dx$

6 $\int_{0}^{1} \frac{e^{\sqrt{x}}}{\sqrt{x}} \, dx$

7 $\int_{0}^{4} \frac{1}{(4-x)^{3/2}} \, dx$

8 $\int_{0}^{-1} \frac{1}{\sqrt[3]{x+1}} \, dx$

9 $\int_{0}^{4} \frac{1}{(4-x)^{2/3}} \, dx$

10 $\int_{1}^{2} \frac{x}{x^2-1} \, dx$

11 $\int_{-2}^{2} \frac{1}{(x+1)^3} \, dx$

12 $\int_{-1}^{1} x^{-4/3} \, dx$

13 $\int_{-2}^{0} \frac{1}{\sqrt{4-x^2}} \, dx$

14 $\int_{-2}^{0} \frac{x}{\sqrt{4-x^2}} \, dx$

15 $\int_{-1}^{2} \frac{1}{x} \, dx$

16 $\int_{0}^{4} \frac{1}{x^2-x-2} \, dx$

17 $\int_{0}^{1} x \ln x \, dx$

18 $\int_{0}^{\pi/2} \tan^2 x \, dx$

19 $\int_{0}^{\pi/2} \tan x \, dx$

20 $\int_{0}^{\pi/2} \frac{1}{1-\cos x} \, dx$

21 $\int_{2}^{4} \frac{x-2}{x^2-5x+4} \, dx$

22 $\int_{1/e}^{e} \frac{1}{x(\ln x)^2} \, dx$

23 $\int_{-1}^{2} \frac{1}{x^2} \cos \frac{1}{x} \, dx$

24 $\int_{0}^{\pi} \sec x \, dx$

25 $\int_{0}^{\pi} \frac{\cos x}{\sqrt{1-\sin x}} \, dx$

26 $\int_{0}^{9} \frac{x}{\sqrt[3]{x-1}} \, dx$

27 $\int_{0}^{4} \frac{1}{x^2-4x+3} \, dx$

28 $\int_{-1}^{3} \frac{x}{\sqrt[3]{x^2-1}} \, dx$

29 $\int_{0}^{\infty} \frac{1}{(x-4)^2} \, dx$

30 $\int_{-\infty}^{0} \frac{1}{x+2} \, dx$

In Exercises 31–34 assign, if possible, a value to (a) the area of the region R, and (b) the volume of the solid obtained by revolving R about the x-axis.

31 $R = \{(x, y): 0 \le x \le 1, 0 \le y \le 1/\sqrt{x}\}$

32 $R = \{(x, y): 0 \le x \le 1, 0 \le y \le 1/\sqrt[3]{x}\}$

33 $R = \{(x, y): -4 \le x \le 4, 0 \le y \le 1/(x+4)\}$

34 $R = \{(x, y): 1 \le x \le 2, 0 \le y \le 1/(x-1)\}$

Find all values of n for which the integrals in Exercises 35 and 36 converge.

35 $\int_{0}^{1} x^n \, dx$

36 $\int_{0}^{1} x^n \ln x \, dx$

Suppose f and g are continuous and $0 \le f(x) \le g(x)$ for all x in $(a, b]$. If f and g are discontinuous at $x = a$, then the following **comparison tests** can be proved.

(i) If $\int_{a}^{b} g(x) \, dx$ converges, then $\int_{a}^{b} f(x) \, dx$ converges.

(ii) If $\int_{a}^{b} f(x) \, dx$ diverges, then $\int_{a}^{b} g(x) \, dx$ diverges.

Analogous tests may be stated for continuity on $[a, b)$ and a discontinuity at $x = b$. Use these tests to determine the convergence or divergence of the improper integrals in Exercises 37–40.

37 $\int_{0}^{\pi} \frac{\sin x}{\sqrt{x}} \, dx$

38 $\int_{0}^{\pi/4} \frac{\sec x}{x^3} \, dx$

39 $\int_{0}^{2} \frac{\cosh x}{(x-2)^2} \, dx$

40 $\int_{0}^{1} \frac{e^{-x}}{x^{2/3}} \, dx$

10.5

TAYLOR'S FORMULA

Recall that f is a polynomial function of degree n if

$$f(x) = a_0 + a_1 x + a_2 x^2 + \cdots + a_n x^n$$

where each a_i is a real number, $a_n \neq 0$, and the exponents are nonnegative integers. Polynomial functions are the simplest functions to use for calculations, in the sense that their values can be found by employing only additions and multiplications of real numbers. More complicated operations are needed to calculate values of logarithmic, exponential, or trigonometric functions; however, sometimes it is possible to *approximate* values by using polynomials. For example, since $\lim_{x \to 0} (\sin x)/x = 1$, it follows that if x is close to 0, then $\sin x \approx x$; that is, the value of the sine function is almost the same as the value of the polynomial x. We say that $\sin x$ *may be approximated by the polynomial* x (provided x is close to 0).

As a second illustration, let f be the natural exponential function, that is, $f(x) = e^x$ for every x. Suppose we are interested in calculating values of f when x is close to 0. Since $f'(x) = e^x$, the slope of the tangent line at the point $(0, 1)$ on the graph of f is $f'(0) = e^0 = 1$. Hence an equation of the tangent line is

$$y - 1 = 1(x - 0) \quad \text{or} \quad y = 1 + x.$$

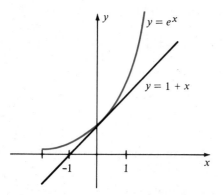

Figure 10.9

Referring to Figure 10.9, it is evident that if x is very close to 0, then the point $(x, 1 + x)$ on the tangent line is close to the point (x, e^x) on the graph of f and hence we may write $e^x \approx 1 + x$. This formula allows us to approximate e^x by means of a polynomial of degree 1. Since the approximation is obviously poor unless x is very close to 0, let us seek a second-degree polynomial

$$g(x) = a + bx + cx^2$$

such that $e^x \approx g(x)$ when x is numerically small. The first and second derivatives of $g(x)$ are

$$g'(x) = b + 2cx$$
$$g''(x) = 2c.$$

If we want the graph of g (a parabola) to have (i) the same y-intercept, (ii) the same tangent line, and (iii) the same concavity, as the graph of f at the point $(0, 1)$, then we must have

$$\text{(i)} \ g(0) = f(0), \quad \text{(ii)} \ g'(0) = f'(0), \quad \text{(iii)} \ g''(0) = f''(0).$$

Since all derivatives of e^x equal e^x, and $e^0 = 1$, these three equations imply that

$$a = 1, \quad b = 1, \quad \text{and} \quad 2c = 1$$

and hence

$$e^x \approx g(x) = 1 + x + \tfrac{1}{2}x^2.$$

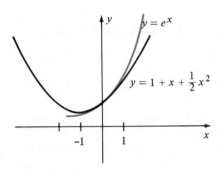

Figure 10.10

The graphs of f and g are sketched in Figure 10.10. Comparing with Figure 10.9, we see that if x is close to 0, then e^x is closer to $1 + x + \frac{1}{2}x^2$ than to $1 + x$.

If we wish to approximate $f(x) = e^x$ by means of a *third*-degree polynomial $h(x)$, it is natural to require that $h(0)$, $h'(0)$, $h''(0)$, and $h'''(0)$ be the same as $f(0)$, $f'(0)$, $f''(0)$, and $f'''(0)$, respectively. The graphs of f and h then have the same tangent line and concavity at $(0, 1)$ and, in addition, their *rates of change of concavity* with respect to x (that is, the third derivatives) are equal. Using the same technique employed previously would give us

$$e^x \approx h(x) = 1 + x + \frac{1}{2}x^2 + \frac{1}{3!}x^3.$$

To get some idea of the accuracy of this approximation, several values of e^x and $h(x)$, approximated to the nearest hundredth, are displayed in the following table.

x	-1.5	-1.0	-0.5	0	0.5	1.0	1.5
e^x	0.22	0.37	0.61	1	1.65	2.72	4.48
$h(x)$	0.06	0.33	0.60	1	1.65	2.67	4.19

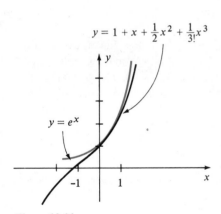

Figure 10.11

Observe that the error in the approximation increases as x increases numerically. The graphs of f and h are sketched in Figure 10.11. Notice the improvement in the approximation near $x = 0$.

If we continued in this manner and determined a polynomial of degree n whose first n derivatives coincide with those of f at $x = 0$, we would arrive at

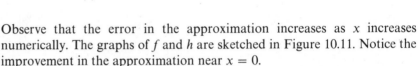

$$e^x \approx 1 + x + \frac{1}{2}x^2 + \frac{1}{3!}x^3 + \cdots + \frac{1}{n!}x^n.$$

It will follow from our later work that this remarkably simple formula can be used to approximate e^x to any desired degree of accuracy.

We now ask the following two questions:

1. Does there exist a *general* formula that may be used to obtain polynomial approximations for arbitrary exponential functions, logarithmic functions, trigonometric functions, inverse trigonometric functions, and other transcendental or algebraic functions?

2. Can the previous discussion be generalized to the case where x is close to an arbitrary number $a \neq 0$?

The answer to both questions is "yes" provided the function under consideration has a sufficient number of derivatives. Specifically, Formula (10.8) of this section may be used to obtain polynomial approximations of a wide variety of functions. To see why this is true, suppose f is a function that has many derivatives, and consider a number a in the domain of f. Let us proceed as we did for the special case of e^x discussed previously, but with a in place of 0. First we note that the only polynomial of degree 0 that coincides with f at a is the constant $f(a)$. To approximate $f(x)$ near a by a polynomial of degree 1, we choose the polynomial whose graph is the tangent line to the graph of f at $(a, f(a))$. Since the equation of this tangent line is

$$y - f(a) = f'(a)(x - a), \quad \text{or} \quad y = f(a) + f'(a)(x - a)$$

the desired first-degree polynomial is

$$f(a) + f'(a)(x - a).$$

A better approximation should be obtained by using a polynomial whose graph has the same *concavity* and tangent line as that of f at $(a, f(a))$. It is left to the reader to verify that these conditions are fulfilled by the second-degree polynomial

$$f(a) + f'(a)(x - a) + \frac{f''(a)}{2}(x - a)^2.$$

If we next consider the third-degree polynomial

$$g(x) = f(a) + f'(a)(x - a) + \frac{f''(a)}{2!}(x - a)^2 + \frac{f'''(a)}{3!}(x - a)^3$$

then it is easy to show that in addition to $g'(a) = f'(a)$ and $g''(a) = f''(a)$, we also have $g'''(a) = f'''(a)$. It appears that if this pattern is continued, we should get better approximations to $f(x)$ when x is near a. This leads to the first $n + 1$ terms on the right side of Formula (10.8). The form of the last term is a consequence of the next result, which bears the name of the English mathematician Brook Taylor (1685–1731).

Taylor's Formula (10.7)

> Let f be a function and n a positive integer such that the derivative $f^{(n+1)}(x)$ exists for every x in an interval I. If a and b are distinct numbers in I, then there is a number z between a and b such that
>
> $$f(b) = f(a) + \frac{f'(a)}{1!}(b - a) + \frac{f''(a)}{2!}(b - a)^2 + \cdots$$
>
> $$+ \frac{f^{(n)}(a)}{n!}(b - a)^n + \frac{f^{(n+1)}(z)}{(n+1)!}(b - a)^{n+1}.$$

Proof   There exists a number R_n (depending on a, b, and n) such that

$$f(b) = f(a) + \frac{f'(a)}{1!}(b - a) + \frac{f''(a)}{2!}(b - a)^2 + \cdots + \frac{f^{(n)}(a)}{n!}(b - a)^n + R_n.$$

Indeed, to find R_n we merely subtract the sum of the first $n + 1$ terms on the right side of this equation from $f(b)$. We wish to show that R_n is the same as the last term of the formula given in the statement of the theorem.

Let g be the function defined by

$$g(x) = f(b) - f(x) - \frac{f'(x)}{1!}(b - x) - \frac{f''(x)}{2!}(b - x)^2 - \cdots$$
$$- \frac{f^{(n)}(x)}{n!}(b - x)^n - R_n \frac{(b - x)^{n+1}}{(b - a)^{n+1}}$$

for all x in I. If we differentiate each side of this equation, then many terms on the right cancel. As a matter of fact, it can be shown (see Exercise 41) that

$$g'(x) = -\frac{f^{(n+1)}(x)}{n!}(b - x)^n + R_n(n + 1)\frac{(b - x)^n}{(b - a)^{n+1}}.$$

It is easy to see that $g(b) = 0$. Moreover, substituting a for x in the expression for $g(x)$ and making use of the first equation of the proof gives us $g(a) = 0$. According to Rolle's Theorem, there is a number z between a and b such that $g'(z) = 0$. Evaluating $g'(x)$ at z and solving for R_n we see that

$$R_n = \frac{f^{(n+1)}(z)}{(n + 1)!}(b - a)^{n+1}$$

which is what we wished to prove.   □

Replacement of b by x in Formula (10.7) leads to the following.

Taylor's Formula with the Remainder (10.8)

Let f have $n + 1$ derivatives throughout an interval containing a. If x is any number in the interval, then

$$f(x) = f(a) + \frac{f'(a)}{1!}(x - a) + \frac{f''(a)}{2!}(x - a)^2 + \cdots$$
$$+ \frac{f^{(n)}(a)}{n!}(x - a)^n + \frac{f^{(n+1)}(z)}{(n + 1)!}(x - a)^{n+1}$$

where z is a number between a and x.

For convenience we shall denote the sum of the first $n + 1$ terms in Taylor's Formula by $P_n(x)$ and give it a special name, as in (i) of the next definition.

Definition (10.9)

(i)   The ***n*th-degree Taylor Polynomial** $P_n(x)$ of f at a is

$$P_n(x) = f(a) + \frac{f'(a)}{1!}(x - a) + \frac{f''(a)}{2!}(x - a)^2 + \cdots + \frac{f^{(n)}(a)}{n!}(x - a)^n.$$

(ii)   The **Taylor Remainder** $R_n(x)$ of f at a is

$$R_n(x) = \frac{f^{(n+1)}(z)}{(n+1)!}(x - a)^{n+1}$$

where z is between a and x.

We may now state the following result.

Theorem (10.10)

Let f have $n + 1$ derivatives throughout an interval containing a. If x is any number in the interval, then

(i) $f(x) = P_n(x) + R_n(x)$.

(ii) $f(x) \approx P_n(x)$ if $x \approx a$, where the error involved in using this approximation is $|R_n(x)|$.

Proof   Part (i) is merely a restatement of Formula (10.8) using the notation of (10.9). Part (ii) follows from the fact that $|f(x) - P_n(x)| = |R_n(x)|$.
□

Example 1   If $f(x) = \ln x$, find Taylor's Formula with the Remainder for $n = 3$ and $a = 1$.

Solution   If $n = 3$ in Taylor's Formula (10.8), then we need the first four derivatives of f. It is convenient to arrange our work as follows.

$$
\begin{aligned}
f(x) &= \ln x & f(1) &= 0 \\
f'(x) &= x^{-1} & f'(1) &= 1 \\
f''(x) &= -x^{-2} & f''(1) &= -1 \\
f'''(x) &= 2x^{-3} & f'''(1) &= 2 \\
f^{(4)}(x) &= -3!x^{-4} & f^{(4)}(z) &= -6z^{-4}
\end{aligned}
$$

By (10.8),

$$\ln x = 0 + \frac{1}{1!}(x - 1) - \frac{1}{2!}(x - 1)^2 + \frac{2}{3!}(x - 1)^3 - \frac{6z^{-4}}{4!}(x - 1)^4$$

where z is between 1 and x. This simplifies to

$$\ln x = (x - 1) - \frac{1}{2}(x - 1)^2 + \frac{1}{3}(x - 1)^3 - \frac{1}{4z^4}(x - 1)^4.$$ ■

In the next two examples Taylor Polynomials are used to approximate values of functions. To discuss the accuracy of an approximation, it is necessary to agree on what is meant by 1-decimal-place accuracy, 2-decimal-place accuracy, etc. Let us adopt the following convention. If E is the error in an approximation, then the approximation will be considered accurate to k decimal places if $|E| < 0.5 \times 10^{-k}$. For example, we have

1-decimal-place accuracy if $|E| < 0.5 \times 10^{-1} = 0.05$
2-decimal-place accuracy if $|E| < 0.5 \times 10^{-2} = 0.005$
3-decimal-place accuracy if $|E| < 0.5 \times 10^{-3} = 0.0005.$

If we are interested in k-decimal-place accuracy in the approximation of a sum, we shall approximate each term of the sum to $k + 1$ decimal places and then round off the final result to k decimal places. In certain cases this may fail to produce the required degree of accuracy; however, it is customary to proceed in this way for elementary approximations. More precise techniques may be found in texts on *numerical analysis*.

Example 2   Use the formula obtained in Example 1 to approximate ln (1.1), and estimate the accuracy of this approximation.

Solution   Substituting 1.1 for x in the formula of Example 1 gives us

$$\ln (1.1) = 0.1 - \frac{1}{2}(0.1)^2 + \frac{1}{3}(0.1)^3 - \frac{1}{4z^4}(0.1)^4$$

where $1 < z < 1.1$. Summing the first three terms we obtain ln $(1.1) \approx 0.0953$. Since $z > 1$, $1/z < 1$ and, therefore, $1/z^4 < 1$. Consequently,

$$|R_3(1.1)| = \left| \frac{(0.1)^4}{4z^4} \right| < \left| \frac{0.0001}{4} \right| = 0.000025.$$

Since $0.000025 < 0.00005 = 0.5 \times 10^{-4}$, it follows from (ii) of Theorem (10.10) and our convention concerning accuracy, that the approximation ln $(1.1) \approx 0.0953$ is accurate to four decimal places.   ■

If we wish to approximate a functional value $f(x)$ for some x, it is desirable to choose the number a in (10.8) such that the remainder $R_n(x)$ is very close to 0 when n is relatively small (say $n = 3$ or $n = 4$). This will be true if we choose a close to x. In addition, a should be chosen in such a way that the values of the first $n + 1$ derivatives of f at a are easy to calculate. This was done in Example 2, where to approximate ln x for $x = 1.1$ we selected $a = 1$ (see Example 1). The next example provides another illustration of a suitable choice for a.

Example 3 Use a Taylor Polynomial to approximate cos 61°, and estimate the accuracy of the approximation.

Solution We wish to approximate $f(x) = \cos x$ if $x = 61°$. Let us begin by observing that 61° is close to 60°, or $\pi/3$ radians, and that it is easy to calculate values of trigonometric functions at $\pi/3$. This suggests that we choose $a = \pi/3$ in (10.8). The choice of n will depend on the accuracy we wish to attain. Let us try $n = 2$. In this event the first three derivatives of f are required and we arrange our work as follows:

$$f(x) = \cos x \qquad f(\pi/3) = 1/2$$
$$f'(x) = -\sin x \qquad f'(\pi/3) = -\sqrt{3}/2$$
$$f''(x) = -\cos x \qquad f''(\pi/3) = -1/2$$
$$f'''(x) = \sin x \qquad f'''(z) = \sin z$$

By (i) of Definition (10.9), the second-degree Taylor Polynomial of f at $\pi/3$ is

$$P_2(x) = \frac{1}{2} - \frac{\sqrt{3}/2}{1!}\left(x - \frac{\pi}{3}\right) - \frac{1/2}{2!}\left(x - \frac{\pi}{3}\right)^2.$$

Since x represents a real number, 61° must be converted to radian measure before substitution on the right side. Writing

$$61° = 60° + 1° = \frac{\pi}{3} + \frac{\pi}{180}$$

and substituting in $P_2(x)$, we obtain

$$P_2\left(\frac{\pi}{3} + \frac{\pi}{180}\right) = \frac{1}{2} - \left(\frac{\sqrt{3}}{2}\right)\left(\frac{\pi}{180}\right) - \frac{1}{4}\left(\frac{\pi}{180}\right)^2 \approx 0.48481.$$

Thus, by (i) of Theorem (10.10),

$$\cos 61° \approx 0.48481.$$

To estimate the accuracy of this approximation, we see from (ii) of Definition (10.9) that

$$|R_2(x)| = \left|\frac{\sin z}{3!}\left(x - \frac{\pi}{3}\right)^3\right|.$$

Substituting $x = (\pi/3) + (\pi/180)$ and using the fact that $|\sin z| \le 1$,

$$\left|R_2\left(\frac{\pi}{3} + \frac{\pi}{180}\right)\right| = \left|\frac{\sin z}{3!}\left(\frac{\pi}{180}\right)^3\right| \le \left|\frac{1}{3!}\left(\frac{\pi}{180}\right)^3\right| \le 0.000001.$$

Thus, by (ii) of Definition (10.9), the approximation $\cos 61° \approx 0.48481$ is accurate to five decimal places. If more accuracy is desired, then it is necessary to find a value of n such that the maximum value of $|R_n((\pi/3) + (\pi/180))|$ is within the desired range. ∎

If we let $a = 0$ in Formula (10.8), we obtain the following important formula, named after the Scottish mathematician Colin Maclaurin (1698–1746).

Maclaurin's Formula (10.11)

> Let f have $n + 1$ derivatives throughout an interval containing 0. If x is any number in the interval, then
>
> $$f(x) = f(0) + \frac{f'(0)}{1!} x + \frac{f''(0)}{2!} x^2 + \cdots + \frac{f^{(n)}(0)}{n!} x^n + \frac{f^{(n+1)}(z)}{(n+1)!} x^{n+1}$$
>
> where z is between 0 and x.

Example 4 Find Maclaurin's Formula for $f(x) = e^x$ if n is any positive integer.

Solution   For every positive integer k we have $f^{(k)}(x) = e^x$, and hence $f^{(k)}(0) = e^0 = 1$. Substituting in (10.11) gives us

$$e^x = 1 + x + \frac{x^2}{2!} + \cdots + \frac{x^n}{n!} + \frac{e^z}{(n+1)!} x^{n+1}$$

where z is between 0 and x. Note that the first $n + 1$ terms are the same as those obtained in our discussion at the beginning of this section. ■

The formula derived in Example 4 may be used to approximate values of the natural exponential function. Another important application will be discussed in the next chapter in conjunction with representations of functions by means of infinite series.

Example 5 Find Maclaurin's Formula for $f(x) = \sin x$ and $n = 8$.

Solution   We need the first nine derivatives of $f(x)$. Let us begin as follows:

$$f(x) = \sin x \qquad f(0) = 0$$
$$f'(x) = \cos x \qquad f'(0) = 1$$
$$f''(x) = -\sin x \qquad f''(0) = 0$$
$$f'''(x) = -\cos x \qquad f'''(0) = -1$$

Since $f^{(4)}(x) = \sin x$, the remaining derivatives follow the same pattern, and we arrive at

$$f^{(9)}(x) = \cos x \qquad f^{(9)}(z) = \cos z.$$

Substituting in (10.11) and noting that the constant term and the coefficients of x^2, x^4, x^6, and x^8 are 0, we obtain

$$\sin x = \frac{1}{1!} x + \frac{(-1)}{3!} x^3 + \frac{1}{5!} x^5 + \frac{(-1)}{7!} x^7 + \frac{\cos z}{9!} x^9$$

which may be written

$$\sin x = x - \frac{x^3}{3!} + \frac{x^5}{5!} - \frac{x^7}{7!} + \frac{\cos z}{9!} x^9. \qquad \blacksquare$$

Incidentally, if we used the first four nonzero terms of the formula found in Example 5 to approximate $\sin (0.1)$, then the error would be less than $|R_8(0.1)|$. Since $|\cos z| \le 1$,

$$|R_8(0.1)| = \left| \frac{\cos z}{9!} (0.1)^9 \right| \le \frac{(0.1)^9}{9!} < 2.7 \times 10^{-15} < 0.5 \times 10^{-14}.$$

According to our convention concerning accuracy, this means that the approximation would be correct to 14 decimal places!

EXERCISES 10.5

In Exercises 1–12, find Taylor's Formula with the Remainder for the given value of a and n.

1 $f(x) = \sin x, a = \pi/2, n = 3$

2 $f(x) = \cos x, a = \pi/4, n = 3$

3 $f(x) = \sqrt{x}, a = 4, n = 3$

4 $f(x) = e^{-x}, a = 1, n = 3$

5 $f(x) = \tan x, a = \pi/4, n = 4$

6 $f(x) = 1/(x-1)^2, a = 2, n = 5$

7 $f(x) = 1/x, a = -2, n = 5$

8 $f(x) = \sqrt[3]{x}, a = -8, n = 3$

9 $f(x) = \tan^{-1} x, a = 1, n = 2$

10 $f(x) = \ln \sin x, a = \pi/6, n = 3$

11 $f(x) = xe^x, a = -1, n = 4$

12 $f(x) = \log x, a = 10, n = 2$

In Exercises 13–24, find Maclaurin's Formula for the given values of n.

13 $f(x) = \ln (x + 1), n = 4$

14 $f(x) = \sin x, n = 7$

15 $f(x) = \cos x, n = 8$

16 $f(x) = \tan^{-1} x, n = 3$

17 $f(x) = e^{2x}, n = 5$

18 $f(x) = \sec x, n = 3$

19 $f(x) = \dfrac{1}{(x-1)^2}, n = 5$

20 $f(x) = \sqrt{4 - x}, n = 3$

21 $f(x) = \arcsin x, n = 2$

22 $f(x) = e^{-x^2}, n = 3$

23 $f(x) = 2x^4 - 5x^3 + x^2 - 3x + 7, n = 4$ and $n = 5$

24 $f(x) = \cosh x, n = 4$ and $n = 5$

Approximate the numbers in Exercises 25–28 to four decimal places by using the indicated exercise. In each case prove that your answer is correct by showing that $|R_n(x)| < 0.5 \times 10^{-4}$.

25 $\sin 89°$ (Use Exercise 1 and $\pi/180 \approx 0.0175$.)

26 $\cos 47°$ (Use Exercise 2 and $\pi/180 \approx 0.0175$.)

27 $\sqrt{4.03}$ (Exercise 3)

28 $e^{-1.02}$ (Exercise 4)

In Exercises 29–34 approximate the number by using the indicated exercise, and estimate the error in the approximation by means of $R_n(x)$.

29 $-1/2.2$; Exercise 7

30 $\sqrt[3]{-8.5}$; Exercise 8

31 $\ln(1.25)$; Exercise 13

32 $\sin 0.1$; Exercise 14

33 $\cos 30°$; Exercise 15

34 $\log 10.01$; Exercise 12

Use Maclaurin's Formula to establish the approximation formulas in Exercises 35–40 and state, in terms of decimal places, the accuracy of the approximation if $|x| \leq 0.1$.

35 $\cos x \approx 1 - \dfrac{x^2}{2}$

36 $\sqrt[3]{1 + x} \approx 1 + \dfrac{1}{3}x$

37 $e^x \approx 1 + x + \dfrac{x^2}{2}$

38 $\sin x \approx x - \dfrac{x^3}{6}$

39 $\ln(1 + x) \approx x - \dfrac{x^2}{2} + \dfrac{x^3}{3}$

40 $\cosh x \approx 1 + \dfrac{x^2}{2}$

41 In the proof of Taylor's Formula (10.7) verify the formula for $g'(x)$.

42 If $f(x)$ is a polynomial of degree n and a is a real number, prove that $f(x) = P_n(x)$, where $P_n(x)$ is given by (i) of Definition (10.9).

10.6

REVIEW

Define or discuss each of the following.

1 Indeterminate forms

2 L'Hopital's Rule

3 Cauchy's Formula

4 Improper integrals

5 Taylor's Formula

6 Maclaurin's Formula

EXERCISES 10.6

Find the limits in Exercises 1–14, if they exist.

1 $\lim\limits_{x \to 0} \dfrac{\ln(2 - x)}{1 + e^{2x}}$

2 $\lim\limits_{x \to 0} \dfrac{\sin 2x - \tan 2x}{x^2}$

3 $\lim\limits_{x \to \infty} \dfrac{x^2 + 2x + 3}{\ln(x + 1)}$

4 $\lim\limits_{x \to 0} \dfrac{\tan^{-1} x}{\sin^{-1} x}$

5 $\lim\limits_{x \to 0} \dfrac{e^{2x} - e^{-2x} - 4x}{x^3}$

6 $\lim\limits_{x \to (\pi/2)^-} \dfrac{\tan x}{\sec x}$

7 $\lim\limits_{x \to \infty} \dfrac{x^e}{e^x}$

8 $\lim\limits_{x \to (\pi/2)^-} \cos x \ln \cos x$

9 $\lim\limits_{x \to \infty} (1 - 2e^{1/x})x$

10 $\lim\limits_{x \to 0} \tan^{-1} x \csc x$

11 $\lim\limits_{x \to 0} (1 + 8x^2)^{1/x^2}$

12 $\lim\limits_{x \to 1} (\ln x)^{x-1}$

13 $\lim\limits_{x \to \infty} (e^x + 1)^{1/x}$

14 $\lim\limits_{x \to 0} \left(\dfrac{1}{\tan x} - \dfrac{1}{x} \right)$

Determine whether the integrals in Exercises 15–26 converge or diverge, and evaluate those that converge.

15 $\displaystyle\int_4^\infty \dfrac{1}{\sqrt{x}}\, dx$

16 $\displaystyle\int_4^\infty \dfrac{1}{x\sqrt{x}}\, dx$

17 $\displaystyle\int_{-\infty}^0 \dfrac{1}{x + 2}\, dx$

18 $\displaystyle\int_0^\infty \sin x\, dx$

19 $\displaystyle\int_{-8}^1 \dfrac{1}{\sqrt[3]{x}}\, dx$

20 $\displaystyle\int_{-4}^0 \dfrac{1}{x + 4}\, dx$

21 $\displaystyle\int_0^2 \dfrac{x}{(x^2 - 1)^2}\, dx$

22 $\displaystyle\int_1^2 \dfrac{1}{x\sqrt{x^2 - 1}}\, dx$

23 $\displaystyle\int_{-\infty}^\infty \dfrac{1}{e^x + e^{-x}}\, dx$

24 $\displaystyle\int_{-\infty}^0 x e^x\, dx$

25 $\displaystyle\int_0^1 \dfrac{\ln x}{x}\, dx$

26 $\displaystyle\int_0^{\pi/2} \csc x\, dx$

27 Find Taylor's Formula with the Remainder for the following.

(a) $f(x) = \ln \cos x$, $a = \pi/6$, $n = 3$

(b) $f(x) = \sqrt{x - 1}$, $a = 2$, $n = 4$

28 Find Maclaurin's Formula for the following.

(a) $f(x) = e^{-x^2}$, $n = 3$

(b) $f(x) = 1/(1 - x)$, $n = 6$

29 Use Taylor's Formula to approximate $\sin^2 43°$ to four decimal places. (*Hint:* $\sin^2 x = (1 - \cos 2x)/2$.)

30 Establish the approximation formula $\sin x \approx x - (x^3/6)$, and state the accuracy of the approximation if $|x| \le 0.2$.

11

INFINITE SERIES

Infinite series are useful in advanced courses in mathematics, physics, and engineering because they may be employed to represent functions in a new way. In this chapter we shall discuss some of the fundamental results associated with this important mathematical concept.

11.1

INFINITE SEQUENCES

A function f from a set X to a set Y is a correspondence that associates with each element x of X a unique element $f(x)$ of Y (see Definition (1.21)). Until now the domain X has usually been an interval of real numbers. In this section we shall consider another type of function.

Definition (11.1)

> An **infinite sequence** is a function whose domain is the set of positive integers.

For convenience we sometimes refer to infinite sequences merely as *sequences*. In this book the range of an infinite sequence will be a set of real numbers.

If f is an infinite sequence, then to each positive integer n there corresponds a real number $f(n)$. These numbers in the range of f may be listed by writing

$$f(1), f(2), f(3), \ldots, f(n), \ldots$$

where the dots at the end indicate that the sequence does not terminate. The number $f(1)$ is called the **first term** of the sequence, $f(2)$ the **second term** and, in general, $f(n)$ the **nth term** of the sequence. It is customary to use a subscript notation and write these numbers as follows:

(11.2)
$$a_1, a_2, a_3, \ldots, a_n, \ldots$$

In (11.2) it is understood that for each positive integer n, the symbol a_n denotes the real number $f(n)$. In this way we obtain an infinite collection of real numbers that is *ordered* in the sense that there is a first number, a second number, a forty-fifth number, and so on. Although sequences are functions, a collection such as (11.2) will also be referred to as an infinite sequence. If we wish to convert (11.2) to a function f, we let $f(n) = a_n$ for each positive integer n.

From the definition of equality of functions we see that a sequence

$$a_1, a_2, a_3, \ldots, a_n, \ldots$$

is **equal** to a sequence

$$b_1, b_2, b_3, \ldots, b_n, \ldots$$

if and only if $a_i = b_i$ for every positive integer i. Infinite sequences are often defined by stating a formula for the nth term, as in the next example.

Example 1   List the first four terms and the tenth term of the sequence whose nth term a_n is as follows.

(a) $a_n = \dfrac{n}{n + 1}$ (b) $a_n = 2 + (0.1)^n$

(c) $a_n = (-1)^{n+1} \dfrac{n^2}{3n - 1}$ (d) $a_n = 4$

Solution   To find the first four terms we substitute, successively, $n = 1, 2, 3,$ and 4 in the formula for a_n. The tenth term is found by substituting 10 for n. Doing this and simplifying gives us the following:

	nth term	First four terms	Tenth term
(a)	$\dfrac{n}{n + 1}$	$\dfrac{1}{2}, \dfrac{2}{3}, \dfrac{3}{4}, \dfrac{4}{5}$	$\dfrac{10}{11}$
(b)	$2 + (0.1)^n$	$2.1, 2.01, 2.001, 2.0001$	2.0000000001
(c)	$(-1)^{n+1} \dfrac{n^2}{3n - 1}$	$\dfrac{1}{2}, -\dfrac{4}{5}, \dfrac{9}{8}, -\dfrac{16}{11}$	$-\dfrac{100}{29}$
(d)	4	$4, 4, 4, 4$	4 ∎

We sometimes denote the sequence (11.2) by $\{a_n\}$. For example, the sequence $\{2^n\}$ has nth term $a_n = 2^n$. According to Definition (11.1), the sequence $\{2^n\}$ is the function f such that $f(n) = 2^n$ for every positive integer n.

Some infinite sequences $\{a_n\}$ have the property that as n increases, a_n gets very close to some real number L. Another way of stating this is to say that $|a_n - L|$ is almost zero when n is large. As an illustration, consider the sequence $\{a_n\}$ where

$$a_n = 2 + \left(-\frac{1}{2}\right)^n.$$

The first few terms are

$$2 - \frac{1}{2}, 2 + \frac{1}{4}, 2 - \frac{1}{8}, 2 + \frac{1}{16}, 2 - \frac{1}{32}, \cdots$$

and it appears that the terms get closer to 2 as n increases. As a matter of fact, for every positive integer n,

$$|a_n - 2| = \left|2 + \left(-\frac{1}{2}\right)^n - 2\right| = \left|\left(-\frac{1}{2}\right)^n\right| = \left(\frac{1}{2}\right)^n = \frac{1}{2^n}$$

and the number $1/2^n$, and hence $|a_n - 2|$, *can be made arbitrarily close to zero by choosing n sufficiently large.* According to the next definition, the given sequence *has the limit* 2, and we write

$$\lim_{n \to \infty}\left[2 + \left(-\frac{1}{2}\right)^n\right] = 2.$$

The situation here is almost identical to that in Chapter 4, where $\lim_{x \to \infty} f(x) = L$ was defined. The only difference is that if $f(n) = a_n$, then the domain of f is the set of positive integers and not an infinite interval of real numbers. As in Definition (4.19), but using a_n instead of $f(x)$, we state the following definition.

Definition (11.3)

A sequence $\{a_n\}$ **has the limit** L, written

$$\lim_{n \to \infty} a_n = L$$

if for every $\varepsilon > 0$, there exists a positive number N such that

$$|a_n - L| < \varepsilon \quad \text{whenever } n > N.$$

If $\lim_{n \to \infty} a_n$ does not exist in the sense of Definition (11.3), then the sequence $\{a_n\}$ has no limit.

A geometric interpretation similar to that shown in Figure 4.29 can be given for the limit of a sequence. The only difference is that the coordinate x of the point shown on the x-axis is always a positive integer. In Figure 11.1 we have illustrated the behavior of the points (k, a_k) for a special case in which $\lim_{n \to \infty} a_n = L$. Note that for any $\varepsilon > 0$, the points (n, a_n) lie between the lines $y = L \pm \varepsilon$, provided n is sufficiently large. Of course, the approach to L may vary from that illustrated in the figure.

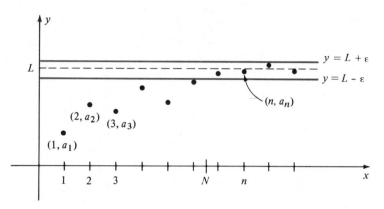

Figure 11.1

Another geometric interpretation of Definition (11.3) may be obtained by plotting the points corresponding to a_k on a coordinate line, as shown in Figure 11.2. In this case if we consider any open interval $(L - \varepsilon, L + \varepsilon)$, then a_n is in the interval whenever n is sufficiently large.

Figure 11.2

If a_n can be made as large as desired by choosing n sufficiently large, then the sequence $\{a_n\}$ has no limit, but we still write $\lim_{n \to \infty} a_n = \infty$. A more precise way of specifying this behavior is as follows.

Definition (11.4)

> The statement $\lim_{n \to \infty} a_n = \infty$ means that for every positive real number P, there exists a number N such that $a_n > P$ whenever $n > N$.

The proof of the next theorem illustrates the use of the preceding definitions.

Theorem (11.5)

> (i) $\lim_{n \to \infty} r^n = 0$ if $|r| < 1$
>
> (ii) $\lim_{n \to \infty} |r^n| = \infty$ if $|r| > 1$

Proof   If $r = 0$ it follows trivially that the limit is 0. Let us assume that $0 < |r| < 1$. To prove (i) by means of Definition (11.3) we must show that for every $\varepsilon > 0$ there exists a positive number N such that

$$\text{if } \quad n > N, \quad \text{then} \quad |r^n - 0| < \varepsilon.$$

The inequality $|r^n - 0| < \varepsilon$ is equivalent to each inequality in the following list:

$$|r|^n < \varepsilon, \quad n \ln |r| < \ln \varepsilon, \quad n > \frac{\ln \varepsilon}{\ln |r|}$$

where the final inequality sign is reversed because $\ln |r|$ is negative if $0 < |r| < 1$. The last inequality in the list gives us a clue to the choice of N. Specifically, if $\varepsilon < 1$, then $\ln \varepsilon < 0$ and we let $N = \ln \varepsilon / \ln |r| > 0$. In this event if $n > N$, then the last inequality in the list is true, and hence so is the first, which is what we wished to prove. If $\varepsilon \geq 1$, then $\ln \varepsilon \geq 0$ and hence $\ln \varepsilon / \ln |r| \leq 0$. In this case if N is *any* positive number, then whenever $n > N$ the last inequality in the list is again true.

To prove (ii) let $|r| > 1$ and consider any positive real number P. The following inequalities are equivalent:

$$|r|^n > P, \quad n \ln |r| > \ln P, \quad n > \frac{\ln P}{\ln |r|}.$$

If we choose $N = \ln P / \ln |r|$, then whenever $n > N$ we have $|r|^n > P$. Hence by Definition (11.4), $\lim_{n \to \infty} |r|^n = \infty$.   □

Example 2   List the first four terms and, if possible, find the limits of the following sequences.

(a) $\{(-\frac{2}{3})^n\}$ (b) $\{(1.01)^n\}$

Solution

(a)   The first four terms of $\{(-\frac{2}{3})^n\}$ are

$$-\frac{2}{3}, \quad \frac{4}{9}, \quad -\frac{8}{27}, \quad \frac{16}{81}.$$

According to Theorem (11.5),

$$\lim_{n \to \infty} \left(-\frac{2}{3}\right)^n = 0.$$

(b)   The first four terms of $\{(1.01)^n\}$ are

$$1.01, \quad 1.0201, \quad 1.030301, \quad 1.04060401.$$

According to Theorem (11.5),

$$\lim_{n \to \infty} (1.01)^n = \infty.$$ ■

Theorems that are analogous to some of the limit theorems in Chapter 2 may be established for infinite sequences. In particular, given two infinite sequences $\{a_n\}$ and $\{b_n\}$, corresponding terms may be added to form the infinite sequence $\{a_n + b_n\}$, which is called the **sum** of the given sequences. Similarly, we could form the **difference**, **product**, or **quotient** by subtracting, multiplying, or dividing corresponding terms (provided no zero denominators occur in the division process).

If $\lim_{n \to \infty} a_n = L$ and $\lim_{n \to \infty} b_n = M$ where L and M are real numbers, then it can be shown that

$$\lim_{n \to \infty} (a_n + b_n) = L + M, \qquad \lim_{n \to \infty} (a_n - b_n) = L - M,$$

$$\lim_{n \to \infty} a_n b_n = LM, \qquad \lim_{n \to \infty} \frac{a_n}{b_n} = \frac{L}{M}.$$

In the last formula we must have $M \neq 0$ and $b_n \neq 0$ for all n.

If $a_n = c$ for all n, so that the infinite sequence is $c, c, \ldots, c, \ldots$, then

$$\lim_{n \to \infty} c = c.$$

Similarly, if c is a real number and k is a positive rational number, then, as in the proof of Theorem (4.21),

$$\lim_{n \to \infty} \frac{c}{n^k} = 0.$$

Example 3 Find the limit of the sequence $\left\{ \dfrac{2n}{5n - 3} \right\}$.

Solution We wish to find $\lim_{n \to \infty} a_n$, where $a_n = 2n/(5n - 3)$. Dividing numerator and denominator of a_n by n and applying limit theorems, we obtain

$$\lim_{n \to \infty} \frac{2n}{5n - 3} = \lim_{n \to \infty} \frac{2}{5 - (3/n)} = \frac{\lim_{n \to \infty} 2}{\lim_{n \to \infty} [5 - (3/n)]}$$

$$= \frac{2}{\lim_{n \to \infty} 5 - \lim_{n \to \infty} (3/n)} = \frac{2}{5 - 0} = \frac{2}{5}.$$

Hence the sequence has the limit $\frac{2}{5}$. ∎

Let $\{a_n\}$ be a sequence and consider the function f, where $f(n) = a_n$. It is often true that $f(x)$ is defined for every *real* number $x \geq 1$. In this case it follows from Definitions (11.3) and (4.19) that

$$\text{if} \quad \lim_{x \to \infty} f(x) = L, \quad \text{then} \quad \lim_{n \to \infty} f(n) = L.$$

This fact enables us to apply theorems about limits of functions (as $x \to \infty$) to limits of sequences. Of special importance is L'Hôpital's Rule, as illustrated in the next example.

Example 4   Find the limit of the sequence $\{5n/e^{2n}\}$.

Solution   Let $f(x) = 5x/e^{2x}$ where x is real. Since f takes on the indeterminate form ∞/∞ as $x \to \infty$, we may use L'Hôpital's Rule, obtaining

$$\lim_{x \to \infty} \frac{5x}{e^{2x}} = \lim_{x \to \infty} \frac{5}{2e^{2x}} = 0.$$

Hence, by the discussion preceding this example,

$$\lim_{n \to \infty} \frac{5n}{e^{2n}} = 0. \qquad \blacksquare$$

The next theorem is analogous to Theorem (2.14). It states that if the terms of an infinite sequence are always sandwiched between corresponding terms of two sequences that have the same limit L, then the given sequence also has the limit L.

The Sandwich Theorem for Infinite Sequences (11.6)

> If $\{a_n\}$, $\{b_n\}$, and $\{c_n\}$ are infinite sequences such that $a_n \leq b_n \leq c_n$ for all n, and if
> $$\lim_{n \to \infty} a_n = L = \lim_{n \to \infty} c_n$$
> then $\lim_{n \to \infty} b_n = L$.

Proof   For every $\varepsilon > 0$ there corresponds a number M such that if $n > M$, then both $|a_n - L| < \varepsilon$ and $|c_n - L| < \varepsilon$; that is,

$$L - \varepsilon < a_n < L + \varepsilon \quad \text{and} \quad L - \varepsilon < c_n < L + \varepsilon.$$

Consequently, if $n > M$, then

$$L - \varepsilon < a_n \quad \text{and} \quad c_n < L + \varepsilon.$$

Since $a_n \leq b_n \leq c_n$, it follows that if $n > M$, then $L - \varepsilon < b_n < L + \varepsilon$ or, equivalently, $|b_n - L| < \varepsilon$. This completes the proof. $\square$

Example 5   Find the limit of the sequence $\left\{\dfrac{\cos^2 n}{3^n}\right\}$.

Solution   Since $0 < \cos^2 n < 1$ for every positive integer n, we may write

$$0 < \frac{\cos^2 n}{3^n} < \frac{1}{3^n}.$$

Applying Theorem (11.5) with $r = \frac{1}{3}$,

$$\lim_{n \to \infty} \frac{1}{3^n} = \lim_{n \to \infty} \left(\frac{1}{3}\right)^n = 0.$$

Moreover $\lim_{n\to\infty} 0 = 0$. It follows from the Sandwich Theorem (11.6) with $a_n = 0$, $b_n = (\cos^2 n)/3^n$, and $c_n = (\frac{1}{3})^n$, that

$$\lim_{n\to\infty} \frac{\cos^2 n}{3^n} = 0.$$

Hence the limit of the sequence is 0. ∎

The proof of the next theorem is left as an exercise.

Theorem (11.7)

Let $\{a_n\}$ be a sequence. If $\lim_{n\to\infty} |a_n| = 0$, then $\lim_{n\to\infty} a_n = 0$.

Example 6 Suppose the nth term of a sequence is $a_n = (-1)^{n+1}(1/n)$. Prove that $\lim_{n\to\infty} a_n = 0$.

Solution The terms of the sequence are alternately positive and negative. For example, the first five terms are

$$1, \quad -\frac{1}{2}, \quad \frac{1}{3}, \quad -\frac{1}{4}, \quad \frac{1}{5}.$$

Since

$$\lim_{n\to\infty} |a_n| = \lim_{n\to\infty} \frac{1}{n} = 0,$$

it follows from Theorem (11.7) that $\lim_{n\to\infty} a_n = 0$. ∎

A sequence is said to be **monotonic** if successive terms are nondecreasing in the sense that

$$a_1 \le a_2 \le \cdots \le a_n \le \cdots$$

or nonincreasing in the sense that

$$a_1 \ge a_2 \ge \cdots \ge a_n \ge \cdots$$

A sequence is **bounded** if there is a positive real number M such that $|a_k| \le M$ for all k. The next theorem is fundamental for later developments.

Theorem (11.8)

A bounded, monotonic, infinite sequence has a limit.

To prove Theorem (11.8), it is necessary to use an important property of real numbers. Let us first state several definitions. If S is a nonempty set of real numbers, then a real number u is called an **upper bound** of S if $x \le u$ for every x in S. A number v is a **least upper bound** of S if v is an upper bound and no number less than v is an upper bound of S. The least upper bound is, therefore, the smallest real number that is greater than or equal to every number in S. To illustrate, if S is the open interval (a, b), then any number greater than b is an upper bound of S; however, the least upper bound of S is unique, and equals b (see Exercise 50).

The following statement is an axiom for the real number system.

The Completeness Property (11.9)

> If a nonempty set S of real numbers has an upper bound, then S has a least upper bound.

Proof of Theorem (11.8)　Let $\{a_n\}$ be a bounded monotonic sequence with nondecreasing terms. Thus

$$a_1 \le a_2 \le \cdots \le a_n \le \cdots$$

and there is a number M such that $a_k \le M$ for every positive integer k. Since the set

$$S = \{a_k : k = 1, 2, \ldots\}$$

has an upper bound, it follows from the Completeness Property that S has a least upper bound L, where $L \le M$. If $\varepsilon > 0$, then $L - \varepsilon$ is not an upper bound of S and hence at least one term of $\{a_n\}$ is greater than $L - \varepsilon$; that is,

$$L - \varepsilon < a_N \quad \text{for some positive integer } N.$$

Since the terms of $\{a_n\}$ are nondecreasing, we have

$$a_N \le a_{N+1} \le a_{N+2} \le \cdots$$

and, therefore,

$$L - \varepsilon < a_n \quad \text{for every } n > N.$$

It follows that if $n > N$, then

$$0 \le L - a_n < \varepsilon, \quad \text{or} \quad |L - a_n| < \varepsilon.$$

By Definition (11.3) this implies that

$$\lim_{n \to \infty} a_n = L \le M;$$

that is, $\{a_n\}$ has a limit. The proof for a sequence $\{a_n\}$ of nonincreasing terms may be obtained by a similar proof or by considering the sequence $\{-a_n\}$. ☐

We shall conclude this section with an application of sequences to the investigation of the time course of an $S \rightarrow I \rightarrow S$ epidemic.* Let us suppose that physicians issue daily reports indicating the number of persons who have become infected with the disease and those who have been cured. We shall label the reporting days as $1, 2, \ldots, n, \ldots$ and let I_n denote the number of individuals in a population of size N who have the disease on the nth day. Let F_n and C_n denote the number of newly infected and the number cured, respectively, on day n. Thus, for $n \geq 1$,

$$I_{n+1} = I_n + F_{n+1} - C_{n+1}.$$

Suppose health officials decide that the number of new cases on a given day is directly proportional to the product of the number ill and the number not infected on the previous day. (This is known as the *Law of Mass Action*, and is typical of a population consisting of students on a college campus.) In addition, assume that the number cured each day is directly proportional to the number ill the previous day. Hence,

$$F_{n+1} = aI_n(N - I_n) \quad \text{and} \quad C_{n+1} = bI_n$$

where a and b are constants. Substituting in the preceding formula for I_{n+1}, we obtain

$$I_{n+1} = I_n + aI_n(N - I_n) - bI_n.$$

In the early stages of an epidemic I_n will be very small compared to N, and from the point of view of public health, it is better to *overestimate* the number ill than it is to underestimate and be unprepared for the spread of the disease. With this in mind, the early dynamics of the epidemic may be realistically investigated by examining the equation

$$I_{n+1} = I_n + aNI_n - bI_n = (1 + aN - b)I_n$$

where I_1 is the initial number of infected individuals. If we let $r = 1 + aN - b$, then $I_{n+1} = rI_n$ and, therefore,

$$I_2 = rI_1, \quad I_3 = rI_2 = r^2I_1, \quad I_4 = rI_3 = r^3I_1, \quad \ldots, \quad I_n = r^{n-1}I_1, \quad \ldots$$

This gives us the following sequence of infected individuals:

$$I_1, rI_1, r^2I_1, r^3I_1, \ldots, r^{n-1}I_1, \ldots$$

The number $r = 1 + aN - b$ (which can be approximated from early data) is of critical import. If $r > 1$, then by Theorem (11.5), $\lim_{n \to \infty} I_n = \infty$ and an epidemic will be in progress. In this case, when n is large, I_n is no longer small compared to N, and the formula for I_{n+1} becomes invalid. If $r < 1$, then $\lim_{n \to \infty} I_n = 0$ and health officials need not be concerned. The case $r = 1$ results in the constant sequence $I_1, I_1, \ldots, I_1, \ldots$

---

* The symbol $S \rightarrow I \rightarrow S$ is an abbreviation for *Susceptible $\rightarrow$ Infected $\rightarrow$ Susceptible*, and is used to signify that an infected person who becomes cured is not immune to the disease, but may contract it again. Examples of such diseases are gonorrhea and strep throat.

EXERCISES 11.1

Given the nth term a_n of an infinite sequence as in Exercises 1–16, find the first four terms and $\lim_{n\to\infty} a_n$, if it exists.

1 $a_n = \dfrac{n}{3n + 2}$

2 $a_n = \dfrac{6n - 5}{5n + 1}$

3 $a_n = \dfrac{7 - 4n^2}{3 + 2n^2}$

4 $a_n = \dfrac{4}{8 - 7n}$

5 $a_n = -5$

6 $a_n = \sqrt{2}$

7 $a_n = \dfrac{(2n - 1)(3n + 1)}{n^3 + 1}$

8 $a_n = 8n + 1$

9 $a_n = \dfrac{2}{\sqrt{n^2 + 9}}$

10 $a_n = \dfrac{100n}{n^{3/2} + 4}$

11 $a_n = (-1)^{n+1} \dfrac{3n}{n^2 + 4n + 5}$

12 $a_n = (-1)^{n+1} \dfrac{\sqrt{n}}{n + 1}$

13 $a_n = 1 + (0.1)^n$

14 $a_n = 1 - (1/2^n)$

15 $a_n = 1 + (-1)^{n+1}$

16 $a_n = (n + 1)/\sqrt{n}$

In Exercises 17–42, find the limit of the sequence, if it exists.

17 $\{6(-5/6)^n\}$

18 $\{8 - (7/8)^n\}$

19 $\{\arctan n\}$

20 $\{(\tan^{-1} n)/n\}$

21 $\{1000 - n\}$

22 $\{(1.0001)^n/1000\}$

23 $\left\{(-1)^n \dfrac{\ln n}{n}\right\}$

24 $\left\{\dfrac{n^2}{\ln (n + 1)}\right\}$

25 $\left\{\dfrac{4n^4 + 1}{2n^2 - 1}\right\}$

26 $\left\{\dfrac{\cos n}{n}\right\}$

27 $\{e^n/n^4\}$

28 $\{e^{-n} \ln n\}$

29 $\left\{\left(1 + \dfrac{1}{n}\right)^n\right\}$

30 $\{(-1)^n n^3 3^{-n}\}$

31 $\{2^{-n} \sin n\}$

32 $\left\{\dfrac{4n^3 + 5n + 1}{2n^3 - n^2 + 5}\right\}$

33 $\left\{\dfrac{n^2}{2n - 1} - \dfrac{n^2}{2n + 1}\right\}$

34 $\left\{n \sin \dfrac{1}{n}\right\}$

35 $\{\cos n\pi\}$

36 $\{4 + \sin (n\pi/2)\}$

37 $\{n^{1/n}\}$

38 $\{n^2/2^n\}$

39 $\left\{\dfrac{n^{-10}}{\sec n}\right\}$

40 $\left\{(-1)^n \dfrac{n^2}{1 + n^2}\right\}$

41 $\{2^n/(2n)!\}$

42 $\{n[\ln (n + 1) - \ln n]\}$

43 A test question lists the first four terms of a sequence as 2, 4, 6, and 8 and asks the person being tested to list the fifth term. Show that the fifth term can be any real number a by finding the nth term of a sequence that has for its first five terms 2, 4, 6, 8, and a.

44 If k is an integer greater than 1, prove that $\lim_{n\to\infty} 1/k^n = 0$. (*Hint*: Use Theorem (11.6).)

45 Prove that if $\{a_n\}$ has the limit L, then L is unique.

46 Prove, directly from Definition (11.3), that if $\lim_{n\to\infty} a_n = L$, then $\lim_{n\to\infty} |a_n| = |L|$.

47 Define, in a manner analogous to Definition (11.4), what is meant by $\lim_{n\to\infty} a_n = -\infty$.

48 Prove Theorem (11.7).

49 If $\{a_n\}$ has a limit and c is a real number, prove that $\lim_{n\to\infty} ca_n = c \lim_{n\to\infty} a_n$.

50 Prove that b is the least upper bound of the open interval (a, b).

11.2

CONVERGENT OR DIVERGENT INFINITE SERIES

Every rational number can be represented by a nonterminating, repeating decimal. For example, applying long division to $\frac{2}{3}$ we obtain the representation

$$\frac{2}{3} = 0.66666\ldots$$

This decimal may also be regarded as the limit of the sequence whose first few terms are

$$0.6, 0.66, 0.666, 0.6666, 0.66666, \ldots$$

As we shall see, it is also possible to express $\frac{2}{3}$ as the *infinite sum* (or *series*)

$$0.6 + 0.06 + 0.006 + 0.0006 + 0.00006 + \cdots$$

Since only finite sums may be added algebraically, it is necessary to introduce concepts that give meaning to an infinite sum of this type. In this section we shall state the required definitions; however, our development will take us far beyond this elementary application. Indeed, the major objective of this chapter is to represent *functions* by means of infinite sums.

Let us begin by introducing some terminology that will be used throughout the remainder of the chapter.

Definition (11.10)

> Let $\{a_n\}$ be an infinite sequence. An expression of the form
>
> $$a_1 + a_2 + \cdots + a_n + \cdots$$
>
> is called an **infinite series**, or simply a **series**.

In summation notation, the series in Definition (11.10) will be denoted by either

$$\sum_{n=1}^{\infty} a_n \quad \text{or} \quad \sum a_n$$

where it is understood that the summation variable in the last sum is n. Each number a_i is called a **term** of the series, and a_n is called the **nth term.**

Definition (11.11)

> (i) The **nth partial sum S_n** of the infinite series $\sum a_n$ is
>
> $$S_n = a_1 + a_2 + \cdots + a_n.$$
>
> (ii) The **sequence of partial sums** associated with the infinite series $\sum a_n$ is
>
> $$S_1, S_2, S_3, \ldots, S_n, \ldots$$

Thus, from (i) of Definition (11.11),

$$S_1 = a_1$$
$$S_2 = a_1 + a_2$$
$$S_3 = a_1 + a_2 + a_3$$
$$S_4 = a_1 + a_2 + a_3 + a_4.$$

To illustrate, consider the series

$$0.6 + 0.06 + 0.006 + \cdots + \frac{6}{10^n} + \cdots$$

The first few terms of the sequence of partial sums are

$$0.6, \quad 0.66, \quad 0.666, \quad 0.6666, \quad 0.66666, \quad \ldots$$

which is the sequence considered at the beginning of this section. In Example 2 we will show that this sequence has the limit $\frac{2}{3}$. Since the partial sums S_n of the series approach $\frac{2}{3}$ as n increases, it is natural to refer to $\frac{2}{3}$ as the *sum* of the infinite series and write

$$\frac{2}{3} = 0.6 + 0.06 + 0.006 + \cdots + \frac{6}{10^n} + \cdots$$

The following definition extends these remarks to any infinite series.

Definition (11.12)

An infinite series

$$a_1 + a_2 + \cdots + a_n + \cdots$$

with sequence of partial sums $S_1, S_2, \ldots, S_n, \ldots$ **is convergent** (or **converges**) if $\lim_{n \to \infty} S_n = S$ for some real number S. The series **is divergent** (or **diverges**) if this limit does not exist.

If $a_1 + a_2 + \cdots + a_n + \cdots$ is a convergent infinite series and $\lim_{n \to \infty} S_n = S$, then S is called the **sum of the series**, and we write

$$S = a_1 + a_2 + \cdots + a_n + \cdots$$

If a series diverges, it has no sum.

Intuitively, if a series converges and has the sum S, then as we add more and more terms, the resulting partial sums get closer and closer to S.

Example 1 Prove that the infinite series

$$\frac{1}{1 \cdot 2} + \frac{1}{2 \cdot 3} + \frac{1}{3 \cdot 4} + \cdots + \frac{1}{n(n + 1)} + \cdots$$

converges and find its sum.

Solution The partial fraction decomposition of a_n is

$$a_n = \frac{1}{n(n + 1)} = \frac{1}{n} - \frac{1}{n + 1}.$$

Consequently, the nth partial sum of the series may be written

$$S_n = a_1 + a_2 + a_3 + \cdots + a_n$$

$$= \left(1 - \frac{1}{2}\right) + \left(\frac{1}{2} - \frac{1}{3}\right) + \left(\frac{1}{3} - \frac{1}{4}\right) + \cdots + \left(\frac{1}{n} - \frac{1}{n+1}\right)$$

$$= 1 - \frac{1}{n+1} = \frac{n}{n+1}.$$

Since

$$\lim_{n \to \infty} S_n = \lim_{n \to \infty} \frac{n}{n+1} = 1,$$

the series converges and has the sum 1. ∎

The series $\sum 1/[n(n+1)]$ of Example 1 is called a **telescoping series**, since writing S_n as shown in the solution of Example 1 causes the terms to collapse to $1 - 1/(n+1)$.

Certain types of infinite series arise frequently in applications. One example is a **geometric series**

$$a + ar + ar^2 + \cdots + ar^{n-1} + \cdots$$

where a and r are real numbers and $a \neq 0$. The next result is extremely important for work later in this chapter.

Theorem (11.13)

> Let $a \neq 0$. The geometric series
>
> $$a + ar + ar^2 + \cdots + ar^{n-1} + \cdots$$
>
> (i)   converges and has the sum $\dfrac{a}{1-r}$ if $|r| < 1$.
>
> (ii) diverges if $|r| \geq 1$.

Proof   If $r = 1$, then $S_n = a + a + \cdots + a = na$ and the series diverges, since $\lim_{n \to \infty} S_n$ does not exist.

If $r \neq 1$, consider

$$S_n = a + ar + ar^2 + \cdots + ar^{n-1}$$

and

$$rS_n = ar + ar^2 + ar^3 + \cdots + ar^n.$$

Subtracting corresponding sides of these equations we obtain

$$(1 - r)S_n = a - ar^n.$$

Dividing both sides by $1 - r$ gives us

$$S_n = \frac{a}{1-r} - \frac{ar^n}{1-r}.$$

Consequently,

$$\lim_{n \to \infty} S_n = \lim_{n \to \infty} \left(\frac{a}{1 - r} - \frac{ar^n}{1 - r} \right)$$

$$= \lim_{n \to \infty} \frac{a}{1 - r} - \lim_{n \to \infty} \frac{ar^n}{1 - r}$$

$$= \frac{a}{1 - r} - \frac{a}{1 - r} \lim_{n \to \infty} r^n.$$

If $|r| < 1$, then $\lim_{r \to \infty} r^n = 0$ (see Theorem (11.5)) and hence

$$\lim_{n \to \infty} S_n = \frac{a}{1 - r}.$$

If $|r| > 1$, then $\lim_{n \to \infty} r^n$ does not exist (see Theorem (11.5)) and hence $\lim_{n \to \infty} S_n$ does not exist. In this case the series diverges.   □

Example 2   Prove that the infinite series

$$0.6 + 0.06 + 0.006 + \cdots + \frac{6}{10^n} + \cdots$$

converges and find its sum.

Solution   The series is geometric with $a = 0.6$ and $r = 0.1$. By (i) of Theorem (11.13) it converges and has the sum

$$S = \frac{0.6}{1 - 0.1} = \frac{0.6}{0.9} = \frac{2}{3}.$$ ■

Example 3   Prove that the following series converges and find its sum:

$$2 + \frac{2}{3} + \frac{2}{3^2} + \cdots + \frac{2}{3^{n-1}} + \cdots$$

Solution   The series converges since it is geometric with $r = \frac{1}{3} < 1$. By (i) of Theorem (11.13), the sum is

$$S = \frac{2}{1 - \frac{1}{3}} = \frac{2}{\frac{2}{3}} = 3.$$ ■

Theorem (11.14)

> If an infinite series $\sum a_n$ is convergent, then $\lim_{n \to \infty} a_n = 0$.

Proof   The nth term a_n of the infinite series can be expressed as

$$a_n = S_n - S_{n-1}.$$

If $\lim_{n \to \infty} S_n = S$, then also $\lim_{n \to \infty} S_{n-1} = S$ and

$$\lim_{n \to \infty} a_n = \lim_{n \to \infty} (S_n - S_{n-1}) = \lim_{n \to \infty} S_n - \lim_{n \to \infty} S_{n-1} = 0. \qquad \square$$

The preceding theorem states that *if* a series converges, *then* its nth term a_n has the limit 0 as $n \to \infty$. The converse is false, that is, *if* $\lim_{n \to \infty} a_n = 0$ *it does not necessarily follow* that the series $\sum a_n$ is convergent. (See Example 5 of this section.)

The next result is an immediate corollary of Theorem (11.14).

Test for Divergence (11.15)

> If $\lim_{n \to \infty} a_n \neq 0$, then the infinite series $\sum a_n$ is divergent.

Example 4   Determine whether the following series converges or diverges:

$$\frac{1}{3} + \frac{2}{5} + \frac{3}{7} + \cdots + \frac{n}{2n + 1} + \cdots$$

Solution   Since

$$\lim_{n \to \infty} a_n = \lim_{n \to \infty} \frac{n}{2n + 1} = \frac{1}{2} \neq 0,$$

the series diverges by (11.15).   ∎

Theorem (11.16)

> If an infinite series $\sum a_n$ is convergent, then for every $\varepsilon > 0$ there exists an integer N such that $|S_k - S_l| < \varepsilon$ whenever $k > N$ and $l > N$.

Proof   If the sum of the series is S, then

$$|S_k - S_l| = |(S_k - S) + (S - S_l)| \leq |S_k - S| + |S - S_l|.$$

Since $\lim_{k \to \infty} S_k = S = \lim_{l \to \infty} S_l$, for every $\varepsilon > 0$ there is an integer N such that

$$|S_k - S| < \frac{\varepsilon}{2} \quad \text{and} \quad |S - S_l| < \frac{\varepsilon}{2}$$

whenever $k > N$ and $l > N$. Thus

$$|S_k - S_l| < \frac{\varepsilon}{2} + \frac{\varepsilon}{2} = \varepsilon. \qquad \square$$

Example 5   Prove that the series $1 + \dfrac{1}{2} + \dfrac{1}{3} + \cdots + \dfrac{1}{n} + \cdots$ is divergent.

Solution   If $n > 1$, then

$$S_{2n} - S_n = \frac{1}{n+1} + \frac{1}{n+2} + \cdots + \frac{1}{2n}$$

$$> \frac{1}{2n} + \frac{1}{2n} + \cdots + \frac{1}{2n} = \frac{1}{2}.$$

If the series is convergent, then by Theorem (11.16) with $\varepsilon = \frac{1}{2}$, $k = 2n$, and $l = n$, $|S_{2n} - S_n| < \frac{1}{2}$ provided n is sufficiently large. Since the last inequality is *never* true, it follows that the series diverges.   ∎

The series in Example 5 will be important in later developments. Because of this we give it the following special name.

Definition (11.17)

> The **harmonic series** is the divergent infinite series
>
> $$1 + \frac{1}{2} + \frac{1}{3} + \cdots + \frac{1}{n} + \cdots$$

The harmonic series is an illustration of a divergent series $\sum a_n$ for which $\lim_{n \to \infty} a_n = 0$. This shows that the converse of Theorem (11.14) is false. Consequently, to establish convergence of an infinite series *it is not enough* to prove that $\lim_{n \to \infty} a_n = 0$, since that may be true for divergent as well as convergent series.

The next theorem states that if corresponding terms of two infinite series are identical after a certain term, then both series converge or both series diverge.

Theorem (11.18)

> If $\sum a_n$ and $\sum b_n$ are infinite series such that $a_i = b_i$ for all $i > k$, where k is a positive integer, then both series converge or both series diverge.

Proof   By hypothesis we may write

$$\sum a_n = a_1 + a_2 + \cdots + a_k + a_{k+1} + \cdots + a_n + \cdots$$

and

$$\sum b_n = b_1 + b_2 + \cdots + b_k + a_{k+1} + \cdots + a_n + \cdots$$

Let S_n and T_n denote the nth partial sums of $\sum a_n$ and $\sum b_n$, respectively. It follows that if $n \geq k$, then

$$S_n - S_k = T_n - T_k$$

or
$$S_n = T_n + (S_k - T_k).$$

Consequently,

$$\lim_{n \to \infty} S_n = \lim_{n \to \infty} T_n + (S_k - T_k)$$

and hence either both of the limits exist or both do not exist. This gives us the desired conclusion. Evidently, if both series converge, then their sums differ by $S_k - T_k$. $\qquad \square$

Theorem (11.18) implies that changing a finite number of terms of an infinite series has no effect on its convergence or divergence (although it does change the sum of a convergent series). In particular, if we replace the first k terms of $\sum a_n$ by 0, convergence is unaffected. It follows that the series

$$a_{k+1} + a_{k+2} + \cdots + a_n + \cdots$$

converges or diverges depending on whether $\sum a_n$ converges or diverges. The series $a_{k+1} + a_{k+2} + \cdots$ is said to have been obtained from $\sum a_n$ by **deleting the first k terms**.

Example 6 Show that the following series converges:

$$\frac{1}{3 \cdot 4} + \frac{1}{4 \cdot 5} + \cdots + \frac{1}{(n+2)(n+3)} + \cdots$$

Solution The series converges since it can be obtained by deleting the first two terms of the convergent telescoping series of Example 1. ∎

The proof of the next theorem follows directly from Definition (11.12) and is left as an exercise.

Theorem (11.19)

> If $\sum a_n$ and $\sum b_n$ are convergent series with sums A and B, respectively, then
>
> (i) $\sum (a_n + b_n)$ converges and has sum $A + B$.
>
> (ii) if c is a real number, $\sum c a_n$ converges and has sum cA.
>
> (iii) $\sum (a_n - b_n)$ converges and has sum $A - B$.

It is also easy to show that if $\sum a_n$ diverges, then so does $\sum c a_n$ for every $c \neq 0$.

Example 7   Prove that the following series converges and find its sum:

$$\sum_{n=1}^{\infty} \left[\frac{7}{n(n+1)} + \frac{2}{3^{n-1}} \right].$$

Solution   The telescoping series $\sum 1/[n(n+1)]$ was considered in Example 1, where it was shown that it converges and has the sum 1. Using (ii) of Theorem (11.19) with $c = 7$ and $a_n = 1/[n(n+1)]$, we see that $\sum 7/[n(n+1)]$ converges and has the sum $7(1) = 7$.

The geometric series $\sum 2/3^{n-1}$ converges and has the sum 3 (see Example 3). Hence by (i) of Theorem (11.19), the given series converges and has the sum $7 + 3$, or 10. ∎

Theorem (11.20)

> If $\sum a_n$ is a convergent series and $\sum b_n$ is divergent, then $\sum (a_n + b_n)$ is divergent.

Proof   We shall give an indirect proof. Thus suppose $\sum (a_n + b_n)$ is convergent. Applying (iii) of Theorem (11.19), $\sum [(a_n + b_n) - a_n] = \sum b_n$ is convergent, a contradiction. Hence our supposition is false, that is, $\sum (a_n + b_n)$ is divergent. □

Example 8   Determine the convergence or divergence of

$$\sum_{n=1}^{\infty} \left(\frac{1}{5^n} + \frac{1}{n} \right).$$

Solution   Since $\sum (1/5^n)$ is a convergent geometric series and $\sum (1/n)$ is the divergent harmonic series, the given series diverges by Theorem (11.20). ∎

We shall conclude this section by returning to the discussion of the epidemic given at the end of Section 11.1. Suppose that instead of I_n (the number ill on day n) we are interested in the total number S_n of individuals who have been ill at some time between the first and nth days. As in our earlier discussion, let us overestimate S_n by approximating the number F_{n+1} of new cases on day $n + 1$ by aNI_n. Thus

$$S_n = I_1 + F_2 + F_3 + F_4 + \cdots + F_n$$
$$= I_1 + aNI_1 + aNI_2 + aNI_3 + \cdots + aNI_{n-1}.$$

Recalling that $I_n = r^{n-1}I_1$, where $r = 1 + aN - b$, we obtain

$$S_n = I_1 + aNI_1 + aNrI_1 + aNr^2I_1 + \cdots + aNr^{n-2}I_1$$
$$= I_1 + aNI_1(1 + r + r^2 + \cdots + r^{n-2}).$$

As in the proof of Theorem (11.13), this may be written

$$S_n = I_1 + aNI_1\left(\frac{1}{1-r} - \frac{r^{n-1}}{1-r}\right).$$

If $r < 1$, then

$$\lim_{n \to \infty} S_n = I_1 + aNI_1\left(\frac{1}{1-r}\right)$$

$$= I_1\left(1 + \frac{aN}{1-r}\right)$$

$$= I_1\left(1 + \frac{aN}{b-aN}\right)$$

$$= I_1\left(\frac{b}{b-aN}\right).$$

If a and b are approximated from early data, this result enables health officials to determine an upper bound for the total number of individuals who will be ill at some stage of the epidemic.

EXERCISES 11.2

In Exercises 1–20 determine whether the infinite series converges or diverges, and if it converges find its sum.

1 $\quad 3 + \dfrac{3}{4} + \cdots + \dfrac{3}{4^{n-1}} + \cdots$

2 $\quad 3 + \dfrac{3}{(-4)} + \cdots + \dfrac{3}{(-4)^{n-1}} + \cdots \qquad \dfrac{12}{5}$

3 $\quad 1 + \dfrac{(-1)}{\sqrt{5}} + \cdots + \left(\dfrac{-1}{\sqrt{5}}\right)^{n-1} + \cdots$

4 $\quad 1 + \left(\dfrac{e}{3}\right) + \cdots + \left(\dfrac{e}{3}\right)^{n-1} + \cdots \qquad \dfrac{3}{3-e}$

5 $\quad 0.37 + 0.0037 + \cdots + \dfrac{37}{(100)^n} + \cdots$

6 $\quad 0.628 + 0.000628 + \cdots + \dfrac{628}{(1000)^n} + \cdots \qquad \dfrac{628}{999}$

7 $\quad \displaystyle\sum_{n=1}^{\infty} 2^{-n}3^{n-1}$

8 $\quad \displaystyle\sum_{n=1}^{\infty} (-5)^{n-1}4^{-n} \qquad \infty$

9 $\quad \displaystyle\sum_{n=1}^{\infty} (-1)^{n-1}$

10 $\quad \displaystyle\sum (\sqrt{2})^{n-1} \qquad D$

11 $\quad \dfrac{1}{4 \cdot 5} + \dfrac{1}{5 \cdot 6} + \cdots + \dfrac{1}{(n+3)(n+4)} + \cdots$

12 $\quad \dfrac{-1}{1 \cdot 2} + \dfrac{-1}{2 \cdot 3} + \cdots + \dfrac{-1}{n(n+1)} + \cdots \qquad C_j - 1$

13 $\quad \dfrac{5}{1 \cdot 2} + \dfrac{5}{2 \cdot 3} + \cdots + \dfrac{5}{n(n+1)} + \cdots$

14 $\quad \dfrac{1}{4} + \dfrac{1}{5} + \cdots + \dfrac{1}{n+3} + \cdots \qquad$ DIVERGES

15 $\quad 3 + \dfrac{3}{2} + \cdots + \dfrac{3}{n} + \cdots$

16 $\quad \dfrac{1}{2} + \dfrac{2}{3} + \cdots + \dfrac{n}{n+1} + \cdots \qquad$ DIVERGES

17 $\quad \displaystyle\sum_{n=1}^{\infty} \dfrac{3n}{5n-1}$

18 $\quad \displaystyle\sum_{n=1}^{\infty} \dfrac{1}{1+(0.3)^n} \qquad$ DIVERGES

19 $\quad \displaystyle\sum_{n=1}^{\infty} \left(\dfrac{1}{8^n} + \dfrac{1}{n(n+1)}\right)$

20 $\quad \displaystyle\sum_{n=1}^{\infty} \left(\dfrac{1}{3^n} - \dfrac{1}{4^n}\right)$

Use the examples and theorems discussed in this section to determine whether the series in Exercises 21–28 converge or diverge.

21 $\displaystyle\sum_{n=1}^{\infty} \frac{1}{\sqrt[n]{e}}$

22 $\displaystyle\sum_{n=1}^{\infty} \frac{n}{\ln(n+1)}$

23 $\displaystyle\sum_{n=1}^{\infty} \left(\frac{5}{n+2} - \frac{5}{n+3} \right)$

24 $\displaystyle\sum_{n=1}^{\infty} \ln\left(\frac{2n}{7n-5} \right)$

25 $\displaystyle\sum_{n=1}^{\infty} \left[\left(\frac{3}{2}\right)^n + \left(\frac{2}{3}\right)^n \right]$

26 $\displaystyle\sum_{n=1}^{\infty} \left[\frac{1}{n(n+1)} - \frac{4}{n} \right]$

27 $\displaystyle\sum_{n=1}^{\infty} n \sin \frac{1}{n}$

28 $\displaystyle\sum_{n=1}^{\infty} (2^{-n} - 2^{-3n})$

Use the method illustrated in Example 1 to prove that the series in Exercises 29 and 30 converge and find their sums.

29 $\displaystyle\sum_{n=1}^{\infty} \frac{1}{4n^2 - 1}$

30 $\displaystyle\sum \frac{-1}{9n^2 + 3n - 2}$

31 Prove that if $\sum a_n$ diverges, then so does $\sum c a_n$ for every $c \neq 0$.

32 Prove Theorem (11.19).

33 Prove or disprove: If $\sum a_n$ and $\sum b_n$ both diverge, then $\sum (a_n + b_n)$ diverges.

34 What is wrong with the following "proof" that the divergent geometric series $\sum_{n=1}^{\infty} (-1)^{n+1}$ has the sum 0?

$$\sum_{n=1}^{\infty} (-1)^{n+1} = [1 + (-1)] + [1 + (-1)] \\ + [1 + (-1)] + \cdots \\ = 0 + 0 + 0 + \cdots = 0$$

In Exercises 35–38, the bar indicates that the digits underneath it repeat indefinitely. Express each repeating decimal as an infinite series and find the rational number it represents.

35 $0.\overline{23}$

36 $5.\overline{146}$

37 $3.2\overline{394}$

38 $2.7\overline{1828}$

39 A rubber ball is dropped from a height of 10 meters. If it rebounds approximately one-half the distance after each fall, use an infinite geometric series to approximate the total distance the ball travels before coming to rest.

40 The bob of a pendulum swings through an arc 24 cm long on its first swing. If each successive swing is approximately five-sixths the length of the preceding swing, use an infinite geometric series to approximate the total distance it travels before coming to rest.

41 If a dosage of Q units of a certain drug is administered to an individual, then the amount remaining in the bloodstream at the end of t minutes is given by Qe^{-ct}, where c is a positive constant. Suppose this same dosage is given at successive T-minute intervals.

(a) Show that the amount $A(k)$ of the drug in the bloodstream immediately after the kth dose is given by $A(k) = \sum_{n=0}^{k-1} Qe^{-ncT}$.

(b) Find an upper bound for the amount of the drug in the bloodstream after any number of doses.

(c) Find the smallest time between doses that will ensure that $A(k)$ does not exceed a certain level M.

42 Suppose that each dollar introduced into the economy recirculates as follows: 85% of the original dollar is spent, then 85% of the remaining $0.85 is spent, and so on. Find the economic impact (the total amount spent) if $1,000,000 is introduced into the economy.

Prove that each of the following is true for every real number x in the indicated interval.

43 $1 - x + x^2 - x^3 + \cdots + (-1)^{n+1} x^n + \cdots = \dfrac{1}{1+x}$, if $-1 < x < 1$

44 $1 + x^2 + x^4 + \cdots + x^{2n} + \cdots = \dfrac{1}{1-x^2}$, if $-1 < x < 1$

45 $\dfrac{1}{2} + \dfrac{(x-3)}{4} + \dfrac{(x-3)^2}{8} + \cdots + \dfrac{(x-3)^n}{2^{n+1}} + \cdots = \dfrac{1}{5-x}$, if $1 < x < 5$

46 $3 + (x-1) + \dfrac{(x-1)^2}{3} + \cdots + \dfrac{(x-1)^n}{3^{n-1}} + \cdots = \dfrac{9}{4-x}$, if $-2 < x < 4$

11.3

POSITIVE TERM SERIES

It is difficult to apply the definition of convergence or divergence (11.12) to an infinite series $\sum a_n$, since in most cases a simple formula for S_n cannot be found. We can, however, develop techniques for using the nth term a_n to test a series for convergence or divergence. When applying these tests we shall be concerned not with the *sum* of the series but with whether the series converges or diverges.

In this section, we shall consider only **positive term series**, that is, series such that $a_n > 0$ for every n.

Theorem (11.21)

> If $\sum a_n$ is a positive term series and if there exists a number M such that $S_n < M$ for every n, then the series converges and has a sum $S \leq M$. If no such M exists the series diverges.

Proof   If $\{S_n\}$ is the sequence of partial sums of the positive term series $\sum a_n$, then

$$S_1 < S_2 < \cdots < S_n < \cdots$$

and therefore $\{S_n\}$ is monotonic. If there exists a number M such that $S_n < M$ for every n, then as in the proof of Theorem (11.8),

$$\lim_{n \to \infty} S_n = S \leq M$$

for some S, and hence the series converges. If no such M exists, then $\lim_{n \to \infty} S_n = \infty$ and the series diverges. $\square$

Suppose a function f is defined for every real number $x \geq 1$. We may then consider the infinite series

$$\sum_{n=1}^{\infty} f(n) = f(1) + f(2) + \cdots + f(n) + \cdots$$

For example, if $f(x) = 1/x^2$, then

$$\sum_{n=1}^{\infty} f(n) = \sum_{n=1}^{\infty} \frac{1}{n^2} = \frac{1}{1^2} + \frac{1}{2^2} + \cdots + \frac{1}{n^2} + \cdots$$

A series of this type may be tested for convergence or divergence by means of an improper integral, as indicated in the following result.

The Integral Test (11.22)

> If a function f is positive-valued, continuous, and decreasing for $x \geq 1$, then the infinite series
> $$f(1) + f(2) + \cdots + f(n) + \cdots$$
> (i)   converges if $\displaystyle\int_1^\infty f(x)\,dx$ converges.
> (ii)   diverges if $\displaystyle\int_1^\infty f(x)\,dx$ diverges.

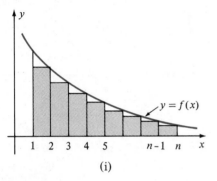

(i)

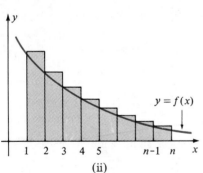

(ii)

Figure 11.3

Proof   If n is a positive integer greater than 1, then the area of the inscribed rectangular polygon illustrated in (i) of Figure 11.3 is

$$\sum_{k=2}^{n} f(k) = f(2) + f(3) + \cdots + f(n).$$

Similarly, the area of the circumscribed rectangular polygon illustrated in (ii) of the figure is

$$\sum_{k=1}^{n-1} f(k) = f(1) + f(2) + \cdots + f(n-1).$$

Since $\int_1^n f(x)\,dx$ is the area under the graph of f from 1 to n,

$$\sum_{k=2}^{n} f(k) \leq \int_1^n f(x)\,dx \leq \sum_{k=1}^{n-1} f(k).$$

If S_n denotes the nth partial sum of the series $f(1) + f(2) + \cdots + f(n) + \cdots$, then this inequality may be written

$$S_n - f(1) \leq \int_1^n f(x)\,dx \leq S_{n-1}.$$

The preceding inequality implies that if the integral $\int_1^\infty f(x)\,dx$ converges and equals $K > 0$, then

$$S_n - f(1) \leq K, \quad \text{or} \quad S_n \leq K + f(1)$$

for every positive integer n. Hence, by Theorem (11.21) the series $\sum f(n)$ converges. If the improper integral diverges, then

$$\lim_{n \to \infty} \int_1^n f(x)\,dx = \infty$$

and since $\int_1^n f(x)\,dx \leq S_{n-1}$ we also have $\lim_{n \to \infty} S_{n-1} = \infty$; that is, the series $\sum f(n)$ diverges. $\square$

Example 1   Use the Integral Test to prove that the harmonic series

$$1 + \frac{1}{2} + \frac{1}{3} + \cdots + \frac{1}{n} + \cdots$$

diverges.

Solution   If we let $f(x) = 1/x$, then f is positive-valued, continuous, and decreasing for $x \geq 1$ and, therefore, the Integral Test may be applied. Since

$$\int_1^{\infty} \frac{1}{x} \, dx = \lim_{t \to \infty} \int_1^t \frac{1}{x} \, dx = \lim_{t \to \infty} \left[\ln x \right]_1^t$$

$$= \lim_{t \to \infty} [\ln t - \ln 1] = \infty,$$

the series diverges by (ii) of (11.22).   ∎

Example 2   Determine whether the infinite series $\sum ne^{-n^2}$ converges or diverges.

Solution   If we let $f(x) = xe^{-x^2}$, then the given series is the same as $\sum f(n)$. If $x \geq 1$, f is positive-valued and continuous. The first derivative may be used to determine whether f is decreasing. Since

$$f'(x) = e^{-x^2} - 2x^2 e^{-x^2} = e^{-x^2}(1 - 2x^2) < 0,$$

f is decreasing on $[1, \infty)$. We may, therefore, apply the Integral Test as follows:

$$\int_1^{\infty} xe^{-x^2} \, dx = \lim_{t \to \infty} \int_1^t xe^{-x^2} \, dx = \lim_{t \to \infty} \left(-\frac{1}{2} \right) e^{-x^2} \bigg]_1^t$$

$$= \left(-\frac{1}{2} \right) \lim_{t \to \infty} \left[\frac{1}{e^{t^2}} - \frac{1}{e} \right] = \frac{1}{2e}.$$

Hence the series converges by (i) of (11.22).   ∎

An integral test may also be used if the function f satisfies the conditions of (11.22) for all $x \geq m$, where m is any positive integer. In this case we merely replace the integral in (11.22) by $\int_m^{\infty} f(x) \, dx$. This corresponds to deleting the first $m - 1$ terms of the series.

If $f(x) = 1/x^p$ where $p > 0$, then the series $\sum f(n)$ has the form

$$1 + \frac{1}{2^p} + \frac{1}{3^p} + \cdots + \frac{1}{n^p} + \cdots$$

and is called the **p-series**, or **hyperharmonic series**. This series will be very useful when we apply comparison tests later in this section. The following theorem provides information about convergence or divergence.

Theorem (*11.23*)

> The *p*-series
>
> $$\sum_{n=1}^{\infty} \frac{1}{n^p} = 1 + \frac{1}{2^p} + \frac{1}{3^p} + \cdots + \frac{1}{n^p} + \cdots$$
>
> converges if $p > 1$ or diverges if $p \le 1$.

Proof   First we note that the special case $p = 1$ is the divergent harmonic series. Next suppose that p is any positive real number different from 1. If we let $f(x) = 1/x^p = x^{-p}$, then f is positive-valued and continuous for $x \ge 1$. Moreover, for these values of x we see that $f'(x) = -px^{-p-1} < 0$, and hence f is decreasing. Thus f satisfies the conditions stated in the Integral Test (11.22) and we consider

$$\int_1^{\infty} \frac{1}{x^p} \, dx = \lim_{t \to \infty} \int_1^t x^{-p} \, dx = \lim_{t \to \infty} \frac{x^{1-p}}{1-p} \Big]_1^t$$

$$= \frac{1}{1-p} \lim_{t \to \infty} [t^{1-p} - 1].$$

If $p > 1$, then $p - 1 > 0$ and the last expression may be written

$$\frac{1}{1-p} \lim_{t \to \infty} \left[\frac{1}{t^{p-1}} - 1 \right] = \frac{1}{1-p}(0 - 1) = \frac{1}{p-1}.$$

Thus by (i) of (11.22), the *p*-series converges if $p > 1$.
    If $0 < p < 1$, then $1 - p > 0$ and

$$\frac{1}{1-p} \lim_{t \to \infty} [t^{1-p} - 1] = \infty.$$

Hence by (ii) of (11.22), the *p*-series diverges.
    If $p \le 0$, then $\lim_{n \to \infty} (1/n^p) \ne 0$, and by (11.15) the series diverges.   □

Example 3   Determine whether the following series converge or diverge.

(a) $1 + \dfrac{1}{2^2} + \dfrac{1}{3^2} + \cdots + \dfrac{1}{n^2} + \cdots$

(b) $5 + \dfrac{5}{\sqrt{2}} + \dfrac{5}{\sqrt{3}} + \cdots + \dfrac{5}{\sqrt{n}} + \cdots$

Solution

(a)   The series $\sum 1/n^2$ converges since it is the *p*-series with $p = 2 > 1$.

(b)   The series $\sum 1/\sqrt{n}$ diverges since it is the *p*-series with $p = \frac{1}{2} < 1$. Consequently $\sum 5/\sqrt{n}$ diverges. (See Exercise 31 of Section 11.2.)   ∎

The next theorem allows us to use known convergent (divergent) series to establish the convergence (divergence) of other series.

Basic Comparison Test (11.24)

Suppose $\sum a_n$ and $\sum b_n$ are positive term series.

(i) If $\sum b_n$ converges and $a_n \leq b_n$ for every positive integer n, then $\sum a_n$ converges.

(ii) If $\sum b_n$ diverges and $a_n \geq b_n$ for every positive integer n, then $\sum a_n$ diverges.

Proof   Let S_n and T_n denote the nth partial sums of $\sum a_n$ and $\sum b_n$, respectively. Suppose $\sum b_n$ converges and has the sum T. If $a_n \leq b_n$ for every n, then $S_n \leq T_n < T$ and hence by Theorem (11.21) $\sum a_n$ converges. This proves part (i).

To prove (ii), suppose $\sum b_n$ diverges and $a_n \geq b_n$ for every n. Then $S_n \geq T_n$ and, since T_n increases without bound as n becomes infinite, so does S_n. Consequently, $\sum a_n$ diverges.   □

Since convergence or divergence of a series is not affected by deleting a finite number of terms, the conditions $a_n \geq b_n$ or $a_n \leq b_n$ of (11.24) are only required from the kth term on, where k is some fixed positive integer.

A series $\sum d_n$ is said to **dominate** a series $\sum c_n$ if $0 < c_n \leq d_n$ for every positive integer n. Using this terminology, part (i) of (11.24) states that a positive term series that is dominated by a convergent series is also convergent. Part (ii) states that a series that dominates a divergent positive term series also diverges.

Example 4 Determine whether the following series converge or diverge.

(a) $\displaystyle\sum_{n=1}^{\infty} \frac{1}{2 + 5^n}$ (b) $\displaystyle\sum_{n=2}^{\infty} \frac{3}{\sqrt{n} - 1}$

Solution

(a) For every $n \geq 1$,

$$\frac{1}{2 + 5^n} < \frac{1}{5^n} = \left(\frac{1}{5}\right)^n.$$

Since $\sum (1/5)^n$ is a convergent geometric series, the given series converges, by (i) of (11.24).

(b) The p-series $\sum 1/\sqrt{n}$ diverges and hence so does the series obtained by discarding the first term $1/\sqrt{1}$. Since

$$\frac{3}{\sqrt{n} - 1} > \frac{1}{\sqrt{n}} \quad \text{if } n \geq 2,$$

it follows from (ii) of (11.24) that the given series diverges.   ∎

Limit Comparison Test (11.25)

> If $\sum a_n$ and $\sum b_n$ are positive term series and if
>
> $$\lim_{n \to \infty} \frac{a_n}{b_n} = k > 0$$
>
> then either both series converge or both diverge.

Proof If $\lim_{n \to \infty} (a_n/b_n) = k > 0$, then there exists a number N such that

$$\frac{k}{2} < \frac{a_n}{b_n} < \frac{3k}{2} \quad \text{whenever } n > N.$$

This is equivalent to

$$\frac{k}{2} b_n < a_n < \frac{3k}{2} b_n \quad \text{whenever } n > N.$$

If the series $\sum a_n$ converges, then $\sum (k/2)b_n$ also converges, since it is dominated by $\sum a_n$. Applying (ii) of (11.19),

$$\sum b_n = \sum \left(\frac{2}{k}\right)\left(\frac{k}{2}\right) b_n$$

converges. Conversely, if $\sum b_n$ converges, then so does $\sum a_n$, since it is dominated by the convergent series $\sum (3k/2)b_n$. We have proved that $\sum a_n$ converges if and only if $\sum b_n$ converges. Consequently, $\sum a_n$ diverges if and only if $\sum b_n$ diverges. □

Two other comparison tests are stated in Exercises 49 and 50.

Example 5 Determine whether the following series converge or diverge.

(a) $\displaystyle\sum_{n=1}^{\infty} \frac{1}{\sqrt[3]{n^2 + 1}}$ (b) $\displaystyle\sum_{n=1}^{\infty} \frac{3n^2 + 5n}{2^n(n^2 + 1)}$

Solution

(a) We shall apply (11.25) with

$$a_n = \frac{1}{\sqrt[3]{n^2 + 1}} \quad \text{and} \quad b_n = \frac{1}{\sqrt[3]{n^2}} = \frac{1}{n^{2/3}}.$$

The series $\sum b_n$ is the p-series with $p = \frac{2}{3} < 1$ and, therefore, diverges. Since

$$\lim_{n \to \infty} \frac{a_n}{b_n} = \lim_{n \to \infty} \frac{\sqrt[3]{n^2}}{\sqrt[3]{n^2 + 1}} = \lim_{n \to \infty} \sqrt[3]{\frac{n^2}{n^2 + 1}} = 1 > 0,$$

$\sum a_n$ also diverges.

(b) If we let

$$a_n = \frac{3n^2 + 5n}{2^n(n^2 + 1)} \quad \text{and} \quad b_n = \frac{1}{2^n}$$

then
$$\lim_{n \to \infty} \frac{a_n}{b_n} = \lim_{n \to \infty} \frac{3n^2 + 5n}{n^2 + 1} = 3 > 0.$$

Since $\sum b_n$ is a convergent geometric series, the series $\sum a_n$ is also convergent by (11.25).   ∎

When we seek a suitable series $\sum b_n$ to be used for comparison with $\sum a_n$, where a_n is a fraction, a good procedure is to discard all terms in the numerator and denominator except those that have the most effect on the magnitude. For example, if we employ this technique in part (b) of Example 5, where

$$a_n = \frac{3n^2 + 5n}{2^n n^2 + 2^n},$$

we are led to
$$\frac{3n^2}{2^n n^2} = \frac{3}{2^n}.$$

This suggests taking $b_n = 1/2^n$, as was done in the solution.

Example 6 If
$$a_n = \frac{8n + \sqrt{n}}{5 + n^2 + n^{7/2}}$$

determine whether $\sum a_n$ converges or diverges.

Solution   To find a suitable comparison series $\sum b_n$ we discard all but the highest powers of n in the numerator and denominator, obtaining

$$\frac{8n}{n^{7/2}} = \frac{8}{n^{5/2}}.$$

This suggests that we choose $b_n = 1/n^{5/2}$. In this event

$$\lim_{n \to \infty} \frac{a_n}{b_n} = \lim_{n \to \infty} \frac{8n^{7/2} + n^3}{5 + n^2 + n^{7/2}} = 8 > 0.$$

Since $\sum b_n$ is a convergent p-series with $p = \frac{5}{2} > 1$, it follows from (11.25) that $\sum a_n$ is also convergent.   ∎

We shall conclude this section with several general remarks about positive term series. Suppose $\sum a_n$ is a positive term series and the terms are grouped in some manner. For example, we could have

$$(a_1 + a_2) + a_3 + (a_4 + a_5 + a_6 + a_7) + \cdots$$

If we denote the last series by $\sum b_n$, so that

$$b_1 = a_1 + a_2, \quad b_2 = a_3, \quad b_3 = a_4 + a_5 + a_6 + a_7, \quad \ldots$$

then any partial sum of the series $\sum b_n$ is also a partial sum of $\sum a_n$. It follows that if $\sum a_n$ converges, then $\sum b_n$ converges and has the same sum. A similar argument may be used for any grouping of the terms of $\sum a_n$. Thus, *if a positive term series converges, then the series obtained by grouping the terms in any manner also converges and has the same sum.*

This result is not necessarily true for a series that contains both positive and negative terms. Also, we cannot make a similar statement about arbitrary divergent series. For example, the terms of the divergent series $\sum (-1)^n$ may be grouped to produce a convergent series (see Exercise 34 of Section 11.2).

Next, suppose that a convergent positive term series $\sum a_n$ has the sum S and that a new series $\sum b_n$ is formed by rearranging the terms in some way. For example, $\sum b_n$ could be the series

$$a_2 + a_8 + a_1 + a_5 + a_7 + a_3 + \cdots$$

If T_n is the nth partial sum of $\sum b_n$, then it is a sum of terms of $\sum a_n$. If m is the largest subscript associated with the terms a_i in T_n, then $T_n \le S_m < S$. Consequently, $T_n < S$ for every n. Applying Theorem (11.21), $\sum b_n$ converges and has a sum $T \le S$. The preceding proof is independent of the particular rearrangement of terms. We may also regard the series $\sum a_n$ as having been obtained by rearranging the terms of $\sum b_n$ and hence, by the same argument, $S \le T$. We have proved that *if the terms of a convergent positive term series $\sum a_n$ are rearranged in any manner, then the resulting series converges and has the same sum.*

EXERCISES 11.3

Use the Integral Test to determine whether the series in Exercises 1–16 converge or diverge.

1 $\displaystyle\sum_{n=1}^{\infty} \frac{1}{(3 + 2n)^2}$

2 $\displaystyle\sum_{n=1}^{\infty} \frac{1}{(4 + n)^{3/2}}$

3 $\displaystyle\sum_{n=1}^{\infty} \frac{1}{4n + 7}$

4 $\displaystyle\sum_{n=2}^{\infty} \frac{1}{n (\ln n)^2}$

5 $\displaystyle\sum_{n=1}^{\infty} \frac{\ln n}{n}$

6 $\displaystyle\sum_{n=1}^{\infty} \frac{n}{n^2 + 1}$

7 $\displaystyle\sum_{n=1}^{\infty} \frac{1}{\sqrt[3]{2n + 1}}$

8 $\displaystyle\sum_{n=1}^{\infty} \frac{1}{1 + 16n^2}$

9 $\displaystyle\sum_{n=1}^{\infty} \frac{\arctan n}{1 + n^2}$

10 $\displaystyle\sum_{n=1}^{\infty} ne^{-n}$

11 $\displaystyle\sum_{n=3}^{\infty} \frac{1}{n(2n - 5)}$

12 $\displaystyle\sum_{n=1}^{\infty} \frac{1}{n(n + 1)(n + 2)}$

13 $\displaystyle\sum_{n=1}^{\infty} n2^{-n^2}$

14 $\displaystyle\sum_{n=1}^{\infty} \frac{1}{\sqrt{n + 9}}$

15 $\displaystyle\sum_{n=2}^{\infty} \frac{1}{n\sqrt[3]{\ln n}}$

16 $\displaystyle\sum_{n=2}^{\infty} \frac{1}{n\sqrt{n^2 - 1}}$

Use comparison tests to determine whether the series in Exercises 17–38 converge or diverge.

17 $\displaystyle\sum_{n=1}^{\infty} \frac{1}{n^4 + n^2 + 1}$

18 $\displaystyle\sum_{n=1}^{\infty} \frac{\sqrt{n}}{n^2 + 1}$

19 $\displaystyle\sum_{n=1}^{\infty} \frac{1}{n3^n}$

20 $\displaystyle\sum_{n=1}^{\infty} \frac{n^2}{n^3 + 1}$

21 $\displaystyle\sum_{n=1}^{\infty} \frac{2n + n^2}{n^3 + 1}$

22 $\displaystyle\sum_{n=1}^{\infty} \frac{2}{3 + \sqrt{n}}$

23 $\displaystyle\sum_{n=1}^{\infty} \frac{\sqrt{n}}{n + 4}$

24 $\displaystyle\sum_{n=1}^{\infty} \frac{n^5 + 4n^3 + 1}{2n^8 + n^4 + 2}$

25 $\sum\limits_{n=2}^{\infty} \dfrac{1}{\sqrt{4n^3 - 5n}}$

26 $\sum\limits_{n=4}^{\infty} \dfrac{3n}{2n^2 - 7}$

27 $\sum\limits_{n=1}^{\infty} \dfrac{8n^2 - 7}{e^n(n + 1)^2}$

28 $\sum\limits_{n=1}^{\infty} \dfrac{\sin^2 n}{2^n}$

29 $\sum\limits_{n=1}^{\infty} \dfrac{1 + 2^n}{1 + 3^n}$

30 $\sum\limits_{n=1}^{\infty} \dfrac{\ln n}{n^4}$ (Hint: $\ln n < n$)

31 $\sum\limits_{n=1}^{\infty} \dfrac{2 + \cos n}{n^2}$

32 $\sum\limits_{n=1}^{\infty} \dfrac{\arctan n}{n^2}$

33 $\sum\limits_{n=1}^{\infty} \dfrac{(2n + 1)^3}{(n^3 + 1)^2}$

34 $\sum\limits_{n=1}^{\infty} \dfrac{1}{\sqrt{n(n + 1)(n + 2)}}$

35 $\sum\limits_{n=1}^{\infty} \dfrac{1}{\sqrt[3]{5n^2 + 1}}$

36 $\sum\limits_{n=1}^{\infty} \dfrac{3n + 5}{n \cdot 2^n}$

37 $\sum\limits_{n=1}^{\infty} \dfrac{1}{n^n}$

38 $\sum\limits_{n=1}^{\infty} \dfrac{1}{n!}$

Determine whether the series in Exercises 39–46 converge or diverge.

39 $\sum\limits_{n=1}^{\infty} \dfrac{n + \ln n}{n^3 + n + 1}$

40 $\sum\limits_{n=1}^{\infty} \dfrac{n + \ln n}{n^2 + 1}$

41 $\sum\limits_{n=1}^{\infty} \sin \dfrac{1}{n^2}$

42 $\sum\limits_{n=1}^{\infty} \tan \dfrac{1}{n}$

43 $\sum\limits_{n=1}^{\infty} \dfrac{\ln n}{n^3}$

44 $\sum\limits_{n=1}^{\infty} \ln\left(1 + \dfrac{1}{2^n}\right)$

45 $\sum\limits_{n=1}^{\infty} \dfrac{n^2 + 2^n}{n + 3^n}$

46 $\sum\limits_{n=1}^{\infty} \dfrac{\sin n + 2^n}{n + 5^n}$

In Exercises 47 and 48 find all real numbers k for which series converge.

47 $\sum\limits_{n=2}^{\infty} \dfrac{1}{n^k \ln n}$

48 $\sum\limits_{n=2}^{\infty} \dfrac{1}{n(\ln n)^k}$

49 Suppose $\sum a_n$ and $\sum b_n$ are positive term series. Prove that if $\lim_{n \to \infty} (a_n/b_n) = 0$ and $\sum b_n$ converges, then $\sum a_n$ converges. (This is not necessarily true for series that contain negative terms.)

50 Prove that if $\lim_{n \to \infty} (a_n/b_n) = \infty$ and $\sum b_n$ diverges, then $\sum a_n$ diverges.

51 Let $\sum a_n$ be a convergent positive series. Let $f(x) = a_x$, and suppose f is continuous and decreasing for $x \geq N$, where N is an integer. Prove that the error in approximating the given series by $\sum_{n=1}^{N} a_n$ is less than $\int_N^{\infty} f(x)\, dx$.

Use Exercise 51 to determine the smallest number of terms that can be added to approximate the sum of the series in Exercises 52–54 with an error less than E.

52 $\sum\limits_{n=1}^{\infty} \dfrac{1}{n^2}$, $E = 0.001$

53 $\sum\limits_{n=1}^{\infty} \dfrac{1}{n^3}$, $E = 0.01$

54 $\sum\limits_{n=2}^{\infty} \dfrac{1}{n(\ln n)^2}$, $E = 0.05$

55 Prove that if a positive term series $\sum a_n$ converges, then $\sum (1/a_n)$ diverges.

56 Prove that if a positive terms series $\sum a_n$ converges, then $\sum \sqrt{a_n a_{n+1}}$ converges.
(Hint: First show that $\sqrt{a_n a_{n+1}} \leq (a_n + a_{n+1})/2$.)

11.4

ALTERNATING SERIES

An infinite series whose terms are alternately positive and negative is called an **alternating series**. It is customary to express an alternating series in one of the forms

$$a_1 - a_2 + a_3 - a_4 + \cdots + (-1)^{n-1} a_n + \cdots$$

or

$$-a_1 + a_2 - a_3 + a_4 - \cdots + (-1)^n a_n + \cdots$$

where each $a_i > 0$. The next theorem provides the main test for convergence of these series.

Alternating Series Test (11.26)

> If $a_k \geq a_{k+1} > 0$ for every positive integer k and $\lim_{n \to \infty} a_n = 0$, then the alternating series $\sum (-1)^{n-1} a_n$ is convergent.

Proof   Let us first consider the partial sums

$$S_2, S_4, S_6, \ldots, S_{2n}, \ldots$$

that contain an even number of terms of the series. Since

$$S_{2n} = (a_1 - a_2) + (a_3 - a_4) + \cdots + (a_{2n-1} - a_{2n})$$

and $a_k - a_{k+1} \geq 0$ for every k, we see that

$$0 \leq S_2 \leq S_4 \leq \cdots \leq S_{2n} \leq \cdots,$$

that is, $\{S_{2n}\}$ is a monotonic sequence. The formula for S_{2n} can also be written

$$S_{2n} = a_1 - (a_2 - a_3) - (a_4 - a_5) - \cdots - (a_{2n-2} - a_{2n-1}) - a_{2n}$$

and hence $S_{2n} \leq a_1$ for every positive integer n. As in the proof of Theorem (11.8),

$$\lim_{n \to \infty} S_{2n} = S \leq a_1$$

for some number S. If we consider a partial sum S_{2n+1} having an *odd* number of terms of the series, then $S_{2n+1} = S_{2n} + a_{2n+1}$ and, since $\lim_{n \to \infty} a_{2n+1} = 0$,

$$\lim_{n \to \infty} S_{2n+1} = \lim_{n \to \infty} S_{2n} = S.$$

It follows that

$$\lim_{n \to \infty} S_n = S \leq a_1$$

that is, the series converges.   □

Example 1   Determine whether the following alternating series converge or diverge.

(a) $\displaystyle\sum_{n=1}^{\infty} (-1)^{n-1} \frac{2n}{4n^2 - 3}$ (b) $\displaystyle\sum_{n=1}^{\infty} (-1)^{n-1} \frac{2n}{4n - 3}$

Solution

(a)   Let $a_n = f(n) = \dfrac{2n}{4n^2 - 3}$.

To apply the Alternating Series Test we must show that:

(i) $a_k \geq a_{k+1}$ for every positive integer k.

(ii) $\lim_{n \to \infty} a_n = 0$.

One method of proving (i) is to show that $f(x) = 2x/(4x^2 - 3)$ is decreasing for $x \geq 1$. By the Quotient Rule,

$$f'(x) = \frac{(4x^2 - 3)(2) - (2x)(8x)}{(4x^2 - 3)^2}$$

$$= \frac{-8x^2 - 6}{(4x^2 - 3)^2} < 0.$$

It follows that f is decreasing and, therefore, $a_k \geq a_{k+1}$ for every positive integer k.

We can obtain the same result directly by proving that $a_k - a_{k+1} \geq 0$. Specifically, if $a_n = 2n/(4n^2 - 3)$, then

$$a_k - a_{k+1} = \frac{2k}{4k^2 - 3} - \frac{2(k + 1)}{4(k + 1)^2 - 3}$$

$$= \frac{8k^2 + 8k + 6}{(4k^2 - 3)(4k^2 + 8k + 1)} \geq 0$$

for every positive integer k. Another technique for proving that $a_k \geq a_{k+1}$ is to show that $a_{k+1}/a_k \leq 1$.

To prove (ii) we see that

$$\lim_{n \to \infty} a_n = \lim_{n \to \infty} \frac{2n}{4n^2 - 3} = 0.$$

Thus the alternating series converges.

(b)   It can be shown that $a_k \geq a_{k+1}$ for every k; however,

$$\lim_{n \to \infty} \frac{2n}{4n - 3} = \frac{1}{2} \neq 0$$

and hence the series diverges by (11.15).   ∎

If an infinite series converges, then the nth partial sum S_n can be used to approximate the sum S of the series. In most cases it is difficult to determine the accuracy of the approximation. However, for an *alternating* series, the next theorem provides a simple way of estimating the error that is involved.

Theorem (11.27)

> If $\sum (-1)^{n-1} a_n$ is an alternating series such that $a_k > a_{k+1} > 0$ for every positive integer k and if $\lim_{n \to \infty} a_n = 0$, then the error involved in approximating the sum S of the series by the nth partial sum S_n is numerically less than a_{n+1}.

Proof   Note that the alternating series $\sum (-1)^{n-1} a_n$ satisfies the conditions of the Alternating Series Test and hence has a sum S. The series obtained by deleting the first n terms, namely

$$(-1)^n a_{n+1} + (-1)^{n+1} a_{n+2} + (-1)^{n+2} a_{n+3} + \cdots$$

also satisfies the conditions of (11.26) and, therefore, has a sum R_n. Thus

$$S - S_n = R_n = (-1)^n(a_{n+1} - a_{n+2} + a_{n+3} - \cdots)$$

and

$$|R_n| = a_{n+1} - a_{n+2} + a_{n+3} - \cdots$$

Employing the same argument used in the proof of the Alternating Series Test (but replacing $\leq$ by $<$) we see that $|R_n| < a_{n+1}$. Consequently,

$$|S - S_n| = |R_n| < a_{n+1}$$

which is what we wished to prove.   □

Example 2 Prove that the series

$$1 - \frac{1}{3!} + \frac{1}{5!} - \cdots + (-1)^{n-1} \frac{1}{(2n-1)!} + \cdots$$

is convergent and approximate its sum S to five decimal places.

Solution   Evidently $a_n = 1/(2n-1)!$ has the limit 0 as $n \to \infty$, and $a_k > a_{k+1}$ for every positive integer k. Hence the series converges by the Alternating Series Test. If we use S_4 to approximate the sum S of the series, then

$$S \approx 1 - \frac{1}{3!} + \frac{1}{5!} - \frac{1}{7!}$$

$$= 1 - \frac{1}{6} + \frac{1}{120} - \frac{1}{5040} \approx 0.84147.$$

By Theorem (11.27) the error involved in the approximation is less than

$$a_5 = \frac{1}{9!} < 0.000005.$$

Thus the approximation 0.84147 is accurate to five decimal places. It will follow from (11.42) that the sum of the given series equals sin 1 and hence sin 1 ≈ 0.84147. This illustrates one technique for constructing trigonometric tables.   ■

EXERCISES 11.4

Determine whether the series in Exercises 1–14 converge or diverge.

1 $\displaystyle\sum_{n=1}^{\infty} (-1)^{n-1} \frac{1}{\sqrt{2n+1}}$

2 $\displaystyle\sum_{n=1}^{\infty} (-1)^{n-1} \frac{1}{n^{2/3}}$

3 $\displaystyle\sum_{n=1}^{\infty} (-1)^{n-1} \frac{1}{\ln(n+1)}$

4 $\displaystyle\sum_{n=1}^{\infty} (-1)^{n-1} \frac{n}{n^2+4}$

5 $\displaystyle\sum_{n=2}^{\infty} (-1)^n \frac{n}{\ln n}$

6 $\displaystyle\sum_{n=1}^{\infty} (-1)^n \frac{\ln n}{n}$

7 $\displaystyle\sum_{n=1}^{\infty} (-1)^{n-1} \frac{n^2+1}{n^3+1}$

8 $\displaystyle\sum_{n=1}^{\infty} (-1)^{n-1} \frac{3n+4}{5n+7}$

9 $\displaystyle\sum_{n=1}^{\infty} (-1)^n \frac{e^n}{n^4}$

10 $\displaystyle\sum_{n=0}^{\infty} (-1)^n \frac{\sqrt{n+1}}{8n+5}$

11 $\displaystyle\sum_{n=1}^{\infty} (-1)^n \frac{\sqrt[3]{n}}{2n+5}$ 12 $\displaystyle\sum_{n=0}^{\infty} (-1)^n \frac{1+4^n}{1+3^n}$

13 $\displaystyle\sum_{n=1}^{\infty} (-1)^n \frac{(\ln n)^k}{n}$, where k is any positive integer.

14 $\displaystyle\sum_{n=1}^{\infty} (-1)^n \frac{1}{\sqrt[k]{n}}$, where k is any positive real number.

In Exercises 15–20, approximate the sum of each series to three decimal places.

15 $\displaystyle\sum_{n=0}^{\infty} (-1)^n \frac{1}{n!}$ 16 $\displaystyle\sum_{n=0}^{\infty} (-1)^{n+1} \frac{1}{(2n)!}$

17 $\displaystyle\sum_{n=1}^{\infty} (-1)^{n-1} \frac{1}{n^3}$ 18 $\displaystyle\sum_{n=1}^{\infty} (-1)^{n-1} \frac{1}{n^5}$

19 $\displaystyle\sum_{n=1}^{\infty} (-1)^{n-1} \frac{n+1}{5^n}$ 20 $\displaystyle\sum_{n=1}^{\infty} (-1)^n \frac{1}{n} \left(\frac{1}{2}\right)^n$

In each of Exercises 21–24, use Theorem (11.27) to find a positive integer n such that S_n approximates the sum of the given series to four decimal places.

21 $\displaystyle\sum_{n=1}^{\infty} (-1)^n \frac{1}{n^2}$ 22 $\displaystyle\sum_{n=1}^{\infty} (-1)^n \frac{1}{\sqrt{n}}$

23 $\displaystyle\sum_{n=1}^{\infty} (-1)^n \frac{1}{n^n}$ 24 $\displaystyle\sum_{n=1}^{\infty} (-1)^n \frac{1}{n^3+1}$

25 If $\sum a_n$ and $\sum b_n$ are both convergent series, is $\sum a_n b_n$ convergent? Explain.

26 If $\sum a_n$ and $\sum b_n$ are both divergent series, is $\sum a_n b_n$ divergent? Explain.

11.5

ABSOLUTE CONVERGENCE

The following concept plays an important role in work with infinite series.

Definition (11.28)

> An infinite series $\sum a_n$ is **absolutely convergent** if the series
>
> $$\sum |a_n| = |a_1| + |a_2| + \cdots + |a_n| + \cdots$$
>
> obtained by taking the absolute value of each term is convergent.

Note that if $\sum a_n$ is a positive term series, then $|a_n| = a_n$, and in this case absolute convergence is the same as convergence.

Example 1 Prove that the following alternating series is absolutely convergent:

$$1 - \frac{1}{2^2} + \frac{1}{3^2} - \frac{1}{4^2} + \cdots + (-1)^n \frac{1}{n^2} + \cdots$$

Solution Taking the absolute value of each term gives us

$$1 + \frac{1}{2^2} + \frac{1}{3^2} + \frac{1}{4^2} + \cdots + \frac{1}{n^2} + \cdots$$

which is a convergent p-series. Hence by Definition (11.28), the alternating series is absolutely convergent. ∎

The next theorem tells us that absolute convergence implies convergence.

Theorem (11.29)

> If an infinite series $\sum a_n$ is absolutely convergent, then $\sum a_n$ is convergent.

Proof   If we let $b_n = a_n + |a_n|$ and use the property $-|a_n| \le a_n \le |a_n|$, then

$$0 \le a_n + |a_n| \le 2|a_n|, \quad \text{or} \quad 0 \le b_n \le 2|a_n|.$$

If $\sum a_n$ is absolutely convergent, then $\sum |a_n|$ is convergent, and hence by (ii) of Theorem (11.19), $\sum 2|a_n|$ is convergent. Applying the Comparison Test (11.24) it follows that $\sum b_n$ is convergent. Using (iii) of (11.19), $\sum (b_n - |a_n|)$ is convergent. Since $b_n - |a_n| = a_n$, this completes the proof.   □

Example 2 Determine whether the following series is convergent or divergent:

$$\sin 1 + \frac{\sin 2}{2^2} + \frac{\sin 3}{3^2} + \cdots + \frac{\sin n}{n^2} + \cdots$$

Solution   The series contains both positive and negative terms, but it is not an alternating series since, for example, the first three terms are positive and the next three are negative. The series of absolute values is

$$\sum_{n=1}^{\infty} \left| \frac{\sin n}{n^2} \right| = \sum_{n=1}^{\infty} \frac{|\sin n|}{n^2}.$$

Since

$$\frac{|\sin n|}{n^2} \le \frac{1}{n^2}$$

the series of absolute values $\sum |(\sin n)/n^2|$ is dominated by the convergent p-series $\sum 1/n^2$ and hence is convergent. Thus the given series is absolutely convergent and, therefore, is convergent by Theorem (11.29).   ■

Series that are convergent, but are not absolutely convergent, are given a special name, as indicated in the next definition.

Definition (11.30)

> An infinite series $\sum a_n$ is **conditionally convergent** if $\sum a_n$ is convergent and $\sum |a_n|$ is divergent.

Example 3 Show that the following alternating series is conditionally convergent:

$$1 - \frac{1}{2} + \frac{1}{3} - \frac{1}{4} + \cdots + (-1)^{n-1}\frac{1}{n} + \cdots$$

Solution   The series is convergent by the Alternating Series Test. Taking the absolute value of each term, we obtain

$$1 + \frac{1}{2} + \frac{1}{3} + \frac{1}{4} + \cdots + \frac{1}{n} + \cdots$$

which is the divergent harmonic series. Hence by Definition (11.30), the alternating series is conditionally convergent.   ∎

We see from the preceding discussion that an arbitrary infinite series may be classified in exactly *one* of the following ways: (i) absolutely convergent, (ii) conditionally convergent, (iii) divergent. Of course, for positive term series we need only determine convergence or divergence.

One of the most important tests for absolute convergence is the following.

The Ratio Test (11.31)

Let $\sum a_n$ be an infinite series of nonzero terms.

(i)   If $\lim\limits_{n \to \infty} \left| \dfrac{a_{n+1}}{a_n} \right| = L < 1$, the series is absolutely convergent.

(ii) If $\lim\limits_{n \to \infty} \left| \dfrac{a_{n+1}}{a_n} \right| = L > 1$, or $\lim\limits_{n \to \infty} \left| \dfrac{a_{n+1}}{a_n} \right| = \infty$, the series is divergent.

(iii) If $\lim\limits_{n \to \infty} \left| \dfrac{a_{n+1}}{a_n} \right| = 1$, the series may be absolutely convergent, conditionally convergent, or divergent.

Proof   (i)   Suppose $\lim_{n \to \infty} |a_{n+1}/a_n| = L < 1$. If r is any number such that $0 \le L < r < 1$, then there exists an integer N such that

$$\left| \frac{a_{n+1}}{a_n} \right| < r \quad \text{whenever } n \ge N.$$

This implies that

$$|a_{N+1}| < |a_N|r$$
$$|a_{N+2}| < |a_{N+1}|r < |a_N|r^2$$
$$|a_{N+3}| < |a_{N+2}|r < |a_N|r^3$$

and, in general,

$$|a_{N+m}| < |a_N|r^m \quad \text{whenever } m > 0.$$

It follows from the Comparison Test (11.24) that the series

$$|a_{N+1}| + |a_{N+2}| + \cdots + |a_{N+m}| + \cdots$$

converges, since its terms are less than the corresponding terms of the convergent geometric series

$$|a_N|r + |a_N|r^2 + \cdots + |a_N|r^n + \cdots$$

Since convergence or divergence is unaffected by discarding a finite number of terms, the series $\sum_{n=1}^{\infty} |a_n|$ also converges, that is, $\sum a_n$ is absolutely convergent.

(ii)   Suppose next that $\lim_{n \to \infty} |a_{n+1}/a_n| = L > 1$. If r is a number such that $1 < r < L$, then there exists an integer N such that

$$\left| \frac{a_{n+1}}{a_n} \right| > r > 1 \quad \text{whenever } n \geq N.$$

Consequently, $|a_{n+1}| > |a_n|$ if $n \geq N$ and therefore $\lim_{n \to \infty} |a_n| \neq 0$. Thus $\lim_{n \to \infty} a_n \neq 0$ and, by (11.15), the series $\sum a_n$ diverges. The proof for $\lim_{n \to \infty} |a_{n+1}/a_n| = \infty$ is similar and is left as an exercise.

(iii)   The Ratio Test provides no useful information if

$$\lim_{n \to \infty} \left| \frac{a_{n+1}}{a_n} \right| = 1.$$

Indeed, it is easy to verify that the limit is 1 for the absolutely convergent series $\sum (-1)^n/n^2$, for the conditionally convergent series $\sum (-1)^n/n$, and for the divergent series $\sum 1/n$. Consequently, *if the limit is 1, then a different test should be employed.*   □

Example 4 Determine whether the following series are absolutely convergent, conditionally convergent, or divergent.

(a) $\displaystyle\sum_{n=1}^{\infty} (-1)^n \frac{3^n}{n!}$ (b) $\displaystyle\sum_{n=1}^{\infty} \frac{3^n}{n^2}$

Solution

(a)   Since

$$\lim_{n \to \infty} \left| \frac{a_{n+1}}{a_n} \right| = \lim_{n \to \infty} \left| \frac{3^{n+1}}{(n+1)!} \cdot \frac{n!}{3^n} \right|$$

$$= \lim_{n \to \infty} \frac{3}{n+1} = 0 < 1$$

the series is absolutely convergent by the Ratio Test.

(b)   Since the series contains only positive terms the absolute value signs in Theorem (11.31) may be deleted. Thus

$$\lim_{n \to \infty} \frac{a_{n+1}}{a_n} = \lim_{n \to \infty} \frac{3^{n+1}}{(n+1)^2} \cdot \frac{n^2}{3^n}$$

$$= \lim_{n \to \infty} \frac{3n^2}{n^2 + 2n + 1} = 3 > 1$$

and hence the series diverges by the Ratio Test.   ∎

Example 5   Determine the convergence or divergence of $\sum_{n=1}^{\infty} \frac{n^n}{n!}$.

Solution   Since all terms of the series are positive, we may drop the absolute value signs in the Ratio Test (11.31). Thus

$$\lim_{n \to \infty} \frac{a_{n+1}}{a_n} = \lim_{n \to \infty} \frac{(n+1)^{n+1}}{(n+1)!} \cdot \frac{n!}{n^n} = \lim_{n \to \infty} \frac{(n+1)^{n+1}}{(n+1)} \cdot \frac{1}{n^n}$$

$$= \lim_{n \to \infty} \frac{(n+1)^n}{n^n} = \lim_{n \to \infty} \left(\frac{n+1}{n}\right)^n$$

$$= \lim_{n \to \infty} \left(1 + \frac{1}{n}\right)^n = e$$

where the last equality is a consequence of Theorem (7.24). Since $e > 1$, the series diverges by the Ratio Test.   ∎

The following test is useful if a_n contains only powers of n. It is not as versatile as the Ratio Test because it cannot be applied if a_n contains factorials.

The Root Test (11.32)

Let $\sum a_n$ be an infinite series.

(i)   If $\lim_{n \to \infty} \sqrt[n]{|a_n|} = L < 1$, the series is absolutely convergent.

(ii)   If $\lim_{n \to \infty} \sqrt[n]{|a_n|} = L > 1$, or $\lim_{n \to \infty} \sqrt[n]{|a_n|} = \infty$, the series is divergent.

(iii)   If $\lim_{n \to \infty} \sqrt[n]{|a_n|} = 1$, the series may be absolutely convergent, conditionally convergent, or divergent.

Proof   The proof is similar to that used for the Ratio Test. If $L < 1$ as in (i), let us consider any number r such that $L < r < 1$. By the definition of limit, there exists a positive integer N such that if $n \geq N$,

$$\sqrt[n]{|a_n|} < r, \quad \text{or} \quad |a_n| < r^n.$$

Since $0 < r < 1$, $\sum_{n=N}^{\infty} r^n$ is a convergent geometric series, and hence by the Comparison Test (11.24), $\sum_{n=N}^{\infty} |a_n|$ converges. Consequently, $\sum_{n=1}^{\infty} |a_n|$

converges; that is, $\sum a_n$ is absolutely convergent. This proves (i). The remainder of the proof is left as an exercise. □

Example 6 Determine the convergence or divergence of $\displaystyle\sum_{n=1}^{\infty} \frac{2^{3n+1}}{n^n}$.

Solution Since all terms are positive we may delete the absolute value signs in the Root Test (11.32). Thus

$$\lim_{n\to\infty} \sqrt[n]{\frac{2^{3n+1}}{n^n}} = \lim_{n\to\infty} \left(\frac{2^{3n+1}}{n^n}\right)^{1/n}$$

$$= \lim_{n\to\infty} \frac{2^{3+(1/n)}}{n} = 0 < 1.$$

Hence the series converges by the Root Test. ■

It can be proved that if a series $\sum a_n$ is absolutely convergent and if the terms are rearranged in any manner, then the resulting series converges and has the same sum as the given series. This is not true for conditionally convergent series. Indeed, if $\sum a_n$ is conditionally convergent, then by suitably rearranging terms one can obtain either a divergent series, or a series that converges and has any desired sum S.*

We now have at our disposal a variety of tests that can be used to investigate an infinite series for convergence or divergence. It takes considerable skill to determine which test is best suited for a particular series. This skill can only be obtained by working many exercises involving different types of series. The following guidelines may be helpful in deciding which test to apply. There are, however, infinite series that are not readily handled by the guidelines, even by experts. In those cases it may be necessary to use methods developed in advanced mathematics courses.

Guidelines for investigating an infinite series $\sum a_n$

I　If $\lim_{n\to\infty} a_n \neq 0$, the series diverges.

II　If $\lim_{n\to\infty} a_n = 0$, proceed as follows.

　A　If $\sum a_n$ is a positive term series, use one of the following tests.
　　(1) Basic Comparison Test
　　(2) Limit Comparison Test
　　(3) Integral Test
　　(4) Ratio Test
　　(5) Root Test

---

* See, for example, R. C. Buck, *Advanced Calculus*, Third Edition (New York: McGraw-Hill, 1978), pp. 238–239.

Remarks: When applying (1) or (2), consider a suitable *p*-series or geometric series. Of the two comparison tests, (1) is seldom used to investigate a series, since it is usually more difficult to apply than (2).

Use (3) if $f(x) = a_x$ is easy to integrate.

Use (4) or (5) if factorials, or exponents involving n, occur in a_n.

B If $\sum a_n$ is an alternating series, either

(1) use the Alternating Series Test, or

(2) apply a test from Part A to $\sum |a_n|$. If $\sum |a_n|$ is convergent, then so is $\sum a_n$.

C If $\sum a_n$ is neither a positive term series nor an alternating series, apply a test from Part A to $\sum |a_n|$. If $\sum |a_n|$ is convergent, then so is $\sum a_n$.

EXERCISES 11.5

In Exercises 1–38, determine whether (a) a series that contains both positive and negative terms is absolutely convergent, conditionally convergent, or divergent; or (b) a positive term series is convergent or divergent.

1 $\displaystyle\sum_{n=1}^{\infty} (-1)^{n+1} \frac{1}{\sqrt{n}}$

2 $\displaystyle\sum_{n=1}^{\infty} (-1)^{n} \frac{n}{n^2 + 1}$

3 $\displaystyle\sum_{n=2}^{\infty} (-1)^{n} \frac{5}{n^4 - 1}$

4 $\displaystyle\sum_{n=1}^{\infty} (-1)^{n} e^{-n}$

5 $\displaystyle\sum_{n=1}^{\infty} \frac{1000 - n}{n!}$

6 $\displaystyle\sum_{n=1}^{\infty} \frac{n!}{e^n}$

7 $\displaystyle\sum_{n=1}^{\infty} (-1)^{n-1} \frac{3n + 1}{2^n}$

8 $\displaystyle\sum_{n=1}^{\infty} (-1)^{n-1} \frac{3^n}{n^2 + 4}$

9 $\displaystyle\sum_{n=1}^{\infty} \frac{5^n}{n(3^{n+1})}$

10 $\displaystyle\sum_{n=1}^{\infty} \frac{2^{n-1}}{5^n(n + 1)}$

11 $\displaystyle\sum_{n=1}^{\infty} \frac{(-100)^n}{n!}$

12 $\displaystyle\sum_{n=1}^{\infty} \frac{10 - n^2}{n!}$

13 $\displaystyle\sum_{n=1}^{\infty} (-1)^{n-1} \frac{\sqrt{n}}{n^2 + 1}$

14 $\displaystyle\sum_{n=1}^{\infty} (-1)^{n-1} \frac{\sqrt{n}}{n + 1}$

15 $\displaystyle\sum_{n=1}^{\infty} (-1)^{n} \frac{n^2 + 1}{n^3 + 1}$

16 $\displaystyle\sum_{n=1}^{\infty} (-1)^{n} \frac{2n + 1}{n^2 + n^3}$

17 $\displaystyle\sum_{n=2}^{\infty} (-1)^{n} \frac{1}{n(\ln n)^5}$

18 $\displaystyle\sum_{n=1}^{\infty} (-1)^{n} \frac{n!}{(n + 1)^5}$

19 $\displaystyle\sum_{n=1}^{\infty} (-1)^{n-1} \frac{2}{n^3 + e^n}$

20 $\displaystyle\sum_{n=1}^{\infty} (-1)^{n-1} \frac{n3^{2n}}{5^{n-1}}$

21 $\displaystyle\sum_{n=1}^{\infty} (-1)^{n} \frac{\arctan n}{n^2}$

22 $\displaystyle\sum_{n=1}^{\infty} \frac{\cos n - 1}{n^{3/2}}$

23 $\displaystyle\sum_{n=1}^{\infty} \frac{n^n}{10^n}$

24 $\displaystyle\sum_{n=1}^{\infty} \frac{10 - 2^n}{n!}$

25 $\displaystyle\sum_{n=1}^{\infty} \frac{n!}{(-5)^n}$

26 $\displaystyle\sum_{n=1}^{\infty} \frac{\sec^{-1} n}{\tan^{-1} n}$

27 $\displaystyle\sum_{n=1}^{\infty} \frac{(n!)^2}{(2n)!}$

28 $\displaystyle\sum_{n=1}^{\infty} (-1)^n \frac{n^2 + 3}{(2n - 5)^2}$

29 $\displaystyle\sum_{n=1}^{\infty} \frac{\sin \sqrt{n}}{\sqrt{n^3 + 1}}$

30 $\displaystyle\sum_{n=1}^{\infty} \frac{n!}{n^n}$

31 $1 + \dfrac{1 \cdot 3}{2!} + \dfrac{1 \cdot 3 \cdot 5}{3!} + \cdots + \dfrac{1 \cdot 3 \cdot 5 \cdots (2n - 1)}{n!} + \cdots$

32 $\dfrac{1}{2} + \dfrac{1 \cdot 4}{2 \cdot 4} + \dfrac{1 \cdot 4 \cdot 7}{2 \cdot 4 \cdot 6} + \cdots + \dfrac{1 \cdot 4 \cdot 7 \cdots (3n - 2)}{2 \cdot 4 \cdot 6 \cdots (2n)} + \cdots$

33 $\displaystyle\sum_{n=2}^{\infty} \frac{1}{(\ln n)^n}$

34 $\displaystyle\sum_{n=1}^{\infty} \frac{(2n)^n}{(5n + 3n^{-1})^n}$

35 $\displaystyle\sum_{n=1}^{\infty} (-1)^n \frac{\ln n}{(1.01)^n}$

36 $\displaystyle\sum_{n=1}^{\infty} (-1)^n 3^{1/n}$

37 $\displaystyle\sum_{n=1}^{\infty} (-1)^n n \tan \frac{1}{n}$

38 $\displaystyle\sum_{n=1}^{\infty} \frac{\cos (n\pi/6)}{n^2}$

39 Complete the proof of the Ratio Test (11.31) by showing that if $\lim_{n \to \infty} |a_{n+1}/a_n| = \infty$, then $\sum a_n$ is divergent.

40 Complete the proof of the Root Test (11.32).

41 (a) Prove that if $\sum a_n$ is absolutely convergent, then $\sum a_n^2$ is convergent.

   (b) Show that the conclusion of part (a) is not necessarily true if $\sum a_n$ is convergent.

   (c) Show that if $\sum a_n^2$ is convergent, it does not necessarily follow that $\sum a_n$ is convergent.

42 Prove that if $\sum a_n$ is absolutely convergent and $a_n \neq 0$ for all n, then $\sum 1/|a_n|$ is divergent.

11.6

POWER SERIES

In the previous sections we concentrated on infinite series with constant terms. Of major importance in applications are series whose terms contain variables, as in the next definition.

Definition (11.33)

> Let x be a variable. A **power series in x** is a series of the form
>
> $$\sum_{n=0}^{\infty} a_n x^n = a_0 + a_1 x + a_2 x^2 + \cdots + a_n x^n + \cdots$$
>
> where each a_i is a real number.

If a number is substituted for x in Definition (11.33) we obtain a series of constant terms that may be tested for convergence or divergence. To simplify the general term of the power series it is assumed that $x^0 = 1$ even if $x = 0$. The main objective of this section is to determine all values of x for which a power series converges. Evidently, every power series in x converges if $x = 0$. To find other numbers that produce a convergent series we often employ the Ratio Test, as illustrated in the following examples.

Example 1 Find all values of x for which the following power series is absolutely convergent:

$$1 + \frac{1}{5}x + \frac{2}{5^2}x^2 + \cdots + \frac{n}{5^n}x^n + \cdots$$

Solution If we let $u_n = (n/5^n)x^n = nx^n/5^n$, then

$$\lim_{n \to \infty}\left|\frac{u_{n+1}}{u_n}\right| = \lim_{n \to \infty}\left|\frac{(n+1)x^{n+1}}{5^{n+1}} \cdot \frac{5^n}{nx^n}\right|$$

$$= \lim_{n \to \infty}\left|\frac{(n+1)x}{5n}\right| = \lim_{n \to \infty}\left(\frac{n+1}{5n}\right)|x| = \frac{1}{5}|x|.$$

By the Ratio Test, the series is absolutely convergent if the following equivalent inequalities are true:

$$\frac{1}{5}|x| < 1, \quad |x| < 5, \quad -5 < x < 5.$$

The series diverges if $\frac{1}{5}|x| > 1$, that is, if $x > 5$ or $x < -5$. If $\frac{1}{5}|x| = 1$, the Ratio Test gives no information and hence the numbers 5 and -5 require special consideration. Substituting 5 for x in the power series we obtain

$$1 + 1 + 2 + 3 + \cdots + n + \cdots$$

which is divergent. If we let $x = -5$ we obtain

$$1 - 1 + 2 - 3 + \cdots + n(-1)^n + \cdots$$

which is also divergent. Consequently, the power series is absolutely convergent for every x in the open interval $(-5, 5)$ and diverges elsewhere. ■

Example 2 Find all values of x for which the following power series is absolutely convergent:

$$1 + \frac{1}{1!}x + \frac{1}{2!}x^2 + \cdots + \frac{1}{n!}x^n + \cdots$$

Solution We shall employ the same technique used in Example 1. If we let

$$u_n = \frac{1}{n!}x^n = \frac{x^n}{n!}$$

then

$$\lim_{n \to \infty}\left|\frac{u_{n+1}}{u_n}\right| = \lim_{n \to \infty}\left|\frac{x^{n+1}}{(n+1)!} \cdot \frac{n!}{x^n}\right|$$

$$= \lim_{n \to \infty}\left|\frac{x}{n+1}\right| = \lim_{n \to \infty}\frac{1}{n+1}|x| = 0.$$

Since the limit 0 is less than 1 for every value of x, it follows from the Ratio Test that the power series is absolutely convergent for all real numbers. ∎

Example 3   Find all values of x for which $\sum n!x^n$ is convergent.

Solution   Let $u_n = n!x^n$. If $x \neq 0$, then

$$\lim_{n \to \infty} \left| \frac{u_{n+1}}{u_n} \right| = \lim_{n \to \infty} \left| \frac{(n+1)!x^{n+1}}{n!x^n} \right|$$

$$= \lim_{n \to \infty} |(n+1)x| = \lim_{n \to \infty} (n+1)|x| = \infty$$

and, by the Ratio Test, the series diverges. Hence the power series is convergent only if $x = 0$. ∎

It will be shown in Theorem (11.35) that the solutions of the preceding examples are typical, in the sense that if a power series converges for nonzero values of x, then either it is absolutely convergent for all real numbers, or it is absolutely convergent throughout some open interval $(-r, r)$ and diverges outside of the closed interval $[-r, r]$. The proof of this fact depends on the next theorem.

Theorem (11.34)

> (i)   If a power series $\sum a_n x^n$ converges for a nonzero number c, then it is absolutely convergent whenever $|x| < |c|$.
>
> (ii) If a power series $\sum a_n x^n$ diverges for a nonzero number d, then it diverges whenever $|x| > |d|$.

Proof   If $\sum a_n c^n$ converges and $c \neq 0$, then by Theorem (11.14), $\lim_{n \to \infty} a_n c^n = 0$. Employing Definition (11.3) with $\varepsilon = 1$, there is a positive integer N such that

$$|a_n c^n| < 1 \quad \text{whenever } n \geq N$$

and, therefore,

$$|a_n x^n| = \left| \frac{a_n c^n x^n}{c^n} \right| = |a_n c^n| \left| \frac{x}{c} \right|^n < \left| \frac{x}{c} \right|^n$$

provided $n \geq N$. If $|x| < |c|$, then $|x/c| < 1$ and $\sum |x/c|^n$ is a convergent geometric series. Hence, by the Comparison Test (11.24), the series obtained by deleting the first N terms of $\sum |a_n x^n|$ is convergent. It follows that the series $\sum |a_n x^n|$ is also convergent, which proves (i).

To prove (ii), suppose the series diverges for $x = d \neq 0$. If the series converges for some c_1, where $|c_1| > |d|$, then by (i) it converges whenever $|x| < |c_1|$. In particular it converges for $x = d$, contrary to our supposition. Hence the series diverges whenever $|x| > |d|$. □

Theorem (11.35)

> If $\sum a_n x^n$ is a power series, then precisely one of the following is true.
>
> (i)   The series converges only if $x = 0$.
>
> (ii) The series is absolutely convergent for all x.
>
> (iii) There is a positive number r such that the series is absolutely convergent if $|x| < r$ and divergent if $|x| > r$.

Proof   If neither (i) nor (ii) is true, then there exist nonzero numbers c and d such that the series converges if $x = c$ and diverges if $x = d$. Let S denote the set of all real numbers for which the series is absolutely convergent. By Theorem (11.34), the series diverges if $|x| > |d|$ and hence every number in S is less than $|d|$. By the Completeness Property (11.9), S has a least upper bound r. It follows that the series is absolutely convergent if $|x| < r$ and diverges if $|x| > r$.   ☐

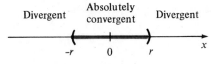

Divergent · Absolutely convergent · Divergent

Figure 11.4 $\sum a_n x^n$ with radius of convergence r.

If (iii) of Theorem (11.35) occurs, then the power series $\sum a_n x^n$ is absolutely convergent throughout the open interval $(-r, r)$ and diverges outside of the closed interval $[-r, r]$, as illustrated in Figure 11.4. The number r is called the **radius of convergence** of the series. Either convergence or divergence may occur at $-r$ or r, depending on the nature of the series.

The totality of numbers for which a power series converges is called its **interval of convergence**. If the radius of convergence r is positive, then the interval of convergence is one of the following:

$$(-r, r), \quad (-r, r], \quad [-r, r), \quad \text{or} \quad [-r, r].$$

If (i) or (ii) of Theorem (11.35) occurs, then the radius of convergence is denoted by 0 or ∞, respectively. In Example 1 of this section, the interval of convergence is $(-5, 5)$ and the radius of convergence is 5. In Example 2 the interval of convergence is $(-\infty, \infty)$ and we write $r = \infty$. In Example 3, $r = 0$. The next example illustrates the case of a half-open interval of convergence.

Example 4   Find the interval of convergence of the power series $\sum\limits_{n=1}^{\infty} \dfrac{1}{\sqrt{n}} x^n$.

Solution   Note that in this example the coefficient of x^0 is 0 and the summation begins with $n = 1$. We let $u_n = x^n/\sqrt{n}$ and consider

$$\lim_{n\to\infty} \left| \frac{u_{n+1}}{u_n} \right| = \lim_{n\to\infty} \left| \frac{x^{n+1}}{\sqrt{n+1}} \cdot \frac{\sqrt{n}}{x^n} \right| = \lim_{n\to\infty} \left| \frac{\sqrt{n}}{\sqrt{n+1}} x \right|$$

$$= \lim_{n\to\infty} \sqrt{\frac{n}{n+1}} |x| = (1)|x| = |x|.$$

It follows from the Ratio Test that the power series is absolutely convergent if $|x| < 1$, that is, if $-1 < x < 1$. The series diverges if $|x| > 1$, that is, if $x > 1$

or $x < -1$. The numbers 1 and -1 must be checked by direct substitution in the power series. If we let $x = 1$ we obtain

$$\sum_{n=1}^{\infty} \frac{1}{\sqrt{n}} (1)^n = 1 + \frac{1}{\sqrt{2}} + \frac{1}{\sqrt{3}} + \cdots + \frac{1}{\sqrt{n}} + \cdots$$

which is a divergent p-series. If we substitute $x = -1$ the result is

$$\sum_{n=1}^{\infty} \frac{1}{\sqrt{n}} (-1)^n = -1 + \frac{1}{\sqrt{2}} - \frac{1}{\sqrt{3}} + \cdots + \frac{(-1)^n}{\sqrt{n}} + \cdots$$

which converges by the Alternating Series Test. Hence the power series converges if $-1 \le x < 1$, and the interval of convergence is $[-1, 1)$. ∎

Definition (11.36)

Let c be a real number and x a variable. A **power series in $x - c$** is a series of the form

$$\sum_{n=0}^{\infty} a_n(x - c)^n = a_0 + a_1(x - c) + a_2(x - c)^2 + \cdots + a_n(x - c)^n + \cdots$$

where each a_i is a real number.

To simplify the nth term in (11.36), it is assumed that $(x - c)^0 = 1$ even if $x = c$. If we employ the same reasoning used to prove Theorem (11.35), and replace x by $x - c$, then it can be shown that precisely one of the following is true.

(i)   The series converges only if $x - c = 0$, that is, if $x = c$.

(ii) The series is absolutely convergent for all x.

(iii) There is a positive number r such that the series is absolutely convergent if $|x - c| < r$ and divergent if $|x - c| > r$.

If (iii) occurs, then the series $\sum a_n(x - c)^n$ is absolutely convergent if

$$-r < x - c < r, \quad \text{or} \quad c - r < x < c + r$$

that is, if x is the interval $(c - r, c + r)$ as illustrated in Figure 11.5. The endpoints of the interval must be checked separately. As before, the totality of numbers for which the series converges is called the **interval of convergence**, and r is called the **radius of convergence**.

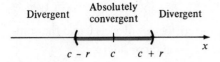

Figure 11.5 $\sum a_n(x - c)^n$ with radius of convergence r.

Example 5   Find the interval of convergence of

$$1 - \frac{1}{2}(x - 3) + \frac{1}{3}(x - 3)^2 + \cdots + (-1)^n \frac{1}{n + 1}(x - 3)^n + \cdots$$

Solution   If we let $u_n = (-1)^n(x-3)^n/(n+1)$, then

$$\lim_{n\to\infty}\left|\frac{u_{n+1}}{u_n}\right| = \lim_{n\to\infty}\left|\frac{(x-3)^{n+1}}{n+2}\cdot\frac{n+1}{(x-3)^n}\right|$$

$$= \lim_{n\to\infty}\left|\frac{n+1}{n+2}(x-3)\right|$$

$$= \lim_{n\to\infty}\left(\frac{n+1}{n+2}\right)|x-3|$$

$$= (1)|x-3| = |x-3|.$$

By the Ratio Test the series is absolutely convergent if $|x-3| < 1$, that is, if

$$-1 < x - 3 < 1 \quad\text{or}\quad 2 < x < 4.$$

The series diverges if $x < 2$ or $x > 4$. The numbers 2 and 4 must be checked separately. If $x = 4$ the resulting series is

$$1 - \frac{1}{2} + \frac{1}{3} - \cdots + (-1)^n\frac{1}{n+1} + \cdots$$

which converges by the Alternating Series Test. For $x = 2$ the series becomes

$$1 + \frac{1}{2} + \frac{1}{3} + \cdots + \frac{1}{n+1} + \cdots$$

which is the divergent harmonic series. Hence the interval of convergence is $(2, 4]$. ■

EXERCISES 11.6

Find the intervals of convergence of the power series in Exercises 1–26.

1 $\displaystyle\sum_{n=0}^{\infty} \frac{1}{n+4} x^n$

2 $\displaystyle\sum_{n=0}^{\infty} \frac{1}{n^2+4} x^n$

3 $\displaystyle\sum_{n=0}^{\infty} \frac{n^2}{2^n} x^n$

4 $\displaystyle\sum_{n=1}^{\infty} \frac{(-3)^n}{n} x^{n+1}$

5 $\displaystyle\sum_{n=1}^{\infty} (-1)^{n-1} \frac{1}{\sqrt{n}} x^n$

6 $\displaystyle\sum_{n=1}^{\infty} \frac{1}{\ln(n+1)} x^n$

7 $\displaystyle\sum_{n=2}^{\infty} \frac{n}{n^2+1} x^n$

8 $\displaystyle\sum_{n=1}^{\infty} \frac{1}{4^n\sqrt{n}} x^n$

9 $\displaystyle\sum_{n=2}^{\infty} \frac{\ln n}{n^3} x^n$

10 $\displaystyle\sum_{n=0}^{\infty} \frac{10^{n+1}}{3^{2n}} x^n$

11 $\displaystyle\sum_{n=0}^{\infty} \frac{n+1}{10^n} (x-4)^n$

12 $\displaystyle\sum_{n=1}^{\infty} \frac{1}{n(n+1)} (x-2)^n$

13 $\displaystyle\sum_{n=0}^{\infty} \frac{n!}{100^n} x^n$

14 $\displaystyle\sum_{n=0}^{\infty} \frac{(3n)!}{(2n)!} x^n$

15 $\displaystyle\sum_{n=0}^{\infty} \frac{1}{(-4)^n} x^{2n+1}$

16 $\displaystyle\sum_{n=1}^{\infty} (-1)^{n-1} \frac{1}{\sqrt[3]{n3^n}} x^n$

17 $\displaystyle\sum_{n=0}^{\infty} \frac{2^n}{(2n)!} x^{2n}$

18 $\displaystyle\sum_{n=0}^{\infty} \frac{10^n}{n!} x^n$

19 $\displaystyle\sum_{n=0}^{\infty} \frac{3^{2n}}{n+1} (x-2)^n$

20 $\displaystyle\sum_{n=1}^{\infty} \frac{1}{n5^n} (x-5)^n$

21 $\displaystyle\sum_{n=0}^{\infty} \frac{n^2}{2^{3n}} (x+4)^n$

22 $\displaystyle\sum_{n=0}^{\infty} \frac{1}{2n+1} (x+3)^n$

23 $\displaystyle\sum_{n=1}^{\infty} \frac{\ln n}{e^n} (x-e)^n$

24 $\displaystyle\sum_{n=0}^{\infty} \frac{n}{3^{2n-1}} (x-1)^{2n}$

25 $\displaystyle\sum_{n=1}^{\infty} (-1)^n \frac{1}{n6^n} (2x-1)^n$

26 $\displaystyle\sum_{n=0}^{\infty} \frac{1}{\sqrt{3n+4}} (3x+4)^n$

Find the radius of convergence of the power series in Exercises 27–30.

27 $\displaystyle\sum_{n=1}^{\infty} (-1)^n \frac{1 \cdot 3 \cdot 5 \cdot \cdots \cdot (2n-1)}{3 \cdot 6 \cdot 9 \cdot \cdots \cdot (3n)} x^n$

28 $\displaystyle\sum_{n=1}^{\infty} \frac{2 \cdot 4 \cdot 6 \cdot \cdots \cdot (2n)}{4 \cdot 7 \cdot 10 \cdot \cdots \cdot (3n+1)} x^n$

29 $\displaystyle\sum_{n=1}^{\infty} \frac{n^n}{n!} x^n$

30 $\displaystyle\sum_{n=0}^{\infty} \frac{(n+1)!}{10^n} (x-5)^n$

Find the radius of convergence of the power series in Exercises 31 and 32, where c and d are positive integers.

31 $\displaystyle\sum_{n=0}^{\infty} \frac{(n+c)!}{n!(n+d)!} x^n$ **32** $\displaystyle\sum_{x=0}^{\infty} \frac{(cn)!}{(n!)^c} x^n$

33 If $\lim_{n \to \infty} |a_{n+1}/a_n| = k$, where $k \neq 0$, prove that the radius of convergence of $\sum a_n x^n$ is $1/k$.

34 If $\lim_{n \to \infty} \sqrt[n]{|a_n|} = k$, where $k \neq 0$, prove that the radius of convergence of $\sum a_n x^n$ is $1/k$.

35 If $\sum a_n x^n$ has radius of convergence r, prove that $\sum a_n x^{2n}$ has radius of convergence $\sqrt{r}$.

36 If $\sum a_n$ is absolutely convergent, prove that $\sum a_n x^n$ is absolutely convergent for all x in the interval $[-1, 1]$.

37 If the interval of convergence of $\sum a_n x^n$ is $(-r, r]$, prove that the series is conditionally convergent at r.

38 If $\sum a_n x^n$ is absolutely convergent at one endpoint of its interval of convergence, prove that it is also absolutely convergent at the other endpoint.

11.7

POWER SERIES REPRESENTATIONS OF FUNCTIONS

A power series $\sum a_n x^n$ may be used to define a function f whose domain is the interval of convergence of the series. Specifically, for each x in this interval we let $f(x)$ equal the sum of the series, that is,

$$f(x) = a_0 + a_1 x + a_2 x^2 + \cdots + a_n x^n + \cdots$$

If a function f is defined in this way we say that $\sum a_n x^n$ is a **power series representation for $f(x)$** (or *of $f(x)$*). We also use the phrase *f is represented by the power series*.

A power series representation $f(x) = \sum a_n x^n$ enables us to find functional values in a new way. Specifically, if c is in the interval of convergence, then $f(c)$ can be found (or approximated) by finding (or approximating) the sum of the series

$$a_0 + a_1 c + a_2 c^2 + \cdots + a_n c^n + \cdots$$

Power series representations will also enable us to solve problems involving differentiation and integration by using techniques different from those considered earlier in the text.

Example 1 Find a function f that is represented by the power series

$$\sum_{n=0}^{\infty} (-1)^n x^n.$$

Solution If $|x| < 1$, then by Theorem (11.13), the geometric series

$$1 - x + x^2 - x^3 + \cdots + (-1)^n x^n + \cdots$$

converges and has the sum $1/[1 - (-x)] = 1/(1 + x)$, and hence we may write

$$\frac{1}{1 + x} = 1 - x + x^2 - x^3 + \cdots + (-1)^n x^n + \cdots$$

This gives us a power series representation for $f(x) = 1/(1 + x)$, if $|x| < 1$. ■

Most of the examples and exercises in this section are concerned with geometric series, since the sum can be determined by means of a specific formula. In the next section we shall consider the following more difficult problem: Given a function f, find a power series representation for f.

A function f defined by a power series has properties similar to those of a polynomial. In particular, it can be shown that f has a derivative whose power series representation may be found by differentiating each term of the given series. Similarly, definite integrals of f may be obtained by integrating each term of the series. These facts are consequences of the next theorem, which is stated without proof.

Theorem (11.37)

Suppose a power series $\sum a_n x^n$ has a nonzero radius of convergence r and let the function f be defined by

$$f(x) = \sum_{n=0}^{\infty} a_n x^n = a_0 + a_1 x + a_2 x^2 + \cdots + a_n x^n + \cdots$$

for every x in the interval of convergence. If $-r < x < r$, then

(i)
$$f'(x) = \sum_{n=0}^{\infty} D_x(a_n x^n) = \sum_{n=1}^{\infty} n a_n x^{n-1}$$
$$= a_1 + 2a_2 x + 3a_3 x^2 + \cdots + n a_n x^{n-1} + \cdots$$

(ii)
$$\int_0^x f(t) \, dt = \sum_{n=0}^{\infty} \int_0^x (a_n t^n) \, dt = \sum_{n=0}^{\infty} \frac{a_n}{n+1} x^{n+1}$$
$$= a_0 x + \frac{1}{2} a_1 x^2 + \frac{1}{3} a_2 x^3 + \cdots + \frac{1}{n+1} a_n x^{n+1} + \cdots$$

It can be shown that the series in (i) and (ii) of Theorem (11.37) have the same radius of convergence as $\sum a_n x^n$. As a corollary of (i), *a function that is represented by a power series in an interval $(-r, r)$ is continuous throughout $(-r, r)$*. (See Theorem (3.8).) Similar results are true for functions represented by power series of the form $\sum a_n(x - c)^n$.

Example 2   Find a power series representation for $1/(1 + x)^2$.

Solution   It was shown in Example 1 that

$$\frac{1}{1 + x} = 1 - x + x^2 - x^3 + \cdots + (-1)^n x^n + \cdots$$

If we differentiate each term of this series, then by (i) of Theorem (11.37),

$$-\frac{1}{(1 + x)^2} = -1 + 2x - 3x^2 - \cdots + (-1)^n n x^{n-1} + \cdots$$

Consequently, if $|x| < 1$, then

$$\frac{1}{(1 + x)^2} = 1 - 2x + 3x^2 + \cdots + (-1)^{n+1} n x^{n-1} + \cdots \qquad \blacksquare$$

Example 3   Find a power series representation for $\ln(1 + x)$ if $|x| < 1$.

Solution   If $|x| < 1$, then

$$\ln(1 + x) = \int_0^x \frac{1}{1 + t}\, dt$$

$$= \int_0^x [1 - t + t^2 - \cdots + (-1)^n t^n + \cdots]\, dt$$

where the last equality follows from Example 1. By (ii) of Theorem (11.37), we may integrate each term of the series as follows:

$$\ln(1 + x) = \int_0^x 1\, dt - \int_0^x t\, dt + \int_0^x t^2\, dt - \cdots + (-1)^n \int_0^x t^n\, dt + \cdots$$

$$= t\Big]_0^x - \frac{t^2}{2}\Big]_0^x + \frac{t^3}{3}\Big]_0^x - \cdots + (-1)^n \frac{t^{n+1}}{n + 1}\Big]_0^x + \cdots$$

Hence

$$\ln(1 + x) = x - \frac{x^2}{2} + \frac{x^3}{3} - \cdots + (-1)^n \frac{x^{n+1}}{n + 1} + \cdots$$

if $|x| < 1$. $\qquad \blacksquare$

Example 4   Calculate $\ln(1.1)$ to five decimal places.

Solution   In Example 3 we found a series representation for $\ln(1 + x)$ if $|x| < 1$. Substituting 0.1 for x in that series gives us

$$\ln(1.1) = 0.1 - \frac{(0.1)^2}{2} + \frac{(0.1)^3}{3} - \frac{(0.1)^4}{4} + \frac{(0.1)^5}{5} - \cdots$$

$$= 0.1 - 0.005 + 0.000333 - 0.000025 + 0.000002 - \cdots$$

If we sum the first four terms on the right and round off to five decimal places, then

$$\ln(1.1) \approx 0.09531.$$

By Theorem (11.27), the error is less than the absolute value 0.000002 of the fifth term and, therefore, the number 0.09531 is accurate to five decimal places. The reader should compare this solution to that of Example 2 in Section 10.5. $\qquad \blacksquare$

Example 5 Find a power series representation for arctan x.

Solution   We first observe that

$$\arctan x = \int_0^x \frac{1}{1 + t^2}\, dt.$$

Next we note that if $|t| < 1$, then by Theorem (11.13), with $a = 1$ and $r = -t^2$,

$$\frac{1}{1 + t^2} = 1 - t^2 + t^4 - \cdots + (-1)^n t^{2n} + \cdots$$

By (ii) of Theorem (11.37) we may integrate each term of the series to obtain

$$\arctan x = x - \frac{x^3}{3} + \frac{x^5}{5} - \cdots + (-1)^n \frac{x^{2n+1}}{2n+1} + \cdots$$

provided $|x| < 1$. It can be proved, by more advanced methods, that this series representation is also valid if $|x| = 1$. ∎

Example 6 Prove that e^x has the power series representation

$$e^x = 1 + x + \frac{x^2}{2!} + \frac{x^3}{3!} + \cdots + \frac{x^n}{n!} + \cdots$$

Solution   The indicated power series was considered in Example 2 of the preceding section, where it was shown to be absolutely convergent for every real number x. If we let f denote the function represented by the series, then

$$f(x) = \sum_{n=0}^{\infty} \frac{x^n}{n!}.$$

Applying (i) of Theorem (11.37),

$$f'(x) = \sum_{n=1}^{\infty} \frac{nx^{n-1}}{n!} = \sum_{n=1}^{\infty} \frac{x^{n-1}}{(n-1)!}$$

$$= 1 + x + \frac{x^2}{2!} + \frac{x^3}{3!} + \cdots + \frac{x^n}{n!} + \cdots$$

that is,

$$f'(x) = f(x) \quad \text{for every } x.$$

If, in Theorem (7.25), we let $y = f(t)$, $t = x$, and $c = 1$, we obtain

$$f(x) = f(0)e^x.$$

However,

$$f(0) = 1 + 0 + \frac{0^2}{2!} + \cdots + \frac{0^n}{n!} + \cdots = 1$$

and hence $$f(x) = e^x$$

which is what we wished to prove. ∎

Note that Example 6 allows us to express the number e as the sum of a convergent positive term series, namely,

$$e = 1 + 1 + \frac{1}{2!} + \frac{1}{3!} + \cdots + \frac{1}{n!} + \cdots$$

Example 7 Approximate $\int_0^{0.1} e^{-x^2} \, dx$.

Solution   Letting $x = -t^2$ in the series of Example 6 gives us

$$e^{-t^2} = 1 - t^2 + \frac{t^4}{2!} - \cdots + \frac{(-1)^n t^{2n}}{n!} + \cdots$$

which is true for all t. Applying (ii) of Theorem (11.37),

$$\int_0^{0.1} e^{-x^2} \, dx = \int_0^{0.1} e^{-t^2} \, dt$$

$$= t \Big]_0^{0.1} - \frac{t^3}{3} \Big]_0^{0.1} + \frac{t^5}{10} \Big]_0^{0.1} - \cdots$$

$$= 0.1 - \frac{(0.1)^3}{3} + \frac{(0.1)^5}{10} - \cdots$$

If we use the first two terms to approximate the sum of this convergent alternating series, then by Theorem (11.27), the error is less than the third term $(0.1)^5/10 = 0.000001$. Hence

$$\int_0^{0.1} e^{-x^2} \, dx \approx 0.1 - \frac{0.001}{3} \approx 0.99667$$

which is accurate to five decimal places.   ∎

Thus far, the techniques used to obtain power series representations of functions were *indirect*, in the sense that we started with known series and then differentiated or integrated, and perhaps used other knowledge about functional behavior, to obtain series representations. In the next section we shall discuss a *direct* method that can be used to find power series representations for a large variety of functions.

EXERCISES 11.7

In Exercises 1–10, find a power series representation for $f(x)$ and specify the interval of convergence.

1 $f(x) = 1/(1 - x)$

2 $f(x) = 1/(1 + x^2)$

3 $f(x) = 1/(1 - x)^2$

4 $f(x) = 1/(1 - 4x)$

5 $f(x) = x^2/(1 - x^2)$

6 $f(x) = x/(1 - x^4)$

7 $f(x) = x/(2 - 3x)$

8 $f(x) = x^3/(4 - x^3)$

9 $f(x) = (x^2 + 1)/(x - 1)$

10 $f(x) = 3/(2x + 5)$

11 Prove that

$$\ln(1 - x) = \sum_{n=1}^{\infty} \frac{-x^n}{n} \quad \text{if } |x| < 1,$$

and use this fact to approximate $\ln(1.2)$ to three decimal places.

12 Use Example 5 to prove that the sum of the series $\sum_{n=0}^{\infty} (-\frac{1}{3})^n / \sqrt{3}(2n + 1)$ is $\pi/6$.

In Exercises 13–18 use the result obtained in Example 6 to find a power series representation for $f(x)$.

13 $f(x) = e^{-x}$

14 $f(x) = e^{2x}$

15 $f(x) = \cosh x$

16 $f(x) = \sinh x$

17 $f(x) = xe^{3x}$

18 $f(x) = x^2 e^{x^2}$

Use infinite series to approximate each of the integrals in Exercises 19–24 to four decimal places.

19 $\displaystyle\int_0^{1/3} \frac{1}{1 + x^6} \, dx$

20 $\displaystyle\int_0^{1/2} \arctan x^2 \, dx$

21 $\displaystyle\int_{0.1}^{0.2} \frac{\arctan x}{x} \, dx$

22 $\displaystyle\int_0^{0.2} \frac{x^3}{1 + x^5} \, dx$

23 $\displaystyle\int_0^1 e^{-x^2/10} \, dx$

24 $\displaystyle\int_0^{0.5} e^{-x^3} \, dx$

25 Use the power series representation for $(1 - x^2)^{-1}$ to find a power series representation for $2x(1 - x^2)^{-2}$.

26 Use the method of Example 3 to find a power series representation for $\ln(3 + 2x)$.

In Exercises 27–30 find a power series representation for $f(x)$. If the integrand is denoted by $g(t)$, assume that the value of $g(0)$ is $\lim_{t \to 0} g(t)$. (Why is a condition of this type necessary?)

27 $f(x) = \displaystyle\int_0^x \frac{e^{-t^2} - 1}{t} \, dt$

28 $f(x) = \displaystyle\int_0^x \frac{e^t - 1}{t} \, dt$

29 $f(x) = \displaystyle\int_0^x \frac{\ln(1 - t)}{t} \, dt$

30 $f(x) = \displaystyle\int_0^x \ln(1 + t^2) \, dt$

11.8

TAYLOR AND MACLAURIN SERIES

Suppose a function f is represented by a power series in $x - c$, such that

$$f(x) = \sum_{n=0}^{\infty} a_n(x - c)^n$$
$$= a_0 + a_1(x - c) + a_2(x - c)^2 + a_3(x - c)^3 + a_4(x - c)^4 + \cdots$$

where the domain of f is an open interval containing c. As in the preceding section, power series representations may be found for $f'(x), f''(x), \ldots,$ by differentiating the terms of the series. Thus

$$f'(x) = \sum_{n=1}^{\infty} na_n(x - c)^{n-1}$$
$$= a_1 + 2a_2(x - c) + 3a_3(x - c)^2 + 4a_4(x - c)^3 + \cdots$$

$$f''(x) = \sum_{n=2}^{\infty} n(n - 1)a_n(x - c)^{n-2}$$
$$= 2a_2 + (3 \cdot 2)a_3(x - c) + (4 \cdot 3)a_4(x - c)^2 + \cdots$$

$$f'''(x) = \sum_{n=3}^{\infty} n(n - 1)(n - 2)a_n(x - c)^{n-3}$$
$$= (3 \cdot 2)a_3 + (4 \cdot 3 \cdot 2)a_4(x - c) + \cdots$$

and, for every positive integer k,

$$f^{(k)}(x) = \sum_{n=k}^{\infty} n(n-1)\cdots(n-k+1)a_n(x-c)^{n-k}.$$

Moreover, each series obtained by differentiation has the same radius of convergence as the original series. Substituting c for x in each of these series representations, we obtain

$$f(c) = a_0, \quad f'(c) = a_1, \quad f''(c) = 2a_2, \quad f'''(c) = (3 \cdot 2)a_3$$

and, for every positive integer n,

$$f^{(n)}(c) = n!a_n, \quad \text{or} \quad a_n = \frac{f^{(n)}(c)}{n!}.$$

We have proved the following result.

Theorem (11.38)

> If f is a function and
>
> $$f(x) = \sum_{n=0}^{\infty} a_n(x-c)^n$$
>
> for all x in an open interval containing c, then
>
> $$f(x) = f(c) + f'(c)(x-c) + \frac{f''(c)}{2!}(x-c)^2 + \cdots$$
>
> $$+ \frac{f^{(n)}(c)}{n!}(x-c)^n + \cdots$$

The series that appears in the conclusion of Theorem (11.38) is called the **Taylor series for $f(x)$ at c**. The special case $c = 0$, stated in the following corollary, is extremely important.

Corollary (11.39)

> If f is a function and $f(x) = \sum a_n x^n$ for all x in an open interval $(-r, r)$, then
>
> $$f(x) = f(0) + f'(0)x + \frac{f''(0)}{2!}x^2 + \cdots + \frac{f^{(n)}(0)}{n!}x^n + \cdots$$

The series in this corollary is called the **Maclaurin series for $f(x)$**. Each example in the preceding section involves a Maclaurin series.

Theorem (11.38) states that *if* a function f is represented by a power series in $x - c$, then the power series *must* be the Taylor series. However, the theorem does not state conditions that guarantee that a power series representation actually exists. We shall now obtain such conditions. Let us begin by noting that the $(n + 1)$st partial sum of the Taylor series stated in (11.38) is the nth-degree Taylor Polynomial $P_n(x)$ of f at c (see Definition (10.9)). Moreover, by Theorem (10.10),

$$P_n(x) = f(x) - R_n(x)$$

where
$$R_n(x) = \frac{f^{(n+1)}(z)}{(n + 1)!} (x - c)^{n+1}$$

for some number z between c and x. In the next theorem we use $R_n(x)$ to specify sufficient conditions for the existence of a power series representation of $f(x)$.

Theorem (11.40)

> If a function f has derivatives of all orders throughout an interval containing c, and if
> $$\lim_{n \to \infty} R_n(x) = 0$$
> for every x in that interval, then $f(x)$ is represented by the Taylor series for $f(x)$ at c.

Proof   The polynomial $P_n(x)$ is a general term for the sequence of partial sums of the Taylor series for $f(x)$ at c. Moreover, since $P_n(x) = f(x) - R_n(x)$,

$$\lim_{n \to \infty} P_n(x) = f(x) - \lim_{n \to \infty} R_n(x) = f(x).$$

Hence the sequence of partial sums converges to $f(x)$. This proves the theorem.   □

In Example 2 of Section 11.6 we proved that the power series $\sum x^n/n!$ is absolutely convergent for all real numbers. Using Theorem (11.14) gives us the following result.

Theorem (11.41)

> For every real number x,
> $$\lim_{n \to \infty} \left| \frac{x^n}{n!} \right| = 0.$$

We shall use Theorem (11.41) in the solution of the following example.

Example 1 Find the Maclaurin series for $\sin x$ and prove that it represents $\sin x$ for every real number x.

Solution   Let us arrange our work as follows:

$$
\begin{array}{ll}
f(x) = \sin x & f(0) = 0 \\
f'(x) = \cos x & f'(0) = 1 \\
f''(x) = -\sin x & f''(0) = 0 \\
f'''(x) = -\cos x & f'''(0) = -1
\end{array}
$$

Successive derivatives follow this same pattern. Substitution in Corollary (11.39) yields the Maclaurin series

$$
\sin x = x - \frac{x^3}{3!} + \frac{x^5}{5!} - \frac{x^7}{7!} + \cdots + (-1)^n \frac{x^{2n+1}}{(2n+1)!} + \cdots
$$

At this stage all we know is that *if* $\sin x$ is represented by a power series in x, then it is given by the preceding series. To prove that the representation is actually true for every number x, let us employ Theorem (11.40). If n is a positive integer, then either

$$
|f^{(n+1)}(x)| = |\cos x| \quad \text{or} \quad |f^{(n+1)}(x)| = |\sin x|.
$$

Hence $|f^{(n+1)}(z)| \leq 1$ for every number z and, using the formula for $R_n(x)$ with $c = 0$,

$$
|R_n(x)| = \frac{|f^{(n+1)}(z)|}{(n+1)!} |x|^{n+1} \leq \frac{|x|^{n+1}}{(n+1)!}.
$$

It follows from Theorem (11.41) and the Sandwich Theorem (11.6) that $\lim_{n \to \infty} |R_n(x)| = 0$. Consequently, $\lim_{n \to \infty} R_n(x) = 0$, and the Maclaurin series representation for $\sin x$ is true for every x. ∎

Example 2 Find the Maclaurin series for $\cos x$.

Solution   We could proceed directly as in Example 1; however, let us obtain the desired series by differentiating the series for $\sin x$ obtained in Example 1. This gives us

$$
\cos x = 1 - \frac{x^2}{2!} + \frac{x^4}{4!} - \frac{x^6}{6!} + \cdots + (-1)^n \frac{x^{2n}}{(2n)!} + \cdots
$$

This series can also be obtained by integrating the terms of the series for $\sin x$. ∎

The Maclaurin series for e^x was obtained in Example 6 of the preceding section by using an indirect technique. We shall next give a direct derivation of this important formula.

Example 3 Find a Maclaurin series that represents e^x for every real number x.

Solution   If $f(x) = e^x$, then $f^{(n)}(x) = e^x$ for every positive integer n. Hence $f^{(n)}(0) = 1$ and substitution in Corollary (11.39) gives us

$$e^x = 1 + x + \frac{x^2}{2!} + \frac{x^3}{3!} + \cdots + \frac{x^n}{n!} + \cdots$$

As in the solution of Example 1, we now employ Theorem (11.40) to prove that this power series representation of e^x is true for every real number x. Using the formula for $R_n(x)$ with $c = 0$,

$$R_n(x) = \frac{f^{(n+1)}(z)}{(n+1)!} x^{n+1} = \frac{e^z}{(n+1)!} x^{n+1}$$

where z is a number between 0 and x. If $0 < x$, then $e^z < e^x$ since the natural exponential function is increasing, and hence, for every positive integer n,

$$0 < R_n(x) < \frac{e^x}{(n+1)!} x^{n+1}.$$

Using Theorem (11.41),

$$\lim_{n \to \infty} \frac{e^x}{(n+1)!} x^{n+1} = e^x \lim_{n \to \infty} \frac{x^{n+1}}{(n+1)!} = 0$$

and, by the Sandwich Theorem (11.6),

$$\lim_{n \to \infty} R_n(x) = 0.$$

If $x < 0$, then also $z < 0$ and hence $e^z < e^0 = 1$. Consequently,

$$0 < |R_n(x)| < \left| \frac{x^{n+1}}{(n+1)!} \right|$$

and we again see that $R_n(x)$ has the limit 0 as n increases without bound. It follows from Theorem (11.40) that the power series representation for e^x is valid for all nonzero x. Finally, note that if $x = 0$, then the series reduces to $e^0 = 1$. ∎

Example 4 Find the Taylor series for $\sin x$ in powers of $x - \pi/6$.

Solution   The derivatives of $f(x) = \sin x$ are listed in Example 1. If we evaluate them at $c = \pi/6$, we obtain

$$f\left(\frac{\pi}{6}\right) = \frac{1}{2}, \quad f'\left(\frac{\pi}{6}\right) = \frac{\sqrt{3}}{2}, \quad f''\left(\frac{\pi}{6}\right) = -\frac{1}{2}, \quad f'''\left(\frac{\pi}{6}\right) = -\frac{\sqrt{3}}{2},$$

and this pattern of four numbers repeats itself indefinitely. Substitution in Theorem (11.38) gives us

$$\sin x = \frac{1}{2} + \frac{\sqrt{3}}{2}\left(x - \frac{\pi}{6}\right) - \frac{1}{2(2!)}\left(x - \frac{\pi}{6}\right)^2 - \frac{\sqrt{3}}{2(3!)}\left(x - \frac{\pi}{6}\right)^3 + \cdots$$

The general term u_n of this series is given by

$$u_n = \begin{cases} (-1)^{n/2}\dfrac{1}{2n!}\left(x-\dfrac{\pi}{6}\right)^n & \text{if } n = 0, 2, 4, 6, \ldots \\[3mm] (-1)^{(n-1)/2}\dfrac{\sqrt{3}}{2n!}\left(x-\dfrac{\pi}{6}\right)^n & \text{if } n = 1, 3, 5, 7, \ldots \end{cases}$$

The proof that the series represents $\sin x$ for all x is similar to that given in Example 1 and is, therefore, omitted. ∎

The next example brings out the fact that a function f may have derivatives of all orders at some number c, but may not have a Taylor series representation at that number. This shows that an additional condition, such as $\lim_{n\to\infty} R_n(x) = 0$, is required to guarantee the existence of a Taylor series.

Example 5   Let f be the function defined by

$$f(x) = \begin{cases} e^{-1/x^2} & \text{if } x \neq 0 \\ 0 & \text{if } x = 0. \end{cases}$$

Show that $f(x)$ cannot be represented by a Maclaurin series.

Solution   Using Theorem (3.7), the derivative of f at 0 is

$$f'(0) = \lim_{x\to 0}\frac{f(x) - f(0)}{x - 0} = \lim_{x\to 0}\frac{e^{-1/x^2}}{x} = \lim_{x\to 0}\frac{(1/x)}{e^{1/x^2}}.$$

The last expression has the indeterminate form ∞/∞. Applying L'Hôpital's Rule we see that

$$f'(0) = \lim_{x\to 0}\frac{(-1/x^2)}{(-2/x^3)e^{1/x^2}} = \lim_{x\to 0}\frac{x}{2e^{1/x^2}} = 0.$$

It can be proved that $f''(0) = 0$, $f'''(0) = 0$ and, in general, $f^{(n)}(0) = 0$ for every positive integer n. According to Corollary (11.39), if $f(x)$ has a Maclaurin series representation, then it is given by

$$f(x) = 0 + 0x + \frac{0}{2!}x^2 + \cdots + \frac{0}{n!}x^n + \cdots$$

which implies that $f(x) = 0$ throughout an interval containing 0. However, this contradicts the definition of f. Consequently, $f(x)$ does not have a Maclaurin series representation. ∎

As a by-product of Example 5, it follows from Theorem (11.40) that for the given function f, $\lim_{x\to\infty} R_n(x) \neq 0$ at $c = 0$.

In previous examples we established some of the Maclaurin series given in the following list. Verifications of the remaining series are left as exercises.

(11.42)   *Important Maclaurin Series*

(a) $\sin x = x - \dfrac{x^3}{3!} + \dfrac{x^5}{5!} - \dfrac{x^7}{7!} + \cdots + (-1)^n \dfrac{x^{2n+1}}{(2n+1)!} + \cdots$ $(-\infty, \infty)$

(b) $\cos x = 1 - \dfrac{x^2}{2!} + \dfrac{x^4}{4!} - \dfrac{x^6}{6!} + \cdots + (-1)^n \dfrac{x^{2n}}{(2n)!} + \cdots$ $(-\infty, \infty)$

(c) $e^x = 1 + x + \dfrac{x^2}{2!} + \dfrac{x^3}{3!} + \cdots + \dfrac{x^n}{n!} + \cdots$ $(-\infty, \infty)$

(d) $\ln (1 + x) = x - \dfrac{x^2}{2} + \dfrac{x^3}{3} - \dfrac{x^4}{4} + \cdots + (-1)^n \dfrac{x^{n+1}}{n+1} + \cdots$ $(-1, 1]$

(e) $\tan^{-1} x = x - \dfrac{x^3}{3} + \dfrac{x^5}{5} - \dfrac{x^7}{7} + \cdots + (-1)^n \dfrac{x^{2n+1}}{2n+1} + \cdots$ $[-1, 1]$

(f) $\sinh x = x + \dfrac{x^3}{3!} + \dfrac{x^5}{5!} + \dfrac{x^7}{7!} + \cdots + \dfrac{x^{2n+1}}{(2n+1)!} + \cdots$ $(-\infty, \infty)$

(g) $\cosh x = 1 + \dfrac{x^2}{2!} + \dfrac{x^4}{4!} + \cdots + \dfrac{x^{2n}}{(2n)!} + \cdots$ $(-\infty, \infty)$

The series in (11.42) can be used to obtain power series representations for other functions. To illustrate, since (c) is true for every x, a power series representation for e^{-x^2} can be found by substituting $-x^2$ for x. This gives us

$$e^{-x^2} = 1 - x^2 + \frac{x^4}{2!} - \cdots + (-1)^n \frac{x^{2n}}{n!} + \cdots$$

Similarly, replacing x by $-x$ in (c) leads to

$$e^{-x} = 1 - x + \frac{x^2}{2!} - \cdots + (-1)^n \frac{x^n}{n!} + \cdots$$

Since $\cosh x = \frac{1}{2}(e^x + e^{-x})$, a power series for $\cosh x$ can be found by adding corresponding terms of the series for e^x and e^{-x}, and then multiplying by $\frac{1}{2}$. This produces

$$\cosh x = 1 + \frac{x^2}{2!} + \frac{x^4}{4!} + \cdots + \frac{x^{2n}}{(2n)!} + \cdots$$

It is interesting to compare this with series (b) for $\cos x$.

Taylor and Maclaurin series may be used to approximate values of functions in a manner similar to those used in the preceding section and in Section 10.5. For example, to find $\sin (0.1)$ we could use (a) of (11.42) to write

$$\sin (0.1) = 0.1 - \frac{0.001}{6} + \frac{0.00001}{120} - \cdots$$

By Theorem (11.27) the error involved in using the sum of the first two terms as an approximation is less than $0.00001/120 \approx 0.00000008$. More generally, we have the following polynomial approximation formula:

$$\sin x \approx x - \frac{x^3}{6}$$

where the error is less than $|x^5|/5!$. (Why?)

The next example illustrates the use of infinite series in approximating values of definite integrals.

Example 6 Approximate $\int_0^1 \sin x^2 \, dx$ to four decimal places.

Solution   From (a) of (11.42),

$$\sin x^2 = x^2 - \frac{x^6}{3!} + \frac{x^{10}}{5!} - \frac{x^{14}}{7!} + \cdots$$

Integrating each term of the series, we obtain

$$\int_0^1 \sin x^2 \, dx = \frac{1}{3} - \frac{1}{42} + \frac{1}{1320} - \frac{1}{75600} + \cdots$$

Summing the first three terms,

$$\int_0^1 \sin x^2 \, dx \approx 0.3103$$

where the error is less than $1/75600 \approx 0.000013$.   ∎

Note that in the preceding example we achieved accuracy to four decimal places by summing only *three* terms of the integrated series for $\sin x^2$. To obtain this degree of accuracy by means of the Trapezoidal Rule or Simpson's Rule, it would be necessary to use an extremely fine partition of the interval [0, 1]. This brings out an important point. For numerical applications, in addition to analyzing a given problem we should also strive to find the most efficient method for computing the answer.

To obtain a Taylor or Maclaurin series representation for a function f by means of (11.38) or (11.39), respectively, it is necessary to find a general formula for $f^{(n)}(x)$ and, in addition, investigate $\lim_{n \to \infty} R_n(x)$. Because of this, our examples were restricted to expressions such as $\sin x$, $\cos x$, and e^x. The method cannot be used if, for example, $f(x)$ equals $\tan x$ or $\sin^{-1} x$, since $f^{(n)}(x)$ becomes very complicated as n increases. Most of the exercises that follow are based on functions whose nth derivatives can be determined easily, or on series representations that we have already established. In more complicated cases we shall restrict our attention to only the first few terms of a Taylor or Maclaurin series representation.

EXERCISES 11.8

In Exercises 1–4 use Corollary (11.39) to find the Maclaurin series for $f(x)$ and prove that the series is valid for all x by means of Theorem (11.40).

1 $f(x) = \cos x$ **2** $f(x) = e^{-x}$

3 $f(x) = e^{2x}$ **4** $f(x) = \cosh x$

In Exercises 5–12 use (11.42) to find a Maclaurin series for $f(x)$ and state the radius of convergence.

5 $f(x) = x^2 e^x$

6 $f(x) = xe^{-2x}$

7 $f(x) = \sinh x$

8 $f(x) = x^2 \sin x$

9 $f(x) = x \sin 3x$

10 $f(x) = \cos x^2$

11 $f(x) = \cos^2 x$ (*Hint*: Use $\cos^2 x = (1 + \cos 2x)/2$.)

12 $f(x) = \sin^2 x$

In Exercises 13–18 find the Taylor Series for $f(x)$ at the indicated number c. (Do not verify that $\lim_{n \to \infty} R_n(x) = 0$.)

13 $f(x) = \sin x; c = \pi/4$

14 $f(x) = \cos x; c = \pi/3$

15 $f(x) = 1/x; c = 2$

16 $f(x) = e^x; c = -3$

17 $f(x) = 10^x; c = 0$

18 $f(x) = \ln (1 + x); c = 0$

19 Find a series representation for e^{2x} in powers of $x + 1$.

20 Find a series representation for $\ln x$ in powers of $x - 1$.

In Exercises 21–26, find the first four terms of the Taylor series for $f(x)$ at the indicated value of c.

21 $f(x) = \sec x; c = \pi/3$ **22** $f(x) = \tan x; c = \pi/4$

23 $f(x) = \sin^{-1} x; c = 1/2$ **24** $f(x) = \csc x; c = 2\pi/3$

25 $f(x) = xe^x; c = -1$ **26** $f(x) = \operatorname{sech} x; c = 0$

In Exercises 27–40, use an infinite series to approximate the given number to four decimal places.

27 $\sqrt{e}$ **28** e^{-1}

29 $\cos 3°$ **30** $\sin 1°$

31 $\tan^{-1} 0.1$ **32** $\ln 1.5$

33 $\sinh 0.5$ **34** $\cosh 0.1$

35 $\displaystyle\int_0^1 e^{-x^2} \, dx$ **36** $\displaystyle\int_0^{1/2} x \cos x^3 \, dx$

37 $\displaystyle\int_0^{0.5} \cos x^2 \, dx$ **38** $\displaystyle\int_0^{0.1} \tan^{-1} x^2 \, dx$

39 $\displaystyle\int_0^{0.25} \sqrt{x} \ln (1 + x) \, dx$ **40** $\displaystyle\int_0^{0.2} \cos \sqrt{x} \, dx$

Approximate the integrals in Exercises 41–46 to four decimal places. Assume that if the integrand is $f(x)$, then $f(0) = \lim_{x \to 0} f(x)$.

41 $\displaystyle\int_0^1 \frac{1 - \cos x}{x^2} \, dx$ **42** $\displaystyle\int_0^1 \frac{\sin x}{x} \, dx$

43 $\displaystyle\int_0^{1/2} \frac{\ln (1 + x)}{x} \, dx$ **44** $\displaystyle\int_0^{0.3} \frac{\tan^{-1} x}{x} \, dx$

45 $\displaystyle\int_{-1}^0 \frac{e^{3x} - 1}{10x} \, dx$ **46** $\displaystyle\int_0^1 \frac{1 - e^{-x}}{x} \, dx$

47 Use (d) of (11.42) to find the Maclaurin series for $\ln [(1 + x)/(1 - x)]$.

48 Use the first five terms of the series in Exercise 47 to calculate $\ln 2$, and compare your answer to the value obtained using a table or calculator.

49 Use (e) of (11.42), with $x = 1$, to represent π as the sum of an infinite series. What accuracy is obtained by using the first five terms of the series to approximate π? Approximately how many terms of the series are required to obtain four decimal-place accuracy for π?

50 Use the identity $\tan^{-1}(1/2) + \tan^{-1}(1/3) = \pi/4$ (see Exercise 30 of Section 8.4) to express π as the sum of two infinite series. Use the first five terms of each series to approximate π and compare the result with that obtained in Exercise 49.

11.9

THE BINOMIAL SERIES

The Binomial Theorem (3.12) states that if k is a positive integer, then for all numbers a and b,

$$(a + b)^k = a^k + ka^{k-1}b + \frac{k(k-1)}{2!} a^{k-2}b^2 + \cdots$$

$$+ \frac{k(k-1)\cdots(k-n+1)}{n!} a^{k-n}b^n + \cdots + b^k.$$

If we let $a = 1$ and $b = x$, then

$$(1 + x)^k = 1 + kx + \frac{k(k-1)}{2!} x^2 + \cdots$$

$$+ \frac{k(k-1)\cdots(k-n+1)}{n!} x^n + \cdots + x^k.$$

If k is not a positive integer (or 0), it is useful to study the power series $\sum a_n x^n$, where $a_0 = 1$ and $a_n = k(k-1)\cdots(k-n+1)/n!$ for $n \geq 1$. This series has the form

$$1 + kx + \frac{k(k-1)}{2!} x^2 + \cdots + \frac{k(k-1)\cdots(k-n+1)}{n!} x^n + \cdots$$

and is called the **Binomial Series**. If k is a nonnegative integer, the series reduces to the finite sum given in the Binomial Theorem. Otherwise the series does not terminate. It is left as an exercise to show that

$$\lim_{n\to\infty} \left| \frac{a_{n+1} x^{n+1}}{a_n x^n} \right| = \lim_{n\to\infty} \left| \frac{k-n}{n+1} \right| |x| = |x|.$$

Hence, by the Ratio Test, the series is absolutely convergent if $|x| < 1$ and is divergent if $|x| > 1$. Thus the Binomial Series represents a function f, where

$$f(x) = \sum_{n=0}^{\infty} a_n x^n \quad \text{if } |x| < 1.$$

It has already been noted that if k is a nonnegative integer, then $f(x) = (1 + x)^k$. We shall now prove that the same is true for *every* real number k. Differentiating the Binomial Series gives us

$$f'(x) = k + k(k-1)x + \cdots + \frac{nk(k-1)\cdots(k-n+1)}{n!} x^{n-1} + \cdots$$

and, therefore,

$$xf'(x) = kx + k(k-1)x^2 + \cdots + \frac{nk(k-1)\cdots(k-n+1)}{n!} x^n + \cdots$$

If we add corresponding terms of the last two power series, then the coefficient of x^n is

$$\frac{(n + 1)k(k - 1)\cdots(k - n)}{(n + 1)!} + \frac{nk(k - 1)\cdots(k - n + 1)}{n!}$$

which simplifies to

$$[(k - n) + n]\frac{k(k - 1)\cdots(k - n + 1)}{n!} = ka_n.$$

Consequently,

$$f'(x) + xf'(x) = \sum_{n=0}^{\infty} ka_n x^n = kf(x)$$

and, therefore,

$$f'(x)(1 + x) - kf(x) = 0.$$

If we define the function g by $g(x) = f(x)/(1 + x)^k$, then

$$g'(x) = \frac{(1 + x)^k f'(x) - f(x)k(1 + x)^{k-1}}{(1 + x)^{2k}}$$

$$= \frac{(1 + x)f'(x) - kf(x)}{(1 + x)^{k+1}} = 0.$$

It follows from Theorem (4.32) that $g(x) = c$ for some constant c, that is,

$$\frac{f(x)}{(1 + x)^k} = c.$$

Since $f(0) = 1$, we see that $c = 1$ and hence $f(x) = (1 + x)^k$, which is what we wished to prove. The next statement summarizes this discussion.

Theorem (11.43)

> If $|x| < 1$, then for every real number k,
>
> $$(1 + x)^k = 1 + kx + \frac{k(k - 1)}{2!}x^2 + \cdots$$
>
> $$+ \frac{k(k - 1)\cdots(k - n + 1)}{n!}x^n + \cdots$$

Example 1 Find a power series representation for $\sqrt[3]{1 + x}$.

Solution Using Theorem (11.43) with $k = \frac{1}{3}$,

$$\sqrt[3]{1 + x} = 1 + \frac{1}{3}x + \frac{\frac{1}{3}(\frac{1}{3} - 1)}{2!}x^2 + \frac{\frac{1}{3}(\frac{1}{3} - 1)(\frac{1}{3} - 2)}{3!}x^3 + \cdots$$

$$+ \frac{\frac{1}{3}(\frac{1}{3} - 1)\cdots(\frac{1}{3} - n + 1)}{n!}x^n + \cdots$$

which may be written

$$\sqrt[3]{1 + x} = 1 + \frac{1}{3}x - \frac{2}{3^2 \cdot 2!}x^2 + \frac{1 \cdot 2 \cdot 5}{3^3 \cdot 3!}x^3 + \cdots$$

$$+ (-1)^{n+1}\frac{1 \cdot 2 \cdots (3n - 4)}{3^n \cdot n!}x^n + \cdots$$

where $|x| < 1$. The formula for the nth term of this series is valid provided $n \geq 2$. ∎

Example 2 Find a power series representation for $\sqrt[3]{1 + x^4}$.

Solution   The desired series can be obtained by substituting x^4 for x in the series of Example 1. Hence, if $|x| < 1$, then

$$\sqrt[3]{1 + x^4} = 1 + \frac{1}{3}x^4 - \frac{2}{3^2 \cdot 2!}x^8 + \cdots$$

$$+ (-1)^{n+1}\frac{1 \cdot 2 \cdots (3n - 4)}{3^n \cdot n!}x^{4n} + \cdots$$ ∎

Example 3 Approximate $\displaystyle\int_0^{0.3} \sqrt[3]{1 + x^4}\, dx$.

Solution   Integrating the terms of the series obtained in Example 2 gives us

$$\int_0^{0.3} \sqrt[3]{1 + x^4}\, dx = 0.3 + 0.000162 - 0.000000243 + \cdots$$

Consequently, the integral may be approximated by 0.300162, which is accurate to six decimal places since the error is less than 0.000000243. (Why?) ∎

The Binomial Series can be used to obtain polynomial approximation formulas for $(1 + x)^k$. To illustrate, if $|x| < 1$, then from Example 1

$$\sqrt[3]{1 + x} \approx 1 + \frac{1}{3}x,$$

where the error involved in this approximation is less than the next term, $x^2/9$.

EXERCISES 11.9

In Exercises 1–6 find a power series representation for $f(x)$ and state the radius of convergence.

1 (a) $f(x) = \sqrt{1 + x}$

   (b) $f(x) = \sqrt{1 - x^3}$

2 (a) $f(x) = \dfrac{1}{\sqrt[3]{1 + x}}$

   (b) $f(x) = \dfrac{1}{\sqrt[3]{1 - x^2}}$

3 $f(x) = (1 + x)^{-3}$

4 $f(x) = x(1 + 2x)^{-2}$

5 $f(x) = \sqrt[3]{8 + x}$

6 $f(x) = (4 + x)^{3/2}$

7 Obtain a power series representation for $\sin^{-1} x$ by using $\sin^{-1} x = \int_0^x 1/\sqrt{1 - t^2}\, dt$. What is the radius of convergence?

8 Obtain a power series representation for $\sinh^{-1} x$ by using $\sinh^{-1} x = \int_0^x 1/\sqrt{1 + t^2}\, dt$. What is the radius of convergence?

Approximate the integrals in Exercises 9 and 10 to three decimal places.

9 $\displaystyle\int_0^{1/2} \sqrt{1 + x^3}\, dx$ (see Exercise 1)

10 $\displaystyle\int_0^{1/2} \dfrac{1}{\sqrt[3]{1 + x^2}}\, dx$ (see Exercise 2)

11.10

REVIEW

Define or discuss each of the following.

1 Infinite sequence

2 Limit of an infinite sequence

3 Theorems on limits of sequences

4 Monotonic sequence

5 Sequence of partial sums of an infinite series

6 Convergent infinite series

7 Divergent infinite series

8 Geometric series

9 The harmonic series

10 The p-series

11 The Integral Test

12 Comparison tests

13 The Alternating Series Test

14 Absolute convergence

15 Conditional convergence

16 The Ratio Test

17 The Root Test

18 Power series

19 Radius of convergence

20 Interval of convergence

21 Differentiation and integration of power series

22 Taylor series

23 Maclaurin series

24 Binomial series

25 Approximations by series

EXERCISES 11.10

In Exercises 1–6 find the limit of the sequence, if it exists.

1 $\left\{ \dfrac{\ln (n^2 + 1)}{n} \right\}$

2 $\{100(0.99)^n\}$

3 $\{10^n/n^{10}\}$

4 $\left\{ \dfrac{1}{n} + (-2)^n \right\}$

5 $\left\{ \dfrac{n}{\sqrt{n + 4}} - \dfrac{n}{\sqrt{n + 9}} \right\}$

6 $\left\{ \left(1 + \dfrac{2}{n}\right)^{2n} \right\}$

In Exercises 7–32, determine whether (a) a positive term series is convergent or divergent, or (b) a series that contains negative terms is absolutely convergent, conditionally convergent, or divergent.

7 $\displaystyle\sum_{n=1}^{\infty} \dfrac{1}{\sqrt[3]{n(n + 1)(n + 2)}}$

8 $\displaystyle\sum_{n=0}^{\infty} \dfrac{(2n + 3)^2}{(n + 1)^3}$

9 $\displaystyle\sum_{n=1}^{\infty} (-2/3)^{n-1}$

10 $\displaystyle\sum_{n=0}^{\infty} \frac{1}{2 + (1/2)^n}$

11 $\displaystyle\sum_{n=1}^{\infty} \frac{3^{2n+1}}{n5^{n-1}}$

12 $\displaystyle\sum_{n=1}^{\infty} \frac{1}{3^n + 2}$

13 $\displaystyle\sum_{n=1}^{\infty} \frac{n!}{\ln(n+1)}$

14 $\displaystyle\sum_{n=1}^{\infty} \frac{n^2 - 1}{n^2 + 1}$

15 $\displaystyle\sum_{n=1}^{\infty} (n^2 + 9)(-2)^{1-n}$

16 $\displaystyle\sum_{n=1}^{\infty} \frac{n + \cos n}{n^3 + 1}$

17 $\displaystyle\sum_{n=1}^{\infty} \frac{e^n}{n^e}$

18 $\displaystyle\sum_{n=1}^{\infty} (-1)^{n-1} \frac{n}{n^2 + 1}$

19 $\displaystyle\sum_{n=1}^{\infty} (-1)^n \frac{1}{\sqrt[n]{n}}$

20 $\displaystyle\sum_{n=2}^{\infty} (-1)^n \frac{(0.9)^n}{\ln n}$

21 $\displaystyle\sum_{n=1}^{\infty} \frac{\sin(n5\pi/3)}{n^{5\pi/3}}$

22 $\displaystyle\sum_{n=2}^{\infty} (-1)^n \frac{\sqrt[3]{n} - 1}{n^2 - 1}$

23 $\displaystyle\sum_{n=1}^{\infty} (-1)^{n-1} \frac{\sqrt{n}}{n + 1}$

24 $\displaystyle\sum_{n=1}^{\infty} (-1)^n \frac{2n + 3}{n!}$

25 $\displaystyle\sum_{n=1}^{\infty} \frac{1 - \cos n}{n^2}$

26 $\displaystyle\frac{2}{1!} - \frac{2 \cdot 4}{2!} + \cdots + (-1)^{n-1} \frac{2 \cdot 4 \cdots (2n)}{n!} + \cdots$

27 $\displaystyle\sum_{n=1}^{\infty} \frac{(2n)^n}{n^{2n}}$

28 $\displaystyle\sum_{n=1}^{\infty} \frac{3^{n-1}}{n^2 + 9}$

29 $\displaystyle\sum_{n=1}^{\infty} \frac{e^{2n}}{(2n - 1)!}$

30 $\displaystyle\sum_{n=1}^{\infty} \left(\frac{1}{3^n} - \frac{5}{\sqrt{n}} \right)$

31 $\displaystyle\sum_{n=2}^{\infty} (-1)^n \frac{\sqrt{\ln n}}{n}$

32 $\displaystyle\sum_{n=1}^{\infty} \frac{\tan^{-1} n}{\sqrt{1 + n^2}}$

Use the Integral Test to determine the convergence or divergence of the series in Exercises 33–38.

33 $\displaystyle\sum_{n=1}^{\infty} \frac{1}{(3n + 2)^3}$

34 $\displaystyle\sum_{n=2}^{\infty} \frac{n}{\sqrt{n^2 - 1}}$

35 $\displaystyle\sum_{n=1}^{\infty} n^{-2} e^{1/n}$

36 $\displaystyle\sum_{n=2}^{\infty} \frac{1}{n(\ln n)^3}$

37 $\displaystyle\sum_{n=1}^{\infty} \frac{10}{\sqrt[3]{n + 8}}$

38 $\displaystyle\sum_{n=5}^{\infty} \frac{1}{n^2 - 4n}$

Approximate the sums of the series in Exercises 39 and 40 to three decimal places.

39 $\displaystyle\sum_{n=1}^{\infty} (-1)^{n-1} \frac{1}{(2n + 1)!}$

40 $\displaystyle\sum_{n=1}^{\infty} (-1)^{n-1} \frac{1}{n^2(n^2 + 1)}$

Find the interval of convergence of the series in Exercises 41–44.

41 $\displaystyle\sum_{n=0}^{\infty} \frac{n + 1}{(-3)^n} x^n$

42 $\displaystyle\sum_{n=0}^{\infty} (-1)^n \frac{4^{2n}}{\sqrt{n + 1}} x^n$

43 $\displaystyle\sum_{n=1}^{\infty} \frac{1}{n \cdot 2^n} (x + 10)^n$

44 $\displaystyle\sum_{n=2}^{\infty} \frac{1}{n(\ln n)^2} (x - 1)^n$

Find the radius of convergence of the series in Exercises 45 and 46.

45 $\displaystyle\sum_{n=0}^{\infty} \frac{(2n)!}{(n!)^2} x^n$

46 $\displaystyle\sum_{n=0}^{\infty} \frac{1}{(n + 5)!} (x + 5)^n$

In Exercises 47–52 find the Maclaurin series for $f(x)$ and state the radius of convergence.

47 $f(x) = (1 - \cos x)/x$ if $x \neq 0$ and $f(0) = 0$

48 $f(x) = xe^{-2x}$

49 $f(x) = \sin x \cos x$

50 $f(x) = \ln(2 + x)$

51 $f(x) = (1 + x)^{2/3}$

52 $f(x) = 1/\sqrt{1 - x^2}$

53 Find a series representation for e^{-x} in powers of $x + 2$.

54 Find a series representation for $\cos x$ in powers of $x - \pi/2$.

55 Find a series representation for $\sqrt{x}$ in powers of $x - 4$.

Use infinite series to approximate the numbers in Exercises 56–60 to three decimal places.

56 $1/\sqrt[3]{3e}$

57 $\displaystyle\int_0^1 x^2 e^{-x^2}\, dx$

58 $\displaystyle\int_0^1 f(x)\, dx$, where $f(x) = (\sin x)/\sqrt{x}$ if $x \neq 0$ and $f(0) = 0$.

59 $\sqrt[5]{1.01}$

60 $e^{-0.25}$

12

TOPICS IN ANALYTIC GEOMETRY

Plane geometry includes the study of figures such as lines, circles, and triangles that lie in a plane. Theorems are proved by reasoning deductively from certain postulates. In *analytic* geometry, plane geometric figures are investigated by introducing a coordinate system and then using equations and formulas of various types. If the study of analytic geometry were to be summarized by means of one statement, perhaps the following would be appropriate: "Given an equation, find its graph and, conversely, given a graph, find its equation." In this chapter we shall apply coordinate methods to several basic plane figures.

12.1

CONIC SECTIONS

Each of the geometric figures to be discussed in this chapter can be obtained by intersecting a double-napped right circular cone with a plane. For this reason they are called **conic sections** or simply **conics**. If, as in (i) of Figure 12.1, the plane cuts entirely across one nappe of the cone and is not per-

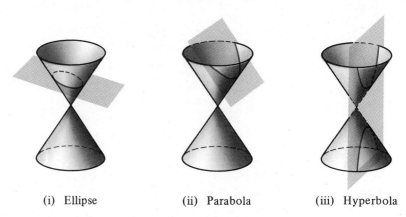

(i) Ellipse   (ii) Parabola   (iii) Hyperbola

Figure 12.1

pendicular to the axis, then the curve of intersection is called an **ellipse**. If the plane is perpendicular to the axis of the cone, a **circle** results. If the plane does not cut across one entire nappe and does not intersect both nappes, as illustrated in (ii) of Figure 12.1, then the curve of intersection is a **parabola**. If the plane cuts through both nappes of the cone, as in (iii) of Figure 12.1, then the resulting figure is called a **hyperbola**.

By changing the position of the plane and the shape of the cone, conics can be made to vary considerably. For certain positions of the plane there result what are called **degenerate conics**. For example, if the plane intersects the cone only at the vertex, then the conic consists of one point. If the axis of the cone lies on the plane, then a pair of intersecting lines is obtained. Finally, if we begin with the parabolic case, as in (ii) of Figure 12.1, and move the plane parallel to its initial position until it coincides with one of the generators of the cone, we obtain a line.

The conic sections were studied extensively by the early Greek mathematicians, who discovered the properties that enable us to define conics in terms of points (foci) and lines (directrices) in the plane of the conic. Reconciliation of these definitions with the previous discussion requires proofs that we shall not go into here.

A remarkable fact about conic sections is that although they were studied thousands of years ago, they are far from obsolete. Indeed, they are important tools for present-day investigations in outer space and for the study of the behavior of atomic particles. It is shown in physics that if a particle moves under the influence of what is called an *inverse square force field*, then its path may be described by means of a conic section. Examples of inverse square fields are gravitational and electromagnetic fields. Planetary orbits are elliptical. If the ellipse is very "flat," the curve resembles the path of a comet. The hyperbola is useful for describing the path of an alpha particle in the electric field of the nucleus of an atom. The interested person can find many other applications of conic sections.

12.2

PARABOLAS

Parabolas are very useful in applications of mathematics to the physical world. For example, it can be shown that if a projectile is fired and it is assumed that it is acted upon only by the force of gravity (that is, air resistance and other outside factors are ignored), then the path of the projectile is parabolic. Properties of parabolas are used in the design of mirrors for telescopes and searchlights, and in the construction of radar antenna. These are only a few of many physical applications.

Definition (12.1)

A **parabola** is the set of all points in a plane equidistant from a fixed point F (the **focus**) and a fixed line l (the **directrix**) in the plane.

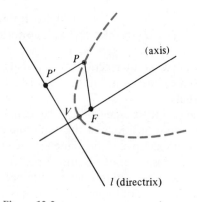

Figure 12.2

We shall assume that F is not on l, for otherwise the parabola degenerates into a line. If P is any point in the plane and P' is the point on l determined by a line through P that is perpendicular to l, then according to Definition (12.1), P is on the parabola if and only if $d(P, F) = d(P, P')$. A typical situation is illustrated in Figure 12.2, where the dashes indicate possible positions of P. The line through F, perpendicular to the directrix, is called the **axis** of the parabola. The point V on the axis, half-way from F to l, is called the **vertex** of the parabola.

To obtain a simple equation for a parabola, let us choose the y-axis along the axis of the parabola, with the origin at the vertex V, as illustrated in Figure 12.3. In this case, the focus F has coordinates $(0, p)$ for some real number $p \neq 0$, and the equation of the directrix is $y = -p$. By the Distance Formula, a point $P(x, y)$ is on the parabola if and only if

$$\sqrt{(x - 0)^2 + (y - p)^2} = \sqrt{(x - x)^2 + (y + p)^2}.$$

Squaring both sides gives us

$$(x - 0)^2 + (y - p)^2 = (y + p)^2,$$

or
$$x^2 + y^2 - 2py + p^2 = y^2 + 2py + p^2.$$

This simplifies to

$$x^2 = 4py.$$

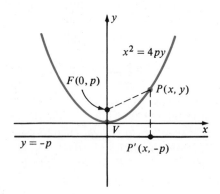

Figure 12.3

We have shown that the coordinates of every point (x, y) on the parabola satisfy $x^2 = 4py$. Conversely, if (x, y) is a solution of this equation, then by reversing the previous steps we see that the point (x, y) is on the parabola. If $p > 0$, the parabola **opens upward**, as in Figure 12.3, whereas if $p < 0$, the parabola **opens downward**. Note that the graph is symmetric with respect to the y-axis, since the solutions of the equation $x^2 = 4py$ are unchanged if $-x$ is substituted for x.

An analogous situation exists if the axis of the parabola is taken along the x-axis. If the vertex is $V(0, 0)$, the focus $F(p, 0)$, and the directrix has equation $x = -p$ (see Figure 12.4), then using the same type of argument we obtain the equation $y^2 = 4px$. If $p > 0$, the parabola opens to the right, whereas if $p < 0$, it opens to the left. In this case the graph is symmetric with respect to the x-axis.

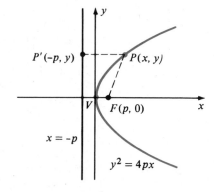

Figure 12.4

The following theorem summarizes our discussion.

Theorem **(12.2)**

> The graph of each of the following equations is a parabola with vertex at the origin and having the indicated focus and directrix.
>
> (i) $x^2 = 4py$: focus $F(0, p)$, directrix $y = -p$.
>
> (ii) $y^2 = 4px$: focus $F(p, 0)$, directrix $x = -p$.

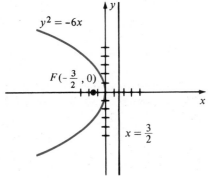

Figure 12.5

Example 1   Find the focus and directrix of the parabola having equation $y^2 = -6x$, and sketch the graph.

Solution   The equation has form (ii) of Theorem (12.2) with $4p = -6$, and hence $p = -\frac{3}{2}$. Thus the focus is $F(p, 0)$, that is, $F(-\frac{3}{2}, 0)$. The equation of the directrix is $x = -p$, or $x = \frac{3}{2}$. The graph is sketched in Figure 12.5. ∎

Example 2   Find an equation of the parabola that has its vertex at the origin, opens upward, and passes through the point $P(-3, 7)$.

Solution   The general form of the equation is given by $x^2 = 4py$ (see (i) of Theorem (12.2)). If P is on the parabola, then $(-3, 7)$ is a solution of the equation. Hence we must have $(-3)^2 = 4p(7)$, or $p = \frac{9}{28}$. Substituting for p in (12.2) leads to the desired equation $x^2 = \frac{9}{7}y$, or $7x^2 = 9y$. ∎

We may use a technique called **translation of axes** to extend the preceding discussion to the case in which the vertex of the parabola is not at the origin. Recall that if a and b are the coordinates of two points A and B, respectively, on a coordinate line l, then the distance between A and B is $d(A, B) = |b - a|$. If we wish to take into account the direction of l, then we use the **directed distance** $\overline{AB}$ from A to B where, by definition,

$$\overline{AB} = b - a.$$

Since $\overline{BA} = a - b$, we have $\overline{AB} = -\overline{BA}$. If the positive direction on l is to the right, then B is to the right of A if and only if $\overline{AB} > 0$, and is to the left of A if and only if $\overline{AB} < 0$. If C is any other point on l with coordinate c, it follows that

$$\overline{AC} = \overline{AB} + \overline{BC}$$

since $c - a = (b - a) + (c - b)$. We shall use this formula in the following discussion to develop formulas for translation of axes.

Suppose that $C(h, k)$ is an arbitrary point in an xy-coordinate plane. Let us introduce a new $x'y'$-coordinate system with origin O' at C such that the x'- and y'-axes are parallel to, and have the same unit lengths and positive directions as, the x- and y-axes, respectively. A typical situation of this type

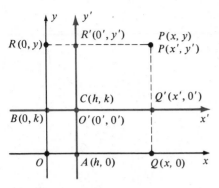

Figure 12.6

is illustrated in Figure 12.6 where, for simplicity, we have placed C in the first quadrant. We shall use primes on letters to denote coordinates of points in the $x'y'$-coordinate system to distinguish them from coordinates with respect to the xy-coordinate system. Thus the point $P(x, y)$ in the xy-system will be denoted by $P(x', y')$ in the $x'y'$-system. If we label projections of P on the various axes as indicated in Figure 12.6, and let A and B denote projections of C on the x- and y-axes, respectively, then using directed distances we obtain

$$x = \overline{OQ} = \overline{OA} + \overline{AQ} = \overline{OA} + \overline{O'Q'} = h + x'$$
$$y = \overline{OR} = \overline{OB} + \overline{BR} = \overline{OB} + \overline{O'R'} = k + y'.$$

This proves the following result.

Theorem (12.3)

> If (x, y) are the coordinates of a point P relative to an xy-coordinate system, and if (x', y') are the coordinates of P relative to an $x'y'$-coordinate system with origin at the point $C(h, k)$ of the xy-system, then
>
> (i) $x = x' + h,$ $y = y' + k$
>
> (ii) $x' = x - h,$ $y' = y - k.$

The formulas in Theorem (12.3) enable us to go from either coordinate system to the other. Their major use is to change the form of equations of graphs. To be specific, if, in the xy-plane, a certain collection of points is the graph of an equation in x and y, then to find an equation in x' and y' that has the same graph in the $x'y'$-plane, we may substitute $x' + h$ for x and $y' + k$ for y in the given equation. Conversely, if a set of points in the $x'y'$-plane is the graph of an equation in x' and y', then to find the corresponding equation in x and y we substitute $x - h$ for x' and $y - k$ for y'.

As a simple illustration of the preceding remarks, the equation

$$(x')^2 + (y')^2 = r^2$$

has, for its graph in the $x'y'$-plane, a circle of radius r with center at the origin O'. Using Theorem (12.3), an equation for this circle in the xy-plane is

$$(x - h)^2 + (y - k)^2 = r^2,$$

which is in agreement with the formula for a circle of radius r with center at $C(h, k)$ in the xy-plane.

As another illustration, we know that

$$(x')^2 = 4py'$$

is an equation of a parabola with a vertex at the origin O' of the $x'y'$-plane. Using (12.3), we see that

$$(x - h)^2 = 4p(y - k)$$

is an equation of the same parabola in the xy-plane. Since the vertex is $V(h, k)$, comparison with (i) of Theorem (12.2) shows that the focus is $F(h, k + p)$ and directrix is $y = k - p$. Similarly, starting with the equation $(y')^2 = 4px'$, and using Theorem (12.3) gives us $(y - k)^2 = 4p(x - h)$. The next theorem summarizes this discussion.

Theorem (12.4)

> The graph of each of the following equations is a parabola with vertex $V(h, k)$ and having the indicated focus and directrix.
>
> (i) $(x - h)^2 = 4p(y - k)$: focus $F(h, k + p)$, directrix $y = k - p$.
>
> (ii) $(y - k)^2 = 4p(x - h)$: focus $F(h + p, k)$, directrix $x = h - p$.

In each case the axis of the parabola is parallel to a coordinate axis. The parabola having equation (i) opens upward or downward, whereas the parabola in (ii) opens to the right or left. Typical graphs are sketched in Figure 12.7.

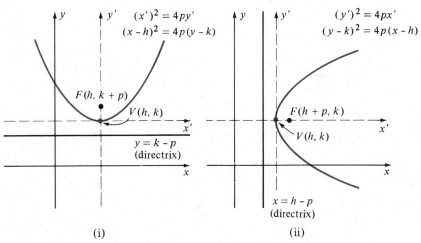

(i)   (ii)

Figure 12.7

Squaring the left side of the equation in (i) of Theorem (12.4) and simplifying leads to an equation of the form

$$y = ax^2 + bx + c$$

where a, b, and c are real numbers. Conversely, given such an equation, we may complete the square in x to arrive at the form in (12.4). This technique will be illustrated in Example 3. Consequently, if $a \neq 0$, then *the graph of $y = ax^2 + bx + c$ is a parabola with a vertical axis*.

Example 3 Discuss and sketch the graph of $y = 2x^2 - 6x + 4$.

Solution　The graph is a parabola with vertical axis. To obtain the proper form we begin by writing the equation as

$$y - 4 = 2x^2 - 6x = 2(x^2 - 3x).$$

Next we complete the square for the expression $x^2 - 3x$. Recall that to complete the square for *any* expression of the form $x^2 + qx$, we add the square of half the coefficient of x, that is, $(q/2)^2$. Thus, for $x^2 - 3x$ we must add $(-\frac{3}{2})^2$, or $\frac{9}{4}$. However, if we add $\frac{9}{4}$ to $x^2 - 3x$ in the equation $y - 4 = 2(x^2 - 3x)$, then because there is a factor 2 outside the parentheses, this amounts to adding $\frac{9}{2}$ to the right side of the equation. Hence we must compensate by adding $\frac{9}{2}$ to the left side. This gives us

$$y - 4 + \frac{9}{2} = 2\left(x^2 - 3x + \frac{9}{4}\right)$$

$$y + \frac{1}{2} = 2\left(x - \frac{3}{2}\right)^2$$

or equivalently

$$\left(x - \frac{3}{2}\right)^2 = \frac{1}{2}\left(y + \frac{1}{2}\right).$$

The last equation is in form (i) of (12.4) with $h = \frac{3}{2}$, $k = -\frac{1}{2}$, and $4p = \frac{1}{2}$, or $p = \frac{1}{8}$. Hence the vertex of the parabola is $V(\frac{3}{2}, -\frac{1}{2})$. The focus is $F(h, k + p)$, or $F(\frac{3}{2}, -\frac{3}{8})$, and the directrix is

$$y = k - p = -\frac{1}{2} - \frac{1}{8}, \quad \text{or} \quad y = -\frac{5}{8}.$$

As an aid to sketching the graph we note that the y-intercept is 4. To find the x-intercepts we solve $2x^2 - 6x + 4 = 0$ or the equivalent equation $(2x - 2)(x - 2) = 0$, obtaining $x = 2$ and $x = 1$. The graph is sketched in Figure 12.8.　∎

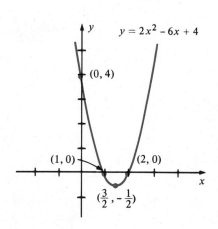

$y = 2x^2 - 6x + 4$

$(0, 4)$

$(1, 0)$　$(2, 0)$

$(\frac{3}{2}, -\frac{1}{2})$

Figure 12.8

The equation in (ii) of (12.4) may be written as

$$x = ay^2 + by + c$$

where a, b, and c are real numbers. Conversely, the preceding equation can be expressed in form (12.4) by completing the square in y as illustrated next, in Example 4. Hence, if $a \neq 0$, the graph of $x = ay^2 + by + c$ is a parabola with a horizontal axis.

Example 4 Discuss and sketch the graph of the equation

$$2x = y^2 + 8y + 22.$$

Solution   By our previous remarks, the graph is a parabola with a horizontal axis. Writing

$$y^2 + 8y = 2x - 22$$

we complete the square on the left by adding 16 to both sides. This gives us

$$y^2 + 8y + 16 = 2x - 6.$$

The last equation may be written

$$(y + 4)^2 = 2(x - 3)$$

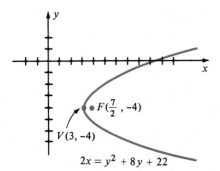

$$F(\tfrac{7}{2}, -4)$$

$$V(3, -4)$$

$$2x = y^2 + 8y + 22$$

Figure 12.9

which is in form (ii) of (12.4) with $h = 3$, $k = -4$, and $4p = 2$, or $p = \tfrac{1}{2}$. Hence the vertex is $V(3, -4)$. Since $p = \tfrac{1}{2} > 0$, the parabola opens to the right with focus at $F(h + p, k)$, that is, $F(\tfrac{7}{2}, -4)$. The equation of the directrix is $x = h - p$, or $x = \tfrac{5}{2}$. The parabola is sketched in Figure 12.9.   ■

Example 5 Find an equation of the parabola with vertex $(4, -1)$, with axis parallel to the y-axis, and that passes through the origin.

Solution   By (12.4) the equation is of the form

$$(x - 4)^2 = 4p(y + 1).$$

If the origin is on the parabola, then $(0, 0)$ is a solution of this equation, and hence $(0 - 4)^2 = 4p(0 + 1)$. Consequently $16 = 4p$ and $p = 4$. The desired equation is, therefore,

$$(x - 4)^2 = 16(y + 1).$$ ■

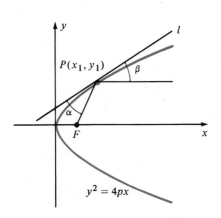

$$P(x_1, y_1)$$

$$\beta$$

$$\alpha$$

$$F$$

$$y^2 = 4px$$

Figure 12.10

We shall conclude this section by pointing out an important property of parabolas. Suppose l is the tangent line at a point $P(x_1, y_1)$ on the graph of $y^2 = 4px$, and let F be the focus. As in Figure 12.10, let α denote the angle between l and the line segment FP, and let β denote the angle between l and the indicated horizontal half-line with endpoint P. In Exercise 34, you are asked to prove that $\alpha = \beta$. This property has many practical applications. For example, the shape of the mirror in a searchlight is obtained by revolving a parabola about its axis. If a light source is placed at F, then by a law of physics a beam of light will be reflected along a line parallel to the axis. The same principle is employed in the construction of mirrors for telescopes or solar ovens, where a beam of light coming toward the parabolic mirror, and parallel to the axis, will be reflected into the focus. Antennas for radar systems and field microphones used at football games also make use of this property.

EXERCISES 12.2

In Exercises 1–16 find the vertex, focus, and directrix of the parabola with the given equation and sketch the graph.

1 $x^2 = -12y$

2 $y^2 = \frac{1}{2}x$

3 $2y^2 = -3x$

4 $x^2 = -3y$

5 $8x^2 = y$

6 $y^2 = -100x$

7 $y^2 - 12 = 12x$

8 $y = 40x - 97 - 4x^2$

9 $y = x^2 - 4x + 2$

10 $y = 8x^2 + 16x + 10$

11 $y^2 - 4y - 2x - 4 = 0$

12 $y^2 + 14y + 4x + 45 = 0$

13 $4x^2 + 40x + y + 106 = 0$

14 $y^2 - 20y + 100 = 6x$

15 $x^2 + 20y = 10$

16 $4x^2 + 4x + 4y + 1 = 0$

17   Describe a method for using a derivative to locate the vertex of a parabola having the equation $y = ax^2 + bx + c$. Use this technique to find the vertices in Exercises 8–10. Describe a similar method for $x = ay^2 + by + c$.

18   Describe how a second derivative may be used to determine whether the parabola $y = ax^2 + bx + c$ opens upward or downward. Illustrate this technique with Exercises 8–10.

In Exercises 19–24 find an equation for the parabola that satisfies the given conditions.

19   Focus $(2, 0)$, directrix $x = -2$

20   Focus $(0, -4)$, directrix $y = 4$

21   Focus $(6, 4)$, directrix $y = -2$

22   Focus $(-3, -2)$, directrix $y = 1$

23   Vertex at the origin, symmetric to the y-axis, and passing through the point $A(2, -3)$

24   Vertex $V(-3, 5)$, axis parallel to the x-axis, and passing through $A(5, 9)$

25   A searchlight reflector is designed so that a cross section through its axis is a parabola and the light source is at the focus. Find the focus if the reflector is 3 ft across at the opening and 1 ft deep.

26   One section of a suspension bridge has its weight uniformly distributed between twin towers that are 400 ft apart and that rise 90 ft above the horizontal roadway (see figure). A cable strung between the tops of the towers has the shape of a parabola and its center point is 10 ft above the roadway. Suppose coordinate axes are introduced as shown in the figure.

(a) Find an equation for the parabola.

(b) Set up an integral that gives the length of the cable.

(c) If nine equispaced vertical cables are used to support the bridge (see figure), find the total length of these supports.

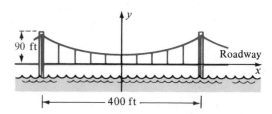

Figure for Exercise 26

27   Find an equation of the parabola with a vertical axis that passes through the points $A(2, 3)$, $B(-1, 6)$, and $C(1, 0)$.

28   Prove that the point on a parabola that is closest to the focus is the vertex.

29   Prove that the equation of the tangent line to the parabola $x^2 = 4py$ at the point $P(x_1, y_1)$ is $x_1x - 2py - 2py_1 = 0$.

30   Prove that the parabola with equation $y = ax^2 + bx + c$ has no points of inflection.

In Exercises 31 and 32 let R denote the region in the xy-plane that is bounded by the given equation.

(a) Find the area of R.

(b) If R is revolved about the y-axis, find the volume of the resulting solid.

(c) If R is revolved about the x-axis, find the volume of the resulting solid.

31   The parabola $x^2 = 4y$ and the line l through the focus that is perpendicular to the axis of the parabola.

32 The graphs of $y^2 = 2x - 6$ and $x = 5$.

33 A vertical gate in a dam has the shape of the parabolic region of Exercise 31. If the line l lies along the surface of the water, find the force exerted on the gate by the water.

34 Prove that $\alpha = \beta$ in Figure 12.10.

35 A line segment that passes through the focus of a parabola and has its endpoints on the parabola is called a **focal chord**. If AB is a focal chord, prove that the tangent lines at A and B are perpendicular.

36 If AB is a focal chord of a parabola (see Exercise 35), prove that the tangent lines at A and B intersect on the directrix.

37 Prove that there is exactly one line of a given slope m that is tangent to the parabola $x^2 = 4py$, and that its equation is $y = mx - pm^2$.

38 Prove that there is exactly one line of a given slope $m \neq 0$ that is tangent to the parabola $y^2 = 4px$, and that its equation is $y = mx + (p/m)$.

39 Prove that any tangent line to a parabola at a point P intersects the axis of the parabola at a point Q such that the midpoint of the line segment PQ is on the line tangent to the parabola at its vertex.

40 Suppose a tangent line to a parabola at a point P intersects the directrix at a point Q. If F is the focus, prove that angle PFQ is a right angle.

12.3

ELLIPSES

An ellipse may be defined as follows.

Definition (12.5)

> An **ellipse** is the set of all points in a plane, the sum of whose distances from two fixed points in the plane (the **foci**) is constant.

It is known that the orbits of planets in the solar system are elliptical, with the sun at one of the foci. This is only one of many important applications of ellipses.

There is an easy way to construct an ellipse on paper. We may begin by inserting two thumbtacks in the paper at points labeled F and F' and fastening the ends of a piece of string to the thumbtacks. If the string is now looped around a pencil and drawn taut at point P, as in Figure 12.11, then moving the pencil and at the same time keeping the string taut, the sum of the distances $d(F, P)$ and $d(F', P)$ is the length of the string, and hence is constant. The pencil will, therefore, trace out an ellipse with foci at F and F'. By varying the positions of F and F' but keeping the length of string fixed, the shape of the ellipse can be made to change considerably. If F and F' are far apart, in the sense that $d(F, F')$ is almost the same as the length of the string, then the ellipse is quite flat. On the other hand, if $d(F, F')$ is close to zero, the ellipse is almost circular. Indeed, if $F = F'$, a circle is obtained.

By introducing suitable coordinate systems we may derive simple equations for ellipses. Let us choose the x-axis as the line through the two foci F and F', with the origin at the midpoint of the segment $F'F$. This point is called the **center** of the ellipse. If F has coordinates $(c, 0)$, where $c > 0$,

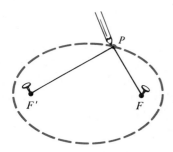

Figure 12.11

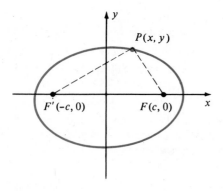

Figure 12.12

then, as shown in Figure 12.12, F' has coordinates $(-c, 0)$ and hence the distance between F and F' is $2c$. Let the constant sum of the distances of P from F and F' be denoted by $2a$, where in order to get points that are not on the x-axis we must have $2a > 2c$, that is, $a > c$. (Why?) By Definition (12.5), $P(x, y)$ is on the ellipse if and only if

$$d(P, F) + d(P, F') = 2a$$

or, by the Distance Formula,

$$\sqrt{(x - c)^2 + (y - 0)^2} + \sqrt{(x + c)^2 + (y - 0)^2} = 2a.$$

Writing the preceding equation as

$$\sqrt{(x - c)^2 + y^2} = 2a - \sqrt{(x + c)^2 + y^2}$$

and squaring both sides, we obtain

$$x^2 - 2cx + c^2 + y^2 = 4a^2 - 4a\sqrt{(x + c)^2 + y^2} + x^2 + 2cx + c^2 + y^2$$

which simplifies to

$$a\sqrt{(x + c)^2 + y^2} = a^2 + cx.$$

Squaring both sides gives us

$$a^2(x^2 + 2cx + c^2 + y^2) = a^4 + 2a^2cx + c^2x^2$$

which may be written in the form

$$x^2(a^2 - c^2) + a^2y^2 = a^2(a^2 - c^2).$$

Dividing both sides by $a^2(a^2 - c^2)$ leads to

$$\frac{x^2}{a^2} + \frac{y^2}{a^2 - c^2} = 1.$$

For convenience, we let

$$b^2 = a^2 - c^2 \quad \text{where } b > 0$$

in the preceding equation, obtaining

$$\frac{x^2}{a^2} + \frac{y^2}{b^2} = 1.$$

Since $c > 0$ and $b^2 = a^2 - c^2$ it follows that $a^2 > b^2$ and hence $a > b$.

We have shown that the coordinates of every point (x, y) on the ellipse in Figure 12.12 satisfy the equation $x^2/a^2 + y^2/b^2 = 1$. Conversely, if (x, y) is a solution of this equation, then by reversing the preceding steps we see that the point (x, y) is on the ellipse.

The x-intercepts may be found by setting $y = 0$. Doing so gives us $x^2/a^2 = 1$, or $x^2 = a^2$, and consequently the x-intercepts are a and $-a$. The corresponding points $V(a, 0)$ and $V'(-a, 0)$ on the graph are called the

vertices of the ellipse, and the line segment $V'V$ is referred to as the **major axis**. Similarly, letting $x = 0$ in the equation of the ellipse, we obtain $y^2/b^2 = 1$, or $y^2 = b^2$. Hence the y-intercepts are b and $-b$. The segment from $M'(0, -b)$ to $M(0, b)$ is called the **minor axis** of the ellipse. Note that the major axis is longer than the minor axis, since $a > b$.

Applying tests for symmetry we see that the ellipse is symmetric to both the x-axis and the y-axis. It is also symmetric with respect to the origin since substitution of $-x$ for x and $-y$ for y does not change the equation.

The preceding discussion may be summarized as follows.

Theorem (12.6)

> The graph of the equation
>
> $$\frac{x^2}{a^2} + \frac{y^2}{b^2} = 1$$
>
> where $a^2 > b^2$ is an ellipse with vertices $(\pm a, 0)$. The endpoints of the minor axis are $(0, \pm b)$. The foci are $(\pm c, 0)$, where $c^2 = a^2 - b^2$.

Example 1 Discuss and sketch the graph of the equation

$$4x^2 + 18y^2 = 36.$$

Solution To obtain the form in Theorem (12.6) we divide both sides of the given equation by 36 and simplify. This leads to

$$\frac{x^2}{9} + \frac{y^2}{2} = 1$$

which is in the proper form with $a^2 = 9$ and $b^2 = 2$. Thus $a = 3, b = \sqrt{2}$, and hence the endpoints of the major axis are $(\pm 3, 0)$ and the endpoints of the minor axis are $(0, \pm\sqrt{2})$. Since

$$c^2 = a^2 - b^2 = 9 - 2 = 7, \quad \text{or} \quad c = \sqrt{7},$$

the foci are $(\pm\sqrt{7}, 0)$. The graph is sketched in Figure 12.13. ∎

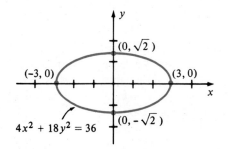

Figure 12.13

Example 2 Find an equation of the ellipse with vertices $(\pm 4, 0)$ and foci $(\pm 2, 0)$.

Solution Using the notation of Theorem (12.6), $a = 4$ and $c = 2$. Since $c^2 = a^2 - b^2$, we see that $b^2 = a^2 - c^2 = 16 - 4 = 12$. This gives us the equation

$$\frac{x^2}{16} + \frac{y^2}{12} = 1.$$

Multiplying both sides by 48 leads to $3x^2 + 4y^2 = 48$. ∎

It is sometimes convenient to choose the major axis of the ellipse along the *y*-axis. If the foci are $(0, \pm c)$, then by the same type of argument used to prove Theorem (12.6) we obtain the following.

Theorem (12.7)

The graph of the equation

$$\frac{x^2}{b^2} + \frac{y^2}{a^2} = 1$$

where $a^2 > b^2$ is an ellipse with vertices $(0, \pm a)$. The endpoints of the minor axis are $(\pm b, 0)$. The foci are $(0, \pm c)$, where $c^2 = a^2 - b^2$.

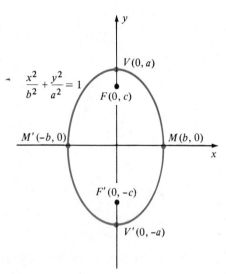

Figure 12.14

A typical graph is sketched in Figure 12.14.

The preceding discussion shows that an equation of an ellipse with center at the origin and foci on a coordinate axis can always be written in the form

$$\frac{x^2}{p} + \frac{y^2}{q} = 1 \quad \text{or} \quad qx^2 + py^2 = pq$$

where p and q are positive. If $p > q$ the major axis lies on the *x*-axis, whereas if $q > p$ the major axis is on the *y*-axis. It is unnecessary to memorize these facts, since in any given problem the major axis can be determined by examining the *x*- and *y*-intercepts.

Example 3 Sketch the graph of the equation $9x^2 + 4y^2 = 25$.

Solution   The graph is an ellipse with center at the origin and foci on one of the coordinate axes. To find the *x*-intercepts, we let $y = 0$, obtaining $9x^2 = 25$, or $x = \pm\frac{5}{3}$. Similarly, to find the *y*-intercepts, we let $x = 0$, obtaining $4y^2 = 25$, or $y = \pm\frac{5}{2}$. This enables us to sketch the ellipse (see Figure 12.15). Since $\frac{5}{3} < \frac{5}{2}$, the major axis is on the *y*-axis. ■

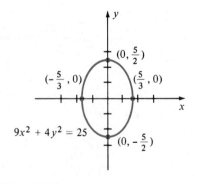

Figure 12.15

Example 4 Find the area of the region bounded by an ellipse whose major and minor axes have lengths $2a$ and $2b$, respectively.

Solution   By (12.6), an equation for the ellipse is $x^2/a^2 + y^2/b^2 = 1$. Solving for *y* gives us

$$y = (\pm b/a)\sqrt{a^2 - x^2}.$$

The graph of the ellipse has the general shape shown in Figure 12.12 and hence, by symmetry, it is sufficient to find the area of the region in the first quadrant and multiply the result by 4. Using 5.12,

$$A = 4\frac{b}{a}\int_0^a \sqrt{a^2 - x^2}\,dx.$$

If we make the trigonometric substitution $x = a \sin \theta$, then

$$\sqrt{a^2 - x^2} = a \cos \theta \quad \text{and} \quad dx = a \cos \theta \, d\theta.$$

Moreover, the values of θ that correspond to $x = 0$ and $x = a$ are $\theta = 0$ and $\theta = \pi/2$, respectively. Consequently,

$$A = 4\frac{b}{a} \int_0^{\pi/2} a^2 \cos^2 \theta \, d\theta = 4ab \int_0^{\pi/2} \frac{1 + \cos 2\theta}{2} \, d\theta$$

$$= 2ab\left[\theta + \tfrac{1}{2}\sin 2\theta\right]_0^{\pi/2} = 2ab[\pi/2] = \pi ab.$$

As a special case, if $b = a$ the ellipse is a circle and $A = \pi a^2$.   ■

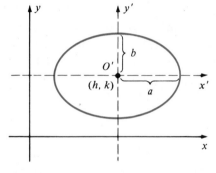

Figure 12.16

By using the translation of axes formulas stated in Theorem (12.3) we can extend our work to an ellipse with center at any point $C(h, k)$ in the xy-plane. For example, since the graph of

$$\frac{(x')^2}{a^2} + \frac{(y')^2}{b^2} = 1$$

is an ellipse with center at O' in an $x'y'$-plane (see Figure 12.16), then according to (12.3), its equation relative to the xy-coordinate system is

$$\frac{(x - h)^2}{a^2} + \frac{(y - k)^2}{b^2} = 1.$$

Squaring the indicated terms in the last equation and simplifying gives us an equation of the form

$$Ax^2 + Cy^2 + Dx + Ey + F = 0$$

where the coefficients are real numbers and A and C are both positive. Conversely, if we start with such an equation, then by completing squares we can obtain a form that displays the center of the ellipse and the lengths of the major and minor axes. This technique is illustrated in the next example.

Example 5 Discuss and sketch the graph of the equation

$$16x^2 + 9y^2 + 64x - 18y - 71 = 0.$$

Solution   We begin by writing the equation in the form

$$16(x^2 + 4x) + 9(y^2 - 2y) = 71.$$

Next, we complete the squares for the expressions within parentheses, obtaining

$$16(x^2 + 4x + 4) + 9(y^2 - 2y + 1) = 71 + 64 + 9.$$

Note that by adding 4 to the expression within the first parentheses we have added 64 to the left side of the equation and hence must compensate by adding 64 to the right side. Similarly, by adding 1 to the expression within the second parentheses, 9 is added to the left side and consequently 9 must also be added to the right side. The last equation may be written

$$16(x + 2)^2 + 9(y - 1)^2 = 144.$$

Dividing by 144 we obtain

$$\frac{(x + 2)^2}{9} + \frac{(y - 1)^2}{16} = 1$$

which is of the form

$$\frac{(x')^2}{9} + \frac{(y')^2}{16} = 1$$

with $x' = x + 2$ and $y' = y - 1$. This corresponds to letting $h = -2$ and $k = 1$ in the translation of axes formulas (12.3). Since the graph of $(x')^2/9 + (y')^2/16 = 1$ is an ellipse with center at the origin O' in the $x'y'$-plane, it follows that the given equation is an ellipse with center $C(-2, 1)$ in the xy-plane and with axes parallel to the coordinate axes. The graph is sketched in Figure 12.17. ■

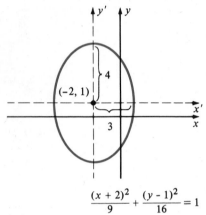

$$\frac{(x + 2)^2}{9} + \frac{(y - 1)^2}{16} = 1$$

Figure 12.17

EXERCISES 12.3

In Exercises 1–14 sketch the graph of the equation and find coordinates of the vertices and foci.

1 $\dfrac{x^2}{9} + \dfrac{y^2}{4} = 1$

2 $\dfrac{x^2}{25} + \dfrac{y^2}{16} = 1$

3 $4x^2 + y^2 = 16$

4 $y^2 + 9x^2 = 9$

5 $5x^2 + 2y^2 = 10$

6 $\frac{1}{2}x^2 + 2y^2 = 8$

7 $4x^2 + 25y^2 = 1$

8 $10y^2 + x^2 = 5$

9 $4x^2 + 9y^2 - 32x - 36y + 64 = 0$

10 $x^2 + 2y^2 + 2x - 20y + 43 = 0$

11 $9x^2 + 16y^2 + 54x - 32y - 47 = 0$

12 $4x^2 + 9y^2 + 24x + 18y + 9 = 0$

13 $25x^2 + 4y^2 - 250x - 16y + 541 = 0$

14 $4x^2 + y^2 = 2y$

In Exercises 15–20 find an equation for the ellipse satisfying the given conditions.

15 Vertices $V(\pm 8, 0)$, foci $F(\pm 5, 0)$

16 Vertices $V(0, \pm 7)$, foci $F(0, \pm 2)$

17 Vertices $V(0, \pm 5)$, length of minor axis 3

18 Foci $F(\pm 3, 0)$, length of minor axis 2

19 Vertices $V(0, \pm 6)$, passing through $(3, 2)$

20 Center at the origin, symmetric with respect to both axes, passing through $A(2, 3)$ and $B(6, 1)$

In Exercises 21 and 22 find the points of intersection of the graphs of the given equations. Sketch both graphs on the same coordinate axes, showing points of intersection.

21 $\begin{cases} x^2 + 4y^2 = 20 \\ x\ + 2y\ = 6 \end{cases}$ **22** $\begin{cases} x^2 + 4y^2 = 36 \\ x^2 +\ y^2 = 12 \end{cases}$

23 An arch of a bridge is semi-elliptical with major axis horizontal. The base of the arch is 30 ft across and the highest part of the arch is 10 ft above the horizontal roadway. Find the height of the arch 6 ft from the center of the base.

24 The **eccentricity** of an ellipse is defined as the ratio $(\sqrt{a^2 - b^2})/a$. If a is fixed and b varies, describe the general shape of the ellipse when the eccentricity is close to 1 and when it is close to zero.

25 Find an equation of the tangent line to the ellipse $5x^2 + 4y^2 = 56$ at the point $P(-2, 3)$.

26 Prove that an equation of the tangent line to the ellipse $x^2/a^2 + y^2/b^2 = 1$ at the point $P(x_1, y_1)$ is

$$xx_1/a^2 + yy_1/b^2 = 1.$$

27 If tangent lines to the ellipse $9x^2 + 4y^2 = 36$ intersect the y-axis at the point $(0, 6)$, find the points of tangency.

28 If tangent lines to the ellipse $b^2x^2 + a^2y^2 = a^2b^2$ intersect the y-axis at the point $(0, d)$, where $d > b$, find the points of tangency.

29 Find the volume of the solid obtained by revolving the region bounded by the ellipse $b^2x^2 + a^2y^2 = a^2b^2$ about the x-axis.

30 Find the volume of the solid obtained by revolving the region bounded by the ellipse $b^2x^2 + a^2y^2 = a^2b^2$ about the y-axis.

31 The base of a solid is a plane region bounded by an ellipse with major and minor axes of lengths 16 and 9, respectively. Find the volume of the solid if every cross section by a plane perpendicular to the major axis is (a) a square; (b) an equilateral triangle.

32 Generalize Exercise 31 to the case where the major and minor axes of the ellipse have lengths $2a$ and $2b$, respectively.

33 Find the dimensions of the rectangle of maximum area that can be inscribed in an ellipse of semi-axes a and b, if two sides of the rectangle are parallel to the major axis.

34 A cylindrical tank whose cross sections are elliptical with axes of lengths 6 ft and 4 ft, respectively, is lying on its side. If the tank is half full of water, find the force exerted by the water on one end of the tank.

35 Let l denote the tangent line at the point P on an ellipse with foci F' and F, as illustrated in the figure. If α is the angle between $F'P$ and l, and if β is the angle between FP and l, prove that $\alpha = \beta$. (This is analogous to the reflective property of the parabola illustrated in Figure 12.10.)

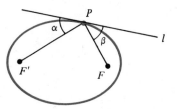

Figure for Exercise 35

36 The base of a right elliptic cone has major and minor axes of lengths $2a$ and $2b$, respectively. Find the volume if the altitude is h.

37 Find the centroid of the semi-elliptical region bounded by the graph of $y = (b/a)\sqrt{a^2 - x^2}$ and the x-axis.

38 The region bounded by the upper half of the ellipse $b^2x^2 + a^2y^2 = a^2b^2$ and the x-axis is revolved about the y-axis. Find the centroid of the resulting solid.

39 A line segment of length $a + b$ moves with its endpoints A and B attached to the coordinate axes, as illustrated in the figure. Prove that if $a \neq b$, then the point P traces an ellipse.

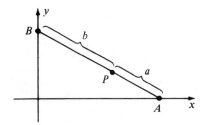

Figure for Exercise 39

40 Consider the ellipse $px^2 + qy^2 = pq$, where $p > 0$ and $q > 0$. Prove that if m is any real number, there are exactly two lines of slope m that are tangent to the ellipse, and that their equations are $y = mx \pm \sqrt{p + qm^2}$.

12.4

HYPERBOLAS

The definition of a hyperbola is similar to that of an ellipse. The only change is that instead of using the *sum* of distances from two fixed points we use the *difference*.

Definition (12.8)

> A **hyperbola** is the set of all points in a plane, the difference of whose distances from two fixed points in the plane (the **foci**) is a positive constant.

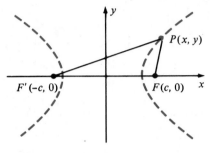

Figure 12.18

To find a simple equation for a hyperbola, we choose a coordinate system with foci at $F(c, 0)$ and $F'(-c, 0)$, and denote the (constant) distance by $2a$. Referring to Figure 12.18, we see that a point $P(x, y)$ is on the hyperbola if and only if either one of the following is true:

$$d(P, F') - d(P, F) = 2a$$
$$d(P, F) - d(P, F') = 2a.$$

For hyperbolas (unlike ellipses) we need $a < c$ in order to obtain points on the hyperbola that are not on the x-axis, for if P is such a point, then from Figure 12.18 we see that

$$d(P, F) < d(F', F) + d(P, F')$$

because the length of one side of a triangle is always less than the sum of the lengths of the other two sides. Similarly,

$$d(P, F') < d(F', F) + d(P, F).$$

Equivalent forms for the previous two inequalities are

$$d(P, F) - d(P, F') < d(F', F)$$
$$d(P, F') - d(P, F) < d(F', F).$$

Since the differences on the left both equal $2a$, and since $d(F', F) = 2c$, the last two inequalities imply that $2a < 2c$, or $a < c$.

The equations $d(P, F') - d(P, F) = 2a$ and $d(P, F) - d(P, F') = 2a$ may be replaced by the single equation

$$|d(P, F) - d(P, F')| = 2a.$$

It then follows from the Distance Formula that an equation of the hyperbola is

$$\left|\sqrt{(x - c)^2 + (y - 0)^2} - \sqrt{(x + c)^2 + (y - 0)^2}\right| = 2a.$$

Employing the type of simplification procedure used to derive an equation for an ellipse, we arrive at the equivalent equation

$$\frac{x^2}{a^2} - \frac{y^2}{c^2 - a^2} = 1.$$

For convenience, let

$$b^2 = c^2 - a^2 \quad \text{where } b > 0$$

in the preceding equation, obtaining

$$\frac{x^2}{a^2} - \frac{y^2}{b^2} = 1.$$

We have shown that the coordinates of every point (x, y) on the hyperbola in Figure 12.18 satisfy the last equation. Conversely, if (x, y) is a solution of that equation, then by reversing the preceding steps we see that the point (x, y) is on the hyperbola.

By the Tests for Symmetry this hyperbola is symmetric with respect to both axes and the origin. The x-intercepts are $\pm a$. The corresponding points $V(a, 0)$ and $V'(-a, 0)$ are called the **vertices**, and the line segment $V'V$ is known as the **transverse axis** of the hyperbola. The origin is called the **center** of the hyperbola. There are no y-intercepts, since the equation $-y^2/b^2 = 1$ has no solutions.

The preceding discussion may be summarized as follows.

Theorem (12.9)

> The graph of the equation
>
> $$\frac{x^2}{a^2} - \frac{y^2}{b^2} = 1$$
>
> is a hyperbola with vertices $(\pm a, 0)$. The foci are $(\pm c, 0)$, where $c^2 = a^2 + b^2$.

If the equation in Theorem (12.9) is solved for y, we obtain

$$y = \pm \frac{b}{a}\sqrt{x^2 - a^2}.$$

Hence there are no points (x, y) on the graph if $x^2 - a^2 < 0$, that is, if $-a < x < a$. However, there *are* points $P(x, y)$ on the graph if $x \geq a$ or $x \leq -a$.

The line $y = (b/a)x$ is an asymptote for the hyperbola in the sense that the distance $d(x)$ between the point $P(x, y)$ on the hyperbola and the corresponding point $P'(x, y_1)$ on the line approaches zero as x increases without bound. To prove this we note that if $x > 0$, then

$$d(x) = \frac{b}{a}x - \frac{b}{a}\sqrt{x^2 - a^2} = \frac{b}{a}(x - \sqrt{x^2 - a^2}).$$

Since

$$x - \sqrt{x^2 - a^2} = \frac{a^2}{x + \sqrt{x^2 - a^2}}$$

we have

$$\lim_{x \to \infty} d(x) = \lim_{x \to \infty} \frac{b}{a} \frac{a^2}{x + \sqrt{x^2 - a^2}} = 0.$$

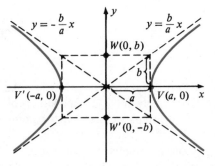

Figure 12.19

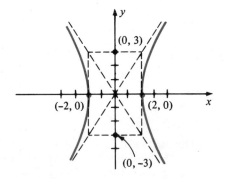

Figure 12.20

Similarly, $d(x)$ has the limit zero as x becomes negatively infinite. It can also be shown that the line $y = (-b/a)x$ is an asymptote for the hyperbola in Theorem (12.9).

The asymptotes serve as excellent guides for sketching the graph. A convenient way to sketch the asymptotes is first to plot the vertices $V(a, 0)$, $V'(-a, 0)$ and the points $W(0, b)$, $W'(0, -b)$ (see Figure 12.19). The line segment $W'W$ of length $2b$ is called the **conjugate axis** of the hyperbola. If horizontal and vertical lines are drawn through the endpoints of the conjugate and transverse axes, respectively, then the diagonals of the resulting rectangle have slopes b/a and $-b/a$. Hence, by extending these diagonals we obtain lines with equations $y = (\pm b/a)x$. The hyperbola is then sketched as in Figure 12.19, using the asymptotes as guides. The two curves that make up the hyperbola are called the **branches** of the hyperbola.

Example 1   Discuss and sketch the graph of the equation $9x^2 - 4y^2 = 36$.

Solution   Dividing both sides by 36, we have

$$\frac{x^2}{4} - \frac{y^2}{9} = 1,$$

which is in the form of Theorem (12.9) with $a^2 = 4$ and $b^2 = 9$. Hence $a = 2$ and $b = 3$. The vertices $(\pm 2, 0)$ and the endpoints $(0, \pm 3)$ of the conjugate axis determine a rectangle whose diagonals (extended) give us the asymptotes. The graph of the equation is sketched in Figure 12.20. The equations of the asymptotes, $y = \pm \frac{3}{2}x$, can be found by referring to the graph or to the equations $y = (\pm b/a)x$. From (12.9) we have $c^2 = a^2 + b^2 = 4 + 9 = 13$, and consequently the foci are $(\pm \sqrt{13}, 0)$.   ∎

The preceding example indicates that for hyperbolas it is not always true that $a < b$, as was the case for ellipses. Indeed, we may have $a < b$, $a > b$, or $a = b$.

Example 2   Find an equation, the foci, and the asymptotes of a hyperbola that has vertices $(\pm 3, 0)$ and passes through the point $P(5, 2)$.

Solution   Substituting $a = 3$ in Theorem (12.9), we obtain the equation

$$\frac{x^2}{9} - \frac{y^2}{b^2} = 1.$$

If $(5, 2)$ is a solution of this equation, then

$$\frac{25}{9} - \frac{4}{b^2} = 1.$$

This gives us $b^2 = \frac{9}{4}$ and hence the desired equation is

$$\frac{x^2}{9} - \frac{4y^2}{9} = 1,$$

or equivalently, $x^2 - 4y^2 = 9$.

From (12.9), $c^2 = a^2 + b^2 = 9 + \frac{9}{4} = \frac{45}{4}$ and, therefore, the foci are $(\pm\frac{3}{2}\sqrt{5}, 0)$. Substituting for b and a in $y = \pm(b/a)x$ and simplifying, we obtain equations $y = \pm\frac{1}{2}x$ for the asymptotes. ∎

If the foci of a hyperbola are the points $(0, \pm c)$ on the y-axis, then by the same type of argument used to prove Theorem (12.9) we obtain the following.

Theorem (12.10)

> The graph of the equation
>
> $$\frac{y^2}{a^2} - \frac{x^2}{b^2} = 1$$
>
> is a hyperbola with vertices $(0, \pm a)$. The foci are $(0, \pm c)$, where $c^2 = a^2 + b^2$.

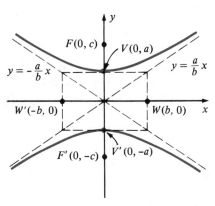

Figure 12.21

In Theorem (12.10) the endpoints of the conjugate axis are $W(b, 0)$ and $W'(-b, 0)$. The asymptotes are found, as before, by using the diagonals of the rectangle determined by these points, the vertices, and lines parallel to the coordinate axes. The graph is sketched in Figure 12.21. The equations of the asymptotes are $y = (\pm a/b)x$. Note the difference between these equations and the equations $y = (\pm b/a)x$ for the asymptotes of the hyperbola given by (12.9).

Example 3 Discuss and sketch the graph of the equation $4y^2 - 2x^2 = 1$.

Solution The form in Theorem (12.10) may be obtained by writing the equation as

$$\frac{y^2}{\frac{1}{4}} - \frac{x^2}{\frac{1}{2}} = 1.$$

Thus $a^2 = \frac{1}{4}$, $b^2 = \frac{1}{2}$, and $c^2 = \frac{1}{4} + \frac{1}{2} = \frac{3}{4}$. Consequently $a = \frac{1}{2}$, $b = \sqrt{2}/2$, and $c = \sqrt{3}/2$. The vertices are $(0, \pm\frac{1}{2})$ and the foci are $(0, \pm\sqrt{3}/2)$. The graph has the general appearance of the graph shown in Figure 12.21. ∎

As was the case for ellipses, we may use translations of axes to generalize our work. The following example illustrates this technique.

Example 4 Discuss and sketch the graph of the equation

$$9x^2 - 4y^2 - 54x - 16y + 29 = 0.$$

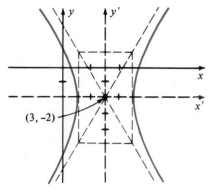

Figure 12.22

Solution   We arrange our work as follows:

$$9(x^2 - 6x) - 4(y^2 + 4y) = -29$$

$$9(x^2 - 6x + 9) - 4(y^2 + 4y + 4) = -29 + 81 - 16$$

$$9(x - 3)^2 - 4(y + 2)^2 = 36$$

$$\frac{(x - 3)^2}{4} - \frac{(y + 2)^2}{9} = 1$$

which is of the form

$$\frac{(x')^2}{4} - \frac{(y')^2}{9} = 1$$

with $x' = x - 3$ and $y' = y + 2$. By translating the x- and y-axes to the new origin $C(3, -2)$ we obtain the sketch shown in Figure 12.22. ∎

The results of the last three sections indicate that the graph of every equation of the form

$$Ax^2 + Cy^2 + Dx + Ey + F = 0$$

is a conic, except for certain degenerate cases in which points, lines, or no graphs are obtained. Although we have only considered special examples, our methods are perfectly general. If A and C are equal and not zero, then the graph, when it exists, is a circle or, in exceptional cases, a point. If A and C are unequal but have the same sign, then by completing squares and properly translating axes we obtain an equation whose graph, when it exists, is an ellipse (or a point). If A and C have opposite signs, an equation of a hyperbola is obtained, or possibly, in the degenerate case, two intersecting straight lines. Finally, if either A or C (but not both) is zero, the graph is a parabola or, in certain cases, a pair of parallel straight lines.

EXERCISES 12.4

In Exercises 1–18 sketch the graph of the equation, find the coordinates of the vertices and foci, and write equations for the asymptotes.

1 $\dfrac{x^2}{9} - \dfrac{y^2}{4} = 1$ **2** $\dfrac{y^2}{49} - \dfrac{x^2}{16} = 1$

3 $\dfrac{y^2}{9} - \dfrac{x^2}{4} = 1$ **4** $\dfrac{x^2}{49} - \dfrac{y^2}{16} = 1$

5 $y^2 - 4x^2 = 16$ **6** $x^2 - 2y^2 = 8$

7 $x^2 - y^2 = 1$ **8** $y^2 - 16x^2 = 1$

9 $x^2 - 5y^2 = 25$ **10** $4y^2 - 4x^2 = 1$

11 $3x^2 - y^2 = -3$ **12** $16x^2 - 36y^2 = 1$

13 $25x^2 - 16y^2 + 250x + 32y + 109 = 0$

14 $y^2 - 4x^2 - 12y - 16x + 16 = 0$

15 $4y^2 - x^2 + 40y - 4x + 60 = 0$

16 $25x^2 - 9y^2 - 100x - 54y + 10 = 0$

17 $9y^2 - x^2 - 36y + 12x - 36 = 0$

18 $4x^2 - y^2 + 32x - 8y + 49 = 0$

In Exercises 19–26 find an equation for the hyperbola satisfying the given conditions.

19 Foci $F(0, \pm 4)$, vertices $V(0, \pm 1)$

20 Foci $F(\pm 8, 0)$, vertices $V(\pm 5, 0)$

21 Foci $F(\pm 5, 0)$, vertices $V(\pm 3, 0)$

22 Foci $F(0, \pm 3)$, vertices $V(0, \pm 2)$

23 Foci $F(0, \pm 5)$, length of conjugate axis 4

24 Vertices $V(\pm 4, 0)$, passing through $P(8, 2)$

25 Vertices $V(\pm 3, 0)$, equations of asymptotes $y = \pm 2x$

26 Foci $F(0, \pm 10)$, equations of asymptotes $y = \pm \frac{1}{3}x$

In Exercises 27 and 28 find the points of intersection of the graphs of the given equations and sketch both graphs on the same coordinate axes, showing points of intersection.

27 $\begin{cases} y^2 - 4x^2 = 16 \\ y \;\; - \;\; x \;\; = \;\; 4 \end{cases}$ **28** $\begin{cases} x^2 - y^2 = 4 \\ y^2 - 3x = 0 \end{cases}$

29 The graphs of the equations

$$\frac{x^2}{a^2} - \frac{y^2}{b^2} = 1 \quad \text{and} \quad \frac{x^2}{a^2} - \frac{y^2}{b^2} = -1$$

are called **conjugate hyperbolas**. Sketch the graphs of both equations on the same coordinate system with $a = 2$ and $b = 5$. Describe the relationship between the two graphs.

30 Find an equation of the hyperbola with foci $(h \pm c, k)$ and vertices $(h \pm a, k)$, where $0 < a < c$ and $c^2 = a^2 + b^2$.

31 Find an equation of the tangent line to the hyperbola $2x^2 - 5y^2 = 3$ at the point $P(-2, 1)$.

32 Prove that an equation of the tangent line to the graph of the hyperbola $x^2/a^2 - y^2/b^2 = 1$ at the point $P(x_1, y_1)$ is

$$\frac{x_1 x}{a^2} - \frac{y_1 y}{b^2} = 1.$$

33 If tangent lines to the hyperbola $9x^2 - y^2 = 36$ intersect the y-axis at the point $(0, 6)$, find the points of tangency.

34 If tangent lines to the hyperbola $b^2x^2 - a^2y^2 = a^2b^2$ intersect the y-axis at the point $(0, d)$, find the points of tangency.

35 Find an equation of a line through $P(2, -1)$ that is tangent to the hyperbola $x^2 - 4y^2 = 16$.

36 Find equations of the tangent lines to the hyperbola $8x^2 - 3y^2 = 48$ that are parallel to the line $2x - y = 10$.

37 The region bounded by the hyperbola $b^2x^2 - a^2y^2 = a^2b^2$ and a vertical line through a focus is revolved about the x-axis. Find the volume of the resulting solid.

38 If the region described in Exercise 37 is revolved about the y-axis, find the volume of the resulting solid.

39 Find the area of the region bounded by the hyperbola $b^2x^2 - a^2y^2 = a^2b^2$ and a line through a focus perpendicular to the transverse axis.

40 The region in the xy-plane that is bounded by the hyperbola $b^2y^2 - a^2x^2 = a^2b^2$ and the line $y = c$, where $c > a$, is revolved about the y-axis. Find the centroid of the resulting solid.

41 Let l denote the tangent line at the point P on the hyperbola illustrated in Figure 12.18. If α is the angle between $F'P$ and l, and if β is the angle between FP and l, prove that $\alpha = \beta$. (This is analogous to the reflective property of the ellipse described in Exercise 35 of Section 12.3.)

42 Let l denote the tangent line at the point P on the hyperbola illustrated in Figure 12.18. If Q and R are the points where l intersects the asymptotes, prove that P is the midpoint of the line segment QR.

12.5

ROTATION OF AXES

The $x'y'$-coordinate system used in a translation of axes may be thought of as having been obtained by moving the origin O of the xy-system to a new position $C(h, k)$ while, at the same time, not changing the positive directions of the axes or the units of length. We shall next introduce a new coordinate system by keeping the origin O fixed and rotating the x- and y-axes about O to another position denoted by x' and y'. A transformation of this type will be referred to as a **rotation of axes**.

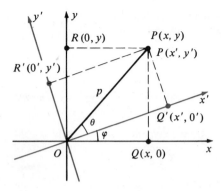

Figure 12.23

Consider a rotation of axes as shown in Figure 12.23 and let φ denote the angle through which the positive x-axis must be rotated in order to coincide with the positive x'-axis. If (x, y) are the coordinates of a point P relative to the xy-plane, then (x', y') will denote its coordinates relative to the new $x'y'$-coordinate system. Let the projections of P on the various axes be denoted as in the figure and let θ denote angle POQ'. If $p = d(O, P)$, then

$$x' = p \cos \theta, \qquad y' = p \sin \theta,$$
$$x = p \cos (\theta + \varphi), \qquad y = p \sin (\theta + \varphi).$$

Applying the addition formulas for the sine and cosine we see that

$$x = p \cos \theta \cos \varphi - p \sin \theta \sin \varphi$$
$$y = p \sin \theta \cos \varphi + p \cos \theta \sin \varphi.$$

Using the fact that $x' = p \cos \theta$ and $y' = p \sin \theta$ gives us (i) of the next theorem. The formulas in (ii) may be obtained from (i) by solving for x' and y'.

Theorem (12.11)

> If the x- and y-axes are rotated about the origin O, through an angle φ, then the coordinates (x, y) and (x', y') of a point P in the two systems are related as follows:
>
> (i) $x = x' \cos \varphi - y' \sin \varphi, \quad y = x' \sin \varphi + y' \cos \varphi$
>
> (ii) $x' = x \cos \varphi + y \sin \varphi, \quad y' = -x \sin \varphi + y \cos \varphi.$

Example 1 The graph of the equation $xy = 1$, or equivalently $y = 1/x$, is sketched in Figure 12.24. If the coordinate axes are rotated through an angle of $45°$, find the equation of the graph relative to the new $x'y'$-coordinate system.

Solution   Letting $\varphi = 45°$ in (i) of Theorem (12.11), we obtain

$$x = x'\left(\frac{\sqrt{2}}{2}\right) - y'\left(\frac{\sqrt{2}}{2}\right) = \left(\frac{\sqrt{2}}{2}\right)(x' - y')$$

$$y = x'\left(\frac{\sqrt{2}}{2}\right) + y'\left(\frac{\sqrt{2}}{2}\right) = \left(\frac{\sqrt{2}}{2}\right)(x' + y').$$

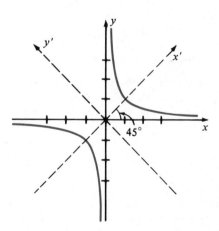

Figure 12.24

Substituting for x and y in the equation $xy = 1$ gives us

$$\left(\frac{\sqrt{2}}{2}\right)(x' - y')\left(\frac{\sqrt{2}}{2}\right)(x' + y') = 1.$$

This reduces to

$$\frac{(x')^2}{2} - \frac{(y')^2}{2} = 1$$

which is an equation of a hyperbola with vertices $(\pm\sqrt{2}, 0)$ on the x'-axis.

Note that the asymptotes for the hyperbola have equations $y' = \pm x'$ in the new system. These correspond to the original x- and y-axes. ∎

Example 1 illustrates a method for eliminating a term of an equation that contains the product xy. This method can be used to transform any equation of the form

$$Ax^2 + Bxy + Cy^2 + Dx + Ey + F = 0$$

where $B \neq 0$, into an equation in x' and y' that contains no $x'y'$ term. Let us prove that this may always be done. If we rotate the axes through an angle φ, then using (i) of Theorem (12.11) to substitute for x and y gives us

$$A(x' \cos \varphi - y' \sin \varphi)^2$$
$$+ B(x' \cos \varphi - y' \sin \varphi)(x' \sin \varphi + y' \cos \varphi)$$
$$+ C(x' \sin \varphi + y' \cos \varphi)^2 + D(x' \cos \varphi - y' \sin \varphi)$$
$$+ E(x' \sin \varphi + y' \cos \varphi) + F = 0.$$

This equation may be written in the form

$$A'(x')^2 + B'x'y' + C'(y')^2 + D'x' + E'y' + F = 0$$

where the coefficient B' of $x'y'$ is given by

$$B' = 2(C - A) \sin \varphi \cos \varphi + B(\cos^2 \varphi - \sin^2 \varphi).$$

To eliminate the $x'y'$ term we must select φ such that

$$2(C - A) \sin \varphi \cos \varphi + B(\cos^2 \varphi - \sin^2 \varphi) = 0.$$

Using double-angle formulas, the last equation may be written

$$(C - A) \sin 2\varphi + B \cos 2\varphi = 0$$

which is equivalent to

$$\cot 2\varphi = \frac{A - C}{B}.$$

This gives us the next result.

Theorem (*12.12*)

> To eliminate the xy-term from the equation
>
> $$Ax^2 + Bxy + Cy^2 + Dx + Ey + F = 0$$
>
> where $B \neq 0$, choose an angle φ such that
>
> $$\cot 2\varphi = \frac{A - C}{B}$$
>
> and use the rotation of axes formulas in Theorem (12.11).

It follows that the graph of any equation in x and y of the type displayed in Theorem (12.12) is a conic, except for certain degenerate cases. (Why?)

Example 2 Discuss and sketch the graph of the equation

$$41x^2 - 24xy + 34y^2 - 25 = 0.$$

Solution   Using the notation of Theorem (12.12) we have $A = 41$, $B = -24$, $C = 34$, and

$$\cot 2\varphi = \frac{41 - 34}{-24} = -\frac{7}{24}.$$

Since $\cot 2\varphi$ is negative we may choose 2φ such that $90° < 2\varphi < 180°$, and consequently $\cos 2\varphi = -\frac{7}{25}$. (Why?) We now use the half-angle formulas to obtain

$$\sin \varphi = \sqrt{\frac{1 - \cos 2\varphi}{2}} = \sqrt{\frac{1 - (-7/25)}{2}} = \frac{4}{5}$$

$$\cos \varphi = \sqrt{\frac{1 + \cos 2\varphi}{2}} = \sqrt{\frac{1 + (-7/25)}{2}} = \frac{3}{5}.$$

It follows that the desired rotation formulas (i) of (12.11) are

$$x = \tfrac{3}{5}x' - \tfrac{4}{5}y', \qquad y = \tfrac{4}{5}x' + \tfrac{3}{5}y'.$$

We leave it to the reader to show that after substituting for x and y in the given equation and simplifying, we obtain the equation

$$(x')^2 + 2(y')^2 = 1.$$

Thus the graph is an ellipse with vertices at $(\pm 1, 0)$ on the x'-axis. Since $\tan \varphi = \sin \varphi / \cos \varphi = (\tfrac{4}{5})/(\tfrac{3}{5}) = \tfrac{4}{3}$, we obtain $\varphi = \tan^{-1}(\tfrac{4}{3})$. To the nearest minute, $\varphi \approx 53°8'$. The graph is sketched in Figure 12.25. ■

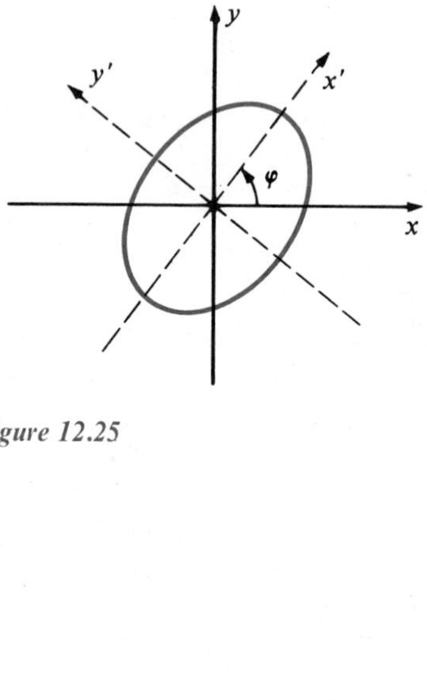

Figure 12.25

In some cases, after elimination of the xy-term in Theorem (12.12), the resulting equation will contain an x' or y' term. It is then necessary to translate the axes of the $x'y'$-coordinate system to obtain the graph. Several problems of this type are contained in the following exercises.

EXERCISES 12.5

After a suitable rotation of axes, describe and sketch the graph of each of the equations in Exercises 1–14.

1   $32x^2 - 72xy + 53y^2 = 80$

2   $7x^2 - 48xy - 7y^2 = 225$

3   $11x^2 + 10\sqrt{3}xy + y^2 = 4$

4   $x^2 - xy + y^2 = 3$

5   $5x^2 - 8xy + 5y^2 = 9$

6   $11x^2 - 10\sqrt{3}xy + y^2 = 20$

7 $16x^2 - 24xy + 9y^2 - 60x - 80y + 100 = 0$

8 $x^2 + 2\sqrt{3}xy + 3y^2 + 8\sqrt{3}x - 8y + 32 = 0$

9 $5x^2 + 6\sqrt{3}xy - y^2 + 8x - 8\sqrt{3}y - 12 = 0$

10 $18x^2 - 48xy + 82y^2 + 6\sqrt{10}x + 2\sqrt{10}y - 80 = 0$

11 $x^2 + 4xy + 4y^2 + 6\sqrt{5}x - 18\sqrt{5}y + 45 = 0$

12 $15x^2 + 20xy - 4\sqrt{5}x + 8\sqrt{5}y - 100 = 0$

13 $40x^2 - 36xy + 25y^2 - 8\sqrt{13}x - 12\sqrt{13}y = 0$

14 $64x^2 - 240xy + 225y^2 + 1020x - 544y = 0$

15 Prove that, except for degenerate cases, the graph of $Ax^2 + Bxy + Cy^2 + Dx + Ey + F = 0$ is

   (a) a parabola if $B^2 - 4AC = 0$.

   (b) an ellipse if $B^2 - 4AC < 0$.

   (c) a hyperbola if $B^2 - 4AC > 0$.

16 Use the results of Exercise 15 to determine the nature of the graphs in Exercises 1–14.

12.6

REVIEW

Define or discuss each of the following.

1 Conic sections

2 Parabola

3 Focus, directrix, vertex, and axis of a parabola

4 Translation of axes

5 Ellipse

6 Major and minor axes of an ellipse

7 Foci and vertices of an ellipse

8 Hyperbola

9 Transverse and conjugate axes of a hyperbola

10 Foci and vertices of a hyperbola

11 Asymptotes of a hyperbola

12 Rotation of axes

EXERCISES 12.6

In Exercises 1–10, find the foci and vertices and sketch the graph of the conic that has the given equation.

1 $y^2 = 64x$

2 $y - 1 = 8(x + 2)^2$

3 $9y^2 = 144 - 16x^2$

4 $9y^2 = 144 + 16x^2$

5 $x^2 - y^2 - 4 = 0$

6 $25x^2 + 36y^2 = 1$

7 $25y = 100 - x^2$

8 $3x^2 + 4y^2 - 18x + 8y + 19 = 0$

9 $x^2 - 9y^2 + 8x + 90y - 210 = 0$

10 $x = 2y^2 + 8y + 3$

Find equations for the conics in Exercises 11–18.

11 The hyperbola with vertices $V(0, \pm 7)$ and endpoints of conjugate axes $(\pm 3, 0)$

12 The parabola with focus $F(-4, 0)$ and directrix $x = 4$

13 The parabola with focus $(0, -10)$ and directrix $y = 10$

14 The parabola with vertex at the origin, symmetric to the x-axis, and passing through the point $(5, -1)$

15 The ellipse with vertices $V(0, \pm 10)$ and foci $F(0, \pm 5)$

16 The hyperbola with foci $F(\pm 10, 0)$ and vertices $V(\pm 5, 0)$

17 The hyperbola with vertices $V(0, \pm 6)$ and asymptotes that have equations $y = \pm 9x$

18 The ellipse with foci $F(\pm 2, 0)$ and passing through the point $(2, \sqrt{2})$

Discuss and sketch the graphs of the equations in Exercises 19–23 after making suitable translations of axes.

19 $4x^2 + 9y^2 + 24x - 36y + 36 = 0$

20 $4x^2 - y^2 - 40x - 8y + 88 = 0$

21 $y^2 - 8x + 8y + 32 = 0$

22 $4x^2 + y^2 - 24x + 4y + 36 = 0$

23 $x^2 - 9y^2 + 8x + 7 = 0$

24 Find equations of the tangent and normal lines to the hyperbola $4x^2 - 9y^2 - 8x + 6y - 36 = 0$ at the point $P(-3, 2)$.

25 Tangent lines to the parabola $y = 2x^2 + 3x + 1$ pass through the point $P(2, -1)$. Find the x-coordinates of the points of tangency.

26 Let R be the region that is bounded by the parabola $9y = 5x^2 - 39x + 90$ and an asymptote of the hyperbola $9y^2 - 4x^2 = 36$.

(a) Find the area of R.

(b) Find the volume of the solid obtained by revolving R about the y-axis.

27 An ellipse having axes of lengths 8 and 4 is revolved about its major axis. Find the volume of the resulting solid.

28 A solid has, for its base, the region in the xy-plane bounded by the graph of $x^2 + y^2 = r^2$. Find the volume of the solid if every cross section by a plane perpendicular to the x-axis is half of an ellipse with one axis always of length c.

After making a suitable rotation of axes, describe and sketch the graphs of the equations in Exercises 29 and 30.

29 $x^2 - 8xy + 16y^2 - 12\sqrt{17}x - 3\sqrt{17}y = 0$

30 $8x^2 + 12xy + 17y^2 - 16\sqrt{5}x - 12\sqrt{5}y = 0$

13

PLANE CURVES AND POLAR COORDINATES

In this chapter parametric and polar equations of curves are discussed.
Applications include tangent lines, areas, arc length, and surfaces of revolution.

13.1

PLANE CURVES

The graph of an equation $y = f(x)$, where f is a function, is often called a *plane curve*. However, to use this as a definition is unnecessarily restrictive, since it rules out most of the conic sections and many other useful graphs. The following statement is satisfactory for most applications.

Definition (13.1)

> A **plane curve** is a set C of ordered pairs of the form $(f(t), g(t))$ where the functions f and g are continuous on an interval I.

For simplicity, we shall often refer to a plane curve as a **curve**. The **graph** of the curve C in Definition (13.1) is the set of all points $P(t) = (f(t), g(t))$ in a rectangular coordinate system that corresponds to the ordered pairs. Each $P(t)$ is referred to as a *point* on the curve. We shall use the term *curve* interchangeably with *graph of a curve*. Sometimes it is convenient to imagine that the point $P(t)$ traces the curve C as t varies through the interval I. This is especially true in applications where t represents time and $P(t)$ is the position of a moving particle at time t.

The graphs of several curves are sketched in Figure 13.1 for the case where I is a closed interval $[a, b]$. If, as in (i) of the figure, $P(a) \neq P(b)$, then $P(a)$ and

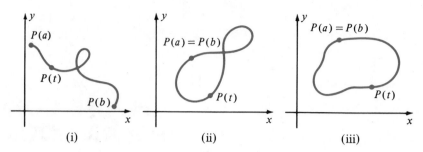

Figure 13.1

$P(b)$ are called the **endpoints** of C. Note that the curve illustrated in (i) intersects itself in the sense that two different values of t give rise to the same point. If $P(a) = P(b)$, as illustrated in (ii) of Figure 13.1, then C is called a **closed curve**. If $P(a) = P(b)$ and C does not intersect itself at any other point, as illustrated in (iii) of the figure, then C is called a **simple closed curve**.

It is convenient to represent curves as indicated in the next definition.

Definition (13.2)

> Let C be the curve consisting of all ordered pairs $(f(t), g(t))$, where f and g are continuous on an interval I. The equations
>
> $$x = f(t), \quad y = g(t),$$
>
> where t is in I, are called **parametric equations** for C, and t is called a **parameter**.

If we are given the parametric representation (13.2), then as t varies through I, the point $P(x, y)$ traces the curve. Sometimes it is possible to eliminate the parameter and obtain an equation for C that involves the variables x and y. The continuity of f and g implies that a small change in t produces a small change in the position of the point $(f(t), g(t))$ on C. This fact may be used to obtain a rough sketch of the graph by plotting many points and connecting them in the order of increasing t, as illustrated in Example 1.

Example 1 Sketch and identify the graph of the curve C given parametrically by

$$x = 2t, \quad y = t^2 - 1, \quad \text{where } -1 \le t \le 2.$$

Solution The parametric equations can be used to tabulate coordinates for points $P(x, y)$ on C as in the following table.

t	-1	$-\frac{1}{2}$	0	$\frac{1}{2}$	1	$\frac{3}{2}$	2
x	-2	-1	0	1	2	3	4
y	0	$-\frac{3}{4}$	-1	$-\frac{3}{4}$	0	$\frac{5}{4}$	3

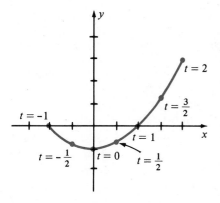

Figure 13.2

Plotting points and using the continuity of f and g gives us the sketch in Figure 13.2.

A precise description of the graph may be obtained by eliminating the parameter. To illustrate, solving the first parametric equation for t we obtain $t = x/2$. If we next substitute for t in the second equation the result is

$$y = \left(\frac{x}{2}\right)^2 - 1, \quad \text{or} \quad y + 1 = \frac{1}{4}x^2.$$

The graph of this equation is a parabola with vertical axis and vertex at the point $(0, -1)$. The curve C is that part of the parabola shown in Figure 13.2.   ∎

Parametric equations of curves are never unique. The reader should verify that the curve C in Example 1 is given by any of the following parametric representations:

$$x = 2t, \quad y = t^2 - 1; \qquad -1 \le t \le 2$$

$$x = t, \quad y = \frac{1}{4}t^2 - 1; \qquad -2 \le t \le 4$$

$$x = t^3, \quad y = \frac{1}{4}t^6 - 1; \quad \sqrt[3]{-2} \le t \le \sqrt[3]{4}.$$

The next example illustrates the fact that it is often useful to eliminate the parameter *before* plotting points.

Example 2 Sketch the graph of the curve C that is given parametrically by

$$x = -2 + t^2, \quad y = 1 + 2t^2$$

where t is in $\mathbb{R}$.

Solution To eliminate the parameter we note, from the first equation, that $t^2 = x + 2$. Substituting for t^2 in the second equation gives us

$$y = 1 + 2(x + 2), \quad \text{or} \quad y - 1 = 2(x + 2).$$

This is an equation of the line of slope 2 through the point $(-2, 1)$, as indicated by the dashes in (i) of Figure 13.3. However, since $t^2 \ge 0$,

$$x = -2 + t^2 \ge -2 \quad \text{and} \quad y = 1 + 2t^2 \ge 1.$$

It follows that the graph of C is that part of the line to the right of $(-2, 1)$, as shown in (ii) of the figure. This fact may also be verified by plotting several points.   ∎

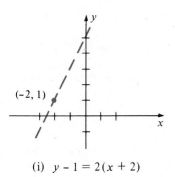

(i) $y - 1 = 2(x + 2)$

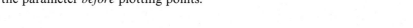

(ii) $x = -2 + t^2, \ y = 1 + 2t^2$

Figure 13.3

Example 3   Describe the graph of the curve C having parametric equations

$$x = \cos t, \quad y = \sin t, \quad 0 \le t \le 2\pi.$$

Solution   We may use the identity $\cos^2 t + \sin^2 t = 1$ to eliminate the parameter. This gives us $x^2 + y^2 = 1$, and hence points on C are on the unit circle with center at the origin. As t increases from 0 to 2π, $P(t)$ starts at the point $A(1, 0)$ and traverses the circle once in the counterclockwise direction. In this example the parameter may be interpreted geometrically as the length of arc from A to P, as illustrated in Figure 8.1. ∎

If a curve C is described by means of an equation $y = f(x)$, where f is a continuous function, then an easy way to obtain parametric equations is to let

$$x = t, \quad y = f(t)$$

where t is in the domain of f. For example, if $y = x^3$, then parametric equations are

$$x = t, \quad y = t^3, \quad \text{where } t \text{ is in } \mathbb{R}.$$

We can use many different substitutions for x, provided that as t varies through some interval, x takes on all values in the domain of f. Thus the graph of $y = x^3$ is also given by

$$x = t^{1/3}, \quad y = t, \quad \text{where } t \text{ is in } \mathbb{R}.$$

Note, however, that the parametric equations

$$x = \sin t, \quad y = \sin^3 t, \quad \text{where } t \text{ is in } \mathbb{R}$$

give only that part of the graph of $y = x^3$ that lies between the points $(-1, -1)$ and $(1, 1)$.

Example 4   Find three different parametric representations for the line of slope m through the point (x_1, y_1).

Solution   By the Point-Slope Form, an equation for the line is

$$y - y_1 = m(x - x_1).$$

If we let $x = t$, then $y - y_1 = m(t - x_1)$ and we obtain the parametric equations

$$x = t, \quad y = y_1 + m(t - x_1), \quad \text{where } t \text{ is in } \mathbb{R}.$$

Another pair of parametric equations results if we let $x - x_1 = t$. In this case $y - y_1 = mt$ and hence the line is given parametrically by

$$x = x_1 + t, \quad y = y_1 + mt, \quad \text{where } t \text{ is in } \mathbb{R}.$$

As a third illustration, if we let $x - x_1 = \tan t$ we obtain

$$x = x_1 + \tan t, \quad y = y_1 + m \tan t, \quad \text{where } -\frac{\pi}{2} < t < \frac{\pi}{2}.$$

There are many other ways to represent the line parametrically. ∎

A curve C is called **smooth** if it has a parametric representation $x = f(t)$, $y = g(t)$ on an interval I such that the derivatives f' and g' are continuous on I and are not simultaneously zero, except possibly at any endpoints of I. The curve C is **piecewise smooth** if the interval I can be partitioned into sub-intervals such that C is smooth on each subinterval. The graph of a smooth curve has no corners. The curves given in Examples 1–4 are smooth. The curve described in the next example is piecewise smooth.

Example 5 The curve traced by a fixed point P on the circumference of a circle as the circle rolls along a straight line in a plane is called a **cycloid**. Find parametric equations for a cycloid.

Solution Suppose the circle has radius a and that it rolls along (and above) the x-axis in the positive direction. If one position of P is the origin, then Figure 13.4 displays part of the curve and a possible position of the circle.

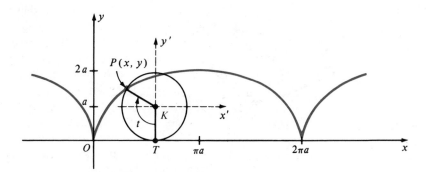

Figure 13.4

Let K denote the center of the circle and T the point of tangency with the x-axis. We introduce a parameter t as the radian measure of angle TKP. Since $\overline{OT}$ is the distance the circle has rolled, $\overline{OT} = at$. Consequently, the coordinates of K are (at, a). If we consider an $x'y'$-coordinate system with origin at $K(at, a)$ and if $P(x', y')$ denotes the point P relative to this system, then by the translation of axes formulas (12.3), with $h = at$ and $k = a$,

$$x = at + x', \quad y = a + y'.$$

If, as in Figure 13.5, θ denotes an angle in standard position on the $x'y'$-system, then $\theta = 3\pi/2 - t$. Hence

$$x' = a \cos \theta = a \cos (3\pi/2 - t) = -a \sin t$$

$$y' = a \sin \theta = a \sin (3\pi/2 - t) = -a \cos t,$$

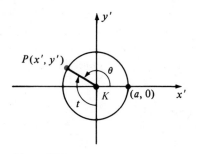

Figure 13.5

and substitution in $x = at + x'$, $y = a + y'$ gives us parametric equations for the cycloid, namely

$$x = a(t - \sin t), \quad y = a(1 - \cos t)$$

where t is in $\mathbb{R}$.   ∎

Differentiating the parametric equations of the cycloid that were obtained in Example 5 gives us

$$\frac{dx}{dt} = a(1 - \cos t), \quad \frac{dy}{dt} = a \sin t.$$

These derivatives are continuous for all t, but are simultaneously 0 if $t = 2\pi n$ for every integer n. Hence the cycloid is not smooth. Note that the points corresponding to $t = 2\pi n$ are the x-intercepts, and the cycloid has a corner at each such point (see Figure 13.4). The graph is piecewise smooth, since it is smooth in each t-interval $[2\pi n, 2\pi(n + 1)]$ where n is an integer.

If $a < 0$, then the graph of $x = a(t - \sin t)$, $y = a(1 - \cos t)$ is the inverted cycloid that results if the circle of Example 5 rolls *below* the x-axis. This curve has a number of important physical properties. In particular, suppose a thin wire passes through two fixed points A and B as illustrated in Figure 13.6, and that the shape of the wire can be changed by bending it in any manner. Suppose further, that a bead is allowed to slide along the wire and the only force acting on the bead is gravity. We now ask which of all the possible paths will allow the bead to slide from A to B in the least amount of time. It is natural to conjecture that the desired path is the straight line segment from A to B; however, this is not the correct answer. The path that requires the least time coincides with the graph of the inverted cycloid with A at the origin and B the lowest point on the curve. The proof of this result will not be given in this text.

To cite another interesting property of this curve, suppose that A is the origin and B is the point with x-coordinate $\pi|a|$, that is, the lowest point on the cycloid occurring in the first arc to the right of A. It can be shown that if the bead is released at *any* point between A and B, the time required for it to reach B is always the same!

Variations of the cycloid occur in practical problems. For example, if a motorcycle wheel rolls along a straight road, then the curve traced by a fixed point on one of the spokes is a cycloid-like curve. In this case, the curve does not have sharp corners, nor does it intersect the road (the x-axis) as does the graph of a cycloid. In like manner, if the wheel of a train rolls along a railroad track, then the curve traced by a fixed point on the circumference of the wheel (which extends below the track) contains loops at regular intervals (see Exercise 38). Several other cycloids are defined in Exercises 33 and 34.

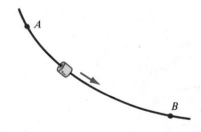

Figure 13.6

EXERCISES 13.1

In Exercises 1–20, (a) sketch the graph of the curve C having the indicated parametric equations, and (b) find a rectangular equation of a graph that contains the points on C.

1 $x = t - 2, y = 2t + 3; 0 \le t \le 5$

2 $x = 1 - 2t, y = 1 + t; -1 \le t \le 4$

3 $x = t^2 + 1, y = t^2 - 1; -2 \le t \le 2$

4 $x = t^3 + 1, y = t^3 - 1; -2 \le t \le 2$

5 $x = 4t^2 - 5, y = 2t + 3; t$ in $\mathbb{R}$

6 $x = t^3, y = t^2; t$ in $\mathbb{R}$

7 $x = e^t, y = e^{-2t}; t$ in $\mathbb{R}$

8 $x = \sqrt{t}, y = 3t + 4; t \ge 0$

9 $x = 2 \sin t, y = 3 \cos t; 0 \le t \le 2\pi$

10 $x = \cos t - 2, y = \sin t + 3; 0 \le t \le 2\pi$

11 $x = \sec t, y = \tan t; -\pi/2 < t < \pi/2$

12 $x = \cos 2t, y = \sin t; -\pi \le t \le \pi$

13 $x = t^2, y = 2 \ln t; t > 0$

14 $x = \cos^3 t, y = \sin^3 t; 0 \le t \le 2\pi$

15 $x = \sin t, y = \csc t; 0 < t \le \pi/2$

16 $x = e^t, y = e^{-t}; t$ in $\mathbb{R}$

17 $x = \cosh t, y = \sinh t; t$ in $\mathbb{R}$

18 $x = 3 \cosh t, y = 2 \sinh t; t$ in $\mathbb{R}$

19 $x = t, y = \sqrt{t^2 - 1}; |t| \ge 1$

20 $x = -2\sqrt{1 - t^2}, y = t; |t| \le 1$

In Exercises 21–24, sketch the graph of the curve C having the indicated parametric equations.

21 $x = t, y = \sqrt{t^2 - 2t + 1}; 0 \le t \le 4$

22 $x = 2t, y = 8t^3; -1 \le t \le 1$

23 $x = (t + 1)^3, y = (t + 2)^2; 0 \le t \le 2$

24 $x = \tan t, y = 1; -\pi/2 < t < \pi/2$

In Exercises 25 and 26 curves C_1, C_2, C_3, and C_4 are given parametrically, where t is in $\mathbb{R}$. Sketch the graphs of C_1, C_2, C_3, C_4, and discuss their similarities and differences.

25 $C_1: x = t^2, y = t$
$C_2: x = t^4, y = t^2$
$C_3: x = \sin^2 t, y = \sin t$
$C_4: x = e^{2t}, y = -e^t$

26 $C_1: x = t, y = 1 - t$
$C_2: x = 1 - t^2, y = t^2$
$C_3: x = \cos^2 t, y = \sin^2 t$
$C_4: x = \ln t - t, y = 1 + t - \ln t$

The parametric representations in Exercises 27 and 28 give the position of a point $P(x, y)$, where t represents time. Describe the motion of the point during the indicated time interval.

27 (a) $x = \cos t, y = \sin t; 0 \le t \le \pi$
(b) $x = \sin t, y = \cos t; 0 \le t \le \pi$
(c) $x = t, y = \sqrt{1 - t^2}; -1 \le t \le 1$

28 (a) $x = t^2, y = 1 - t^2; 0 \le t \le 1$
(b) $x = 1 - \ln t, y = \ln t; 1 \le t \le e$
(c) $x = \cos^2 t, y = \sin^2 t; 0 \le t \le 2\pi$

29 If $P_1(x_1, y_1)$ and $P_2(x_2, y_2)$ are distinct points, show that

$$x = (x_2 - x_1)t + x_1, \quad y = (y_2 - y_1)t + y_1,$$

where t is in $\mathbb{R}$, are parametric equations of the line l through P_1 and P_2. Find three other pairs of parametric equations for the line l. Show that there is an infinite number of different pairs of parametric equations for l.

30 What is the difference between the graph of the hyperbola $b^2x^2 - a^2y^2 = a^2b^2$ and the graph of $x = a \cosh t, y = b \sinh t$, where t is in $\mathbb{R}$?

31 Show that

$$x = a \cos t + h, \quad y = b \sin t + k,$$

where $0 \le t \le 2\pi$ are parametric equations of an ellipse with center at the point (h, k) and semi-axes of lengths a and b.

32 Find parametric equations for the parabola that has (a) vertex $V(0, 0)$ and focus $F(0, p)$; (b) vertex $V(h, k)$ and focus $F(h, k + p)$.

33 A circle C of radius b rolls on the inside of a second circle having equation $x^2 + y^2 = a^2$, where $b < a$. Let P be a fixed point on C and let the initial position of P be $A(a, 0)$ as shown in the figure. If the parameter t is the angle from the positive x-axis to the line segment from O to the center of C, show that parametric equations for the curve traced by P (called a **hypocycloid**) are

$$x = (a - b) \cos t + b \cos \frac{a - b}{b} t$$

$$y = (a - b) \sin t - b \sin \frac{a - b}{b} t$$

where $0 \le t \le 2\pi$. If $b = a/4$ show that

$$x = a \cos^3 t, \quad y = a \sin^3 t$$

and sketch the graph of the curve.

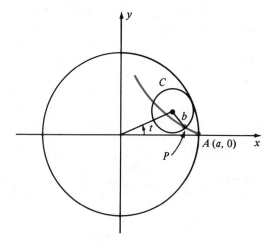

Figure for Exercise 33

34 If the circle C of Exercise 33 rolls on the outside of the second circle (see figure) then the curve traced by P is called an **epicycloid**. Show that parametric equations for this curve are

$$x = (a + b) \cos t - b \cos \frac{a + b}{b} t$$

$$y = (a + b) \sin t - b \sin \frac{a + b}{b} t.$$

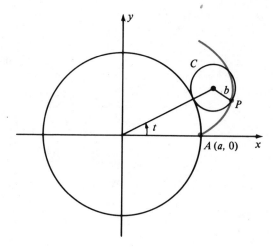

Figure for Exercise 34

35 If $b = a/3$ in Exercise 34, find parametric equations for the epicycloid and sketch the graph.

36 Problem 17 of the 1982 SAT examination was as follows: The radius of circle A is one-third that of circle B. How many revolutions will circle A make as it rolls around circle B until it reaches its starting point? Use Exercise 35 to show that the answer is 4. (To the embarassment of the Educational Testing Service of Princeton, NJ, this was not one of the choices given as an answer.)

37 If a string is unwound from around a circle and is kept taut in the plane of the circle, then a fixed point P on the string will trace a curve called the **involute of the circle**. If the circle is chosen as in Exercise 33 and the parameter t is the angle from the positive x-axis to the point of tangency of the string, show that parametric equations for the involute are

$$x = a(\cos t + t \sin t), \quad y = a(\sin t - t \cos t).$$

38 Generalize the cycloid of Example 5 to the case where P is any point on a fixed line through the center C of the circle. If $b = d(C, P)$, derive the parametric equations

$$x = at - b \sin t, \quad y = a - b \cos t.$$

Sketch a typical graph if $b < a$ (a **curtate cycloid**) and if $b > a$ (a **prolate cycloid**). The term **trochoid** is sometimes used for either of these curves.

13.2

TANGENT LINES
TO CURVES

In Example 1 of the preceding section we saw that the curve C having parametric equations

$$x = 2t, \quad y = t^2 - 1, \quad \text{where } -1 \leq t \leq 2$$

can also be described by using an equation of the form $y = k(x)$, where k is a function defined on a suitable interval. As a matter of fact we may take

$$y = k(x) = \frac{1}{4}x^2 - 1, \quad \text{where } -2 \leq x \leq 4.$$

It follows that the slope of the tangent line at any point $P(x, y)$ on C is

$$k'(x) = \frac{1}{2}x \quad \text{or} \quad k'(x) = \frac{1}{2}(2t) = t.$$

We shall now introduce a technique for finding the slope directly from the parametric equations.

Let $x = f(t)$, $y = g(t)$ be parametric equations of a curve C and suppose C can also be represented by a rectangular equation of the form $y = k(x)$, where k is a function. Let us also assume that f, g, and k are differentiable throughout their domains. If $P(f(t), g(t))$ is a point on C, then the coordinates satisfy the equation $y = k(x)$, that is,

$$g(t) = k(f(t)).$$

Applying the Chain Rule,

$$g'(t) = k'(f(t))f'(t) = k'(x)f'(t)$$

and hence the slope of the tangent line at P is

$$k'(x) = \frac{g'(t)}{f'(t)}, \quad \text{provided } f'(t) \neq 0.$$

If $f'(t) = 0$ and $g'(t) \neq 0$ for some t, then C has a vertical tangent line at the point corresponding to t. If $f'(t)$ and $g'(t)$ are simultaneously zero, then further information is required.

The conditions imposed in the preceding discussion are fulfilled if C is smooth and $f'(t) \neq 0$ for all t in a closed interval $[a, b]$. In this case either $f'(t) > 0$ or $f'(t) < 0$ throughout $[a, b]$ (Why?), and it follows from Theorem (7.28) or Theorem (7.29) that f has an inverse function f^{-1}. Since $x = f(t)$ we may write $t = f^{-1}(x)$, where x is in the interval $[f(a), f(b)]$. Hence

$$y = g(t) = g(f^{-1}(x)) = k(x)$$

where k is the composite function $g \circ f^{-1}$.

An easy way to remember the formula for $k'(x)$ is to use the notation given in the next theorem.

Theorem (13.3)

> If a smooth curve C is given parametrically by $x = f(t)$, $y = g(t)$, then the slope dy/dx of the tangent line to C at $P(x, y)$ is
>
> $$\frac{dy}{dx} = \frac{dy/dt}{dx/dt}, \quad \text{provided } \frac{dx}{dt} \neq 0.$$

Example 1 Find the slopes of the tangent and normal lines at the point $P(t)$ on the curve having parametric equations $x = 2t$, $y = t^2 - 1$, where $-1 \leq t \leq 2$.

Solution The curve is the same as that considered in Example 1 of the preceding section (see Figure 13.2). Using Theorem (13.3) with $x = 2t$ and $y = t^2 - 1$, the slope of the tangent line at $P(t)$ is

$$\frac{dy}{dx} = \frac{dy/dt}{dx/dt} = \frac{2t}{2} = t.$$

Note that this agrees with the discussion at the beginning of this section. The slope of the normal line is $-1/t$, provided $t \neq 0$. ∎

Example 2 If C has parametric equations $x = t^3 - 3t$, $y = t^2 - 5t - 1$, where t is in $\mathbb{R}$, find an equation of the tangent line to C at the point corresponding to $t = 2$. For what values of t is the tangent line horizontal or vertical?

Solution Applying (13.3),

$$\frac{dy}{dx} = \frac{dy/dt}{dx/dt} = \frac{2t - 5}{3t^2 - 3}.$$

Thus at the point P corresponding to $t = 2$, the slope m of the tangent line is

$$m = \frac{2(2) - 5}{3(2^2) - 3} = -\frac{1}{9}.$$

Substituting $t = 2$ in the parametric equations gives us the coordinates $(2, -7)$ of P. Hence an equation of the tangent line is

$$y + 7 = -\frac{1}{9}(x - 2)$$

or, equivalently,

$$x + 9y + 61 = 0.$$

The tangent line is horizontal if $dy/dt = 0$, that is, if $2t - 5 = 0$ or $t = 5/2$.

The tangent line is vertical if $3t^2 - 3 = 0$. (Why?) Thus there are vertical tangent lines at the points corresponding to $t = 1$ and $t = -1$. ∎

If the graph of a curve C coincides with the graph of $y = k(x)$, then

$$y' = \frac{dy}{dx} = \frac{dy/dt}{dx/dt}.$$

If y' is a differentiable function of t we can find d^2y/dx^2 by applying Theorem (13.3) to y' as follows:

$$\frac{d^2y}{dx^2} = \frac{d}{dx}(y') = \frac{dy'/dt}{dx/dt}.$$

Example 3 Suppose C has parametric equations

$$x = e^{-t}, \quad y = e^{2t}, \quad \text{where } t \text{ is in } \mathbb{R}.$$

(a) Find d^2y/dx^2 directly from the parametric equations.

(b) Define a function k that has the same graph as C and check the answer to part (a).

(c) Discuss the concavity of C.

Solution

(a) As in the preceding discussion,

$$y' = \frac{dy}{dx} = \frac{dy/dt}{dx/dt} = \frac{2e^{2t}}{-e^{-t}} = -2e^{3t},$$

and

$$\frac{d^2y}{dx^2} = \frac{dy'}{dx} = \frac{dy'/dt}{dx/dt} = \frac{-6e^{3t}}{-e^{-t}} = 6e^{4t}.$$

(b) Using the fact that $x = e^{-t} = 1/e^t$, we obtain $e^t = 1/x = x^{-1}$. Since $y = e^{2t} = (e^t)^2$, it follows that C is the graph of the equation

$$y = (x^{-1})^2 = x^{-2}, \quad \text{where } x > 0.$$

The restriction $x > 0$ is necessary because $e^{-t} > 0$ for all t. Thus a function k whose graph coincides with the graph of C may be defined by

$$k(x) = x^{-2}, \quad \text{where } x > 0.$$

Since $k'(x) = -2x^{-3}$,

$$k''(x) = 6x^{-4} = 6(e^{-t})^{-4} = 6e^{4t}$$

which is in agreement with part (a).

(c) Since $d^2y/dx^2 = 6e^{4t} > 0$ for all t, the curve C is concave upward at every point. ∎

EXERCISES 13.2

Find the slopes of the tangent lines at the point corresponding to $t = 1$ on the curves defined in Exercises 1–10. (See the graphs in Exercises 1–10 of Section 13.1).

1 $x = t - 2, y = 2t + 3; 0 \le t \le 5$

2 $x = 1 - 2t, y = 1 + t; -1 \le t \le 4$

3 $x = t^2 + 1, y = t^2 - 1; -2 \le t \le 2$

4 $x = t^3 + 1, y = t^3 - 1; -2 \le t \le 2$

5 $x = 4t^2 - 5, y = 2t + 3; t$ in $\mathbb{R}$

6 $x = t^3, y = t^2; t$ in $\mathbb{R}$

7 $x = e^t, y = e^{-2t}; t$ in $\mathbb{R}$

8 $x = \sqrt{t}, y = 3t + 4; t \ge 0$

9 $x = 2 \sin t, y = 3 \cos t; 0 \le t \le 2\pi$

10 $x = \cos t - 2, y = \sin t + 3; 0 \le t \le 2\pi$

11 Let l be the line with parametric equations $x = t/2$, $y = 2t - 5$. If a curve C is given by $x = t^4 - 7, y = 8t + 3$, where t is in $\mathbb{R}$, find a rectangular equation of the normal line to C that is parallel to l.

12 If C has parametric equations $x = 6t + 1$, $y = t^3 - 2t$, where t is in $\mathbb{R}$, find the points on C at which the tangent line is perpendicular to the line $3x + 5y - 8 = 0$.

13 If C is given parametrically by $x = 4t^2$, $y = 2t - 5$, where t is in $\mathbb{R}$, find a rectangular equation of the line through the point $(4, 1)$ that is tangent to C.

14 Find a rectangular equation of the line through the origin that is normal to the curve having parametric equations $x = 3t^2 - 2, y = 2t^3$, where t is in $\mathbb{R}$.

In Exercises 15–18 find the points on the indicated curve at which the tangent line is either horizontal or vertical. Sketch the graph, showing the horizontal and vertical tangent lines. Find d^2y/dx^2 in each case.

15 $x = 4t^2, y = t^3 - 12t; t$ in $\mathbb{R}$

16 $x = t^3 + 1, y = t^2 - 2t; t$ in $\mathbb{R}$

17 $x = 3t^2 - 6t, y = \sqrt{t}; t \ge 0$

18 $x = \sqrt[3]{t}, y = \sqrt[3]{t} - t; t$ in $\mathbb{R}$

19 What is the slope of the tangent line at the point $P(t)$ on the cycloid $x = a(t - \sin t), y = a(1 - \cos t)$, where t is in $\mathbb{R}$ and $a > 0$? At what points is the tangent line horizontal or vertical? Where does the slope of the tangent line equal 1? Find d^2y/dx^2 and discuss concavity if $0 < t < 2\pi$.

20 Answer the questions of Exercise 19 for the hypocycloid $x = a \cos^3 t, y = a \sin^3 t$ where $0 \le t \le 2\pi$ and $a > 0$.

13.3

POLAR COORDINATE SYSTEMS

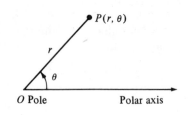

Figure 13.7

We have previously specified points in a plane in terms of rectangular coordinates, using the ordered pair (a, b) to denote the point whose directed distances from the x- and y-axes are b and a, respectively. Another important method for representing points is by means of *polar coordinates*. To introduce a system of polar coordinates in a plane we begin with a fixed point O (called the **origin**, or **pole**) and a directed half-line (called the **polar axis**) with endpoint O. Next we consider any point P in the plane different from O. If, as illustrated in Figure 13.7, $r = d(O, P)$ and θ denotes the measure of any angle determined by the polar axis and OP, then r and θ are called **polar coordinates** of P and the symbols (r, θ) or $P(r, \theta)$ are used to denote P. As usual θ is considered positive if the angle is generated by a counterclockwise rotation of the polar axis and negative if the rotation is clockwise. Either radians or degrees may be used for the measure of θ.

If the polar axis is the initial side, then there are many angles with the same terminal side. Hence the polar coordinates of a point are not unique. For example, $(3, \pi/4)$, $(3, 9\pi/4)$, and $(3, -7\pi/4)$ all represent the same point (see Figure 13.8). We shall also allow r to be negative. In this event, instead of measuring $|r|$ units along the terminal side of the angle θ, we measure along

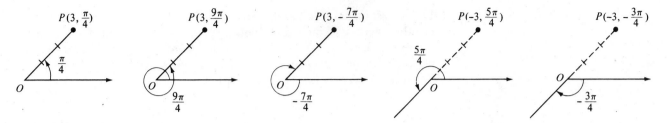

Figure 13.8

the half-line with endpoint O that has direction opposite to that of the terminal side. The points corresponding to the pairs $(-3, 5\pi/4)$ and $(-3, -3\pi/4)$ are also plotted in Figure 13.8.

Finally, we agree that the pole O has polar coordinates $(0, \theta)$ for *any* θ. An assignment of ordered pairs of the form (r, θ) to points in a plane will be referred to as a **polar coordinate system** and the plane will be called an *rθ-plane*.

A **polar equation** is an equation in r and θ. A **solution** of a polar equation is an ordered pair (a, b) that leads to equality if a is substituted for r and b for θ. The **graph** of a polar equation is the set of all points (in an $r\theta$-plane) that correspond to the solutions.

Example 1 Sketch the graph of the polar equation $r = 4 \sin \theta$.

Solution The following table contains some solutions of the equation.

θ	0	$\pi/6$	$\pi/4$	$\pi/3$	$\pi/2$	$2\pi/3$	$3\pi/4$	$5\pi/6$	π
r	0	2	$2\sqrt{2}$	$2\sqrt{3}$	4	$2\sqrt{3}$	$2\sqrt{2}$	2	0

In rectangular coordinates, the graph of the given equation consists of sine waves of amplitude 4 and period 2π. However, if polar coordinates are used, then the points that correspond to the pairs in the table lie on a circle of radius 2 and we draw the graph accordingly (see Figure 13.9). As an aid to plotting points, we have extended the polar axis in the negative direction and introduced a vertical line through the pole. The proof that the graph is a circle will be given in Example 5. Additional points obtained by letting θ vary from π to 2π lie on the same circle. For example, the solution $(-2, 7\pi/6)$ gives us the same point as $(2, \pi/6)$; the point corresponding to $(-2\sqrt{2}, 5\pi/4)$ is the same as that obtained from $(2\sqrt{2}, \pi/4)$, etc. If we let θ increase through all real numbers we obtain the same points again and again because of the periodicity of the sine function. ∎

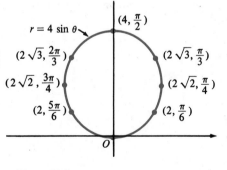

Figure 13.9

Example 2 Sketch the graph of the equation $r = 2 + 2 \cos \theta$.

Solution Since the cosine function decreases from 1 to -1 as θ varies from 0 to π, it follows that r decreases from 4 to 0 in this θ-interval. The following table exhibits some solutions of the equation.

θ	0	$\pi/6$	$\pi/4$	$\pi/3$	$\pi/2$	$2\pi/3$	$3\pi/4$	$5\pi/6$	π
r	4	$2 + \sqrt{3}$	$2 + \sqrt{2}$	3	2	1	$2 - \sqrt{2}$	$2 - \sqrt{3}$	0

If θ increases from π to 2π, then $\cos \theta$ increases from -1 to 1, and consequently r increases from 0 to 4. Plotting points leads to the sketch shown in Figure 13.10, where we have used polar coordinate graph paper, which displays lines through O at various angles and circles with centers at the pole. The same graph may be obtained by taking other intervals for θ.

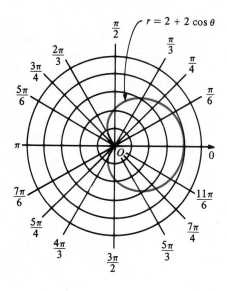

Figure 13.10

The heart-shaped graph in Example 2 is called a **cardioid**. In general, the graph of any polar equation of the form

$$r = a(1 + \cos \theta), \qquad r = a(1 + \sin \theta)$$

$$r = a(1 - \cos \theta), \qquad r = a(1 - \sin \theta)$$

where a is a real number, is a cardioid.

The graph of an equation of the form

$$r = a + b \cos \theta \quad \text{or} \quad r = a + b \sin \theta$$

where $a \neq b$, is called a **limaçon**. The graph is similar in shape to a cardioid; however, there may be an added "loop" as shown in the next example.

Example 3   Sketch the graph of $r = 2 + 4 \cos \theta$.

Solution   Some points corresponding to $0 \leq \theta \leq \pi$ are exhibited in the following table.

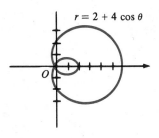

$r = 2 + 4 \cos \theta$

Figure 13.11

θ	0	$\pi/6$	$\pi/4$	$\pi/3$	$\pi/2$	$2\pi/3$	$3\pi/4$	$5\pi/6$	π
r	6	$2 + 2\sqrt{3}$	$2 + 2\sqrt{2}$	4	2	0	$2 - 2\sqrt{2}$	$2 - 2\sqrt{3}$	-2

Note that $r = 0$ at $\theta = 2\pi/3$. The values of r are negative if $2\pi/3 < \theta \le \pi$, and this leads to the lower half of the small loop in Figure 13.11. (Verify this fact.) Letting θ range from π to 2π gives us the upper half of the small loop and the lower half of the large loop. ■

Example 4 Sketch the graph of the equation $r = a \sin 2\theta$, where $a > 0$.

Solution Instead of tabulating solutions, let us reason as follows. If θ increases from 0 to $\pi/4$, then 2θ varies from 0 to $\pi/2$ and hence $\sin 2\theta$ increases from 0 to 1. It follows that r increases from 0 to a in the θ-interval $[0, \pi/4]$. If we next let θ increase from $\pi/4$ to $\pi/2$, then 2θ changes from $\pi/2$ to π. Consequently, r decreases from a to 0 in the θ-interval $[\pi/4, \pi/2]$. (Why?) The corresponding points on the graph constitute the first-quadrant loop illustrated in Figure 13.12. Note that the point $P(r, \theta)$ traces the loop in the *counterclockwise* direction as θ increases from 0 to $\pi/2$.

If we let θ vary from $\pi/2$ to π, then 2θ varies from π to 2π and, therefore, $r = a \sin 2\theta \le 0$. Thus, the points $P(r, \theta)$ are in the fourth quadrant if $\pi/2 < \theta < \pi$. We shall leave it to the reader to verify that if θ increases from $\pi/2$ to π, then $P(r, \theta)$ traces (in a counterclockwise direction) the loop shown in the fourth quadrant.

Similar loops are obtained for the θ-intervals $[\pi, 3\pi/2]$ and $[3\pi/2, 2\pi]$. In Figure 13.12 we have plotted only those points on the graph that correspond to the largest numerical values of r. ■

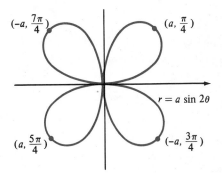

$\left(-a, \dfrac{7\pi}{4}\right)$ $\left(a, \dfrac{\pi}{4}\right)$

$r = a \sin 2\theta$

$\left(a, \dfrac{5\pi}{4}\right)$ $\left(-a, \dfrac{3\pi}{4}\right)$

Figure 13.12

The graph in Example 4 is called a **four-leafed rose**. In general, any equation of the form

$$r = a \sin n\theta \quad \text{or} \quad r = a \cos n\theta$$

where n is a positive integer greater than 1 and a is a real number, has a graph that consists of a number of loops attached to the origin. If n is even there are $2n$ loops, whereas if n is odd there are n loops (see, for example, Exercises 9, 10, and 22).

Many other interesting graphs result from polar equations. Some are included in the exercises at the end of this section. Polar coordinates are very useful in applications involving circles with centers at the origin or lines that pass through the origin, since equations that have these graphs may be written in the simple forms $r = k$ or $\theta = k$ for some fixed number k. (Verify this fact.)

Let us now superimpose an xy-plane on an $r\theta$-plane such that the positive x-axis coincides with the polar axis. Any point P in the plane may then be assigned rectangular coordinates (x, y) or polar coordinates (r, θ). It is not difficult to obtain formulas that specify the relationship between the two coordinate systems. If $r > 0$ we have a situation similar to that illustrated in (i) of Figure 13.13. If $r < 0$ we have that shown in (ii) of the figure where, for later purposes, we have also plotted the point P' having polar coordinates $(|r|, \theta)$ and rectangular coordinates $(-x, -y)$.

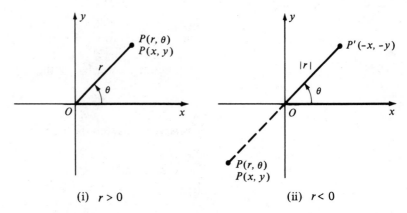

(i) $r > 0$ (ii) $r < 0$

Figure 13.13

The following result specifies relationships between (x, y) and (r, θ). In the statement of the theorem it is assumed that the positive x-axis coincides with the polar axis.

Theorem (13.4)

> The rectangular coordinates (x, y) and polar coordinates (r, θ) of a point P are related as follows:
>
> (i) $x = r \cos \theta, \quad y = r \sin \theta.$
>
> (ii) $\tan \theta = \dfrac{y}{x}, \quad r^2 = x^2 + y^2.$

Proof   Although we have pictured θ as an acute angle in Figure 13.13, the discussion that follows is valid for all angles. On the one hand, if $r > 0$ as in (i) of Figure 13.13, then

$$x = r \cos \theta, \quad y = r \sin \theta.$$

On the other hand, if $r < 0$ then $|r| = -r$, and from (ii) of Figure 13.13 we see that

$$\cos \theta = \frac{-x}{|r|} = \frac{-x}{-r} = \frac{x}{r},$$

$$\sin \theta = \frac{-y}{|r|} = \frac{-y}{-r} = \frac{y}{r}.$$

Multiplication by r produces (i) of (13.4) and, therefore, these formulas hold whether r is positive or negative. If $r = 0$, then the point is the pole and we again see that the formulas in (i) are true.

The formulas in (ii) follow readily from Figure 13.13. $\square$

We may use Theorem (13.4) to change from one system of coordinates to the other. A more important use is for transforming a polar equation to an equation in x and y, and vice versa. This is illustrated in the next two examples.

Example 5 Find an equation in x and y that has the same graph as the polar equation $r = 4 \sin \theta$.

Solution   The equation $r = 4 \sin \theta$ was considered in Example 1. It is convenient to multiply both sides by r, obtaining $r^2 = 4r \sin \theta$. Applying Theorem (13.4) gives us $x^2 + y^2 = 4y$. The last equation is equivalent to $x^2 + (y - 2)^2 = 4$ whose graph is a circle of radius 2 with center at $(0, 2)$ in the xy-plane.   ∎

Example 6 Find a polar equation of an arbitrary line.

Solution   We know that every line in an xy-coordinate system is the graph of a linear equation $ax + by + c = 0$. Substituting for x and y from (i) of (13.4) leads to the polar equation

$$r(a \cos \theta + b \sin \theta) + c = 0.$$ ∎

If we superimpose an xy-plane on an $r\theta$-plane, then the graph of a polar equation may be symmetric with respect to the x-axis, the y-axis, or the origin. It is left to the reader to show that if a substitution listed in the following table does not change the solutions of a polar equation, then the graph has the indicated symmetry.

Tests for Symmetry **(13.5)**

Substitution	Symmetry
$-\theta$ for θ	the line $\theta = 0$ (x-axis)
$-r$ for r	the pole (origin)
$\pi + \theta$ for θ	the pole (origin)
$\pi - \theta$ for θ	the line $\theta = \pi/2$ (y-axis)

To illustrate, since $\cos(-\theta) = \cos \theta$, the graph of the equation in Example 2 (see Figure 13.10) is symmetric with respect to the x-axis. Since $\sin(\pi - \theta) = \sin \theta$, the graph in Example 1 is symmetric with respect to the y-axis. The graph in Example 4 is symmetric to both axes and the origin. Other tests for symmetry may be stated; however, those listed in (13.5) are among the easiest to apply.

Tangent lines to graphs of polar equations may be found by means of the next theorem.

Theorem (13.6)

The slope m of the tangent line to the graph of $r = f(\theta)$ at the point $P(r, \theta)$ is

$$m = \frac{\dfrac{dr}{d\theta} \sin \theta + r \cos \theta}{\dfrac{dr}{d\theta} \cos \theta - r \sin \theta}.$$

Proof   If (x, y) are the rectangular coordinates of $P(r, \theta)$, then by Theorem (13.4),

$$x = r \cos \theta = f(\theta) \cos \theta$$
$$y = r \sin \theta = f(\theta) \sin \theta.$$

These may be considered as parametric equations for the graph, where the parameter is θ. Applying Theorem (13.3), the slope of the tangent line at (x, y) is

$$\frac{dy}{dx} = \frac{(dy/d\theta)}{(dx/d\theta)} = \frac{f'(\theta) \sin \theta + f(\theta) \cos \theta}{f'(\theta) \cos \theta - f(\theta) \sin \theta}.$$

This is equivalent to the formula in the statement of the theorem.   □

Horizontal tangent lines occur if the numerator in the formula for m is 0 and the denominator is not 0. Vertical tangent lines occur if the denominator is 0 and the numerator is not 0. The case 0/0 requires special attention.

To find the slopes of the tangent lines at the pole, it is necessary to determine the values of θ for which $r = 0$. For such values (and with $r = 0$) the formula in Theorem (13.6) reduces to $m = \tan \theta$. These remarks are illustrated in the next example.

Example 7   Given the cardioid $r = 2 + 2 \cos \theta$, find

(a) the slope of the tangent line at $\theta = \pi/6$.
(b) the slope of the tangent line at the pole.
(c) the points at which the tangent line is horizontal.
(d) the points at which the tangent line is vertical.

Solution   The graph of $r = 2 + 2 \cos \theta$ is sketched in Figure 13.10. If we apply Theorem (13.6), the slope m of the tangent line is

$$m = \frac{(-2 \sin \theta) \sin \theta + (2 + 2 \cos \theta) \cos \theta}{(-2 \sin \theta) \cos \theta - (2 + 2 \cos \theta) \sin \theta}$$

$$= \frac{2(\cos^2 \theta - \sin^2 \theta) + 2 \cos \theta}{-2(2 \sin \theta \cos \theta) - 2 \sin \theta}$$

$$= -\frac{\cos 2\theta + \cos \theta}{\sin 2\theta + \sin \theta}.$$

(a) At $\theta = \pi/6$,

$$m = -\frac{\cos(\pi/3) + \cos(\pi/6)}{\sin(\pi/3) + \sin(\pi/6)} = -\frac{(1/2) + (\sqrt{3}/2)}{(\sqrt{3}/2) + (1/2)} = -1.$$

(b) To find the slope of the tangent line at the pole we need the values of θ such that $r = 2 + 2\cos\theta = 0$. This gives us $\cos\theta = -1$, or $\theta = \pi$. Substitution in the formula for m produces the meaningless expression $0/0$, and hence we let $r = 0$ in Theorem (13.6), obtaining $m = \tan\theta$. Hence at the pole, $m = \tan\pi = 0$.

(c) To find horizontal tangents we let

$$\cos 2\theta + \cos\theta = 0.$$

This equation may be written as

$$2\cos^2\theta - 1 + \cos\theta = 0,$$

or $$(2\cos\theta - 1)(\cos\theta + 1) = 0.$$

From $\cos\theta = 1/2$ we obtain $\theta = \pi/3$ and $\theta = 5\pi/3$. The corresponding points are $(3, \pi/3)$ and $(3, 5\pi/3)$. Using $\cos\theta = -1$ gives us $\theta = \pi$. Since the denominator in the formula for m is 0 at $\theta = \pi$, further investigation is required. Indeed, we saw in (b) that there is a horizontal tangent line at the point $(0, \pi)$.

(d) To find vertical tangent lines we let

$$\sin 2\theta + \sin\theta = 0.$$

Equivalent equations are

$$2\sin\theta\cos\theta + \sin\theta = 0,$$

and $$\sin\theta(2\cos\theta + 1) = 0.$$

Letting $\sin\theta = 0$ and $\cos\theta = -1/2$ leads to the following values of θ: 0, π, $2\pi/3$, and $4\pi/3$. We found, in (c), that π gives us a horizontal tangent. The remaining values result in the points $(4, 0)$, $(1, 2\pi/3)$, and $(1, 4\pi/3)$, at which the graph has vertical tangent lines. ∎

EXERCISES 13.3

Sketch the graphs of the polar equations in Exercises 1–24.

1 $r = 5$

2 $\theta = \pi/4$

3 $\theta = -\pi/6$

4 $r = -2$

5 $r = 4\cos\theta$

6 $r = -2\sin\theta$

7 $r = 4(1 - \sin\theta)$

8 $r = 1 + 2\cos\theta$

9 $r = 8\cos 3\theta$

10 $r = 2\sin 4\theta$

11 $r^2 = 4\cos 2\theta$ (lemniscate)

12 $r = 6 \sin^2 (\frac{1}{2}\theta)$

13 $r = 4 \csc \theta$

14 $r = -3 \sec \theta$

15 $r = 2 - \cos \theta$

16 $r = 2 + 2 \sec \theta$ (conchoid)

17 $r = 2^\theta, \theta \geq 0$ (spiral)

18 $r\theta = 1, \theta > 0$ (spiral)

19 $r = -6(1 + \cos \theta)$

20 $r = e^{2\theta}$ (logarithmic spiral)

21 $r = 2 + 4 \sin \theta$

22 $r = 8 \cos 5\theta$

23 $r^2 = -16 \sin 2\theta$

24 $4r = \theta$

In Exercises 25–32 find a polar equation that has the same graph as the given equation.

25 $x = -3$

26 $y = 2$

27 $x^2 + y^2 = 16$

28 $x^2 = 8y$

29 $y = 6$

30 $y = 6x$

31 $x^2 - y^2 = 16$

32 $9x^2 + 4y^2 = 36$

In Exercises 33–48 find an equation in x and y that has the same graph as the given polar equation and use it as an aid in sketching the graph in a polar coordinate system.

33 $r \cos \theta = 5$

34 $r \sin \theta = -2$

35 $r - 6 \sin \theta = 0$

36 $r = 6 \cot \theta$

37 $r = 2$

38 $\theta = \pi/4$

39 $r = \tan \theta$

40 $r = 4 \sec \theta$

41 $r^2(4 \sin^2 \theta - 9 \cos^2 \theta) = 36$

42 $r^2(\cos^2 \theta + 4 \sin^2 \theta) = 16$

43 $r^2 \cos 2\theta = 1$

44 $r^2 \sin 2\theta = 4$

45 $r(\sin \theta - 2 \cos \theta) = 6$

46 $r(\sin \theta + r \cos^2 \theta) = 1$

47 $r = 1/(1 + \cos \theta)$

48 $r = 4/(2 + \sin \theta)$

49 If $P_1(r_1, \theta_1)$ and $P_2(r_2, \theta_2)$ are points in an $r\theta$-plane, use the Law of Cosines to prove that
$$[d(P_1, P_2)]^2 = r_1^2 + r_2^2 - 2r_1r_2 \cos (\theta_2 - \theta_1).$$

50 Prove that the graphs of the following polar equations are circles and find the center and radius in each case.

(a) $r = a \sin \theta, a \neq 0$

(b) $r = b \cos \theta, b \neq 0$

(c) $r = a \sin \theta + b \cos \theta, ab \neq 0$

In Exercises 51–60 find the slope of the tangent line to the graph of the equation at the indicated value of θ. (Refer to the graphs in Exercises 5–12 and 17–18.)

51 $r = 4 \cos \theta, \theta = \pi/3$

52 $r = -2 \sin \theta, \theta = \pi/6$

53 $r = 4(1 - \sin \theta), \theta = 0$

54 $r = 1 + 2 \cos \theta, \theta = \pi/2$

55 $r = 8 \cos 3\theta, \theta = \pi/4$

56 $r = 2 \sin 4\theta, \theta = \pi/4$

57 $r^2 = 4 \cos 2\theta, \theta = \pi/6$

58 $r = 6 \sin^2 (\frac{1}{2}\theta), \theta = 2\pi/3$

59 $r = 2^\theta, \theta = \pi$

60 $r\theta = 1, \theta = 2\pi$

61 If $\cos \theta \neq 0$, show that the slope of the tangent line to the graph of $r = f(\theta)$ is
$$m = \frac{(dr/d\theta) \tan \theta + r}{(dr/d\theta) - r \tan \theta}.$$

62 Let $a > 0$ and let C be the set of points in a coordinate plane, the product of whose distances from $P(a, 0)$ and $Q(-a, 0)$ is equal to a^2. Show that a polar equation for C is $r^2 = 2a^2 \cos 2\theta$. (*Hint*: Use Exercise 49.)

63 If the graphs of the polar equations $r = f(\theta)$ and $r = g(\theta)$ intersect at $P(r, \theta)$, prove that the tangent lines at P are perpendicular if and only if
$$f'(\theta)g'(\theta) + f(\theta)g(\theta) = 0.$$

(The graphs are said to be *orthogonal* at P.)

64 Use Exercise 63 to prove that the graphs of each of the following pairs of equations are orthogonal at their points of intersection.

(a) $r = a \sin \theta, r = a \cos \theta$

(b) $r = a\theta, r\theta = a$

13.4

POLAR EQUATIONS OF CONICS

The following theorem provides another method for describing conic sections.

Theorem (13.7)

> Let F be a fixed point and l a fixed line in a plane. The set of all points P in the plane such that the ratio $d(P, F)/d(P, Q)$ is a positive constant e, where $d(P, Q)$ is the distance from P to l, is a conic section. Moreover, the conic is a parabola if $e = 1$, an ellipse if $0 < e < 1$, and a hyperbola if $e > 1$.

The constant e is called the **eccentricity** of the conic and should not be confused with the base of the natural logarithms. It will be seen that the point F is a focus of the conic. The line l is called a **directrix**. A typical situation is illustrated in Figure 13.14, where the curve indicates possible positions of P.

We shall prove the theorem if $e \leq 1$ and leave the case $e > 1$ to the reader.

If $e = 1$ in Theorem (13.7), then $d(P, F) = d(P, Q)$ and, according to Definition (12.1), a parabola with focus F and directrix l is obtained.

Suppose next that $0 < e < 1$. It is convenient to introduce a polar coordinate system in the plane with F as the pole and with l perpendicular to the polar axis at the point $D(d, 0)$, where $d > 0$. If $P(r, \theta)$ is a point in the plane such that $d(P, F)/d(P, Q) = e < 1$, then from Figure 13.15 we see that P lies to the left of l. Let C be the projection of P on the polar axis. Since

$$d(P, F) = r \quad \text{and} \quad d(P, Q) = \overline{FD} - \overline{FC} = d - r\cos\theta,$$

it follows that P satisfies the condition in (13.7) if and only if

$$\frac{r}{d - r\cos\theta} = e$$

or
$$r = de - er\cos\theta.$$

Solving for r gives us

$$r = \frac{de}{1 + e\cos\theta}$$

which is a polar equation of the graph. Actually, the same equation is obtained if $e = 1$; however, there is no point (r, θ) on the graph if $1 + \cos\theta = 0$.

The rectangular equation corresponding to $r = de - er\cos\theta$ is

$$\pm\sqrt{x^2 + y^2} = de - ex.$$

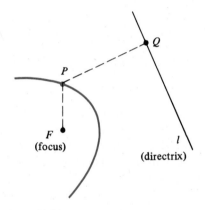

Figure 13.14

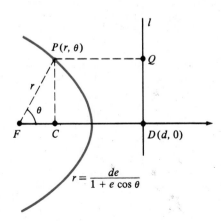

$$r = \frac{de}{1 + e\cos\theta}$$

Figure 13.15

Squaring both sides and rearranging terms leads to

$$(1 - e^2)x^2 + 2de^2x + y^2 = d^2e^2.$$

Completing the square in the previous equation and simplifying, we obtain

$$\left(x + \frac{de^2}{1 - e^2}\right)^2 + \frac{y^2}{1 - e^2} = \frac{d^2e^2}{(1 - e^2)^2}.$$

Finally, dividing both sides by $d^2e^2/(1 - e^2)^2$ leads to an equation of the form

$$\frac{(x - h)^2}{a^2} + \frac{y^2}{b^2} = 1$$

where $h = -de^2/(1 - e^2)$. Consequently, the graph is an ellipse with center at the point $(-de^2/(1 - e^2), 0)$ on the x-axis and where

$$a^2 = \frac{d^2e^2}{(1 - e^2)^2}, \qquad b^2 = \frac{d^2e^2}{1 - e^2}.$$

By Theorem (12.6),

$$c^2 = a^2 - b^2 = \frac{d^2e^4}{(1 - e^2)^2}$$

and hence $c = de^2/(1 - e^2)$. This proves that F is a focus of the ellipse. It also follows that $e = c/a$. A similar proof may be given for the case $e > 1$.

It can be shown, conversely, that every conic that is not a circle may be described by means of the statement in Theorem (13.7). This gives us a formulation of conic sections that is equivalent to the approach used in Chapter 12. Since (13.7) includes all three types of conics, it is sometimes regarded as a definition for the conic sections.

If we had chosen the focus F to the *right* of the directrix, as illustrated in Figure 13.16 (where $d > 0$), then the equation $r = de/(1 - e \cos \theta)$ would have resulted. (Note the minus sign in place of the plus sign.) Other sign changes occur if d is allowed to be negative.

If l is taken *parallel* to the polar axis through one of the points $(d, \pi/2)$ or $(d, 3\pi/2)$, as illustrated in Figure 13.17, then the corresponding equations would contain $\sin \theta$ instead of $\cos \theta$. The proofs of these facts are left to the reader.

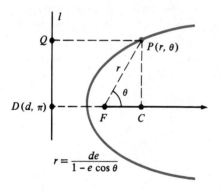

$$r = \frac{de}{1 - e \cos \theta}$$

Figure 13.16

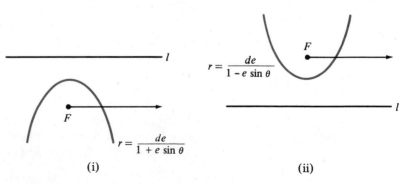

(i)   (ii)

Figure 13.17

The following theorem summarizes our discussion.

Theorem (*13.8*)

A polar equation having one of the four forms

$$r = \frac{de}{1 \pm e \cos \theta}, \qquad r = \frac{de}{1 \pm e \sin \theta}$$

is a conic section. Moreover, the conic is a parabola if $e = 1$, an ellipse if $0 < e < 1$, or a hyperbola if $e > 1$.

Example 1 Describe and sketch the graph of the equation

$$r = \frac{10}{3 + 2 \cos \theta}.$$

Solution   Dividing numerator and denominator of the given fraction by 3 gives us

$$r = \frac{\frac{10}{3}}{1 + \frac{2}{3} \cos \theta}$$

which has one of the forms in Theorem (13.8) with $e = \frac{2}{3}$. Thus the graph is an ellipse with focus F at the pole and major axis along the polar axis. The endpoints of the major axis may be found by setting θ equal to 0 and π. This gives us $V(2, 0)$ and $V'(10, \pi)$. Hence

$$2a = d(V', V) = 12, \quad \text{or} \quad a = 6.$$

The center of the ellipse is the midpoint of the segment $V'V$, namely $(4, \pi)$. Using the fact that $e = c/a$, we obtain

$$c = ae = 6(\tfrac{2}{3}) = 4.$$

Hence

$$b^2 = a^2 - c^2 = 36 - 16 = 20,$$

that is, the semiminor axis has length $\sqrt{20}$. The graph is sketched in Figure 13.18 where, for reference, we have superimposed a rectangular coordinate system on the polar system. ∎

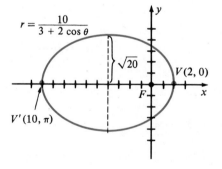

$$r = \frac{10}{3 + 2 \cos \theta}$$

Figure 13.18

Example 2 Describe and sketch the graph of the equation

$$r = \frac{10}{2 + 3 \sin \theta}.$$

Solution   To express the equation in one of the forms in Theorem (13.8) we divide numerator and denominator of the given fraction by 2, obtaining

$$r = \frac{5}{1 + \frac{3}{2} \sin \theta}.$$

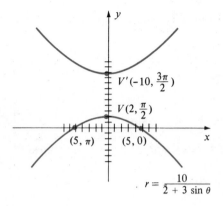

Figure 13.19

Thus $e = \frac{3}{2}$ and the graph is a hyperbola with a focus at the pole. The expression $\sin \theta$ tells us that the transverse axis of the hyperbola is perpendicular to the polar axis. To find the vertices we let θ equal $\pi/2$ and $3\pi/2$ in the given equation. This gives us the points $V(2, \pi/2)$, $V'(-10, 3\pi/2)$, and hence

$$2a = d(V, V') = 8, \quad \text{or} \quad a = 4.$$

The points $(5, 0)$ and $(5, \pi)$ on the graph can be used to get a rough estimate of the lower branch of the hyperbola. The upper branch is obtained by symmetry, as illustrated in Figure 13.19. If more accuracy or additional information is desired, we may calculate

$$c = ae = 4(\tfrac{3}{2}) = 6$$

and
$$b^2 = c^2 - a^2 = 36 - 16 = 20.$$

Asymptotes may then be constructed in the usual way.   ∎

Example 3   Sketch the graph of the equation

$$r = \frac{15}{4 - 4 \cos \theta}.$$

Solution   To obtain one of the forms in (13.8) we divide numerator and denominator of the given fraction by 4, obtaining

$$r = \frac{\frac{15}{4}}{1 - \cos \theta}.$$

Consequently, $e = 1$ and the graph is a parabola with focus at the pole. A rough sketch can be found by plotting the points that correspond to the x- and y- intercepts. These are indicated in the following table.

θ	0	$\pi/2$	π	$3\pi/2$
r	—	15/4	15/8	15/4

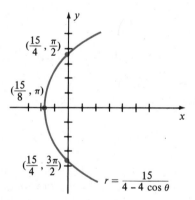

Figure 13.20

Note that there is no point on the graph corresponding to $\theta = 0$. (Why?) Plotting the three points and using the fact that the graph is a parabola with focus at the pole gives us the sketch in Figure 13.20.   ∎

If only a rough sketch of a conic is desired, then the technique employed in Example 3 is recommended. To use this method we plot (if possible) points corresponding to $\theta = 0$, $\pi/2$, π, and $3\pi/2$. These points, together with the type of conic (obtained from the value of e), readily lead to the sketch.

EXERCISES 13.4

In Exercises 1–10, identify and sketch the graph of the polar equation.

1 $r = \dfrac{12}{6 + 2\sin\theta}$

2 $r = \dfrac{12}{6 - 2\sin\theta}$

3 $r = \dfrac{12}{2 - 6\cos\theta}$

4 $r = \dfrac{12}{2 + 6\cos\theta}$

5 $r = \dfrac{3}{2 + 2\cos\theta}$

6 $r = \dfrac{3}{2 - 2\sin\theta}$

7 $r = \dfrac{4}{\cos\theta - 2}$

8 $r = \dfrac{4\sec\theta}{2\sec\theta - 1}$

9 $r = \dfrac{6\csc\theta}{2\csc\theta + 3}$

10 $r = \csc\theta(\csc\theta - \cot\theta)$

11–20 Find rectangular equations for the graphs in Exercises 1–10.

In Exercises 21–26 use Theorem (13.7) to find a polar equation of the conic with focus at the pole and the given eccentricity and equation of directrix.

21 $e = 1/3, r = 2\sec\theta$

22 $e = 2/5, r = 4\csc\theta$

23 $e = 4, r = -3\csc\theta$

24 $e = 3, r = -4\sec\theta$

25 $e = 1, r\cos\theta = 5$

26 $e = 1, r\sin\theta = -2$

27 Find a polar equation of the parabola with focus at the pole and vertex $(4, \pi/2)$.

28 Find a polar equation of an ellipse with eccentricity 2/3, a vertex at $(1, 3\pi/2)$, and a focus at the pole.

29 Prove Theorem (13.7) for the case $e > 1$.

30 (a) Use Figure 3.16 to derive the formula
$$r = de/(1 - e\cos\theta).$$
(b) Derive the formulas $r = de/(1 \pm e\sin\theta)$ in Theorem (13.8) by using Figure 13.17.

In Exercises 31–36, express the given equation in one of the forms in Theorem (13.8) and then find the eccentricity and an equation for the directrix.

31 $y^2 = 4 - 4x$

32 $x^2 = 1 - 2y$

33 $3y^2 - 16y - x^2 + 16 = 0$

34 $5x^2 + 9y^2 = 32x + 64$

35 $8x^2 + 9y^2 + 4x = 4$

36 $4x^2 - 5y^2 + 36y - 36 = 0$

In Exercises 37–40, find the slope of the tangent line to the graph of the equation at the point corresponding to the indicated value of θ. (Refer to the graphs in Exercises 1–4.)

37 $r = \dfrac{12}{6 + 2\sin\theta}, \theta = \pi/6$

38 $r = \dfrac{12}{6 - 2\sin\theta}, \theta = 0$

39 $r = \dfrac{12}{2 - 6\cos\theta}, \theta = \pi/2$

40 $r = \dfrac{12}{2 + 6\cos\theta}, \theta = \pi/3$

41 Given the hyperbola $r = ep/(1 - e\cos\theta)$, where $e > 1$, show that the angles between the line through the polar axis and the asymptotes are given by $\arccos(\pm 1/e)$.

42 A **focal chord** of a conic is a line segment through a focus with endpoints on the conic. Let d_1 and d_2 be the lengths into which a focal chord is divided by the focus. Show that the sum of the reciprocals of d_1 and d_2 is the same for all focal chords of a given conic.

13.5

AREAS IN POLAR COORDINATES

Areas of certain regions that are bounded by graphs of polar equations can be found by employing limits of sums of the areas of circular sectors. Suppose R is a region in an $r\theta$-plane bounded by the lines through O with equations $\theta = a$ and $\theta = b$, where $0 \le a < b \le 2\pi$, and by the graph of $r = f(\theta)$, where f is continuous and $f(\theta) \ge 0$ on $[a, b]$. A region of this type is illustrated in (i) of Figure 13.21.

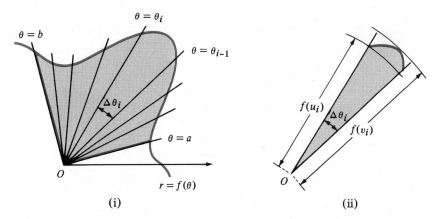

(i)   **(ii)**

Figure 13.21

Let P denote a partition of $[a, b]$ determined by

$$a = \theta_0 < \theta_1 < \theta_2 < \cdots < \theta_n = b$$

and let $\Delta\theta_i = \theta_i - \theta_{i-1}$ for $i = 1, 2, \ldots, n$. The lines with equations $\theta = \theta_i$ divide R into wedge-shaped subregions. If $f(u_i)$ is the minimum value and $f(v_i)$ is the maximum value of f on $[\theta_{i-1}, \theta_i]$, then as illustrated in (ii) of Figure 13.21, the area ΔA_i of the ith subregion is between the areas of the inscribed and circumscribed circular sectors having central angle $\Delta\theta_i$ and radii $f(u_i)$ and $f(v_i)$, respectively. Hence, by (8.3),

$$\tfrac{1}{2}[f(u_i)]^2 \, \Delta\theta_i \le \Delta A_i \le \tfrac{1}{2}[f(v_i)]^2 \, \Delta\theta_i.$$

Summing from $i = 1$ to $i = n$, and using the fact that the sum of the ΔA_i is the area A of R, we obtain

$$\sum_{i=1}^{n} \tfrac{1}{2}[f(u_i)]^2 \, \Delta\theta_i \le A \le \sum_{i=1}^{n} \tfrac{1}{2}[f(v_i)]^2 \, \Delta\theta_i.$$

The limit of each sum, as the norm $\|P\|$ of the subdivision approaches zero, equals the integral $\int_a^b \tfrac{1}{2}[f(\theta)]^2 \, d\theta$. This gives us the following result.

Theorem (13.9)

If f is continuous and $f(\theta) \geq 0$ on $[a, b]$, where $0 \leq a < b \leq 2\pi$, then the area A of the region bounded by the graphs of $r = f(\theta)$, $\theta = a$, and $\theta = b$ is

$$A = \int_a^b \tfrac{1}{2}[f(\theta)]^2 \, d\theta = \int_a^b \tfrac{1}{2} r^2 \, d\theta.$$

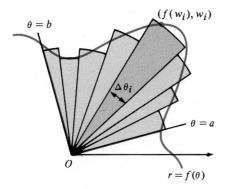

Figure 13.22

The integral in Theorem (13.9) may be interpreted as a limit of a sum by writing

$$A = \int_a^b \tfrac{1}{2}[f(\theta)]^2 \, d\theta = \lim_{\|P\| \to 0} \sum_{i=1}^{n} \tfrac{1}{2}[f(w_i)]^2 \, \Delta\theta_i$$

where w_i is *any* number in the subinterval $[\theta_{i-1}, \theta_i]$ of $[a, b]$. The sketch in Figure 13.22 provides a geometric illustration of a typical Riemann sum. It is convenient to think of starting with the area of a circular sector of the type shown in Figure 13.21, and then sweeping out the region R by letting θ vary from a to b while at the same time each $\Delta\theta_i$ approaches zero.

Example 1 Find the area of the region that is bounded by the cardioid $r = 2 + 2 \cos \theta$.

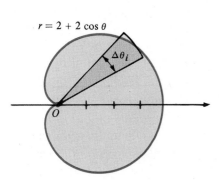

Figure 13.23

Solution The graph was sketched in Figure 13.10. It is sketched again in Figure 13.23 together with a typical circular sector of the type discussed previously. Making use of symmetry we shall find the area of the upper half of the region and double the result. The function f of Theorem (13.9) is given by $f(\theta) = 2 + 2 \cos \theta$. To find a and b we observe that the circular sectors will sweep out the upper half of the cardioid if θ varies from 0 to π. Consequently, $a = 0$, $b = \pi$, and

$$A = 2 \int_0^\pi \tfrac{1}{2}(2 + 2 \cos \theta)^2 \, d\theta$$

$$= \int_0^\pi (4 + 8 \cos \theta + 4 \cos^2 \theta) \, d\theta.$$

Using the fact that $\cos^2 \theta = \tfrac{1}{2}(1 + \cos 2\theta)$,

$$A = \int_0^\pi (6 + 8 \cos \theta + 2 \cos 2\theta) \, d\theta$$

$$= \left[6\theta + 8 \sin \theta + \sin 2\theta \right]_0^\pi = 6\pi.$$

The area could also have been found by using (13.9) with $a = 0$ and $b = 2\pi$.

The next result generalizes Theorem (13.9).

Theorem (13.10)

> Let f and g be continuous functions such that $f(\theta) \geq g(\theta) \geq 0$ for every θ in $[a, b]$, where $0 \leq a < b \leq 2\pi$. Let R denote the region bounded by the graphs of $r = f(\theta)$, $r = g(\theta)$, $\theta = a$, and $\theta = b$. The area A of R is
>
> $$A = \frac{1}{2} \int_a^b ([f(\theta)]^2 - [g(\theta)]^2)\, d\theta.$$

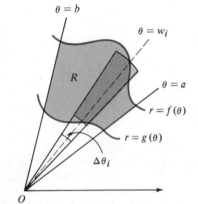

Figure 13.24

A region of the type described in Theorem (13.10) is sketched in Figure 13.24. The area A can be found by subtracting the areas of two regions of the type described in (13.9). Specifically,

$$A = \int_a^b \frac{1}{2}[f(\theta)]^2 \, d\theta - \int_a^b \frac{1}{2}[g(\theta)]^2 \, d\theta.$$

Writing this as one integral gives us Theorem (13.10).

The integral in (13.10) can be expressed as a limit of a sum as follows:

$$A = \lim_{\|P\| \to 0} \sum_i \frac{1}{2}([f(w_i)]^2 - [g(w_i)]^2)\, \Delta\theta_i.$$

The sum may be thought of as the process of starting with the area of a region of the type illustrated by the truncated wedge in Figure 13.24 and then sweeping out R by letting θ vary from a to b.

Example 2 Find the area A of the region that is inside the cardioid $r = 2 + 2\cos\theta$ and outside the circle $r = 3$.

Solution   The region is sketched in Figure 13.25 together with a typical polar element of area. Since the two graphs intersect at the points $(3, -\pi/3)$ and $(3, \pi/3)$, the region will be swept out by the indicated element if θ varies from $-\pi/3$ to $\pi/3$. Using Theorem (13.10),

$$A = \frac{1}{2} \int_{-\pi/3}^{\pi/3} [(2 + 2\cos\theta)^2 - 3^2]\, d\theta$$

$$= \frac{1}{2} \int_{-\pi/3}^{\pi/3} (8\cos\theta + 4\cos^2\theta - 5)\, d\theta.$$

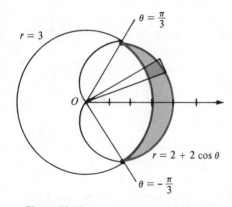

Figure 13.25

As in Example 1, the integral may be evaluated by letting $\cos^2\theta = \frac{1}{2}(1 + \cos 2\theta)$. It is left to the reader to show that $A = (9\sqrt{3}/2) - \pi$.

EXERCISES 13.5

In Exercises 1–8 sketch the graph of the equation and find the area of the region bounded by the graph.

1 $r = 2 \cos \theta$ **2** $r = 5 \sin \theta$

3 $r = 1 - \cos \theta$ **4** $r = 6 - 6 \sin \theta$

5 $r = \sin 2\theta$ **6** $r^2 = 9 \cos 2\theta$

7 $r = 4 + \sin \theta$ **8** $r = 3 + 2 \cos \theta$

In Exercises 9 and 10 find the area of region R.

9 $R = \{(r, \theta): 0 \le \theta \le \pi/2, 0 \le r \le e^{\theta}\}$.

10 $R = \{(r, \theta): 0 \le \theta \le \pi, 0 \le r \le 2\theta\}$.

In Exercises 11–14, find the area of the region bounded by one loop of the graph of the given equation.

11 $r^2 = 4 \cos 2\theta$ **12** $r = 2 \cos 3\theta$

13 $r = 3 \cos 5\theta$ **14** $r = \sin 6\theta$

In Exercises 15–18 find the area of the region that is outside the graph of the first equation and inside the graph of the second equation.

15 $r = 2 + 2 \cos \theta, r = 3$

16 $r = 2, r = 4 \cos \theta$

17 $r = 2, r^2 = 8 \sin 2\theta$

18 $r = 1 - \sin \theta, r = 3 \sin \theta$

In Exercises 19–22, find the area of the region that is inside the graphs of *both* of the given equations.

19 $r = \sin \theta, r = \sqrt{3} \cos \theta$

20 $r = 2(1 + \sin \theta), r = 1$

21 $r = 1 + \sin \theta, r = 5 \sin \theta$

22 $r^2 = 4 \cos 2\theta, r = 1$

In Exercises 23–26, find the area of the region bounded by the graphs of the equations.

23 $r = 2 \sec \theta, \theta = \pi/6, \theta = \pi/3$

24 $r = \csc \theta \cot \theta, \theta = \pi/6, \theta = \pi/4$

25 $r = 4/(1 - \cos \theta), \theta = \pi/4$

26 $r(1 + \sin \theta) = 2, \theta = \pi/3$

27 A line crosses a circular disc of radius a at a distance b from its center, where $0 < b < a$. Find the area of the smaller of the two circular segments cut from the disc by the line.

28 Let OP be the ray from the pole to the point $P(r, \theta)$ on the spiral $r = a\theta$, where $a > 0$. If the ray makes two revolutions (starting from $\theta = 0$), find the area of the region swept out in its second revolution that was not swept out in the first revolution (see figure).

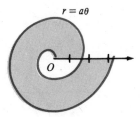

$r = a\theta$

Figure for Exercise 28

13.6

LENGTHS OF CURVES

If a curve is the graph of an equation $y = f(x)$, where f is smooth on an interval $[a, b]$, then its arc length may be found by means of Definition (6.15). If the curve is given parametrically or in terms of polar coordinates, then (6.15) is not necessarily applicable. In this section we shall develop a formula that can be used for more general curves.

Suppose a smooth curve C is given parametrically by

$$x = f(t), \quad y = g(t)$$

where $a \le t \le b$. Furthermore, suppose C does not intersect itself, that is, different values of t between a and b determine different points on C. Consider a

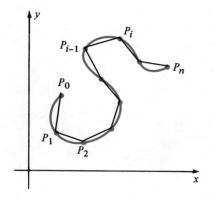

Figure 13.26

partition P of $[a, b]$ given by $a = t_0 < t_1 < t_2 < \cdots < t_n = b$. Let $\Delta t_i = t_i - t_{i-1}$ and let $P_i = (f(t_i), g(t_i))$ be the point on C determined by t_i. If $d(P_{i-1}, P_i)$ is the length of the line segment $P_{i-1}P_i$, then the length L_P of the broken line shown in Figure 13.26 is

$$L_P = \sum_{i=1}^{n} d(P_{i-1}, P_i).$$

In a manner similar to Definition (5.7) we write

$$L = \lim_{\|P\| \to 0} L_P$$

and call L the **length of C from P_0 to P_n** if, for every $\varepsilon > 0$, there exists a $\delta > 0$ such that $|L_P - L| < \varepsilon$ for all partitions P with $\|P\| < \delta$.

By the Distance Formula,

$$d(P_{i-1}, P_i) = \sqrt{[f(t_i) - f(t_{i-1})]^2 + [g(t_i) - g(t_{i-1})]^2}.$$

Applying the Mean Value Theorem (4.12), there exist numbers w_i and z_i in the open interval (t_{i-1}, t_i) such that

$$f(t_i) - f(t_{i-1}) = f'(w_i)\,\Delta t_i,$$
$$g(t_i) - g(t_{i-1}) = g'(z_i)\,\Delta t_i.$$

Substituting in the formula for $d(P_{i-1}, P_i)$ and removing the common factor $(\Delta t_i)^2$ from the radicand gives us

$$d(P_{i-1}, P_i) = \sqrt{[f'(w_i)]^2 + [g'(z_i)]^2}\,\Delta t_i.$$

Consequently,

$$L = \lim_{\|P\| \to 0} L_P = \lim_{\|P\| \to 0} \sum_{i=1}^{n} \sqrt{[f'(w_i)]^2 + [g'(z_i)]^2}\,\Delta t_i,$$

provided the limit exists. If $w_i = z_i$ for all i, then the sum is a Riemann sum for the function k defined by $k(t) = \sqrt{[f'(t)]^2 + [g'(t)]^2}$. The limit of this sum is

$$L = \int_a^b \sqrt{[f'(t)]^2 + [g'(t)]^2}\,dt.$$

It can be shown that the limit exists even if $w_i \neq z_i$; however, the proof requires advanced methods and hence is omitted. The next theorem summarizes this discussion.

Theorem (13.11)

> If a smooth curve C is given parametrically by $x = f(t)$, $y = g(t)$, where $a \leq t \leq b$, and if C does not intersect itself, except possibly at the endpoints of $[a, b]$, then the length L of C is
>
> $$L = \int_a^b \sqrt{[f'(t)]^2 + [g'(t)]^2}\,dt = \int_a^b \sqrt{\left(\frac{dx}{dt}\right)^2 + \left(\frac{dy}{dt}\right)^2}\,dt.$$

If a curve C is described in rectangular coordinates by an equation of the form $y = k(x)$, where k' is continuous on $[a, b]$, then parametric equations for C are

$$x = t, \quad y = k(t), \quad \text{where } a \le t \le b.$$

In this case

$$\frac{dx}{dt} = 1, \quad \frac{dy}{dt} = k'(t) = k'(x), \quad dt = dx$$

and from Theorem (13.11),

$$L = \int_a^b \sqrt{1 + [k'(x)]^2} \, dx.$$

This is in agreement with the arc length formula given in Definition (6.15).

Example 1 Find the length of one arch of the cycloid that has parametric equations $x = t - \sin t, \, y = 1 - \cos t$.

Solution The graph has the shape illustrated in Figure 13.4, where the radius a of the circle is 1. One arch is obtained if t varies from 0 to 2π. Applying Theorem (13.11),

$$L = \int_0^{2\pi} \sqrt{(1 - \cos t)^2 + (\sin t)^2} \, dt.$$

Since $\cos^2 t + \sin^2 t = 1$, the integrand reduces to

$$\sqrt{2 - 2 \cos t} = \sqrt{2}\sqrt{1 - \cos t}.$$

Thus

$$L = \int_0^{2\pi} \sqrt{2}\sqrt{1 - \cos t} \, dt.$$

By a half-angle formula, $\sin^2 (t/2) = (1 - \cos t)/2$, or equivalently,

$$1 - \cos t = 2 \sin^2 (t/2).$$

Hence

$$\sqrt{1 - \cos t} = \sqrt{2 \sin^2 (t/2)} = \sqrt{2}|\sin (t/2)|.$$

The absolute value sign may be deleted since if $0 \le t \le 2\pi$, then $0 \le t/2 \le \pi$ and hence $\sin (t/2) \ge 0$. Consequently,

$$L = \int_0^{2\pi} \sqrt{2}\sqrt{2} \sin (t/2) \, dt = 2 \int_0^{2\pi} \sin (t/2) \, dt$$

$$= -4 \cos (t/2) \Big]_0^{2\pi} = (-4)(-1) - (-4)(1) = 8.$$ ∎

The next theorem provides a formula for arc length in polar coordinates.

Theorem (13.12)

If a curve C is the graph of a polar equation $r = f(\theta)$, where f' is continuous on $[a, b]$, then the length L of C is

$$L = \int_a^b \sqrt{[f(\theta)]^2 + [f'(\theta)]^2} \, d\theta.$$

Proof   As in the proof of Theorem (13.6), parametric equations for C are

$$x = f(\theta) \cos \theta, \quad y = f(\theta) \sin \theta$$

where $a \le \theta \le b$. Hence

$$\frac{dx}{d\theta} = -f(\theta) \sin \theta + f'(\theta) \cos \theta$$

$$\frac{dy}{d\theta} = f(\theta) \cos \theta + f'(\theta) \sin \theta.$$

It is left as an exercise to show that

$$\left(\frac{dx}{d\theta}\right)^2 + \left(\frac{dy}{d\theta}\right)^2 = [f(\theta)]^2 + [f'(\theta)]^2.$$

Substitution in Theorem (13.11) with $\theta = t$ gives us the desired formula.   □

Example 2   Find the length of the spiral having polar equation $r = e^{\theta/2}$, from $(\sqrt{e}, 1)$ to $(e, 2)$.

Solution   A portion of the spiral is sketched in Figure 13.27, together with the points corresponding to $\theta = 1$ and $\theta = 2$. Using Theorem (13.12) with $f(\theta) = e^{\theta/2}$,

$$L = \int_1^2 \sqrt{(e^{\theta/2})^2 + (\tfrac{1}{2}e^{\theta/2})^2} \, d\theta$$

$$= \int_1^2 \sqrt{\tfrac{5}{4}e^\theta} \, d\theta = \frac{\sqrt{5}}{2} \int_1^2 e^{\theta/2} \, d\theta$$

$$= \sqrt{5} \, e^{\theta/2} \Big]_1^2 = \sqrt{5}(e - \sqrt{e}).$$

If an approximation is desired, $L \approx 2.4$.   ∎

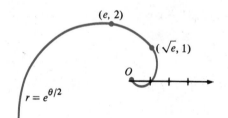

Figure 13.27

EXERCISES 13.6

Find the lengths of the curves defined in Exercises 1–7.

1    $x = 5t^2, y = 2t^3; 0 \le t \le 1$

2    $x = 3t, y = 2t^{3/2}; 0 \le t \le 4$

3    $x = e^t \cos t, y = e^t \sin t; 0 \le t \le \pi/2$

4    $x = 3t^2 - 5, y = 2t^2 + 1; 0 \le t \le 2$

5    $x = \ln \cos t, y = t; 0 \le t \le \pi/3$

6    $x = \cos^3 t, y = \sin^3 t; 0 \le t \le \pi/2$

7    $x = 2t, y = t^4 + \tfrac{1}{8}t^{-2}; 1 \le t \le 2$

8   Find the length of the curve having polar equation $r = \sin^3 (\theta/3)$.

9   Find the length of the spiral $r = e^{-\theta}$ from $\theta = 0$ to $\theta = 2\pi$.

10   Find the length of the spiral $r = \theta$ from $\theta = 0$ to $\theta = 4\pi$.

11   Find the length of the curve $r = \cos^2 (\theta/2)$ from $\theta = 0$ to $\theta = \pi$.

12   Find the length of the spiral $r = 2^\theta$ from $\theta = 0$ to $\theta = \pi$.

The parametric equations in Exercises 13–16 give the position (x, y) of a particle at time t. Find the distance the particle travels during the indicated time interval.

13 $x = 4t + 3$, $y = 2t^2$; $0 \leq t \leq 5$

14 $x = \cos 2t$, $y = \sin^2 t$; $0 \leq t \leq \pi$

15 $x = t \cos t - \sin t$, $y = t \sin t + \cos t$; $0 \leq t \leq \pi/2$

16 $x = t^2$, $y = 2t$; $0 \leq t \leq 4$

17 Verify the formula for $(dx/d\theta)^2 + (dy/d\theta)^2$ in the proof of Theorem (13.12).

18 In a manner analogous to the work with plane areas, define the *centroid* $(\bar{x}, \bar{y})$ of a curve and state integral formulas for finding $(\bar{x}, \bar{y})$ when the curve is given parametrically.

19 Find the length of the curve defined by $x = 1/t$, $y = \ln t$; $1 \leq t \leq 2$.

20 Find the length of the cardioid $r = 1 + \cos \theta$.

13.7

SURFACES OF REVOLUTION

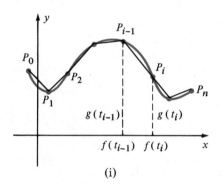

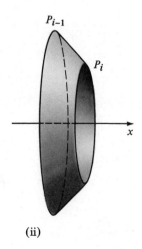

(ii)

Figure 13.28

If a plane curve C is revolved about a line in the plane, a **surface of revolution** is generated. For example, if a circle is revolved about a diameter, the surface of a sphere is obtained. Formulas for surface area may be developed provided C is sufficiently well behaved. In particular, suppose a smooth curve C is given parametrically by $x = f(t)$, $y = g(t)$, where $a \leq t \leq b$ and $g(t) \geq 0$ for all t. The last condition implies that the graph of C is above the x-axis. We shall use the notation of the previous section, letting P denote a partition of $[a, b]$ and $P_i = (f(t_i), g(t_i))$ denote the point on C that corresponds to t_i, as illustrated in (i) of Figure 13.28. Reasoning intuitively, we observe that if the norm $\|P\|$ is close to zero, then the broken line C' obtained by connecting P_{i-1} to P_i for each i is an approximation to C, and hence the area of the surface generated by revolving C' about the x-axis should approximate the area of the surface generated by C. As illustrated in (ii) of Figure 13.28, each segment $P_{i-1}P_i$ generates the lateral surface of a frustum of a right circular cone having base radii $g(t_{i-1})$ and $g(t_i)$ and slant height $d(P_{i-1}, P_i)$. It can be shown that the lateral area of a frustum of a right circular cone of base radii r_1 and r_2 and slant height s is given by $\pi(r_1 + r_2)s$. Hence the area of the surface generated by $P_{i-1}P_i$ is

$$\pi[g(t_{i-1}) + g(t_i)]d(P_{i-1}, P_i).$$

Summing terms of this form from $i = 1$ to $i = n$ gives us the area S_P of the surface generated by the broken line C'.

If we use the expression obtained for $d(P_{i-1}, P_i)$ in Section 13.6, then

$$S_P = \sum_{i=1}^{n} \pi[g(t_i) + g(t_{i-1})]\sqrt{[f'(w_i)]^2 + [g'(z_i)]^2} \, \Delta t_i$$

where w_i and z_i are in the interval (t_{i-1}, t_i). Since the length of C is the limit of the length L_P of C' as $\|P\| \to 0$, it is natural to define the area S of the surface generated by C as

$$S = \lim_{\|P\| \to 0} S_P.$$

From the form of S_P it is reasonable to expect that this limit is given by the definite integral in the following theorem; however, the proof is beyond the scope of this book.

Theorem (13.13)

Let a smooth curve C be given by $x = f(t)$, $y = g(t)$, where $a \le t \le b$, and suppose that $g(t) \ge 0$ for all t. The area S of the surface of revolution obtained by revolving C about the x-axis is

$$S = \int_a^b 2\pi g(t)\sqrt{[f'(t)]^2 + [g'(t)]^2}\, dt = \int_a^b 2\pi y \sqrt{\left(\frac{dx}{dt}\right)^2 + \left(\frac{dy}{dt}\right)^2}\, dt.$$

As an aid to remembering Theorem (13.13), recall that if ds is the differential of arc length, then by Theorem (6.18),

$$(ds)^2 = (dx)^2 + (dy)^2.$$

Assuming that ds and dt are positive, we may write

$$ds = \sqrt{(dx)^2 + (dy)^2} = \sqrt{\left(\frac{dx}{dt}\right)^2 + \left(\frac{dy}{dt}\right)^2}\, dt.$$

If we regard ds as the slant height of a typical frustum, and if (x, y) represents a point in the corresponding subarc, then *the integrand $2\pi y\, ds$ in Theorem (13.13) may be regarded as the product of the slant height and the circumference $2\pi y$ of the circle traced by (x, y).*

It follows in similar fashion that *if the curve C is revolved about the y-axis, and $a \ge 0$, then S is given by*

$$S = \int_a^b 2\pi x \sqrt{\left(\frac{dx}{dt}\right)^2 + \left(\frac{dy}{dt}\right)^2}\, dt,$$

where again we may regard $2\pi x$ as the circumference of a circle traced by a point (x, y) on C.

Example 1 Verify that the surface area of a sphere of radius a is $4\pi a^2$.

Solution If C is the upper half of the circle $x^2 + y^2 = a^2$, then the desired surface may be obtained by revolving C about the x-axis. Parametric equations for C are

$$x = a \cos t, \quad y = a \sin t$$

where $0 \le t \le \pi$. Applying Theorem (13.13) and using the trigonometric identity $\sin^2 t + \cos^2 t = 1$,

$$S = \int_0^\pi 2\pi a \sin t \sqrt{a^2 \sin^2 t + a^2 \cos^2 t}\, dt = 2\pi a^2 \int_0^\pi \sin t\, dt$$

$$= -2\pi a^2 \left[\cos t\right]_0^\pi = -2\pi a^2[-1 - 1] = 4\pi a^2. \qquad \blacksquare$$

If C is the graph of an equation $y = f(x)$, where f is smooth on an interval $[a, b]$, then parametric equations for C are

$$x = t, \quad y = f(t)$$

where $a \le t \le b$. Using (13.13) and the fact that $t = x$, the area of the surface generated by revolving C around the x-axis is given by (i) of the next theorem.

Theorem (13.14)

If C is the graph of $y = f(x)$, where f' is continuous on $[a, b]$, then

(i) the area S of the surface generated by revolving C about the x-axis is

$$S = \int_a^b 2\pi f(x)\sqrt{1 + [f'(x)]^2}\, dx.$$

(ii) if $a \ge 0$, the area S of the surface generated by revolving C about the y-axis is

$$S = \int_a^b 2\pi x\sqrt{1 + [f'(x)]^2}\, dx.$$

Formulas analogous to those in Theorem (13.14) may be stated if C is the graph of an equation $x = g(y)$, where $c \le y \le d$.

Example 2 The arc of the parabola $y^2 = x$ from $(1, 1)$ to $(4, 2)$ is revolved about the x-axis. Find the area of the resulting surface.

Solution The surface is illustrated in Figure 13.29. Using (i) of Theorem (13.14) with $y = f(x) = x^{1/2}$ we obtain

$$S = \int_1^4 2\pi x^{1/2}\sqrt{1 + \left(\frac{1}{2x^{1/2}}\right)^2}\, dx$$

$$= \int_1^4 2\pi x^{1/2}\sqrt{\frac{4x + 1}{4x}}\, dx = \pi \int_1^4 \sqrt{4x + 1}\, dx$$

$$= \frac{\pi}{6}(4x + 1)^{3/2}\Big]_1^4 = \frac{\pi}{6}(17^{3/2} - 5^{3/2}). \qquad \blacksquare$$

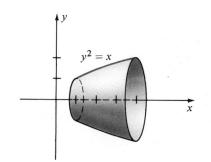

Figure 13.29

Formulas for S may also be stated if C is the graph of a polar equation $r = f(\theta)$, where $a \le \theta \le b$. In this case parametric equations for C are

$$x = f(\theta)\cos\theta, \quad y = f(\theta)\sin\theta$$

where $a \le \theta \le b$. If C is revolved about the polar axis, then substituting in Theorem (13.13) and using the arc length differential obtained in the proof of (13.12) leads to (i) of the next theorem.

Surface Area (Polar Coordinates)
(13.15)

$$\text{(i)} \quad S = \int_a^b 2\pi f(\theta) \sin \theta \sqrt{[f(\theta)]^2 + [f'(\theta)]^2} \, d\theta \quad \text{(around polar axis)}$$

$$\text{(ii)} \quad S = \int_a^b 2\pi f(\theta) \cos \theta \sqrt{[f(\theta)]^2 + [f'(\theta)]^2} \, d\theta \quad \text{(around } \theta = \pi/2)$$

The formula in (ii) is found by using $2\pi x$ instead of $2\pi y$. When (13.15) is used, a and b must be chosen such that the surface does not retrace itself when C is revolved, as would be the case of revolving the graph of $r = \cos \theta$, with $0 \le \theta \le \pi$, about the polar axis.

Example 3 The part of the spiral $r = e^{\theta/2}$ from $\theta = 0$ to $\theta = \pi$ is revolved about the polar axis. Find the area of the resulting surface.

Solution The surface is illustrated in Figure 13.30. Using (i) of (13.15) with $f(\theta) = e^{\theta/2}$,

$$S = \int_0^\pi 2\pi e^{\theta/2} \sin \theta \sqrt{(e^{\theta/2})^2 + (\tfrac{1}{2}e^{\theta/2})^2} \, d\theta$$

$$= \int_0^\pi 2\pi e^{\theta/2} \sin \theta \sqrt{\tfrac{5}{4}e^{\theta}} \, d\theta = \sqrt{5}\pi \int_0^\pi e^{\theta} \sin \theta \, d\theta.$$

Integration by parts leads to

$$S = \frac{\sqrt{5\pi}}{2} e^{\theta}(\sin \theta - \cos \theta) \Big]_0^\pi = \frac{\sqrt{5\pi}}{2} (e^\pi + 1). \quad \blacksquare$$

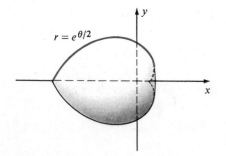

$r = e^{\theta/2}$

Figure 13.30

EXERCISES 13.7

In Exercises 1–8 find the area of the surface generated by revolving the given curve about the x-axis.

1 $x = t^2, y = 2t; 0 \le t \le 4$

2 $x = 4t, y = t^3; 1 \le t \le 2$

3 $x = t^2, y = t - \tfrac{1}{3}t^3; 0 \le t \le 1$

4 $x = 4t^2 + 1, y = 3 - 2t; -2 \le t \le 0$

5 $y = x^3$ from $x = 1$ to $x = 2$

6 $y = e^{-x}$ from $x = 0$ to $x = 1$

7 $x = a(t - \sin t), y = a(1 - \cos t); 0 \le t \le 2\pi$

8 $12y = 4x^3 + (3/x)$ from $x = 1$ to $x = 2$

In Exercises 9–16 find the area of the surface generated by revolving the given arc about the y-axis.

9 $x = 4t^{1/2}, y = \tfrac{1}{2}t^2 + t^{-1}; 1 \le t \le 4$

10 $x = 3t, y = t + 1; 0 \le t \le 5$

11 $x = e^t \sin t, y = e^t \cos t; 0 \le t \le \pi/2$

12 $x = 3t^2, y = 2t^3; 0 \le t \le 1$

13 $x^2 = 16y$ from $(0, 0)$ to $(8, 4)$

14 $8x = y^3$ from $(1, 2)$ to $(8, 4)$

15 $y = \ln x$ from $x = 1$ to $x = 2$

16 $y = \cosh x$ from $x = 1$ to $x = 2$

In Exercises 17–20 find the area of the surface generated by revolving the given curve about the polar axis.

17 $r = 2 + 2\cos\theta$ **18** $r^2 = 4\cos 2\theta$

19 $r = 2a\sin\theta$ **20** $r = 2a\cos\theta$

21 The smaller arc of the circle $x^2 + y^2 = 25$ between the points $(-3, 4)$ and $(3, 4)$ is revolved about the y-axis. Find the area of the resulting surface.

22 If the arc described in Exercise 21 is revolved about the x-axis, find the area of the resulting surface.

23 Prove that the surface area of a right circular cone of altitude a and base radius b is $\pi b\sqrt{a^2 + b^2}$.

24 If the ellipse $b^2x^2 + a^2y^2 = a^2b^2$ is revolved about the y-axis, find the area of the resulting surface.

25 A **torus** is the surface generated by revolving a circle about a nonintersecting line in its plane. Find the surface area of the torus generated by revolving the circle $x^2 + y^2 = a^2$ about the line $x = b$, where $0 < a < b$.

26 The shape of a reflector in a searchlight is obtained by revolving a parabola about its axis. If the reflector is 4 feet across at the opening and 1 foot deep, find its surface area.

27 Parametric equations for the four-cusped hypocycloid $x^{2/3} + y^{2/3} = a^{2/3}$ are $x = a\cos^3 t$, $y = a\sin^3 t$, where $0 \le t \le 2\pi$. Find the area of the surface generated by revolving the hypocycloid about the x-axis.

28 Prove that the area of the surface of a sphere that lies between two parallel planes depends only on the distance between the planes and not on their relative positions along a diameter of the sphere.

13.8

REVIEW

Define or discuss each of the following.

1 Plane curves

2 Closed curves

3 Parametric equations of a curve

4 Polar coordinate systems

5 Graphs of polar equations

6 Areas in polar coordinates

7 Length of a curve

8 Surfaces of revolution

EXERCISES 13.8

In Exercises 1–3 sketch the graph of the curve, find a rectangular equation of a graph that contains the points on the curve, and find the slope of the tangent line at the point corresponding to $t = 1$.

1 $x = (1/t) + 1$, $y = (2/t) - t$; $0 < t \le 4$

2 $x = \cos^2 t - 2$, $y = \sin t + 1$; $0 \le t \le 2\pi$

3 $x = \sqrt{t}$, $y = e^{-t}$; $t \ge 0$

4 Let C be the curve with parametric equations $x = t^2$, $y = 2t^3 + 4t - 1$, where t is in $\mathbb{R}$. Find the x-coordinates of the points on C at which the tangent line passes through the origin.

5 Compare the graphs of the following curves, where t is in $\mathbb{R}$.

    (a) $x = t^2$, $y = t^3$ (b) $x = t^4$, $y = t^6$

    (c) $x = e^{2t}$, $y = e^{3t}$ (d) $x = 1 - \sin^2 t$, $y = \cos^3 t$

In Exercises 6–18, sketch the graph of the equation and find an equation in x and y that has the same graph.

6 $r = 3 + 2\cos\theta$ **7** $r = 6\cos 2\theta$

8 $r = 4 - 4\cos\theta$ **9** $r^2 = -4\sin 2\theta$

10 $r = 2\sin 3\theta$ **11** $r(3\cos\theta - 2\sin\theta) = 6$

12 $r = e^{-\theta}$, $\theta \ge 0$ **13** $r^2 = \sec 2\theta$

14 $r = 8\sec\theta$ **15** $r = 4\cos^2(\theta/2)$

16 $r = 6 - r\cos\theta$ **17** $r = 8/(3 + \cos\theta)$

18 $r = 8/(1 - 3\sin\theta)$

19 Find the slope of the tangent line to the graph of $r = 3/(2 + 2\cos\theta)$ at the point corresponding to $\theta = \pi/2$.

Find polar equations for the graphs in Exercises 20 and 21.

20 $x^2 + y^2 = 2xy$ **21** $y^2 = x^2 - 2x$

22 Find a polar equation of the hyperbola that has focus at the pole, eccentricity 2, and equation of directrix $r = 6\sec\theta$.

23 Find the area of the region bounded by one loop of the graph of $r^2 = 4\sin 2\theta$.

24 Find the area of the region that is inside the graph of $r = 3 + 2\sin\theta$ and outside the graph of $r = 4$.

25 The position (x, y) of a particle at time t is given by $x = 2\sin t$, $y = \sin^2 t$. Find the distance the particle travels from $t = 0$ to $t = \pi/2$.

26 Find the length of the spiral $r = 1/\theta$ from $(1, 1)$ to $(\frac{1}{2}, 2)$.

27 Find the area of the surface generated by revolving the graph of $y = \cosh x$ from $x = 0$ to $x = 1$ about the x-axis.

28 The curve $x = 2t^2 + 1$, $y = 4t - 3$ where $0 \le t \le 1$ is revolved about the y-axis. Find the area of the resulting surface.

29 The arc of the spiral $r = e^\theta$ from $(1, 0)$ to $(e, 1)$ is revolved about the line $\theta = \pi/2$. Find the area of the resulting surface.

30 Find the area of the surface generated by revolving the lemniscate $r^2 = a^2\cos 2\theta$ about the polar axis.

14

VECTORS AND
SOLID ANALYTIC GEOMETRY

Vectors have two different natures, one geometric and the other algebraic.
For applications it is necessary to have an understanding of both aspects.
Because of this, our development will oscillate between the two approaches,
sometimes emphasizing geometry and other times algebra. For simplicity we
shall begin with vectors in a plane. The extension to three dimensions is then
easily accomplished by simply taking a third component into account. We shall
also discuss some basic topics from solid analytic geometry including lines,
planes, cylinders, quadric surfaces, and space curves.

14.1

VECTORS IN TWO DIMENSIONS

In previous work we assigned directions to certain lines, such as the x- and
y-axes. In similar fashion, a **directed line segment** is a line segment to which a
direction has been assigned. Another name for a directed line segment is a
vector. If a vector extends from a point A (called the **initial point**) to a point B
(called the **terminal point**), it is customary to place an arrowhead at B and
use $\overrightarrow{AB}$ to represent the vector (see Figure 14.1). The length of the directed line
segment is called the **magnitude** of the vector $\overrightarrow{AB}$ and is denoted by $|\overrightarrow{AB}|$.
The vectors $\overrightarrow{AB}$ and $\overrightarrow{CD}$ are considered **equal**, and we write $\overrightarrow{AB} = \overrightarrow{CD}$ if and

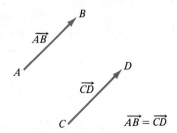

Figure 14.1

617

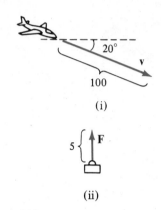

Figure 14.2

only if they have the same magnitude and direction, as illustrated in Figure 14.1. Consequently, vectors may be translated from one position to another, provided neither the magnitude nor direction is changed. Vectors of this type are often referred to as **free vectors**.

Many physical concepts may be represented by vectors. To illustrate, suppose an airplane is descending at a constant speed of 100 mph and the line of flight makes an angle of 20° with the horizontal. Both of these facts are represented by the vector in (i) of Figure 14.2, where it is assumed that units have been chosen so that the magnitude is 100. As shown in the figure, we will use a boldface letter such as **v** to denote a vector whose endpoints are not specified. For written work, where it is difficult to represent boldface type, a notation such as $\vec{v}$ may be employed. The vector **v** in this illustration is called a **velocity vector**.

As a second illustration, suppose a person pulls directly upward on an object with a force of 5 lb, as would be the case in lifting a 5-lb weight. We may indicate this fact by the vector **F** in (ii) of Figure 14.2. A vector that represents a pull or push of some type is called a **force vector**.

Another use for vectors is to let $\overrightarrow{AB}$ represent the path of a point as it moves along the line segment from A to B. We then refer to $\overrightarrow{AB}$ as a **displacement** of the point. As illustrated in (i) of Figure 14.3, a displacement $\overrightarrow{AB}$ followed by a displacement $\overrightarrow{BC}$ will lead to the same point as the single displacement $\overrightarrow{AC}$. By definition, the vector $\overrightarrow{AC}$ is called the **sum** of $\overrightarrow{AB}$ and $\overrightarrow{BC}$, and we write

$$\overrightarrow{AC} = \overrightarrow{AB} + \overrightarrow{BC}.$$

Since we are working with free vectors, *any* two vectors may be added by placing the initial point of one on the terminal point of the other and then proceeding as in (i) of Figure 14.3.

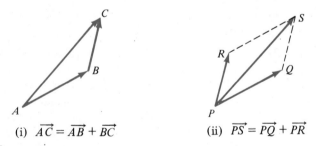

(i) $\overrightarrow{AC} = \overrightarrow{AB} + \overrightarrow{BC}$ (ii) $\overrightarrow{PS} = \overrightarrow{PQ} + \overrightarrow{PR}$

Figure 14.3

Another way to find the sum of two vectors is to consider vectors that are equal to the given ones and have the same initial point, as illustrated by $\overrightarrow{PQ}$ and $\overrightarrow{PR}$ in (ii) of Figure 14.3. If we construct the parallelogram $RPQS$, then since $\overrightarrow{PR} = \overrightarrow{QS}$, it follows that $\overrightarrow{PS} = \overrightarrow{PQ} + \overrightarrow{PR}$. If $\overrightarrow{PQ}$ and $\overrightarrow{PR}$ are two forces acting at P, then $\overrightarrow{PS}$ is the **resultant force**, that is, the single force that produces the same effect as the two combined forces.

If c is a real number and $\overrightarrow{AB}$ is a vector, then $c\overrightarrow{AB}$ is defined as a vector whose magnitude is $|c|$ times the magnitude of $\overrightarrow{AB}$ and whose direction is the same as $\overrightarrow{AB}$ if $c > 0$, and opposite that of $\overrightarrow{AB}$ if $c < 0$. Geometric illustrations are given in Figure 14.4. We refer to c as a **scalar** and $c\overrightarrow{AB}$ as a **scalar multiple** of $\overrightarrow{AB}$.

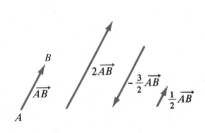

Figure 14.4

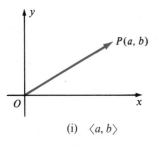

(i) $\langle a, b \rangle$

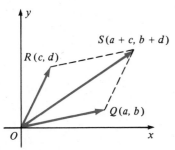

(ii) $\langle a, b \rangle + \langle c, d \rangle = \langle a + c, b + d \rangle$

Figure 14.5

Let us next introduce a coordinate plane and assume that all vectors under discussion are in that plane. Since the position of a vector may be changed, provided the magnitude and direction are not altered, we may place the initial point of each vector at the origin. The terminal point of a typical vector $\overrightarrow{OP}$ may then be assigned coordinates (a, b), as shown in (i) of Figure 14.5. Conversely, every ordered pair (a, b) determines the vector $\overrightarrow{OP}$, where P has coordinates (a, b). We thus obtain a one-to-one correspondence between vectors and ordered pairs. This allows us to regard a vector in a plane as an ordered pair of real numbers instead of a directed line segment. To avoid confusion with the notation for open intervals or points, we shall use the symbol $\langle a, b \rangle$ for an ordered pair that represents a vector.

If, as in (ii) of Figure 14.5, we consider vectors $\overrightarrow{OQ}$ and $\overrightarrow{OR}$ corresponding to $\langle a, b \rangle$ and $\langle c, d \rangle$, respectively, and if we let $\overrightarrow{OS}$ be the vector corresponding to $\langle a + c, b + d \rangle$, it is easy to show, using slopes, that $O, Q, S,$ and R are vertices of a parallelogram. It follows that $\overrightarrow{OQ} + \overrightarrow{OR} = \overrightarrow{OS}$ or, in terms of ordered pairs,

$$\langle a, b \rangle + \langle c, d \rangle = \langle a + c, b + d \rangle.$$

Although we shall not prove it at this time, the ordered-pair formula that corresponds to a scalar multiple $k\overrightarrow{OP}$ of the vector in (i) of the figure is

$$k\langle a, b \rangle = \langle ka, kb \rangle.$$

In this way, vectors take on a dual role, with the algebraic viewpoint of ordered pairs emerging from the previous work involving geometry. It is possible to reverse the order of this development and *begin* with an algebraic definition of vectors and *then* obtain directed line segments as an alternate interpretation. There are several advantages to this approach. One is that proofs of properties of vectors become simple exercises in algebraic manipulations instead of problems in geometry. Another advantage is that the algebraic approach to vectors is easily generalized to vectors in three or more dimensions. With these remarks in mind we shall next state an algebraic definition for vectors. Geometric representations that agree with our previous discussion will be introduced after the definition. We shall continue to use $\langle x, y \rangle$ for an ordered pair that refers to a vector. Such ordered pairs will also be denoted by symbols such as **a** or $\vec{a}$. As before, real numbers will be called *scalars*.

Definition (14.1)

> The **two-dimensional vector space V_2** is the set of all ordered pairs $\langle x, y \rangle$ of real numbers, called **vectors**, subject to the following axioms.
>
> (i) **Addition of vectors.** If $\mathbf{a} = \langle a_1, a_2 \rangle$ and $\mathbf{b} = \langle b_1, b_2 \rangle$ are vectors, then
>
> $$\mathbf{a} + \mathbf{b} = \langle a_1 + b_1, a_2 + b_2 \rangle.$$
>
> (ii) **Multiplication of vectors by scalars.** If $\mathbf{a} = \langle a_1, a_2 \rangle$ and c is a scalar, then
>
> $$c\mathbf{a} = \langle ca_1, ca_2 \rangle.$$

It should be clear that (i) and (ii) of Definition (14.1) were motivated by our representation of directed line segments as ordered pairs. The numbers a_1 and a_2 in $\langle a_1, a_2 \rangle$ are called the **components** of the vector. Thus, to add two vectors we add corresponding components.

The **zero vector 0** and the **negative** $-\mathbf{a}$ of a vector $\mathbf{a} = \langle a_1, a_2 \rangle$ are defined as follows:

(14.2) $$\mathbf{0} = \langle 0, 0 \rangle; \qquad -\mathbf{a} = -\langle a_1, a_2 \rangle = \langle -a_1, -a_2 \rangle.$$

A nonzero vector $\mathbf{a} = \langle a_1, a_2 \rangle$ will be represented in a coordinate plane by a directed line segment $\overrightarrow{PQ}$ with any initial point $P(x, y)$ and terminal point $Q(x + a_1, y + a_2)$. Several representations are shown in Figure 14.6, where the symbol $\mathbf{a}$ is placed next to any directed line segment that represents the vector $\mathbf{a}$ of V_2. The zero vector $\mathbf{0} = \langle 0, 0 \rangle$ will be represented by any point in the plane. Strictly speaking, a directed line segment $\overrightarrow{PQ}$ *represents* a vector; however, we shall also refer to $\overrightarrow{PQ}$ as a *vector*. It should always be clear from the discussion whether the term "vector" refers to an ordered pair or a directed line segment.

The geometric representation for the sum of two vectors $\mathbf{a} = \langle a_1, a_2 \rangle$ and $\mathbf{b} = \langle b_1, b_2 \rangle$ is illustrated in Figure 14.7, where we have chosen an arbitrary point $P(x, y)$ for the initial point of $\mathbf{a}$.

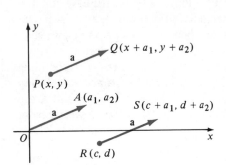

Figure 14.6

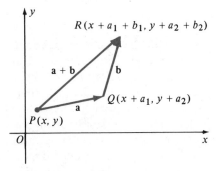

Figure 14.7

Example 1   If $\mathbf{a} = \langle -1, 3 \rangle$ and $\mathbf{b} = \langle 4, 2 \rangle$,

(a)   find $\mathbf{a} + \mathbf{b}$, $-\frac{3}{2}\mathbf{b}$, and $2\mathbf{a} + 3\mathbf{b}$.

(b)   represent $\mathbf{a}, \mathbf{b}, \mathbf{a} + \mathbf{b}$, and $-\frac{3}{2}\mathbf{b}$ geometrically.

Solution

(a)   Applying Definition (14.1),

$$\mathbf{a} + \mathbf{b} = \langle -1, 3 \rangle + \langle 4, 2 \rangle = \langle -1 + 4, 3 + 2 \rangle = \langle 3, 5 \rangle$$
$$-\tfrac{3}{2}\mathbf{b} = -\tfrac{3}{2}\langle 4, 2 \rangle = \langle -6, -3 \rangle$$
$$2\mathbf{a} + 3\mathbf{b} = 2\langle -1, 3 \rangle + 3\langle 4, 2 \rangle = \langle -2, 6 \rangle + \langle 12, 6 \rangle = \langle 10, 12 \rangle.$$

(b)   The vectors $\mathbf{a}, \mathbf{b}, \mathbf{a} + \mathbf{b}$, and $-\frac{3}{2}\mathbf{b}$ are represented geometrically in Figure 14.8, where we have chosen the origin for the initial point of each vector. Note that $-\frac{3}{2}\mathbf{b}$ has a direction opposite to that of $\mathbf{b}$ and is $\frac{3}{2}$ as long.

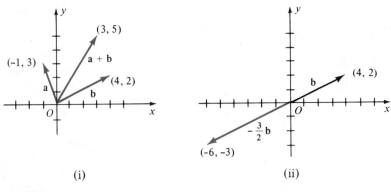

(i)   (ii)

Figure 14.8

Definition (14.3)

> The **magnitude** $|\mathbf{a}|$ of a vector $\mathbf{a} = \langle a_1, a_2 \rangle$ is $|\mathbf{a}| = \sqrt{a_1^2 + a_2^2}$.

It follows from the Distance Formula that $|\mathbf{a}|$ is the length of any directed line segment used to represent $\mathbf{a}$. Note also that $|\mathbf{a}| \geq 0$, where $|\mathbf{a}| = 0$, if and only if $\mathbf{a} = \mathbf{0}$. The symbol for the magnitude of a vector should not be confused with the symbol $|c|$ for the absolute value of a real number c. As an illustration, if $\mathbf{a} = \langle 3, -2 \rangle$, then

$$|\mathbf{a}| = |\langle 3, -2 \rangle| = \sqrt{9 + 4} = \sqrt{13}.$$

If P is any point in a coordinate plane and O is the origin, then $\overrightarrow{OP}$ is called the **position vector** of P. If $\mathbf{a} = \langle a_1, a_2 \rangle$, then the position vector $\overrightarrow{OA}$ of the point $A(a_1, a_2)$ (see (i) of Figure 14.6) is called the **position vector corresponding to a**. Although $\mathbf{a}$ has many geometric representations, there is precisely one position vector corresponding to $\mathbf{a}$.

Theorem (14.4)

> If $P_1(x_1, y_1)$ and $P_2(x_2, y_2)$ are points, then the vector $\mathbf{a}$ in V_2 that has geometric representation $\overrightarrow{P_1 P_2}$ is
>
> $$\mathbf{a} = \langle x_2 - x_1, y_2 - y_1 \rangle.$$

Proof   If $P_1(x_1, y_1)$ is the initial point of a vector corresponding to $\mathbf{a} = \langle a_1, a_2 \rangle$, then the terminal point is $Q(x_1 + a_1, y_1 + a_2)$. If Q is to coincide with $P_2(x_2, y_2)$, then we must have $x_2 = x_1 + a_1$ and $y_2 = y_1 + a_2$. Consequently, $\mathbf{a} = \langle x_2 - x_1, y_2 - y_1 \rangle$. One illustration of $\overrightarrow{P_1 P_2}$ is shown in Figure 14.9, together with the position vector corresponding to $\mathbf{a}$.

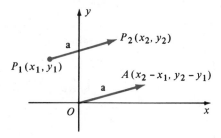

Figure 14.9   □

Example 2   Given the points $P(-2, 3)$ and $Q(4, 5)$, find vectors $\mathbf{a}$ and $\mathbf{b}$ that have geometric representations $\overrightarrow{PQ}$ and $\overrightarrow{QP}$, respectively. Sketch $\overrightarrow{PQ}$, $\overrightarrow{QP}$, and the position vectors corresponding to $\mathbf{a}$ and $\mathbf{b}$.

Solution   Applying Theorem (14.4), the vectors $\mathbf{a}$ and $\mathbf{b}$ are

$$\mathbf{a} = \langle 4 - (-2), 5 - 3 \rangle = \langle 6, 2 \rangle,$$

$$\mathbf{b} = \langle -2 - 4, 3 - 5 \rangle = \langle -6, -2 \rangle.$$

The sketches are shown in Figure 14.10. Observe that $\mathbf{b} = -\mathbf{a}$.

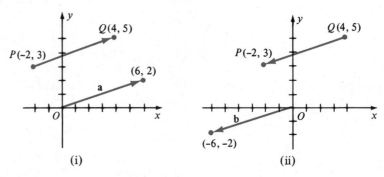

Figure 14.10   ■

Four important properties are stated in the next theorem, where $\mathbf{a}$, $\mathbf{b}$, and $\mathbf{c}$ are any vectors in V_2.

Theorem (14.5)

> (i) $\mathbf{a} + \mathbf{b} = \mathbf{b} + \mathbf{a}$
> (ii) $\mathbf{a} + (\mathbf{b} + \mathbf{c}) = (\mathbf{a} + \mathbf{b}) + \mathbf{c}$
> (iii) $\mathbf{a} + \mathbf{0} = \mathbf{a}$
> (iv) $\mathbf{a} + (-\mathbf{a}) = \mathbf{0}$

Proof   For part (i) let $\mathbf{a} = \langle a_1, a_2 \rangle$ and $\mathbf{b} = \langle b_1, b_2 \rangle$. Since $a_1 + b_1 = b_1 + a_1$ and $a_2 + b_2 = b_2 + a_2$,

$$\mathbf{a} + \mathbf{b} = \langle a_1 + b_1, a_2 + b_2 \rangle = \langle b_1 + a_1, b_2 + a_2 \rangle = \mathbf{b} + \mathbf{a}.$$

The remainder of the proof is left as an exercise.   □

The operation of **subtraction** of vectors, denoted by $-$, is defined as follows.

Definition (14.6)

> Let $\mathbf{a} = \langle a_1, a_2 \rangle$ and $\mathbf{b} = \langle b_1, b_2 \rangle$. The **difference a $-$ b** is
>
> $$\mathbf{a} - \mathbf{b} = \mathbf{a} + (-\mathbf{b}) = \langle a_1 - b_1, a_2 - b_2 \rangle.$$

The last equality in (14.6) is true because $-\mathbf{b} = \langle -b_1, -b_2 \rangle$. Thus, to find $\mathbf{a} - \mathbf{b}$ we merely subtract the components of $\mathbf{b}$ from the corresponding components of $\mathbf{a}$.

Example 3   If $\mathbf{a} = \langle 5, -4 \rangle$ and $\mathbf{b} = \langle -3, 2 \rangle$, find $\mathbf{a} - \mathbf{b}$ and $2\mathbf{a} - 3\mathbf{b}$.

Solution

$$\mathbf{a} - \mathbf{b} = \langle 5, -4 \rangle - \langle -3, 2 \rangle = \langle 5 - (-3), -4 - 2 \rangle = \langle 8, -6 \rangle$$

$$2\mathbf{a} - 3\mathbf{b} = 2\langle 5, -4 \rangle - 3\langle -3, 2 \rangle = \langle 10, -8 \rangle - \langle -9, 6 \rangle = \langle 19, -14 \rangle$$

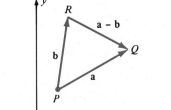

Figure 14.11

It is easy to show that if $\mathbf{a}$ and $\mathbf{b}$ are arbitrary vectors, then

$$\mathbf{b} + (\mathbf{a} - \mathbf{b}) = \mathbf{a},$$

that is, $\mathbf{a} - \mathbf{b}$ is the vector that, when added to $\mathbf{b}$, gives us $\mathbf{a}$. This fact leads to a simple geometric interpretation for $\mathbf{a} - \mathbf{b}$. Specifically, if we represent $\mathbf{a}$ and $\mathbf{b}$ by vectors $\overrightarrow{PQ}$ and $\overrightarrow{PR}$ *with the same initial point*, as illustrated in Figure 14.11, then $\overrightarrow{RQ}$ represents $\mathbf{a} - \mathbf{b}$.

At the beginning of this section the symbol $c\overrightarrow{AB}$ was used to denote a vector having the same direction as $\overrightarrow{AB}$ if $c > 0$ or the opposite direction if $c < 0$. The next definition is the analogue of this concept for vectors in V_2.

Definition (14.7)

> Nonzero vectors $\mathbf{a}$ and $\mathbf{b}$ of V_2 have
>
> (i)   the **same direction** if $\mathbf{b} = c\mathbf{a}$ for some scalar $c > 0$.
>
> (ii) **opposite directions** if $\mathbf{b} = c\mathbf{a}$ for some scalar $c < 0$.

It is not difficult to prove that if $\mathbf{b} = c\mathbf{a}$, then position vectors corresponding to $\mathbf{a}$ and $\mathbf{b}$ lie on the same line through the origin. Moreover, if $c > 0$, then the terminal points A and B of the position vectors lie in the same quadrant (or on the same positive or negative coordinate axis). If $c < 0$, and A is not on a coordinate axis, then A and B lie in diagonally opposite quadrants.

It is convenient to assume that the zero vector $\mathbf{0}$ has *all* directions.

We say that two vectors $\mathbf{a}$ and $\mathbf{b}$ are **parallel** if and only if $\mathbf{b} = c\mathbf{a}$ for some scalar c. In particular, nonzero vectors $\mathbf{a}$ and $\mathbf{b}$ are parallel if they have the same, or opposite, directions. The next theorem brings out the relationship between the magnitudes of $\mathbf{a}$ and $c\mathbf{a}$.

Theorem (14.8)

> If $\mathbf{a}$ is a vector and c is a scalar, then $|c\mathbf{a}| = |c|\,|\mathbf{a}|$.

Proof   If $\mathbf{a} = \langle a_1, a_2 \rangle$, then $c\mathbf{a} = \langle ca_1, ca_2 \rangle$. Using Definition (14.3) and properties of real numbers,

$$|c\mathbf{a}| = \sqrt{(ca_1)^2 + (ca_2)^2} = \sqrt{c^2(a_1^2 + a_2^2)}$$
$$= \sqrt{c^2}\sqrt{a_1^2 + a_2^2} = |c|\,|\mathbf{a}|. \qquad \square$$

Theorem (14.8) implies that the length of a directed line segment that represents $c\mathbf{a}$ is $|c|$ times the length of a directed line segment that represents $\mathbf{a}$. This agrees with the geometric interpretation discussed at the beginning of this section and illustrated in Figure 14.4.

Some properties of scalar multiples of vectors in V_2 are stated in the following theorem.

Theorem (14.9)

> (i) $c(\mathbf{a} + \mathbf{b}) = c\mathbf{a} + c\mathbf{b}$
>
> (ii) $(c + d)\mathbf{a} = c\mathbf{a} + d\mathbf{a}$
>
> (iii) $(cd)\mathbf{a} = c(d\mathbf{a}) = d(c\mathbf{a})$
>
> (iv) $1\mathbf{a} = \mathbf{a}$
>
> (v) $0\mathbf{a} = \mathbf{0} = c\mathbf{0}$

Proof   We shall prove (i) and leave proofs of the remaining properties as exercises. Letting $\mathbf{a} = \langle a_1, a_2 \rangle$ and $\mathbf{b} = \langle b_1, b_2 \rangle$, we have

$$
\begin{aligned}
c(\mathbf{a} + \mathbf{b}) &= c\langle a_1 + b_1, a_2 + b_2 \rangle \\
&= \langle ca_1 + cb_1, ca_2 + cb_2 \rangle \\
&= \langle ca_1, ca_2 \rangle + \langle cb_1, cb_2 \rangle \\
&= c\mathbf{a} + c\mathbf{b}. \qquad \square
\end{aligned}
$$

There should be no difficulty in remembering the properties in Theorem (14.9) since they resemble familiar properties of real numbers.

The following special vectors $\mathbf{i}$ and $\mathbf{j}$ will be important in future developments.

(14.10)
$$
\mathbf{i} = \langle 1, 0 \rangle, \qquad \mathbf{j} = \langle 0, 1 \rangle
$$

We may use $\mathbf{i}$ and $\mathbf{j}$ to obtain another way of denoting vectors in two dimensions, as indicated by the next theorem.

Theorem (14.11)

> If $\mathbf{a} = \langle a_1, a_2 \rangle$ is any vector in V_2, then $\mathbf{a} = a_1\mathbf{i} + a_2\mathbf{j}$.

Proof   Using (i) and (ii) of Definition (14.1),

$$
\langle a_1, a_2 \rangle = \langle a_1, 0 \rangle + \langle 0, a_2 \rangle = a_1\langle 1, 0 \rangle + a_2\langle 0, 1 \rangle = a_1\mathbf{i} + a_2\mathbf{j}. \qquad \square
$$

The formula $\mathbf{a} = a_1\mathbf{i} + a_2\mathbf{j}$ for the vector $\mathbf{a} = \langle a_1, a_2 \rangle$ has an interesting geometric interpretation. Position vectors corresponding to $\mathbf{i}$, $\mathbf{j}$, and $\mathbf{a}$ are

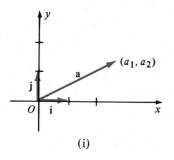

(i)

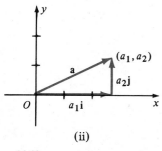

(ii)

Figure 14.12

illustrated in (i) of Figure 14.12. Since $\mathbf{i}$ and $\mathbf{j}$ are unit vectors, $a_1\mathbf{i}$ and $a_2\mathbf{j}$ may be represented by horizontal and vertical vectors of magnitudes $|a_1|$ and $|a_2|$, respectively, as illustrated in (ii) of Figure 14.12. The position vector for $\mathbf{a}$ may be regarded as the sum of these vectors. For this reason a_1 is called the **horizontal component** and a_2 the **vertical component** of the vector $\mathbf{a}$.

The vector sum $a_1\mathbf{i} + a_2\mathbf{j}$ is called a **linear combination** of $\mathbf{i}$ and $\mathbf{j}$. Note that if $\mathbf{b} = \langle b_1, b_2 \rangle = b_1\mathbf{i} + b_2\mathbf{j}$, and if c is a scalar, then

$$(a_1\mathbf{i} + a_2\mathbf{j}) + (b_1\mathbf{i} + b_2\mathbf{j}) = (a_1 + b_1)\mathbf{i} + (a_2 + b_2)\mathbf{j},$$
$$(a_1\mathbf{i} + a_2\mathbf{j}) - (b_1\mathbf{i} + b_2\mathbf{j}) = (a_1 - b_1)\mathbf{i} + (a_2 - b_2)\mathbf{j},$$

and
$$c(a_1\mathbf{i} + a_2\mathbf{j}) = (ca_1)\mathbf{i} + (ca_2)\mathbf{j}.$$

These formulas show that linear combinations of $\mathbf{i}$ and $\mathbf{j}$ may be regarded as ordinary algebraic sums.

Example 4 If $\mathbf{a} = 5\mathbf{i} + \mathbf{j}$ and $\mathbf{b} = 4\mathbf{i} - 7\mathbf{j}$, express $3\mathbf{a} - 2\mathbf{b}$ as a linear combination of $\mathbf{i}$ and $\mathbf{j}$.

Solution
$$3\mathbf{a} - 2\mathbf{b} = 3(5\mathbf{i} + \mathbf{j}) - 2(4\mathbf{i} - 7\mathbf{j})$$
$$= (15\mathbf{i} + 3\mathbf{j}) - (8\mathbf{i} - 14\mathbf{j})$$
$$= 7\mathbf{i} + 17\mathbf{j} \qquad \blacksquare$$

A **unit vector** is a vector of magnitude 1. The vectors $\mathbf{i}$ and $\mathbf{j}$ are unit vectors.

Theorem (14.12)

> If $\mathbf{a}$ is a nonzero vector, then a unit vector $\mathbf{u}$ having the same direction as $\mathbf{a}$ is
> $$\mathbf{u} = \frac{1}{|\mathbf{a}|}\mathbf{a}.$$

Proof If we let $c = 1/|\mathbf{a}|$, then $c > 0$ and it follows from Definition (14.7) that the vector $\mathbf{u} = c\mathbf{a}$ has the same direction as $\mathbf{a}$. Moreover, by Theorem (14.8),

$$|\mathbf{u}| = |c\mathbf{a}| = |c|\,|\mathbf{a}| = \frac{1}{|\mathbf{a}|}\,|\mathbf{a}| = 1. \qquad \square$$

Example 5 Find a unit vector $\mathbf{u}$ having the same direction as $3\mathbf{i} - 4\mathbf{j}$.

Solution Since $|\mathbf{a}| = \sqrt{9 + 16} = 5$ we have, from Theorem (14.12),

$$\mathbf{u} = \tfrac{1}{5}(3\mathbf{i} - 4\mathbf{j}) = \tfrac{3}{5}\mathbf{i} - \tfrac{4}{5}\mathbf{j}. \qquad \blacksquare$$

EXERCISES 14.1

1 (a) Prove (ii)–(iv) of Theorem (14.5).

(b) Give geometric arguments that justify the conclusions.

2 Prove (ii)–(v) of Theorem (14.9).

Prove the properties in Exercises 3–10, where $\mathbf{a}$, $\mathbf{b}$ are vectors and c is a scalar.

3 $(-1)\mathbf{a} = -\mathbf{a}$

4 $(-c)\mathbf{a} = -c\mathbf{a}$

5 $-(\mathbf{a} + \mathbf{b}) = -\mathbf{a} - \mathbf{b}$

6 $c(\mathbf{a} - \mathbf{b}) = c\mathbf{a} - c\mathbf{b}$

7 If $\mathbf{a} + \mathbf{b} = \mathbf{0}$, then $\mathbf{b} = -\mathbf{a}$.

8 If $\mathbf{a} + \mathbf{b} = \mathbf{a}$, then $\mathbf{b} = \mathbf{0}$.

9 If $c\mathbf{a} = \mathbf{0}$ and $c \neq 0$, then $\mathbf{a} = \mathbf{0}$.

10 If $c\mathbf{a} = \mathbf{0}$ and $\mathbf{a} \neq \mathbf{0}$, then $c = 0$.

In Exercises 11–20 find $\mathbf{a} + \mathbf{b}$, $\mathbf{a} - \mathbf{b}$, $4\mathbf{a} + 5\mathbf{b}$, and $4\mathbf{a} - 5\mathbf{b}$.

11 $\mathbf{a} = \langle 2, -3 \rangle$, $\mathbf{b} = \langle 1, 4 \rangle$

12 $\mathbf{a} = \langle -2, 6 \rangle$, $\mathbf{b} = \langle 2, 3 \rangle$

13 $\mathbf{a} = -\langle 7, -2 \rangle$, $\mathbf{b} = 4\langle -2, 1 \rangle$

14 $\mathbf{a} = 2\langle 5, -4 \rangle$, $\mathbf{b} = -\langle 6, 0 \rangle$

15 $\mathbf{a} = \mathbf{i} + 2\mathbf{j}$, $\mathbf{b} = 3\mathbf{i} - 5\mathbf{j}$

16 $\mathbf{a} = -3\mathbf{i} + \mathbf{j}$, $\mathbf{b} = -3\mathbf{i} + \mathbf{j}$

17 $\mathbf{a} = -(4\mathbf{i} - \mathbf{j})$, $\mathbf{b} = 2(\mathbf{i} - 3\mathbf{j})$

18 $\mathbf{a} = 8\mathbf{j}$, $\mathbf{b} = (-3)(-2\mathbf{i} + \mathbf{j})$

19 $\mathbf{a} = 2\mathbf{j}$, $\mathbf{b} = -3\mathbf{i}$

20 $\mathbf{a} = \mathbf{0}$, $\mathbf{b} = \mathbf{i} + \mathbf{j}$

In Exercises 21–26 find a vector $\mathbf{a}$ that has geometric representation $\overrightarrow{PQ}$. Sketch $\overrightarrow{PQ}$ and a position vector corresponding to $\mathbf{a}$.

21 $P(1, -4)$, $Q(5, 3)$

22 $P(7, -3)$, $Q(-2, 4)$

23 $P(2, 5)$, $Q(-4, 5)$

24 $P(-4, 6)$, $Q(-4, -2)$

25 $P(-3, -1)$, $Q(6, -4)$

26 $P(2, 3)$, $Q(-6, 0)$

In Exercises 27–30 sketch vectors corresponding to $\mathbf{a}$, $\mathbf{b}$, $\mathbf{a} + \mathbf{b}$, $\mathbf{a} - \mathbf{b}$, $2\mathbf{a}$, and $-3\mathbf{b}$.

27 $\mathbf{a} = 3\mathbf{i} + 2\mathbf{j}$, $\mathbf{b} = -\mathbf{i} + 5\mathbf{j}$

28 $\mathbf{a} = -5\mathbf{i} + 2\mathbf{j}$, $\mathbf{b} = \mathbf{i} - 3\mathbf{j}$

29 $\mathbf{a} = \langle -4, 6 \rangle$, $\mathbf{b} = \langle -2, 3 \rangle$

30 $\mathbf{a} = \langle 2, 0 \rangle$, $\mathbf{b} = \langle -2, 0 \rangle$

In Exercises 31–38 find the magnitude of $\mathbf{a}$.

31 $\mathbf{a} = \langle 3, -3 \rangle$

32 $\mathbf{a} = \langle -2, -2\sqrt{3} \rangle$

33 $\mathbf{a} = \langle -5, 0 \rangle$

34 $\mathbf{a} = \langle 0, 10 \rangle$

35 $\mathbf{a} = -4\mathbf{i} + 5\mathbf{j}$

36 $\mathbf{a} = 10\mathbf{i} - 10\mathbf{j}$

37 $\mathbf{a} = -18\mathbf{j}$

38 $\mathbf{a} = 0\mathbf{i} + 0\mathbf{j}$

In Exercises 39–42 find a unit vector having (a) the same direction as $\mathbf{a}$; (b) the direction opposite to $\mathbf{a}$.

39 $\mathbf{a} = -8\mathbf{i} + 15\mathbf{j}$

40 $\mathbf{a} = 5\mathbf{i} - 3\mathbf{j}$

41 $\mathbf{a} = \langle 2, -5 \rangle$

42 $\mathbf{a} = \langle 0, 6 \rangle$

43 Find a vector that has the same direction as $\langle -6, 3 \rangle$ and (a) twice the magnitude; (b) one-half the magnitude.

44 Find a vector that has the opposite direction of $8\mathbf{i} - 5\mathbf{j}$ and (a) three times the magnitude; (b) one-third the magnitude.

45 Find a vector of magnitude 6 that has the same direction as $\mathbf{a} = 4\mathbf{i} - 7\mathbf{j}$.

46 Find a vector of magnitude 4 whose direction is opposite that of $\mathbf{a} = \langle 2, -5 \rangle$.

47 Demonstrate graphically that $|\mathbf{a} + \mathbf{b}| \leq |\mathbf{a}| + |\mathbf{b}|$. Under what conditions is $|\mathbf{a} + \mathbf{b}| = |\mathbf{a}| + |\mathbf{b}|$?

48 (a) Let $\mathbf{a} = \langle a_1, a_2 \rangle$ be any nonzero vector and let $\overrightarrow{OA}$ be the position vector corresponding to $\mathbf{a}$. If θ is the smallest nonnegative angle from the positive x-axis to $\overrightarrow{OA}$, show that $\mathbf{a} = |\mathbf{a}|(\cos\theta\,\mathbf{i} + \sin\theta\,\mathbf{j})$.

(b) Show that every unit vector in V_2 can be expressed in the form $\cos\theta\,\mathbf{i} + \sin\theta\,\mathbf{j}$ for some θ.

49 If $\mathbf{a} = \langle 1, -4 \rangle$, $\mathbf{b} = \langle -2, 6 \rangle$, and $\mathbf{c} = \langle -2, 3 \rangle$, find scalars p and q such that $\mathbf{c} = p\mathbf{a} + q\mathbf{b}$.

50 Generalize Exercise 49 as follows. If $\mathbf{a} = \langle a_1, a_2 \rangle$ and $\mathbf{b} = \langle b_1, b_2 \rangle$ are any nonzero, nonparallel vectors, and if $\mathbf{c}$ is any other vector, prove that there exist scalars p and q such that $\mathbf{c} = p\mathbf{a} + q\mathbf{b}$. Interpret this fact geometrically.

51 If $\mathbf{a} = \langle 4, -2 \rangle$, $\mathbf{b} = \langle -6, 3 \rangle$, and $\mathbf{c} = \langle 2, 5 \rangle$, prove that $\mathbf{c}$ cannot be expressed in the form $\mathbf{c} = p\mathbf{a} + q\mathbf{b}$, where p and q are scalars.

52 If $\mathbf{a} = 3\mathbf{i} - 4\mathbf{j}$, find all real numbers c such that

(a) $|c\mathbf{a}| = 3$;   (b) $|c\mathbf{a}| = -3$;   (c) $|c\mathbf{a}| = 0$.

53 Let $\mathbf{r}_0 = \langle x_0, y_0 \rangle$, $\mathbf{r} = \langle x, y \rangle$, and $c > 0$. Describe the set of all points $P(x, y)$ such that $|\mathbf{r} - \mathbf{r}_0| = c$.

54 Let $\mathbf{r}_0 = \langle x_0, y_0 \rangle$, $\mathbf{r} = \langle x, y \rangle$, and $\mathbf{a} = \langle a_1, a_2 \rangle \neq \mathbf{0}$. Describe the set of all points $P(x, y)$ such that $\mathbf{r} - \mathbf{r}_0$ is a scalar multiple of $\mathbf{a}$.

55 If $P_1, P_2, \ldots, P_n$ are arbitrary points in a coordinate plane, show that $\sum_{i=1}^{n-1} \overrightarrow{P_i P_{i+1}} + \overrightarrow{P_n P_1}$ is the zero vector. Illustrate this fact graphically if $n = 5$.

56 If $P_1, P_2, \ldots, P_n$ are vertices of a regular polygon and if O is the center of the polygon, prove that $\sum_{i=1}^{n} \overrightarrow{OP_i}$ is the zero vector.

14.2

RECTANGULAR COORDINATE SYSTEMS IN THREE DIMENSIONS

If we wish to discuss geometric aspects of vectors in space, it is necessary to introduce a three-dimensional coordinate system. This is accomplished by means of ordered triples of real numbers. An **ordered triple** (a, b, c) is a set $\{a, b, c\}$ of three numbers in which a is considered as the first number of the set, b as the second number, and c as the third number. The totality of all such ordered triples is denoted by $\mathbb{R} \times \mathbb{R} \times \mathbb{R}$. Two ordered triples (a_1, a_2, a_3) and (b_1, b_2, b_3) are said to be **equal**, and we write $(a_1, a_2, a_3) = (b_1, b_2, b_3)$, if and only if $a_1 = b_1$, $a_2 = b_2$, and $a_3 = b_3$.

To specify points in space we choose a fixed point O (called the **origin**) and consider three mutually perpendicular coordinate lines (called the x-, y-, and z-axes) with common origin O, as illustrated in (i) of Figure 14.13.

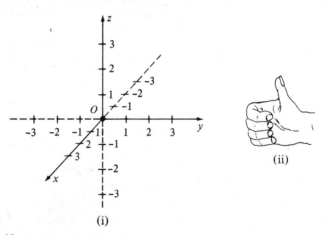

(i)

(ii)

Figure 14.13 Right-handed coordinate system

To visualize this configuration, we may imagine that the y- and z-axes lie in the plane of the paper and that the x-axis projects out from the paper. A coordinate system of this type is called **right-handed**. If the x- and y-axes are interchanged, the resulting coordinate system is said to be **left-handed**. The coordinate plane determined by the x- and y-axes is called the **xy-plane**. Similarly, the coordinate plane determined by the y- and z-axes is called the **yz-plane** and that determined by the x- and z-axes is called the **xz-plane**. The terminology *right-handed* may be justified as follows. If, as in (ii) of Figure 14.13, the fingers of the right hand are curled in the direction of a 90° rotation of axes in the xy-plane (so that the positive x-axis is transformed into the positive y-axis), then the extended thumb points in the direction of the positive z-axis.

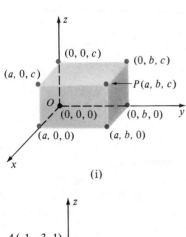

Figure 14.14

If P is a point, then the projection of P on the x-axis has some coordinate a, called the **x-coordinate** of P. The x-coordinate of P may also be thought of as a directed distance from the yz-plane to P. Similarly, the coordinates b and c of the projections of P on the y- and z-axes, respectively, are called the **y-coordinate** and **z-coordinate** of P. We shall use the ordered triple (a, b, c) to denote the coordinates of P, as well as P itself. The notation $P(a, b, c)$ will also be used for the point P with coordinates a, b, and c. If P is not on a coordinate plane, then the three planes through P that are parallel to the coordinate planes, together with the coordinate planes, determine a rectangular parallelepiped. This is illustrated in (i) of Figure 14.14, where the eight vertices are labeled in the manner just described. Conversely, to each ordered triple (a, b, c) of real numbers, there corresponds a point P having coordinates a, b, and c. The concept of plotting points is similar to that used in two dimensions. As an aid to plotting, it is sometimes convenient to construct a parallelepiped as shown in (i) of Figure 14.14. The points $A(-1, -3, 1)$ and $B(3, 4, -2)$ are plotted in (ii) of the figure.

The one-to-one correspondence between the points in space and ordered triples of real numbers we have described is called a **rectangular coordinate system in three dimensions**. The three coordinate planes partition space into eight parts, called **octants**. The part consisting of all points $P(a, b, c)$ whose three coordinates a, b, and c are positive is called the **first octant**.

It is not difficult to find a formula for the distance $d(P_1, P_2)$ between two points P_1 and P_2. If P_1 and P_2 are on a line parallel to the z-axis, as illustrated in (i) of Figure 14.15, and if their projections on the z-axis are $A_1(0, 0, z_1)$ and $A_2(0, 0, z_2)$, respectively, then $d(P_1, P_2) = d(A_1, A_2) = |z_2 - z_1|$. Similar formulas hold if the line through P_1 and P_2 is parallel to the x- or y-axis.

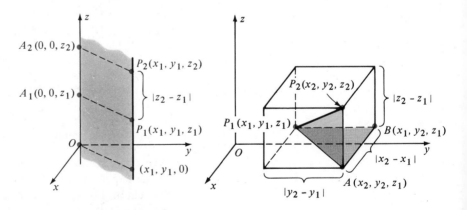

Figure 14.15

If we have a situation of the type illustrated in (ii) of Figure 14.15, then triangle P_1AP_2 is a right triangle, and hence by the Pythagorean Theorem,

$$d(P_1, P_2) = \sqrt{[d(P_1, A)]^2 + [d(A, P_2)]^2}.$$

Since P_1 and A are in a plane parallel to the xy-plane, it follows from the Distance Formula (1.10) that $[d(P_1, A)]^2 = (x_2 - x_1)^2 + (y_2 - y_1)^2$, whereas from our previous remarks $[d(A, P_2)]^2 = (z_2 - z_1)^2$. Substituting in the preceding formula for $d(P_1, P_2)$ gives us the following.

Distance Formula (14.13)

The distance between any two points $P_1(x_1, y_1, z_1)$ and $P_2(x_2, y_2, z_2)$ is

$$d(P_1, P_2) = \sqrt{(x_2 - x_1)^2 + (y_2 - y_1)^2 + (z_2 - z_1)^2}.$$

Note that if P_1 and P_2 are on the xy-plane, so that $z_1 = z_2 = 0$, then (14.13) reduces to the two-dimensional Distance Formula.

Example 1 Find the distance between $A(-1, -3, 1)$ and $B(3, 4, -2)$.

Solution Points A and B are plotted in Figure 14.14. Using the Distance Formula,

$$d(A, B) = \sqrt{(3 + 1)^2 + (4 + 3)^2 + (-2 - 1)^2}$$
$$= \sqrt{16 + 49 + 9} = \sqrt{74}. \qquad \blacksquare$$

By referring to Figure 14.15 and using similar triangles, the following result can be proved.

Midpoint Formula (14.14)

The midpoint of the line segment from $P_1(x_1, y_1, z_1)$ to $P_2(x_2, y_2, z_2)$ is

$$\left(\frac{x_1 + x_2}{2}, \frac{y_1 + y_2}{2}, \frac{z_1 + z_2}{2} \right).$$

As an illustration of (14.14), if A and B are the points in Example 1, then the midpoint of segment AB has coordinates $(1, \frac{1}{2}, -\frac{1}{2})$.

The **graph of an equation** in three variables x, y, and z is defined as the set of all points $P(a, b, c)$ in a rectangular coordinate system such that the ordered triple (a, b, c) is a solution of the equation; that is, equality is obtained when a, b, and c are substituted for x, y, and z, respectively. The graph of such an equation is a **surface**. It is easy to derive an equation that has, as its graph, a sphere of radius r with center at the point $C(x_0, y_0, z_0)$. As illustrated in

Figure 14.16 a point $P(x, y, z)$ is on the sphere if and only if $[d(C, P)]^2 = r^2$. Applying the Distance Formula gives us the next theorem.

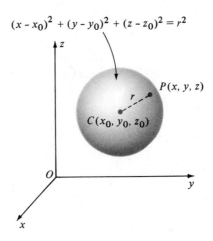

$$(x - x_0)^2 + (y - y_0)^2 + (z - z_0)^2 = r^2$$

Figure 14.16

Theorem (14.15)

> An equation of a sphere of radius r and center $C(x_0, y_0, z_0)$ is
>
> $$(x - x_0)^2 + (y - y_0)^2 + (z - z_0)^2 = r^2.$$

If we square the indicated expressions and simplify, then the equation in (14.15) may be written in the form

$$x^2 + y^2 + z^2 + ax + by + cz + d = 0$$

where the coefficients are real numbers. Conversely, if we begin with an equation of this form and if the graph exists, then by completing squares we can obtain the form in (14.15), and hence the graph is a sphere or a point.

Example 2   Discuss the graph of the equation

$$x^2 + y^2 + z^2 - 6x + 8y + 4z + 4 = 0.$$

Solution   We complete squares as follows:

$$(x^2 - 6x) + (y^2 + 8y) + (z^2 + 4z) = -4$$

$$(x^2 - 6x + 9) + (y^2 + 8y + 16) + (z^2 + 4z + 4) = -4 + 9 + 16 + 4$$

$$(x - 3)^2 + (y + 4)^2 + (z + 2)^2 = 25.$$

Comparing the last equation with (14.15) it follows that the graph is a sphere of radius 5 with center $C(3, -4, -2)$.   ∎

EXERCISES 14.2

In Exercises 1–6 plot the points A and B and find (a) $d(A, B)$; (b) the midpoint of AB.

1 $A(2, 4, -5), B(4, -2, 3)$ **2** $A(1, -2, 7), B(2, 4, -1)$

3 $A(-4, 0, 1), B(3, -2, 1)$ **4** $A(0, 5, -4), B(1, 1, 0)$

5 $A(1, 0, 0), B(0, 1, 1)$ **6** $A(0, 0, 0), B(-8, -1, 4)$

In Exercises 7 and 8 suppose A and B are opposite vertices of a parallelepiped having its faces parallel to the coordinate planes. Find the coordinates of the other vertices.

7 $A(2, 5, -3), B(-4, 2, 1)$ **8** $A(-3, 2, 6), B(1, 5, -1)$

In Exercises 9 and 10 prove that A, B, and C are vertices of a right triangle and find its area.

9 $A(2, 0, 1), B(3, 1, 2), C(1, 2, 0)$

10 $A(4, -3, 2), B(6, -2, 1), C(7, -6, 5)$

In Exercises 11–14 find an equation of the sphere with center C and radius r.

11 $C(3, -1, 2), r = 3$ **12** $C(4, -5, 1), r = 5$

13 $C(-5, 0, 1), r = 1/2$ **14** $C(0, -3, -6), r = \sqrt{3}$

15 Find an equation of the sphere with center $(-2, 4, -6)$ that is tangent to (a) the yz-plane; (b) the xz-plane; (c) the xy-plane.

16 Find an equation of the sphere having endpoints of a diameter at $A(1, 4, -2)$ and $B(-7, 1, 2)$.

In Exercises 17–22 find the center and radius of the sphere having the given equation.

17 $x^2 + y^2 + z^2 + 4x - 2y + 2z + 2 = 0$

18 $x^2 + y^2 + z^2 - 6x - 10y + 6z + 34 = 0$

19 $x^2 + y^2 + z^2 - 8x + 8z + 16 = 0$

20 $4x^2 + 4y^2 + 4z^2 - 4x + 8y - 3 = 0$

21 $x^2 + y^2 + z^2 + 4y = 0$

22 $x^2 + y^2 + z^2 - z = 0$

23 Find an equation for the set of all points equidistant from $A(2, -1, 3)$ and $B(-1, 5, 1)$. Describe the graph of the equation.

24 Work Exercise 23 using $A(5, 0, -4)$ and $B(2, -1, 7)$.

25 Describe the graphs of the equations (a) $z = 5$; (b) $y = 2$; and (c) $x = 0$.

26 Find an equation for the set of all points $P(x, y, z)$ that are three times as far from $A(2, -1, 3)$ as they are from $B(-1, 5, 1)$. Describe the graph of the equation.

In Exercises 27–34 describe the given region R in a three-dimensional coordinate system.

27 $R = \{(x, y, z): x^2 + y^2 + z^2 \le 1\}$

28 $R = \{(x, y, z): x^2 + y^2 + z^2 > 1\}$

29 $R = \{(x, y, z): |x| \le 1, |y| \le 2, |z| \le 3\}$

30 $R = \{(x, y, z): x^2 + y^2 = 1\}$

31 $R = \{(x, y, z): 0 < x^2 + y^2 \le 1\}$

32 $R = \{(x, y, z): 4 < x^2 + y^2 + z^2 < 9\}$

33 $R = \{(x, y, z): x^2 + y^2 + z^2 - 4x - 2y + 1 < 0\}$

34 $R = \{(x, y, z): y = x\}$

14.3

VECTORS IN THREE DIMENSIONS

The next definition extends the concept of vectors in V_2 to three dimensions.

Definition (14.16)

> The **three-dimensional vector space** V_3 is the set of all ordered triples $\langle x, y, z \rangle$ of real numbers, called **vectors**, such that if $\mathbf{a} = \langle a_1, a_2, a_3 \rangle$, $\mathbf{b} = \langle b_1, b_2, b_3 \rangle$, and c is a scalar, then
>
> (i) $\mathbf{a} + \mathbf{b} = \langle a_1 + b_1, a_2 + b_2, a_3 + b_3 \rangle$
>
> (ii) $c\mathbf{a} = \langle ca_1, ca_2, ca_3 \rangle$.

The numbers a_1, a_2, and a_3 are called the **components** of the vector $\langle a_1, a_2, a_3 \rangle$. Following the same procedure used in V_2, we define the **zero vector 0**, the **negative** $-\mathbf{a}$ of $\mathbf{a}$, the **magnitude** $|\mathbf{a}|$ of $\mathbf{a}$, and the **difference** $\mathbf{a} - \mathbf{b}$ as follows.

Definition (14.17)

> (i) $\mathbf{0} = \langle 0, 0, 0 \rangle$
>
> (ii) $-\mathbf{a} = -\langle a_1, a_2, a_3 \rangle = \langle -a_1, -a_2, -a_3 \rangle$
>
> (iii) $|\mathbf{a}| = \sqrt{a_1^2 + a_2^2 + a_3^2}$
>
> (iv) $\mathbf{a} - \mathbf{b} = \mathbf{a} + (-\mathbf{b}) = \langle a_1 - b_1, a_2 - b_2, a_3 - b_3 \rangle$

Properties of vectors in two dimensions may be extended without difficulty to V_3 by simply taking the third component into account. In particular, the properties listed in Theorems (14.5) and (14.9) are readily proved.

A vector $\mathbf{a} = \langle a_1, a_2, a_3 \rangle$ in V_3 may be represented in a rectangular coordinate system by a directed line segment $\overrightarrow{PQ}$ with arbitrary initial point $P(x, y, z)$ and terminal point $Q(x + a_1, y + a_2, z + a_3)$, as illustrated in Figure 14.17.

As before, directed line segments are also referred to as *vectors*. The vector $\overrightarrow{OP}$ from the origin to a point P is called the **position vector of P**. If P has coordinates (a_1, a_2, a_3), then $\overrightarrow{OP}$ (see Figure 14.17) is called the **position vector corresponding to $\mathbf{a} = \langle a_1, a_2, a_3 \rangle$**. The geometric interpretation of vector addition in three dimensions is exactly the same as that in two dimensions.

Note that the magnitude of $\mathbf{a}$ is the length of any of its geometric representations. Observe also that $|\mathbf{a}| \geq 0$, where $|\mathbf{a}| = 0$ if and only if $\mathbf{a} = \mathbf{0}$. As in two dimensions, it can be proved that $|c\mathbf{a}| = |c|\,|\mathbf{a}|$.

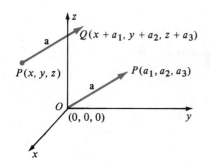

Figure 14.17

Example 1 If $\mathbf{a} = \langle 2, 5, -3 \rangle$ and $\mathbf{b} = \langle -4, 1, 7 \rangle$, find the vectors $\mathbf{a} + \mathbf{b}$ and $2\mathbf{a} - 3\mathbf{b}$, and the scalar $|\mathbf{a}|$.

Solution

$$\mathbf{a} + \mathbf{b} = \langle 2, 5, -3 \rangle + \langle -4, 1, 7 \rangle = \langle -2, 6, 4 \rangle$$

$$2\mathbf{a} - 3\mathbf{b} = 2\langle 2, 5, -3 \rangle - 3\langle -4, 1, 7 \rangle = \langle 4, 10, -6 \rangle - \langle -12, 3, 21 \rangle$$
$$= \langle 16, 7, -27 \rangle$$

$$|\mathbf{a}| = \sqrt{(2)^2 + (5)^2 + (-3)^2} = \sqrt{38} \qquad \blacksquare$$

Exactly as for vectors in two dimensions, we say that two nonzero vectors $\mathbf{a}$ and $\mathbf{b}$ of V_3 have the *same direction* if $\mathbf{b} = c\mathbf{a}$ for some scalar $c > 0$, or *opposite directions* if $\mathbf{b} = c\mathbf{a}$ for some $c < 0$. The vectors $\mathbf{a}$ and $\mathbf{b}$ are *parallel* if $\mathbf{b} = c\mathbf{a}$ for some scalar c. To illustrate, given the vectors

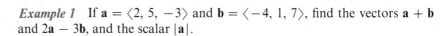

$$\mathbf{a} = \langle 15, -6, 24 \rangle, \quad \mathbf{b} = \langle 5, -2, 8 \rangle, \quad \mathbf{c} = \langle -\tfrac{15}{2}, 3, -12 \rangle$$

the vectors **a** and **b** have the same direction since $\mathbf{a} = 3\mathbf{b}$ or, equivalently, $\mathbf{b} = \frac{1}{3}\mathbf{a}$; whereas **c** and **a** have opposite directions since $\mathbf{c} = -\frac{1}{2}\mathbf{a}$, or $\mathbf{a} = -2\mathbf{c}$. Each of these vectors is parallel to the other two.

Theorem (14.18)

> If $P_1(x_1, y_1, z_1)$ and $P_2(x_2, y_2, z_2)$ are points, then the vector **a** in V_3 that has geometric representation $\overrightarrow{P_1P_2}$ is
>
> $$\mathbf{a} = \langle x_2 - x_1, y_2 - y_1, z_2 - z_1 \rangle.$$

The proof is similar to that of Theorem (14.4).

Two vectors $\overrightarrow{PQ}$ and $\overrightarrow{RS}$ are said to have the same (or the opposite) direction if their corresponding vectors in V_3 have the same (or the opposite) direction. If the magnitude $|\overrightarrow{PQ}|$ is defined as the distance between P and Q, then the notation $\overrightarrow{PQ} = \overrightarrow{RS}$ means that the indicated vectors have the same magnitude and direction. If **a** is the vector in V_3 corresponding to $\overrightarrow{PQ}$, and c is a scalar, then a geometric representation of $c\mathbf{a}$ is denoted by $c\overrightarrow{PQ}$. Such scalar multiples have the same geometric meaning as their counterparts in two dimensions. Moreover, the vectors $\overrightarrow{PQ}$ and $\overrightarrow{RS}$ have the same direction if $\overrightarrow{PQ} = c\overrightarrow{RS}$ for some $c > 0$, and opposite directions if $\overrightarrow{PQ} = c\overrightarrow{RS}$ for some $c < 0$. In either case, we say that $\overrightarrow{PQ}$ and $\overrightarrow{RS}$ are *parallel*.

Example 2 Given the points $P_1(5, 6, -2)$ and $P_2(-3, 8, 7)$,

(a) find the position vector corresponding to $\overrightarrow{P_1P_2}$.

(b) find $|\overrightarrow{P_1P_2}|$.

Solution

(a) By (14.18), the vector $\mathbf{a} = \langle -3 - 5, 8 - 6, 7 + 2 \rangle = \langle -8, 2, 9 \rangle$ has geometric representation $\overrightarrow{P_1P_2}$. Thus $A(-8, 2, 9)$ is the terminal point of the position vector $\overrightarrow{OA}$ corresponding to $\overrightarrow{P_1P_2}$.

(b) $|\overrightarrow{P_1P_2}| = |\mathbf{a}| = \sqrt{64 + 4 + 81} = \sqrt{149}$ ∎

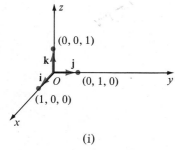

(i)

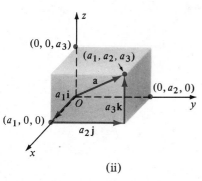

(ii)

Figure 14.18

A vector **a** is a **unit vector** if $|\mathbf{a}| = 1$. The special unit vectors

$$\mathbf{i} = \langle 1, 0, 0 \rangle, \quad \mathbf{j} = \langle 0, 1, 0 \rangle, \quad \mathbf{k} = \langle 0, 0, 1 \rangle$$

are important since any vector $\mathbf{a} = \langle a_1, a_2, a_3 \rangle$ can be expressed as a linear combination of **i**, **j**, and **k**. Specifically,

$$\mathbf{a} = \langle a_1, a_2, a_3 \rangle = a_1\mathbf{i} + a_2\mathbf{j} + a_3\mathbf{k}.$$

Geometric representations of **i**, **j**, and **k** are sketched in (i) of Figure 14.18. Part (ii) of the figure illustrates how the position vector for $\mathbf{a} = \langle a_1, a_2, a_3 \rangle$ may be regarded as the sum of three vectors corresponding to $a_1\mathbf{i}$, $a_2\mathbf{j}$, and $a_3\mathbf{k}$.

As in Section 14.1, rules for addition, subtraction, and multiplication by scalars may be easily translated into the **i, j, k** notation. It is often convenient to regard V_2 as a subset of V_3 by identifying the vector $\langle a_1, a_2 \rangle$ with $\langle a_1, a_2, 0 \rangle$. If this is done, then there is no essential difference between the vectors $\mathbf{i} = \langle 1, 0 \rangle$, $\mathbf{j} = \langle 0, 1 \rangle$, and the vectors **i, j** defined in V_3.

The next concept to be introduced has many mathematical and physical applications. In the manner of our previous work we shall begin with an algebraic definition and give geometric and physical interpretations later in the section.

Definition (14.19)

> The **dot product a · b** of $\mathbf{a} = \langle a_1, a_2, a_3 \rangle$ and $\mathbf{b} = \langle b_1, b_2, b_3 \rangle$ is
>
> $$\mathbf{a} \cdot \mathbf{b} = a_1 b_1 + a_2 b_2 + a_3 b_3.$$

The symbol **a · b** is read "**a** dot **b**." The dot product is also referred to as the **scalar product**, or **inner product**. It is important to note that **a · b** is a scalar, not a vector. For example,

$$\langle 2, 4, -3 \rangle \cdot \langle -1, 5, 2 \rangle = (2)(-1) + (4)(5) + (-3)(2) = 12$$

$$(3\mathbf{i} - 2\mathbf{j} + \mathbf{k}) \cdot (4\mathbf{i} + 5\mathbf{j} - 2\mathbf{k}) = (3)(4) + (-2)(5) + (1)(-2) = 0.$$

Some properties of the dot product are listed in the following theorem, where **a**, **b**, and **c** are any vectors, and c is a scalar.

Theorem (14.20)

> (i) $\mathbf{a} \cdot \mathbf{a} = |\mathbf{a}|^2$
>
> (ii) $\mathbf{a} \cdot \mathbf{b} = \mathbf{b} \cdot \mathbf{a}$
>
> (iii) $\mathbf{a} \cdot (\mathbf{b} + \mathbf{c}) = \mathbf{a} \cdot \mathbf{b} + \mathbf{a} \cdot \mathbf{c}$
>
> (iv) $(c\mathbf{a}) \cdot \mathbf{b} = c(\mathbf{a} \cdot \mathbf{b}) = \mathbf{a} \cdot (c\mathbf{b})$
>
> (v) $\mathbf{0} \cdot \mathbf{a} = 0$

Proof   If $\mathbf{a} = \langle a_1, a_2, a_3 \rangle$, $\mathbf{b} = \langle b_1, b_2, b_3 \rangle$, and $\mathbf{c} = \langle c_1, c_2, c_3 \rangle$, then

$$\begin{aligned}
\mathbf{a} \cdot (\mathbf{b} + \mathbf{c}) &= \langle a_1, a_2, a_3 \rangle \cdot \langle b_1 + c_1, b_2 + c_2, b_3 + c_3 \rangle \\
&= a_1(b_1 + c_1) + a_2(b_2 + c_2) + a_3(b_3 + c_3) \\
&= (a_1 b_1 + a_2 b_2 + a_3 b_3) + (a_1 c_1 + a_2 c_2 + a_3 c_3) \\
&= \mathbf{a} \cdot \mathbf{b} + \mathbf{a} \cdot \mathbf{c}.
\end{aligned}$$

This proves property (iii). Proofs of the remaining properties are left as exercises.   □

There is a close connection between dot products and the angle between two vectors, as defined below.

Definition (14.21)

> Let **a** and **b** be nonzero vectors. If **b** is not a scalar multiple of **a**, and if $\overrightarrow{OA}$ and $\overrightarrow{OB}$ are the position vectors corresponding to **a** and **b**, respectively, then **the angle θ between a and b** (or between $\overrightarrow{OA}$ and $\overrightarrow{OB}$) is angle AOB of the triangle determined by points A, O, and B (see Figure 14.19). If $\mathbf{b} = c\mathbf{a}$ for some scalar c, then $\theta = 0$ or $\theta = \pi$ according as $c > 0$ or $c < 0$.

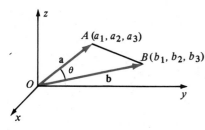

Figure 14.19

Note that if **a** and **b** are parallel then, by (14.21), $\theta = 0$ or $\theta = \pi$. The vectors **a** and **b** are said to be **orthogonal** if $\theta = \pi/2$. For convenience we shall assume that the zero vector **0** is parallel and orthogonal to every vector **a**.

Theorem (14.22)

> If θ is the angle between nonzero vectors **a** and **b**, then
>
> $$\mathbf{a} \cdot \mathbf{b} = |\mathbf{a}|\,|\mathbf{b}| \cos \theta.$$

Proof　　If $\mathbf{b} \neq c\mathbf{a}$ we have a situation similar to that illustrated in Figure 14.19. Applying the Law of Cosines to triangle AOB gives us

$$|\overrightarrow{AB}|^2 = |\mathbf{a}|^2 + |\mathbf{b}|^2 - 2|\mathbf{a}|\,|\mathbf{b}| \cos \theta.$$

Consequently,

$$(b_1 - a_1)^2 + (b_2 - a_2)^2 + (b_3 - a_3)^2$$
$$= a_1^2 + a_2^2 + a_3^2 + b_1^2 + b_2^2 + b_3^2 - 2|\mathbf{a}|\,|\mathbf{b}| \cos \theta$$

which reduces to

$$-2a_1 b_1 - 2a_2 b_2 - 2a_3 b_3 = -2|\mathbf{a}|\,|\mathbf{b}| \cos \theta.$$

Dividing both sides of the last equation by -2 gives us the desired conclusion.

If $\mathbf{b} = c\mathbf{a}$, then by properties (iv) and (i) of Theorem (14.20),

$$\mathbf{a} \cdot \mathbf{b} = \mathbf{a} \cdot (c\mathbf{a}) = c(\mathbf{a} \cdot \mathbf{a}) = c|\mathbf{a}|^2.$$

Also, 　　　　$|\mathbf{a}|\,|\mathbf{b}| \cos \theta = |\mathbf{a}|\,|c\mathbf{a}| \cos \theta = |c|\,|\mathbf{a}|^2 \cos \theta.$

If $c > 0$, then $|c| = c$, $\theta = 0$, and $|c|\,|\mathbf{a}|^2 \cos \theta$ reduces to $c|\mathbf{a}|^2$. Thus $|\mathbf{a}|\,|\mathbf{b}| \cos \theta = \mathbf{a} \cdot \mathbf{b}$. If $c < 0$, then $|c| = -c$, $\theta = \pi$, and again $|c|\,|\mathbf{a}|^2 \cos \theta$ reduces to $c|\mathbf{a}|^2$. This completes the proof of the theorem. □

Corollary (14.23)

> If θ is the angle between nonzero vectors $\mathbf{a}$ and $\mathbf{b}$, then
> $$\cos \theta = \frac{\mathbf{a} \cdot \mathbf{b}}{|\mathbf{a}||\mathbf{b}|}.$$

Example 3   Find the angle between $\mathbf{a} = \langle 4, -3, 1 \rangle$ and $\mathbf{b} = \langle -1, -2, 2 \rangle$.

Solution   Applying Corollary (14.23),

$$\cos \theta = \frac{\mathbf{a} \cdot \mathbf{b}}{|\mathbf{a}||\mathbf{b}|} = \frac{(4)(-1) + (-3)(-2) + (1)(2)}{\sqrt{16 + 9 + 1}\sqrt{1 + 4 + 4}} = \frac{4}{3\sqrt{26}} = \frac{4\sqrt{26}}{78} = \frac{2\sqrt{26}}{39},$$

or
$$\theta = \arccos(2\sqrt{26}/39).$$

Using a table or calculator gives us the approximations

$$\theta \approx 74.84° \approx 1.31 \text{ radians.} \quad \blacksquare$$

The following theorem is an immediate consequence of Theorem (14.22).

Theorem (14.24)

> Two vectors $\mathbf{a}$ and $\mathbf{b}$ are orthogonal if and only if $\mathbf{a} \cdot \mathbf{b} = 0$.

Example 4   Prove that the following pairs of vectors are orthogonal.

(a) $\mathbf{i}, \mathbf{j}$ (b) $3\mathbf{i} - 7\mathbf{j} + 2\mathbf{k}, 10\mathbf{i} + 4\mathbf{j} - \mathbf{k}$

Solution   The proof follows from Theorem (14.24). Thus

(a) $\mathbf{i} \cdot \mathbf{j} = \langle 1, 0, 0 \rangle \cdot \langle 0, 1, 0 \rangle = (1)(0) + (0)(1) + (0)(0) = 0.$

(b) $(3\mathbf{i} - 7\mathbf{j} + 2\mathbf{k}) \cdot (10\mathbf{i} + 4\mathbf{j} - \mathbf{k}) = 30 - 28 - 2 = 0.$ $\quad \blacksquare$

Each of the next two results is true for all vectors $\mathbf{a}$ and $\mathbf{b}$.

Cauchy–Schwarz Inequality (14.25)

> $$|\mathbf{a} \cdot \mathbf{b}| \leq |\mathbf{a}||\mathbf{b}|$$

Proof   The result is trivial if either $\mathbf{a}$ or $\mathbf{b}$ is $\mathbf{0}$. If $\mathbf{a}$ and $\mathbf{b}$ are nonzero vectors, then by Theorem (14.22), $|\mathbf{a} \cdot \mathbf{b}| = |\mathbf{a}||\mathbf{b}||\cos \theta|$, where θ is the angle between $\mathbf{a}$ and $\mathbf{b}$. Since $|\cos \theta| \leq 1$, it follows that $|\mathbf{a} \cdot \mathbf{b}| \leq |\mathbf{a}||\mathbf{b}|$. $\quad \square$

Triangle Inequality (14.26)

> $$|\mathbf{a} + \mathbf{b}| \leq |\mathbf{a}| + |\mathbf{b}|$$

Proof   Using properties of the dot product we may write

$$|\mathbf{a} + \mathbf{b}|^2 = (\mathbf{a} + \mathbf{b}) \cdot (\mathbf{a} + \mathbf{b}) = \mathbf{a} \cdot \mathbf{a} + 2\mathbf{a} \cdot \mathbf{b} + \mathbf{b} \cdot \mathbf{b}$$
$$= |\mathbf{a}|^2 + 2\mathbf{a} \cdot \mathbf{b} + |\mathbf{b}|^2.$$

Since $\mathbf{a} \cdot \mathbf{b} \le |\mathbf{a} \cdot \mathbf{b}| \le |\mathbf{a}||\mathbf{b}|$, we have

$$|\mathbf{a} + \mathbf{b}|^2 \le |\mathbf{a}|^2 + 2|\mathbf{a}||\mathbf{b}| + |\mathbf{b}|^2 = (|\mathbf{a}| + |\mathbf{b}|)^2.$$

Taking square roots gives us

$$|\mathbf{a} + \mathbf{b}| \le |\mathbf{a}| + |\mathbf{b}|. \qquad \square$$

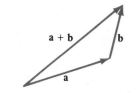

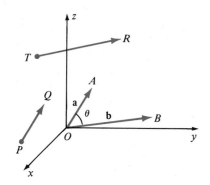

Figure 14.20

Figure 14.21

The reason for the name "Triangle Inequality" is apparent if we represent **a** and **b** geometrically as in Figure 14.20. In this case (14.26) states that the length of a side of a triangle is not greater than the sum of the lengths of the other two sides.

The dot product of two vectors $\overrightarrow{PQ}$ and $\overrightarrow{TR}$ may be obtained by first defining the angle between these vectors as the angle θ between their corresponding vectors **a** and **b** in V_3. If $\theta = \pi/2$, then $\overrightarrow{PQ}$ and $\overrightarrow{TR}$ are **orthogonal**, or **perpendicular**. A typical angle is illustrated in Figure 14.21, where $\overrightarrow{OA}$ and $\overrightarrow{OB}$ are position vectors corresponding to **a** and **b**, respectively. The **dot product** of $\overrightarrow{PQ}$ and $\overrightarrow{TR}$ is defined by

$$\overrightarrow{PQ} \cdot \overrightarrow{TR} = \mathbf{a} \cdot \mathbf{b} = |\mathbf{a}||\mathbf{b}| \cos \theta.$$

Since $|\overrightarrow{PQ}| = |\mathbf{a}|$ and $|\overrightarrow{TR}| = |\mathbf{b}|$, this may be written

$$\overrightarrow{PQ} \cdot \overrightarrow{TR} = |\overrightarrow{PQ}||\overrightarrow{TR}| \cos \theta.$$

If $\overrightarrow{PQ}$ and $\overrightarrow{PR}$ have the same initial point, and if S is the projection of Q on the line through P and R (see Figure 14.22), then the scalar $|\overrightarrow{PQ}| \cos \theta$ is called the **component of $\overrightarrow{PQ}$ along $\overrightarrow{PR}$** abbreviated **comp$_{\overrightarrow{PR}}$ $\overrightarrow{PQ}$**.

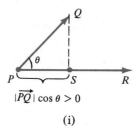
$|\overrightarrow{PQ}| \cos \theta > 0$
(i)

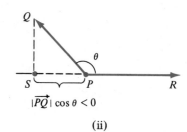
$|\overrightarrow{PQ}| \cos \theta < 0$
(ii)

Figure 14.22   comp$_{\overrightarrow{PR}}$ $\overrightarrow{PQ}$

Note that $|\overrightarrow{PQ}| \cos \theta$ is positive or negative if $0 \le \theta < \pi/2$ or $\pi/2 < \theta \le \pi$, respectively. If $\theta = \pi/2$, the component is 0. It follows that

$$\text{comp}_{\overrightarrow{PR}} \overrightarrow{PQ} = |\overrightarrow{PQ}| \cos \theta = \overrightarrow{PQ} \cdot \frac{1}{|\overrightarrow{PR}|} \overrightarrow{PR}.$$

One way to remember this fact is to use the following statement.

Theorem (14.27)

> The component of $\overrightarrow{PQ}$ along $\overrightarrow{PR}$ equals the dot product of $\overrightarrow{PQ}$ with a unit vector having the same direction as $\overrightarrow{PR}$.

The preceding concept may be applied to vectors **a** and **b** of V_3 by representing them geometrically as $\overrightarrow{PQ}$ and $\overrightarrow{PR}$, respectively. This gives us the following definition.

Definition (14.28)

> Let **a** and **b** be vectors in V_3. The **component of a along b**, denoted by comp$_b$ **a**, is
>
> $$\text{comp}_b\, \mathbf{a} = \mathbf{a} \cdot \frac{1}{|\mathbf{b}|}\mathbf{b}.$$

If $\mathbf{a} = a_1\mathbf{i} + a_2\mathbf{j} + a_3\mathbf{k}$, then by Definition (14.28)

$$\text{comp}_i\, \mathbf{a} = \mathbf{a} \cdot \mathbf{i} = a_1, \quad \text{comp}_j\, \mathbf{a} = \mathbf{a} \cdot \mathbf{j} = a_2, \quad \text{and} \quad \text{comp}_k\, \mathbf{a} = \mathbf{a} \cdot \mathbf{k} = a_3.$$

Thus, the components of **a** along **i**, **j**, and **k** are the same as the components a_1, a_2, and a_3 of **a**.

Example 5 If $\mathbf{a} = 4\mathbf{i} - \mathbf{j} + 5\mathbf{k}$ and $\mathbf{b} = 6\mathbf{i} + 3\mathbf{j} - 2\mathbf{k}$, find
(a) comp$_b$ **a**.   (b) comp$_a$ **b**.

Solution   (a)   Using Definition (14.28),

$$\text{comp}_b\, \mathbf{a} = \mathbf{a} \cdot \frac{1}{|\mathbf{b}|}\mathbf{b} = (4\mathbf{i} - \mathbf{j} + 5\mathbf{k}) \cdot \frac{1}{7}(6\mathbf{i} + 3\mathbf{j} - 2\mathbf{k})$$

$$= \frac{24 - 3 - 10}{7} = \frac{11}{7}.$$

(b)   Interchanging the roles of **a** and **b** in (14.28),

$$\text{comp}_a\, \mathbf{b} = \mathbf{b} \cdot \frac{1}{|\mathbf{a}|}\mathbf{a} = (6\mathbf{i} + 3\mathbf{j} - 2\mathbf{k}) \cdot \frac{1}{\sqrt{42}}(4\mathbf{i} - \mathbf{j} + 5\mathbf{k})$$

$$= \frac{24 - 3 - 10}{\sqrt{42}} = \frac{11}{\sqrt{42}}.$$

We shall conclude this section with an important physical interpretation for the dot product. Recall from Definition (6.9) that the work done if a constant force F is exerted through a distance d is given by $W = Fd$. This formula is very restrictive since it can only be used if the force is applied along the line of motion. More generally, suppose a vector $\overrightarrow{PQ}$ represents a force, and that its point of application moves along a vector $\overrightarrow{PR}$. This is illustrated in Figure 14.23, where a force $\overrightarrow{PQ}$ is used to pull an object along a level path from P to R. The vector $\overrightarrow{PQ}$ is the sum of the vectors $\overrightarrow{PS}$ and $\overrightarrow{SQ}$. Since $\overrightarrow{SQ}$ does not contribute to the horizontal movement, we may assume that the motion from P to R is caused by $\overrightarrow{PS}$ alone. Applying (6.9), the work W is found by multiplying the component of $\overrightarrow{PQ}$ in the direction of $\overrightarrow{PR}$ by the distance $|\overrightarrow{PR}|$, that is,

$$W = (|\overrightarrow{PQ}|\cos\theta)|\overrightarrow{PR}| = \overrightarrow{PQ} \cdot \overrightarrow{PR}.$$

This leads to the following definition.

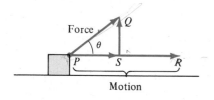

Figure 14.23

Definition (14.29)

> The **work done by a constant force** $\overrightarrow{PQ}$ as its point of application moves along the vector $\overrightarrow{PR}$ is $\overrightarrow{PQ} \cdot \overrightarrow{PR}$.

Example 6   The magnitude and direction of a constant force are given by $\mathbf{a} = 5\mathbf{i} + 2\mathbf{j} + 6\mathbf{k}$. Find the work done if the point of application of the force moves from $P(1, -1, 2)$ to $R(4, 3, -1)$.

Solution   By Theorem (14.18) the vector in V_3 that corresponds to $\overrightarrow{PR}$ is $\mathbf{b} = \langle 3, 4, -3\rangle$. If $\overrightarrow{PQ}$ is a geometric representation for $\mathbf{a}$, then by Definition (14.29) the work done is given by

$$\overrightarrow{PQ} \cdot \overrightarrow{PR} = \mathbf{a} \cdot \mathbf{b} = 15 + 8 - 18 = 5.$$

If, for example, the unit of length is feet and the magnitude of the force is measured in pounds, then the work done is 5 ft-lb. If the length is in meters and the force is in Newtons, then the work done is 5 joules.   ∎

EXERCISES 14.3

1 Complete the proof of Theorem (14.20).

2 Prove Theorems (14.5) and (14.9) for V_3.

In Exercises 3–8 find $\mathbf{a} + \mathbf{b}$, $5\mathbf{a} - 4\mathbf{b}$, $\mathbf{a} \cdot \mathbf{b}$, $|\mathbf{a}|$, $|3\mathbf{a}|$, $|-3\mathbf{a}|$, $\mathbf{a} - \mathbf{b}$, and $|\mathbf{a} - \mathbf{b}|$.

3 $\mathbf{a} = \langle -2, 6, 1\rangle$, $\mathbf{b} = \langle 3, -3, -1\rangle$

4 $\mathbf{a} = \langle 1, 2, -3\rangle$, $\mathbf{b} = \langle -4, 0, 1\rangle$

5 $\mathbf{a} = 3\mathbf{i} - 4\mathbf{j} + 2\mathbf{k}$, $\mathbf{b} = \mathbf{i} + 2\mathbf{j} - 5\mathbf{k}$

6 $\mathbf{a} = 2\mathbf{i} - \mathbf{j} + 4\mathbf{k}$, $\mathbf{b} = \mathbf{i} - \mathbf{k}$

7 $\mathbf{a} = \mathbf{i} + \mathbf{j}$, $\mathbf{b} = -\mathbf{j} + \mathbf{k}$

8 $\mathbf{a} = 2\mathbf{i}$, $\mathbf{b} = 3\mathbf{k}$

In Exercises 9 and 10 sketch geometric representations for $\mathbf{a}$, $\mathbf{b}$, $2\mathbf{a}$, $-3\mathbf{b}$, $\mathbf{a} + \mathbf{b}$, and $\mathbf{a} - \mathbf{b}$.

9 $\mathbf{a} = \langle 2, 3, 4\rangle$, $\mathbf{b} = \langle 1, -2, 2\rangle$

10 $\mathbf{a} = -\mathbf{i} + 2\mathbf{j} + 3\mathbf{k}$, $\mathbf{b} = -2\mathbf{j} + \mathbf{k}$

If $\mathbf{a} = \langle -2, 3, 1 \rangle$, $\mathbf{b} = \langle 7, 4, 5 \rangle$, and $\mathbf{c} = \langle 1, -5, 2 \rangle$ find the numbers in Exercises 11–20.

11 $\mathbf{a} \cdot \mathbf{b}$

12 $\mathbf{b} \cdot \mathbf{c}$

13 $\mathbf{a} \cdot (\mathbf{b} + \mathbf{c})$

14 $\mathbf{b} \cdot (\mathbf{a} - \mathbf{c})$

15 $(2\mathbf{a} + \mathbf{b}) \cdot 3\mathbf{c}$

16 $(\mathbf{a} - \mathbf{b}) \cdot (\mathbf{b} + \mathbf{c})$

17 $\text{comp}_{\mathbf{c}}\mathbf{b}$

18 $\text{comp}_{\mathbf{a}}\mathbf{c}$

19 $\text{comp}_{\mathbf{b}}(\mathbf{a} + \mathbf{c})$

20 $\text{comp}_{\mathbf{c}}\mathbf{c}$

In Exercises 21–24 find the cosine of the angle between $\mathbf{a}$ and $\mathbf{b}$.

21 $\mathbf{a} = -4\mathbf{i} + 8\mathbf{j} - 3\mathbf{k}, \mathbf{b} = 2\mathbf{i} + \mathbf{j} + \mathbf{k}$

22 $\mathbf{a} = \mathbf{i} - 7\mathbf{j} + 4\mathbf{k}, \mathbf{b} = 5\mathbf{i} - \mathbf{k}$

23 $\mathbf{a} = -2\mathbf{i} - 3\mathbf{j}, \mathbf{b} = -6\mathbf{i} + 4\mathbf{k}$

24 $\mathbf{a} = \langle 3, -5, -1 \rangle, \mathbf{b} = \langle 2, 1, -3 \rangle$

Given the points $P(3, -2, -1)$, $Q(1, 5, 4)$, $R(2, 0, -6)$, and $S(-4, 1, 5)$, find the quantities in Exercises 25–30.

25 $\overrightarrow{PQ} \cdot \overrightarrow{RS}$

26 $\overrightarrow{QS} \cdot \overrightarrow{RP}$

27 The angle between $\overrightarrow{PQ}$ and $\overrightarrow{RS}$

28 The angle between $\overrightarrow{QS}$ and $\overrightarrow{RP}$

29 The component of $\overrightarrow{PS}$ along $\overrightarrow{QR}$

30 The component of $\overrightarrow{QR}$ along $\overrightarrow{PS}$

31 Given $P(8, -3, 5)$, $Q(6, 1, -7)$, and $R(x, y, z)$, state an equation in x, y, and z that guarantees that $\overrightarrow{PR}$ is orthogonal to $\overrightarrow{PQ}$. Give a geometric description of all such points $R(x, y, z)$.

32 Prove, by means of vectors, that $P(2, -3, 1)$, $Q(-5, 1, 7)$, and $R(6, 1, 3)$ are vertices of a right triangle and find its area.

In Exercises 33 and 34, the magnitude and direction of a force are given by the vector $\mathbf{a}$. Find the work done if the point of application moves from P to Q.

33 $\mathbf{a} = -\mathbf{i} + 5\mathbf{j} - 3\mathbf{k}; P(4, 0, -7), Q(2, 4, 0)$

34 $\mathbf{a} = \langle 8, 0, -4 \rangle; P(-1, 2, 5), Q(4, 1, 0)$

35 A constant force of magnitude 4 lb has the same direction as the vector $\mathbf{a} = \mathbf{i} + \mathbf{j} + \mathbf{k}$. If distance is measured in feet, find the work done if the point of application moves along the y-axis from $(0, 2, 0)$ to $(0, -1, 0)$.

36 A constant force of magnitude 5 Newtons has the same direction as the positive z-axis. If distance is measured in meters, find the work done if the point of application moves along a straight line from the origin to the point $P(1, 2, 3)$.

37 Determine c such that the vectors $3\mathbf{i} - \mathbf{j} + c\mathbf{k}$ and $2c\mathbf{i} + 3\mathbf{j} + 4\mathbf{k}$ are orthogonal.

38 Determine all values of c such that the vectors $\langle 3c, 1, -4c \rangle$ and $\langle c, 4, 2 \rangle$ are orthogonal.

39 Find a unit vector orthogonal to both $\langle 7, -10, 1 \rangle$ and $\langle -2, 2, 3 \rangle$.

40 Find two unit vectors in V_2 orthogonal to $\mathbf{a} = \langle -2, -1 \rangle$.

41 If $\mathbf{a} = 14\mathbf{i} - 15\mathbf{j} + 6\mathbf{k}$,
   (a) find a vector having the same direction and twice the magnitude.
   (b) find a vector having the opposite direction and one-third the magnitude.

42 Work Exercise 41 if $\mathbf{a} = \langle -6, -3, 6 \rangle$.

Work Exercises 43–56 *without introducing components* for the vectors.

43 Under what conditions are the following true?
   (a) $|\mathbf{a} \cdot \mathbf{b}| = |\mathbf{a}| \, |\mathbf{b}|$ (b) $|\mathbf{a} + \mathbf{b}| = |\mathbf{a}| + |\mathbf{b}|$

44 Prove that $|\mathbf{a} - \mathbf{b}| \geq |\mathbf{a}| - |\mathbf{b}|$. (*Hint*: Let $\mathbf{a} = \mathbf{b} + (\mathbf{a} - \mathbf{b})$ and use the Triangle Inequality.)

45 Prove that $(\mathbf{a} + \mathbf{b}) \cdot (\mathbf{a} - \mathbf{b}) = \mathbf{a} \cdot \mathbf{a} - \mathbf{b} \cdot \mathbf{b}$.

46 Prove that $|\mathbf{a} + \mathbf{b}|^2 = |\mathbf{a}|^2 + 2\mathbf{a} \cdot \mathbf{b} + |\mathbf{b}|^2$.

47 Prove that $|\mathbf{a} + \mathbf{b}|^2 + |\mathbf{a} - \mathbf{b}|^2 = 2(|\mathbf{a}|^2 + |\mathbf{b}|^2)$.

48 Prove that $\mathbf{a} \cdot \mathbf{b} = \frac{1}{4}(|\mathbf{a} + \mathbf{b}|^2 - |\mathbf{a} - \mathbf{b}|^2)$.

49 Prove that if $\mathbf{c}$ is orthogonal to both $\mathbf{a}$ and $\mathbf{b}$, then $\mathbf{c}$ is orthogonal to $p\mathbf{a} + q\mathbf{b}$ for all scalars p and q.

50 Extend (14.26) to any finite sum of vectors.

51 If A, B, C are any three points in space and P is the midpoint of the line segment BC, prove that $\overrightarrow{AP} = \frac{1}{2}(\overrightarrow{AB} + \overrightarrow{AC})$.

52 Generalize Exercise 51 as follows. If A, B, and C are points in space, and if a point P divides the line segment BC in the ratio $d(B, P)/d(B, C) = m$, prove that
$$\overrightarrow{AP} = (1 - m)\overrightarrow{AB} + m\overrightarrow{AC}.$$

53 If $\mathbf{a}$ and $\mathbf{b}$ are nonzero, nonparallel vectors, and $c\mathbf{a} + d\mathbf{b} = p\mathbf{a} + q\mathbf{b}$ where c, d, p, and q are scalars, prove that $c = p$ and $d = q$.

54 Use vectors to prove that the line segments that join the midpoints of consecutive sides of a quadrilateral form a parallelogram. (*Hint*: See Exercise 51.)

55 Prove that a quadrilateral is a parallelogram if and only if the diagonals bisect one another.

56 Prove that the diagonals of a rhombus are perpendicular.

57 The **direction angles** of a nonzero vector $\mathbf{a} = \langle a_1, a_2, a_3 \rangle$ are defined as the angles α, β, and γ between the vectors $\mathbf{i}$, $\mathbf{j}$, and $\mathbf{k}$, respectively, and the vector $\mathbf{a}$. The **direction cosines** of $\mathbf{a}$ are $\cos \alpha$, $\cos \beta$, and $\cos \gamma$. Prove that

(a) $\cos \alpha = \dfrac{a_1}{|\mathbf{a}|}$, $\cos \beta = \dfrac{a_2}{|\mathbf{a}|}$, $\cos \gamma = \dfrac{a_3}{|\mathbf{a}|}$.

(b) $\cos^2 \alpha + \cos^2 \beta + \cos^2 \gamma = 1$.

58 Refer to Exercise 57.

(a) Find the direction cosines of $\mathbf{a} = \langle -2, 1, 5 \rangle$.

(b) Find the direction angles and direction cosines of $\mathbf{i}$, $\mathbf{j}$, and $\mathbf{k}$.

(c) Find two unit vectors that satisfy the condition

$$\cos \alpha = \cos \beta = \cos \gamma.$$

59 Three nonzero numbers l, m, and n are called **direction numbers** of a nonzero vector $\mathbf{a}$ if they are proportional to the direction cosines, that is, if there exists a positive number k such that

$$l = k \cos \alpha, \quad m = k \cos \beta, \quad n = k \cos \gamma.$$

If l, m, n are direction numbers of $\mathbf{a}$ and $d = (l^2 + m^2 + n^2)^{1/2}$, prove that $\cos \alpha = l/d$, $\cos \beta = m/d$, $\cos \gamma = n/d$.

60 Refer to Exercise 59. If l_1, m_1, n_1 and l_2, m_2, n_2 are direction numbers of $\mathbf{a}$ and $\mathbf{b}$, respectively, prove the following.

(a) $\mathbf{a}$ and $\mathbf{b}$ are orthogonal if and only if

$$l_1 l_2 + m_1 m_2 + n_1 n_2 = 0.$$

(b) $\mathbf{a}$ and $\mathbf{b}$ are parallel if and only if there is a number k such that $l_1 = k l_2$, $m_1 = k m_2$ and $n_1 = k n_2$.

14.4

THE VECTOR PRODUCT

In this section we introduce the *vector product* (or *cross product*) $\mathbf{a} \times \mathbf{b}$ of two vectors $\mathbf{a}$ and $\mathbf{b}$. Unlike the dot product, which is a scalar, this new operation produces another vector. The vector product was first used as a tool for physical problems involving moments of forces. It is possible to define $\mathbf{a} \times \mathbf{b}$ geometrically and then obtain an algebraic form by introducing a rectangular coordinate system. We shall reverse this process and begin with an algebraic definition. This approach disguises the geometric nature of $\mathbf{a} \times \mathbf{b}$; however, it leads to simpler proofs of properties.

It is convenient to use *determinants* when working with vector products. A **determinant of order 2** is defined by

$$\begin{vmatrix} a_1 & a_2 \\ b_1 & b_2 \end{vmatrix} = a_1 b_2 - a_2 b_1$$

where all letters represent real numbers. For example,

$$\begin{vmatrix} 2 & -3 \\ 4 & 5 \end{vmatrix} = (2)(5) - (-3)(4) = 10 + 12 = 22.$$

A **determinant of order 3** is given by

$$\begin{vmatrix} c_1 & c_2 & c_3 \\ a_1 & a_2 & a_3 \\ b_1 & b_2 & b_3 \end{vmatrix} = \begin{vmatrix} a_2 & a_3 \\ b_2 & b_3 \end{vmatrix} c_1 - \begin{vmatrix} a_1 & a_3 \\ b_1 & b_3 \end{vmatrix} c_2 + \begin{vmatrix} a_1 & a_2 \\ b_1 & b_2 \end{vmatrix} c_3.$$

This is sometimes called the *expansion of the determinant by the first row*. The numerical value can be found by evaluating the second-order determinants on the right side of the equation.

Example 1 Find the value of $\begin{vmatrix} 2 & -1 & 3 \\ -2 & 5 & 1 \\ 1 & 2 & -4 \end{vmatrix}$.

Solution By definition,

$$\begin{vmatrix} 2 & -1 & 3 \\ -2 & 5 & 1 \\ 1 & 2 & -4 \end{vmatrix} = \begin{vmatrix} 5 & 1 \\ 2 & -4 \end{vmatrix}(2) - \begin{vmatrix} -2 & 1 \\ 1 & -4 \end{vmatrix}(-1) + \begin{vmatrix} -2 & 5 \\ 1 & 2 \end{vmatrix}(3)$$

$$= (-20 - 2)(2) - (8 - 1)(-1) + (-4 - 5)(3)$$

$$= -44 + 7 - 27 = -64. \qquad \blacksquare$$

Definition (14.30)

The **vector product** (or **cross product**) $\mathbf{a} \times \mathbf{b}$ of $\mathbf{a} = \langle a_1, a_2, a_3 \rangle$ and $\mathbf{b} = \langle b_1, b_2, b_3 \rangle$ is

$$\mathbf{a} \times \mathbf{b} = \begin{vmatrix} a_2 & a_3 \\ b_2 & b_3 \end{vmatrix}\mathbf{i} - \begin{vmatrix} a_1 & a_3 \\ b_1 & b_3 \end{vmatrix}\mathbf{j} + \begin{vmatrix} a_1 & a_2 \\ b_1 & b_2 \end{vmatrix}\mathbf{k}.$$

The symbol $\mathbf{a} \times \mathbf{b}$ is read "**a** cross **b**". Note that the formula for $\mathbf{a} \times \mathbf{b}$ can be obtained by replacing c_1, c_2, c_3 in our definition of a determinant of order 3 by the unit vectors $\mathbf{i}, \mathbf{j}, \mathbf{k}$. This suggests the following notation for the formula in Definition (14.30):

(14.31)
$$\mathbf{a} \times \mathbf{b} = \begin{vmatrix} \mathbf{i} & \mathbf{j} & \mathbf{k} \\ a_1 & a_2 & a_3 \\ b_1 & b_2 & b_3 \end{vmatrix}$$

The symbol on the right side of this equation is not a determinant, since the first row contains vectors instead of scalars. Consequently, properties of determinants may not be applied to (14.31). The determinant notation is used as a mnemonic device for remembering the more cumbersome formula (14.30). With this warning in mind we shall use (14.31) to find vector products, as in the following example.

Example 2 Find $\mathbf{a} \times \mathbf{b}$ if $\mathbf{a} = \langle 2, -1, 6 \rangle$ and $\mathbf{b} = \langle -3, 5, 1 \rangle$.

Solution Writing

$$\mathbf{a} \times \mathbf{b} = \begin{vmatrix} \mathbf{i} & \mathbf{j} & \mathbf{k} \\ 2 & -1 & 6 \\ -3 & 5 & 1 \end{vmatrix}$$

we obtain

$$\mathbf{a} \times \mathbf{b} = \begin{vmatrix} -1 & 6 \\ 5 & 1 \end{vmatrix}\mathbf{i} - \begin{vmatrix} 2 & 6 \\ -3 & 1 \end{vmatrix}\mathbf{j} + \begin{vmatrix} 2 & -1 \\ -3 & 5 \end{vmatrix}\mathbf{k}$$

$$= (-1 - 30)\mathbf{i} - (2 + 18)\mathbf{j} + (10 - 3)\mathbf{k}$$

$$= -31\mathbf{i} - 20\mathbf{j} + 7\mathbf{k}. \qquad \blacksquare$$

If **a** is any vector in V_3, then

$$\mathbf{a} \times \mathbf{0} = \mathbf{0} = \mathbf{0} \times \mathbf{a}$$

for if one of the vectors in Definition (14.30) is **0**, then each determinant has a row of zeros. Also, $\mathbf{a} \times \mathbf{a} = \mathbf{0}$ for every **a**, since in this case each determinant in (14.30) has equal rows.

The next theorem brings out an important property of vector products.

Theorem (14.32)

> The vector $\mathbf{a} \times \mathbf{b}$ is orthogonal to both **a** and **b**.

Proof   By Theorem (14.24) it is sufficient to show that

$$(\mathbf{a} \times \mathbf{b}) \cdot \mathbf{a} = 0 \quad \text{and} \quad (\mathbf{a} \times \mathbf{b}) \cdot \mathbf{b} = 0.$$

If we apply the definition of dot product to (14.30) and if $\mathbf{a} = \langle a_1, a_2, a_3 \rangle$, then

$$(\mathbf{a} \times \mathbf{b}) \cdot \mathbf{a} = \begin{vmatrix} a_2 & a_3 \\ b_2 & b_3 \end{vmatrix} a_1 - \begin{vmatrix} a_1 & a_3 \\ b_1 & b_3 \end{vmatrix} a_2 + \begin{vmatrix} a_1 & a_2 \\ b_1 & b_2 \end{vmatrix} a_3$$

$$= (a_2 b_3 - a_3 b_2)a_1 - (a_1 b_3 - a_3 b_1)a_2 + (a_1 b_2 - a_2 b_1)a_3$$

$$= a_2 b_3 a_1 - a_3 b_2 a_1 - a_1 b_3 a_2 + a_3 b_1 a_2 + a_1 b_2 a_3 - a_2 b_1 a_3$$

$$= 0.$$

Hence $\mathbf{a} \times \mathbf{b}$ is orthogonal to **a**. The verification that $(\mathbf{a} \times \mathbf{b}) \cdot \mathbf{b} = 0$ is left to the reader.   □

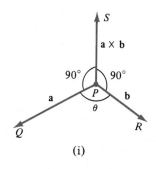

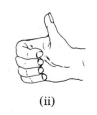

(ii)

Figure 14.24

In geometric terms, Theorem (14.32) implies that if nonzero vectors **a** and **b** are represented by vectors $\overrightarrow{PQ}$ and $\overrightarrow{PR}$ with the same initial point P, then $\mathbf{a} \times \mathbf{b}$ may be represented by a vector $\overrightarrow{PS}$ that is perpendicular to the plane determined by P, Q, and R, as illustrated in (i) of Figure 14.24. We shall write

$$\overrightarrow{PS} = \overrightarrow{PQ} \times \overrightarrow{PR}.$$

It can be shown that the direction of $\overrightarrow{PS}$ may be obtained using the right-hand rule illustrated in (ii) of Figure 14.24. Specifically, if θ denotes the angle between $\overrightarrow{PQ}$ and $\overrightarrow{PR}$, and if the fingers of the right hand are curled such that a rotation through θ will transform $\overrightarrow{PQ}$ into a vector having the same direction as $\overrightarrow{PR}$, then the extended thumb points in the direction of $\overrightarrow{PQ} \times \overrightarrow{PR}$.

The following result provides information about the magnitude of $\mathbf{a} \times \mathbf{b}$.

Theorem (14.33)

> If θ is the angle between nonzero vectors **a** and **b**, then
>
> $$|\mathbf{a} \times \mathbf{b}| = |\mathbf{a}| |\mathbf{b}| \sin \theta.$$

Proof   Applying Definitions (14.2) and (14.30),

$$|\mathbf{a} \times \mathbf{b}|^2 = \begin{vmatrix} a_2 & a_3 \\ b_2 & b_3 \end{vmatrix}^2 + \begin{vmatrix} a_1 & a_3 \\ b_1 & b_3 \end{vmatrix}^2 + \begin{vmatrix} a_1 & a_2 \\ b_1 & b_2 \end{vmatrix}^2$$

$$= (a_2 b_3 - a_3 b_2)^2 + (a_1 b_3 - a_3 b_1)^2 + (a_1 b_2 - a_2 b_1)^2$$

$$= a_2^2 b_3^2 - 2a_2 a_3 b_2 b_3 + a_3^2 b_2^2 + a_1^2 b_3^2 - 2a_1 a_3 b_1 b_3$$
$$\quad + a_3^2 b_1^2 + a_1^2 b_2^2 - 2a_1 a_2 b_1 b_2 + a_2^2 b_1^2$$

$$= (a_1^2 + a_2^2 + a_3^2)(b_1^2 + b_2^2 + b_3^2) - (a_1 b_1 + a_2 b_2 + a_3 b_3)^2.$$

The last equality may be verified by multiplying the indicated expressions. The vector form of this identity is

$$|\mathbf{a} \times \mathbf{b}|^2 = (|\mathbf{a}||\mathbf{b}|)^2 - (\mathbf{a} \cdot \mathbf{b})^2$$

or, since $\mathbf{a} \cdot \mathbf{b} = |\mathbf{a}||\mathbf{b}| \cos \theta$,

$$|\mathbf{a} \times \mathbf{b}|^2 = (|\mathbf{a}||\mathbf{b}|)^2 - (|\mathbf{a}||\mathbf{b}|)^2 \cos^2 \theta$$
$$= (|\mathbf{a}||\mathbf{b}|)^2 (1 - \cos^2 \theta) = (|\mathbf{a}||\mathbf{b}|)^2 \sin^2 \theta.$$

Finally, taking square roots, we obtain

$$|\mathbf{a} \times \mathbf{b}| = |\mathbf{a}||\mathbf{b}| \sin \theta. \qquad \square$$

Corollary (14.34)

> Two vectors $\mathbf{a}$ and $\mathbf{b}$ are parallel if and only if $\mathbf{a} \times \mathbf{b} = \mathbf{0}$.

Proof   Suppose $\mathbf{a}$ and $\mathbf{b}$ are nonzero vectors. If θ is the angle between $\mathbf{a}$ and $\mathbf{b}$, then the vectors are parallel if and only if $\theta = 0$ or $\theta = \pi$, or equivalently $\sin \theta = 0$. By Theorem (14.33) the last statement is equivalent to $\mathbf{a} \times \mathbf{b} = \mathbf{0}$. If either $\mathbf{a}$ or $\mathbf{b}$ is the zero vector the proof is trivial. $\square$

To interpret $|\mathbf{a} \times \mathbf{b}|$ geometrically let us represent $\mathbf{a}$ and $\mathbf{b}$ by vectors $\overrightarrow{PQ}$ and $\overrightarrow{PR}$ having the same initial point P. Let S be the point such that segments PQ and PR are adjacent sides of a parallelogram with vertices $P, Q, R,$ and S, as illustrated in Figure 14.25. An altitude of the parallelogram is $|\mathbf{b}| \sin \theta$ and hence its area is $|\mathbf{a}||\mathbf{b}| \sin \theta$. Comparing this with Corollary (14.33) we see that *the magnitude of the vector product* $\mathbf{a} \times \mathbf{b}$ *equals the area of the parallelogram determined by* $\mathbf{a}$ *and* $\mathbf{b}$. Physical interpretations of the vector product will be given in the next chapter.

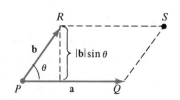

Figure 14.25

Example 3   Find the area of the triangle determined by $P(4, -3, 1)$, $Q(6, -4, 7)$, and $R(1, 2, 2)$.

Solution   Applying Theorem (14.18), the vectors in V_3 corresponding to $\overrightarrow{PQ}$ and $\overrightarrow{PR}$ are $\mathbf{a} = \langle 2, -1, 6 \rangle$ and $\mathbf{b} = \langle -3, 5, 1 \rangle$, respectively. From

Example 2, $\mathbf{a} \times \mathbf{b} = -31\mathbf{i} - 20\mathbf{j} + 7\mathbf{k}$. Hence the area of the parallelogram with adjacent sides PQ and PR is

$$|\mathbf{a} \times \mathbf{b}| = \sqrt{961 + 400 + 49} = \sqrt{1410}.$$

It follows that the area of the triangle is $\frac{1}{2}\sqrt{1410}$. ∎

The vector products of the special unit vectors $\mathbf{i}, \mathbf{j}$, and $\mathbf{k}$ are of interest. For example, using Definition (14.30) with $\mathbf{a} = \mathbf{i} = \langle 1, 0, 0 \rangle$ and $\mathbf{b} = \mathbf{j} = \langle 0, 1, 0 \rangle$,

$$\mathbf{i} \times \mathbf{j} = \begin{vmatrix} 0 & 0 \\ 1 & 0 \end{vmatrix} \mathbf{i} - \begin{vmatrix} 1 & 0 \\ 0 & 0 \end{vmatrix} \mathbf{j} + \begin{vmatrix} 1 & 0 \\ 0 & 1 \end{vmatrix} \mathbf{k} = \mathbf{k}.$$

In general, each of the following is true:

(14.35)

$$\begin{aligned} \mathbf{i} \times \mathbf{j} &= \mathbf{k}, & \mathbf{j} \times \mathbf{k} &= \mathbf{i}, & \mathbf{k} \times \mathbf{i} &= \mathbf{j} \\ \mathbf{j} \times \mathbf{i} &= -\mathbf{k}, & \mathbf{k} \times \mathbf{j} &= -\mathbf{i}, & \mathbf{i} \times \mathbf{k} &= -\mathbf{j} \\ \mathbf{i} \times \mathbf{i} &= \mathbf{j} \times \mathbf{j} = \mathbf{k} \times \mathbf{k} = \mathbf{0}. \end{aligned}$$

The fact that $\mathbf{i} \times \mathbf{j} \neq \mathbf{j} \times \mathbf{i}$ shows that the vector product is not commutative. The associative law does not hold either since, for example,

$$\mathbf{i} \times (\mathbf{j} \times \mathbf{j}) = \mathbf{i} \times \mathbf{0} = \mathbf{0},$$

whereas

$$(\mathbf{i} \times \mathbf{j}) \times \mathbf{j} = \mathbf{k} \times \mathbf{j} = -\mathbf{i}.$$

The following theorem lists some important properties of the vector product, where $\mathbf{a}, \mathbf{b}$, and $\mathbf{c}$ are any vectors, and m is a scalar.

Theorem (14.36)

(i)	$\mathbf{a} \times \mathbf{b} = -\mathbf{b} \times \mathbf{a}$
(ii)	$(m\mathbf{a}) \times \mathbf{b} = m(\mathbf{a} \times \mathbf{b}) = \mathbf{a} \times (m\mathbf{b})$
(iii)	$\mathbf{a} \times (\mathbf{b} + \mathbf{c}) = (\mathbf{a} \times \mathbf{b}) + (\mathbf{a} \times \mathbf{c})$
(iv)	$(\mathbf{a} + \mathbf{b}) \times \mathbf{c} = (\mathbf{a} \times \mathbf{c}) + (\mathbf{b} \times \mathbf{c})$
(v)	$(\mathbf{a} \times \mathbf{b}) \cdot \mathbf{c} = \mathbf{a} \cdot (\mathbf{b} \times \mathbf{c})$
(vi)	$\mathbf{a} \times (\mathbf{b} \times \mathbf{c}) = (\mathbf{a} \cdot \mathbf{c})\mathbf{b} - (\mathbf{a} \cdot \mathbf{b})\mathbf{c}$

Proof   All of the properties may be established by straightforward (but sometimes lengthy) applications of Definition (14.30). For example, if $\mathbf{a} = \langle a_1, a_2, a_3 \rangle$ and $\mathbf{b} = \langle b_1, b_2, b_3 \rangle$, then

$$\mathbf{b} \times \mathbf{a} = \begin{vmatrix} b_2 & b_3 \\ a_2 & a_3 \end{vmatrix} \mathbf{i} - \begin{vmatrix} b_1 & b_3 \\ a_1 & a_3 \end{vmatrix} \mathbf{j} + \begin{vmatrix} b_1 & b_2 \\ a_1 & a_2 \end{vmatrix} \mathbf{k}.$$

Since interchanging two rows of a determinant changes its sign, we have

$$\mathbf{b} \times \mathbf{a} = - \begin{vmatrix} a_2 & a_3 \\ b_2 & b_3 \end{vmatrix} \mathbf{i} + \begin{vmatrix} a_1 & a_3 \\ b_1 & b_3 \end{vmatrix} \mathbf{j} - \begin{vmatrix} a_1 & a_2 \\ b_1 & b_2 \end{vmatrix} \mathbf{k}$$

$$= -\mathbf{a} \times \mathbf{b}.$$

This proves property (i).

If $\mathbf{c} = \langle c_1, c_2, c_3 \rangle$, then the $\mathbf{i}$ component of $\mathbf{a} \times (\mathbf{b} + \mathbf{c})$ is

$$\begin{vmatrix} a_2 & a_3 \\ b_2 + c_2 & b_3 + c_3 \end{vmatrix} = a_2(b_3 + c_3) - a_3(b_2 + c_2)$$

$$= (a_2 b_3 - a_3 b_2) + (a_2 c_3 - a_3 c_2)$$

$$= \begin{vmatrix} a_2 & a_3 \\ b_2 & b_3 \end{vmatrix} + \begin{vmatrix} a_2 & a_3 \\ c_2 & c_3 \end{vmatrix}$$

which is equal to the $\mathbf{i}$ component of $(\mathbf{a} \times \mathbf{b}) + (\mathbf{a} \times \mathbf{c})$. A similar calculation can be used to prove that the $\mathbf{j}$ and $\mathbf{k}$ components of $\mathbf{a} \times (\mathbf{b} + \mathbf{c})$ are the same as those of $(\mathbf{a} \times \mathbf{b}) + (\mathbf{a} \times \mathbf{c})$. This establishes (iii). The proofs of the remaining properties are left to the reader.   □

The formula in (vi) of Theorem (14.36) is called the **triple vector product** of $\mathbf{a}$, $\mathbf{b}$, and $\mathbf{c}$.

We shall conclude this section with two geometric applications of the vector product.

Example 4   Find a formula for the distance d from a point R to a line l.

Solution   As illustrated in Figure 14.26, let P and Q be points on l, and let θ be the angle between $\overrightarrow{PQ}$ and $\overrightarrow{PR}$. Using the fact that $d = |\overrightarrow{PR}| \sin \theta$ and $|\overrightarrow{PQ} \times \overrightarrow{PR}| = |\overrightarrow{PQ}||\overrightarrow{PR}| \sin \theta$ gives us

$$d = \frac{|\overrightarrow{PQ} \times \overrightarrow{PR}|}{|\overrightarrow{PQ}|}.$$ ■

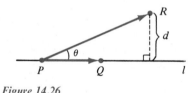

Figure 14.26

Example 5   Suppose $\overrightarrow{PQ}$, $\overrightarrow{PR}$, and $\overrightarrow{PS}$ represent adjacent sides of a rectangular parallelepiped. If $\mathbf{a}$, $\mathbf{b}$, and $\mathbf{c}$ are the corresponding vectors in V_3, show that $|(\mathbf{a} \times \mathbf{b}) \cdot \mathbf{c}|$ is the volume of the parallelepiped.

Solution   One illustration of the parallelepiped is given in Figure 14.27.

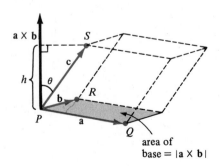

Figure 14.27

The area of the base is $|\mathbf{a} \times \mathbf{b}|$. Let θ be the angle between $\mathbf{c}$ and $\mathbf{a} \times \mathbf{b}$. Since the vector corresponding to $\mathbf{a} \times \mathbf{b}$ is perpendicular to the base, it follows that the altitude h of the parallelepiped is given by $h = |\mathbf{c}||\cos \theta|$. It is necessary to use the absolute value $|\cos \theta|$ since θ may be an obtuse angle. Hence the volume V of the parallelepiped is given by

$$V = \text{(area of base) (altitude)}$$
$$= |\mathbf{a} \times \mathbf{b}||\mathbf{c}||\cos \theta|$$
$$= |(\mathbf{a} \times \mathbf{b}) \cdot \mathbf{c}|.$$ ∎

EXERCISES 14.4

In Exercises 1–10 find $\mathbf{a} \times \mathbf{b}$.

1 $\mathbf{a} = \langle 1, -2, 3 \rangle, \mathbf{b} = \langle 2, 1, -4 \rangle$

2 $\mathbf{a} = \langle -5, 1, -1 \rangle, \mathbf{b} = \langle 3, 6, -2 \rangle$

3 $\mathbf{a} = 5\mathbf{i} - 6\mathbf{j} - \mathbf{k}, \mathbf{b} = 3\mathbf{i} + \mathbf{k}$

4 $\mathbf{a} = 2\mathbf{i} + \mathbf{j}, \mathbf{b} = -5\mathbf{j} + 2\mathbf{k}$

5 $\mathbf{a} = \langle 0, 1, 2 \rangle, \mathbf{b} = \langle 1, 2, 0 \rangle$

6 $\mathbf{a} = -3\mathbf{i} + \mathbf{j} + 2\mathbf{k}, \mathbf{b} = 9\mathbf{i} - 3\mathbf{j} - 6\mathbf{k}$

7 $\mathbf{a} = 3\mathbf{i} - \mathbf{j} + 8\mathbf{k}, \mathbf{b} = 5\mathbf{j}$

8 $\mathbf{a} = \langle 0, 0, 4 \rangle, \mathbf{b} = \langle -7, 1, 0 \rangle$

9 $\mathbf{a} = 4\mathbf{i} - 6\mathbf{j} + 2\mathbf{k}, \mathbf{b} = -2\mathbf{i} + 3\mathbf{j} - \mathbf{k}$

10 $\mathbf{a} = 3\mathbf{i}, \mathbf{b} = 4\mathbf{k}$

11 If $\mathbf{a} = \langle 2, 0, -1 \rangle$, $\mathbf{b} = \langle -3, 1, 0 \rangle$, and $\mathbf{c} = \langle 1, -2, 4 \rangle$, find $\mathbf{a} \times (\mathbf{b} \times \mathbf{c})$ and $(\mathbf{a} \times \mathbf{b}) \times \mathbf{c}$.

12 If $\mathbf{a}, \mathbf{b}$, and $\mathbf{c}$ are the vectors of Exercise 11, find $\mathbf{a} \times (\mathbf{b} - \mathbf{c})$ and $(\mathbf{a} \times \mathbf{b}) - (\mathbf{a} \times \mathbf{c})$.

13 Prove that $(\mathbf{a} \times \mathbf{b}) \cdot \mathbf{b} = 0$ for all vectors $\mathbf{a}$ and $\mathbf{b}$.

14 Prove (14.35).

15 Complete the proof of Theorem (14.36).

16 (a) If $\mathbf{a} \times \mathbf{b} = \mathbf{a} \times \mathbf{c}$ and $\mathbf{a} \neq \mathbf{0}$, does it follow that $\mathbf{b} = \mathbf{c}$? Explain.

　(b) Let $\mathbf{a} \neq \mathbf{0}$. If $\mathbf{a} \times \mathbf{b} = \mathbf{a} \times \mathbf{c}$ *and* $\mathbf{a} \cdot \mathbf{b} = \mathbf{a} \cdot \mathbf{c}$, prove that $\mathbf{b} = \mathbf{c}$.

In Exercises 17–20 find (a) a vector perpendicular to the plane determined by the points P, Q, and R; (b) the area of the triangle determined by P, Q, and R.

17 $P(1, -1, 2), Q(0, 3, -1), R(3, -4, 1)$

18 $P(-3, 0, 5), Q(2, -1, -3), R(4, 1, -1)$

19 $P(4, 0, 0), Q(0, 5, 0), R(0, 0, 2)$

20 $P(-1, 2, 0), Q(0, 2, -3), R(5, 0, 1)$

21 If $\mathbf{a} = \langle a_1, a_2, a_3 \rangle$, $\mathbf{b} = \langle b_1, b_2, b_3 \rangle$, and $\mathbf{c} = \langle c_1, c_2, c_3 \rangle$, prove that

$$\mathbf{a} \cdot (\mathbf{b} \times \mathbf{c}) = (\mathbf{a} \times \mathbf{b}) \cdot \mathbf{c} = \begin{vmatrix} a_1 & a_2 & a_3 \\ b_1 & b_2 & b_3 \\ c_1 & c_2 & c_3 \end{vmatrix}.$$

(This number is called the **triple scalar product** of $\mathbf{a}, \mathbf{b}$, and $\mathbf{c}$.)

22 If $\mathbf{a}, \mathbf{b}$, and $\mathbf{c}$ are represented by vectors with a common initial point, show that $\mathbf{a} \cdot (\mathbf{b} \times \mathbf{c}) = 0$ if and only if the vectors are coplanar.

23 Given $P(1, -1, 2), Q(0, 3, -1)$, and $R(3, -4, 1)$, use Exercise 21 to find the volume of the parallelepiped having adjacent sides OP, OQ, and OR.

24 Given $A(2, 1, -1), B(3, 0, 2), C(4, -2, 1)$, and $D(5, -3, 0)$, use Exercise 21 to find the volume of the parallelepiped having adjacent sides AB, AC, and AD.

Without using components, verify the identities in Exercises 25–30, where $\mathbf{a}, \mathbf{b}, \mathbf{c}$, and $\mathbf{d}$ are arbitrary vectors.

25 $(\mathbf{a} + \mathbf{b}) \times (\mathbf{a} - \mathbf{b}) = 2\mathbf{b} \times \mathbf{a}$

26 $\mathbf{a} \times (\mathbf{b} \times \mathbf{c}) + \mathbf{b} \times (\mathbf{c} \times \mathbf{a}) + \mathbf{c} \times (\mathbf{a} \times \mathbf{b}) = \mathbf{0}$ (*Hint*: Use (vi) of Theorem (14.36).)

27 $(\mathbf{a} \times \mathbf{b}) \times \mathbf{c} = (\mathbf{a} \cdot \mathbf{c})\mathbf{b} - (\mathbf{b} \cdot \mathbf{c})\mathbf{a}$

28 $(\mathbf{a} \times \mathbf{b}) \cdot (\mathbf{c} \times \mathbf{d}) = \begin{vmatrix} \mathbf{a} \cdot \mathbf{c} & \mathbf{b} \cdot \mathbf{c} \\ \mathbf{a} \cdot \mathbf{d} & \mathbf{b} \cdot \mathbf{d} \end{vmatrix}$

29 $(\mathbf{a} \times \mathbf{b}) \times (\mathbf{c} \times \mathbf{d}) = (\mathbf{a} \times \mathbf{b} \cdot \mathbf{d})\mathbf{c} - (\mathbf{a} \times \mathbf{b} \cdot \mathbf{c})\mathbf{d}$

30 $(\mathbf{a} \times \mathbf{b}) \cdot (\mathbf{b} \times \mathbf{c}) \times (\mathbf{c} \times \mathbf{a}) = (\mathbf{a} \cdot \mathbf{b} \times \mathbf{c})^2$

14.5

LINES IN SPACE

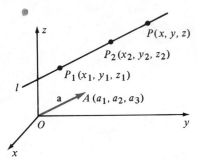

Figure 14.28

If $P_1(x_1, y_1, z_1)$ and $P_2(x_2, y_2, z_2)$ are distinct points, then by Theorem (14.18), the vector $\mathbf{a} = \langle a_1, a_2, a_3 \rangle$ in V_3 with geometric representation $\overrightarrow{P_1 P_2}$ is given by $\mathbf{a} = \langle x_2 - x_1, y_2 - y_1, z_2 - z_1 \rangle$. To find equations for the line l through P_1 and P_2 we observe that a point $P(x, y, z)$ is on l if and only if $\overrightarrow{P_1 P}$ and $\overrightarrow{P_1 P_2}$ are parallel. This is illustrated in Figure 14.28, where the position vector $\overrightarrow{OA}$ corresponding to $\mathbf{a}$ is also shown.

We know from our work in Section 14.3 that $\overrightarrow{P_1 P}$ and $\overrightarrow{P_1 P_2}$ are parallel if and only if

$$\overrightarrow{P_1 P} = t\overrightarrow{P_1 P_2} \quad \text{for some scalar } t.$$

This condition may be stated algebraically as follows:

$$\langle x - x_1, y - y_1, z - z_1 \rangle = t\langle a_1, a_2, a_3 \rangle = \langle a_1 t, a_2 t, a_3 t \rangle.$$

Equating components and solving for x, y, and z gives us

$$x = x_1 + a_1 t, \quad y = y_1 + a_2 t, \quad z = z_1 + a_3 t$$

where t is a real number. These equations are called **parametric equations for the line l**, and the variable t is called a **parameter**. The points $P(x, y, z)$ on l are obtained by letting t take on all real values. For example, P_1 corresponds to $t = 0$, the midpoint of $P_1 P_2$ to $t = \frac{1}{2}$, and P_2 to $t = 1$.

We may summarize the preceding discussion as follows.

Theorem (14.37)

> If $P_1(x_1, y_1, z_1)$ and $P_2(x_2, y_2, z_2)$ are two distinct points, and if $\mathbf{a} = \langle a_1, a_2, a_3 \rangle$ is the vector corresponding to $\overrightarrow{P_1 P_2}$, then parametric equations for the line through P_1 and P_2 are
>
> $$x = x_1 + a_1 t, \quad y = y_1 + a_2 t, \quad z = z_1 + a_3 t.$$

Example 1 Find parametric equations for the line l through $P_1(3, 1, -2)$ and $P_2(-2, 7, -4)$. At what point does l intersect the xy-plane?

Solution   The vector $\mathbf{a}$ in V_3 corresponding to $\overrightarrow{P_1 P_2}$ is

$$\mathbf{a} = \langle -2 - 3, 7 - 1, -4 + 2 \rangle = \langle -5, 6, -2 \rangle.$$

Applying Theorem (14.37), parametric equations for l are

$$x = 3 - 5t, \quad y = 1 + 6t, \quad z = -2 - 2t.$$

The line intersects the xy-plane at the point $R(x, y, z)$ if $z = -2 - 2t = 0$, that is, if $t = -1$. Thus R is the point with coordinates $(8, -5, 0)$. ∎

If, in Figure 14.28, we let

$$\mathbf{r} = x\mathbf{i} + y\mathbf{j} + z\mathbf{k} \quad \text{and} \quad \mathbf{r}_1 = x_1\mathbf{i} + y_1\mathbf{j} + z_1\mathbf{k}$$

then $\overrightarrow{OP}$ and $\overrightarrow{OP_1}$ are position vectors of $\mathbf{r}$ and $\mathbf{r}_1$, respectively, and $\overrightarrow{P_1P}$ represents the vector $\mathbf{r} - \mathbf{r}_1$. Thus we may write $\mathbf{r} - \mathbf{r}_1 = t\mathbf{a}$. This leads to the following vector equation for the line in Theorem (14.37):

$$\mathbf{r} = \mathbf{r}_1 + t\mathbf{a}.$$

If $\mathbf{a} = \langle a_1, a_2, a_3 \rangle$ is a nonzero vector in V_3 and $P_1(x_1, y_1, z_1)$ is any point, then there is a unique line l through P_1 parallel to the position vector $\overrightarrow{OA}$ corresponding to $\mathbf{a}$. We shall refer to l as **the line through P_1 parallel to a.** Parametric equations for l are given in Theorem (14.37). If $\mathbf{b}$ is any nonzero vector that is parallel to $\mathbf{a}$, then the same line is determined, for in this case $\mathbf{b} = c\mathbf{a} = \langle ca_1, ca_2, ca_3 \rangle$ and parametric equations for the line through P_1 determined by $\mathbf{b}$ are

$$x = x_1 + (ca_1)v, \quad y = y_1 + (ca_2)v, \quad z = z_1 + (ca_3)v$$

where the parameter v ranges through all real numbers. These equations determine the same line, since the point given by t can be obtained by letting $v = t/c$.

Example 2   Find parametric equations for the line l through $P(5, -6, 2)$ parallel to $\mathbf{a} = \langle \frac{1}{2}, 2, -\frac{4}{3} \rangle$.

Solution   To avoid fractions we shall use the vector $\mathbf{b} = 6\mathbf{a} = \langle 3, 12, -8 \rangle$ instead of $\mathbf{a}$. Applying Theorem (14.37), parametric equations for l are

$$x = 5 + 3t, \quad y = -6 + 12t, \quad z = 2 - 8t. \quad\blacksquare$$

If l_1 and l_2 are lines parallel to vectors $\mathbf{a} = \langle a_1, a_2, a_3 \rangle$ and $\mathbf{b} = \langle b_1, b_2, b_3 \rangle$, respectively, then the angles between l_1 and l_2 are defined as θ and $\pi - \theta$, where θ is the angle between $\mathbf{a}$ and $\mathbf{b}$. The lines are **orthogonal** if $\mathbf{a} \cdot \mathbf{b} = 0$ or, equivalently, if $a_1b_1 + a_2b_2 + a_3b_3 = 0$. The lines are **parallel** if $\mathbf{b} = c\mathbf{a}$ for some scalar c, that is, if $b_1 = ca_1, b_2 = ca_2, b_3 = ca_3$.

In the next section we shall discuss representations of lines as intersections of planes.

EXERCISES 14.5

In Exercises 1–4 find parametric equations for the line through P_1 and P_2. Determine (if possible) the points at which the line intersects each of the coordinate planes.

1   $P_1(5, -2, 4), P_2(2, 6, 1)$

2   $P_1(-3, 1, -1), P_2(7, 11, -8)$

3   $P_1(2, 0, 5), P_2(-6, 0, 3)$

4   $P_1(2, -2, 4), P_2(2, -2, -3)$

In Exercises 5–8 find parametric equations for the line through P parallel to $\mathbf{a}$.

5   $P(4, 2, -3), \mathbf{a} = \langle \frac{1}{3}, 2, \frac{1}{2} \rangle$

6 $P(5, 0, -2)$, $\mathbf{a} = \langle -1, -4, 1 \rangle$

7 $P(0, 0, 0)$, $\mathbf{a} = \mathbf{j}$

8 $P(1, 2, 3)$, $\mathbf{a} = \mathbf{i} + 2\mathbf{j} + 3\mathbf{k}$

9 Suppose a line l has parametric equations $x = 5 - 3t$, $y = -2 + t$, $z = 1 + 9t$. Find parametric equations for a line through $P(-6, 4, -3)$ that is parallel to l.

10 If l_1 is the line through $P(5, -2, 4)$ and $Q(2, 6, 1)$, and if l_2 is the line through $R(-3, 1, -1)$ and $S(7, 11, -8)$, find the angles between l_1 and l_2.

11 Find the angles between the two lines having parametric equations $x = 7 - 2t, y = 4 + 3t, z = 5t$ and $x = -1 + t$, $y = 3 + 4t, z = 1 + t$, respectively.

12 Find parametric equations for the line through $P(4, -1, 0)$ that is parallel to the line through $P_1(-3, 9, -2)$ and $P_2(5, 7, -3)$.

In Exercises 13–16 determine whether the two lines intersect and, if so, find the point of intersection.

13 $x = 1 + 2t, y = 1 - 4t, z = 5 - t$;
$x = 4 - v, y = -1 + 6v, z = 4 + v$

14 $x = 1 - 6t, y = 3 + 2t, z = 1 - 2t$;
$x = 2 + 2v, y = 6 + v, z = 2 + v$

15 $x = 3 + t, y = 2 - 4t, z = t$;
$x = 4 - v, y = 3 + v, z = -2 + 3v$

16 $x = 2 - 5t, y = 6 + 2t, z = -3 - 2t$;
$x = 4 - 3v, y = 7 + 5v, z = 1 + 4v$

In Exercises 17 and 18 show that the two lines intersect orthogonally.

17 $x = 2 + 3t, y = -4 - 2t, z = -1 + 4t$;
$x = 6 + 4v, y = -2 + 2v, z = -3 - 2v$

18 $x = 4 - t, y = -1 - t, z = 4 + 3t$;
$x = 3 + 2v, y = -1 + v, z = v$

19 Use the dot product to find the distance d from the point $A(2, -6, 1)$ to the line l through $B(3, 4, -2)$ and $C(7, -1, 5)$.

20 Work Exercise 19 for the points $A(1, 5, 0)$, $B(-2, 1, -4)$, and $C(0, -3, 2)$.

21 Find the distance from the point $C(2, 1, -2)$ to the line having parametric equations $x = 3 - 2t$, $y = -4 + 3t$, $z = 1 + 2t$.

22 Suppose a line l has parametric equations $x = 2t + 1$, $y = -t + 3$, $z = 5t$. Find the distance from $A(3, 1, -1)$ to l.

23 Find parametric equations for the line through $P(3, 1, -2)$ that is orthogonal to, and intersects, the line $x = -1 + t$, $y = -2 + t, z = -1 + t$.

24 Find the point on the line through $A(2, 1, -2)$ and $B(1, -3, 2)$ that is equidistant from $C(0, 1, 1)$ and $D(1, 2, 3)$.

14.6

PLANES

If P_1 and P_2 are distinct points, then all points P such that $\overline{P_1P}$ is orthogonal to $\overrightarrow{P_1P_2}$ lie on a plane Γ through P_1, as illustrated in Figure 14.29.

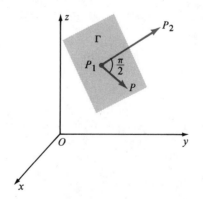

Figure 14.29

We shall call $\overrightarrow{P_1 P_2}$, or the vector $\mathbf{a} = \langle a_1, a_2, a_3 \rangle$ with geometric representation $\overrightarrow{P_1 P_2}$, a **normal vector** to Γ. It follows from Theorem (14.24) that P is on Γ if and only if $\overrightarrow{P_1 P_2} \cdot \overrightarrow{P_1 P} = 0$ or, equivalently,

$$\langle a_1, a_2, a_3 \rangle \cdot \langle x - x_1, y - y_1, z - z_1 \rangle = 0.$$

Applying the definition of dot product gives us the next result.

Theorem (14.38)

An equation of the plane through $P_1(x_1, y_1, z_1)$ with normal vector $\mathbf{a} = \langle a_1, a_2, a_3 \rangle$ is

$$a_1(x - x_1) + a_2(y - y_1) + a_3(z - z_1) = 0.$$

If we let

$$\mathbf{r} = x\mathbf{i} + y\mathbf{j} + z\mathbf{k} \quad \text{and} \quad \mathbf{r}_1 = x_1\mathbf{i} + y_1\mathbf{j} + z_1\mathbf{k}$$

then $\overrightarrow{OP}$ and $\overrightarrow{OP_1}$ are position vectors of $\mathbf{r}$ and $\mathbf{r}_1$, respectively, and $\overrightarrow{P_1 P}$ represents the vector $\mathbf{r} - \mathbf{r}_1$. This leads to the following **vector equation for a plane**:

$$\mathbf{a} \cdot (\mathbf{r} - \mathbf{r}_1) = 0.$$

Example 1 Find an equation of the plane through the point $(5, -2, 4)$ with normal vector $\mathbf{a} = \langle 1, 2, 3 \rangle$.

Solution Applying Theorem (14.38), we obtain

$$1(x - 5) + 2(y + 2) + 3(z - 4) = 0$$

which reduces to

$$x + 2y + 3z - 13 = 0. \qquad \blacksquare$$

Example 2 Find an equation of the plane determined by the points $P(4, -3, 1)$, $Q(6, -4, 7)$, and $R(1, 2, 2)$.

Solution Vectors $\mathbf{a}$ and $\mathbf{b}$ corresponding to $\overrightarrow{PQ}$ and $\overrightarrow{PR}$ are

$$\mathbf{a} = \langle 2, -1, 6 \rangle \quad \text{and} \quad \mathbf{b} = \langle -3, 5, 1 \rangle.$$

As illustrated in Figure 14.24, the vector $\mathbf{a} \times \mathbf{b}$ is normal to the plane determined by P, Q, and R. From Example 2 of Section 14.4,

$$\mathbf{a} \times \mathbf{b} = -31\mathbf{i} - 20\mathbf{j} + 7\mathbf{k}.$$

Using Theorem (14.38) with $P_1 = P$ gives us the equation

$$-31(x - 4) - 20(y + 3) + 7(z - 1) = 0,$$

or

$$-31x - 20y + 7z + 57 = 0. \qquad \blacksquare$$

The equation of the plane in Theorem (14.38) may be written in the form

$$ax + by + cz + d = 0$$

where $a = a_1$, $b = b_1$, $c = c_1$, and $d = -a_1x_1 - a_2y_1 - a_3z_1$. Conversely, given $ax + by + cz + d = 0$, where $a, b,$ and c are not all zero, we may choose numbers x_1, y_1, and z_1 such that $ax_1 + by_1 + cz_1 + d = 0$. Consequently, $d = -ax_1 - by_1 - cz_1$, and hence

$$ax + by + cz - ax_1 - by_1 - cz_1 = 0,$$

or
$$a(x - x_1) + b(y - y_1) + c(z - z_1) = 0.$$

According to Theorem (14.38), the graph of the last equation is a plane through $P(x_1, y_1, z_1)$ with normal vector $\langle a, b, c \rangle$. An equation of the form $ax + by + cz + d = 0$, where a, b, and c are not all zero, is called a **linear equation in three variables** x, y, and z. We have proved the following theorem.

Theorem (14.39)

> The graph of every linear equation $ax + by + cz + d = 0$ is a plane with normal vector $\langle a, b, c \rangle$.

For simplicity we often use the phrase "the plane $ax + by + cz + d = 0$" instead of the more accurate statement "the plane that has equation $ax + by + cz + d = 0$."

To sketch the graph of a linear equation we often find, if possible, the **trace** of the graph in each coordinate plane, that is, the line in which the graph intersects the coordinate plane. To find the trace in the xy-plane we substitute 0 for z, since this will lead to all points of the graph that lie on the xy-plane. Similarly, to find the trace in the yz-plane or the xz-plane we let $x = 0$ or $y = 0$, respectively, in the equation $ax + by + cz + d = 0$.

Example 3 Sketch the graph of the equation $2x + 3y + 4z = 12$.

Solution   There are three points on the plane that are easily found, namely the points of intersection of the plane with the coordinate axes. Substituting 0 for both y and z in the equation, we obtain $2x = 12$, or $x = 6$. Thus the point $(6, 0, 0)$ is on the graph. As in two dimensions, 6 is called the x-*intercept* of the graph. Similarly, substitution of 0 for x and z gives us the y-*intercept* 4, and hence the point $(0, 4, 0)$ is on the graph. The point $(0, 0, 3)$ (or z-*intercept* 3) is obtained in like manner. The trace in the xy-plane is found by substituting 0 for z in the given equation. This leads to $2x + 3y = 12$, which has as its graph in the xy-plane a line with x-intercept 6 and y-intercept 4. This trace and the traces of the graph in the xz- and yz-planes are illustrated in Figure 14.30. ∎

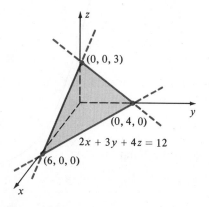

Figure 14.30

Definition (14.40)

> Two planes with normal vectors **a** and **b** are
>
> (i) **parallel** if **a** and **b** are parallel.
>
> (ii) **orthogonal** if **a** and **b** are orthogonal.

Example 4 Prove that the graphs of the equations $2x - 3y - z - 5 = 0$ and $-6x + 9y + 3z + 2 = 0$ are parallel planes.

Solution By Theorem (14.39), the graphs are planes with normal vectors $\mathbf{a} = \langle 2, -3, -1 \rangle$ and $\mathbf{b} = \langle -6, 9, 3 \rangle$. Since $\mathbf{b} = -3\mathbf{a}$, the vectors **a** and **b** are parallel and hence, by Definition (14.40), so are the planes. ∎

Example 5 Find an equation of the plane Γ through $P(5, -2, 4)$ that is parallel to the plane $3x + y - 6z + 8 = 0$.

Solution The vector $\mathbf{a} = \langle 3, 1, -6 \rangle$ may be used as a normal vector for Γ and hence the desired equation may be written $3x + y - 6z + d = 0$ for some number d. If $P(5, -2, 4)$ is on Γ, then its coordinates must satisfy this equation, that is, $3(5) + (-2) - 6(4) + d = 0$ or $d = 11$. Hence an equation for Γ is $3x + y - 6z + 11 = 0$. ∎

The vector $\mathbf{i} = \langle 1, 0, 0 \rangle$ is a normal vector for the yz-plane. A plane that has an equation of the form $x - x_1 = 0$ (or $x = x_1$) also has normal vector **i** and hence is parallel to the yz-plane (and orthogonal to both the xy- and xz-planes). A portion of the graph of $x = a$ is sketched in (i) of Figure 14.31. Similarly, the graph of $y = b$ is a plane parallel to the xz-plane with y-intercept b, whereas the graph of $z = c$ is a plane parallel to the xy-plane with z-intercept c (see (ii) and (iii) of Figure 14.31).

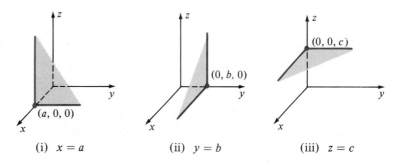

(i) $x = a$ (ii) $y = b$ (iii) $z = c$

Figure 14.31

A plane with an equation of the form $by + cz + d = 0$ has normal vector $\mathbf{a} = \langle 0, b, c \rangle$ and is orthogonal to the yz-plane since $\mathbf{a} \cdot \mathbf{i} = 0$. Similarly, graphs of $ax + by + d = 0$ and $ax + cz + d = 0$ are planes that are orthogonal to the xy-plane and xz-plane, respectively.

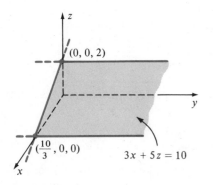

 shows axes labeled with points $(0, 0, 2)$, $(\frac{10}{3}, 0, 0)$, and the equation $3x + 5z = 10$.

Figure 14.32

Example 6   Sketch the graph of the equation $3x + 5z = 10$.

Solution   The graph is a plane orthogonal to the xz-plane with x-intercept $10/3$ and z-intercept 2. Note that the trace in the yz-plane has equation $5z = 10$ and hence is a line parallel to the y-axis with z-intercept 2. Similarly, the trace in the xy-plane has equation $3x = 10$ and is a line parallel to the y-axis with x-intercept $10/3$. A portion of the graph showing traces in the three coordinate planes is sketched in Figure 14.32.   ∎

Lines may be described as intersections of planes. If a line l is given parametrically as in Theorem (14.37), and if a_1, a_2, a_3 are different from zero, we may solve each equation for t, obtaining

$$t = \frac{x - x_1}{a_1}, \quad t = \frac{y - y_1}{a_2}, \quad t = \frac{z - z_1}{a_3}.$$

It follows that a point $P(x, y, z)$ is on l if and only if

(14.41)
$$\frac{x - x_1}{a_1} = \frac{y - y_1}{a_2} = \frac{z - z_1}{a_3}.$$

The equations in (14.41) are called a **symmetric form** for the line l. A symmetric form is not unique, since in (14.41) we may use any three numbers b_1, b_2, b_3 that are proportional to a_1, a_2, a_3, or any point on l other than (x_1, y_1, z_1).

If, in (14.41), we take the indicated expressions in pairs, say

$$\frac{x - x_1}{a_1} = \frac{y - y_1}{a_2} \quad \text{and} \quad \frac{x - x_1}{a_1} = \frac{z - z_1}{a_3},$$

we obtain a description of l as an intersection of two planes, the first orthogonal to the xy-plane and the second orthogonal to the xz-plane. If one of the numbers $a_1, a_2,$ or a_3 is zero, we cannot solve each equation in (14.37) for t. For example, if $a_3 = 0$ and $a_1 a_2 \neq 0$, then the third equation reduces to $z = z_1$, and a symmetric form may be written as

$$\frac{x - x_1}{a_1} = \frac{y - y_1}{a_2}, \quad z = z_1$$

which again expresses l as an intersection of two planes. A similar situation exists if $a_1 = 0$ or $a_2 = 0$.

Example 7   Find a symmetric form for the line through $P_1(3, 1, -2)$ and $P_2(-2, 7, -4)$.

Solution   As in Example 1 of the preceding section, a vector **a** corresponding to $\overrightarrow{P_1 P_2}$ is

$$\mathbf{a} = \langle -2 - 3, 7 - 1, -4 + 2 \rangle = \langle -5, 6, -2 \rangle$$

and by Theorem (14.37) a parametric representation for the line is

$$x = 3 - 5t, \quad y = 1 + 6t, \quad z = -2 - 2t.$$

Solving each equation for t and equating the results we obtain the symmetric form

$$\frac{x-3}{-5} = \frac{y-1}{6} = \frac{z+2}{-2}. \qquad \blacksquare$$

Example 8 Find a formula for the distance h from a point $P(x_0, y_0, z_0)$ to the plane $ax + by + cz + d = 0$.

Solution   Let $R(x_1, y_1, z_1)$ be any point on the plane and let $\mathbf{n}$ be a normal vector to the plane. The vector $\mathbf{p} = \langle x_0 - x_1, y_0 - y_1, z_0 - z_1 \rangle$ has a geometric representation $\overrightarrow{RP}$. As illustrated in Figure 14.33, the distance h is given by

$$h = |\text{comp}_{\mathbf{n}}\, \mathbf{p}| = \left| \mathbf{p} \cdot \frac{1}{|\mathbf{n}|} \mathbf{n} \right|.$$

Since $\langle a, b, c \rangle$ is a normal vector to the plane we may let

$$\frac{1}{|\mathbf{n}|}\mathbf{n} = \frac{1}{\sqrt{a^2 + b^2 + c^2}} \langle a, b, c \rangle.$$

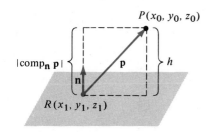

$P(x_0, y_0, z_0)$

$|\text{comp}_{\mathbf{n}}\, \mathbf{p}|$

$\mathbf{p}$

h

$\mathbf{n}$

$R(x_1, y_1, z_1)$

Figure 14.33

Consequently,

$$h = \left| \mathbf{p} \cdot \frac{1}{|\mathbf{n}|} \mathbf{n} \right| = \left| \frac{a(x_0 - x_1) + b(y_0 - y_1) + c(z_0 - z_1)}{\sqrt{a^2 + b^2 + c^2}} \right|$$

$$= \frac{|(ax_0 + by_0 + cz_0) + (-ax_1 - by_1 - cz_1)|}{\sqrt{a^2 + b^2 + c^2}}.$$

Since R is on the plane, $ax_1 + by_1 + cz_1 + d = 0$ and, therefore, $d = -ax_1 - by_1 - cz_1$. Hence the preceding formula may be written

$$h = \frac{|ax_0 + by_0 + cz_0 + d|}{\sqrt{a^2 + b^2 + c^2}}. \qquad \blacksquare$$

Example 9 Find the distance from the point $P(-6, 2, 3)$ to the plane $4x - 5y + 8z - 7 = 0$.

Solution   Using the formula obtained in Example 8,

$$h = \frac{|4(-6) - 5(2) + 8(3) - 7|}{\sqrt{16 + 25 + 64}} = \frac{17}{\sqrt{105}}. \qquad \blacksquare$$

Example 10 Find a formula for the shortest distance d between two skew lines l_1 and l_2.

Solution   Two lines are skew if they are not parallel and do not intersect. Choose points P_1, Q_1, on l_1 and P_2, Q_2 on l_2, as illustrated in Figure 14.34. It follows that the vector

$$\mathbf{n} = \frac{1}{|\overrightarrow{P_1Q_1} \times \overrightarrow{P_2Q_2}|} (\overrightarrow{P_1Q_1} \times \overrightarrow{P_2Q_2})$$

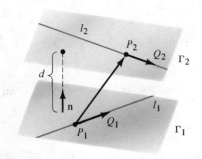

Figure 14.34

is a unit vector orthogonal to both $\overrightarrow{P_1Q_1}$ and $\overrightarrow{P_2Q_2}$. (Why?)

Let us consider planes Γ_1 and Γ_2 through P_1 and P_2, respectively, each having normal vector $\mathbf{n}$. Thus Γ_1 and Γ_2 are parallel planes containing l_1 and l_2, respectively. The distance d between the planes is measured along a line parallel to the common normal $\mathbf{n}$, as shown in the figure. It follows that d is the shortest distance between l_1 and l_2. Moreover,

$$d = |\text{comp}_{\mathbf{n}} \, \overrightarrow{P_1P_2}| = |\mathbf{n} \cdot \overrightarrow{P_1P_2}|$$
$$= \frac{1}{|\overrightarrow{P_1Q_1} \times \overrightarrow{P_2Q_2}|} |(\overrightarrow{P_1Q_1} \times \overrightarrow{P_2Q_2}) \cdot \overrightarrow{P_1P_2}|.$$ ∎

EXERCISES 14.6

Sketch the graphs of the equations in Exercises 1–8.

1 (a) $x = 3$
(b) $y = -2$
(c) $z = 5$

2 (a) $x = -4$
(b) $y = 0$
(c) $z = -2/3$

3 $2x + y - 6 = 0$

4 $3x - 2z - 24 = 0$

5 $4y - 2z - 15 = 0$

6 $5x + y - 4z + 20 = 0$

7 $2x - y + 5z + 10 = 0$

8 $x + y + z = 0$

In Exercises 9–18 find an equation of the plane that satisfies the stated conditions.

9 Through $P(6, -7, 4)$, parallel to (a) the xy-plane; (b) the yz-plane; (c) the xz-plane

10 Through $P(-2, 5, -8)$ with normal vector (a) $\mathbf{i}$; (b) $\mathbf{j}$; (c) $\mathbf{k}$

11 Through $P(-11, 4, -2)$ with normal vector $\mathbf{a} = 6\mathbf{i} - 5\mathbf{j} - \mathbf{k}$

12 Through $P(4, 2, -9)$ with normal vector $\overrightarrow{OP}$

13 Through $P(2, 5, -6)$ and parallel to the plane $3x - y + 2z = 10$

14 Through the origin and parallel to the plane $x - 6y + 4z = 7$

15 Through the origin and the points $P(0, 2, 5)$ and $Q(1, 4, 0)$

16 Through the points $P(3, 2, 1)$, $Q(-1, 1, -2)$, and $R(3, -4, 1)$

17 Through $P(5, 2, -3)$ and orthogonal to the planes $2x - y + 4z = 7$ and $3y - z = 8$

18 Through $P(-4, 1, 6)$ and having the same trace in the xz-plane as $x + 4y - 5z = 8$

In Exercises 19–22, find a symmetric form for the line through P_1 and P_2.

19 $P_1(5, -2, 4)$, $P_2(2, 6, 1)$

20 $P_1(-3, 1, -1)$, $P_2(7, 11, -8)$

21 $P_1(4, 2, -3)$, $P_2(-3, 2, 5)$

22 $P_1(5, -7, 4)$, $P_2(-2, -1, 4)$

In Exercises 23–26 determine whether the given lines (a) are parallel; (b) are orthogonal; (c) intersect.

23 $(x + 1)/4 = (y + 7)/(-2) = (z - 5)/3$;
$(x - 3)/3 = (y + 2)/3 = z/(-2)$

24 $(2x - 3)/6 = (y - 4)/(-1) = (z + 1)/2$;
$(x + 3)/(-6) = (y + 1)/2 = (z - 5)/(-4)$

25 $(x + 1)/3 = (y - 3)/(-1) = (z + 5)/2$;
   $x/(-4) = (y - 4)/2 = (z - 9)/4$

26 $(x - 3)/5 = (y + 6)/(-2) = z - 2$;
   $x/7 = (y - 4)/3 = (1 - z)/5$.

In Exercises 27–30 find the distance from the point P to the given plane.

27 $3x - 7y + z - 5 = 0, P(1, -1, 2)$

28 $2x + 4y - 5z + 1 = 0, P(3, 1, -2)$

29 $4x - 3z = 2, P(5, -8, 1)$

30 $6y - z = 10, P(4, 1, -3)$

In Exercises 31 and 32 show that the two planes are parallel and find the distance between the planes.

31 $4x - 2y + 6z = 3, -6x + 3y - 9z = 4$

32 $3x + 12y - 6z = -2, 5x + 20y - 10z = 7$

33 If a line l has parametric equations $x = 3t + 1$, $y = -2t + 4$, $z = t - 3$, find an equation of the plane that contains l and the point $P(5, 0, 2)$.

34 Find parametric equations for the line of intersection of the planes $2x + y + 4z = 8$ and $x + 3y - z = -1$.

35 Find an equation of the plane through the point $A(1, 2, -3)$ and the line of intersection of the planes $x + 3y - 5z = 10$ and $6x - y + z = 0$.

36 Let $\mathbf{a}$, $\mathbf{b}$, $\mathbf{c}$ be vectors in V_3. Prove that their position vectors $\overrightarrow{OA}, \overrightarrow{OB}, \overrightarrow{OC}$ are coplanar if and only if

$$\mathbf{c} = m\mathbf{a} + n\mathbf{b}$$

for some scalars m and n.

In Exercises 37 and 38 let l_1 be the line through A, B and let l_2 be the line through C, D. Find the distance between l_1 and l_2.

37 $A(1, -2, 3), B(2, 0, 5), C(4, 1, -1), D(-2, 3, 4)$

38 $A(1, 3, 0), B(0, 4, 5), C(-2, -1, 2), D(5, 1, 0)$

39 If a plane has nonzero x-, y-, and z-intercepts a, b, and c, respectively, show that an equation of the plane may be written $(x/a) + (y/b) + (z/c) = 1$. (This is called the **intercept form** for the equation of a plane.)

40 Prove that the planes with equation $a_1x + b_1y + c_1z = d_1$ and $a_2x + b_2y + c_2z = d_2$ are
   (a) orthogonal if and only if $a_1a_2 + b_1b_2 + c_1c_2 = 0$.
   (b) parallel if and only if there is a real number m such that $a_1 = ma_2, b_1 = mb_2$, and $c_1 = mc_2$.

The following concept is used in Exercises 41–44. If S_1 and S_2 are two spheres with equations

$$x^2 + y^2 + z^2 + a_1x + b_1y + c_1z + d_1 = 0,$$

and $x^2 + y^2 + z^2 + a_2x + b_2y + c_2z + d_2 = 0,$

respectively, then their **radical plane** is the plane whose equation is obtained by subtracting the equations of the spheres:

$$(a_1 - a_2)x + (b_1 - b_2)y + (c_1 - c_2)z + (d_1 - d_2) = 0.$$

41 If P_1 and P_2 are the centers of the spheres S_1 and S_2, respectively, show that $\overrightarrow{P_1P_2}$ is normal to the radical plane of the spheres.

42 Show that if two spheres are tangent to each other, then their radical plane is tangent to both spheres.

43 Show that if two spheres intersect in a circle C, then their radical plane contains C.

44 Show that the three radical planes obtained from three intersecting spheres are either parallel or intersect in a common line.

14.7

CYLINDERS AND SURFACES OF REVOLUTION

In this section and the next we shall consider equations in x, y, and z whose graphs are fundamental in the study of analytic geometry. We previously defined the concept of *trace* in a coordinate plane. More generally, the trace of a surface in *any* plane is the intersection of the surface and the plane. To find the shape of a surface from its equation, we often make considerable use of traces in planes that are parallel to coordinate planes.

Example 1 Find traces, in various planes, of the surface having equation $z = x^2 + y^2$, and sketch the graph of the equation.

Solution   Substitution of 0 for x in the equation gives us $z = y^2$, and hence the trace of the surface in the yz-plane is a parabola with vertex at the origin and opening upward, as shown in Figure 14.35.

Similarly, substituting 0 for y leads to $z = x^2$, and hence the trace in the xz-plane is also a parabola. It is instructive to find traces in planes parallel to the xy-plane, that is, planes with equations of the form $z = c$. Substituting c for z in the given equation produces $x^2 + y^2 = c$. Thus, if $c > 0$, then the trace in the plane $z = c$ is a circle of radius $\sqrt{c}$. Three such circles are sketched in Figure 14.35. If $c < 0$, then $x^2 + y^2 = c$ has no graph and consequently none of the points beneath the xy-plane is on the surface. The trace in the xy-plane has equation $x^2 + y^2 = 0$ and, therefore, consists of only one point, the origin. Although traces in planes parallel to the xz- or yz-planes could be determined, those we have obtained are sufficient for an accurate description of the graph. It will follow from the discussion at the end of this section that the surface in this example may be regarded as having been generated by revolving the graph of the parabola $z = y^2$ in the yz-plane about the z-axis. This surface is called a **circular paraboloid**, or **paraboloid of revolution**. ■

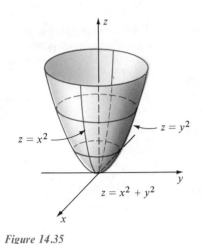

$z = x^2$

$z = y^2$

$z = x^2 + y^2$

Figure 14.35

Definition (14.42)

> If C is a curve in a plane and l is a line not in the plane, then the set of points on all lines that intersect C and are parallel to l is called a **cylinder**.

The curve C in Definition (14.42) is called a **directrix** for the cylinder, and each line through C parallel to l is a **ruling** of the cylinder. The most familiar type of cylinder is a **right circular cylinder**, obtained by letting C be a circle in a plane and l a line perpendicular to the plane, as illustrated in (i) of Figure 14.36. Although we have cut off the cylinder in the figure, it is to be understood that the rulings extend indefinitely. It is not required that the directrix C in (14.42) be a closed curve. This is illustrated in (ii) of Figure 14.36, where C is a parabola.

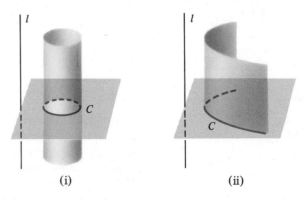

(i)   (ii)

Figure 14.36

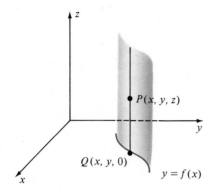

Figure 14.37

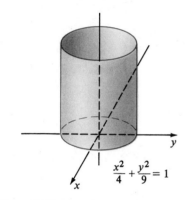

Figure 14.38

In the discussion to follow we shall only consider the case in which the directrix C is on a coordinate plane and the line l is parallel to the coordinate axis that is not on the plane. Suppose that C is on the xy-plane and has equation $y = f(x)$, where f is a function, and that the rulings are parallel to the z-axis. As illustrated in Figure 14.37, a point $P(x, y, z)$ is on the cylinder if and only if $Q(x, y, 0)$ is on C; that is, if and only if the first two coordinates x and y of P satisfy the equation $y = f(x)$. It follows that the equation of the cylinder is $y = f(x)$, and hence is the same as the equation of the directrix in the xy-plane.

Example 2 Sketch the graph of $\dfrac{x^2}{4} + \dfrac{y^2}{9} = 1$.

Solution From our previous remarks, the graph is a cylinder with rulings parallel to the z-axis. We begin by sketching the graph of $x^2/4 + y^2/9 = 1$ in the xy-plane. This ellipse is a directrix for the cylinder. All traces in planes parallel to the xy-plane are ellipses congruent to this directrix. A portion of the graph is shown in Figure 14.38. This surface is called an **elliptic cylinder**.

It can be shown that the graph of an equation that contains only the variables y and z is a cylinder with rulings parallel to the x-axis and whose trace (directrix) in the yz-plane is the graph of the given equation. Similarly, the graph of an equation that does not contain the variable y is a cylinder with rulings parallel to the y-axis and whose directrix is the graph of the given equation in the xz-plane.

Example 3 Sketch the graphs of the following equations.

(a) $y^2 = 9 - z$ (b) $z = \sin x$

Solution (a) The graph is a cylinder with rulings parallel to the x-axis. A directrix in the yz-plane is the graph of $y^2 = 9 - z$. Part of the graph is sketched in (i) of Figure 14.39. This surface is called a **parabolic cylinder**.

(b) The graph is a cylinder with rulings parallel to the y-axis and whose directrix in the xz-plane is the graph of the equation $z = \sin x$. A portion of the graph is sketched in (ii) of Figure 14.39.

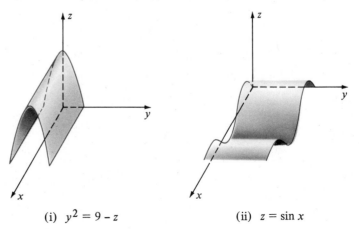

(i) $y^2 = 9 - z$ (ii) $z = \sin x$

Figure 14.39

A surface of revolution was defined in Chapter 13 as the surface generated by revolving a plane curve C about a line in the plane, called the axis of revolution. In the discussion to follow, C will always lie in a coordinate plane and the axis of revolution will be one of the coordinate axes. It will be convenient to use the symbol $f(x, y)$ to denote an expression in the variables x and y. In this case $f(a, b)$ will denote the number obtained by substituting a for x and b for y. This notation will be discussed further in Chapter 16.

The graph of the equation $f(x, y) = 0$ in the xy-plane is a curve C. (We are interested here only in the graph in the xy-plane and not in the three-dimensional graph, which is a cylinder.) Suppose, for simplicity, that x and y are nonnegative for all points (x, y) on C and let S denote the surface obtained by revolving C about the y-axis, as illustrated in Figure 14.40. A point $P(x, y, z)$ is on S if and only if $Q(x_1, y, 0)$ is on C, where $x_1 = \sqrt{x^2 + z^2}$. Consequently, $P(x, y, z)$ is on S if and only if $f(\sqrt{x^2 + z^2}, y) = 0$. Thus, to find an equation for S we replace the variable x in the equation for C by $\sqrt{x^2 + z^2}$. Similarly, if the graph of $f(x, y) = 0$ is revolved about the x-axis, then an equation for the resulting surface may be found by replacing y by $\sqrt{y^2 + z^2}$. For some curves that contain points (x, y) where x or y is negative, the preceding discussion can be extended by substituting $\pm\sqrt{x^2 + z^2}$ for x or $\pm\sqrt{y^2 + z^2}$ for y. If, as in the next example, x or y appear only to even powers, then there is no need to make this distinction since the radical disappears when the equation is simplified.

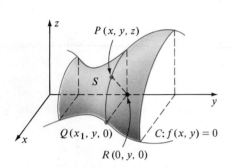

Figure 14.40

Example 4   The graph of $9x^2 + 4y^2 = 36$ is revolved about the y-axis. Find an equation for the resulting surface.

Solution   The desired equation may be found by substituting $x^2 + z^2$ for x^2. This gives us

$$9(x^2 + z^2) + 4y^2 = 36.$$

The surface is called an **ellipsoid of revolution**.   ∎

A similar discussion can be given for curves that lie in the yz-plane or the xz-plane. For example, if a suitable curve C in the xz-plane is revolved about the z-axis, then an equation for the resulting surface may be found by replacing x by $\sqrt{x^2 + y^2}$. If C is revolved about the x-axis we replace z by $\sqrt{y^2 + z^2}$.

Finally, note that equations for surfaces of revolution are characterized by the fact that two of the variables appear in combinations such as $x^2 + y^2$, $y^2 + z^2$, or $x^2 + z^2$.

Example 5   Describe the graph of the equation $y^2 + z^2 - 4x^2 = 16$ as a surface of revolution.

Solution   The given equation may be obtained from the equation $y^2 - 4x^2 = 16$ by substituting $\sqrt{y^2 + z^2}$ for y. Hence the graph is a surface generated by revolving the hyperbola in the xy-plane having equation $y^2 - 4x^2 = 16$ about the x-axis. We could also describe the surface as that generated by revolving the hyperbola in the xz-plane that has the equation $z^2 - 4x^2 = 16$ about the x-axis. ∎

EXERCISES 14.7

In Exercises 1–16 sketch the graph of the equation in three dimensions.

1 $x^2 + y^2 = 9$

2 $y^2 + z^2 = 16$

3 $4y^2 + 9z^2 = 36$

4 $x^2 + 5z^2 = 25$

5 $x^2 = 9z$

6 $x^2 - 4y = 0$

7 $y^2 - x^2 = 16$

8 $xz = 1$

9 $4x - 3y = 12$

10 $2z + y = 5$

11 $z = x^3 - 4x$

12 $z = \cos(y/2)$

13 $yz = 0$

14 $|x + y| = 2$

15 $z = e^y$

16 $z = \log x$

In Exercises 17–22 find an equation of the surface obtained by revolving the graph of the given equation about the indicated axis. Sketch the graph of the surface.

17 $x^2 + 4y^2 = 16$; y-axis

18 $y^2 = 4x$; x-axis

19 $z = 4 - y^2$; z-axis

20 $z = e^{-y^2}$; y-axis

21 $z^2 - x^2 = 1$; x-axis

22 $xz = 1$; z-axis

In Exercises 23–26 sketch the graph of the given equation.

23 $x^2 + 4y^2 + z^2 = 16$

24 $x^2 - 4y^2 + z^2 = 16$

25 $y - 4z^2 = 4x^2$

26 $x^2 - y^2 + z^2 = 1$

14.8

QUADRIC SURFACES

In Chapter 12 it was shown that, in two dimensions, the graph of a second-degree equation in x and y is a conic section. In three-dimensional analytic geometry, the graph of a second-degree equation in x, y, and z is referred to as a **quadric surface**. In this section we shall investigate equations for such surfaces. Let us begin with the following, where a, b, and c are positive real numbers. The graph of the equation is given the name listed in the margin.

Ellipsoid (*14.43*)

$$\frac{x^2}{a^2} + \frac{y^2}{b^2} + \frac{z^2}{c^2} = 1$$

A typical graph of an ellipsoid is sketched in Figure 14.41. Note that traces of this surface in planes parallel to coordinate planes are ellipses. For example, the trace in the xy-plane is the ellipse $x^2/a^2 + y^2/b^2 = 1$. Similarly, the traces in the yz- and xz-planes are ellipses, as indicated in the figure. Let us find the trace in an arbitrary plane parallel to the xy-plane, that is, in a plane having an equation of the form $z = k$. Substituting k for z in Definition (14.43) gives us the equation

$$\frac{x^2}{a^2} + \frac{y^2}{b^2} = 1 - \frac{k^2}{c^2}.$$

If $|k| > c$, then $1 - k^2/c^2 < 0$ and there is no graph. Thus the graph of (14.43) lies between the planes $z = -c$ and $z = c$. If $|k| < c$, then $1 - k^2/c^2 > 0$ and hence the trace in the plane $z = k$ is an ellipse. Similarly, traces in planes parallel to the other two coordinate planes are ellipses, provided they do not

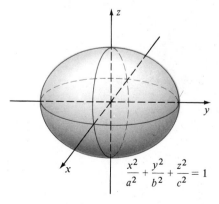

$$\frac{x^2}{a^2} + \frac{y^2}{b^2} + \frac{z^2}{c^2} = 1$$

Figure 14.41

intersect the x-axis outside of the open interval $(-a, a)$ or the y-axis outside of $(-b, b)$. If $a = b = c$ then the graph is a sphere of radius a with center at the origin.

Another quadric surface is given by the following equation.

Hyperboloid of One Sheet (*14.44*)

$$\frac{x^2}{a^2} + \frac{y^2}{b^2} - \frac{z^2}{c^2} = 1$$

Observe that the traces in the xz- and yz-planes are hyperbolas with equations

$$\frac{x^2}{a^2} - \frac{z^2}{c^2} = 1 \quad \text{and} \quad \frac{y^2}{b^2} - \frac{z^2}{c^2} = 1,$$

respectively. Traces on planes parallel to the xy-plane have equations of the form

$$\frac{x^2}{a^2} + \frac{y^2}{b^2} = 1 + \frac{k^2}{c^2},$$

where k is a real number and, therefore, are ellipses. The graph of (14.44) is sketched in Figure 14.42. The z-axis is called the **axis of the hyperboloid**.

The graphs of

$$\frac{x^2}{a^2} - \frac{y^2}{b^2} + \frac{z^2}{c^2} = 1 \quad \text{and} \quad -\frac{x^2}{a^2} + \frac{y^2}{b^2} + \frac{z^2}{c^2} = 1$$

are also hyperboloids of one sheet; however, in the first case the axis of the hyperboloid is the y-axis, whereas in the second the axis of the hyperboloid coincides with the x-axis.

Hyperboloid of Two Sheets (*14.45*)

$$\frac{x^2}{a^2} - \frac{y^2}{b^2} - \frac{z^2}{c^2} = 1$$

Traces in the xy- and xz-planes are hyperbolas, whereas traces in planes with equations of the form $x = k$ where $|k| > a$ are ellipses. We leave it to the reader to show that the graph of (14.45) has the appearance shown in Figure 14.43. The x-axis is the axis of the hyperboloid.

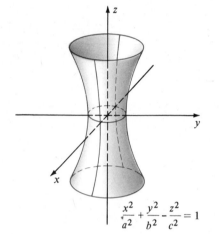

$$\frac{x^2}{a^2} + \frac{y^2}{b^2} - \frac{z^2}{c^2} = 1$$

Figure 14.42

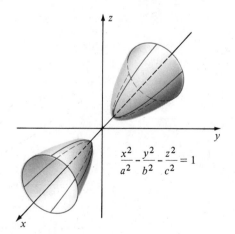

$$\frac{x^2}{a^2} - \frac{y^2}{b^2} - \frac{z^2}{c^2} = 1$$

Figure 14.43

By using minus signs on different terms, we can obtain a hyperboloid of two sheets whose axis is the y-axis or the z-axis.

Cone (14.46)

$$\frac{x^2}{a^2} + \frac{y^2}{b^2} - \frac{z^2}{c^2} = 0$$

A typical graph of a cone is sketched in Figure 14.44. The trace in the yz-plane has equation

$$\frac{y^2}{b^2} - \frac{z^2}{c^2} = 0.$$

Solving for y we obtain $y = \pm(b/c)z$, which gives us the equations of two lines through the origin. Similarly, the trace in the xz-plane is a pair of lines that intersect at the origin. Traces in planes parallel to the xy-plane are ellipses. (Why?) The z-axis is the **axis of the cone**. By changing signs of the terms in (14.46), we obtain a cone whose axis is either the x- or the y-axis.

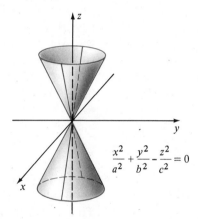

$$\frac{x^2}{a^2} + \frac{y^2}{b^2} - \frac{z^2}{c^2} = 0$$

Figure 14.44

Paraboloid (14.47)

$$\frac{x^2}{a^2} + \frac{y^2}{b^2} = cz$$

Example 1 in the previous section is the special case of (14.47) in which $a = b = c = 1$. If $c > 0$, then the graph of (14.47) is similar to that shown in Figure 14.35, except that if $a \neq b$, then traces in planes parallel to the xy-plane are ellipses instead of circles. If $c < 0$, then the paraboloid opens downward. The z-axis is called the **axis of the paraboloid**. The graphs of the equations.

$$\frac{x^2}{a^2} + \frac{z^2}{b^2} = cy \quad \text{and} \quad \frac{y^2}{a^2} + \frac{z^2}{b^2} = cx$$

are paraboloids whose axes are the y-axis and x-axis, respectively.

Hyperbolic Paraboloid (14.48)

$$\frac{y^2}{a^2} - \frac{x^2}{b^2} = cz$$

A typical sketch of this saddle-shaped surface for the case $c > 0$ is shown in Figure 14.45. Variations are obtained by interchanging x, y, and z in (14.48).

It is possible to obtain formulas for translation or rotation of axes in three dimensions that are analogous to those in two dimensions. These can be used to show that the graph of an equation of degree two in x, y, and z, is one of the surfaces discussed in this chapter, except for degenerate cases.

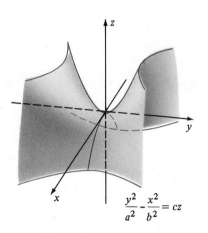

$$\frac{y^2}{a^2} - \frac{x^2}{b^2} = cz$$

Figure 14.45

Example 1 Sketch the graph of $16x^2 - 9y^2 + 36z^2 = 144$ and identify the surface.

Solution   Dividing both sides of the equation by 144 leads to

$$\frac{x^2}{9} - \frac{y^2}{16} + \frac{z^2}{4} = 1.$$

The trace in the xy-plane is the hyperbola $x^2/9 - y^2/16 = 1$, with vertices at $(\pm 3, 0, 0)$. The trace in the yz-plane is another hyperbola $z^2/4 - y^2/16 = 1$, with vertices at $(0, 0, \pm 2)$. The trace in the xz-plane is the ellipse with the equation $x^2/9 + z^2/4 = 1$. These traces are illustrated in Figure 14.46. To find traces on planes that are parallel to the xz-plane we substitute $y = k$ in the given equation, obtaining

$$\frac{x^2}{9} + \frac{z^2}{4} = 1 + \frac{k^2}{16}.$$

This shows that every cross section in a plane parallel to the xz-plane is an ellipse. For example, if $k = \pm 8$ we obtain $x^2/9 + z^2/4 = 5$ or, equivalently, $x^2/45 + z^2/20 = 1$. Several of these traces are illustrated in Figure 14.46. Traces on planes parallel to the xy-plane or the yz-plane may be found by substituting $z = k$ or $x = k$, respectively, in the given equation. We leave it to the reader to verify that in either case the trace is a hyperbola. The surface is a hyperboloid of one sheet with the y-axis as its axis.

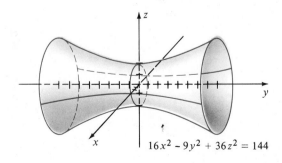

$$16x^2 - 9y^2 + 36z^2 = 144$$

Figure 14.46

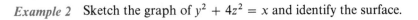

Example 2   Sketch the graph of $y^2 + 4z^2 = x$ and identify the surface.

Solution   The trace in the xy-plane is the parabola $y^2 = x$. The trace in the xz-plane is the parabola $4z^2 = x$. These traces are illustrated in Figure 14.47. The trace in the yz-plane is the graph of $y^2 + 4z^2 = 0$ and hence consists of only one point, the origin. To find traces in planes parallel to the yz-plane we let $x = k$ in the given equation, obtaining

$$y^2 + 4z^2 = k.$$

If $k < 0$ there is no graph. (Why?) If $k > 0$ the trace is an ellipse. For example, letting $k = 9$, we obtain

$$\frac{y^2}{9} + \frac{z^2}{\frac{9}{4}} = 1$$

which is an ellipse with semi-axes of lengths 3 and $\frac{3}{2}$, as illustrated in Figure 14.47. The reader should verify that traces on planes parallel to either the xz- or xy-planes are parabolas. The surface is a paraboloid having the x-axis as its axis. ∎

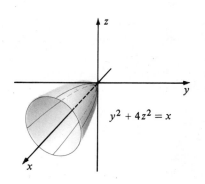

Figure 14.47

EXERCISES 14.8

Sketch the graphs of the equations in Exercises 1–20 and identify the surfaces.

1 $4x^2 + 9y^2 = 36z$

2 $8x^2 + 4y^2 + z^2 = 16$

3 $16x^2 + 100y^2 - 25z^2 = 400$

4 $25x^2 - 225y^2 + 9z^2 = 225$

5 $3x^2 - 4y^2 - z^2 = 12$

6 $4x^2 + y^2 = 9z^2$

7 $x^2 - 16y^2 = 4z^2$

8 $4y^2 - 25z^2 = 100x$

9 $9x^2 + 4y^2 + z^2 = 36$

10 $16x^2 - 25y^2 + 100z^2 = 200$

11 $16x^2 - 4y^2 - z^2 + 1 = 0$

12 $36x = 9y^2 + z^2$

13 $y^2 - 9x^2 - z^2 - 9 = 0$

14 $z^2 - x^2 - y^2 = 1$

15 $4y^2 + 25z^2 + 100x = 0$

16 $16y = x^2 + 4z^2$

17 $36x^2 - 16y^2 + 9z^2 = 0$

18 $4y^2 + 9z^2 = 9x^2$

19 $4y = x^2 - z^2$

20 $4x^2 + 16y = z^2$

In Exercises 21–24, find an equation for the set of all points $P(x, y, z)$ satisfying the given condition, and identify the graph of the equation.

21   Equally distant from the point $(0, 0, -1)$ and the plane $z = 1$

22   Equally distant from the z-axis and the xy-plane

23   Half as far from the xy-plane as from the origin

24   The sum of the squares of the distances from P to the origin and to the xy-plane is 4

25   Show that all cross sections of the ellipsoid $x^2/a^2 + y^2/b^2 + z^2/c^2 = 1$ by planes parallel to the xy-plane are ellipses with the same eccentricity.

26   Identify the surface $z = xy$ by making a suitable rotation of axes in the xy-plane.

14.9

CYLINDRICAL AND SPHERICAL COORDINATE SYSTEMS

The system of polar coordinates can be extended to three dimensions. We merely represent a point P by an ordered triple (r, θ, z) where z is the usual (third) rectangular coordinate of P and (r, θ) are polar coordinates for the projection P' of P onto the xy-plane (see Figure 14.48). This system of coordinates is called the **cylindrical coordinate system**.

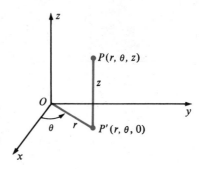

Figure 14.48

Since we have changed to polar coordinates in the xy-plane, the formulas in Theorem (13.4) give relationships between the rectangular coordinates and the cylindrical coordinates of P. Let us restate these formulas as follows.

Theorem (14.49)

> The rectangular coordinates (x, y, z) and cylindrical coordinates (r, θ, z) of a point P are related as follows:
>
> $$x = r \cos \theta, \quad y = r \sin \theta, \quad \tan \theta = \frac{y}{x}, \quad r^2 = x^2 + y^2.$$

The reason for using the term *cylindrical* for this system is that if $k > 0$, then the graph of the equation $r = k$, or equivalently of $x^2 + y^2 = k^2$, is a circular cylinder of radius k with axis along the z-axis. Note that the graph of $\theta = k$ is a plane containing the z-axis, whereas the graph of $z = k$ is a plane perpendicular to the z-axis.

Example 1 Find a rectangular equation for, and describe the graph of, each of the following equations.

(a) $z = r^2$ (b) $r = 4 \sin \theta$

Solution

(a) By Theorem (14.49), a rectangular equation for the graph of $z = r^2$ is $z = x^2 + y^2$. The graph is the paraboloid sketched in Figure 14.35.

(b)   Multiplying both sides of $r = 4 \sin \theta$ by r gives us $r^2 = 4r \sin \theta$. Using (14.49) we obtain the rectangular equation $x^2 + y^2 = 4y$, or equivalently,

$$x^2 + (y - 2)^2 = 4.$$

The graph is a cylinder with rulings parallel to the z-axis. The directrix of the cylinder is a circle of radius 2 in the xy-plane whose center in rectangular coordinates is $(0, 2, 0)$. ∎

Example 2 For each of the following equations, find an equation in cylindrical coordinates and describe the graph.

(a) $z^2 = x^2 - y^2$ (b) $z^2 = x^2 + y^2$

Solution

(a)   Applying the first two equations in Theorem (14.49),

$$z^2 = r^2 \cos^2 \theta - r^2 \sin^2 \theta$$
$$= r^2(\cos^2 \theta - \sin^2 \theta),$$

or $z^2 = r^2 \cos 2\theta.$

The graph is a circular cone with axis along the x-axis.

(b)   By the last equation in (14.49),

$$z^2 = r^2 \quad \text{or} \quad z = r.$$

The graph is a circular cone with axis along the z-axis. ∎

A system called **spherical coordinates** may also be introduced in three dimensions. In this system a point P different from the origin is represented by an ordered triple (ρ, ϕ, θ), where $\rho = |\overrightarrow{OP}|$, ϕ is the angle between the positive z-axis and $\overrightarrow{OP}$, and θ is a polar angle associated with the projection P' of P onto the xy-plane (see Figure 14.49). The origin is represented by any triple of the form $(0, \phi, \theta)$. The term *spherical* arises from the fact that the graph of the equation $\rho = k$, where $k > 0$, is a sphere with center at O. As in cylindrical coordinates, the graph of $\theta = k$ is a half-plane containing the z-axis. The graph of $\phi = k$ is usually a half-cone with vertex at O.

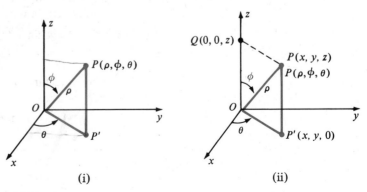

(i)   (ii)

Figure 14.49

The relationship between spherical coordinates (ρ, ϕ, θ) and rectangular coordinates (x, y, z) of a point P can be found by referring to (ii) of Figure 14.49. Using the fact that

$$x = |\overrightarrow{OP'}| \cos \theta \quad \text{and} \quad y = |\overrightarrow{OP'}| \sin \theta$$

together with

$$|\overrightarrow{OP'}| = |\overrightarrow{QP}| = \rho \sin \phi \quad \text{and} \quad |\overrightarrow{OQ}| = \rho \cos \phi$$

gives us (i) of the next theorem. Part (ii) is a consequence of the Distance Formula.

Theorem (14.50)

> The rectangular coordinates (x, y, z) and spherical coordinates (ρ, ϕ, θ) of a point P are related as follows:
>
> (i) $x = \rho \sin \phi \cos \theta, \quad y = \rho \sin \phi \sin \theta, \quad z = \rho \cos \phi$
>
> (ii) $\rho^2 = x^2 + y^2 + z^2$

Example 3 Find a rectangular equation for $\rho = 2 \sin \phi \cos \theta$ and describe the graph of this equation.

Solution   Multiplying both sides of the given equation by ρ we obtain $\rho^2 = 2\rho \sin \phi \cos \theta$. Applying (14.50) gives us $x^2 + y^2 + z^2 = 2x$ or, equivalently, $(x - 1)^2 + y^2 + z^2 = 1$. By (14.15), the graph is a sphere of radius 1 whose center in rectangular coordinates is $(1, 0, 0)$.   ■

Example 4 If a point P has spherical coordinates $(4, \pi/6, \pi/3)$, find rectangular and cylindrical coordinates for P.

Solution   Using (i) of Theorem (14.50) with $\rho = 4$, $\phi = \pi/6$, and $\theta = \pi/3$, we obtain the rectangular coordinates

$$x = 4 \sin \frac{\pi}{6} \cos \frac{\pi}{3} = 4\left(\frac{1}{2}\right)\left(\frac{1}{2}\right) = 1$$

$$y = 4 \sin \frac{\pi}{6} \sin \frac{\pi}{3} = 4\left(\frac{1}{2}\right)\left(\frac{\sqrt{3}}{2}\right) = \sqrt{3}$$

$$z = 4 \cos \frac{\pi}{6} = 4\left(\frac{\sqrt{3}}{2}\right) = 2\sqrt{3}.$$

To find the cylindrical coordinates we note that $r^2 = x^2 + y^2 = 1 + 3 = 4$. Hence cylindrical coordinates for P are (r, θ, z) or $(2, \pi/3, 2\sqrt{3})$.   ■

EXERCISES 14.9

1 Change the following from cylindrical coordinates to rectangular coordinates.

(a) $(5, \pi/2, 3)$ (b) $(6, \pi/3, -5)$

2 Change the following from spherical coordinates to rectangular coordinates.

(a) $(4, \pi/6, \pi/2)$ (b) $(1, 3\pi/4, 2\pi/3)$

3 Change the following from rectangular coordinates to spherical coordinates.

(a) $(1, 1, -2\sqrt{2})$ (b) $(1, \sqrt{3}, 0)$

4 Change the following from rectangular coordinates to cylindrical coordinates.

(a) $(2\sqrt{3}, 2, -2)$ (b) $(\sqrt{2}, -\sqrt{2}, 1)$

5 Change the following from spherical coordinates to cylindrical coordinates.

(a) $(4, \pi/3, \pi/3)$ (b) $(2, 5\pi/6, \pi/4)$

6 Change the following from cylindrical coordinates to spherical coordinates.

(a) $(\sqrt{2}, \pi/4, 1)$ (b) $(3, \pi/3, 1)$

Describe the graphs of the equations in Exercises 7–20.

7 $r = 4$ **8** $\theta = \pi/4$

9 $\phi = \pi/6$ **10** $\rho = 3$

11 $r = 4\cos\theta$ **12** $r = \sin 2\theta$

13 $\rho = 4\cos\phi$ **14** $\rho = \sec\phi$

15 $\phi = 0$ **16** $\phi = \pi/2$

17 $r^2 = \cos 2\theta$ **18** $z = 1 - r^2$

19 $\rho^2 - 3\rho + 2 = 0$ **20** $\phi^2 - \pi\phi + (3\pi^2/16) = 0$

Find equations in cylindrical coordinates and in spherical coordinates for the graphs of the equations in Exercises 21–30.

21 $x^2 + y^2 + z^2 = 4$ **22** $x^2 + y^2 = 4z$

23 $3x + y - 4z = 12$ **24** $x^2 - y^2 - z^2 = 1$

25 $y^2 + z^2 = 9$ **26** $x^2 + z^2 = 9$

27 $6x = x^2 + y^2$ **28** $x^2 - 4z^2 + y^2 = 0$

29 $y = xz$ **30** $y = x$

14.10

REVIEW

Define or discuss each of the following.

1 Vectors in V_2 and V_3

2 Scalar

3 Scalar multiples of vectors

4 Addition of vectors

5 Components of a vector

6 Zero vector

7 Magnitude of a vector

8 Position vector

9 Subtraction of vectors

10 Parallel vectors

11 Unit vector

12 Rectangular coordinate system in three dimensions

13 Distance Formula

14 Midpoint Formula

15 Equation of a sphere

16 The dot product and its properties

17 The angle between two vectors

18 Orthogonal vectors

19 Cauchy–Schwarz Inequality

20 Triangle Inequality

21 Work done by a force

22 The vector product and its properties

23 Equations of a line

24 Equations of a plane

25 Cylinders

26 Quadric surfaces

27 Cylindrical coordinates

28 Spherical coordinates

EXERCISES 14.10

Find the vectors or scalars indicated in Exercises 1–10 if $\mathbf{a} = 2\mathbf{i} + 5\mathbf{j}$ and $\mathbf{b} = 4\mathbf{i} - \mathbf{j}$.

1 $4\mathbf{a} + \mathbf{b}$

2 $2\mathbf{a} - 3\mathbf{b}$

3 $(\mathbf{a} + \mathbf{b}) \cdot (\mathbf{a} - 2\mathbf{b})$

4 $3\mathbf{a} \cdot (5\mathbf{b} + \mathbf{i})$

5 $|\mathbf{a}| - |\mathbf{b}|$

6 $|\mathbf{a} - \mathbf{b}|$

7 The angle between $\mathbf{a}$ and $\mathbf{i}$

8 A unit vector having the same direction as $\mathbf{b}$

9 A unit vector orthogonal to $\mathbf{a}$

10 The cosine of the angle between $\mathbf{a}$ and $\mathbf{b}$

11 Given the points $A(5, -3, 2)$ and $B(-1, -4, 3)$, find

(a) $d(A, B)$.

(b) the coordinates of the midpoint of the line segment AB.

(c) an equation of the sphere with center B and tangent to the xz-plane.

(d) an equation of the plane through B parallel to the xz-plane.

(e) parametric equations for the line through A and B.

(f) an equation of the plane through A with normal vector $\overrightarrow{AB}$.

12 Find an equation for the plane through $A(0, 4, 9)$ and $B(0, -3, 7)$ that is perpendicular to the yz-plane.

13 Find an equation for the plane that has x-intercept 5, y-intercept -2, and z-intercept 6.

14 Find an equation for the cylinder that is perpendicular to the xy-plane and has, for its directrix, the circle in the xy-plane with center $C(4, -3, 0)$ and radius 5.

15 Find an equation for an ellipsoid with center O and having x-intercept 8, y-intercept 3, and z-intercept 1.

16 Find an equation for the surface obtained by revolving the graph of the equation $z = x$ about the z-axis.

In Exercises 17–27 sketch the graph of the given equation.

17 $x^2 + y^2 + z^2 - 14x + 6y - 8z + 10 = 0$

18 $4y - 3z - 15 = 0$

19 $3x - 5y + 2z = 10$

20 $y = z^2 + 1$

21 $9x^2 + 4z^2 = 36$

22 $x^2 + 4y + 9z^2 = 0$

23 $z^2 - 4x^2 = 9 - 4y^2$

24 $2x^2 + 4z^2 - y^2 = 0$

25 $z^2 - 4x^2 - y^2 = 4$

26 $x^2 + 2y^2 + 4z^2 = 16$

27 $x^2 - 4y^2 = 4z$

If $\mathbf{a} = 3\mathbf{i} - \mathbf{j} - 4\mathbf{k}$, $\mathbf{b} = 2\mathbf{i} + 5\mathbf{j} - 2\mathbf{k}$, and $\mathbf{c} = -\mathbf{i} + 6\mathbf{k}$, find the vectors or scalars in Exercises 28–44.

28 $3\mathbf{a} - 2\mathbf{b}$

29 $\mathbf{a} \cdot (\mathbf{b} - \mathbf{c})$

30 $|\mathbf{b} + \mathbf{c}|$

31 $|\mathbf{b}| + |\mathbf{c}|$

32 A unit vector having the same direction as $\mathbf{a}$

33 The direction cosines of $\mathbf{a}$

34 The cosine of the angle between $\mathbf{a}$ and $\mathbf{c}$

35 $\mathbf{a} \times \mathbf{b}$

36 $\mathbf{a} \cdot (\mathbf{b} \times \mathbf{c})$

37 $\text{comp}_{\mathbf{b}} \, \mathbf{a}$

38 $\text{comp}_{\mathbf{a}} \, (\mathbf{b} \times \mathbf{c})$

39 $\mathbf{a} \cdot \mathbf{a}$

40 $\mathbf{a} \times \mathbf{a}$

41 $(\mathbf{a} \times \mathbf{c}) + (\mathbf{c} \times \mathbf{a})$

42 $\mathbf{a} \times (\mathbf{b} + \mathbf{c})$

43 A vector having the opposite direction of $\mathbf{b}$ and twice the magnitude of $\mathbf{b}$

44 Two unit vectors orthogonal to both $\mathbf{b}$ and $\mathbf{c}$

45 Given the points $P(2, -1, 1)$, $Q(-3, 2, 0)$, and $R(4, -5, 3)$, find

(a) a unit normal vector orthogonal to the plane determined by P, Q, and R.

(b) an equation for the plane determined by P, Q, and R.

(c) parametric equations for a line through P that is parallel to the line through Q and R.

(d) $\overrightarrow{QP} \cdot \overrightarrow{QR}$.

(e) the angle between $\overrightarrow{QP}$ and $\overrightarrow{QR}$.

(f) the area of the triangle with vertices P, Q, and R.

46 Find the angle between the two lines

$$(x - 3)/2 = (y + 1)/(-4) = (z - 5)/8$$

and

$$(x + 1)/7 = (6 - y)/2 = (2z + 7)/(-4).$$

47 Find parametric equations for each of the lines in Exercise 46.

48 Determine whether the following lines intersect and, if so, find the point of intersection:

$$x = 2 + t, y = 1 + t, z = 4 + 7t;$$
$$x = -4 + 5t, y = 2 - 2t, z = 1 - 4t.$$

49 Find the angle between the lines in Exercise 48.

50 The position of a particle at time t is $(t, t \sin t, t \cos t)$. Find the distance the particle moves during the time interval $[0, 5]$.

51 If the rectangular coordinates of a point P are $(2, -2, 1)$, find cylindrical and spherical coordinates for P.

52 If spherical coordinates of a point P are $(12, \pi/6, 3\pi/4)$, find cylindrical and rectangular coordinates for P.

Find a rectangular equation and describe the graph of each equation in Exercises 53–56.

53 $\phi = 3\pi/4$

54 $r = \cos 2\theta$

55 $\rho \sin \phi \cos \theta = 1$

56 $\rho^2 - 3\rho = 0$

Find equations in cylindrical coordinates and in spherical coordinates for the graphs of the equations in Exercises 57–60.

57 $x^2 + y^2 = 1$

58 $z = x^2 - y^2$

59 $x^2 + y^2 + z^2 - 2z = 0$

60 $2x + y - 3z = 4$

15

VECTOR-VALUED FUNCTIONS

In this chapter we shall extend the theory of limits, derivatives, and integrals to functions whose ranges consist of vectors instead of real numbers. Our principal application of these concepts will be to the study of motion. Further applications will be given in later chapters.

15.1

DEFINITIONS AND GRAPHS

A function was defined in (1.21) as a correspondence that associates with each element of a set X a unique element of a set Y. If X is a set of real numbers and Y is a set of vectors, then the function will be called a **vector-valued function** and denoted by symbols such as $\mathbf{r}$ or $\vec{r}$. Thus, if $\mathbf{r}$ is a vector-valued function, then for each real number t in the domain X there exists a unique vector

$$\mathbf{r}(t) = x\mathbf{i} + y\mathbf{j} + z\mathbf{k}.$$

If, with each t, we associate the first component of $\mathbf{r}(t)$, we obtain a function f from X to $\mathbb{R}$, where $x = f(t)$. Similarly, there are functions g and h such that $y = g(t)$ and $z = h(t)$, respectively. Hence for every t in X we have the following:

$$(15.1) \qquad \mathbf{r}(t) = f(t)\mathbf{i} + g(t)\mathbf{j} + h(t)\mathbf{k}$$

Conversely, if f, g, and h are functions from X to $\mathbb{R}$, then we may define a vector-valued function $\mathbf{r}$ by means of (15.1). Consequently, $\mathbf{r}$ *is a vector-valued function if and only if* $\mathbf{r}(t)$ *is expressible in the form* (15.1). If it is not stated explicitly, the domain of $\mathbf{r}$ is assumed to be the intersection of the domains of f, g, and h.

Example 1 If $\mathbf{r}(t) = 3t^2\mathbf{i} + 2t^3\mathbf{j} + (t - 2)\mathbf{k}$, find $\mathbf{r}(1)$, $\mathbf{r}(2)$, and $\mathbf{r}(0)$.

Solution The specified vectors may be found by substituting 1, 2 and 0 for t. Thus

$$\mathbf{r}(1) = 3\mathbf{i} + 2\mathbf{j} - \mathbf{k}$$
$$\mathbf{r}(2) = 12\mathbf{i} + 16\mathbf{j} + 0\mathbf{k} = 12\mathbf{i} + 16\mathbf{j}$$
$$\mathbf{r}(0) = 0\mathbf{i} + 0\mathbf{j} - 2\mathbf{k} = -2\mathbf{k}.$$ ∎

There is a close connection between vector-valued functions and curves in space. As in Definition (13.1) for plane curves, we define a **space curve**, or a **curve in three dimensions**, as a set C of ordered triples of the form

$$(f(t), g(t), h(t))$$

where the functions f, g, and h are continuous on an interval I. The graph of C in a rectangular coordinate system consists of all points $P(f(t), g(t), h(t))$ that correspond to the ordered triples in C. The equations

$$x = f(t), \quad y = g(t), \quad z = h(t)$$

where t is in I, are called **parametric equations** for C. The notions of **endpoints**, **closed curve**, **simple closed curve**, and **length of a curve** are defined exactly as in the two-dimensional case. A curve C is **smooth** if it has a parametric representation $x = f(t), y = g(t), z = h(t)$ on an interval I such that f', g', and h' are continuous and not simultaneously zero, except possibly at endpoints. In this case the parametric equations are called a **smooth parametric representation** of C.

The next result is analogous to Theorem (13.11).

Theorem (15.2)

> If a smooth curve C is given parametrically by $x = f(t)$, $y = g(t)$, $z = h(t)$, where $a \le t \le b$, and if C does not intersect itself, except possibly at the endpoints of $[a, b]$, then the length L of C is
>
> $$L = \int_a^b \sqrt{[f'(t)]^2 + [g'(t)]^2 + [h'(t)]^2}\, dt$$
>
> $$= \int_a^b \sqrt{\left(\frac{dx}{dt}\right)^2 + \left(\frac{dy}{dt}\right)^2 + \left(\frac{dz}{dt}\right)^2}\, dt.$$

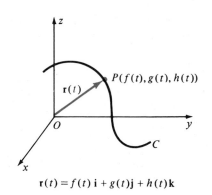

$$\mathbf{r}(t) = f(t)\,\mathbf{i} + g(t)\mathbf{j} + h(t)\mathbf{k}$$

Figure 15.1

Consider a vector function $\mathbf{r}$, where $\mathbf{r}(t) = f(t)\mathbf{i} + g(t)\mathbf{j} + h(t)\mathbf{k}$. As illustrated in Figure 15.1, if $\overrightarrow{OP}$ is the position vector corresponding to $\mathbf{r}(t)$, then as t varies, the terminal point P traces the curve having parametric equations $x = f(t)$, $y = g(t)$, $z = h(t)$. This geometric representation of $\mathbf{r}(t)$ is especially useful if t is time and P is the position of a particle in motion.

Example 2 Let $\mathbf{r}(t) = a \cos t\mathbf{i} + a \sin t\mathbf{j} + bt\mathbf{k}$ and let $\overrightarrow{OP}$ be the position vector of $\mathbf{r}(t)$.

(a)   Sketch the graph of the curve C traced by P as t varies.

(b)   Find the length of C between the points corresponding to $t = 0$ and $t = 2\pi$.

Solution

(a)   Parametric equations for C are $x = a \cos t, y = a \sin t, z = bt$. If the point $P(x, y, z)$ is on C, then P is on the circular cylinder having equation $x^2 + y^2 = a^2$. (Why?) As t varies from 0 to 2π, the point P starts at $(a, 0, 0)$ and winds around this cylinder one time. Other intervals of length 2π lead to similar loops (see Figure 15.2). This curve is called a **circular helix**.

(b)   The length of C between the points corresponding to $t = 0$ and $t = 2\pi$ can be found by means of Theorem (15.2). Thus

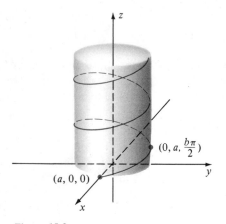

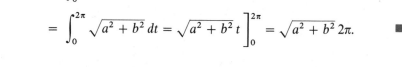

$$L = \int_0^{2\pi} \sqrt{a^2 \sin^2 t + a^2 \cos^2 t + b^2} \, dt$$

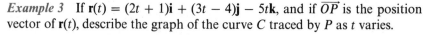

$$= \int_0^{2\pi} \sqrt{a^2 + b^2} \, dt = \sqrt{a^2 + b^2}\, t \;\Big]_0^{2\pi} = \sqrt{a^2 + b^2}\, 2\pi. \qquad\blacksquare$$

Figure 15.2

Example 3 If $\mathbf{r}(t) = (2t + 1)\mathbf{i} + (3t - 4)\mathbf{j} - 5t\mathbf{k}$, and if $\overrightarrow{OP}$ is the position vector of $\mathbf{r}(t)$, describe the graph of the curve C traced by P as t varies.

Solution   Parametric equations for C are

$$x = 2t + 1, \quad y = 3t - 4, \quad z = -5t.$$

By Theorem (14.37) the graph is the line through the point $(1, -4, 0)$ that is parallel to the vector $\langle 2, 3, -5 \rangle$. $\qquad\blacksquare$

As in the preceding chapter, we may regard V_2 as the subset of V_3 consisting of all vectors whose third component is 0. In this case (15.1) takes on the form

$$\mathbf{r}(t) = f(t)\mathbf{i} + g(t)\mathbf{j}$$

and as t varies, the curve traced by the terminal point of the position vector $\overrightarrow{OP}$ lies in the xy-plane.

Example 4 Let $\mathbf{r}(t) = (\tfrac{1}{2}t \cos t)\mathbf{i} + (\tfrac{1}{2}t \sin t)\mathbf{j}$, and let $\overrightarrow{OP}$ be the position vector of $\mathbf{r}(t)$.

(a)   Sketch the graph of the curve traced by P as t varies from 0 to 3π.

(b)   Find $\mathbf{r}(6)$ and sketch its position vector.

Solution

(a)   The curve is in the xy-plane and has parametric equations

$$x = \tfrac{1}{2}t \cos t, \quad y = \tfrac{1}{2}t \sin t, \quad \text{where } 0 \le t \le 3\pi.$$

There is no loss of generality in assuming that t is the radian measure of an angle θ. Indeed, we may write

$$x = \tfrac{1}{2}\theta \cos \theta, \quad y = \tfrac{1}{2}\theta \sin \theta, \quad \text{where } 0 \le \theta \le 3\pi.$$

Moreover, since

$$\frac{y}{x} = \frac{\tfrac{1}{2}\theta \sin \theta}{\tfrac{1}{2}\theta \cos \theta} = \tan \theta$$

it follows from Theorem (13.4) that θ is an angle that can be used to express (x, y) in polar coordinates. We next observe that

$$x^2 + y^2 = \tfrac{1}{4}\theta^2 \cos^2 \theta + \tfrac{1}{4}\theta^2 \sin^2 \theta$$
$$= \tfrac{1}{4}\theta^2 (\cos^2 \theta + \sin^2 \theta) = \tfrac{1}{4}\theta^2.$$

Changing $x^2 + y^2$ to polar coordinates gives us

$$r^2 = \tfrac{1}{4}\theta^2 \quad \text{or} \quad r = \tfrac{1}{2}\theta.$$

The graph of the polar equation $r = \tfrac{1}{2}\theta$, where $0 \le \theta \le 3\pi$, is the portion of the spiral illustrated in Figure 15.3.

(b)   Substitution of 6 for t gives us

$$\mathbf{r}(6) = (3 \cos 6)\mathbf{i} + (3 \sin 6)\mathbf{j} \approx 2.9\mathbf{i} - 0.8\mathbf{j}$$

and we sketch $\mathbf{r}(6)$ as indicated in Figure 15.3.   ∎

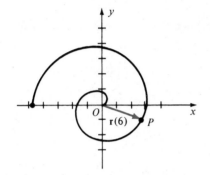

Figure 15.3

EXERCISES 15.1

In Exercises 1–14 sketch the graph of the curve C traced by the terminal point of the position vector of $\mathbf{r}(t)$ if t varies as indicated.

1   $\mathbf{r}(t) = 3t\mathbf{i} + (1 - 9t^2)\mathbf{j}; t$ in $\mathbb{R}$

2   $\mathbf{r}(t) = (1 - t^3)\mathbf{i} + t\mathbf{j}; t \ge 0$

3   $\mathbf{r}(t) = (t^3 - 1)\mathbf{i} + (t^2 + 2)\mathbf{j}; -2 \le t \le 2$

4   $\mathbf{r}(t) = (2 + \cos t)\mathbf{i} - (3 - \sin t)\mathbf{j}; 0 \le t \le 2\pi$

5   $\mathbf{r}(t) = e^t \cos t\mathbf{i} + e^t \sin t\mathbf{j}; 0 \le t \le \pi$

6   $\mathbf{r}(t) = 2 \cosh t\mathbf{i} + 3 \sinh t\mathbf{j}; t$ in $\mathbb{R}$

7   $\mathbf{r}(t) = t\mathbf{i} + 4 \cos t\mathbf{j} + 9 \sin t\mathbf{k}; t \ge 0$

8   $\mathbf{r}(t) = \tan t\mathbf{i} + \sec t\mathbf{j} + 2\mathbf{k}; -\pi/2 < t < \pi/2$

9   $\mathbf{r}(t) = t\mathbf{i} + t^2\mathbf{j} + t^3\mathbf{k}; t$ in $\mathbb{R}$

10   $\mathbf{r}(t) = t^3\mathbf{i} + t^2\mathbf{j} + t\mathbf{k}; 0 \le t \le 4$

11   $\mathbf{r}(t) = (t^2 + 1)\mathbf{i} + t\mathbf{j} + 3\mathbf{k}; t$ in $\mathbb{R}$

12   $\mathbf{r}(t) = 6 \sin t\mathbf{i} + 4\mathbf{j} + 25 \cos t\mathbf{k}; -2\pi \le t \le 2\pi$

13   $\mathbf{r}(t) = t\mathbf{i} + t\mathbf{j} + \sin t\mathbf{k}; t$ in $\mathbb{R}$

14   $\mathbf{r}(t) = t\mathbf{i} + 2t\mathbf{j} + e^t\mathbf{k}; t$ in $\mathbb{R}$

In Exercises 15–20 find the length of the curve.

15   $C = \{(5t, 4t^2, 3t^2): 0 \le t \le 2\}$

16   $C = \{(t^2, t \sin t, t \cos t): 0 \le t \le 1\}$

17   $x = e^t \cos t, y = e^t, z = e^t \sin t; 0 \le t \le 2\pi$

18   $x = 2t, y = 4 \sin 3t, z = 4 \cos 3t; 0 \le t \le 2\pi$

19   $x = 3t^2, y = t^3, z = 6t; 0 \le t \le 1$

20   $x = 1 - 2t^2, y = 4t, z = 3 + 2t^2; 0 \le t \le 2$

15.2

LIMITS, DERIVATIVES, AND INTEGRALS

The limit of a vector-valued function may be defined as follows.

Definition (15.3)

If $\mathbf{r}(t) = f(t)\mathbf{i} + g(t)\mathbf{j} + h(t)\mathbf{k}$, then

$$\lim_{t \to a} \mathbf{r}(t) = \left[\lim_{t \to a} f(t)\right]\mathbf{i} + \left[\lim_{t \to a} g(t)\right]\mathbf{j} + \left[\lim_{t \to a} h(t)\right]\mathbf{k}.$$

provided f, g, and h have limits as t approaches a.

The preceding definition may be extended to one-sided limits as was done in Chapter 2. If, in Definition (15.3),

$$\lim_{t \to a} f(t) = a_1, \quad \lim_{t \to a} g(t) = a_2, \quad \text{and} \quad \lim_{t \to a} h(t) = a_3$$

then

$$\lim_{t \to a} \mathbf{r}(t) = a_1\mathbf{i} + a_2\mathbf{j} + a_3\mathbf{k} = \mathbf{a}.$$

Intuitively, we have a situation of the type illustrated in Figure 15.4, where if $\overrightarrow{OP}$ and $\overrightarrow{OA}$ are position vectors corresponding to $\mathbf{r}(t)$ and $\mathbf{a}$, respectively, then as t approaches a, $\overrightarrow{OP}$ approaches $\overrightarrow{OA}$, in the sense that P gets closer and closer to A.

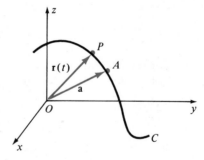

Figure 15.4

Theorems on limits of vector-valued functions similar to those in Chapter 2 can be established and are left as exercises. Limits of vector-valued functions may also be formulated using an ε-δ approach similar to that in Definition (2.3) (see Exercise 35).

Definition (15.4)

A vector-valued function $\mathbf{r}$ is **continuous** at a if

$$\lim_{t \to a} \mathbf{r}(t) = \mathbf{r}(a).$$

It follows that if $\mathbf{r}(t) = f(t)\mathbf{i} + g(t)\mathbf{j} + h(t)\mathbf{k}$, then $\mathbf{r}$ is continuous at a if and only if f, g, and h are continuous at a. Continuity on an interval is defined in the usual way.

Definition (15.5)

If $\mathbf{r}$ is a vector-valued function, then the **derivative** of $\mathbf{r}$ is the vector-valued function $\mathbf{r}'$ defined by

$$\mathbf{r}'(t) = \lim_{\Delta t \to 0} \frac{1}{\Delta t}[\mathbf{r}(t + \Delta t) - \mathbf{r}(t)]$$

for all t such that the limit exists.

If we adopt the convention that

$$\frac{1}{\Delta t}[\mathbf{r}(t + \Delta t) - \mathbf{r}(t)] = \frac{\mathbf{r}(t + \Delta t) - \mathbf{r}(t)}{\Delta t}$$

then Definition (15.5) takes on the familiar form for derivatives introduced in Chapter 3. We shall use this notation in the proof of the next theorem, which states that to find $\mathbf{r}'(t)$ we may differentiate each component of $\mathbf{r}(t)$.

Theorem (15.6)

If $\quad \mathbf{r}(t) = f(t)\mathbf{i} + g(t)\mathbf{j} + h(t)\mathbf{k}$

where f, g, and h are differentiable, then

$$\mathbf{r}'(t) = f'(t)\mathbf{i} + g'(t)\mathbf{j} + h'(t)\mathbf{k}.$$

Proof

$$\mathbf{r}'(t) = \lim_{\Delta t \to 0} \frac{\mathbf{r}(t + \Delta t) - \mathbf{r}(t)}{\Delta t}$$

$$= \lim_{\Delta t \to 0} \frac{[f(t + \Delta t)\mathbf{i} + g(t + \Delta t)\mathbf{j} + h(t + \Delta t)\mathbf{k}] - [f(t)\mathbf{i} + g(t)\mathbf{j} + h(t)\mathbf{k}]}{\Delta t}$$

$$= \lim_{\Delta t \to 0} \left[\frac{f(t + \Delta t) - f(t)}{\Delta t}\mathbf{i} + \frac{g(t + \Delta t) - g(t)}{\Delta t}\mathbf{j} + \frac{h(t + \Delta t) - h(t)}{\Delta t}\mathbf{k}\right].$$

Taking the limit of each component gives us the desired result. □

If $\mathbf{r}'(t)$ exists we say that the function $\mathbf{r}$ is **differentiable** at t. We also denote derivatives as follows:

$$\mathbf{r}'(t) = D_t\mathbf{r}(t) = \frac{d}{dt}\mathbf{r}(t).$$

Higher derivatives may be obtained in like manner. For example,

$$\mathbf{r}''(t) = f''(t)\mathbf{i} + g''(t)\mathbf{j} + h''(t)\mathbf{k}.$$

Example 1 If $\mathbf{r}(t) = (\ln t)\mathbf{i} + e^{-3t}\mathbf{j} + t^2\mathbf{k}$,

(a) find the domain of $\mathbf{r}$ and determine where $\mathbf{r}$ is continuous.

(b) find $\mathbf{r}'(t)$ and $\mathbf{r}''(t)$.

Solution

(a) Since $\ln t$ is undefined if $t \le 0$, the domain of $\mathbf{r}$ is the set of positive real numbers. Moreover, $\mathbf{r}$ is continuous throughout its domain since each component determines a continuous function.

(b) By Theorem (15.6),

$$\mathbf{r}'(t) = (1/t)\mathbf{i} - 3e^{-3t}\mathbf{j} + 2t\mathbf{k},$$

and

$$\mathbf{r}''(t) = (-1/t^2)\mathbf{i} + 9e^{-3t}\mathbf{j} + 2\mathbf{k}. \quad \blacksquare$$

Since the definition of $\mathbf{r}'(t)$ in (15.5) has the same form as that given for $f'(x)$ in Chapter 3, it is reasonable to expect that differentation formulas for vector-valued functions are similar to those for real-valued functions. The next theorem brings out this fact. Other differentiation formulas may be found in the Exercises.

Theorem (15.7)

> If $\mathbf{u}$ and $\mathbf{v}$ are differentiable vector-valued functions and c is a scalar, then
>
> (i) $D_t[\mathbf{u}(t) + \mathbf{v}(t)] = \mathbf{u}'(t) + \mathbf{v}'(t)$.
>
> (ii) $D_t[c\mathbf{u}(t)] = c\mathbf{u}'(t)$.
>
> (iii) $D_t[\mathbf{u}(t) \cdot \mathbf{v}(t)] = \mathbf{u}(t) \cdot \mathbf{v}'(t) + \mathbf{u}'(t) \cdot \mathbf{v}(t)$.
>
> (iv) $D_t[\mathbf{u}(t) \times \mathbf{v}(t)] = \mathbf{u}(t) \times \mathbf{v}'(t) + \mathbf{u}'(t) \times \mathbf{v}(t)$.

Proof   We shall prove (iii) and leave the other parts as exercises. If we write

$$\mathbf{u}(t) = f_1(t)\mathbf{i} + f_2(t)\mathbf{j} + f_3(t)\mathbf{k}$$

$$\mathbf{v}(t) = g_1(t)\mathbf{i} + g_2(t)\mathbf{j} + g_3(t)\mathbf{k}$$

where each f_i and g_i is a differentiable function of t, then

$$\mathbf{u}(t) \cdot \mathbf{v}(t) = \sum_{i=1}^{3} f_i(t)g_i(t).$$

Consequently,

$$D_t[\mathbf{u}(t) \cdot \mathbf{v}(t)] = D_t \sum_{i=1}^{3} f_i(t)g_i(t) = \sum_{i=1}^{3} D_t[f_i(t)g_i(t)]$$

$$= \sum_{i=1}^{3} [f_i(t)g_i'(t) + f_i'(t)g_i(t)]$$

$$= \sum_{i=1}^{3} f_i(t)g_i'(t) + \sum_{i=1}^{3} f_i'(t)g_i(t)$$

$$= \mathbf{u}(t) \cdot \mathbf{v}'(t) + \mathbf{u}'(t) \cdot \mathbf{v}(t). \qquad \square$$

There is an interesting geometric interpretation for the derivative $\mathbf{r}'(t)$. Given $\mathbf{r}(t) = f(t)\mathbf{i} + g(t)\mathbf{j} + h(t)\mathbf{k}$ where f, g, and h are differentiable, let C denote the curve with parametric equations $x = f(t)$, $y = g(t)$, $z = h(t)$. As in (i) of Figure 15.5, if $\overrightarrow{OP}$ and $\overrightarrow{OQ}$ are the position vectors corresponding to $\mathbf{r}(t)$ and $\mathbf{r}(t + \Delta t)$, respectively, then

$$\overrightarrow{PQ} = \overrightarrow{OQ} - \overrightarrow{OP}$$

is a geometric vector corresponding to $\mathbf{r}(t + \Delta t) - \mathbf{r}(t)$. It follows that $(1/\Delta t)\overrightarrow{PQ}$ is a vector corresponding to $(1/\Delta t)[\mathbf{r}(t + \Delta t) - \mathbf{r}(t)]$. If $\Delta t > 0$, then $(1/\Delta t)\overrightarrow{PQ}$ is a vector $\overrightarrow{PR}$ having the same direction as $\overrightarrow{PQ}$. Moreover, if $0 < \Delta t < 1$, then $(1/\Delta t) > 1$ and $|\overrightarrow{PR}| > |\overrightarrow{PQ}|$ as illustrated in (ii) of Figure 15.5.

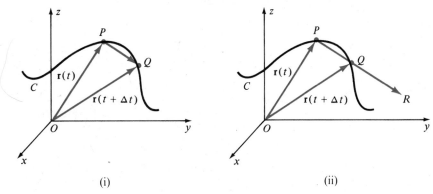

(i)   (ii)

Figure 15.5

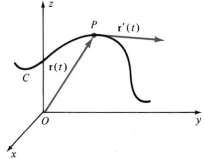

Figure 15.6

If we let $\Delta t \to 0^+$, then Q approaches P and it appears that the vector $\overrightarrow{PR}$ approaches a vector that lies on the tangent line to C at P, as illustrated in Figure 15.6. A similar discussion may be given if $\Delta t < 0$. For this reason we refer to $\mathbf{r}'(t)$, or any of its geometric representations, as a **tangent vector** to C at P. The vector $\mathbf{r}'(t)$ points in the direction of increasing values of t. By definition, the tangent line to C at P is the line through P parallel to $\mathbf{r}'(t)$.

For the two-dimensional case $\mathbf{r}(t) = f(t)\mathbf{i} + g(t)\mathbf{j}$ we have

$$\mathbf{r}'(t) = f'(t)\mathbf{i} + g'(t)\mathbf{j}.$$

If $f'(t) \neq 0$, then the slope of the line parallel to $\mathbf{r}'(t)$ is $g'(t)/f'(t)$. This is in agreement with the formula in Theorem (13.3) obtained for tangent lines to plane curves. These remarks are illustrated in the next example.

Example 2 If $\mathbf{r}(t) = 2t\mathbf{i} + (4 - t^2)\mathbf{j}$, where $-2 \leq t \leq 2$, find $\mathbf{r}'(t)$. Sketch the curve C with parametric equations $x = 2t$, $y = 4 - t^2$, where $-2 \leq t \leq 2$. Illustrate $\mathbf{r}(1)$ and $\mathbf{r}'(1)$ geometrically.

Solution   Since we are dealing with V_2, it is sufficient to use an xy-plane to represent vectors. Eliminating the parameter from $x = 2t$, $y = 4 - t^2$, we obtain $y = 4 - (x/2)^2$, which is an equation of a parabola. Coordinates of points on C that correspond to several values of t are listed below.

t	-2	-1	0	1	2
x	-4	-2	0	2	4
y	0	3	4	3	0

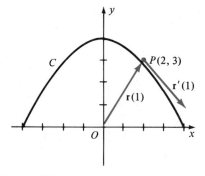

Figure 15.7

The curve C is that part of the parabola illustrated in Figure 15.7. Since $\mathbf{r}(1) = 2\mathbf{i} + 3\mathbf{j}$, the position vector corresponding to $\mathbf{r}(1)$ is $\overrightarrow{OP}$, where P is the point on C with coordinates $(2, 3)$. By differentiation we obtain $\mathbf{r}'(t) = 2\mathbf{i} - 2t\mathbf{j}$. In the figure we have represented $\mathbf{r}'(1) = 2\mathbf{i} - 2\mathbf{j}$ by a vector with initial point P to illustrate that it is a tangent vector to C.   ∎

Example 3 Let C be the curve with parametric equations

$$x = t^2 - 5, \quad y = t^3, \quad z = 3t + 1$$

where t is in $\mathbb{R}$. Find parametric equations for the tangent line to C at the point corresponding to $t = 2$.

Solution   The point $P(t)$ on C corresponding to t is $(t^2 - 5, t^3, 3t + 1)$. If we let

$$\mathbf{r}(t) = (t^2 - 5)\mathbf{i} + t^3\mathbf{j} + (3t + 1)\mathbf{k}$$

then from the previous discussion, a tangent vector to C at $P(t)$ is

$$\mathbf{r}'(t) = 2t\mathbf{i} + 3t^2\mathbf{j} + 3\mathbf{k}.$$

In particular, the point corresponding to $t = 2$ is $P(2) = (-1, 8, 7)$ and a tangent vector is

$$\mathbf{r}'(2) = 4\mathbf{i} + 12\mathbf{j} + 3\mathbf{k}.$$

Applying Theorem (14.37), parametric equations of the tangent line are

$$x = 4t - 1, \quad y = 12t + 8, \quad z = 3t + 7.$$ ∎

Example 4 Prove that if $|\mathbf{r}(t)|$ is constant, then $\mathbf{r}'(t)$ is orthogonal to $\mathbf{r}(t)$ for every t.

Solution   By hypothesis

$$\mathbf{r}(t) \cdot \mathbf{r}(t) = |\mathbf{r}(t)|^2 = c$$

for some real number c and, therefore,

$$D_t \left[\mathbf{r}(t) \cdot \mathbf{r}(t) \right] = D_t c = 0.$$

Using (iii) of Theorem (15.7) gives us

$$\mathbf{r}(t) \cdot \mathbf{r}'(t) + \mathbf{r}'(t) \cdot \mathbf{r}(t) = 0,$$
or
$$2\mathbf{r}(t) \cdot \mathbf{r}'(t) = 0.$$

Hence, by Theorem (14.24), $\mathbf{r}(t)$ and $\mathbf{r}'(t)$ are orthogonal. ∎

The result stated in Example 4 is also geometrically evident, for if $|\mathbf{r}(t)|$ is constant, then as t varies, the endpoint of the position vector corresponding to $\mathbf{r}(t)$ moves on a curve C that lies on the surface of a sphere with center at O. The tangent vector $\mathbf{r}'(t)$ to C is always orthogonal to the position vector and hence to $\mathbf{r}(t)$.

Definite integrals of vector-valued functions may also be considered. Specifically, if

$$\mathbf{r}(t) = f(t)\mathbf{i} + g(t)\mathbf{j} + h(t)\mathbf{k}$$

where f, g, and h are integrable on $[a, b]$, then by definition

$$\int_a^b \mathbf{r}(t)\, dt = \left[\int_a^b f(t)\, dt \right] \mathbf{i} + \left[\int_a^b g(t)\, dt \right] \mathbf{j} + \left[\int_a^b h(t)\, dt \right] \mathbf{k}.$$

In this event we say that $\mathbf{r}$ is **integrable** on $[a, b]$.

If $\mathbf{R}'(t) = \mathbf{r}(t)$, then $\mathbf{R}(t)$ is called an **antiderivative** of $\mathbf{r}(t)$. The next result is analogous to the Fundamental Theorem of Calculus.

Theorem (15.8)

If $\mathbf{R}(t)$ is an antiderivative of $\mathbf{r}(t)$ on $[a, b]$, then

$$\int_a^b \mathbf{r}(t)\, dt = \mathbf{R}(t) \Big]_a^b = \mathbf{R}(b) - \mathbf{R}(a).$$

The proof is left as an exercise.

Example 5   Find $\int_0^2 \mathbf{r}(t)\,dt$ if $\mathbf{r}(t) = 12t^3\mathbf{i} + 4e^{2t}\mathbf{j} + (t+1)^{-1}\mathbf{k}$.

Solution   By finding an antiderivative of each component of $\mathbf{r}(t)$, we obtain

$$\mathbf{R}(t) = 3t^4\mathbf{i} + 2e^{2t}\mathbf{j} + \ln(t+1)\mathbf{k}.$$

Since $\mathbf{R}'(t) = \mathbf{r}(t)$ it follows from Theorem (15.8) that

$$\int_0^2 \mathbf{r}(t)\,dt = \mathbf{R}(2) - \mathbf{R}(0)$$
$$= (48\mathbf{i} + 2e^4\mathbf{j} + \ln 3\mathbf{k}) - (0\mathbf{i} + 2\mathbf{j} + 0\mathbf{k})$$
$$= 48\mathbf{i} + 2(e^4 - 1)\mathbf{j} + \ln 3\mathbf{k}. \qquad \blacksquare$$

Indefinite integrals of vector-valued functions have a theory similar to that developed for real-valued functions. The proofs of theorems require only minor modifications of those given earlier and are omitted. If $\mathbf{R}(t)$ is an antiderivative of $\mathbf{r}(t)$, then every antiderivative has the form $\mathbf{R}(t) + \mathbf{c}$ for some (constant) vector $\mathbf{c}$ and we write

$$\int \mathbf{r}(t)\,dt = \mathbf{R}(t) + \mathbf{c}, \qquad \text{where } \mathbf{R}'(t) = \mathbf{r}(t).$$

If boundary conditions are known, then $\mathbf{c}$ may be determined (see Exercises 27–30).

EXERCISES 15.2

In Exercises 1–4 find (a) the domain of $\mathbf{r}$ and determine where $\mathbf{r}$ is continuous; (b) $\mathbf{r}'(t)$ and $\mathbf{r}''(t)$.

1 $\mathbf{r}(t) = \sqrt{t-1}\,\mathbf{i} + \sqrt{2-t}\,\mathbf{j}$

2 $\mathbf{r}(t) = (1/t)\mathbf{i} + \sin 3t\mathbf{j}$

3 $\mathbf{r}(t) = \tan t\mathbf{i} + (t^2 + 8t)\mathbf{j}$

4 $\mathbf{r}(t) = e^{t^2}\mathbf{i} + \sin^{-1} t\mathbf{j}$

In Exercises 5–12 (a) sketch the curve in the xy-plane determined by the components of $\mathbf{r}(t)$; (b) find $\mathbf{r}'(t)$ and sketch vectors corresponding to $\mathbf{r}(t)$ and $\mathbf{r}'(t)$ at the indicated value of t.

5 $\mathbf{r}(t) = (-t^4/4)\mathbf{i} + t^2\mathbf{j}; t = 2$

6 $\mathbf{r}(t) = e^{2t}\mathbf{i} + e^{-4t}\mathbf{j}; t = 0$

7 $\mathbf{r}(t) = 4\cos t\mathbf{i} + 2\sin t\mathbf{j}; t = 3\pi/4$

8 $\mathbf{r}(t) = 2\sec t\mathbf{i} + 3\tan t\mathbf{j}; t = \pi/4$

9 $\mathbf{r}(t) = t^3\mathbf{i} + t^{-3}\mathbf{j}; t = 1$

10 $\mathbf{r}(t) = t^2\mathbf{i} + t^3\mathbf{j}; t = -1$

11 $\mathbf{r}(t) = (2t-1)\mathbf{i} + (4-t)\mathbf{j}; t = 3$

12 $\mathbf{r}(t) = 5\mathbf{i} + t^3\mathbf{j}; t = 2$

In Exercises 13–16 (a) find the domain of $\mathbf{r}$ and determine where $\mathbf{r}$ is continuous; (b) find $\mathbf{r}'(t)$ and $\mathbf{r}''(t)$.

13 $\mathbf{r}(t) = t^2\mathbf{i} + \tan t\mathbf{j} + 3\mathbf{k}$

14 $\mathbf{r}(t) = \sqrt[3]{t}\,\mathbf{i} + (1/t)\mathbf{j} + e^{-t}\mathbf{k}$

15 $\mathbf{r}(t) = \sqrt{t}\,\mathbf{i} + e^{2t}\mathbf{j} + t\mathbf{k}$

16 $\mathbf{r}(t) = \ln(1-t)\mathbf{i} + \sin t\mathbf{j} + t^2\mathbf{k}$

Parametric equations of a curve C are given in each of Exercises 17–20. Find parametric equations for the tangent line to C at the point P.

17 $x = 2t^3 - 1, y = -5t^2 + 3, z = 8t + 2; P(1, -2, 10)$

18 $x = 4\sqrt{t}, y = t^2 - 10, z = 4/t; P(8, 6, 1)$

19 $x = e^t, y = te^t, z = t^2 + 4; P(1, 0, 4)$

20 $x = t\sin t, y = t\cos t, z = t; P(\pi/2, 0, \pi/2)$

In Exercises 21 and 22 find two different unit tangent vectors to the given curve at the indicated point P.

21 $x = e^{2t}, y = e^{-t}, z = t^2 + 4; P(1, 1, 4)$

22 $x = 2 + \sin t, y = \cos t, z = t; P(2, 1, 0)$

Evaluate the integrals in Exercises 23–26.

23 $\int_0^2 (6t^2\mathbf{i} - 4t\mathbf{j} + 3\mathbf{k}) \, dt$

24 $\int_{-1}^1 (-5t\mathbf{i} + 8t^3\mathbf{j} - 3t^2\mathbf{k}) \, dt$

25 $\int_0^{\pi/4} (\sin t\mathbf{i} - \cos t\mathbf{j} + \tan t\mathbf{k}) \, dt$

26 $\int_0^1 [te^{t^2}\mathbf{i} + \sqrt{t}\mathbf{j} + (t^2 + 1)^{-1}\mathbf{k}] \, dt$

27 If $\mathbf{u}'(t) = t^2\mathbf{i} + (6t + 1)\mathbf{j} + 8t^3\mathbf{k}$ and $\mathbf{u}(0) = 2\mathbf{i} - 3\mathbf{j} + \mathbf{k}$, find $\mathbf{u}(t)$.

28 If $\mathbf{r}'(t) = 2\mathbf{i} - 4t^3\mathbf{j} + 6\sqrt{t}\mathbf{k}$ and $\mathbf{r}(0) = \mathbf{i} + 5\mathbf{j} + 3\mathbf{k}$, find $\mathbf{r}(t)$.

29 If $\mathbf{u}''(t) = 6t\mathbf{i} - 12t^2\mathbf{j} + \mathbf{k}, \mathbf{u}'(0) = \mathbf{i} + 2\mathbf{j} - 3\mathbf{k}$, and $\mathbf{u}(0) = 7\mathbf{i} + \mathbf{k}$, find $\mathbf{u}(t)$.

30 If $\mathbf{r}''(t) = 6t\mathbf{i} + 3\mathbf{j}, \mathbf{r}'(0) = 4\mathbf{i} - \mathbf{j} + \mathbf{k}$, and $\mathbf{r}(0) = 5\mathbf{j}$, find $\mathbf{r}(t)$.

31 If a curve C has a tangent vector $\mathbf{a}$ at a point P, then the **normal plane** to C at P is the plane through P with normal vector $\mathbf{a}$. Find an equation of the normal plane to the curve of Exercise 19 at the point $P(1, 0, 4)$.

32 Refer to Exercise 31. Find an equation for the normal plane to the curve of Exercise 20 at the point $P(\pi/2, 0, \pi/2)$.

33 If $\mathbf{r}$ and $\mathbf{s}$ are vector-valued functions that have limits as $t \to a$, prove the following.

(a) $\lim_{t \to a} [\mathbf{r}(t) + \mathbf{s}(t)] = \lim_{t \to a} \mathbf{r}(t) + \lim_{t \to a} \mathbf{s}(t)$

(b) $\lim_{t \to a} [\mathbf{r}(t) \cdot \mathbf{s}(t)] = \lim_{t \to a} \mathbf{r}(t) \cdot \lim_{t \to a} \mathbf{s}(t)$

(c) $\lim_{t \to a} c\mathbf{r}(t) = c \lim_{t \to a} \mathbf{r}(t)$ where c is a scalar

34 If a function f and a vector-valued function $\mathbf{u}$ have limits as $t \to a$, prove that

$$\lim_{t \to a} f(t)\mathbf{u}(t) = \left[\lim_{t \to a} f(t)\right]\left[\lim_{t \to a} \mathbf{u}(t)\right].$$

35 Prove that $\lim_{t \to a} \mathbf{u}(t) = \mathbf{b}$ if and only if for every $\varepsilon > 0$ there is a $\delta > 0$ such that $|\mathbf{u}(t) - \mathbf{b}| < \varepsilon$ whenever $0 < |t - a| < \delta$. Interpret this fact geometrically.

36 If $\mathbf{u}$ and $\mathbf{v}$ are vector-valued functions that have limits as $t \to a$, prove that

$$\lim_{t \to a} [\mathbf{u}(t) \times \mathbf{v}(t)] = \lim_{t \to a} \mathbf{u}(t) \times \lim_{t \to a} \mathbf{v}(t).$$

37 Complete the proof of Theorem (15.7).

38 Prove Theorem (15.8).

39 Prove that if f and $\mathbf{u}$ are differentiable functions of t, then

$$D_t[f(t)\mathbf{u}(t)] = f(t)\mathbf{u}'(t) + f'(t)\mathbf{u}(t).$$

40 Prove the Chain Rule for vector-valued functions:

$$D_t\mathbf{u}(f(t)) = f'(t)\mathbf{u}'(f(t))$$

provided $\mathbf{u}$ and f are differentiable and the domains are suitably restricted.

41 Prove that if $\mathbf{u}, \mathbf{v},$ and $\mathbf{w}$ are differentiable vector-valued functions, then

$D_t[\mathbf{u}(t) \cdot \mathbf{v}(t) \times \mathbf{w}(t)] =$
$\quad \mathbf{u}'(t) \cdot \mathbf{v}(t) \times \mathbf{w}(t) + \mathbf{u}(t) \cdot \mathbf{v}'(t) \times \mathbf{w}(t) + \mathbf{u}(t) \cdot \mathbf{v}(t) \times \mathbf{w}'(t).$

42 Prove that if $\mathbf{u}'(t)$ and $\mathbf{u}''(t)$ exist, then

$$D_t[\mathbf{u}(t) \times \mathbf{u}'(t)] = \mathbf{u}(t) \times \mathbf{u}''(t).$$

43 If $\mathbf{u}$ and $\mathbf{v}$ are integrable on $[a, b]$ and if c is a scalar, prove the following.

(a) $\int_a^b [\mathbf{u}(t) + \mathbf{v}(t)] \, dt = \int_a^b \mathbf{u}(t) \, dt + \int_a^b \mathbf{v}(t) \, dt$

(b) $\int_a^b c\mathbf{u}(t) \, dt = c \int_a^b \mathbf{u}(t) \, dt$

44 If $\mathbf{u}$ is integrable on $[a, b]$ and $\mathbf{c}$ is any vector in V_3, prove that

$$\int_a^b \mathbf{c} \cdot \mathbf{u}(t) \, dt = \mathbf{c} \cdot \int_a^b \mathbf{u}(t) \, dt.$$

In Exercises 45 and 46, find $D_t[\mathbf{u}(t) \cdot \mathbf{v}(t)]$ and $D_t[\mathbf{u}(t) \times \mathbf{v}(t)]$.

45 $\mathbf{u}(t) = t\mathbf{i} + t^2\mathbf{j} + t^3\mathbf{k}$
$\quad \mathbf{v}(t) = \sin t\mathbf{i} + \cos t\mathbf{j} + 2 \sin t\mathbf{k}$

46 $\mathbf{u}(t) = 2t\mathbf{i} + 6t\mathbf{j} + t^2\mathbf{k}$
$\quad \mathbf{v}(t) = e^{-t}\mathbf{i} - e^{-t}\mathbf{j} + \mathbf{k}$

15.3

MOTION

In this section we introduce concepts that are needed for the study of motion. If we assume that the mass of an object is concentrated at its center of gravity, then we can reduce our study to the motion of a point (or a particle). Motion often takes place in a plane. For example, although the earth moves through space, its orbit lies in a plane. (This will be proved in Section 15.6.) Let us begin, therefore, by studying a particle moving in a coordinate plane. To analyze the behavior of the particle, it is essential to know its position (x, y) at every instant. Since at time t the particle has a unique position, it follows that the coordinates x and y are functions of t. Thus, we shall assume that at time t the particle is at the point $P(x, y)$, where $x = f(t)$ and $y = g(t)$ for certain functions f and g. If we let

$$\mathbf{r}(t) = f(t)\mathbf{i} + g(t)\mathbf{j}$$

then as t varies, the endpoint of the position vector $\overrightarrow{OP}$ corresponding to $\mathbf{r}(t)$ traces the path C of the particle. From the discussion in the preceding section, the tangent line to C at P is parallel to

$$\mathbf{r}'(t) = f'(t)\mathbf{i} + g'(t)\mathbf{j}.$$

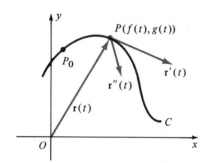

Figure 15.8

(See Figure 15.8.) The magnitude of this vector is

$$|\mathbf{r}'(t)| = \sqrt{[f'(t)]^2 + [g'(t)]^2}.$$

If P_0 is the point corresponding to $t = t_0$, and if C is smooth, then by Theorem (13.11), the arc length $s(t)$ of C from P_0 to P is

$$s(t) = \int_{t_0}^{t} \sqrt{[f'(t)]^2 + [g'(t)]^2}\, dt = \int_{t_0}^{t} |\mathbf{r}'(t)|\, dt.$$

Applying Theorem (5.24) we obtain $s'(t) = |\mathbf{r}'(t)|$; that is, $|\mathbf{r}'(t)|$ *is the rate of change of arc length with respect to time.* For this reason we refer to $|\mathbf{r}'(t)|$ as the **speed** of the particle. The tangent vector $\mathbf{r}'(t)$, whose magnitude is the speed, is defined as the **velocity** of the particle at time t. In analogy with Theorem (4.28) we call the vector $\mathbf{r}''(t)$ the **acceleration** of the particle at time t and usually represent it by a vector with initial point P. It can be shown that in most cases $\mathbf{r}''(t)$ is directed toward the concave side of C, as illustrated in Figure 15.8.

Example 1 The position of a particle moving in a plane is given by

$$\mathbf{r}(t) = (t^2 + t)\mathbf{i} + t^3\mathbf{j}$$

where $0 \le t \le 2$. Find the velocity and acceleration at time t. Sketch the path C of the particle and represent $\mathbf{r}'(1)$ and $\mathbf{r}''(1)$ geometrically.

Solution   The velocity and acceleration are

$$\mathbf{r}'(t) = (2t + 1)\mathbf{i} + 3t^2\mathbf{j} \quad \text{and} \quad \mathbf{r}''(t) = 2\mathbf{i} + 6t\mathbf{j}.$$

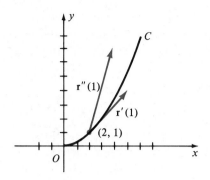

Figure 15.9

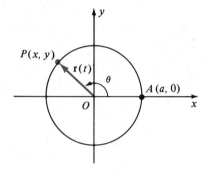

Figure 15.10

In particular, at $t = 1$

$$\mathbf{r}'(1) = 3\mathbf{i} + 3\mathbf{j} \quad \text{and} \quad \mathbf{r}''(1) = 2\mathbf{i} + 6\mathbf{j}.$$

These vectors and the path of the particle are sketched in Figure 15.9. ∎

Example 2 Show that if a point P moves around a circle of radius a at a constant speed v, then the acceleration vector has a constant magnitude v^2/a and is directed from P toward the center of the circle. (This vector is called the **centripetal acceleration vector**.)

Solution There is no loss of generality in assuming that the center of the circle is at the origin O of a rectangular coordinate system. Suppose that the radius of the circle is a, the initial position of P is $A(a, 0)$, and θ is the angle generated by $\overline{OP}$ at the end of time t (see Figure 15.10). The fact that P moves around the circle at a constant speed is equivalent to stating that the rate at which the angle θ changes per unit time (called the **angular speed**) is a constant ω. This fact may be expressed by means of the differential equation $d\theta/dt = \omega$, or equivalently by $\theta = \omega t + c$ for some number c. Since $\theta = 0$ when $t = 0$, the number c is 0; that is $\theta = \omega t$. It follows that the coordinates (x, y) of P are

$$x = a \cos \omega t, \quad y = a \sin \omega t$$

and the motion of P is given by

$$\mathbf{r}(t) = a \cos \omega t \mathbf{i} + a \sin \omega t \mathbf{j}.$$

Consequently,

$$\mathbf{r}'(t) = -\omega a \sin \omega t \mathbf{i} + \omega a \cos \omega t \mathbf{j}$$
$$\mathbf{r}''(t) = -\omega^2 a \cos \omega t \mathbf{i} - \omega^2 a \sin \omega t \mathbf{j}$$

and hence $$\mathbf{r}''(t) = -\omega^2 \mathbf{r}(t).$$

This shows that the direction of the acceleration vector is opposite that of $\mathbf{r}(t)$, and hence is directed from P toward O.

The magnitude of $\mathbf{r}''(t)$ is

$$|\mathbf{r}''(t)| = |-\omega^2 \mathbf{r}(t)| = |-\omega^2| \, |\mathbf{r}(t)| = \omega^2 a$$

which is a constant. Since

$$v = |\mathbf{r}'(t)| = \sqrt{(\omega a)^2 \sin^2 \omega + (\omega a)^2 \cos^2 \omega} = \sqrt{\omega^2 a^2} = \omega a$$

we see that $\omega = v/a$. Substitution in the formula $|\mathbf{r}''(t)| = \omega^2 a$ gives us

$$|\mathbf{r}''(t)| = \frac{v^2}{a}.$$

Note that $|\mathbf{r}''(t)|$ will increase if we either increase v or decrease a, a fact that is evident to anyone who has attached an object to a string or rope and twirled it about in a circular path.

For an application of Example 2, see Exercises 23 and 24. ∎

Our discussion of motion may be extended to a particle moving in a three-dimensional coordinate system. Thus, suppose the position of a particle at time t is given by $P(t) = (f(t), g(t), h(t))$ where the functions f, g, and h are defined on an interval I. If we let

$$\mathbf{r}(t) = f(t)\mathbf{i} + g(t)\mathbf{j} + h(t)\mathbf{k}$$

then as t varies, the endpoint of the position vector $\overrightarrow{OP}$ traces the path of the particle. As before, the derivative $\mathbf{r}'(t)$, if it exists, is called the **velocity** of the particle at time t and is a tangent vector to C at $P(t)$. The vector $\mathbf{r}''(t)$ is called the **acceleration** of the particle at time t. The **speed** of the particle is the magnitude $|\mathbf{r}'(t)|$ of the velocity and, as in two dimensions, equals the rate of change of arc length with respect to time.

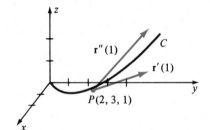

Figure 15.11

Example 3   The position of a particle is given by $\mathbf{r}(t) = 2t\mathbf{i} + 3t^2\mathbf{j} + t^3\mathbf{k}$, where $0 \leq t \leq 2$. Find the velocity and acceleration at time t. Sketch the path of the particle and illustrate $\mathbf{r}'(1)$ and $\mathbf{r}''(1)$ geometrically.

Solution   The velocity and acceleration are

$$\mathbf{r}'(t) = 2\mathbf{i} + 6t\mathbf{j} + 3t^2\mathbf{k} \quad \text{and} \quad \mathbf{r}''(t) = 6\mathbf{j} + 6t\mathbf{k}.$$

In particular, at $t = 1$ the particle is at point $P(2, 3, 1)$ and

$$\mathbf{r}'(1) = 2\mathbf{i} + 6\mathbf{j} + 3\mathbf{k} \quad \text{and} \quad \mathbf{r}''(1) = 6\mathbf{j} + 6\mathbf{k}.$$

These vectors and the path of the particle are sketched in Figure 15.11.   ∎

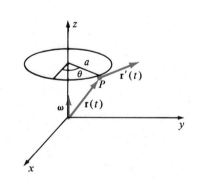

Figure 15.12

Suppose a particle P is rotating about the z-axis on a circle of radius a that lies in a plane parallel to the xy-plane, as illustrated in Figure 15.12. In addition, suppose the angular speed $d\theta/dt$ is a constant ω. The vector $\boldsymbol{\omega} = \omega\mathbf{k}$ directed along the z-axis and having magnitude ω is called the **angular velocity** of P. As in Example 2, the motion of P is given by

$$\mathbf{r}(t) = a \cos \omega t\mathbf{i} + a \sin \omega t\mathbf{j} + h\mathbf{k},$$

where h is the distance from the xy-plane to the particle. A direct computation shows that

$$\boldsymbol{\omega} \times \mathbf{r}(t) = -\omega a \sin \omega t\mathbf{i} + \omega a \cos \omega t\mathbf{j} = \mathbf{r}'(t).$$

Thus the velocity vector is the cross product of the angular velocity and the position vector of P. This fact can be generalized to the rotation of a particle P about any line.

As a final illustration of how vectors may be applied to problems involving motion, suppose a projectile is fired from a cannon, with an angle of elevation α. If the initial speed is v_0, then we may introduce a coordinate plane as in Figure 15.13, where the muzzle of the cannon is at the origin O and the initial velocity $\mathbf{v}_0$ has direction α and magnitude v_0. Let $P(x, y)$ denote the position of the projectile after time t and let $\mathbf{r}(t)$ correspond to $\overrightarrow{OP}$. Our objective is to find an explicit form for $\mathbf{r}(t)$.

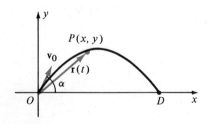

Figure 15.13

We shall assume that air resistance is negligible and that the only force acting on the projectile is the force **g** of gravitational acceleration. Since **g** acts in the downward direction we may write $\mathbf{g} = -g\mathbf{j}$, where $|\mathbf{g}| = g \approx 32$ ft/sec^2. According to Newton's Second Law of Motion, the force **F** acting on an object of mass m is related to its acceleration **a** by the formula $\mathbf{F} = m\mathbf{a}$. In the present situation, $m\mathbf{a} = m\mathbf{g}$, or $\mathbf{a} = \mathbf{g}$, which leads to the vector differential equation

$$\mathbf{r}''(t) = \mathbf{g}.$$

Indefinite integration gives us

$$\mathbf{r}'(t) = t\mathbf{g} + \mathbf{c}$$

where **c** is some vector. Since $\mathbf{r}'(t)$ is the velocity at time t,

$$\mathbf{v}_0 = \mathbf{r}'(0) = \mathbf{c}$$

and hence

$$\mathbf{r}'(t) = t\mathbf{g} + \mathbf{v}_0.$$

Integrating both sides of the last equation we obtain

$$\mathbf{r}(t) = \tfrac{1}{2}t^2\mathbf{g} + t\mathbf{v}_0 + \mathbf{d}$$

for some vector **d**. Since $\mathbf{r}(0) = \mathbf{0}$, it follows that $\mathbf{d} = \mathbf{0}$. Consequently,

$$\mathbf{r}(t) = \tfrac{1}{2}t^2\mathbf{g} + t\mathbf{v}_0$$

which may also be written

$$x\mathbf{i} + y\mathbf{j} = -\tfrac{1}{2}t^2 g\mathbf{j} + t(v_0 \cos \alpha\mathbf{i} + v_0 \sin \alpha\mathbf{j}).$$

Equating components we see that

(∗) $$x = (v_0 \cos \alpha)t, \quad y = -\tfrac{1}{2}gt^2 + (v_0 \sin \alpha)t.$$

These are parametric equations for the path of the projectile. Eliminating the parameter gives us the rectangular equation

$$y = \frac{-g}{2v_0^2 \cos^2 \alpha}x^2 + (\tan \alpha)x.$$

This shows that the path of the projectile is parabolic. To find the range, that is, the distance $d(O, D)$ from O to D in Figure 15.13, we let $y = 0$ in (∗), obtaining

$$t(-\tfrac{1}{2}gt + v_0 \sin \alpha) = 0.$$

Thus the point D corresponds to $t = (2v_0 \sin \alpha)/g$. Using the first equation in (∗) gives us

$$d(O, D) = (v_0 \cos \alpha)\left(\frac{2v_0 \sin \alpha}{g}\right) = \frac{v_0^2 \sin 2\alpha}{g}.$$

In particular, the range will have its maximum value v_0^2/g if $\sin 2\alpha = 1$, or $\alpha = 45°$. The maximum altitude h occurs when $t = (v_0 \sin \alpha)/g$. (Why?)

Substituting in the second equation of (∗), we obtain

$$h = -\frac{1}{2}g\left(\frac{v_0 \sin \alpha}{g}\right)^2 + (v_0 \sin \alpha)\left(\frac{v_0 \sin \alpha}{g}\right) = \frac{v_0^2 \sin^2 \alpha}{2g}.$$

EXERCISES 15.3

In Exercises 1–8 the position of a particle moving in a plane is given by $\mathbf{r}(t)$. Find its velocity, acceleration, and speed at time t. Sketch the path of the particle together with vectors corresponding to the velocity and acceleration at the indicated time t.

1    $\mathbf{r}(t) = 2t\mathbf{i} + (4t^2 + 1)\mathbf{j}, t = 1$

2    $\mathbf{r}(t) = (4 - 9t^2)\mathbf{i} + 3t\mathbf{j}, t = 1$

3    $\mathbf{r}(t) = (2/t)\mathbf{i} + [3/(t + 1)]\mathbf{j}, t = 2$

4    $\mathbf{r}(t) = \sqrt{t}\mathbf{i} + (1 + \sqrt{t})\mathbf{j}, t = 4$

5    $\mathbf{r}(t) = \sin t\mathbf{i} + 4 \cos 2t\mathbf{j}, t = \pi/6$

6    $\mathbf{r}(t) = \cos^2 t\mathbf{i} + 2 \sin t\mathbf{j}, t = 3\pi/4$

7    $\mathbf{r}(t) = e^{2t}\mathbf{i} + e^{-t}\mathbf{j}, t = 0$

8    $\mathbf{r}(t) = 2t\mathbf{i} + e^{-t^2}\mathbf{j}, t = 1$

In Exercises 9–16 $\mathbf{r}(t)$ is the position vector of a particle moving in space. Find the velocity, acceleration, and speed at time t. Sketch the path of the particle together with vectors corresponding to $\mathbf{r}'(t)$ and $\mathbf{r}''(t)$ for the indicated values of t.

9    $\mathbf{r}(t) = \cos t\mathbf{i} + \sin t\mathbf{j} + t\mathbf{k}; t = 0, \pi/4, \pi/2$

10   $\mathbf{r}(t) = t^2\mathbf{i} + t^3\mathbf{j} + t\mathbf{k}; t = 0, 1, 2$

11   $\mathbf{r}(t) = t^2\mathbf{i} + 2\sqrt{t}\mathbf{j} + 4\sqrt{t^3}\mathbf{k}; t = 1, 4, 9$

12   $\mathbf{r}(t) = 4 \sin t\mathbf{i} + 2t\mathbf{j} + 9 \cos t\mathbf{k}; t = 0, \pi/2, 3\pi/4$

13   $\mathbf{r}(t) = e^t(\cos t\mathbf{i} + \sin t\mathbf{j} + \mathbf{k}), t = 0, \pi/4, \pi/2$

14   $\mathbf{r}(t) = t(\cos t\mathbf{i} + \sin t\mathbf{j} + t\mathbf{k}), t = 0, \pi/4, \pi/2$

15   $\mathbf{r}(t) = (1 + t)\mathbf{i} + 2t\mathbf{j} + (2 + 3t)\mathbf{k}, t = 0, 1, 2$

16   $\mathbf{r}(t) = 2t\mathbf{i} + \mathbf{j} + 9t^2\mathbf{k}, t = 0, 1, 2$

17   Prove that if a particle moves at a constant speed, then the velocity and acceleration vectors are orthogonal.

18   Prove that if the acceleration of a moving particle is always $\mathbf{0}$, then the motion is rectilinear.

19   A projectile is fired with an initial speed of 1500 ft/sec and angle of elevation 30°. Find (a) the velocity at time t; (b) the maximum altitude; (c) the range; and (d) the speed at which the projectile strikes level ground.

20   Work Exercise 19 if the angle of elevation is 60°.

21   An outfielder on a baseball team threw a ball a distance of 250 ft. If he released the ball at an angle of 45° to the ground, what was the initial speed?

22   A projectile is fired horizontally with a velocity of 1800 ft/sec from an altitude of 1000 ft above level ground. When and where will it strike the ground?

Use the results of Example 2 to solve Exercises 23 and 24. Take the radius of the earth as 4,000 miles.

23   If a space shuttle is in a circular orbit 150 mi above the surface of the earth, approximate (a) its speed; (b) the length of time for one complete revolution around the earth.

24   A satellite that is in a circular orbit makes one revolution around the earth every 88 minutes. Approximate its altitude.

15.4

CURVATURE

For many applications involving vector-valued functions it is convenient to employ unit tangent vectors to curves. Let us begin by considering a vector-valued function $\mathbf{r}$ in two dimensions such that $\mathbf{r}(t) = f(t)\mathbf{i} + g(t)\mathbf{j}$. Let C be the curve determined by the endpoint of the position vector corresponding to $\mathbf{r}(t)$. From our work in Section 15.3 we know that $\mathbf{r}'(t)$ is a tangent vector to C. If $\mathbf{r}'(t) \neq \mathbf{0}$, then by Theorem (14.12), a **unit tangent vector** $\mathbf{T}(t)$ to C is given by the following formula.

(15.9)
$$\mathbf{T}(t) = \frac{1}{|\mathbf{r}'(t)|}\,\mathbf{r}'(t)$$

Since $|\mathbf{T}(t)| = 1$, it follows from Example 4 of Section 15.2 that $\mathbf{T}'(t)$ is orthogonal to $\mathbf{T}(t)$ for every t. Letting $\mathbf{N}(t) = (1/|\mathbf{T}'(t)|)\mathbf{T}'(t)$ gives us a *unit vector orthogonal to* $\mathbf{T}(t)$. We shall refer to $\mathbf{N}(t)$ as a **unit normal vector** to C. Let us state this for reference as follows.

(15.10)
$$\mathbf{N}(t) = \frac{1}{|\mathbf{T}'(t)|}\,\mathbf{T}'(t)$$

The next example illustrates these concepts.

Example 1 Let C be the curve determined by $\mathbf{r}(t) = t^2\mathbf{i} + t\mathbf{j}$.
(a) Find the unit tangent and normal vectors $\mathbf{T}(t)$ and $\mathbf{N}(t)$.
(b) Sketch C, together with $\mathbf{T}(1)$ and $\mathbf{N}(1)$.

Solution Using (15.9) with $\mathbf{r}'(t) = 2t\mathbf{i} + \mathbf{j}$, we obtain

$$\mathbf{T}(t) = \frac{1}{(4t^2 + 1)^{1/2}}\,(2t\mathbf{i} + \mathbf{j}) = \frac{2t}{(4t^2 + 1)^{1/2}}\,\mathbf{i} + \frac{1}{(4t^2 + 1)^{1/2}}\,\mathbf{j}.$$

Differentiating the components of $\mathbf{T}(t)$ gives us

$$\mathbf{T}'(t) = \frac{2}{(4t^2 + 1)^{3/2}}\,\mathbf{i} - \frac{4t}{(4t^2 + 1)^{3/2}}\,\mathbf{j} = \frac{2}{(4t^2 + 1)^{3/2}}\,(\mathbf{i} - 2t\mathbf{j}).$$

It is easy to verify that $|\mathbf{T}'(t)| = 2/(4t^2 + 1)$. Applying (15.10) and simplifying leads to

$$\mathbf{N}(t) = \frac{1}{(4t^2 + 1)^{1/2}}\,(\mathbf{i} - 2t\mathbf{j}).$$

Note that

$$\mathbf{T}(t) \cdot \mathbf{N}(t) = \frac{1}{4t^2 + 1}\,(2t - 2t) = 0$$

and hence $\mathbf{T}(t)$ and $\mathbf{N}(t)$ are, indeed, orthogonal.
At $t = 1$ we have

$$\mathbf{T}(1) = \frac{1}{\sqrt{5}}\,(2\mathbf{i} + \mathbf{j}) \quad \text{and} \quad \mathbf{N}(1) = \frac{1}{\sqrt{5}}\,(\mathbf{i} - 2\mathbf{j}).$$

These vectors and the curve (a parabola) are sketched in Figure 15.14. ∎

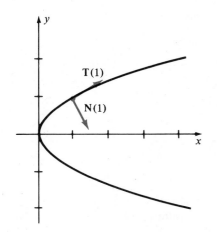

Figure 15.14

Note that the unit normal vector $\mathbf{N}(t)$ in Example 1 points toward the concave side of the curve. It can be shown that this is always the case.

A plane curve may be represented parametrically in many different ways. Sometimes it is convenient to use arc length as a parameter. Thus, suppose a curve C in the xy-plane is given by

$$x = f(s), \quad y = g(s)$$

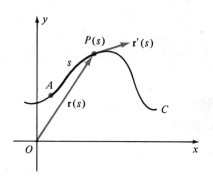

Figure 15.15

where f' and g' are continuous on an interval I and where the parameter s is arc length measured along C from some fixed point A. Geometrically we have a situation similar to that illustrated in Figure 15.15, where for each s in I there corresponds a unique point $P(s) = (f(s), g(s))$ that is s units from A (measured along C). The positive direction along C is determined by increasing values of s. We shall call s the **arc length parameter** for the curve C. The position of the fixed point A is irrelevant.

If, as illustrated in Figure 15.15, $\mathbf{r}(s) = f(s)\mathbf{i} + g(s)\mathbf{j}$ is the position vector of $P(s)$, then we know that $\mathbf{r}'(s)$ is a tangent vector to C. However, if s is the arc length parameter for C, it can be shown that $\mathbf{r}'(s)$ is a *unit* tangent vector. To establish this fact first note that by Theorem (13.11)

$$s = \int_0^S \sqrt{[f'(t)]^2 + [g'(t)]^2} \, dt.$$

Differentiating with respect to s gives us

$$1 = \sqrt{[f'(s)]^2 + [g'(s)]^2} = |\mathbf{r}'(s)|.$$

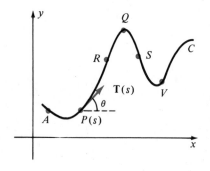

Figure 15.16

We shall denote the unit tangent vector $\mathbf{r}'(s)$ by $\mathbf{T}(s)$ and study its variation as $P(s)$ moves along C. For each s let θ be the angle between $\mathbf{T}(s)$ and $\mathbf{i}$, as illustrated in Figure 15.16. Observe that θ is a function of s, because for each s there corresponds a point $P(s)$ on C that, in turn, determines a value of θ. The rate of change $d\theta/ds$ of θ with respect to s is the basis for the next definition.

Definition (15.11)

> Let a smooth curve C be given by $x = f(s)$, $y = g(s)$, where s is the arc length parameter. The **curvature** K of C at the point $P(s) = P(x, y)$ is
>
> $$K = \left| \frac{d\theta}{ds} \right|.$$

Thus, curvature is the absolute value of the rate at which θ changes with respect to arc length s. The curvature is small for points such as R and S in Figure 15.16, since θ changes slowly as $P(s)$ moves along C. The curvature is large for the points Q and V, since θ changes rapidly near those points. Thus, roughly speaking, K provides information about the sharpness of the curve at various points.

The next two examples give important illustrations of Definition 15.11.

Example 2 Prove that the curvature of a line is 0 at every point on the line.

Solution   If C is a line, then the angle θ is the same for every point $P(s)$ on the line, that is, θ is constant. Hence

$$K = |d\theta/ds| = |0| = 0.$$ ∎

Example 3 Prove that the curvature at any point on a circle of radius a is $1/a$.

Solution   There is no loss of generality in assuming, as in Figure 15.17, that the circle has its center at O, and that the point P is in the first quadrant. If we let $A(a, 0)$ be fixed, and let s be the length of arc $\overset{\frown}{AP}$, then

$$s = a\alpha \quad \text{or} \quad \alpha = s/a,$$

where α is angle POA (see Appendix III). Referring to Figure 15.17 we see that

$$\theta = \alpha + (\pi/2) = (s/a) + (\pi/2)$$

and, therefore, $$\dfrac{d\theta}{ds} = \dfrac{1}{a} \quad \text{and} \quad K = \dfrac{1}{a}.$$ ∎

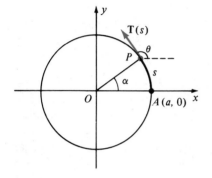

Figure 15.17

Note, from Example 3, that as the radius a of a circle increases, its curvature $K = 1/a$ decreases. Indeed, if we let a increase without bound, K approaches 0, the curvature of a line. As the radius a approaches 0, the curvature K increases without bound.

Suppose a curve C is the graph of a rectangular equation $y = h(x)$, where h' is continuous on some interval. Since y' is the slope of the tangent line at P we see, as in Figure 15.16, that

(a) $$\tan \theta = y' \quad \text{or} \quad \theta = \tan^{-1} y'.$$

From Definition (6.17), the arc length function s may be defined by

(b) $$s(x) = \int_a^x \sqrt{1 + (y')^2}\, dx$$

where a is the x-coordinate of the fixed point A on C. If y'' exists, then by the Chain Rule

$$\frac{d\theta}{dx} = \frac{d\theta}{ds}\frac{ds}{dx}.$$

Consequently,

(c) $$K = \left|\frac{d\theta}{ds}\right| = \left|\frac{d\theta/dx}{ds/dx}\right|.$$

Referring to (a) and (b),

$$\frac{d\theta}{dx} = \frac{y''}{1 + (y')^2} \quad \text{and} \quad \frac{ds}{dx} = \sqrt{1 + (y')^2}.$$

Substitution in (c) gives us the following theorem.

Theorem (15.12)

If C is the graph of $y = f(x)$, then the curvature K at $P(x, y)$ is

$$K = \frac{|y''|}{[1 + (y')^2]^{3/2}}.$$

Example 4 Sketch the graph of $y = 1 - x^2$ and find the curvature at the points (x, y), $(0, 1)$, $(1, 0)$, and $(2, -3)$.

Solution The graph (a parabola) is sketched in Figure 15.18. Since $y' = -2x$ and $y'' = -2$ we have, from Theorem (15.12),

$$K = \frac{2}{(1 + 4x^2)^{3/2}}.$$

In particular, at $(0, 1)$ we see that $K = 2$; that is, the direction of the tangent vector changes at the rate of 2 radians per unit change in arc length. At the point $(1, 0)$, $K = 2/(5)^{3/2} \approx 0.18$, which shows that the curve is less sharp here than at $(0, 1)$. This can also be seen from the graph. Finally, at $(2, -3)$, $K = 2/(17)^{3/2} \approx 0.03$. Observe that as x increases without bound the curvature approaches 0. ∎

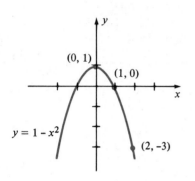

(0, 1)
(1, 0)
$y = 1 - x^2$
(2, -3)

Figure 15.18

In the next example we shall employ Theorem (15.12) to work Example 3.

Example 5 Prove that the curvature at any point on a circle of radius a is $1/a$.

Solution Let us assume, as in Figure 15.17, that the circle has center at the origin and equation $x^2 + y^2 = a^2$. Implicit differentiation yields

$$2x + 2yy' = 0, \quad \text{or} \quad y' = -x/y.$$

Consequently,

$$y'' = -\frac{y - xy'}{y^2} = -\frac{y + (x^2/y)}{y^2}$$

$$= -\frac{y^2 + x^2}{y^3} = -\frac{a^2}{y^3}.$$

Substituting in (15.12), and using the fact that $x^2 + y^2 = a^2$,

$$K = \frac{|a^2/y^3|}{[1 + x^2/y^2]^{3/2}} = \frac{|a^2/y^3|}{[(y^2 + x^2)/y^2]^{3/2}} = \frac{|a^2/y^3|}{(a^2/y^2)^{3/2}} = \frac{1}{a}. \quad ∎$$

Let us next derive a formula for finding K if C is described in terms of any parameter t. Suppose C is given parametrically by

$$x = f(t), \quad y = g(t)$$

and that f'' and g'' exist for all t under consideration. As in the proof of Theorem (15.12)

$$K = \left| \frac{d\theta}{ds} \right| = \left| \frac{d\theta/dt}{ds/dt} \right|.$$

Since the slope of the tangent line is given by $g'(t)/f'(t)$, we have

$$\tan \theta = \frac{g'(t)}{f'(t)}, \quad \text{or} \quad \theta = \tan^{-1} \frac{g'(t)}{f'(t)}$$

provided $f'(t) \neq 0$. Hence

$$\frac{d\theta}{dt} = \frac{1}{1 + [g'(t)/f'(t)]^2} \frac{f'(t)g''(t) - g'(t)f''(t)}{[f'(t)]^2} = \frac{f'(t)g''(t) - g'(t)f''(t)}{[f'(t)]^2 + [g'(t)]^2}.$$

Moreover,

$$\left| \frac{ds}{dt} \right| = \sqrt{[f'(t)]^2 + [g'(t)]^2}.$$

Substituting in the formula $K = |(d\theta/dt)/(ds/dt)|$ gives us the following.

Theorem (15.13)

> If a curve C is given parametrically by $x = f(t)$, $y = g(t)$, where f'' and g'' exist, then the curvature K at $P(x, y)$ is
>
> $$K = \frac{|f'(t)g''(t) - g'(t)f''(t)|}{[(f'(t))^2 + (g'(t))^2]^{3/2}}.$$

If the curvature at a point P on a curve C is not zero, then the circle of radius $\rho = 1/K$ whose center lies on the concave side of C and which has the same tangent line at P as C is called the **circle of curvature** for P. Its radius ρ and center are called the **radius of curvature** and **center of curvature**, respectively, for P. According to Examples 3 and 5 the curvature of the circle of curvature is $1/\rho$, or K, and hence is the same as the curvature of C. For this reason the circle of curvature may be thought of as the circle that best coincides with C at P.

Example 6 Let C be the curve with parametric equations $x = t^2$, $y = t^3$. Find the curvature at the point P corresponding to $t = \frac{1}{2}$. Sketch the graph of C and the circle of curvature for P.

Solution Letting $f(t) = t^2$ and $g(t) = t^3$, we obtain $f'(t) = 2t$, $f''(t) = 2$, $g'(t) = 3t^2$, and $g''(t) = 6t$. Substituting in Theorem (15.13),

$$K = \frac{|(2t)(6t) - (3t^2)(2)|}{[(2t)^2 + (3t^2)^2]^{3/2}} = \frac{6t^2}{(4t^2 + 9t^4)^{3/2}}.$$

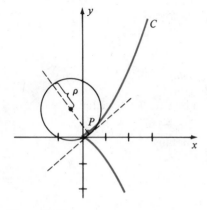

Figure 15.19

If $t = \frac{1}{2}$, then $K = (\frac{6}{4})/(\frac{25}{16})^{3/2} = \frac{96}{125} \approx 0.77$. The point corresponding to $t = \frac{1}{2}$ has coordinates $(\frac{1}{4}, \frac{1}{8})$ and the radius of curvature ρ at that point is $\frac{1}{K}$ or $\frac{125}{96} \approx 1.3$. The graph of C and the circle of curvature are sketched in Figure 15.19. ∎

The definition of curvature $K = |d\theta/ds|$ introduced in (15.11) has no immediate analogue in three dimensions, because the unit tangent vector $\mathbf{T}(s)$ cannot be specified in terms of a single angle θ. Thus it is necessary to use a different approach for space curves. Of course, we then must show that if we specialize to vectors in a plane, the new definition of curvature agrees with (15.11).

To find a clue to a suitable definition, we first observe that in two dimensions,

$$\mathbf{T}(s) = \cos\theta\mathbf{i} + \sin\theta\mathbf{j}$$

where θ is the angle considered in Theorem (15.11). Regarding θ as a function of s and differentiating gives us

$$\mathbf{T}'(s) = \left(-\sin\theta\frac{d\theta}{ds}\right)\mathbf{i} + \left(\cos\theta\frac{d\theta}{ds}\right)\mathbf{j} = \frac{d\theta}{ds}(-\sin\theta\mathbf{i} + \cos\theta\mathbf{j}).$$

Hence $$|\mathbf{T}'(s)| = \left|\frac{d\theta}{ds}\right||-\sin\theta\mathbf{i} + \cos\theta\mathbf{j}| = \left|\frac{d\theta}{ds}\right| = K.$$

We shall use this fact to introduce the concept of curvature in three dimensions. Indeed, our plan is to describe the vector $\mathbf{T}(s)$ *without* referring to the angle θ, and then *define* K as $|\mathbf{T}'(s)|$. Let us proceed as follows.

Suppose a curve C is given by

$$x = f(s), \quad y = g(s), \quad z = h(s)$$

where s is arc length measured along C from a fixed point A to the point $P(s) = (f(s), g(s), h(s))$ and where $f''(s)$, $g''(s)$, and $h''(s)$ exist. As before, if $\mathbf{r}(s) = f(s)\mathbf{i} + g(s)\mathbf{j} + h(s)\mathbf{k}$ is the position vector of $P(s)$, then $\mathbf{r}'(s)$ is a unit tangent vector to C at $P(s)$ which we shall denote by $\mathbf{T}(s)$. Since $|\mathbf{T}(s)|$ is a constant, it follows that $\mathbf{T}'(s)$ is orthogonal to $\mathbf{T}(s)$ (see Example 4 of Section 15.2). If $\mathbf{T}'(s) \neq \mathbf{0}$, let

$$\mathbf{N}(s) = \frac{1}{|\mathbf{T}'(s)|}\mathbf{T}'(s).$$

The vector $\mathbf{N}(s)$ is a unit vector orthogonal to $\mathbf{T}(s)$ and is referred to as the **principal unit normal vector** to C at the point $P(s)$. These concepts are illustrated in Figure 15.20.

Having suitably arrived at the unit tangent vector $\mathbf{T}(s)$, we now follow our plan and define curvature in three dimensions as follows, where it is assumed that $\mathbf{T}'(s)$ exists.

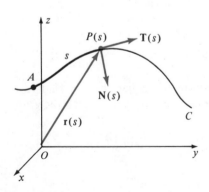

Figure 15.20

Definition (15.14)

> Let a curve C be given by
>
> $$x = f(s), \quad y = g(s), \quad z = h(s)$$
>
> where s is the arc length parameter. Let $\mathbf{r}(s) = f(s)\mathbf{i} + g(s)\mathbf{j} + h(s)\mathbf{k}$ and let $\mathbf{T}(s) = \mathbf{r}'(s)$. The **curvature** K of C at the point $P(x, y, z)$ is
>
> $$K = |\mathbf{T}'(s)|.$$

As we have already proved, the preceding definition reduces to (15.11) if C is a plane curve. Note that

$$\mathbf{N}(s) = \frac{1}{K}\mathbf{T}'(s), \quad \text{or} \quad \mathbf{T}'(s) = K\mathbf{N}(s).$$

In the next section we shall derive a formula that can be used to find the curvature of a space curve.

EXERCISES 15.4

In Exercises 1–12 find the curvature of the curve at the point P.

1 $y = 2 - x^3$; $P(1, 1)$

2 $y = x^4$; $P(1, 1)$

3 $y = e^{x^2}$; $P(0, 1)$

4 $y = \ln(x - 1)$; $P(2, 0)$

5 $y = \cos 2x$; $P(0, 1)$

6 $y = \sec x$; $P(\pi/3, 2)$

7 $x = t - 1, y = \sqrt{t}$; $P(3, 2)$

8 $x = t + 1, y = t^2 + 4t + 3$; $P(1, 3)$

9 $x = t - t^2, y = 1 - t^3$; $P(0, 1)$

10 $x = t - \sin t, y = 1 - \cos t$; $P(\pi/2 - 1, 1)$

11 $x = 2 \sin t, y = 3 \cos t$; $P(1, 3\sqrt{3}/2)$

12 $x = \cos^3 t, y = \sin^3 t$; $P(\sqrt{2}/4, \sqrt{2}/4)$

For the curves specified in Exercises 13–16 (a) find the radius of curvature for P; (b) find the center of curvature for P; and (c) sketch the graph and the circle of curvature for P.

13 $y = \sin x$; $P(\pi/2, 1)$ **14** $y = \sec x$; $P(0, 1)$

15 $y = e^x$; $P(0, 1)$ **16** $xy = 1$; $P(1, 1)$

For the curves described in Exercises 17–22 find the points at which the curvature is a maximum.

17 $y = e^{-x}$ **18** $y = \cosh x$

19 $9x^2 + 4y^2 = 36$ **20** $9x^2 - 4y^2 = 36$

21 $y = \ln x$ **22** $y = \sin x$

23 Use the equation $y = mx + b$ to prove that the curvature at every point on a line is 0 (cf. Example 2).

24 Prove that the maximum curvature of a parabola occurs at the vertex.

25 Prove that for an ellipse, the maximum and minimum curvature occur at the ends of the major and minor axes, respectively.

26 Prove that for a hyperbola, the maximum curvature occurs at the ends of the transverse axis.

In Exercises 27–30 find the points on the indicated curve at which the curvature is 0.

27 $y = x^4 - 12x^2$ **28** $y = \tan x$

29 $y = \sinh x$ **30** $y = e^{-x^2}$

31 Suppose that a curve C is the graph of a polar equation $r = f(\theta)$. If $r' = dr/d\theta$ and $r'' = d^2r/d\theta^2$, show that the curvature K at $P(r, \theta)$ is

$$K = \frac{|2(r')^2 - rr'' + r^2|}{[(r')^2 + r^2]^{3/2}}.$$

(*Hint:* Use $x = r \cos \theta$ and $y = r \sin \theta$ to express C in parametric form.)

In Exercises 32–34 find the curvature of the polar curve at $P(r, \theta)$ by means of the formula in Exercise 31.

32 The cardioid $r = a(1 - \cos \theta)$, where $0 < \theta < 2\pi$

33 The four-leafed rose $r = \sin 2\theta$, where $0 < \theta < \pi/2$

34 The spiral $r = e^{a\theta}$

35 Let $P(x, y)$ be a point on the graph of $y = f(x)$ at which $K \neq 0$. If (h, k) is the center of curvature for P show that

$$h = x - \frac{y'[1 + (y')^2]}{y''}, \quad k = y + \frac{[1 + (y')^2]}{y''}.$$

36–40 Use the formulas from Exercise 35 to find the center of curvature for P in Exercises 1–5.

41 A highway that coincides with the x-axis has an exit ramp at the origin O that follows along the curve $y = -x^3/27$ from O to the point $P(3, -1)$ and then along the

arc of a circle as shown in the figure. Find the center of the circular arc that makes the curvature at $P(3, -1)$ continuous.

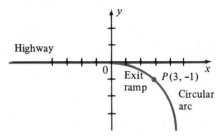

Highway

$P(3,-1)$

Exit ramp

Circular arc

Figure for Exercise 41

42 Prove that straight lines and circles are the only *plane* curves that have a constant curvature. (Exercise 25 of Section 15.5 shows that this result is not true for space curves.)

43 Prove that if k is a nonnegative function that is nonzero and continuous on an interval $[0, a]$, then there is a plane curve C such that k represents the curvature of C as a function of arc length. (*Hint:* If s is in $[0, a]$, define

$$h(s) = \int_0^s k(t)\, dt,$$

$$x = f(s) = \int_0^s \cos h(t)\, dt, \quad \text{and} \quad y = g(s) = \int_0^s \sin h(t)\, dt.)$$

44 Use Exercise 43 to rework Exercise 42.

15.5

TANGENTIAL AND NORMAL COMPONENTS OF ACCELERATION

The results of the previous section may be applied to the motion of a particle. Let us suppose that at time t the particle is at the point $P(x, y, z)$ on a curve C, and that C is given parametrically by

$$x = f(t), \quad y = g(t), \quad z = h(t)$$

where f'', g'', and h'' exist. As usual, let $\mathbf{r}(t)$ represent the position vector $\overline{OP}$ and let s denote arc length measured along C. We shall also assume that s increases as t increases. The unit tangent vector $\mathbf{T}(s)$ introduced in the previous section may then be expressed as

$$\mathbf{T}(s) = \frac{1}{|\mathbf{r}'(t)|} \mathbf{r}'(t)$$

and hence

$$\mathbf{r}'(t) = |\mathbf{r}'(t)| \mathbf{T}(s) = \frac{ds}{dt} \mathbf{T}(s).$$

Differentiating with respect to t and using Exercises 39 and 40 of Section 15.2 gives us

$$\mathbf{r}''(t) = \frac{d^2s}{dt^2}\,\mathbf{T}(s) + \frac{ds}{dt}\frac{d}{dt}\,\mathbf{T}(s)$$

$$= \frac{d^2s}{dt^2}\,\mathbf{T}(s) + \frac{ds}{dt}\frac{ds}{dt}\,\mathbf{T}'(s).$$

From the remark following Definition (15.14), we may write $\mathbf{T}'(s) = K\mathbf{N}(s)$ where K is the curvature of C and $\mathbf{N}(s)$ is the unit normal. Consequently,

$$\mathbf{r}''(t) = \frac{d^2s}{dt^2}\,\mathbf{T}(s) + \left(\frac{ds}{dt}\right)^2 K\mathbf{N}(s).$$

If we denote the speed ds/dt by v and write $K = 1/\rho$, where ρ is the radius of curvature of C, then the last formula for $\mathbf{r}''(t)$ may be written

$$\mathbf{r}''(t) = \frac{dv}{dt}\,\mathbf{T}(s) + \frac{v^2}{\rho}\,\mathbf{N}(s).$$

This formula expresses the acceleration $\mathbf{r}''(t)$ in terms of a **tangential component** dv/dt (the rate of change of speed with respect to time) and a **normal component** v^2/ρ. The sketch in Figure 15.21 illustrates one possibility that could occur.

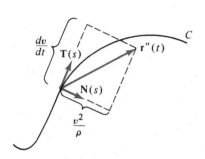

Figure 15.21

The next theorem summarizes the preceding discussion.

Theorem (15.15)

> Suppose the position of a point P on a curve C is given by
>
> $$\mathbf{r}(t) = f(t)\mathbf{i} + g(t)\mathbf{j} + h(t)\mathbf{k}$$
>
> where t represents time. If the speed of P is $v = ds/dt$, then the acceleration of P is
>
> $$\mathbf{r}''(t) = \frac{dv}{dt}\,\mathbf{T}(s) + \frac{v^2}{\rho}\,\mathbf{N}(s)$$
>
> where ρ is the radius of curvature of C.

Note that the normal component v^2/ρ depends only on the speed and the curvature (or radius of curvature) of the curve C. If the speed or curvature is large (or the radius of curvature is small), then the normal component of acceleration is large. This result, obtained theoretically, proves the well-known fact that an automobile driver should slow down when attempting to negotiate a sharp turn.

Let us next find formulas for the tangential and normal components of acceleration that depend only on $\mathbf{r}(t)$. First, recall from the discussion at beginning of this section, that

$$\mathbf{r}'(t) = v\mathbf{T}(s).$$

Taking the dot product with $\mathbf{r}''(t)$ as given in Theorem (15.15) we obtain

$$\mathbf{r}'(t) \cdot \mathbf{r}''(t) = \frac{dv}{dt} v\mathbf{T}(s) \cdot \mathbf{T}(s) + \frac{v^3}{\rho} \mathbf{T}(s) \cdot \mathbf{N}(s).$$

Since $\mathbf{T}(s)$ has magnitude 1 and is orthogonal to $\mathbf{N}(s)$, the last equality reduces to

$$\mathbf{r}'(t) \cdot \mathbf{r}''(t) = \frac{dv}{dt} v = \frac{dv}{dt} |\mathbf{r}'(t)|.$$

If $|\mathbf{r}'(t)| \neq 0$, we obtain the following formula for the **tangential component of acceleration**:

(15.16)
$$\frac{dv}{dt} = \frac{\mathbf{r}'(t) \cdot \mathbf{r}''(t)}{|\mathbf{r}'(t)|}$$

In like manner, if we take the cross product of $\mathbf{r}'(t)$ with $\mathbf{r}''(t)$, then

$$\mathbf{r}'(t) \times \mathbf{r}''(t) = v\frac{dv}{dt}\mathbf{T}(s) \times \mathbf{T}(s) + \frac{v^3}{\rho}\mathbf{T}(s) \times \mathbf{N}(s)$$

$$= \frac{v^3}{\rho}\mathbf{T}(s) \times \mathbf{N}(s).$$

Since $\mathbf{T}(s)$ and $\mathbf{N}(s)$ are orthogonal unit vectors, $|\mathbf{T}(s) \times \mathbf{N}(s)| = 1$ and therefore

$$|\mathbf{r}'(t) \times \mathbf{r}''(t)| = \frac{v^3}{\rho}.$$

This gives us the formula for the **normal component of acceleration**:

(15.17)
$$\frac{v^2}{\rho} = \frac{|\mathbf{r}'(t) \times \mathbf{r}''(t)|}{|\mathbf{r}'(t)|}.$$

There is an alternative way of finding the normal component of acceleration. To simplify the notation, denote $\mathbf{T}(s)$, $\mathbf{N}(s)$, and $\mathbf{r}''(t)$ in Theorem (15.15) by $\mathbf{T}$, $\mathbf{N}$, and $\mathbf{a}$, respectively. If the tangential and normal components of $\mathbf{a}$ are denoted by $a_{\mathbf{T}}$ and $a_{\mathbf{N}}$, then we may write

$$\mathbf{a} = a_{\mathbf{T}}\mathbf{T} + a_{\mathbf{N}}\mathbf{N}.$$

Since $\mathbf{T}$ and $\mathbf{N}$ are mutually orthogonal unit vectors, $\mathbf{T} \cdot \mathbf{N} = 0$, $\mathbf{T} \cdot \mathbf{T} = 1$, and $\mathbf{N} \cdot \mathbf{N} = 1$. Consequently,

$$|\mathbf{a}|^2 = \mathbf{a} \cdot \mathbf{a} = (a_{\mathbf{T}}\mathbf{T} + a_{\mathbf{N}}\mathbf{N}) \cdot (a_{\mathbf{T}}\mathbf{T} + a_{\mathbf{N}}\mathbf{N}) = a_{\mathbf{T}}^2 + a_{\mathbf{N}}^2.$$

Thus,

(15.18)
$$a_{\mathbf{N}} = \sqrt{|\mathbf{a}|^2 - a_{\mathbf{T}}^2}.$$

Formula (15.18) should be used if it is difficult to find the normal component by means of (15.17).

Example 1 The position of a particle at time t is (t, t^2, t^3). Find the tangential and normal components of acceleration at time t.

Solution We may write

$$\mathbf{r}(t) = t\mathbf{i} + t^2\mathbf{j} + t^3\mathbf{k}$$
$$\mathbf{r}'(t) = \mathbf{i} + 2t\mathbf{j} + 3t^2\mathbf{k}$$
$$\mathbf{r}''(t) = 2\mathbf{j} + 6t\mathbf{k}$$
$$|\mathbf{r}'(t)| = (1 + 4t^2 + 9t^4)^{1/2}.$$

Using (15.16), the tangential component of acceleration is

$$a_{\mathbf{T}} = \frac{dv}{dt} = \frac{\mathbf{r}'(t) \cdot \mathbf{r}''(t)}{|\mathbf{r}'(t)|} = \frac{4t + 18t^3}{(1 + 4t^2 + 9t^4)^{1/2}}.$$

Let us find the normal component by means of (15.17). First,

$$\mathbf{r}'(t) \times \mathbf{r}''(t) = \begin{vmatrix} \mathbf{i} & \mathbf{j} & \mathbf{k} \\ 1 & 2t & 3t^2 \\ 0 & 2 & 6t \end{vmatrix} = 6t^2\mathbf{i} - 6t\mathbf{j} + 2\mathbf{k}.$$

Applying (15.17) gives us

$$a_{\mathbf{N}} = \frac{v^2}{\rho} = \frac{|\mathbf{r}'(t) \times \mathbf{r}''(t)|}{|\mathbf{r}'(t)|} = \frac{(36t^4 + 36t^2 + 4)^{1/2}}{(1 + 4t^2 + 9t^4)^{1/2}} = 2\left(\frac{9t^4 + 9t^2 + 1}{9t^4 + 4t^2 + 1}\right)^{1/2}$$

After determining $a_{\mathbf{T}}$ we could have used (15.18) to find the normal component as follows

$$a_{\mathbf{N}} = \sqrt{(4 + 36t^2) - \frac{(4t + 18t^3)^2}{1 + 4t^2 + 9t^4}}.$$

This simplifies to the preceding expression for $a_{\mathbf{N}}$. ∎

Example 2 If a point P moves around a circle of radius a with a constant speed v, find the tangential and normal components of acceleration.

Solution Since $v = ds/dt$ is a constant, the tangential component dv/dt is 0. By Example 3 of the previous section, the curvature of the circle is $1/a$. Hence the normal component of acceleration (see (15.17)) is v^2/a. This shows that the acceleration is a vector of constant magnitude v^2/a directed from P toward the center of the circle (see Example 2 in Section 15.3). ∎

We may use (15.17) to obtain a formula for curvature of a space curve C. Specifically, if C is given parametrically by

$$x = f(t), \quad y = g(t), \quad z = h(t)$$

let

$$\mathbf{r}(t) = f(t)\mathbf{i} + g(t)\mathbf{j} + h(t)\mathbf{k}.$$

We may then regard C as the curve traced by the endpoint of $\mathbf{r}(t)$ as t varies. Since $v = |\mathbf{r}'(t)|$ and $K = 1/\rho$ we have, from (15.17),

$$\frac{|\mathbf{r}'(t)|^2}{(1/K)} = \frac{|\mathbf{r}'(t) \times \mathbf{r}''(t)|}{|\mathbf{r}'(t)|}.$$

Solving for K gives us the following result.

Theorem (15.19)

> Let a curve C be given by $x = f(t)$, $y = g(t)$, $z = h(t)$, where f'', g'', and h'' exist. The curvature K at the point $P(x, y, z)$ on C is
>
> $$K = \frac{|\mathbf{r}'(t) \times \mathbf{r}''(t)|}{|\mathbf{r}'(t)|^3}.$$

This formula may also be used for plane curves (see Exercise 24).

Example 3 Find the curvature of the twisted cubic $x = t$, $y = t^2$, $z = t^3$ at the point (x, y, z).

Solution   If we let

$$\mathbf{r}(t) = t\mathbf{i} + t^2\mathbf{j} + t^3\mathbf{k}$$

then the curve is the same as that considered in Example 1. Substituting the expressions obtained there for $\mathbf{r}'(t)$ and $\mathbf{r}'(t) \times \mathbf{r}''(t)$ into Theorem (15.19), we obtain

$$K = \frac{2(9t^4 + 9t^2 + 1)^{1/2}}{(1 + 4t^2 + 9t^4)^{3/2}}. \qquad \blacksquare$$

EXERCISES 15.5

In Exercises 1–8, $\mathbf{r}(t)$ is the position vector of a particle at time t. Find the tangential and normal components of acceleration at time t.

1 $\mathbf{r}(t) = t^2\mathbf{i} + (3t + 2)\mathbf{j}$

2 $\mathbf{r}(t) = (2t^2 - 1)\mathbf{i} + 5t\mathbf{j}$

3 $\mathbf{r}(t) = 3t\mathbf{i} + t^3\mathbf{j} + 3t^2\mathbf{k}$

4 $\mathbf{r}(t) = 4t\mathbf{i} + t^2\mathbf{j} + 2t^2\mathbf{k}$

5 $\mathbf{r}(t) = t(\cos t\mathbf{i} + \sin t\mathbf{j})$

6 $\mathbf{r}(t) = \cosh t\mathbf{i} + \sinh t\mathbf{j}$

7 $\mathbf{r}(t) = 4\cos t\mathbf{i} + 9\sin t\mathbf{j} + t\mathbf{k}$

8 $\mathbf{r}(t) = e^t(\sin t\mathbf{i} + \cos t\mathbf{j} + \mathbf{k})$

9–16 Use (15.19) to find the curvature, at the point corresponding to t, of the curve traced by the endpoint of $\mathbf{r}(t)$ in each of Exercises 1–8.

17 A particle moves along the parabola $y = x^2$ such that the horizontal component of velocity is always 3. Find the tangential and normal components of acceleration at the point $P(1, 1)$. Sketch the path and illustrate the acceleration as a sum of two vectors as in Theorem (15.15).

18 Work Exercise 17 if the particle moves along the graph of $y = 2x^3 - x$.

19 Prove that if a particle moves along a curve C with a constant speed, then the acceleration is always normal to C.

20 Use Theorem (15.19) to prove that if a particle moves through space with an acceleration that is always **0**, then the motion is rectilinear.

21 If a particle P moves along a curve C with a constant speed, show that the magnitude of the acceleration is directly proportional to the curvature of the curve; that is, $|\mathbf{a}| = cK$ for some real number c.

22 If, in Exercise 21, a second particle Q moves along C with a speed twice that of P, show that the magnitude of the acceleration is four times greater than that of P.

23 Show that if a particle moves along the graph of $y = f(x)$, where $a \leq x \leq b$, then the normal component of accelera-

tion is 0 at a point of inflection. Illustrate this fact by using $\mathbf{r}(t) = t\mathbf{i} + t^3\mathbf{j}$.

24 If a plane curve is given parametrically by $x = f(t)$, $y = g(t)$, where f'' and g'' exist, use Theorem (15.19) to prove that the curvature at the point K is given by Theorem (15.13).

25 Show that the curvature at every point on the circular helix $x = a\cos t$, $y = a\sin t$, $z = bt$ where $a > 0$, is given by $K = a/(a^2 + b^2)$.

26 Find the curvature, at (x, y, z), of the elliptic helix that has parametric equations $x = a\cos t$, $y = b\sin t$, $z = ct$, where a, b, and c are positive.

15.6

KEPLER'S LAWS

It is fitting to conclude this chapter with a display of the power and beauty of vector methods when applied to the derivation of three classical physical laws. The discussion in this section is not simple, for it is not a simple problem that we intend to consider. However, the reader should gain considerable insight and be justly rewarded by studying the material that follows very carefully.

After many years of analyzing an enormous amount of empirical data, the German astronomer Johannes Kepler (1571–1630) formulated three laws that describe the motion of planets about the sun. These laws may be stated as follows.

Kepler's Laws (*15.20*)

First Law.   The orbit of each planet is an ellipse with the sun at one focus.

Second Law.   The vector from the sun to a moving planet sweeps out area at a constant rate.

Third Law.   If the time required for a planet to travel once around its elliptical orbit is T, and if the major axis of the ellipse is $2a$, then $T^2 = ka^3$ for some constant k.

Approximately 50 years later, Sir Isaac Newton (1642–1727) proved that Kepler's Laws were consequences of his Law of Universal Gravitation and Second Law of Motion. The achievements of both men were monumental, because these laws clarified all astronomical observations that had been made up to that time.

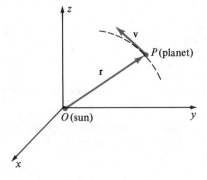

Figure 15.22

In this section, we shall prove Kepler's Laws by the use of vector techniques. Since the force of gravity that the sun exerts on a planet far exceeds that exerted by other celestial bodies, we shall neglect all other forces acting on a planet. From this point of view there are only two objects to consider: the sun and a planet revolving around it.

It is convenient to introduce a coordinate system with the center of mass of the sun at the origin, O, as illustrated in Figure 15.22. The point P represents the center of mass of the planet. To simplify the notation we shall denote the position vector of P by $\mathbf{r}$ instead of $\mathbf{r}(t)$, and use $\mathbf{v}$ and $\mathbf{a}$ to denote the velocity $\mathbf{r}'(t)$ and acceleration $\mathbf{r}''(t)$, respectively. Throughout our discussion we shall not distinguish between vectors represented by directed line segments and vectors represented by triples of real numbers.

Before proving Kepler's Laws, let us show that the motion of the planet takes place in one plane. If we let $r = |\mathbf{r}|$, then $\mathbf{u} = (1/r)\mathbf{r}$ is a unit vector having the same direction as $\mathbf{r}$. According to Newton's Law of Gravitation, the force $\mathbf{F}$ of gravitational attraction on the planet is given by

$$\mathbf{F} = -G\frac{Mm}{r^2}\mathbf{u}$$

where M is the mass of the sun, m is the mass of the planet, and G is the gravitational constant. Newton's Second Law of Motion states that

$$\mathbf{F} = m\mathbf{a}.$$

If we equate these two expressions for $\mathbf{F}$ and solve for $\mathbf{a}$ we obtain

(a) $$\mathbf{a} = -\frac{GM}{r^2}\mathbf{u}.$$

This shows that $\mathbf{a}$ is parallel to $\mathbf{r} = r\mathbf{u}$ and hence $\mathbf{r} \times \mathbf{a} = \mathbf{0}$. In addition, since $\mathbf{v} \times \mathbf{v} = \mathbf{0}$ we see that

$$\frac{d}{dt}(\mathbf{r} \times \mathbf{v}) = \mathbf{r} \times \frac{d\mathbf{v}}{dt} + \frac{d\mathbf{r}}{dt} \times \mathbf{v}$$

$$= \mathbf{r} \times \mathbf{a} + \mathbf{v} \times \mathbf{v} = \mathbf{0}.$$

It follows that

(b) $$\mathbf{r} \times \mathbf{v} = \mathbf{c}$$

where $\mathbf{c}$ is a constant vector. The vector $\mathbf{c}$ will play an important role in the proofs of Kepler's Laws.

Since $\mathbf{r} \times \mathbf{v} = \mathbf{c}$, the vector $\mathbf{r}$ is orthogonal to $\mathbf{c}$ for every value of t. This implies that the curve traced by P lies in one plane, that is, *the orbit of the planet is a plane curve.*

Let us now prove Kepler's First Law. There is no loss of generality in assuming that the motion of the planet takes place in the xy-plane. In this case the vector $\mathbf{c}$ is perpendicular to the xy-plane, and we may assume that $\mathbf{c}$ has the same direction as the positive z-axis, as illustrated in Figure 15.23.

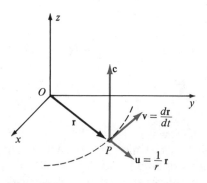

Figure 15.23

Since $\mathbf{r} = r\mathbf{u}$ we see that

$$\mathbf{v} = \frac{d\mathbf{r}}{dt} = r\frac{d\mathbf{u}}{dt} + \frac{dr}{dt}\mathbf{u}.$$

Substitution in $\mathbf{c} = \mathbf{r} \times \mathbf{v}$, and use of properties of vector products gives us

$$\mathbf{c} = r\mathbf{u} \times \left(r\frac{d\mathbf{u}}{dt} + \frac{dr}{dt}\mathbf{u}\right)$$

$$= r^2\left(\mathbf{u} \times \frac{d\mathbf{u}}{dt}\right) + r\frac{dr}{dt}(\mathbf{u} \times \mathbf{u}).$$

Since $\mathbf{u} \times \mathbf{u} = \mathbf{0}$, this reduces to

(c) $$\mathbf{c} = r^2\left(\mathbf{u} \times \frac{d\mathbf{u}}{dt}\right).$$

Using (c) and (a) together with (ii) and (vi) of Theorem 14.36, we see that

$$\mathbf{a} \times \mathbf{c} = \left(-\frac{GM}{r^2}\mathbf{u}\right) \times \left[r^2\left(\mathbf{u} \times \frac{d\mathbf{u}}{dt}\right)\right]$$

$$= -GM\left[\mathbf{u} \times \left(\mathbf{u} \times \frac{d\mathbf{u}}{dt}\right)\right]$$

$$= -GM\left[\left(\mathbf{u} \cdot \frac{d\mathbf{u}}{dt}\right)\mathbf{u} - (\mathbf{u} \cdot \mathbf{u})\frac{d\mathbf{u}}{dt}\right].$$

Since $|\mathbf{u}| = 1$ it follows from Example 4 of Section 15.2 that $\mathbf{u} \cdot (d\mathbf{u}/dt) = 0$. In addition, $\mathbf{u} \cdot \mathbf{u} = |\mathbf{u}|^2 = 1$, and hence the last formula for $\mathbf{a} \times \mathbf{c}$ reduces to

$$\mathbf{a} \times \mathbf{c} = GM\frac{d\mathbf{u}}{dt} = \frac{d}{dt}(GM\mathbf{u}).$$

We may also write

$$\mathbf{a} \times \mathbf{c} = \frac{d\mathbf{v}}{dt} \times \mathbf{c} = \frac{d}{dt}(\mathbf{v} \times \mathbf{c})$$

and, consequently,

$$\frac{d}{dt}(\mathbf{v} \times \mathbf{c}) = \frac{d}{dt}(GM\mathbf{u}).$$

Integrating both sides of this equation gives us

(d) $$\mathbf{v} \times \mathbf{c} = GM\mathbf{u} + \mathbf{b}$$

where $\mathbf{b}$ is a constant vector.

The vector $\mathbf{v} \times \mathbf{c}$ is orthogonal to $\mathbf{c}$ and, therefore, is in the xy-plane. Since $\mathbf{u}$ is also in the xy-plane it follows from (d) that $\mathbf{b}$ is in the xy-plane.

Up to this point, our proof has been independent of the positions of the x- and y-axes. Let us now choose a coordinate system such that the positive x-axis has the direction of the constant vector $\mathbf{b}$, as illustrated in Figure 15.24.

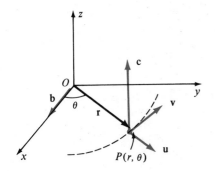

Figure 15.24

Let (r, θ) be polar coordinates for the point P, where $r = |\mathbf{r}|$. It follows that

$$\mathbf{u} \cdot \mathbf{b} = |\mathbf{u}|\,|\mathbf{b}| \cos \theta = b \cos \theta$$

where $b = |\mathbf{b}|$. If we let $c = |\mathbf{c}|$, then using (b) together with properties of the dot and vector products, and also (d),

$$c^2 = \mathbf{c} \cdot \mathbf{c} = (\mathbf{r} \times \mathbf{v}) \cdot \mathbf{c} = \mathbf{r} \cdot (\mathbf{v} \times \mathbf{c})$$

$$= (r\mathbf{u}) \cdot (GM\mathbf{u} + \mathbf{b})$$

$$= rGM(\mathbf{u} \cdot \mathbf{u}) + r(\mathbf{u} \cdot \mathbf{b})$$

$$= rGM + rb \cos \theta.$$

Solving the last equation for r gives us

$$r = \frac{c^2}{GM + b \cos \theta}.$$

Dividing numerator and denominator of this fraction by GM we obtain

(e) $$r = \frac{p}{1 + e \cos \theta}$$

where $p = c^2/GM$ and $e = b/GM$. From Theorem (13.8), the graph of this polar equation is a conic with eccentricity e and focus at the origin. Since the orbit is a closed curve, it follows that $0 < e < 1$ and that the conic is an ellipse. This completes the proof of Kepler's First Law.

Let us next prove Kepler's Second Law. It may be assumed that the orbit of the planet is an ellipse in the xy-plane. Let $r = f(\theta)$ be a polar equation of the orbit, with the sun centered at the focus O. Let P_0 denote the position of the planet at time t_0, and P its position at any time $t \geq t_0$. As illustrated in Figure 15.25, θ_0 and θ will denote the angles measured from the positive x-axis to $\overrightarrow{OP_0}$ and $\overrightarrow{OP}$, respectively.

By Theorem (13.9), the area A swept out by $\overrightarrow{OP}$ in the time interval $[t_0, t]$ is

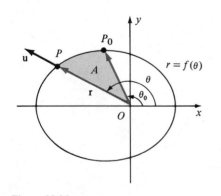

Figure 15.25

$$A = \int_{\theta_0}^{\theta} \frac{1}{2} r^2 \, d\theta$$

and hence

$$\frac{dA}{d\theta} = \frac{d}{d\theta} \int_{\theta_0}^{\theta} \frac{1}{2} r^2 \, d\theta = \frac{1}{2} r^2.$$

Using this fact and the Chain Rule gives us

(f) $$\frac{dA}{dt} = \frac{dA}{d\theta} \frac{d\theta}{dt} = \frac{1}{2} r^2 \frac{d\theta}{dt}.$$

Next we observe that since $\mathbf{r} = r \cos \theta \mathbf{i} + r \sin \theta \mathbf{j} + 0\mathbf{k}$, the unit vector $\mathbf{u} = (1/r)\mathbf{r}$ may be expressed in the form

$$\mathbf{u} = \cos \theta \mathbf{i} + \sin \theta \mathbf{j} + 0\mathbf{k}.$$

Consequently,

$$\frac{d\mathbf{u}}{dt} = -\sin \theta \frac{d\theta}{dt}\mathbf{i} + \cos \theta \frac{d\theta}{dt}\mathbf{j} + 0\mathbf{k}.$$

A direct calculation may be used to show that

$$\mathbf{u} \times \frac{d\mathbf{u}}{dt} = \frac{d\theta}{dt}\mathbf{k}.$$

If $\mathbf{c}$ is the vector obtained in the proof of Kepler's First Law, then by (c) and the last equation,

$$\mathbf{c} = r^2\left(\mathbf{u} \times \frac{d\mathbf{u}}{dt}\right) = r^2\frac{d\theta}{dt}\mathbf{k}$$

and hence

(g) $$c = |\mathbf{c}| = r^2\frac{d\theta}{dt}.$$

Combining (f) and (g) we see that

(h) $$\frac{dA}{dt} = \frac{1}{2}c$$

that is, the rate at which A is swept out by $\overrightarrow{OP}$ is a constant. This establishes Kepler's Second Law.

To prove Kepler's Third Law we shall retain the notation used in the proofs of the first two laws. In particular, we assume that a polar equation of the planetary orbit is given by

$$r = \frac{p}{1 + e \cos \theta}$$

where $p = c^2/GM$ and $e = b/GM$ (see (e)).

Let T denote the time required for the planet to make one complete revolution about the sun. By (h) the area swept out in the time interval $[0, T]$ is given by

$$A = \int_0^T \frac{dA}{dt} dt = \int_0^T \frac{1}{2}c \, dt = \frac{1}{2}cT.$$

This also equals the area of the plane region bounded by the ellipse. However, it was shown in Example 4 of Section 12.3 that $A = \pi ab$, where $2a$ and $2b$ are lengths of the major and minor axes, respectively, of the ellipse. Consequently,

$$\frac{1}{2}cT = \pi ab \quad \text{or} \quad T = \frac{2\pi ab}{c}.$$

From our work in Section 13.4, we know that

$$e = \frac{\sqrt{a^2 - b^2}}{a}$$

and hence

$$a^2 e^2 = a^2 - b^2, \quad \text{or} \quad b^2 = a^2(1 - e^2).$$

Thus
$$T^2 = \frac{4\pi^2 a^2 b^2}{c^2} = \frac{4\pi^2 a^4 (1 - e^2)}{c^2}.$$

It was shown in the proof of Theorem 13.7 (where we had $p = de$) that

$$a^2 = \frac{p^2}{(1 - e^2)^2}, \quad \text{or} \quad a = \frac{p}{1 - e^2}$$

and hence
$$T^2 = \frac{4\pi^2 a^4}{c^2} \left(\frac{p}{a}\right).$$

Since $p = c^2/GM$, this reduces to

$$T^2 = \frac{4\pi^2}{GM} a^3 = k a^3$$

where $k = 4\pi^2/GM$. This completes the proof.

In our proofs of Kepler's Laws, remember that we assumed that the only gravitational force acting on a planet was that of the sun. If forces exerted by other planets are taken into account, then irregularities in the elliptical orbits may occur. Indeed, it was because of observed irregularities in the motion of Uranus that the British astronomer J. Adams (1819–1892) and the French astronomer U. LeVerrier (1811–1877) were able to predict the orbit of an unknown planet that was causing the irregularities. Using their predictions, this planet, later named Neptune, was first observed with a telescope by the German astronomer J. Galle in 1846.

15.7

REVIEW

Define or discuss each of the following.

1 Vector-valued function

2 Limit of a vector-valued function

3 Continuity of a vector-valued function

4 Derivative of a vector-valued function

5 Differentiation formulas for vector-valued functions

6 Tangent vector to a curve

7 Integrals of vector-valued functions

8 Velocity and acceleration

9 Normal vector to a curve

10 Curvature of a plane curve

11 Circle of curvature

12 Radius of curvature

13 Curvature of a space curve

14 Tangential component of acceleration

15 Normal component of acceleration

16 Kepler's Laws

EXERCISES 15.7

1 If $\mathbf{r}(t) = t^2\mathbf{i} + (4t^2 - t^4)\mathbf{j}$,

(a) sketch the curve determined by the components of $\mathbf{r}(t)$.

(b) find $\mathbf{r}'(t)$ and $\mathbf{r}''(t)$.

(c) sketch vectors corresponding to $\mathbf{r}'(t)$ and $\mathbf{r}''(t)$ if $t = 0$, $t = 1$, and $t = 2$.

2 The position of a particle moving in a plane is given by

$$\mathbf{r}(t) = (t - \sin t)\mathbf{i} + (1 - \cos t)\mathbf{j}.$$

Find its velocity, acceleration, and speed at time t. Sketch the path of the particle together with vectors corresponding to the velocity and acceleration for the following values of t.

(a) $t = 0$ (f) $t = 5\pi/4$

(b) $t = \pi/4$ (g) $t = 3\pi/2$

(c) $t = \pi/2$ (h) $t = 7\pi/4$

(d) $t = 3\pi/4$ (i) $t = 2\pi$

(e) $t = \pi$

3 If the curve C is given by $x = e^t \sin t$, $y = e^t \cos t$, $z = e^t$, where $0 \le t \le 1$, find

(a) a unit tangent vector to C at the point P corresponding to $t = 0$.

(b) the length of C.

4 The position of a particle at time t is given by

$$\mathbf{r}(t) = 3t\mathbf{i} + t^3\mathbf{j} + t^4\mathbf{k}.$$

(a) Find the velocity and acceleration at time t.

(b) Sketch the path of the particle together with vectors corresponding to the velocity and acceleration at $t = 1$.

(c) Find the speed at $t = 1$.

5 If the curve C is given by $x = 3t^2 + 1$, $y = 4t$, $z = e^{t-1}$, find an equation of the tangent line at the point $P(4, 4, 1)$.

6 If $\mathbf{u}(t) = t^2\mathbf{i} + 6t\mathbf{j} + t\mathbf{k}$ and $\mathbf{v}(t) = t\mathbf{i} - 5t\mathbf{j} + 4t^2\mathbf{k}$ find the values of t for which $\mathbf{u}(t)$ and $\mathbf{v}'(t)$ are orthogonal.

7 If $\mathbf{u}(t)$ and $\mathbf{v}(t)$ are as in Exercise 6, find $D_t[\mathbf{u}(t) \times \mathbf{v}(t)]$ and $D_t[\mathbf{u}(t) \cdot \mathbf{v}(t)]$.

8 Find $\int (\sin 3t\mathbf{i} + e^{-2t}\mathbf{j} + \cos t\mathbf{k})\, dt$.

9 Evaluate $\int_0^1 (4t\mathbf{i} + t^3\mathbf{j} - \mathbf{k})\, dt$.

10 Find $\mathbf{u}(t)$ if $\mathbf{u}'(t) = e^{-t}\mathbf{i} - 4\sin 2t\mathbf{j} + 3\sqrt{t}\mathbf{k}$ and $\mathbf{u}(0) = -\mathbf{i} + 2\mathbf{j}$.

Verify the identities in Exercises 11 and 12 without using components.

11 $D_t|\mathbf{u}(t)|^2 = 2\mathbf{u}(t) \cdot \mathbf{u}'(t)$

12 $D_t(\mathbf{u}(t) \cdot \mathbf{u}(t) \times \mathbf{u}''(t)) = \mathbf{u}(t) \cdot \mathbf{u}'(t) \times \mathbf{u}'''(t)$

In Exercises 13–15 find the curvature of the given curve at the point P.

13 $y = xe^x$; $P(0, 0)$

14 $x = 1/(1 + t)$, $y = 1/(1 - t)$; $P(\tfrac{2}{3}, 2)$

15 $x = 2t^2$, $y = t^4$, $z = 4t$; $P(x, y, z)$

16 Find the x-coordinates of the points on the graph of $y = x^3 - 3x$ at which the curvature is a maximum.

17 If C is the graph of $y = \cosh x$,

    (a) find an equation of the circle of curvature for the point $P(0, 1)$.

    (b) sketch C and the circle of curvature for P.

18 Given the limaçon $r = 2 + \sin \theta$, find the radius of curvature for the point $P(\pi, 2)$.

In Exercises 19 and 20 find the tangential and normal components of acceleration at time t if the position vector of a particle is as indicated.

19 $\mathbf{r}(t) = \sin 2t\,\mathbf{i} + \cos t\,\mathbf{j}$ **20** $\mathbf{r}(t) = 3t\,\mathbf{i} + t^3\,\mathbf{j} + t\,\mathbf{k}$

16

PARTIAL DIFFERENTIATION

In this chapter the concept of derivative is generalized to functions of more than one variable. Applications include finding rates of change, extrema, and approximations by differentials.

16.1

FUNCTIONS OF SEVERAL VARIABLES

The functions considered in previous chapters involved only one independent variable. As we have seen, such functions can be used to solve a variety of problems; however, there are numerous applications in which *several* independent variables occur. For example, the area of a rectangle depends on *two* quantities, length and width. If an object is located in space, then the temperature at a point P in the object may depend on *three* rectangular coordinates x, y, and z of P. If, in addition, the temperature changes with time t, then *four* variables x, y, z, and t are involved. As a final illustration, a manufacturer may find that the cost C of producing a certain item depends on material, labor, equipment, overhead, and maintenance charges. Thus C is influenced by *five* different variables.

In this section we shall define functions of several variables and discuss some of their properties. Recall that a function f is a correspondence that associates with each element of a set X a unique element of a set Y (see Definition (1.21)). If both X and Y are subsets of $\mathbb{R}$, then f is called a function of one (real) variable. If X is a subset of $\mathbb{R} \times \mathbb{R}$ and Y is a subset of $\mathbb{R}$, then f is called a *function of two (real) variables.* Another way of stating this is as follows.

Definition (16.1)

Let D be a set of ordered pairs of real numbers. A **function f of two variables** is a correspondence that associates with each pair (x, y) in D a unique real number, denoted by $f(x, y)$. The set D is called the **domain** of f. The **range** of f consists of all real numbers $f(x, y)$, where (x, y) is in D.

In Chapter 1 the domain and range of a function of one variable were represented geometrically by points on two real lines (see Figure 1.33). For functions of *two* variables we may represent the domain D by points in an xy-plane and the range by points on a real line, say a w-axis. This is illustrated in Figure 16.1, where several curved arrows are drawn from ordered pairs in D to the corresponding numbers in the range. To obtain a physical interpretation of this situation we could imagine a flat metal plate having the shape of D. To each point (x, y) on the plate there corresponds a temperature $f(x, y)$ that can be recorded on a thermometer, represented by the w-axis. As another illustration, we could regard D as the surface of a lake and let $f(x, y)$ denote the depth of the water under the point (x, y).

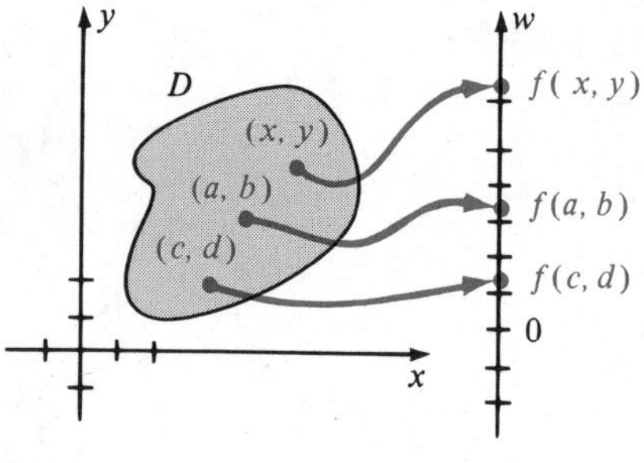

A function f of two variables is often defined by using an expression in x and y to specify $f(x, y)$. In this event, it is assumed that the domain is the set of all pairs (x, y) for which the given expression is meaningful, and we call f **a function of x and y**.

Figure 16.1

Example 1 If
$$f(x, y) = \frac{xy - 5}{2\sqrt{y - x^2}},$$
find the domain D of f. Illustrate D and the numbers $f(2, 5)$, $f(1, 2)$, and $f(-1, 2)$ in the manner of Figure 16.1.

Solution   The domain D is the set of all pairs (x, y) such that $y - x^2$ is positive, that is, $y > x^2$. (Why?) Thus the graph of D is that part of the xy-plane that lies above the parabola $y = x^2$, as shown in Figure 16.2. By substitution,

$$f(2, 5) = \frac{(2)(5) - 5}{2\sqrt{5 - 4}} = \frac{5}{2}.$$

Similarly, $f(1, 2) = -\frac{3}{2}$ and $f(-1, 2) = -\frac{7}{2}$. These functional values are indicated on the w-axis in Figure 16.2. ∎

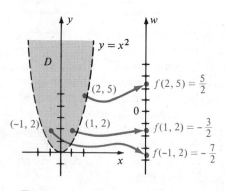

Figure 16.2

Formulas are sometimes used to define functions of two variables. For example, the formula $V = \pi r^2 h$ expresses the volume V of a right circular cylinder as a function of the altitude h and base radius r. The symbols r and h are referred to as **independent variables** and V is the **dependent variable**.

A function f of three (real) variables is defined as in (16.1), except that the domain D is a subset of $\mathbb{R} \times \mathbb{R} \times \mathbb{R}$. In this case, with each ordered triple (x, y, z) in D there is associated a unique real number $f(x, y, z)$. If we represent

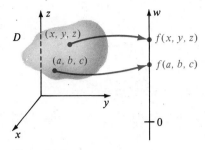

Figure 16.3

D by a region in a three-dimensional rectangular coordinate system, then as illustrated in Figure 16.3, to each point (x, y, z) in D there corresponds a unique point on the w-axis with coordinate $f(x, y, z)$. To obtain a physical illustration we could regard D as a solid object and, as before, let $f(x, y, z)$ be the temperature at (x, y, z).

As in the two-variable case, functions of three variables are often defined by means of expressions. For example,

$$f(x, y, z) = \frac{xe^y + \sqrt{z^2 + 1}}{xy \sin z}$$

determines a function of $x, y,$ and z. Formulas such as $V = lwh$ for the volume of a rectangular parallelepiped of dimensions $l, w,$ and h also illustrate how functions of three variables may arise.

Let us now return to a function f of two variables x and y. The **graph** of f is, by definition, the graph of the equation $z = f(x, y)$ in a three-dimensional rectangular coordinate system, and hence is usually a surface S of some type. If we represent D by a region in the xy-plane, then the pair (x, y) in D is represented by the point $(x, y, 0)$. Functional values $f(x, y)$ are the (directed) distances from the xy-plane to S as illustrated in Figure 16.4.

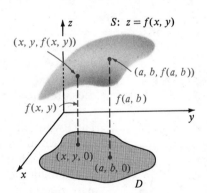

Figure 16.4

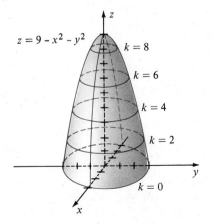

Figure 16.5 Circular traces are on the planes $z = k$.

Example 2 Sketch the graph of f if $f(x, y) = 9 - x^2 - y^2$ and the domain is $D = \{(x, y): x^2 + y^2 \leq 9\}$.

Solution The domain D of f may be represented geometrically by the collection of all points within and on the circle $x^2 + y^2 = 9$ in the xy-plane. The graph of f is the portion of the graph of

$$z = 9 - x^2 - y^2$$

shown in Figure 16.5. For future reference we have sketched circular traces on the planes $z = k$ where $k = 0, 2, 4, 6,$ and 8.

There is another useful graphical technique for describing a function f of two variables. The method consists of sketching, in an xy-plane, the

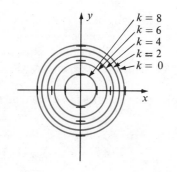

Figure 16.6 Level curves:
$9 - x^2 - y^2 = k$

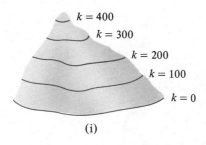

(i)

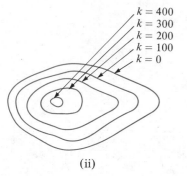

(ii)

Figure 16.7

graphs of the equations $f(x, y) = k$ for various values of k. The graphs obtained in this way are called **level curves** of the function f. It is important to note that as a point (x, y) moves on a level curve, the values $f(x, y)$ of the function are constant.

Example 3   Sketch some level curves of the function f in Example 2.

Solution   The level curves are graphs, in the xy-plane, of equations of the form $f(x, y) = k$, that is,

$$9 - x^2 - y^2 = k$$

or

$$x^2 + y^2 = 9 - k.$$

These are circles, provided $0 \leq k < 9$. In Figure 16.6 we have sketched the level curves corresponding to $k = 0, 2, 4, 6,$ and 8. These level curves are the projections, in the xy-plane, of the circular traces shown in Figure 16.5. ∎

If we sketch the level curves $f(x, y) = k$ for equispaced values of k, such as $k = 0, 2, 4, 6,$ and 8 in Example 3, then the nearness of successive curves gives information about the steepness of the surface. In Figure 16.6 the level curves corresponding to $k = 0$ and $k = 2$ are closer than those corresponding to $k = 6$ and $k = 8$, which indicates that the surface shown in Figure 16.5 is steeper at points near the xy-plane than at points farther away.

Level curves are often used in making **topographic maps** of rough terrain. For example, suppose $f(x, y)$ denotes the altitude (in feet) at a point (x, y) of latitude x and longitude y. On the hill pictured in (i) of Figure 16.7 we have sketched curves (in three dimensions) corresponding to altitudes of $k = 0, 100, 200, 300,$ and 400 feet. We may regard these curves as having been obtained by "slicing" the hill with planes parallel to the base. A person walking along one of these curves would always remain at the same altitude. The (two-dimensional) level curves corresponding to the same values of k are shown in (ii) of the figure. They represent the view obtained by looking down on the hill from an airplane.

Configurations of the type illustrated in (ii) of Figure 16.7 also occur on maps that indicate the depth of the water in a lake. One example is sketched in Figure 16.8, where $f(x, y)$ is the depth under the point (x, y).

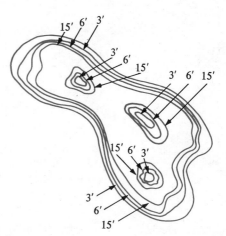

Figure 16.8   Depth of water in a lake

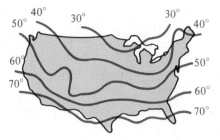

Figure 16.9   Isothermal curves

As another illustration of level curves, Figure 16.9 shows a weather map of the United States, where $f(x, y)$ denotes the high temperature at (x, y) during a certain day. The level curves are called **isothermal curves** to indicate that the temperature is constant on each curve. Another weather map could be drawn in which $f(x, y)$ represents the barometric pressure at (x, y). In this case, the level curves are called **isobars**.

If f is a function of three variables x, y, and z, then by definition the **level surfaces** of f are the graphs of $f(x, y, z) = k$, where k takes on suitable real values. If we let $k = w_0$, w_1, and w_2, there may result surfaces S_0, S_1, and S_2, respectively, as illustrated in Figure 16.10. As a point (x, y, z) moves along one of these surfaces, $f(x, y, z)$ does not change. In applications, if $f(x, y, z)$ is the temperature at (x, y, z), then the level surfaces are called **isothermal surfaces**. If $f(x, y, z)$ represents electrical potential, the level surfaces are called **equipotential surfaces**, since the voltage does not change if (x, y, z) remains on such a surface.

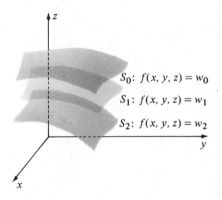

Figure 16.10   Level surfaces of $f(x, y, z)$

Example 4   If $f(x, y, z) = z - \sqrt{x^2 + y^2}$, sketch some level surfaces of f.

Solution   The level surfaces are graphs of

$$z - \sqrt{x^2 + y^2} = k$$

or equivalently,

$$z = \sqrt{x^2 + y^2} + k$$

where k is any real number. For each k we obtain a right circular cone with its axis along the z-axis. The special cases $k = -1$, $k = 0$, $k = 1$, and $k = 2$ are sketched in Figure 16.11.   ∎

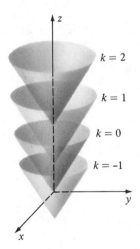

Figure 16.11   Level surfaces of
$z - \sqrt{x^2 + y^2} = k$

Computers can be programmed to exhibit level curves and traces of surfaces on various planes. It is also possible to represent surfaces from various perspectives. Sketches of this type are sometimes referred to as *computer graphics*, or *computer generated graphs*. Several illustrations obtained by means of a computer are shown in Figure 16.12. Students who are interested in computer programming should look for details of this technique in other books and courses.

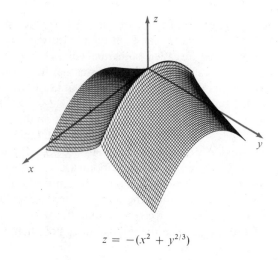

$$z = -(x^2 + y^{2/3})$$

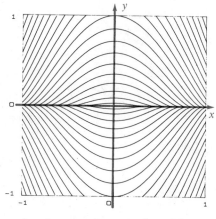

Level curves of $z = -(x^2 + y^{2/3})$

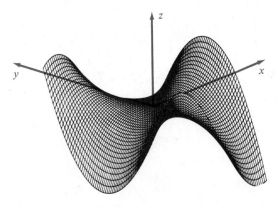

$$z = x^3 - 2xy^2 - x^2 - y^2$$

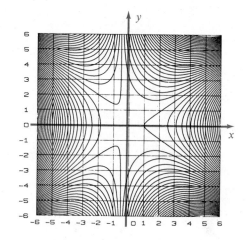

Level curves of $z = x^3 - 2xy^2 - x^2 - y^2$

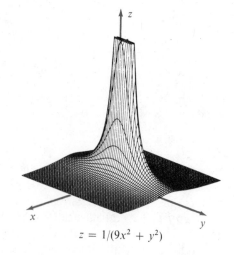

$$z = 1/(9x^2 + y^2)$$

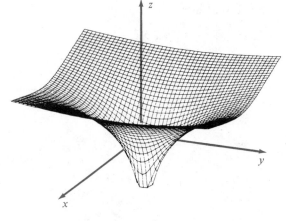

$$z = \ln(2x^2 + y^2)$$

Figure 16.12

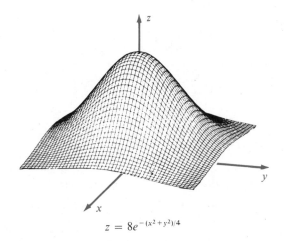

$$z = 8e^{-(x^2 + y^2)/4}$$

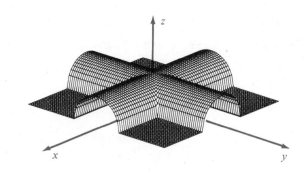

$$z = \begin{cases} \sqrt{1 - y^2} & \text{if } |x| \geq 1 \text{ and } |y| \leq 1, \text{ or if } |x| \leq 1 \text{ and } |y| \leq |x| \\ \sqrt{1 - x^2} & \text{if } |y| \geq 1 \text{ and } |x| \leq 1, \text{ or if } |y| \leq 1 \text{ and } |x| \leq |y| \\ 0 & \text{if } |x| \geq 1 \text{ and } |y| \geq 1 \end{cases}$$

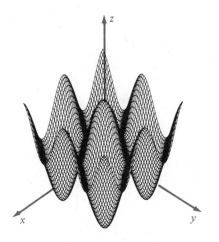

$$z = \cos x + \cos y$$

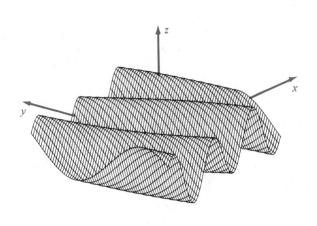

$$z = \cos x + 3 \cos (3x + y)$$

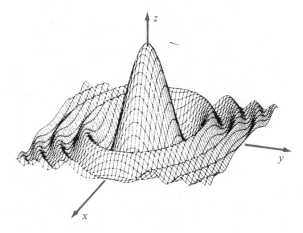

$$z = \frac{\frac{1}{2} \cos (2x^2 + y^2)}{1 + 2x^2 + y^2}$$

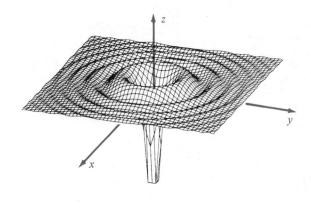

$$z = \frac{1 - 2 \cos (x^2 + y^2)}{x^2 + y^2}$$

EXERCISES 16.1

In Exercises 1–8 determine the domain of f and the value of f at the indicated points.

1 $f(x, y) = 2x - y^2; (-2, 5), (5, -2), (0, -2)$

2 $f(x, y) = (y + 2)/x; (3, 1), (1, 3), (2, 0)$

3 $f(u, v) = uv/(u - 2v); (2, 3), (-1, 4), (0, 1)$

4 $f(r, s) = \sqrt{1 - r} - e^{r/s}; (1, 1), (0, 4), (-3, 3)$

5 $f(x, y, z) = \sqrt{25 - x^2 - y^2 - z^2}; (1, -2, 2), (-3, 0, 2)$

6 $f(x, y, z) = 2 + \tan x + y \sin z; (\pi/4, 4, \pi/6), (0, 0, 0)$

7 $f(r, s, v, p) = rs^2 \tan v + 4sv \ln p; (3, -1, \pi/4, e)$

8 $f(x, y, u, v, w) = w \ln (x - y) - xue^{v/w}; (2, 1, 3, 4, -1)$

In Exercises 9–12 find

$$\frac{f(x + h, y) - f(x, y)}{h} \quad \text{and} \quad \frac{f(x, y + h) - f(x, y)}{h}$$

where $h \neq 0$.

9 $f(x, y) = x^2 + y^2$

10 $f(x, y) = y^2 + 3x$

11 $f(x, y) = xy^2 + 3x$

12 $f(x, y) = xy^3 + 4x^2 - 2$

In Exercises 13–22 describe the graph of f.

13 $f(x, y) = \sqrt{1 - x^2 - y^2}$

14 $f(x, y) = x^2 + y^2 - 1$

15 $f(x, y) = 6 - 2x - 3y$

16 $f(x, y) = \sqrt{y^2 - 4x^2 - 16}$

17 $f(x, y) = 5$

18 $f(x, y) = 4 - x$

19 $f(x, y) = \sqrt{72 + 4x^2 - 9y^2}$

20 $f(x, y) = \sqrt{x^2 + 4y^2 + 25}$

21 $f(x, y) = 16y^2 - x^2$

22 $f(x, y) = -\sqrt{16 - 9x^2}$

In Exercises 23–30 sketch some level curves of f.

23 $f(x, y) = y^2 - x^2$

24 $f(x, y) = 3x - 2y$

25 $f(x, y) = x^2 - y$

26 $f(x, y) = xy$

27 $f(x, y) = y - \sin x$

28 $f(x, y) = 4x^2 + y^2$

29 $f(x, y) = x^2 + y^2 - 4x + 6y + 13$

30 $f(x, y) = e^x - y$

31   If $f(x, y) = y \arctan x$, find an equation of the level curve of f that passes through the point $P(1, 4)$.

32   If $f(x, y, z) = x^2 + 4y^2 - z^2$, find an equation of the level surface of f that passes through the point $P(2, -1, 3)$.

In Exercises 33–38 describe the level surfaces of f.

33 $f(x, y, z) = x^2 + y^2 + z^2$

34 $f(x, y, z) = z + x^2 + 4y^2$

35 $f(x, y, z) = x + 2y + 3z$

36 $f(x, y, z) = x^2 + y^2 - z^2$

37 $f(x, y, z) = x^2 + y^2$

38 $f(x, y, z) = z$

39   A flat metal plate is situated on an xy-plane such that the temperature T (in degrees Celsius) at the point (x, y) is inversely proportional to the distance from the origin.

     (a) Describe the isothermal curves.

     (b) If the temperature at the point $P(4, 3)$ is 40, find an equation of the isothermal curve where the temperature is 20.

40   If the voltage V at the point $P(x, y, z)$ is given by $V = 6/(x^2 + 4y^2 + 9z^2)^{1/2}$,

     (a) describe the equipotential surfaces.

     (b) find an equation of the equipotential surface where $V = 120$.

41   According to Newton's *Law of Universal Gravitation*, if a particle of mass M is at the origin of a rectangular coordinate system, then the magnitude F of the force exerted on a particle of mass m located at the point (x, y, z) is given by

$$F = \frac{gMm}{x^2 + y^2 + z^2}$$

where g is a gravitational constant. How many independent variables are present? If M and m are constant, describe the level surfaces of the resulting function of x, y, and z. What is the physical significance of these level surfaces?

42 According to the *Ideal Gas Law*, the pressure P, volume V, and temperature T of a confined gas are related by the formula $PV = kT$, where k is a constant. Express P as a function of V and T and describe the level curves associated with this function. What is the physical significance of these level curves?

43 Define, in a manner similar to (16.1), a function of (a) four variables; (b) five variables; (c) n variables, where n is any positive integer.

44 If f and g are functions of two variables define, in a manner that is analogous to the single variable case, the *sum* $f + g$, *difference* $f - g$, *product* fg, and *quotient* f/g. What restrictions are required for the domains of these functions?

16.2

LIMITS AND CONTINUITY

If f is a function of two variables, it is often important to study the variation of $f(x, y)$ as (x, y) varies through the domain D of f. As a physical illustration, consider a thin metal plate that has the shape of the region D in Figure 16.1. To each point (x, y) on the plate there corresponds a temperature $f(x, y)$, which is recorded on a thermometer represented by the w-axis. As the point (x, y) moves on the plate the temperature may increase, decrease, or remain constant, and hence the point on the w-axis that corresponds to $f(x, y)$ will move in a positive direction, negative direction, or remain fixed, respectively. If the temperature $f(x, y)$ gets closer to a fixed value L as (x, y) gets closer and closer to a fixed point (a, b), we use the following notation.

(16.2)
$$\lim_{(x, y) \to (a, b)} f(x, y) = L$$

This may be read as follows: *f has the limit L as (x, y) approaches (a, b)*. To make our remarks mathematically precise, let us employ a geometric device similar to that used for functions of one variable in Chapter 2. For any $\varepsilon > 0$, consider the open interval $(L - \varepsilon, L + \varepsilon)$ on the w-axis. If (16.2) is true, then as illustrated in Figure 16.13 there is a $\delta > 0$ such that for every point (x, y) inside the circle of radius δ with center (a, b), except possibly (a, b) itself, the number corresponding to $f(x, y)$ is in the interval $(L - \varepsilon, L + \varepsilon)$. This is equivalent to the following statement:

If $0 < \sqrt{(x - a)^2 + (y - b)^2} < \delta$, then $|f(x, y) - L| < \varepsilon$.

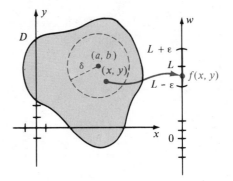

Figure 16.13

The preceding discussion is summarized in the next definition.

Definition (16.3)

Let a function f of two variables be defined throughout the interior of a circle with center (a, b), except possibly at (a, b) itself. The **limit of $f(x, y)$ as (x, y) approaches (a, b) is L**, written

$$\lim_{(x, y) \to (a, b)} f(x, y) = L$$

means that for every $\varepsilon > 0$ there corresponds a $\delta > 0$ such that

if $\quad 0 < \sqrt{(x - a)^2 + (y - b)^2} < \delta, \quad$ then $\quad |f(x, y) - L| < \varepsilon.$

If we consider the graph of f illustrated in Figure 16.4, then intuitively, Definition (16.3) means that as the point $(x, y, 0)$ approaches $(a, b, 0)$ on the xy-plane, the corresponding point $(x, y, f(x, y))$ on the graph S of f approaches (a, b, L) (which may or may not be on S). It can be shown that if the limit L exists, it is unique.

If f and g are functions of two variables, then $f + g, f - g, fg$, and f/g are defined in the usual way, and Theorem (2.8) concerning limits of sums, products, and quotients can be extended. For example, if f and g have limits as (x, y) approaches (a, b), then it can be proved that

$$\lim_{(x, y) \to (a, b)} [f(x, y) + g(x, y)] = \lim_{(x, y) \to (a, b)} f(x, y) + \lim_{(x, y) \to (a, b)} g(x, y).$$

A function f of two variables is a **polynomial function** if $f(x, y)$ can be expressed as a sum of terms of the form $cx^m y^n$, where c is a real number and m and n are nonnegative integers. A **rational function** is a quotient of two polynomial functions. As in the single variable case, it can be shown that limits of such functions may be found by substituting for x and y. For example,

$$\lim_{(x, y) \to (2, -3)} (x^3 - 4xy^2 + 5y - 7) = 2^3 - 4(2)(-3)^2 + 5(-3) - 7$$

$$= 8 - 72 - 15 - 7 = -86.$$

The next example is an illustration of a function that has no limit as (x, y) approaches $(0, 0)$.

Example 1 Show that $\lim_{(x, y) \to (0, 0)} \dfrac{x^2 - y^2}{x^2 + y^2}$ does not exist.

Solution If $f(x, y) = (x^2 - y^2)/(x^2 + y^2)$, then f is defined everywhere except the origin $(0, 0)$. If we consider any point $(x, 0)$ on the x-axis, then $f(x, 0) = x^2/x^2 = 1$, provided $x \neq 0$. For points $(0, y)$ on the y-axis, we have $f(0, y) = -y^2/y^2 = -1$, if $y \neq 0$. Consequently, as illustrated in Figure 16.14,

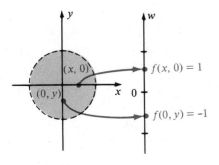

Figure 16.14

every circle with center $(0, 0)$ contains points at which the value of f is 1 and points at which the value of f is -1. It follows that the limit does not exist, for if we take $\varepsilon = 1$, there is no open interval $(L - \varepsilon, L + \varepsilon)$ on the w-axis containing both 1 and -1, and hence it is impossible to find a $\delta > 0$ that satisfies the conditions of Definition (16.3). ∎

For certain functions in Chapter 2 we proved that $\lim_{x \to a} f(x)$ did not exist by showing that $\lim_{x \to a^-} f(x)$ and $\lim_{x \to a^+} f(x)$ were not equal. When considering such one-sided limits we may regard the point on the x-axis with coordinate x as "approaching" the point with coordinate a either from the left or from the right, respectively. The analogous situation for functions of two variables is more complicated, since in a coordinate plane there are an infinite number of different curves (which we shall call **paths**) along which (x, y) can approach (a, b). However, if the limit in Definition (16.3) exists, then $f(x, y)$ must have the limit L regardless of the path taken. This illustrates the following important rule for investigating certain limits.

(16.4) If two different paths to a point $P(a, b)$ produce two different limiting values for f, then the limit of $f(x, y)$ as (x, y) approaches (a, b) does not exist.

Example 2 Rework Example 1 using (16.4).

Solution If, on the one hand, (x, y) approaches $(0, 0)$ along the x-axis, then the y-coordinate is always zero and the expression $(x^2 - y^2)/(x^2 + y^2)$ reduces to x^2/x^2 or 1. Hence the limiting value is 1. On the other hand, if we let (x, y) approach $(0, 0)$ along the y-axis, then the x-coordinate is 0 and $(x^2 - y^2)/(x^2 + y^2)$ reduces to $-y^2/y^2$ or -1. Since two different limits are obtained, the given limit does not exist by (16.4). We could, of course, have chosen other paths to the origin $(0, 0)$. For example, if we let (x, y) approach $(0, 0)$ along the line $y = 2x$, then

$$\frac{x^2 - y^2}{x^2 + y^2} = \frac{x^2 - 4x^2}{x^2 + 4x^2} = \frac{-3x^2}{5x^2} = -\frac{3}{5}$$

and hence the limiting value would be $-\frac{3}{5}$. ∎

Example 3 If $f(x, y) = \dfrac{x^2 y}{x^4 + y^2}$ show that $\lim\limits_{(x, y) \to (0, 0)} f(x, y)$ does not exist.

Solution If we let (x, y) approach $(0, 0)$ along any line $y = mx$ that passes through the origin, we see that if $m \neq 0$,

$$\lim_{(x, y) \to (0, 0)} \frac{x^2 y}{x^4 + y^2} = \lim_{(x, y) \to (0, 0)} \frac{x^2 (mx)}{x^4 + (mx)^2} = \lim_{(x, y) \to (0, 0)} \frac{mx^3}{x^4 + m^2 x^2}$$

$$= \lim_{(x, y) \to (0, 0)} \frac{mx}{x^2 + m^2} = \frac{0}{0 + m^2} = 0.$$

Because of this fact it is tempting to conclude that $f(x, y)$ has the limit 0 as (x, y) approaches $(0, 0)$. However, if we let (x, y) approach $(0, 0)$ along the parabola $y = x^2$, then

$$\lim_{(x, y) \to (0, 0)} \frac{x^2 y}{x^4 + y^2} = \lim_{(x, y) \to (0, 0)} \frac{x^2(x^2)}{x^4 + (x^2)^2}$$

$$= \lim_{(x, y) \to (0, 0)} \frac{x^4}{2x^4} = \lim_{(x, y) \to (0, 0)} \frac{1}{2} = \frac{1}{2}.$$

Thus not every path through $(0, 0)$ leads to the same limiting value of $f(x, y)$, and hence by (16.4), the limit does not exist. ∎

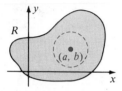

(i) Interior point of R

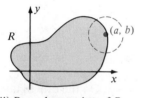

(ii) Boundary point of R

Figure 16.15

If R is a region in the xy-plane, then a point (a, b) is called an **interior point** of R if there exists a circle with center (a, b) that contains only points of R. An illustration of an interior point is shown in (i) of Figure 16.15. A point (a, b) is called a **boundary point** of R if *every* circle with center (a, b) contains points that are in R and points that are not in R, as illustrated in (ii) of Figure 16.15.

A region is called **closed** if it contains all of its boundary points. A region is **open** if it contains *none* of its boundary points; that is, every point of the region is an interior point. A region that contains some, but not all, of its boundary points is neither open nor closed. These concepts are analogous to closed intervals, open intervals, and half-open intervals of real numbers.

Suppose a function f of two variables is defined for all (x, y) in a region R, except possibly at (a, b). If (a, b) is an interior point, then Definition (16.3) can be used to investigate the limit of f as (x, y) approaches (a, b). If (a, b) is a *boundary point we shall use Definition* (16.3) *with the added restriction that* (x, y) *must be both in R and in the circle of radius δ.* Theorems on limits may then be extended to boundary points.

A function f of two variables is **continuous** at an interior point (a, b) of a region R if $f(a, b)$ exists, $f(x, y)$ has a limit as (x, y) approaches (a, b), and

$$\lim_{(x, y) \to (a, b)} f(x, y) = f(a, b).$$

This notion can be extended to a boundary point (a, b) of R by applying the previous remarks on limits concerning boundary points. If R is in the domain D of f, then f is said to be **continuous on R** if it is continuous at every pair (a, b) in R. If f is continuous on R, then a small change in (x, y) produces a small change in $f(x, y)$. Referring to the graph S of f, if (x, y) is close to (a, b), then the point $(x, y, f(x, y))$ on S is close to $(a, b, f(a, b))$. Thus there are no holes or vertical steps in the graph of a continuous function.

Theorems on continuity that are analogous to those for functions of one variable may be proved. In particular, polynomial functions are continuous everywhere and rational functions are continuous except at points where the denominator vanishes.

The preceding discussion on limits and continuity can be extended to functions of three or more variables. For example, if f is a function of three variables we may state the following definition.

Definition (16.5)

Let a function f of three variables be defined throughout the interior of a sphere with center (a, b, c), except possibly at (a, b, c) itself. The **limit of $f(x, y, z)$ as (x, y, z) approaches (a, b, c) is L**, written

$$\lim_{(x, y, z) \to (a, b, c)} f(x, y, z) = L$$

means that for every $\varepsilon > 0$ there corresponds a $\delta > 0$ such that if

$$0 < \sqrt{(x - a)^2 + (y - b)^2 + (z - c)^2} < \delta,$$

then
$$|f(x, y, z) - L| < \varepsilon.$$

We may give a graphical interpretation for Definition (16.5) that is similar to that illustrated in Figure 16.13. The difference is that in place of the circle in the xy-plane, we use a sphere of radius δ with center at the point (a, b, c) in three dimensions (see Exercise 25).

An *interior point* or *boundary point* P of a three-dimensional region R is defined as in two dimensions, except that we use a *sphere* with center P instead of a circle. The limit definition (16.5) may be extended to the case where (a, b, c) is a boundary point by adding the restriction that (x, y, z) is both in R and in the sphere of radius δ.

A function f of three variables is *continuous* at an interior point (a, b, c) of a region if $f(a, b, c)$ exists, $f(x, y, z)$ has a limit as (x, y, z) approaches (a, b, c), and

$$\lim_{(x, y, z) \to (a, b, c)} f(x, y, z) = f(a, b, c).$$

Continuity at a boundary point is defined by using the extension of Definition (16.5) to boundary points.

There is no essential difference in defining limits for functions of four or more variables; however, in these cases we no longer give geometric interpretations (see Exercise 27).

In Section 16.5 we shall discuss composite functions of several variables. As a simple illustration, if f is a function of two variables and g is a function of one variable, then a function h of two variables may be defined by letting $h(x, y) = g(f(x, y))$, provided the range of f is in the domain of g.

Example 4 In each of the following, express $g(f(x, y))$ in terms of x and y, and find the domain of the resulting composite function.

(a) $f(x, y) = xe^y$, $g(t) = 3t^2 + t + 1$

(b) $f(x, y) = y - 4x^2$, $g(t) = \sin \sqrt{t}$

Solution In each case we substitute $f(x, y)$ for t. Thus:

(a) $g(f(x, y)) = g(xe^y) = 3x^2 e^{2y} + xe^y + 1.$

(b) $g(f(x, y)) = g(y - 4x^2) = \sin\sqrt{y - 4x^2}.$

The domain in part (a) is $\mathbb{R} \times \mathbb{R}$, and the domain in part (b) consists of all ordered pairs (x, y) such that $y \geq 4x^2$. ∎

The following analogue of (2.28) may be proved for functions of the type illustrated in Example 4.

Theorem (16.6)

> If a function f of two variables is continuous at (a, b), and a function g of one variable is continuous at $f(a, b)$, then the function h defined by $h(x, y) = g(f(x, y))$ is continuous at (a, b).

Theorem (16.6) is important because it provides a technique for establishing the continuity of certain functions by using previously proved results, as illustrated in the next example.

Example 5 If $h(x, y) = e^{x^2 + 5xy + y^3}$, show that h is continuous at every pair (a, b).

Solution   If we let $f(x, y) = x^2 + 5xy + y^3$ and $g(t) = e^t$, it follows that $h(x, y) = g(f(x, y))$. Since f is a polynomial function, it is continuous at every pair (a, b). Moreover, g is continuous at every $t = f(a, b)$. Thus, by Theorem (16.6), h is continuous at (a, b). ∎

EXERCISES 16.2

In Exercises 1–10 find the limits, if they exist.

1 $\displaystyle\lim_{(x, y) \to (0, 0)} \frac{x^2 - 2}{3 + xy}$

2 $\displaystyle\lim_{(x, y) \to (0, 0)} \frac{x^3 - x^2y + xy^2 - y^3}{x^2 + y^2}$

3 $\displaystyle\lim_{(x, y) \to (0, 0)} \frac{2x^2 - y^2}{x^2 + 2y^2}$

4 $\displaystyle\lim_{(x, y) \to (0, 0)} \frac{x^2 - 2xy + 5y^2}{3x^2 + 4y^2}$

5 $\displaystyle\lim_{(x, y) \to (0, 0)} \frac{4x^2y}{x^3 + y^3}$

6 $\displaystyle\lim_{(x, y) \to (0, 0)} \frac{x^4 - y^4}{x^2 + y^2}$

7 $\displaystyle\lim_{(x, y) \to (1, 2)} \frac{xy - 2x - y + 2}{x^2 + y^2 - 2x - 4y + 5}$

8 $\displaystyle\lim_{(x, y) \to (0, 0)} \frac{3xy}{5x^4 + 2y^4}$

9 $\displaystyle\lim_{(x, y) \to (0, 0)} \frac{3x^3 - 2x^2y + 3y^2x - 2y^3}{x^2 + y^2}$

10 $\displaystyle\lim_{(x, y, z) \to (0, 0, 0)} \frac{xy + yz + xz}{x^2 + y^2 + z^2}$

In Exercises 11–16 discuss the continuity of the function f.

11 $f(x, y) = \ln(x + y - 1)$

12 $f(x, y) = xy/(x^2 - y^2)$

13 $f(x, y, z) = 1/(x^2 + y^2 - z^2)$

14 $f(x, y, z) = \sqrt{xy} \tan z$

15 $f(x, y) = \sqrt{x} \, e^{\sqrt{1 - y^2}}$

16 $f(x, y) = \sqrt{25 - x^2 - y^2}$

In Exercises 17–20 find $h(x, y) = g(f(x, y))$ and use Theorem (16.6) to determine where h is continuous.

17 $f(x, y) = x^2 - y^2, g(t) = (t^2 - 4)/t$

18 $f(x, y) = 3x + 2y - 4, g(t) = \ln(t + 5)$

19 $f(x, y) = x + \tan y, g(z) = z^2 + 1$

20 $f(x, y) = y \ln x, g(w) = e^w$

21 If $f(x, y) = x^2 + 2y$, $g(t) = e^t$, and $h(t) = t^2 - 3t$, find $g(f(x, y)), h(f(x, y))$, and $f(g(t), h(t))$.

22 If $f(x, y, z) = 2x + ye^z$ and $g(t) = t^2$, find $g(f(x, y, z))$.

23 If $f(u, v) = uv - 3u + v$, $g(x, y) = x - 2y$, and $k(x, y) = 2x + y$, find $f(g(x, y), k(x, y))$.

24 If $f(x, y) = 2x + y$, find $f(f(x, y), f(x, y))$.

25 Give a graphical interpretation of Definition (16.5).

26 Prove that if the limit L in Definition (16.3) exists, then it is unique.

27 Extend Definition (16.3) to functions of four variables.

28 Prove that if f is a continuous function of two variables and $f(a, b) > 0$, then there is a circle C in the xy-plane with center (a, b) such that $f(x, y) > 0$ for all pairs (x, y) that are in the domain of f and within C.

29 Prove, directly from Definition (16.3), that

(a) $\lim\limits_{(x, y) \to (a, b)} x = a.$　　(b) $\lim\limits_{(x, y) \to (a, b)} y = b.$

30 If $\lim\limits_{(x, y) \to (a, b)} f(x, y) = L$ and c is any real number, prove, directly from Definition (16.3), that

$$\lim_{(x, y) \to (a, b)} cf(x, y) = cL.$$

16.3

PARTIAL DERIVATIVES

In this section we begin the study of derivatives of functions of several variables.

Definition (16.7)

> If f is a function of two variables, then the **first partial derivatives of f with respect to x and y** are the functions f_x and f_y defined as follows:
>
> $$f_x(x, y) = \lim_{h \to 0} \frac{f(x + h, y) - f(x, y)}{h}$$
>
> $$f_y(x, y) = \lim_{h \to 0} \frac{f(x, y + h) - f(x, y)}{h}$$
>
> provided the limits exist.

Note that in the definition of $f_x(x, y)$, the variable y is fixed. It is for this reason that the notation for limits of functions of one variable is used instead of that introduced in the previous section. Indeed, if we let $y = b$ and define g by $g(x) = f(x, b)$, then $g'(x) = f_x(x, b) = f_x(x, y)$. Thus, to find $f_x(x, y)$ we may regard y as a constant and differentiate $f(x, y)$ with respect to x in the usual way. Similarly, to find $f_y(x, y)$ the variable x is regarded as a constant and $f(x, y)$ is differentiated with respect to y.

Other common notations for partial derivatives are

$$f_x = \frac{\partial f}{\partial x} \quad \text{and} \quad f_y = \frac{\partial f}{\partial y}.$$

For brevity we often speak of $\partial f/\partial x$ or $\partial f/\partial y$ as *the partial of f with respect to x or y*, respectively.

If $w = f(x, y)$ we write

$$f_x(x, y) = \frac{\partial}{\partial x} f(x, y) = \frac{\partial w}{\partial x} = w_x$$

$$f_y(x, y) = \frac{\partial}{\partial y} f(x, y) = \frac{\partial w}{\partial y} = w_y.$$

Example 1 If $f(x, y) = x^3 y^2 - 2x^2 y + 3x$, find

(a) $f_x(x, y)$ and $f_y(x, y)$.   (b) $f_x(2, -1)$ and $f_y(2, -1)$.

Solution

(a) Regarding y as a constant and differentiating with respect to x, we obtain

$$f_x(x, y) = 3x^2 y^2 - 4xy + 3.$$

Regarding x as a constant and differentiating with respect to y gives us

$$f_y(x, y) = 2x^3 y - 2x^2.$$

(b) Using the results of (a),

$$f_x(2, -1) = 3(2)^2(-1)^2 - 4(2)(-1) + 3 = 23$$

$$f_y(2, -1) = 2(2)^3(-1) - 2(2)^2 = -24. \qquad\blacksquare$$

Formulas such as the Product Rule, Quotient Rule, and Power Rule are true for partial derivatives. For example, if u and v are functions of x and y, and n is a real number,

$$\frac{\partial}{\partial x}(uv) = u\frac{\partial v}{\partial x} + v\frac{\partial u}{\partial x}, \qquad \frac{\partial}{\partial x}\left(\frac{u}{v}\right) = \frac{v\dfrac{\partial u}{\partial x} - u\dfrac{\partial v}{\partial x}}{v^2},$$

$$\frac{\partial}{\partial x}(u^n) = nu^{n-1}\frac{\partial u}{\partial x}, \qquad \frac{\partial}{\partial x}\cos u = -\sin u\frac{\partial u}{\partial x}.$$

Example 2 Find $\dfrac{\partial w}{\partial y}$ if $w = xy^2 e^{xy}$.

Solution   We may proceed as follows.

$$\frac{\partial w}{\partial y} = xy^2 \frac{\partial}{\partial y}(e^{xy}) + e^{xy}\frac{\partial}{\partial y}(xy^2)$$

$$= xy^2(xe^{xy}) + e^{xy}(2xy) = (xy + 2)xye^{xy} \qquad\blacksquare$$

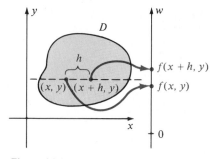

Figure 16.16

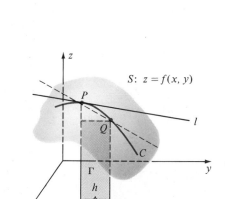

Figure 16.17

It is instructive to interpret Definition (16.7) by means of the illustration given at the beginning of the previous section, where $f(x, y)$ is the temperature of a flat metal plate at the point (x, y). Note that the points (x, y) and $(x + h, y)$ lie on a horizontal line, as indicated in Figure 16.16. In this case the difference $f(x + h, y) - f(x, y)$ is the change in temperature as a point moves from (x, y) to $(x + h, y)$, and the ratio

$$\frac{f(x + h, y) - f(x, y)}{h}$$

is the *average* change in temperature. For example, if the temperature change is 2 degrees and h is 4, then the average change in temperature is $\frac{2}{4}$, or $\frac{1}{2}$. Thus, *on the average*, the temperature changes at a rate of $\frac{1}{2}$ degree per unit change in distance. Taking the limit of the average change as h approaches 0 gives us the *rate of change* of temperature with respect to distance as the point (x, y) moves in a horizontal direction. Similarly, $f_y(x, y)$ measures the rate of change of $f(x, y)$ as (x, y) moves in a vertical direction.

As another illustration, suppose D represents the surface of a lake and $f(x, y)$ is the depth of the water under the point (x, y) on the surface. In this case $f_x(x, y)$ is the rate at which the depth changes as a point moves away from (x, y) parallel to the x-axis. Similarly, $f_y(x, y)$ is the rate of change of the depth in the direction of the y-axis.

There is an interesting geometric interpretation for Definition (16.7) in terms of the graph of f. In order to illustrate the meaning of $f_y(x, y)$, consider the points $M(x, y, 0)$ and $N(x, y + h, 0)$. The plane Γ, parallel to the yz-plane and passing through M and N, intersects S in a curve C. If we introduce a coordinate system on Γ, and if P and Q are the points on C that have projections M and N on the xy-plane, then as illustrated in Figure 16.17, $\overline{MP} = f(x, y)$, $\overline{NQ} = f(x, y + h)$, and $\overline{MN} = h$. It follows that the ratio $[f(x, y + h) - f(x, y)]/h$ is the slope of the secant line through P and Q in the plane Γ. The limit of this ratio as h approaches 0, namely $f_y(x, y)$, is the slope of the tangent line l to C at P. Similarly, if C' is the trace of S on the plane Γ', parallel to the xz-plane and passing through M, then $f_x(x, y)$ is the slope of the tangent line to C' at P.

First partial derivatives of functions of three or more variables are defined in the same manner as that used in Definition (16.7). Specifically, all variables except one are held constant and differentiation takes place with respect to the remaining variable. Thus, given $f(x, y, z)$ we may find f_x, f_y, and f_z (or equivalently $\partial f/\partial x$, $\partial f/\partial y$, and $\partial f/\partial z$). For example,

$$f_z(x, y, z) = \lim_{h \to 0} \frac{f(x, y, z + h) - f(x, y, z)}{h}.$$

In a manner similar to our discussion in two dimensions, if $f(x, y, z)$ is the temperature at the point $P(x, y, z)$, then $f_x(x, y, z)$ is the rate of change of temperature with respect to distance along a line through P that is parallel to the x-axis. The partial derivatives $f_y(x, y, z)$ and $f_z(x, y, z)$ give the rates of change in the directions of the y- and z-axes, respectively.

Example 3   If $w = x^2 y^3 \sin z + e^{xz}$, find $\partial w/\partial x$, $\partial w/\partial y$, and $\partial w/\partial z$.

Solution

$$\frac{\partial w}{\partial x} = 2xy^3 \sin z + ze^{xz}$$

$$\frac{\partial w}{\partial y} = 3x^2 y^2 \sin z$$

$$\frac{\partial w}{\partial z} = x^2 y^3 \cos z + xe^{xz} \qquad\blacksquare$$

If f is a function of two variables x and y, then f_x and f_y are also functions of two variables and we may consider *their* first partial derivatives. These are called the **second partial derivatives** of f and are denoted as follows.

$$\frac{\partial}{\partial x} f_x = (f_x)_x = f_{xx} = \frac{\partial}{\partial x} \left(\frac{\partial f}{\partial x} \right) = \frac{\partial^2 f}{\partial x^2}$$

$$\frac{\partial}{\partial y} f_x = (f_x)_y = f_{xy} = \frac{\partial}{\partial y} \left(\frac{\partial f}{\partial x} \right) = \frac{\partial^2 f}{\partial y \partial x}$$

$$\frac{\partial}{\partial x} f_y = (f_y)_x = f_{yx} = \frac{\partial}{\partial x} \left(\frac{\partial f}{\partial y} \right) = \frac{\partial^2 f}{\partial x \partial y}$$

$$\frac{\partial}{\partial y} f_y = (f_y)_y = f_{yy} = \frac{\partial}{\partial y} \left(\frac{\partial f}{\partial y} \right) = \frac{\partial^2 f}{\partial y^2}$$

If $w = f(x, y)$ we write

$$\frac{\partial^2}{\partial x^2} f(x, y) = f_{xx}(x, y) = \frac{\partial^2 w}{\partial x^2} = w_{xx}$$

and likewise for the other second partial derivatives.

It is customary to refer to f_{xy} and f_{yx} as the **mixed second partial derivatives of f** (or simply *mixed partials of f*). The next theorem brings out the fact that under suitable conditions, these mixed partials are equal. The proof may be found in texts on advanced calculus.

Theorem (16.8)

Let f be a function of x and y. If f, f_x, f_y, f_{xy}, and f_{yx} are continuous on an open region R, then

$$f_{xy} = f_{yx}$$

throughout R.

The hypothesis of Theorem (16.8) is satisfied for most functions encountered in calculus and its applications. Similarly, if $w = f(x, y, z)$ and f

has continuous second partial derivatives, then it can be proved that the following equalities hold for mixed partials:

$$\frac{\partial^2 w}{\partial y \, \partial x} = \frac{\partial^2 w}{\partial x \, \partial y}; \qquad \frac{\partial^2 w}{\partial z \, \partial x} = \frac{\partial^2 w}{\partial x \, \partial z}; \qquad \frac{\partial^2 w}{\partial z \, \partial y} = \frac{\partial^2 w}{\partial y \, \partial z}.$$

Example 4   Find the second partial derivatives of f if

$$f(x, y) = x^3 y^2 - 2x^2 y + 3x.$$

Solution   This function was considered in Example 1. Since the first partials of f are $f_x(x, y) = 3x^2 y^2 - 4xy + 3$ and $f_y(x, y) = 2x^3 y - 2x^2$,

$$f_{xx}(x, y) = \frac{\partial}{\partial x}(3x^2 y^2 - 4xy + 3) = 6xy^2 - 4y$$

$$f_{xy}(x, y) = \frac{\partial}{\partial y}(3x^2 y^2 - 4xy + 3) = 6x^2 y - 4x$$

$$f_{yx}(x, y) = \frac{\partial}{\partial x}(2x^3 y - 2x^2) = 6x^2 y - 4x$$

$$f_{yy}(x, y) = \frac{\partial}{\partial y}(2x^3 y - 2x^2) = 2x^3. \qquad \blacksquare$$

Third and higher partial derivatives are defined in like manner. For example,

$$\frac{\partial}{\partial x} f_{xx} = f_{xxx} = \frac{\partial}{\partial x}\left(\frac{\partial^2 f}{\partial x^2}\right) = \frac{\partial^3 f}{\partial x^3}$$

$$\frac{\partial}{\partial x} f_{xy} = f_{xyx} = \frac{\partial}{\partial x}\left(\frac{\partial^2 f}{\partial y \, \partial x}\right) = \frac{\partial^3 f}{\partial x \, \partial y \, \partial x},$$

and so on. If first, second, and third partial derivatives are continuous, then again it can be shown that the order of differentiation is immaterial, that is,

$$f_{xyx} = f_{yxx} = f_{xxy} \quad \text{and} \quad f_{yxy} = f_{xyy} = f_{yyx}.$$

Similar notations and results apply to functions of more than two variables. Of course, letters other than x and y may be used. To illustrate, if f is a function of r and s, then symbols such as

$$f_r(r, s), \quad f_s(r, s), \quad f_{rs}, \quad \frac{\partial f}{\partial r}, \quad \frac{\partial^2 f}{\partial r^2}$$

are employed for partial derivatives.

EXERCISES 16.3

In Exercises 1–18 find the first partial derivatives of f.

1 $f(x, y) = 2x^4y^3 - xy^2 + 3y + 1$

2 $f(x, y) = (x^3 - y^2)^2$

3 $f(r, s) = \sqrt{r^2 + s^2}$

4 $f(s, t) = (t/s) - (s/t)$

5 $f(x, y) = xe^y + y \sin x$

6 $f(x, y) = e^x \ln xy$

7 $f(t, v) = \ln \sqrt{(t + v)/(t - v)}$

8 $f(u, w) = \arctan (u/w)$

9 $f(x, y) = x \cos (x/y)$

10 $f(x, y) = \sqrt{4x^2 - y^2} \sec x$

11 $f(r, s, t) = r^2 e^{2s} \cos t$

12 $f(x, y, t) = (x^2 - t^2)/(1 + \sin 3y)$

13 $f(x, y, z) = (y^2 + z^2)^x$

14 $f(r, s, v) = (2r + 3s)^{\cos v}$

15 $f(x, y, z) = xe^z - ye^x + ze^{-y}$

16 $f(r, s, v, p) = r^3 \tan s + \sqrt{s}e^{v^2} - v \cos 2p$

17 $f(q, v, w) = \sin^{-1} \sqrt{qv} + \sin vw$

18 $f(x, y, z) = xyze^{xyz}$

In Exercises 19–24 verify that $w_{xy} = w_{yx}$.

19 $w = xy^4 - 2x^2y^3 + 4x^2 - 3y$

20 $w = x^2/(x + y)$

21 $w = x^3e^{-2y} + y^{-2} \cos x$

22 $w = y^2e^{x^2} + (1/x^2y^3)$

23 $w = x^2 \cosh (z/y)$

24 $w = \sqrt{x^2 + y^2 + z^2}$

25 If $w = 3x^2y^3z + 2xy^4z^2 - yz$, find w_{xyz}.

26 If $w = u^4vt^2 - 3uv^2t^3$, find w_{tut}.

27 If $u = v \sec rt$, find u_{rvr}.

28 If $v = y \ln (x^2 + z^4)$, find v_{zzy}.

29 If $w = \sin xyz$ find $\partial^3w/\partial z \, \partial y \, \partial x$.

30 If $w = x^2/(y^2 + z^2)$, find $\partial^3w/\partial z \, \partial y^2$.

31 If $w = r^4s^3t - 3s^2e^{rt}$, verify that $w_{rrs} = w_{rsr} = w_{srr}$.

32 If $w = \tan uv + 2 \ln (u + v)$, verify that $w_{uvv} = w_{vuv} = w_{vvu}$.

A function f of x and y is **harmonic** if $\partial^2f/\partial x^2 + \partial^2f/\partial y^2 = 0$. Prove that the functions defined in Exercises 33–36 are harmonic.

33 $f(x, y) = \ln \sqrt{x^2 + y^2}$

34 $f(x, y) = \arctan (y/x)$

35 $f(x, y) = \cos x \sinh y + \sin x \cosh y$

36 $f(x, y) = e^{-x} \cos y + e^{-y} \cos x$

37 If $w = \cos (x - y) + \ln (x + y)$, show that $\partial^2w/\partial x^2 - \partial^2w/\partial y^2 = 0$.

38 If $w = (y - 2x)^3 - \sqrt{y - 2x}$, show that $w_{xx} - 4w_{yy} = 0$.

39 If $w = e^{-c^2t} \sin cx$, show that $w_{xx} = w_t$ for every real number c.

40 The Ideal Gas Law may be stated as $PV = knT$, where n is the number of moles of gas, V is the volume, T is the temperature, P is the pressure, and k is a constant. Show that

$$\frac{\partial V}{\partial T} \frac{\partial T}{\partial P} \frac{\partial P}{\partial V} = -1.$$

In Exercises 41 and 42 show that v satisfies the *wave equation*

$$\frac{\partial^2v}{\partial t^2} = a^2 \frac{\partial^2v}{\partial x^2}.$$

41 $v = (\sin akt)(\sin kx)$

42 $v = (x - at)^4 + \cos (x + at)$

Two functions u and v of x and y are said to satisfy the *Cauchy–Riemann equations* if $u_x = v_y$ and $u_y = -v_x$. Show that the pairs of functions in Exercises 43–46 satisfy these equations.

43 $u(x, y) = x^2 - y^2, v(x, y) = 2xy$

44 $u(x, y) = y/(x^2 + y^2), v(x, y) = x/(x^2 + y^2)$

45 $u(x, y) = e^x \cos y, v(x, y) = e^x \sin y$

46 $u(x, y) = \cos x \cosh y + \sin x \sinh y$
$v(x, y) = \cos x \cosh y - \sin x \sinh y$

47 List all possible second partial derivatives of $w = f(x, y, z)$.

48 If $w = f(x, y, z, t, v)$, define w_t as a limit.

49 If $w = f(x, y, z)$ and f has continuous third partial derivatives, how many of them can be different?

50 If $w = f(x, y)$ and f has continuous fourth partial derivatives, how many of them can be different?

51 A flat metal plate is situated on an xy-plane such that the temperature T at the point (x, y) is given by $T = 10(x^2 + y^2)^2$, where T is in degrees and x, y are in centimeters. Find the rate of change of T with respect to distance at $P(1, 2)$ in the direction of (a) the x-axis; (b) the y-axis.

52 The surface of a certain lake is represented by a region D in the xy-plane such that the depth under the point corresponding to (x, y) is given by $f(x, y) = 300 - 2x^2 - 3y^2$, where x, y and $f(x, y)$ are in feet. If a boy in the water is at the point $(4, 9)$, find the rate at which the depth changes if he swims in the direction of (a) the x-axis; (b) the y-axis.

53 Suppose the electrical potential V at the point (x, y, z) is given by $V = 100/(x^2 + y^2 + z^2)$, where V is in volts and x, y, z are in inches. Find the rate of change of V with respect to distance at $P(2, -1, 1)$ in the direction of (a) the x-axis; (b) the y-axis; (c) the z-axis.

54 An object is situated in a rectangular coordinate system such that the temperature T at the point $P(x, y, z)$ is given by $T = 4x^2 - y^2 + 16z^2$, where T is in degrees and x, y, z are in centimeters. Find the rate of change of T at the point $P(4, -2, 1)$ in the direction of (a) the x-axis; (b) the y-axis; (c) the z-axis.

55 Let C be the trace of the paraboloid $z = 9 - x^2 - y^2$ on the plane $x = 1$. Find parametric equations of the tangent line l to C at the point $P(1, 2, 4)$. Sketch the paraboloid, C, and l. (*Hint*: See Figure 16.17.)

56 Let C be the trace of the graph of $z = \sqrt{36 - 9x^2 - 4y^2}$ on the plane $y = 2$. Find parametric equations of the tangent line l to C at the point $(1, 2, \sqrt{11})$. Sketch the surface, C, and l.

16.4

INCREMENTS AND DIFFERENTIALS

If f is a function of two variables x and y, then the symbols Δx and Δy will denote increments of x and y. It should be observed that in the present situation Δy is an increment of the *independent* variable y and is not the same as that defined in (3.20), where y was a *dependent* variable. In terms of this increment notation, Definition (16.7) may be written

$$f_x(x, y) = \lim_{\Delta x \to 0} \frac{f(x + \Delta x, y) - f(x, y)}{\Delta x}$$

$$f_y(x, y) = \lim_{\Delta y \to 0} \frac{f(x, y + \Delta y) - f(x, y)}{\Delta y}.$$

The increment of $w = f(x, y)$ is defined as follows.

Definition (16.9)

> Let $w = f(x, y)$ and let Δx and Δy be increments of x and y, respectively. The **increment** Δw of $w = f(x, y)$ is
>
> $$\Delta w = f(x + \Delta x, y + \Delta y) - f(x, y).$$

Note that the increment Δw represents the change in the value of f if (x, y) changes to $(x + \Delta x, y + \Delta y)$.

Example 1 Let $w = f(x, y) = 3x^2 - xy$.

(a) Find Δw.

(b) Use Δw to calculate the change in $f(x, y)$ if (x, y) changes from $(1, 2)$ to $(1.01, 1.98)$.

Solution

(a) From Definition (16.9),

$$\begin{aligned}
\Delta w &= [3(x + \Delta x)^2 - (x + \Delta x)(y + \Delta y)] - [3x^2 - xy] \\
&= [3x^2 + 6x\,\Delta x + 3(\Delta x)^2 - (xy + x\,\Delta y + y\,\Delta x + \Delta x\,\Delta y)] - 3x^2 + xy \\
&= 6x\,\Delta x + 3(\Delta x)^2 - x\,\Delta y - y\,\Delta x - \Delta x\,\Delta y.
\end{aligned}$$

(b) To find the change in $f(x, y)$ we substitute $x = 1$, $y = 2$, $\Delta x = 0.01$, $\Delta y = -0.02$ in the formula for Δw, obtaining

$$\begin{aligned}
\Delta w &= 6(1)(0.01) + 3(0.01)^2 - (1)(-0.02) - 2(0.01) - (0.01)(-0.02) \\
&= 0.0605.
\end{aligned}$$

This number can also be found by calculating $f(1.01, 1.98) - f(1, 2)$.   ■

Our first objective in this section is to obtain an alternative formula for the increment Δw in Definition (16.9). The reason for doing so will become apparent later, when we prove several important results that have many applications, both in mathematics and in the real world. To gain some insight into the type of formula we should expect, let us consider a differentiable function f of *one* variable x, and set $u = f(x)$. Let x_0 be a fixed value of x. If Δx is an increment of x, then by Definition (3.20),

$$\Delta u = f(x_0 + \Delta x) - f(x_0).$$

Since

$$\lim_{\Delta x \to 0} \frac{\Delta u}{\Delta x} = f'(x_0),$$

it follows that

$$\lim_{\Delta x \to 0} \left[\frac{\Delta u}{\Delta x} - f'(x_0) \right] = 0.$$

If we define ε by

$$\varepsilon = \frac{\Delta u}{\Delta x} - f'(x_0)$$

then ε is a function of Δx, and $\varepsilon \to 0$ as $\Delta x \to 0$. (Strictly speaking, we should use the symbol $\varepsilon(\Delta x)$ to denote the fact that ε depends on Δx.) It follows from the definition of ε that

$$\Delta u = f'(x_0)\,\Delta x + \varepsilon\,\Delta x.$$

This gives us an alternative formula for the increment of a function of *one* variable.

Suppose next that f is a function of *two* variables, and let $w = f(x, y)$. Suppose further that f_x and f_y exist at (x_0, y_0). Consider increments Δx, Δy, and, as in Definition (16.9), let

$$\Delta w = f(x_0 + \Delta x, y_0 + \Delta y) - f(x_0, y_0).$$

From the alternative formula for Δu obtained in the preceding paragraph, it is not unreasonable to expect that Δw may be written in the form

$$\Delta w = f_x(x_0, y_0)\, \Delta x + \varepsilon_1\, \Delta x + f_y(x_0, y_0)\, \Delta y + \varepsilon_2\, \Delta y$$

where the symbols ε_1 and ε_2 denote functions of Δx and Δy that have the limit 0 as $(\Delta x, \Delta y) \to (0, 0)$. The next theorem establishes this fact, provided that suitable restrictions are placed on f. As indicated previously, we shall use the abbreviations ε_1 and ε_2 in place of the more cumbersome notation $\varepsilon_1(\Delta x, \Delta y)$ and $\varepsilon_2(\Delta x, \Delta y)$.

Theorem (16.10)

> Let $w = f(x, y)$, where the function f is defined on a rectangular region $R = \{(x, y): a < x < b, c < y < d\}$. Suppose f_x and f_y exist throughout R and are continuous at the pair (x_0, y_0) in R. If $(x_0 + \Delta x, y_0 + \Delta y)$ is in R and $\Delta w = f(x_0 + \Delta x, y_0 + \Delta y) - f(x_0, y_0)$, then
>
> $$\Delta w = f_x(x_0, y_0)\, \Delta x + f_y(x_0, y_0)\, \Delta y + \varepsilon_1\, \Delta x + \varepsilon_2\, \Delta y$$
>
> where ε_1 and ε_2 are functions of Δx and Δy that have the limit 0 as $(\Delta x, \Delta y) \to (0, 0)$.

Proof   The graph of R and the pairs we wish to consider are illustrated in Figure 16.18.

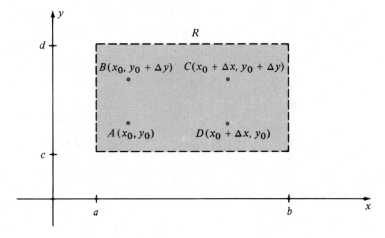

Figure 16.18

We first note that the increment Δw of (16.9) may be expressed as

$$\Delta w = [f(x_0 + \Delta x, y_0 + \Delta y) - f(x_0, y_0 + \Delta y)]$$
$$+ [f(x_0, y_0 + \Delta y) - f(x_0, y_0)].$$

If we next define a function g of *one* variable x by $g(x) = f(x, y_0 + \Delta y)$, then $g'(x) = f_x(x, y_0 + \Delta y)$ for all x under consideration. Applying the Mean Value Theorem (4.12) to g we obtain

$$g(x_0 + \Delta x) - g(x_0) = g'(u)\,\Delta x$$

where u is between x_0 and $x_0 + \Delta x$. Using the definition of g, the last equation can be written

$$f(x_0 + \Delta x, y_0 + \Delta y) - f(x_0, y_0 + \Delta y) = f_x(u, y_0 + \Delta y)\,\Delta x.$$

Similarly, if we let $h(y) = f(x_0, y)$ and apply the Mean Value Theorem to the function h, we obtain

$$h(y_0 + \Delta y) - h(y_0) = h'(v)\,\Delta y$$

where v is between y_0 and $y_0 + \Delta y$. Thus,

$$f(x_0, y_0 + \Delta y) - f(x_0, y_0) = f_y(x_0, v)\,\Delta y.$$

Substitution in the expression for Δw obtained at the beginning of the proof gives us

$$\Delta w = f_x(u, y_0 + \Delta y)\,\Delta x + f_y(x_0, v)\,\Delta y.$$

Let us define ε_1 and ε_2 by

$$\varepsilon_1 = f_x(u, y_0 + \Delta y) - f_x(x_0, y_0)$$

$$\varepsilon_2 = f_y(x_0, v) - f_y(x_0, y_0).$$

Using the fact that f_x and f_y are continuous and noting that $u \to x_0$ and $v \to y_0$ as $\Delta x \to 0$ and $\Delta y \to 0$, respectively, it follows that ε_1 and ε_2 have the limit 0 as $(\Delta x, \Delta y) \to (0, 0)$. If we rewrite the preceding equations as

$$f_x(u, y_0 + \Delta y) = f_x(x_0, y_0) + \varepsilon_1$$

$$f_y(x_0, v) = f_y(x_0, y_0) + \varepsilon_2$$

and substitute into the last expression for Δw, we see that

$$\Delta w = [f_x(x_0, y_0) + \varepsilon_1]\,\Delta x + [f_y(x_0, y_0) + \varepsilon_2]\,\Delta y$$

which leads to the conclusion of the theorem. $\square$

Example 2 If $w = 3x^2 - xy$, find expressions for ε_1 and ε_2 that satisfy the conclusion of Theorem 16.10.

Solution   From the solution of Example 1,

$$\Delta w = 6x\,\Delta x + 3(\Delta x)^2 - x\,\Delta y - y\,\Delta x - \Delta x\,\Delta y$$

which may also be written

$$\Delta w = (6x - y)\,\Delta x + (-x)\,\Delta y + (3\,\Delta x)(\Delta x) + (-\Delta x)\,\Delta y.$$

If we define $\varepsilon_1 = 3\,\Delta x$ and $\varepsilon_2 = (-\Delta x)$, then the form in Theorem (16.10) is obtained. Note that ε_1 and ε_2 are not unique since we may also write

$$\Delta w = (6x - y)\,\Delta x + (-x)\,\Delta y + (3\,\Delta x - \Delta y)\,\Delta x + 0\,\Delta y$$

in which case $\varepsilon_1 = 3\,\Delta x - \Delta y$ and $\varepsilon_2 = 0$.   ∎

In Chapter 3 we defined differentials for functions of one variable. Recall that if $u = f(x)$ and Δx is an increment of x, then by (3.22) $dx = \Delta x$ and $du = f'(x)\,dx$. The next definition extends this concept to functions of two variables.

Definition (16.11)

Let $w = f(x, y)$, and let Δx and Δy be increments of x and y.

(i) The **differentials dx and dy of the independent variables** x and y are

$$dx = \Delta x \quad \text{and} \quad dy = \Delta y.$$

(ii) The **differential dw of the dependent variable** w is

$$dw = f_x(x, y)\,dx + f_y(x, y)\,dy = \frac{\partial w}{\partial x}\,dx + \frac{\partial w}{\partial y}\,dy.$$

If f satisfies the hypothesis of Theorem (16.10), then using the conclusion of that theorem, with the pair (x_0, y_0) replaced by (x, y), we see that

$$\Delta w - dw = \varepsilon_1\,\Delta x + \varepsilon_2\,\Delta y$$

where ε_1 and ε_2 approach 0 as $(\Delta x, \Delta y) \to (0, 0)$. It follows that if Δx and Δy are small, then $\Delta w - dw \approx 0$, that is, $dw \approx \Delta w$. This fact may be used to approximate the change in w corresponding to small changes in x and y.

Example 3   If $w = 3x^2 - xy$, find dw and use it to approximate the change in w if (x, y) changes from $(1, 2)$ to $(1.01, 1.98)$. How does this compare with the exact change in w?

Solution   Applying Definition (16.11),

$$dw = \frac{\partial w}{\partial x}\,dx + \frac{\partial w}{\partial y}\,dy$$

$$= (6x - y)\,dx + (-x)\,dy.$$

Substituting $x = 1, y = 2, dx = \Delta x = 0.01,$ and $dy = \Delta y = -0.02,$ we obtain

$$dw = (6 - 2)(0.01) + (-1)(-0.02) = 0.06.$$

It was shown in Example 1 that $\Delta w = 0.0605$. Hence the error involved in using dw is 0.0005. ∎

Example 4 The radius and altitude of a right circular cylinder are measured as 3 in. and 8 in., respectively, with a possible error in measurement of ± 0.05 in. Use differentials to approximate the maximum error in the calculated volume of the cylinder.

Solution   The volume V of a cylinder of radius r and altitude h is $V = \pi r^2 h$. Consequently,

$$dV = \frac{\partial V}{\partial r}\, dr + \frac{\partial V}{\partial h}\, dh = 2\pi r h\, dr + \pi r^2\, dh.$$

Substituting $r = 3$, $h = 8$, and $dr = dh = \pm 0.05$ gives us the following approximation to the maximum error:

$$dV = 48\pi(\pm 0.05) + 9\pi(\pm 0.05) = \pm 2.85\pi \approx \pm 8.95 \text{ in.}^3. \qquad ∎$$

For a function of one variable, the term *differentiable* means that the derivative exists. For functions of two variables we use the much stronger condition stated in the next definition.

Definition (16.12)

> If $w = f(x, y)$, then f is **differentiable** at (x_0, y_0) provided Δw can be expressed in the form
>
> $$\Delta w = f_x(x_0, y_0)\, \Delta x + f_y(x_0, y_0)\, \Delta y + \varepsilon_1\, \Delta x + \varepsilon_2\, \Delta y$$
>
> where ε_1 and ε_2 have the limit 0 as $(\Delta x, \Delta y) \to (0, 0)$.

A function f of two variables is said to be **differentiable on a region R** if it is differentiable at each point of R. The next theorem follows directly from Theorem (16.10) and Definition (16.12). The terminology *rectangular region* means a region of the type described in (16.10).

Theorem (16.13)

> If $w = f(x, y)$, and f_x and f_y are continuous on a rectangular region R, then f is differentiable on R.

The following result shows that a differentiable function is continuous, and is one of the reasons for using Definition (16.12) as the definition of differentiability.

Theorem (16.14)

> If a function f of two variables is differentiable at (x_0, y_0), then f is continuous at (x_0, y_0).

Proof We begin by writing the formula for Δw in Definition (16.12) as follows:

$$\Delta w = [f_x(x_0, y_0) + \varepsilon_1]\,\Delta x + [f_y(x_0, y_0) + \varepsilon_2]\,\Delta y.$$

If we let $x = x_0 + \Delta x$ and $y = y_0 + \Delta y$, then

$$\Delta w = f(x, y) - f(x_0, y_0)$$
$$= [f_x(x_0, y_0) + \varepsilon_1](x - x_0) + [f_y(x_0, y_0) + \varepsilon_2](y - y_0).$$

It follows that

$$\lim_{(x, y) \to (x_0, y_0)} [f(x, y) - f(x_0, y_0)] = 0$$

or

$$\lim_{(x, y) \to (x_0, y_0)} f(x, y) = f(x_0, y_0).$$

Hence f is continuous at (x_0, y_0). $\qquad\square$

Applying Theorem (16.13) gives us the following corollary.

Corollary (16.15)

> If f is a function of two variables, and if f_x and f_y are continuous on a rectangular region R, then f is continuous on R.

It can be shown by means of examples that the mere *existence* of f_x and f_y is not enough to ensure continuity of f (see Exercise 39). This brings out a difference from the single variable case, where the existence of f' implies continuity of f.

The discussion in this section can be extended to functions of more than two variables. For example, suppose $w = f(x, y, z)$ where f is defined on a suitable region R (such as a rectangular parallelepiped) and f_x, f_y, f_z exist in R and are continuous at (x, y, z). If x, y, and z are given increments Δx, Δy, and Δz, respectively, then the corresponding increment

$$\Delta w = f(x + \Delta x, y + \Delta y, z + \Delta z) - f(x, y, z)$$

can be written in the form

$$\Delta w = f_x(x, y, z)\, \Delta x + f_y(x, y, z)\, \Delta y + f_z(x, y, z)\, \Delta z + \varepsilon_1\, \Delta x + \varepsilon_2\, \Delta y + \varepsilon_3\, \Delta z$$

where ε_1, ε_2, and ε_3 are functions of Δx, Δy, and Δz that have the limit 0 as $(\Delta x, \Delta y, \Delta z) \to (0, 0, 0)$. Results that are analogous to (16.13)–(16.15) may be established for functions of three variables.

The following definition is an extension of (16.11) to functions of three variables.

Definition (16.16)

Let $w = f(x, y, z)$ and let Δx, Δy, and Δz be increments of x, y, and z, respectively.

(i)   The **differentials of the independent variables x, y, and z** are

$$dx = \Delta x, \quad dy = \Delta y, \quad dz = \Delta z.$$

(ii)   The **differential of the dependent variable w** is

$$dw = \frac{\partial w}{\partial x}\, dx + \frac{\partial w}{\partial y}\, dy + \frac{\partial w}{\partial z}\, dz.$$

We can use dw to approximate Δw if the increments of x, y, and z are small. Differentiability is defined as in (16.12). The extension to four or more variables is made in like manner.

Example 5   Suppose the dimensions (in inches) of a rectangular parallelepiped change from 9, 6, and 4 to 9.02, 5.97, and 4.01, respectively. Use differentials to approximate the change in volume. What is the exact change in volume?

Solution   If we denote the dimensions by x, y, and z, then the volume is $V = xyz$ and, by Definition (16.16),

$$dV = yz\, dx + xz\, dy + xy\, dz.$$

Substituting $x = 9$, $y = 6$, $z = 4$, $dx = 0.02$, $dy = -0.03$, and $dz = 0.01$ gives us

$$\Delta V \approx dV = 24(0.02) + 36(-0.03) + 54(0.01)$$
$$= 0.48 - 1.08 + 0.54 = -0.06.$$

Thus the volume decreases by approximately 0.06 in.[3] The exact change in volume is

$$\Delta V = (9.02)(5.97)(4.01) - (9)(6)(4) = -0.063906. \qquad \blacksquare$$

EXERCISES 16.4

In Exercises 1–4 find values of ε_1 and ε_2 that satisfy Definition (16.12).

1 $f(x, y) = 4y^2 - 3xy + 2x$

2 $f(x, y) = (2x - y)^2$

3 $f(x, y) = x^3 + y^3$

4 $f(x, y) = 2x^2 - xy^2 + 3y$

In Exercises 5–16 find dw.

5 $w = x^3 - x^2y + 3y^2$

6 $w = 5x^2 + 4y - 3xy^3$

7 $w = x^2 \sin y + 2y^{3/2}$

8 $w = ye^{-2x} - 3x^4$

9 $w = x^2e^{xy} + (1/y^2)$

10 $w = \ln(x^2 + y^2) + x \tan^{-1} y$

11 $w = x^2 \ln(y^2 + z^2)$

12 $w = x^2y^3z + e^{-2z}$

13 $w = xyz/(x + y + z)$

14 $w = x^2e^{yz} + y \ln z$

15 $w = x^2z + 4yt^3 - xz^2t$

16 $w = x^2y^3zt^{-1}v^4$

17 Use differentials to approximate the change in

$$f(x, y) = x^2 - 3x^3y^2 + 4x - 2y^3 + 6$$

if (x, y) changes from $(-2, 3)$ to $(-2.02, 3.01)$.

18 Use differentials to approximate the change in

$$f(x, y) = x^2 - 2xy + 3y$$

if (x, y) changes from $(1, 2)$ to $(1.03, 1.99)$.

19 Use differentials to approximate the change in

$$f(x, y, z) = x^2z^3 - 3yz^2 + x^{-3} + 2y^{1/2}z$$

if (x, y, z) changes from $(1, 4, 2)$ to $(1.02, 3.97, 1.96)$.

20 Use differentials to approximate the change in

$$w = r^2 + 3sv + 2p^3$$

if r changes from 1 to 1.02, s from 2 to 1.99, v from 4 to 4.01, and p from 3 to 2.97.

21 Use differentials to approximate $\sqrt[3]{26.98}\sqrt{36.04}$.

22 Use differentials to approximate $(32.03)^{2/5}/(1.95)^4$.

23 The dimensions of a closed rectangular box are measured as 3 ft, 4 ft, and 5 ft with a possible error of $\frac{1}{16}$ in. Use differentials to approximate the maximum error in the calculated value of (a) the surface area of the box; (b) the volume of the box.

24 The two shortest sides of a right triangle are measured as 3 cm and 4 cm, respectively, with a possible error of 0.02 cm. Use differentials to approximate the maximum error in the calculated value of (a) the hypotenuse; (b) the area of the triangle.

25 An open cylindrical tin can has diameter 3 in. and altitude 4 in. Use differentials to approximate the amount of tin in the can if the tin is 0.015 in. thick.

26 The total resistance R of three resistances $R_1, R_2,$ and R_3, connected in parallel (see figure) is given by

$$\frac{1}{R} = \frac{1}{R_1} + \frac{1}{R_2} + \frac{1}{R_3}.$$

If measurements of $R_1, R_2,$ and R_3 are 100, 200, and 400 ohms, respectively, with a maximum error of 1% in each measurement, approximate the maximum error in the calculated value of R.

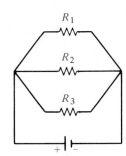

Figure for Exercise 26

27 The specific gravity of an object is given by $s = A/(A - W)$, where A and W are the weights (in pounds) of the object in air and water, respectively. If measurements are $A = 12$ lb and $W = 5$ lb with maximum errors of $\frac{1}{2}$ oz in air and 1 oz in water, what is the maximum error in the calculated value of s?

28 The pressure P, volume V, and temperature T of a confined gas are related by the *Ideal Gas Law* $PV = kT$, where k is a constant. If $P = 0.5$ lb/in.^{2} when $V = 64$ in.^{3} and $T = 80°$, approximate the change in P if V and T change to 70 in.^{3} and 76°, respectively.

29 In using the specific gravity formula $s = A/(A - W)$ (see Exercise 27), suppose that there are percentage errors of 2% and 4% in the measurements A and W, respectively. Express the maximum possible percentage error in the calculated value of s in terms of A and W.

30 In using the Ideal Gas Law $PV = kT$ (see Exercise 28), suppose that there are percentage errors of 0.8% and 0.5% in the measurements of T and P, respectively. Approximate the maximum percentage error in the calculated value of V.

31 The electrical resistance R of a wire is directly proportional to its length and inversely proportional to the square of its diameter. If the length is measured with a possible error of 1% and the diameter is measured with a possible error of 3%, what is the maximum percentage error in the calculated value of R?

32 It was shown in Section 6.8 that the flow of blood through an arteriole is given by $F = \pi PR^4/8vl$, where l is the length of the arteriole, R is the radius, P is the pressure difference between the two ends, and v is the viscosity of the blood. Suppose that v and l are constant. If the radius decreases by 2% and the pressure increases by 3%, use differentials to find the percentage change in the flow.

33 The temperature T at the point $P(x, y, z)$ in a rectangular coordinate system is given by $T = 8(2x^2 + 4y^2 + 9z^2)^{1/2}$. Use differentials to approximate the temperature difference between the points $(6, 3, 2)$ and $(6.1, 3.3, 1.98)$.

34 Approximate the change in area of an isosceles triangle if each of the two equal sides increases from 100 to 101 and the angle between them decreases from $120°$ to $119°$.

35 If the cylinder described in Example 4 of this section has a closed top and bottom, use differentials to approximate the maximum error in the calculated total surface area.

36 Use differentials to approximate the change in surface area of the parallelepiped described in Example 5 of this section. What is the exact change?

In Exercises 37 and 38 prove that f is differentiable at every point in its domain.

37 $f(x, y) = (x^2 - y^2)/(x^2 + y^2)$

38 $f(x, y, z) = (x + y + z)/(x^2 + y^2 + z^2)$

39 Suppose $f(x, y) = \begin{cases} \dfrac{xy}{x^2 + y^2} & \text{if } (x, y) \neq (0, 0) \\ 0 & \text{if } (x, y) = (0, 0). \end{cases}$

Prove that

(a) $f_x(0, 0)$ and $f_y(0, 0)$ exist. (*Hint*: Use (16.7).)

(b) f is not continuous at $(0, 0)$.

(c) f is not differentiable at $(0, 0)$.

40 Suppose

$$f(x, y, z) = \begin{cases} \dfrac{xyz}{x^3 + y^3 + z^3} & \text{if } (x, y, z) \neq (0, 0, 0) \\ 0 & \text{if } (x, y, z) = (0, 0, 0). \end{cases}$$

Prove that f_x, f_y, and f_z exist at $(0, 0, 0)$, but that f is not differentiable at $(0, 0, 0)$.

16.5

THE CHAIN RULE

If f, g, and k are functions of two variables such that $w = f(u, v)$, $u = g(x, y)$, and $v = k(x, y)$, and if for each pair (x, y) in a subset D of $\mathbb{R} \times \mathbb{R}$ the corresponding pair (u, v) is in the domain of f, then $w = f(g(x, y), k(x, y))$ defines w as a (composite) function of x and y. For example, if

$$w = u^2 + u \sin v, \quad u = xe^{2y}, \quad \text{and} \quad v = xy$$

then

$$w = x^2 e^{4y} + xe^{2y} \sin xy.$$

The next theorem provides formulas for expressing $\partial w/\partial x$ and $\partial w/\partial y$ in terms of the first partial derivatives of the functions g, k, and f. In the statement of the theorem it is assumed that domains are chosen such that the composite function is defined on a suitable domain D. Each of the formulas stated in Theorem (16.17) is referred to as a *Chain Rule*.

The Chain Rule (16.17)

> If $w = f(u, v)$, $u = g(x, y)$, and $v = k(x, y)$ where f is differentiable and g and k have continuous first partial derivatives, then
>
> $$\frac{\partial w}{\partial x} = \frac{\partial w}{\partial u}\frac{\partial u}{\partial x} + \frac{\partial w}{\partial v}\frac{\partial v}{\partial x}$$
>
> $$\frac{\partial w}{\partial y} = \frac{\partial w}{\partial u}\frac{\partial u}{\partial y} + \frac{\partial w}{\partial v}\frac{\partial v}{\partial y}.$$

Proof   If x is given an increment Δx and y is held constant (that is, $\Delta y = 0$), then

(a)
$$\Delta w = f(g(x + \Delta x, y), k(x + \Delta x, y)) - f(g(x, y), k(x, y)).$$

Also,

(b)
$$\Delta u = g(x + \Delta x, y) - g(x, y)$$
$$\Delta v = k(x + \Delta x, y) - k(x, y)$$

and hence

$$g(x + \Delta x, y) = g(x, y) + \Delta u = u + \Delta u$$

$$k(x + \Delta x, y) = k(x, y) + \Delta v = v + \Delta v.$$

Substituting into equation (a),

$$\Delta w = f(u + \Delta u, v + \Delta v) - f(u, v).$$

Since f is differentiable, we may write

(c)
$$\Delta w = \frac{\partial w}{\partial u}\Delta u + \frac{\partial w}{\partial v}\Delta v + \varepsilon_1 \Delta u + \varepsilon_2 \Delta v$$

where the symbols ε_1 and ε_2 denote functions of Δu and Δv that have the limit 0 as $(\Delta u, \Delta v) \to (0, 0)$. Moreover, we may assume that both ε_1 and ε_2 are 0 if $(\Delta u, \Delta v) = (0, 0)$, for if they are not, they can be replaced by functions μ_1 and μ_2 that have this property and are equal to ε_1 and ε_2 elsewhere. With this agreement, the functions ε_1 and ε_2 in (c) are continuous at $(0, 0)$. Dividing both sides of that equation by Δx gives us

(d)
$$\frac{\Delta w}{\Delta x} = \frac{\partial w}{\partial u}\frac{\Delta u}{\Delta x} + \frac{\partial w}{\partial v}\frac{\Delta v}{\Delta x} + \varepsilon_1 \frac{\Delta u}{\Delta x} + \varepsilon_2 \frac{\Delta v}{\Delta x}.$$

From equations (a) and (b),

$$\lim_{\Delta x \to 0} \frac{\Delta w}{\Delta x} = \frac{\partial w}{\partial x}, \quad \lim_{\Delta x \to 0} \frac{\Delta u}{\Delta x} = \frac{\partial u}{\partial x}, \quad \lim_{\Delta x \to 0} \frac{\Delta v}{\Delta x} = \frac{\partial v}{\partial x}.$$

If Δx approaches 0, then Δu and Δv also approach 0 and hence so do ε_1 and ε_2. Consequently, if we take the limit in equation (d) as $\Delta x \to 0$, we obtain

$$\frac{\partial w}{\partial x} = \frac{\partial w}{\partial u}\frac{\partial u}{\partial x} + \frac{\partial w}{\partial v}\frac{\partial v}{\partial x}.$$

The second formula in the statement of the theorem is established in similar fashion.
□

It is of interest to note that (16.17) is a generalization of the Chain Rule (3.28) for functions of one variable. Specifically, if we let $w = f(u)$, $u = g(x)$, and $v = 0$, then the partial derivatives become ordinary derivatives, and the first formula in (16.17) takes on the form

$$\frac{dw}{dx} = \frac{dw}{du}\frac{du}{dx}$$

which is in agreement with (3.28).

Example 1   If $w = u^3 + v^2$, $u = xy^2$, and $v = x^2 \sin y$, use the Chain Rule to find $\partial w/\partial x$ and $\partial w/\partial y$.

Solution   Note that w is a function of x and y. Specifically,

$$w = (xy^2)^3 + (x^2 \sin y)^2.$$

We could find $\partial w/\partial x$ and $\partial w/\partial y$ directly from this expression; however, our objective is to illustrate the Chain Rule. Thus, applying (16.17),

$$\frac{\partial w}{\partial x} = (3u^2)y^2 + (2v)(2x \sin y)$$
$$= 3(xy^2)^2 y^2 + (2x^2 \sin y)(2x \sin y)$$
$$= 3x^2 y^6 + 4x^3 \sin^2 y.$$

Similarly,

$$\frac{\partial w}{\partial y} = (3u^2)(2xy) + (2v)(x^2 \cos y)$$
$$= 3(xy^2)^2(2xy) + 2(x^2 \sin y)(x^2 \cos y)$$
$$= 6x^3 y^5 + 2x^4 \sin y \cos y. \qquad \blacksquare$$

Note that after applying the Chain Rule in Example 1 we substituted for u and v, thereby expressing $\partial w/\partial x$ and $\partial w/\partial y$ in terms of x and y. This was done to emphasize the fact that w is a (composite) function of the *two* variables x and y.

The Chain Rule can be extended to any number of variables, as stated in the following theorem.

General Chain Rule (16.18)

If w is a differentiable function of n variables $u_1, u_2, \ldots, u_n$ where each u_i is a function of m variables $x_1, x_2, \ldots, x_m$ having continuous first partial derivatives, and if w is a (composite) function of $x_1, x_2, \ldots, x_m$, then

$$\frac{\partial w}{\partial x_i} = \frac{\partial w}{\partial u_1} \frac{\partial u_1}{\partial x_i} + \frac{\partial w}{\partial u_2} \frac{\partial u_2}{\partial x_i} + \cdots + \frac{\partial w}{\partial u_n} \frac{\partial u_n}{\partial x_i}$$

for $i = 1, 2, \ldots, m$.

Example 2 If $w = r^2 + sv + t^3$, where $r = x^2 + y^2 + z^2$, $s = xyz$, $v = xe^y$, and $t = yz^2$, use the Chain Rule to find $\partial w/\partial z$.

Solution Note that w is a function of r, s, v, t and that each of these four variables is a function of x, y, and z. Applying Theorem (16.18),

$$\frac{\partial w}{\partial z} = \frac{\partial w}{\partial r} \frac{\partial r}{\partial z} + \frac{\partial w}{\partial s} \frac{\partial s}{\partial z} + \frac{\partial w}{\partial v} \frac{\partial v}{\partial z} + \frac{\partial w}{\partial t} \frac{\partial t}{\partial z}$$

$$= (2r)(2z) + v(xy) + s(0) + (3t^2)(2yz)$$

$$= 4z(x^2 + y^2 + z^2) + xe^y(xy) + 0 + 3(yz^2)^2(2yz)$$

$$= 4z(x^2 + y^2 + z^2) + x^2ye^y + 6y^3z^5. \qquad \blacksquare$$

The following special case of the Chain Rule follows directly from Theorem (16.18) by letting $x_i = t$ for each i.

Theorem (16.19)

If w is a function of $u_1, u_2, \ldots, u_n$ and each u_i is a function of *one* variable t such that du_i/dt is continuous, then w is a function of t and

$$\frac{dw}{dt} = \frac{\partial w}{\partial u_1} \frac{du_1}{dt} + \frac{\partial w}{\partial u_2} \frac{du_2}{dt} + \cdots + \frac{\partial w}{\partial u_n} \frac{du_n}{dt}.$$

In Theorem (16.19) the notations dw/dt and du_i/dt are used because w and each u_i is a function of *one* variable t. This situation could occur in applications where w is a function of several variables and each variable is a function of time t. In this case the derivative dw/dt would give us the rate of change of w with respect to time.

Example 3 If $w = x^2 + yz$ and $x = 3t^2 + 1$, $y = 2t - 4$, $z = t^3$, find dw/dt.

Solution The problem could be solved without the Chain Rule by writing

$$w = (3t^2 + 1)^2 + (2t - 4)t^3$$

and then finding dw/dt by single variable methods. However, if we apply Theorem (16.19), then

$$
\begin{aligned}
\frac{dw}{dt} &= \frac{\partial w}{\partial x}\frac{dx}{dt} + \frac{\partial w}{\partial y}\frac{dy}{dt} + \frac{\partial w}{\partial z}\frac{dz}{dt} \\
&= (2x)(6t) + z(2) + y(3t^2) \\
&= 2(3t^2 + 1)6t + t^3(2) + (2t - 4)3t^2 \\
&= 44t^3 - 12t^2 + 12t.
\end{aligned}
$$
∎

Partial derivatives can be used to find derivatives of functions that are determined implicitly. Suppose, as in Section 3.6, an equation $F(x, y) = 0$ determines a differentiable function f of one variable x such that $F(x, f(x)) = 0$ for all x in the domain D of f. If we let

$$w = F(u, y), \quad \text{where } u = x \text{ and } y = f(x)$$

then applying Theorem (16.19) and using the fact that $w = F(x, f(x)) = 0$ for all x in D,

$$\frac{dw}{dx} = \frac{\partial w}{\partial u}\frac{du}{dx} + \frac{\partial w}{\partial y}\frac{dy}{dx} = 0,$$

that is,

$$\frac{\partial w}{\partial u}(1) + \frac{\partial w}{\partial y}f'(x) = 0.$$

If $\partial w/\partial y \neq 0$, then (since $u = x$)

$$f'(x) = -\frac{\partial w/\partial x}{\partial w/\partial y} = -\frac{F_x(x, y)}{F_y(x, y)}.$$

We may summarize this discussion as follows.

Theorem (16.20)

> If an equation $F(x, y) = 0$ determines, implicitly, a differentiable function f of one variable x such that $y = f(x)$, then
> $$\frac{dy}{dx} = -\frac{F_x(x, y)}{F_y(x, y)}.$$

Example 4 Find y' if $y = f(x)$ satisfies the equation

$$y^4 + 3y - 4x^3 - 5x - 1 = 0.$$

Solution If $F(x, y)$ is the expression on the left side of the equation, then by Theorem (16.20),

$$y' = -\frac{-12x^2 - 5}{4y^3 + 3} = \frac{12x^2 + 5}{4y^3 + 3}.$$

Compare this solution with that of Example 2 in Section 3.6, which was obtained using single-variable methods. ∎

Given an equation such as

$$x^2 - 4y^3 + 2z - 7 = 0$$

we can solve for z, obtaining

$$z = \tfrac{1}{2}(-x^2 + 4y^3 + 7)$$

which is of the form　　　　　　$z = f(x, y).$

In analogy with the single variable case, we say that the function f of two variables x and y is determined *implicitly* by the given equation. The next theorem gives us formulas for finding f_x and f_y or, equivalently, $\partial z/\partial x$ and $\partial z/\partial y$ without actually solving the equation for z.

Theorem (16.21)

> If an equation $F(x, y, z) = 0$ determines an implicit differentiable function f of two variables x and y such that $z = f(x, y)$ for all (x, y) in the domain of f, then
>
> $$\frac{\partial z}{\partial x} = -\frac{F_x(x, y, z)}{F_z(x, y, z)}, \qquad \frac{\partial z}{\partial y} = -\frac{F_y(x, y, z)}{F_z(x, y, z)}.$$

Proof　　　The statement that $F(x, y, z) = 0$ determines the function f, where $z = f(x, y)$, means that $F(x, y, f(x, y)) = 0$ for all (x, y) in the domain of f. If we let

$$w = F(u, v, z), \quad \text{where } u = x, v = y, z = f(x, y)$$

then by the Chain Rule and the fact that $w = 0$ for all (x, y) in D,

$$\frac{\partial w}{\partial x} = \frac{\partial w}{\partial u}\frac{\partial u}{\partial x} + \frac{\partial w}{\partial v}\frac{\partial v}{\partial x} + \frac{\partial w}{\partial z}\frac{\partial z}{\partial x} = 0.$$

Since $u = x$ and $v = y$, we have

$$\frac{\partial w}{\partial x}(1) + \frac{\partial w}{\partial y}(0) + \frac{\partial w}{\partial z}\frac{\partial z}{\partial x} = 0$$

and, if $\partial w/\partial z \neq 0$,

$$\frac{\partial z}{\partial x} = -\frac{\partial w/\partial x}{\partial w/\partial z} = -\frac{F_x(x, y, z)}{F_z(x, y, z)}.$$

The formula for $\partial z/\partial y$ may be obtained in similar fashion. □

Example 5 If $z = f(x, y)$ satisfies the equation

$$x^2z^2 + xy^2 - z^3 + 4yz - 5 = 0$$

find $\partial z/\partial x$ and $\partial z/\partial y$.

Solution   If we let $F(x, y, z)$ denote the expression on the left of the given equation, then by Theorem (16.21),

$$\frac{\partial z}{\partial x} = -\frac{2xz^2 + y^2}{2x^2z - 3z^2 + 4y}$$

$$\frac{\partial z}{\partial y} = -\frac{2xy + 4z}{2x^2z - 3z^2 + 4y}.$$ ■

EXERCISES 16.5

Use the Chain Rule to solve Exercises 1–14.

In Exercises 1 and 2 find $\partial w/\partial x$ and $\partial w/\partial y$.

1 $w = u \sin v, u = x^2 + y^2, v = xy$

2 $w = uv + v^2, u = x \sin y, v = y \sin x$

In Exercises 3 and 4 find $\partial w/\partial r$ and $\partial w/\partial s$.

3 $w = u^2 + 2uv, u = r \ln s, v = 2r + s$

4 $w = e^{tv}, t = r + s, v = rs$

In Exercises 5 and 6 find $\partial z/\partial x$ and $\partial z/\partial y$.

5 $z = r^3 + s + v^2, r = xe^y, s = ye^x, v = x^2y$

6 $z = pq + qw, p = 2x - y, q = x - 2y, w = -2x + 2y$

In Exercises 7 and 8 find $\partial r/\partial u, \partial r/\partial v,$ and $\partial r/\partial t$.

7 $r = x \ln y, x = 3u + vt, y = uvt$

8 $r = w^2 \cos z, w = u^2vt, z = ut^2$

9 Find $\partial p/\partial r$ if $p = u^2 + 3v^2 - 4w^2, u = x - 3y + 2r - s,$ $v = 2x + y - r + 2s, w = -x + 2y + r + s.$

10 Find $\partial s/\partial \partial$ if $s = tr + ue^v, t = xy^2z, r = x^2yz,$ $u = xyz^2, v = xyz.$

In Exercises 11–14 find dw/dt.

11 $w = x^3 - y^3, x = 1/(t + 1), y = t/(t + 1)$

12 $w = \ln (u + v), u = e^{-2t}, v = t^3 - t^2$

13 $w = r^2 - s \tan v, r = \sin^2 t, s = \cos t, v = 4t$

14 $w = x^2y^3z^4, x = 2t + 1, y = 3t - 2, z = 5t + 4$

In Exercises 15–18 find y' under the assumption that $y = f(x)$ satisfies the given equation.

15 $2x^3 + x^2y + y^3 = 1$

16 $x^4 + 2x^2y^2 - 3xy^3 + 2x = 0$

17 $6x + \sqrt{xy} = 3y - 4$

18 $x^{2/3} + y^{2/3} = 4$

In Exercises 19–22 find $\partial z/\partial x$ and $\partial z/\partial y$ under the assumption that $z = f(x, y)$ satisfies the given equation.

19 $2xz^3 - 3yz^2 + x^2y^2 + 4z = 0$

20 $xz^2 + 2x^2y - 4y^2z + 3y - 2 = 0$

21 $xe^{yz} - 2ye^{xz} + 3ze^{xy} = 1$

22 $yx^2 + z^2 + \cos xyz = 4$

23 If $w = f(x, y), x = r \cos \theta,$ and $y = r \sin \theta$ show that

$$\left(\frac{\partial w}{\partial x}\right)^2 + \left(\frac{\partial w}{\partial y}\right)^2 = \left(\frac{\partial w}{\partial r}\right)^2 + \frac{1}{r^2}\left(\frac{\partial w}{\partial \theta}\right)^2.$$

24 If $w = f(x, y)$ and $x = e^r \cos \theta, y = e^r \sin \theta$, show that

$$\frac{\partial^2 w}{\partial x^2} + \frac{\partial^2 w}{\partial y^2} = e^{-2r}\left(\frac{\partial^2 w}{\partial r^2} + \frac{\partial^2 w}{\partial \theta^2}\right).$$

25 If $w = f(x, y)$ and $x = r \cos \theta, y = r \sin \theta$, show that

$$\frac{\partial^2 w}{\partial x^2} + \frac{\partial^2 w}{\partial y^2} = \frac{\partial^2 w}{\partial r^2} + \frac{1}{r^2}\frac{\partial^2 w}{\partial \theta^2} + \frac{1}{r}\frac{\partial w}{\partial r}.$$

26 If $v = f(x - at) + g(x + at)$ where f and g have second partial derivatives, show that v satisfies the wave equation

$$\frac{\partial^2 v}{\partial t^2} = a^2 \frac{\partial^2 v}{\partial x^2}.$$

(Compare with Exercise 42 of Section 16.3.)

27 If $w = \cos(x + y) + \cos(x - y)$, show, without using addition formulas, that $w_{xx} - w_{yy} = 0$.

28 If $w = f(x^2 + y^2)$, show that $y(\partial w/\partial x) - x(\partial w/\partial y) = 0$. (*Hint*: Let $u = x^2 + y^2$.)

29 If $w = f(u, v)$, $u = g(x, y)$, and $v = k(x, y)$, show that

$$\frac{\partial^2 w}{\partial x^2} = \frac{\partial^2 w}{\partial u^2}\left(\frac{\partial u}{\partial x}\right)^2 + \left(\frac{\partial^2 w}{\partial v \, \partial u} + \frac{\partial^2 w}{\partial u \, \partial v}\right)\frac{\partial u}{\partial x}\frac{\partial v}{\partial x}$$

$$+ \frac{\partial^2 w}{\partial v^2}\left(\frac{\partial v}{\partial x}\right)^2 + \frac{\partial w}{\partial u}\frac{\partial^2 u}{\partial x^2} + \frac{\partial w}{\partial v}\frac{\partial^2 v}{\partial x^2}.$$

30 Given w, u, and v as in Exercise 25, show that

$$\frac{\partial^2 w}{\partial y \, \partial x} = \frac{\partial^2 w}{\partial u^2}\frac{\partial u}{\partial x}\frac{\partial u}{\partial y} + \frac{\partial^2 w}{\partial v \, \partial u}\frac{\partial u}{\partial x}\frac{\partial v}{\partial y} + \frac{\partial^2 w}{\partial u \, \partial v}\frac{\partial u}{\partial y}\frac{\partial v}{\partial x}$$

$$+ \frac{\partial^2 w}{\partial v^2}\frac{\partial v}{\partial x}\frac{\partial v}{\partial y} + \frac{\partial w}{\partial u}\frac{\partial^2 u}{\partial y \, \partial x} + \frac{\partial w}{\partial v}\frac{\partial^2 v}{\partial y \, \partial x}.$$

31 If n resistances $R_1, R_2, \ldots, R_n$ are connected in parallel, then the total resistance R is given by

$$\frac{1}{R} = \sum_{i=1}^{n} \frac{1}{R_i}.$$

Prove that for $i = 1, 2, \ldots, n$,

$$\frac{\partial R}{\partial R_i} = \left(\frac{R}{R_i}\right)^2.$$

32 A function f of two variables is said to be **homogeneous of degree n** if $f(tx, ty) = t^n f(x, y)$ for every t such that (tx, ty) is in the domain of f. Show that for such functions, $x f_x(x, y) + y f_y(x, y) = nf(x, y)$. (*Hint*: Differentiate $f(tx, ty)$ with respect to t.)

33 The radius r and altitude h of a right circular cylinder are increasing at rates of 0.01 in./min and 0.02 in./min, respectively. Use the Chain Rule to find the rate at which the volume is increasing at the time when $r = 4$ in. and $h = 7$ in. At what rate is the curved surface area changing at this time?

34 The equal sides and included angle of an isosceles triangle are increasing at rates of 0.1 ft/hr and 2°/hr, respectively.

Use the Chain Rule to find the rate at which the area of the triangle is increasing at the time when the length of each of the equal sides is 20 ft and the included angle is 60°.

35 The pressure P, volume V, and temperature T of a confined gas are related by the Ideal Gas Law $PV = kT$, where k is a constant. If P and V are changing at the rates dP/dt and dV/dt, respectively, use the Chain Rule to find a formula for dT/dt. Check your answer by using the product rule for functions of one variable.

36 If the base radius r and altitude h of a right circular cylinder are changing at the rates dr/dt and dh/dt, respectively, use the Chain Rule to find a formula for dV/dt, where V is the volume of the cylinder. Check your answer by using single variable methods.

37 A certain gas obeys the Ideal Gas Law $PV = 8T$. Suppose that the gas is being heated at a rate of 2°/min and the pressure is increasing at a rate of $\frac{1}{2}$ (lb/in²)/min. If, at a certain instant, the temperature is 200° and the pressure is 10 lb/in.², find the rate at which the volume is changing.

38 Sand is leaking out of a hole in a container at a rate of 6 in.³/min. As it leaks out it forms a pile in the shape of a right circular cone whose base radius is increasing at a rate of $\frac{1}{4}$ in./min. If, at the instant that 40 in.³ has leaked out, the radius is 5 in., find the rate at which the height of the pile is increasing.

39 Prove the following Mean Value Theorem for a function f of two variables x and y:

If f has first partial derivatives that are continuous on a rectangular region $R = \{(x, y): a < x < b, \ c < y < d\}$, and if $A(x_1, y_1)$ and $B(x_2, y_2)$ are points in R, then there is a point $P(x^*, y^*)$ on the line segment AB such that

$$f(x_2, y_2) - f(x_1, y_1) =$$
$$f_x(x^*, y^*)(x_2 - x_1) + f_y(x^*, y^*)(y_2 - y_1).$$

40 Use Exercise 39 to prove the following extension of Theorem (4.32): If $f_x(x, y) = 0$ and $f_y(x, y) = 0$ for all (x, y) in a rectangular region R, then $f(x, y)$ is constant on R.

41 Extend Exercise 39 to a function of three variables.

42 Extend Exercise 40 to a function of three variables.

43 Suppose $u = f(x, y)$ and $v = g(x, y)$ satisfy the *Cauchy-Riemann equations* $u_x = v_y$ and $u_y = -v_x$. If $x = r \cos \theta$, $y = r \sin \theta$, show that

$$\frac{\partial u}{\partial r} = \frac{1}{r}\frac{\partial v}{\partial \theta} \quad \text{and} \quad \frac{\partial v}{\partial r} = -\frac{1}{r}\frac{\partial u}{\partial \theta}.$$

16.6

DIRECTIONAL DERIVATIVES

Let us return to the illustration in which $w = f(x, y)$ is the temperature of a flat metal plate at the point $P(x, y)$. In Section 16.3 the partial derivatives $f_x(x, y)$ and $f_y(x, y)$ were interpreted as the rates of change of w with respect to distance at P in the horizontal or vertical directions, respectively. This can be generalized to any direction as follows. Let Q be any other point on the plate and let $R(x + \Delta x, y + \Delta y)$ be a point on the vector $\overrightarrow{PQ}$, as illustrated in Figure 16.19. The increment

$$\Delta w = f(x + \Delta x, y + \Delta y) - f(x, y)$$

is the difference in temperature between points P and R.

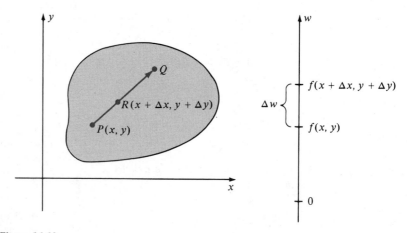

Figure 16.19

The ratio

$$\frac{\Delta w}{d(P, R)} = \frac{\Delta w}{\sqrt{(\Delta x)^2 + (\Delta y)^2}}$$

is the *average change in temperature* as a point moves along $\overrightarrow{PQ}$ from P to R. For example, if the temperature at P is 50° and the temperature at R is 51.5°, then $\Delta w = 1.5°$. If $d(P, R) = 3$ in., then the average rate of change in temperature as a point moves from P to R is

$$\frac{\Delta w}{d(P, R)} = 0.5°/\text{in.}$$

If the ratio $\Delta w/d(P, R)$ has a limit as R approaches P along $\overrightarrow{PQ}$, that is, as $(\Delta x, \Delta y) \to (0, 0)$, then the limit is called *the rate of change of w with respect to distance in the direction of the vector* $\overrightarrow{PQ}$.

If f is any function of two variables, then the preceding discussion may be used to motivate the concept of the *directional derivative* of f at $P(x, y)$ in any direction. Let us specify the direction from P by means of a unit vector $\mathbf{u} = u_1\mathbf{i} + u_2\mathbf{j}$ and consider the line l through P and parallel to $\mathbf{u}$, as illustrated in (i) of Figure 16.20.

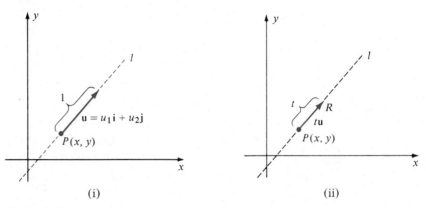

(i)   (ii)

Figure 16.20

If R is another point on l and $\overrightarrow{PR}$ has the same direction as $\mathbf{u}$, then

$$\overrightarrow{PR} = t\mathbf{u} = (tu_1)\mathbf{i} + (tu_2)\mathbf{j}$$

where $t > 0$ (see (ii) of Figure 16.20). The coordinates of R are

$$(x + tu_1, y + tu_2)$$

and hence

$$\Delta w = f(x + tu_1, y + tu_2) - f(x, y).$$

Since $d(P, R) = |\overrightarrow{PR}| = t$, the average change in w corresponding to the displacement $\overrightarrow{PR}$ is

$$\frac{\Delta w}{d(P, R)} = \frac{f(x + tu_1, y + tu_2) - f(x, y)}{t}.$$

Taking the limit as $t \to 0^+$ gives us the rate of change of f at P in the direction of $\mathbf{u}$. This is the motivation for the next definition.

Definition (16.22)

If f is a function of x and y and if $\mathbf{u} = u_1\mathbf{i} + u_2\mathbf{j}$ is a unit vector, then **the directional derivative of f at $P(x, y)$ in the direction of u,** denoted by $D_{\mathbf{u}} f(x, y)$, is

$$D_{\mathbf{u}} f(x, y) = \lim_{t \to 0} \frac{f(x + tu_1, y + tu_2) - f(x, y)}{t}.$$

It is easy to see that the first partial derivatives of f are special cases of the directional derivative. As a matter of fact, if $\mathbf{u} = \mathbf{i}$, then $u_1 = 1$, $u_2 = 0$, and the limit in Definition (16.22) reduces to

$$D_{\mathbf{i}} f(x, y) = \lim_{t \to 0} \frac{f(x + t, y) - f(x, y)}{t} = f_x(x, y).$$

Similarly, if $\mathbf{u} = \mathbf{j}$, then $u_1 = 0$, $u_2 = 1$, and we obtain

$$D_{\mathbf{j}} f(x, y) = \lim_{t \to 0} \frac{f(x, y + t) - f(x, y)}{t} = f_y(x, y).$$

If $\mathbf{a}$ is any vector having the same direction as $\mathbf{u}$ we shall also refer to $D_{\mathbf{u}} f(x, y)$ as **the directional derivative of f in the direction of a**.

The following theorem provides a formula for finding directional derivatives.

Theorem (16.23)

> If f is a differentiable function of two variables and $\mathbf{u} = u_1\mathbf{i} + u_2\mathbf{j}$ is a unit vector, then
>
> $$D_{\mathbf{u}} f(x, y) = f_x(x, y)u_1 + f_y(x, y)u_2.$$

Proof   If g is the function of one variable t defined by

$$g(t) = f(x + tu_1, y + tu_2),$$

then from Theorem (3.7) and Definition (16.22)

$$\begin{aligned} g'(0) &= \lim_{t \to 0} \frac{g(t) - g(0)}{t - 0} \\ &= \lim_{t \to 0} \frac{f(x + tu_1, y + tu_2) - f(x, y)}{t} \\ &= D_{\mathbf{u}} f(x, y). \end{aligned}$$

If we write

$$g(t) = f(r, v), \quad \text{where } r = x + tu_1, v = y + tu_2,$$

then from Theorem (16.19)

$$\begin{aligned} g'(t) &= f_r(r, v)\frac{dr}{dt} + f_v(r, v)\frac{dv}{dt} \\ &= f_r(r, v)u_1 + f_v(r, v)u_2. \end{aligned}$$

As we saw in the first part of the proof, the directional derivative equals $g'(0)$. However, if $t = 0$ then $r = x$, $v = y$, and we obtain the formula given in the statement of the theorem. $\square$

Example 1 If $f(x, y) = x^3y^2$, find the directional derivative of f at the point $P(-1, 2)$ in the direction of the vector $\mathbf{a} = 4\mathbf{i} - 3\mathbf{j}$.

Solution We wish to find $D_{\mathbf{u}}f(-1, 2)$, where $\mathbf{u}$ is a *unit* vector having the direction of $\mathbf{a}$. Since

$$\mathbf{u} = \frac{1}{|\mathbf{a}|}\mathbf{a} = \frac{1}{5}(4\mathbf{i} - 3\mathbf{j}) = \frac{4}{5}\mathbf{i} - \frac{3}{5}\mathbf{j}$$

and $f_x(x, y) = 3x^2y^2$ and $f_y(x, y) = 2x^3y,$

it follows from (16.23) that

$$D_{\mathbf{u}}f(x, y) = 3x^2y^2\left(\frac{4}{5}\right) + 2x^3y\left(-\frac{3}{5}\right).$$

Hence

$$D_{\mathbf{u}}f(-1, 2) = 6(-1)^2(2)^2\left(\frac{4}{5}\right) + 2(-1)^3(2)\left(-\frac{3}{5}\right)$$

$$= \frac{96}{5} + \frac{12}{5} = \frac{108}{5}. \quad \blacksquare$$

We may use Theorem (16.23) to express a directional derivative as a dot product of two vectors as follows:

$$D_{\mathbf{u}}f(x, y) = \langle f_x(x, y), f_y(x, y) \rangle \cdot \langle u_1, u_2 \rangle = \langle f_x(x, y), f_y(x, y) \rangle \cdot \mathbf{u}.$$

The vector in V_2 whose components are the first partial derivatives of $f(x, y)$ is designated by the following special name and notation.

Definition (16.24)

> If f is a function of two variables, then the **gradient of f** (or of $f(x, y)$) is
>
> $$\nabla f(x, y) = f_x(x, y)\mathbf{i} + f_y(x, y)\mathbf{j}.$$

In applications, the gradient $\nabla f(x, y)$ is sometimes denoted by $\text{grad}\, f(x, y)$. From the previous discussion we have the following.

(16.25) $$D_{\mathbf{u}}f(x, y) = \nabla f(x, y) \cdot \mathbf{u}$$

Thus, *to find the directional derivative of f in the direction of the unit vector* $\mathbf{u}$, *we may dot the gradient of f with* $\mathbf{u}$. The symbol ∇ (called "**del**") is a *vector differential operator* and is symbolized by

$$\nabla = \mathbf{i}\frac{\partial}{\partial x} + \mathbf{j}\frac{\partial}{\partial y}.$$

It has properties similar to the operator d/dx (see Exercises 31–36). Standing alone it is meaningless; however, if it operates on $f(x, y)$ it produces the two-dimensional vector function given by Definition (16.24).

Example 2 If $f(x, y) = x^2 - 4xy$,

(a) find the gradient of f at the point $P(3, -2)$.

(b) use the gradient to find the directional derivative of f at $P(3, -2)$ in the direction from $P(3, -2)$ to $Q(4, 1)$.

(c) discuss the significance of the answer to part (b) if $f(x, y)$ is the temperature at (x, y).

Solution

(a) By Definition (16.24),

$$\nabla f(x, y) = (2x - 4y)\mathbf{i} - 4x\mathbf{j}.$$

Hence, at $P(3, -2)$,

$$\nabla f(3, -2) = (6 + 8)\mathbf{i} - 12\mathbf{j} = 14\mathbf{i} - 12\mathbf{j}.$$

(b) If we let $\mathbf{a} = \overrightarrow{PQ}$, then

$$\mathbf{a} = \langle 4 - 3, 1 - (-2) \rangle = \langle 1, 3 \rangle = \mathbf{i} + 3\mathbf{j}.$$

A unit vector having the direction of $\overrightarrow{PQ}$ is

$$\mathbf{u} = \frac{1}{|\mathbf{a}|}\mathbf{a} = \frac{1}{\sqrt{10}}(\mathbf{i} + 3\mathbf{j}).$$

Applying (16.25) gives us

$$D_{\mathbf{u}} f(3, -2) = \nabla f(3, -2) \cdot \mathbf{u}$$

$$= (14\mathbf{i} - 12\mathbf{j}) \cdot \frac{1}{\sqrt{10}}(\mathbf{i} + 3\mathbf{j})$$

$$= \frac{1}{\sqrt{10}}(14 - 36) = -\frac{22}{\sqrt{10}} \approx -6.96.$$

(c) If $f(x, y)$ is the temperature (in degrees) at (x, y), then the fact that $D_{\mathbf{u}} f(3, -2) \approx -6.96$ indicates that if a point moves in the direction of $\overrightarrow{PQ}$, the temperature at P is decreasing at approximately $6.96°$ per unit change in distance. ∎

Let $P(x, y)$ be a fixed point, and consider the directional derivative $D_{\mathbf{u}} f(x, y)$ as $\mathbf{u} = \langle u_1, u_2 \rangle$ varies. For certain unit vectors $\mathbf{u}$ the directional derivative may be positive (that is, $f(x, y)$ may increase), or negative ($f(x, y)$ may decrease), or it may be 0. In many applications it is important to find the direction in which $f(x, y)$ increases most rapidly and also to find the maximum rate of change. The next theorem provides this information.

Theorem (16.26)

> The maximum value of $D_{\mathbf{u}} f(x, y)$ at the point $P(x, y)$ occurs in the direction of $\nabla f(x, y)$. Moreover, this maximum value is $|\nabla f(x, y)|$.

Proof   Let γ denote the angle between $\mathbf{u}$ and $\nabla f(x, y)$. Applying (16.25) and Theorem (14.22),

$$D_{\mathbf{u}} f(x, y) = \nabla f(x, y) \cdot \mathbf{u}$$
$$= |\nabla f(x, y)| \, |\mathbf{u}| \cos \gamma = |\nabla f(x, y)| \cos \gamma.$$

The maximum value occurs if $\cos \gamma = 1$; that is, if $\gamma = 0$, in which case $\mathbf{u}$ has the same direction as $\nabla f(x, y)$. Moreover, in this direction we see that $D_{\mathbf{u}} f(x, y) = |\nabla f(x, y)|$. This completes the proof.   □

In terms of rates of change, Theorem (16.26) tells us that $f(x, y)$ increases most rapidly in the direction of $\nabla f(x, y)$. It follows that $f(x, y)$ *decreases* most rapidly in the direction of $-\nabla f(x, y)$.

Example 3   If $f(x, y) = x^2 - 4xy$, find the direction in which $f(x, y)$ increases most rapidly at the point $P(3, -2)$. What is this maximum rate of increase?

Solution   The function f was considered in Example 2, where we found that $\nabla f(3, -2) = 14\mathbf{i} - 12\mathbf{j}$. Hence, by Theorem (16.26), $f(x, y)$ increases most rapidly at $P(3, -2)$ in the direction of the vector $14\mathbf{i} - 12\mathbf{j}$. The maximum rate is

$$|14\mathbf{i} - 12\mathbf{j}| = \sqrt{196 + 144} = \sqrt{340} \approx 18.4.$$ ■

The directional derivative of a function f of three variables is defined in a manner similar to Definition (16.22). Specifically, if $\mathbf{u} = u_1\mathbf{i} + u_2\mathbf{j} + u_3\mathbf{k}$ is a unit vector, then

$$D_{\mathbf{u}} f(x, y, z) = \lim_{t \to 0} \frac{f(x + tu_1, y + tu_2, z + tu_3) - f(x, y, z)}{t}.$$

As in the two variable case, $D_{\mathbf{u}} f(x, y, z)$ is the rate of change of f with respect to distance at $P(x, y, z)$ in the direction of $\mathbf{u}$. If f is a function of three variables, then in a manner analogous to Definition (16.24), the **gradient of** f (or of $f(x, y, z)$), denoted by $\nabla f(x, y, z)$ or grad $f(x, y, z)$, is defined as follows.

(16.27) $$\nabla f(x, y, z) = f_x(x, y, z)\mathbf{i} + f_y(x, y, z)\mathbf{j} + f_z(x, y, z)\mathbf{k}$$

The next result is the three-dimensional version of Theorem (16.23).

Theorem (16.28)

If f is a function of three variables and $\mathbf{u} = u_1\mathbf{i} + u_2\mathbf{j} + u_3\mathbf{k}$ is a unit vector, then

$$D_{\mathbf{u}} f(x, y, z) = \nabla f(x, y, z) \cdot \mathbf{u}$$
$$= f_x(x, y, z)u_1 + f_y(x, y, z)u_2 + f_z(x, y, z)u_3.$$

It is easy to show, as in Theorem (16.26), that of all possible directional derivatives $D_{\mathbf{u}} f(x, y, z)$ at the point $P(x, y, z)$, the one in the direction of $\nabla f(x, y, z)$ has the largest value. Moreover, this maximum value is $|\nabla f(x, y, z)|$.

Example 4 Suppose a rectangular coordinate system is located in space such that the temperature T at the point (x, y, z) is given by the formula $T = 100/(x^2 + y^2 + z^2)$.

(a) Find the rate of change of T with respect to distance at the point $P(1, 3, -2)$ in the direction of the vector $\mathbf{a} = \mathbf{i} - \mathbf{j} + \mathbf{k}$.

(b) In what direction from P does T increase most rapidly? What is the maximum rate of change of T at P?

Solution

(a) By definition, the gradient of T is

$$\nabla T = \frac{\partial T}{\partial x}\mathbf{i} + \frac{\partial T}{\partial y}\mathbf{j} + \frac{\partial T}{\partial z}\mathbf{k}.$$

Since

$$\frac{\partial T}{\partial x} = \frac{-200x}{(x^2 + y^2 + z^2)^2}, \quad \frac{\partial T}{\partial y} = \frac{-200y}{(x^2 + y^2 + z^2)^2}, \quad \frac{\partial T}{\partial z} = \frac{-200z}{(x^2 + y^2 + z^2)^2}$$

we have

$$\nabla T = \frac{-200}{(x^2 + y^2 + z^2)^2} (x\mathbf{i} + y\mathbf{j} + z\mathbf{k}).$$

If we let $\nabla T]_P$ denote the value of ∇T at the point $P(1, 3, -2)$, then

$$\nabla T]_P = \frac{-200}{196} (\mathbf{i} + 3\mathbf{j} - 2\mathbf{k}).$$

A unit vector $\mathbf{u}$ having the same direction as $\mathbf{a} = \mathbf{i} - \mathbf{j} + \mathbf{k}$ is

$$\mathbf{u} = \frac{1}{\sqrt{3}} (\mathbf{i} - \mathbf{j} + \mathbf{k}).$$

Using Theorem (16.28), the rate of change of T at P in the direction of **a** is

$$D_{\mathbf{u}} T]_P = \nabla T]_P \cdot \mathbf{u} = \frac{-200}{196} \frac{(1 - 3 - 2)}{\sqrt{3}} = \frac{200}{49\sqrt{3}} \approx 2.4.$$

If, for example, T is measured in degrees and distance is in inches, then T is increasing at a rate of 2.4 degrees per inch at P in the direction of **a**.

(b) The maximum rate of change of T at P occurs in the direction of the gradient, that is, in the direction of the vector $-\mathbf{i} - 3\mathbf{j} + 2\mathbf{k}$. The maximum rate of change equals the magnitude of the gradient, that is,

$$|\nabla T]_P| = \frac{200}{196} \sqrt{1 + 9 + 4} \approx 3.8. \qquad \blacksquare$$

EXERCISES 16.6

In Exercises 1–6 find the gradient of f at the indicated point.

1 $f(x, y) = \sqrt{x^2 + y^2}$; $P(-4, 3)$

2 $f(x, y) = 7y - 5x$; $P(2, 6)$

3 $f(x, y) = e^{3x} \tan y$; $P(0, \pi/4)$

4 $f(x, y) = x \ln (x - y)$; $P(5, 4)$

5 $f(x, y, z) = yz^3 - 2x^2$; $P(2, -3, 1)$

6 $f(x, y, z) = xy^2 e^z$; $P(2, -1, 0)$

In Exercises 7–20 find the directional derivative of f at the point P in the indicated direction.

7 $f(x, y) = x^2 - 5xy + 3y^2$; $P(3, -1)$, $\mathbf{u} = (\sqrt{2}/2)(\mathbf{i} + \mathbf{j})$

8 $f(x, y) = x^3 - 3x^2 y - y^3$; $P(1, -2)$, $\mathbf{u} = \frac{1}{2}(-\mathbf{i} + \sqrt{3}\mathbf{j})$

9 $f(x, y) = \arctan y/x$; $P(4, -4)$, $\mathbf{a} = 2\mathbf{i} - 3\mathbf{j}$

10 $f(x, y) = x^2 \ln y$; $P(5, 1)$, $\mathbf{a} = -\mathbf{i} + 4\mathbf{j}$

11 $f(x, y) = \sqrt{9x^2 - 4y^2 - 1}$; $P(3, -2)$, $\mathbf{a} = \mathbf{i} + 5\mathbf{j}$

12 $f(x, y) = (x - y)/(x + y)$; $P(2, -1)$, $\mathbf{a} = 3\mathbf{i} + 4\mathbf{j}$

13 $f(x, y) = x \cos^2 y$; $P(2, \pi/4)$, $\mathbf{a} = \langle 5, 1 \rangle$

14 $f(x, y) = xe^{3y}$; $P(4, 0)$, $\mathbf{a} = \langle -1, 3 \rangle$

15 $f(x, y, z) = xy^3 z^2$; $P(2, -1, 4)$, $\mathbf{a} = \mathbf{i} + 2\mathbf{j} - 3\mathbf{k}$

16 $f(x, y, z) = x^2 + 3yz + 4xy$; $P(1, 0, -5)$, $\mathbf{a} = 2\mathbf{i} - 3\mathbf{j} + \mathbf{k}$

17 $f(x, y, z) = z^2 e^{xy}$; $P(-1, 2, 3)$, $\mathbf{a} = 3\mathbf{i} + \mathbf{j} - 5\mathbf{k}$

18 $f(x, y, z) = \sqrt{xy} \sin z$; $P(4, 9, \pi/4)$, $\mathbf{a} = 2\mathbf{i} + 3\mathbf{j} - 2\mathbf{k}$

19 $f(x, y, z) = (x + y)(y + z)$; $P(5, 7, 1)$, $\mathbf{a} = \langle -3, 0, 1 \rangle$

20 $f(x, y, z) = z^2 \tan^{-1} (x + y)$; $P(0, 0, 4)$, $\mathbf{a} = \langle 6, 0, 1 \rangle$

In Exercises 21–24

(a) find the directional derivative of f at P in the direction from P to Q.

(b) find a unit vector in the direction in which f increases most rapidly at P and find the rate of change of f in that direction.

(c) find a unit vector in the direction in which f decreases most rapidly at P and find the rate of change of f in that direction.

21 $f(x, y) = x^2 e^{-2y}$; $P(2, 0)$, $Q(-3, 1)$

22 $f(x, y) = \sin (2x - y)$; $P(-\pi/3, \pi/6)$, $Q(0, 0)$

23 $f(x, y, z) = \sqrt{x^2 + y^2 + z^2}$; $P(-2, 3, 1)$, $Q(0, -5, 4)$

24 $f(x, y, z) = (x/y) - (y/z)$; $P(0, -1, 2)$, $Q(3, 1, -4)$

25 A metal plate is situated on an xy-plane such that the temperature T is inversely proportional to the distance from the origin. If the temperature at $P(3, 4)$ is $100°$, find the rate of change of T at P in the direction of the vector $\mathbf{i} + \mathbf{j}$. In what direction does T increase most rapidly at P? In what direction does T decrease most rapidly at P? In what direction is the rate of change 0?

26 The surface of a certain lake is represented by a region D in the xy-plane such that the depth (in feet) under the point corresponding to (x, y) is $f(x, y) = 300 - 2x^2 - 3y^2$. If a boy in the water is at the point $(4, 9)$, in what direction should he swim in order for the depth to decrease most rapidly? In what direction will the depth remain the same?

27 The electrical potential V at the point $P(x, y, z)$ in a rectangular coordinate system is $V = x^2 + 4y^2 + 9z^2$. Find the rate of change of V at $P(2, -1, 3)$ in the direction from P to the origin. Find the direction that produces the maximum rate of change of V at P. What is the maximum rate of change at P?

28 An object is situated in a rectangular coordinate system such that the temperature T at the point $P(x, y, z)$ is given by $T = 4x^2 - y^2 + 16z^2$. Find the rate of change of T at the point $P(4, -2, 1)$ in the direction of the vector $2\mathbf{i} + 6\mathbf{j} - 3\mathbf{k}$. In what direction does T increase most rapidly at P? What is this maximum rate of change? In what direction does T decrease most rapidly at P? What is this rate of change?

29 Prove Theorem (16.28).

30 Prove that $D_{\mathbf{k}} f(x, y, z) = f_z(x, y, z)$.

If $u = f(x, y)$ and $v = g(x, y)$ where f and g are differentiable, prove the identities in Exercises 31–36.

31 $\nabla(cu) = c\nabla u$, where c is a constant

32 $\nabla(u + v) = \nabla u + \nabla v$

33 $\nabla(uv) = u\nabla v + v\nabla u$

34 $\nabla\left(\dfrac{u}{v}\right) = \dfrac{v\nabla u - u\nabla v}{v^2}, \quad v \neq 0$

35 $\nabla u^n = nu^{n-1}\nabla u$ for every real number n

36 If $w = h(u)$, then $\nabla w = \dfrac{dw}{du}\nabla u$.

37 Let $\mathbf{u}$ be a unit vector and let θ be the angle, measured in the counterclockwise direction, from the positive x-axis to the position vector corresponding to $\mathbf{u}$.

(a) Show that

$$D_{\mathbf{u}} f(x, y) = f_x(x, y) \cos\theta + f_y(x, y) \sin\theta.$$

(b) If $f(x, y) = x^2 + 2xy - y^2$ and $\theta = 5\pi/6$, find $D_{\mathbf{u}} f(2, -3)$.

38 Refer to Exercise 37. If $f(x, y) = (xy + y^2)^4$ and $\theta = \pi/3$, find $D_{\mathbf{u}} f(2, -1)$.

39 If f, f_x, and f_y are continuous and $\nabla f(x, y) = \mathbf{0}$ throughout a rectangular region $R = \{(x, y): a < x < b, c < y < d\}$, prove that $f(x, y)$ is constant on R (cf. Exercise 40 of Section 16.5).

40 Suppose $w = f(x, y)$ and $x = g(t)$, $y = h(t)$, where all functions are differentiable. If $\mathbf{r}(t) = x\mathbf{i} + y\mathbf{j}$, prove that $dw/dt = \nabla w \cdot \mathbf{r}'(t)$.

16.7

TANGENT PLANES AND NORMAL LINES TO SURFACES

Suppose a surface S is the graph of an equation $F(x, y, z) = 0$, where F has continuous first partial derivatives. Let $P_0(x_0, y_0, z_0)$ be a point on S at which F_x, F_y, and F_z are not all zero. A **tangent line** to S at P_0 is, by definition, a tangent line l to any curve C that lies on S and contains P_0, as illustrated in Figure 16.21. Suppose parametric equations for C are $x = f(t)$, $y = g(t)$, $z = h(t)$ and let t_0 be the value of t that corresponds to P_0. For each t the point $(f(t), g(t), h(t))$ on C is also on S and, therefore,

$$F(f(t), g(t), h(t)) = 0.$$

If we let

$$w = F(x, y, z), \quad \text{where } x = f(t), y = g(t), z = h(t),$$

then using Theorem (16.19) and the fact that $w = 0$ for all t,

$$\frac{dw}{dt} = \frac{\partial w}{\partial x}\frac{dx}{dt} + \frac{\partial w}{\partial y}\frac{dy}{dt} + \frac{\partial w}{\partial z}\frac{dz}{dt} = 0,$$

that is,

$$F_x(x, y, z)f'(t) + F_y(x, y, z)g'(t) + F_z(x, y, z)h'(t) = 0.$$

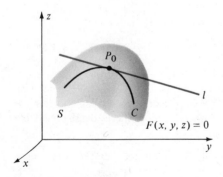

Figure 16.21

In particular, at the point P_0,

$$F_x(x_0, y_0, z_0)f'(t_0) + F_y(x_0, y_0, z_0)g'(t_0) + F_z(x_0, y_0, z_0)h'(t_0) = 0.$$

The preceding equation can be written compactly as

$$\nabla F]_{P_0} \cdot \mathbf{r}'(t_0) = 0,$$

where the symbol $\nabla F]_{P_0}$ denotes $\nabla F(x_0, y_0, z_0)$, and $\mathbf{r}(t) = \langle f(t), g(t), h(t) \rangle$. Since $\mathbf{r}'(t_0)$ is a tangent vector to C at P_0, this implies that *the vector $\nabla F]_{P_0}$ is orthogonal to every tangent line l to S at P_0*. The plane Γ through P_0 with normal vector $\nabla F]_{P_0}$ is called the **tangent plane to S at P_0**. We have shown that every tangent line l to S at P_0 lies in the tangent plane at P_0. A sketch illustrating these concepts appears in Figure 16.22. The following theorem summarizes our discussion.

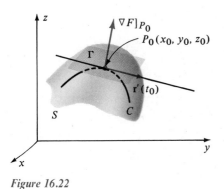

Figure 16.22

Theorem (16.29)

Suppose $F(x, y, z)$ has continuous first partial derivatives and P_0 is a point on the graph S of $F(x, y, z) = 0$. If F_x, F_y, and F_z are not all 0 at P_0, then the vector $\nabla F]_{P_0}$ is normal to the tangent plane to S at P_0.

We shall refer to the vector $\nabla F]_{P_0}$ in Theorem (16.29) as a vector that is *normal to the surface S at P_0*. Applying Theorem (14.38) gives us the following corollary, where it is assumed that F_x, F_y, and F_z are not all 0 at (x_0, y_0, z_0).

Corollary (16.30)

An equation for the tangent plane to the graph of $F(x, y, z) = 0$ at the point (x_0, y_0, z_0) is

$$F_x(x_0, y_0, z_0)(x - x_0) + F_y(x_0, y_0, z_0)(y - y_0)$$
$$+ F_z(x_0, y_0, z_0)(z - z_0) = 0.$$

Example 1 Find an equation for the tangent plane to the ellipsoid given by $4x^2 + 9y^2 + z^2 - 49 = 0$ at the point $P_0(1, -2, 3)$.

Solution If $F(x, y, z)$ denotes the left side of the given equation, then

$$F_x(x, y, z) = 8x, \quad F_y(x, y, z) = 18y, \quad F_z(x, y, z) = 2z$$

and hence at P_0

$$F_x(1, -2, 3) = 8, \quad F_y(1, -2, 3) = -36, \quad F_z(1, -2, 3) = 6.$$

Applying Corollary (16.30) we obtain the equation

$$8(x - 1) - 36(y + 2) + 6(z - 3) = 0$$

which simplifies to

$$4x - 18y + 3z - 49 = 0. \qquad \blacksquare$$

If $z = f(x, y)$ is an equation for S, then letting $F(x, y, z) = f(x, y) - z$, the equation in Corollary (16.30) becomes

$$f_x(x_0, y_0)(x - x_0) + f_y(x_0, y_0)(y - y_0) + (-1)(z - z_0) = 0.$$

We have proved the following result.

Theorem (16.31)

> An equation for the tangent plane to the graph of $z = f(x, y)$ at the point (x_0, y_0, z_0) is
>
> $$z - z_0 = f_x(x_0, y_0)(x - x_0) + f_y(x_0, y_0)(y - y_0).$$

The line perpendicular to the tangent plane at a point P_0 on a surface S is called a **normal line** to S at P_0. If S is the graph of $F(x, y, z)$, then the normal line is determined by the vector $\nabla F]_{P_0}$.

Example 2 Find an equation of the normal line to the ellipsoid given by $4x^2 + 9y^2 + z^2 - 49 = 0$ at the point $P_0(1, -2, 3)$.

Solution This surface is the same as the one considered in Example 1. Consequently, the vector $\nabla F]_{P_0} = \langle 8, -36, 6 \rangle$ determines the desired normal line. Using Theorem (14.37) we obtain the following parametric equations for the normal line:

$$x = 1 + 8t, \quad y = -2 - 36t, \quad z = 3 + 6t$$

where t is a real number. A symmetric form for the line (see (14.41)) is

$$\frac{x - 1}{8} = \frac{y + 2}{-36} = \frac{z - 3}{6}. \qquad \blacksquare$$

Suppose f is a function of x and y, and S is the graph of the equation $z = f(x, y)$. The tangent plane Γ to S at $P_0(x_0, y_0, z_0)$ may be used to obtain a geometric interpretation for the differential

$$dz = f_x(x_0, y_0)\, \Delta x + f_y(x_0, y_0)\, \Delta y.$$

Since $\qquad\qquad \Delta z = f(x_0 + \Delta x, y_0 + \Delta y) - f(x_0, y_0)$

the point $P(x_0 + \Delta x, y_0 + \Delta y, z_0 + \Delta z)$ is on S. Let $D(x_0 + \Delta x, y_0 + \Delta y, z)$ be on the tangent plane Γ and consider the points $A(x_0 + \Delta x, y_0 + \Delta y, 0)$ and $B(x_0 + \Delta x, y_0 + \Delta y, z_0)$ as illustrated in Figure 16.23. Since D is on Γ its coordinates satisfy the equation in Theorem (16.31), that is,

$$z - z_0 = f_x(x_0, y_0)(x_0 + \Delta x - x_0) + f_y(x_0, y_0)(y_0 + \Delta y - y_0)$$
$$= f_x(x_0, y_0)\, \Delta x + f_y(x_0, y_0)\, \Delta y = dz. \quad \cdot$$

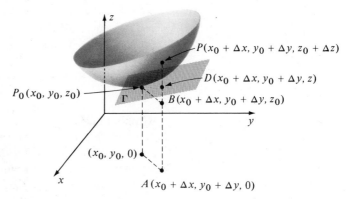

Figure 16.23

In terms of directed distances, we have shown that $dz = \overline{BD}$, that is, dz *is the directed distance from B to the point on the tangent plane that lies above (or below) B.* This is analogous to the single variable case $y = f(x)$ discussed in Section 3.4, where dy was interpreted as a distance to a point on a tangent line to the graph of f (see Figure 3.10). If follows that when dz is used as an approximation to z, it is assumed that the surface S almost coincides with its tangent plane at points close to P_0.

As an application of the discussion in this section, suppose a solid, liquid, or gas is situated in a three-dimensional coordinate system such that the temperature at the point $P(x, y, z)$ is $w = F(x, y, z)$. If $P_0(x_0, y_0, z_0)$ is a fixed point, then the graph of the equation

$$F(x, y, z) = F(x_0, y_0, z_0)$$

is the level (isothermal) surface S_0 that passes through P_0. For every point on S_0 the temperature is $w_0 = F(x_0, y_0, z_0)$. It is convenient to visualize all possible level surfaces that can be obtained by choosing different points. Several surfaces S_0, S_1, and S_2 through P_0, P_1, and P_2, respectively, corresponding to temperatures w_0, w_1, and w_2, are sketched in Figure 16.24. By Theorem (16.29), $\nabla F]_{P_0}$, $\nabla F]_{P_1}$, and $\nabla F]_{P_2}$ are normal vectors to S_0, S_1, and S_2 at the points P_0, P_1, and P_2, respectively.

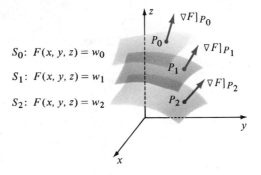

Figure 16.24

The preceding discussion may be summarized as in the next theorem.

Theorem (16.32)

> If a function F of three variables is differentiable at $P_0(x_0, y_0, z_0)$ and $\nabla F]_{P_0} \neq \mathbf{0}$, then $\nabla F]_{P_0}$ is a normal vector to the level surface of $F(x, y, z)$ that contains P_0.

Combining Theorem (16.32) with our work in the previous section gives us the fact that *the direction for the maximum rate of change of $F(x, y, z)$ at P_0 is along a normal vector to the level surface through P_0*, as illustrated by the isothermal surfaces in Figure 16.24. Another illustration of (16.32) occurs in electrical theory where $F(x, y, z)$ is the potential at $P(x, y, z)$. If a particle moves on one of the level (equipotential) surfaces, the potential remains constant. The potential changes most rapidly when a particle moves in a direction that is normal to an equipotential surface.

Example 3   If $F(x, y, z) = x^2 + y^2 + z$, sketch the level surface for F that passes through the point $P(1, 2, 4)$ and draw a vector with initial point P that corresponds to $\nabla F]_P$.

Solution   The level surfaces for F are graphs of equations of the form $F(x, y, z) = c$, where c is a constant. In particular, the level surface that passes through $P(1, 2, 4)$ is the graph of

$$F(x, y, z) = F(1, 2, 4).$$

Using the formula for $F(x, y, z)$ gives us

$$x^2 + y^2 + z = (1)^2 + (2)^2 + 4 = 9$$

or equivalently, $z = 9 - x^2 - y^2.$

The graph of this level surface is a paraboloid of revolution and is sketched in Figure 16.25. Since

$$\nabla F(x, y, z) = 2x\mathbf{i} + 2y\mathbf{j} + \mathbf{k}$$

we see that

$$\nabla F]_P = 2(1)\mathbf{i} + 2(2)\mathbf{j} + \mathbf{k} = 2\mathbf{i} + 4\mathbf{j} + \mathbf{k}.$$

The vector with initial point P that represents $\nabla F]_P$ is shown in Figure 16.25. By Theorem (16.32), $\nabla F]_P$ is orthogonal to the level surface. ∎

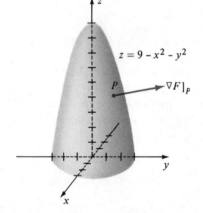

Figure 16.25

A result similar to Theorem (16.32) may be proved for functions of two variables. Specifically, given $f(x, y)$, we may show that the direction for the maximum rate of change of f at $P_0(x_0, y_0)$ is orthogonal to the level curve C through P_0; that is, orthogonal to the tangent line to C at P_0. For example, if $f(x, y) = 9 - x^2 - y^2$, the level curves are circles (see Figure 16.6) and hence the maximum rate of change of f takes place along lines through the origin. (Why?) This corresponds to going up or down the steepest part of the graph of f sketched in Figure 16.25.

Example 4 If $f(x, y) = x^2 + 2y^2$, sketch the level curve of f that passes through the point $P(3, 1)$ and draw a vector with initial point P that corresponds to $\nabla f]_P$.

Solution The level curve of f that passes through the point $P(3, 1)$ is given by $f(x, y) = f(3, 1)$, or

$$x^2 + 2y^2 = (3)^2 + 2(1)^2 = 11.$$

The graph is the ellipse sketched in Figure 16.26. The gradient of f is

$$\nabla f(x, y) = 2x\mathbf{i} + 4y\mathbf{j}$$

and hence $\nabla f]_P = 2(3)\mathbf{i} + 4(1)\mathbf{j} = 6\mathbf{i} + 4\mathbf{j}.$

The vector with initial point P that represents $\nabla f]_P$ is shown in Figure 16.26. Note that this vector is orthogonal to the level curve through P. ∎

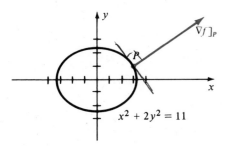

Figure 16.26

EXERCISES 16.7

In Exercises 1–10 find equations for the tangent plane and the normal line to the graph of the given equation at the indicated point P.

1 $4x^2 - y^2 + 3z^2 = 10$; $P(2, -3, 1)$

2 $9x^2 - 4y^2 - 25z^2 = 40$; $P(4, 1, -2)$

3 $z = 4x^2 + 9y^2$; $P(-2, -1, 25)$

4 $z = 4x^2 - y^2$; $P(5, -8, 36)$

5 $xy + 2yz - xz^2 + 10 = 0$; $P(-5, 5, 1)$

6 $x^3 - 2xy + z^3 + 7y + 6 = 0$; $P(1, 4, -3)$

7 $z = 2e^{-x} \cos y$; $P(0, \pi/3, 1)$

8 $z = \ln xy$; $P(\frac{1}{2}, 2, 0)$

9 $x = \ln(y/2z)$; $P(0, 2, 1)$

10 $xyz - 4xz^3 + y^3 = 10$; $P(-1, 2, 1)$

In Exercises 11–14 prove that an equation of the tangent plane to the given quadric surface at the point $P_0(x_0, y_0, z_0)$ may be written in the indicated form.

11 $\dfrac{x^2}{a^2} + \dfrac{y^2}{b^2} + \dfrac{z^2}{c^2} = 1$; $\dfrac{xx_0}{a^2} + \dfrac{yy_0}{b^2} + \dfrac{zz_0}{c^2} = 1$

12 $\dfrac{x^2}{a^2} - \dfrac{y^2}{b^2} + \dfrac{z^2}{c^2} = 1$; $\dfrac{xx_0}{a^2} - \dfrac{yy_0}{b^2} + \dfrac{zz_0}{c^2} = 1$

13 $\dfrac{x^2}{a^2} - \dfrac{y^2}{b^2} - \dfrac{z^2}{c^2} = 1$; $\dfrac{xx_0}{a^2} - \dfrac{yy_0}{b^2} - \dfrac{zz_0}{c^2} = 1$

14 $\dfrac{x^2}{a^2} + \dfrac{y^2}{b^2} = cz$; $\dfrac{2xx_0}{a^2} + \dfrac{2yy_0}{b^2} = c(z + z_0)$

15 Find the points on the hyperboloid $x^2 - 2y^2 - 4z^2 = 16$ at which the tangent plane is parallel to the plane $4x - 2y + 4z = 5$.

16 Prove that the sum of the squares of the x-, y-, and z-intercepts of every tangent plane to the graph of the equation $x^{2/3} + y^{2/3} + z^{2/3} = a^{2/3}$ is a constant.

17 Prove that every normal line to a sphere passes through the center of the sphere.

18 Find the points on the paraboloid $z = 4x^2 + 9y^2$ at which the normal line is parallel to the line through $P(-2, 4, 3)$ and $Q(5, -1, 2)$.

19 Two surfaces are said to be **orthogonal** at a point of intersection $P(x, y, z)$ if their normal lines at P are orthogonal. Show that the graphs of $F(x, y, z) = 0$ and $G(x, y, z) = 0$ (where F and G have partial derivatives) are orthogonal at P if and only if

$$F_x G_x + F_y G_y + F_z G_z = 0.$$

20 Prove that the sphere $x^2 + y^2 + z^2 = a^2$ and the cone $x^2 + y^2 - z^2 = 0$ are orthogonal at every point of intersection (see Exercise 19).

In Exercises 21–24 sketch the level curve C for f that passes through the point P and draw a vector (with initial point P) corresponding to $\nabla f]_P$ (see Exercises 23–26 of Section 16.1).

21 $f(x, y) = y^2 - x^2$; $P(2, 1)$

22 $f(x, y) = 3x - 2y$; $P(-2, 1)$

23 $f(x, y) = x^2 - y$; $P(-3, 5)$

24 $f(x, y) = xy$; $P(3, 2)$

In Exercises 25–30 sketch the level surface S for F that passes through the point P and draw a vector (with initial point P) corresponding to $\nabla F]_P$ (see Exercises 33–38 of Section 16.1).

25 $F(x, y, z) = x^2 + y^2 + z^2$; $P(1, 5, 2)$

26 $F(x, y, z) = z - x^2 - y^2$; $P(2, -2, 1)$

27 $F(x, y, z) = x + 2y + 3z$; $P(3, 4, 1)$

28 $F(x, y, z) = x^2 + y^2 - z^2$; $P(3, -1, 1)$

29 $F(x, y, z) = x^2 + y^2$; $P(2, 0, 3)$

30 $F(x, y, z) = z$; $P(2, 3, 4)$

16.3

EXTREMA OF FUNCTIONS OF SEVERAL VARIABLES

Throughout the remainder of this chapter, the term **rectangular region** will be used for points in a coordinate plane that lie inside a rectangle whose sides are parallel to the coordinate axes. If we wish to include the points on the boundary, the term **closed rectangular region** will be used. A region R is **bounded** if it is a subregion of a closed rectangular region.

In a manner analogous to the single variable case, a function f of two variables is said to have a **local maximum** at (a, b) if there is a rectangular region R containing (a, b) such that $f(x, y) \le f(a, b)$ for all other pairs (x, y) in R. Geometrically, if a surface S is the graph of f, then the local maxima correspond to the high points on S. If f_y exists, then as pointed out in Section 16.3, $f_y(a, b)$ is the slope of the tangent line to the trace C of S on the plane $x = a$ (see Figure 16.17). It follows that if $f(a, b)$ is a local maximum, then $f_y(a, b) = 0$. Similarly $f_x(a, b) = 0$.

The function f has a **local minimum** at (c, d) if there is a rectangular region R containing (c, d) such that $f(x, y) \ge f(c, d)$ for all other (x, y) in R. If f has first partial derivatives, they must be zero at (c, d). The local minima correspond to the low points on the graph of f.

It can be shown that if a function f of two variables is continuous on a closed and bounded region R, then f has an **absolute maximum** $f(a, b)$ and an **absolute minimum** $f(c, d)$ for some (a, b) and (c, d) in R. This means that

$$f(c, d) \le f(x, y) \le f(a, b)$$

for all (x, y) in R. The proof may be found in texts on advanced calculus.

The local maxima and minima are called the **local extrema** of f. The **extrema** include the absolute maximum and minimum (if they exist). If f has continuous first partial derivatives, then from the preceding discussion, the pairs that give rise to local extrema are solutions of *both* of the equations

$$f_x(x, y) = 0 \quad \text{and} \quad f_y(x, y) = 0.$$

As was the case for functions of one variable, local extrema can also occur at pairs where f_x or f_y does not exist. Since all of these pairs are crucial for finding local extrema, we shall give them a special name.

Definition (16.33)

Let f be a function of two variables. A pair (a, b) is a **critical point** of f if either

(i) $f_x(a, b) = 0$ and $f_y(a, b) = 0$, or

(ii) $f_x(a, b)$ or $f_y(a, b)$ does not exist.

If f has continuous first partial derivatives, and if (a, b) is a critical point, then the graph of f has a horizontal tangent plane at the point $(a, b, f(a, b))$. (Why?) When searching for local extrema of a function we usually begin by finding the critical points and then, in some way, we test each pair to determine whether it corresponds to a local maximum or minimum.

An absolute maximum or minimum of a function of two variables may occur at a boundary point of its domain R. The investigation of such **boundary extrema** usually requires a separate procedure, as was the case with endpoint extrema for functions of one variable. If there are no boundary points as, for example, when R is the entire xy-plane or the interior of a circle, then there can be no boundary extrema.

Example 1 Let $f(x, y) = 1 + x^2 + y^2$, where $0 \leq x^2 + y^2 \leq 4$. Find the extrema of f.

Solution The domain R of f is the region in the xy-plane consisting of the circle $x^2 + y^2 = 4$ and its interior. Since $f_x(x, y) = 2x$ and $f_y(x, y) = 2y$, the critical points are the (simultaneous) solutions of

$$2x = 0 \quad \text{and} \quad 2y = 0.$$

The only pair that satisfies both equations is $(0, 0)$ and hence $f(0, 0) = 1$ is the only possible local extremum. Moreover, since

$$f(x, y) = 1 + x^2 + y^2 > 1 \quad \text{if } (x, y) \neq (0, 0),$$

f has the local minimum 1 at $(0, 0)$. This fact may also be seen by referring to the graph of $z = f(x, y)$, sketched in Figure 16.27. It should be clear that the number $f(0, 0) = 1$ is also the absolute minimum of f.

To find possible boundary extrema, we investigate points (a, b) that lie on the boundary of the domain of f, that is, on the circle $x^2 + y^2 = 4$. Referring to Figure 16.27, we see that any such point, for example $(0, 2)$, leads to the absolute maximum $f(0, 2) = 5$. ∎

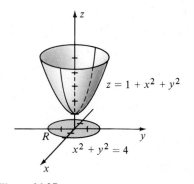

$z = 1 + x^2 + y^2$

$x^2 + y^2 = 4$

Figure 16.27

As illustrated in the next example, not every critical point may lead to an extremum.

Example 2 If $f(x, y) = y^2 - x^2$ and the domain of f is $\mathbb{R} \times \mathbb{R}$, find the extrema of f.

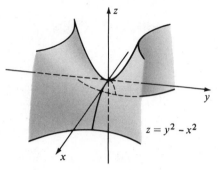

Figure 16.28

Solution   Since $f_x(x, y) = -2x$ and $f_x(x, y) = 2y$, the only possible local extremum is $f(0, 0) = 0$. (Why?) However, if $y \neq 0$, then $f(0, y) = y^2 > 0$; and if $x \neq 0$, then $f(x, 0) = -x^2 < 0$. Thus every rectangular region in the xy-plane containing $(0, 0)$ contains pairs at which functional values are greater than $f(0, 0)$ as well as pairs at which values are less than $f(0, 0)$. Consequently f has no local extrema. This is also evident from the graph of $z = f(x, y)$ sketched in Figure 16.28. Because of the shape of the surface near $(0, 0, 0)$, this point is called a **saddle point** of the graph. Since the domain of f has no boundary, there are no boundary extrema.   ∎

The following test, which we state without proof, is useful for more complicated functions. In a sense, it is the counterpart of the Second Derivative Test for functions of one variable. The interested reader may consult advanced calculus books for a proof.

***Test for Extrema* (16.34)**

> Let f be a function of two variables that has continuous second partial derivatives on a rectangular region Q and let
>
> $$g(x, y) = f_{xx}(x, y) f_{yy}(x, y) - [f_{xy}(x, y)]^2$$
>
> for all (x, y) in Q. If (a, b) is in Q and $f_x(a, b) = 0$, $f_y(a, b) = 0$, then
> (i) $f(a, b)$ is a local maximum of f if $g(a, b) > 0$ and $f_{xx}(a, b) < 0$.
> (ii) $f(a, b)$ is a local minimum of f if $g(a, b) > 0$ and $f_{xx}(a, b) > 0$.
> (iii) $f(a, b)$ is not an extremum of f if $g(a, b) < 0$.

Note that if $g(a, b) > 0$, then $f_{xx}(a, b) f_{yy}(a, b)$ is positive and hence $f_{xx}(a, b)$ and $f_{yy}(a, b)$ agree in sign. Thus, we may replace $f_{xx}(a, b)$ in (i) and (ii) by $f_{yy}(a, b)$. If $g(a, b) = 0$ the test gives no information and in this case it is necessary to use a more direct approach such as that illustrated in Examples 1 and 2.

Example 3   If $f(x, y) = x^2 - 4xy + y^3 + 4y$, find the local extrema of f.

Solution   Since $f_x(x, y) = 2x - 4y$ and $f_y(x, y) = -4x + 3y^2 + 4$, the critical points are solutions of the equations

$$2x - 4y = 0 \quad \text{and} \quad -4x + 3y^2 + 4 = 0.$$

Solving these simultaneously we obtain the pairs $(4, 2)$ and $(\frac{4}{3}, \frac{2}{3})$. (Verify this fact.) The second partial derivatives of f are

$$f_{xx}(x, y) = 2, \quad f_{xy}(x, y) = -4, \quad \text{and} \quad f_{yy}(x, y) = 6y$$

and, therefore, $g(x, y) = 12y - 16$. Since $g(\frac{4}{3}, \frac{2}{3}) = -8 < 0$, it follows from (iii) of (16.34) that $f(\frac{4}{3}, \frac{2}{3})$ is not an extremum of f. Since $g(4, 2) = 8 > 0$ and $f_{xx}(4, 2) = 2 > 0$, it follows from (ii) of (16.34) that f has a local minimum $f(4, 2) = 0$.   ∎

Example 4 If $f(x, y) = x^2 - 4xy + y^3 + 4y$, find the absolute extrema of f on the triangular region R that has vertices $(-1, -1)$, $(7, -1)$, and $(7, 7)$.

Solution   The boundary of R consists of the line segments C_1, C_2, and C_3 shown in Figure 16.29. In Example 3 we found that f has a local minimum 0 at the point $(4, 2)$, which is within R. Thus we only need to check for boundary extrema.

On C_1 we have $y = -1$, and hence values of f are given by

$$f(x, -1) = x^2 + 4x - 1 - 4 = x^2 + 4x - 5.$$

This determines a function of one variable whose domain is the interval $[-1, 7]$. The first derivative is $2x + 4$, which equals 0 at $x = -2$, a number outside the interval $[-1, 7]$. Thus, there is no local extremum of $f(x, -1)$ on $[-1, 7]$. Indeed, $f(x, -1)$ is increasing throughout this interval and there is an endpoint minimum $f(-1, -1) = -8$ and an endpoint maximum $f(7, -1) = 72$.

On C_2 we have $x = 7$ and values of f are given by

$$f(7, y) = 49 - 28y + y^3 + 4y = y^3 - 24y + 49,$$

where $-1 \le y \le 7$. The first derivative of this function of y is $3y^2 - 24$, and hence there is a critical number if $3y^2 = 24$, or $y = \sqrt{8} = 2\sqrt{2}$. Using the second derivative $6y$ of $f(7, y)$ we see that $f(7, 2\sqrt{2})$ is a local minimum for f on the segment C_2. However, since $f(7, 2\sqrt{2}) = 49 - 32\sqrt{2} \approx 3.7$ and $f(-1, -1) = -8$, it is not an absolute minimum for f on R. The values of f at the endpoints of C_2 are

$$f(7, -1) = 72 \quad \text{and} \quad f(7, 7) = 224.$$

Finally, on C_3 we have $y = x$ and values of f are given by

$$f(x, x) = x^2 - 4x^2 + x^3 + 4x = x^3 - 3x^2 + 4x.$$

The first derivative $3x^2 - 6x + 4$ has no real roots and hence there are no critical numbers for $f(x, x)$. We have already calculated the endpoint values $f(-1, -1) = -8$ and $f(7, 7) = 224$. It follows that these are the absolute minimum and absolute maximum values for f on the triangular region R. ∎

Example 5 A rectangular box with no top and having a volume of 12 ft³ is to be constructed. The cost per square foot of the material to be used is $4 for the bottom, $3 for two of the opposite sides, and $2 for the remaining pair of opposite sides. Find the dimensions of the box that will minimize the cost.

Solution   Let x and y denote the dimensions (in feet) of the base, and z the altitude (in feet). Since there are two sides of area xz and two sides of area yz, the cost C (in dollars) of the material is given by

$$C = 4xy + 3(2xz) + 2(2yz)$$

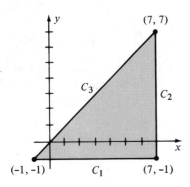

Figure 16.29

where x, y, and z are all different from 0. Since $xyz = 12$, it follows that $z = 12/xy$. Substituting in the formula for C and simplifying, we obtain

$$C = 4xy + \frac{72}{y} + \frac{48}{x}.$$

There are no boundary points (Why?) and hence there are no boundary extrema. To determine the possible local extrema we must find the simultaneous solutions of the equations

$$C_x = 4y - \frac{48}{x^2} = 0 \quad \text{and} \quad C_y = 4x - \frac{72}{y^2} = 0.$$

These equations imply that

$$x^2 y = 12 \quad \text{and} \quad xy^2 = 18.$$

We see from the first equation that $y = 12/x^2$. Substitution in the second equation gives us

$$x\left(\frac{144}{x^4}\right) = 18, \quad \text{or} \quad 144 = 18x^3.$$

Solving for x we obtain $x^3 = 8$, or $x = 2$. Since $y = 12/x^2$, the corresponding value of y is $\frac{12}{4}$, or 3. Theorem 16.34 can be used to show that these values of x and y determine a minimum value of C. Finally, using $z = 12/xy$ we obtain $z = 12/(2 \cdot 3) = 2$. Thus the minimum cost occurs if the dimensions of the base are 3 ft by 2 ft and the altitude is 2 ft. The longer sides should be made out of the \$2 material and the shorter sides out of the \$3 material. ∎

The material in this section can be generalized to functions of more than two variables. For example, given $f(x, y, z)$ we define local maxima and minima in a manner analogous to the two-variable case. If f has first partial derivatives, then a local extremum can occur only at a point where f_x, f_y, and f_z are simultaneously 0. It is difficult to obtain tests for determining whether such a point corresponds to a maximum, a minimum, or neither. However, in applications it is often possible to determine this by analyzing the physical nature of the problem.

EXERCISES 16.8

In Exercises 1–16 find the extrema of f.

1 $f(x, y) = x^2 + 2xy + 3y^2$

2 $f(x, y) = x^2 - 3xy - y^2 + 2y - 6x$

3 $f(x, y) = x^3 + 3xy - y^3$

4 $f(x, y) = 4x^3 - 2x^2y + y^2$

5 $f(x, y) = x^2 + 4y^2 - x + 2y$

6 $f(x, y) = 5 + 4x - 2x^2 + 3y - y^2$

7 $f(x, y) = x^4 + y^3 + 32x - 9y$

8 $f(x, y) = \cos x + \cos y$

9 $f(x, y) = e^x \sin y$

10 $f(x, y) = x \sin y$

11 $f(x, y) = (4y + x^2y^2 + 8x)/xy$

12 $f(x, y) = x/(x + y)$

13 $f(x, y) = \sin x + \sin y + \sin (x + y)$;
$0 \le x \le 2\pi, 0 \le y \le 2\pi$

14 $f(x, y) = (x + y + 1)^2/(x^2 + y^2 + 1)$

15 $f(x, y) = yx^2 + y^3 - 4x^2$

16 $f(x, y) = x^2 - 6x \cos y + 9$

In Exercises 17–22 find the absolute maximum and minimum values of f if the domain is the indicated region R. (Refer to Exercises 1–6 for local extrema.)

17 $f(x, y) = x^2 + 2xy + 3y^2$;
$R = \{(x, y): -2 \le x \le 4, -1 \le y \le 3\}$.

(*Hint*: To determine boundary extrema, consider the following functions of *one* variable: $f(x, -1)$, $f(x, 3)$, $f(-2, y)$, and $f(-4, y)$.)

18 $f(x, y) = x^2 - 3xy - y^2 + 2y - 6x$;
$R = \{(x, y): |x| \le 3, |y| \le 2\}$

19 $f(x, y) = x^3 + 3xy - y^3$; R the triangular region with vertices $(1, 2)$, $(1, -2)$ and $(-1, -2)$.

20 $f(x, y) = 4x^3 - 2x^2y + y^2$; R the region bounded by the graphs of $y = x^2$ and $y = 9$.

21 $f(x, y) = x^2 + 4y^2 - x + 2y$; R the region bounded by the ellipse $x^2 + 4y^2 = 1$.

22 $f(x, y) = 5 + 4x - 2x^2 + 3y - y^2$; R the triangle bounded by the lines $y = x$, $y = -x$, and $y = 2$.

23 Find the shortest distance from the point $P(2, 1, -1)$ to the plane $4x - 3y + z = 5$.

24 Find the shortest distance between the planes $2x + 3y - z = 2$ and $2x + 3y - z = 4$.

25 Find the points on the graph of $xy^3z^2 = 16$ that are closest to the origin.

26 Find three positive real numbers whose sum is 1000 and whose product is a maximum.

27 If an open rectangular box is to have a fixed volume V, what relative dimensions will make the surface area a minimum?

28 If an open rectangular box is to have a fixed surface area A, what relative dimensions will make the volume a maximum?

29 Find the dimensions of the rectangular parallelepiped of maximum volume with faces parallel to the coordinate planes that can be inscribed in the ellipsoid $16x^2 + 4y^2 + 9z^2 = 144$.

30 Generalize Exercise 29 to any ellipsoid $x^2/a^2 + y^2/b^2 + z^2/c^2 = 1$.

31 Find the dimensions of the rectangular parallelepiped of maximum volume that has three of its faces in the coordinate planes, one vertex at the origin, and another vertex in the first octant on the plane $4x + 3y + z = 12$.

32 Generalize Exercise 31 to any plane $x/a + y/b + z/c = 1$ where a, b, and c are positive real numbers.

33 A company plans to manufacture closed rectangular boxes that have a volume of 8 ft^3. If the material for the top and bottom costs twice as much as the material for the sides, find the dimensions that will minimize the cost.

34 A window has the shape of a rectangle surmounted by an isosceles triangle, as illustrated in the figure. If the perimeter of the window is 12 ft, what values of x, y, and θ will maximize the total area?

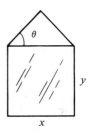

Figure for Exercise 34

35 Suppose a post office will not mail a rectangular box if the sum of its length and girth (the perimeter of a cross section that is perpendicular to the length) is more than 84 in. Find the dimensions of the box of maximum volume that can be mailed.

36 Find a vector in three dimensions having magnitude 8 such that the sum of its components is as large as possible.

37 In scientific experiments, corresponding values of two quantities x and y are often tabulated as follows:

x values	x_1	x_2	$\cdots$	x_n
y values	y_1	y_2	$\cdots$	y_n

Plotting the points (x_i, y_i) may lead the investigator to conjecture that x and y are related *linearly*, that is, $y = mx + b$ for some m and b. Thus it is desirable to find a line l having that equation which "best fits" the data, as illustrated in the figure on the next page. Statisticians call l a **linear regression line**.

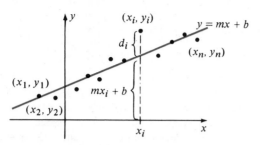

Figure for Exercise 37

One technique for finding l is to employ the **method of least squares**. To use this method consider, for each i, the *vertical deviation* $d_i = y_i - (mx_i + b)$ of the point (x_i, y_i) from the line $y = mx + b$ (see figure). Values of m and b are then determined that minimize the sum of the squares $\sum_{i=1}^{n} d_i^2$ (the squares d_i^2 are used because some of the d_i may be negative). Substituting for d_i produces the following function f of m and b:

$$f(m, b) = \sum_{i=1}^{n} (y_i - mx_i - b)^2.$$

Show that the line $y = mx + b$ of best fit occurs if

$$\left(\sum_{i=1}^{n} x_i \right) m + nb = \sum_{i=1}^{n} y_i$$

and

$$\left(\sum_{i=1}^{n} x_i^2 \right) m + \left(\sum_{i=1}^{n} x_i \right) b = \sum_{i=1}^{n} x_i y_i.$$

Thus the line can be found by solving this system of two equations for the two unknowns m and b.

38 If the equations in Exercise 37 are true, show that the sum $\sum_{i=1}^{n} d_i$ of the deviations is 0. (This means that the positive and negative deviations cancel one another, and it is one reason for using $\sum_{i=1}^{n} d_i^2$ in the method of least squares.)

In Exercises 39 and 40, use the method of least squares (see Exercise 37) to find a line $y = mx + b$ that best fits the given data.

39

x values	1	4	7
y values	3	5	6

40

x values	1	4	6	8
y values	1	3	2	4

41 The following table lists the relationship between semester averages and scores on the final examination for ten students in a mathematics class.

Semester average	40 55 62 68 72 76 80 86 90 94
Final examination	30 45 65 72 60 82 76 92 88 98

Fit these data to a straight line and use it to estimate the final examination grade of a student with a class average of 70.

42 In studying the stress-strain diagram of an elastic material (see page 255), an engineer finds that part of the curve appears to be linear. Experimental values are listed in the following table, where the stress is measured in pounds and the strain in inches.

Stress	2	2.2	2.4	2.6	2.8	3.0
Strain	0.10	0.30	0.40	0.60	0.70	0.90

Fit these data to a straight line and estimate the strain when the stress is 2.5 lb.

16.9

LAGRANGE MULTIPLIERS

The definitions of extrema given in the preceding section can be extended to a function f of any number of variables. In many applications it is necessary to find the extrema of f when the variables are restricted in some manner. As an illustration, suppose we wish to find the volume of the largest rectangular box with faces parallel to the coordinate planes that can be inscribed in the ellipsoid $16x^2 + 4y^2 + 9z^2 = 144$. Note that by symmetry it is sufficient to examine the part in the first octant illustrated in Figure 16.30.

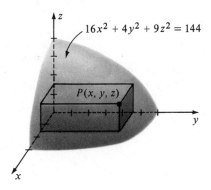

$16x^2 + 4y^2 + 9z^2 = 144$

$P(x, y, z)$

Figure 16.30

If $P(x, y, z)$ is the vertex indicated in the figure, then the volume V of the box is $V = 8xyz$. The problem is to find the maximum value of V subject to the **constraint** (or **side condition**)

$$16x^2 + 4y^2 + 9z^2 - 144 = 0.$$

Solving this equation for z and substituting in the formula for V, we obtain

$$V = (8xy)\tfrac{1}{3}\sqrt{144 - 16x^2 - 4y^2}.$$

The extrema may then be found using (16.34); however, this method is cumbersome because of the manipulations involved in finding partial derivatives and critical points. Another disadvantage of this technique is that for similar problems it may be impossible to solve for z. For these reasons it is sometimes simpler to employ the method of *Lagrange multipliers* discussed in this section.*

As another illustration, let $f(x, y)$ be the temperature at the point $P(x, y)$ on the flat metal plate illustrated in Figure 16.31, and let C be a curve that has the rectangular equation $g(x, y) = 0$. Suppose we wish to find the points on C at which the temperature attains its largest or smallest value. This amounts to finding the extrema of $f(x, y)$ *subject to the constraint* $g(x, y) = 0$. One technique for accomplishing this is to use the following result.

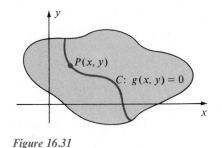

$P(x, y)$

$C: g(x, y) = 0$

Figure 16.31

_____

* This method was discovered by the French mathematician Joseph Lagrange (1736–1813).

Lagrange's Theorem (16.35)

Let $f(x, y)$ and $g(x, y)$ have continuous first partial derivatives, and suppose f has an extremum $f(x_0, y_0)$ when (x, y) is subject to the constraint $g(x, y) = 0$. If $\nabla g(x_0, y_0) \neq \mathbf{0}$, then there is a real number λ such that

$$\nabla f(x_0, y_0) = \lambda \nabla g(x_0, y_0).$$

Proof   The graph of $g(x, y) = 0$ is a curve C in the xy-plane. It can be shown that under the given conditions, C has a smooth parametric representation

$$x = k(t), \quad y = h(t),$$

where t is in some interval I. Let

$$\mathbf{r}(t) = x\mathbf{i} + y\mathbf{j} = k(t)\mathbf{i} + h(t)\mathbf{k}$$

be the position vector of the point $P(x, y)$ on C (see Figure 16.32) and let the point $P_0(x_0, y_0)$ correspond to $t = t_0$; that is,

$$\mathbf{r}(t_0) = x_0\mathbf{i} + y_0\mathbf{j} = k(t_0)\mathbf{i} + h(t_0)\mathbf{j}.$$

If we define a function F of one variable t by

$$F(t) = f(k(t), h(t))$$

then as t varies we obtain functional values $f(x, y)$ that correspond to (x, y) on C; that is, $f(x, y)$ is subject to the constraint $g(x, y) = 0$. Since $f(x_0, y_0)$ is an extremum of f under these conditions, it follows that $F(t_0) = f(k(t_0), h(t_0))$ is an extremum of $F(t)$. Thus $F'(t_0) = 0$. If we regard F as a composite function, then by Theorem (16.19),

$$F'(t) = f_x(x, y) \frac{dx}{dt} + f_y(x, y) \frac{dy}{dt}$$

and, therefore,

$$0 = F'(t_0) = f_x(x_0, y_0)k'(t_0) + f_y(x_0, y_0)h'(t_0) = \nabla f(x_0, y_0) \cdot \mathbf{r}'(t_0).$$

This shows that the vector $\nabla f(x_0, y_0)$ is orthogonal to the tangent vector $\mathbf{r}'(t_0)$ to C. However, $\nabla g(x_0, y_0)$ is also orthogonal to $\mathbf{r}'(t_0)$, because C is a level curve for g. Since $\nabla f(x_0, y_0)$ and $\nabla g(x_0, y_0)$ are orthogonal to the same vector they are parallel, that is $\nabla f(x_0, y_0) = \lambda \nabla g(x_0, y_0)$ for some λ.   □

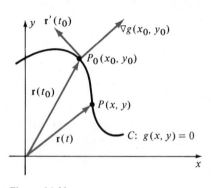

Figure 16.32

The number λ that appears in Theoream (16.35) is called a **Lagrange multiplier**. When applying this method we begin by considering the equations

$$\nabla f(x, y) = \lambda \nabla g(x, y) \quad \text{and} \quad g(x, y) = 0$$

or, equivalently,

(16.36) $f_x(x, y) = \lambda g_x(x, y), \quad f_y(x, y) = \lambda g_y(x, y) \quad \text{and} \quad g(x, y) = 0.$

If f has an extremum at (x_0, y_0) subject to the constraint $g(x, y) = 0$, then (x_0, y_0) must be a solution of each equation in (16.36). Thus, to determine the extrema, find the pairs (a_i, b_i) that, along with a suitable λ, satisfy all three equations in (16.36). If f has a maximum (or minimum) value it will be the largest (or smallest) of the corresponding functional values $f(a_i, b_i)$.

Example 1 Find the extrema of $f(x, y) = xy$ if (x, y) is restricted to the ellipse $4x^2 + y^2 = 4$.

Solution In this example the constraint is $g(x, y) = 4x^2 + y^2 - 4 = 0$. Setting $\nabla f(x, y) = \lambda \nabla g(x, y)$, we obtain

$$y\mathbf{i} + x\mathbf{j} = \lambda(8x\mathbf{i} + 2y\mathbf{j})$$

and equations (16.36) take on the form

$$y = 8x\lambda, \quad x = 2y\lambda \quad \text{and} \quad 4x^2 + y^2 - 4 = 0.$$

There are many ways to solve this system. One is to begin by writing

$$x = 2y\lambda = 2(8x\lambda)\lambda = 16x\lambda^2.$$

This leads to

$$x - 16x\lambda^2 = 0, \quad \text{or} \quad x(1 - 16\lambda^2) = 0$$

and hence either $x = 0$ or $\lambda = \pm\frac{1}{4}$. If $x = 0$, then using $4x^2 + y^2 - 4 = 0$ we obtain $y = \pm 2$. Thus the points $(0, \pm 2)$ are possibilities for extrema of $f(x, y)$. If $\lambda = \pm\frac{1}{4}$, then $y = 8x\lambda = 8x(\pm\frac{1}{4})$ or $y = \pm 2x$. Again using the fact that $4x^2 + y^2 - 4 = 0$ we get

$$4x^2 + 4x^2 - 4 = 0, \quad 8x^2 = 4 \quad \text{or} \quad x = \pm 1/\sqrt{2} = \pm\sqrt{2}/2.$$

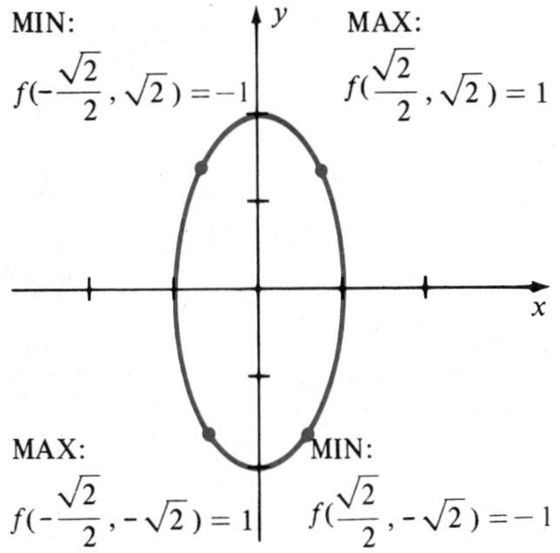

The corresponding y-values satisfy

$$4(\tfrac{1}{2}) + y^2 - 4 = 0, \quad y^2 = 2 \quad \text{or} \quad y = \pm\sqrt{2}.$$

This gives us the points $(\sqrt{2}/2, \pm\sqrt{2})$ and $(-\sqrt{2}/2, \pm\sqrt{2})$. The value of f at each of the points we have found is listed below.

(x, y)	$(0, 2)$	$(0, -2)$	$(\sqrt{2}/2, \sqrt{2})$	$(\sqrt{2}/2, -\sqrt{2})$	$(-\sqrt{2}/2, \sqrt{2})$	$(-\sqrt{2}/2, -\sqrt{2})$
$f(x, y)$	0	0	1	-1	-1	1

It follows that $f(x, y)$ takes on a maximum value 1 at either $(\sqrt{2}/2, \sqrt{2})$ or $(-\sqrt{2}/2, -\sqrt{2})$ and a minimum value -1 at $(\sqrt{2}/2, -\sqrt{2})$ or $(-\sqrt{2}/2, \sqrt{2})$. These facts are indicated on the ellipse in Figure 16.33. ∎

MIN:
$f(-\dfrac{\sqrt{2}}{2}, \sqrt{2}) = -1$

MAX:
$f(\dfrac{\sqrt{2}}{2}, \sqrt{2}) = 1$

MAX:
$f(-\dfrac{\sqrt{2}}{2}, -\sqrt{2}) = 1$

MIN:
$f(\dfrac{\sqrt{2}}{2}, -\sqrt{2}) = -1$

Figure 16.33

Lagrange's Theorem (16.35) can be extended to functions of more than two variables. For example, to determine the extrema of $f(x, y, z)$ subject to the constraint $g(x, y, z) = 0$ we begin by finding the (simultaneous) solutions of the equations

(16.37) $$\nabla f(x, y, z) = \lambda \nabla g(x, y, z) \quad \text{and} \quad g(x, y, z) = 0$$

or, equivalently, of

(16.38)
$$f_x = \lambda g_x, \quad f_y = \lambda g_y, \quad f_z = \lambda g_z, \quad g = 0$$

where we have simplified the notation by deleting the functional value symbols in $f_x(x, y, z), g_x(x, y, z)$, etc. A similar situation exists for functions of four or more variables.

The next example was discussed briefly at the beginning of this section.

Example 2 Find the volume of the largest rectangular box with faces parallel to the coordinate planes that can be inscribed in the ellipsoid $16x^2 + 4y^2 + 9z^2 = 144$.

Solution   A typical position for the box is illustrated in Figure 16.30. We wish to maximize

$$V = f(x, y, z) = 8xyz$$

subject to the constraint

$$g(x, y, z) = 16x^2 + 4y^2 + 9z^2 - 144 = 0.$$

As in (16.37), let us begin by considering $\nabla f(x, y, z) = \lambda \nabla g(x, y, z)$, or

$$8yz\mathbf{i} + 8xz\mathbf{j} + 8xy\mathbf{k} = \lambda(32x\mathbf{i} + 8y\mathbf{j} + 18z\mathbf{k}).$$

This, together with $g(x, y, z) = 0$ gives us the following system of four equations:

$$8yz = 32x\lambda, \quad 8xz = 8y\lambda, \quad 8xy = 18z\lambda, \quad 16x^2 + 4y^2 + 9z^2 - 144 = 0.$$

Multiplying the first equation by x, the second by y, and the third by z and adding gives us

$$24xyz = 32x^2\lambda + 8y^2\lambda + 18z^2\lambda = 2\lambda(16x^2 + 4y^2 + 9z^2).$$

This, together with the fact that $16x^2 + 4y^2 + 9z^2 = 144$, implies that

$$24xyz = 2\lambda(144) \quad \text{or} \quad xyz = 12\lambda.$$

The last equation may be used to find x, y, and z. For example, multiplying both sides of the equation $8yz = 32x\lambda$ by x, we obtain

$$8xyz = 32x^2\lambda$$

$$8(12\lambda) = 32x^2\lambda$$

$$96\lambda - 32x^2\lambda = 0$$

$$32\lambda(3 - x^2) = 0.$$

Consequently, either $\lambda = 0$ or $x = \sqrt{3}$. We may reject $\lambda = 0$ since in this case $xyz = 0$ and hence $V = 8xyz = 0$. Thus the only possibility for x is $\sqrt{3}$.

Similarly, multiplying both sides of the equation $8xz = 8y\lambda$ by y leads to

$$8xyz = 8y^2\lambda$$

$$8(12\lambda) = 8y^2\lambda$$

$$96\lambda - 8y^2\lambda = 0$$

$$8\lambda(12 - y^2) = 0$$

and hence $y = \sqrt{12} = 2\sqrt{3}$.

Finally, multiplying the equation $8xy = 18z\lambda$ by z and using the same technique results in $z = 4\sqrt{3}/3$. It follows that the desired volume is

$$V = 8(\sqrt{3})(2\sqrt{3})\left(\frac{4\sqrt{3}}{3}\right) = 64\sqrt{3}. \qquad \blacksquare$$

Example 3   If $f(x, y, z) = 4x^2 + y^2 + 5z^2$, find the point on the plane $2x + 3y + 4z = 12$ at which $f(x, y, z)$ has its least value.

Solution   We wish to find the minimum of $f(x, y, z) = 4x^2 + y^2 + 5z^2$ subject to the constraint $g(x, y, z) = 2x + 3y + 4z - 12 = 0$. If, as in (16.37),

$$\nabla(4x^2 + y^2 + 5z^2) = \lambda\nabla(2x + 3y + 4z - 12)$$

then

$$8x = 2\lambda, \quad 2y = 3\lambda, \quad \text{and} \quad 10z = 4\lambda.$$

Solving the last three equations for λ gives us

$$\lambda = 4x = \tfrac{2}{3}y = \tfrac{5}{2}z.$$

These conditions imply that

$$y = 6x \quad \text{and} \quad z = \tfrac{8}{5}x.$$

Substituting into the equation $2x + 3y + 4z - 12 = 0$, we obtain

$$2x + 18x + \tfrac{32}{5}x - 12 = 0$$

or $x = \tfrac{5}{11}$. Hence $y = 6(\tfrac{5}{11}) = \tfrac{30}{11}$ and $z = (\tfrac{8}{5})(\tfrac{5}{11}) = \tfrac{8}{11}$. It follows that the minimum value occurs at the point $(\tfrac{5}{11}, \tfrac{30}{11}, \tfrac{8}{11})$. $\blacksquare$

For some applications there may be more than one constraint. In particular, consider the problem of finding the extrema of $f(x, y, z)$ subject to the *two* constraints

$$g(x, y, z) = 0 \quad \text{and} \quad h(x, y, z) = 0.$$

It can be shown that if f has an extremum subject to these constraints, then the following condition must be satisfied for some real numbers λ and μ:

$$\nabla f(x, y, z) = \lambda\nabla g(x, y, z) + \mu\nabla h(x, y, z).$$

A specific illustration of this situation is given in the next example. This method can also be extended to functions of more than three variables and other numbers of side conditions.

Example 4 Let C denote the first octant arc of the curve in which the paraboloid $2z = 16 - x^2 - y^2$ and the plane $x + y = 4$ intersect. Find the points on C that are closest to, and farthest from, the origin. Find the minimum and maximum distances from the origin to C.

Solution   The arc C is sketched in Figure 16.34. If $P(x, y, z)$ is an arbitrary point on C, then we wish to find the largest and smallest values of $d(O, P) = \sqrt{x^2 + y^2 + z^2}$. These may be found by determining the triples (x, y, z) that give the extrema of the radicand

$$f(x, y, z) = x^2 + y^2 + z^2$$

subject to the constraints

$$g(x, y, z) = x^2 + y^2 + 2z - 16 = 0$$

and $$h(x, y, z) = x + y - 4 = 0.$$

As in our discussion, consider

$$\nabla(x^2 + y^2 + z^2) = \lambda\nabla(x^2 + y^2 + 2z - 16) + \mu\nabla(x + y - 4).$$

This leads to the following system of equations:

$$2x = 2x\lambda + \mu$$

$$2y = 2y\lambda + \mu$$

$$2z = 2\lambda$$

$$x^2 + y^2 + 2z - 16 = 0$$

$$x + y - 4 = 0$$

From the first two equations we obtain

$$2x - 2y = (2x\lambda + \mu) - (2y\lambda + \mu) = 2x\lambda - 2y\lambda$$

or $$2(x - y) - 2\lambda(x - y) = 0$$

and therefore $$2(x - y)(1 - \lambda) = 0.$$

Consequently, either $\lambda = 1$ or $x = y$.

If $\lambda = 1$, it follows that $2z = 2\lambda = 2(1)$ or $z = 1$. The first constraint $x^2 + y^2 + 2z - 16 = 0$ then gives us $x^2 + y^2 - 14 = 0$. Solving this equation simultaneously with $x + y - 4 = 0$ we find that either

$$x = 2 + \sqrt{3}, y = 2 - \sqrt{3} \quad \text{or} \quad x = 2 - \sqrt{3}, y = 2 + \sqrt{3}.$$

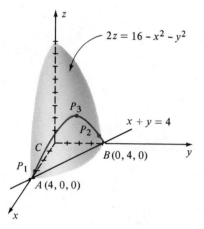

Figure 16.34

$2z = 16 - x^2 - y^2$

P_3

P_2

$x + y = 4$

C

$B(0, 4, 0)$

P_1

$A(4, 0, 0)$

Thus, points on C that may lead to extrema are

$$P_1(2 + \sqrt{3}, 2 - \sqrt{3}, 1) \quad \text{and} \quad P_2(2 - \sqrt{3}, 2 + \sqrt{3}, 1).$$

The corresponding distances from O are

$$d(O, P_1) = \sqrt{15} = d(O, P_2).$$

If $x = y$, then using the constraint $x + y - 4 = 0$ we get $x + x - 4 = 0$, $2x = 4$, or $x = 2$. This gives us $P_3(2, 2, 4)$ and $d(O, P_3) = 2\sqrt{6}$.

Referring to Figure 16.34, we may now make the following observations. As a point moves continuously along C from $A(4, 0, 0)$ to $B(0, 4, 0)$, its distance from the origin starts at $d(O, A) = 4$, decreases to a minimum value $\sqrt{15}$ at P_1 and then increases to a maximum value $2\sqrt{6}$ at P_3. The distance then decreases to $\sqrt{15}$ at P_2 and again increases to 4 at B.

As a check on the solution, note that parametric equations for C are

$$x = 4 - t, \quad y = t, \quad z = 4t - t^2$$

where $0 \leq t \leq 4$. In this case

$$f(x, y, z) = (4 - t)^2 + t^2 + (4t - t^2)^2$$

and the extrema of f may be found using single variable methods. It is left to the reader to show that the same points are obtained. ■

EXERCISES 16.9

In Exercises 1–10 use Lagrange multipliers to find the local extrema of the function f subject to the stated constraints.

1 $f(x, y) = y^2 - 4xy + 4x^2; x^2 + y^2 = 1$

2 $f(x, y) = 2x^2 + xy - y^2 + y; 2x + 3y = 1$

3 $f(x, y, z) = x + y + z; x^2 + y^2 + z^2 = 25$

4 $f(x, y, z) = x^2 + y^2 + z^2; x + y + z = 25$

5 $f(x, y, z) = x^2 + y^2 + z^2; x - y + z = 1$

6 $f(x, y, z) = x + 2y - 3z; z = 4x^2 + y^2$

7 $f(x, y, z) = x^2 + y^2 + z^2; x - y = 1; y^2 - z^2 = 1$

8 $f(x, y, z) = z - x^2 - y^2; x + y + z = 1; x^2 + y^2 = 4$

9 $f(x, y, z, t) = xyzt; x - z = 2; y^2 + t = 4$

10 $f(x, y, z, t) = x^2 + y^2 + z^2 + t^2; x + y + 2z = 1;$
   $2x - z + t = 2; y + 3z + 2t = -1$

11 Find the point on the sphere $x^2 + y^2 + z^2 = 9$ that is closest to the point $(2, 3, 4)$.

12 Let C denote the line of intersection of the planes $3x + 2y + z = 6$ and $x - 4y + 2z = 8$. Find the point on C that is closest to the origin.

13 A closed rectangular box having a volume of $2 \, \text{ft}^3$ is to be constructed. If the cost per square foot of the material for the sides, bottom, and top is $1.00, $2.00, and $1.50, respectively, what dimensions will minimize the cost?

14 Find the volume of the largest rectangular box that has three of its vertices on the positive x-, y-, and z-axes, respectively, and a fourth vertex on the plane $2x + 3y + 4z = 12$.

15 A container with a closed top is to be constructed in the shape of a right circular cylinder. If the surface area has a fixed value S, what relative dimensions will maximize the volume?

16 Prove that the triangle of maximum area having a given perimeter p is equilateral. (*Hint:* If the sides are x, y, z and if $s = p/2$, then the area is $[s(s - x)(s - y)(s - z)]^{1/2}$.)

17–26 Work Exercises 23–32 of Section 16.8 using Lagrange multipliers.

27 Prove that the product of the sines of the angles of a triangle is greatest when the triangle is equilateral.

28 A rectangular box with an open top is to have a fixed volume V. What dimensions will minimize the surface area?

29 The strength of a rectangular beam varies as the product of its width and the square of its depth. Find the dimensions of the strongest rectangular beam that can be cut from a cylindrical log whose cross sections are elliptical with major and minor axes of lengths 24 in. and 16 in., respectively.

30 If x units of capital and y units of labor are required to manufacture $f(x, y)$ units of a certain commodity, then the *Cobb–Douglas production function* is defined by $f(x, y) = kx^a y^b$, where k is a constant and a and b are positive numbers such that $a + b = 1$. Suppose that $f(x, y) = x^{1/5} y^{4/5}$ and that each unit of capital costs C dollars and each unit of labor costs L dollars. If the total amount available for these costs is M dollars, so that $xC + yL = M$, how many units of capital and labor will maximize production?

16.10

REVIEW

Define or discuss each of the following.

1 Function of more than one variable

2 Level curves

3 Level surfaces

4 Limits of functions of more than one variable

5 Continuous function

6 First partial derivatives

7 Higher partial derivatives

8 Differentials

9 Differentiable function

10 The Chain Rule

11 Implicit differentiation

12 Directional derivative

13 Gradient of a function

14 Tangent plane to a surface

15 Extrema of functions of two variables

16 Lagrange multipliers

EXERCISES 16.10

In Exercises 1–4 determine the domain of the function f.

1 $f(x, y) = \sqrt{36 - 4x^2 + 9y^2}$

2 $f(x, y) = \ln xy$

3 $f(x, y, z) = (z^2 - x^2 - y^2)^{-3/2}$

4 $f(x, y, z) = \sec z/(x - y)$

In Exercises 5–10 find the first partial derivatives of f.

5 $f(x, y) = x^3 \cos y - y^2 + 4x$

6 $f(r, s) = r^2 e^{rs}$

7 $f(x, y, z) = (x^2 + y^2)/(y^2 + z^2)$

8 $f(u, v, t) = u \ln (v/t)$

9 $f(x, y, z, t) = x^2 z \sqrt{2y + t}$

10 $f(v, w) = v^2 \cos w + w^2 \cos v$

In Exercises 11 and 12 find the second partial derivatives of f.

11 $f(x, y) = x^3 y^2 - 3xy^3 + x^4 - 3y + 2$

12 $f(x, y, z) = x^2 e^{y^2 - z^2}$

13 If $u = (x^2 + y^2 + z^2)^{-1/2}$ prove that

$$\frac{\partial^2 u}{\partial x^2} + \frac{\partial^2 u}{\partial y^2} + \frac{\partial^2 u}{\partial z^2} = 0.$$

14 (a) Find dw if $w = y^3 \tan^{-1} x^2 + 2x - y$.
(b) Find dw if $w = x^2 \sin yz$.

15 Find Δw and dw if $w = x^2 + 3xy - y^2$. Use Δw to find the exact change and dw to find the approximate change in w if (x, y) changes from $(-1, 2)$ to $(-1.1, 2.1)$.

16 In using Ohm's Law $R = E/I$, suppose that there are percentage errors of 3% and 2% in the measurements of E and I, respectively. Use differentials to estimate the maximum percentage error in the calculated value of R.

Use the Chain Rule to find the solutions in Exercises 17–19.

17 If $s = uv + vw - uw$, $u = 2x + 3y$, $v = 4x - y$, and $w = -x + 2y$, find $\partial s/\partial x$ and $\partial s/\partial y$.

18 If $z = ye^x$, $x = r + st$, $y = 2r + 3s - t$, find $\partial z/\partial r$, $\partial z/\partial s$, and $\partial z/\partial t$.

19 If $w = x \sin yz$, $x = 3e^{-t}$, $y = t^2$, $z = 3t$, find dw/dt.

20 Find the directional derivative of $f(x, y) = 3x^2 - y^2 + 5xy$ at the point $P(2, -1)$ in the direction of the vector $\mathbf{a} = -3\mathbf{i} - 4\mathbf{j}$. What is the maximum rate of increase of $f(x, y)$ at P?

21 Suppose that the temperature at the point (x, y, z) is given by $T(x, y, z) = 3x^2 + 2y^2 - 4z$. Find the rate of change of T at the point $P(-1, -3, 2)$ in the direction from P to the point $Q(-4, 1, -2)$.

22 A curve C is given parametrically by $x = t$, $y = t^2$, $z = t^3$. If $f(x, y, z) = y^2 + xz$, find $D_{\mathbf{u}}f(2, 4, 8)$, where $\mathbf{u}$ is a unit tangent vector to C at $P(2, 4, 8)$.

23 Find equations of the tangent plane and normal line to the graph of $7z = 4x^2 - 2y^2$ at the point $P(-2, -1, 2)$.

24 Show that every plane tangent to the cone

$$\frac{x^2}{a^2} - \frac{y^2}{b^2} + \frac{z^2}{c^2} = 0$$

passes through the origin.

25 If $y = f(x)$ satisfies $x^3 - 4xy^3 - 3y + x - 2 = 0$, use partial derivatives to find $f'(x)$.

26 If $z = f(x, y)$ satisfies $x^2y + z \cos y - xz^3 = 0$, find $\partial z/\partial x$ and $\partial z/\partial y$.

27 Find the extrema of f if $f(x, y) = x^2 + 3y - y^3$.

28 The material for the bottom of a rectangular box costs twice as much per square inch as that for the sides and top. What relative dimensions will minimize the cost if the volume is fixed?

29 Given $f(x, y) = x^2/4 + y^2/25$, sketch several level curves associated with f and represent $\nabla f]_P$ by a vector for a point P on each curve.

30 Given $F(x, y, z) = z + 4x^2 + 9y^2$, sketch several level surfaces associated with F and represent $\nabla F]_P$ by a vector for a point P on each surface.

31 Find the local extrema of $f(x, y, z) = xyz$ subject to the constraint $x^2 + 4y^2 + 2z^2 = 8$.

32 Use Lagrange multipliers to find the local extrema of $f(x, y, z) = 4x^2 + y^2 + z^2$ subject to the constraints $2x - y + z = 4$ and $x + 2y - z = 1$. Check your answer using single variable methods.

33 Find the points on the graph of $1/x + 2/y + 3/z = 1$ that are closest to the origin.

34 A hopper in a grain elevator has the shape of a right circular cone of radius 2 ft, surmounted by a right circular cylinder. If the volume is 100 ft^3, find the altitudes h and k of the cylinder and cone, respectively, that will minimize the curved surface area.

17

MULTIPLE INTEGRALS

The concept of the definite integral, which was introduced in Chapter 5, may be extended to functions of several variables. In this chapter we shall define double and triple integrals and discuss some of their fundamental properties and applications.

17.1

DOUBLE INTEGRALS

Most of the integrals considered previously were of the form $\int_a^b f(x)\, dx$, where f is continuous on the interval $[a, b]$. There are many other types of integrals. Among those we shall study are *double integrals*, *triple integrals*, *surface integrals*, and *line integrals*. Each of these integrals is obtained in similar fashion. The principle difference is the domain of the integrand. Recall that $\int_a^b f(x)\, dx$ may be defined by applying the following four steps.

(i) Partition $[a, b]$ by choosing $a = x_0 < x_1 < x_2 < \cdots < x_n = b$.

(ii) For each i, select any number w_i in the subinterval $[x_{i-1}, x_i]$.

(iii) Consider the Riemann sum $\sum_i f(w_i)\, \Delta x_i$, where $\Delta x_i = x_i - x_{i-1}$.

(iv) If $\lim_{\|\Delta\| \to 0} \sum_i f(w_i)\, \Delta x_i$ exists, where $\|\Delta\|$ is the norm of the partition, then this limit equals $\int_a^b f(x)\, dx$.

Suppose next that f is a function of *two* variables and that $f(x, y)$ exists throughout a region R in the xy-plane. Our objective in this section is to define the *double integral* $\iint_R f(x, y)\, dA$. As will be seen, the procedure parallels steps (i)–(iv). To define this double integral it is essential to place restrictions on the region R. Although the restrictions may appear strong, most regions encountered in applications are included among those we shall consider. Of primary importance are regions in the *xy*-plane of the types illustrated in Figure 17.1, where the functions g_1, g_2 and h_1, h_2 are continuous on the intervals $[a, b]$ and $[c, d]$, respectively, and where $g_1(x) \leq g_2(x)$ for all x in $[a, b]$ and $h_1(y) \leq h_2(y)$ for all y in $[c, d]$. The region illustrated in (i) of Figure 17.1 will be called a **region of Type I**, whereas that illustrated in (ii) of the figure will be called a **region of Type II**.

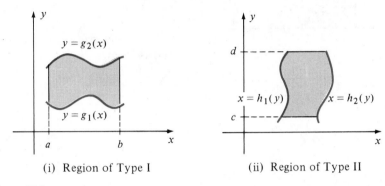

(i) Region of Type I   (ii) Region of Type II

Figure 17.1

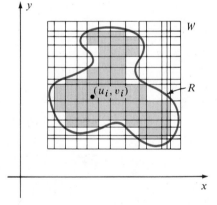

Figure 17.2

Throughout the remainder of this chapter, the symbol R will denote a region that can be decomposed into a finite number of subregions of Type I or II. Every such region R is a subset of a closed rectangular region W. If W is divided into smaller rectangles by means of a finite number of horizontal and vertical lines, then the totality of closed rectangular subregions that lie completely *within* R is called an **inner partition** of R. The shaded rectangles in Figure 17.2 illustrate an inner partition. If these shaded rectangular subregions are labeled $R_1, R_2, \ldots, R_n$, then the inner partition is $P = \{R_i : i = 1, \ldots, n\}$. The length of the longest diagonal of all the R_i will be denoted by $\|P\|$ and called the **norm of the partition** P. The symbol ΔA_i will be used for the area of R_i. Next, for each i, we choose any point (u_i, v_i) in R_i as illustrated in Figure 17.2. This corresponds to step (ii) of the four step process. Riemann sums (cf. step (iii)) are defined as follows.

Definition (17.1)

> Let f be a function of two variables that is defined on a region R and let $P = \{R_i : i = 1, \ldots, n\}$ be an inner partition of R. A **Riemann sum** of f for P is any sum of the form
>
> $$\sum_{i=1}^{n} f(u_i, v_i) \, \Delta A_i$$
>
> where (u_i, v_i) is in R_i and ΔA_i is the area of R_i.

The fourth step is to consider limits of such sums. First, however, let us consider a specific example of Definition (17.1).

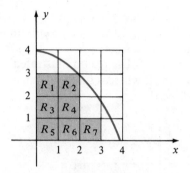

Figure 17.3

Example   Suppose $f(x, y) = 4x + 2y + 1$ and

$$R = \{(x, y): 0 \le x \le 4, 0 \le y \le \tfrac{1}{4}(16 - x^2)\}.$$

Let $P = \{R_i : i = 1, \ldots, 7\}$ be the inner partition of R determined by vertical and horizontal lines with integral intercepts, as shown in Figure 17.3. If the point (u_i, v_i) is the centroid of R_i, calculate the Riemann sum (17.1).

Solution   Referring to Figure 17.3 we see that the points (u_i, v_i) should be chosen as follows: $(\tfrac{1}{2}, \tfrac{5}{2})$ in R_1; $(\tfrac{3}{2}, \tfrac{5}{2})$ in R_2; $(\tfrac{1}{2}, \tfrac{3}{2})$ in R_3; $(\tfrac{3}{2}, \tfrac{3}{2})$ in R_4; $(\tfrac{1}{2}, \tfrac{1}{2})$ in R_5; $(\tfrac{3}{2}, \tfrac{1}{2})$ in R_6; and $(\tfrac{5}{2}, \tfrac{1}{2})$ in R_7. Since the area ΔA_i of each R_i is 1, the Riemann sum is

$$f(\tfrac{1}{2}, \tfrac{5}{2}) \cdot 1 + f(\tfrac{3}{2}, \tfrac{5}{2}) \cdot 1 + f(\tfrac{1}{2}, \tfrac{3}{2}) \cdot 1 + f(\tfrac{3}{2}, \tfrac{3}{2}) \cdot 1$$
$$+ f(\tfrac{1}{2}, \tfrac{1}{2}) \cdot 1 + f(\tfrac{3}{2}, \tfrac{1}{2}) \cdot 1 + f(\tfrac{5}{2}, \tfrac{1}{2}) \cdot 1.$$

This simplifies to

$$8 + 12 + 6 + 10 + 4 + 8 + 12 = 60. \quad \blacksquare$$

In the preceding example, the point (u_i, v_i) was the centroid of R_i. It is important to remember that when working with *arbitrary* Riemann sums, we may choose *any* pair (u_i, v_i) in R_i (see Exercise 1). For brevity we shall often denote inner partitions of R by $P = \{R_i\}$ without specifying the domain of the variable i, and the symbol Σ_i will signify that the summation takes place over all rectangles R_i in the partition. The next definition of **limit of a sum** is patterned after Definition (5.7).

Definition (17.2)

> Let f be a function of two variables that is defined on a region R and let L be a real number. The statement
>
> $$\lim_{\|P\| \to 0} \sum_i f(u_i, v_i)\, \Delta A_i = L$$
>
> means that for every $\varepsilon > 0$ there exists a $\delta > 0$ such that if $P = \{R_i\}$ is an inner partition of R with $\|P\| < \delta$, then
>
> $$\left| \sum_i f(u_i, v_i)\, \Delta A_i - L \right| < \varepsilon$$
>
> for every choice of (u_i, v_i) in R_i.

Intuitively, Definition (17.2) states that the Riemann sums of f get closer and closer to L as the norm $\|P\|$ approaches zero. If the limit L exists, it is unique. The next definition is analogous to (5.8).

Definition (17.3)

> Let f be a function of two variables that is defined on a region R. The **double integral of f over R**, denoted by $\iint_R f(x, y)\, dA$, is
>
> $$\iint\limits_R f(x, y)\, dA = \lim_{\|P\| \to 0} \sum_i f(u_i, v_i)\, \Delta A_i$$
>
> provided the limit exists.

If the double integral of f over R exists, then f is said to be **integrable over R**. It can be proved that if f is continuous on R, then f is integrable over R.

There is a useful geometric interpretation for double integrals whenever f is continuous and $f(x, y) \geq 0$ throughout R. Let S denote the graph of f and T the solid that lies under S and over R, as illustrated in Figure 17.4.

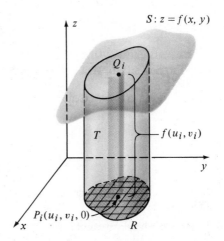

Figure 17.4

If $P_i(u_i, v_i, 0)$ is a point in the subregion R_i of an inner partition P of R, then $f(u_i, v_i)$ is the distance from the xy-plane to the point Q_i on S directly above P_i. The number $f(u_i, v_i)\, \Delta A_i$ is the volume of the prism with the rectangular base illustrated in Figure 17.4. The sum of all such volumes is an approximation to the volume V of T. Since it appears that this approximation improves as $\|P\|$ approaches zero, we define V as the limit of a sum of the numbers $f(u_i, v_i)\, \Delta A_i$. Applying (17.3) gives us the following.

Definition (17.4)

> Let f be a function of two variables such that $f(x, y) \geq 0$ for all (x, y) in a region R. The **volume** V of the solid that lies under the graph of $z = f(x, y)$ and over R is
>
> $$V = \iint\limits_R f(x, y)\, dA.$$

Evidently, if $f(x, y) \le 0$ throughout R, then the double integral of f over R is the *negative* of the volume of the solid that lies *over* the graph of f and *under* the region R.

We shall conclude this section by listing, without proof, some properties of double integrals that correspond to those given for definite integrals in Chapter 5. It is assumed that all regions and functions are suitably restricted so that the indicated integrals exist.

(17.5)   (i) $\displaystyle\iint\limits_{R} cf(x, y)\, dA = c \iint\limits_{R} f(x, y)\, dA$,   if c is any real number.

(ii) $\displaystyle\iint\limits_{R} [f(x, y) + g(x, y)]\, dA = \iint\limits_{R} f(x, y)\, dA + \iint\limits_{R} g(x, y)\, dA.$

(iii)   If R is the union of two nonoverlapping regions R_1 and R_2,

$$\iint\limits_{R} f(x, y)\, dA = \iint\limits_{R_1} f(x, y)\, dA + \iint\limits_{R_2} f(x, y)\, dA.$$

(iv)   If $f(x, y) \ge 0$ throughout R, then $\displaystyle\iint\limits_{R} f(x, y)\, dA \ge 0.$

EXERCISES 17.1

1   If f, R, and P are as in the Example of this section, find the Riemann sum of f for P if (u_i, v_i) is

(a) the point in the lower left-hand corner of R_i.

(b) the point in the upper right-hand corner of R_i.

2   Suppose f and R are as in the Example of this section and $P = \{R_i\}$ is the inner partition determined by horizontal and vertical lines having intercepts $0, \frac{1}{2}, \frac{3}{2}, \frac{5}{2}$, and 4. Find the Riemann sum of f for P if (u_i, v_i) is the centroid of R_i.

3   Let R be the triangular region with vertices $(0, 0)$, $(6, 0)$, $(6, 12)$ and let P be the inner partition of R determined by vertical lines with x-intercepts $0, 1, 3, 4, 6$ and horizontal lines with y-intercepts $0, 2, 5, 7, 8, 12$. If $f(x, y) = x^2 y$, find the Riemann sum of f for P if (u_i, v_i) is the centroid of R_i.

4   Let $R = \{(x, y): 0 \le x \le 5, 0 \le y \le \sqrt{25 - x^2}\}$. Let P be the partition of R determined by vertical lines with x-intercepts $0, 2, 3, 4, 5$ and by horizontal lines with y-intercepts $0, 2, 4, 5$. If $f(x, y) = 4 - xy$, find the Riemann sum of f for P if (u_i, v_i) is the point in the lower right-hand corner of R_i.

5   Let R be the region bounded by the trapezoid with vertices $(0, 0)$, $(4, 4)$, $(8, 4)$, and $(12, 0)$. Let P be the partition of R determined by vertical lines with x-intercepts $0, 2, 4,$

6, 8, 10, 12 and by horizontal lines with y-intercepts $0, 2, 4$. If $f(x, y) = xy$, find the Riemann sum of f for P if (u_i, v_i) is the centroid of R_i.

6   Let R be the region bounded by the triangle with vertices $(-4, 0)$, $(0, 8)$, and $(4, 0)$. Let P be the partition of R determined by vertical lines with x-intercepts $-4, -2, 0, 1, 3, 4$ and by horizontal lines with y-intercepts $0, 2, 4, 6, 8$. If $f(x, y) = x + y$, find the Riemann sum of f for P if (u_i, v_i) is the upper right-hand corner of R_i.

7   If $R = \{(x, y): a \le x \le b, c \le y \le d\}$ and $f(x, y) = k$ for all (x, y) in R, prove that the double integral of f over R equals $k(b - a)(d - c)$ by using (a) Definition (17.3), and (b) Definition (17.4).

8   Suppose that the functions f and g are integrable over R and $f(x, y) \ge g(x, y)$ for all (x, y) in R. Use properties of double integrals to prove that

$$\iint\limits_{R} f(x, y)\, dA \ge \iint\limits_{R} g(x, y)\, dA.$$

9   Prove (i) of (17.5). (*Hint*: See the proof of (5.14) in Appendix II.)

10   Prove (ii) of (17.5). (*Hint*: See the proof of (5.15) in Appendix II.)

17.2

EVALUATION OF DOUBLE INTEGRALS

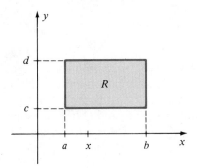

Figure 17.5

Except for elementary cases, it is virtually impossible to find the value of a double integral directly from Definition (17.3). This section contains methods of evaluation that make use of the Fundamental Theorem of Calculus.

Suppose f is a function of two variables that is continuous on a closed rectangular region R of the type illustrated in Figure 17.5. We shall use the symbol $\int_c^d f(x, y)\, dy$ to mean that x is held fixed and the integration is performed with respect to y. This is sometimes called a **partial integration with respect to y**. To each x in the interval $[a, b]$ there corresponds a unique value for this integral and hence a function A is determined, where the value $A(x)$ is given by

$$A(x) = \int_c^d f(x, y)\, dy.$$

As an illustration, if $f(x, y) = x^3 + 4y$, $c = 1$, and $d = 2$, then

$$A(x) = \int_1^2 (x^3 + 4y)\, dy = x^3 y + 2y^2 \Big]_1^2$$

$$= (2x^3 + 8) - (x^3 + 2)$$

$$= x^3 + 6.$$

It can be proved that the function A is continuous on $[a, b]$ and hence has a definite integral that may be written

$$\int_a^b A(x)\, dx = \int_a^b \left[\int_c^d f(x, y)\, dy \right] dx.$$

The expression on the right side of this equation is called an **iterated (double) integral**. Ordinarily the notation is shortened by omitting the brackets, as in the following definition.

Definition (17.6)

$$\int_a^b \int_c^d f(x, y)\, dy\, dx = \int_a^b \left[\int_c^d f(x, y)\, dy \right] dx$$

Example 1 Evaluate $\displaystyle\int_1^4 \int_{-1}^2 (2x + 6x^2 y)\, dy\, dx.$

Solution As in Definition (17.6), the integral equals

$$\int_1^4 \left[\int_{-1}^2 (2x + 6x^2 y)\, dy \right] dx = \int_1^4 \left[2xy + 3x^2 y^2 \right]_{-1}^2 dx$$

$$= \int_1^4 [(4x + 12x^2) - (-2x + 3x^2)]\, dx$$

$$= \int_1^4 (6x + 9x^2)\, dx$$

$$= 3x^2 + 3x^3 \Big]_1^4 = 234.$$ ∎

Similarly, we may consider iterated integrals of the following form.

Definition (17.7)

$$\int_{c}^{d} \int_{a}^{b} f(x, y)\, dx\, dy = \int_{c}^{d} \left[\int_{a}^{b} f(x, y)\, dx \right] dy$$

In Definition (17.7) we integrate first with respect to x (with y held fixed). After substituting the limits b and a for x in the usual way, the resulting expression is integrated with respect to y from c to d. It is proved in more advanced texts that the integrals in Definitions (17.6) and (17.7) are equal. This is illustrated in the next example where we reevaluate the integral in Example 1.

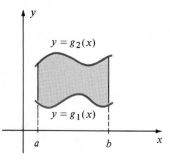

$y = g_2(x)$

$y = g_1(x)$

a b

(i) Region of Type I

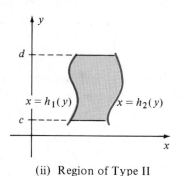

$x = h_1(y)$ $x = h_2(y)$

(ii) Region of Type II

Figure 17.6

Example 2 Evaluate $\displaystyle\int_{-1}^{2} \int_{1}^{4} (2x + 6x^2 y)\, dx\, dy$.

Solution   Applying Definition (17.7),

$$\int_{-1}^{2} \left[\int_{1}^{4} (2x + 6x^2 y)\, dx \right] dy = \int_{-1}^{2} \left[x^2 + 2x^3 y \right]_{1}^{4} dy$$

$$= \int_{-1}^{2} \left[(16 + 128y) - (1 + 2y) \right] dy$$

$$= \int_{-1}^{2} (126y + 15)\, dy$$

$$= 63y^2 + 15y \Big]_{-1}^{2} = 234. \qquad \blacksquare$$

Iterated integrals may be defined over nonrectangular regions. In particular, suppose f is continuous on a region of Type I or II, as shown in Figure 17.6. We define iterated integrals of f over the regions illustrated in (i) and (ii) of the figure as follows.

Definition (17.8)

(i) $\displaystyle\int_{a}^{b} \int_{g_1(x)}^{g_2(x)} f(x, y)\, dy\, dx = \int_{a}^{b} \left[\int_{g_1(x)}^{g_2(x)} f(x, y)\, dy \right] dx$

(ii) $\displaystyle\int_{c}^{d} \int_{h_1(y)}^{h_2(y)} f(x, y)\, dx\, dy = \int_{c}^{d} \left[\int_{h_1(y)}^{h_2(y)} f(x, y)\, dx \right] dy$

In (i) of Definition (17.8) we perform a partial integration with respect to y and then substitute $g_2(x)$ and $g_1(x)$ for y in the usual way. The resulting expression in x is then integrated from a to b. Similar remarks hold for (ii).

Example 3   Evaluate $\int_0^2 \int_{x^2}^{2x} (x^3 + 4y)\,dy\,dx$.

Solution   By (i) of Definition (17.8) the integral equals

$$\int_0^2 \left[\int_{x^2}^{2x} (x^3 + 4y)\,dy \right] dx = \int_0^2 \left[x^3 y + 2y^2 \right]_{x^2}^{2x} dx$$

$$= \int_0^2 [(2x^4 + 8x^2) - (x^5 + 2x^4)]\,dx$$

$$= \frac{8}{3}x^3 - \frac{1}{6}x^6 \Big]_0^2 = \frac{32}{3}. \qquad \blacksquare$$

Example 4   Evaluate $\int_1^3 \int_{\pi/6}^{y^2} 2y \cos x\,dx\,dy$.

Solution   By (ii) of Definition (17.8) the integral equals

$$\int_1^3 \left[\int_{\pi/6}^{y^2} 2y \cos x\,dx \right] dy = \int_1^3 \left[2y \sin x \right]_{\pi/6}^{y^2} dy$$

$$= \int_1^3 (2y \sin y^2 - y)\,dy$$

$$= -\cos y^2 - \frac{1}{2}y^2 \Big]_1^3$$

$$= (-\cos 9 - \tfrac{9}{2}) - (-\cos 1 - \tfrac{1}{2})$$

$$= \cos 1 - \cos 9 - 4 \approx -2.55. \qquad \blacksquare$$

The next theorem indicates that if the region R is of Type I or II, then the double integral defined in Section 17.1 may be evaluated by means of an iterated integral.

Theorem (17.9)

(i)   Let R be a region of Type I that lies between the graphs of $y = g_1(x)$ and $y = g_2(x)$, where g_1 and g_2 are continuous on $[a, b]$. If f is continuous on R, then

$$\iint_R f(x, y)\,dA = \int_a^b \int_{g_1(x)}^{g_2(x)} f(x, y)\,dy\,dx.$$

(ii)   Let R be a region of Type II that lies between the graphs of $x = h_1(y)$ and $x = h_2(y)$, where h_1 and h_2 are continuous on $[c, d]$. If f is continuous on R then

$$\iint_R f(x, y)\,dA = \int_c^d \int_{h_1(y)}^{h_2(y)} f(x, y)\,dx\,dy.$$

It is important to note that in order to use (i) of Theorem 17.9, the region R must have a **lower boundary**, consisting of the graph of an equation $y = g_1(x)$ and an **upper boundary** consisting of the graph of an equation $y = g_2(x)$ (see (i) of Fig. 17.6). To use (ii), R must have a **left-hand boundary** consisting of the graph of an equation $x = h_1(y)$ and a **right-hand boundary** consisting of the graph of an equation $x = h_2(y)$ (see (ii) of Fig. 17.6). If the region is more complicated, then it is often possible to divide R into subregions of the required type and then apply (i) or (ii) to each subregion.

Instead of giving a rigorous proof of Theorem (17.9), we shall present an intuitive discussion that makes the result plausible, at least for nonnegative-valued functions. Suppose $f(x, y) \geq 0$ throughout the region R of Type I illustrated in (i) of Figure 17.6. Let S denote the graph of f, T the solid that lies under S and over R, and V the volume of T. We shall begin by considering the plane Γ that is parallel to the yz-plane and intersects the x-axis at the point $(x, 0, 0)$, where $a \leq x \leq b$. As illustrated in Figure 17.7, C will denote the trace of S on Γ. The points $P(x, g_1(x), 0)$ and $Q(x, g_2(x), 0)$ indicate where Γ intersects the boundaries $y = g_1(x)$ and $y = g_2(x)$ of R. In the figure we have shown the part of Γ that lies above the segment PQ and under the surface S.

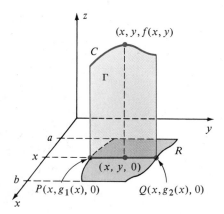

Figure 17.7

From our work in Chapter 6, the area $A(x)$ of this part of Γ is

$$A(x) = \int_{g_1(x)}^{g_2(x)} f(x, y)\, dy.$$

Since $A(x)$ is the area of a typical cross section of T, it follows from (6.8) that

$$V = \int_a^b A(x)\, dx = \int_a^b \int_{g_1(x)}^{g_2(x)} f(x, y)\, dy\, dx.$$

Using the fact that V is also given by Definition (17.4) we obtain the evaluation formula in part (i) of Theorem (17.9). A similar discussion can be given for part (ii).

Example 5   Evaluate $\iint_R (x^3 + 4y)\, dA$, where R is the region in the xy-plane bounded by the graphs of the equations $y = x^2$ and $y = 2x$.

Solution   The graph of R is sketched in Figure 17.8. Note that R is a region both of Type I and Type II. If we regard R as a region of Type I with lower boundary $y = x^2$ and upper boundary $y = 2x$, where $0 \le x \le 2$, then by (i) of Theorem 17.9,

$$\iint_R (x^3 + 4y)\, dA = \int_0^2 \int_{x^2}^{2x} (x^3 + 4y)\, dy\, dx.$$

From Example 3 we know that the last integral equals $\frac{32}{3}$.

If we regard R as a region of Type II, then the left-hand boundary is the graph of $x = \frac{1}{2}y$ and the right-hand boundary is the graph of $x = \sqrt{y}$, where $0 \le y \le 4$. Hence by (ii) of Theorem 17.9,

$$\iint_R f(x, y)\, dA = \int_0^4 \int_{(1/2)y}^{\sqrt{y}} (x^3 + 4y)\, dx\, dy$$

$$= \int_0^4 \left[\frac{1}{4}x^4 + 4yx \right]_{(1/2)y}^{\sqrt{y}} dy$$

$$= \int_0^4 \left[\left(\frac{1}{4}y^2 + 4y^{3/2} \right) - \left(\frac{1}{64}y^4 + 2y^2 \right) \right] dy = \frac{32}{3}. \qquad \blacksquare$$

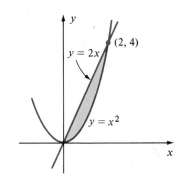

$y = 2x$

$(2, 4)$

$y = x^2$

Figure 17.8

Example 6   Let R be the region bounded by the graphs of the equations $y = \sqrt{x}$, $y = \sqrt{3x - 18}$, and $y = 0$. If f is continuous on R, express the double integral $\iint_R f(x, y)\, dA$ in terms of iterated integrals by

(a)   using only part (i) of Theorem 17.9.

(b)   using only part (ii) of Theorem 17.9.

Solution   The graphs of $y = \sqrt{x}$ and $y = \sqrt{3x - 18}$ are the top halves of the parabolas $y^2 = x$ and $y^2 = 3x - 18$. The region R is sketched in Figure 17.9.

(a)   If we wish to use only part (i) of Theorem 17.9, then it is necessary to employ two iterated integrals, because if $0 \le x \le 6$, then the lower boundary of the region is the graph of $y = 0$, whereas if $6 \le x \le 9$, the lower boundary is the graph of $y = \sqrt{3x - 18}$. If R_1 denotes the part of the region R that lies between $x = 0$ and $x = 6$, and if R_2 denotes the part between $x = 6$ and $x = 9$, then both R_1 and R_2 are regions of Type I. Hence

$$\iint_R f(x, y)\, dA = \iint_{R_1} f(x, y)\, dA + \iint_{R_2} f(x, y)\, dA$$

$$= \int_0^6 \int_0^{\sqrt{x}} f(x, y)\, dy\, dx + \int_6^9 \int_{\sqrt{3x-18}}^{\sqrt{x}} f(x, y)\, dy\, dx.$$

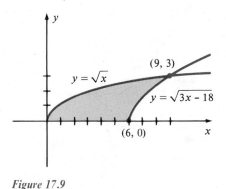

$y = \sqrt{x}$

$(9, 3)$

$y = \sqrt{3x - 18}$

$(6, 0)$

Figure 17.9

(b)   To use (ii) of Theorem 17.9 we must solve each of the given equations for x in terms of y, obtaining

$$x = y^2 \quad \text{and} \quad x = \frac{y^2 + 18}{3} = \frac{1}{3}y^2 + 6.$$

Only one iterated integral is required in this case since R is a region of Type II. Thus

$$\iint\limits_{R} f(x, y)\, dA = \int_0^3 \int_{y^2}^{(1/3)y^2 + 6} f(x, y)\, dx\, dy. \qquad\blacksquare$$

Examples 5 and 6 indicate how certain double integrals may be evaluated by using either (i) or (ii) of Theorem 17.9. Generally, the choice of the *order of integration dy dx* or *dx dy* often depends on the form of $f(x, y)$ and the region R. Sometimes it is extremely difficult, or even impossible, to evaluate a given iterated double integral. However, by **reversing the order of integration** from *dy dx* to *dx dy*, or vice versa, it may be possible to find an equivalent iterated double integral that is easy to evaluate. This technique is illustrated in the next example.

Example 7   Given $\int_0^4 \int_{\sqrt{y}}^2 y \cos x^5\, dx\, dy$, reverse the order of integration and evaluate the resulting integral.

Solution   Since the given order of integration is *dx dy*, the region R is of Type II. As illustrated in Figure 17.10, the left-hand and right-hand boundaries are the graphs of $x = \sqrt{y}$ and $x = 2$, respectively, and $0 \le y \le 4$.

Note that R is also a region of Type I whose lower and upper boundaries are given by $y = 0$ and $y = x^2$, respectively, and where $0 \le x \le 2$. Hence by Theorem 17.9,

$$\int_0^4 \int_{\sqrt{y}}^2 y \cos x^5\, dx\, dy = \iint\limits_{R} y \cos x^5\, dA = \int_0^2 \int_0^{x^2} y \cos x^5\, dy\, dx$$

$$= \int_0^2 \frac{y^2}{2} \cos x^5 \Big]_0^{x^2} dx = \int_0^2 \frac{x^4}{2} \cos x^5\, dx$$

$$= \frac{1}{10} \int_0^2 \cos x^5 (5x^4)\, dx$$

$$= \frac{1}{10} \sin x^5 \Big]_0^2 = \frac{1}{10} \sin 32 \approx 0.055. \qquad\blacksquare$$

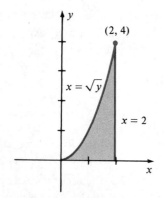

Figure 17.10

On page 784 we saw that the volume V of a solid that lies under the graph of $z = f(x, y)$ and over a region of Type I in the xy-plane can be expressed as

$$V = \int_a^b A(x)\, dx = \int_a^b \int_{g_1(x)}^{g_2(x)} f(x, y)\, dy\, dx$$

where $A(x)$ is the area of the typical cross section of the solid shown in Figure 17.7. It is often useful to regard these two integrals as limits of sums. Thus, by (6.8), we may write

$$V = \int_a^b A(x)\, dx = \lim_{\|P'\| \to 0} \sum_i A(u_i)\, \Delta x_i$$

where P' is a partition of the interval $[a, b]$, u_i is any number in the ith subinterval $[x_{i-1}, x_i]$ of P', and $\Delta x_i = x_i - x_{i-1}$. As illustrated in Figure 17.11, $A(u_i)\, \Delta x_i$ is the volume of a lamina L_i whose face is parallel to the yz-plane and whose base is a rectangle of width Δx_i in the xy-plane. Consequently, the volume V may be regarded as a limit of a sum of volumes of such laminas.

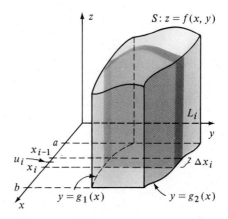

Figure 17.11

The *iterated* integral $V = \int_a^b \int_{g_1(x)}^{g_2(x)} f(x, y)\, dy\, dx$ may also be expressed in terms of limits of sums. First, applying the definition of the definite integral,

$$A(x) = \int_{g_1(x)}^{g_2(x)} f(x, y)\, dy = \lim_{\|P''\| \to 0} \sum_j f(x, v_j)\, \Delta y_j$$

where, for each x, P'' is a partition of the y-interval $[g_1(x), g_2(x)]$, v_j is a number in the jth subinterval $[y_{j-1}, y_j]$ of P'', and $\Delta y_j = y_j - y_{j-1}$. Hence, for each u_i in $[a, b]$,

$$A(u_i) = \lim_{\|P''\| \to 0} \sum_j f(u_i, v_j)\, \Delta y_j$$

and, therefore,

$$V = \int_a^b A(x)\, dx = \lim_{\|P'\| \to 0} \sum_i A(u_i)\, \Delta x_i$$

$$= \lim_{\|P'\| \to 0} \sum_i \left[\lim_{\|P''\| \to 0} \sum_j f(u_i, v_j)\, \Delta y_j \right] \Delta x_i.$$

If we ignore the limits in this formula, there results what is called a **double sum**, written

$$\sum_i \sum_j f(u_i, v_j) \, \Delta y_j \, \Delta x_i.$$

Referring to Figure 17.12, we see that the expression $f(u_i, v_j) \, \Delta y_j \, \Delta x_i$ may be regarded as the volume of a prism having base area $\Delta y_j \, \Delta x_i$ and altitude $f(u_i, v_j)$.

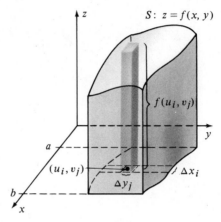

Figure 17.12

The double sum may then be interpreted as follows. *Holding x fixed* we sum the volumes of the prisms in the direction of the y-axis, thereby obtaining the volume of the lamina illustrated in Figure 17.11. The volumes of all such laminas are then added by summing in the direction of the x-axis. In a certain sense, the iterated integral does this and at the same time takes limits as the bases of the prisms shrink to points. This interpretation is often useful in visualizing applications of the double integral. To emphasize this point of view we sometimes write

$$(17.10) \qquad V = \lim_{\|\Delta\| \to 0} \sum_i \sum_j f(u_i, v_j) \, \Delta y_j \, \Delta x_i = \int_a^b \int_{g_1(x)}^{g_2(x)} f(x, y) \, dy \, dx$$

where the symbol $\|\Delta\| \to 0$ is used to signify that all the Δx_i and Δy_j approach 0.

EXERCISES 17.2

Using the values given in Exercises 1 and 2, verify that

$$\int_a^b \int_c^d f(x, y) \, dy \, dx = \int_c^d \int_a^b f(x, y) \, dx \, dy.$$

1 $a = 1, b = 2, c = -1, d = 2; f(x, y) = 12xy^2 - 8x^3$

2 $a = -2, b = -1, c = 0, d = 3; f(x, y) = 4xy^3 + y$

Evaluate the iterated integrals in Exercises 3–12.

3 $\displaystyle\int_1^2 \int_{1-x}^{\sqrt{x}} x^2 y \, dy \, dx$

4 $\displaystyle\int_{-1}^1 \int_{x^3}^{x+1} (3x + 2y) \, dy \, dx$

5 $\displaystyle\int_0^2 \int_{y^2}^{2y} (4x - y) \, dx \, dy$

6 $\displaystyle\int_0^1 \int_{-y-1}^{y-1} (x^2 + y^2)\, dx\, dy$

7 $\displaystyle\int_1^2 \int_{x^3}^{x} e^{y/x}\, dy\, dx$

8 $\displaystyle\int_0^{\pi/6} \int_0^{\pi/2} (x \cos y - y \cos x)\, dy\, dx$

9 $\displaystyle\int_1^e \int_0^x \ln x\, dy\, dx$

10 $\displaystyle\int_0^1 \int_y^1 \frac{1}{1 + y^2}\, dx\, dy$

11 $\displaystyle\int_{\pi/6}^{\pi/4} \int_{\tan x}^{\sec x} (y + \sin x)\, dy\, dx$

12 $\displaystyle\int_{\pi/6}^{\pi/4} \int_0^{\sin x} e^y \cos x\, dy\, dx$

In Exercises 13–18 sketch the region R bounded by the graphs of the given equations and express $\iint_R f(x, y)\, dA$ as an iterated integral by using (a) part (i) of Theorem (17.9) and (b) part (ii) of Theorem (17.9).

13 $y = \sqrt{x},\, x = 4,\, y = 0$ **14** $y = \sqrt{x},\, x = 0,\, y = 2$

15 $y = x^3,\, x = 0,\, y = 8$ **16** $y = x^3,\, x = 2,\, y = 0$

17 $y = \sqrt{x},\, y = x^3$ **18** $y = \sqrt{1 - x^2},\, y = 0$

In Exercises 19–24 express the double integral over R as an iterated integral and find its value.

19 $\iint_R (y + 2x)\, dA$, where R is the rectangular region with vertices $(-1, -1),\, (2, -1),\, (2, 4)$, and $(-1, 4)$.

20 $\iint_R (x - y)\, dA$, where R is the triangular region with vertices $(2, 9),\, (2, 1)$, and $(-2, 1)$.

21 $\iint_R xy^2\, dA$, where R is the triangular region with vertices $(0, 0),\, (3, 1)$, and $(-2, 1)$.

22 $\iint_R (y + 1)\, dA$, where R is the region between the graphs of $y = \sin x$ and $y = \cos x$ from $x = 0$ to $x = \pi/4$.

23 $\iint_R x^3 \cos xy\, dA$, where R is the region bounded by the graphs of $y = x^2,\, y = 0$, and $x = 2$.

24 $\iint_R e^{x/y}\, dA$, where R is the region bounded by the graphs of $y = 2x,\, y = -x$, and $y = 4$.

In Exercises 25–30 sketch the region R bounded by the graphs of the given equations. If f is continuous on R, use Definition (17.4) to express $\iint_R f(x, y)\, dA$ as a sum of two iterated integrals of the type used in (a) part (i) of Theorem (17.9); (b) part (ii) of Theorem (17.9).

25 $8y = x^3,\, y - x = 4,\, 4x + y = 9$

26 $x = 2\sqrt{y},\, \sqrt{3}x = \sqrt{y},\, y = 2x + 5$

27 $x = \sqrt{3 - y},\, y = 2x,\, x + y + 3 = 0$

28 $x + 2y = 5,\, x - y = 2,\, 2x + y = -2$

29 $y = e^x,\, y = \ln x,\, x + y = 1,\, x + y = 1 + e$

30 $y = \sin x,\, \pi y = 2x$

In Exercises 31–38 sketch the region of integration for the iterated integral.

31 $\displaystyle\int_{-1}^2 \int_{-\sqrt{4-x^2}}^{4-x^2} f(x, y)\, dy\, dx$

32 $\displaystyle\int_{-1}^2 \int_{x^2-4}^{x-2} f(x, y)\, dy\, dx$

33 $\displaystyle\int_0^1 \int_{\sqrt{y}}^{\sqrt[3]{y}} f(x, y)\, dx\, dy$

34 $\displaystyle\int_{-2}^{-1} \int_{3y}^{2y} f(x, y)\, dx\, dy$

35 $\displaystyle\int_{-3}^1 \int_{\tan^{-1} x}^{e^x} f(x, y)\, dy\, dx$

36 $\displaystyle\int_{-1}^2 \int_{\sinh x}^{\cosh x} f(x, y)\, dy\, dx$

37 $\displaystyle\int_{-1}^1 \int_{1-y^4}^{\cosh y} f(x, y)\, dx\, dy$

38 $\displaystyle\int_{\pi}^{2\pi} \int_{\sin y}^{\ln y} f(x, y)\, dx\, dy$

In Exercises 39–44 reverse the order of integration and evaluate the resulting integral.

39 $\displaystyle\int_0^1 \int_{2x}^2 e^{(y^2)}\, dy\, dx$

40 $\displaystyle\int_0^9 \int_{\sqrt{y}}^3 \sin x^3\, dx\, dy$

41 $\displaystyle\int_0^2 \int_{y^2}^4 y \cos x^2\, dx\, dy$

42 $\displaystyle\int_1^e \int_0^{\ln x} y\, dy\, dx$

43 $\displaystyle\int_0^8 \int_{\sqrt[3]{y}}^2 \frac{y}{\sqrt{16 + x^7}}\, dx\, dy$

44 $\displaystyle\int_0^1 \int_x^1 \frac{1}{y} \sin y \cos \frac{x}{y}\, dy\, dx$

45   Interpret the iterated integral in part (ii) of Theorem (17.9) as a limit of a double sum in a manner analogous to (17.10).

17.3

AREAS AND VOLUMES

It is customary to denote the double integral $\iint_R 1 \, dA$ by $\iint_R dA$. If R is a region of Type I, as illustrated in (i) of Figure 17.6, then by Theorem (17.9),

$$\iint\limits_R dA = \int_a^b \int_{g_1(x)}^{g_2(x)} 1 \, dy \, dx = \int_a^b y \Big]_{g_1(x)}^{g_2(x)} dx$$

$$= \int_a^b [g_2(x) - g_1(x)] \, dx$$

which, according to Theorem (6.1), equals the area A of R. The same is true if R is a region of Type II, as illustrated in (ii) of Figure 17.6. These facts are also evident from the definition of the double integral, for if $f(x, y) = 1$ throughout R, then the Riemann sum in Definition (17.3) is a sum of areas of rectangles in an inner partition P of R. As the norm $\|P\|$ approaches zero, these rectangles cover more of R and it is apparent that the limit equals the area of R.

If an iterated integral is used to find an area it may be regarded as a limit of a double sum in a manner similar to that used for volumes at the end of the previous section. Specifically, as in (17.10), with $f(x, y) = 1$, we write

$$A = \iint\limits_R dA = \int_a^b \int_{g_1(x)}^{g_2(x)} dy \, dx = \lim_{\|\Delta\| \to 0} \sum_i \sum_j \Delta y_j \, \Delta x_i.$$

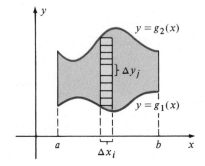

Figure 17.13

The double sum may be interpreted as follows. First, *holding x fixed* we sum over rectangles of areas $\Delta y_j \, \Delta x_i$ in the direction of the y-axis, from the graph of $y = g_1(x)$ to the graph of $y = g_2(x)$. This gives us the area of a rectangular strip that is parallel to the y-axis, as illustrated in Figure 17.13. Second, we sweep out the region R by summing over all such strips from $x = a$ to $x = b$. The limit of the double sum (that is, the iterated integral) does this and, at the same time, forces all the Δx_i and Δy_j to approach 0.

Example 1 Find the area A of the region bounded by the graphs of $2y = 16 - x^2$ and $x + 2y - 4 = 0$.

Solution   The region lies under the parabola $y = 8 - (x^2/2)$ and over the line $y = 2 - (x/2)$, as illustrated in Figure 17.14. Also shown in the figure is a typical rectangle of area $\Delta y_j \, \Delta x_i$ and the strip obtained by summing in the direction of the y-axis. Using an iterated integral we have

$$A = \int_{-3}^4 \int_{2-(x/2)}^{8-(x^2/2)} dy \, dx = \int_{-3}^4 \left[\left(8 - \frac{x^2}{2}\right) - \left(2 - \frac{x}{2}\right) \right] dx$$

$$= 6x - \frac{x^3}{6} + \frac{x^2}{4} \Big]_{-3}^4 = \frac{343}{12}.$$ ■

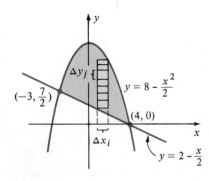

Figure 17.14

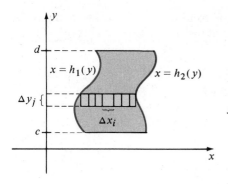

Figure 17.15

If part (ii) of Theorem (17.9) is used to find areas, then

$$A = \iint_R dA = \int_c^d \int_{h_1(y)}^{h_2(y)} dx\, dy = \lim_{\|\Delta\| \to 0} \sum_j \sum_i \Delta x_i\, \Delta y_j.$$

The preceding double sum may be interpreted as follows. First, *holding y fixed* we sum over rectangles of areas $\Delta x_i\, \Delta y_j$ in the direction of the x-axis, from the graph of $x = h_1(y)$ to the graph of $x = h_2(y)$. This gives us the area of a rectangular strip that is parallel to the x-axis, as illustrated in Figure 17.15. Second, we sweep out the region R by summing over all such strips from $y = c$ to $y = d$. As before, the iterated integral accomplishes this and simultaneously forces all the Δx_i and Δy_j to approach 0.

Example 2   Find the area A of the region in the xy-plane bounded by the graphs of $x = y^3$, $x + y = 2$, and $y = 0$.

Solution   The region is sketched in Figure 17.16 together with a typical rectangle of area $\Delta x_i\, \Delta y_j$ and a strip obtained by summing in the direction of the x-axis.

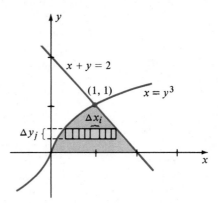

Figure 17.16

Using an iterated integral gives us

$$A = \iint_R dA = \int_0^1 \int_{y^3}^{2-y} dx\, dy = \int_0^1 x\Big]_{y^3}^{2-y} dy$$

$$= \int_0^1 (2 - y - y^3)\, dy = 2y - \frac{y^2}{2} - \frac{y^4}{4}\Big]_0^1 = \frac{5}{4}.$$

The area can also be found by employing part (i) of Theorem (17.9); however, in this case it is necessary to divide R into two parts by means of the vertical line through (1, 1). We then have

$$A = \int_0^1 \int_0^{\sqrt[3]{x}} dy\, dx + \int_1^2 \int_0^{2-x} dy\, dx. \qquad \blacksquare$$

We can use Definition (17.4) to express volumes of certain solids in terms of double integrals that may then be evaluated by means of iterated integrals. This is illustrated in the next example.

Example 3   Find the volume V of the solid in the first octant bounded by the coordinate planes and the graphs of the equations $z = x^2 + y^2 + 1$ and $2x + y = 2$.

Solution   As illustrated in Figure 17.17, the solid lies under the paraboloid $z = x^2 + y^2 + 1$ and over the triangular region R in the xy-plane bounded by the coordinate axes and the line $y = 2 - 2x$.

By Definition (17.4), with $f(x, y) = x^2 + y^2 + 1$,

$$V = \iint\limits_{R} (x^2 + y^2 + 1)\, dA.$$

As in (17.10), this integral may be interpreted as the limit of a double sum

$$\sum_i \sum_j f(u_i, v_j)\, \Delta y_j \, \Delta x_i = \sum_i \sum_j (u_i^2 + v_j^2 + 1)\, \Delta y_j \, \Delta x_i$$

where $(u_i^2 + v_j^2 + 1)\, \Delta y_j \, \Delta x_i$ represents the volume of a prism of the type shown in (i) of Figure 17.17, and where we first sum in the direction of the y-axis from $y = 0$ to $y = 2 - 2x$. The sketch in (ii) of the figure shows the region R in the xy-plane and a strip corresponding to the first summation. This first summation gives us the volume of a lamina whose face is parallel to the yz-plane. We then sum these laminar volumes in the direction of the x-axis from $x = 0$ to $x = 1$. Using an iterated integral,

$$V = \int_0^1 \int_0^{2-2x} (x^2 + y^2 + 1)\, dy \, dx = \int_0^1 \left[x^2 y + \frac{1}{3}y^3 + y \right]_0^{2-2x} dx$$

$$= \int_0^1 \left(-\frac{14}{3}x^3 + 10x^2 - 10x + \frac{14}{3} \right) dx$$

$$= -\frac{7}{6}x^4 + \frac{10}{3}x^3 - 5x^2 + \frac{14}{3}x \bigg]_0^1 = \frac{11}{6}.$$

We may also find V by integrating first with respect to x. The reader should verify that in this case the iterated integral has the form

$$V = \int_0^2 \int_0^{(2-y)/2} (x^2 + y^2 + 1)\, dx \, dy. \qquad \blacksquare$$

Example 4   Find the volume V of the solid bounded by the graphs of $x^2 + y^2 = 9$ and $y^2 + z^2 = 9$.

Solution   The graphs are cylinders of radius 3 whose axes intersect orthogonally. That portion of the solid that lies in the first octant is illustrated in (i) of Figure 17.18. By symmetry it is sufficient to find the volume of this portion and then multiply the result by 8. Referring to (i) of the figure we see that the solid lies under the graph of $z = \sqrt{9 - y^2}$ and over the region R bounded by the x- and y-axes and the graph of $x^2 + y^2 = 9$. Thus, by Definition (17.4),

$$V = 8 \iint\limits_{R} (9 - y^2)^{1/2} \, dA.$$

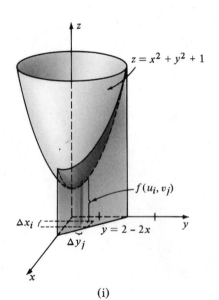

$z = x^2 + y^2 + 1$

$f(u_i, v_j)$

Δx_i

$y = 2 - 2x$

Δy_j

(i)

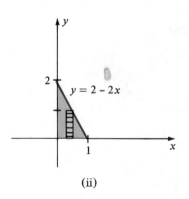

$y = 2 - 2x$

(ii)

Figure 17.17

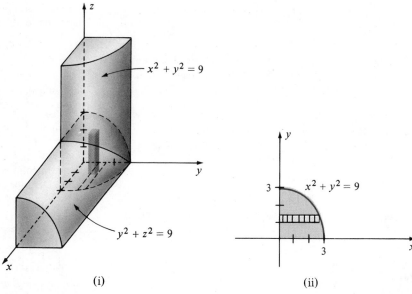

Figure 17.18

We shall evaluate the double integral by using (ii) of Theorem (17.9). In this case we regard the integral as a limit of a double sum of volumes of prisms of the type shown in (i) of Figure 17.18, where we first sum in the direction of the x-axis from $x = 0$ to $x = (9 - y^2)^{1/2}$. The sketch in (ii) of Figure 17.18 indicates the region R in the xy-plane and a strip corresponding to this first summation. This gives us the volume of a lamina whose face is parallel to the xz-plane. We then sum these laminar volumes in the direction of the y-axis from $y = 0$ to $y = 3$. Using an iterated integral, we have

$$V = 8 \int_0^3 \int_0^{(9-y^2)^{1/2}} (9 - y^2)^{1/2} \, dx \, dy = 8 \int_0^3 \left[(9 - y^2)^{1/2} x \right]_0^{(9-y^2)^{1/2}} dy$$

$$= 8 \int_0^3 (9 - y^2) \, dy = 8 \left[9y - \frac{1}{3}y^3 \right]_0^3 = 144.$$

The double integral could also be evaluated by integrating first with respect to y; however, this order of integration is much more complicated.   ∎

EXERCISES 17.3

In Exercises 1–10 sketch the region bounded by the graphs of the given equations and find its area by means of double integrals. Interpret each iterated double integral as a limit of a double sum by sketching a typical rectangle of area $\Delta y_j \, \Delta x_i$ and the rectangular strip that corresponds to the first summation of the double sum.

1   $y = 1/x^2, y = -x^2, x = 1, x = 2$

2   $y = \sqrt{x}, y = -x, x = 1, x = 4$

3   $y^2 = -x, x - y = 4, y = -1, y = 2$

4   $x = y^2, y - x = 2, y = -2, y = 3$

5   $y = x, y = 3x, x + y = 4$

6   $x - y + 1 = 0, 7x - y - 17 = 0, 2x + y + 2 = 0$

7   $y = e^x, y = \sin x, x = -\pi, x = \pi$

8   $y = \ln |x|, y = 0, y = 1$

9 $y = x^2, y = 1/(1 + x^2)$

10 $y = \cosh x, y = \sinh x, x = -1, x = 1$

11 Find the volume of the solid that lies under the graph of $z = 4x^2 + y^2$ and over the rectangular region R in the xy-plane having vertices $(0, 0, 0)$, $(0, 1, 0)$, $(2, 0, 0)$, and $(2, 1, 0)$.

12 Find the volume of the solid that lies under the graph of $z = x^2 + 4y^2$ and over the triangular region R in the xy-plane having vertices $(0, 0, 0)$, $(1, 0, 0)$, and $(1, 2, 0)$.

In Exercises 13–20 sketch the solid in the first octant that is bounded by the graphs of the given equations and find its volume.

13 $x^2 + z^2 = 9, y = 2x, y = 0, z = 0$

14 $z = 4 - x^2, x + y = 2, x = 0, y = 0, z = 0$

15 $2x + y + z = 4, x = 0, y = 0, z = 0$

16 $y^2 = z, y = x, x = 4, z = 0$

17 $z = x^2 + y^2, y = 4 - x^2, x = 0, y = 0, z = 0$

18 $z = y^3, y = x^3, x = 0, z = 0, y = 1$

19 $z = x^3, x = 4y^2, 16y = x^2, z = 0$

20 $x^2 + y^2 = 16, x = z, y = 0, z = 0$

In Exercises 21 and 22 find the volume of the solid bounded by the graphs of the given equations.

21 $z = x^2 + 4, y = 4 - x^2, x + y = 2, z = 0$

22 $y = x^3, y = x^4, z - x - y = 4, z = 0$

23 Find the volume of the solid bounded by the graphs of $x^2 + z^2 = a^2$ and $x^2 + y^2 = a^2$, where $a > 0$.

24 Find the volume of the solid bounded by the graph of $x^{2/3} + y^{2/3} + z^{2/3} = a^{2/3}$, where $a > 0$.

The iterated integrals in Exercises 25–30 represent volumes of solids. Describe each solid.

25 $\displaystyle\int_{-2}^{1} \int_{x-1}^{1-x^2} (x^2 + y^2)\, dy\, dx$

26 $\displaystyle\int_{0}^{1} \int_{3-x}^{3-x^2} \sqrt{25 - x^2 - y^2}\, dy\, dx$

27 $\displaystyle\int_{0}^{4} \int_{y/4}^{\sqrt{y}} (x + y)\, dx\, dy$

28 $\displaystyle\int_{0}^{1} \int_{y^{1/2}}^{y^{1/3}} \sqrt{x^2 + y^2}\, dx\, dy$

29 $\displaystyle\int_{0}^{4} \int_{-1}^{2} 3\, dy\, dx$

30 $\displaystyle\int_{0}^{\pi} \int_{0}^{\sin x} dy\, dx$

17.4

MOMENTS AND CENTER OF MASS

The first moment and center of mass of a homogeneous lamina were discussed in Chapter 9. We shall employ double integrals to extend these concepts to laminas in which the density is not constant.

Let T denote a lamina that has the shape of a region R in the xy-plane and suppose that the area density, that is, the mass per unit area, at the point (x, y) is given by $\rho(x, y)$, where ρ is a continuous function on R. If $P = \{R_i\}$ is an inner partition of R, then (u_i, v_i) will denote a point in R_i (see Figure 17.19). The symbol T_i will represent the part of the lamina T that corresponds to R_i. Since ρ is continuous, a small change in (x, y) produces a small change in the density $\rho(x, y)$, that is, ρ is almost constant on R_i. Hence, if $\|P\| \approx 0$, then the mass T_i may be approximated by $\rho(u_i, v_i)\, \Delta A_i$, where ΔA_i is the area of R_i. The sum $\Sigma_i\, \rho(u_i, v_i)\, \Delta A_i$ is an approximation to the mass of the lamina T. The mass M of T is defined as the limit of such sums as $\|P\| \to 0$. This gives us

$$M = \lim_{\|P\| \to 0} \sum_i \rho(u_i, v_i)\, \Delta A_i = \iint\limits_{R} \rho(x, y)\, dA.$$

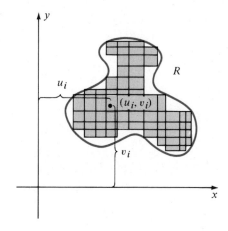

Figure 17.19

If the mass of T_i is assumed to be concentrated at (u_i, v_i), then as in Chapter 9, the moment of T_i with respect to the x-axis is the product $v_i \rho(u_i, v_i) \Delta A_i$. The moment M_x of T with respect to the x-axis is defined as the limit of such sums, that is,

$$M_x = \lim_{\|P\| \to 0} \sum_i v_i \rho(u_i, v_i) \Delta A_i = \iint_R y\rho(x, y) \, dA.$$

Similarly, the moment M_y of T with respect to the y-axis is

$$M_y = \lim_{\|P\| \to 0} \sum_i u_i \rho(u_i, v_i) \Delta A_i = \iint_R x\rho(x, y) \, dA.$$

As usual, the center of mass of the lamina is the point $(\bar{x}, \bar{y})$ such that $\bar{x} = M_y/M$, $\bar{y} = M_x/M$.

We shall list these facts for reference as follows.

Definition (17.11)

Let T be a lamina that has the shape of a region R in the xy-plane. If the area density at (x, y) is $\rho(x, y)$, where ρ is continuous on R, then

(i)   the **mass** of T is $M = \iint_R \rho(x, y) \, dA$.

(ii)   the **moments** of T with respect to the x- and y-axes are

$$M_x = \iint_R y\rho(x, y) \, dA; \quad M_y = \iint_R x\rho(x, y) \, dA.$$

(iii)   the **center of mass** $(\bar{x}, \bar{y})$ of T is given by

$$\bar{x} = \frac{M_y}{M}, \quad \bar{y} = \frac{M_x}{M}.$$

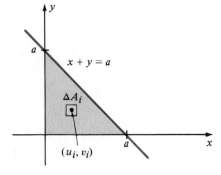

Figure 17.20

Example 1   A lamina T has the shape of an isosceles right triangle with equal sides of length a. Find the center of mass if the density at a point Q is directly proportional to the square of the distance from Q to the vertex V that is opposite the hypotenuse.

Solution   It is convenient to introduce a coordinate system as illustrated in Figure 17.20, with V at the origin and the hypotenuse of the triangle along the line $x + y = a$. Shown in the figure is a typical rectangle of an inner partition P, where (u_i, v_i) is a point in the rectangle.

By hypothesis, the density at $Q(x, y)$ is given by $\rho(x, y) = k(x^2 + y^2)$ for some constant k. Applying (i) of Definition (17.11) and Theorem (17.9), the mass of T is

$$M = \iint_R k(x^2 + y^2)\, dA = \int_0^a \int_0^{a-x} k(x^2 + y^2)\, dy\, dx$$

$$= k \int_0^a \left[x^2 y + \frac{1}{3}y^3 \right]_0^{a-x} dx = k \int_0^a \left[x^2(a - x) + \frac{1}{3}(a - x)^3 \right] dx.$$

It may be verified that the last integral equals $ka^4/6$. Similarly, by (ii) of Definition (17.11),

$$M_y = \iint_R xk(x^2 + y^2)\, dA = \int_0^a \int_0^{a-x} xk(x^2 + y^2)\, dy\, dx$$

which may be shown to equal $ka^5/15$. From (iii) of (17.11),

$$\bar{x} = \frac{ka^5/15}{ka^4/6} = \frac{2a}{5}.$$

By symmetry, $\bar{y} = 2a/5$.   ∎

Example 2   A lamina has the shape of the region R in the xy-plane bounded by the graphs of $x = y^2$ and $x = 4$. Find the center of mass if the density at the point $P(x, y)$ is directly proportional to the distance from the y-axis to P.

Solution   The region is sketched in Figure 17.21 where, by hypothesis, $\rho(x, y) = kx$ for some constant k. It follows from the form of ρ and the symmetry of the region that the center of mass is on the x-axis, that is, $\bar{y} = 0$.

Using (i) of Definition (17.11) and then integrating first with respect to x as indicated by the strip in Figure 17.21,

$$M = \iint_R kx\, dA = k \int_{-2}^2 \int_{y^2}^4 x\, dx\, dy$$

$$= k \int_{-2}^2 \frac{x^2}{2} \bigg]_{y^2}^4 dy = \frac{k}{2} \int_{-2}^2 (16 - y^4)\, dy = \frac{128}{5}k.$$

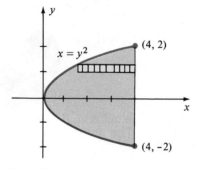

Figure 17.21

By (ii) of Definition (17.11),

$$M_y = \iint_R x(kx)\, dA = k \int_{-2}^2 \int_{y^2}^4 x^2\, dx\, dy$$

$$= k \int_{-2}^2 \frac{x^3}{3} \bigg]_{y^2}^4 dy = \frac{k}{3} \int_{-2}^2 (64 - y^6)\, dy = \frac{512}{7}k.$$

Consequently,

$$\bar{x} = \frac{M_y}{M} = \frac{512k}{7} \cdot \frac{5}{128k} = \frac{20}{7}$$

and hence the centroid is $(\frac{20}{7}, 0)$.   ∎

If n particles of masses $m_1, m_2, \ldots, m_n$ are located at points (x_1, y_1), $(x_2, y_2), \ldots, (x_n, y_n)$, respectively, then the moments of the system with respect to the x- and y-axes were defined in (9.7) by

$$M_x = \sum_{i=1}^{n} y_i m_i \quad \text{and} \quad M_y = \sum_{i=1}^{n} x_i m_i.$$

These are also called the *first* moments of the system with respect to the coordinate axes. If we use the *squares* of the distances from the coordinate axes we obtain the *second moments*, or **moments of inertia** I_x and I_y of the system with respect to the x-axis and y-axis, respectively. Thus, by definition,

$$I_x = \sum_{i=1}^{n} y_i^2 m_i \quad \text{and} \quad I_y = \sum_{i=1}^{n} x_i^2 m_i.$$

This concept may be extended to laminas by employing the limiting processes used for double integrals. In particular, let us consider a lamina T of the type considered at the beginning of this section If the area density at (x, y) is $\rho(x, y)$ where ρ is continuous, then it is natural to define the **moment of inertia I_x of T with respect to the x-axis** as follows:

(17.12)
$$I_x = \lim_{\|P\| \to 0} \sum_i v_i^2 \rho(u_i, v_i) \, \Delta A_i = \iint\limits_R y^2 \rho(x, y) \, dA.$$

In like manner, the **moment of inertia I_y of T with respect to the y-axis** is

(17.13)
$$I_y = \lim_{\|P\| \to 0} \sum_i u_i^2 \, \rho(u_i, v_i) \, \Delta A_i = \iint\limits_R x^2 \, \rho(x, y) \, dA.$$

If we multiply $\rho(u_i, v_i) \, \Delta A_i$ by the square $u_i^2 + v_i^2$ of the distance from the origin to (u_i, v_i) and take the limit of a sum of such terms, we obtain the **moment of inertia I_O of T with respect to the origin**. This leads to

(17.14)
$$I_O = \iint\limits_R (x^2 + y^2) \, \rho(x, y) \, dA.$$

The number I_O is also called the **polar moment of inertia** of T. Observe that $I_O = I_x + I_y$.

Example 3 A lamina T has the semicircular shape illustrated in (i) of Figure 17.22. If the density at a point P is directly proportional to the distance from the diameter AB to P, find the moment of inertia of T with respect to the line through A and B.

Solution Let us introduce a coordinate system as in (ii) of Figure 17.22. Since the density at (x, y) is $\rho(x, y) = ky$ it follows from (17.12) that the desired moment of inertia is

$$I_x = \int_{-a}^{a} \int_0^{\sqrt{a^2 - x^2}} y^2 (ky) \, dy \, dx = k \int_{-a}^{a} \frac{1}{4} y^4 \Big]_0^{\sqrt{a^2 - x^2}} dx$$

$$= \frac{k}{4} \int_{-a}^{a} (a^4 - 2a^2 x^2 + x^4) \, dx.$$

Carrying out the integration we obtain $I_x = 4ka^5/15$. ∎

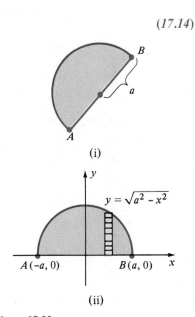

(i)

(ii)

Figure 17.22

Moments of inertia are useful in problems that involve the rotation of an object about a fixed axis. For example, suppose a wheel is rotating about an axle. If a particle P of mass m on the wheel is a distance a from the axis of rotation (see Figure 15.12), then the moment of inertia I of P with respect to the axis is ma^2. If the angular speed is ω, then the speed v of the particle, that is, the distance traveled per unit time, is $a\omega$. By definition, the **kinetic energy** K.E. of P is

$$\text{K.E.} = \tfrac{1}{2}mv^2.$$

Consequently,

$$\text{K.E.} = \tfrac{1}{2}ma^2\omega^2 = \tfrac{1}{2}I\omega^2.$$

If we represent the wheel by a flat disc and introduce a limit of a sum, then the same formula may be derived for the kinetic energy of the wheel. The formula can also be extended to laminas having noncircular shapes. The kinetic energy of a rotating object tells the engineer or physicist the amount of work necessary to bring the object to rest. Since K.E. $= \tfrac{1}{2}I\omega^2$, the kinetic energy of a rotating lamina is directly proportional to the moment of inertia, and hence for a fixed ω, the larger the moment of inertia, the larger the amount of work required to stop the rotation.

If a lamina T of mass M has moments of inertia I_x and I_y, then the **radius of gyration $\bar{\bar{y}}$ of T with respect to the x-axis**, and the **radius of gyration $\bar{\bar{x}}$ of T with respect to the y-axis** are defined by the equations

(17.15)
$$\bar{\bar{y}}^2 M = I_x \quad \text{and} \quad \bar{\bar{x}}^2 M = I_y.$$

These formulas imply that $(\bar{\bar{x}}, \bar{\bar{y}})$ is the point at which all the mass can be concentrated without changing the moments of inertia of T with respect to the coordinate axes.

Example 4 Find the radius of gyration with respect to the x-axis of the lamina described in Example 3.

Solution   In Example 3 we found that $I_x = 4ka^5/15$. The mass of T is

$$M = \int_{-a}^{a} \int_{0}^{\sqrt{a^2 - x^2}} ky \, dy \, dx.$$

Evaluating this integral gives us $M = 2ka^3/3$. Applying (17.15),

$$\bar{\bar{y}} = \sqrt{\frac{I_x}{M}} = \sqrt{\frac{4ka^5/15}{2ka^3/3}} = \sqrt{\frac{2a^2}{5}} = \frac{a\sqrt{10}}{5}. \qquad \blacksquare$$

EXERCISES 17.4

In Exercises 1–14 find the mass and the center of mass of the lamina that has the shape of the region bounded by the graphs of the given equations and having the indicated density.

1 $y = \sqrt{x}, x = 9, y = 0; \rho(x, y) = x + y$

2 $y = \sqrt[3]{x}, x = 8, y = 0; \rho(x, y) = y^2$

3 $y = x^2$, $y = 4$; density at the point $P(x, y)$ is directly proportional to the distance from the y-axis to P

4 $y = x^3$, $y = 2x$; density at the point $P(x, y)$ is directly proportional to the distance from the x-axis to P

5 $y = e^{-x^2}, y = 0, x = -1, x = 1; \rho(x, y) = |xy|$

6 $y = \sin x, y = 0, x = 0, x = \pi; \rho(x, y) = y$

7 $x = y^2, y - x = 2, y = -2, y = 3; \rho(x, y) = 1$

8 $y = x, y = 3x, x + y = 4; \rho(x, y) = 2$

9 $y = \sec x, x = -\pi/4, x = \pi/4, y = 1/2; \rho(x, y) = 4$

10 $xy^2 = 1, y = 1, y = 2; \rho(x, y) = x^2 + y^2$

11 $y = 1/x, y = x, x = 2, y = 0; \rho(x, y) = x$

12 $x = e^y, y = -1, x = 2; \rho(x, y) = 1$

13 $y = e^{-x}, y = 0, x = 0, x = 1; \rho(x, y) = y^2$

14 $y = \ln x, y = 0, x = 2; \rho(x, y) = 1/x$

15 Find I_x, I_y, and I_O for the lamina of Exercise 1.

16 Find I_x, I_y, and I_O for the lamina of Exercise 2.

17 Find I_x, I_y, and I_O for the lamina of Exercise 3.

18 Find I_x, I_y, and I_O for the lamina of Exercise 4.

19 A homogeneous lamina has the shape of a square of side a. Find the moment of inertia with respect to (a) a side; (b) a diagonal; and (c) the center of mass.

20 A homogeneous lamina has the shape of an equilateral triangle of side a. Find the moment of inertia with respect to (a) an altitude; (b) a side; and (c) a vertex.

21 Find the radius of gyration in (a) of Exercise 19.

22 Find the radius of gyration in (a) of Exercise 20.

23 Find the moment of inertia and radius of gyration of a homogeneous circular disc of radius a with respect to a line along a diameter.

24 A lamina T has the shape of an isosceles right triangle with equal sides of length a. The density at a point P is directly proportional to the square of the distance from P to the vertex V that is opposite the hypotenuse. Find the moment of inertia and the radius of gyration with respect to a line along one of the equal sides (see Example 1).

Homogeneous laminas of density ρ are illustrated in Exercises 25–28, where $a < b$. Find the moment of inertia and radius of gyration with respect to the indicated side of length b.

25 Right triangle

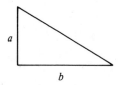

26 Rectangle

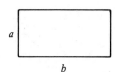

27 Isosceles triangle

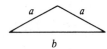

28 Isosceles trapezoid

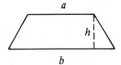

17.5

DOUBLE INTEGRALS IN POLAR COORDINATES

A region in a polar coordinate plane of the type illustrated in Figure 17.23, which is bounded by arcs of circles of radii r_1 and r_2 with centers at the pole and by two rays from the origin, will be called an **elementary polar region**. If $\Delta\theta$ denotes the radian measure of the angle between the rays and $\Delta r = r_2 - r_1$, then the area ΔA of the region is

$$\Delta A = \tfrac{1}{2}r_2^2\,\Delta\theta - \tfrac{1}{2}r_1^2\,\Delta\theta.$$

This formula may also be written

$$\Delta A = \tfrac{1}{2}(r_2^2 - r_1^2)\,\Delta\theta = \tfrac{1}{2}(r_2 + r_1)\,(r_2 - r_1)\,\Delta\theta.$$

If we denote the **average radius** $\tfrac{1}{2}(r_2 + r_1)$ by $\bar{r}$, then

$$\Delta A = \bar{r}\,\Delta r\,\Delta\theta.$$

Next, consider a region R of the type illustrated in (i) of Figure 17.24, bounded by two rays that make positive angles α and β with the polar axis and by the graphs of two polar equations $r = g_1(\theta)$ and $r = g_2(\theta)$, where the functions g_1 and g_2 are continuous and $g_1(\theta) \le g_2(\theta)$ for all θ in the interval $[\alpha, \beta]$. If R is subdivided by means of circular arcs and rays as shown in (ii) of Figure 17.24, then the collection of elementary polar regions $R_1, R_2, \ldots, R_n$ that lie completely within R is called an **inner polar partition** P of R. The norm $\|P\|$ of P is the length of the longest diagonal of the R_i. If we choose a point (r_i, θ_i) in R_i such that r_i is the average radius, then as in Figure 17.23, the area ΔA_i of R_i is $r_i\,\Delta r_i\,\Delta\theta_i$.

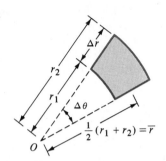

Figure 17.23 $\Delta A = \bar{r}\,\Delta r\,\Delta\theta$

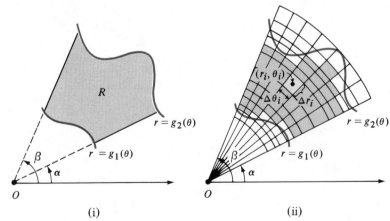

(i)   (ii)

Figure 17.24

If f is a continuous function of the polar variables r and θ, then it can be proved that

$$(17.16) \qquad \lim_{\|P\| \to 0} \sum_i f(r_i, \theta_i)r_i\,\Delta r_i\,\Delta\theta_i = \iint\limits_R f(r, \theta)\,dA = \int_\alpha^\beta \int_{g_1(\theta)}^{g_2(\theta)} f(r, \theta)r\,dr\,d\theta.$$

It is important to note that the integrand on the right in (17.16) is the product of $f(r, \theta)$ and r. This is due to the fact that ΔA_i equals $r_i \Delta r_i \Delta \theta_i$.

It is sometimes convenient to regard the iterated integral in (17.16) in terms of limits of double sums in a manner similar to our discussion for rectangular coordinates at the end of Section 17.2. In the present situation, we first hold θ fixed and sum along one of the wedge-shaped regions shown in (ii) of Figure 17.24, from the graph of g_1 to the graph of g_2. For the second summation we sweep out the region by letting θ vary from α to β. Intuitively, the iterated integral does this and simultaneously takes the limit as $\|P\|$ approaches zero.

If $f(r, \theta) = 1$ throughout R, then the integral in (17.16) equals the area of R. This also follows from our work in Chapter 13 since after the partial integration with respect to r, the formula given in Theorem (13.10) is obtained.

Example 1 A lamina has the shape of the region R that lies outside the graph of $r = a$ and inside the graph of $r = 2a \sin \theta$. Find the mass if the density at a point $P(r, \theta)$ is inversely proportional to the distance from the pole to P.

Solution   The region is sketched in Figure 17.25, together with a typical strip of elementary polar regions involved in summing from one boundary of R to the other. By hypothesis, the density $f(r, \theta)$ at the point (r, θ) is k/r, where k is some real number. The mass M of the lamina is $\iint_R f(r, \theta)\, dA$. Thus, by (17.16),

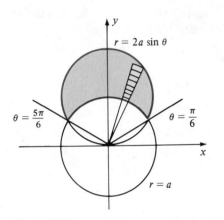

Figure 17.25

$$M = \iint_R f(r, \theta)\, dA = \int_{\pi/6}^{5\pi/6} \int_a^{2a \sin \theta} (k/r) r\, dr\, d\theta = k \int_{\pi/6}^{5\pi/6} \Big[r \Big]_a^{2a \sin \theta} d\theta$$

$$= k \int_{\pi/6}^{5\pi/6} (2a \sin \theta - a)\, d\theta = ka \Big[-2 \cos \theta - \theta \Big]_{\pi/6}^{5\pi/6}$$

$$= ka[(\sqrt{3} - 5\pi/6) - (-\sqrt{3} - \pi/6)]$$

$$= ka\,(2\sqrt{3} - 2\pi/3). \qquad \blacksquare$$

It can be shown that under suitable conditions an iterated double integral in rectangular coordinates can be transformed into a double integral in polar coordinates. First, the variables x and y in the integrand are replaced by $r \cos \theta$ and $r \sin \theta$. Next, $dy\, dx$ or $dx\, dy$ is replaced by $r\, dr\, d\theta$ or $r\, d\theta\, dr$ and the limits of integration are changed to polar coordinates. This gives us the following **change of variable formula**:

(17.17)
$$\iint_R f(x, y)\, dA = \iint_R f(r \cos \theta, r \sin \theta) r\, dr\, d\theta$$

This formula will be justified in Section 18.9. As an illustration, formula (17.13) for the moment of inertia with respect to the y-axis of a lamina having the shape of (i) in Figure 17.24 is

$$I_y = \iint_R x^2 \rho(x, y)\, dA = \int_\alpha^\beta \int_{g_1(\theta)}^{g_2(\theta)} (r \cos \theta)^2 \rho(r \cos \theta, r \sin \theta) r\, dr\, d\theta,$$

where $\rho(x, y)$ is the density at (x, y). For the special case of the lamina in Example 1, the last equation takes on the form

$$I_y = \int_{\pi/6}^{5\pi/6} \int_a^{2a\sin\theta} (r\cos\theta)^2 (k/r)r\, dr\, d\theta = \int_{\pi/6}^{5\pi/6} \int_a^{2a\sin\theta} kr^2 \cos^2\theta\, dr\, d\theta.$$

If the integrand $f(x, y)$ of a double integral $\iint_R f(x, y)\, dy\, dx$ contains the expression $x^2 + y^2$, or if the region R involves circular arcs, then the introduction of polar coordinates often leads to a simpler evaluation. This is illustrated in the following example.

Example 2   Use polar coordinates to evaluate

$$\int_{-a}^a \int_0^{\sqrt{a^2 - x^2}} (x^2 + y^2)^{3/2}\, dy\, dx.$$

Solution   Evidently the region of integration is bounded by the graphs of $y = \sqrt{a^2 - x^2}$ and $y = 0$ illustrated in (ii) of Figure 17.22. Replacing $x^2 + y^2$ in the integrand by r^2, $dy\, dx$ by $r\, dr\, d\theta$, and changing the limits, we have

$$\int_{-a}^a \int_0^{\sqrt{a^2 - x^2}} (x^2 + y^2)^{3/2}\, dy\, dx = \int_0^\pi \int_0^a r^3 r\, dr\, d\theta$$

$$= \int_0^\pi \frac{r^5}{5}\Big]_0^a d\theta = \frac{a^5}{5} \int_0^\pi d\theta$$

$$= \frac{a^5}{5}\theta\Big]_0^\pi = \frac{\pi a^5}{5}. \qquad \blacksquare$$

Polar coordinates can also be used for double integrals over a region R of the type illustrated in (i) of Figure 17.26. In this case R is bounded by the arcs of two circles of radii a and b and by the graphs of polar equations $\theta = h_1(r)$ and $\theta = h_2(r)$, where the functions h_1 and h_2 are continuous and $h_1(r) \le h_2(r)$ for all r in the interval $[a, b]$.

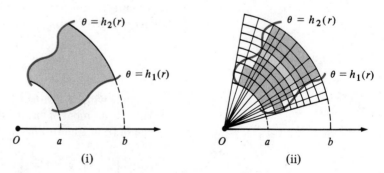

Figure 17.26

If f is a function of r and θ that is continuous on R, then the limit in (17.16) exists and the corresponding evaluation formula is

$$\iint\limits_{R} f(r, \theta) \, dA = \int_{a}^{b} \int_{h_1(r)}^{h_2(r)} f(r, \theta) r \, d\theta \, dr.$$

The preceding iterated integral may be interpreted in terms of limits of double sums. In this case we first hold r fixed and sum along a circular arc, as illustrated by the colored region in (ii) of Figure 17.26. Secondly, we sweep out R by summing terms that correspond to such ring-shaped regions. This technique is illustrated in the next example.

Example 3 Find the polar moment of inertia of a homogeneous lamina that has the shape of the smaller of the regions bounded by the polar axis, the graphs of $r = 1$, $r = 2$, and the part of the spiral $r\theta = 1$ from $\theta = \frac{1}{2}$ to $\theta = 1$.

Solution The region, together with some elementary polar regions, is sketched in Figure 17.27.

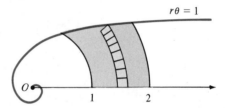

Figure 17.27

By hypothesis, the density at every point (r, θ) is a constant k. From the previous discussion,

$$I_0 = \iint\limits_{R} (x^2 + y^2)\rho(x, y) \, dA = \int_{1}^{2} \int_{0}^{1/r} r^2 k r \, d\theta \, dr$$

$$= k \int_{1}^{2} r^3 \theta \Big]_{0}^{1/r} dr = k \int_{1}^{2} r^2 \, dr = k \frac{r^3}{3}\Big]_{1}^{2} = \frac{7}{3} k. \qquad \blacksquare$$

If $f(r, \theta) \geq 0$ throughout a polar region R, then the double integral $\iint_{R} f(r, \theta) \, dA$ in (17.16) may be regarded as the volume of a solid in a manner completely analogous to our formulation of Definition (17.4) or (17.10) in rectangular coordinates. The principal difference in the present situation is that we consider the graph S of $z = f(r, \theta)$ in *cylindrical* coordinates. The desired solid lies under the graph of S and over the region R. If we use a polar partition of R, then the expression $f(r_i, \theta_i) r_i \, \Delta r_i \, \Delta \theta_i$ in (17.16) may be interpreted as the volume of a prism of height $f(r_i, \theta_i)$ and base area $r_i \, \Delta r_i \, \Delta \theta_i$, as illustrated in Figure 17.28. The limit of a sum of these expressions equals the volume.

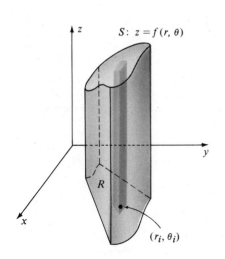

Figure 17.28

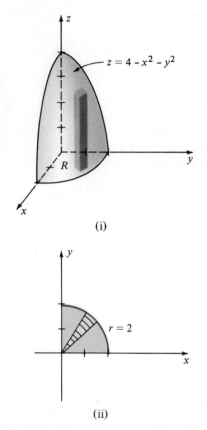

(i)

(ii)

Figure 17.29

Example 4 Find the volume V of the solid bounded by the paraboloid $z = 4 - x^2 - y^2$ and the xy-plane.

Solution   The first octant portion of the solid is sketched in (i) of Figure 17.29. By symmetry, it is sufficient to find the volume of this portion and multiply the result by 4. If we introduce cylindrical coordinates, then an equation for the paraboloid is $z = 4 - r^2$. The region R in the xy-plane is bounded by the coordinate axes and one-fourth of the circle $r = 2$. This is illustrated in (ii) of Figure 17.29 together with a wedge of elements of a polar partition.

By the discussion preceding this example, with $f(r, \theta) = 4 - r^2$, we have

$$V = 4 \iint_R (4 - r^2)\, dA = 4 \int_0^{\pi/2} \int_0^2 (4 - r^2) r\, dr\, d\theta$$

$$= 4 \int_0^{\pi/2} \left[2r^2 - \frac{r^4}{4} \right]_0^2 d\theta = 4 \int_0^{\pi/2} 4\, d\theta = 16\theta \Big]_0^{\pi/2} = 8\pi.$$

The same problem, using rectangular coordinates, leads to the following double integral:

$$V = 4 \iint_R (4 - x^2 - y^2)\, dA = 4 \int_0^2 \int_0^{\sqrt{4 - x^2}} (4 - x^2 - y^2)\, dy\, dx.$$

After evaluating the last integral, the reader should be thoroughly convinced of the advantage of using cylindrical (or polar) coordinates in certain problems.   ∎

EXERCISES 17.5

In Exercises 1–4 use a double integral to find the area of the indicated region.

1 One loop of $r^2 = 9 \sin 2\theta$

2 One loop of $r = 2 \cos 4\theta$

3 Inside $r = 2 - 2 \cos \theta$ and outside $r = 3$

4 Bounded by $r = 3 + 2 \sin \theta$

In Exercises 5–8 find the mass and center of mass of the lamina that has the indicated shape and density.

5 Bounded by $r = 2 \cos \theta$; density at the point $P(r, \theta)$ is directly proportional to the distance from the pole to P.

6 Bounded by the loop of $r^2 = a \cos 2\theta$ from $\theta = -\pi/4$ to $\theta = \pi/4$; density at the point $P(r, \theta)$ is directly proportional to the square of the distance from the pole to P.

7 Inside the graph of $r = 3 \sin \theta$ and outside the graph of $r = 1 + \sin \theta$; density at the point $P(r, \theta)$ is inversely proportional to the distance from the pole to P.

8 Inside the graph of $r = 4 \cos \theta$ and outside the graph of $r = 2$; density at the point $P(r, \theta)$ is directly proportional to the distance from the polar axis to P.

9 Find the polar moment of inertia of a homogeneous lamina that has the shape of the region bounded by $r^2 = a \sin 2\theta$.

10 A homogeneous lamina has the shape of the region $R = \{(r, \theta): r^{-1} \le \theta \le 2r^{-1}, 1 \le r \le 2\}$. Find the polar moment of inertia.

11 Find the moment of inertia and radius of gyration of a homogeneous circular lamina of radius a with respect to a line through the center.

12 Find the moment of inertia and radius of gyration of a homogeneous circular lamina of radius a with respect to a tangent line.

Use polar coordinates to evaluate the integrals in Exercises 13–18.

13 $\iint_R (x^2 + y^2)^{3/2} \, dA$, where R is the region bounded by the graph of $x^2 + y^2 = 4$.

14 $\iint_R x^2(x^2 + y^2)^3 \, dA$, where R is the region bounded by the graphs of $y = \sqrt{1 - x^2}$ and $y = 0$.

15 $\iint_R x^2/(x^2 + y^2) \, dA$, where R is the region between the concentric circles $x^2 + y^2 = a^2$ and $x^2 + y^2 = b^2$, with $0 < a < b$.

16 $\iint_R (x^2 + y^2) \, dA$, where R is the region bounded by the graph of $(x^2 + y^2)^2 = x^2 - y^2$.

17 $\iint_R (x + y) \, dA$, where R is the region bounded by the graph of $x^2 + y^2 - 2y = 0$.

18 $\iint_R \sqrt{x^2 + y^2} \, dA$, where R is the triangular region with vertices $(0, 0)$, $(3, 0)$ and $(3, 3)$.

Evaluate the integrals in Exercises 19–24 by changing to polar coordinates.

19 $\int_{-a}^{a} \int_{0}^{\sqrt{a^2 - x^2}} e^{-(x^2 + y^2)} \, dy \, dx$

20 $\int_{0}^{a} \int_{0}^{\sqrt{a^2 - x^2}} (x^2 + y^2)^{3/2} \, dy \, dx$

21 $\int_{1}^{2} \int_{0}^{x} \frac{1}{\sqrt{x^2 + y^2}} \, dy \, dx$

22 $\int_{0}^{1} \int_{0}^{\sqrt{1 - x^2}} e^{\sqrt{x^2 + y^2}} \, dy \, dx$

23 $\int_{0}^{2} \int_{0}^{\sqrt{4 - y^2}} \cos(x^2 + y^2) \, dx \, dy$

24 $\int_{0}^{2} \int_{-\sqrt{2y - y^2}}^{\sqrt{2y - y^2}} x \, dx \, dy$

25 Find the volume of the solid that lies inside the sphere $x^2 + y^2 + z^2 = 25$ and outside the cylinder $x^2 + y^2 = 9$.

26 Find the volume of the solid that is cut out of the ellipsoid $4x^2 + 4y^2 + z^2 = 16$ by the cylinder $x^2 + y^2 = 1$.

27 Find the volume of the solid bounded by the cone $z = r$ and the cylinder $r = 2 \cos \theta$.

28 Find the volume of the solid bounded by the paraboloid $z = 4r^2$, the cylinder $r = 3 \sin \theta$, and the plane $z = 0$.

29 Find the volume of the solid that lies inside the graphs of both $x^2 + y^2 + z^2 = 16$ and $x^2 + y^2 - 4y = 0$.

30 Find the volume of the solid bounded by the graphs of $z = 9 - x^2 - y^2$ and $z = 5$.

31 (a) Let R_a be the region that is bounded by the circle $x^2 + y^2 = a^2$. If we define

$$\int_{-\infty}^{\infty} \int_{-\infty}^{\infty} e^{-(x^2 + y^2)} \, dx \, dy = \lim_{a \to \infty} \iint_{R_a} e^{-(x^2 + y^2)} \, dA,$$

evaluate this improper integral and interpret the result geometrically.

(b) It can be proved, using advanced methods, that

$$\int_{-\infty}^{\infty} \int_{-\infty}^{\infty} e^{-(x^2 + y^2)} \, dx \, dy = \int_{-\infty}^{\infty} e^{-x^2} \, dx \int_{-\infty}^{\infty} e^{-y^2} \, dy.$$

Use this and part (a) to prove that $\int_{-\infty}^{\infty} e^{-x^2} \, dx = \sqrt{\pi}$. Interpret this result geometrically.

(c) Use (b) to prove that

$$\frac{1}{\sqrt{2\pi}} \int_{-\infty}^{\infty} e^{-x^2/2} \, dx = 1.$$

(This result is important in the field of statistics.)

17.6

TRIPLE INTEGRALS

It is possible to define *triple integrals* for a function f of *three* variables x, y, and z in a manner similar to that used in Section 17.2 for functions of two variables. The simplest case occurs if f is continuous throughout a region of the form

$$Q = \{(x, y, z): a \le x \le b, c \le y \le d, k \le z \le l\},$$

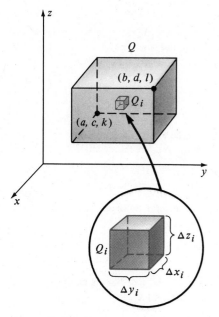

Figure 17.30

as illustrated in Figure 17.30. If Q is divided into subregions $Q_1, Q_2, \ldots, Q_n$ by means of planes parallel to the three coordinate planes, then the collection $\{Q_i : i = 1, \ldots, n\}$ is called a **partition** P of Q. The **norm** $\|P\|$ of the partition is the length of the longest diagonal of all the Q_i. As in Figure 17.30, if Δx_i, Δy_i, and Δz_i are the dimensions of Q_i, then the volume ΔV_i of Q_i is the product $\Delta x_i \, \Delta y_i \, \Delta z_i$.

If (u_i, v_i, w_i) is a point in Q_i, then the sum

$$\sum_{i=1}^{n} f(u_i, v_i, w_i) \, \Delta V_i$$

is called a **Riemann sum** of f for P.

The concept of a limit of Riemann sums for a function of three variables is defined in a manner similar to that given in Definition (17.2) for functions of two variables. If this limit exists it is called the **triple integral of f over Q**, denoted by

$$(17.18) \qquad \iiint\limits_{Q} f(x, y, z) \, dV = \lim_{\|P\| \to 0} \sum_{i} f(u_i, v_i, w_i) \, \Delta V_i.$$

Moreover, it can be shown that

$$\iiint\limits_{Q} f(x, y, z) \, dV = \int_{k}^{l} \int_{c}^{d} \int_{a}^{b} f(x, y, z) \, dx \, dy \, dz,$$

where the **iterated integral** on the right means that the first integration is with respect to x (with y and z fixed), the second is with respect to y (with z fixed), and the third is with respect to z. There are five other iterated integrals that equal the triple integral of f over the parallelepiped Q. For example, if we integrate in the order y, z, x, then

$$\iiint\limits_{Q} f(x, y, z) \, dV = \int_{a}^{b} \int_{k}^{l} \int_{c}^{d} f(x, y, z) \, dy \, dz \, dx.$$

Example 1 Evaluate $\iiint_{Q} 3xy^3 z^2 \, dV$ if

$$Q = \{(x, y, z) : -1 \le x \le 3, 1 \le y \le 4, 0 \le z \le 2\}.$$

Solution Of the six possible iterated integrals, we shall use the following:

$$\int_{1}^{4} \int_{-1}^{3} \int_{0}^{2} 3xy^3 z^2 \, dz \, dx \, dy = \int_{1}^{4} \int_{-1}^{3} xy^3 z^3 \Big]_{0}^{2} \, dx \, dy$$

$$= \int_{1}^{4} \int_{-1}^{3} 8xy^3 \, dx \, dy = \int_{1}^{4} 4x^2 y^3 \Big]_{-1}^{3} \, dy$$

$$= \int_{1}^{4} (36y^3 - 4y^3) \, dy = 8y^4 \Big]_{1}^{4} = 2040.$$

∎

Triple integrals may be defined over regions other than those bounded by rectangular parallelepipeds. Suppose, for example, that R is a region in the xy-plane that can be divided into subregions of Types I and II, and that Q is the region in three dimensions defined by

$$Q = \{(x, y, z): (x, y) \text{ is in } R \text{ and } k_1(x, y) \le z \le k_2(x, y)\}$$

where the functions k_1 and k_2 have continuous first partial derivatives throughout R. Geometrically, Q lies between the graphs of $z = k_1(x, y)$ and $z = k_2(x, y)$ and over or under the region R. If Q is subdivided by means of planes parallel to the three coordinate planes, then the resulting parallelepipeds $Q_1, Q_2, \ldots, Q_n$ that lie completely within Q form an **inner partition** P of Q. The sketch in Figure 17.31 shows a region Q of the type under consideration together with a typical element Q_i of an inner partition.

A **Riemann sum** of f for P is any sum of the form $\Sigma_i f(u_i, v_i, w_i) \Delta V_i$, where (u_i, v_i, w_i) is an arbitrary point in Q_i and ΔV_i is the volume of Q_i. The triple integral of f over Q is again defined as the limit (17.18). It can be shown that if f is continuous throughout Q, then

$$\iiint\limits_Q f(x, y, z) \, dV = \iint\limits_R \left[\int_{k_1(x, y)}^{k_2(x, y)} f(x, y, z) \, dz \right] dA.$$

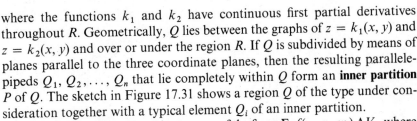

Figure 17.31

In particular, if R is of Type I, as in (i) of Figure 17.1, then

(17.19)
$$\iiint\limits_Q f(x, y, z) \, dV = \int_a^b \int_{g_1(x)}^{g_2(x)} \int_{k_1(x, y)}^{k_2(x, y)} f(x, y, z) \, dz \, dy \, dx.$$

The symbol on the right-hand side of the last equation is called an **iterated triple integral**. It is evaluated by means of partial integrations of $f(x, y, z)$ in the order z, y, and x, substituting the indicated limits in the usual way. Similarly, if R is of Type II as in (ii) of Figure 17.1, then

(17.20)
$$\iiint\limits_Q f(x, y, z) \, dV = \int_c^d \int_{h_1(y)}^{h_2(y)} \int_{k_1(x,y)}^{k_2(x,y)} f(x, y, z) \, dz \, dx \, dy.$$

It is convenient to regard the iterated integral in either (17.19) or (17.20) as a limit of a *triple sum* of terms of the form $f(u_i, v_j, w_k) \Delta z_k \Delta y_j \Delta x_i$, where we first sum along a column of parallelepipeds in the direction of the z-axis from the lower surface (with equation $z = k_1(x, y)$) to the upper surface (with equation $z = k_2(x, y)$). The resulting double integral is then evaluated over R using the techniques discussed in Section 17.2.

Example 2 Express $\iiint_Q f(x, y, z) \, dV$ as an iterated integral, where Q is the region in the first octant bounded by the coordinate planes and the graphs of $z - 2 = x^2 + (y^2/4)$ and $x^2 + y^2 = 1$.

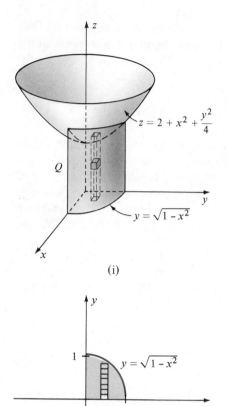

(i)

(ii)

Figure 17.32

Solution   As illustrated in (i) of Figure 17.32, Q lies under the paraboloid $z = 2 + x^2 + (y^2/4)$ and over the graph of $z = 0$ (the xy-plane). The region R in the xy-plane is bounded by the coordinate axes and the graph of $y = \sqrt{1 - x^2}$. Also shown in (i) of the figure is a column that corresponds to a first summation over the parallelepipeds of volumes $\Delta z_k \, \Delta y_j \, \Delta x_i$ in the direction of the z-axis. Since the column extends from the xy-plane ($z = 0$) to the paraboloid, the lower limit on the innermost integral sign in (17.19) is 0 and the upper limit is $2 + x^2 + (y^2/4)$. The second and third integrations are taken over the region R (see (ii) of Figure 17.32). Consequently, the integral in (17.19) has the form

$$\int_0^1 \int_0^{\sqrt{1-x^2}} \int_0^{2+x^2+(y^2/4)} f(x, y, z) \, dz \, dy \, dx. \qquad \blacksquare$$

If $f(x, y, z) = 1$ throughout Q, then the triple integral of f over Q is written $\iiint_Q dV$ and its value is the volume of the region Q. It follows from Example 2 that the value of the integral

$$\int_0^1 \int_0^{\sqrt{1-x^2}} \int_0^{2+x^2+(y^2/4)} dz \, dy \, dx$$

is the volume of the region Q shown in (i) of Figure 17.32.

Example 3   Find the volume of the solid that is bounded by the cylinder $y = x^2$ and by the planes $y + z = 4$ and $z = 0$.

Solution   The solid is sketched in (i) of Figure 17.33 together with a column corresponding to a first summation in the z-direction. Note that the column extends from $z = 0$ to $z = 4 - y$. The region R in the xy-plane is shown in (ii)

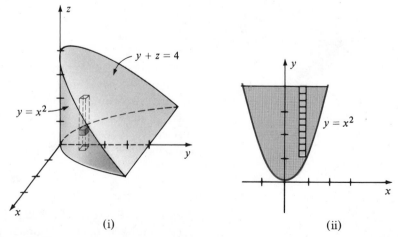

(i)

(ii)

Figure 17.33

of the figure, together with a strip corresponding to the first (of a double) integration with respect to y. Applying (17.19) with $f(x, y, z) = 1$,

$$V = \int_{-2}^{2} \int_{x^2}^{4} \int_{0}^{4-y} dz \, dy \, dx = \int_{-2}^{2} \int_{x^2}^{4} (4 - y) \, dy \, dx$$

$$= \int_{-2}^{2} \left[4y - \frac{1}{2} y^2 \right]_{x^2}^{4} dx = \int_{-2}^{2} \left(8 - 4x^2 + \frac{1}{2} x^4 \right) dx$$

$$= 8x - \frac{4}{3} x^3 + \frac{1}{10} x^5 \bigg]_{-2}^{2} = \frac{256}{15}.$$

If we use (17.20), then the strip in (ii) of Figure 17.33 would be horizontal and

$$V = \int_{0}^{4} \int_{-\sqrt{y}}^{\sqrt{y}} \int_{0}^{4-y} dz \, dx \, dy.$$

The reader may verify that the value of this integral is also $\frac{256}{15}$. ■

For certain regions, a triple integral may be evaluated by means of an iterated triple integral in which the first integration is with respect to y or x. Thus, consider

$$Q = \{(x, y, z): a \le x \le b, h_1(x) \le z \le h_2(x), k_1(x, z) \le y \le k_2(x, z)\}$$

where the given functions are sufficiently well behaved. A graph of this type of region is illustrated in Figure 17.34. It can be shown that

(17.21)
$$\iiint_Q f(x, y, z) \, dV = \int_{a}^{b} \int_{h_1(x)}^{h_2(x)} \int_{k_1(x,z)}^{k_2(x,z)} f(x, y, z) \, dy \, dz \, dx.$$

This iterated integral may be interpreted as a limit of a triple sum obtained by first summing over a row of parallelepipeds in the direction of the y-axis from the left surface (with equation $y = k_1(x, z)$) to the right surface (with equation $y = k_2(x, z)$), as indicated in Figure 17.34. The resulting double integral is then evaluated over the region R in the xz-plane.

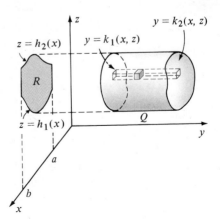

Figure 17.34

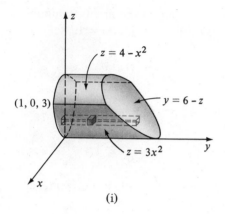

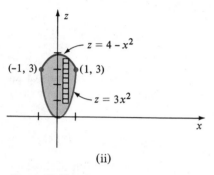

Figure 17.35

Example 4 Find the volume of the region Q bounded by the graphs of $z = 3x^2$, $z = 4 - x^2$, $y = 0$, and $z + y = 6$.

Solution   As illustrated in (i) of Figure 17.35, Q lies under the cylinder $z = 4 - x^2$, over the cylinder $z = 3x^2$, to the right of the xz-plane, and to the left of the plane $z + y = 6$. Hence Q is a region of the type illustrated in Figure 17.34, where $k_1(x, z) = 0$ and $k_2(x, z) = 6 - z$. The region R in the xz-plane is sketched in (ii) of Figure 17.35. Applying (17.21),

$$V = \iiint_Q dV = \int_{-1}^{1} \int_{3x^2}^{4-x^2} \int_{0}^{6-z} dy\, dz\, dx = \int_{-1}^{1} \int_{3x^2}^{4-x^2} (6 - z)\, dz\, dx$$

$$= \int_{-1}^{1} \left[6z - \frac{1}{2}z^2 \right]_{3x^2}^{4-x^2} dx = \int_{-1}^{1} (16 - 20x^2 + 4x^4)\, dx = \frac{304}{15}.$$

If a different order of integration is used it is necessary to use several triple integrals. (Why?) ∎

Finally, if Q is a region of the type illustrated in Figure 17.36, where the functions l_1 and l_2 have continuous first partial derivatives on a suitable region R in the yz-plane, then

$$\iiint_Q f(x, y, z)\, dV = \iint_R \left[\int_{l_1(y,z)}^{l_2(y,z)} f(x, y, z)\, dx \right] dA.$$

In the final iterated double integral, dA will be replaced by $dz\, dy$ or $dy\, dz$.

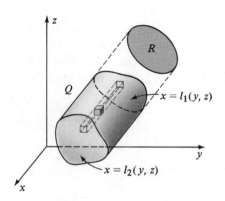

Figure 17.36

Example 5 Rework Example 3 by integrating first with respect to x.

Solution   The solid is bounded by the graphs of $y = x^2$, $y + z = 4$, and $z = 0$. Referring to (i) of Figure 17.37, a first integration with respect to x would correspond to a summation along a row of parallelepipeds extending from the graph of $x = -\sqrt{y}$ to the graph of $x = \sqrt{y}$. The region R in the yz-plane lies between these graphs and is bounded by the y- and z-axes and the line $y + z = 4$.

If the second integration is with respect to y, as indicated by the strip in (ii) of Figure 17.37, then

$$V = \int_{0}^{4} \int_{0}^{4-z} \int_{-\sqrt{y}}^{\sqrt{y}} dx\, dy\, dz.$$

It may be verified that the value of this integral is $\frac{256}{15}$. If we take the second integration with respect to z instead of y, then the iterated integral is

$$V = \int_{0}^{4} \int_{0}^{4-y} \int_{-\sqrt{y}}^{\sqrt{y}} dx\, dz\, dy.$$

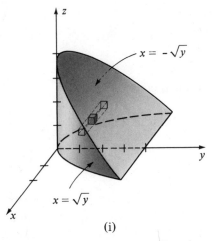

Figure 17.37

EXERCISES 17.6

1 Evaluate the integral of Example 1 using five orders of integration different from that used in the solution.

2 If $Q = \{(x, y, z): 1 \le x \le 2,\ -1 \le y \le 0,\ 0 \le z \le 3\}$, evaluate

$$\iiint\limits_{Q} (x + 2y + 4z)\,dV$$

in six different ways.

Evaluate the iterated integrals in Exercises 3–6.

3 $\displaystyle\int_{0}^{1}\int_{1+x}^{2x}\int_{z}^{x+z} x\,dy\,dz\,dx$

4 $\displaystyle\int_{1}^{2}\int_{0}^{z^2}\int_{x+z}^{x-z} z\,dy\,dx\,dz$

5 $\displaystyle\int_{-1}^{2}\int_{1}^{x^2}\int_{0}^{x+y} 2x^2 y\,dz\,dy\,dx$

6 $\displaystyle\int_{2}^{3}\int_{0}^{3y}\int_{1}^{yz} (2x + y + z)\,dx\,dz\,dy$

In Exercises 7–10 sketch the region Q bounded by the graphs of the given equations and express $\iiint_{Q} f(x, y, z)\,dV$ as an iterated integral in six different ways.

7 $x + 2y + 3z = 6;\, x = 0,\, y = 0,\, z = 0$

8 $x^2 + y^2 = 9,\, z = 0,\, z = 2$

9 $z = 9 - 4x^2 - y^2,\, z = 0$

10 $36x^2 + 9y^2 + 4z^2 = 36$

In Exercises 11–20 sketch the region bounded by the graphs of the given equations and use a triple integral to find its volume.

11 $z + x^2 = 4,\, y + z = 4,\, y = 0,\, z = 0$

12 $x^2 + z^2 = 4,\, y^2 + z^2 = 4$

13 $y = 2 - z^2,\, y = z^2,\, x + z = 4,\, x = 0$

14 $z = 4y^2,\, z = 2,\, x = 2,\, x = 0$

15 $y^2 + z^2 = 1,\, x + y + z = 2,\, x = 0$

16 $z = x^2 + y^2,\, y + z = 2$

17 $z = 9 - x^2,\, z = 0,\, y = -1,\, y = 2$

18 $z = e^{x+y},\, y = 3x,\, x = 2,\, y = 0,\, z = 0$

19 $z = x^2,\, z = x^3,\, y = z^2,\, y = 0$

20 $y = x^2 + z^2,\, z = x^2,\, z = 4,\, y = 0$

Each of the iterated integrals in Exercises 21–26 represents the volume of a region Q. Describe Q.

21 $\displaystyle\int_{0}^{1}\int_{\sqrt{1-z}}^{\sqrt{4-z}}\int_{2}^{3} dx\,dy\,dz$

22 $\displaystyle\int_{0}^{1}\int_{z^3}^{\sqrt{z}}\int_{0}^{4-x} dy\,dx\,dz$

23 $\displaystyle\int_{0}^{2}\int_{x^2}^{2x}\int_{0}^{x+y} dz\,dy\,dx$

24 $\displaystyle\int_{0}^{1}\int_{x}^{3x}\int_{0}^{xy} dz\,dy\,dx$

25 $\displaystyle\int_1^2 \int_{-\sqrt{z}}^{\sqrt{z}} \int_{-\sqrt{z-x^2}}^{\sqrt{z-x^2}} dy\, dx\, dz$

26 $\displaystyle\int_1^4 \int_{-z}^{z} \int_{-\sqrt{z^2-y^2}}^{\sqrt{z^2-y^2}} dx\, dy\, dz$

27 Define the limit of Riemann sums for functions of three variables (see Definition (17.2)).

28 Define the integral $\displaystyle\iiiint_U f(x, y, z, t)\, dW$, where

$$U = \{(x, y, z, t): a \le x \le b, c \le y \le d, m \le z \le n,$$
$$p \le t \le q\}.$$

17.7

APPLICATIONS OF TRIPLE INTEGRALS

If a solid has the shape of a three-dimensional region Q, and the density (mass per unit volume) at (x, y, z) is $\rho(x, y, z)$, where ρ is continuous throughout Q, then as in Definition (17.11) the mass is defined by

(17.22)
$$M = \iiint_Q \rho(x, y, z)\, dV.$$

If a particle of mass m is at the point (x, y, z), then its moments with respect to the xy-, xz-, and yz-planes are defined as zm, ym, and xm, respectively. By employing the usual techniques involving limits of sums, we are led to define the **moments** of the *solid* with respect to the coordinate planes as

$$M_{xy} = \iiint_Q z\rho(x, y, z)\, dV$$

(17.23)
$$M_{xz} = \iiint_Q y\rho(x, y, z)\, dV$$

$$M_{yz} = \iiint_Q x\rho(x, y, z)\, dV.$$

For example, if $\{Q_i\}$ is a partition of the region Q and if T_i is the part of the solid corresponding to Q_i, then the mass of T_i may be approximated by $\rho(x_i, y_i, z_i)\, \Delta V_i$ where (x_i, y_i, z_i) is a point in Q_i. The moment of T_i with respect to the xy-plane may be approximated by $z_i\rho(x_i, y_i, z_i)\, \Delta V_i$. Summing all such moments and taking the limit leads to the first integral in (17.23). The other formulas may be motivated in similar fashion. The **center of mass** is the point $(\bar{x}, \bar{y}, \bar{z})$, such that

(17.24)
$$\bar{x} = \frac{M_{yz}}{M}, \quad \bar{y} = \frac{M_{xz}}{M}, \quad \bar{z} = \frac{M_{xy}}{M}.$$

If Q is homogeneous, then the function ρ is constant and hence cancels in (17.24). Consequently, the center of mass of a homogeneous solid depends only on the shape of Q. As in two dimensions, the corresponding point for geometric solids is called the **centroid** of the solid.

Example 1 A solid has the shape of a right circular cylinder of base radius a and height h. Find the center of mass if the density at a point P is directly proportional to the distance from one of the bases to P.

Solution   If we introduce a coordinate system as in Figure 17.38, then the solid is bounded by the graphs of $x^2 + y^2 = a^2$, $z = 0$, and $z = h$. By hypothesis the density at (x, y, z) is $\rho(x, y, z) = kz$ for some constant k. Evidently, the center of mass is on the z-axis and, therefore, it is sufficient to find $\bar{z} = M_{xy}/M$. Moreover, by the form of ρ and the symmetry of the solid, we may calculate M and M_{xy} for the first octant portion and multiply by 4. Using (17.22),

$$M = 4 \int_0^a \int_0^{\sqrt{a^2-x^2}} \int_0^h kz \, dz \, dy \, dx = 4k \int_0^a \int_0^{\sqrt{a^2-x^2}} \frac{h^2}{2} \, dy \, dx$$

$$= 2kh^2 \int_0^a \sqrt{a^2 - x^2} \, dx = 2kh^2 \left(\frac{\pi a^2}{4}\right) = \frac{k\pi h^2 a^2}{2}.$$

Next using (17.23),

$$M_{xy} = 4 \int_0^a \int_0^{\sqrt{a^2-x^2}} \int_0^h z(kz) \, dz \, dy \, dx = 4k \int_0^a \int_0^{\sqrt{a^2-x^2}} \frac{h^3}{3} \, dy \, dx$$

$$= \frac{4kh^3}{3} \int_0^a \sqrt{a^2 - x^2} \, dx = \frac{4kh^3}{3}\left(\frac{\pi a^2}{4}\right) = \frac{k\pi h^3 a^2}{3}.$$

Finally, by (17.24),

$$\bar{z} = \frac{M_{xy}}{M} = \frac{k\pi h^3 a^2}{3} \cdot \frac{2}{k\pi h^2 a^2} = \frac{2h}{3}.$$

Hence the center of mass is on the axis of the cylinder, $\frac{2}{3}$ of the way from the lower base. ■

In the following example we shall merely *set up* the integrals necessary for the solution; that is, the integrals will be expressed in iterated form, but not evaluated.

Example 2 A solid has the shape of the region in the first octant bounded by the graphs of $4 - z = 9x^2 + y^2$, $y = 4x$, $z = 0$, and $y = 0$. If the density at the point $P(x, y, z)$ is proportional to the distance from the origin to P, set up the integrals needed to find $\bar{x}$.

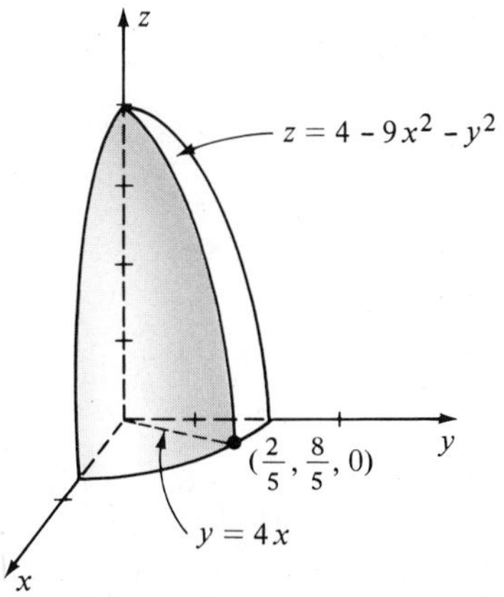

Solution   The region is sketched in Figure 17.39. The density at (x, y, z) is $\rho(x, y, z) = k(x^2 + y^2 + z^2)^{1/2}$ for some k. Using (17.22) and (17.23), M and M_{yz} are given by

$$M = \int_0^{8/5} \int_{y/4}^{\sqrt{4-y^2}/3} \int_0^{4-9x^2-y^2} k(x^2 + y^2 + z^2)^{1/2} \, dz \, dx \, dy$$

$$M_{yz} = \int_0^{8/5} \int_{y/4}^{\sqrt{4-y^2}/3} \int_0^{4-9x^2-y^2} xk(x^2 + y^2 + z^2)^{1/2} \, dz \, dx \, dy.$$

By (17.24), $\bar{x} = M_{yz}/M$. ■

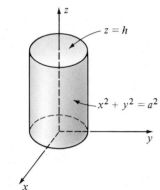

Figure 17.38

Figure 17.39

If a particle of mass m is at the point (x, y, z), then its distance to the z-axis is $(x^2 + y^2)^{1/2}$ and its **moment of inertia I_z with respect to the z-axis** is defined as $(x^2 + y^2)m$. Next, consider a solid T that has the shape of a region Q and density $\rho(x, y, z)$ at (x, y, z). As before, let $\{Q_i\}$ be a partition of Q and choose any point (x_i, y_i, z_i) in Q_i. If T_i is the part of the solid corresponding to Q_i, then the moment of inertia of T_i with respect to the z-axis may be approximated by $(x_i^2 + y_i^2)\rho(x_i, y_i, z_i)\,\Delta V_i$. Taking the limit of a sum of such terms leads to the following definition of the moment of inertia of T with respect to the z-axis:

(17.25)
$$I_z = \iiint_Q (x^2 + y^2)\rho(x, y, z)\,dV.$$

Similarly, the moments of inertia with respect to the x- and y-axis, respectively, are

(17.26)
$$I_x = \iiint_Q (y^2 + z^2)\rho(x, y, z)\,dV, \quad I_y = \iiint_Q (x^2 + z^2)\rho(x, y, z)\,dV.$$

If a solid of mass M has moment of inertia I with respect to a line, then the **radius of gyration** is, by definition, the number d such that $I = Md^2$. This formula implies that the radius of gyration is the distance from the line at which all the mass could be concentrated without changing the moment of inertia of the solid.

Example 3 Find the moment of inertia and radius of gyration, with respect to the axis of symmetry, of the cylindrical solid described in Example 1.

Solution The solid is sketched in Figure 17.38, where $\rho(x, y, z) = kz$. Using (17.25) and symmetry,

$$I_z = 4 \int_0^a \int_0^{\sqrt{a^2 - x^2}} \int_0^h (x^2 + y^2)kz\,dz\,dy\,dx$$

$$= 4k \int_0^a \int_0^{\sqrt{a^2 - x^2}} (x^2 + y^2)\frac{h^2}{2}\,dy\,dx$$

$$= 2kh^2 \int_0^a \left[x^2 y + \frac{y^3}{3} \right]_0^{\sqrt{a^2 - x^2}} dx$$

$$= 2kh^2 \int_0^a \left[x^2\sqrt{a^2 - x^2} + \tfrac{1}{3}\sqrt{(a^2 - x^2)^3} \right] dx.$$

The last integral may be evaluated by using a trigonometric substitution or a table of integrals. This results in $I_z = k\pi h^2 a^4/4$.

If d is the radius of gyration, then $d^2 = I_z/M$. Using the value for M found in Example 1, we obtain

$$d^2 = \frac{k\pi h^2 a^4}{4} \cdot \frac{2}{k\pi h^2 a^2} = \frac{a^2}{2}.$$

Hence $d = a/\sqrt{2} \approx 0.7a$. Thus, the radius of gyration is a distance from the axis of the cylinder that is approximately $\frac{7}{10}$ of the radius of the cylinder. ∎

Example 4 A homogeneous solid has the shape of the region Q bounded by the graphs of $4x^2 - y^2 + z^2 = 0$ and $y = 3$. Set up a triple integral for finding the moment of inertia with respect to the y-axis.

Solution   The region is part of an elliptic cone and is sketched in Figure 17.40 Note that the trace of the cone on the xy-plane is the pair of lines $x = \pm y/2$. If we denote the constant density by k, then applying (17.26),

$$I_y = \int_0^3 \int_{-y/2}^{y/2} \int_{-\sqrt{y^2-4x^2}}^{\sqrt{y^2-4x^2}} (x^2 + z^2)k \, dz \, dx \, dy.$$

We could also find I_y by multiplying by 4 the moment of inertia of that part of the solid that lies in the first octant. Thus

$$I_y = 4\int_0^3 \int_0^{y/2} \int_0^{\sqrt{y^2-4x^2}} (x^2 + z^2)k \, dz \, dx \, dy. \qquad \blacksquare$$

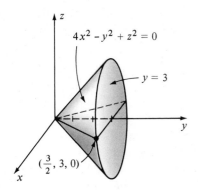

$4x^2 - y^2 + z^2 = 0$

$y = 3$

$\left(\dfrac{3}{2}, 3, 0\right)$

Figure 17.40

EXERCISES 17.7

In Exercises 1 and 2 set up the integrals needed to find the centroid of the given solid, and then find the centroid.

1 The solid in the first octant bounded by the coordinate planes and the graphs of $z = 9 - x^2$ and $2x + y = 6$.

2 The solid bounded by the graphs of $z = x^2$, $y = x^2$, $y = x^3$, and $z = 0$.

In Exercises 3 and 4 set up, but do not evaluate, the integrals needed to find the centroid of the given solid.

3 The solid bounded by the graphs of $y = 4x^2 + 9z^2$, $z = x^2$, $z = 1$, and $y = 0$.

4 The solid in the first octant bounded by the coordinate planes and the graphs of $x^2 + z^2 = 9$ and $y^2 + z^2 = 9$.

5 The density at a point P in a cubical solid of edge a is directly proportional to the square of the distance from P to a fixed corner of the cube. Find the center of mass.

6 Let Q be the tetrahedron bounded by the coordinate planes and the plane $2x + 5y + z = 10$. Find the center of mass if the density at the point $P(x, y, z)$ is directly proportional to the distance from the xz-plane to P.

7 Let Q be bounded by the paraboloid $x = y^2 + 4z^2$ and the plane $x = 4$. If the density at (x, y, z) is $\rho(x, y, z) = x^2 + z^2$, set up, but do not evaluate, the integrals necessary for finding the center of mass of Q.

8 Let Q be bounded by the hyperboloid $y^2 - x^2 - z^2 = 1$ and the plane $y = 2$. If the density at (x, y, z) is given by $\rho(x, y, z) = x^2 y^2 z^2$, set up, but do not evaluate, the integrals necessary for finding the center of mass.

9 Find the moment of inertia with respect to a line through one of its edges of a homogeneous cube of volume a^3. What is the radius of gyration?

10 Find the moment of inertia with respect to the z-axis of the tetrahedron described in Exercise 6. What is the radius of gyration?

In Exercises 11–16 set up the integral for the moment of inertia with respect to the z-axis of the indicated solid.

11 The solid bounded by the graphs of $z = 36 - 4x^2 - 9y^2$ and $z = 0$; $\rho(x, y, z) = z$

12 The cylinder bounded by the graphs of $4x^2 + y^2 = a^2$, $z = 0$, and $z = h$; $\rho(x, y, z) = 1 + z$

13 A sphere of radius a with center at the origin; $\rho(x, y, z) = x^2 + y^2 + z^2$

14 The cone bounded by the graphs of $x^2 + 9y^2 - z^2 = 0$ and $z = 36$; $\rho(x, y, z) = x^2 + y^2$

15 The homogeneous tetrahedron bounded by the coordinate planes and the graph of $x/a + y/b + z/c = 1$, where a, b, and c are positive.

16 The homogeneous solid bounded by the ellipsoid

$$\frac{x^2}{a^2} + \frac{y^2}{b^2} + \frac{z^2}{c^2} = 1.$$

17.8

TRIPLE INTEGRALS IN CYLINDRICAL AND SPHERICAL COORDINATES

Triple integrals may sometimes be expressed in terms of cylindrical coordinates. The simplest case occurs if a function of r, θ, and z is continuous throughout a region of the form

$$Q = \{(r, \theta, z) : a \le r \le b, c \le \theta \le d, k \le z \le l\}.$$

We first divide Q into subregions $Q_1, Q_2, \ldots, Q_n$ that have the same shape as Q by using graphs of equations $r = a_i$, $\theta = c_i$, and $z = k_i$, where a_i, c_i, and k_i are numbers in the intervals $[a, b]$, $[c, d]$, and $[k, l]$, respectively. A typical subregion Q_i, appropriately labeled, is sketched in Figure 17.41, where $\bar{r}_i$ is the average radius of the base of Q_i. The volume ΔV_i of Q_i is the product of the area of the base $\bar{r}_i \, \Delta r_i \, \Delta \theta_i$ and the altitude Δz_i. Thus

$$\Delta V_i = \bar{r}_i \, \Delta r_i \, \Delta \theta_i \, \Delta z_i.$$

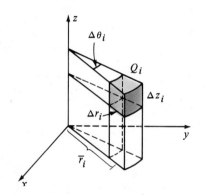

If (r_i, θ_i, z_i) is any point in Q_i, then the triple integral of f over Q is

$$\iiint_Q f(r, \theta, z) \, dV = \lim_{\|P\| \to 0} \sum_i f(r_i, \theta_i, z_i) \, \Delta V_i$$

where the norm $\|P\|$ is the length of the longest diagonal of the Q_i. Furthermore, it can be proved that

$$\iiint_Q f(r, \theta, z) \, dV = \int_k^l \int_c^d \int_a^b f(r, \theta, z) r \, dr \, d\theta \, dz.$$

Figure 17.41

There are five other possible orders of integration for this iterated integral.

Triple integrals in cylindrical coordinates may be defined over more complicated regions by employing inner partitions. For example, it can be shown that if R is a polar region of the type discussed in Section 17.5 and if

$$Q = \{(r, \theta, z) : (r, \theta) \text{ is in } R \text{ and } k_1(r, \theta) \le z \le k_2(r, \theta)\}$$

where the functions k_1 and k_2 have continuous first partial derivatives throughout R, then

$$\iiint_Q f(r, \theta, z) \, dV = \iint_R \left[\int_{k_1(r,\theta)}^{k_2(r,\theta)} f(r, \theta, z) \, dz \right] dA.$$

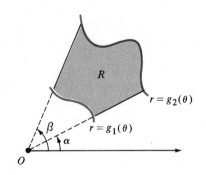

Figure 17.42

In particular, suppose R is a region of the type illustrated in Figure 17.42. We may then evaluate the triple integral of f over Q as follows.

$$(17.27) \qquad \iiint_Q f(r, \theta, z) \, dV = \int_\alpha^\beta \int_{g_1(\theta)}^{g_2(\theta)} \int_{k_1(r,\theta)}^{k_2(r,\theta)} f(r, \theta, z) \, r \, dz \, dr \, d\theta.$$

The iterated integral in (17.27) may be interpreted as a limit of a triple sum, in which the first summation takes place along a column of the subregions

Q_i from the lower surface (with equation $z = k_1(r, \theta)$) to the upper surface (with equation $z = k_2(r, \theta)$) (see Figure 17.43). The second summation extends over a row of these columns, where θ is held fixed and r is allowed to vary. At this stage we have summed over a slice of Q as illustrated in Figure 17.43. Finally, we sweep out Q by letting θ vary from α to β.

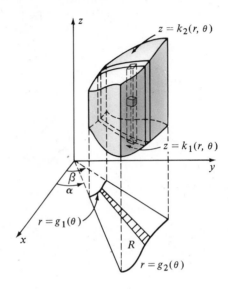

Figure 17.43

Example 1 A solid has the shape of the region Q that lies inside the cylinder $r = a$, within the sphere $r^2 + z^2 = 4a^2$ and above the xy-plane. Find the center of mass and the moment of inertia I_z if the density at a point P is directly proportional to the distance from the xy-plane to P.

Solution The region Q is sketched in Figure 17.44 together with a column that corresponds to a first summation in the direction of the z-axis. Since the density at the point $P(r, \theta, z)$ is given by kz for some constant k, the mass M may be found by applying (17.27) with $f(r, \theta, z) = kz$. Thus

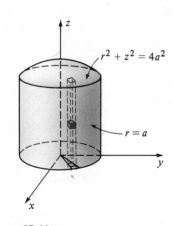

Figure 17.44

$$M = \int_0^{2\pi} \int_0^a \int_0^{\sqrt{4a^2 - r^2}} kzr \, dz \, dr \, d\theta = \frac{k}{2} \int_0^{2\pi} \int_0^a z^2 r \Big]_0^{\sqrt{4a^2 - r^2}} dr \, d\theta$$

$$= \frac{k}{2} \int_0^{2\pi} \int_0^a (4a^2 r - r^3) \, dr \, d\theta = \frac{k}{2} \int_0^{2\pi} 2a^2 r^2 - \frac{1}{4} r^4 \Big]_0^a d\theta$$

$$= \frac{7a^4 k}{8} \int_0^{2\pi} d\theta = \frac{7a^4 \pi k}{4}.$$

To find I_z we employ (17.25) and use cylindrical coordinates, obtaining

$$I_z = \int_0^{2\pi} \int_0^a \int_0^{\sqrt{4a^2 - r^2}} r^2 kzr \, dz \, dr \, d\theta = \frac{k}{2} \int_0^{2\pi} \int_0^a r^3 z^2 \Big]_0^{\sqrt{4a^2 - r^2}} dr \, d\theta$$

$$= \frac{k}{2} \int_0^{2\pi} \int_0^a (4a^2 r^3 - r^5) \, dr \, d\theta = \frac{5a^6 \pi k}{6}. \qquad \blacksquare$$

Triple integrals in spherical coordinates may also be considered. Suppose a function f of ρ, ϕ, and θ is continuous throughout a region of the form

$$Q = \{(\rho, \phi, \theta): a \leq \rho \leq b, c \leq \phi \leq d, k \leq \theta \leq l\}.$$

We may divide Q into subregions $Q_1, Q_2, \ldots, Q_n$ having the same shape as Q by using graphs of equations of the form $\rho = a_i$, $\phi = c_i$, and $\theta = k_i$. A typical subregion Q_i is sketched in (i) of Figure 17.45, where $\Delta\rho_i$, $\Delta\phi_i$, and $\Delta\theta_i$ are the indicated increments.

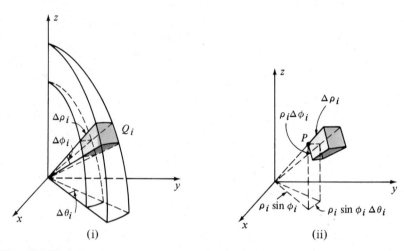

(i)   (ii)

Figure 17.45

The volume of the subregion Q_i may be approximated by regarding it as a rectangular box. If $P(\rho_i, \phi_i, \theta_i)$ is a corner of the box, then as illustrated in (ii) of Figure 17.45, the dimensions of the box are approximately $\Delta\rho_i$, $\rho_i \Delta\phi_i$, and $\rho_i \sin \phi_i \Delta\theta_i$. Hence, if ΔV_i is the volume of Q_i, then

(17.28)
$$\Delta V_i \approx \rho_i^2 \sin \phi_i \, \Delta\rho_i \, \Delta\phi_i \, \Delta\theta_i.$$

If the triple integral of f over Q is defined as a limit of a sum in the usual way, then it can be proved that

(17.29)
$$\iiint\limits_{Q} f(\rho, \phi, \theta) \, dV = \int_k^l \int_c^d \int_a^b f(\rho, \phi, \theta)\rho^2 \sin \phi \, d\rho \, d\phi \, d\theta.$$

There are five other possible orders of integration. Spherical coordinates may also be employed over more complicated regions by using inner partitions. In this case the limits of integration in (17.29) must be suitably adjusted.

Example 2 Find the volume and the centroid of the region Q that is bounded above by the sphere $\rho = a$ and below by the cone $\phi = m$ where $0 < m < \pi/2$.

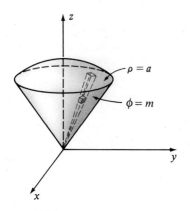

Figure 17.46

Solution The region Q is sketched in Figure 17.46. The volume V is

$$V = \int_0^{2\pi} \int_0^m \int_0^a \rho^2 \sin\phi \, d\rho \, d\phi \, d\theta$$

$$= \int_0^{2\pi} \int_0^m \frac{a^3}{3} \sin\phi \, d\phi \, d\theta = \frac{a^3}{3} \int_0^{2\pi} \left[-\cos\phi \right]_0^m d\theta$$

$$= \frac{a^3}{3} \int_0^{2\pi} (1 - \cos m) \, d\theta = \frac{2\pi a^3}{3} (1 - \cos m).$$

By symmetry, the centroid is on the z-axis. If (x, y, z) are rectangular coordinates of a point, then by Theorem (14.50) $z = \rho \cos\phi$. Consequently,

$$M_{xy} = \iiint_Q z \, dV = \int_0^{2\pi} \int_0^m \int_0^a \rho \cos\phi \, \rho^2 \sin\phi \, d\rho \, d\phi \, d\theta$$

$$= \int_0^{2\pi} \int_0^m \frac{a^4}{4} \sin\phi \cos\phi \, d\phi \, d\theta = \frac{a^4}{4} \int_0^{2\pi} \frac{1}{2} \sin^2\phi \Big]_0^m d\theta$$

$$= \frac{a^4}{8} \sin^2 m \int_0^{2\pi} d\theta = \frac{\pi a^4}{4} \sin^2 m.$$

The centroid, in rectangular coordinates, is $(0, 0, \bar{z})$, where

$$\bar{z} = \frac{M_{xy}}{V} = \frac{3}{8} a(1 + \cos m). \qquad \blacksquare$$

The letter ρ used in spherical coordinates should not be confused with the density function. For problems that involve spherical coordinates *and* a density function, a different symbol should be used to denote density.

EXERCISES 17.8

Use cylindrical coordinates in Exercises 1–10.

1 Find the volume and the centroid of the solid bounded by the graphs of $z = x^2 + y^2$, $x^2 + y^2 = 4$, and $z = 0$.

2 Find the volume and the centroid of the solid bounded by the graphs of $z = \sqrt{x^2 + y^2}$, $x^2 + y^2 = 4$, and $z = 0$.

3 Find the moment of inertia of a homogeneous right circular cylinder of altitude h and radius of base a with respect to (a) the axis of the cylinder; (b) a diameter of the base.

4 A homogeneous solid is bounded by the graphs of $z = \sqrt{x^2 + y^2}$ and $z = x^2 + y^2$. Find (a) the center of mass; (b) the moment of inertia with respect to the z-axis.

5 The density at a point P of a spherical solid of radius a is directly proportional to the distance from P to a fixed line l through the center of the solid. Find the mass of the solid.

6 Find the mass of the conical solid bounded by the graphs of $z = \sqrt{x^2 + y^2}$ and $z = 4$ if the density at a point P is directly proportional to the distance from the z-axis to P.

7 Find the moment of inertia with respect to l for the solid described in Exercise 5. What is the radius of gyration?

8 Find the moment of inertia with respect to the z-axis for the solid described in Exercise 6. What is the radius of gyration?

9 Let $Q = \{(x, y, z): 1 \le z \le 5 - x^2 - y^2, \; x^2 + y^2 \ge 1\}$. If the density at the point $P(x, y, z)$ is directly proportional to the distance from the xy-plane to P, find the mass and the center of mass of Q.

10 Find the mass and center of mass of the solid that lies inside both the cylinder $x^2 + y^2 - 2y = 0$ and the sphere $x^2 + y^2 + z^2 = 4$ if the density at the point $P(x, y, z)$ is directly proportional to the distance from the xy-plane to P.

Use spherical coordinates in Exercises 11–18.

11 Find the mass and center of mass of a solid hemisphere of radius a if the density at a point P is directly proportional to the distance from the center of the base to P.

12 Rework Exercise 2.

13 Find the moment of inertia with respect to the axis for the hemisphere in Exercise 11.

14 Find the moment of inertia with respect to a diameter of the base of a homogeneous solid hemisphere of radius a.

15 Find the volume of the solid that lies above the cone $z^2 = x^2 + y^2$ and inside the sphere $x^2 + y^2 + z^2 = 4z$.

16 Find the volume of the solid that lies outside the cone $z^2 = x^2 + y^2$ and inside the sphere $x^2 + y^2 + z^2 = 1$.

17 Find the mass of the solid that lies outside the sphere $x^2 + y^2 + z^2 = 1$ and inside the sphere $z^2 + y^2 + z^2 = 2$ if the density at a point P is directly proportional to the square of the distance from the center of the spheres to P.

18 A homogeneous spherical shell has inner radius a and outer radius b. Find its moment of inertia with respect to a line through the center.

Evaluate the integrals in Exercises 19 and 20 by changing to cylindrical coordinates.

19 $\displaystyle\int_0^1 \int_0^{\sqrt{1-y^2}} \int_0^{\sqrt{4-x^2-y^2}} z \, dz \, dx \, dy$

20 $\displaystyle\int_0^2 \int_{-\sqrt{2x-x^2}}^{\sqrt{2x-x^2}} \int_0^{x^2+y^2} \sqrt{x^2 + y^2} \, dz \, dy \, dx$

Evaluate the integrals in Exercises 21 and 22 by changing to spherical coordinates.

21 $\displaystyle\int_{-2}^2 \int_{-\sqrt{4-x^2}}^{\sqrt{4-x^2}} \int_{\sqrt{x^2+y^2}}^{\sqrt{8-x^2-y^2}} (x^2 + y^2 + z^2) \, dz \, dy \, dx$

22 $\displaystyle\int_0^{\sqrt{2}} \int_y^{\sqrt{4-y^2}} \int_0^{\sqrt{4-x^2-y^2}} \sqrt{x^2 + y^2 + z^2} \, dz \, dx \, dy$

23 Interpret (17.29) as a limit of a triple sum.

17.9

SURFACE AREA

In Chapter 13 we obtained formulas for the area of a surface of revolution (see (13.13)–(13.15)). We shall now discuss a method for finding areas of surfaces of the type illustrated in Figure 17.47. Suppose that $f(x, y) \ge 0$ throughout a region R in the xy-plane and that f has continuous first partial

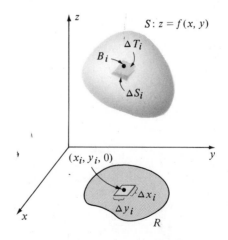

Figure 17.47

derivatives on R. Let S denote the part of the graph of f whose projection on the xy-plane is R. To simplify the discussion we shall assume that no normal vector to S is parallel to the xy-plane. Our objective is to define the area A of S and to find a formula that can be used to calculate A.

Let $P = \{R_i\}$ be an inner partition of R, where the dimensions of the rectangle R_i are Δx_i and Δy_i. We choose a point $(x_i, y_i, 0)$ in each R_i and let $B_i(x_i, y_i, f(x_i, y_i))$ be the corresponding point on S. Next consider the tangent plane to S at B_i. Let ΔT_i and ΔS_i be the areas of the regions on the tangent plane and S, respectively, obtained by projecting R_i vertically upward (see Figure 17.47). If the norm $\|P\|$ of the partition is small, then ΔT_i is an approximation to ΔS_i, and $\sum_i \Delta T_i$ is an approximation to the area A of S. Since this approximation should improve as $\|P\|$ approaches 0, we define the **surface area** by

(17.30)
$$A = \lim_{\|P\| \to 0} \sum_i \Delta T_i.$$

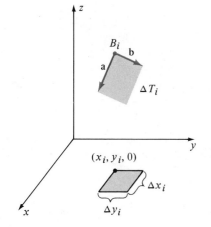

Figure 17.48

To find an integral formula for A, let us choose $(x_i, y_i, 0)$ as the corner of R_i closest to the origin. Let $\mathbf{a}$ and $\mathbf{b}$ be the vectors with initial point $B_i(x_i, y_i, f(x_i, y_i))$ that are tangent to the traces of S on the planes $y = y_i$ and $x = x_i$, respectively, as illustrated in the isolated view shown in Figure 17.48. Using the geometric interpretation for partial derivatives given in Section 16.3, the slopes of the lines determined by $\mathbf{a}$ and $\mathbf{b}$ in these planes are $f_x(x_i, y_i)$ and $f_y(x_i, y_i)$, respectively. It follows that

$$\mathbf{a} = \Delta x_i \mathbf{i} + f_x(x_i, y_i)\, \Delta x_i \mathbf{k}$$
$$\mathbf{b} = \Delta y_i \mathbf{j} + f_y(x_i, y_i)\, \Delta y_i \mathbf{k}.$$

The area ΔT_i of the parallelogram determined by $\mathbf{a}$ and $\mathbf{b}$ is $|\mathbf{a} \times \mathbf{b}|$. Since

$$\mathbf{a} \times \mathbf{b} = \begin{vmatrix} \mathbf{i} & \mathbf{j} & \mathbf{k} \\ \Delta x_i & 0 & f_x(x_i, y_i)\, \Delta x_i \\ 0 & \Delta y_i & f_y(x_i, y_i)\, \Delta y_i \end{vmatrix}$$

we have

$$\mathbf{a} \times \mathbf{b} = -f_x(x_i, y_i)\, \Delta x_i \Delta y_i \mathbf{i} - f_y(x_i, y_i)\, \Delta x_i \Delta y_i \mathbf{j} + \Delta x_i \Delta y_i \mathbf{k}.$$

Consequently,

$$\Delta T_i = |\mathbf{a} \times \mathbf{b}| = \sqrt{(f_x(x_i, y_i))^2 + (f_y(x_i, y_i))^2 + 1}\, \Delta x_i \Delta y_i$$

where $\Delta x_i \Delta y_i = \Delta A_i$. If, as in (17.30), we take the limit of a sum of the ΔT_i and apply the definition of the double integral, we obtain

(17.31)
$$A = \iint\limits_R \sqrt{(f_x(x, y))^2 + (f_y(x, y))^2 + 1}\, dA.$$

This formula may also be used if $f(x, y) \le 0$ on R.

Example 1 Let R be the triangular region in the xy-plane with vertices $(0, 0, 0)$, $(0, 1, 0)$, and $(1, 1, 0)$. Find the surface area of that part of the graph of $z = 3x + y^2$ that lies over R.

Solution   The region R is bounded by the graphs of $y = x$, $x = 0$, and $y = 1$ as shown in Figure 17.49. Letting $f(x, y) = 3x + y^2$ and applying (17.31) gives us

$$A = \iint_R \sqrt{3^2 + (2y)^2 + 1}\, dA = \int_0^1 \int_0^y (10 + 4y^2)^{1/2}\, dx\, dy$$

$$= \int_0^1 (10 + 4y^2)^{1/2}\, x \bigg]_0^y dy = \int_0^1 (10 + 4y^2)^{1/2}\, y\, dy$$

$$= \frac{1}{12}(10 + 4y^2)^{3/2} \bigg]_0^1 = \frac{14^{3/2} - 10^{3/2}}{12} \approx 1.7. \qquad \blacksquare$$

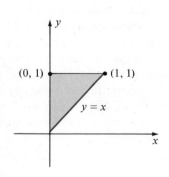

Figure 17.49

Example 2   Find the surface area of the part of the graph of $z = 9 - x^2 - y^2$ that lies above or on the xy-plane.

Solution   The graph is sketched in Figure 16.5. By (17.31),

$$A = \iint_R \sqrt{4x^2 + 4y^2 + 1}\, dx\, dy$$

where R is the region in the xy-plane bounded by the circle $x^2 + y^2 = 9$. If we use polar coordinates to evaluate the double integral, then

$$A = \int_0^{2\pi} \int_0^3 (4r^2 + 1)^{1/2}\, r\, dr\, d\theta.$$

It is left to the reader to show that $A = \pi(37^{3/2} - 1)/6 \approx 117.3$. $\qquad \blacksquare$

Formulas similar to (17.31) may be stated if the surface S has suitable projections on the yz- or xz-planes. Thus, if S is the graph of an equation $y = g(x, z)$ and if the projection on the xz-plane is R_1, then

$$A = \iint_{R_1} \sqrt{(g_x(x, z))^2 + (g_z(x, z))^2 + 1}\, dx\, dz.$$

Another formula of this type may be stated if S is given by $x = h(y, z)$.

EXERCISES 17.9

1 Find the surface area of that part of the graph of $z = y + (x^2/2)$ that lies over the square region in the xy-plane having vertices $(0, 0, 0)$, $(1, 0, 0)$, $(1, 1, 0)$, and $(0, 1, 0)$.

2 Find the area of the region on the plane $z = y + 1$ that lies inside the cylinder $x^2 + y^2 = 1$.

3 Find the area of the part of the plane $x/a + y/b + z/c = 1$ cut out by the cylinder $x^2 + y^2 = d^2$, where a, b, c, and d are positive.

4 Set up an integral for finding the surface area of that part of the sphere $x^2 + y^2 + z^2 = 4$ that lies over the square region in the xy-plane having vertices $(1, 1, 0)$, $(1, -1, 0)$, $(-1, 1, 0)$, and $(-1, -1, 0)$.

5 Find the surface area of that part of the sphere with equation $x^2 + y^2 + z^2 = a^2$ that lies inside the cylinder $x^2 + y^2 - ay = 0$.

6 Find the surface area of that part of the cylinder $y^2 + z^2 = a^2$ that lies inside the cylinder $x^2 + y^2 = a^2$.

7 Find the surface area of the part of the paraboloid $z = x^2 + y^2$ that is cut off by the plane $z = 1$.

8 Let R be the triangular region in the xy-plane with vertices $(0, 0, 0)$, $(0, 2, 0)$, and $(2, 2, 0)$. Find the surface area of that part of the graph of $z = y^2$ that lies over R.

9 Find the surface area of the part of the graph of $z = xy$ that is inside the cylinder $x^2 + y^2 = 4$.

10 Find the surface area of that part of the hyperbolic paraboloid $z = x^2 - y^2$ that lies in the first octant and is inside the cylinder $x^2 + y^2 = 1$.

17.10

REVIEW

Define or discuss each of the following.

1 Inner partitions

2 Riemann sums

3 Double integral

4 Iterated double integral

5 Evaluation of double integrals

6 Areas and volumes using double integrals

7 Center of mass of a lamina

8 Moments of inertia of a lamina

9 Radius of gyration

10 Double integrals in polar coordinates

11 Triple integral

12 Iterated triple integral

13 Evaluation of triple integrals

14 Applications of triple integrals

15 Triple integrals in cylindrical and spherical coordinates

16 Surface area

EXERCISES 17.10

Evaluate the integrals in Exercises 1–6.

1 $\displaystyle\int_{-1}^{0} \int_{x+1}^{x^3} (x^2 - 2y)\, dy\, dx$

2 $\displaystyle\int_{1}^{2} \int_{1}^{\ln y} \frac{1}{y}\, dx\, dy$

3 $\displaystyle\int_{0}^{3} \int_{r}^{r^2+1} r\, d\theta\, dr$

4 $\displaystyle\int_{2}^{0} \int_{0}^{z^2} \int_{x}^{z} (x + z)\, dy\, dx\, dz$

5 $\displaystyle\int_{0}^{2} \int_{\sqrt{y}}^{1} \int_{z^2}^{y} xy^2 z^3\, dx\, dz\, dy$

6 $\displaystyle\int_{0}^{\pi/2} \int_{0}^{\pi/4} \int_{0}^{a\cos\phi} \rho^2 \sin\phi\, d\rho\, d\phi\, d\theta$

In Exercises 7–10 express $\iint_R f(x, y)\, dA$ as an iterated integral if R is the region bounded by the graphs of the given equations.

7 $x^2 - y^2 = 4, x = 4$

8 $x^2 - y^2 = 4, y = 4, y = 0$

9 $y^2 = 4 + x, y^2 = 4 - x$

10 $y = -x^2 + 4, y = 3x^2$

Each of the integrals in Exercises 11 and 12 represents the area of a region R in the xy-plane. Describe R.

11 $\displaystyle\int_{-1}^{1} \int_{e^y}^{y^3} dx\, dy$

12 $\displaystyle\int_{-1}^{0} \int_{x}^{-x^2} dy\, dx$

In Exercises 13 and 14 reverse the order of integration and evaluate the resulting integral.

13 $\displaystyle\int_{0}^{3} \int_{y^2}^{9} ye^{-x^2}\, dx\, dy$

14 $\displaystyle\int_{0}^{1} \int_{x}^{\sqrt{x}} e^{x/y}\, dy\, dx$

In Exercises 15 and 16 find the mass and the center of mass of the lamina that has the shape of the region bounded by the graphs of the given equations and having the indicated density.

15 $y = x, y = 2x, x = 3$; density at the point $P(x, y)$ is directly proportional to the distance from the y-axis to P.

16 $y^2 = x, x = 4$; density at the point $P(x, y)$ is directly proportional to the distance from the line with equation $x = -1$ to P.

17 A lamina has the shape of the region that lies inside the graph of $r = 2 + \sin\theta$ and outside the graph of $r = 1$. Find the mass if the density at the point $P(r, \theta)$ is inversely proportional to the distance from the pole to P.

18 Find the area of the region bounded by the polar axis and the graphs of $r = e^\theta$ and $r = 2$ from $\theta = 0$ to $\theta = \ln 2$.

19 Use polar coordinates to evaluate

$$\int_{-a}^{0} \int_{-\sqrt{a^2-x^2}}^{0} \sqrt{x^2 + y^2}\, dy\, dx.$$

20 Find I_x, I_y, and I_O for the lamina bounded by the graphs of $y = x^2$ and $y = x^3$ if the density at the point $P(x, y)$ is directly proportional to the distance from the y-axis to P.

21 A lamina has the shape of a right triangle with sides of lengths a, b, and $\sqrt{a^2 + b^2}$. The density is directly proportional to the distance from the side of length a. Find the moment of inertia and the radius of gyration with respect to a line along the side of length a.

22 A lamina has the shape of the region between concentric circles of radii a and b, where $a < b$. If the density at a point P is directly proportional to the distance from the center to P, use polar coordinates to find the moment of inertia with respect to a line through the center.

23 Find the volume of the solid that lies under the graph of $z = xy^2$ and over the rectangle in the xy-plane with vertices $(1, 1, 0)$, $(2, 1, 0)$, $(1, 3, 0)$, and $(2, 3, 0)$.

24 Express $\iiint_Q f(x, y, z)\, dV$ as an iterated integral in six different ways, where Q is bounded by the graphs of $y = x^2 + 4z^2$ and $y = 4$.

25 Use triple integrals to find the volume and centroid of the solid bounded by the graphs of $z = x^2$, $z = 4$, $y = 0$, and $y + z = 4$.

26 Set up a triple integral for the moment of inertia with respect to the z-axis of the solid bounded by the graphs of $z = 9x^2 + y^2$ and $z = 9$ if the density at the point $P(x, y, z)$ is inversely proportional to the square of the distance from the point $(0, 0, -1)$ to P.

27 Set up a triple integral for the moment of inertia with respect to the y-axis of the solid bounded by the graphs of $x^2 - y^2 + z^2 = 1$, $y = 0$, and $y = 4$ if the density at the point $P(x, y, z)$ is directly proportional to the distance from the y-axis to P.

28 A homogeneous solid is bounded by the graph of $z = 9 - x^2 - y^2$, the interior of the cylinder $x^2 + y^2 = 4$, and the xy-plane. Use cylindrical coordinates to find (a) the mass; (b) the center of mass; (c) the moment of inertia with respect to the z-axis.

29 A solid has the shape of a sphere of radius a. Use spherical coordinates to find the mass if the density at a point P is directly proportional to the distance from the center to P.

30 Find the surface area of that part of the cone $z = (x^2 + y^2)^{1/2}$ that is inside the cylinder $x^2 + y^2 = 4x$.

18

TOPICS IN VECTOR CALCULUS

The concepts introduced in this chapter have many applications in the physical sciences. After discussing *vector fields* and defining *line integrals*, we consider evaluation theorems and determine conditions for independence of path. *Green's Theorem* is then proved for elementary plane regions. This important result provides a connection between line integrals and double integrals. We also investigate the *divergence* and *curl* of a vector field and discuss two of the major theorems in vector calculus: the *Divergence Theorem* and *Stokes' Theorem*. The chapter ends with a study of *transformations of coordinates, Jacobians*, and *change of variables in multiple integrals*.

18.1

VECTOR FIELDS

If, to each point K in a region, there is associated a unique vector having initial point K, then the totality of such vectors is called a **vector field**. Vector fields are common in everyday life. The diagram in (i) of Figure 18.1 illustrates a vector field determined by a wheel rotating about an axle. To each point on the wheel there corresponds a velocity vector. A vector field of this type is called a **velocity field**.

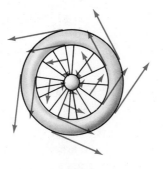

(i) Rotating wheel

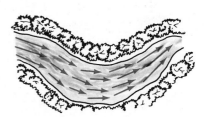

(ii) Flow of water

Figure 18.1 Velocity fields

Other examples of velocity fields are those determined by the flow of water or of wind. Thus, (ii) of Figure 18.1 could indicate the velocity vectors associated with fluid particles moving in a stream. Sketches of this type show only a few vectors of the vector field. It is important to remember that a vector is associated with *every* point in the region under consideration. Other common vector fields are **force fields**, which arise in the study of mechanics or electricity. In each of these illustrations it was assumed that the vectors were independent of time. These are called **steady vector fields** and are the only types we shall consider in this chapter.

If a rectangular coordinate system is introduced, then the vector associated with the point $K(x, y, z)$ may be denoted by $\mathbf{F}(x, y, z)$. Since the components of $\mathbf{F}(x, y, z)$ depend on the coordinates of K, we may write

(18.1)
$$\mathbf{F}(x, y, z) = M(x, y, z)\mathbf{i} + N(x, y, z)\mathbf{j} + P(x, y, z)\mathbf{k}$$

where M, N, and P are functions of x, y, and z. Conversely, every equation of the form (18.1) determines a vector field. It follows that a vector field may be defined as a **vector function F** whose domain D is a subset of $\mathbb{R} \times \mathbb{R} \times \mathbb{R}$ and whose range is a subset of V_3. As a special case, if the domain corresponds to a region in a plane and the range is a subset of V_2, then a vector field (in two dimensions) is given by

(18.2)
$$\mathbf{F}(x, y) = M(x, y)\mathbf{i} + N(x, y)\mathbf{j}$$

where M and N are functions of x and y. For convenience we shall use the terms "vector field" and "vector function" interchangeably. The functions discussed in Chapter 16, which associate a *number* with each point K, will be called **scalar functions** (or **scalar fields**).

Example 1   Describe the vector field $\mathbf{F}$ if $\mathbf{F}(x, y) = -y\mathbf{i} + x\mathbf{j}$.

Solution   The vectors $\mathbf{F}(x, y)$ associated with several points (x, y) are listed in the following table and are sketched in (i) of Figure 18.2.

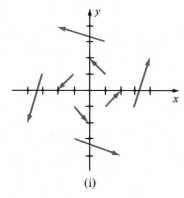

(i)

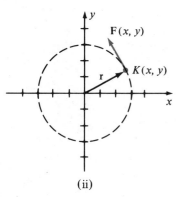

(ii)

Figure 18.2

(x, y)	$\mathbf{F}(x, y)$	(x, y)	$\mathbf{F}(x, y)$
$(1, 1)$	$-\mathbf{i} + \mathbf{j}$	$(1, 3)$	$-3\mathbf{i} + \mathbf{j}$
$(-1, 1)$	$-\mathbf{i} - \mathbf{j}$	$(-3, 1)$	$-\mathbf{i} - 3\mathbf{j}$
$(-1, -1)$	$\mathbf{i} - \mathbf{j}$	$(-1, -3)$	$3\mathbf{i} - \mathbf{j}$
$(1, -1)$	$\mathbf{i} + \mathbf{j}$	$(3, -1)$	$\mathbf{i} + 3\mathbf{j}$

The vectors are much like those associated with the rotating wheel in (i) of Figure 18.1. To arrive at a more general description of the vector field $\mathbf{F}$, consider an arbitrary point $K(x, y)$ and let $\mathbf{r} = x\mathbf{i} + y\mathbf{j}$ be the position vector of $K(x, y)$ (see (ii) of Figure 18.2). Since

$$\mathbf{r} \cdot \mathbf{F}(x, y) = (x\mathbf{i} + y\mathbf{j}) \cdot (-y\mathbf{i} + x\mathbf{j})$$
$$= -xy + yx = 0,$$

it follows that $\mathbf{F}(x, y)$ is orthogonal to the position vector $\mathbf{r}$ and, therefore, is tangent to the circle of radius $|\mathbf{r}|$ with center at the origin, as shown in (ii) of the figure. Note that

$$|\mathbf{F}(x, y)| = \sqrt{y^2 + x^2} = |\mathbf{r}|$$

and hence the magnitude of $\mathbf{F}(x, y)$ equals the radius of this circle. Thus the vector field is, indeed, similar to the velocity field of the rotating wheel illustrated in Figure 18.1, where the origin is at the point of rotation. ∎

The next definition introduces one of the most important vector fields that occurs in the physical sciences.

Definition (18.3)

> Let $\mathbf{r} = x\mathbf{i} + y\mathbf{j} + z\mathbf{k}$ be the position vector of a point $K(x, y, z)$. A vector field $\mathbf{F}$ is called an **inverse square field** if
>
> $$\mathbf{F}(x, y, z) = \frac{c}{|\mathbf{r}|^2} \mathbf{u}$$
>
> where c is a scalar and $\mathbf{u} = (1/|\mathbf{r}|)\mathbf{r}$ is a unit vector having the same direction as $\mathbf{r}$.

Example 2 Describe the inverse square field $\mathbf{F}(x, y, z)$ of (18.3) if $c < 0$.

Solution Since $\mathbf{u} = (1/|\mathbf{r}|)\mathbf{r}$, where $\mathbf{r} = x\mathbf{i} + y\mathbf{j} + z\mathbf{k}$, we have

$$\mathbf{F}(x, y, z) = \frac{c}{|\mathbf{r}|^3} \mathbf{r} = \frac{c}{(x^2 + y^2 + z^2)^{3/2}} (x\mathbf{i} + y\mathbf{j} + z\mathbf{k}).$$

The last expression may be written in form (18.1). It is simpler, however, to analyze the vectors in the field by using the expression containing $\mathbf{r}$. Since $\mathbf{F}(x, y, z)$ is a negative scalar multiple of $\mathbf{r}$, the direction of $\mathbf{F}(x, y, z)$ is toward the origin O. Also,

$$|\mathbf{F}(x, y, z)| = \frac{|c|}{|\mathbf{r}|^2} |\mathbf{u}| = \frac{|c|}{|\mathbf{r}|^2}$$

and hence the magnitude of $\mathbf{F}(x, y, z)$ is inversely proportional to the square of the distance from O to the point (x, y, z). This means that as the point $K(x, y, z)$ moves away from the origin, the length of the associated vector $\mathbf{F}(x, y, z)$ decreases. Some typical vectors in the field are sketched in Figure 18.3. ∎

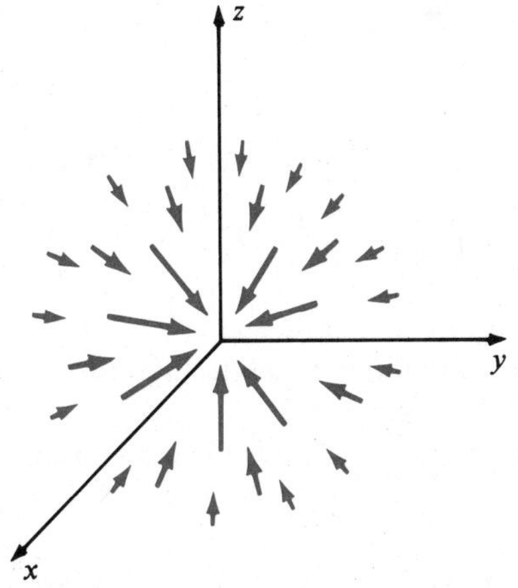

Figure 18.3 Inverse square vector field

The force of gravity determines an inverse square vector field. According to Newton's Law of Universal Gravitation, if a particle of mass M is located at the origin of a rectangular coordinate system, then the force exerted on a particle of mass m located at $K(x, y, z)$ is

$$\mathbf{F}(x, y, z) = -G \frac{Mm}{|\mathbf{r}|^2} \mathbf{u}$$

where G is a gravitational constant, $\mathbf{r}$ is the position vector of the point K, and $\mathbf{u} = (1/|\mathbf{r}|)\mathbf{r}$.

Inverse square fields also occur in electrical theory. In particular, Coulomb's Law states that if a point charge of Q coulombs is at the origin, then the force $\mathbf{F}(x, y, z)$ it exerts on a point charge of q coulombs located at $K(x, y, z)$ is

$$\mathbf{F}(x, y, z) = c \frac{Qq}{|\mathbf{r}|^2} \mathbf{u}$$

for some constant c, where $\mathbf{r}$ and $\mathbf{u}$ are as in Definition (18.3). Note that Coulomb's Law has the same form as Newton's Law of Universal Gravitation.

Important types of vector fields may be obtained by using the gradient of a scalar function f of two or three variables. For example, if $w = f(x, y, z)$, then as in (16.27), the *gradient* ∇w of w is

$$\nabla w = f_x(x, y, z)\mathbf{i} + f_y(x, y, z)\mathbf{j} + f_z(x, y, z)\mathbf{k}.$$

We know from our work in Chapter 16 that the direction of the vector ∇w at any point K is orthogonal to the level surface associated with f that contains K. Moreover, the magnitude of ∇w equals the maximum rate of change of f at the point K.

Definition (18.4)

> A vector field $\mathbf{F}$ is called a **conservative vector field** if it is the gradient of a scalar function, that is, if
>
> $$\mathbf{F}(x, y, z) = \nabla f(x, y, z)$$
>
> for some function f.

If $\mathbf{F}$ is conservative, then the function f in Definition (18.4) is called a **potential function** for $\mathbf{F}$ and $f(x, y, z)$ is called the **potential** at the point $K(x, y, z)$.

Inverse square fields are conservative. In particular, if $\mathbf{F}$ is the field given in Definition (18.3), then

$$\mathbf{F}(x, y, z) = \nabla\left(\frac{-c}{r}\right)$$

where $r = |\mathbf{r}| = (x^2 + y^2 + z^2)^{1/2}$. The verification of this fact is left to the reader (see Exercise 11). In physics, the potential function of a conservative vector field $\mathbf{F}$ is defined as the function p such that $\mathbf{F}(x, y, z) = -\nabla p(x, y, z)$. In this event, the field in Definition (18.3) is $\mathbf{F}(x, y, z) = \nabla(c/r)$. We shall have more to say about conservative fields later in this chapter.

The vector differential operator ∇ in three dimensions is

$$\nabla = \mathbf{i}\,\frac{\partial}{\partial x} + \mathbf{j}\,\frac{\partial}{\partial y} + \mathbf{k}\,\frac{\partial}{\partial z}.$$

It has no practical use standing alone; however, if it operates on a scalar function f it produces the gradient of f:

$$\text{grad } f = \nabla f = \frac{\partial f}{\partial x}\mathbf{i} + \frac{\partial f}{\partial y}\mathbf{j} + \frac{\partial f}{\partial z}\mathbf{k}.$$

We shall next use ∇ as an operator on *vector* functions.

In certain applications a vector function $\mathbf{F}$ is used to obtain another vector function called the **curl of F**, denoted by curl $\mathbf{F}$ or $\nabla \times \mathbf{F}$, and defined as follows.

Definition **(18.5)**

Let a vector function $\mathbf{F}$ in three dimensions be given by

$$\mathbf{F}(x, y, z) = M(x, y, z)\mathbf{i} + N(x, y, z)\mathbf{j} + P(x, y, z)\mathbf{k}$$

where M, N, and P have partial derivatives in some region. The **curl** of $\mathbf{F}$ is

$$\text{curl } \mathbf{F} = \nabla \times \mathbf{F} = \left(\frac{\partial P}{\partial y} - \frac{\partial N}{\partial z}\right)\mathbf{i} + \left(\frac{\partial M}{\partial z} - \frac{\partial P}{\partial x}\right)\mathbf{j} + \left(\frac{\partial N}{\partial x} - \frac{\partial M}{\partial y}\right)\mathbf{k}.$$

As indicated in Definition (18.5), it is customary to use either the symbol curl $\mathbf{F}$ or $\nabla \times \mathbf{F}$ to denote the *value* of this function at (x, y, z). The formula for curl $\mathbf{F}$ may be regarded as the determinant "expansion" (by the first row) of the following expression:

(18.6)

$$\text{curl } \mathbf{F} = \nabla \times \mathbf{F} = \begin{vmatrix} \mathbf{i} & \mathbf{j} & \mathbf{k} \\ \dfrac{\partial}{\partial x} & \dfrac{\partial}{\partial y} & \dfrac{\partial}{\partial z} \\ M & N & P \end{vmatrix}.$$

This determinant notation is ambiguous, since the first row is made up of vectors, the second row of partial derivative operators, and the third row of scalar functions; however, it is an extremely useful device for remembering the cumbersome formula given in Definition (18.5).

Example 3   Find $\nabla \times \mathbf{F}$ if $\mathbf{F}(x, y, z) = xy^2z^4\mathbf{i} + (2x^2y + z)\mathbf{j} + y^3z^2\mathbf{k}$.

Solution   Applying (18.6),

$$\nabla \times \mathbf{F} = \begin{vmatrix} \mathbf{i} & \mathbf{j} & \mathbf{k} \\ \dfrac{\partial}{\partial x} & \dfrac{\partial}{\partial y} & \dfrac{\partial}{\partial z} \\ xy^2z^4 & (2x^2y + z) & y^3z^2 \end{vmatrix}$$

$$= (3y^2z^2 - 1)\mathbf{i} + 4xy^2z^3\mathbf{j} + (4xy - 2xyz^4)\mathbf{k}. \qquad \blacksquare$$

The operator ∇ is also used to obtain a certain *scalar* field from a vector field F as follows.

Definition (18.7)

> Let $\mathbf{F}(x, y, z) = M(x, y, z)\mathbf{i} + N(x, y, z)\mathbf{j} + P(x, y, z)\mathbf{k}$. The **divergence** of $\mathbf{F}$, denoted by div $\mathbf{F}$ or $\nabla \cdot \mathbf{F}$, is defined by
>
> $$\text{div } \mathbf{F} = \nabla \cdot \mathbf{F} = \frac{\partial M}{\partial x} + \frac{\partial N}{\partial y} + \frac{\partial P}{\partial z}.$$

The reason for using the symbol $\nabla \cdot \mathbf{F}$ is evident from the fact that the formula for div $\mathbf{F}$ may be obtained by taking the formal dot product of ∇ and $\mathbf{F}$.

Example 4   Find $\nabla \cdot \mathbf{F}$ if $\mathbf{F}(x, y, z) = xy^2z^4\mathbf{i} + (2x^2y + z)\mathbf{j} + y^3z^2\mathbf{k}$.

Solution   By Definition (18.7),

$$\nabla \cdot \mathbf{F} = \frac{\partial}{\partial x}(xy^2z^4) + \frac{\partial}{\partial y}(2x^2y + z) + \frac{\partial}{\partial z}(y^3z^2)$$

$$= y^2z^4 + 2x^2 + 2y^3z. \qquad \blacksquare$$

We shall discuss the physical significance of curl $\mathbf{F}$ and div $\mathbf{F}$ later in this chapter. In this section let us concentrate on algebraic properties of these functions. A property that is analogous to the product rule for derivatives is given in the next example. Several other properties are stated in exercises at the end of this section.

Example 5   If f and $\mathbf{F}$ are scalar and vector functions, respectively, that possess partial derivatives, show that

$$\nabla \cdot [f\mathbf{F}] = f[\nabla \cdot \mathbf{F}] + [\nabla f] \cdot \mathbf{F}.$$

Solution If we write $\mathbf{F} = M\mathbf{i} + N\mathbf{j} + P\mathbf{k}$, where $M, N,$ and P are functions of $x, y,$ and z, then

$$f\mathbf{F} = fM\mathbf{i} + fN\mathbf{j} + fP\mathbf{k}.$$

Applying Definition (18.7),

$$\nabla \cdot [f\mathbf{F}] = \frac{\partial}{\partial x}(fM) + \frac{\partial}{\partial y}(fN) + \frac{\partial}{\partial z}(fP)$$

$$= f\frac{\partial M}{\partial x} + \frac{\partial f}{\partial x}M + f\frac{\partial N}{\partial y} + \frac{\partial f}{\partial y}N + f\frac{\partial P}{\partial z} + \frac{\partial f}{\partial z}P.$$

Rearranging terms gives us

$$\nabla \cdot [f\mathbf{F}] = f\left[\frac{\partial M}{\partial x} + \frac{\partial N}{\partial y} + \frac{\partial P}{\partial z}\right] + \left[\frac{\partial f}{\partial x}M + \frac{\partial f}{\partial y}N + \frac{\partial f}{\partial z}P\right]$$

$$= f[\nabla \cdot \mathbf{F}] + [\nabla f] \cdot \mathbf{F}. \qquad \blacksquare$$

Finally, if a vector field $\mathbf{F}$ is described as in (18.1), then we may define limits, continuity, partial derivatives, and multiple integrals by using the components of $\mathbf{F}(x, y, z)$ in a fashion similar to that used for vector-valued functions in Chapter 15. For example, if we wish to differentiate or integrate $\mathbf{F}(x, y, z)$, we differentiate or integrate each component. The usual theorems may be established. Thus, $\mathbf{F}$ is continuous if and only if the component functions $M, N,$ and P are continuous, $\mathbf{F}$ has partial derivatives if and only if the same is true for $M, N,$ and P, etc.

EXERCISES 18.1

In Exercises 1–10 sketch a sufficient number of vectors $\mathbf{F}(x, y, z)$ to illustrate the pattern of the vectors in the field $\mathbf{F}$.

1 $\mathbf{F}(x, y) = x\mathbf{i} - y\mathbf{j}$

2 $\mathbf{F}(x, y) = -x\mathbf{i} + y\mathbf{j}$

3 $\mathbf{F}(x, y) = 2x\mathbf{i} + 3y\mathbf{j}$

4 $\mathbf{F}(x, y) = 3\mathbf{i} + x\mathbf{j}$

5 $\mathbf{F}(x, y) = (x^2 + y^2)^{-1/2}(x\mathbf{i} + y\mathbf{j})$

6 $\mathbf{F}(x, y, z) = x\mathbf{i} + z\mathbf{k}$

7 $\mathbf{F}(x, y, z) = -x\mathbf{i} - y\mathbf{j} - z\mathbf{k}$

8 $\mathbf{F}(x, y, z) = x\mathbf{i} + y\mathbf{j} + z\mathbf{k}$

9 $\mathbf{F}(x, y, z) = \mathbf{i} + \mathbf{j} + \mathbf{k}$

10 $\mathbf{F}(x, y, z) = 2\mathbf{k}$

11 If $\mathbf{F}$ is the inverse square vector field of Definition (18.3), prove that $\mathbf{F}(x, y, z) = \nabla(-c/r)$, where $r = |\mathbf{r}|$.

12 Find a potential function for $\mathbf{F}(x, y, z) = -(GMm/|\mathbf{r}|^2)\mathbf{u}$.

In Exercises 13–16 find a conservative vector field that has the given potential.

13 $f(x, y, z) = x^2 - 3y^2 + 4z^2$

14 $f(x, y, z) = \sin(x^2 + y^2 + z^2)$

15 $f(x, y) = \arctan(xy)$

16 $f(x, y) = y^2 e^{-3x}$

17 If $\mathbf{F}$ is given by (18.1), define

$$\lim_{(x, y, z) \to (x_0, y_0, z_0)} \mathbf{F}(x, y, z) = \mathbf{a}$$

in a manner analogous to Definition (15.3). Give an ε-δ definition for this limit. What is the geometric significance of the limit?

18 If **F** is given by (18.1), define the notion of *continuity* at (x_0, y_0, z_0) in a manner similar to Definition (15.4). What is the geometric significance of a continuous vector function?

In Exercises 19–22 find $\nabla \times \mathbf{F}$ and $\nabla \cdot \mathbf{F}$.

19 $\mathbf{F}(x, y, z) = x^2 z\mathbf{i} + y^2 x\mathbf{j} + (y + 2z)\mathbf{k}$

20 $\mathbf{F}(x, y, z) = (3x + y)\mathbf{i} + xy^2 z\mathbf{j} + xz^2\mathbf{k}$

21 $\mathbf{F}(x, y, z) = 3xyz^2\mathbf{i} + y^2 \sin z\mathbf{j} + xe^{2z}\mathbf{k}$

22 $\mathbf{F}(x, y, z) = x^3 \ln z\mathbf{i} + xe^{-y}\mathbf{j} - (y^2 + 2z)\mathbf{k}$

Verify the identities in Exercises 23–26.

23 $\nabla \times (\mathbf{F} + \mathbf{G}) = \nabla \times \mathbf{F} + \nabla \times \mathbf{G}$

24 $\nabla \cdot (\mathbf{F} + \mathbf{G}) = \nabla \cdot \mathbf{F} + \nabla \cdot \mathbf{G}$

25 $\nabla \times (f\mathbf{F}) = f(\nabla \times \mathbf{F}) + (\nabla f) \times \mathbf{F}$

26 $\nabla \cdot (\mathbf{F} \times \mathbf{G}) = (\nabla \times \mathbf{F}) \cdot \mathbf{G} - (\nabla \times \mathbf{G}) \cdot \mathbf{F}$

If f and **F** have continuous second partial derivatives, and **a** is a constant vector, verify the identities in Exercises 27–30.

27 curl grad $f = \mathbf{0}$

28 div curl $\mathbf{F} = 0$

29 curl (grad f + curl **F**) = curl curl **F**

30 curl $\mathbf{a} = \mathbf{0}$

31 If $\mathbf{r} = x\mathbf{i} + y\mathbf{j} + z\mathbf{k}$ prove that $\nabla \cdot \mathbf{r} = 3$, $\nabla \times \mathbf{r} = \mathbf{0}$, and $\nabla|\mathbf{r}| = \mathbf{r}/|\mathbf{r}|$.

32 If $\mathbf{r} = x\mathbf{i} + y\mathbf{j} + z\mathbf{k}$ and **a** is a constant vector, prove that curl $(\mathbf{a} \times \mathbf{r}) = 2\mathbf{a}$ and div $(\mathbf{a} \times \mathbf{r}) = 0$.

33 Prove that both the curl and the divergence of an inverse square vector field are zero.

34 Let $\mathbf{r} = x\mathbf{i} + y\mathbf{j} + z\mathbf{k}$ and $r = |\mathbf{r}|$. If $\mathbf{F}(x, y, z) = (c/r^k)\mathbf{r}$, where c is a constant and k is any positive real number, prove that the curl of **F** is **0**. (*Hint:* Use Exercise 25.)

35 If a vector field **F** is conservative and has continuous partial derivatives, prove that curl $\mathbf{F} = \mathbf{0}$.

36 The differential operator Lap is denoted by $\nabla^2 = \nabla \cdot \nabla = \partial^2/\partial x^2 + \partial^2/\partial y^2 + \partial^2/\partial z^2$. If it operates on $f(x, y, z)$, it produces a scalar function called the **Laplacian** of f, that is,

$$\text{Lap } f = \nabla^2 f = \frac{\partial^2 f}{\partial x^2} + \frac{\partial^2 f}{\partial y^2} + \frac{\partial^2 f}{\partial z^2}.$$

If f and g are scalar functions that have second partial derivatives, prove that

(a) div grad $f = $ Lap f.

(b) Lap $(fg) = $
$$f \text{ Lap } g + g \text{ Lap } f + 2 (\text{grad } f) \cdot (\text{grad } g).$$

Using the notation of Exercise 36, prove that the functions defined in Exercises 37 and 38 satisfy **Laplace's equation** $\nabla^2 f = 0$. (Functions of this type are called **harmonic** and are very important in physical applications.)

37 $f(x, y, z) = (x^2 + y^2 + z^2)^{-1/2}$

38 $f(x, y, z) = ax^2 + by^2 + cz^2$, where $a + b + c = 0$.

18.2

LINE INTEGRALS

To define $\int_a^b f(x)\, dx$ we began by dividing the interval $[a, b]$ into n subintervals of lengths $\Delta x_1, \Delta x_2, \ldots, \Delta x_n$. We then chose a number w_i in each subinterval and took the limit of the Riemann sum $\sum_i f(w_i)\, \Delta x_i$ as all the Δx_i approached 0. A similar process can be carried out for functions of several variables. For example, if $f(x, y)$ is defined on a finite plane curve C, we could divide C into n subarcs of lengths $\Delta s_1, \Delta s_2, \ldots, \Delta s_n$. After choosing a point (u_i, v_i) in each subarc we could then consider the limit of the sum $\sum_i f(u_i, v_i)\, \Delta s_i$ as all the Δs_i approach 0. This technique produces what is known as a *line integral*. Let us now make these remarks more precise. We shall begin by recalling the definition of *smooth curve* that was introduced in Section 13.1.

Definition (18.8)

> A plane curve C is **smooth** if it has a parametric representation
>
> $$x = g(t), \quad y = h(t), \quad \text{where} \quad a \leq t \leq b,$$
>
> such that g' and h' are continuous and not simultaneously 0 on $[a, b]$.

The preceding definition can be extended to space curves in the obvious way, by considering a third function k such that $z = k(t)$.

The parametric representation of C in Definition (18.8) is sometimes called a **smooth parameterization**. For example, if C is the parabola $y = x^2$, then $x = t$, $y = t^2$ is a smooth parameterization; however, $x = t^3$, $y = t^6$ is not, since dx/dt and dy/dt are simultaneously 0 at $t = 0$. Hereafter, whenever we refer to a smooth curve C, it will be assumed that a smooth parameterization $x = f(t)$, $y = g(t)$ has been chosen.

The cycloid $x = a(t - \sin t)$, $y = a(1 - \cos t)$ sketched in Figure 13.4 is not smooth. Note that both $dx/dt = 0$ and $dy/dt = 0$ if $t = 2\pi n$, where n is any integer. Moreover, there is no smooth parameterization, since the cycloid has a corner at every point $(2\pi n, 0)$ on the x-axis.

Consider $f(x, y)$, where f is continuous on a region D containing a smooth curve C. Let A and B be the points on C determined by the parameter values a and b, respectively. We shall assign a positive direction, or **orientation**, along C as that determined by increasing values of t. Let us partition the parameter interval $[a, b]$ by choosing

$$a = t_0 < t_1 < t_2 < \cdots < t_n = b.$$

This leads to a subdivision of C into subarcs $\overset{\frown}{P_{i-1}P_i}$, where $P_i(x_i, y_i)$ is the point on C corresponding to t_i. As in Figure 18.4, let $\Delta x_i = x_i - x_{i-1}$, $\Delta y_i = y_i - y_{i-1}$, and let Δs_i denote the length of the subarc $\overset{\frown}{P_{i-1}P_i}$. The **norm** $\|\Delta\|$ of the subdivision of C is, by definition, the largest of the Δs_i.

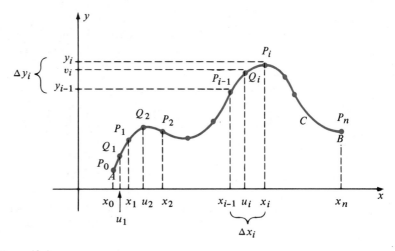

Figure 18.4

We next choose a point $Q_i(u_i, v_i)$ in each subarc $\overset{\frown}{P_{i-1}P_i}$, as illustrated in Figure 18.4. For each i, we evaluate the function f at (u_i, v_i), multiply this number by Δs_i, and form the sum

$$\sum_{i=1}^{n} f(u_i, v_i)\, \Delta s_i.$$

If, in a manner similar to Definition (5.7), this sum has a limit L as $n \to \infty$ and $\|\Delta\| \to 0$ that is independent of the partition of $[a, b]$ and the points Q_i, then L is called the **line integral of f along C from A to B** and is denoted by

(18.9)
$$\int_C f(x, y)\, ds = \lim_{\|\Delta\| \to 0} \sum_i f(u_i, v_i)\, \Delta s_i.$$

The terminology *line integral* is a misnomer. It would be more descriptive to use *curve integral* for (18.9).

It can be shown that if f is continuous on D, then the limit in (18.9) exists and is the same for all parametric representations of C (provided the same orientation is used). Moreover, the integral may be evaluated as follows.

Theorem (18.10)

> If a smooth curve C is given by $x = g(t)$, $y = h(t)$, where $a \le t \le b$, and if $f(x, y)$ is continuous on a region D containing C, then
>
> $$\int_C f(x, y)\, ds = \int_a^b f(g(t), h(t))\sqrt{[g'(t)]^2 + [h'(t)]^2}\, dt.$$

The form of the radical in Theorem (18.10) is a consequence of the discussion of arc length in Chapter 13 (see Theorem (13.11)). It is worth noting that if C is an interval on the x-axis, then $\Delta s_i = \Delta x_i$, all $v_i = 0$, and (18.10) reduces to a definite integral of the type considered in Chapter 5.

The preceding discussion can be extended to more complicated curves. In particular, suppose C is a **piecewise smooth curve**, in the sense that it can be expressed as the union of a finite number of smooth curves $C_1, C_2, \ldots, C_n$, where the terminal point B_i of C_i is the initial point A_{i+1} of C_{i+1} for $i = 1, 2, \ldots, n - 1$. In this case the line integral of f along C is defined as the sum of the line integrals along the individual curves.

Example 1 Evaluate $\int_C xy^2\, ds$ if C has parametric equations $x = \cos t$, $y = \sin t$ where $0 \le t \le \pi/2$.

Solution   The curve C is that part of the unit circle with center O that lies in the first quadrant. Applying Theorem (18.10),

$$\int_C xy^2\, ds = \int_0^{\pi/2} \cos t \sin^2 t \sqrt{\sin^2 t + \cos^2 t}\, dt$$

$$= \int_0^{\pi/2} \sin^2 t \cos t\, dt = \frac{1}{3} \sin^3 t \Big]_0^{\pi/2} = \frac{1}{3}. \qquad \blacksquare$$

It is possible to establish properties of line integrals that are similar to those obtained for the definite integrals of Chapter 5. For example, it can be shown that reversing the direction of integration changes the sign of the integral, that the integral of a sum of two functions is the sum of the integrals of the individual functions, etc.

An elementary physical application of the line integral (18.9) may be obtained by regarding the curve as a thin wire of variable density. If the wire is represented by the curve C in Figure 18.4, and if the **linear density** (the mass per unit length) at the point (x, y) is given by $f(x, y)$, then $f(u_i, v_i) \Delta s_i$ is an approximation to the mass Δm_i of the part of the wire between P_{i-1} and P_i. The sum

$$\sum_i \Delta m_i = \sum_i f(u_i, v_i) \Delta s_i$$

is an approximation to the total mass M of the wire. To define M we take the limit of this sum, obtaining

$$M = \int_C f(x, y) \, ds.$$

Example 2 A thin wire is bent into the shape of a semicircle of radius a. If the density at a point P is directly proportional to its distance from the line through the endpoints, find the mass of the wire.

Solution If we introduce a coordinate system such that the shape of the wire coincides with the upper half of a circle of radius a and center O, then parametric equations for C are

$$x = a \cos t, \quad y = a \sin t; \quad 0 \le t \le \pi.$$

By hypothesis, the density function f is given by $f(x, y) = ky$ for some constant k. Hence, by the previous discussion, the mass of the wire is

$$M = \int_C ky \, ds = \int_0^\pi ka \sin t \sqrt{a^2 \sin^2 t + a^2 \cos^2 t} \, dt$$

$$= \int_0^\pi ka(\sin t)a \, dt = -ka^2 \cos t \Big]_0^\pi = 2ka^2. \quad \blacksquare$$

Two other types of line integrals may be obtained by using Δx_i and Δy_i in place of Δs_i in (18.9). They are called the **line integrals of f along C with respect to x and y**, respectively. Thus, by definition,

(18.11)

$$\int_C f(x, y) \, dx = \lim_{\|\Delta\| \to 0} \sum_i f(u_i, v_i) \Delta x_i$$

$$\int_C f(x, y) \, dy = \lim_{\|\Delta\| \to 0} \sum_i f(u_i, v_i) \Delta y_i.$$

If C is given parametrically by $x = g(t)$, $y = h(t)$, where $a \leq t \leq b$, then the integrals in (18.11) may be evaluated by substituting $g(t)$ and $h(t)$ for x and y, respectively, letting $dx = g'(t)\, dt$, $dy = h'(t)\, dt$, and using a and b for the limits of integration.

Example 3   Evaluate $\int_C xy^2\, dx$ and $\int_C xy^2\, dy$ if C is the part of the parabola $y = x^2$ from $A(0, 0)$ to $B(2, 4)$.

Solution   Parametric equations for C are $x = t$, $y = t^2$, where $0 \leq t \leq 2$. Hence $dx = dt$, $dy = 2t\, dt$, and

$$\int_C xy^2\, dx = \int_0^2 t^5\, dt = \frac{1}{6} t^6 \Big]_0^2 = \frac{32}{3}$$

$$\int_C xy^2\, dy = \int_0^2 t^5(2t)\, dt = \frac{2}{7} t^7 \Big]_0^2 = \frac{256}{7}.$$ ∎

If C is the graph of a rectangular equation $y = g(x)$ where $a \leq x \leq b$, then parametric equations for C can be found by letting

$$x = t, \quad y = g(t) \quad \text{where} \quad a \leq t \leq b.$$

The line integrals in (18.11) may then be evaluated as follows:

$$\int_C f(x, y)\, dx = \int_a^b f(t, g(t))\, dt = \int_a^b f(x, g(x))\, dx$$

$$\int_C f(x, y)\, dy = \int_a^b f(t, g(t))g'(t)\, dt = \int_a^b f(x, g(x))g'(x)\, dx.$$

This shows that for curves given in the rectangular form $y = g(x)$ where $a \leq x \leq b$, we may bypass the parametric equations by substituting $y = g(x)$, $dy = g'(x)\, dx$, and then using a and b for the limits of integration. To illustrate, in Example 3, where $y = x^2$, we could have written

$$\int_C xy^2\, dx = \int_0^2 x(x^4)\, dx = \int_0^2 x^5\, dx = \frac{32}{3}$$

$$\int_C xy^2\, dy = \int_0^2 x(x^4)2x\, dx = \int_0^2 2x^6\, dx = \frac{256}{7}.$$

In applications, line integrals often occur in the combination

$$\int_C M(x, y)\, dx + \int_C N(x, y)\, dy$$

where M and N are continuous functions on a domain D containing C. This sum is usually abbreviated by writing it in the form

$$\int_C M(x, y)\, dx + N(x, y)\, dy.$$

If C is given parametrically by $x = g(t)$, $y = h(t)$, where $a \le t \le b$, then the integral may be evaluated by substituting for x, y, dx, and dy as was done previously.

Example 4   Evaluate $\int_C xy \, dx + x^2 \, dy$ if:

(a) C consists of line segments from $(2, 1)$ to $(4, 1)$ and from $(4, 1)$ to $(4, 5)$.

(b) C is the line segment from $(2, 1)$ to $(4, 5)$.

(c)   parametric equations for C are $x = 3t - 1$, $y = 3t^2 - 2t$; $1 \le t \le \frac{5}{3}$.

Solution   The curves for parts (a)–(c) are sketched in Figure 18.5.

(a)   If C is subdivided into two parts C_1 and C_2 as shown in (i) of Figure 18.5, then parametric equations for these curves are

$$C_1 : x = t, y = 1; \quad 2 \le t \le 4$$

$$C_2 : x = 4, y = t; \quad 1 \le t \le 5.$$

The line integral along C may be expressed as a sum of two line integrals, the first along C_1 and the second along C_2. On C_1 we have $dy = 0$, $dx = dt$, and hence

$$\int_{C_1} xy \, dx + x^2 \, dy = \int_2^4 t(1) \, dt + 0 = \frac{1}{2} t^2 \Big]_2^4 = 6.$$

On C_2 we have $dx = 0$, $dy = dt$, and therefore

$$\int_{C_2} xy \, dx + x^2 \, dy = \int_1^5 0 + 16 \, dt = 16t \Big]_1^5 = 64.$$

Consequently, the line integral along C equals $6 + 64$ or 70.

(b)   The graph of C is sketched in (ii) of Figure 18.5. A rectangular equation for C is $y = 2x - 3$, where $2 \le x \le 4$. In this case $dy = 2 \, dx$ and we may write

$$\int_C xy \, dx + x^2 \, dy = \int_2^4 x(2x - 3) \, dx + x^2 2 \, dx$$

$$= \int_2^4 (4x^2 - 3x) \, dx = \frac{170}{3} = 56\tfrac{2}{3}.$$

(c)   The graph of C is part of a parabola (see (iii) of Figure 18.5). In this case $dx = 3 \, dt$, $dy = (6t - 2) \, dt$, and the given integral equals

$$\int_1^{5/3} (3t - 1)(3t^2 - 2t)3 \, dt + (3t - 1)^2 (6t - 2) \, dt.$$

It is left to the reader to show that the value is 58. Another method of solution is to use the rectangular equation $y = \frac{1}{3}(x^2 - 1)$ for the parabola, where $2 \le x \le 4$. In this event we obtain the following integral:

$$\int_C xy \, dx + x^2 \, dy = \int_2^4 x \frac{1}{3}(x^2 - 1) \, dx + x^2 \left(\frac{2}{3} x \right) dx = 58. \quad ■$$

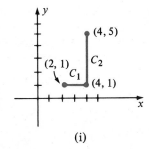

(i)

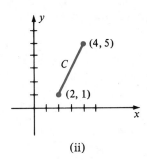

(ii)

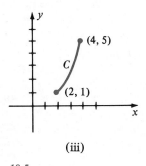

(iii)

Figure 18.5

In the preceding example we obtained three different values for the line integral along three different paths from (2, 1) to (4, 5). In Section 18.3 we shall consider line integrals that have the same value along every curve joining two points A and B. For such integrals we use the phrase *independent of path*.

If a smooth curve C in three dimensions is given parametrically by

$$x = g(t), \quad y = h(t), \quad z = k(t)$$

where $a \leq t \leq b$, then line integrals of a function f of *three* variables are defined in a manner similar to those for two variables. In this case, instead of using (x_i, y_i) and (u_i, v_i) as the coordinates of P_i and Q_i on C (see Figure 18.4), we use (x_i, y_i, z_i) and (u_i, v_i, w_i), respectively. Instead of (18.9) we now have

$$\int_C f(x, y, z)\, ds = \lim_{\|\Delta\| \to 0} \sum_i f(u_i, v_i, w_i)\, \Delta s_i.$$

This integral may be evaluated by using the formula

$$\int_a^b f(g(t), h(t), k(t))\sqrt{(g'(t))^2 + (h'(t))^2 + (k'(t))^2}\, dt.$$

In addition to line integrals $\int_C f(x, y, z)\, dx$ and $\int_C f(x, y, z)\, dy$, there is a line integral with respect to z in three dimensions, given by

$$\int_C f(x, y, z)\, dz = \lim_{\|\Delta\| \to 0} \sum_i f(u_i, v_i, w_i)\, \Delta z_i$$

where $\Delta z_i = z_i - z_{i-1}$. As in two dimensions, these line integrals often occur in the form

$$\int_C M(x, y, z)\, dx + N(x, y, z)\, dy + P(x, y, z)\, dz$$

where M, N, and P are functions of x, y, and z that are continuous throughout a region containing C. If C is given parametrically, then this line integral may be evaluated by substituting for x, y, and z in the same manner as in the two-variable case.

Example 5   Evaluate $\int_C yz\, dx + xz\, dy + xy\, dz$ where C is the twisted cubic

$$x = t, \quad y = t^2, \quad z = t^3; \quad 0 \leq t \leq 2.$$

Solution   Substituting for x, y, and z and using $dx = dt$, $dy = 2t\, dt$, $dz = 3t^2\, dt$, we obtain

$$\int_0^2 t^5\, dt + 2t^5\, dt + 3t^5\, dt = \int_0^2 6t^5\, dt = t^6 \Big]_0^2 = 64. \quad \blacksquare$$

One of the most important physical applications of line integrals has to do with force fields. Let us suppose that the force acting at the point (x, y, z) is

$$\mathbf{F}(x, y, z) = M(x, y, z)\mathbf{i} + N(x, y, z)\mathbf{j} + P(x, y, z)\mathbf{k}$$

where the functions M, N, and P are continuous. Our objective is to define the work done as the point of application of $\mathbf{F}(x, y, z)$ moves along a smooth curve C, where C is given parametrically by $x = g(t)$, $y = h(t)$, $z = k(t)$; $a \le t \le b$. We shall assume that the motion is in the direction determined by *increasing* values of t.

Let us subdivide C by choosing points $P_0, P_1, P_2, \ldots, P_n$ and then, for each i, choose a point Q_i in the subarc $\overparen{P_{i-1}P_i}$, as illustrated in Figure 18.6. Let the coordinates of P_i and Q_i be (x_i, y_i, z_i) and (u_i, v_i, w_i), respectively. If the norm $\|\Delta\|$ is small, then the work done by $\mathbf{F}(x, y, z)$ along the arc $\overparen{P_{i-1}P_i}$ can be approximated by the work Δw_i done by the *constant* force $\mathbf{F}(u_i, v_i, w_i)$ as its point of application moves along $\overrightarrow{P_{i-1}P_i}$.

If $\Delta x_i = x_i - x_{i-1}$, $\Delta y_i = y_i - y_{i-1}$, and $\Delta z_i = z_i - z_{i-1}$, then $\overrightarrow{P_{i-1}P_i}$ corresponds to the vector $\Delta x_i\mathbf{i} + \Delta y_i\mathbf{j} + \Delta z_i\mathbf{k}$ of V_3. By (14.29) the work Δw_i done by $\mathbf{F}(u_i, v_i, w_i)$ along $\overrightarrow{P_{i-1}P_i}$ is

$$
\begin{aligned}
\Delta w_i &= \mathbf{F}(u_i, v_i, w_i) \cdot (\Delta x_i\mathbf{i} + \Delta y_i\mathbf{j} + \Delta z_i\mathbf{k}) \\
&= M(u_i, v_i, w_i)\,\Delta x_i + N(u_i, v_i, w_i)\,\Delta y_i + P(u_i, v_i, w_i)\,\Delta z_i.
\end{aligned}
$$

The work W done by $\mathbf{F}$ along C is, by definition,

$$W = \lim_{\|\Delta\| \to 0} \sum_i \Delta w_i$$

that is,

$$(18.12) \qquad W = \int_C M(x, y, z)\,dx + N(x, y, z)\,dy + P(x, y, z)\,dz.$$

In applications, the preceding integral is usually expressed in vector form. If we let

$$\mathbf{r}(t) = x\mathbf{i} + y\mathbf{j} + z\mathbf{k}$$

where $x = g(t)$, $y = h(t)$, and $z = k(t)$, then $\mathbf{r}(t)$ is the position vector of the point $Q(x, y, z)$ on C. If s denotes arc length measured along C, then as in Section 15.4, a unit tangent vector $\mathbf{T}(s)$ to C at Q is

$$\mathbf{T}(s) = \frac{d}{ds}\mathbf{r}(t) = \frac{dx}{ds}\mathbf{i} + \frac{dy}{ds}\mathbf{j} + \frac{dz}{ds}\mathbf{k}.$$

The vectors $\mathbf{r}(t)$, $\mathbf{F}(x, y, z)$, and $\mathbf{T}(s)$ are illustrated in Figure 18.7.

The **tangential component of F at Q** (see Figure 18.7), is

$$\mathbf{F}(x, y, z) \cdot \mathbf{T}(s) = M(x, y, z)\frac{dx}{ds} + N(x, y, z)\frac{dy}{ds} + P(x, y, z)\frac{dz}{ds}.$$

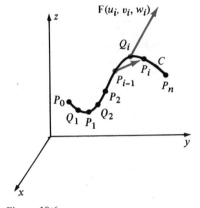

Figure 18.6

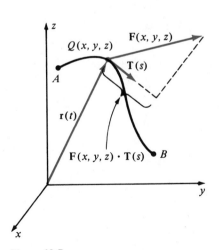

Figure 18.7

Note that this component is positive if the angle between $\mathbf{F}(x, y, z)$ and $\mathbf{T}(s)$ is acute, whereas it is negative if this angle is obtuse. Formula (18.12) may now be rewritten

$$W = \int_C \mathbf{F}(x, y, z) \cdot \mathbf{T}(s)\, ds$$

that is, *the work done as the point of application of $\mathbf{F}(x, y, z)$ moves along C equals the line integral of the tangential component of $\mathbf{F}$ along C.* To simplify the notation we shall denote $\mathbf{F}(x, y, z)$ by $\mathbf{F}$, $\mathbf{T}(s)$ by $\mathbf{T}$, and formally let

$$d\mathbf{r} = dx\mathbf{i} + dy\mathbf{j} + dz\mathbf{k} = \mathbf{T}\, ds.$$

With this understanding, our discussion may be summarized as follows.

Definition (18.13)

> Let C be a smooth space curve, and let $\mathbf{T}$ be a unit tangent vector to C at (x, y, z). If $\mathbf{F}$ is the force acting at (x, y, z), then the **work W done by F along C** is
>
> $$W = \int_C \mathbf{F} \cdot \mathbf{T}\, ds = \int_C \mathbf{F} \cdot d\mathbf{r}$$
>
> where $\mathbf{r} = x\mathbf{i} + y\mathbf{j} + z\mathbf{k}$.

Intuitively, we may regard $\mathbf{F} \cdot d\mathbf{r}$ in Definition (18.13) as representing the work done as the point of application of $\mathbf{F}$ moves along the tangent vector $d\mathbf{r}$ to C. The integral sign represents the limit of a sum of such elements of work.

Example 6   If an inverse force field $\mathbf{F}$ is given by

$$\mathbf{F}(x, y, z) = \frac{k}{|\mathbf{r}|^3}\, \mathbf{r}$$

find the work done by $\mathbf{F}$ as its point of application moves along the x-axis from $A(1, 0, 0)$ to $B(2, 0, 0)$.

Solution   Let C denote the line segment from A to B. Parametric equations for C are $x = t$, $y = 0$, $z = 0$, where $1 \le t \le 2$. Since

$$\mathbf{F}(x, y, z) = \frac{k}{(x^2 + y^2 + z^2)^{3/2}}\, (x\mathbf{i} + y\mathbf{j} + z\mathbf{k}),$$

it follows from Definition (18.13) that

$$W = \int_C \mathbf{F} \cdot d\mathbf{r} = \int_C \frac{k}{(x^2 + y^2 + z^2)^{3/2}}\, (x\, dx + y\, dy + z\, dz).$$

Substituting for x, y, and z from the parametric equations for C gives us

$$W = \int_1^2 \frac{k}{(t^2)^{3/2}} \, t \, dt = \int_1^2 \frac{k}{t^2} \, dt = -\frac{k}{t}\Big]_1^2 = \frac{k}{2}.$$

The units for W depend on those for distance and $|\mathbf{F}(x, y, z)|$. ∎

If we restrict our discussion to two dimensions, then a force field may be expressed in the form

$$\mathbf{F}(x, y) = M(x, y)\mathbf{i} + N(x, y)\mathbf{j}.$$

If C is a finite piecewise smooth plane curve, then the work done as the point of application of $\mathbf{F}(x, y)$ moves along C is

$$W = \int_C M(x, y) \, dx + N(x, y) \, dy = \int_C \mathbf{F} \cdot d\mathbf{r}.$$

Example 7 Let C be the part of the parabola $y = x^2$ between $A(0, 0)$ and $B(3, 9)$. If $\mathbf{F}(x, y) = -y\mathbf{i} + x\mathbf{j}$, find the work done by $\mathbf{F}$ along C from A to B.

Solution The vector field $\mathbf{F}$ was discussed in Example 1 of Section 18.1. Some typical vectors are sketched in (i) of Figure 18.2. Parametric equations for C (from A to B) are $x = t$, $y = t^2$, where $0 \leq t \leq 3$. Hence

$$W = \int_C \mathbf{F} \cdot d\mathbf{r} = \int_C -y \, dx + x \, dy$$

$$= \int_0^3 -t^2 \, dt + t(2t) \, dt = \int_0^3 t^2 \, dt = \frac{t^3}{3}\Big]_0^3 = 9.$$

If, for example, $|\mathbf{F}|$ is measured in pounds and s is in feet, then $W = 9$ ft-lb. ∎

EXERCISES 18.2

In Exercises 1 and 2 evaluate the line integrals $\int_C f(x, y) \, ds$, $\int_C f(x, y) \, dx$, and $\int_C f(x, y) \, dy$ if C is defined parametrically as indicated.

1 $f(x, y) = x^3 + y$; $x = 3t$, $y = t^3$; $0 \leq t \leq 1$

2 $f(x, y) = xy^{2/5}$; $x = \frac{1}{2}t$, $y = t^{5/2}$; $0 \leq t \leq 1$

In Exercises 3 and 4 evaluate the line integral along the given curve C and then verify directly that reversing the direction of integration changes the sign of the integral.

3 $\int_C 6x^2y \, dx + xy \, dy$ where C is the graph of $y = x^3 + 1$ from $(-1, 0)$ to $(1, 2)$.

4 $\int_C y \, dx + (x + y) \, dy$ where C is the graph of $y = x^2 + 2x$ from $(0, 0)$ to $(2, 8)$.

5 Evaluate $\int_C (x - y) \, dx + x \, dy$ where C is the graph of $y^2 = x$ from $(4, -2)$ to $(4, 2)$.

6 Evaluate $\int_C xy \, dx + x^2y^3 \, dy$ where C is the graph of $x = y^3$ from $(0, 0)$ to $(1, 1)$.

7 Evaluate $\int_C xy \, dx + (x + y) \, dy$ for each of the following curves C from $(0, 0)$ to $(1, 3)$.

(a) C consists of line segments from $(0, 0)$ to $(1, 0)$ and from $(1, 0)$ to $(1, 3)$.

(b) C consists of line segments from $(0, 0)$ to $(0, 3)$ and from $(0, 3)$ to $(1, 3)$.

(c) C is the line segment from $(0, 0)$ to $(1, 3)$.

(d) C is the part of the parabola $y = 3x^2$ from $(0, 0)$ to $(1, 3)$.

8 Evaluate $\int_C (x^2 + y^2)\,dx + 2x\,dy$ for each of the following curves C from $(1, 2)$ to $(-2, 8)$.

(a) C consists of line segments from $(1, 2)$ to $(1, 8)$ and from $(1, 8)$ to $(-2, 8)$.

(b) C consists of line segments from $(1, 2)$ to $(-2, 2)$ and from $(-2, 2)$ to $(-2, 8)$.

(c) C is the line segment from $(1, 2)$ to $(-2, 8)$.

(d) C is the graph of $y = 2x^2$ from $(1, 2)$ to $(-2, 8)$.

9 Evaluate $\int_C xz\,dx + (y + z)\,dy + x\,dz$ if C is given by $x = e^t$, $y = e^{-t}$, $z = e^{2t}$; $0 \le t \le 1$.

10 Evaluate $\int_C y\,dx + z\,dy + x\,dz$ if C is given by $x = \sin t$, $y = 2 \sin t$, $z = \sin^2 t$; $0 \le t \le \pi/2$.

11 Evaluate

$$\int_C (x + y + z)\,dx + (x - 2y + 3z)\,dy + (2x + y - z)\,dz$$

for each of the following curves C from $(0, 0, 0)$ to $(2, 3, 4)$.

(a) C consists of three line segments, the first parallel to the x-axis, the second parallel to the y-axis, and the third parallel to the z-axis.

(b) C consists of three line segments, the first parallel to the z-axis, the second parallel to the x-axis, and the third parallel to the y-axis.

(c) C is a line segment.

12 Evaluate $\int_C (x - y)\,dx + (y - z)\,dy + x\,dz$ where the curve C from $(1, -2, 3)$ to $(-4, 5, 2)$ is of the type described in parts (a)–(c) of Exercise 11.

13 Evaluate $\int_C xyz\,ds$, if C is the line segment from $(0, 0, 0)$ to $(1, 2, 3)$.

14 Evaluate $\int_C (xy + z)\,ds$ if C is the helix $x = a \cos t$, $y = a \sin t$, $z = bt$; $0 \le t \le 2\pi$.

15 If a thin wire has the shape of a plane curve C, and if the (linear) density at (x, y) is $f(x, y)$, then the **moments with respect to the x- and y-axes** are defined by

$$M_x = \int_C yf(x, y)\,ds \quad \text{and} \quad M_y = \int_C xf(x, y)\,ds.$$

Use limits of sums to show that these are "natural" definitions for M_x and M_y. The center of mass $(\bar{x}, \bar{y})$ of the wire is defined by $\bar{x} = M_y/M$ and $\bar{y} = M_x/M$, where M is the mass of the wire. Find the center of mass of the wire in Example 2 of this section.

16 Refer to Exercise 15. A thin wire is situated in a coordinate plane such that its shape coincides with the part of the parabola $y = 4 - x^2$ between $(-2, 0)$ and $(2, 0)$. Find the mass and center of mass if the density at the point (x, y) is directly proportional to its distance from the y-axis.

17 Extend the definitions of mass and center of mass of a wire to three dimensions (see Exercise 15).

18 A wire of constant density is bent into the shape of the helix $x = a \cos t$, $y = a \sin t$, $z = bt$, where $0 \le t \le 3\pi$. Find the mass and center of mass of the wire (see Exercise 17).

19 If the force at (x, y) is $\mathbf{F}(x, y) = xy^2\mathbf{i} + x^2y\mathbf{j}$, find the work done by $\mathbf{F}$ along the curves described in parts (a)–(d) of Exercise 7.

20 If the force at (x, y) is $\mathbf{F}(x, y) = (2x + y)\mathbf{i} + (x + 2y)\mathbf{j}$, find the work done by $\mathbf{F}$ along the curves described in parts (a)–(d) of Exercise 8.

21 The force acting at a point (x, y) in a coordinate plane is $\mathbf{F}(x, y) = (4/|\mathbf{r}|^3)\mathbf{r}$, where $\mathbf{r} = x\mathbf{i} + y\mathbf{j}$. Find the work done by $\mathbf{F}$ along the upper half of the circle $x^2 + y^2 = a^2$ from $(-a, 0)$ to $(a, 0)$.

22 The force at a point (x, y) in a coordinate plane is given by $\mathbf{F}(x, y) = (x^2 + y^2)\mathbf{i} + xy\mathbf{j}$. Find the work done by $\mathbf{F}(x, y)$ along the graph of $y = x^3$ from $(0, 0)$ to $(2, 8)$.

23 The force at a point (x, y, z) in three dimensions is given by $\mathbf{F}(x, y, z) = y\mathbf{i} + z\mathbf{j} + x\mathbf{k}$. Find the work done by $\mathbf{F}(x, y, z)$ along the curve $x = t$, $y = t^2$, $z = t^3$ from $(0, 0, 0)$ to $(2, 4, 8)$.

24 Solve Exercise 23 if $\mathbf{F}(x, y, z) = e^x\mathbf{i} + e^y\mathbf{j} + e^z\mathbf{k}$.

25 If a thin wire of variable density is represented by a curve C in the xy-plane, define the moments of inertia I_x and I_y with respect to the x- and y-axes, respectively. Find I_x and I_y for the wire in Example 2.

26 Find the moments of inertia I_x and I_y for the wire in Exercise 16.

27 If a thin wire of variable density is represented by a curve in three dimensions, define the moments of inertia with respect to the x-, y-, and z-axes.

28 Given the wire in Exercise 18, find the moment of inertia with respect to the z-axis (see also Exercise 27).

29 If an object moves through a force field $\mathbf{F}$ such that at each point (x, y, z) its velocity vector is orthogonal to $\mathbf{F}(x, y, z)$, show that the work done by $\mathbf{F}$ on the object is 0.

30 If a constant force $\mathbf{c}$ acts on a moving object as it travels once around a circle, show that the work done by $\mathbf{c}$ on the object is 0.

31 Discuss the significance of the case where the work integral $\int_C \mathbf{F} \cdot d\mathbf{r}$ in Definition (18.13) is negative.

32 Show that (14.29) is a special case of (18.13).

18.3

INDEPENDENCE OF PATH

A piecewise smooth curve that connects two points A and B is sometimes called a **path** from A to B. We shall now obtain conditions under which a line integral is **independent of path** in a region, in the sense that if A and B are arbitrary points, then the same value is obtained for *every* path from A to B. The results will be established for line integrals in two dimensions. Proofs for the three-dimensional case are similar and will be omitted.

If the line integral $\int_C f(x, y)\, ds$ is independent of path we sometimes denote it by $\int_A^B f(x, y)\, ds$, since the value of the integral depends only on the endpoints A and B of the curve C. A similar notation is used for $\int_C f(x, y)\, dx$, $\int_C f(x, y)\, dy$, and for line integrals in three dimensions.

It will be assumed throughout this section that all regions in the plane are **connected**. This means that any two points in a region can be joined by a piecewise smooth curve that lies in the region. We shall also assume that regions are *open*, that is, for every point A in a region D there exists a circle with center A that lies completely in D.

The next theorem gives us the fundamental result that if a vector function $\mathbf{F}$ is continuous on D, then the line integral $\int_C \mathbf{F} \cdot d\mathbf{r}$ *is independent of path if and only if $\mathbf{F}$ is conservative*.

Theorem (18.14)

> If $\mathbf{F}(x, y) = M(x, y)\mathbf{i} + N(x, y)\mathbf{j}$ is continuous on an open connected region, then the line integral $\int_C \mathbf{F} \cdot d\mathbf{r}$ is independent of path if and only if $\mathbf{F}(x, y) = \nabla f(x, y)$ for some function f.

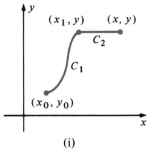

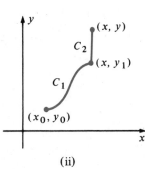

Figure 18.8

Proof   Suppose the integral is independent of path. If (x_0, y_0) is a fixed point in D, let f be defined by

$$f(x, y) = \int_{(x_0, y_0)}^{(x, y)} \mathbf{F} \cdot d\mathbf{r}$$

where (x, y) is any point in D. Since the integral is independent of path, f depends only on x and y, and not on the path C from (x_0, y_0) to (x, y). Choose a circle in D with center (x, y) and let (x_1, y) be a point within the circle such that $x_1 \neq x$. Let C_1 be any path from (x_0, y_0) to (x_1, y) and let C_2 be the horizontal segment from (x_1, y) to (x, y), as illustrated in (i) of Figure 18.8. We may, therefore, write

$$f(x, y) = \int_{C_1} \mathbf{F} \cdot d\mathbf{r} + \int_{C_2} \mathbf{F} \cdot d\mathbf{r} = \int_{(x_0, y_0)}^{(x_1, y)} \mathbf{F} \cdot d\mathbf{r} + \int_{(x_1, y)}^{(x, y)} \mathbf{F} \cdot d\mathbf{r}.$$

Since the first integral does not depend on x,

$$\frac{\partial}{\partial x} f(x, y) = 0 + \frac{\partial}{\partial x} \int_{(x_1, y)}^{(x, y)} \mathbf{F} \cdot d\mathbf{r}.$$

Writing $\mathbf{F} \cdot d\mathbf{r} = M(x, y)\,dx + N(x, y)\,dy$ and using the fact that $dy = 0$ on C_2 gives us

$$\frac{\partial}{\partial x} f(x, y) = \frac{\partial}{\partial x} \int_{(x_1, y)}^{(x, y)} M(x, y)\,dx.$$

Since y is fixed in this partial differentiation we may regard the integrand as a function of one variable x. Applying, (5.24),

$$\frac{\partial}{\partial x} f(x, y) = M(x, y).$$

Similarly, if we choose the path shown in (ii) of Figure 18.8 and differentiate with respect to y, we obtain

$$\frac{\partial}{\partial y} f(x, y) = N(x, y).$$

This proves that $\nabla f(x, y) = M(x, y)\mathbf{i} + N(x, y)\mathbf{j} = \mathbf{F}(x, y)$.

Conversely, if there exists a function f such that $\mathbf{F}(x, y) = \nabla f(x, y)$, then

$$M(x, y)\mathbf{i} + N(x, y)\mathbf{j} = f_x(x, y)\mathbf{i} + f_y(x, y)\mathbf{j}$$

and hence

$$M(x, y) = f_x(x, y) \quad \text{and} \quad N(x, y) = f_y(x, y).$$

Consequently, if $A(x_1, y_1)$ and $B(x_2, y_2)$ are points in D, then

$$\int_C \mathbf{F} \cdot d\mathbf{r} = \int_C f_x(x, y)\,dx + f_y(x, y)\,dy$$

where C is any path from A to B. If C is smooth and is given parametrically by $x = g(t)$, $y = h(t)$ where $t_1 \le t \le t_2$, then substituting in the integrand,

$$\int_C \mathbf{F} \cdot d\mathbf{r} = \int_{t_1}^{t_2} [f_x(g(t), h(t))g'(t) + f_y(g(t), h(t))h'(t)]\,dt.$$

Applying the Chain Rule (16.19) and the Fundamental Theorem of Calculus,

$$\begin{aligned} \int_C \mathbf{F} \cdot d\mathbf{r} &= \int_{t_1}^{t_2} \frac{d}{dt} [f(g(t), h(t))]\,dt \\ &= f(g(t_2), h(t_2)) - f(g(t_1), h(t_1)) \\ &= f(x_2, y_2) - f(x_1, y_1). \end{aligned}$$

Thus the line integral depends only on the coordinates of A and B, not on the path C, which is what we wished to prove. The proof can be extended to piecewise smooth curves by subdividing C into a finite number of smooth curves. $\square$

The proof of the converse of Theorem (18.14) exhibits a technique for evaluating line integrals that are independent of path. We may state this result as follows.

Theorem (18.15)

Let $\mathbf{F}(x, y) = M(x, y)\mathbf{i} + N(x, y)\mathbf{j}$ be continuous on an open connected region D, and let C be a piecewise smooth curve in D with endpoints $A(x_1, y_1)$ and $B(x_2, y_2)$. If $\mathbf{F}(x, y) = \nabla f(x, y)$, then

$$\int_C M(x, y)\, dx + N(x, y)\, dy = \int_{(x_1, y_1)}^{(x_2, y_2)} \mathbf{F} \cdot d\mathbf{r}$$
$$= f(x_2, y_2) - f(x_1, y_1) = f(x, y)\Big]_{(x_1, y_1)}^{(x_2, y_2)}$$

This evaluation theorem for line integrals is analogous to the Fundamental Theorem of Calculus.

In terms of conservative force fields, Theorem (18.15) implies that the work done along any path C from A to B equals the difference in the potentials between A and B. It follows that if C is a closed curve, that is, if $A = B$, then the work done in traversing C is 0. It can be shown that a converse of this result is also true, namely, if $\int_C \mathbf{F} \cdot d\mathbf{r}$ is 0 for every simple closed curve C, then the line integral is independent of path and hence the field is conservative. These facts are important in applications, since many of the vector fields that occur in nature (such as inverse square fields) are conservative. In physical terms, if a unit particle moves completely around a closed curve in a conservative force field, and if there is no loss of energy due to friction, then the work done is 0. More will be said about conservative fields at the end of this section.

Before considering an example, let us note another important byproduct of Theorem (18.14). If the integral $\int_C M(x, y)\, dx + N(x, y)\, dy$ is independent of path, then by Theorem (18.14), there is a function f such that

$$M = \frac{\partial f}{\partial x} \quad \text{and} \quad N = \frac{\partial f}{\partial y}.$$

Consequently,

$$\frac{\partial M}{\partial y} = \frac{\partial^2 f}{\partial y\, \partial x} \quad \text{and} \quad \frac{\partial N}{\partial x} = \frac{\partial^2 f}{\partial x\, \partial y}.$$

If M and N have continuous first partial derivatives, then f has continuous *second* partial derivatives, and hence the order of differentiation is immaterial, that is

$$\frac{\partial M}{\partial y} = \frac{\partial N}{\partial x}.$$

The converse of this result is false, unless we place additional restrictions on the domain D of $\mathbf{F}(x, y)$. In particular, if D is a **simply connected region**, in the sense that every simple closed curve C in D encloses only points in D (that is, there are no "holes" in the region), then the condition $\partial M/\partial y = \partial N/\partial x$ implies that the line integral $\int_C M(x, y)\, dx + N(x, y)\, dy$ is independent of path (a proof may be found in books on advanced calculus). Our discussion can be summarized as follows.

Theorem (18.16)

> If $M(x, y)$ and $N(x, y)$ have continuous first partial derivatives on a simply connected region D, then
>
> $$\int_C M(x, y)\, dx + N(x, y)\, dy$$
>
> is independent of path in D if and only if
>
> $$\frac{\partial M}{\partial y} = \frac{\partial N}{\partial x}.$$

Example 1 If $\mathbf{F}(x, y) = (2x + y^3)\mathbf{i} + (3xy^2 + 4)\mathbf{j}$, show that $\int_C \mathbf{F} \cdot d\mathbf{r}$ is independent of path and evaluate $\displaystyle\int_{(0, 1)}^{(2, 3)} \mathbf{F} \cdot d\mathbf{r}$.

Solution   The vector function $\mathbf{F}$ has continuous first partial derivatives for all (x, y), and hence Theorem (18.16) is applicable. Letting $M = 2x + y^3$ and $N = 3xy^2 + 4$, we see that

$$\frac{\partial M}{\partial y} = 3y^2 = \frac{\partial N}{\partial x}.$$

Thus the line integral is independent of path. It follows from Theorem (18.14) that there exists a (potential) function f such that

$$f_x(x, y) = 2x + y^3 \quad \text{and} \quad f_y(x, y) = 3xy^2 + 4.$$

If we (partially) integrate $f_x(x, y)$ with respect to x we obtain

$$f(x, y) = x^2 + xy^3 + k(y)$$

where k is a function of y alone. Differentiating with respect to y and comparing with $f_y(x, y) = 3xy^2 + 4$ gives us

$$f_y(x, y) = 3xy^2 + k'(y) = 3xy^2 + 4.$$

Consequently, $k'(y) = 4$, or $k(y) = 4y + c$ for some constant c. Thus

$$f(x, y) = x^2 + xy^3 + 4y + c$$

defines a function of the desired type. Applying Theorem (18.15),

$$\int_{(0,1)}^{(2,3)} (2x + y^3)\, dx + (3xy^2 + 4)\, dy = x^2 + xy^3 + 4y \Big]_{(0,1)}^{(2,3)}$$

$$= (4 + 54 + 12) - 4 = 66$$

where we have dropped the constant c, since *any* potential function f may be used to evaluate the integral. ∎

Example 2 Determine whether $\int_C x^2 y\, dx + 3xy^2\, dy$ is independent of path.

Solution   If we let $M = x^2 y$ and $N = 3xy^2$, then

$$\frac{\partial M}{\partial y} = x^2 \quad \text{and} \quad \frac{\partial N}{\partial x} = 3y^2.$$

Since $\partial M / \partial y \neq \partial N / \partial x$, the integral is not independent of path by Theorem (18.16). ∎

In Example 1 we determined a potential function f for a two-dimensional conservative vector field. The solution of the next example illustrates a technique that can be used in three dimensions.

Example 3   If $\mathbf{F}(x, y, z) = y^2 \cos x\, \mathbf{i} + (2y \sin x + e^{2z})\mathbf{j} + 2ye^{2z}\mathbf{k}$, prove that $\int_C \mathbf{F} \cdot d\mathbf{r}$ is independent of path and find a potential function f for $\mathbf{F}$. Find the work done by $\mathbf{F}$ along any curve C from $(0, 1, \frac{1}{2})$ to $(\pi/2, 3, 2)$.

Solution   The integral is independent of path if there exists a differentiable function f of x, y, and z such that $\nabla f(x, y, z) = \mathbf{F}(x, y, z)$; that is, if

(a)
$$f_x(x, y, z) = y^2 \cos x$$
$$f_y(x, y, z) = 2y \sin x + e^{2z}$$
$$f_z(x, y, z) = 2ye^{2z}.$$

If we (partially) integrate $f_x(x, y, z)$ with respect to x we see that

(b)
$$f(x, y, z) = y^2 \sin x + g(y, z)$$

for some function g of y and z. Differentiating with respect to y and comparing with the equation for f_y in (a) gives us

$$f_y(x, y, z) = 2y \sin x + g_y(y, z) = 2y \sin x + e^{2z}$$

and hence

$$g_y(y, z) = e^{2z}.$$

Integrating with respect to y, we obtain

$$g(y, z) = ye^{2z} + k(z)$$

where k is a function of z alone. Consequently, from equation (b),

$$f(x, y, z) = y^2 \sin x + ye^{2z} + k(z).$$

Differentiating with respect to z, and using the equation for f_z in (a),

$$f_z(x, y, z) = 2ye^{2z} + k'(z) = 2ye^{2z}.$$

It follows that $k'(z) = 0$ or $k(z) = c$ for some constant c. Thus

$$f(x, y, z) = y^2 \sin x + ye^{2z} + c$$

defines a potential function for $\mathbf{F}$.

Using the extension of Theorem (18.15) to three dimensions, the work done by $\mathbf{F}$ along any curve C from $(0, 1, \frac{1}{2})$ to $(\pi/2, 3, 2)$ is

$$\int_C \mathbf{F} \cdot d\mathbf{r} = y^2 \sin x + ye^{2z} \Big]_{(0, 1, 1/2)}^{(\pi/2, 3, 2)}$$

$$= (9 + 3e^4) - (0 + e) = 9 + 3e^4 - e. \qquad \blacksquare$$

Let $\mathbf{F}(x, y, z)$ be a conservative vector field with potential function f. In physics, the potential energy $p(x, y, z)$ of a particle at the point (x, y, z) is defined by $p(x, y, z) = -f(x, y, z)$ and, therefore,

$$\mathbf{F}(x, y, z) = -\nabla p(x, y, z).$$

If A and B are two points, then as in Theorem (18.15), the work done by $\mathbf{F}$ along a smooth curve C with endpoints A and B is

$$W = \int_A^B \mathbf{F} \cdot d\mathbf{r} = -p(x, y, z) \Big]_A^B = p(A) - p(B)$$

where $p(A)$ and $p(B)$ denote the potential energy at A and B, respectively. Thus, the work W equals the difference in the potential energies at A and B. In particular, if B is a point at which the potential is 0, then $W = p(A)$. This is in agreement with the classical physical description of potential energy as the type of energy that a body has by virtue of its position.

Now suppose that the particle moves from A to B along the curve C such that its position at time t is given by $x = g(t)$, $y = h(t)$, $z = k(t)$, where $a \leq t \leq b$. If $\mathbf{r} = x\mathbf{i} + y\mathbf{j} + z\mathbf{k}$ is the position vector of the particle, then the velocity and acceleration at time t are

$$\mathbf{v} = \frac{d\mathbf{r}}{dt} \quad \text{and} \quad \mathbf{a} = \frac{d\mathbf{v}}{dt}$$

where we have used the abbreviations $\mathbf{v}$ and $\mathbf{a}$ for $\mathbf{v}(t)$ and $\mathbf{a}(t)$, respectively. The speed of the particle is $v(t) = v = |\mathbf{v}|$. In this event, the work done by $\mathbf{F}(x, y, z)$ along C is

$$W = \int_A^B \mathbf{F} \cdot d\mathbf{r} = \int_a^b \left(\mathbf{F} \cdot \frac{d\mathbf{r}}{dt} \right) dt = \int_a^b (\mathbf{F} \cdot \mathbf{v}) \, dt.$$

By Newton's Second Law of Motion, $\mathbf{F} = m\mathbf{a} = m(d\mathbf{v}/dt)$ and hence

$$W = \int_a^b m\left(\frac{d\mathbf{v}}{dt} \cdot \mathbf{v}\right) dt = \int_a^b m \frac{1}{2} \frac{d}{dt} (\mathbf{v} \cdot \mathbf{v}) \, dt$$

$$= \frac{1}{2} m \int_a^b \frac{d}{dt} (v^2) \, dt = \frac{1}{2} mv^2 \Big]_a^b$$

$$= \tfrac{1}{2}m[v(b)]^2 - \tfrac{1}{2}m[v(a)]^2.$$

Since the kinetic energy of a particle of mass m and velocity v is $\frac{1}{2}mv^2$, the last equality may be written

$$W = K(B) - K(A)$$

where $K(A)$ and $K(B)$ denote the kinetic energy of the particle at A and B, respectively. We have shown that

$$W = p(A) - p(B) = K(B) - K(A)$$

or $\qquad\qquad\qquad p(A) + K(A) = p(B) + K(B).$

The last formula states that if a particle moves from one point to another in a conservative vector field, then the sum of the potential and kinetic energies remains constant; that is, the *total* energy does not change. This is known as the *Law of Conservation of Energy*, and is the reason for calling the vector field *conservative*.

EXERCISES 18.3

In Exercises 1–10 determine whether or not $\int_C \mathbf{F} \cdot d\mathbf{r}$ is independent of path. If it is, find a potential function f for $\mathbf{F}$.

1 $\mathbf{F}(x, y) = (3x^2y + 2)\mathbf{i} + (x^3 + 4y^3)\mathbf{j}$

2 $\mathbf{F}(x, y) = (6x^2 - 2xy^2)\mathbf{i} + (2x^2y + 5)\mathbf{j}$

3 $\mathbf{F}(x, y) = e^x\mathbf{i} + (3 - e^x \sin y)\mathbf{j}$

4 $\mathbf{F}(x, y) = (x^2 + y^2)^{-1/2}(-y\mathbf{i} + x\mathbf{j})$

5 $\mathbf{F}(x, y) = 4xy^3\mathbf{i} + 2xy^3\mathbf{j}$

6 $\mathbf{F}(x, y) = y^3 \cos x\mathbf{i} - 3y^2 \sin x\mathbf{j}$

7 $\mathbf{F}(x, y, z) = (y \sec^2 x - ze^x)\mathbf{i} + \tan x\mathbf{j} - e^x\mathbf{k}$

8 $\mathbf{F}(x, y, z) = (y + z)\mathbf{i} + (x + z)\mathbf{j} + (x + y)\mathbf{k}$

9 $\mathbf{F}(x, y, z) = 8xz\mathbf{i} + (1 - 6yz^3)\mathbf{j} + (4x^2 - 9y^2z^2)\mathbf{k}$

10 $\mathbf{F}(x, y, z) = e^{-z}\mathbf{i} + 2y\mathbf{j} + xe^{-z}\mathbf{k}$

In Exercises 11–14 prove that the line integral is independent of path and find its value.

11 $\displaystyle\int_{(-1, 2)}^{(3, 1)} (y^2 + 2xy) \, dx + (x^2 + 2xy) \, dy$

12 $\displaystyle\int_{(0, 0)}^{(1, \pi/2)} e^x \sin y \, dx + e^x \cos y \, dy$

13 $\displaystyle\int_{(1, 0, 2)}^{(-2, 1, 3)} (6xy^3 + 2z^2) \, dx + 9x^2y^2 \, dy + (4xz + 1) \, dz$

14 $\displaystyle\int_{(4, 0, 3)}^{(-1, 1, 2)} (yz + 1) \, dx + (xz + 1) \, dy + (xy + 1) \, dz$

15 Suppose a force $\mathbf{F}(x, y, z)$ is directed toward the origin with a magnitude that is inversely proportional to the distance from the origin. Prove that $\mathbf{F}$ is conservative and find a potential function for $\mathbf{F}$.

16 Suppose a force $\mathbf{F}(x, y, z)$ is directed away from the origin with a magnitude that is directly proportional to the distance from the origin. Prove that $\mathbf{F}$ is conservative and find a potential function for $\mathbf{F}$.

17 If $\int_C M(x, y, z) \, dx + N(x, y, z) \, dy + P(x, y, z) \, dz$ is independent of path and M, N, and P have continuous first partial derivatives, prove that

$$\frac{\partial M}{\partial y} = \frac{\partial N}{\partial x}, \quad \frac{\partial M}{\partial z} = \frac{\partial P}{\partial x}, \quad \frac{\partial N}{\partial z} = \frac{\partial P}{\partial y}.$$

18 Prove the analogue of Theorem (18.14) for line integrals in three dimensions.

Use Exercise 17 to prove that the line integrals in Exercises 19 and 20 are not independent of path.

19 $\int_C 2xy \, dx + (x^2 + z^2) \, dy + yz \, dz$

20 $\int_C e^y \cos z \, dx + xe^y \cos z \, dy + xe^y \sin z \, dz$

21 If $\mathbf{F}$ is a constant force, prove that the work done along any curve with endpoints P and Q is $\mathbf{F} \cdot \overline{PQ}$.

22 If $\mathbf{F}(x, y, z) = g(x)\mathbf{i} + h(y)\mathbf{j} + k(z)\mathbf{k}$, where g, h, and k are continuous functions of one variable, show that $\mathbf{F}$ is conservative.

23 Let $\mathbf{F}(x, y) = M(x, y)\mathbf{i} + N(x, y)\mathbf{j}$, where

$$M(x, y) = -y/(x^2 + y^2) \quad \text{and} \quad N(x, y) = x/(x^2 + y^2).$$

Show that $\partial M/\partial y = \partial N/\partial x$ for all (x, y) in the domain D of $\mathbf{F}$, but that $\int_C \mathbf{F} \cdot d\mathbf{r}$ is not independent of path in D. (*Hint*: Consider two different semicircles with endpoints $(-1, 0)$ and $(1, 0)$.) Why doesn't this contradict Theorem (18.16)?

24 Suppose a satellite of mass m is orbiting the earth at a constant altitude of h miles, and that it completes one revolution every k minutes. If the radius of the earth is taken as 4,000 miles, find the work done by the gravitational field of the earth during any time interval.

25 Let $\mathbf{F}$ be an inverse square field such that $\mathbf{F}(x, y, z) = (c/|\mathbf{r}|^3)\mathbf{r}$, where $\mathbf{r} = x\mathbf{i} + y\mathbf{j} + z\mathbf{k}$ and c is a constant. Let P_1 and P_2 be any points whose distances from the origin are d_1 and d_2, respectively. Express the work done by $\mathbf{F}$ along any piecewise smooth curve joining P_1 to P_2 in terms of d_1 and d_2.

26 Suppose that as a particle moves through a conservative vector field, its potential energy is decreasing at a rate of k units per second. At what rate is its kinetic energy changing?

18.4

GREEN'S THEOREM

Let a smooth plane curve C have parametric representation $x = g(t)$, $y = h(t)$, where $a \le t \le b$. Recall that if $A = (g(a), h(a)) = (g(b), h(b)) = B$, then C is a smooth *closed* curve. If, in addition, $(g(t_1), h(t_1)) \ne (g(t_2), h(t_2))$ for all other numbers t_1, t_2 in $[a, b]$, then C does not cross itself between A and B and it is referred to as *simple*. (See Section 13.1.) Some common examples of smooth simple closed curves are circles and ellipses. A *piecewise smooth simple closed curve* consists of a finite union of smooth curves C_i such that as t varies from a to b, the point $P(t)$ obtained from parametric representations of the C_i traces C exactly one time, with the exception that $P(a) = P(b)$. A curve of this type forms the boundary of a region R in the plane and, by definition, the positive direction along C is such that R is on the left as $P(t)$ traces C. This is illustrated in Figure 18.9, where the arrows indicate the positive direction along C. The symbol

$$\oint_C M(x, y) \, dx + N(x, y) \, dy$$

will denote a line integral along C once in the positive direction.

The next theorem, named after the English mathematical physicist George Green (1793–1841), indicates the relationship between the line integral around C and a double integral over R. To simplify the statement, the symbols M, N, $\partial M/\partial x$, and $\partial N/\partial y$ are used in the integrands to denote the values of these functions at (x, y).

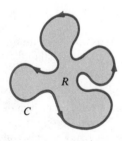

Figure 18.9

Green's Theorem (*18.17*)

Let C be a piecewise smooth simple closed curve and let R be the region consisting of C and its interior. If M and N are functions that are continuous and have continuous first partial derivatives throughout an open region D containing R, then

$$\oint_C M \, dx + N \, dy = \iint_R \left(\frac{\partial N}{\partial x} - \frac{\partial M}{\partial y} \right) dA.$$

Partial proof We shall prove the theorem for a region R that is both of Type I and of Type II. Thus we may write

$$R = \{(x, y): a \le x \le b, g_1(x) \le y \le g_2(x)\}$$

and

$$R = \{(x, y): c \le y \le d, h_1(y) \le x \le h_2(y)\}$$

where the functions g_1, g_2, h_1, and h_2 are smooth. An illustration of a region of this type appears in Figure 18.10. It is sufficient to show that each of the following is true:

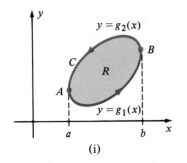

(i)

(a)

$$\oint_C M \, dx = - \iint_R \frac{\partial M}{\partial y} \, dA$$

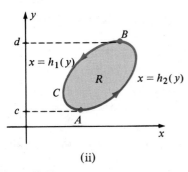

(ii)

Figure 18.10

(b)

$$\oint_C N \, dy = \iint_R \frac{\partial N}{\partial x} \, dA.$$

Note that addition of these integrals produces the desired conclusion.

To prove (a) we refer to (i) of Figure 18.10 and note that C consists of two curves C_1 and C_2 that have equations $y = g_1(x)$ and $y = g_2(x)$, respectively. The line integral $\oint_C M \, dx$ may be written

$$\oint_C M \, dx = \int_{C_1} M(x, y) \, dx + \int_{C_2} M(x, y) \, dx$$

$$= \int_a^b M(x, g_1(x)) \, dx + \int_b^a M(x, g_2(x)) \, dx$$

$$= \int_a^b M(x, g_1(x)) \, dx - \int_a^b M(x, g_2(x)) \, dx.$$

Applying (i) of Theorem (17.9) to the double integral $\iint_R (\partial M/\partial y) \, dA$ gives us

$$\iint_R \frac{\partial M}{\partial y} \, dA = \int_a^b \int_{g_1(x)}^{g_2(x)} \frac{\partial M}{\partial y} \, dy \, dx$$

$$= \int_a^b M(x, y) \Big]_{g_1(x)}^{g_2(x)} dx$$

$$= \int_a^b [M(x, g_2(x)) - M(x, g_1(x))] \, dx.$$

Comparing the last expression with that obtained for $\oint_C M\,dx$ gives us (a). The formula in (b) may be established in similar fashion by referring to (ii) of Figure 18.10. This is left as an exercise.   □

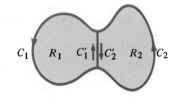

Figure 18.11

Although we shall not prove it, Green's Theorem is true for regions where part of the boundary consists of horizontal or vertical line segments. The theorem may then be extended to the case where R is a finite union of such regions. For example if, as illustrated in Figure 18.11, $R = R_1 \cup R_2$, where the boundary of R_1 is $C_1 \cup C_1'$ and the boundary of R_2 is $C_2 \cup C_2'$, then

$$\iint\limits_{R_1} \left(\frac{\partial N}{\partial x} - \frac{\partial M}{\partial y}\right) dy\,dx = \oint_{C_1 \cup C_1'} M\,dx + N\,dy$$

$$\iint\limits_{R_2} \left(\frac{\partial N}{\partial x} - \frac{\partial M}{\partial y}\right) dy\,dx = \oint_{C_2 \cup C_2'} M\,dx + N\,dy.$$

The sum of the two double integrals equals a double integral over R. We next observe that a line integral along C_1' in the direction indicated in Figure 18.11 is the negative of that along C_2' since the curve is the same but the directions are opposite one another. It follows that the sum of the two line integrals reduces to a line integral along $C_1 \cup C_2$, which is the boundary C of R. Thus

$$\iint\limits_{R} \left(\frac{\partial N}{\partial x} - \frac{\partial M}{\partial y}\right) dy\,dx = \oint_C M\,dx + N\,dy.$$

We may now proceed, by mathematical induction, to any finite union. The proof of Green's Theorem for the most general case is beyond the scope of this book.

Example 1   Use Green's Theorem to evaluate $\oint_C 5xy\,dx + x^3\,dy$, where C is the closed curve consisting of the graphs of $y = x^2$ and $y = 2x$ between the points $(0, 0)$ and $(2, 4)$.

Solution   The region R bounded by C is illustrated in Figure 18.12. Applying Green's Theorem, with $M(x, y) = 5xy$ and $N(x, y) = x^3$,

$$\oint_C 5xy\,dx + x^3\,dy = \iint\limits_{R}\left[\frac{\partial}{\partial x}(x^3) - \frac{\partial}{\partial y}(5xy)\right] dA$$

$$= \int_0^2 \int_{x^2}^{2x} (3x^2 - 5x)\,dy\,dx$$

$$= \int_0^2 \Big[3x^2y - 5xy\Big]_{x^2}^{2x} dx$$

$$= \int_0^2 (11x^3 - 10x^2 - 3x^4)\,dx = -\frac{28}{15}.$$

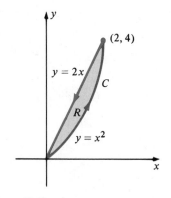

Figure 18.12

Of course, the line integral can also be evaluated directly.   ∎

Example 2 Use Green's Theorem to evaluate the line integral

$$\oint_C 2xy \, dx + (x^2 + y^2) \, dy$$

where C is the ellipse $4x^2 + 9y^2 = 36$.

Solution If R is the region bounded by C, then applying Green's Theorem with $M(x, y) = 2xy$ and $N(x, y) = x^2 + y^2$,

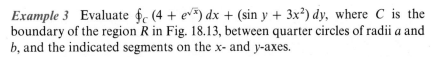

$$\oint_C 2xy \, dx + (x^2 + y^2) \, dy = \iint_R (2x - 2x) \, dA$$

$$= \iint_R 0 \, dA = 0. \qquad \blacksquare$$

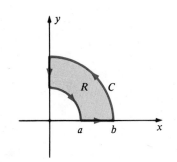

Figure 18.13

Example 3 Evaluate $\oint_C (4 + e^{\sqrt{x}}) \, dx + (\sin y + 3x^2) \, dy$, where C is the boundary of the region R in Fig. 18.13, between quarter circles of radii a and b, and the indicated segments on the x- and y-axes.

Solution Applying Green's Theorem, and then changing the variables in the double integral to polar coordinates, the given line integral equals

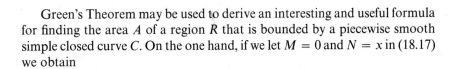

$$\iint_R (6x - 0) \, dA = \int_0^{\pi/2} \int_a^b (6r \cos \theta) r \, dr \, d\theta$$

$$= \int_0^{\pi/2} 2r^3 \cos \theta \Big]_a^b \, d\theta = \int_0^{\pi/2} 2(b^3 - a^3) \cos \theta \, d\theta$$

$$= 2(b^3 - a^3) \sin \theta \Big]_0^{\pi/2} = 2(b^3 - a^3).$$

The reader will gain a deeper appreciation of Green's Theorem by attempting to evaluate the line integral in this example directly. $\blacksquare$

Green's Theorem may be used to derive an interesting and useful formula for finding the area A of a region R that is bounded by a piecewise smooth simple closed curve C. On the one hand, if we let $M = 0$ and $N = x$ in (18.17) we obtain

$$A = \iint_R dA = \oint_C x \, dy.$$

On the other hand, if we let $M = -y$ and $N = 0$ the result is

$$A = \iint_R dA = -\oint_C y \, dx.$$

We may combine these two formulas for A by adding both sides of the equations and dividing by 2. This gives us the following result.

Theorem (18.18)

If a region R in the xy-plane is bounded by a piecewise smooth simple closed curve C, then the area A of R is

$$A = \frac{1}{2} \oint_C x \, dy - y \, dx.$$

Example 4   Use Theorem (18.18) to find the area of the ellipse $\dfrac{x^2}{a^2} + \dfrac{y^2}{b^2} = 1$.

Solution   Parametric equations for the ellipse C are $x = a \cos t, y = b \sin t$, where $0 \le t \le 2\pi$. Hence

$$A = \frac{1}{2} \oint_C x \, dy - y \, dx$$

$$= \frac{1}{2} \int_0^{2\pi} (a \cos t)(b \cos t) \, dt - (b \sin t)(-a \sin t) \, dt$$

$$= \frac{1}{2} \int_0^{2\pi} ab(\cos^2 t + \sin^2 t) \, dt$$

$$= \frac{ab}{2} \int_0^{2\pi} dt = \frac{ab}{2}(2\pi) = \pi ab. \qquad \blacksquare$$

Green's Theorem can be extended to a region R that contains holes, provided we integrate over the entire boundary and always keep the region R to the left. This is illustrated in Figure 18.14, where it can be shown that the double integral over R equals the sum of the line integrals along C_1 and C_2 in the indicated directions. The proof consists of making a slit in R as illustrated in the figure and then noting that the sum of two line integrals in opposite directions along the same curve is zero. A similar argument can be used if the region has several holes. We shall use this observation in the solution of the next example.

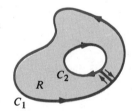

Figure 18.14

Example 5   Let C_1 and C_2 be two nonintersecting piecewise smooth simple closed curves having the origin O as an interior point. If $M = -y/(x^2 + y^2)$ and $N = x/(x^2 + y^2)$, prove that

$$\oint_{C_1} M \, dx + N \, dy = \oint_{C_2} M \, dx + N \, dy.$$

Solution   If R denotes the region between C_1 and C_2, then we have a situation similar to that illustrated in Figure 18.14, where O is inside of C_2.

By the remarks preceding this example,

$$\oint_{C_1} M\,dx + N\,dy + \oint_{C_2} M\,dx + N\,dy = \iint_R \left(\frac{\partial N}{\partial x} - \frac{\partial M}{\partial y}\right) dA,$$

where $\circlearrowright$ indicates the positive direction along C_2 *with respect to R.* Since

$$\frac{\partial N}{\partial x} = \frac{(x^2 + y^2)(1) - x(2x)}{(x^2 + y^2)^2} = \frac{y^2 - x^2}{(x^2 + y^2)^2} = \frac{\partial M}{\partial y}$$

the double integral over R is zero. Consequently,

$$\oint_{C_1} M\,dx + N\,dy = -\oint_{C_2} M\,dx + N\,dy.$$

Since the positive direction along C_2 *with respect to the region in the interior of* C_2 is opposite to that indicated in the last integral we have

$$\oint_{C_1} M\,dx + N\,dy = \oint_{C_2} M\,dx + N\,dy$$

which is what we wished to prove. ∎

Example 6 If $\mathbf{F}(x, y) = \dfrac{1}{x^2 + y^2}(-y\mathbf{i} + x\mathbf{j})$, prove that $\oint_C \mathbf{F} \cdot d\mathbf{r} = 2\pi$ for every piecewise smooth simple closed curve C having the origin in its interior.

Solution The line integral has the same form as those considered in Example 5. Hence, if we choose a circle C_1 of radius a with center at the origin that lies entirely within C, then by Example 5,

$$\oint_C \mathbf{F} \cdot d\mathbf{r} = \oint_{C_1} \mathbf{F} \cdot d\mathbf{r}.$$

Since parametric equations for C_1 are

$$x = a\cos t, \quad y = a\sin t; \quad 0 \le t \le 2\pi,$$

we have

$$\oint_C \mathbf{F} \cdot d\mathbf{r} = \oint_{C_1} \frac{-y}{x^2 + y^2}\,dx + \frac{x}{x^2 + y^2}\,dy$$

$$= \int_0^{2\pi} \frac{-a\sin t}{a^2}(-a\sin t)\,dt + \frac{a\cos t}{a^2}(a\cos t)\,dt$$

$$= \int_0^{2\pi} (\sin^2 t + \cos^2 t)\,dt = \int_0^{2\pi} dt = 2\pi. \qquad ∎$$

The conclusion of Green's Theorem (18.17) may be expressed in vector form. To see how this is accomplished, first note that if **F** is a vector field in two dimensions, then we may write

$$\mathbf{F}(x, y) = M\mathbf{i} + N\mathbf{j} + 0\mathbf{k}$$

where $M = M(x, y)$ and $N = N(x, y)$. The curl of **F** is

$$\nabla \times \mathbf{F} = \begin{vmatrix} \mathbf{i} & \mathbf{j} & \mathbf{k} \\ \dfrac{\partial}{\partial x} & \dfrac{\partial}{\partial y} & \dfrac{\partial}{\partial z} \\ M & N & 0 \end{vmatrix}$$

$$= 0\mathbf{i} + 0\mathbf{j} + \left(\frac{\partial N}{\partial x} - \frac{\partial M}{\partial y} \right) \mathbf{k}.$$

Observe that the coefficient of **k** has the same form as the double integral in (18.17). If we consider the unit tangent vector

$$\mathbf{T} = \frac{dx}{ds}\mathbf{i} + \frac{dy}{ds}\mathbf{j} + \frac{dz}{ds}\mathbf{k}$$

to C, where s represents arc length, then Green's Theorem may be written

(18.19)
$$\oint_C \mathbf{F} \cdot \mathbf{T} \, ds = \iint_R (\nabla \times \mathbf{F}) \cdot \mathbf{k} \, dA.$$

Since $(\nabla \times \mathbf{F}) \cdot \mathbf{k}$ is the component of curl **F** in the direction of the z-axis we shall refer to it as the **normal component** (to R) of curl **F**. In words, (18.19) may be phrased as follows: *the line integral of the tangential component of* **F** *taken along C once in the positive direction is equal to the double integral over R of the normal component of* curl **F**. The three-dimensional analogue of this result is *Stokes' Theorem*, which will be discussed in Section 18.7. Physical interpretations will be given at that time.

Green's Theorem is important in the area of mathematics known as *complex variables*, a subject that is fundamental for many applications in the physical sciences and engineering.

EXERCISES 18.4

Use Green's Theorem to evaluate the line integrals in Exercises 1–14.

1 $\oint_C (x^2 + y)\, dx + xy^2\, dy$, where C is the closed curve determined by $y^2 = x$ and $y = -x$ from $(0, 0)$ to $(1, -1)$.

2 $\oint_C (x + y^2)\, dx + (1 + x^2)\, dy$, where C is the closed curve determined by $y = x^3$ and $y = x^2$ from $(0, 0)$ to $(1, 1)$.

3 $\oint_C x^2 y^2\, dx + (x^2 - y^2)\, dy$ where C is the square with vertices $(0, 0)$, $(1, 0)$, $(1, 1)$, and $(0, 1)$.

4 $\oint_C \sqrt{y}\, dx + \sqrt{x}\, dy$ where C is the triangle with vertices $(1, 1)$, $(3, 1)$, and $(2, 2)$.

5 $\oint_C xy\, dx + (y + x)\, dy$ where C is the unit circle with center at $(0, 0)$.

6 $\oint_C e^x \sin y\, dx + e^x \cos y\, dy$ where C is the ellipse $3x^2 + 8y^2 = 24$.

7 $\oint_C xy\, dx + \sin y\, dy$ where C is the triangle with vertices $(1, 1)$, $(2, 2)$, and $(3, 0)$.

8 $\oint_C \tan^{-1} x\, dx + 3x\, dy$ where C is the rectangle with vertices $(1, 0)$, $(0, 1)$, $(2, 3)$, and $(3, 2)$.

9 $\oint_C y^2(1 + x^2)^{-1}\, dx + 2y \tan^{-1} x\, dy$ where C is the graph of $x^{2/3} + y^{2/3} = 1$.

10 $\oint_C (x^2 + y^2)\, dx + 2xy\, dy$ where C is the boundary of the region bounded by the graphs of $y = \sqrt{x}$, $y = 0$, and $x = 4$.

11 $\oint_C (x^4 + 4)\, dx + xy\, dy$, where C is the cardioid $r = 1 + \cos \theta$.

12 $\oint_C xy\, dx + (x^2 + y^2)\, dy$, where C is the first quadrant loop of $r = \sin 2\theta$.

13 $\oint_C (x + y)\, dx + (y + x^2)\, dy$, where C is the boundary of the region between the graphs of $x^2 + y^2 = 1$ and $x^2 + y^2 = 4$.

14 $\oint_C (1 - x^2 y)\, dx + \sin y\, dy$, where C is the boundary of the region that lies inside the square with vertices $(\pm 2, \pm 2)$ and outside of the square with vertices $(\pm 1, \pm 1)$.

In Exercises 15 and 16 use Theorem (18.18) to find the area of the region bounded by the curve C with the given parametric equations.

15 $x = a \cos^3 t$, $y = a \sin^3 t$; $0 \le t \le 2\pi$

16 $x = a \cos t$, $y = a \sin t$; $0 \le t \le 2\pi$

In Exercises 17 and 18 use (18.18) to find the area of the region bounded by the graphs of the given equations.

17 $y = 4x^2$, $y = 16x$

18 $y = x^3$, $y^2 = x$

19 Prove formula (b) in the partial proof of Green's Theorem.

20 If $\mathbf{F}(x, y)$ is a two-dimensional vector field and $\int_A^B \mathbf{F} \cdot d\mathbf{r}$ is independent of path in a region D, use Green's Theorem to prove that $\oint_C \mathbf{F} \cdot d\mathbf{r} = 0$ for every piecewise smooth simple closed curve in D.

21 Let R be the region bounded by a piecewise smooth simple closed curve C in the xy-plane. If the area of R is A, use Green's Theorem to prove that the centroid $(\bar{x}, \bar{y})$ is

$$\bar{x} = \frac{1}{2A} \oint_C x^2\, dy, \qquad \bar{y} = -\frac{1}{2A} \oint_C y^2\, dx.$$

22 Suppose a homogeneous lamina of density k has the shape of a region in the xy-plane that is bounded by a piecewise smooth simple closed curve C. Prove that the moments of inertia with respect to the x- and y-axes are

$$I_x = -\frac{k}{3} \oint_C y^3\, dx, \qquad I_y = \frac{k}{3} \oint_C x^3\, dy.$$

23 Use Exercise 21 to find the centroid of a semicircular region of radius a.

24 Use Exercise 22 to find the moment of inertia of a homogeneous circular disc of radius a with respect to a diameter.

25 Suppose f and g are differential functions. Prove that $\oint_C f(x)\, dx + g(y)\, dy = 0$ for every piecewise smooth simple closed curve C.

26 Let $M = y/(x^2 + y^2)$ and $N = -x/(x^2 + y^2)$. If R is the region bounded by the unit circle C with center at the origin, show that

$$\oint_C M\, dx + N\, dy \ne \iint_R \left(\frac{\partial N}{\partial x} - \frac{\partial M}{\partial y} \right) dA.$$

Explain why Green's Theorem is not true in this instance.

27 Suppose $\mathbf{F}(x, y)$ has continuous first partial derivatives in a simply connected region R. If curl $\mathbf{F} = \mathbf{0}$ in R, prove that $\oint_C \mathbf{F} \cdot d\mathbf{r} = 0$ for every piecewise smooth simple closed curve C in R.

18.5

SURFACE INTEGRALS

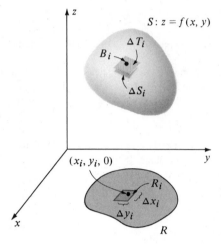

Figure 18.15

Line integrals are evaluated along curves. Double and triple integrals are defined on regions in two and three dimensions, respectively. It is also possible to consider an integral of a function over a surface. For simplicity we shall restrict our discussion to rather well-behaved surfaces, and our demonstrations will be at an intuitive level. Rigorous treatments may be found in texts on advanced calculus.

If the projection of a surface S on a coordinate plane is a region of the type we considered for double integrals, then S is said to have a **regular projection** on the coordinate plane. Suppose that S is the graph of $z = f(x, y)$ and that S has a regular projection R on the xy-plane. Suppose further that f has continuous first partial derivatives on R. In Chapter 17 we defined the area A of S. A similar technique may be used to define an integral of $g(x, y, z)$ over the surface S, where the function g is continuous throughout a region containing S. We shall employ the notation used in Section 17.9. In particular, as illustrated in Figure 18.15, ΔS_i and ΔT_i will denote areas of parts of S and the tangent plane to S at $B_i(x_i, y_i, z_i)$, respectively, that project onto the rectangle R_i of an inner partition P of R. As in the definitions of all previous integrals, we evaluate g at B_i for each i and form the sum $\sum_i g(x_i, y_i, z_i)\, \Delta T_i$. By definition, the **surface integral of g over S** is

$$(18.20) \qquad \iint_S g(x, y, z)\, dS = \lim_{\|P\| \to 0} \sum_i g(x_i, y_i, z_i)\, \Delta T_i$$

where the limit of the sum is defined in the usual way. In a manner similar to the development of formula (17.31) for surface area, this integral may be evaluated by means of the formula

$$(18.21) \qquad \iint_S g(x, y, z)\, dS = \iint_R g(x, y, f(x, y))\sqrt{(f_x(x, y))^2 + (f_y(x, y))^2 + 1}\, dA.$$

If S is the union of several surfaces of the type we are considering, then the surface integral is defined as the sum of the individual surface integrals. Observe that if $g(x, y, z) = 1$ for all (x, y, z), then (18.21) reduces to (17.31) and hence the surface integral equals the surface area of S.

An elementary physical application may be obtained by considering a thin metal sheet, such as tinfoil, that has the shape of S. If the area density at (x, y, z) is $g(x, y, z)$, then (18.21) is the mass of the sheet. The center of gravity and moments of inertia may be obtained by employing the methods used for solids in the preceding chapter.

Example 1   Evaluate $\iint_S x^2 z\, dS$ where S is the portion of the circular cone $z^2 = x^2 + y^2$ that lies between the planes $z = 1$ and $z = 4$.

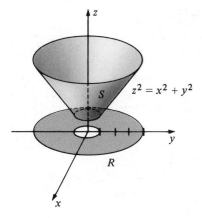

Figure 18.16

Solution   As shown in Figure 18.16, the projection R of S onto the xy-plane is the annular region bounded by circles of radii 1 and 4 with centers at the origin. If we write the equation for S in the form

$$z = (x^2 + y^2)^{1/2} = f(x, y)$$

then

$$f_x(x, y) = \frac{x}{(x^2 + y^2)^{1/2}} \quad \text{and} \quad f_y(x, y) = \frac{y}{(x^2 + y^2)^{1/2}}.$$

Applying (18.21) and noting that the radical reduces to $\sqrt{2}$, we obtain

$$\iint_S x^2 z \, dS = \iint_R x^2 (x^2 + y^2)^{1/2} \sqrt{2} \, dx \, dy.$$

Using polar coordinates to evaluate the double integral,

$$\iint_S x^2 z \, dS = \int_0^{2\pi} \int_1^4 (r^2 \cos^2 \theta) r \sqrt{2} \, r \, dr \, d\theta = \sqrt{2} \int_0^{2\pi} \cos^2 \theta \left. \frac{r^5}{5} \right]_1^4 d\theta$$

$$= \frac{1023\sqrt{2}}{5} \int_0^{2\pi} \frac{1 + \cos 2\theta}{2} \, d\theta = \frac{1023\sqrt{2}}{10} \left[\theta + \frac{1}{2} \sin 2\theta \right]_0^{2\pi}$$

$$= \frac{1023\sqrt{2}\,\pi}{5} \approx 909.0. \qquad \blacksquare$$

As at the beginning of this section, suppose S is the graph of $z = f(x, y)$, where S has a regular projection R on the xy-plane, and f has continuous first partial derivatives on R. Consider a unit normal vector $\mathbf{n}$ to S, that is, a unit vector that is normal to the tangent plane to S at B_i, as illustrated in (i) of Figure 18.17. Let α, β, and γ be the angles between $\mathbf{n}$ and the vectors $\mathbf{i}$, $\mathbf{j}$, and $\mathbf{k}$, respectively. As in the figure, we shall choose the upper normal to S, in the sense that $0 \le \gamma \le \pi/2$.

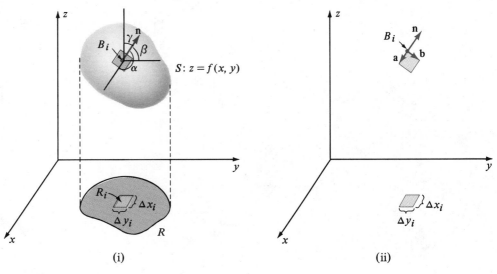

(i)   (ii)

Figure 18.17

Since the components of $\mathbf{n}$ on $\mathbf{i}$, $\mathbf{j}$, and $\mathbf{k}$ are

$$\mathbf{n} \cdot \mathbf{i} = \cos \alpha, \quad \mathbf{n} \cdot \mathbf{j} = \cos \beta, \quad \mathbf{n} \cdot \mathbf{k} = \cos \gamma$$

we have

(18.22)
$$\mathbf{n} = \cos \alpha \mathbf{i} + \cos \beta \mathbf{j} + \cos \gamma \mathbf{k}.$$

If B_i, $\mathbf{a}$ and $\mathbf{b}$ are as shown in (ii) of Figure 18.17, then

$$\mathbf{n} = \frac{1}{|\mathbf{a} \times \mathbf{b}|} \mathbf{a} \times \mathbf{b}.$$

Using the formulas for $\mathbf{a} \times \mathbf{b}$ and $|\mathbf{a} \times \mathbf{b}|$ obtained on page 821, and abbreviating $f_x(x_i, y_i)$ and $f_y(x_i, y_i)$ by f_x and f_y gives us

(18.23)
$$\mathbf{n} = \frac{1}{\sqrt{f_x^2 + f_y^2 + 1}} (-f_x \mathbf{i} - f_y \mathbf{j} + \mathbf{k}).$$

Comparing this formula for $\mathbf{n}$ with (18.22) we see that

$$\cos \gamma = \frac{1}{\sqrt{f_x^2 + f_y^2 + 1}}$$

and hence, if γ is acute, (18.21) may be written

(18.24)
$$\iint_S g(x, y, z) \, dS = \iint_R g(x, y, f(x, y)) \sec \gamma \, dx \, dy.$$

In like manner, if an equation for S is $y = h(x, z)$ where h has continuous first partial derivatives, and S has a regular projection R_1 on the xz-plane (see Figure 18.18), then

(18.25)
$$\iint_S g(x, y, z) \, dS = \iint_{R_1} g(x, h(x, z), z) \sec \beta \, dx \, dz$$

$$= \iint_{R_1} g(x, h(x, z), z) \sqrt{(h_x(x, z))^2 + (h_z(x, z))^2 + 1} \, dx \, dz.$$

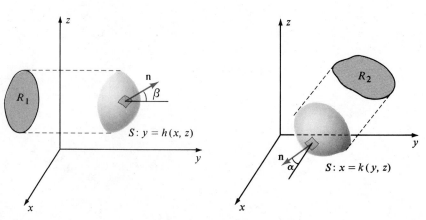

Figure 18.18   *Figure 18.19*

Similarly, if S is given by $x = k(y, z)$ where k has continuous first partial derivatives and if S has a regular projection R_2 on the yz-plane (see Figure 18.19), then

(18.26)
$$\iint_S g(x, y, z) \, dS = \iint_{R_2} g(k(y, z), y, z) \sec \alpha \, dy \, dz$$
$$= \iint_{R_2} g(k(y, z), y, z) \sqrt{(k_y(y, z))^2 + (k_z(y, z))^2 + 1} \, dy \, dz.$$

Example 2 Evaluate $\iint_S (xz/y) \, dS$, where S is the part of the cylinder $x = y^2$ that lies in the first octant between the planes $z = 0, z = 5, y = 1$, and $y = 4$.

Solution The surface S is sketched in Figure 18.20, where for clarity we have used a different unit of length on the x-axis. Since the projection of S on the yz-plane is the rectangle with vertices $(0, 1, 0), (0, 4, 0), (0, 4, 5),$ and $(0, 1, 5)$, we may use formula (18.26) with $k(y, z) = y^2$, as follows:

$$\iint_S \frac{xz}{y} \, dS = \int_1^4 \int_0^5 \frac{y^2 z}{y} \sqrt{(2y)^2 + 0^2 + 1} \, dz \, dy$$
$$= \int_1^4 \int_0^5 yz \sqrt{4y^2 + 1} \, dz \, dy = \int_1^4 y \sqrt{4y^2 + 1} \left[\frac{z^2}{2} \right]_0^5 dy$$
$$= \frac{25}{2} \int_1^4 y \sqrt{4y^2 + 1} \, dy = \frac{25}{24} (4y^2 + 1)^{3/2} \Big]_1^4$$
$$= \frac{25}{24} [65^{3/2} - 5^{3/2}] \approx 534.2. \qquad \blacksquare$$

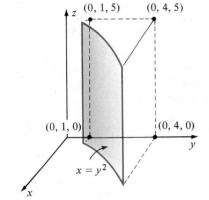

Figure 18.20

It follows from our work in Section 17.9 (see page 821) and the previous discussion, that

$$\Delta T_i = \sec \gamma \, \Delta x_i \, \Delta y_i$$

that is, the area on the tangent plane in (i) of Figure 18.17 can be found by multiplying the area $\Delta x_i \, \Delta y_i$ of R_i by $\sec \gamma$. If we represent the area of R_i by $dx \, dy$, then we are motivated to define the **differential dS of surface area** by

$$dS = \sec \gamma \, dx \, dy.$$

In applications we may regard dS as the element of surface area on S that projects onto a rectangular region in the xy-plane of area $dx \, dy$. Similarly, for the surface integrals corresponding to Figures 18.18 and 18.19 we write

$$dS = \sec \beta \, dx \, dz \quad \text{and} \quad dS = \sec \alpha \, dy \, dz$$

respectively.

In certain cases we may consider surface integrals that involve a vector function $\mathbf{F}$, where

$$\mathbf{F}(x, y, z) = M(x, y, z)\mathbf{i} + N(x, y, z)\mathbf{j} + P(x, y, z)\mathbf{k}$$

and M, N, and P are continuous functions. If $\mathbf{n}$ is the unit normal (18.22), then $\mathbf{F}(x, y, z) \cdot \mathbf{n}$ is called the **normal component** of $\mathbf{F}$ at the point (x, y, z). By definition, the **surface integral of the normal component of F over S** is

$$(18.27) \qquad \iint_S \mathbf{F} \cdot \mathbf{n} \, dS = \iint_S (M \cos \alpha + N \cos \beta + P \cos \gamma) \, dS$$

where we have used the symbols M, N, and P for the functions to denote their values at (x, y, z).

If S has regular projections R, R_1, and R_2 on the xy-, xz-, and yz-planes, respectively, and if α, β, and γ are acute, then using the three forms for dS obtained previously we may write (18.27) as

$$(18.28) \qquad \iint_S \mathbf{F} \cdot \mathbf{n} \, dS = \iint_{R_2} M \, dy \, dz + \iint_{R_1} N \, dx \, dz + \iint_R P \, dx \, dy.$$

To attach a physical significance to (18.27), suppose that S is submerged in a fluid having a velocity field $\mathbf{F}(x, y, z)$. Let dS represent a small element of area on S. If $\mathbf{F}$ is continuous, then it is almost constant on dS and, as illustrated in Figure 18.21, the amount of fluid crossing dS in a unit of time may be approximated by the volume of a cylinder of base area dS and altitude $\mathbf{F} \cdot \mathbf{n}$.

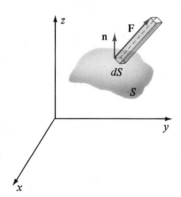

Figure 18.21

If dV denotes the volume of the cylinder in the figure, then $dV = \mathbf{F} \cdot \mathbf{n} \, dS$. Since dV represents the amount of fluid crossing dS per unit time, the surface integral (18.27) is the volume of fluid crossing S per unit time. In this context it is called the *flux* of $\mathbf{F}$ *through* (or *over*) S. Thus we have the following, where it is assumed that $\mathbf{F}$ and S satisfy the conditions we have imposed.

Definition (18.29) | The **flux** of a vector field $\mathbf{F}$ through (or over) a surface S is $\displaystyle\iint_S \mathbf{F} \cdot \mathbf{n} \, dS$.

If the fluid in the previous discussion has density $\rho = \rho(x, y, z)$, then the value of the flux integral $\iint_S \rho \mathbf{F} \cdot \mathbf{n} \, dS$ is the *mass* of the fluid crossing S.

Surface integrals may be defined over closed surfaces, such as spheres, ellipsoids, etc. In this case if we choose $\mathbf{n}$ as the *outer normal*, then the flux measures the *net outward flow* per unit time. If the integral in Definition (18.29) is positive, then the flow out of S exceeds the flow into S and we say there is a **source** of $\mathbf{F}$ within S. If the integral is negative, then the flow into S exceeds the flow out of S and we say there is a **sink** within S. If the integral is 0, then the flow into and the flow out of S are equal, that is, the sources and sinks balance one another.

Example 3   Let S be the part of the graph of $z = 9 - x^2 - y^2$ such that $z \geq 0$. If $\mathbf{F}(x, y, z) = 3x\mathbf{i} + 3y\mathbf{j} + z\mathbf{k}$, find the flux of $\mathbf{F}$ through S.

Solution   The graph is sketched in Figure 18.22, together with a typical unit upper normal $\mathbf{n}$ at $P(x, y, z)$ and the vector $\mathbf{F}(x, y, z)$. According to (18.23), a unit upper normal to S at the point (x, y, z) is

$$\mathbf{n} = \frac{2x\mathbf{i} + 2y\mathbf{j} + \mathbf{k}}{\sqrt{4x^2 + 4y^2 + 1}}.$$

Hence by (18.29) the flux of $\mathbf{F}$ through S is

$$\iint_S \mathbf{F} \cdot \mathbf{n} \, dS = \iint_S \frac{6x^2 + 6y^2 + z}{\sqrt{4x^2 + 4y^2 + 1}} \, dS.$$

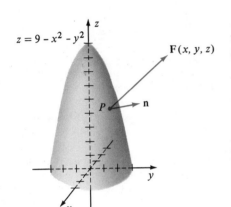

$z = 9 - x^2 - y^2$

$F(x, y, z)$

P $\mathbf{n}$

Figure 18.22

Applying (18.21) gives us

$$\iint_S \mathbf{F} \cdot \mathbf{n} \, dS = \iint_R \frac{6x^2 + 6y^2 + 9 - x^2 - y^2}{\sqrt{4x^2 + 4y^2 + 1}} \sqrt{4x^2 + 4y^2 + 1} \, dx \, dy$$

$$= \iint_R (5x^2 + 5y^2 + 9) \, dx \, dy$$

where R is the circular region in the xy-plane bounded by the graph of $x^2 + y^2 = 9$. Changing to polar coordinates,

$$\iint_S \mathbf{F} \cdot \mathbf{n} \, dS = \int_0^{2\pi} \int_0^3 (5r^2 + 9) r \, dr \, d\theta = \frac{567\pi}{2} \approx 890.6.$$

If, for example, $\mathbf{F}$ is the velocity field of an expanding gas, and $|\mathbf{F}|$ is measured in cm/sec, then the unit of flux would be cm^3/sec. Hence the flow of gas across the surface S in Figure 18.22 is approximately 890.6 cm^3/sec.   ∎

EXERCISES 18.5

In Exercises 1–4 evaluate $\iint_S f(x, y, z)\, dS$.

1 $f(x, y, z) = x^2$; S the upper half of the sphere $x^2 + y^2 + z^2 = a^2$.

2 $f(x, y, z) = x^2 + y^2 + z^2$; S the part of the plane $z = y + 4$ that is inside the cylinder $x^2 + y^2 = 4$.

3 $f(x, y, z) = x + y$; S the first octant portion of the plane $2x + 3y + z = 6$.

4 $f(x, y, z) = (x^2 + y^2 + 1)^{1/2}$; S the part of the paraboloid $2z = x^2 + y^2$ that lies inside the cylinder $x^2 + y^2 = 2y$.

In Exercises 5–8 set up, but do not evaluate, the given surface integral by using a projection of S on (a) the yz-plane; (b) the xz-plane.

5 $\iint_S xy^2z^3\, dS$; S the part of the plane $2x + 3y + 4z = 12$ that is in the first octant.

6 $\iint_S (xz + 2y)\, dS$; S the part of the graph of $y = x^3$ between the planes $y = 0$, $y = 8$, $z = 2$, and $z = 0$.

7 $\iint_S (x^2 - 2y + z)\, dS$; S the part of the graph of $4x + y = 8$ bounded by the coordinate planes and the plane $z = 6$.

8 $\iint_S (x^2 + y^2 + z^2)\, dS$; S the first octant part of the graph of $x^2 + y^2 = 4$ bounded by the coordinate planes and the plane $x + z = 2$.

9 Interpret $\iint_S f(x, y, z)\, dS$ geometrically if f is the constant function $f(x, y, z) = c$, where $c > 0$, and S has a regular projection on the xy-plane.

10 Show that a double integral $\iint_R f(x, y)\, dA$ of the type considered in Theorem 17.9 is a special case of a surface integral.

In Exercises 11–14 find $\iint_S \mathbf{F} \cdot \mathbf{n}\, dS$, where $\mathbf{n}$ is a unit upper normal to S.

11 $\mathbf{F} = x\mathbf{i} + y\mathbf{j} + z\mathbf{k}$; S the upper half of the sphere $x^2 + y^2 + z^2 = a^2$.

12 $\mathbf{F} = x\mathbf{i} - y\mathbf{j}$; S the part of the sphere $x^2 + y^2 + z^2 = a^2$ lies in the first octant.

13 $\mathbf{F} = 2\mathbf{i} + 5\mathbf{j} + 3\mathbf{k}$; S the part of the cone $z = (x^2 + y^2)^{1/2}$ that is inside the cylinder $x^2 + y^2 = 1$.

14 $\mathbf{F} = x\mathbf{i} + y\mathbf{j} + z\mathbf{k}$; S the part of the plane $3x + 2y + z = 12$ cut out by the planes $x = 0$, $y = 0$, $x = 1$, and $y = 2$.

In Exercises 15 and 16 use (18.28) to find the flux of F through S.

15 $\mathbf{F}(x, y, z) = x\mathbf{i} + y\mathbf{j} + z\mathbf{k}$; S the part of the plane $2x + 3y + z = 6$ that lies in the first octant.

16 $\mathbf{F}(x, y, z) = (x^2 + z)\mathbf{i} + y^2z\mathbf{j} + (x^2 + y^2 + z)\mathbf{k}$; S the part of the paraboloid $z = x^2 + y^2$ in the first octant that is cut off by the plane $z = 4$.

In Exercises 17 and 18 find the flux of $\mathbf{F}$ over the closed surface S. (Be sure to use the *outer* normal to S.)

17 $\mathbf{F}(x, y, z) = (x + y)\mathbf{i} + z\mathbf{j} + xz\mathbf{k}$; S the surface of the cube having vertices $(\pm 1, \pm 1, \pm 1)$.

18 $\mathbf{F}(x, y, z) = x\mathbf{i} - y\mathbf{j} + z\mathbf{k}$; S the surface of the solid bounded by the graphs of $z = x^2 + y^2$ and $z = 4$.

19 Prove that if S is given by $z = f(x, y)$, then (18.29) can be written

$$\iint_S \mathbf{F} \cdot \mathbf{n}\, dS = \iint_R (-Mf_x - Nf_y + P)\, dx\, dy.$$

20 By Coulomb's Law (see page 828), if a point charge of q coulombs is located at the origin, then the force exerted on a unit charge at (x, y, z) is given by $\mathbf{F}(x, y, z) = (cq/|\mathbf{r}|^3)\mathbf{r}$, where $\mathbf{r} = x\mathbf{i} + y\mathbf{j} + z\mathbf{k}$. If S is any sphere with center at the origin, show that the flux of $\mathbf{F}$ through S is $4\pi cq$.

21 If a thin metal sheet T has the shape of a surface S and has area density $f(x, y, z)$, then the *moments* of T with respect to the coordinate planes are defined by

$$M_{xy} = \iint_S zf(x, y, z)\, dS$$

$$M_{xz} = \iint_S yf(x, y, z)\, dS$$

$$M_{yz} = \iint_S xf(x, y, z)\, dS.$$

Use limits of sums to show that these are "natural" definitions. The *center of mass* $(\bar{x}, \bar{y}, \bar{z})$ is given by

$$\bar{x}M = M_{yz}, \quad \bar{y}M = M_{xz}, \quad \bar{z}M = M_{xy}$$

where M is the mass of T. Suppose a metal funnel has the shape of the surface S described in Example 1 (see Figure 18.16). If the unit of length is centimeters, and the area density at the point (x, y, z) is z^2 gm/cm², find the mass and the center of mass of the funnel. Find the moment of inertia of the funnel with respect to the z-axis.

22 Refer to Exercise 21. A thin metal sheet of constant density k has the shape of the hemisphere $z = \sqrt{a^2 - x^2 - y^2}$. Find its center of mass.

18.6

THE DIVERGENCE THEOREM

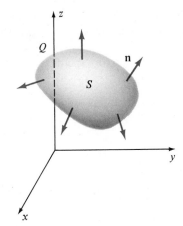

Figure 18.23

One of the most important theorems of vector calculus is the *Divergence Theorem*. It is also called *Gauss' Theorem*, in honor of Karl Friedrich Gauss (1777–1855), whom many consider the greatest mathematician of all time. The theorem has to do with a surface S that completely encloses a region Q in three dimensions. A surface of this type will be called a **closed surface**. For example, S could be the surface of a sphere, an ellipsoid, a cube, or a tetrahedron. The Divergence Theorem can be proved for very complicated regions; however, the proof would take us into the field of advanced calculus. Instead, *throughout this section*, we shall assume that Q is a region over which triple integrals can be evaluated using the techniques developed in Chapter 17. It will also be assumed that the resulting surface integrals can be evaluated over S. Most of the regions that are encountered in applications, such as quadric surfaces, polyhedra, closed cylinders, etc., are of this type. In this section **n** will denote a unit *outer* normal to S. Illustrations of a region Q and a surface S of the types to be considered, together with some typical positions of **n** are illustrated in Figure 18.23.

Under the restrictions we have placed on Q and S, the Divergence Theorem may be stated as follows.

The Divergence Theorem (18.30)

> Let Q be a region in three dimensions bounded by a closed surface S, and let **n** denote the unit outer normal to S at (x, y, z). If **F** is a vector function that has continuous partial derivatives on Q, then
>
> $$\iint_S \mathbf{F} \cdot \mathbf{n} \, dS = \iiint_Q \nabla \cdot \mathbf{F} \, dV.$$

In physical applications the formula in Theorem (18.30) states that *the flux of* **F** *over S equals the triple integral of the divergence of* **F** *over Q*. If $\mathbf{F}(x, y, z) = M(x, y, z)\mathbf{i} + N(x, y, z)\mathbf{j} + P(x, y, z)\mathbf{k}$, then using (18.27) and the definition of $\nabla \cdot \mathbf{F}$, the formula may be written

$$\iint_S (M \cos \alpha + N \cos \beta + P \cos \gamma) \, dS = \iiint_Q \left(\frac{\partial M}{\partial x} + \frac{\partial N}{\partial y} + \frac{\partial P}{\partial z} \right) dV.$$

To prove this equality it is sufficient to show that

$$\iint_S M \cos \alpha \, dS = \iiint_Q \frac{\partial M}{\partial x} \, dV$$

$$\iint_S N \cos \beta \, dS = \iiint_Q \frac{\partial N}{\partial y} \, dV$$

$$\iint_S P \cos \gamma \, dS = \iiint_Q \frac{\partial P}{\partial z} \, dV.$$

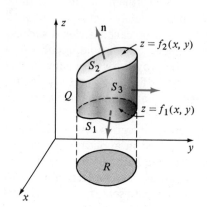

Figure 18.24

Since the proofs for all three formulas are basically similar, we shall prove only the third. It is quite difficult to prove the theorem for the most general case, and hence we shall specialize Q and S. Specifically, let R be a region in the xy-plane that satisfies the conditions of Green's Theorem, and suppose Q lies over R and between two surfaces having the equations $z = f_1(x, y)$ and $z = f_2(x, y)$, where the functions f_1 and f_2 have continuous first partial derivatives and $f_1(x, y) \le f_2(x, y)$ for all (x, y) in R. As illustrated in Figure 18.24, the surface S that bounds Q consists of a bottom surface S_1, a top surface S_2, and a lateral surface S_3.

On S_3 we have $\gamma = 90°$ and, therefore, $\cos \gamma = 0$. Consequently,

$$\iint_S P \cos \gamma \, dS = \iint_{S_1} P \cos \gamma \, dS + \iint_{S_2} P \cos \gamma \, dS.$$

Applying (18.24),

$$\iint_{S_2} P \cos \gamma \, dS = \iint_R P(x, y, f_2(x, y)) \, dx \, dy.$$

Since the formulas for surface integrals in Section 18.5 were derived under the assumption that $\mathbf{n}$ is always an *upper* normal, and since the outer normal $\mathbf{n}$ on S_1 is a *lower* normal, it is necessary to change the sign on the right side of (18.24) when working with S_1. Indeed, a formula for $\mathbf{n}$ on S_1 may be found by taking the negative of the vector in (18.23). Thus

$$\iint_{S_1} P \cos \gamma \, dS = - \iint_R P(x, y, f_1(x, y)) \, dx \, dy$$

and hence

$$\iint_S P \cos \gamma \, dS = \iint_R [P(x, y, f_2(x, y)) - P(x, y, f_1(x, y))] \, dx \, dy$$

$$= \iint_R \left[\int_{f_1(x, y)}^{f_2(x, y)} \frac{\partial P}{\partial z} \, dz \right] dx \, dy$$

$$= \iiint_Q \frac{\partial P}{\partial z} \, dV$$

which is what we wished to prove. It is not difficult to extend our proof to finite unions of regions of the type we have considered. The proofs of the formulas for $\iint_S M \cos \alpha \, dS$ and $\iint_S N \cos \beta \, dS$ are similar, provided Q and S are suitably restricted.

Example 1 Let Q be the region bounded by the graphs of $x^2 + y^2 = 4$, $z = 0$, and $z = 3$. Let S denote the surface of Q. If $\mathbf{F}(x, y, z) = x^3\mathbf{i} + y^3\mathbf{j} + z^3\mathbf{k}$, use the Divergence Theorem to evaluate $\iint_S \mathbf{F} \cdot \mathbf{n} \, dS$.

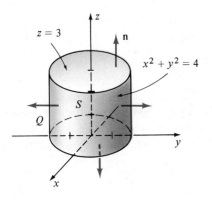

$z = 3$

$\mathbf{n}$

$x^2 + y^2 = 4$

S

Q

x

Figure 18.25

Solution The surface S and some typical positions of $\mathbf{n}$ are sketched in Figure 18.25. Since

$$\nabla \cdot \mathbf{F} = 3x^2 + 3y^2 + 3z^2 = 3(x^2 + y^2 + z^2)$$

we have, by the Divergence Theorem,

$$\iint_S \mathbf{F} \cdot \mathbf{n} \, dS = 3 \iiint_Q (x^2 + y^2 + z^2) \, dV.$$

It is convenient to use cylindrical coordinates to evaluate the triple integral. Thus

$$\iint_S \mathbf{F} \cdot \mathbf{n} \, dS = 3 \int_0^{2\pi} \int_0^2 \int_0^3 (r^2 + z^2) r \, dz \, dr \, d\theta$$

$$= 3 \int_0^{2\pi} \int_0^2 \left[r^2 z + \frac{1}{3} z^3 \right]_0^3 r \, dr \, d\theta = 3 \int_0^{2\pi} \int_0^2 (3r^2 + 9) r \, dr \, d\theta$$

$$= 3 \int_0^{2\pi} \left[\frac{3}{4} r^4 + \frac{9}{2} r^2 \right]_0^2 d\theta = 3 \int_0^{2\pi} 30 \, d\theta = 90\theta \Big]_0^{2\pi} = 180\pi. \quad \blacksquare$$

Example 2 Let Q be the region bounded by the cylinder $z = 4 - x^2$, the plane $y + z = 5$, and the xy- and xz-planes. Evaluate $\iint_S \mathbf{F} \cdot \mathbf{n} \, dS$ if $\mathbf{F}(x, y, z) = (x^3 + \sin z)\mathbf{i} + (x^2 y + \cos z)\mathbf{j} + e^{x^2 + y^2}\mathbf{k}$.

Solution The region Q is sketched in Figure 18.26. It would be an incredibly difficult job to evaluate the integral directly; however, using the Divergence Theorem we can obtain the value from the triple integral

$$\iiint_Q (3x^2 + x^2) \, dV.$$

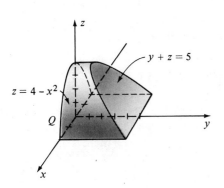

z

$y + z = 5$

$z = 4 - x^2$

Q

y

x

Figure 18.26

Referring to Figure 18.26 and using (17.21), we see that this integral equals

$$\int_{-2}^2 \int_0^{4-x^2} \int_0^{5-z} 4x^2 \, dy \, dz \, dx = \int_{-2}^2 \int_0^{4-x^2} 4x^2 (5 - z) \, dz \, dx$$

$$= \int_{-2}^2 \left[20x^2 z - 2x^2 z^2 \right]_0^{4-x^2} dx$$

$$= \int_{-2}^2 (48x^2 - 4x^4 - 2x^6) \, dx$$

$$= \frac{4608}{35} \approx 131.7. \quad \blacksquare$$

We may use the Divergence Theorem to obtain a physical interpretation for the divergence of a vector field. Let us begin by recalling from the Mean Value Theorem for Definite Integrals (5.20) that if a function f of one variable is continuous on a closed interval $[a, b]$, then there is a number c in (a, b) such that

$$\int_a^b f(x)\, dx = f(c)L$$

where $L = b - a$ is the length of $[a, b]$. An analogous result may be proved for triple integrals. Specifically, if a function f of three variables is continuous throughout a spherical region Q, then there is a point $A(u, v, w)$ in the interior of Q such that

$$\iiint\limits_Q f(x, y, z)\, dV = \left[f(x, y, z) \right]_A V$$

where V is the volume of Q and $[f(x, y, z)]_A$ denotes $f(u, v, w)$. This result is sometimes called the **Mean Value Theorem for Triple Integrals**. It follows that if $\mathbf{F}$ is a continuous vector function, then

$$\iiint\limits_Q \nabla \cdot \mathbf{F}\, dV = \left[\nabla \cdot \mathbf{F} \right]_A V$$

or

$$\left[\nabla \cdot \mathbf{F} \right]_A = \frac{1}{V} \iiint\limits_Q \nabla \cdot \mathbf{F}\, dV.$$

Applying the Divergence Theorem,

$$\left[\nabla \cdot \mathbf{F} \right]_A = \frac{1}{V} \iint\limits_S \mathbf{F} \cdot \mathbf{n}\, dS$$

where S is the surface of Q. The ratio on the right may be interpreted as *the flux of $\mathbf{F}$ per unit volume over the sphere.*

Next let P be an arbitrary point and suppose $\mathbf{F}$ is continuous throughout a region containing P in its interior. Let S_k be the surface of a sphere of radius k with center at P. From the previous discussion, for each k there is a point P_k within S_k such that

$$\left[\nabla \cdot \mathbf{F} \right]_{P_k} = \frac{1}{V_k} \iint\limits_{S_k} \mathbf{F} \cdot \mathbf{n}\, dS$$

where V_k is the volume of the sphere (see Figure 18.27).

If we let $k \to 0$, then $P_k \to P$ and we obtain

(18.31)

$$\left[\operatorname{div} \mathbf{F} \right]_P = \lim_{k \to 0} \frac{1}{V_k} \iint\limits_{S_k} \mathbf{F} \cdot \mathbf{n}\, dS$$

that is, *the divergence of $\mathbf{F}$ at P is the limiting value of the flux per unit volume over a sphere with center P, as the radius of the sphere approaches 0.* In particular, if $\mathbf{F}$ represents the velocity of a fluid, then div $\mathbf{F}]_P$ can be interpreted as the rate of loss or gain of fluid per unit volume at P. If $\rho = \rho(x, y, z)$ is the density

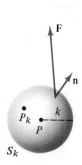

Figure 18.27

at (x, y, z), then $[\text{div } \rho\mathbf{F}]_P$ is the change in *mass* per unit volume. It follows from the remarks at the end of Section 18.5 that there is a source or sink at P if div $\mathbf{F}]_P > 0$ or div $\mathbf{F}]_P < 0$, respectively. If the fluid is incompressible and no sources or sinks are present, then there can be no gain or loss within the volume element V_k and hence at every point P,

$$\text{div } \mathbf{F} = \frac{\partial M}{\partial x} + \frac{\partial N}{\partial y} + \frac{\partial P}{\partial z} = 0.$$

This formula is called the **equation of continuity** for incompressible fluids.

The limit formulation of divergence given in (18.31) may also be applied to physical concepts such as magnetic or electric flux, since these entities have many characteristics that are similar to those of fluids. In like manner, if $\mathbf{F}$ represents the flow of heat in a region, and if $\nabla \cdot \mathbf{F}]_P > 0$, then there is a source of heat at P (or heat is *leaving* P and, therefore, the temperature at P is decreasing). If $\nabla \cdot \mathbf{F}]_P < 0$, then heat is being *absorbed* at P (or the temperature at P is increasing).

Example 3 Suppose a closed surface S forms the boundary of a region Q and the origin O is an interior point of Q. If an inverse square field is $\mathbf{F} = (q/r^3)\mathbf{r}$ where q is a constant, $\mathbf{r} = x\mathbf{i} + y\mathbf{j} + z\mathbf{k}$, and $r = |\mathbf{r}|$, prove that the flux of $\mathbf{F}$ over S is $4\pi q$ regardless of the shape of Q.

Solution Since $\mathbf{F}$ is not continuous at O the Divergence Theorem cannot be applied directly; however, we may proceed as follows. Let S_1 be a sphere of radius a and center O that lies completely within S (see Figure 18.28), and let Q_1 denote the region that lies outside of S_1 and inside of S. Since $\mathbf{F}$ is continuous throughout Q_1 we may apply the Divergence Theorem, obtaining

$$\iiint\limits_{Q_1} \nabla \cdot \mathbf{F} \, dV = \iint\limits_{S} \mathbf{F} \cdot \mathbf{n} \, dS + \iint\limits_{S_1} \mathbf{F} \cdot \mathbf{n} \, dS.$$

It can be shown that if $\mathbf{F}$ is an inverse square field, then $\nabla \cdot \mathbf{F} = 0$ (see Exercise 33 in Section 18.1) and, therefore,

$$\iint\limits_{S} \mathbf{F} \cdot \mathbf{n} \, dS = -\iint\limits_{S_1} \mathbf{F} \cdot \mathbf{n} \, dS.$$

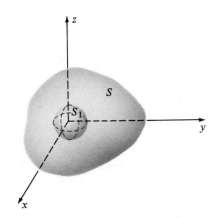

Figure 18.28

Since the unit normal $\mathbf{n}$ on S_1 is an *outer* normal, it points *away* from the region Q_1 that has S_1 as part of its boundary. Hence, on S_1 the unit normal $\mathbf{n}$ points *toward* the origin. Thus $\mathbf{n} = (-1/r)\mathbf{r}$. where $r = a$, and

$$\iint\limits_{S} \mathbf{F} \cdot \mathbf{n} \, dS = -\iint\limits_{S_1} \left(\frac{q}{r^3}\right)\mathbf{r} \cdot \left(-\frac{1}{r}\mathbf{r}\right) dS$$

$$= \iint\limits_{S_1} \frac{q}{r^4}(\mathbf{r} \cdot \mathbf{r}) \, dS = \iint\limits_{S_1} \frac{q}{r^2} \, dS$$

$$= \frac{q}{a^2} \iint\limits_{S_1} dS = \frac{q}{a^2}(4\pi a^2) = 4\pi q. \quad \blacksquare$$

Example 3 has an important application in the theory of electricity. By Coulomb's Law (see page 828), if a point charge of q coulombs is located at the origin, then the force exerted on a unit charge at (x, y, z) is $\mathbf{F}(x, y, z) = (cq/|\mathbf{r}|^3)\mathbf{r}$, where $\mathbf{r} = x\mathbf{i} + y\mathbf{j} + z\mathbf{k}$ and c is a constant. It follows that the *electric flux* of F over any closed surface containing the origin is $4\pi cq$. Thus, the electric flux is independent of the size or shape of S and depends only on q. It is not difficult to extend this to the case where there is more than one point charge within S. This result is called *Gauss' Law*, and has far-reaching consequences in the study of electric fields.

EXERCISES 18.6

In Exercises 1–4 use the Divergence Theorem to find $\iint_S \mathbf{F} \cdot \mathbf{n} \, dS$.

1 $\mathbf{F} = y \sin x\mathbf{i} + y^2 z\mathbf{j} + (x + 3z)\mathbf{k}$; S the surface of the region bounded by the planes $x = \pm 1, y = \pm 1, z = \pm 1$.

2 $\mathbf{F} = y^3 e^z\mathbf{i} - xy\mathbf{j} + x \arctan y\mathbf{k}$; S the surface of the region bounded by the coordinate planes and the plane $x + y + z = 1$.

3 $\mathbf{F} = (x^2 + \sin yz)\mathbf{i} + (y - xe^{-z})\mathbf{j} + z^2\mathbf{k}$; S the surface of the region bounded by the graphs of $x^2 + y^2 = 4$, $x + z = 2$, and $z = 0$.

4 $\mathbf{F} = 2xy\mathbf{i} + z \cosh x\mathbf{j} + (z^2 + y \sin^{-1} x)\mathbf{k}$; S the surface of the region bounded by the graphs of $z = x^2 + y^2$ and $z = 9$.

In Exercises 5–10 use the Divergence Theorem to find the flux of $\mathbf{F}$ through S.

5 $\mathbf{F}(x, y, z) = yz\mathbf{i} + xz\mathbf{j} + xy\mathbf{k}$; S the graph of $x^{2/3} + y^{2/3} + z^{2/3} = 1$.

6 $\mathbf{F}(x, y, z) = (x^2 + z^2)\mathbf{i} + (y^2 - 2xy)\mathbf{j} + (4z - 2yz)\mathbf{k}$; S the part of the cone $x = \sqrt{y^2 + z^2}$ between the planes $x = 0$ and $x = 9$.

7 $\mathbf{F}(x, y, z) = 3x\mathbf{i} + xz\mathbf{j} + z^2\mathbf{k}$; S the surface of the region bounded by the paraboloid $z = 4 - x^2 - y^2$ and the xy-plane.

8 $\mathbf{F}(x, y, z) = xy^2\mathbf{i} + yz^2\mathbf{j} + zx^2\mathbf{k}$; S the surface of the region that lies between the cylinders $x^2 + y^2 = 4$ and $x^2 + y^2 = 9$, and the planes $z = -1$ and $z = 2$.

9 $\mathbf{F}(x, y, z) = 2xz\mathbf{i} + xyz\mathbf{j} + yz\mathbf{k}$; S the surface of the region bounded by the coordinate planes and the graphs of $x + 2z = 4$ and $y = 2$.

10 $\mathbf{F}(x, y, z) = x^3\mathbf{i} + y^3\mathbf{j} + z^3\mathbf{k}$; S the surface of the region that is inside both the cone $z = \sqrt{x^2 + y^2}$ and the sphere $x^2 + y^2 + z^2 = 25$.

In Exercise 11–14 verify the Divergence Theorem (18.30) by evaluating both the surface integral and the triple integral.

11 $\mathbf{F} = x\mathbf{i} + y\mathbf{j} + z\mathbf{k}$; S the graph of $x^2 + y^2 + z^2 = a^2$.

12 $\mathbf{F} = x^2\mathbf{i} + y^2\mathbf{j} + z^2\mathbf{k}$; S the surface of the cube bounded by the coordinate planes and the planes $x = a, y = a, z = a$, where $a > 0$.

13 $\mathbf{F} = (x + z)\mathbf{i} + (y + z)\mathbf{j} + (x + y)\mathbf{k}$; S the surface of the region

$$Q = \{(x, y, z): 0 \le y^2 + z^2 \le 1, 0 \le x \le 2\}.$$

14 $\mathbf{F} = |\mathbf{r}|^2\mathbf{r}$, where $\mathbf{r} = x\mathbf{i} + y\mathbf{j} + z\mathbf{k}$; S the graph of $x^2 + y^2 + z^2 = a^2$.

15 If $\iint_S \mathbf{F} \cdot \mathbf{n} \, dS = 0$ for every closed surface of the type considered in the Divergence Theorem, prove that div $\mathbf{F} = 0$.

16 Use the Divergence Theorem to prove that if a scalar function f has continuous second partial derivatives, then $\iiint_Q \nabla^2 f \, dV = \iint_S D_\mathbf{n} f \, dS$ where $D_\mathbf{n} f$ is the directional derivative of f in the direction of an outer normal $\mathbf{n}$ to S.

In Exercises 17–22 assume that S and Q satisfy the conditions of the Divergence Theorem.

17 If f and g are scalar functions that have continuous second partial derivatives, prove that

$$\iiint_Q (f\nabla^2 g + \nabla f \cdot \nabla g) \, dV = \iint_S (f\nabla g) \cdot \mathbf{n} \, dS.$$

(*Hint*: Let $\mathbf{F} = f\nabla g$ in (18.30).)

18 If f and g are scalar functions that have continuous second partial derivatives, prove that

$$\iiint_Q (f\nabla^2 g - g\nabla^2 f) \, dV = \iint_S (f\nabla g - g\nabla f) \cdot \mathbf{n} \, dS.$$

(*Hint*: Use the identity in Exercise 17 together with that obtained by interchanging f and g.)

19 If $\mathbf{F}$ is a conservative vector field with potential function f, and if div $\mathbf{F} = 0$, prove that $\iiint_Q \mathbf{F} \cdot \mathbf{F} \, dV = \iint_S f\mathbf{F} \cdot \mathbf{n} \, dS$.

20 If $\mathbf{r} = x\mathbf{i} + y\mathbf{j} + z\mathbf{k}$ and V is the volume of Q, prove that $V = \frac{1}{3} \iint_S \mathbf{r} \cdot \mathbf{n} \, dS$.

21 If $\mathbf{F}$ has continuous second partial derivatives prove that $\iint_S \text{curl } \mathbf{F} \cdot \mathbf{n} \, dS = 0$.

22 If $\mathbf{a}$ is a constant vector prove that $\iint_S \mathbf{a} \cdot \mathbf{n} \, dS = 0$.

A surface (or triple) integral of a vector function is defined as the sum of the surface (or triple) integrals of each component of the function. With this understanding, establish the results in

Exercises 23 and 24 ,where S and Q satisfy the conditions of the Divergence Theorem.

23 $\iint_S \mathbf{F} \times \mathbf{n} \, dS = -\iiint_Q \nabla \times \mathbf{F} \, dV$ (*Hint*: Apply the Divergence Theorem and Exercise 26 of Section 18.1 to $\mathbf{c} \times \mathbf{F}$, where $\mathbf{c}$ is an arbitrary constant vector.)

24 $\iint_S f\mathbf{n} \, dS = \iiint_Q \nabla f \, dV$ (*Hint*: Apply the Divergence Theorem to $f\mathbf{c}$ where $\mathbf{c}$ is an arbitrary constant vector.)

25 If $\mathbf{F}$ is orthogonal to S at each point (x, y, z), prove that $\iiint_Q \text{curl } \mathbf{F} \, dV = 0$.

26 If $\mathbf{r}$ is the position vector of (x, y, z) and $r = |\mathbf{r}|$, prove that $\iiint_Q \mathbf{r} \, dV = \frac{1}{2} \iint_S r^2 \mathbf{n} \, dS$.

18.7

STOKES' THEOREM

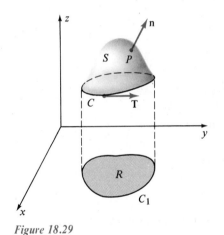

Figure 18.29

In (18.19) we stated the following vector form for Green's Theorem:

$$\oint_C \mathbf{F} \cdot \mathbf{T} \, ds = \iint_R \text{curl } \mathbf{F} \cdot \mathbf{k} \, dA$$

where the plane curve C is the boundary of R. This result may be extended to a piecewise smooth simple closed curve C in three dimensions that forms the boundary of a surface S. The sketch in Figure 18.29 illustrates a special case in which S is the graph of $z = f(x, y)$, where f has continuous first partial derivatives and the projection C_1 of C on the xy-plane is a curve that bounds a region R of the type considered in Green's Theorem. In the figure, $\mathbf{n}$ represents a unit *upper* normal to S. We take the positive direction along C as that which corresponds to the positive direction along C_1. The vector $\mathbf{T}$ is a unit tangent vector to C. If $\mathbf{F}$ is a vector function that has continuous partial derivatives in a region containing S, then we have the following theorem, named after the English mathematical physicist George G. Stokes (1819–1903).

Stokes' Theorem **(18.32)**

$$\oint_C \mathbf{F} \cdot \mathbf{T} \, ds = \iint_S \text{curl } \mathbf{F} \cdot \mathbf{n} \, dS$$

The formula in Theorem (18.32) may be stated as follows: *the line integral of the tangential component of* $\mathbf{F}$ *taken along* C *once in the positive direction equals the surface integral of the normal component of curl* $\mathbf{F}$ *over* S. Of course, the line integral may also be written $\oint_C \mathbf{F} \cdot d\mathbf{r}$, where $\mathbf{r}$ is the position vector of the point (x, y, z) on C. To consider more general situations than the one pictured in Figure 18.29 it is necessary to consider what is called an *oriented surface* S; that is, a surface on which a unit normal $\mathbf{n}$ and positive direction along C may be defined in a suitable manner. The interested reader should consult more advanced texts for a proof of Stokes' Theorem.

Example 1   Let S be the part of the paraboloid $z = 9 - x^2 - y^2$ where $z \geq 0$, and let C be the trace of S on the xy-plane. Verify Stokes' Theorem for the vector field $\mathbf{F} = 3z\mathbf{i} + 4x\mathbf{j} + 2y\mathbf{k}$.

Solution   We wish to show that the two integrals in Theorem (18.32) have the same value.

The surface is the same as that considered in Example 3 of Section 18.5 (see Figure 18.22) where we found that

$$\mathbf{n} = \frac{2x\mathbf{i} + 2y\mathbf{j} + \mathbf{k}}{\sqrt{4x^2 + 4y^2 + 1}}.$$

By (18.6)

$$\text{curl } \mathbf{F} = \begin{vmatrix} \mathbf{i} & \mathbf{j} & \mathbf{k} \\ \dfrac{\partial}{\partial x} & \dfrac{\partial}{\partial y} & \dfrac{\partial}{\partial z} \\ 3z & 4x & 2y \end{vmatrix} = 2\mathbf{i} + 3\mathbf{j} + 4\mathbf{k}.$$

Consequently,

$$\iint_S \text{curl } \mathbf{F} \cdot \mathbf{n} \, dS = \iint_S \frac{4x + 6y + 4}{\sqrt{4x^2 + 4y^2 + 1}} \, dS.$$

Using (18.21) to evaluate this surface integral gives us

$$\iint_S \text{curl } \mathbf{F} \cdot \mathbf{n} \, dS = \iint_R (4x + 6y + 4) \, dx \, dy$$

where R is the region in the xy-plane bounded by the circle of radius 3 with center at the origin. Changing to polar coordinates, we obtain

$$\iint_S \text{curl } \mathbf{F} \cdot \mathbf{n} \, dS = \int_0^{2\pi} \int_0^3 (4r \cos \theta + 6r \sin \theta + 4)r \, dr \, d\theta$$

$$= \int_0^{2\pi} \int_0^3 [r^2(4 \cos \theta + 6 \sin \theta) + 4r] \, dr \, d\theta$$

$$= \int_0^{2\pi} (36 \cos \theta + 54 \sin \theta + 18) \, d\theta$$

$$= \left[36 \sin \theta - 54 \cos \theta + 18\theta \right]_0^{2\pi} = 36\pi.$$

The line integral in Theorem (18.32) may be written

$$\oint_C \mathbf{F} \cdot \mathbf{T} \, ds = \oint_C \mathbf{F} \cdot d\mathbf{r} = \oint_C 3z \, dx + 4x \, dy + 2y \, dz$$

where C is the circle in the xy-plane having equation $x^2 + y^2 = 9$. Since $z = 0$ on C, this reduces to

$$\oint_C \mathbf{F} \cdot d\mathbf{r} = \oint_C 4x \, dy.$$

Using the parametric equations $x = 3 \cos t$, $y = 3 \sin t$ for C, we obtain

$$\oint_C \mathbf{F} \cdot d\mathbf{r} = \int_0^{2\pi} 4(3 \cos t) \, 3 \cos t \, dt = \int_0^{2\pi} 36 \cos^2 dt$$

$$= 18 \int_0^{2\pi} (1 + \cos 2t) \, dt = 36\pi$$

which is the same as the value of the surface integral. ∎

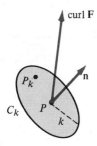

Figure 18.30

We may use Stokes' Theorem to obtain a physical interpretation for curl $\mathbf{F}$ in a manner similar to that in which the Divergence Theorem was used for div $\mathbf{F}$ in (18.31). If P is any point, let S_k be a circular disc of radius k with center at P and let C_k denote the circumference of S_k (see Figure 18.30). Applying Stokes' Theorem and a mean value theorem for double integrals leads to

$$\oint_{C_k} \mathbf{F} \cdot \mathbf{T} \, ds = \iint_{S_k} \text{curl } \mathbf{F} \cdot \mathbf{n} \, dS = \left[\text{curl } \mathbf{F} \cdot \mathbf{n} \right]_{P_k} (\pi k^2)$$

where P_k is some point in S_k and πk^2 is the area of S_k. Thus

$$\left[\text{curl } \mathbf{F} \cdot \mathbf{n} \right]_{P_k} = \frac{1}{\pi k^2} \oint_{C_k} \mathbf{F} \cdot \mathbf{T} \, ds.$$

If we let $k \to 0$, then $P_k \to P$ and hence

(18.33)
$$\left[\text{curl } \mathbf{F} \cdot \mathbf{n} \right]_P = \lim_{k \to 0} \frac{1}{\pi k^2} \oint_{C_k} \mathbf{F} \cdot \mathbf{T} \, ds.$$

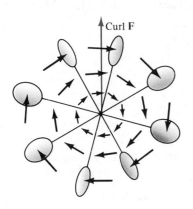

Figure 18.31

If $\mathbf{F}$ is the velocity field of a fluid, then the line integral $\oint_C \mathbf{F} \cdot \mathbf{T} \, ds$ is called the **circulation around C**. It measures the average tendency of the fluid to move, or *circulate*, around the curve. We see from (18.33) that $[\text{curl } \mathbf{F} \cdot \mathbf{n}]_P$ provides information about the motion of the fluid around the circumference of a circular disc that is perpendicular to $\mathbf{n}$, as the disc shrinks to a point. Since $[\text{curl } \mathbf{F} \cdot \mathbf{n}]_P$ will have its maximum value when $\mathbf{n}$ is parallel to curl $\mathbf{F}$, it follows that the direction of curl $\mathbf{F}$ at P is that for which the circulation around the boundary of a disc perpendicular to curl $\mathbf{F}$ will have its maximum value as the disc shrinks to a point. If we replace the disc S_k in Figure 18.30 by a miniature paddlewheel of the type illustrated in Figure 18.31, then the wheel will rotate most rapidly when $\mathbf{n}$ is parallel to curl $\mathbf{F}$. Occasionally curl $\mathbf{F} \cdot \mathbf{n}$ is called the **rotation of F about n** and is denoted by **rot F**, since it measures the tendency of the vector field to *rotate* about P.

Example 2 Let $\mathbf{F}(x, y, z) = 3(y^2 + 1)^{-1}\mathbf{i} + 0\mathbf{j} + 0\mathbf{k}$, where $|y| \le 4$. Describe the vector field $\mathbf{F}$ and discuss the circulation of $\mathbf{F}$ around several circles C_k that lie in the xy-plane. Where does curl $\mathbf{F} \cdot \mathbf{n}$ achieve its maximum value?

Solution   If we restrict our attention to the xy-plane, the vectors in $\mathbf{F}$ form the pattern shown in Figure 18.32. (Verify this fact.) The same pattern is repeated in any plane parallel to the xy-plane.

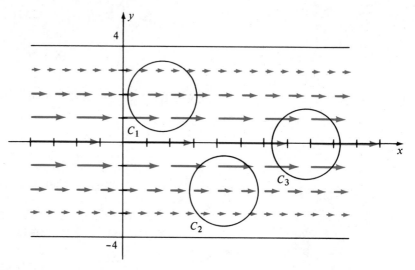

Figure 18.32

Let us regard $\mathbf{F}$ as the velocity field of a fluid that is flowing in a culvert. Note that the speed of the fluid is greatest along the x-axis and decreases as the point (x, y) approaches either of the lines $y = \pm 4$. If we consider the circle C_1, then evidently $\oint_{C_1} \mathbf{F} \cdot \mathbf{T}\, ds > 0$, since as C_1 is traversed in the positive direction, the tangential component $\mathbf{F} \cdot \mathbf{T}$ is large and positive on the lower half of C_1 and negative (but close to 0) on the upper half. If C_1 represents a small paddle wheel, it would rotate in the counterclockwise direction. By again considering values of $\mathbf{F} \cdot \mathbf{T}$ it is evident that $\oint_{C_2} \mathbf{F} \cdot \mathbf{T}\, ds < 0$ and $\oint_{C_3} \mathbf{F} \cdot \mathbf{T}\, ds = 0$. The paddlewheel corresponding to C_2 would rotate in a clockwise direction, and that for C_3 would not rotate at all. It is easy to show that

$$\text{curl } \mathbf{F} = \frac{6y}{(y^2 + 1)^2}\, \mathbf{k}.$$

Since the unit normal $\mathbf{n}$ is $\mathbf{k}$ we have

$$\text{curl } \mathbf{F} \cdot \mathbf{n} = \text{curl } \mathbf{F} \cdot \mathbf{k} = \frac{6y}{(y^2 + 1)^2}.$$

Observe that curl $\mathbf{F} \cdot \mathbf{n}$ is positive for points above the x-axis, negative for points below the x-axis, and 0 for points on the x-axis. Differentiating,

$$D_y(\text{curl } \mathbf{F} \cdot \mathbf{n}) = \frac{6(1 - 3y^2)}{(y^2 + 1)^3}$$

and hence critical numbers for curl $\mathbf{F} \cdot \mathbf{n}$ are $y = \pm 1/\sqrt{3} \approx \pm 0.577$. It follows from the first derivative test that these lead to maximum values. Thus the maximum values for curl $\mathbf{F} \cdot \mathbf{n}$ occur at any point on either of the horizontal lines $y = \pm 1/\sqrt{3}$. $\blacksquare$

To bring out another connection between curl $\mathbf{F}$ and rotational aspects of motion, consider the special case where a fluid is rotating uniformly about an axis, as if it were a rigid body. If we concentrate on a single fluid particle at the point $P(x, y, z)$, where x, y, and z are functions of time t, then we have a situation similar to that illustrated in Figure 15.12, where $\mathbf{r} = x\mathbf{i} + y\mathbf{j} + z\mathbf{k}$ is the position vector of P and $\boldsymbol{\omega} = \omega_1\mathbf{i} + \omega_2\mathbf{j} + \omega_3\mathbf{k}$ is the (constant) angular velocity. If we denote the velocity $d\mathbf{r}/dt$ of P by $\mathbf{F}$, then as in Section 15.3,

$$\mathbf{F} = \boldsymbol{\omega} \times \mathbf{r} = (\omega_2 z - \omega_3 y)\mathbf{i} + (\omega_3 x - \omega_1 z)\mathbf{j} + (\omega_1 y - \omega_2 x)\mathbf{k}.$$

By (18.6),

$$\text{curl } \mathbf{F} = \begin{vmatrix} \mathbf{i} & \mathbf{j} & \mathbf{k} \\ \dfrac{\partial}{\partial x} & \dfrac{\partial}{\partial y} & \dfrac{\partial}{\partial z} \\ \omega_2 z - \omega_3 y & \omega_3 x - \omega_1 z & \omega_1 y - \omega_2 x \end{vmatrix}.$$

Using the fact that $\boldsymbol{\omega}$ is a constant vector it may be verified that

$$\text{curl } \mathbf{F} = 2\omega_1\mathbf{i} + 2\omega_2\mathbf{j} + 2\omega_3\mathbf{k} = 2\boldsymbol{\omega}.$$

This shows that the magnitude of curl $\mathbf{F}$ is twice the angular velocity, and the direction of curl $\mathbf{F}$ is along the axis of rotation.

In our discussion of line integrals that are independent of path we introduced the notion of a simply connected region in the plane (see page 845). This concept can be extended to three dimensions; however, the usual definition requires properties of curves and surfaces studied in more advanced courses. For our purposes a region D in three dimensions will be called **simply connected** if every simple closed curve C in D is the boundary of a surface S that satisfies the conditions of Stokes' Theorem. In regions of this type any simple closed curve can be continuously shrunk to a point in D without crossing the boundary of D. For example, the region inside of a sphere or a rectangular parallelepiped is simply connected. The region inside of a torus (a doughnut-shaped surface) is not simply connected. Using this rather restrictive definition we can establish the following result.

Theorem (**18.34**)

If $\mathbf{F}(x, y, z)$ has continuous partial derivatives throughout a simply connected region D, then curl $\mathbf{F} = \mathbf{0}$ in D if and only if $\oint_C \mathbf{F} \cdot d\mathbf{r} = 0$ for every simple closed curve C in D.

Proof   If curl $\mathbf{F} = \mathbf{0}$, then by Stokes' Theorem,

$$\oint_C \mathbf{F} \cdot d\mathbf{r} = \iint_S \text{curl } \mathbf{F} \cdot \mathbf{n} \, dS = 0.$$

Conversely, suppose $\oint_C \mathbf{F} \cdot d\mathbf{r} = 0$ for every simple closed curve C. If at some point P, curl $\mathbf{F} \neq \mathbf{0}$, then by continuity there is a subregion of D containing P throughout which curl $\mathbf{F} \neq \mathbf{0}$. If, in this subregion, we choose a circular disc of the type illustrated in Figure 18.30, where $\mathbf{n}$ is parallel to curl $\mathbf{F}$, then

$$\oint_{C_k} \mathbf{F} \cdot d\mathbf{r} = \iint_{S_k} \text{curl } \mathbf{F} \cdot \mathbf{n} \, dS > 0,$$

a contradiction. Consequently, curl $\mathbf{F} = \mathbf{0}$ throughout D.   □

If curl $\mathbf{F} = \mathbf{0}$ in D, then the circulation around every closed curve C is 0. (Why?) This also means that the *rotation* of $\mathbf{F}$ about any unit vector $\mathbf{n}$ is 0. For this reason a vector field $\mathbf{F}$ such that curl $\mathbf{F} = \mathbf{0}$ is often called **irrotational**.

It can be shown that if $\mathbf{F}$ is continuous on a suitable region, then the vanishing of the line integral $\oint_C \mathbf{F} \cdot d\mathbf{r}$ around every simple closed curve is equivalent to independence of path. Moreover, as in Theorem (18.14), independence of path is equivalent to $\mathbf{F} = \nabla f$ for some function f (that is, $\mathbf{F}$ is conservative). Combining these facts with Theorem (18.34), we obtain the following.

Theorem (18.35)

> If $\mathbf{F}(x, y, z)$ has continuous first partial derivatives throughout a simply connected region, then the following conditions are equivalent.
>
> (i) $\mathbf{F}$ is conservative.
>
> (ii) $\displaystyle\int_C \mathbf{F} \cdot d\mathbf{r}$ is independent of path.
>
> (iii) $\displaystyle\oint_C \mathbf{F} \cdot d\mathbf{r} = 0$ for every simple closed curve C.
>
> (iv) curl $\mathbf{F} = \mathbf{0}$.

Example 3 Prove that if $\mathbf{F}(x, y, z) = (3x^2 + y^2)\mathbf{i} + 2xy\mathbf{j} - 3z^2\mathbf{k}$, then $\mathbf{F}$ is conservative.

Solution   The function $\mathbf{F}$ has continuous partial derivatives for all (x, y, z). Since

$$\mathbf{F} = \begin{vmatrix} \mathbf{i} & \mathbf{j} & \mathbf{k} \\ \dfrac{\partial}{\partial x} & \dfrac{\partial}{\partial y} & \dfrac{\partial}{\partial z} \\ 3x^2 + y^2 & 2xy & -3z^2 \end{vmatrix}$$

$$= (0 - 0)\mathbf{i} + (0 - 0)\mathbf{j} + (2y - 2y)\mathbf{k} = \mathbf{0},$$

$\mathbf{F}$ is conservative by (iv) of Theorem (18.35).   ∎

In advanced mathematics and applications it is useful to study pairs of vector fields **F** and **G** that have the form

$$\mathbf{F}(x, y) = v(x, y)\mathbf{i} + u(x, y)\mathbf{j}$$
$$\mathbf{G}(x, y) = u(x, y)\mathbf{i} - v(x, y)\mathbf{j}$$

where the functions u and v have continuous partial derivatives. Note that $\mathbf{F}(x, y)$ and $\mathbf{G}(x, y)$ are orthogonal at every point since

$$\mathbf{F} \cdot \mathbf{G} = vu - uv = 0.$$

Moreover, under suitable restrictions (see Theorem 18.16) it can be shown that **F** and **G** are both conservative if and only if

(18.36)
$$\frac{\partial u}{\partial x} = \frac{\partial v}{\partial y} \quad \text{and} \quad \frac{\partial u}{\partial y} = -\frac{\partial v}{\partial x}.$$

These conditions on u and v are called the **Cauchy–Riemann equations** and are exploited extensively in the study of complex variables. We see from (18.36) that

$$\frac{\partial^2 u}{\partial x^2} = \frac{\partial^2 v}{\partial x \, \partial y} = \frac{\partial^2 v}{\partial y \, \partial x} = -\frac{\partial^2 u}{\partial y^2}$$

and hence
$$\nabla^2 u = \frac{\partial^2 u}{\partial x^2} + \frac{\partial^2 u}{\partial y^2} = 0.$$

Similarly, we can show that

$$\nabla^2 v = \frac{\partial^2 v}{\partial x^2} + \frac{\partial^2 v}{\partial y^2} = 0.$$

Consequently, u and v satisfy Laplace's equation and hence are harmonic. (See Exercise 36 of Section 18.1).

EXERCISES 18.7

In Exercises 1–4 verify Stokes' Theorem for **F** and S.

1 $\mathbf{F} = y^2\mathbf{i} + z^2\mathbf{j} + x^2\mathbf{k}$; S the first octant portion of the plane $x + y + z = 1$.

2 $\mathbf{F} = 2y\mathbf{i} - z\mathbf{j} + 3\mathbf{k}$; S the part of the paraboloid $z = 4 - x^2 - y^2$ that lies inside the cylinder $x^2 + y^2 = 1$.

3 $\mathbf{F} = z\mathbf{i} + x\mathbf{j} + y\mathbf{k}$; S the hemisphere $z = (a^2 - x^2 - y^2)^{1/2}$.

4 $\mathbf{F} = x^2\mathbf{i} + y^2\mathbf{j} + z^2\mathbf{k}$; S the part of the cone $z = (x^2 + y^2)^{1/2}$ cut off by the plane $z = 1$.

5 If $\mathbf{F} = (3z - \sin x)\mathbf{i} + (x^2 + e^y)\mathbf{j} + (y^3 - \cos z)\mathbf{k}$, use Stokes' Theorem to evaluate $\oint_C \mathbf{F} \cdot d\mathbf{r}$ where C is the curve $x = \cos t, y = \sin t, z = 1; 0 \le t \le 2\pi$.

6 If $\mathbf{F} = yz\mathbf{i} + xy\mathbf{j} + xz\mathbf{k}$, use Stokes' Theorem to evaluate $\oint_C \mathbf{F} \cdot d\mathbf{r}$, where C is the square with vertices $(0, 0, 2)$, $(1, 0, 2), (1, 1, 2), (0, 1, 2)$.

7 If $\mathbf{F} = 2y\mathbf{i} + e^z\mathbf{j} - \arctan x\mathbf{k}$, use Stokes' Theorem to evaluate $\iint_S \text{curl } \mathbf{F} \cdot \mathbf{n} \, dS$, where S is the part of the paraboloid $z = 4 - x^2 - y^2$ cut off by the xy-plane.

8 If $\mathbf{F} = xy^2\mathbf{i} + yz^2\mathbf{j} + zx^2\mathbf{k}$, use Stokes' Theorem to evaluate $\iint_S \text{curl } \mathbf{F} \cdot \mathbf{n}\, dS$, where S is the triangle with vertices $(1, 0, 0)$, $(0, 2, 0)$, $(0, 0, 3)$.

Establish the identities in Exercises 9–11 under the assumption that C and S satisfy the conditions of Stokes' Theorem.

9 $\oint_C f\nabla g \cdot d\mathbf{r} = \iint_S (\nabla f \times \nabla g) \cdot \mathbf{n}\, dS$, where f and g are scalar functions.

10 $\oint_C \mathbf{a} \times \mathbf{r} \cdot d\mathbf{r} = 2\mathbf{a} \cdot \iint_S \mathbf{n}\, dS$, where $\mathbf{a}$ is a constant vector.

11 $\oint_C f\, d\mathbf{r} = \iint_S \mathbf{n} \times \nabla f\, dS$, where f is a scalar function. (*Hint*: Compare with Exercise 18 of Section 15.7.)

12 If $\mathbf{F}(x, y, z) = y\mathbf{i} + (x + e^z)\mathbf{j} + (1 + ye^z)\mathbf{k}$, use (ii) of Theorem (18.35) to prove that $\mathbf{F}$ is conservative.

13 If u and v have continuous partial derivatives and satisfy the Cauchy–Riemann equations (18.36), prove that v satisfies Laplace's equation $\nabla^2 v = 0$.

14 Let $\mathbf{n}$ denote the unit outer normal at any point P on the surface of a sphere S. If $\mathbf{F}$ has continuous first partial derivatives within and on S, prove that $\iint_S \text{curl } \mathbf{F} \cdot \mathbf{n}\, dS = 0$ by using (a) the Divergence Theorem; (b) Stokes' Theorem.

15 If M, N, and P are functions of x, y, and z that have continuous first partial derivatives on a simply connected region, prove that

$$\int_C M(x, y, z)\, dx + N(x, y, z)\, dy + P(x, y, z)\, dz$$

is independent of path if and only if

$$\frac{\partial M}{\partial y} = \frac{\partial N}{\partial x}, \quad \frac{\partial M}{\partial z} = \frac{\partial P}{\partial x}, \quad \frac{\partial N}{\partial z} = \frac{\partial P}{\partial y}.$$

16 If S and C satisfy the conditions of Stoke's Theorem, and if $\mathbf{F}$ is a constant vector function, use (18.32) to prove that $\oint_C \mathbf{F} \cdot \mathbf{T}\, ds = 0$.

18.8

TRANSFORMATIONS OF COORDINATES

In previous sections of this text we studied many different types of scalar and vector functions. Let us now consider a function T whose domain and range are subsets of $\mathbb{R} \times \mathbb{R}$. Thus, to each ordered pair (x, y) in the domain there corresponds a unique ordered pair (u, v) in the range such that $T(x, y) = (u, v)$. We may represent T geometrically as illustrated in Figure 18.33, where the point (u, v) in the uv-plane corresponds to the point (x, y) in the xy-plane. As usual, (u, v) is referred to as the **image** of (x, y) under T.

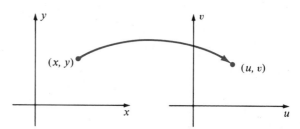

Figure 18.33 $T(x, y) = (u, v)$

Since each pair (u, v) is determined by (x, y), it follows that u and v are functions of x and y; that is,

(18.37) $$u = f(x, y), \qquad v = g(x, y)$$

where f and g have the same domain as T. We shall refer to equations (18.37), or to the function T, as a **transformation of coordinates** from the xy-plane to the uv-plane.

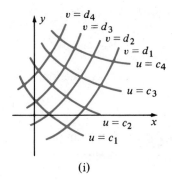

(i)

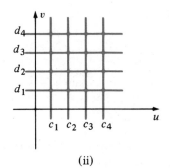

(ii)

Figure 18.34

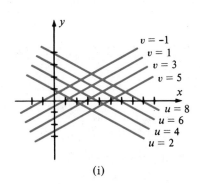

(i)

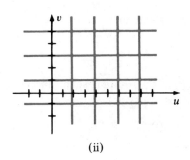

(ii)

Figure 18.35

Given the transformation of coordinates (18.37), let us partition a region in the uv-plane by means of vertical lines $u = c_1$, $u = c_2$, $u = c_3, \ldots$, and horizontal lines $v = d_1$, $v = d_2$, $v = d_3, \ldots$. The corresponding level curves for f and g, that is

$$u = f(x, y) = c_i, \quad v = g(x, y) = d_j$$

where $i = 1, 2, 3, \ldots$ and $j = 1, 2, 3, \ldots$ determine a corresponding "curvilinear" partition of a region in the xy-plane. Figure 18.34 illustrates four curves of each type associated with a certain transformation T.

We shall refer to the curves $u = f(x, y) = c_i$ and $v = g(x, y) = d_j$ in the xy-plane as **u-curves** and **v-curves**, respectively. Of course, the types of curves obtained depend on the nature of the functions f and g. The next example illustrates a case where each u-curve and v-curve is a line.

Example 1 Let T be the transformation of coordinates defined by

$$u = x + 2y, \quad v = x - 2y.$$

(a) Find the images of $(0, 1)$, $(1, 2)$, and $(2, -3)$.

(b) Sketch, in the uv-plane, the vertical lines $u = 2, u = 4, u = 6, u = 8$, and the horizontal lines $v = -1, v = 1, v = 3, v = 5$. Sketch the corresponding u-curves and v-curves in the xy-plane.

Solution

(a) To find the images of the given ordered pairs we may use the formula

$$T(x, y) = (u, v) = (x + 2y, x - 2y).$$

This gives us

$$T(0, 1) = (2, -2), \quad T(1, 2) = (5, -3), \quad T(2, -3) = (-4, 8).$$

(b) The desired vertical and horizontal lines in the uv-plane are sketched in (ii) of Figure 18.35. The u-curves in the xy-plane are the lines

$$x + 2y = 2, \quad x + 2y = 4, \quad x + 2y = 6, \quad x + 2y = 8$$

and the v-curves are the lines

$$x - 2y = -1, \quad x - 2y = 1, \quad x - 2y = 3, \quad x - 2y = 5$$

as illustrated in (i) of the figure. ∎

The transformation T in Example 1 is one-to-one, in the sense that different ordered pairs in the xy-plane have different images in the uv-plane (see Exercise 13). In general, if T is a one-to-one transformation of coordinates, then by reversing the correspondence we obtain a transformation T^{-1} from the *uv*-plane to the *xy*-plane called the **inverse** of T. We may specify T^{-1} by means of equations of the form

$$x = \varphi(u, v), \quad y = \psi(u, v).$$

Note that T, followed by T^{-1}, or vice versa, is an identity function.

Example 2

(a) Find the inverse of the transformation T defined in Example 1.

(b) Find the curve in the *uv*-plane that maps into the ellipse $x^2 + 4y^2 = 1$ under T^{-1}.

Solution

(a) The transformation T is given by

$$u = x + 2y, \quad v = x - 2y.$$

If we add corresponding sides of these equations we obtain $u + v = 2x$. Subtracting corresponding sides leads to $u - v = 4y$. Thus T^{-1} is given by

$$x = \tfrac{1}{2}(u + v), \quad y = \tfrac{1}{4}(u - v).$$

(b) Since x and y are related by means of the last two equations in part (a), the points (u, v) that map into $x^2 + 4y^2 = 1$ must satisfy the equation

$$[\tfrac{1}{2}(u + v)]^2 + 4[\tfrac{1}{4}(u - v)]^2 = 1.$$

This simplifies to

$$u^2 + v^2 = 2.$$

It follows that the circle of radius $\sqrt{2}$ with center at the origin in the *uv*-plane maps into the ellipse $x^2 + 4y^2 = 1$ under T^{-1}. ∎

Example 3 If $P(x, y)$ is any point other than $(0, 0)$ in the *xy*-plane, let r and θ be defined as follows:

$$r = \sqrt{x^2 + y^2}$$
$\theta = $ the smallest nonnegative angle from the
positive *x*-axis to the position vector $\overrightarrow{OP}$.

Thus, r and θ are polar coordinates for P, and the preceding equalities define a transformation of coordinates T from the *xy*-plane to the *rθ*-plane.

(a) Sketch, in the *rθ*-plane, the graphs of $r = 1$, $r = 2$, $r = 3$ and $\theta = \tfrac{1}{2}$, $\theta = \tfrac{3}{2}, \theta = \tfrac{5}{2}$. Sketch the corresponding *r*-curves and *θ*-curves in the *xy*-plane.

(b) Find T^{-1}.

Solution

(a) Note that we must exclude points (r, θ) in the *rθ*-plane such that $r \le 0$, $\theta < 0$, or $\theta \ge 2\pi$. (Why?) The desired graphs are sketched in Figure 18.36.

(b) It is not difficult to show that T is one-to-one (see Exercise 14). The familiar polar coordinate formulas

$$x = r \cos \theta, \quad y = r \sin \theta$$

define the inverse transformation T^{-1}. ∎

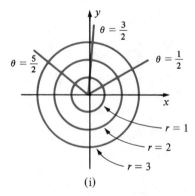

(i)

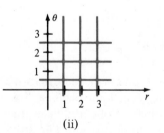

(ii)

Figure 18.36

The discussion in this section can be extended to three dimensions. Thus, a transformation of coordinates T from an xyz-coordinate system to a uvw-coordinate system is determined by three equations of the form

$$u = f(x, y, z), \quad v = g(x, y, z), \quad w = h(x, y, z)$$

where the functions f, g, and h are defined on some region R. In this case the planes $u = c_i$, $v = d_j$, $z = e_k$, where c_i, d_j, e_k are real numbers, determine **u-surfaces**, **v-surfaces**, and **w-surfaces** in the xyz-coordinate system (see Exercises 15 and 16).

EXERCISES 18.8

In Exercises 1–10 let T be the transformation from the xy-plane to the uv-plane defined by the given equations.

(a) Describe the u-curves and the v-curves.

(b) If T is one-to-one, find equations $x = \varphi(u, v)$, $y = \psi(u, v)$ that specify T^{-1}.

1 $u = 3x, v = 5y$

2 $u = e^x, v = e^y$

3 $u = x - y, v = 3y + 2x$

4 $u = 5x - 4y, v = 2x + 3y$

5 $u = 2x + y, v = xy$

6 $u = x^3, v = x + y$

7 $u = x^2 + 4y^2, v = 4x^2 - y^2$

8 $u = x + y, v = x^2 - y$

9 $u = x^2 - y^2, v = 2xy$

10 $u = y/(x^2 + y^2), v = x/(x^2 + y^2), x^2 + y^2 \neq 0$

11 If T is the transformation defined in Exercise 1, find the image of the rectangle with vertices $(0, 0)$, $(0, 1)$, $(2, 1)$, and $(2, 0)$. What is the image of the unit circle $x^2 + y^2 = 1$?

12 If T is the transformation defined in Exercise 3, find the image of the triangle with vertices $(0, 0)$, $(0, 1)$, $(2, 0)$. What is the image of the line $x + 2y = 1$?

13 Prove that the transformation T of Example 1 is one-to-one.

14 Prove that the transformation T of Example 3 is one-to-one.

In Exercises 15 and 16 let T be the indicated transformation from the xyz-coordinate system to the uvw-coordinate system. Find the u-surfaces, the v-surfaces, and the w-surfaces. If T is one-to-one, determine T^{-1}.

15 $u = 2x + y, v = y + z, w = x + y + z$

16 $u = x^2 + y^2 + z^2, v = x^2 + y^2 - z^2, w = x^2 - y^2 - z^2$.

18.9

CHANGE OF VARIABLES IN MULTIPLE INTEGRALS

In Section 5.6 we developed a technique for making a change of variables in a definite integral $\int_a^b f(x)\, dx$. Under suitable conditions, if we let $x = g(u)$, then $dx = g'(u)\, du$ and

$$\int_a^b f(x)\, dx = \int_c^d f(g(u))g'(u)\, du$$

where $a = g(c)$ and $b = g(d)$. In this section we shall obtain an analogous formula for a change of variables in a double integral $\iint_R F(x, y)\, dA$. In particular, suppose we let

$$x = f(u, v), \quad y = g(u, v)$$

where f and g have continuous second partial derivatives. These equations define a transformation of coordinates W from the uv-plane to the xy-plane. If we use these equations to substitute for x and y in the double integral, then the integrand becomes a function of u and v. Our objective is to find a region S in the uv-plane that maps onto R under W, as illustrated in Figure 18.37, and such that

$$\iint_R F(x, y)\, dA = \iint_S F(f(u, v), g(u, v))\, dA.$$

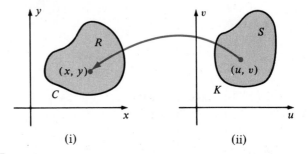

Figure 18.37

To change variables in a double integral $\iint_R F(x, y)\, dA$ it is necessary to impose suitable restrictions on the region R and the function F. Specifically, we shall assume that R consists of all points that are either inside or on a piecewise smooth simple closed curve C, and that F has continuous first partial derivatives throughout an open region containing R. We shall also require that the transformation W maps a region S of the uv-plane in a one-to-one manner onto R, and that S is bounded by a piecewise smooth simple closed curve K that W maps onto C. Finally, we impose the condition that as (u, v) traces K once in the positive direction, the corresponding point (x, y) traces C once, either in the positive or negative direction. It is possible to weaken these conditions; however, that is beyond the scope of our work.

The function of u and v introduced in the next definition will be used in the change of variable process. It is named after the mathematician C. G. Jacobi (1804–51).

Definition (18.38)

If $x = f(u, v)$ and $y = g(u, v)$, then the **Jacobian** of x and y with respect to u and v, denoted by $\partial(x, y)/\partial(u, v)$, is

$$\frac{\partial(x, y)}{\partial(u, v)} = \begin{vmatrix} \dfrac{\partial x}{\partial u} & \dfrac{\partial x}{\partial v} \\[2mm] \dfrac{\partial y}{\partial u} & \dfrac{\partial y}{\partial v} \end{vmatrix} = \frac{\partial x}{\partial u}\frac{\partial y}{\partial v} - \frac{\partial y}{\partial u}\frac{\partial x}{\partial v}.$$

In the next theorem all symbols have the same meanings as before, and it is assumed that the regions and functions satisfy the conditions we have discussed. In the statement of the theorem the notations $dx\, dy$ and $du\, dv$ are used in place of dA so that there is no misunderstanding about the region of integration.

Theorem (18.39)

If $x = f(u, v)$, $y = g(u, v)$ is a transformation of coordinates, then

$$\iint_R F(x, y) \, dx \, dy = \pm \iint_S F(f(u, v), g(u, v)) \frac{\partial(x, y)}{\partial(u, v)} \, du \, dv.$$

The + sign or the − sign is chosen if, as (u, v) traces the boundary K of S once in the positive direction, the corresponding point (x, y) traces the boundary C of R once in the positive or negative direction, respectively.

Proof   Let us begin by choosing $G(x, y)$ such that $\partial G/\partial x = F$. Applying Green's Theorem (18.17) with $G = N$ gives us

(a)
$$\iint_R F(x, y) \, dx \, dy = \iint_R \frac{\partial}{\partial x} [G(x, y)] \, dx \, dy = \oint_C G(x, y) \, dy.$$

Suppose the curve K in the uv-plane is given parametrically by

$$u = \varphi(t), v = \psi(t), \quad \text{where } a \le t \le b.$$

From our assumptions on the transformation, parametric equations for the curve C in the xy-plane are

(b)
$$x = f(u, v) = f(\varphi(t), \psi(t))$$
$$y = g(u, v) = g(\varphi(t), \psi(t))$$

where $a \le t \le b$. We may, therefore, evaluate the line integral $\oint_C G(x, y) \, dy$ in (a) by formal substitutions for x and y. To simplify the notation, let

$$H(t) = G[f(\varphi(t), \psi(t)), g(\varphi(t), \psi(t))].$$

Applying the Chain Rule to y in (b) gives us

$$\frac{dy}{dt} = \frac{\partial y}{\partial u} \frac{du}{dt} + \frac{\partial y}{\partial v} \frac{dv}{dt} = \frac{\partial y}{\partial u} \varphi'(t) + \frac{\partial y}{\partial v} \psi'(t).$$

Consequently,

$$\oint_C G(x, y) \, dy = \oint_C H(t) \frac{dy}{dt} \, dt$$
$$= \int_a^b H(t) \left[\frac{\partial y}{\partial u} \varphi'(t) + \frac{\partial y}{\partial v} \psi'(t) \right] dt.$$

Since $du = \varphi'(t) \, dt$ and $dv = \psi'(t) \, dt$, we may regard the last line integral as a line integral around the curve K in the uv-plane. Thus

(c)
$$\oint_C G(x, y) \, dy = \pm \oint_K G \frac{\partial y}{\partial u} \, du + G \frac{\partial y}{\partial v} \, dv$$

where for simplicity we have used G as an abbreviation for $G(f(u, v), g(u, v))$. The choice of the + sign or the − sign is determined by letting t vary from a to b and noting whether (x, y) traces C in the same or opposite direction, respectively, as (u, v) traces K.

The line integral on the right in (c) has the form

$$\oint_K M \, du + N \, dv$$

where

$$M = G \frac{\partial y}{\partial u} \quad \text{and} \quad N = G \frac{\partial y}{\partial v}.$$

Applying Green's Theorem, we obtain

$$\oint_K M \, du + N \, dv$$

$$= \iint_S \left(\frac{\partial N}{\partial u} - \frac{\partial M}{\partial v} \right) du \, dv$$

$$= \iint_S \left(G \frac{\partial^2 y}{\partial u \, \partial v} + \frac{\partial G}{\partial u} \frac{\partial y}{\partial v} - G \frac{\partial^2 y}{\partial v \, \partial u} - \frac{\partial G}{\partial v} \frac{\partial y}{\partial u} \right) du \, dv$$

$$= \iint_S \left[\left(\frac{\partial G}{\partial x} \frac{\partial x}{\partial u} + \frac{\partial G}{\partial y} \frac{\partial y}{\partial u} \right) \frac{\partial y}{\partial v} - \left(\frac{\partial G}{\partial x} \frac{\partial x}{\partial v} + \frac{\partial G}{\partial y} \frac{\partial y}{\partial v} \right) \frac{\partial y}{\partial u} \right] du \, dv$$

$$= \iint_S \frac{\partial G}{\partial x} \left(\frac{\partial x}{\partial u} \frac{\partial y}{\partial v} - \frac{\partial y}{\partial u} \frac{\partial x}{\partial v} \right) du \, dv.$$

Using the fact that $\partial G / \partial x = F(x, y)$, together with the definition of Jacobian (18.38), gives us

$$\oint_K M \, du + N \, dv = \iint_S F(f(u, v), g(u, v)) \frac{\partial(x, y)}{\partial(u, v)} \, du \, dv.$$

Combining this formula with (a) and (c) leads to the desired result.   □

Example 1   Evaluate $\iint_R e^{(y-x)/(y+x)} \, dx \, dy$ where R is the region in the xy-plane bounded by the trapezoid with vertices $(0, 1)$, $(0, 2)$, $(2, 0)$, and $(1, 0)$.

Solution   Since $y - x$ and $y + x$ appear in the integrand, it is convenient to let

$$u = y - x, \quad v = y + x.$$

These equations define a transformation T from the xy-plane to the uv-plane. To apply Theorem (18.39), it is necessary to use the inverse transformation T^{-1}. To find T^{-1} we solve the preceding equations for x and y in terms of u and v, obtaining

$$x = \tfrac{1}{2}(v - u), \quad y = \tfrac{1}{2}(v + u).$$

The Jacobian of this transformation from the uv-plane to the xy-plane is

$$\frac{\partial(x, y)}{\partial(u, v)} = \begin{vmatrix} -\tfrac{1}{2} & \tfrac{1}{2} \\ \tfrac{1}{2} & \tfrac{1}{2} \end{vmatrix} = -\frac{1}{4} - \frac{1}{4} = -\frac{1}{2}.$$

The region R is sketched in (i) of Figure 18.38.

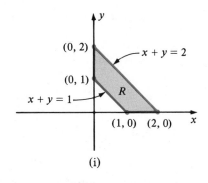

(i)

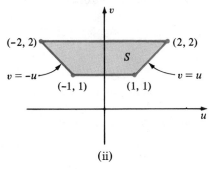

(ii)

Figure 18.38

To determine the region S in the uv-plane that corresponds to R, we note that the sides of R lie on the lines

$$x = 0, \quad y = 0, \quad x + y = 1, \quad x + y = 2.$$

Using the equations for the transformations T and T^{-1} we see that the corresponding curves in the uv-plane are

$$v = u, \quad v = -u, \quad v = 1, \quad v = 2,$$

respectively. These lines form the boundary of the trapezoidal region S shown in (ii) of Figure 18.38. It is left to the reader to verify that interior points of S correspond to interior points of R, and that as (u, v) traces the boundary of S once in the *positive* (counterclockwise) direction, the corresponding point (x, y) traces the boundary of R once in the *negative* (clockwise) direction. Applying Theorem 18.39,

$$\iint_R e^{(y-x)/(y+x)} \, dx \, dy = -\int_1^2 \int_{-v}^v e^{u/v}\left(-\frac{1}{2}\right) du \, dv$$

$$= \frac{1}{2}\int_1^2 \left[v e^{u/v} \right]_{-v}^v dv = \frac{1}{2}\int_1^2 v(e - e^{-1}) \, dv$$

$$= \frac{1}{2}(e - e^{-1})\left[\frac{v^2}{2}\right]_1^2 = \frac{3}{4}(e - e^{-1}). \qquad \blacksquare$$

Example 2   Evaluate $\iint_R e^{-(x^2+y^2)} \, dx \, dy$ where R is the region in the xy-plane bounded by the circle $x^2 + y^2 = a^2$.

Solution   The polar coordinate substitutions

$$x = r \cos \theta, \quad y = r \sin \theta$$

determine a transformation from the $r\theta$-plane to the xy-plane, where

$$\frac{\partial(x, y)}{\partial(r, \theta)} = \begin{vmatrix} \cos \theta & -r \sin \theta \\ \sin \theta & r \cos \theta \end{vmatrix} = r.$$

A transformation similar to that defined in Example 3 of Section 18.8 maps the rectangle S in the $r\theta$-plane bounded by $r = 0, r = a, \theta = 0$, and $\theta = 2\pi$ onto R. Moreover, as (r, θ) traces the boundary of S once in the positive direction, the corresponding point (x, y) traces R once in the positive direction. Hence, from Theorem (18.39),

$$\iint_R e^{-(x^2+y^2)} \, dx \, dy = \int_0^{2\pi} \int_0^a e^{-r^2} r \, dr \, d\theta = \int_0^{2\pi} \left[-\frac{1}{2} e^{-r^2} \right]_0^a d\theta$$

$$= -\frac{1}{2}(e^{-a^2} - 1)2\pi = \pi(1 - e^{-a^2}). \qquad \blacksquare$$

The theory we have developed can be extended to triple integrals. Given a transformation

$$x = f(u, v, w), \quad y = g(u, v, w), \quad z = h(u, v, w)$$

from a uvw-coordinate system to an xyz-coordinate system, we define the **Jacobian** $\partial(x, y, z)/\partial(u, v, w)$ of the transformation by

$$(18.40) \qquad \frac{\partial(x, y, z)}{\partial(u, v, w)} = \begin{vmatrix} \dfrac{\partial x}{\partial u} & \dfrac{\partial x}{\partial v} & \dfrac{\partial x}{\partial w} \\[2mm] \dfrac{\partial y}{\partial u} & \dfrac{\partial y}{\partial v} & \dfrac{\partial y}{\partial w} \\[2mm] \dfrac{\partial z}{\partial u} & \dfrac{\partial z}{\partial v} & \dfrac{\partial z}{\partial w} \end{vmatrix}.$$

If R and S are corresponding regions in the xyz- and uvw-coordinate systems, then under suitable restrictions,

$$(18.41) \qquad \iiint_R F(x, y, z)\, dx\, dy\, dz = \pm \iiint_S G(u, v, w) \frac{\partial(x, y, z)}{\partial(u, v, w)}\, du\, dv\, dw$$

where $G(u, v, w)$ is the expression obtained by substituting for x, y, and z in $F(x, y, z)$.

To illustrate, if we use the spherical coordinate formulas (14.50), then

$$x = \rho \sin \phi \cos \theta$$
$$y = \rho \sin \phi \sin \theta$$
$$z = \rho \cos \phi.$$

It can be shown (see Exercise 17), that

$$\frac{\partial(x, y, z)}{\partial(\rho, \phi, \theta)} = \rho^2 \sin \phi.$$

Hence (18.41) takes on the form

$$\iiint_R F(x, y, z)\, dx\, dy\, dz = \pm \iiint_S G(\rho, \theta, \phi)\rho^2 \sin \phi \, d\rho\, d\phi\, d\theta$$

which is in agreement with formula (17.29), obtained in an intuitive manner.

EXERCISES 18.9

In Exercises 1–4 find the Jacobian $\partial(x, y)/\partial(u, v)$.

1 $x = u^2 - v^2$, $y = 2uv$

2 $x = e^u \sin v$, $y = e^u \cos v$

3 $x = ve^{-2u}$, $y = u^2 e^{-v}$

4 $x = u/(u^2 + v^2)$, $y = v/(u^2 + v^2)$

In Exercises 5 and 6 find $\partial(x, y, z)/\partial(u, v, w)$.

5 $x = 2u + 3v - w$, $y = v - 5w$, $z = u + 4w$

6 $x = u^2 + vw$, $y = 2v + u^2 w$, $z = uvw$

In Exercises 7–10 use the indicated change of variables to express the integral as a double integral over a region S in the uv-plane. (Do not evaluate.)

7 $\iint_R (y - x)\, dx\, dy$; R the region bounded by $y = 2x$, $y = 0$, $x = 2$; $x = u + v$, $y = 2v$

8 $\iint_R (3x - 4y)\, dx\, dy$; R the region bounded by $y = 3x$, $2y = x$, $x = 4$; $x = u - 2v$, $y = 3u - v$

9 $\iint_R (\frac{1}{4}x^2 + \frac{1}{9}y^2)\, dx\, dy$; $R = \{(x, y): \frac{1}{4}x^2 + \frac{1}{9}y^2 \le 1\}$; $x = 2u$, $y = 3v$

10 $\iint_R xy\, dx\, dy$; R the region bounded by $y = 2\sqrt{1 - x}$, $x = 0$, $y = 0$; $x = u^2 - v^2$, $y = 2uv$

In Exercises 11–16 evaluate the given integral by making the indicated change of variable.

11 $\iint_R (x - y)^2 \cos^2 (x + y)\, dx\, dy$; R the region bounded by the square with vertices $(0, 1)$, $(1, 2)$, $(2, 1)$, $(1, 0)$; $u = x - y$, $v = x + y$.

12 $\iint_R \sin [(y - x)/(y + x)]\, dx\, dy$; R the trapezoid with vertices $(1, 1)$, $(2, 2)$, $(4, 0)$, $(2, 0)$; $u = y - x$, $v = y + x$.

13 $\iint_R (x^2 + 2y^2)\, dx\, dy$; R the region in the first quadrant bounded by the graphs of $xy = 1$, $xy = 2$, $y = x$, $y = 2x$; $x = u/v$, $y = v$.

14 $\iint_R (4x - 4y + 1)^{-2}\, dx\, dy$; R the region bounded by $x = \sqrt{-y}$, $x = y$, $x = 1$; $x = u + v$, $y = v - u^2$.

15 $\iint_R (2y + x)/(y - 2x)\, dx\, dy$; R the trapezoid with vertices $(-1, 0)$, $(-2, 0)$, $(0, 4)$, $(0, 2)$; $u = y - 2x$, $v = 2y + x$.

16 $\iint_R (\sqrt{x - 2y} + \frac{1}{4}y^2)\, dx\, dy$; R the triangle with vertices $(0, 0)$, $(4, 0)$, $(4, 2)$; $u = \frac{1}{2}y$, $v = x - 2y$.

17 Verify that for the spherical coordinate transformation,

$$\frac{\partial(x, y, z)}{\partial(\rho, \phi, \theta)} = \rho^2 \sin \phi.$$

18 Use Theorem (18.39) to derive a formula for

$$\iiint_R F(x, y, z)\, dx\, dy\, dz$$

for a transformation from rectangular to cylindrical coordinates.

19 If the transformation of coordinates $x = f(u, v)$, $y = g(u, v)$ is one-to-one, show that

$$\frac{\partial(x, y)}{\partial(u, v)} \frac{\partial(u, v)}{\partial(x, y)} = 1.$$

(*Hint*: Use the following property of determinants:

$$\begin{vmatrix} a & b \\ c & d \end{vmatrix} \begin{vmatrix} p & q \\ r & s \end{vmatrix} = \begin{vmatrix} ap + br & aq + bs \\ cp + dr & cq + ds \end{vmatrix}.)$$

20 Given the transformation $x = f(u, v)$, $y = g(u, v)$ and the transformation $u = h(r, s)$, $v = k(r, s)$ show that

$$\frac{\partial(x, y)}{\partial(u, v)} \frac{\partial(u, v)}{\partial(r, s)} = \frac{\partial(x, y)}{\partial(r, s)}.$$

(*Hint*: Use the hint given in Exercise 19.)

18.10

REVIEW

Define or discuss each of the following.

1 Vector field

2 Vector function

3 Scalar function

4 Inverse square field

5 Conservative vector field

6 Divergence

7 Curl

8 Line integral

9 Applications of line integrals

10 Independence of path

11 Green's Theorem

12 Surface integrals

13 The Divergence Theorem (Gauss' Theorem)

14 Stokes' Theorem

15 Transformations of coordinates

16 Jacobians

17 Change of variables in multiple integrals

EXERCISES 18.10

In Exercises 1–4 give a geometric description of the vector field **F**.

1 $\mathbf{F}(x, y) = 2x\mathbf{i} + y\mathbf{j}$

2 $\mathbf{F}(x, y, z) = x\mathbf{i} + y\mathbf{j} + \mathbf{k}$

3 $\mathbf{F}(x, y, z) = -\mathbf{k}$

4 $\mathbf{F}(x, y, z) = \nabla(x^2 + y^2 + z^2)^{-1/2}$

5 Find a two-dimensional conservative vector field that has the potential function $f(x, y) = y^2 \tan x$.

6 Find a three-dimensional conservative vector field that has the potential function $f(x, y, z) = \ln (x + y + z)$.

Given the points $A(1, 0)$ and $B(-1, 4)$, evaluate the line integral $\int_C y^2\,dx + xy\,dy$ if C is the curve described in each of Exercises 7–10.

7 C is given parametrically by $x = 1 - t$, $y = t^2$; $0 \leq t \leq 2$.

8 C consists of the line segments from A to the point $D(-1, 0)$, and from D to B.

9 C is the line segment from A to B.

10 C is the graph of $y = 2 - 2x^3$ from A to B.

11 Evaluate $\int_C xy\,ds$, where C is the part of the graph of $y = x^4$ between $(-1, 1)$ and $(2, 16)$.

Given $A(0, 0, 0)$ and $B(2, 4, 8)$ evaluate

$$\int_C x\,dx + (x + y)\,dy + (x + y + z)\,dz$$

if C is the curve described in each of Exercises 12–14.

12 C consists of three line segments, the first parallel to the z-axis, the second parallel to the x-axis, and the third parallel to the y-axis.

13 C consists of the line segment from A to B.

14 C is given parametrically by $x = t$, $y = t^2$, $z = t^3$.

15 If $\mathbf{F}(x, y) = (x + y)\mathbf{i} + (x - y)\mathbf{j}$, evaluate $\int_C \mathbf{F} \cdot d\mathbf{r}$, where C has parametric equations $x = \cos t$, $y = \sin t$, $-\pi \leq t \leq 0$.

16 The force $\mathbf{F}$ at a point (x, y, z) in three dimensions is given by $\mathbf{F}(x, y, z) = xy\mathbf{i} + y^2 z\mathbf{j} + xz^2\mathbf{k}$. If C is the square with vertices $A_1(1, 1, 1)$, $A_2(-1, 1, 1)$, $A_3(-1, -1, 1)$, and $A_4(1, -1, 1)$, find the work done by $\mathbf{F}$ as its point of application moves around C once in the direction determined by increasing subscripts on the A_i.

Prove that the line integrals in Exercises 17 and 18 are independent of path and find their values.

17 $\displaystyle\int_{(1, -1)}^{(2, 3)} (x + y)\,dx + (x + y)\,dy$

18 $\displaystyle\int_{(0, 0, 0)}^{(2, 1, 3)} (8x^3 + z^2)\,dx - 3z\,dy + (2xz - 3y)\,dz$

19 If $\mathbf{F}(x, y, z) = 2xe^{2y}\mathbf{i} + 2(x^2 e^{2y} + y\cot z)\mathbf{j} - y^2 \csc^2 z\mathbf{k}$, prove that $\int_C \mathbf{F} \cdot d\mathbf{r}$ is independent of path and find a potential function f for $\mathbf{F}$.

Use Green's Theorem to evaluate the line integral

$$\oint_C xy\,dx + (x^2 + y^2)\,dy$$

if C is the curve described in each of Exercises 20–22.

20 C is the closed curve determined by $y = x^2$ and $y - x = 2$ between $(-1, 1)$ and $(2, 4)$.

21 C is the triangle with vertices $(0, 0)$, $(1, 0)$, and $(0, 1)$.

22 C is the circle with equation $x^2 + y^2 - 2x = 0$.

23 Find div $\mathbf{F}$ and curl $\mathbf{F}$ if $\mathbf{F}(x, y, z) = x^3 z^4\mathbf{i} + xyz^2\mathbf{j} + x^2 y^2\mathbf{k}$.

24 If f and g are scalar functions of three variables prove that $\nabla \cdot (f\nabla g) = f\nabla^2 g + \nabla f \cdot \nabla g$.

25 Evaluate $\iint_S xyz\,dS$, where S is the part of the plane $z = x + y$ that lies over the triangular region in the xy-plane having vertices $(0, 0, 0)$, $(1, 0, 0)$, and $(0, 2, 0)$.

26 Evaluate $\iint_S x^2 z^2\,dS$ where S is the top half of the cylinder $y^2 + z^2 = 4$ between $x = 0$ and $x = 1$.

27 Let Q be the region bounded by the cylinder $x^2 + y^2 = 1$ and the planes $z = 0$ and $z = 1$. If $\mathbf{F} = x^3\mathbf{i} + y^3\mathbf{j} + z^3\mathbf{k}$, use the Divergence Theorem to find $\iint_S \mathbf{F} \cdot \mathbf{n}\,dS$, where S is the surface of Q and $\mathbf{n}$ is the unit outer normal to S.

28 Verify the Divergence Theorem if $\mathbf{F} = 2x\mathbf{i} + y\mathbf{j} - z\mathbf{k}$ and S is the surface of the parallelepiped bounded by the planes $x = \pm 1$, $y = \pm 2$, $z = \pm 3$.

Verify Stokes' Theorem for the given $\mathbf{F}$ and S in Exercises 29 and 30.

29 $\mathbf{F} = y^2\mathbf{i} + 2x\mathbf{j} + 5y\mathbf{k}$ and S is the hemisphere $z = (4 - x^2 - y^2)^{1/2}$.

30 $\mathbf{F} = (x + y)\mathbf{i} + (y + z)\mathbf{j} + (x + z)\mathbf{k}$ and S is the region bounded by the triangle with vertices $(1, 0, 0)$, $(0, 1, 0)$, and $(0, 0, 1)$.

31 If a transformation T of the xy-plane to the uv-plane is given by $u = 2x + 5y$, $v = 3x - 4y$, find:

(a) the u-curves.

(b) the v-curves.

(c) T^{-1}.

(d) $\partial(x, y)/\partial(u, v)$.

(e) the image of the line $ax + by + c = 0$.

(f) the image of the circle $x^2 + y^2 = a^2$.

32 Evauate the integral $\int_0^1 \int_y^{2-y} e^{(x-y)/(x+y)}\,dx\,dy$ by means of the change of variables $u = x - y$, $v = x + y$.

19

DIFFERENTIAL EQUATIONS

A *differential equation* is an equation that involves derivatives or differentials. If only derivatives of a function of one variable occur, then a differential equation is called *ordinary*. A *partial differential equation* contains partial derivatives. The primary objective of this chapter is to develop techniques for solving basic types of ordinary differential equations. The discussion is not intended to be a treatise on the subject, but rather to serve as an introduction to this vast and important area of mathematics.

19.1

SEPARABLE AND FIRST-ORDER LINEAR DIFFERENTIAL EQUATIONS

Separable and first-order linear differential equations were discussed in Sections 7.6 and 7.7. At that time our choice of examples and exercises was limited, because integration formulas for trigonometric and inverse trigonometric functions were not available, and we had not developed techniques of integration such as partial fractions. We shall begin this section with two examples and then introduce several new topics.

Example 1 Solve the differential equation

$$(1 + y^2) + (1 + x^2)\frac{dy}{dx} = 0.$$

Solution Writing the equation in differential form and then separating the variables gives us

$$(1 + y^2)\, dx + (1 + x^2)\, dy = 0$$

$$\frac{1}{1 + x^2}\, dx + \frac{1}{1 + y^2}\, dy = 0.$$

Integrating each term we obtain the (implicit) solution

$$\tan^{-1} x + \tan^{-1} y = C.$$

To find an explicit solution we may solve for y as follows:

$$\tan^{-1} y = C - \tan^{-1} x$$

$$y = \tan (C - \tan^{-1} x).$$

The form of this solution may be changed by using the subtraction formula for the tangent function (see Appendix III). Thus

$$y = \frac{\tan C - \tan (\tan^{-1} x)}{1 + \tan C \tan (\tan^{-1} x)}.$$

If we let $k = \tan C$ and use the fact that $\tan (\tan^{-1} x) = x$, this may be written

$$y = \frac{k - x}{1 + kx}. \qquad \blacksquare$$

Example 2 Solve the differential equation

$$y' + y \tan x = \sec x + 2x \cos x.$$

Solution   The equation is a first-order linear differential equation (see Definition (7.26)). By Theorem (7.27), an integrating factor is

$$e^{\int \tan x \, dx} = e^{\ln |\sec x|} = |\sec x|.$$

Multiplying both sides of the differential equation by $|\sec x|$ and discarding the absolute value sign gives us

$$y' \sec x + y \sec x \tan x = \sec^2 x + 2x \cos x \sec x$$

or

$$D_x(y \sec x) = \sec^2 x + 2x.$$

Integrating both sides yields the (implicit) solution

$$y \sec x = \tan x + x^2 + C.$$

Finally, multiplying both sides of the last equation by $1/\sec x = \cos x$ we obtain the explicit solution

$$y = \sin x + (x^2 + C) \cos x. \qquad \blacksquare$$

A generalization of a first-order linear differential equation is the following **Bernoulli equation**

(19.1) $$y' + P(x)y = Q(x)y^n$$

where $n \neq 0$. Evidently $y = 0$ is a solution. If $y \neq 0$ we may divide both sides by y^n, obtaining

(*) $$y^{-n}y' + P(x)y^{1-n} = Q(x).$$

If we let $w = y^{1-n}$, then

$$w' = D_x w = (1 - n)y^{-n}y'$$

or

$$y^{-n}y' = \frac{1}{1 - n} w'.$$

Replacing $y^{-n}y'$ in (∗) by the last expression gives us

$$\frac{1}{1-n}\,w' + P(x)w = Q(x).$$

This first-order linear differential equation may be solved for w using the integrating factor technique. After w has been found, the solution of (19.1) is given by $y^{1-n} = w$ (and $y = 0$).

Example 3   Solve the differential equation $y' + 2x^{-1}y = x^6y^3$.

Solution   The equation has the Bernoulli form (19.1) with $n = 3$. If, as in the preceding discussion, we multiply both sides by y^{-3} and substitute $w = y^{1-n} = y^{-2}$ we obtain

$$y^{-3}y' + 2x^{-1}y^{-2} = x^6$$

$$-\tfrac{1}{2}w' + 2x^{-1}w = x^6$$

$$w' - 4x^{-1}w = -2x^6.$$

By Theorem (7.27), an integrating factor for the last equation is

$$e^{\int(-4/x)\,dx} = e^{-4\ln|x|} = e^{\ln|x|^{-4}} = |x|^{-4} = x^{-4}$$

and we proceed as follows:

$$x^{-4}w' - 4x^{-5}w = -2x^2,$$

$$x^{-4}w = -\tfrac{2}{3}x^3 + C$$

$$w = -\tfrac{2}{3}x^7 + Cx^4.$$

Finally, since $w = y^{-2}$, the solution of the given equation is

$$y^{-2} = -\tfrac{2}{3}x^7 + Cx^4$$

or $$(-\tfrac{2}{3}x^7 + Cx^4)y^2 = 1. \qquad\blacksquare$$

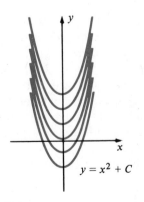

Figure 19.1

$y = x^2 + C$

Since general solutions of differential equations contain arbitrary constants, the graphs of these solutions are families of curves. For example, the solution $y = x^2 + C$ of the differential equation $y' = 2x$ leads to a family of parabolas of the type illustrated in Figure 19.1. Each member of the family is obtained by assigning a specific value to C. Conversely, if we *begin* with the equation of a family of curves, then differentiation may be used to obtain a differential equation called a **differential equation of the family**.

Example 4   Find a differential equation of the family of parabolas defined by $y = Cx^2$ where C is a nonzero real number.

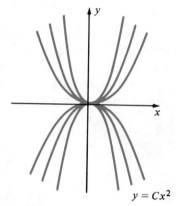

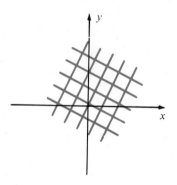

Figure 19.2

Solution   The family consists of all parabolas with vertex at the origin and focus on the y-axis, as illustrated in Figure 19.2. Differentiating $y = Cx^2$ we obtain $y' = 2Cx$ and therefore

$$y = Cx^2 = (Cx)x = (\tfrac{1}{2}y')x.$$

Hence a differential equation for the family is

$$2y - y'x = 0. \qquad \blacksquare$$

An equation in x and y that contains two arbitrary constants is said to define a **two-parameter family of curves**. Similar terminology is used if there are more than two constants. To find a differential equation when several constants are involved, it is necessary to differentiate the given equation several times.

Example 5 Find a differential equation for the two-parameter family of curves

$$y = C_1 \cos x + C_2 \sin x.$$

Solution   Differentiating twice, we obtain

$$y' = -C_1 \sin x + C_2 \cos x$$

$$y'' = -C_1 \cos x - C_2 \sin x.$$

Comparing with the given equation we see that

$$y'' = -y, \quad \text{or} \quad y'' + y = 0.$$

This is a differential equation for the family. $\blacksquare$

(i) $y = 2x + b,\ y = -\tfrac{1}{2}x + c$

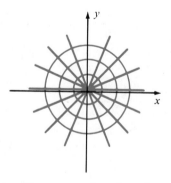

(ii) $y = mx,\ x^2 + y^2 = a^2$

Figure 19.3 Mutually orthogonal families of curves

An **orthogonal trajectory** of a family of curves is a curve that intersects each curve of the family orthogonally. We shall restrict our discussion to curves in a coordinate plane. To illustrate, given the family $y = 2x + b$ of all lines of slope 2, every line $y = (-1/2)x + c$ of slope $-1/2$ is an orthogonal trajectory. Several curves from each family are sketched in (i) of Figure 19.3. We sometimes call two such families of curves **mutually orthogonal**. As another example, the family $y = mx$ of all lines through the origin and the family $x^2 + y^2 = a^2$ of all concentric circles with centers at the origin are mutually orthogonal (see (ii) of Figure 19.3).

Pairs of mutually orthogonal families of curves occur frequently in physical applications of mathematics. In the theory of electricity and magnetism, the lines of force associated with a given field are orthogonal trajectories of the corresponding equipotential curves. Similarly, the stream lines studied in aerodynamics and hydrodynamics are orthogonal trajectories of the so-called *velocity-equipotential* curves. As a final illustration, in the study of thermodynamics the flow of heat across a plane surface is shown to be orthogonal to the isothermal curves.

The next example illustrates a technique for finding the orthogonal trajectories of a family of curves.

Example 6 Find the orthogonal trajectories of the family of ellipses $x^2 + 3y^2 = c$ and sketch several members of each family.

Solution   Differentiating the given equation implicitly we obtain

$$2x + 6yy' = 0, \quad \text{or} \quad y' = -\frac{x}{3y}.$$

Hence the slope of the tangent line at any point (x, y) on one of the ellipses is $y' = -x/3y$. If dy/dx is the slope of the tangent line on a corresponding orthogonal trajectory, then it must equal the negative reciprocal of y'. This gives us the following differential equation for the family of orthogonal trajectories:

$$\frac{dy}{dx} = \frac{3y}{x}.$$

Separating the variables,

$$\frac{dy}{y} = 3\frac{dx}{x}.$$

Integrating, and writing the constant of integration as $\ln |k|$ gives us

$$\ln |y| = 3 \ln |x| + \ln |k| = \ln |kx^3|.$$

It follows that $y = kx^3$ is an equation for the family of orthogonal trajectories. Several members of the given family and corresponding orthogonal trajectories are sketched (with dashes) in Figure 19.4. ∎

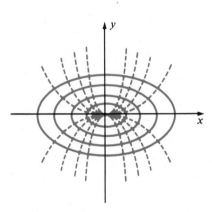

Figure 19.4   Orthogonal trajectories

EXERCISES 19.1

Solve the differential equations in Exercises 1–22.

1 $x \tan y - y' \sec x = 0$

2 $\cos x \, dy - y \, dx = 0$

3 $\sin y \cos x \, dx + (1 + \sin^2 x) \, dy = 0$

4 $e^y \sin x \, dx - \cos^2 x \, dy = 0$

5 $\sqrt{1 - y^2} + x(x - 1)y' = 0$

6 $xy\sqrt{x^2 - 1}\, y' + \sqrt{y^2 - 1} = 0$

7 $y' + y \cot x = \csc x$

8 $y' - y \tan x = e^{-x} \sec x$

9 $y' + y \cot x = 4x^2 \csc x$

10 $y' + y \tan x = \sin x$

11 $(y \sin x - 2) \, dx + \cos x \, dy = 0$

12 $(x^2 y - 1) \, dx + x^3 \, dy = 0$

13 $(x^2 \cos x + y) \, dx - x \, dy = 0$

14 $y' + y = \sin x$

15 $xy' - y = x^3 y^4$

16 $xy' + 2y = 4x^4 y^4$

17 $(2xy - x^2 y^2) \, dx + (1 + x^2) \, dy = 0$

18 $y' + y = y^2 e^x$

19 $x^{-1} y' + 2y = 3$

20 $\cos x \, dy - (y \sin x + e^{-x}) \, dx = 0$

21 $\tan x \, dy + (y - \sin x) \, dx = 0$

22 $y' + y \tan x = \cos^3 x$

In Exercises 23–28 find the particular solution of the differential equation that satisfies the given boundary conditions.

23 $\sec 2y \, dx - \cos^2 x \, dy = 0$; $y = \pi/6$ when $x = \pi/4$

24 $x \, dy - \sqrt{1 - y^2} \, dx = 0$; $y = 1/2$ when $x = 1$

25 $\cot x \, dy - (1 + y^2) \, dx = 0$; $y = 1$ when $x = 0$

26 $\csc y \, dx - e^x \, dy = 0$; $y = 0$ when $x = 0$

27 $y' + y \tan x = 2x \cos x$; $y = 3$ when $x = 0$

28 $y' + y = \cos x$; $y = 2$ when $x = 0$

In Exercises 29–34 find a differential equation that has the given general solution.

29 $y^2 = x^3 + C$

30 $x^2 + y^2 = C^2$

31 $y = e^{Cx}$

32 $y = C_1 x^3 + C_2 x$

33 $y = C_1 e^x + C_2 x e^x$

34 $y = C_1 e^x + C_2 x e^x + C_3 e^{-x}$

In Exercises 35–38 find a differential equation of the indicated family of curves.

35 All nonvertical lines through the origin

36 All circles with centers at the origin

37 All circles with centers on the x-axis

38 All parabolas having focus and vertex on the x-axis

Find the orthogonal trajectories of the families of curves in Exercises 39–46 and sketch several members of each family.

39 $x^2 - y^2 = c$

40 $xy = c$

41 $y^2 = cx$

42 $y = cx^2$

43 $y^2 = cx^3$

44 $y = ce^{-x}$

45 $x^2 + y^2 = 2cx$

46 $x^2 + y^2 - 2cy = 0$

47 An object of mass m is released from a balloon. Find the distance it falls in t seconds if the force of resistance due to the air is directly proportional to the square of the speed of the object (cf. Example 3 of Section 7.7).

48 The equation

$$R \frac{di}{dt} + \frac{i}{C} = \frac{dE}{dt}$$

describes an electrical circuit consisting of a resistance R and capacity C in series. Express i as a function of t if $i = i_0$ when $t = 0$, and $E = E_0 \sin \omega t$ where E_0 and ω are constants.

49 In chemistry, the notation $A + B \rightarrow Y$ is used to denote the production of a substance Y due to the interaction of two substances A and B. Let a and b be the initial amounts of A and B, respectively. If, at time t, the concentration of Y is $y = f(t)$, then the concentrations of A and B are $a - f(t)$ and $b - f(t)$, respectively. If the rate at which the production of Y takes place is given by $dy/dt = k(a - y)(b - y)$, where k is a positive constant, and if $f(0) = 0$, find $f(t)$.

50 Let $y = f(t)$ be the population, at time t, of a collection of objects such as human beings, insects, or bacteria. If the rate of growth dy/dt is proportional to y, that is, if $dy/dt = cy$ for some positive constant c, then $f(t) = f(0)e^{ct}$ (see Theorem (7.25)). In most cases the rate of growth is dependent on available resources such as food supplies, and as t becomes large, $f'(t)$ begins to decrease. An equation that is often used to describe this type of variation in population is the **Law of Logistic Growth** $dy/dt = y(c - by)$ where c and b are positive constants (cf. Exercise 33 of Section 7.6).

(a) If $f(0) = k$, find $f(t)$.

(b) Find $\lim_{t \to \infty} f(t)$.

(c) Show that $f'(t)$ is increasing if $f(t) < c/2b$ and is decreasing if $f(t) > c/2b$.

(d) Sketch a typical graph of f. (A graph of this type is called a **logistic curve**.)

19.2

EXACT DIFFERENTIAL EQUATIONS

Separable differential equations are special cases of the first-order equation

$$P(x, y) + Q(x, y)\frac{dy}{dx} = 0$$

where P and Q are functions of x and y. This equation may be expressed in terms of differentials as

$$P(x, y)\,dx + Q(x, y)\,dy = 0.$$

In this section we shall restrict our attention to differential equations that are *exact*, in the sense of the following definition.

Definition (19.2)

> Let P and Q have continuous first partial derivatives. The differential equation
>
> $$P(x, y) + Q(x, y)\frac{dy}{dx} = 0, \quad \text{or} \quad P(x, y)\,dx + Q(x, y)\,dy = 0$$
>
> is an **exact differential equation** provided
>
> $$\frac{\partial P}{\partial y} = \frac{\partial Q}{\partial x}.$$

Example 1 Show that the following differential equation is exact:

$$(3x^2y - 2y^3 + 3)\,dx + (x^3 - 6xy^2 + 2y)\,dy = 0.$$

Solution If $P(x, y) = 3x^2y - 2y^3 + 3$ and $Q(x, y) = x^3 - 6xy^2 + 2y$, we see that

$$P_y(x, y) = 3x^2 - 6y^2 = Q_x(x, y).$$

Hence the equation is exact by Definition (19.2). ∎

Theorem (19.3)

> If $P(x, y) + Q(x, y)y' = 0$ is an exact differential equation, then there exists a function F of x and y such that
>
> $$\frac{\partial F}{\partial x} = P \quad \text{and} \quad \frac{\partial F}{\partial y} = Q.$$

Proof If (x_1, y_1) is in the domain of P and Q, let us define the function F by

$$F(x, y) = \int_{x_1}^{x} P(t, y)\,dt + \int_{y_1}^{y} Q(x_1, t)\,dt.$$

It is left as an exercise to prove that $F_x(x, y) = P(x, y)$ and $F_y(x, y) = Q(x, y)$. ☐

The next result is the fundamental existence theorem for solutions of exact differential equations.

Theorem (19.4)

> An exact differential equation $P(x, y) + Q(x, y)y' = 0$ has a solution of the form $F(x, y) = C$, where C is a constant and F is any function of x and y such that $F_x = P$ and $F_y = Q$.

Proof   By Theorem (19.3) there exists at least one function F such that $F_x = P$ and $F_y = Q$. We shall complete the proof by showing that $y = f(x)$ is a solution of the exact differential equation if and only if $F(x, f(x)) = C$ for some constant C.

By the Chain Rule,

$$D_x F(x, f(x)) = F_x(x, f(x)) + F_y(x, f(x))f'(x)$$

or

$$D_x F(x, f(x)) = P(x, f(x)) + Q(x, f(x))f'(x).$$

If $y = f(x)$ is a solution of the differential equation, then $D_x F(x, f(x)) = 0$ and, therefore, $F(x, f(x)) = C$ for some C. Conversely, if $F(x, f(x)) = C$, then $D_x F(x, f(x)) = 0$, that is,

$$0 = P(x, f(x)) + Q(x, f(x))f'(x)$$

which means that $f(x)$ is a solution of the given differential equation.   □

The following examples illustrate a technique for finding the solution $F(x, y) = C$ in Theorem (19.4).

Example 2   Solve the differential equation

$$(3x^2y - 2y^3 + 3) \, dx + (x^3 - 6xy^2 + 2y) \, dy = 0.$$

Solution   It was shown in Example 1 that the equation is exact. Consequently, by Theorem (19.4), there is a function F such that

$$F_x(x, y) = 3x^2y - 2y^3 + 3$$

$$F_y(x, y) = x^3 - 6xy^2 + 2y.$$

Integrating $F_x(x, y)$ with respect to x gives us

$$F(x, y) = x^3y - 2xy^3 + 3x + g(y)$$

where g is a function of y. It follows that

$$F_y(x, y) = x^3 - 6xy^2 + g'(y).$$

Comparing this equation with the previous formula for $F_y(x, y)$ we see that $g'(y) = 2y$, and hence $g(y) = y^2 + C_1$. Substituting for $g(y)$ in the formula for $F(x, y)$ gives us

$$F(x, y) = x^3y - 2xy^3 + 3x + y^2 + C_1.$$

Hence, by Theorem (19.4), a solution of the differential equation is

$$x^3y - 2xy^3 + 3x + y^2 = C$$

where the constant C_1 has been absorbed by the constant C. This solution may be checked by implicit differentiation.   ∎

Example 3   Solve the differential equation

$$2x \sin y - y \sin x + (x^2 \cos y + \cos x)y' = 0$$

subject to the boundary conditions $y = \pi/6$ if $x = \pi/2$.

Solution   If $P(x, y) = 2x \sin y - y \sin x$ and $Q(x, y) = x^2 \cos y + \cos x$ we see that

$$P_y(x, y) = 2x \cos y - \sin x = Q_x(x, y)$$

and hence the differential equation is exact. By Theorem (19.3) there is a function F such that

$$F_x(x, y) = 2x \sin y - y \sin x$$

$$F_y(x, y) = x^2 \cos y + \cos x.$$

Integrating $F_x(x, y)$ with respect to x gives us

$$F(x, y) = x^2 \sin y + y \cos x + g(y)$$

where g is a function of y. Consequently,

$$F_y(x, y) = x^2 \cos y + \cos x + g'(y).$$

Comparing with the other form of $F_y(x, y)$ we see that $g'(y) = 0$ and therefore $g(y) = C_1$. Applying Theorem (19.4), a solution of the given differential equation is

$$x^2 \sin y + y \cos x = C$$

where again C_1 has been absorbed by C. Finally, the boundary conditions imply that

$$\left(\frac{\pi}{2}\right)^2\left(\frac{1}{2}\right) + \left(\frac{\pi}{6}\right)\cdot 0 = C$$

or $C = \pi^2/8$. Thus the required particular solution is

$$x^2 \sin y + y \cos x = \frac{\pi^2}{8}.$$ ∎

EXERCISES 19.2

Solve the differential equations in Exercises 1–14.

1 $(2x + y) dx + (2y + x) dy = 0$

2 $(2xy + y^3) dx + (x^2 + 3y^2x) dy = 0$

3 $(1 - 2xy + 3x^2y^2) dx - (x^2 + 3y^2 - 2x^3y) dy = 0$

4 $(2x - 3x^2y) dx + (1 + 2y - x^3) dy = 0$

5 $(x^2 \cos y + 4y)y' + 2x \sin y = -5$

6 $y^2e^x + 4y + (2ye^x + 4x)y' = 0$

7 $e^{xy}(xy + 1) dx + (x^2e^{xy} + 2y) dy = 0$

8 $(v \ln v + 2u \ln u) du + (u \ln v + u) dv = 0$

9 $(\sin^2 \theta + 2r \cos \theta) dr + r(\sin 2\theta - r \sin \theta) d\theta = 0$

10 $(x^2 + 2xy - y^2)y' = -y^2 - 2xy + x^2$

11 $(e^{-x} \cos y - 2x) dx + e^{-x} \sin y \, dy = 0$

12 $y \cos x \, dx + (\sin x - \sin y) dy = 0$

13 $(x - yx^{-1} + \ln y) + (xy^{-1} - \ln x)y' = 0$

14 $\left(\dfrac{x}{1 + y^2} + y^3 - 1\right)y' + \left(\dfrac{1}{1 + x^2} + \tan^{-1} y\right) = 0$

In Exercises 15–18 find a solution of the differential equation that satisfies the given boundary conditions.

15 $(2xy^3 + 8x) dx + (3x^2y^2 + 5) dy = 0$;
$y = -1$ when $x = 2$

16 $(x^2e^y + 3e^x)y' + (2xe^y + 3ye^x) = 0$; $y = 0$ when $x = 1$

17 $\left(ye^{2x} + \dfrac{y}{1 + 4y^2}\right) dy + (y^2e^{2x} - 1) dx = 0$; $y = 1/2$ when $x = 0$

18 $(x \cos x + \sin x + y) dx + (\sin y + x) dy = 0$;
$y = \pi/3$ when $x = \pi/6$

19 Complete the proof of Theorem (19.3).

20 Prove that the separable differential equation $M(x) + N(y)y' = 0$ is a special case of

$$P(x, y) + Q(x, y)\frac{dy}{dx} = 0.$$

19.3

HOMOGENEOUS DIFFERENTIAL EQUATIONS

Definition (19.5)

> A function f of two variables is said to be **homogeneous of degree n** if
>
> $$f(tx, ty) = t^n f(x, y)$$
>
> for every $t > 0$ such that (tx, ty) is in the domain of f.

Example 1 Show that the following define homogeneous functions, and find their degrees.

(a) $f(x, y) = 2x^4 - x^2y^2 + 5xy^3$ (b) $g(x, y) = \dfrac{1}{x^2 + y^2} e^{x/y}$

Solution

(a) Since

$$f(tx, ty) = 2(tx)^4 - (tx)^2(ty)^2 + 5(tx)(ty)^3$$
$$= t^4(2x^4 - x^2y^2 + 5xy^3) = t^4f(x, y),$$

it follows from Definition (19.5) that f is homogeneous of degree 4.

(b) Since

$$g(tx, ty) = \frac{1}{t^2x^2 + t^2y^2}\, e^{tx/ty} = t^{-2}g(x, y),$$

we see from (19.5) that g is homogeneous of degree -2. ■

Definition (19.6)

> A **homogeneous differential equation** is an equation the form
>
> $$P(x, y)dx + Q(x, y)dy = 0$$
>
> where P and Q are homogeneous functions of the same degree.

The next result is the principal theorem for homogeneous differential equations.

Theorem (19.7)

> Let $P(x, y)\, dx + Q(x, y)\, dy = 0$ be a homogeneous differential equation. If g is a differentiable function, then either of the following substitutions leads to a separable differential equation:
>
> (i) $y = xv$,   where $v = g(x)$.
>
> (ii) $x = yu$,   where $u = g(y)$.

Proof   We shall only prove (i), since the proof of (ii) is similar. If $y = xv$, where $v = g(x)$, then

$$dy = v\, dx + x\, dv.$$

Thus, substitution of xv for y in $P(x, y)\, dx + Q(x, y)\, dy = 0$ yields

$$P(x, xv)\, dx + Q(x, xv)(v\, dx + x\, dv) = 0.$$

If P and Q are homogeneous of degree n, then

$$P(x, xv) = x^n P(1, v) \quad \text{and} \quad Q(x, xv) = x^n Q(1, v).$$

Substituting in the preceding differential equation and dividing both sides by x^n, we obtain

$$P(1, v)\, dx + Q(1, v)(v\, dx + x\, dv) = 0.$$

This equation may be written in the separable form

$$\frac{1}{x}\, dx + \frac{Q(1, v)}{P(1, v) + vQ(1, v)}\, dv = 0$$

provided no zero denominators occur. We have proved that if $y = xv$ is a solution of the homogeneous differential equation, then v is a solution of the preceding separable differential equation. Conversely, if v is a solution of the separable differential equation, then we may reverse the argument to show that $y = xv$ is a solution of $P(x, y) \, dx + Q(x, y) \, dy = 0$. It is not advisable to memorize the final form of the separable differential equation. Instead, remember the substitution $y = xv$. □

Example 2 Solve the differential equation

$$(y^2 - xy) \, dx + x^2 \, dy = 0.$$

Solution   If $P(x, y) = y^2 - xy$ and $Q(x, y) = x^2$, then the functions P and Q are both homogeneous of degree 2. (Why?) Hence the differential equation is homogeneous and as in (i) of Theorem (19.7) we substitute

$$y = xv, \qquad dy = v \, dx + x \, dv.$$

This leads to the following chain of equations.

$$(x^2v^2 - x^2v) \, dx + x^2(v \, dx + x \, dv) = 0$$
$$x^2(v^2 - v) \, dx + x^2(v \, dx + x \, dv) = 0$$
$$(v^2 - v) \, dx + v \, dx + x \, dv = 0$$
$$v^2 \, dx + x \, dv = 0$$
$$\frac{1}{x} \, dx + \frac{1}{v^2} \, dv = 0$$

Integrating each term we obtain

$$\ln |x| - \frac{1}{v} = C_1.$$

Since $v = y/x$, this gives us

$$\ln |x| - \frac{x}{y} = C_1.$$

If we let $C_1 = \ln |C|$, then

$$\ln |x| - \ln |C| = \frac{x}{y}, \quad \text{or} \quad \ln \left| \frac{x}{C} \right| = \frac{x}{y}.$$

This may be written

$$\frac{x}{C} = e^{x/y}, \quad \text{or} \quad x = Ce^{x/y}.$$

An explicit form for the solution may be obtained by solving for y. ■

Example 3 Solve the differential equation

$$\left(y - x \cot \frac{x}{y} \right) dy + y \cot \frac{x}{y} \, dx = 0.$$

Solution   The equation is homogeneous of degree 1. (Why?) It is convenient to let $x = vy$, since in this case cot (x/y) reduces to cot v. Thus we substitute

$$x = vy, \qquad dx = v\,dy + y\,dv$$

obtaining

$$(y - vy \cot v)\,dy + y \cot v(v\,dy + y\,dv) = 0.$$

Dividing both sides by y and simplifying leads to

$$(1 - v \cot v)\,dy + \cot v(v\,dy + y\,dv) = 0$$

$$dy + y \cot v\,dv = 0.$$

Separating the variables and integrating, we obtain

$$\frac{1}{y}\,dy + \cot v\,dv = 0$$

$$\ln|y| + \ln|\sin v| = C_1 = \ln|C|$$

$$\ln|y \sin v| = \ln|C|$$

$$y \sin v = C.$$

Since $v = x/y$, a solution to the given equation is

$$y \sin \frac{x}{y} = C. \qquad\blacksquare$$

EXERCISES 19.3

Solve the differential equations in Exercises 1–10.

1 $(x + 3y)\,dx + x\,dy = 0$

2 $(2x - y)\,dx + (x + 2y)\,dy = 0$

3 $\left(y \sin \dfrac{y}{x} + x \cos \dfrac{y}{x}\right) dx - x \sin \dfrac{y}{x}\,dy = 0$

4 $x^2\,dy + (y^2 - xy)\,dx = 0$

5 $(x^2 + y^2)\,dx - x^2\,dy = 0$

6 $(y^2 - xy + x^2)\,dx + xy\,dy = 0$

7 $y' = x^3/(4x^3 - 3x^2y)$

8 $y' = x^{-1}y - \cos x^{-1}\,y$

9 $(y + \sqrt{x^2 - y^2})\,dx = x\,dy$

10 $(x + y)\,dx = (x - y)\,dy$

Solve the differential equations in Exercises 11–14 by using the substitution $x = vy$.

11 $2xy\,dx + (y^2 - x^2)\,dy = 0$

12 $y^2\,dx + x(x - y)\,dy = 0$

13 $3y\,dx + (x + 2y)\,dy = 0$

14 $dx - (y^{-1}x + \tan y^{-1}\,x)\,dy = 0$

In Exercises 15 and 16 show that the differential equation is both homogeneous and exact and solve it by the corresponding method.

15 $(x + 3y)\,dx + (3x - 2y)\,dy = 0$

16 $(x^2 + y^2)\,dx + 2xy\,dy = 0$

19.4

SECOND-ORDER LINEAR DIFFERENTIAL EQUATIONS

Definition (19.8)

> Let $f_1, f_2, \ldots, f_n$ and k be functions of one variable that have the same domain. A **linear differential equation of order n** is an equation of the form
>
> $$y^{(n)} + f_1(x)y^{(n-1)} + \cdots + f_{n-1}(x)y' + f_n(x)y = k(x).$$
>
> If $k(x) = 0$ for all x, the equation is said to be **homogeneous**.

Notice that the meaning of the word *homogeneous* is different from that in Section 19.2. If $k(x) \neq 0$ for some x, the equation is said to be **nonhomogeneous**. A thorough analysis of (19.8) may be found in textbooks on differential equations. We shall restrict our work to second-order equations in which f_1 and f_2 are constant functions. In this section we shall consider the homogeneous case. Nonhomogeneous equations will be discussed in the next section.

The general second-order homogeneous linear differential equation with constant coefficients has the form

$$y'' + by' + cy = 0$$

where b and c are constants. Before attempting to find particular solutions let us establish the following result.

Theorem (19.9)

> If $y = f(x)$ and $y = g(x)$ are solutions of $y'' + by' + cy = 0$, then
>
> $$y = C_1 f(x) + C_2 g(x)$$
>
> is a solution for all real numbers C_1 and C_2.

Proof   By hypothesis,

$$f''(x) + bf'(x) + cf(x) = 0$$

$$g''(x) + bg'(x) + cg(x) = 0.$$

If we multiply the first of these equations by C_1, the second by C_2, and add, the result is

$$[C_1 f''(x) + C_2 g''(x)] + b[C_1 f'(x) + C_2 g'(x)] + c[C_1 f(x) + C_2 g(x)] = 0.$$

Thus $C_1 f(x) + C_2 g(x)$ is a solution.   □

It can be shown that if the solutions f and g in Theorem (19.9) have the property that $f(x) \neq Cg(x)$ for every real number C, and if $g(x)$ is not identically 0, then $y = C_1 f(x) + C_2 g(x)$ is the general solution of the differential equation $y'' + by' + cy = 0$. Thus, to determine the general solution it is sufficient to find two such functions f and g and employ (19.9).

In our search for a solution of $y'' + by' + cy = 0$ we shall use $y = e^{mx}$ as a trial solution. Since $y' = me^{mx}$ and $y'' = m^2 e^{mx}$, it follows that $y = e^{mx}$ is a solution if and only if

$$m^2 e^{mx} + bme^{mx} + ce^{mx} = 0$$

or, since $e^{mx} \neq 0$, if and only if

$$m^2 + bm + c = 0.$$

The last equation is very important in finding solutions of $y'' + by' + cy = 0$, and is given the following special name.

Definition (19.10)

> The **auxiliary equation** of the differential equation $y'' + by' + c = 0$ is $m^2 + bm + c = 0$.

Note that the auxiliary equation can be obtained from the differential equation by replacing y'' by m^2, y' by m, and y by 1. In simple cases the roots of the auxiliary equation can be found by factoring. If a factorization is not evident, then applying the quadratic formula, we see that the roots are

$$m = \frac{-b \pm \sqrt{b^2 - 4c}}{2}.$$

Thus the auxiliary equation has unequal real roots m_1 and m_2, a double real root m, or two complex conjugate roots depending on whether $b^2 - 4c$ is positive, zero, or negative, respectively. The next theorem is a consequence of the remark following the proof of Theorem (19.9).

Theorem (19.11)

> If the roots m_1, m_2 of the auxiliary equation are real and unequal, then the general solution of $y'' + by' + cy = 0$ is
>
> $$y = C_1 e^{m_1 x} + C_2 e^{m_2 x}.$$

Example 1   Solve the differential equation $y'' - 3y' - 10y = 0$.

Solution   The auxiliary equation is $m^2 - 3m - 10 = 0$, or equivalently, $(m - 5)(m + 2) = 0$. Since the roots $m_1 = 5$ and $m_2 = -2$ are real and unequal, it follows from Theorem (19.11) that the general solution is

$$y = C_1 e^{5x} + C_2 e^{-2x}. \qquad \blacksquare$$

Theorem (19.12)

> If the auxiliary equation has a double root m, then the general solution of $y'' + by' + cy = 0$ is
>
> $$y = C_1 e^{mx} + C_2 x e^{mx}.$$

Proof   The roots of $m^2 + bm + c = 0$ are $m = (-b \pm \sqrt{b^2 - 4c})/2$. If $b^2 - 4c = 0$, we obtain $m = -b/2$ or $2m + b = 0$. Since m satisfies the auxiliary equation, $y = e^{mx}$ is a solution of the differential equation. According to the remark following the proof of Theorem (19.9), it is sufficient to show that $y = xe^{mx}$ is also a solution. Substitution of xe^{mx} for y in the equation $y'' + by' + cy = 0$ gives us

$$(2me^{mx} + m^2 xe^{mx}) + b(mxe^{mx} + e^{mx}) + cxe^{mx}$$
$$= (m^2 + bm + c)xe^{mx} + (2m + b)e^{mx}$$
$$= 0xe^{mx} + 0e^{mx} = 0$$

which is what we wished to show.   □

Example 2 Solve the differential equation $y'' - 6y' + 9y = 0$.

Solution   The auxiliary equation $m^2 - 6m + 9 = 0$, or equivalently $(m - 3)^2 = 0$, has a double root 3. Hence, by Theorem (19.12), the general solution is

$$y = C_1 e^{3x} + C_2 xe^{3x} = e^{3x}(C_1 + C_2 x). \qquad \blacksquare$$

We may also consider second-order differential equations of the form

$$ay'' + by' + cy = 0$$

where $a \neq 1$. It is possible to obtain the form stated in Theorems (19.11) and (19.12) by dividing both sides by a; however, it is usually simpler to employ the auxiliary equation

$$am^2 + bm + c = 0$$

as illustrated in the next example.

Example 3 Solve the differential equation $6y'' - 7y' + 2y = 0$.

Solution   The auxiliary equation $6m^2 - 7m + 2 = 0$ can be factored as follows:

$$(2m - 1)(3m - 2) = 0.$$

Hence the roots are $m_1 = \frac{1}{2}$ and $m_2 = \frac{2}{3}$. Applying Theorem (19.11), the general solution of the given equation is

$$y = C_1 e^{x/2} + C_2 e^{2x/3}. \qquad \blacksquare$$

The final case to consider is that in which the roots of the auxiliary equation $m^2 + bm + c = 0$ of $y'' + by' + cy = 0$ are complex numbers. Recall that complex numbers may be represented by expressions of the form $a + bi$, where a and b are real numbers, and i is a symbol that may be manipulated in the same manner as a real number, but has the additional property that $i^2 = -1$. Two complex numbers $a + bi$ and $c + di$ are said to be **equal**, and we write $a + bi = c + di$, if and only if $a = c$ and $b = d$. Operations of addition, subtraction, multiplication, and division are defined *just as though* all letters denote real numbers, with the additional stipulation that whenever i^2 occurs, it may be replaced by -1. For example, the formulas for addition and multiplication of two complex numbers $a + bi$ and $c + di$ are

$$(a + bi) + (c + di) = (a + c) + (b + d)i$$

$$(a + bi)(c + di) = (ac - bd) + (ad + bc)i.$$

We may regard the real numbers as a subset of the complex numbers by identifying the real number a with the complex number $a + 0i$. A complex number of the form $0 + bi$ is abbreviated bi.

Complex numbers are often required for solving equations of the form $f(x) = 0$, where $f(x)$ is a polynomial. For example, if only real numbers are allowed, then the equation $x^2 = -4$ has no solutions. However, if complex numbers are available, then the equation has a solution $2i$, since

$$(2i)^2 = 2^2 i^2 = 4(-1) = -4.$$

Similarly, $-2i$ is a solution of $x^2 = -4$.

Since $i^2 = -1$, we sometimes use the symbol $\sqrt{-1}$ in place of i and write

$$\sqrt{-13} = \sqrt{13}\, i, \qquad 2 + \sqrt{-25} = 2 + \sqrt{25}\, i = 2 + 5i,$$

etc. A quadratic equation $ax^2 + bx + c = 0$, where a, b, c are real numbers and $a \neq 0$, has roots given by the quadratic formula

$$x = \frac{-b \pm \sqrt{b^2 - 4ac}}{2a}.$$

If $b^2 - 4ac < 0$, then the roots are complex numbers. To illustrate, if we apply the quadratic formula to the equation $x^2 - 4x + 13 = 0$, we obtain

$$x = \frac{4 \pm \sqrt{16 - 52}}{2} = \frac{4 \pm \sqrt{-36}}{2} = \frac{4 \pm 6i}{2} = 2 \pm 3i,$$

Thus the equation has the two complex roots $2 + 3i$ and $2 - 3i$.

The complex number $a - bi$ is called the **conjugate** of the complex number $a + bi$. We see from the quadratic formula that if a quadratic equation with real coefficients has complex roots, then they are necessarily conjugates of one another.

It follows from the preceding discussion that if the auxiliary equation $m^2 + bm + c = 0$ of $y'' + by' + cy = 0$ has complex roots, then they are of the form

$$z_1 = s + ti \quad \text{and} \quad z_2 = s - ti$$

where s and t are real numbers. We may anticipate, from Theorem (19.11), that the general solution of the differential equation is $y = C_1 e^{z_1 x} + C_2 e^{z_2 x}$, or

(19.13)
$$y = C_1 e^{(s+ti)x} + C_2 e^{(s-ti)x}.$$

To handle such complex exponents it is necessary to extend some of the concepts of calculus to include functions whose domains include complex numbers. Since a complete development is beyond the scope of our work, we shall merely outline the main ideas.

In Section 11.8 we discussed how certain functions can be represented by power series. It is not difficult to extend the definitions and theorems of Chapter 11 to infinite series that involve complex numbers. Since this is true, we shall use the power series representations (11.42) to *define* e^z, $\sin z$, and $\cos z$ for every complex number z as follows:

(19.14)
$$e^z = 1 + z + \frac{z^2}{2!} + \cdots + \frac{z^n}{n!} + \cdots$$
$$\sin z = z - \frac{z^3}{3!} + \frac{z^5}{5!} - \cdots + (-1)^n \frac{z^{2n+1}}{(2n+1)!} + \cdots$$
$$\cos z = 1 - \frac{z^2}{2!} + \frac{z^4}{4!} - \cdots + (-1)^n \frac{z^{2n}}{(2n)!} + \cdots$$

Using the first formula in (19.14) gives us

$$e^{iz} = 1 + (iz) + \frac{(iz)^2}{2!} + \frac{(iz)^3}{3!} + \frac{(iz)^4}{4!} + \frac{(iz)^5}{5!} + \cdots$$
$$= 1 + iz + i^2 \frac{z^2}{2!} + i^3 \frac{z^3}{3!} + i^4 \frac{z^4}{4!} + i^5 \frac{z^5}{5!} + \cdots$$

Since $i^2 = -1, i^3 = -i, i^4 = 1, i^5 = i$, etc., we see that

$$e^{iz} = 1 + iz - \frac{z^2}{2!} - i\frac{z^3}{3!} + \frac{z^4}{4!} + i\frac{z^5}{5!} - \cdots$$

which may also be written in the form

$$e^{iz} = \left(1 - \frac{z^2}{2!} + \frac{z^4}{4!} - \cdots\right) + i\left(z - \frac{z^3}{3!} + \frac{z^5}{5!} - \cdots\right).$$

If we now use the formulas for $\cos z$ and $\sin z$ in (19.14), we obtain the following important result.

Euler's Formula **(19.15)**

> For every complex number z,
>
> $$e^{iz} = \cos z + i \sin z.$$

It can be shown that the Laws of Exponents are true for complex numbers. In addition, formulas for derivatives developed earlier in this text may be extended to functions of a *complex* variable z. For example, $D_z e^{kz} = k e^{kz}$, where k is a complex number. It can then be proved that the general solution of $y'' + by' + c = 0$, where the roots of the auxiliary equation are the complex numbers $s \pm ti$, is given by (19.13). The form of this solution may be changed as follows:

$$
\begin{aligned}
y &= C_1 e^{(s+ti)x} + C_2 e^{(s-ti)x} \\
&= C_1 e^{sx+txi} + C_2 e^{sx-txi} \\
&= C_1 e^{sx} e^{txi} + C_2 e^{sx} e^{-txi}
\end{aligned}
$$

or equivalently,

$$(19.16) \qquad y = e^{sx}(C_1 e^{itx} + C_2 e^{-itx}).$$

This can be further simplified by using Euler's Formula. Specifically, we see from (19.15) that

$$
\begin{aligned}
e^{itx} &= \cos tx + i \sin tx, \\
e^{-itx} &= \cos tx - i \sin tx
\end{aligned}
$$

from which it follows that

$$\cos tx = \frac{e^{itx} + e^{-itx}}{2}, \qquad \sin tx = \frac{e^{itx} - e^{-itx}}{2i}.$$

If we let $C_1 = C_2 = \tfrac{1}{2}$ in (19.16) we obtain

$$y = \tfrac{1}{2} e^{sx}(e^{itx} + e^{-itx}).$$

Applying the preceding formula for $\cos tx$ gives us

$$y = \tfrac{1}{2} e^{sx}(2 \cos tx) = e^{sx} \cos tx.$$

Thus, $y = e^{sx} \cos tx$ is a particular solution of $y'' + by' + cy = 0$. Letting $C_1 = -C_2 = i/2$ leads to the particular solution $y = e^{sx} \sin tx$. This is a partial proof of the next theorem.

Theorem (19.17)

> If the auxiliary equation $m^2 + bm + c = 0$ has distinct complex roots $s \pm ti$, then the general solution of $y'' + by' + cy = 0$ is
>
> $$y = e^{sx}(C_1 \cos tx + C_2 \sin tx).$$

Example 4 Solve the differential equation $y'' - 10y' + 41y = 0$.

Solution The roots of the auxiliary equation $m^2 - 10m + 41 = 0$ are

$$m = \frac{10 \pm \sqrt{100 - 164}}{2} = \frac{10 \pm \sqrt{-64}}{2} = \frac{10 \pm 8i}{2} = 5 \pm 4i.$$

Hence by Theorem (19.17), the general solution of the differential equation is

$$y = e^{5x}(C_1 \cos 4x + C_2 \sin 4x). \qquad \blacksquare$$

EXERCISES 19.4

Solve the differential equations in Exercises 1–22.

1 $y'' - 5y' + 6y = 0$

2 $y'' - y' - 2y = 0$

3 $y'' - 3y' = 0$

4 $y'' + 6y' + 8y = 0$

5 $y'' + 4y' + 4y = 0$

6 $y'' - 4y' + 4y = 0$

7 $y'' - 4y' + y = 0$

8 $6y'' - 7y' - 3y = 0$

9 $y'' + 2\sqrt{2}y' + 2y = 0$

10 $4y'' + 20y' + 25y = 0$

11 $8y'' + 2y' - 15y = 0$

12 $y'' + 4y' + y = 0$

13 $9y'' - 24y' + 16y = 0$

14 $4y'' - 8y' + 7y = 0$

15 $2y'' - 4y' + y = 0$

16 $2y'' + 7y' = 0$

17 $y'' - 2y' + 2y = 0$

18 $y'' - 2y' + 5y = 0$

19 $y'' - 4y' + 13y = 0$

20 $y'' + 4 = 0$

21 $\dfrac{d^2y}{dx^2} + 6\dfrac{dy}{dx} + 2y = 0$

22 $\dfrac{d^2y}{dx^2} + 2\dfrac{dy}{dx} + 6y = 0$

In Exercises 23–30 find the particular solution of the differential equation that satisfies the stated boundary conditions.

23 $y'' - 3y' + 2y = 0$; $y = 0$ and $y' = 2$ when $x = 0$

24 $y'' - 2y' + y = 0$; $y = 1$ and $y' = 2$ when $x = 0$

25 $y'' + y = 0$; $y = 1$ and $y' = 2$ when $x = 0$

26 $y'' - y' - 6y = 0$; $y = 0$ and $y' = 1$ when $x = 0$

27 $y'' + 8y' + 16y = 0$; $y = 2$ and $y' = 1$ when $x = 0$

28 $y'' + 5y = 0$; $y = 4$ and $y' = 2$ when $x = 0$

29 $\dfrac{d^2y}{dx^2} - 2\dfrac{dy}{dx} + 5y = 0$; $y = 0$ and $\dfrac{dy}{dx} = 1$ when $x = 0$

30 $\dfrac{d^2y}{dx^2} - 6\dfrac{dy}{dx} + 13y = 0$; $y = 2$ and $\dfrac{dy}{dx} = 3$ when $x = 0$

19.5

NONHOMOGENEOUS LINEAR DIFFERENTIAL EQUATIONS

In this section we shall consider second-order nonhomogeneous linear differential equations with constant coefficients, that is, equations of the form

$$y'' + by' + cy = k(x)$$

where b and c are constants and the function k is continuous.

It is convenient to use the differential operator symbols D and D^2 where if $y = f(x)$, then

$$Dy = y' = f'(x) \quad \text{and} \quad D^2y = y'' = f''(x).$$

Definition (19.18)

> If $y = f(x)$ and f'' exists, then the **linear differential operator** $L = D^2 + bD + c$ is defined by
>
> $$L(y) = (D^2 + bD + c)y = D^2y + bDy + cy = y'' + by' + cy.$$

Using L, the differential equation $y'' + by' + cy = k(x)$ may be written in the compact form $L(y) = k(x)$. In Exercises 19 and 20 you are asked to verify that for every real number C,

$$L(Cy) = CL(y)$$

and that if $y_1 = f_1(x)$ and $y_2 = f_2(x)$, then

$$L(y_1 \pm y_2) = L(y_1) \pm L(y_2).$$

Given the differential equation $y'' + by' + cy = k(x)$, that is, $L(y) = k(x)$, the corresponding homogeneous equation $L(y) = 0$ is called the **complementary equation**.

Theorem (19.19)

> Let $y'' + by' + cy = k(x)$ be a second-order nonhomogeneous linear differential equation. If y_p is a particular solution of $L(y) = k(x)$ and if y_c is the general solution of the complementary equation $L(y) = 0$, then the general solution of $L(y) = k(x)$ is $y = y_p + y_c$.

Proof   Since $L(y_p) = k(x)$ and $L(y_c) = 0$,

$$L(y_p + y_c) = L(y_p) + L(y_c) = k(x) + 0 = k(x)$$

which means that $y_p + y_c$ is a solution of $L(y) = k(x)$. Moreover, if $y = f(x)$ is any other solution, then

$$L(y - y_p) = L(y) - L(y_p) = k(x) - k(x) = 0.$$

Consequently, $y - y_p$ is a solution of the complementary equation. Thus $y - y_p = y_c$ for some y_c, which is what we wished to prove. ☐

If we use the results of the previous section to find the general solution y_c of $L(y) = 0$, then according to Theorem (19.19) all that is needed to determine the general solution of $L(y) = k(x)$ is *one* particular solution y_p.

Example 1   Solve the differential equation $y'' - 4y = 6x - 4x^3$.

Solution   We see by inspection that $y_p = x^3$ is a particular solution of the given equation. The complementary equation is $y'' - 4y = 0$, which by Theorem (19.11) has the general solution

$$y_c = C_1 e^{2x} + C_2 e^{-2x}.$$

Applying Theorem (19.19), the general solution of the given nonhomogeneous equation is

$$y = C_1 e^{2x} + C_2 e^{-2x} + x^3. \qquad \blacksquare$$

If a particular solution of $y'' + by' + cy = k(x)$ cannot be found by inspection, then the following technique, called *variation of parameters*, may be employed. Given the differential equation $L(y) = k(x)$, let y_1 and y_2 be the expressions that appear in the general solution $y = C_1 y_1 + C_2 y_2$ of the complementary equation $L(y) = 0$. For example we might have, as in (19.11), $y_1 = e^{m_1 x}$ and $y_2 = e^{m_2 x}$. Let us now attempt to find a particular solution of $L(y) = k(y)$ that has the form

$$y_p = uy_1 + vy_2$$

where $u = g(x)$ and $v = h(x)$ for some functions g and h. The first and second derivatives of y_p are

$$y_p' = (uy_1' + vy_2') + (u'y_1 + v'y_2)$$

$$y_p'' = (uy_1'' + vy_2'') + (u'y_1' + v'y_2') + (u'y_1 + v'y_2)'.$$

Substituting these into $L(y_p) = y_p'' + by_p' + cy_p$ and rearranging terms, we obtain

$$L(y_p) = u(y_1'' + by_1' + cy_1) + v(y_2'' + by_2' + cy_2)$$
$$+ b(u'y_1 + v'y_2) + (u'y_1 + v'y_2)' + (u'y_1' + v'y_2').$$

Since y_1 and y_2 are solutions of $y'' + by' + cy = 0$, the first two terms on the right are 0. Hence, to obtain $L(y_p) = k(x)$, it is sufficient to choose u and v such that

$$u'y_1 + v'y_2 = 0$$
$$u'y_1' + v'y_2' = k(x).$$

It can be shown that this system of equations always has a unique solution u' and v'. We may then determine u and v by integration and use the fact that $y_p = uy_1 + vy_2$ to find a particular solution of the differential equation. Our discussion may be summarized as follows.

Variation of Parameters **(19.20)**

> If $y = C_1 y_1 + C_2 y_2$ is the general solution of the complementary equation $L(y) = 0$ of $y'' + by' + cy = k(x)$, then a particular solution of $L(y) = k(x)$ is $y_p = uy_1 + vy_2$, where $u = g(x)$ and $v = h(x)$ satisfy the following system of equations:
>
> $$u'y_1 + v'y_2 = 0$$
>
> $$u'y_1' + v'y_2' = k(x).$$

Example 2 Solve the differential equation $y'' + y = \cot x$.

Solution   The complementary equation is $y'' + y = 0$. Since the auxiliary equation $m^2 + 1 = 0$ has roots $\pm i$, we see from Theorem (19.17) that the general solution of the homogeneous differential equation $y'' + y = 0$ is $y = C_1 \cos x + C_2 \sin x$. As in the preceding discussion, we let $y_1 = \cos x$ and $y_2 = \sin x$. The system of equations in Theorem (19.20) is, therefore,

$$u' \cos x + v' \sin x = 0$$
$$-u' \sin x + v' \cos x = \cot x.$$

Solving for u' and v' gives us

$$u' = -\cos x, \qquad v' = \csc x - \sin x.$$

If we integrate each of these expressions (and drop the constants of integration) we obtain

$$u = -\sin x, \qquad v = \ln|\csc x - \cot x| + \cos x.$$

Applying Theorem (19.20), a particular solution of the given equation is

$$y_p = -\sin x \cos x + \sin x \ln|\csc x - \cot x| + \sin x \cos x$$

or $\qquad y_p = \sin x \ln|\csc x - \cot x|.$

Finally, by Theorem (19.19), the general solution of $y'' + y = \cot x$ is

$$y = C_1 \cos x + C_2 \sin x + \sin x \ln|\csc x - \cot x|. \qquad \blacksquare$$

Given the differential equation

$$L(y) = y'' + by' + cy = e^{nx}$$

where e^{nx} *is not a solution of* $L(y) = 0$, it is reasonable to expect that there exists a particular solution of the form $y_p = Ae^{nx}$, since e^{nx} is the result of finding $L(Ae^{nx})$. This suggests that we use Ae^{nx} as a trial solution in the given equation and attempt to find the value of the coefficient A. This technique is called the **method of undetermined coefficients**, and is illustrated in the next example.

Example 3 Solve the differential equation $y'' + 2y' - 8y = e^{3x}$.

Solution   Since the auxiliary equation $m^2 + 2m - 8 = 0$ of the differential equation $y'' + 2y' - 8y = 0$ has roots 2 and -4, it follows from the previous section that the general solution of the complementary equation is

$$y_c = C_1 e^{2x} + C_2 e^{-4x}.$$

From the preceding remarks we seek a particular solution of the form $y_p = Ae^{3x}$. Since $y_p' = 3Ae^{3x}$ and $y_p'' = 9Ae^{3x}$, substitution in the given equation leads to

$$9Ae^{3x} + 6Ae^{3x} - 8Ae^{3x} = e^{3x}.$$

Dividing both sides by e^{3x}, we obtain

$$9A + 6A - 8A = 1, \quad \text{or} \quad A = \tfrac{1}{7}.$$

Thus $y_p = \tfrac{1}{7}e^{3x}$ and by Theorem (19.19), the general solution is

$$y = C_1 e^{2x} + C_2 e^{-4x} + \tfrac{1}{7}e^{3x}. \qquad \blacksquare$$

Three rules for arriving at trial solutions of second-order nonhomogeneous differential equations with constant coefficients are stated without proof in the next theorem. The reader is referred to texts on differential equations for a more extensive treatment of this topic.

Theorem (19.21)

(i) If $y'' + by' + cy = e^{nx}$, and n is not a root of the auxiliary equation $m^2 + bm + c = 0$, then there is a particular solution of the form $y_p = Ae^{nx}$.

(ii) If $y'' + by' + cy = xe^{nx}$, and n is not a solution of the auxiliary equation $m^2 + bm + c = 0$, then there is a particular solution of the form $y_p = (A + Bx)e^{nx}$.

(iii) If either

$$y'' + by' + cy = e^{sx} \sin tx$$

or
$$y'' + by' + cy = e^{sx} \cos tx$$

and the complex number $s + ti$ is not a solution of the auxiliary equation $m^2 + bm + c = 0$, then there is a particular solution of the form

$$y_p = Ae^{sx} \cos tx + Be^{sx} \sin tx.$$

Rule (i) of Theorem (19.21) was used in the solution of Example 3. Illustrations of rules (ii) and (iii) are given in the following examples.

Example 4 Solve $y'' - 3y' - 18y = xe^{4x}$.

Solution   Since the auxiliary equation $m^2 - 3m - 18 = 0$ has roots 6 and -3, it follows from the preceding section that the general solution of $y'' - 3y' - 18y = 0$ is

$$y = C_1 e^{6x} + C_2 e^{-3x}.$$

Since 4 is not a root of the auxiliary equation, we see from (ii) of Theorem (19.21) that there is a particular solution of the form

$$y_p = (A + Bx)e^{4x}.$$

Differentiating, we obtain

$$y_p' = (4A + 4Bx + B)e^{4x}$$

$$y_p'' = (16A + 16Bx + 8B)e^{4x}.$$

Substitution in the given differential equation produces

$$(16A + 16Bx + 8B)e^{4x} - 3(4A + 4Bx + B)e^{4x} - 18(A + Bx)e^{4x} = xe^{4x}$$

which reduces to

$$-14A + 5B - 14Bx = x.$$

Thus y_p is a solution provided

$$-14A + 5B = 0 \quad \text{and} \quad -14B = 1.$$

This gives us $B = -\frac{1}{14}$ and $A = -\frac{5}{196}$. Consequently,

$$y_p = \left(-\frac{5}{196} - \frac{1}{14}x\right)e^{4x} = -\frac{1}{196}(5 + 14x)e^{4x}.$$

Applying Theorem (19.19), the general solution is

$$y = C_1 e^{6x} + C_2 e^{-3x} - \tfrac{1}{196}(5 + 14x)e^{4x}. \qquad \blacksquare$$

Example 5 Solve $y'' - 10y' + 41y = \sin x$.

Solution The general solution $y_c = e^{5x}(C_1 \cos 4x + C_2 \sin 4x)$ of the complementary equation was found in Example 4 of the preceding section. Referring to (iii) of Theorem (19.21) with $s = 0$ and $t = 1$, we seek a particular solution of the form

$$y_p = A \cos x + B \sin x.$$

Since

$$y_p' = -A \sin x + B \cos x \quad \text{and} \quad y_p'' = -A \cos x - B \sin x,$$

substitution in the given equation produces

$$-A \cos x - B \sin x + 10A \sin x - 10B \cos x$$
$$+ 41A \cos x + 41B \sin x = \sin x$$

which can be written

$$(40A - 10B) \cos x + (10A + 40B) \sin x = \sin x.$$

Consequently, y_p is a solution provided

$$40A - 10B = 0 \quad \text{and} \quad 10A + 40B = 1.$$

The solution of this system of equations is $A = \frac{1}{170}$ and $B = \frac{4}{170}$. Hence

$$y_p = \frac{1}{170} \cos x + \frac{4}{170} \sin x = \frac{1}{170}(\cos x + 4 \sin x)$$

and the general solution is

$$y = e^{5x}(C_1 \cos 4x + C_2 \sin 4x) + \frac{1}{170}(\cos x + 4 \sin x). \qquad \blacksquare$$

EXERCISES 19.5

Solve the differential equations in Exercises 1–10 by using the method of variation of parameters.

1 $y'' + y = \tan x$

2 $y'' + y = \sec x$

3 $y'' - 6y' + 9y = x^2 e^{3x}$

4 $y'' + 3y' = e^{-3x}$

5 $y'' - y = e^x \cos x$

6 $y'' - 4y' + 4y = x^{-2} e^{2x}$

7 $y'' - 9y = e^{3x}$

8 $y'' + y = \sin x$

9 $\dfrac{d^2 y}{dx^2} - 3\dfrac{dy}{dx} - 4y = 2$

10 $\dfrac{d^2 y}{dx^2} - \dfrac{dy}{dx} = x + 1$

Solve the differential equations in Exercises 11–18 by using undetermined coefficients.

11 $y'' - 3y' + 2y = 4e^{-x}$

12 $y'' + 6y' + 9y = e^{2x}$

13 $y'' + 2y' = \cos 2x$

14 $y'' + y = \sin 5x$

15 $y'' - y = xe^{2x}$

16 $y'' + 3y' - 4y = xe^{-x}$

17 $\dfrac{d^2 y}{dx^2} - 6\dfrac{dy}{dx} + 13y = e^x \cos x$

18 $\dfrac{d^2 y}{dx^2} - 2\dfrac{dy}{dx} + 2y = e^{-x} \sin 2x$

Prove the identities in Exercises 19 and 20, where L is the linear differential operator (19.18) and y, y_1, y_2 represent functions of x.

19 $L(Cy) = CL(y)$

20 $L(y_1 \pm y_2) = L(y_1) \pm L(y_2)$

19.6

VIBRATIONS

Vibrations in mechanical systems are caused by external forces. Thus, a violin string vibrates if bowed, a steel beam vibrates if struck by a hammer, and a bridge vibrates if a marching band crosses it in cadence. In this section we shall use differential equations to analyze vibrations that may occur in a spring.

According to Hooke's Law, the force required to stretch a spring y units beyond its natural length is ky, here k is a positive real number called the **spring constant** (see (6.11)). The **restoring force** of the spring is $-ky$. Suppose that a weight W is attached to the spring and, at the equilibrium position, the spring is stretched a distance l_1 beyond its natural length l_0, as illustrated in Figure 19.5. If g is the gravitational constant and m is the mass of the weight, then $W = mg$, and at the equilibrium position,

$$mg = kl_1, \quad \text{or} \quad mg - kl_1 = 0$$

where we have assumed that the mass of the spring is negligible compared to m.

Suppose the weight is pulled down and released. Consider the coordinate line shown in Figure 19.6, where y denotes the directed distance from the equilibrium point to the center of mass of the weight after t seconds. By

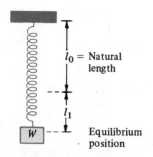

$l_0 =$ Natural length

l_1

W — Equilibrium position

Figure 19.5

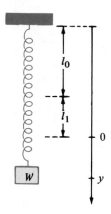

Figure 19.6

Newton's Second Law of Motion, the force acting on the weight is $F = ma$, where a is its acceleration. Assuming there is no external retarding force and the weight moves in a frictionless medium, we see that

$$F = mg - k(l_1 + y) = mg - kl_1 - ky = -ky.$$

Since $a = d^2y/dt^2$, this implies that

$$m\frac{d^2y}{dt^2} = -ky, \quad \text{or} \quad \frac{d^2y}{dt^2} + \frac{k}{m}y = 0.$$

If, as is customary, we denote k/m by ω^2, the preceding equation may be written

$$\textbf{(19.22)} \qquad \frac{d^2y}{dt^2} + \omega^2 y = 0.$$

Since the solutions of the auxiliary equation are $\pm\omega i$ (Why?), it follows from Theorem (19.17) that the general solution of the differential equation is

$$y = C_1 \cos \omega t + C_2 \sin \omega t.$$

Thus the weight is in simple harmonic motion (see Exercise 86 of Section 8.2).

Suppose, for example, that the weight is pulled down a distance l_2 and is released with zero velocity. In this case, at $t = 0$,

$$l_2 = C_1(1) + C_2(0) \quad \text{or} \quad C_1 = l_2.$$

Since

$$\frac{dy}{dt} = -\omega C_1 \sin \omega t + \omega C_2 \cos \omega t$$

we also have (at $t = 0$),

$$0 = -\omega C_1(0) + \omega C_2(1) \quad \text{or} \quad C_2 = 0.$$

Hence the displacement y of the weight at time t is

$$y = l_2 \cos \omega t.$$

This type of motion was discussed earlier in the text (see Definition (8.7)). The **amplitude** (the maximum displacement of W) is l_2, and the **period** (the time for one complete vibration) is $2\pi/\omega = 2\pi\sqrt{m/k}$. A typical graph indicating this type of motion is illustrated in Figure 19.7.

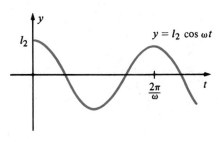

Figure 19.7 Simple harmonic motion

Example 1 An 8-pound weight stretches a spring 2 ft beyond its natural length. The weight is then pulled down another $\frac{1}{2}$ foot and released with an initial upward velocity of 6 ft/sec. Find a formula for the displacement of the weight at any time t.

Solution   From Hooke's Law, $8 = k(2)$, or $k = 4$. If y is the displacement of the weight from the equilibrium position at time t, then

$$m\frac{d^2y}{dt^2} = -4y.$$

Since $W = mg$ it follows that $m = W/g = \frac{8}{32} = \frac{1}{4}$. Hence

$$\frac{1}{4}\frac{d^2y}{dt^2} = -4y, \quad \text{or} \quad \frac{d^2y}{dt^2} + 16y = 0$$

which has the form of (19.22). As in our discussion, this implies that

$$y = C_1 \cos 4t + C_2 \sin 4t.$$

At $t = 0$ we have $y = \frac{1}{2}$ and therefore

$$\tfrac{1}{2} = C_1(1) + C_2(0), \quad \text{or} \quad C_1 = \tfrac{1}{2}.$$

Since

$$\frac{dy}{dt} = -4C_1 \sin 4t + 4C_2 \cos 4t$$

and $dy/dt = -6$ at $t = 0$, we obtain

$$-6 = -4C_1(0) + 4C_2(1), \quad \text{or} \quad C_2 = -\tfrac{3}{2}.$$

Hence the displacement at time t is given by

$$y = \tfrac{1}{2} \cos 4t - \tfrac{3}{2} \sin 4t.$$

Let us next consider the motion of the spring if a damping (or frictional) force is present, as is the case when the weight moves through a fluid (see Figure 19.8). A shock absorber in an automobile is a good illustration of this situation. We shall assume that the direction of the damping force is opposite to that of the motion and that it is directly proportional to the velocity of the weight. Thus, the damping force is given by $-c(dy/dt)$, where c is a positive constant. According to Newton's Second Law, the differential equation that describes the motion is

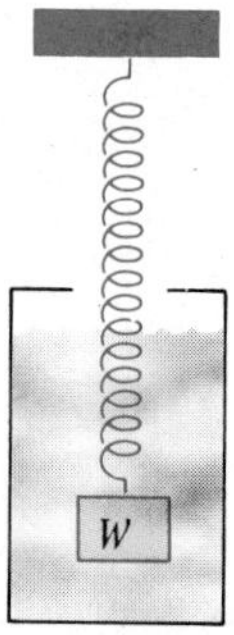

$$m\frac{d^2y}{dt^2} = -ky - c\frac{dy}{dt}, \quad \text{or} \quad \frac{d^2y}{dt^2} + \frac{c}{m}\frac{dy}{dt} + \frac{k}{m}y = 0.$$

This equation, called the differential equation of **free, damped vibration**, is often written in the form

(*19.23*)
$$\frac{d^2y}{dt^2} + 2p\frac{dy}{dt} + \omega^2 y = 0$$

where $2p = c/m$ and $\omega^2 = k/m$. The roots of the auxiliary equation are

$$\frac{-2p \pm \sqrt{4p^2 - 4\omega^2}}{2} = -p \pm \sqrt{p^2 - \omega^2}.$$

Figure 19.8   Damping force

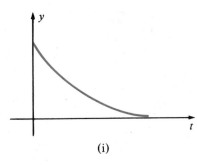

(i)

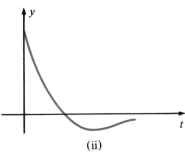

(ii)

Figure 19.9 Overdamped

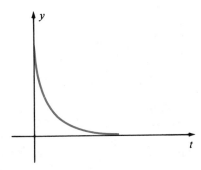

Figure 19.10 Critically damped

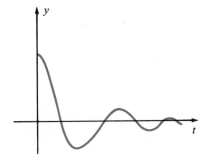

Figure 19.11 Underdamped

The following three possibilities may arise:

$$p^2 - \omega^2 > 0, \quad p^2 - \omega^2 = 0, \quad \text{or} \quad p^2 - \omega^2 < 0.$$

Let us classify the corresponding types of motion as follows:

Case (i) $p^2 - \omega^2 > 0$: **Overdamped**.

By Theorem (19.11), the general solution of the differential equation may be written

$$y = e^{-pt}(C_1 e^{\sqrt{p^2 - \omega^2}\,t} + C_2 e^{-\sqrt{p^2 - \omega^2}\,t}).$$

In the case under consideration,

$$p^2 - \omega^2 = \frac{c^2}{4m^2} - \frac{k}{m} = \frac{c^2 - 4mk}{4m^2} > 0$$

so that $c^2 > 4mk$. This shows that c is usually large compared to k, that is, the damping force dominates the restoring force of the spring, and the weight returns to the equilibrium position relatively quickly. In particular, this happens if the fluid has a high viscosity, as is true for heavy oil or grease.

The manner in which y approaches 0 depends on the constants C_1 and C_2 in the general solution. If the constants are both positive, the graph has the general shape shown in (i) of Figure 19.9, whereas if they have opposite signs, the graph resembles that shown in (ii) of the figure.

Case (ii) $p^2 - \omega^2 = 0$: **Critically damped**.

In this case the auxiliary equation has a double root $-p$, and by Theorem (19.12), the general solution of the differential equation is

$$y = e^{-pt}(C_1 + C_2 t).$$

The typical graph sketched in Figure 19.10 is similar to that in (i) of Figure 19.9; however, any decrease in the damping force leads to the oscillatory motion discussed in the following case.

Case (iii) $p^2 - \omega^2 < 0$: **Underdamped**.

The roots of the auxiliary equation are complex conjugates $a \pm bi$, and by Theorem (19.17) the general solution of the differential equation is

$$y = e^{-at}(C_1 \sin bt + C_2 \cos bt).$$

In this case c is usually smaller than k, and the spring oscillates while returning to the equilibrium position, as illustrated in Figure 19.11. This type of motion could occur in a worn shock absorber in an automobile.

Example 2 A 24-pound weight stretches a spring 1 ft beyond its natural length. If the weight is pushed downward from its equilibrium position with an initial velocity of 2 ft/sec, and if the damping force is $-9(dy/dt)$, find a formula for the displacement y of the weight at any time t.

Solution   By Hooke's Law, $24 = k(1)$, or $k = 24$. The mass of the weight is $m = W/g = \frac{24}{32} = \frac{3}{4}$. Using the notation introduced in (19.23),

$$2p = \frac{c}{m} = \frac{9}{\frac{3}{4}} = 12 \quad \text{and} \quad \omega^2 = \frac{k}{m} = \frac{24}{\frac{3}{4}} = 32.$$

Substitution in (19.23) gives us

$$\frac{d^2y}{dt^2} + 12\frac{dy}{dt} + 32y = 0.$$

It may be verified that the roots of the auxiliary equation are -8 and -4. Thus the motion is overdamped and the general solution of the differential equation is

$$y = C_1e^{-4t} + C_2e^{-8t}.$$

Letting $t = 0$,

$$0 = C_1 + C_2 \quad \text{or} \quad C_2 = -C_1.$$

Since

$$\frac{dy}{dt} = -4C_1e^{-4t} - 8C_2e^{-8t}$$

we have, at $t = 0$,

$$2 = -4C_1 - 8C_2 = -4C_1 - 8(-C_1) = 4C_1.$$

Consequently, $C_1 = \frac{1}{2}$ and $C_2 = -C_1 = -\frac{1}{2}$. Thus the displacement y of the weight at time t is

$$y = \tfrac{1}{2}e^{-4t} - \tfrac{1}{2}e^{-8t} = \tfrac{1}{2}e^{-4t}(1 - e^{-4t}).$$

The graph is similar in appearance to that in (i) of Figure 19.9. In Exercise 9 you are asked to find the damping force that will produce critical damping.   ∎

EXERCISES 19.6

1 A 5-pound weight stretches a spring 6 in. beyond its natural length. If the weight is raised 4 in. and then released with zero velocity, find a formula for the displacement of the weight at any time t.

2 A 10-pound weight stretches a spring 8 in. beyond its natural length. If the weight is pulled down another 3 in. and then released with a downward velocity of 6 in./sec, find a formula for the displacement at any time t.

3 A 10-pound weight stretches a spring 1 ft beyond its natural length. If the weight is relased from its equilibrium position with a downward velocity of 2 ft/sec, and if the

frictional force is $-5(dy/dt)$, find a formula for the displacement of the weight.

4 A 6-pound weight is suspended from a spring whose spring constant is 48 lb/ft. Initially the weight has a downward velocity of 4 ft/sec from a position 5 in. below the equilibrium point. Find a formula for the displacement.

5 A 4-pound weight stretches a spring 3 in. beyond its natural length. If the weight is pulled down another 4 in. and released with zero velocity in a medium where the damping force is $-2(dy/dt)$, find the displacement y at any time t.

6 A spring stretches $\frac{1}{2}$ ft when an 8-pound weight is attached. If the weight is released from the equilibrium position with an upward velocity of 1 ft/sec, and if the damping force is $-4(dy/dt)$, find a formula for the displacement y at any time t.

7 Describe a spring/weight system that leads to the following differential equation and initial values:

$$\frac{1}{4}\frac{d^2y}{dt^2} + \frac{dy}{dt} + 6y = 0;$$

$$y = -2 \quad \text{and} \quad \frac{dy}{dt} = -1 \quad \text{at } t = 0.$$

8 A 4-pound weight stretches a spring 1 ft. If the weight moves in a medium whose damping force is $-c(dy/dt)$ where $c > 0$, for what values of c is the motion (a) overdamped? (b) critically damped? (c) underdamped?

9 In Example 2 of this section, find the damping force that will produce critical damping.

10 Suppose an undamped weight of mass m is attached to a spring having spring constant k and that an external periodic force given by $F \sin \alpha t$ is applied to the weight.

(a) Show that the differential equation that describes the motion of the weight is

$$\frac{d^2y}{dt^2} + \omega^2 y = \frac{F}{m}\sin \alpha t$$

where $\omega^2 = k/m$.

(b) Prove that the displacement y of the weight is given by

$$y = C_1 \cos \omega t + C_2 \sin \omega t + C \sin \alpha t,$$

where $C = F/m(\omega^2 - \alpha^2)$. This type of motion is called **forced vibration**.

19.7

SERIES SOLUTIONS OF DIFFERENTIAL EQUATIONS

As shown in Chapter 11, a power series $\sum a_n x^n$ determines a function f, such that

$$y = f(x) = a_0 + a_1 x + a_2 x^2 + a_3 x^3 + a_4 x^4 + \cdots$$

for every x in the interval of convergence. Moreover, series representations for the derivatives of f may be obtained by differentiating each term. Thus

$$y' = a_1 + 2a_2 x + 3a_3 x^2 + 4a_4 x^3 + \cdots = \sum_{n=1}^{\infty} na_n x^{n-1},$$

$$y'' = 2a_2 + 3 \cdot 2a_3 x + 4 \cdot 3a_4 x^2 + \cdots = \sum_{n=2}^{\infty} n(n-1)a_n x^{n-2},$$

etc. Power series may be used to solve certain differential equations. In this case, the solution is often expressed as an infinite series, and is called a **series solution** of the differential equation.

Example 1   Find a series solution of the differential equation $y' = 2xy$.

Solution   If the solution is given by $y = \sum a_n x^n$, then $y' = \sum na_n x^{n-1}$, and substitution in the differential equation gives us

$$\sum_{n=1}^{\infty} na_n x^{n-1} = 2x \sum_{n=0}^{\infty} a_n x^n = \sum_{n=0}^{\infty} 2a_n x^{n+1}.$$

It is convenient to change the summation on the left so that the same power of x appears as in the summation on the right. This may be accomplished by replacing n by $n + 2$ and beginning the summation at $n = -1$. Thus

$$\sum_{n=-1}^{\infty} (n + 2)a_{n+2}x^{n+1} = \sum_{n=0}^{\infty} 2a_n x^{n+1}, \quad \text{or}$$

$$a_1 + 2a_2 x + \cdots + (n + 2)a_{n+2}x^{n+1} + \cdots = 2a_0 x + \cdots + 2a_n x^{n+1} + \cdots$$

Comparing coefficients we see that $a_1 = 0$ and $(n + 2)a_{n+2} = 2a_n$ if $n \geq 0$. Consequently, the coefficients are given by

$$a_1 = 0 \quad \text{and} \quad a_{n+2} = \frac{2}{n + 2} a_n \quad \text{if} \quad n \geq 0.$$

In particular,

$$a_1 = 0, \; a_2 = a_0, \; a_3 = \tfrac{2}{3}a_1 = 0, \; a_4 = \tfrac{1}{2}a_2 = \tfrac{1}{2}a_0, \; a_5 = \tfrac{2}{5}a_3 = 0,$$

$$a_6 = \tfrac{1}{3}a_4 = \frac{1}{2 \cdot 3} a_0, \; a_7 = \tfrac{2}{7}a_5 = 0, \; a_8 = \tfrac{1}{4}a_6 = \frac{1}{2 \cdot 3 \cdot 4} a_0,$$

etc. It can be shown that if n is odd, then $a_n = 0$, whereas $a_{2n} = (1/n!)a_0$ for every positive integer n. The series solution is, therefore,

$$y = \sum_{n=0}^{\infty} a_n x^n = a_0 \left(1 + x^2 + \frac{1}{2!} x^4 + \cdots + \frac{1}{n!} x^{2n} + \cdots \right). \qquad \blacksquare$$

It follows from (19.14) that the series solution in Example 1 can be written as $y = a_0 e^{(x^2)}$. Indeed, this form may be found directly from $y' = 2xy$ by using the separation of variables technique. The objective in Example 1, however, was to illustrate series solutions and not to find the solution in the simplest manner. In many instances it is impossible to find the sum of $\sum a_n x^n$ and the solution must be left in series form.

Example 2 Solve the differential equation $y'' - xy' - 2y = 0$.

Solution   Substituting

$$y = \sum_{n=0}^{\infty} a_n x^n, \quad y' = \sum_{n=1}^{\infty} na_n x^{n-1}, \quad \text{and} \quad y'' = \sum_{n=2}^{\infty} n(n - 1)a_n x^{n-2}$$

we obtain

$$\sum_{n=2}^{\infty} n(n - 1)a_n x^{n-2} - x\sum_{n=1}^{\infty} na_n x^{n-1} - 2\sum_{n=0}^{\infty} a_n x^n = 0$$

or

$$\sum_{n=2}^{\infty} n(n - 1)a_n x^{n-2} = \sum_{n=0}^{\infty} na_n x^n + \sum_{n=0}^{\infty} 2a_n x^n.$$

We next adjust the summation on the left so that the power x^n appears instead of x^{n-2}. This can be accomplished by replacing n by $n + 2$ and starting the summation at $n = 0$. This gives us

$$\sum_{n=0}^{\infty} (n + 2)(n + 1)a_{n+2}x^n = \sum_{n=0}^{\infty} (n + 2)a_n x^n.$$

Comparing coefficients we see that $(n + 2)(n + 1)a_{n+2} = (n + 2)a_n$, that is,

$$a_{n+2} = \frac{1}{n + 1} a_n.$$

Letting $n = 0, 1, 2, \ldots, 7$ leads to the following form for the coefficients a_k:

$$a_2 = a_0 \qquad\qquad a_3 = \frac{1}{2} a_1$$

$$a_4 = \frac{1}{3} a_2 = \frac{1}{3} a_0 \qquad a_5 = \frac{1}{4} a_3 = \frac{1}{2 \cdot 4} a_1$$

$$a_6 = \frac{1}{5} a_4 = \frac{1}{3 \cdot 5} a_0 \qquad a_7 = \frac{1}{6} a_5 = \frac{1}{2 \cdot 4 \cdot 6} a_1$$

$$a_8 = \frac{1}{7} a_6 = \frac{1}{3 \cdot 5 \cdot 7} a_0 \qquad a_9 = \frac{1}{8} a_7 = \frac{1}{2 \cdot 4 \cdot 6 \cdot 8} a_1$$

In general,

$$a_{2n} = \frac{1}{1 \cdot 3 \cdots (2n - 1)} a_0, \qquad a_{2n+1} = \frac{1}{2 \cdot 4 \cdots (2n)} a_1 = \frac{1}{2^n n!} a_1.$$

The solution $y = \sum a_n x^n$ may, therefore, be expressed as a sum of two infinite series:

$$y = a_0 \left[1 + \sum_{n=1}^{\infty} \frac{1}{1 \cdot 3 \cdots (2n - 1)} x^{2n} \right] + a_1 \sum_{n=0}^{\infty} \frac{1}{2^n n!} x^{2n+1}. \qquad\blacksquare$$

EXERCISES 19.7

Find series solutions for the differential equations in Exercises 1–12.

1 $y'' + y = 0$

2 $y'' - 4y = 0$

3 $y'' - 2xy = 0$

4 $y'' + 2xy' + y = 0$

5 $\dfrac{d^2y}{dx^2} - x\dfrac{dy}{dx} + 2y = 0$

6 $\dfrac{d^2y}{dx^2} + x^2 y = 0$

7 $(x + 1)y' = 3y$

8 $y' = 4x^3 y$

9 $y'' - y = 5x$

10 $y'' - xy = x^4$

11 $(x^2 - 1)y'' + 6xy' + 4y = -4$

12 $y'' + y = e^x$

19.8

REVIEW

Discuss methods for solving the following types of differential equations.

1 Separable

2 First-order linear

3 Exact

4 Homogeneous

5 Bernoulli

6 Second-order linear

EXERCISES 19.8

Solve the differential equations in Exercises 1–40.

1 $xe^y\,dx - \csc x\,dy = 0$

2 $(2xy - 1)\,dx + (x^2 + 2y)\,dy = 0$

3 $(3x - y)\,dx + (x + y)\,dy = 0$

4 $y' + 4y = e^{-x}$

5 $y^2 - ye^{-x} + (e^{-x} + 2xy + 3)y' = 0$

6 $(x^2y + x^2)\,dy + y\,dx = 0$

7 $y\tan x + y' = 2\sec x$

8 $(x^2 + y^2) - xyy' = 0$

9 $y\sqrt{1 - x^2}\,y' = \sqrt{1 - y^2}$

10 $(2y + x^3)\,dx - x\,dy = 0$

11 $\left(2x\sin\dfrac{y}{x} - y\cos\dfrac{y}{x}\right)dx + x\cos\dfrac{y}{x}\,dy = 0$

12 $(y\cos x - 2x) + (\sin x + 2y)y' = 0$

13 $xy' - 2y = x^3y^3$

14 $y'' + y' - 6y = 0$

15 $y'' - 8y' + 16y = 0$

16 $y'' - 6y' + 25y = 0$

17 $\dfrac{d^2y}{dx^2} - 2\dfrac{dy}{dx} = 0$

18 $\dfrac{d^2y}{dx^2} = y + \sin x$

19 $y'' - y = e^x\sin x$

20 $y'' - y' - 6y = e^{2x}$

21 $\sec^2 y\,dx = \sqrt{1 - x^2}\,dy - x\sec^2 y\,dx$

22 $(2x - yx^{-1} + \ln y)\,dx + (xy^{-1} - \ln x + 1)\,dy = 0$

23 $y' + y = e^{4x}$

24 $y'' + 2y' = 0$

25 $y'' - 3y' + 2y = e^{5x}$

26 $xe^y\,dx - (x + 1)y\,dy = 0$

27 $xy' + y = (x - 2)^2$

28 $(3x^2 - 2xy^2 + 1)\,dx + (y^2 - 2x^2y)\,dy = 0$

29 $y'' - y' - 20y = xe^{-x}$

30 $(x^2 - y^2)y' + 3xy = 0$

31 $\dfrac{d^2y}{dx^2} + 5\dfrac{dy}{dx} + 7y = 0$

32 $\dfrac{d^2y}{dx^2} + y = \csc x$

33 $e^{x+y}\,dx - \csc x\,dy = 0$

34 $y'' + 10y' + 25y = 0$

35 $\cot x\,dy = (y - \cos x)\,dx$

36 $y'' + y' + y = e^x\cos x$

37 $(y - 2e^{-2x}\sin y)\,dx + (e^{-2x}\cos y + x)\,dy = 0$

38 $y''' = 0$

39 $y' + y\csc x = \tan x$

40 $xy^2y' = x^3 + y^3$

APPENDIX I

MATHEMATICAL INDUCTION

The method of proof known as **mathematical induction** may be used to show that certain statements or formulas are true for all positive integers. For example, if n is a positive integer, let P_n denote the statement

$$(xy)^n = x^n y^n$$

where x and y are real numbers. Thus, P_1 represents the statement $(xy)^1 = x^1 y^1$, P_2 denotes $(xy)^2 = x^2 y^2$, P_3 is $(xy)^3 = x^3 y^3$, etc. It is easy to show that P_1, P_2, and P_3 are *true* statements. However, since the set of positive integers is infinite, it is impossible to check the validity of P_n for every positive integer n. To give a proof, the method provided by (I.2) is required. This method is based on the following fundamental axiom.

Axiom of Mathematical Induction
(I.1)

Suppose a set S of positive integers has the following two properties:

(i) S contains the integer 1.

(ii) Whenever S contains a positive integer k, S also contains $k + 1$.

Then S contains every positive integer.

The reader should have little reluctance about accepting (I.1). If S is a set of positive integers satisfying property (ii), then whenever S contains an arbitrary positive integer k, it must also contain the next positive integer, $k + 1$. If S also satisfies property (i), then S contains 1 and hence by (ii), S contains $1 + 1$, or 2. Applying (ii) again, we see that S contains $2 + 1$, or 3. Once again, S must contain $3 + 1$, or 4. If we continue in this manner, it can be argued that if n is any *specific* positive integer, then n is in S, since we can proceed a step at a time as above, eventually reaching n. Although this argument does not *prove* (I.1), it certainly makes it plausible.

We shall use (I.1) to establish the following fundamental principle.

Principle of Mathematical Induction
(I.2)

> If with each positive integer n there is associated a statement P_n, then all the statements P_n are true provided the following two conditions hold:
>
> (i) P_1 is true.
> (ii) Whenever k is a positive integer such that P_k is true, then P_{k+1} is also true.

Proof   Assume that (i) and (ii) of (I.2) hold and let S denote the set of all positive integers n such that P_n is true. By assumption, P_1 is true and consequently 1 is in S. Thus S satisfies property (i) of (I.1). Whenever S contains a positive integer k, then by the definition of S, P_k is true and hence from assumption (ii) of (I.2), P_{k+1} is also true. This means that S contains $k + 1$. We have shown that whenever S contains a positive integer k, then S also contains $k + 1$. Consequently, property (ii) of (I.1) is true. Hence by (I.1), S contains every positive integer; that is, P_n is true for every positive integer n.

$\square$

There are other variations of the principle of mathematical induction. One variation appears in (I.5). In most of our work the statement P_n will usually be given in the form of an equation involving the arbitrary positive integer n, as in our illustration $(xy)^n = x^n y^n$.

When applying (I.2), the following two steps should always be used.

(I.3)

> Step (i)   Prove that P_1 is true.
> Step (ii) Assume that P_k is true and prove that P_{k+1} is true.

Step (ii) is usually the most confusing for the beginning student. We do not *prove* that P_k is true (except for $k = 1$). Instead, we show that *if* P_k is true, then the statement P_{k+1} is true. That is all that is necessary according to (I.2). The assumption that P_k is true is referred to as the **induction hypothesis**.

Example 1 Prove that for every positive integer n, the sum of the first n positive integers is $n(n + 1)/2$.

Solution   For any positive integer n, let P_n denote the statement

$$1 + 2 + 3 + \cdots + n = \frac{n(n + 1)}{2}$$

where by convention, when $n \leq 4$, the left side is adjusted so that there are precisely n terms in the sum. The following are some special cases of P_n:

If $n = 2$, then P_2 is

$$1 + 2 = \frac{2(2 + 1)}{2}, \quad \text{or} \quad 3 = 3.$$

If $n = 3$, then P_3 is

$$1 + 2 + 3 = \frac{3(3 + 1)}{2}, \quad \text{or} \quad 6 = 6.$$

If $n = 5$, then P_5 is

$$1 + 2 + 3 + 4 + 5 = \frac{5(5 + 1)}{2}, \quad \text{or} \quad 15 = 15.$$

We wish to show that P_n is true for every n. Although it is instructive to check P_n for several values of n as we have done, it is unnecessary to do so. We need only follow steps (i) and (ii) of (I.3).

(i) If we substitute $n = 1$ in P_n then by convention the left side collapses to 1 and the right side is $\dfrac{1(1 + 1)}{2}$, which also equals 1. This proves that P_1 is true.

(ii) *Assume* that P_k is true. Thus the induction hypothesis is

$$1 + 2 + 3 + \cdots + k = \frac{k(k + 1)}{2}.$$

Our goal is to prove that P_{k+1} is true, that is,

$$1 + 2 + 3 + \cdots + (k + 1) = \frac{(k + 1)[(k + 1) + 1]}{2}.$$

By the induction hypothesis we already have a formula for the sum of the first k positive integers. Hence a formula for the sum of the first $k + 1$ positive integers may be found simply by adding $(k + 1)$ to both sides of the induction hypothesis. Doing so and simplifying, we obtain

$$1 + 2 + 3 + \cdots + k + (k + 1) = \frac{k(k + 1)}{2} + (k + 1)$$

$$= \frac{k(k + 1) + 2(k + 1)}{2}$$

$$= \frac{k^2 + 3k + 2}{2}$$

$$= \frac{(k + 1)(k + 2)}{2}$$

$$= \frac{(k + 1)[(k + 1) + 1]}{2}.$$

We have shown that P_{k+1} is true, and therefore the proof by mathematical induction is complete. ∎

Consider a positive integer j and suppose that with each integer $n \geq j$ there is associated a statement P_n. For example, if $j = 6$, then the statements are numbered P_6, P_7, P_8, and so on. The principle of mathematical induction may be extended to cover this situation. Just as before, two steps are used. Specifically, to prove that the statements S_n are true for $n \geq j$, we use the following two steps.

(I.4)

> (i′) Prove that S_j is true,
>
> (ii′) *Assume* that S_k is true for $k \geq j$ and *prove* that S_{k+1} is true.

Example 2 Let a be a nonzero real number such that $a > -1$. Prove that $(1 + a)^n > 1 + na$ for every integer $n \geq 2$.

Solution   For each positive integer n, let P_n denote $(1 + a)^n > 1 + na$. Note that P_1 is *false* since $(1 + a)^1 = 1 + (1)(a)$. However, we can show that P_n is true for $n \geq 2$ by using (I.5) with $j = 2$.

(i′) We first note that $(1 + a)^2 = 1 + 2a + a^2$. Since $a \neq 0$, we have $a^2 > 0$ and therefore $1 + 2a + a^2 > 1 + 2a$. Thus $(1 + a)^2 > 1 + 2a$, and hence P_2 is true.

(ii′) Assume that P_k is true for $k \geq 2$. Thus the induction hypothesis is

$$(1 + a)^k > 1 + ka.$$

We wish to show that P_{k+1} is true, that is,

$$(1 + a)^{k+1} > 1 + (k + 1)a.$$

Since $a > -1$, we have $a + 1 > 0$, and hence multiplying both sides of the induction hypothesis by $1 + a$ will not change the inequality sign. Consequently,

$$(1 + a)^k(1 + a) > (1 + ka)(1 + a)$$

which may be rewritten as

$$(1 + a)^{k+1} > 1 + ka + a + ka^2$$

or as

$$(1 + a)^{k+1} > 1 + (k + 1)a + ka^2.$$

Since $ka^2 > 0$, we have

$$1 + (k + 1)a + ka^2 > 1 + (k + 1)a$$

and therefore

$$(1 + a)^{k+1} > 1 + (k + 1)a.$$

Thus, P_{k+1} is true and the proof is complete.   ∎

EXERCISES A.I

In Exercises 1–18 prove that the given formula is true for every positive integer n.

1 $2 + 4 + 6 + \cdots + 2n = n(n + 1)$

2 $1 + 4 + 7 + \cdots + (3n - 2) = \dfrac{n(3n - 1)}{2}$

3 $1 + 3 + 5 + \cdots + (2n - 1) = n^2$

4 $3 + 9 + 15 + \cdots + (6n - 3) = 3n^2$

5 $2 + 7 + 12 + \cdots + (5n - 3) = \dfrac{n}{2}(5n - 1)$

6 $2 + 6 + 18 + \cdots + 2 \cdot 3^{n-1} = 3^n - 1$

7 $1 + 2 \cdot 2 + 3 \cdot 2^2 + 4 \cdot 2^3 + \cdots + n \cdot 2^{n-1} = 1 + (n - 1) \cdot 2^n$

8 $(-1)^1 + (-1)^2 + (-1)^3 + \cdots + (-1)^n = \dfrac{(-1)^n - 1}{2}$

9 $1^2 + 2^2 + 3^2 + \cdots + n^2 = \dfrac{n(n + 1)(2n + 1)}{6}$

10 $1^3 + 2^3 + 3^3 + \cdots + n^3 = \left[\dfrac{n(n + 1)}{2}\right]^2$

11 $\dfrac{1}{1 \cdot 2} + \dfrac{1}{2 \cdot 3} + \dfrac{1}{3 \cdot 4} + \cdots + \dfrac{1}{n(n + 1)} = \dfrac{n}{n + 1}$

12 $\dfrac{1}{1 \cdot 2 \cdot 3} + \dfrac{1}{2 \cdot 3 \cdot 4} + \dfrac{1}{3 \cdot 4 \cdot 5} + \cdots + \dfrac{1}{n(n + 1)(n + 2)}$

$$= \dfrac{n(n + 3)}{4(n + 1)(n + 2)}$$

13 $3 + 3^2 + 3^3 + \cdots + 3^n = \tfrac{3}{2}(3^n - 1)$

14 $1^3 + 3^3 + 5^3 + \cdots + (2n - 1)^3 = n^2(2n^2 - 1)$

15 $n < 2^n$

16 $1 + 2n \le 3^n$

17 $1 + 2 + 3 + \cdots + n < \tfrac{1}{8}(2n + 1)^2$

18 If $0 < a < b$, then $\left(\dfrac{a}{b}\right)^{n+1} < \left(\dfrac{a}{b}\right)^n$.

Prove that the statements in Exercises 19–22 are true for every positive integer n.

19 3 is a factor of $n^3 - n + 3$.

20 2 is a factor of $n^2 + n$.

21 4 is a factor of $5^n - 1$.

22 9 is a factor of $10^{n+1} + 3 \cdot 10^n + 5$.

23 Use mathematical induction to prove that if a is any real number greater than 1, then $a^n > 1$ for every positive integer n.

24 If $a \ne 1$, prove that

$$1 + a + a^2 + \cdots + a^{n-1} = \dfrac{a^n - 1}{a - 1}$$

for every positive integer n.

25 Use mathematical induction to prove that $a - b$ is a factor of $a^n - b^n$ for every positive integer n.
(*Hint*: $a^{k+1} - b^{k+1} = a^k(a - b) + (a^k - b^k)b$.)

26 Prove that $a + b$ is a factor of $a^{2n-1} + b^{2n-1}$ for every positive integer n.

27 Prove that

$$\log (a_1 a_2 \cdots a_n) = \log a_1 + \log a_2 + \cdots + \log a_n$$

for all $n \ge 2$, where each a_i is a positive real number.

28 Prove the **Generalized Distributive Law**

$$a(b_1 + b_2 + \cdots + b_n) = ab_1 + ab_2 + \cdots + ab_n$$

for all $n \ge 2$, where a and each b_i are real numbers.

29 Prove that

$$a + ar + ar^2 + \cdots + ar^{n-1} = \dfrac{a(1 - r^n)}{1 - r}$$

where n is any positive integer and a and r are real numbers with $r \ne 1$.

30 Prove that

$$a + (a + d) + (a + 2d) + \cdots + [a + (n - 1)d]$$
$$= (n/2)[2a + (n - 1)d]$$

where n is any positive integer and a and d are real numbers.

APPENDIX II

THEOREMS ON LIMITS AND DEFINITE INTEGRALS

This appendix contains proofs for some theorems stated in Chapters 2, 3, and 5. Part of the numbering system corresponds to that given in those chapters.

Uniqueness Theorem for Limits (II.0)

> If $f(x)$ has a limit as x approaches a, then the limit is unique.

Proof   Suppose that $\lim_{x \to a} f(x) = L_1$ and $\lim_{x \to a} f(x) = L_2$ where $L_1 \neq L_2$. It may be assumed, without loss of generality, that $L_1 < L_2$. Choose ε such that $\varepsilon < \frac{1}{2}(L_2 - L_1)$ and consider the open intervals $(L_1 - \varepsilon, L_1 + \varepsilon)$ and $(L_2 - \varepsilon, L_2 + \varepsilon)$ on the coordinate line l' (see Figure II.1). Since $\varepsilon < \frac{1}{2}(L_2 - L_1)$ these two intervals do not intersect. By Definition (2.4) there is a $\delta_1 > 0$ such that whenever x is in $(a - \delta_1, a + \delta_1)$, but $x \neq a$, then $f(x)$ is in $(L_1 - \varepsilon, L_1 + \varepsilon)$. Similarly, there is a $\delta_2 > 0$ such that whenever x is in $(a - \delta_2, a + \delta_2)$, but $x \neq a$, then $f(x)$ is in $(L_2 - \varepsilon, L_2 + \varepsilon)$. This is illustrated in Figure II.1, where the case $\delta_1 < \delta_2$ is shown. If an x is selected that is in *both* $(a - \delta_1, a + \delta_1)$ and $(a - \delta_2, a + \delta_2)$, then $f(x)$ is in $(L_1 - \varepsilon, L_1 + \varepsilon)$ and also in $(L_2 - \varepsilon, L_2 + \varepsilon)$, contrary to the fact that these two intervals do not intersect. Hence our original supposition is false and consequently $L_1 = L_2$.

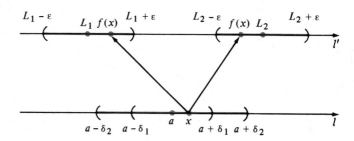

Figure II.1   □

Theorem (2.8)

> If $\lim_{x \to a} f(x) = L$ and $\lim_{x \to a} g(x) = M$, then
>
> (i) $\lim_{x \to a} [f(x) + g(x)] = L + M$.
>
> (ii) $\lim_{x \to a} [f(x) \cdot g(x)] = L \cdot M$.
>
> (iii) $\lim_{x \to a} \left[\dfrac{f(x)}{g(x)} \right] = \dfrac{L}{M}$,   provided $M \neq 0$.

Proof (i) According to Definition (2.3) we must show that for every $\varepsilon > 0$ there corresponds a $\delta > 0$ such that

(II.1)
$$\text{if} \quad 0 < |x - a| < \delta, \quad \text{then} \quad |f(x) + g(x) - (L + M)| < \varepsilon.$$

We begin by writing

(II.2)
$$|f(x) + g(x) - (L + M)| = |(f(x) - L) + (g(x) - M)|.$$

Employing the Triangle Inequality (1.5),

$$|(f(x) - L) + (g(x) - M)| \le |f(x) - L| + |g(x) - M|.$$

Combining the last inequality with (II.2) gives us

(II.3)
$$|f(x) + g(x) - (L + M)| \le |f(x) - L| + |g(x) - M|.$$

Since $\lim_{x \to a} f(x) = L$ and $\lim_{x \to a} g(x) = M$, the numbers $|f(x) - L|$ and $|g(x) - M|$ can be made arbitrarily small by choosing x sufficiently close to a. In particular, they can be made less than $\varepsilon/2$. Thus there exist $\delta_1 > 0$ and $\delta_2 > 0$ such that

(II.4)
$$\text{if} \quad 0 < |x - a| < \delta_1 \quad \text{then} \quad |f(x) - L| < \varepsilon/2, \quad \text{and}$$
$$\text{if} \quad 0 < |x - a| < \delta_2 \quad \text{then} \quad |g(x) - M| < \varepsilon/2.$$

If δ denotes the *smaller* of δ_1 and δ_2, then whenever $0 < |x - a| < \delta$, the inequalities in (II.4) involving $f(x)$ and $g(x)$ are both true. Consequently, if $0 < |x - a| < \delta$, then from (II.4) and (II.3),

$$|f(x) + g(x) - (L + M)| < \varepsilon/2 + \varepsilon/2 = \varepsilon$$

which is the desired statement (II.1).

(ii) We first show that if k is a function and

(II.5)
$$\text{if} \quad \lim_{x \to a} k(x) = 0, \quad \text{then} \quad \lim_{x \to a} f(x)k(x) = 0.$$

Since $\lim_{x \to a} f(x) = L$, it follows from Definition (2.3) (with $\varepsilon = 1$) that there is a $\delta_1 > 0$ such that if $0 < |x - a| < \delta_1$, then $|f(x) - L| < 1$ and hence also

$$|f(x)| = |f(x) - L + L| \le |f(x) - L| + |L| < 1 + |L|.$$

Consequently,

(II.6)
$$\text{if} \quad 0 < |x - a| < \delta_1, \quad \text{then} \quad |f(x)k(x)| < (1 + |L|)|k(x)|.$$

Since $\lim_{x \to a} k(x) = 0$, for every $\varepsilon > 0$ there corresponds a $\delta_2 > 0$ such that

(II.7)
$$\text{if} \quad 0 < |x - a| < \delta_2, \quad \text{then} \quad |k(x) - 0| < \frac{\varepsilon}{1 + |L|}.$$

If δ denotes the smaller of δ_1 and δ_2, then whenever $0 < |x - a| < \delta$ both inequalities (II.6) and (II.7) are true and consequently,

$$|f(x)k(x)| < (1 + |L|) \cdot \frac{\varepsilon}{1 + |L|}.$$

Therefore,

$$\text{if} \quad 0 < |x - a| < \delta, \quad \text{then} \quad |f(x)k(x) - 0| < \varepsilon$$

which proves (II.5).

Next consider the identity

(II.8) $$f(x)g(x) - LM = f(x)[g(x) - M] + M[f(x) - L].$$

Since $\lim_{x \to a} [g(x) - M] = 0$ it follows from (II.5), with $k(x) = g(x) - M$, that $\lim_{x \to a} f(x)[g(x) - M] = 0$. In addition, $\lim_{x \to a} M[f(x) - L] = 0$ and hence, from (II.8), $\lim_{x \to a} [f(x)g(x) - LM] = 0$. The last statement is equivalent to $\lim_{x \to a} f(x)g(x) = LM$.

(iii) It is sufficient to show that $\lim_{x \to a} 1/(g(x)) = 1/M$, for once this is done, the desired result may be obtained by applying (ii) to the product $f(x) \cdot 1/(g(x))$. Consider

(II.9) $$\left| \frac{1}{g(x)} - \frac{1}{M} \right| = \left| \frac{M - g(x)}{g(x)M} \right| = \frac{1}{|M||g(x)|} |g(x) - M|.$$

Since $\lim_{x \to a} g(x) = M$, there exists a $\delta_1 > 0$ such that if $0 < |x - a| < \delta_1$ then $|g(x) - M| < |M|/2$. Consequently, for all such x,

$$|M| = |g(x) + (M - g(x))|$$
$$\leq |g(x)| + |M - g(x)|$$
$$< |g(x)| + |M|/2$$

and, therefore,

$$\frac{|M|}{2} < |g(x)|, \quad \text{or} \quad \frac{1}{|g(x)|} < \frac{2}{|M|}.$$

Substitution in (II.9) leads to

(II.10) $$\left| \frac{1}{g(x)} - \frac{1}{M} \right| < \frac{2}{|M|^2} |g(x) - M|, \quad \text{provided } 0 < |x - a| < \delta_1.$$

Again using the fact that $\lim_{x \to a} g(x) = M$, it follows that for every $\varepsilon < 0$ there corresponds a $\delta_2 > 0$ such that

(II.11) $$\text{if} \quad 0 < |x - a| < \delta_2, \quad \text{then} \quad |g(x) - M| < \frac{|M|^2}{2} \varepsilon.$$

If δ denotes the smaller of δ_1 and δ_2, then both inequalities (II.10) and (II.11) are true. Thus

$$\text{if} \quad 0 < |x - a| < \delta, \quad \text{then} \quad \left| \frac{1}{g(x)} - \frac{1}{M} \right| < \varepsilon$$

which means that $\lim_{x \to a} 1/(g(x)) = 1/M$. $\qquad \square$

Theorem (2.12)

> If $a > 0$ and n is a positive integer, or if $a \leq 0$ and n is an odd positive integer, then $\lim_{x \to a} \sqrt[n]{x} = \sqrt[n]{a}$.

Proof   Suppose $a > 0$ and n is any positive integer. It must be shown that for every $\varepsilon > 0$ there corresponds a $\delta > 0$ such that

$$\text{if} \quad 0 < |x - a| < \delta, \quad \text{then} \quad |\sqrt[n]{x} - \sqrt[n]{a}| < \varepsilon$$

or, equivalently,

(II.12) $$\text{if} \quad -\delta < x - a < \delta, \, x \neq a, \quad \text{then} \quad -\varepsilon < \sqrt[n]{x} - \sqrt[n]{a} < \varepsilon.$$

It is sufficient to prove (II.12) if $\varepsilon < \sqrt[n]{a}$, for if a δ exists under this condition then the same δ can be used for any *larger* value of ε. Thus, in the remainder of the proof $\sqrt[n]{a} - \varepsilon$ is considered as a positive number less than ε. The inequalities in the following list are all equivalent:

$$-\varepsilon < \sqrt[n]{x} - \sqrt[n]{a} < \varepsilon$$
$$\sqrt[n]{a} - \varepsilon < \sqrt[n]{x} < \sqrt[n]{a} + \varepsilon$$
$$(\sqrt[n]{a} - \varepsilon)^n < x < (\sqrt[n]{a} + \varepsilon)^n$$
$$(\sqrt[n]{a} - \varepsilon)^n - a < x - a < (\sqrt[n]{a} + \varepsilon)^n - a$$
$$-[a - (\sqrt[n]{a} - \varepsilon)^n] < x - a < (\sqrt[n]{a} + \varepsilon)^n - a.$$

If δ denotes the smaller of the two positive numbers $a - (\sqrt[n]{a} - \varepsilon)^n$ and $(\sqrt[n]{a} + \varepsilon)^n - a$, then whenever $-\delta < x - a < \delta$ the last inequality in the list is true and hence so is the first. This gives us (II.12).

Next suppose $a < 0$ and n is an odd positive integer. In this case $-a$ and $\sqrt[n]{-a}$ are positive and, by the first part of the proof, we may write

$$\lim_{-x \to -a} \sqrt[n]{-x} = \sqrt[n]{-a}.$$

Thus for every $\varepsilon > 0$, there corresponds a $\delta > 0$ such that

$$\text{if} \quad 0 < |-x - (-a)| < \delta, \quad \text{then} \quad |\sqrt[n]{-x} - \sqrt[n]{-a}| < \varepsilon$$

or equivalently

$$\text{if} \quad 0 < |x - a| < \delta, \quad \text{then} \quad |\sqrt[n]{x} - \sqrt[n]{a}| < \varepsilon.$$

The last inequalities imply that $\lim_{x \to a} \sqrt[n]{x} = \sqrt[n]{a}$. □

The Sandwich Theorem (2.14)

> If $f(x) \leq h(x) \leq g(x)$ for all x in an open interval containing a, except possibly at a, and if $\lim_{x \to a} f(x) = L = \lim_{x \to a} g(x)$, then $\lim_{x \to a} h(x) = L$.

Proof   For every $\varepsilon > 0$, there correspond $\delta_1 > 0$ and $\delta_2 > 0$ such that

(II.13)

$$\text{if} \quad 0 < |x - a| < \delta_1, \quad \text{then} \quad |f(x) - L| < \varepsilon,$$
$$\text{if} \quad 0 < |x - a| < \delta_2, \quad \text{then} \quad |g(x) - L| < \varepsilon.$$

If δ denotes the smaller of δ_1 and δ_2, then whenever $0 < |x - a| < \delta$, both inequalities in (II.13) that involve ε are true, that is,

$$-\varepsilon < f(x) - L < \varepsilon \quad \text{and} \quad -\varepsilon < g(x) - L < \varepsilon.$$

Consequently, if $0 < |x - a| < \delta$, then $L - \varepsilon < f(x)$ and $g(x) < L + \varepsilon$. Since $f(x) \le h(x) \le g(x)$, it follows that if $0 < |x - a| < \delta$, then $L - \varepsilon < h(x) < L + \varepsilon$ or, equivalently, $|h(x) - L| < \varepsilon$, which is what we wished to prove.

Theorem (3.7)

> If f is defined on an open interval containing a, then
>
> $$f'(a) = \lim_{x \to a} \frac{f(x) - f(a)}{x - a}$$
>
> provided the limit exists.

Proof   Suppose

$$\lim_{x \to a} \frac{f(x) - f(a)}{x - a} = L$$

for some number L. According to the definition of limit (2.3), this means that for every $\varepsilon > 0$ there exists a $\delta > 0$ such that

$$\text{if} \quad 0 < |x - a| < \delta, \quad \text{then} \quad \left| \frac{f(x) - f(a)}{x - a} - L \right| < \varepsilon.$$

If we let $h = x - a$, then $x = a + h$ and the last statement may be written

$$\text{if} \quad 0 < |h| < \delta, \quad \text{then} \quad \left| \frac{f(a + h) - f(a)}{h} - L \right| < \varepsilon$$

which means that

$$\lim_{h \to 0} \frac{f(a + h) - f(a)}{h} = L.$$

However, by Definition (3.4) this limit must equal $f'(a)$, and consequently $L = f'(a)$. This gives us the formula in the statement of the theorem. Conversely, if $f'(a)$ exists then by reversing the steps in the previous proof we arrive at the desired limit.   □

The Chain Rule (3.28)

> If $y = f(u)$, $u = g(x)$, and the derivatives dy/du and du/dx both exist, then the composite function defined by $y = f(g(x))$ has a derivative given by
>
> $$\frac{dy}{dx} = \frac{dy}{du}\frac{du}{dx} = f'(u)g'(x) = f'(g(x))g'(x).$$

Proof   If $y = f(x)$ and $\Delta x \approx 0$, then by (3.21), the difference between the derivative $f'(x)$ and the ratio $\Delta y/\Delta x$ is numerically small. Since this difference depends on the size of Δx, we shall represent it by means of the functional notation $\eta(\Delta x)$. Thus, for each $\Delta x \neq 0$,

(a)
$$\eta(\Delta x) = \frac{\Delta y}{\Delta x} - f'(x).$$

It should be noted that $\eta(\Delta x)$ does *not* represent the product of η and Δx, but rather than η *is a function of* Δx, whose values are given by (a). Moreover, applying (3.21) we see that

(b)
$$\lim_{\Delta x \to 0} \eta(\Delta x) = \lim_{\Delta x \to 0} \left[\frac{\Delta y}{\Delta x} - f'(x) \right] = 0.$$

The function η has been defined only for nonzero values of Δx. It is convenient to extend the definition of η to include $\Delta x = 0$ by letting $\eta(0) = 0$. It then follows from (b) that η *is continuous at* 0.

Multiplying both sides of (a) by Δx and rearranging terms gives us

(c)
$$\Delta y = f'(x)\,\Delta x + \eta(\Delta x) \cdot \Delta x$$

which is true whether $\Delta x \neq 0$ or $\Delta x = 0$. Since $f'(x)\,\Delta x = dy$, it follows from (c) that

(d)
$$\Delta y - dy = \eta(\Delta x) \cdot \Delta x.$$

Let us now consider the situation stated in the hypothesis of the theorem, where

$$y = f(u) \quad \text{and} \quad u = g(x).$$

If $g(x)$ is in the domain of f, then we may write

$$y = f(u) = f(g(x)),$$

that is, y is a function of x. If we give x an increment Δx there corresponds an increment Δu in u and, in turn, an increment Δy in $y = f(u)$. Thus

$$\Delta u = g(x + \Delta x) - g(x)$$

$$\Delta y = f(u + \Delta u) - f(u).$$

Since dy/du exists we may use (c) with u as the independent variable to write

(e)
$$\Delta y = f'(u)\,\Delta u + \eta(\Delta u) \cdot \Delta u$$

where η is a function of Δu and where, by (b),

(f)
$$\lim_{\Delta u \to 0} \eta(\Delta u) = 0.$$

Moreover, η is continuous at $\Delta u = 0$ and (e) is true if $\Delta u = 0$. Dividing both sides of (e) by Δx gives us

$$\frac{\Delta y}{\Delta x} = f'(u)\frac{\Delta u}{\Delta x} + \eta(\Delta u) \cdot \frac{\Delta u}{\Delta x}.$$

If we now take the limit as Δx approaches zero and use the fact that

$$\lim_{\Delta x \to 0} \frac{\Delta y}{\Delta x} = \frac{dy}{dx} \quad \text{and} \quad \lim_{\Delta x \to 0} \frac{\Delta u}{\Delta x} = \frac{du}{dx}$$

we see that

$$\frac{dy}{dx} = f'(u)\frac{du}{dx} + \lim_{\Delta x \to 0} \eta(\Delta u) \cdot \frac{du}{dx}.$$

Since $f'(u) = dy/du$ we may complete the proof by showing that the limit indicated in the last equation is 0. To accomplish this we first observe that since g is differentiable it is continuous, and hence

$$\lim_{\Delta x \to 0} [g(x + \Delta x) - g(x)] = 0$$

or equivalently
$$\lim_{\Delta x \to 0} \Delta u = 0.$$

In other words, Δu approaches 0 as Δx approaches 0. Using this fact, together with (f), gives us

(g)
$$\lim_{\Delta x \to 0} \eta(\Delta u) = \lim_{\Delta u \to 0} \eta(\Delta u) = 0$$

and the theorem is proved. (The fact that $\lim_{\Delta x \to 0} \eta(\Delta u) = 0$ can also be established by means of an $\varepsilon - \delta$ argument using (2.3).)

Theorem (5.14)

> If f is integrable on $[a, b]$ and k is any number, then kf is integrable on $[a, b]$ and
>
> $$\int_a^b kf(x)\,dx = k\int_a^b f(x)\,dx.$$

Proof   If $k = 0$ the result follows from Theorem (5.13). Assume, therefore, that $k \neq 0$. Since f is integrable, $\int_a^b f(x)\,dx = I$ for some number I. If P is a partition of $[a, b]$, then every Riemann sum R_P for the function kf has the form $\Sigma_i k f(w_i)\,\Delta x_i$, where for each i, w_i is in the ith subinterval $[x_{i-1}, x_i]$

of P. We wish to show that for every $\varepsilon > 0$ there corresponds a $\delta > 0$ such that whenever $\|P\| < \delta$, then

(II.14)
$$\left| \sum_i k f(w_i) \, \Delta x_i - kI \right| < \varepsilon$$

for all w_i in $[x_{i-1}, x_i]$. If we let $\varepsilon' = \varepsilon/|k|$, then since f is integrable there exists a $\delta > 0$ such that whenever $\|P\| < \delta$,

$$\left| \sum_i f(w_i) \, \Delta x_i - I \right| < \varepsilon' = \frac{\varepsilon}{|k|}.$$

Multiplying both sides of this inequality by $|k|$ leads to (II.14). Hence

$$\lim_{\|P\| \to 0} \sum_i k f(w_i) \, \Delta x_i = kI = k \int_a^b f(x) \, dx. \qquad \square$$

Theorem (5.15)

> If f and g are integrable on $[a, b]$, then $f + g$ is integrable on $[a, b]$ and
> $$\int_a^b [f(x) + g(x)] \, dx = \int_a^b f(x) \, dx + \int_a^b g(x) \, dx.$$

Proof By hypothesis there exist real numbers I_1 and I_2 such that

$$\int_a^b f(x) \, dx = I_1 \quad \text{and} \quad \int_a^b g(x) \, dx = I_2.$$

Let P denote a partition of $[a, b]$ and let R_P denote an arbitrary Riemann sum for $f + g$ associated with P, that is,

(II.15)
$$R_P = \sum_i [f(w_i) + g(w_i)] \, \Delta x_i$$

where w_i is in $[x_{i-1}, x_i]$ for each i. We wish to show that for every $\varepsilon > 0$ there corresponds a $\delta > 0$ such that whenever $\|P\| < \delta$, then $|R_P - (I_1 + I_2)| < \varepsilon$. Using (i) of Theorem (5.3), we may write (II.15) in the form

$$R_P = \sum_i f(w_i) \, \Delta x_i + \sum_i g(w_i) \, \Delta x_i.$$

Rearranging terms and using the Triangle Inequality (1.5),

(II.16)
$$|R_P - (I_1 + I_2)| = \left| \left(\sum_i f(w_i) \, \Delta x_i - I_1 \right) + \left(\sum_i g(w_i) \, \Delta x_i - I_2 \right) \right|$$
$$\leq \left| \sum_i f(w_i) \, \Delta x_i - I_1 \right| + \left| \sum_i g(w_i) \, \Delta x_i - I_2 \right|.$$

By the integrability of f and g, if $\varepsilon' = \varepsilon/2$, then there exist $\delta_1 > 0$ and $\delta_2 > 0$ such that whenever $\|P\| < \delta_1$ and $\|P\| < \delta_2$,

(II.17)

$$\left| \sum_i f(w_i)\, \Delta x_i - I_1 \right| < \varepsilon' = \varepsilon/2,$$

and

$$\left| \sum_i g(w_i)\, \Delta x_i - I_2 \right| < \varepsilon' = \varepsilon/2$$

for all w_i in $[x_{i-1}, x_i]$. If δ denotes the smaller of δ_1 and δ_2, then whenever $\|P\| < \delta$, both inequalities in (II.17) are true and hence, from (II.16),

$$|R_P - (I_1 + I_2)| < \varepsilon/2 + \varepsilon/2 = \varepsilon$$

which is what we wished to prove.   □

Theorem (5.16)

> If $a < c < b$, and if f is integrable on both $[a, c]$ and $[c, b]$, then f is integrable on $[a, b]$ and
>
> $$\int_a^b f(x)\, dx = \int_a^c f(x)\, dx + \int_c^b f(x)\, dx.$$

Proof   By hypothesis there exist real numbers I_1 and I_2 such that

(II.18)

$$\int_a^c f(x)\, dx = I_1 \quad \text{and} \quad \int_c^b f(x)\, dx = I_2.$$

Let us denote a partition of $[a, c]$ by P_1, of $[c, b]$ by P_2, and of $[a, b]$ by P. Arbitrary Riemann sums associated with P_1, P_2, and P will be denoted by R_{P_1}, R_{P_2}, and R_P, respectively. It must be shown that for every $\varepsilon > 0$ there corresponds a $\delta > 0$ such that if $\|P\| < \delta$, then $|R_P - (I_1 + I_2)| < \varepsilon$.

If we let $\varepsilon' = \varepsilon/4$, then by (II.18) there exist positive numbers δ_1 and δ_2 such that if $\|P_1\| < \delta_1$ and $\|P_2\| < \delta_2$, then

(II.19)

$$|R_{P_1} - I_1| < \varepsilon' = \varepsilon/4 \quad \text{and} \quad |R_{P_2} - I_2| < \varepsilon' = \varepsilon/4.$$

If δ denotes the smaller of δ_1 and δ_2, then both inequalities in (II.19) are true whenever $\|P\| < \delta$. Moreover, since f is integrable on $[a, c]$ and $[c, b]$ it is bounded on both intervals and hence there exists a number M such that $|f(x)| \le M$ for all x in $[a, b]$. We shall now assume that δ has been chosen so that in addition to the previous requirement we also have $\delta < \varepsilon/4M$.

Let P be a partition of $[a, b]$ such that $\|P\| < \delta$. If the numbers that determine P are

$$a = x_0, x_1, x_2, \ldots, x_n = b,$$

then there is a unique half-open interval of the form $(x_{k-1}, x_k]$ that contains c. If $R_P = \sum_{i=1}^n f(w_i)\, \Delta x_i$ we may write

(II.20)

$$R_P = \sum_{i=1}^{k-1} f(w_i)\, \Delta x_i + f(w_k)\, \Delta x_k + \sum_{i=k+1}^{n} f(w_i)\, \Delta x_i.$$

Let P_1 denote the partition of $[a, c]$ determined by $\{a, x_1, \ldots, x_{k-1}, c\}$, let P_2 denote the partition of $[c, b]$ determined by $\{c, x_k, \ldots, x_{n-1}, b\}$, and consider the Riemann sums

$$R_{P_1} = \sum_{i=1}^{k-1} f(w_i) \Delta x_i + f(c)(c - x_{k-1})$$

(II.21)

$$R_{P_2} = f(c)(x_k - c) + \sum_{i=k+1}^{n} f(w_i) \Delta x_i.$$

Using the triangle inequality and (II.19),

$$
\begin{aligned}
|(R_{P_1} + R_{P_2}) - (I_1 + I_2)| &= |(R_{P_1} - I_1) + (R_{P_2} - I_2)| \\
&\leq |R_{P_1} - I_1| + |R_{P_2} - I_2| \\
&< \frac{\varepsilon}{4} + \frac{\varepsilon}{4} = \frac{\varepsilon}{2}.
\end{aligned}
$$

(II.22)

It follows from (II.20) and (II.21) that

$$|R_P - (R_{P_1} + R_{P_2})| = |f(w_k) - f(c)| \Delta x_k.$$

Employing the triangle inequality and the choice of δ gives us

$$
\begin{aligned}
|R_P - (R_{P_1} + R_{P_2})| &\leq \{|f(w_k)| + |f(c)|\} \Delta x_k \\
&\leq (M + M)(\varepsilon/4M) = \varepsilon/2
\end{aligned}
$$

(II.23)

provided $\|P\| < \delta$. If we write

$$
\begin{aligned}
|R_P - (I_1 + I_2)| &= |R_P - (R_{P_1} + R_{P_2}) + (R_{P_1} + R_{P_2}) - (I_1 + I_2)| \\
&\leq |R_P - (R_{P_1} + R_{P_2})| + |(R_{P_1} + R_{P_2}) - (I_1 + I_2)|
\end{aligned}
$$

then it follows from (II.23) and (II.22) that whenever $\|P\| < \delta$

$$|R_P - (I_1 + I_2)| < \varepsilon/2 + \varepsilon/2 = \varepsilon$$

for every Riemann sum R_P. This completes the proof. □

Theorem (5.18)

If f is integrable on $[a, b]$ and if $f(x) \geq 0$ for all x in $[a, b]$, then

$$\int_a^b f(x) \, dx \geq 0.$$

Proof   We shall give an indirect proof. Thus let $\int_a^b f(x) \, dx = I$ and suppose that $I < 0$. Consider any partition P of $[a, b]$ and let $R_P = \sum_i f(w_i) \Delta x_i$ be an arbitrary Riemann sum associated with P. Since $f(w_i) \geq 0$ for all w_i in $[x_{i-1}, x_i]$ it follows that $R_P \geq 0$. If we let $\varepsilon = -(I/2)$, then according to Definition (5.7), whenever $\|P\|$ is sufficiently small,

$$|R_P - I| < \varepsilon = -(I/2).$$

It follows that $R_P < I - (I/2) = I/2 < 0$, a contradiction. Therefore, the supposition $I < 0$ is false and hence $I \geq 0$.

APPENDIX III

THE TRIGONOMETRIC FUNCTIONS

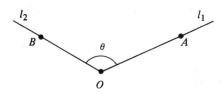

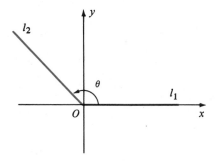

Figure III.1

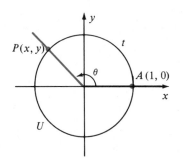

Figure III.2

Figure III.3 $\theta = t$ radians

An angle θ is often regarded as the set of all points on two rays, or half-lines, l_1 and l_2, having the same initial point O. If A and B are points on l_1 and l_2, respectively (see Figure III.1), then we may refer to **angle** AOB. For trigonometric purposes it is convenient to regard angle AOB as generated by starting with the fixed ray l_1 and rotating it about O, in a plane, to a position specified by ray l_2. We call l_1 the **initial side**, l_2 the **terminal side**, and O the **vertex** of the angle. The amount or direction of rotation is not restricted in any way, that is, we may let l_1 make several rotations in either direction about O before coming to the position l_2. Thus, different angles may have the same initial and terminal sides.

If a rectangular coordinate system is introduced, then the **standard position** of an angle is obtained by taking the vertex at the origin and letting l_1 coincide with the positive x-axis (see Figure III.2). If l_1 is rotated in a counterclockwise direction to position l_2, then the angle is considered **positive**; whereas if l_1 is rotated in a clockwise direction, the angle is **negative**. We sometimes specify the direction of rotation by using a curved arrow, as illustrated in Figure III.2.

The magnitude of an angle may be expressed in terms of either degrees or radians. An angle of degree measure $1°$ is obtained by $1/360$ of a complete revolution in the counterclockwise direction. In calculus, the most important unit of angular measure is the *radian*. To define radian measure, let us consider a unit circle U with center at the origin of a rectangular system, and let θ be an angle in standard position. We regard θ as generated by rotating the positive x-axis about O. As the x-axis rotates to the terminal side of θ, its point of intersection with U travels a certain distance t before arriving at its final position $P(x, y)$, as illustrated in Figure III.3. If t is considered positive for a counterclockwise rotation and negative for a clockwise rotation, then a natural way of assigning a measure to θ is to use the number t. When this is done, we say that θ **is an angle of t radians** and we write $\theta = t$ or $\theta = t$ *radians*. Note that it is customary to let θ denote either the angle or the angular measure of the angle.

According to the preceding discussion, if $\theta = 1$, then θ is an angle that subtends an arc of unit length on the unit circle U. The notation $\theta = -7.5$ means that θ is the angle generated by a clockwise rotation in which the point of intersection of the x-axis with the unit circle U travels 7.5 units. Since the circumference of U is 2π, we see that if $\theta = \pi/2$, then θ is obtained by $\frac{1}{4}$ of a complete revolution in the counterclockwise direction. Similarly, if $\theta = -\pi/4$, then θ is generated by $\frac{1}{8}$ of a revolution in the clockwise direction. These angles, measured in radians, are sketched in Figure III.4.

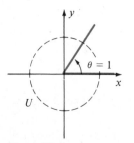

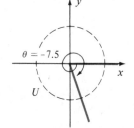

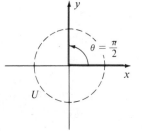

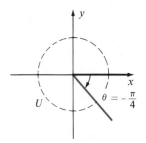

Figure III.4

If an angle in standard position is generated by $\frac{1}{2}$ of a complete counter-clockwise rotation, then the degree measure is $180°$ and the radian measure is π. This gives us the basic relation

$$180° = \pi \text{ radians.}$$

Equivalent formulas are

$$1° = \frac{\pi}{180} \text{ radians} \quad \text{and} \quad 1 \text{ radian} = \left(\frac{180}{\pi}\right)°.$$

Thus, to change degrees to radians, multiply by $\pi/180$. To change radians to degrees, multiply by $180/\pi$. The following table gives the relationships between the radian and degree measures of several common angles.

Radians	0	$\pi/6$	$\pi/4$	$\pi/3$	$\pi/2$	$2\pi/3$	$3\pi/4$	$5\pi/6$	π
Degrees	$0°$	$30°$	$45°$	$60°$	$90°$	$120°$	$135°$	$150°$	$180°$

By division we obtain the following:

$$1° \approx 0.01745 \text{ radians}; \qquad 1 \text{ radian} \approx 57.296°.$$

Example 1

(a)   Find the radian measure of θ if $\theta = -150°$ and if $\theta = 225°$.

(b)   Find the degree measure of θ if $\theta = 7\pi/4$ and if $\theta = -\pi/3$.

Solution

(a)   Since there are $\pi/180$ radians in each degree, the number of radians in $-150°$ can be found by multiplying -150 by $\pi/180$. Thus,

$$-150° = -150\left(\frac{\pi}{180}\right) = -\frac{5\pi}{6} \text{ radians.}$$

Similarly, $\qquad 225° = 225\left(\frac{\pi}{180}\right) = \frac{5\pi}{4} \text{ radians.}$

(b)   The number of degrees in 1 radian is $180/\pi$. Consequently, to find the number of degrees in $7\pi/4$ radians, we multiply by $180/\pi$, obtaining

$$\frac{7\pi}{4} \text{ radians} = \frac{7\pi}{4}\left(\frac{180}{\pi}\right) = 315°.$$

In like manner, $\qquad -\frac{\pi}{3} \text{ radians} = -\frac{\pi}{3}\left(\frac{180}{\pi}\right) = -60°.$ ∎

The radian measure of an angle can be found by using a circle of *any* radius. In the following discussion, the terminology **central angle** of a circle refers to an angle whose vertex is at the center of the circle. Suppose that θ is a central angle of a circle of radius r, and that θ subtends an arc of length s, where $0 \le s < 2\pi r$. To find the radian measure of θ, let us place θ in standard position on a rectangular coordinate system and superimpose a unit circle U,

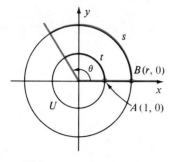

Figure III.5

as shown in Figure III.5. If θ is subtended by an arc of length t on U, then by definition we may write $\theta = t$. From plane geometry, the ratio of the arcs in Figure III.5 is the same as the ratio of the radii; that is,

$$\frac{t}{s} = \frac{1}{r} \quad \text{or} \quad t = \frac{s}{r}.$$

Substituting θ for t gives us the following result.

Theorem (III.1)

If a central angle θ of a circle of radius r is subtended by an arc of length s, then the radian measure of θ is given by

$$\theta = \frac{s}{r}.$$

The radian measure of an angle is independent of the size of the circle. For example, if the radius of the circle is $r = 4$ cm and an arc of length 8 cm subtends a central angle θ, then using $\theta = s/r$, the radian measure is

$$\theta = \frac{8 \text{ cm}}{4 \text{ cm}} = 2.$$

If the radius of the circle is 4 km and the arc is 8 km, then

$$\theta = \frac{8 \text{ km}}{4 \text{ km}} = 2.$$

These calculations indicate that the radian measure of an angle is dimensionless and hence may be regarded as a real number. Indeed, it is for this reason that we usually employ the notation $\theta = t$ instead of $\theta = t$ radians.

There are two standard techniques for introducing the trigonometric functions—one through the use of a unit circle, and the other by means of right triangles. Both lead to the same result. We shall begin with the unit circle approach. Descriptions of trigonometric functions in terms of right triangles are stated in (III.7).

Given any real number t, let θ denote the angle (in standard position) of radian measure t. The point $P(x, y)$ at which the terminal side of θ intersects the unit circle U (see Figure III.3) will be called **the point on U that corresponds to t**. The coordinates of $P(x, y)$ may be used to define the six **trigonometric or (circular) functions**. These functions are referred to as the **sine, cosine, tangent, cotangent, secant,** and **cosecant functions**, and are designated by the symbols **sin, cos, tan, cot, sec,** and **csc,** respectively. If t is a real number, then the real number that the sine function associates with t will be denoted by either sin (t) or sin t, and similarly for the other five functions.

Definition (III.2)

> If t is any real number and $P(x, y)$ is the point on the unit circle U that corresponds to t, then the **trigonometric functions** are given by
>
> $$\sin t = y \qquad\qquad \csc t = \frac{1}{y} \quad (\text{if } y \neq 0)$$
>
> $$\cos t = x \qquad\qquad \sec t = \frac{1}{x} \quad (\text{if } x \neq 0)$$
>
> $$\tan t = \frac{y}{x} \quad (\text{if } x \neq 0) \qquad \cot t = \frac{x}{y} \quad (\text{if } y \neq 0).$$

Example 2 Find the values of the trigonometric functions at

(a) $t = 0$;   (b) $t = \pi/4$;   (c) $t = \pi/2$.

Solution   The points $P(x, y)$ corresponding to the given values of t are plotted in Figure III.6.

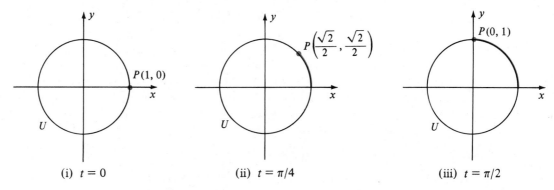

(i) $t = 0$ (ii) $t = \pi/4$ (iii) $t = \pi/2$

Figure III.6

Thus, for $t = 0$ we let $x = 1$ and $y = 0$ in Definition (III.2), obtaining the values in the first line of the following table. Note that since $y = 0$, $\csc 0$ and $\cot 0$ are undefined, as indicated by the dashes in the table.

Part (b) may be solved by taking $x = \sqrt{2}/2$ and $y = \sqrt{2}/2$ in Definition (III.2). Finally, for (c), let $x = 0$ and $y = 1$ in the definition.

t	$\sin t$	$\cos t$	$\tan t$	$\csc t$	$\sec t$	$\cot t$
0	0	1	0	—	1	—
$\dfrac{\pi}{4}$	$\dfrac{\sqrt{2}}{2}$	$\dfrac{\sqrt{2}}{2}$	1	$\sqrt{2}$	$\sqrt{2}$	1
$\dfrac{\pi}{2}$	1	0	—	1	—	0

Values corresponding to $t = \pi/6$ and $t = \pi/3$ will be determined in Example 3. By using methods developed in this text, values for every real number t may be approximated to any degree of accuracy. It will be assumed that the reader knows how to use trigonometric tables (see Appendix IV) or a calculator to approximate values of the trigonometric functions.

If, in Definition (III.2), $P(x, y)$ is in quadrant I, then x and y are both positive, and hence all values of the trigonometric functions are positive. If $P(x, y)$ is in quadrant II, then x is negative, y is positive, and hence $\sin t$ and $\csc t$ are positive, whereas the other four functions are negative. Similar remarks can be made for the remaining quadrants.

The domain of the sine and cosine functions is $\mathbb{R}$. However, in the definitions of $\tan t$ and $\sec t$, x appears in the denominator, and hence we must exclude the points $P(x, y)$ on the y-axis, that is, $P(0, 1)$ and $P(0, -1)$. It follows that the domain of the tangent and secant functions consists of all numbers t *except* those of the form $(\pi/2) + n\pi$, where n is an integer. Similarly, for the cotangent and cosecant functions we must exclude all numbers of the form $t = n\pi$, where n is an integer.

Since $|x| \leq 1$ and $|y| \leq 1$ for all points $P(x, y)$ on the unit circle U, we see that $-1 \leq \sin t \leq 1$ and $-1 \leq \cos t \leq 1$. It follows from the discussion in Chapter 8 that $\sin t$ and $\cos t$ take on *every* value between -1 and 1. It can also be shown that the range of the tangent and cotangent functions is $\mathbb{R}$, and that the range of the cosecant and secant functions is $(-\infty, -1] \cup [1, \infty)$.

Since the circumference of the unit circle U is 2π, the same point $P(x, y)$ is obtained for $t + 2\pi n$, where n is any integer. Hence the values of the trigonometric functions repeat in successive intervals of length 2π. A function f with domain X is said to be **periodic** if there exists a positive real number k such that $f(t + k) = f(t)$ for every t in X. Geometrically, this means that the graph of f repeats itself as x-coordinates of points vary over successive intervals of length k. If a least such positive real number k exists, it is called the **period** of f. It can be shown that the sine, cosine, cosecant, and secant functions have period 2π, whereas the tangent and cotangent functions have period π.

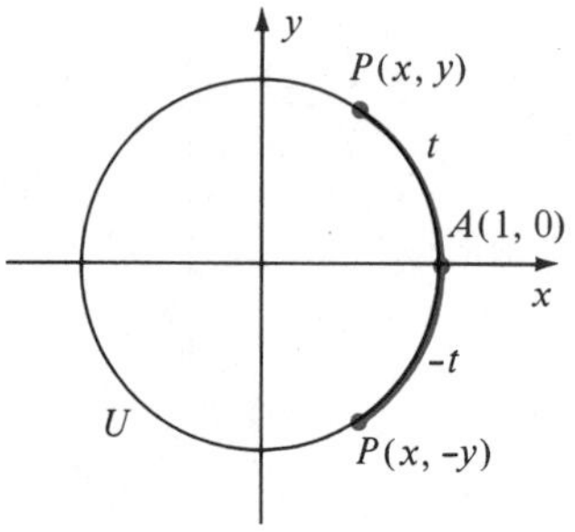

If $P(x, y)$ is the point on U corresponding to t, then as illustrated in Figure III.7, $P(x, -y)$ corresponds to $-t$. Consequently, $\sin(-t) = -y = -\sin t$ and $\cos(-t) = x = \cos t$. Similarly, $\tan(-t) = -\tan t$. This gives us the following formulas for negatives:

Figure III.7

$$(III.3) \qquad \sin(-t) = -\sin t \qquad \cos(-t) = \cos t \qquad \tan(-t) = -\tan t$$

Each of the formulas in (III.3) is an *identity*, that is, each is true for every t in the domain of the indicated function.

The formulas listed in (III.4) are, without doubt, the most important identities in trigonometry, because they may be used to simplify and unify many different aspects of the subject. Three of these identities involve squares, such as $(\sin t)^2$ and $(\cos t)^2$. In general, if n is an integer different from -1, then powers such as $(\sin t)^n$ are written in the form $\sin^n t$. The case $n = -1$ is reserved for the inverse trigonometric functions.

The Fundamental Identities (III.4)

$$\csc t = \frac{1}{\sin t} \qquad \sec t = \frac{1}{\cos t} \qquad \cot t = \frac{1}{\tan t}$$

$$\tan t = \frac{\sin t}{\cos t} \qquad \cot t = \frac{\cos t}{\sin t}$$

$$\sin^2 t + \cos^2 t = 1 \qquad 1 + \tan^2 t = \sec^2 t \qquad 1 + \cot^2 t = \csc^2 t$$

Proof   The proofs follow directly from the definition of the trigonometric functions. Thus,

$$\csc t = \frac{1}{y} = \frac{1}{\sin t}, \qquad \sec t = \frac{1}{x} = \frac{1}{\cos t}, \qquad \cot t = \frac{x}{y} = \frac{1}{(y/x)} = \frac{1}{\tan t},$$

$$\tan t = \frac{y}{x} = \frac{\sin t}{\cos t}, \qquad \cot t = \frac{x}{y} = \frac{\cos t}{\sin t}$$

where we assume that no denominator is zero.

If (x, y) is a point on the unit circle U, then

$$y^2 + x^2 = 1.$$

Since $y = \sin t$ and $x = \cos t$, this gives us

$$(\sin t)^2 + (\cos t)^2 = 1$$

or, equivalently,

$$\sin^2 t + \cos^2 t = 1.$$

If $\cos t \neq 0$, then, dividing both sides of the last equation by $\cos^2 t$, we obtain

$$\frac{\sin^2 t}{\cos^2 t} + 1 = \frac{1}{\cos^2 t}$$

or

$$\left(\frac{\sin t}{\cos t}\right)^2 + 1 = \left(\frac{1}{\cos t}\right)^2.$$

Since $\tan t = \sin t/\cos t$ and $\sec t = 1/\cos t$, we see that

$$\tan^2 t + 1 = \sec^2 t.$$

The final Fundamental Identity is left as an exercise.   □

In certain applications it is convenient to change the domain of a trigonometric function from a subset of $\mathbb{R}$ to a set of angles. This may be accomplished by means of the following definition.

Definition (III.5)

If θ is an angle and if the radian measure of θ is t, then **the value of a trigonometric function at θ** is its value at the real number t.

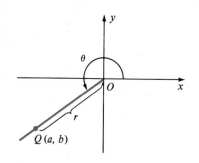

Figure III.8

It follows from Definition (III.5) that $\sin \theta = \sin t$, $\cos \theta = \cos t$, etc., where t is the radian measure of θ. To make the unit of angular measure clear, we shall use the degree symbol and write $\sin 65°$, $\tan 150°$, etc., whenever the angle is measured in degrees. Numerals without any symbol attached, such as $\cos 3$ and $\csc (\pi/6)$, will indicate that radian measure is being used. This is not in conflict with our previous work, where, for example, $\cos 3$ meant the value of the cosine function at the real number 3, since by definition the cosine of an angle of measure 3 radians is identical with the cosine of the real number 3.

Let θ be an angle in standard position and let $Q(a, b)$ be an arbitrary point on the terminal side of θ, as illustrated in Figure III.8. The next theorem specifies how the coordinates of the point Q may be used to determine the values of the trigonometric functions of θ.*

Theorem (III.6)

> Let θ be an angle in standard position on a rectangular coordinate system and let $Q(a, b)$ be any point other than O on the terminal side of θ. If $d(O, Q) = r$, then
>
> $$\sin \theta = \frac{b}{r} \qquad\qquad \csc \theta = \frac{r}{b} \quad (\text{if } b \neq 0)$$
>
> $$\cos \theta = \frac{a}{r} \qquad\qquad \sec \theta = \frac{r}{a} \quad (\text{if } a \neq 0)$$
>
> $$\tan \theta = \frac{b}{a} \quad (\text{if } a \neq 0) \qquad\qquad \cot \theta = \frac{a}{b} \quad (\text{if } b \neq 0).$$

Note that if $r = 1$, then Theorem (III.6) reduces to (III.2), with $a = x$, $b = y$, and $\theta = t$.

For acute angles, values of the trigonometric functions can be interpreted as ratios of the lengths of the sides of a right triangle. Recall that a triangle is called a **right triangle** if one of its angles is a right angle. If θ is an acute angle, then it can be regarded as an angle of a right triangle and we may refer to the lengths of the **hypotenuse**, the **opposite side**, and the **adjacent side** in the usual way. For convenience, we shall use **hyp, opp,** and **adj,** respectively, to denote these numbers. Let us introduce a rectangular coordinate system as in Figure III.9. Referring to the figure, we see that the lengths of the adjacent side and the opposite side for θ are the x-coordinate and y-coordinate, respectively, of a point Q on the terminal side of θ. By Theorem (III.6) we have the following.

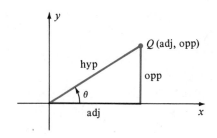

Figure III.9

---

* For a proof see E. W. Swokowski, *Fundamentals of Algebra and Trigonometry*, Fifth Edition (Boston: Prindle, Weber & Schmidt, 1981).

Right Triangle Trigonometry (III.7)

$$\sin \theta = \frac{\text{opp}}{\text{hyp}} \qquad \csc \theta = \frac{\text{hyp}}{\text{opp}}$$

$$\cos \theta = \frac{\text{adj}}{\text{hyp}} \qquad \sec \theta = \frac{\text{hyp}}{\text{adj}}$$

$$\tan \theta = \frac{\text{opp}}{\text{adj}} \qquad \cot \theta = \frac{\text{adj}}{\text{opp}}$$

These formulas are very important in work with right triangles. The next example illustrates how they may be used.

Example 3 Find $\sin \theta$, $\cos \theta$, and $\tan \theta$ for the following values of θ:

(a) $\theta = 60°$;   (b) $\theta = 30°$;   (c) $\theta = 45°$.

Solution   Let us consider an equilateral triangle having sides of length 2. The median from one vertex to the opposite side bisects the angle at that vertex, as illustrated in (i) of Figure III.10. By the Pythagorean Theorem, the length of this median is $\sqrt{3}$. Using the colored triangle and (III.7), we obtain the following.

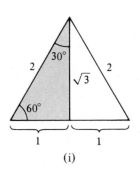

(i)

(a) $\sin 60° = \dfrac{\sqrt{3}}{2}$, $\cos 60° = \dfrac{1}{2}$, $\tan 60° = \dfrac{\sqrt{3}}{1} = \sqrt{3}$

(b) $\sin 30° = \dfrac{1}{2}$, $\cos 30° = \dfrac{\sqrt{3}}{2}$, $\tan 30° = \dfrac{1}{\sqrt{3}} = \dfrac{\sqrt{3}}{3}$

(c)   To find the functional values for $\theta = 45°$, let us consider an isosceles right triangle whose two equal sides have length 1, as illustrated in (ii) of Figure III.10. Thus,

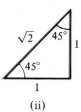

(ii)

Figure III.10

$$\sin 45° = \frac{1}{\sqrt{2}} = \frac{\sqrt{2}}{2} = \cos 45°, \quad \tan 45° = \frac{1}{1} = 1. \qquad \blacksquare$$

Some of the following trigonometric identities will be useful later in the text. Proofs may be found in books on trigonometry.

Addition and Subtraction Formulas

$$\sin (u + v) = \sin u \cos v + \cos u \sin v$$

$$\cos (u + v) = \cos u \cos v - \sin u \sin v$$

$$\tan (u + v) = \frac{\tan u + \tan v}{1 - \tan u \tan v}$$

$$\sin (u - v) = \sin u \cos v - \cos u \sin v$$

$$\cos (u - v) = \cos u \cos v + \sin u \sin v$$

$$\tan (u - v) = \frac{\tan u - \tan v}{1 + \tan u \tan v}$$

Double Angle Formulas

$$\sin 2u = 2 \sin u \cos u$$
$$\cos 2u = \cos^2 u - \sin^2 u = 1 - 2 \sin^2 u = 2 \cos^2 u - 1$$
$$\tan 2u = \frac{2 \tan u}{1 - \tan^2 u}$$

Half-Angle Formulas

$$\sin^2 \frac{u}{2} = \frac{1 - \cos u}{2} \qquad \cos^2 \frac{u}{2} = \frac{1 + \cos u}{2}$$
$$\tan \frac{u}{2} = \frac{1 - \cos u}{\sin u} = \frac{\sin u}{1 + \cos u}$$

Product Formulas

$$\sin u \cos v = \tfrac{1}{2}[\sin (u + v) + \sin (u - v)]$$
$$\cos u \sin v = \tfrac{1}{2}[\sin (u + v) - \sin (u - v)]$$
$$\cos u \cos v = \tfrac{1}{2}[\cos (u + v) + \cos (u - v)]$$
$$\sin u \sin v = \tfrac{1}{2}[\cos (u - v) - \cos (u + v)]$$

Factoring Formulas

$$\sin u + \sin v = 2 \cos \frac{u - v}{2} \sin \frac{u + v}{2}$$
$$\sin u - \sin v = 2 \cos \frac{u + v}{2} \sin \frac{u - v}{2}$$
$$\cos u + \cos v = 2 \cos \frac{u + v}{2} \cos \frac{u - v}{2}$$
$$\cos u - \cos v = 2 \sin \frac{v + u}{2} \sin \frac{v - u}{2}$$

EXERCISES A.III

1 Verify the entries in the table of radians and degrees on page A17.

2 Prove that $1 + \cot^2 t = \csc^2 t$.

3 Find the quadrant containing θ if
(a) $\sec \theta < 0$ and $\sin \theta > 0$.
(b) $\cot \theta > 0$ and $\csc \theta < 0$.
(c) $\cos \theta > 0$ and $\tan \theta < 0$.

4 Find the values of the remaining trigonometric functions if
(a) $\sin t = -\frac{4}{5}$ and $\cos t = \frac{3}{5}$.
(b) $\csc t = \sqrt{13}/2$ and $\cot t = -\frac{3}{2}$.

5 Without the use of tables or calculators, find the values of the trigonometric functions corresponding to each of the following real numbers.
(a) $9\pi/2$ (b) $-5\pi/4$ (c) 0   (d) $11\pi/6$

6 Find, without using a calculator, the radian measures that correspond to the following degree measures: $330°$, $405°$, $-150°$, $240°$, $36°$.

7 Find, without using a calculator, the degree measures that correspond to the following radian measures: $9\pi/2$, $-2\pi/3$, $7\pi/4$, 5π, $\pi/5$.

8 A central angle θ is subtended by an arc 20 cm long on a circle of radius 2 meters. What is the radian measure of θ?

9 Find the values of the six trigonometric functions of θ if θ is in standard position and satisfies the stated condition.

(a) The point $(30, -40)$ is on the terminal side of θ.

(b) The terminal side of θ is in quadrant II and is parallel to the line $2x + 3y + 6 = 0$.

(c) $\theta = -90°$.

10 Find each of the following without the use of tables or calculators.

(a) $\cos 225°$ (b) $\tan 150°$ (c) $\sin(-\pi/6)$

(d) $\sec(4\pi/3)$ (e) $\cot(7\pi/4)$ (f) $\csc(300°)$

Verify the identities in Exercises 11–30.

11 $\cos\theta \sec\theta = 1$

12 $\tan\alpha \cot\alpha = 1$

13 $\sin\theta \sec\theta = \tan\theta$

14 $\sin\alpha \cot\alpha = \cos\alpha$

15 $\dfrac{\csc x}{\sec x} = \cot x$

16 $\cot\beta \sec\beta = \csc\beta$

17 $(1 + \cos\alpha)(1 - \cos\alpha) = \sin^2\alpha$

18 $\cos^2 x(\sec^2 x - 1) = \sin^2 x$

19 $\cos^2 t - \sin^2 t = 2\cos^2 t - 1$

20 $(\tan\theta + \cot\theta)\tan\theta = \sec^2\theta$

21 $\dfrac{\sin t}{\csc t} + \dfrac{\cos t}{\sec t} = 1$

22 $1 - 2\sin^2 x = 2\cos^2 x - 1$

23 $(1 + \sin\alpha)(1 - \sin\alpha) = 1/\sec^2\alpha$

24 $(1 - \sin^2 t)(1 + \tan^2 t) = 1$

25 $\sec\beta - \cos\beta = \tan\beta \sin\beta$

26 $\dfrac{\sin w + \cos w}{\cos w} = 1 + \tan w$

27 $\dfrac{\csc^2\theta}{1 + \tan^2\theta} = \cot^2\theta$

28 $\sin x + \cos x \cot x = \csc x$

29 $\sin t(\csc t - \sin t) = \cos^2 t$

30 $\cot t + \tan t = \csc t \sec t$

In Exercises 31–42 find the solutions of the given equation that are in the interval $[0, 2\pi)$, and also find the degree measure of each solution.

31 $2\cos^3\theta - \cos\theta = 0$

32 $2\cos\alpha + \tan\alpha = \sec\alpha$

33 $\sin\theta = \tan\theta$

34 $\csc^5\theta - 4\csc\theta = 0$

35 $2\cos^3 t + \cos^2 t - 2\cos t - 1 = 0$

36 $\cos x \cot^2 x = \cos x$

37 $\sin\beta + 2\cos^2\beta = 1$

38 $2\sec u \sin u + 2 = 4\sin u + \sec u$

39 $\cos 2x + 3\cos x + 2 = 0$

40 $\sin 2u = \sin u$

41 $2\cos^2\frac{1}{2}\theta - 3\cos\theta = 0$

42 $\sec 2x \csc 2x = 2\csc 2x$

If θ and φ are acute angles such that $\csc\theta = \frac{5}{3}$ and $\cos\varphi = \frac{8}{17}$, find the numbers in Exercises 43–51.

43 $\sin(\theta + \varphi)$

44 $\cos(\theta + \varphi)$

45 $\tan(\theta - \varphi)$

46 $\sin(\varphi - \theta)$

47 $\sin 2\varphi$

48 $\cos 2\varphi$

49 $\tan 2\theta$

50 $\sin\theta/2$

51 $\tan\theta/2$

52 Express $\cos(\alpha + \beta + \gamma)$ in terms of functions of α, β, and γ.

53 Is there a real number t such that $7\sin t = 9$? Explain.

54 Is there a real number t such that $3\csc t = 1$? Explain.

55 If $f(t) = \cos t$ and $g(t) = t/4$, find (a) $f(g(\pi))$; (b) $g(f(\pi))$.

56 If $f(t) = \tan t$ and $g(t) = t/4$, find (a) $f(g(\pi))$; (b) $g(f(\pi))$.

APPENDIX IV

TABLES

TABLE A

TRIGONOMETRIC FUNCTIONS

Degrees	Radians	sin	tan	cot	cos		
0	0.000	0.000	0.000		1.000	1.571	90
1	0.017	0.017	0.017	57.29	1.000	1.553	89
2	0.035	0.035	0.035	28.64	0.999	1.536	88
3	0.052	0.052	0.052	19.081	0.999	1.518	87
4	0.070	0.070	0.070	14.301	0.998	1.501	86
5	0.087	0.087	0.087	11.430	0.996	1.484	85
6	0.105	0.105	0.105	9.514	0.995	1.466	84
7	0.122	0.122	0.123	8.144	0.993	1.449	83
8	0.140	0.139	0.141	7.115	0.990	1.431	82
9	0.157	0.156	0.158	6.314	0.988	1.414	81
10	0.175	0.174	0.176	5.671	0.985	1.396	80
11	0.192	0.191	0.194	5.145	0.982	1.379	79
12	0.209	0.208	0.213	4.705	0.978	1.361	78
13	0.227	0.225	0.231	4.331	0.974	1.344	77
14	0.244	0.242	0.249	4.011	0.970	1.326	76
15	0.262	0.259	0.268	3.732	0.966	1.309	75
16	0.279	0.276	0.287	3.487	0.961	1.292	74
17	0.297	0.292	0.306	3.271	0.956	1.274	73
18	0.314	0.309	0.325	3.078	0.951	1.257	72
19	0.332	0.326	0.344	2.904	0.946	1.239	71
20	0.349	0.342	0.364	2.747	0.940	1.222	70
21	0.367	0.358	0.384	2.605	0.934	1.204	69
22	0.384	0.375	0.404	2.475	0.927	1.187	68
23	0.401	0.391	0.424	2.356	0.921	1.169	67
24	0.419	0.407	0.445	2.246	0.914	1.152	66
25	0.436	0.423	0.466	2.144	0.906	1.134	65
26	0.454	0.438	0.488	2.050	0.899	1.117	64
27	0.471	0.454	0.510	1.963	0.891	1.100	63
28	0.489	0.469	0.532	1.881	0.883	1.082	62
29	0.506	0.485	0.554	1.804	0.875	1.065	61
30	0.524	0.500	0.577	1.732	0.866	1.047	60
31	0.541	0.515	0.601	1.664	0.857	1.030	59
32	0.559	0.530	0.625	1.600	0.848	1.012	58
33	0.576	0.545	0.649	1.540	0.839	0.995	57
34	0.593	0.559	0.675	1.483	0.829	0.977	56
35	0.611	0.574	0.700	1.428	0.819	0.960	55
36	0.628	0.588	0.727	1.376	0.809	0.942	54
37	0.646	0.602	0.754	1.327	0.799	0.925	53
38	0.663	0.616	0.781	1.280	0.788	0.908	52
39	0.681	0.629	0.810	1.235	0.777	0.890	51
40	0.698	0.643	0.839	1.192	0.766	0.873	50
41	0.716	0.656	0.869	1.150	0.755	0.855	49
42	0.733	0.669	0.900	1.111	0.743	0.838	48
43	0.750	0.682	0.933	1.072	0.731	0.820	47
44	0.768	0.695	0.966	1.036	0.719	0.803	46
45	0.785	0.707	1.000	1.000	0.707	0.785	45
		cos	cot	tan	sin	Radians	Degrees

TABLE B
EXPONENTIAL FUNCTIONS

x	e^x	e^{-x}	x	e^x	e^{-x}
0.00	1.0000	1.0000	2.50	12.182	0.0821
0.05	1.0513	0.9512	2.60	13.464	0.0743
0.10	1.1052	0.9048	2.70	14.880	0.0672
0.15	1.1618	0.8607	2.80	16.445	0.0608
0.20	1.2214	0.8187	2.90	18.174	0.0550
0.25	1.2840	0.7788	3.00	20.086	0.0498
0.30	1.3499	0.7408	3.10	22.198	0.0450
0.35	1.4191	0.7047	3.20	24.533	0.0408
0.40	1.4918	0.6703	3.30	27.113	0.0369
0.45	1.5683	0.6376	3.40	29.964	0.0334
0.50	1.6487	0.6065	3.50	33.115	0.0302
0.55	1.7333	0.5769	3.60	36.598	0.0273
0.60	1.8221	0.5488	3.70	40.447	0.0247
0.65	1.9155	0.5220	3.80	44.701	0.0224
0.70	2.0138	0.4966	3.90	49.402	0.0202
0.75	2.1170	0.4724	4.00	54.598	0.0183
0.80	2.2255	0.4493	4.10	60.340	0.0166
0.85	2.3396	0.4274	4.20	66.686	0.0150
0.90	2.4596	0.4066	4.30	73.700	0.0136
0.95	2.5857	0.3867	4.40	81.451	0.0123
1.00	2.7183	0.3679	4.50	90.017	0.0111
1.10	3.0042	0.3329	4.60	99.484	0.0101
1.20	3.3201	0.3012	4.70	109.95	0.0091
1.30	3.6693	0.2725	4.80	121.51	0.0082
1.40	4.0552	0.2466	4.90	134.29	0.0074
1.50	4.4817	0.2231	5.00	148.41	0.0067
1.60	4.9530	0.2019	6.00	403.43	0.0025
1.70	5.4739	0.1827	7.00	1096.6	0.0009
1.80	6.0496	0.1653	8.00	2981.0	0.0003
1.90	6.6859	0.1496	9.00	8103.1	0.0001
2.00	7.3891	0.1353	10.00	22026.0	0.00005
2.10	8.1662	0.1225			
2.20	9.0250	0.1108			
2.30	9.9742	0.1003			
2.40	11.0232	0.0907			

TABLE C
NATURAL LOGARITHMS

n	0.0	0.1	0.2	0.3	0.4	0.5	0.6	0.7	0.8	0.9
0*		7.697	8.391	8.796	9.084	9.307	9.489	9.643	9.777	9.895
1	0.000	0.095	0.182	0.262	0.336	0.405	0.470	0.531	0.588	0.642
2	0.693	0.742	0.788	0.833	0.875	0.916	0.956	0.993	1.030	1.065
3	1.099	1.131	1.163	1.194	1.224	1.253	1.281	1.308	1.335	1.361
4	1.386	1.411	1.435	1.459	1.482	1.504	1.526	1.548	1.569	1.589
5	1.609	1.629	1.649	1.668	1.686	1.705	1.723	1.740	1.758	1.775
6	1.792	1.808	1.825	1.841	1.856	1.872	1.887	1.902	1.917	1.932
7	1.946	1.960	1.974	1.988	2.001	2.015	2.028	2.041	2.054	2.067
8	2.079	2.092	2.104	2.116	2.128	2.140	2.152	2.163	2.175	2.186
9	2.197	2.208	2.219	2.230	2.241	2.251	2.262	2.272	2.282	2.293
10	2.303	2.313	2.322	2.332	2.342	2.351	2.361	2.370	2.380	2.389

* Subtract 10 if $n < 1$; for example, $\ln 0.3 \approx 8.796 - 10 = -1.204$.

APPENDIX V

FORMULAS FROM GEOMETRY

The most frequently used notation is as follows.

$$r = \text{radius}$$
$$h = \text{altitude}$$
$$b \text{ (or } a) = \text{length of base}$$
$$A = \text{area}$$
$$C = \text{circumference}$$
$$V = \text{volume}$$
$$S = \text{curved surface area}$$
$$B = \text{area of base}$$

Formulas for areas, circumference of a circle, volumes, and curved surface area are as follows.

TRIANGLE	$A = \frac{1}{2}bh$
CIRCLE	$A = \pi r^2, \quad C = 2\pi r$
PARALLELOGRAM	$A = bh$
TRAPEZOID	$A = \frac{1}{2}(a + b)h$
RIGHT CIRCULAR CYLINDER	$V = \pi r^2 h, \quad S = 2\pi rh$
RIGHT CIRCULAR CONE	$V = \frac{1}{3}\pi r^2 h, \quad S = \pi r\sqrt{r^2 + h^2}$
SPHERE	$V = \frac{4}{3}\pi r^3, \quad S = 4\pi r^2$
PRISM	$V = Bh$
PYRAMID	$V = \frac{1}{3}Bh$

ANSWERS TO ODD-NUMBERED EXERCISES

CHAPTER 1

Exercises 1.1, page 9

1 (a) > (b) < (c) = (d) > (e) = (f) <
3 (a) 3 (b) 7 (c) 7 (d) 3 (e) $\frac{22}{7} - \pi$ (f) -1
   (g) 0 (h) 9 (i) $x - 5$ (j) $b - a$
5 (a) 4 (b) 8 (c) 8 (d) 12   **7** $\left(\frac{17}{5}, \infty\right)$
9 $[-2, \infty)$ **11** $(-\infty, -3) \cup (2, \infty)$ **13** $(5, \infty)$
15 $\left(-\frac{4}{5}, 3\right]$ **17** $[-3, 1)$ **19** $\left(-\infty, \frac{7}{2}\right)$
21 $(9.7, 10.3)$ **23** $\left[\frac{5}{3}, 3\right]$ **25** $\left(-\infty, \frac{1}{25}\right) \cup \left(\frac{3}{5}, \infty\right)$
27 $\left(-2, \frac{1}{3}\right)$ **29** $(-\infty, -4] \cup \left[-\frac{1}{2}, \infty\right)$
31 $\left(-\infty, -\frac{1}{10}\right) \cup \left(\frac{1}{10}, \infty\right)$ **33** $\left[-\frac{2}{3}, \frac{7}{2}\right)$
35 $\frac{140}{9} \le C \le \frac{80}{3}$ **37** $\frac{20}{9} \le x \le 4$ **39** $\frac{1}{2} \le t \le 4$
41 $|a| = |(a - b) + b| \le |a - b| + |b|$. Hence
   $|a| - |b| \le |a - b|$.

Exercises 1.2, page 18

1 $5, \left(4, -\frac{1}{2}\right)$ **3** $\sqrt{26}, \left(-\frac{1}{2}, -\frac{9}{2}\right)$ **5** $5, \left(-\frac{11}{2}, -2\right)$
7 35   **9** $d(A, D) = d(B, C)$ and $d(A, B) = d(C, D)$
11 $a > 4$ or $a < \frac{2}{5}$

15   **17**   **19**

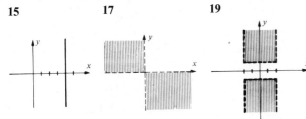

21   **23**   **25** Symmetry:
                              y-axis

27 Symmetry: **29** Symmetry: **31**
   y-axis   origin

33   **35**

37 Circle of radius 4, center at the origin
39 $(x - 3)^2 + (y + 2)^2 = 16$ **41** $x^2 + y^2 = 34$
43 $(x + 4)^2 + (y - 2)^2 = 4$
45 $(x - 1)^2 + (y - 2)^2 = 34$ **47** $(-2, 3), 3$
49 $(-3, 0), 3$ **51** $\left(\frac{1}{4}, -\frac{1}{4}\right), \sqrt{26}/4$

Exercises 1.3, page 24

1 4   **3** The slope does not exist.
5 The slopes of opposite sides are equal.
7 *Hint*: Show that opposite sides are parallel and two adjacent sides are perpendicular.
9 $(-12, 0)$ **11** $x - 2y - 14 = 0$
13 $3x - 8y - 41 = 0$ **15** $x - 8y - 24 = 0$
17 (a) $x = 10$ (b) $y = -6$ **19** $5x + 2y - 29 = 0$
21 $5x - 7y + 15 = 0$
23 $x + 6y - 9 = 0; 3x - 5y + 5 = 0; 4x + y - 4 = 0;$
    $(\frac{15}{23}, \frac{32}{23})$
25 $m = \frac{3}{4}, b = 2$ **27** $m = -\frac{1}{2}, b = 0$

   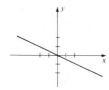

29 $m = 0, b = 4$ **31** $m = -\frac{5}{4}, b = 5$

   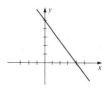

33 $m = \frac{1}{3}, b = -\frac{7}{3}$ **35** $k = -3$

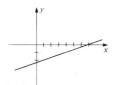

37 $x/(3/2) + y/(-3) = 1$ **39** $r < 1$ or $r > 2$
41 $y = 59,000 + 6000x$, where y is the value of the house and x is the number of years after the purchase date; 2 years and 4 months after the purchase date.

Exercises 1.4, page 32

1 $2, -8, -3, 6\sqrt{2} - 3$
3 (a) $3a^2 - a + 2$ (b) $3a^2 + a + 2$ (c) $-3a^2 + a - 2$
    (d) $3a^2 + 6ah + 3h^2 - a - h + 2$
    (e) $3a^2 + 3h^2 - a - h + 4$ (f) $6a + 3h - 1$
5 (a) $a^2/(1 + 4a^2)$ (b) $a^2 + 4$ (c) $1/(a^4 + 4)$
    (d) $1/(a^2 + 4)^2$ (e) $1/(a + 4)$ (f) $1/\sqrt{a^2 + 4}$
7 $[\frac{5}{3}, \infty)$ **9** $[-2, 2]$
11 All real numbers except 0, 3, and -3
13 $\frac{9}{7}, (a + 5)/7, \mathbb{R}$
15 $19, a^2 + 3$, all nonnegative real numbers
17 $\sqrt[3]{4}, \sqrt[3]{a}, \mathbb{R}$ **19** Yes   **21** No   **23** Yes
25 No   **27** Odd   **29** Even   **31** Even

33 Neither   **35** Neither
37 $(-\infty, \infty); (-\infty, \infty)$ **39** $(-\infty, \infty); \{-3\}$

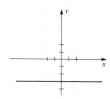   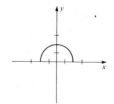

41 $(-\infty, \infty); (-\infty, 4]$ **43** $[-2, 2]; [0, 2]$

   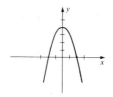

45 $(-\infty, 4) \cup (4, \infty);$ **47** $(-\infty, 4) \cup (4, \infty); (0, \infty)$
    $(-\infty, 0) \cup (0, \infty)$

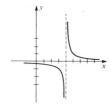   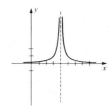

49 $(-\infty, \infty); [0, \infty)$ **51** $(-\infty, 0) \cup (0, \infty); \{-1, 1\}$

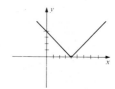   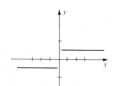

53 $(-\infty, 4]; [0, \infty)$ **55** $(-\infty, \infty);$
    $(-\infty, 4) \cup (4, \infty)$

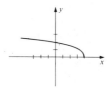   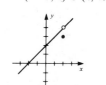

57 $(-\infty, \infty); [-5, 5]$ **59**

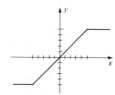   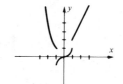

61 (a) 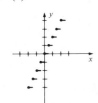   (b)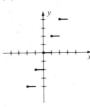

63 If $-1 < x < 1$, then there are two different points on the graph with x-coordinate x.

67 $r = C/2\pi$; $6/\pi \approx 1.9$ in.

69 $V = 4x^3 - 100x^2 + 600x$ **71** $P = 4\sqrt{A}$

73 $d = 2\sqrt{t^2 + 2500}$

75 $A = 20x$ if $x < 50$ and $A = 20x - 0.02x^2$ if $50 \le x \le 600$

Exercises 1.5, page 38

1 $3x^2 + 1/(2x - 3)$; $3x^2 - 1/(2x - 3)$; $3x^2/(2x - 3)$; $3x^2(2x - 3)$

3 $2x$; $2/x$, $x^2 - 1/x^2$, $(x^2 + 1)/(x^2 - 1)$

5 $2x^3 + x^2 + 7$; $2x^3 - x^2 - 2x + 3$; $2x^5 + 2x^4 + 3x^3 + 4x^2 + 3x + 10$, $(2x^3 - x + 5)/(x^2 + x + 2)$

7 $98x^2 - 112x + 37$; $-14x^2 - 31$ **9** $(x + 1)^3$; $x^3 + 1$

11 $3/(3x^2 + 2)^2 + 2$; $1/(27x^4 + 36x^2 + 14)$

13 $\sqrt{2x^2 + 7}$; $2x + 4$ **15** $5, -5$ **17** $1/x^4$, $1/x^4$

19 x, x

21 No; the coefficients of x^5 may be the negatives of one another.

25 *Hint*: Consider
$$f(x) = \tfrac{1}{2}[f(x) + f(-x)] + \tfrac{1}{2}[f(x) - f(-x)].$$

Exercises 1.6, page 40

1 $(-\infty, -\tfrac{3}{5})$ **3** $[3.495, 3.505]$ **5** $(1, \tfrac{3}{2})$

7 $(-5, \tfrac{1}{3}) \cup (\tfrac{7}{5}, \infty)$ **9** (a) 12 (b) $(\tfrac{1}{2}, \tfrac{5}{2})$ (c) 7

11 Symmetry: y-axis   **13** Symmetry: origin

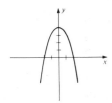   

15   **17**

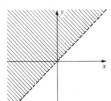   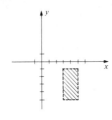

19 $(x + 4)^2 + (y + 3)^2 = 81$ **21** $(5, -7)$; 9

23 $6x - 7y + 24 = 0$ **25** $x = -4$

27 $(-\infty, 0) \cup (0, 1) \cup (1, \infty)$

29 The open interval $(5, 7)$

31 (a) $1/\sqrt{2}$ (b) $\tfrac{1}{2}$ (c) 1 (d) $1/\sqrt[4]{2}$ (e) $1/\sqrt{1 - x}$
   (f) $-1/\sqrt{x + 1}$ (g) $1/\sqrt{x^2 + 1}$ (h) $1/(x + 1)$

33   **35**

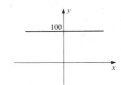   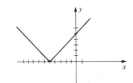

37 $x^2 + 4 + \sqrt{2x + 5}$; $x^2 + 4 - \sqrt{2x + 5}$;
   $(x^2 + 4)\sqrt{2x + 5}$; $(x^2 + 4)/\sqrt{2x + 5}$; $2x + 9$;
   $\sqrt{2x^2 + 13}$

39 If $a \ne b$, then $5 - 7a \ne 5 - 7b$; that is, $f(a) \ne f(b)$.

CHAPTER 2

Exercises 2.1, page 45

1 4   **3** $\tfrac{1}{9}$ **5** $\tfrac{17}{13}$ **7** 32   **9** $2x$ **11** 12

13 Does not exist   **15** Does not exist

17 (a) $10a - 4$ (b) $y = 16x - 20$

19 (a) $3a^2$ (b) $y = 12x - 16$

21 (a) 3   **23** (a) $-1/a^2$
   (b) $y = 3x + 2$ (b) $x + 4y - 4 = 0$

          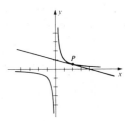

29 $(3, 9)$

Exercises 2.2, page 52

1 Given any ε, choose $\delta \le \varepsilon/5$.

3 Given any ε, choose $\delta \le \varepsilon/9$.

5 Given any ε, choose $\delta \le 4\varepsilon$.

7 Given any ε, let δ be any positive number.

9 Given any ε, choose $\delta \le \varepsilon$.

11 Every interval $(3 - \delta, 3 + \delta)$ contains numbers for which the quotient equals 1, and other numbers for which the quotient equals -1.

13 $1/(x + 5)$ can be made as large as desired by choosing x sufficiently close to -5.

15 There are many examples. One is $f(x) = (x^2 - 1)/(x - 1)$ if $x \neq 1$ and $f(1) = 3$.

17 Every interval $(a - \delta, a + \delta)$ contains numbers such that $f(x) = 0$ and other numbers such that $f(x) = 1$.

Exercises 2.3, page 60

1 -13 **3** $5\sqrt{2} - 20$ **5** -2 **7** 0 **9** 15
11 -7 **13** $\frac{1}{12}$ **15** 8 **17** -23 **19** 2
21 $\frac{72}{7}$ **23** -2 **25** $-\frac{1}{8}$ **27** 0 **29** -810
31 -64 **33** -108 **35** $-\frac{1}{4}$
37 If $a < 0$ and $r = m/n$ is in lowest terms, then n must be an odd integer.

Exercises 2.4, page 64

1 (a) 0 (b) Does not exist (c) Does not exist
3 (a) 0 (b) 0 (c) 0
5 (a) -1 (b) 1 (c) Does not exist **7** 4 **9** -6
11 3 **13** 4 **15** 1 **17** $\frac{1}{8}$ **19** 1
21 Does not exist **23** $6; 4$

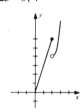

25 $-1; \frac{1}{11}$; does not exist
27 **29** **31**

 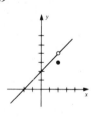 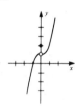

33 $(-1)^{n-1}; (-1)^n$ **35** $0; 0$

 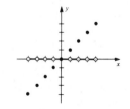

37 $1; 0$

Exercises 2.5, page 72

11 $\{x : x \neq \frac{3}{2}, x \neq -1\}$ **13** $[\frac{3}{2}, \infty)$
15 $(-\infty, -1) \cup (1, \infty)$ **17** $\{x : x \neq -9\}$
19 $\{x : x \neq 0, x \neq 1\}$ **21** $[-5, -3] \cup [3, 4) \cup (4, 5]$
23, 25, 27 Discontinuous at a; continuous at every other real number **29** $\frac{5}{2}$
31 $c = d = 8$
33 $f \circ g$ is continuous at 0; $g \circ f$ is not continuous at 0.
35 f is discontinuous on any open interval containing the origin.
37 No. $\lim_{x \to 3} f(x)$ does not exist.
39 Yes. All conditions of (2.18) are fulfilled.
43 $c = \sqrt[3]{w - 1}$ **45** $c = \sqrt{w - 2}$

Exercises 2.6, page 74

1 13 **3** $-4 - \sqrt{14}$ **5** $\frac{7}{8}$ **7** $\frac{32}{3}$
9 Does not exist **11** 3 **13** -1 **15** $4a^3$
17 $-\frac{3}{16}$ **19** -1 **21** -6 **25** $\mathbb{R}$
27 $[-3, -2) \cup (-2, 2) \cup (2, 3]$
29 Discontinuous at 4 and -4
31 Discontinuous at 0 and 2 **33** $c = 1/\sqrt{w}$

CHAPTER 3

Exercises 3.1, page 79

1 $-3a^2$ **3** $1/2\sqrt{a}$
5 (a) 11.8, 11.4, 11.04, 11.004 cm/sec (b) 11 cm/sec
  (c) $(-\frac{3}{8}, \infty)$ (d) $(-\infty, -\frac{3}{8})$
7 48 ft/sec, 16 ft/sec, -16 ft/sec. Maximum height attained at $t = \frac{7}{2}$ sec. Ground struck at $t = 7$ sec with velocity -112 ft/sec.
9 If $f(t) = at + b$, then $v(t) = a$.

Exercises 3.2, page 85

1 0 **3** 9 **5** $8 - 10x$ **7** $-1/(x - 2)^2$
9 $3/2\sqrt{3x + 1}$ **11** $-7/2\sqrt{x^3}$ **13** $6x^2 - 4$
15 $2a$ **17** $-12/a^3$ **19** $-1/(a + 5)^2$
23 *Hint:* Use (3.7).
25 If $f(x) = ax + b$, then $f'(x) = a$ has degree 0. If $f(x)$ has degree 2, then $f'(x)$ has degree 1. If $f(x)$ has degree 3, then $f'(x)$ has degree 2.
27 Domain of **29** Domain of
  $f' = \{x : x \neq 0\}$ $f' = \{x : x \neq 0, x \neq \pm 1\}$

 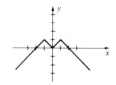

Exercises 3.3, page 93

1 $f'(x) = 20x + 9$　　**3** $f'(s) = -1 + 8s - 20s^3$
5 $g'(x) = 10x^4 + 9x^2 - 28x$
7 $h'(r) = 18r^5 - 21r^2 + 4r$　　**9** $f'(x) = 23/(3x + 2)^2$
11 $h'(z) = (-27z^2 + 12z + 70)/(2 - 9z)^2$
13 $f'(x) = 9x^2 - 4x + 4$　　**15** $F'(t) = 2t - (2/t^3)$
17 $g'(x) = 416x^3 - 195x^2 + 64x - 20$
19 $G'(v) = 6v^2/(v^3 + 1)^2$
21 $f'(x) = -(1 + 2x + 3x^2)/(1 + x + x^2 + x^3)^2$
23 $g'(z) = 72z^5 - 64z^3 - 18z^2 - 70z - 7$
25 $K'(s) = -\frac{4}{81}s^{-5}$　　**27** $h'(x) = 10(5x - 4)$
29 $f'(t) = (6 - 20t - 21t^2)/5(2 + 7t^2)^2$ where $t \neq 0$
31 $M'(x) = 2 - 4/x^2 - 6/x^3$　　**33** $y' = (-3x + 2)/x^3$
35 $y' = 120x - 49$
39 $y' = (8x - 1)(x^2 + 4x + 7)(3x^2) +$
　　$(8x - 1)(2x + 4)(x^3 - 5) + 8(x^2 + 4x + 7)(x^3 - 5)$
41 (a) $y = 5$　(b) $5x + 2y - 10 = 0$
　　(c) $4x - 5y + 13 = 0$
43 (a) $\frac{2}{3}, -2$　(b) $0, -\frac{4}{3}$
45 (a) $(-\infty, -5)$ and $(2, \infty)$
　　(b) $(-5, 2)$. The velocity is 0 at $t = -5$ and $t = 2$.
47 (a) 1　(b) -3　(c) -4　(d) 11　(e) $-\frac{1}{25}$
49 $(0, 1)$; $(1, 2)$; $(3, 22)$　　**51** $16y - 40x + 99 = 0$
53 $y - 9 = 2(x - 5)$, point of tangency is $(1, 1)$;
　　$y - 9 = 18(x - 5)$, point of tangency is $(9, 81)$.
55 (a) In ft/sec: 4, 10, 18　(b) $6\sqrt{5} \approx 13.4$ ft/sec

Exercises 3.4, page 101

1 (a) $(4x - 4)\Delta x + 2(\Delta x)^2$　(b) -0.72
3 (a) $-[(2x + \Delta x)\Delta x]/x^2(x + \Delta x)^2$
　　(b) $-\frac{189}{9801} \approx -0.019$
5 (a) $(6x + 5)\Delta x + 3(\Delta x)^2$　(b) $(6x + 5)\,dx$
　　(c) $-3(\Delta x)^2$
7 (a) $-\Delta x/x(x + \Delta x)$　(b) $-dx/x^2$
　　(c) $-(\Delta x)^2/x^2(x + \Delta x)$
9 (a) $-9\Delta x$　(b) $-9\Delta x$　(c) 0
11 (a) $4x^3(\Delta x) + 6x^2(\Delta x)^2 + 4x(\Delta x)^3 + (\Delta x)^4$
　　(b) $4x^3\Delta x$　(c) $-6x^2(\Delta x)^2 - 4x(\Delta x)^3 - (\Delta x)^4$
13 0.06　　**15** $dw = (3z^2 - 6z + 2)\,dz, \Delta w \approx -1.30$
17 $1.92\pi \approx 6.03$ in.^{2}; 0.0075; 0.75 %
19 30 in.^{3}; 30.301 in.^{3}
21 Area ≈ 3301.661; maximum error ≈ 11.459;
　　average error ≈ 0.00347; percentage error $\approx 0.347 \%$
23 $1/50\pi \approx 0.00637$　　**25** -1　　**27** 0.92; 0.92236816
29 dA is the shaded region in the figure.

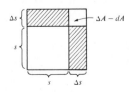

31 40 % increase in the pressure difference

Exercises 3.5, page 108

1 $f'(x) = 3(x^2 - 3x + 8)^2(2x - 3)$
3 $g'(x) = -40(8x - 7)^{-6}$
5 $f'(x) = -(7x^2 + 1)/(x^2 - 1)^5$
7 $f'(x) = 5(8x^3 - 2x^2 + x - 7)^4(24x^2 - 4x + 1)$
9 $F'(v) = 17,000(17v - 5)^{999}$
11 $s'(t) = -2(4t^5 - 3t^3 + 2t)^{-3}(20t^4 - 9t^2 + 2)$
13 $N'(x) = 32x(6x - 7)^3(8x^2 + 9) + 18(8x^2 + 9)^2(6x - 7)^2$
　　which reduces to $(6x - 7)^2(8x^2 + 9)(336x^2 - 224x + 162)$
15 $g'(z) = 6(z^2 - 1/z^2)^5(2z + 2/z^3)$
17 $k'(u) = -20(u^2 + 1)^3/(4u - 5)^6 + 6u(u^2 + 1)^2/(4u - 5)^5$
　　which reduces to $(u^2 + 1)^2(4u^2 - 30u - 20)/(4u - 5)^6$
19 $f'(x) = 124x(3x^2 - 5)/(2x^2 + 7)^3$
21 $G'(s) = -6(s^{-4} + 3s^{-2} + 2)^{-7}(-4s^{-5} - 6s^{-3})$
23 $h'(x) = 200(2x + 1)^9[(2x + 1)^{10} + 1]^9$
25 $F'(t) = 2(2t + 1)(2t + 3)^2(24t^2 + 26t + 3)$
27 (a) $y - 81 = 864(x - 2)$　(b) $(1, 1), (\frac{1}{2}, 0), (\frac{3}{2}, 0)$
29 (a) $y - 1 = 20(x - 1)$　(b) $(\frac{1}{2}, 0)$
31 $dy = 10(x^4 - 3x^2 + 1)^9(4x^3 - 6x)\,dx, \Delta y \approx 0.2$
33 (a) $\dfrac{dw}{ds} = \dfrac{dw}{dz}\dfrac{dz}{ds}$　(b) $dw/ds = (3z^2 + 2/z^2)5(s^2 + 1)^4 2s$
35 $dK/dt = mv(dv/dt)$　　**37** -0.02
39 15　　**45** $x + 2y - 5 = 0$　　**47** $4x - 2y - 3 = 0$
49 $64x - 16y - 7 = 0$　　**51** $(4x^3 + 4x)/(x^4 + 2x^2 + 5)$

Exercises 3.6, page 113

1 $f(x) = -\frac{2}{5}x^2 + 2x + \frac{4}{5}, \mathbb{R}$
3 $f(x) = \sqrt{16 - x^2}, [-4, 4]$. There are other answers.
5 $f(x) = x + \sqrt{x}, [0, \infty)$
7 $f(x) = 1 - 2\sqrt{x} + x, [0, 1]$　　**9** $-8x/y$
11 $-(6x^2 + 2xy)/(x^2 + 3y^2)$　　**13** $(10x - y)/(x + 8y)$
15 $-y^3/x^3$　　**17** $-(2xy^3 + 4y + 1)/(3x^2y^2 + 4x - 6)$
19 $(4x^2 + 3x - 1)(8x + 3)/4y(y^2 - 9)^3$
　　　　　　　　　　$= (8x + 3)/4y(4x^2 + 3x - 1)^{1/2}$
21 $4x - y + 16 = 0$　　**23** $y + 3 = -\frac{36}{23}(x - 2)$
25 If it did, then $[f(x)]^2 + x^2 = -1$, an impossibility.
27 (a) Infinitely many
　　(b) One, $f(x) = 0$, with domain $x = 0$
　　(c) None
29 0.09

Exercises 3.7, page 115

1 $f'(x) = \frac{2}{3}x^{-1/3} + 6x^{1/2}$
3 $k'(r) = 8r^2(8r^3 + 27)^{-2/3}$
5 $F'(v) = -5v^4(v^5 - 32)^{-6/5}$　　**7** $f'(x) = 1/\sqrt{2x}$
9 $F'(z) = 15\sqrt{z} - 1/z\sqrt[3]{z}$
11 $g'(w) = (w^2 + 4w - 9)/2w^{5/2}$
13 $M'(x) = (8x - 7)/2\sqrt{4x^2 - 7x + 4}$
15 $f'(t) = -48t/(9t^2 + 16)^{5/3}$
17 $H'(u) = -1/2\sqrt{(3u + 8)(2u + 5)^3}$

19 $k'(s) = 16(s^2 + 9)^{1/4}(4s + 5)^3 + \frac{1}{2}s(4s + 5)^4(s^2 + 9)^{-3/4}$

21 $h'(x) = \frac{9}{5}x^2(x^2 + 4)^{5/3}(x^3 + 1)^{-2/5}$
$$+ \tfrac{10}{3}x(x^2 + 4)^{2/3}(x^3 + 1)^{3.5}$$

23 $g'(z) = -3\sqrt[3]{2z + 3}/2\sqrt{(3z + 2)^3}$
$$+ 2/(3\sqrt{3z + 2} \sqrt[3]{(2z + 3)^2})$$

25 $f'(x) = 6(7x + \sqrt{x^2 + 3})^5(7 + x/\sqrt{x^2 + 3})$

27 $2x + \sqrt{3}y - 1 = 0$ **29** $y - 5 = -\frac{33}{24}(x + 8)$

31 $(4, 2)$ **33** $-\sqrt{y/x}$ **35** $(12\sqrt{xy} + y)/(6\sqrt{xy} - x)$

37 $(18x^{5/3}y^{2/3} + y)/(12x^{2/3}y^{5/3} - x)$

39 $dy = 2(6x + 11)^{-2/3}\,dx$ **41** $4 + \frac{1}{48} \approx 4.02$

43 -0.24

45 60π cm^2; maximum error ≈ 1.508;
percentage error $\approx 0.8\%$

47 $f'(x) = (x - 1)/|x - 1|$; domain of $f' = \{x: x \neq 1\}$

49 $f'(x) = 2x(x^2 - 1)/|x^2 - 1|$; domain of $f' = \{x: x \neq \pm 1\}$

47   **49**

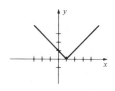   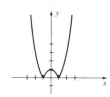

Exercises 3.8, page 118

1 $f'(x) = 12x^3 - 8x + 1$, $f''(x) = 36x^2 - 8$

3 $H'(s) = \frac{1}{3}\sqrt[3]{s^2} - 4/s^3$, $H''(s) = -\frac{2}{9}\sqrt[3]{s^5} + 12/s^4$

5 $g'(z) = 3/2\sqrt{3z + 1}$, $g''(z) = -9/4\sqrt{(3z + 1)^3}$

7 $k'(r) = 20(4r + 7)^4$, $k''(r) = 320(4r + 7)^3$

9 $f'(x) = x(x^2 + 4)^{-1/2}$, $f''(x) = 4(x^2 + 4)^{-3/2}$

11 $120x^2 + 18$ **13** $594/(3x + 1)^4$

15 $-270(2 - 9x)^{-8/3}$ **17** $(2xy^3 - 2x^4)/y^5 = -2x/y^5$

19 $10(y^2 - 3xy + x^2)/(2y - 3x)^3 = 40/(2y - 3x)^3$

21 $f'(x) = 6x^5 - 8x^3 + 9x^2 - 1$, $f''(x) = 30x^4 - 24x^2 + 18x$,
$f'''(x) = 120x^3 - 48x + 18$, $f^{(4)}(x) = 360x^2 - 48$,
$f^{(5)}(x) = 720x$, $f^{(6)}(x) = 720$

23 $f^{(n)}(x) = (-1)^n n!/x^{n+1}$, $f^{(n)}(1) = (-1)^n n!$

25   The degree of f' is $n - 1$, of f'' is $n - 2$, ..., of $f^{(n)}$ is 0. Since $f^{(n)}$ is a constant function, all higher derivatives are zero.

27 $f''(1) = 18$, $f''(-2) = -18$ **29** 0.14

31 $D_x^2 y = f''(g(x))(g'(x))^2 + f'(g(x))g''(x)$

Exercises 3.9, page 121

1 1.2599 **3** 0.5641 **5** 1.3315 **7** -1.7321

9 4.64575 **11** ± 3.34 **13** $-1, 1.35$

15 $-1.88, 0.35, 1.53$ **17** $-2.62, -0.38, 0.27, 3.73$

Exercises 3.10, page 122

1 $-24x/(3x^2 + 2)^2$ **3** $6x^2 - 7$ **5** $3/\sqrt{6t + 5}$

7 $\frac{1}{3}(7z^2 - 4z + 3)^{-2/3}(14z - 4)$ **9** $-144x/(3x^2 - 1)^5$

---

11 $-2(y^2 - y^{-2})^{-3}(2y + 2y^{-3})$ **13** $\frac{12}{5}(3x + 2)^{-1/5}$

15 $4(8s^2 - 4)^3(72s^4 - 108s^2 + 16s)/(1 - 9s^3)^5$

17 $(x^6 + 1)^4(3x + 2)^2(99x^6 + 60x^5 + 9)$

19 $\frac{4}{3}(7y - 2)^{-2}(2y + 1)^{-1/3} - 14(7y - 2)^{-3}(2y + 1)^{2/3}$

21 $2x[(x^2 + 2)(x^2 + 3) + (x^2 + 1)(x^2 + 3) + (x^2 + 1)(x^2 + 2)]$

23 $\dfrac{1}{2\sqrt{x + \sqrt{x + \sqrt{x}}}}\left(1 + \dfrac{2\sqrt{x} + 1}{4\sqrt{x}\sqrt{x + \sqrt{x}}}\right)$

25 $\dfrac{[\frac{2}{3}(3x + 2)^{1/2}(2x + 3)^{-2/3} - \frac{3}{2}(2x + 3)^{1/3}(3x + 2)^{-1/2}]}{3x + 2}$

27 $3(9z^{5/3} - 5z^{3/5})^2(15z^{2/3} - 3z^{-2/5})$

29 $(9s - 1)^3(108s^2 - 139s + 39)$

31 $(4xy^2 - 15x^2)/(12y^2 - 4x^2y)$

33 $1/\sqrt{x}(3\sqrt{y} + 2)$ **35** $9x - 4y - 12 = 0$

37 $x - y + 4 = 0$ **39** $(4 \pm \sqrt{10})/6$

41 $y' = 15x^2 + 2/\sqrt{x}$; $y'' = 30x - 1/\sqrt{x^3}$;
$y''' = 30 + 3/2\sqrt{x^5}$

43 $y'' = 5(y^2 - 4xy - x^2)/(y - 2x)^3 = -40/(y - 2x)^3$

45 $f^{(n)}(x) = n!/(1 - x)^{n+1}$

47   maximum error $\approx 0.06\sqrt{3}$; percentage error $\approx 1.5\%$

49 -0.57

51   (a) 2   (b) -7 (c) -14 (d) 21   (e) $-\frac{10}{9}$ (f) $-\frac{19}{27}$

53 2% **55** -0.7560

CHAPTER 4

Exercises 4.1, page 131

1 $5; -3$ **3** $1; -3$

5   (a) Since $f'(x) = 1/(3x^{2/3})$, $f'(0)$ does not exist. If $a \neq 0$, then $f'(a) \neq 0$. Hence 0 is the only critical number of f. The number $f(0) = 0$ is not a local extremum since $f(x) < 0$ if $x < 0$ and $f(x) > 0$ if $x > 0$.
(b) The facts that 0 is the only critical number and that there is a vertical tangent line at $(0, 0)$ follow as in part (a). The number $f(0) = 0$ is a local minimum since $f(x) > 0$ if $x \neq 0$.

7   There is a critical number 0, but $f(0)$ is not a local extremum since $f(x) < f(0)$ if $x < 0$ and $f(x) > f(0)$ if $x > 0$. The function is continuous at every number a since $\lim_{x \to a} f(x) = f(a)$. If $0 < x_1 < x_2 < 1$, then $f(x_1) < f(x_2)$ and hence there is neither a maximum nor minimum on $(0, 1)$. This does not contradict (4.3) because the interval $(0, 1)$ is open.

9 $\frac{3}{8}$ **11** $\frac{5}{3}$ and -2 **13**   2   **15**   4 and -4 (not 0)

17 $0, \frac{15}{7}$, and $\frac{5}{2}$ **19**   None   **21** $0, \pm\sqrt{3}, \pm 3$

23 $0, \pm\frac{3}{2}$ **25** $-5, -\frac{9}{7}, \frac{3}{2}$

27   If $f(x) = cx + d$ and $c \neq 0$, then $f'(x) = c \neq 0$. Hence there are no critical numbers. On $[a, b]$ the function has absolute extrema at a and b.

29   If $x = n$ is an integer, then $f'(n)$ does not exist. Otherwise, $f'(x) = 0$ for all $x \neq n$.

31 If $f(x) = ax^2 + bx + c$ and $a \neq 0$, then $f'(x) = 2ax + b$. Hence $-b/2a$ is the only critical number of f.

33 Since $f'(x) = nx^{n-1}$, the only possible critical number is $x = 0$, and $f(0) = 0$. If n is even then $f(x) > 0$ if $x \neq 0$ and hence 0 is a local minimum. If n is odd, then 0 is not an extremum since $f(x) < 0$ if $x < 0$ and $f(x) > 0$ if $x > 0$.

Exercises 4.2, page 135

1 f is not differentiable at the number 0 in the interval $(-1, 1)$.

3 f is not continuous on the interval $[-1, 4]$.

5 $c = 2$ **7** $c = 0$ **9** $c = 2$ **11** $c = 2$

13 Hypotheses not satisfied   **15** $c = 2$

17 $c = (2 - \sqrt{7})/3$

19 If $f(x) = cx + d$, then $f'(x) = c$ for all x. Moreover,
$$f(b) - f(a) = (cb + d) - (ca + d)$$
$$= c(b - a) = f'(x)(b - a).$$

21 If f has degree 3, then $f'(x)$ is a polynomial of degree 2. Consequently, the equation $f(b) - f(a) = f'(x)(b - a)$ has at most two solutions x_1 and x_2. If f has degree 4, there are at most three such solutions. If f has degree n, there are at most $n - 1$ solutions.

Exercises 4.3, page 141

1 Max: $f(-\frac{7}{8}) = \frac{129}{16}$; increasing on $(-\infty, -\frac{7}{8}]$; decreasing on $[-\frac{7}{8}, \infty)$

3 Max: $f(-2) = 29$; min: $f(\frac{5}{3}) = -\frac{548}{27}$; increasing on $(-\infty, -2]$ and $[\frac{5}{3}, \infty)$; decreasing on $[-2, \frac{5}{3}]$

5 Max: $f(0) = 1$; min: $f(-2) = -15$; min: $f(2) = -15$; increasing on $[-2, 0]$ and $[2, \infty)$; decreasing on $(-\infty, -2]$ and $[0, 2]$

1   **3**   **5**

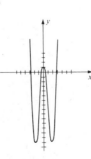

7 Min: $f(-1) = -3$; increasing on $[-1, \infty)$; decreasing on $(-\infty, -1]$ (*see graph*)

9 Max: $f(0) = 0$; min: $f(-\sqrt{3}) = f(\sqrt{3}) = -3$; increasing on $[-\sqrt{3}, 0]$ and $[\sqrt{3}, \infty)$; decreasing on $(-\infty, -\sqrt{3}]$ and $[0, \sqrt{3}]$ (*see graph*)

11 Max: $f(\frac{7}{4}) \approx 42$; min: $f(0) = 2$; min: $f(7) = 2$; increasing on $[0, \frac{7}{4}]$ and $[7, \infty)$; decreasing on $(-\infty, 0]$ and $[\frac{7}{4}, 7]$ (*see graph*)

7   **9**   **11**

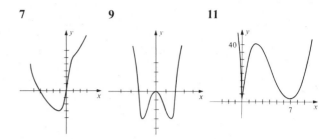

13 Max: $f(-1) = -4$; min: $f(1) = 4$; increasing on $(-\infty, -1]$ and $[1, \infty)$; decreasing on $[-1, 0)$ and $(0, 1]$.

15 Max: $f(\frac{3}{5}) \approx 0.346$; min: $f(1) = 0$; increasing on $(-\infty, \frac{3}{5}]$ and $[1, \infty)$; decreasing on $[\frac{3}{5}, 1]$

13   **15**

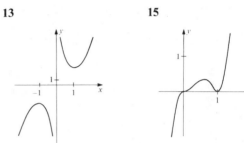

17 Max: $f(-\sqrt{3}) = (6\sqrt{3})^{1/3}$; min: $f(\sqrt{3}) = -(6\sqrt{3})^{1/3}$

19 Max: $f(-1) = 0$; min: $f(\frac{5}{7}) = -9^3 12^4/7^7$ **21** None

In Exercises 23 and 25 the absolute maximum is given first and the absolute minimum second.

23 (a) $f(-\frac{7}{8}) = \frac{129}{16}$; $f(1) = -6$
(b) $f(-\frac{7}{8}) = \frac{129}{16}$; $f(-4) = -31$
(c) $f(0) = 5$; $f(5) = -130$

25 (a) $f(-1) = 20$; $f(1) = -16$
(b) $f(-2) = 29$; $f(-4) = -31$
(c) $f(5) = 176$; $f(\frac{5}{3}) = -\frac{548}{27}$

27 (*graph below*)

29 Max: $f'(-1) = 8$; min: $f'(1) = -8$; f' is increasing on $(-\infty, -1]$ and $[1, \infty)$; decreasing on $[-1, 1]$.

27   **29**

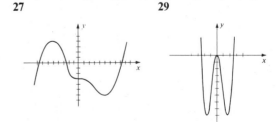

31 $a = \frac{3}{4}$, $b = 0$, $c = -\frac{9}{4}$, $d = \frac{1}{2}$

Exercises 4.4, page 149

In Exercises 1–11 the notations CU and CD mean that the graph is concave upward or downward, respectively, in the interval that follows. PI denotes point(s) of inflection.

1 Max: $f(\frac{1}{3}) = \frac{31}{27}$; min: $f(1) = 1$. CD on $(-\infty, \frac{2}{3})$; CU on $(\frac{2}{3}, \infty)$; x-coordinate of PI is $\frac{2}{3}$.

3 Min: $f(1) = 5$; CU on $(-\infty, 0)$ and $(\frac{2}{3}, \infty)$; CD on $(0, \frac{2}{3})$; x-coordinates of PI are 0 and $\frac{2}{3}$.

1   **3**

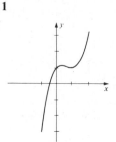 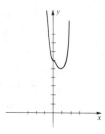

5 Max: $f(0) = 0$ (by first derivative test); min: $f(-\sqrt{2}) = f(\sqrt{2}) = -8$; CU on $(-\infty, -\sqrt{\frac{6}{5}})$ and $(\sqrt{\frac{6}{5}}, \infty)$; CD on $(-\sqrt{\frac{6}{5}}, \sqrt{\frac{6}{5}})$; x-coordinates of PI are $\pm\sqrt{\frac{6}{5}}$.

7 Max: $f(0) = 1$; min: $f(-1) = f(1) = 0$. CU on $(-\infty, -1/\sqrt{3})$ and $(1/\sqrt{3}, \infty)$; CD on $(-1/\sqrt{3}, 1/\sqrt{3})$; x-coordinates of PI are $\pm 1/\sqrt{3}$.

5   **7**

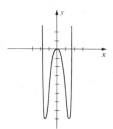 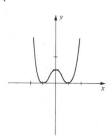

9 No local extrema; CU on $(-\infty, 0)$; CD on $(0, \infty)$; PI $(0, -1)$

11 No max or min. CU on $(-\infty, -3)$ and $(3, \infty)$; CD on $(-3, 0)$ and $(0, 3)$; x-coordinates of PI are ± 3.

9   **11**

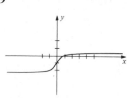 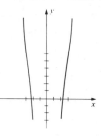

13 Min: $f(-1) = -\frac{1}{2}$; max: $f(1) = \frac{1}{2}$. CU on $(-\sqrt{3}, 0)$ and $(\sqrt{3}, \infty)$; CD on $(-\infty, -\sqrt{3})$ and $(0, \sqrt{3})$; x-coordinates of PI are 0, $\pm\sqrt{3}$.

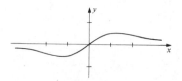

15 Max: $f(-\frac{4}{3}) \approx 7.27$; min: $f(0) = 0$; CD on $(-\infty, 0)$ and $(0, \frac{2}{3})$; CU on $(\frac{2}{3}, \infty)$; PI is $(\frac{2}{3}, \frac{10}{3}\sqrt[3]{12})$.

17 Min: $f(-2) \approx -7.55$; CU on $(-\infty, 0)$ and $(4, \infty)$; CD on $(0, 4)$; x-coordinates of PI are 0 and 4.

15   **17**

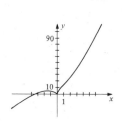 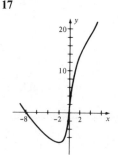

19   **21**

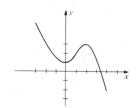 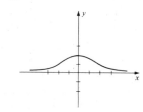

23   **25**

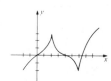 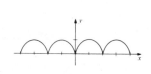

27 If $f(x) = ax^2 + bx + c$, then $f''(x) = 2a$. (a) CU if $a > 0$. (b) CD if $a < 0$.

29 If f has degree n, then f'' has degree $n - 2$.

Exercises 4.5, page 161

1 $\frac{5}{2}$ **3** $-\frac{7}{3}$ **5** 1   **7** 0   **9** $\frac{5}{2}$ **11** -1

13 $y = 4$

15 Both limits equal the quotient of the coefficients of x^n in $f(x)$ and $g(x)$. If f has lower degree than g, then both limits are 0.

21 $\infty; -\infty; x = 4; y = 0$.   **23** $\infty; -\infty; x = -\frac{5}{2};$
No max or min. $y = 0$. No max or min.

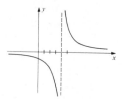 

25 $-\infty -\infty; x = -8; y = 0$. Max: $f(8) = \frac{3}{32}$ (*see graph*)

27 $-\infty, \infty$ for $a = -1$; $\infty, -\infty$ for $a = 2$; $x = -1$, $x = 2$, $y = 2$. Max: $f(0) = 0$; min: $f(-4) = \frac{16}{9}$ (*see graph*)

25

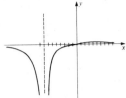

27

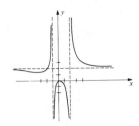

29 ∞, $-\infty$ for $a = 0$; ∞, ∞ for $a = 3$; $x = 0$, $x = 3$, $y = 0$.
31 $x = 2$, $x = -2$, $y = 0$

29

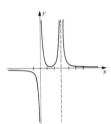

31

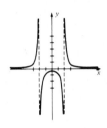

33 $y = 2$ **35** $x = -3$, $x = 0$, $x = 2$,
                                $y = 0$

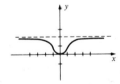   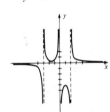

37 $x = -3$, $x = 1$, $y = 1$ **39** $x = 4$, $y = 0$

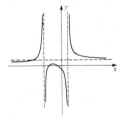   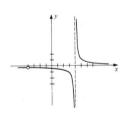

41 Vertical: $x = -1$ **43** Vertical: $x = 0$;
     oblique: $y = x - 2$ oblique: $y = -\frac{1}{2}x$

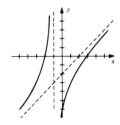   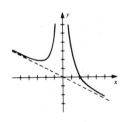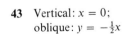

45 Vertical: $x = 1$; oblique: $y = x$

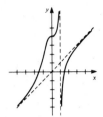

47 ∞ or $-\infty$ depending on whether the leading coefficients of f and g have the same sign or opposite signs, respectively.
49 $(-2, 0)$ **51** $(\frac{4}{3}, 3)$ and $(-\frac{4}{3}, 3)$

Exercises 4.6, page 169

1 20, -20 **3** Side of base $= 2$ ft; height $= 1$ ft
5 $5\sqrt{5}$ ft ≈ 11.2 ft
7 Length of base $= \sqrt{2}a$, height $= a/\sqrt{2}$
9 $\frac{32}{81}\pi a^3$ **11** Radius of base $=$ height $= 1/\sqrt[3]{\pi}$
13 Length $= 2\sqrt[3]{300} \approx 13.38$ ft;
     width $= \frac{3}{2}\sqrt[3]{300} \approx 10.04$ ft;
     height $= \sqrt[3]{300} \approx 6.69$ ft
17 55 **19** 3 in.
23 Width $= 2a/\sqrt{3}$; depth $= 2\sqrt{2}a/\sqrt{3}$
25 $(1, 2)$ **27** 500 **29** $d\sqrt[3]{S_1}/(\sqrt[3]{S_1} + \sqrt[3]{S_2})$
31 25 ft by $\frac{50}{7}$ ft
33 Radius $= \sqrt[3]{15}/2$; length of cylinder $= 2\sqrt[3]{15}$
35 (a) Use $36\sqrt{3}/(2 + \sqrt{3}) \approx 16.71$ cm for the rectangle.
     (b) Use all the wire for the rectangle.
37 17.20 ft
39 Width $= 12/(6 - \sqrt{3}) \approx 2.81$ ft;
     height $= (18 - 6\sqrt{3})/(6 - \sqrt{3}) \approx 1.78$ ft
41 37 **43** 16 in., 16 in., 32 in.
45 $4/(1 + \sqrt[4]{\frac{1}{2}}) \approx 2.17$ miles from A

Exercises 4.7, page 177

1 (a) $16^{-2/3}$ cm/min (b) 36π cm³/min
     (c) $(6)4^{1/3}\pi$ cm²/min
3 In (beats/min)/sec: (a) 7 (b) 15 (c) 23
5 (a) 3200π cm²/sec (b) 6400π cm²/sec
     (c) 9600π cm²/sec

In Exercises 7–15 the symbols $[a, b)$, $(a, b]$, and (a, b) denote intervals of time.

7 $v(t) = 6t - 12$, $a(t) = 6$. The point moves to the left in $[0, 2)$ and to the right in $(2, 5]$.
9 $v(t) = 3t^2 - 9$, $a(t) = 6t$; to the right in $[-3, -\sqrt{3})$; to the left in $(-\sqrt{3}, \sqrt{3})$; to the right in $(\sqrt{3}, 3]$

11 $v(t) = 1 - 4/t^2$, $a(t) = 8/t^3$; to the left in $[1, 2)$; to the right in $(2, 4]$

13 $v(t) = 8t^3 - 12t$, $a(t) = 24t^2 - 12$; to the left in $[-2, -\sqrt{3/2})$; to the right in $(-\sqrt{3/2}, 0)$; to the left in $(0, \sqrt{3/2})$; to the right in $(\sqrt{3/2}, 2]$

15 $v(t) = 3t^2$, $a(t) = 6t$; to the right in $(0, 4]$

17 $v(t) = 144 - 32t$, $a(t) = -32$; $v(3) = 48$, $a(3) = -32$. Maximum height is 324 ft. Strikes ground at $t = 9$.

19 $a(2) = -6$, $a(3) = 6$; $v(\frac{10}{3}) = \frac{44}{3}$

21 -0.12 units/ft   **23** $dr/dc = 1/2\pi$ **25** $\frac{9}{5}$

27 $dT/dl = C/\sqrt{l}$, where C is a constant

29 $V = 2400s - 200s^2 + 4s^3$; $dV/ds = 2400 - 400s + 12s^2$

Exercises 4.8, page 182

1 $-3\sqrt{336}/8 \approx -6.9$ ft/sec   **3** $20/9\pi \approx 0.71$ ft/min

5 $\frac{64}{11}$ ft/sec, $\frac{20}{11}$ ft/sec   **7** $-7442\pi \approx -23{,}368$ in.^{3}/hr

9 $\frac{10}{3}$ ft/sec   **11**   Increasing at a rate of 5 in.^{3}/min

13 $15\sqrt{3}/32 \approx 0.8$ ft/min   **15**   10 units/sec

17 $-4\sqrt{\sqrt{3}/600} \approx -0.215$ cm/min   **19** π m/sec

21 $\frac{11}{1600}$ ohms/sec   **23** $13.37/112\pi \approx 0.38$ ft/min

25   64 ft/sec

27 $(6 + \sqrt{2})180/\sqrt{10 + 3\sqrt{2}} \approx 353.6$ mi/hr

29 $-27/25\pi \approx 0.344$ in./hr

Exercises 4.9, page 189

1 $3x^3 - 2x^2 + 3x + C$

3 $\frac{1}{2}x^4 - \frac{1}{3}x^3 + \frac{3}{2}x^2 - 7x + C$

5 $-\frac{1}{2}x^{-2} + 3x^{-1} + C$ **7** $2x^{3/2} + 2x^{1/2} + C$

9 $9x^{2/3} - \frac{1}{8}x^{4/3} + 7x + C$

11 $\frac{8}{9}x^{9/4} + \frac{24}{5}x^{5/4} - x^{-3} + C$

13 $-(27x^2 + 36x + 16)/(3x^3) + C$

15 $3x^3 - 3x^2 + x + C$ **17** $\frac{24}{5}x^{5/3} - \frac{15}{2}x^{2/3} + C$

19 $\frac{10}{9}x^{9/5} + C$ **21** $\frac{1}{3}x^3 + \frac{1}{2}x^2 + x + C \; (x \neq 1)$

23 $f(x) = 4x^3 - 3x^2 + x + 3$

25 $f(x) = \frac{2}{3}x^3 - \frac{1}{2}x^2 - 8x + \frac{65}{6}$

27 $f(t) = \frac{8}{5}t^{5/2} + t^2 - 4t + \frac{27}{5}$

29 $s(t) = t^2 - t^3 - 5t + 4$

31 $s(t) = -16t^2 + 80t + 240$

33 $s(t) = -16t^2 + 1600t$. Maximum height is $s(50) = 40{,}000$.

35   (a) $s(t) = -16t^2 - 16t + 96$ (b) $t = 2$ (c) 80 ft/sec

39 $a(t) = 10$ ft/sec^2 **41** $F = \frac{9}{5}C + 32$

43 $V = 2t^{3/2} + \frac{1}{8}t^2 + 2$

Exercises 4.10, page 196

1   (a) 806

(b) $c(x) = (800/x) + 0.04 + 0.0002x$;

$C'(x) = 0.04 + 0.0004x$; $c(100) = 8.06$;

$C'(100) = 0.08$

(c) 0.84   (d) $x = 0$

3   (a) 11,250

(b) $c(x) = (250/x) + 100 + 0.001x^2$;

$C'(x) = 100 + 0.003x^2$; $c(100) = 112.50$;

$C'(100) = 130$

(c) 107.5   (d) $x = 0$

5   16

Answers for 7–13 indicate functional values at x.

7 $\frac{800}{3} - \frac{4}{3}x$; $-\frac{4}{3}$; $\frac{800}{3}x - \frac{4}{3}x^2$; $\frac{800}{3} - \frac{8}{3}x$;

100 units at 133.33

9 $4 - \sqrt{x}$; $-1/2\sqrt{x}$; $4x - x^{3/2}$; $4 - \frac{3}{2}\sqrt{x}$; 7 units at 1.35

11   (a) $(800 - 3x)/\sqrt{400 - x}$ (b) 6158.39   (c) $2\sqrt{400 - x}$

13   (a) $300 - 3x^2$ (b) 2000   (c) $300 - x^2$

15   (a) $-\frac{1}{10}$ (b) $50x - \frac{1}{10}x^2$ (c) $48x - \frac{1}{10}x^2 - 10$

(d) $48 - \frac{1}{5}x$ (e) 5750   (f) 2

17   (a) $1800x - 2x^2$ (b) $1799x - 2.01x^2 - 1000$ (c) 100

(d) \$158,800

19   3990 units; \$15,420.10

21 $p(x) = 150 - \frac{1}{10}x$; $R(x) = 150x - \frac{1}{10}x^2$

25 $C(x) = 20x - (0.015/2)x^2 + 5.0075$; 986.26

27 $R(x) = \frac{1}{3}x^3 - 3x^2 + 15x$; $g'(x) = \frac{2}{3}x - 3$

Exercises 4.11, page 198

1   Tangent: $y + 3 = -\frac{30}{23}(x - 2)$;

normal: $y + 3 = \frac{23}{30}(x - 2)$

3   Horizontal at $x = -\frac{3}{2}$; vertical at $x = -1$, $x = -2$

5 $(\sqrt{61} - 1)/3$

7   Local maximum $f(0) = 1$; increasing on $(-\infty, 0]$, decreasing on $[0, \infty)$

9   Local maximum $f((4 + \sqrt{7})/3)$ and local minimum $f((4 - \sqrt{7})/3)$; increasing on $[(4 - \sqrt{7})/3, (4 + \sqrt{7})/3]$, decreasing on $(-\infty, (4 - \sqrt{7})/3]$ and $[(4 + \sqrt{7})/3, \infty)$; x-coordinate of PI is $x = \frac{4}{3}$; CU on $(-\infty, \frac{4}{3})$ and CD on $(\frac{4}{3}, \infty)$.

7   **9**

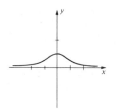   

11   Local maximum $f(\sqrt[3]{20}) = 400$; increasing on $(-\infty, \sqrt[3]{20}]$, decreasing on $[\sqrt[3]{20}, \infty)$; x-coordinates of PIs are 0 and 2; CU on $(0, 2)$; CD on $(-\infty, 0)$ and $(2, \infty)$. (see graph)

13   Local max: $f(\sqrt{2}) = f(-\sqrt{2}) = 4\sqrt{2}$;

local min: $f(-\sqrt{6}) = f(0) = f(\sqrt{6}) = 0$;

CD on $(-\sqrt{6}, 0)$ and $(0, \sqrt{6})$; CU on $(-\infty, -\sqrt{6})$ and $(\sqrt{6}, \infty)$. (see graph)

11   **13**

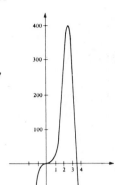
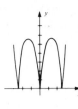

15 $v(t) = 12t^3 - 12t^2 - 24t$; $a(t) = 36t^2 - 24t - 24$; to the left in $[-3, -1)$ and $(0, 2)$; to the right in $(-1, 0)$ and $(2, 3)]$.

17 $s(t) = -16t^2 - 30t + 900$; $v(5) = -190$ ft/sec; $t = 15(-1 + \sqrt{65})/16 \approx 6.6$ sec

19 $(x^2/2)(x^4 + x^2 - 1) + C$

21 $\frac{10}{13}x^{13/5} - \frac{10}{21}x^{21/10} + C$ **23** C

25 125 yd by 250 yd   **27** $\sqrt{2}a$

29 Radius of semicircle is $1/8\pi$ mi, length of rectangle is $\frac{1}{8}$ mi.

31 $-18/5\pi \approx -1.15$ ft/min   **33** $20\sqrt{5} \approx 44.7$ mi/hr

35 $dp/dv = -p/v$ **37** $\frac{3}{2}$ **39** 0   **41** $-\infty$

43 $-\infty$

45 $y = \frac{1}{3}, x = \frac{5}{3}, x = -\frac{5}{3}$ **47**

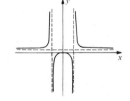
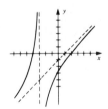

CHAPTER 5

Exercises 5.1, page 208

1 -5 **3** 34   **5** 40   **7** 510   **9** 500

13 $\frac{1}{3}(n^3 + 6n^2 + 20n)$ **15** $\frac{1}{3}(4n^3 - 12n^2 + 11n)$

17 28   **19** $\frac{125}{3}$ **21** 78   **23** 18   **25** $\frac{19}{4}$

Exercises 5.2, page 215

1 1.1, 1.5, 1.1, 0.4, 0.9; $\|P\| = 1.5$

3 0.3, 1.7, 1.4, 0.5, 0.1; $\|P\| = 1.7$

5 (a) 40 (b) 32 (c) 36   **7** $\frac{49}{4}$ **9** 79

11 $\int_{-1}^2 (3x^2 - 2x + 5)\, dx$ **13** $\int_0^4 2\pi x(1 + x^3)\, dx$

15 $-\frac{14}{3}$ **17** 30   **19** 25   **21** $9\pi/4$ **23** $\frac{625}{4}$

25 Any unbounded function. For example, $f(x) = 1/x$, $g(x) = 1/\sqrt{1 - x}$, $h(x) = \csc x$. There is no contradiction since the interval in (5.11) is closed.

Exercises 5.3, page 220

1 30   **3** -12 **5** 2   **7** 78   **9** $-\frac{291}{2}$

11 $-14\sqrt{5}/3$ **13** $\frac{215}{6}$ **19** $\int_{-3}^1 f(x)\, dx$

21 $\int_h^{c+h} f(x)\, dx$ **29** Hint: $-|f(x)| \le f(x) \le |f(x)|$

Exercises 5.4, page 223

1 $\sqrt{3}$ **3** 3   **5** $\sqrt[3]{15/4}$ **7** $\sqrt{50} - 3$

9 $a\sqrt{16 - \pi^2/4}$

11 If $f(x) = k$ on $[a, b]$, then $\int_a^b f(x)\, dx = k(b - a) = f(c)(b - a)$ for every c in $[a, b]$, since $f(c) = k$.

Exercises 5.5, page 229

1 -18 **3** $\frac{265}{2}$ **5** 5   **7** $\frac{31}{256}$ **9** $\frac{20}{3}$

11 $\frac{352}{5}$ **13** $-\frac{37}{6}$ **15** $\frac{13}{3}$ **37** $-\frac{7}{2}$ **19** 0

21 $\frac{10}{3}$ **23** $\frac{53}{2}$ **25** $8\sqrt{3} + 16$ **29** $1/(x + 1)$

31 6   **37** $z = \frac{16}{9}$ **39** $\sqrt[3]{\frac{5}{4}}$ **43** $\frac{3}{2}$

47 $4x^7/\sqrt{x^{12} + 2}$ **49** $3x^2(x^9 + 1)^{10} - 3(27x^3 + 1)^{10}$

Exercises 5.6, page 237

1 $\frac{1}{15}(3x + 1)^5 + C$ **3** $\frac{2}{9}(t^3 - 1)^{3/2} + C$

5 $-1/(4(x^2 - 4x + 3)^2) + C$ **7** $-\frac{3}{8}(1 - 2s^2)^{2/3} + C$

9 $\frac{2}{5}(\sqrt{u} + 3)^5 + C$ **11** $\frac{14}{3}$ **13** 0   **15** $\frac{1}{3}$

17 $\frac{1}{7}x^7 + \frac{3}{5}x^5 + x^3 + x + C$ **19** $\frac{5}{19}(8x + 5)^{3/2} + C$

21 $\frac{5}{36}$

23 (a) $\frac{1}{3}(x + 4)^3 + C_1$
(b) $\frac{1}{3}x^3 + 4x^2 + 16x + C_2$, where $C_2 = C_1 + \frac{64}{3}$

25 (a) $\frac{2}{3}(\sqrt{x} + 3)^3 + C_1$
(b) $\frac{2}{3}x^{3/2} + 6x + 18x^{1/2} + C_2$, $18 + C_1 = C_2$

29 $1/\sqrt{x^3 + x + 5}$ **31** 1   **33** $\frac{14}{3}$ **35** $z = \sqrt{3}$

37 $\frac{544}{225}$

Exercises 5.7, page 245

1 (a) 1.41 (b) 1.39   **3** (a) 0.88 (b) 0.88

5 (a) 0.39 (b) 0.39   **7** (a) 3.35 (b) 3.35

9 1.38   **11** 0.88   **15** (a) 8.65 (b) 8.59

17 (a) 127.5 (b) 131.7 (c) 128   **19** (a) 41 (b) 7

Exercises 5.8, page 247

1 70   **3** $\frac{11}{4}$ **5** -10 **7** $\frac{3}{5}$ **9** $\frac{1}{6}$

11 $-\frac{1}{16}(1 - 2x^2)^4 + C$ **13** $\sqrt{8} - \sqrt{3} \approx 1.10$

15 $-2/(1 + \sqrt{x}) + C$ **17** $3x - x^2 - \frac{5}{4}x^4 + C$

19 $\frac{52}{9}$ **21** $\frac{1}{6}(4t^2 + 2t - 7)^3 + C$

23 $-x^{-2} + 3x^{-1} + C$ **25** $\sqrt{3}/2$

27 $\sqrt[5]{y^4 + 2y^2 + 1} + C$ **29** 0

31 (a) 341.36 (b) 334.42   **33** $\frac{15}{136}$

CHAPTER 6

Exercises 6.1, page 256

1 $\frac{17}{6}$ **3** $\frac{33}{2}$ **5** $\frac{32}{3}$

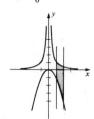 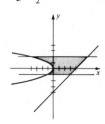 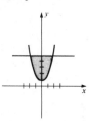

7 $\frac{32}{3}$ **9** $\frac{9}{2}$ **11** $8\sqrt{3}$

 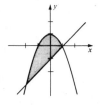 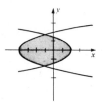

13 2   **15** $\frac{1}{2}$ **17** 8

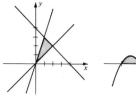 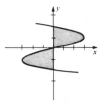

19 $\frac{16}{3}$ **21**

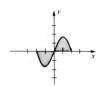 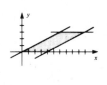

21 If $g(y) = (2y + 4) - 2y = 4$ on $[0, 3]$, then
$A = \lim_{\|P\| \to 0} \sum_i 4 \Delta y_i$.
(a) $A = \int_0^3 4 \, dy = 12$
(b) The figure is a parallelogram with altitude 3 and base 4, and hence $A = (3)(4) = 12$.

23 Let $f(x) = \sqrt{9 - (x - 4)^2}$ on $[1, 7]$. Then
$A = \lim_{\|P\| \to 0} \sum_i 2f(w_i) \Delta x_i$. Since the region is bounded by a circle of radius 3, $A = 9\pi$.

25 The limit equals the area of the region under the graph of $y = 4x + 1$ from 0 to 1. $A = 3$.

27 The limit equals the area of the region to the left of the graph of $x = 4 - y^2$ and to the right of the y-axis from $y = 0$ to $y = 1$. $A = \frac{11}{3}$.

29 The area A of $\{(x, y): 2 \le x \le 5, 0 \le y \le x(x^2 + 1)^{-2}\}$, $A = \frac{21}{260}$

31 The area A of $\{(x, y): 1 \le y \le 4, 0 \le x \le (5 + \sqrt{y})/\sqrt{y}\}$, $A = 13$

33 9   **35** 12   **37** (a) 4.25   (b) 4.50

Exercises 6.2, page 264

1 $2\pi/3$ **3** 2π **5** $512\pi/15$

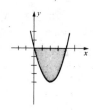 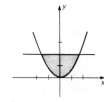

7 $64\pi/15$ **9** $64\sqrt{2}\pi/3$ **11** $72\pi/5$

 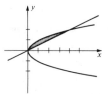 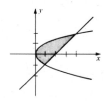

13 $576\pi/7$ **15** $264\pi/5$

17 (a) $512\pi/15$ (b) $832\pi/15$ (c) $128\pi/3$

19 $V = \pi \int_{-2}^0 [(8 - 4x)^2 - (8 - x^3)^2] \, dx + \pi \int_0^2 [(8 - x^3)^2 - (8 - 4x)^2] \, dx$

21 $V = \pi \int_2^3 [(y - 1)^2 - (2 - \sqrt{3 - y})^2] \, dy$

23 $V = \pi \int_{-1}^1 [(5 + \sqrt{1 - y^2})^2 - (5 - \sqrt{1 - y^2})^2] \, dy$
$\quad = 20\pi \int_{-1}^1 \sqrt{1 - y^2} \, dy$

25 $V = \frac{1}{3}\pi r^2 h$ **27** $V = \frac{1}{3}\pi h(r_1^2 + r_2^2 + r_1 r_2)$

29 The limit equals the volume of the solid obtained by revolving the region between $y = x^2$ and $y = x^3$, $0 \le x \le 1$, about the x-axis. $V = 2\pi/35$.

Exercises 6.3, page 269

1 $128\pi/5$ **3** $24\pi/5$ **5** $135\pi/2$

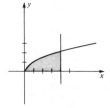 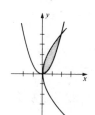 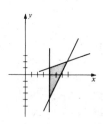

7 $512\pi/5$ **9** 72π **11** $64\pi/3$

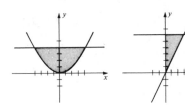

13 (a) 16π (b) $64\pi/3$
15 (a) $512\pi/15$ (b) $832\pi/15$ (c) $128\pi/3$
17 $V = 2\pi \int_{-8}^{0} (8 - y)(y/4 - y^{1/3}) \, dy +$
$\qquad 2\pi \int_{0}^{8} (8 - y)(y^{1/3} - y/4) \, dy$
19 $V = 2\pi \int_{0}^{1} (2 - x)(x - x^2) \, dx$
21 $V = 4\pi \int_{-1}^{1} (5 - x)\sqrt{1 - x^2} \, dx$ **23** $V = \frac{1}{3}\pi r^2 h$
25 $V = \frac{1}{3}\pi h(r_1^2 + r_2^2 + r_1 r_2)$
27 The limit of the sum equals the volume of the solid
obtained by revolving the region between $y = x$ and
$y = x^2$, $0 \le x \le 1$, about the y-axis. The volume is $\pi/6$.

Exercises 6.4, page 272

1 $16a^3/3$ **3** $\frac{128}{15}$ **5** $2a^2h/3$ **7** $128\pi/15$
9 $2a^3/3$ **11** $\pi a^2 b/2$ **13** 4 **15** $\pi a^3/24$

Exercises 6.5, page 279

1 (a) $\frac{128}{3}$ in.-lb (b) $\frac{64}{3}$ in.-lb **3** $F_2 = 3F_1$
5 2250 ft-lb **7** 44,660 joules
9 (a) $81(62.5)\pi/2 \approx 7952$ ft-lb
$\qquad$ (b) $(189)(62.5)\pi/2 \approx 18,555$ ft-lb
11 500 ft-lb
13 (a) $3k/10$ ergs (b) $9k/40$ ergs (k a constant)
15 276 ft-lb **17** $575(\frac{1}{2} - 1/\sqrt[5]{40}) \approx 12.55$ in.-lb
19 $W = gm_1 m_2 h/4000(4000 + h)$ **21** 36.85 ft-lb

Exercises 6.6, page 284

1 (a) 31.25 lb (b) 93.75 lb
3 (a) $62.5/\sqrt{3}$ lb (b) $62.5\sqrt{3}/24$ lb **5** 320 lb
7 303,356.25 lb **9** $(592)(62.5)/3$ lb $\approx 12,333.3$ lb
11 $\frac{3200}{3}$ lb **13** 4500 lb **15** (a) 1516 (b) 1614.6

Exercises 6.7, page 290

1 $(4 + \frac{16}{81})^{3/2} - (1 + \frac{16}{81})^{3/2} \approx 7.29$
3 $\frac{8}{27}(10^{3/2} - (13^{3/2}/8)) \approx 7.63$ **7** $\frac{13}{12}$ **9** $\frac{353}{240}$
11 $s = \int_{0}^{2} \sqrt{\frac{53}{4} - 21y^2 + 9y^4} \, dy$ **13** 6
15 $s(x) = (x^{2/3} + \frac{4}{9})^{3/2} - (1 + \frac{4}{9})^{3/2}$;
$\qquad \Delta s = \frac{1}{27}[(9(1.1)^{2/3} + 4)^{3/2} - 13^{3/2}] \approx 0.1196$;
$\qquad ds = \sqrt{13/30} \approx 0.1202$
17 $ds = \sqrt{17}(0.1) \approx 0.412$, $d(A, B) = \sqrt{0.1781} \approx 0.422$
19 1.44 **21** 8.61

Exercises 6.8, page 296

1 (a) \$49.54 (b) \$96.30 (c) \$137.50 (d) \$170.37
3 $10^{3/5}$ yr ≈ 3.98 yr ≈ 4 yr (to the nearest month)
5 (a) $9((601)^{2/3} - 1) \approx 632$ min
$\qquad$ (b) $9((301)^{2/3} - 1) \approx 395$ min
7 In minutes: (a) 18.16 (b) 66.22 (c) 115.24 (d) 197.12
9 $(27 - 5\sqrt{5})/3$ gal **11** 11 (by the Trapezoidal Rule)

Exercises 6.9, page 298

1 $\frac{64}{3}$ **3** $5\sqrt{5}/6$ **5** $\frac{253}{12}$

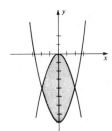

7 10π **9** $3\pi/5$

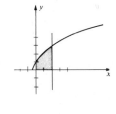

11 (a) $1152\pi/5$ (b) 54π (c) $1728\pi/5$
13 $(37^{3/2} - 10^{3/2})/27 \approx 7.16$
15 $432(62.5)\pi$ ft-lb $\approx 84,823$ ft-lb **17** 6,000 lb **19** $\pi/5$
21 Two possibilities exist. The solid could be obtained by
revolving the region under $y = x^2$ from $x = 0$ to $x = 1$
around the x-axis, or by revolving the region under
$y = \frac{1}{2}x^3$ from $x = 0$ to $x = 1$ around the y-axis.

CHAPTER 7

Exercises 7.1, page 302

5 $f^{-1}(x) = \frac{1}{11}(x - 8)$
7 $f^{-1}(x) = \sqrt{6 - x}$, where $0 \le x \le 6$
9 $f^{-1}(x) = \frac{1}{7}(x^2 + 2)$, where $x \ge 0$
11 $f^{-1}(x) = \sqrt[3]{(7 - x)/3}$

13 $f^{-1}(x) = (x^{1/5} - 8)^{1/3}$

15

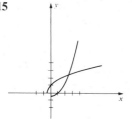

17 $f^{-1}(x) = (x - b)/a$; a constant function has no inverse. The identity function is its own inverse.

19 If $f(x) = x^2$, then f is not one-to-one and hence has no inverse function.

Exercises 7.2, page 310

1 $9/(9x + 4)$ **3** $-15/(2 - 3x)$

5 $-3x^2/(7 - 2x^3)$ **7** $(6x - 2)/(3x^2 - 2x + 1)$

9 $(8x + 7)/3(4x^2 + 7x)$ **11** $1 + \ln x$

13 $\dfrac{1}{2x}\left(1 + \dfrac{1}{\sqrt{\ln x}}\right)$ **15** $-\dfrac{1}{x}\left(\dfrac{1}{(\ln x)^2} + 1\right)$

17 $\dfrac{20}{5x - 7} + \dfrac{6}{2x + 3}$ **19** $\dfrac{x}{x^2 + 1} - \dfrac{18}{9x - 4}$

21 $\dfrac{2x}{3(x^2 - 1)} - \dfrac{2x}{3(x^2 + 1)}$ **23** $\dfrac{1}{\sqrt{x^2 - 1}}$

25 $[x/(x^2 + 1)] \ln(x^2 + 1)$ **27** $x/(x^2 + 1)\sqrt{\ln(x^2 + 1)}$

29 $\dfrac{(2x^2 - 1)y}{x(3y + 1)}$ **31** $\dfrac{y^2 - xy \ln y}{x^2 - xy \ln x}$

33 $1/(x + y - 1)$ **35** $y = 8x - 15$

37 The graphs coincide if $x > 0$; however, the graph of $y = \ln(x^2)$ contains points with negative x-coordinates.

39 $(1, 1), (2, 4 + 4 \ln 2)$ **41** 0.6982

43 $v(t) = 2t - 4/(t - 1)$; $a(t) = 2 + 4/(t + 1)^2$. Moves to the left in $[0, 1)$ and to the right in $(1, 4]$.

45 $1, \frac{1}{5}, \frac{1}{10}, \frac{1}{100}, \frac{1}{1000}$; the slope approaches 0; the slope increases without bound.

Exercises 7.3, page 318

1 $-5e^{-5x}$ **3** $6xe^{3x^2}$ **5** $e^{2x}/\sqrt{1 + e^{2x}}$

7 $e^{\sqrt{x+1}}/(2\sqrt{x + 1})$ **9** $(-2x^2 + 2x)e^{-2x}$

11 $\dfrac{e^x(x^2 + 1) - 2xe^x}{(x^2 + 1)^2}$ or $\dfrac{e^x(x - 1)^2}{(x^2 + 1)^2}$

13 $12(e^{4x} - 5)^2 e^{4x}$ **15** $-e^{1/x}/x^2 - e^{-x}$

17 $\dfrac{(e^x + e^{-x})^2 - (e^x - e^{-x})^2}{(e^x + e^{-x})^2}$ or $4/(e^x + e^{-x})^2$.

19 $e^{-2x}\left(\dfrac{1}{x} - 2 \ln x\right)$ **21** $\dfrac{e^x}{e^x + 1} - \dfrac{e^x}{e^x - 1}$

23 $\dfrac{e^{2x} - e^{-2x}}{(e^{2x} + e^{-2x})\sqrt{\ln(e^{2x} + e^{-2x})}}$ **25** $\dfrac{3x^2 - ye^{xy}}{xe^{xy} + 6y}$

27 $\dfrac{6x - e^y}{3y^2 + xe^y}$ **29** $y = (e + 3)x - e - 1$

31 $c = \ln((e^b - e^a)/(b - a))$ **33** $(\frac{1}{2}, e)$

35 Min: $f(-1) = -e^{-1}$; decreasing on $(-\infty, -1)$, increasing on $(-1, \infty)$; CU on $(-2, \infty)$, CD on $(-\infty, -2)$; PI at $(-2, -2e^{-2})$

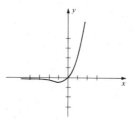

37 No local extrema; decreasing on $(-\infty, \infty)$; CU on $(-\infty, \infty)$; no PI

39 Min: $f(1/e) = -1/e$; decreasing on $(0, 1/e]$, increasing on $[1/e, \infty)$; CU on $(0, \infty)$; no PI

37   **39**

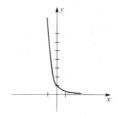

43 (a) $(\ln a - \ln b)/(a - b)$ (b) $\lim_{t \to \infty} C(t) = 0$

45 $0.03e \approx 0.082, 2.800$ **47** $1, 2$ **49** $1, 2, -2$

53 Max: $1/\sigma\sqrt{2\pi}$; increasing on $(-\infty, \mu]$, decreasing on $[\mu, \infty)$; CU on $(-\infty, \mu - \sigma]$ and $[\mu + \sigma, \infty)$, CD on $[\mu - \sigma, \mu + \sigma]$; PI at $(\mu \pm \sigma, c)$ where $c = e^{-1/2}/\sigma\sqrt{2\pi}$; $\lim_{x \to \infty} f(x) = 0$, $\lim_{x \to -\infty} f(x) = 0$

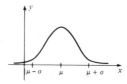

57 $r/R = e^{-1/2} \approx 0.607$

Exercises 7.4, page 325

1 $\frac{1}{2} \ln(x^2 + 1) + C$ **3** $-\frac{1}{5} \ln |7 - 5x| + C$

5 $\frac{1}{2} \ln |x^2 - 4x + 9| + C$ **7** $\frac{1}{3} \ln |x^3 + 1| + C$

9 $\frac{1}{2}(\ln 9 - \ln 3)$, or $\ln \sqrt{3}$

11 $4(\ln 6 - \ln 5)$, or $4 \ln \frac{6}{5}$ **13** $\frac{1}{2}x^2 + \frac{1}{5}e^{5x} + C$

15 $\frac{1}{2}(\ln x)^2 + C$ **17** $-\frac{1}{4}(e^{-12} - e^{-4})$

19 $2e^{\sqrt{x}} + C$ **21** $e^x + 2x - e^{-x} + C$

23 $\ln(e^x + e^{-x}) + C$ **25** $-1/(x + 1) + C$

27 $x^2 + x - 4 \ln|x - 3| + C$ **29** 4
31 $\ln 2 + e^{-2} - e^{-1} \approx 0.46$ **33** $\pi(1 - e^{-1})$
35 $(1 - e^{-6})/3$
37 (a) $\int_0^1 \sqrt{1 + e^{2x}}\, dx$
  (b) If $f(x) = \sqrt{1 + e^{2x}}$, then $L \approx \frac{1}{10}[f(0) + 2f(0.2) + 2f(0.4) + 2f(0.6) + 2f(0.8) + f(1)]$
  (c) $L \approx 2.0096$
39 $(5x + 2)^2(6x + 1)(150x + 39)$
41 $\dfrac{(19x^2 + 20x - 3)(x^2 + 3)^4}{2(x + 1)^{3/2}}$
43 $\sqrt[3]{2x + 1}\,(4x - 1)^2(3x + 5)^4\left[\dfrac{2}{3(2x + 1)} + \dfrac{8}{(4x - 1)} + \dfrac{12}{(3x + 5)}\right]$
45 $\left[\dfrac{3x}{3x^2 + 2} + \dfrac{3}{2(6x - 7)}\right]\sqrt{(3x^2 + 2)}\sqrt{6x - 7}$
47 $-2/(3 - 2x)$ **49** $6e^{-2x}/(1 - e^{-2x})$

Exercises 7.5, page 332

1 $7^x \ln 7$ **3** $8^{x^2 + 1}(2x \ln 8)$
5 $(4x^3 + 6x)/\ln 10(x^4 + 3x^2 + 1)$ **7** $5^{3x - 4}3 \ln 5$
9 $\dfrac{-(x^2 + 1)10^{1/x}\ln 10}{x^2} + 2x10^{1/x}$ **11** $\dfrac{2x^3(\ln 7)7^{\sqrt{x^4 + 9}}}{\sqrt{x^4 + 9}}$
13 $\dfrac{30x}{(3x^2 + 2)\ln 10}$ **15** $\left(\dfrac{6}{6x + 4} - \dfrac{2}{2x - 3}\right)/\ln 5$
17 $1/(x \ln x \ln 10)$ **19** $ex^{e - 1} + e^x$
21 $(x + 1)^x\left(\dfrac{x}{x + 1} + \ln(x + 1)\right)$
23 $(1/(3 \ln 10))10^{3x} + C$ **25** $(-1/(2 \ln 3))3^{-x^2} + C$
27 $(1/\ln 2)\ln(2^x + 1) + C$ **29** $24/1250 \ln 5$
31 $2^{x^3}/(3 \ln 2) + C$ **33** $2x + (3^{2x} - 3^{-2x})/2 \ln 3 + C$
35 $\ln 10 \ln|\log x| + C$ **37** $1/\ln 2 - \frac{1}{2}$
39 Tangent: $y - 3 = 3(1 + \ln 3)(x - 1)$;
  normal: $y - 3 = -(3(1 + \ln 3))^{-1}(x - 1)$
43 pH ≈ 2.201; $d(\text{pH}) \approx \pm 0.002$
45 (a) $1/I_0 \ln 10$ (b) $1/100I_0 \ln 10$ (c) $1/1000I_0 \ln 10$
47 They are symmetric with respect to the line $y = x$.

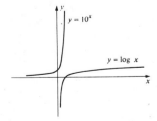

Exercises 7.6, page 341

7 $y = mx$ **9** $y = -1 + Ce^{(x^2/2) - x}$
11 $y = -\frac{1}{3}\ln 3(C + e^{-x})$ **13** $y = Ce^{x - (1/x)}$

15 $y^2 + \ln y = 3x - 8$ **17** $y = \ln(2x + \ln x + e^2 - 2)$
19 $q(t) = 5000(3)^{t/10}$; 45,000; $10 \ln 10/\ln 3 \approx 21$ hr
21 $30(\frac{29}{30})^5 \approx 25.3$
23 Approximately 109.24 yr after Jan. 1, 1980 (March 29, 2089)
25 $5 \ln 6/\ln 3 \approx 8.2$ min
27 We must determine i such that at $t = 0$, $Ri + L(di/dt) = 0$ and $i = I$. Proceeding in a manner similar to the solution of Example 8 we obtain $(1/i)di = (-R/L)dt$; $\ln i = (-R/L)t + C$ and $i = e^C e^{(-R/L)t}$. Since $i = I$ at $t = 0$, $i = Ie^{-Rt/L}$.
29 Write $P(1 + r/m)^{mt} = P((1 + r/m)^{m/r})^{rt}$. If we let $h = r/m$, then $h \to 0$ as $m \to \infty$ and hence
$$\lim_{m \to \infty} P((1 + r/m)^{m/r})^{rt} = \lim_{h \to 0} P((1 + h)^{1/h})^{rt} = Pe^{rt}.$$
35 (a) $(\ln 2)/100$ (b) $(\ln 2)/1000$ (c) $(\ln 2)/c$
37 \$14,310.84   **39** 683.3 mg   **41** 13,235 yr

Exercises 7.7, page 348

1 $y = \frac{1}{4}e^{2x} + Ce^{-2x}$ **3** $y = \frac{1}{2}x^5 + Cx^3$
5 $y = (1/x)e^x - \frac{1}{2}x + C/x$ **7** $y = (e^x + C)/x^2$
9 $y = (\frac{1}{3}x + C/x^2)e^{-3x}$ **11** $y = (x + C)e^{5x}$
13 $y = x(x + \ln x + 1)$ **15** $y = e^{-x}(1 - x^{-1})$
17 $Q = CE(1 - e^{-t/RC})$
19 $f(t) = \frac{80}{3}(1 - e^{-0.075t}) + Ke^{-0.075}$ lb
21 (a) $f(t) = M + (A - M)e^{k(1 - t)}$ where k is a constant.
  (b) 28

Exercises 7.8, page 353

1 $(x^2 - 3)/2$, $[\sqrt{5}, 5]$; x
3 $\sqrt{4 - x}$, $[-45, 4]$; $-1/2\sqrt{4 - x}$
5 $1/x$, $(0, \infty)$; $-1/x^2$
7 $\sqrt{-\ln x}$, $(0, 1]$; $-1/2x\sqrt{-\ln x}$
9 $\ln(x + \sqrt{x^2 + 4}) - \ln 2$; $\mathbb{R}$; $1/\sqrt{x^2 + 4}$
11 f is increasing since $f'(x) = 5x^4 + 9x^2 + 2 \geq 2 > 0$.
  Slope is $\frac{1}{16}$.
13 f is increasing since $f'(x) = 4e^{2x}/(e^{2x} + 1)^2 > 0$.
  Slope is 1.
15 Since f decreases on $(-\infty, 0]$ and increases on $[0, \infty)$, there is no inverse function. If the domain is restricted to a subset of one of these intervals, then f will have an inverse function.
17 f increases on $(-\infty, 0]$ and decreases on $[0, \infty)$. If the domain is restricted to a subset of one of these intervals, then f^{-1} will exist.

Exercises 7.9, page 353

1 $-2(1 + \ln|1 - 2x|)$
3 $12/(3x + 2) + 3/(6x - 5) - 8/(8x - 7)$
5 $-4x/(2x^2 + 3)[\ln(2x^2 + 3)]^2$ **7** $2x$
9 $(\ln 10)10^x \log x + 10^x/x \ln 10$ **11** $(1/x)(2 \ln x)x^{\ln x}$
13 $2x(1 - x^2)e^{1 - x^2}$
15 $2^{-1/x}[(x^3 + 4)\ln 2 - 3x^4]/x^2(x^3 + 4)^2$

17 $e(1 + \sqrt{x})^{e-1}/2\sqrt{x}$ **19** $(10^{\ln x} \ln 10)/x$

21 $(2x - 5)/(x^2 - 5x - 3)$

23 $(1/x)(1 + \ln(\ln x))(\ln x)^{\ln x}$

25 $y(e^{xy} - 1)/x(1 - e^{xy})$

29 $-\frac{1}{2}e^{-2x} - 2e^{-x} + x + C$ **31** $2(e^{-1} - e^{-2})$

33 $-\ln|1 - \ln x| + C$

35 $\frac{1}{6}x^2 - \frac{2}{9}x + \frac{4}{27}\ln|3x + 2| + C$ **37** $3/(2\ln 4)$

39 $2 \ln 10 \sqrt{\log x} + C$

41 $\frac{1}{2}x^2 - x + 2\ln|x + 1| + C$

43 $(5e)^x/(1 + \ln 5) + C$ **45** $x^{e+1}/(e + 1) + C$

47 $\ln|x^3 + 1| + y^3 = C$ **49** $y = \frac{1}{3} + Ce^{3/x}$

51 $4 \ln|y^3 + 1| - \ln|1 - x^2| = 8 \ln 3$

53 $4e^2 + 12$ cm   **55** $y - e = -2(1 + e)(x - 1)$

57 $(\pi/8)(e^{-16} - e^{-24})$

59 (a) $f'(x) > 0$, so that f is increasing and f^{-1} exists with
domain $[4, \infty)$.
(b) $\ln(\sqrt{x} - 1)$, $1/2\sqrt{x}(\sqrt{x} - 1)$ (c) $4, \frac{1}{4}$

61 The amount in solution at any time t hours past 1:00 P.M.
is $10(1 - 2^{-t/3})$.
(a) 2.21 hours (approximately 6:14 P.M.)
(b) $10(1 - 2^{-7/3}) \approx 8.016$ lb

63 6,400,000

CHAPTER 8

Exercises 8.1, page 360

1 1   **3** $\frac{1}{8}$ **5** $\frac{2}{3}$ **7** 0   **9** $-\frac{3}{4}$ **11** 0

13 7   **15** 1   **17** 0   **19** 2   **21** 1   **23** 1

25 -1

Exercises 8.2, page 369

3 $2x \cos(x^2 + 2)$ **5** $-15 \cos^4 3x \sin 3x$

7 $x \sin(1/x) + 3x^2 \cos(1/x)$

9 $8 \sec^2(8x + 3)$

11 $(1/2\sqrt{x} - 1) \sec(\sqrt{x} - 1) \tan(\sqrt{x} - 1)$

13 $-(3x^2 - 2) \csc^2(x^3 - 2x)$

15 $-6x \sin 3x^2 - 6 \cos 3x \sin 3x$

17 $-4 \csc^2 2x \cot 2x$

19 $-5x^2 \csc 5x \cot 5x + 2x \csc 5x$

21 $3 \tan^3 x \sec^3 x + 2 \tan x \sec^5 x$

23 $5(\sin 5x - \cos 5x)^4(5 \cos 5x + 5 \sin 5x)$

25 $-9 \cot^2(3x + 1) \csc^2(3x + 1)$

27 $\dfrac{-4(1 - \sin 4x)\sin 4x + 4 \cos^2 4x}{(1 - \sin 4x)^2}$ or $\dfrac{4}{1 - \sin 4x}$

29 $\dfrac{-\csc x \cot x - \csc^2 x}{\csc x + \cot x}$ or $-\csc x$

31 $\dfrac{e^{-3x} \sec^2 \sqrt{x}}{2\sqrt{x}} - 3e^{-3x} \tan \sqrt{x}$ **33** $\dfrac{2 \tan 2x}{\ln \sec 2x}$

35 $6 \tan^2 2x \sec^2 2x - 6 \tan 2x \sec^3 2x$

37 $-3 \csc 3x[(x^3 + 1) \cot 3x + x^2]/(x^3 + 1)^2$

39 $x^{\cot x}(\cot x/x - \csc^2 x \ln x)$

41 $y' = 6 \sec^2 3x \tan 3x$, $y'' = 18 \sec^4 3x + 36 \sec^2 3x \tan^2 3x$

43 $y' = x \sin x$, $y'' = x \cos x + \sin x$

45 $y' = \dfrac{\sec^2 x}{2\sqrt{\tan x}}$, $y'' = \sec^2 x \sqrt{\tan x} - \dfrac{1}{4} \dfrac{\sec^4 x}{\sqrt{\tan^3 x}}$

47 $\cos y/(2y + x \sin y)$

49 $(e^x \cot y - e^{2y})/(2xe^{2y} + e^x \csc^2 y)$

51 Max: $f(\pi/4) = \sqrt{2}$; min: $f(5\pi/4) = -\sqrt{2}$; increasing on $[0, \pi/4]$ and $[5\pi/4, 2\pi]$; decreasing on $[\pi/4, 5\pi/4]$

53 Min: $f(\pi/3) = (\pi - 3\sqrt{3})/6$; max: $f(5\pi/3) = (5\pi + 3\sqrt{3})/6$; decreasing on $[0, \pi/3]$ and $[5\pi/3, 2\pi]$; increasing on $[\pi/3, 5\pi/3]$

51   **53**

 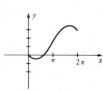

55 Max: $f(\pi/6) = 3\sqrt{3}/2$; min: $f(5\pi/6) = -3\sqrt{3}/2$; increasing on $[0, \pi/6]$ and $[5\pi/6, 2\pi]$; decreasing on $[\pi/6, 5\pi/6]$

57 Max: $f(\pi/4) = -e^{-\pi/4}/\sqrt{2} \approx 0.32$,
$f(\pi/4 + 2\pi) = e^{-9\pi/4}/\sqrt{2} \approx 6 \times 10^{-4}$;
min: $f(5\pi/4) = -e^{-5\pi/4}/\sqrt{2} \approx -0.014$,
$f(5\pi/4 + 2\pi) = e^{-3.25\pi}/\sqrt{2} \approx -3 \times 10^{-5}$

55   **57**

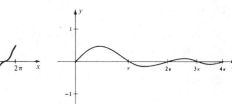

59 Min: $f(0) = -\sqrt{2}$; max: $f(\pi/4) = -1$

61 Tangent: $y - 1 = -3\sqrt{3}(x - \pi/6)$;
normal: $y - 1 = (\sqrt{3}/9)(x - \pi/6)$

63 $\pm\pi/9$ **65** $\frac{2000}{9}\pi \approx 698$ ft/sec   **67** $40\sqrt{3}$ ft

69 -0.31 rad/sec   **71** $dF \approx 0.27$ lb

73 $L = (4/\sin\theta) + (3/\cos\theta)$, where $\tan\theta = \sqrt[3]{4}/3$.
$L = (4^{2/3} + 3^{2/3})^{3/2} \approx 9.87$

75 $2\pi(1 - \sqrt{\frac{2}{3}})$ **77** $\pi/180 \approx 0.2$

79 (a) $\sqrt{3} - 2, 0$ (b) $(0, \pi/6), (\pi/3, 7\pi/6), (4\pi/3, 2\pi)$

81 $5, 8, \frac{1}{8}$. The particle oscillates on a coordinate line between 5 and -5, completing one such oscillation every 8 seconds.

83 $6, 3, \frac{1}{3}$. The particle oscillates on a coordinate line between 6 and -6, completing one such oscillation every 3 seconds.

87 1.4958   **89** 2.71

Exercises 8.3, page 376

1 $\frac{1}{4}\sin 4x + C$ **3** $\frac{1}{3}\sec 3x + C$
5 $\frac{1}{3}(\ln|\sec 3x| + \ln|\sec 3x + \tan 3x|) + C$
7 $\frac{1}{2}\ln|\sec 2x + \tan 2x| + C$
9 $-\frac{1}{2}\cot(x^2 + 1) + C$ **11** $\frac{1}{6}\sin 6x + C$
13 $\frac{1}{4}(\sin 3x)^{4/3} + C$
15 $-\cos x - \frac{4}{3}(\cos x)^{3/2} + \frac{1}{2}\sin^2 x + C$
17 $(1/\cos x) + C = \sec x + C$ **19** $\frac{1}{2}\tan^2 x]_0^{\pi/4} = \frac{1}{2}$
21 $\frac{1}{2}(\ln|\sec 2x + \tan 2x| - \sin 2x) + C$
23 $\frac{1}{2}\sin^2 x]_{\pi/6}^{\pi} = -\frac{1}{8}$ **25** $\ln|x + \cos x| + C$
27 $\tan x + \sec x]_{\pi/4}^{\pi/3} = \sqrt{3} - \sqrt{2} + 1$
29 $e^x + \ln|\sec e^x| + C$ **31** $-e^{\cos x} + C$
33 $\frac{1}{2}\ln|2\tan x + 1| + C$ **35** 2
37 $\ln\left(\dfrac{\sqrt{2}+1}{\sqrt{2}-1}\right) = \ln(3 + 2\sqrt{2}) = 2\ln(1 + \sqrt{2})$

39 $2\pi\sqrt{3}$
45 (a) $L = \frac{1}{2}\int_0^{\pi/2}\sqrt{4 + \sec^2(x/2)\tan^2(x/2)}\,dx$
   (b) If $f(x) = \frac{1}{2}\sqrt{4 + \sec^2(x/2)\tan^2(x/2)}$, then
   $L \approx (\pi/24)[f(0) + 4f(\pi/8) + 2f(\pi/4)$
   $\qquad\qquad\qquad + 4f(3\pi/8) + f(\pi/2)]$
   (c) $L \approx 1.65$
47 (a) $750\int_1^2 (2 - y)\operatorname{arcsec} y\,dy$
   (b) $375\int_{-\pi/3}^{\pi/3}(2 - \sec x)x\sec x\tan x\,dx \approx 227$ ft-lb
49 (a) 2.24 (b) 2.34
51 (b) $q(t) = q_0 e^u$, where $u = (k/2\pi)(1 - \cos 2\pi t)$

Exercises 8.4, page 381

1 (a) $\pi/3$ (b) $-\pi/3$ **3** (a) $\pi/4$ (b) $3\pi/4$
5 (a) $\pi/3$ (b) $-\pi/3$ **7** $\frac{1}{2}$ **9** $\frac{4}{5}$ **11** $\pi - \sqrt{5}$
13 0 **15** Undefined   **17** $-\frac{24}{25}$ **19** $x/\sqrt{x^2 + 1}$
21 $\sqrt{2 + 2x/2}$
31 $\cot^{-1} x = y$ if and only if $\cot y = x$, where $0 < y < \pi$
33   **35**

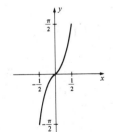   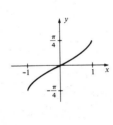

37   **39**

41

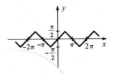

Exercises 8.5, page 387

1 $\dfrac{3}{9x^2 - 30x + 26}$ **3** $\dfrac{1}{2\sqrt{x}\sqrt{1 - x}}$
5 $\dfrac{-e^{-x}}{\sqrt{e^{-2x} - 1}} - e^{-x}\operatorname{arcsec} e^{-x}$
7 $\dfrac{2x^3}{1 + x^4} + 2x\arctan x^2$ **9** $-\dfrac{9(1 + \cos^{-1}3x)^2}{\sqrt{1 - 9x^2}}$
11 $\dfrac{2x}{(1 + x^4)\arctan x^2}$ **13** $\dfrac{-1}{(\sin^{-1}x)^2\sqrt{1 - x^2}}$
15 $\dfrac{x}{(x^2 - 1)\sqrt{x^2 - 2}}$ **17** $\dfrac{1 - 2x\arctan x}{(x^2 + 1)^2}$
19 $(1/2\sqrt{x})\sec^{-1}\sqrt{x} + 1/2\sqrt{x}\sqrt{x - 1}$
21 $(1 - x^6)^{-1/2}(3\ln 3)x^2 3^{\arcsin x^3}$
23 $(\tan x)^{\arctan x}[\cot x\sec^2 x\arctan x + (\ln\tan x)/(1 + x^2)]$
25 $(ye^x - \sin^{-1} y - 2x)/(x/\sqrt{1 - y^2} - e^x)$ **27** $\pi/16$
29 $\pi/12$ **31** $-\arctan(\cos x) + C$
33 $2\arctan\sqrt{x} + C$ **35** $\sin^{-1}(e^x/4) + C$
37 $\frac{1}{6}\sec^{-1}(x^3/2) + C$ **39** $\frac{1}{2}\ln(x^2 + 9) + C$
41 $\frac{1}{5}\sec^{-1}(e^x/5) + C$ **43** $4\pi/3$
45 $\pm\frac{7}{3576} \approx \pm 0.002$ **47** $-\frac{25}{1044}$ rad/sec   **49** $40\sqrt{3}$
51 $f'(x) = \cos x/|\cos x|$; $f''(x) = 0$ if $x \neq (\pi/2) + n\pi$, where n is any integer; local max of $\pi/2$ at $x = (\pi/2) + 2\pi n$; local min of $-\pi/2$ at $x = (3\pi/2) + 2\pi n$; for graph, see Exercise 41 of Section 8.4.
53 Tangent: $y - \pi/6 = (2/\sqrt{3})(x - \frac{3}{2})$;
   normal: $y - \pi/6 = (-\sqrt{3}/2)(x - \frac{3}{2})$
55 CU on $(-\infty, 0)$; CD on $(0, \infty)$
57 $(\pi e^2/2) - (\pi^2/4) - (\pi/2) \approx 7.57$
59 $2\pi/27$ mi/sec ≈ 0.233 mi/sec
63 $l_1 = a - b\cot\theta, l_2 = b\csc\theta; \theta = \arccos(r_2/r_1)^4$

Exercises 8.6, page 393

15 $5\cosh 5x$ **17** $(1/2\sqrt{x})(\sqrt{x}\operatorname{sech}^2\sqrt{x} + \tanh\sqrt{x})$
19 $-2x\operatorname{sech} x^2[(x^2 + 1)\tanh x^2 + 1]/(x^2 + 1)^2$
21 $3x^2\sinh x^3$ **23** $3\cosh^2 x\sinh x$ **25** $2\coth 2x$
27 $-e^{3x}\operatorname{sech} x\tanh x + 3e^{3x}\operatorname{sech} x$
29 $-\operatorname{sech}^2 x/(\tanh x + 1)^2$ **31** $\dfrac{y(e^x - \cosh xy)}{(x\cosh xy - e^x)}$
33 $2\cosh\sqrt{x} + C$ **35** $\ln|\sinh x| + C$
37 $\frac{1}{2}\sinh^2 x + C$ (or $\frac{1}{2}\cosh^2 x + C$, or $\frac{1}{4}\sinh 2x + C$)
39 $-\frac{1}{3}\operatorname{sech} 3x + C$ **41** $\frac{1}{9}\tanh^3 3x + C$
43 $-\frac{1}{2}\ln|1 - 2\tanh x| + C$ **45** $\frac{1}{3}(-1 + \cosh 3)$

47 $(\ln (2 + \sqrt{3}), \sqrt{3}), (\ln (2 - \sqrt{3}), -\sqrt{3})$

51

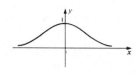

53 Let $A = \frac{1}{2} \sin t \cos t - \int_1^{\cosh t} \sqrt{x^2 - 1}\, dx$ and show that $dA/dt = \frac{1}{2}$.

Exercises 8.7, page 398

7   **9**

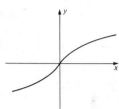   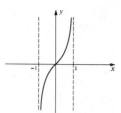

11 $5/\sqrt{25x^2 + 1}$ **13** $1/(2\sqrt{x}\sqrt{x - 1})$

15 $2x/(2x^2 - x^4)$ **17** $-|x|/x\sqrt{x^2 + 1} + \sinh^{-1}(1/x)$

19 $4/(\sqrt{16x^2 - 1}\cosh^{-1} 4x)$ **21** $\frac{1}{3}\sinh^{-1}\frac{3}{5}x + C$

23 $\frac{1}{14}\tanh^{-1}\frac{2}{7}x + C$ **25** $\cosh^{-1}(e^x/4) + C$

27 $-\frac{1}{6}\operatorname{sech}^{-1}(x^2/3) + C$

Exercises 8.8, page 398

1 0   **3** $\frac{2}{3}$ **5** $\frac{3}{5}$

7 $-(6x + 1)(\sin\sqrt{3x^2 + x})/2\sqrt{3x^2 + x}$

9 $5(\sec x + \tan x)^4(\sec x \tan x + \sec^2 x)$

11 $2x \operatorname{arcsec} x^2 + 2x/\sqrt{x^4 - 1}$

13 $\dfrac{12(3x + 7)^3 \sin^{-1} 5x - 5(3x + 7)^4(1 - 25x^2)^{-1/2}}{(\sin^{-1} 5x)^2}$

15 $(\cos x)^{x+1}[\ln \cos x - (x + 1)\tan x]$

17 $(2x - 2\operatorname{sech} 4x \tanh 4x)/\sqrt{2x^2 + \operatorname{sech} 4x}$

19 $-6\cot 2x$ **21** $-(2 + 2\sec^2 x \tan x)/(2x + \sec^2 x)^2$

23 $(-\sin x)e^{\cos x} - e \sin x(\cos x)^{e-1}$

25 $-5e^{-5x}\sinh e^{-5x}$ **27** $\frac{1}{3}x^{-2/3}$

29 $2^{\arctan 2x}(2\ln 2)/(1 + 4x^2)$

31 $-6e^{-2x}\sin^2 e^{-2x}\cos e^{-2x}$

33 $-2xe^{-x^2}(\csc^2 x^2 + \cot x^2)$

35 $[-(\cot x + 1)\csc x \cot x + \csc^2 x(\csc x + 1)]/(\cot x + 1)^2$

37 $-x/\sqrt{x^2(1 - x^2)}$ **39** $3\sec^2(\sin 3x)\cos 3x$

41 $4(\tan x + \tan^{-1} x)^3[\sec^2 x + 1/(1 + x^2)]$

43 $1/(1 + x^2)[1 + (\tan^{-1} x)^2]$

45 $-e^{-x}(e^{-x}\cosh e^{-x} + \sinh e^{-x})$

47 $(\cosh x - \sinh x)^{-2}$, or e^{2x} **49** $2x/\sqrt{x^4 + 1}$

51 $-\frac{1}{3}\sin(5 - 3x) + C$ **53** $2\tan\sqrt{x} + C$

55 $\frac{1}{9}\ln|\sin 9x| + \frac{1}{9}\ln|\csc 9x - \cot 9x| + C$

57 $-\ln|\cos e^x| + C$

59 $-\frac{1}{3}\cot 3x + \frac{2}{3}\ln|\csc 3x - \cot 3x| + x + C$

61 $\frac{1}{4}\sin 4x + C$ **63** $\frac{1}{6}\sin(2x^3) + C$

65 $\frac{2}{15}(16\sqrt{2} - 3\sqrt{3})$ **67** $-\frac{1}{6}\csc^2 3x + C$

69 $\frac{1}{18}\ln(4 + 9x^2) + C$ **71** $-\sqrt{1 - e^{2x}} + C$

73 $\frac{1}{2}\sinh x^2 + C$ **75** $\pi/3$ **77** $\frac{1}{3}(1 + \tan x)^3 + C$

79 $-\ln|2 + \cot x| + C$ **81** $\cosh(\ln x) + C$

83 $\frac{1}{2}\sin^{-1}(2x/3) + C$ **85** $-\frac{1}{3}\operatorname{sech}^{-1}|2x/3| + C$

87 $\frac{1}{25}\sqrt{25x^2 + 36} + C$

89 $(\frac{4}{15}, \sin^{-1}(\frac{4}{5})), (-\frac{4}{15}, \sin^{-1}(-\frac{4}{5}))$

91 Min: $f(\tan^{-1}\frac{1}{2}) = 5\sqrt{5}$; decreasing on $(0, c)$ and increasing on $(c, \pi/2)$, where $c = \tan^{-1}\frac{1}{2}$.

93 $\pi(4 - \pi)/4$ **95** $(\pi/4) - 0.01(1 + \pi)$

97 $\frac{1}{260}$ rad/sec

CHAPTER 9

Exercises 9.1, page 407

1 $-(x + 1)e^{-x} + C$ **3** $e^{3x}(\frac{1}{3}x^2 - \frac{2}{9}x + \frac{2}{27}) + C$

5 $\frac{1}{25}\cos 5x + \frac{1}{5}x \sin 5x + C$

7 $x \sec x - \ln|\sec x + \tan x| + C$

9 $(x^2 - 2)\sin x + 2x \cos x + C$

11 $x \tan^{-1} x - \frac{1}{2}\ln(1 + x^2) + C$

13 $\frac{2}{9}x^{3/2}(3\ln x - 2) + C$ **15** $-x \cot x + \ln|\sin x| + C$

17 $-\frac{1}{2}e^{-x}(\cos x + \sin x) + C$

19 $\cos x(1 - \ln \cos x) + C$

21 $\frac{1}{2}(\sec x \tan x + \ln|\sec x + \tan x|) + C$

23 $\frac{1}{3}(2 - \sqrt{2})$ **25** $\pi/4$

27 $\frac{1}{40400}(2x + 3)^{100}(200x - 3) + C$

29 $\frac{1}{41}e^{4x}(4\sin 5x - 5\cos 5x) + C$

31 $x(\ln x)^2 - 2x \ln x + 2x + C$

33 $x^3 \cosh x - 3x^2 \sinh x + 6x \cosh x - 6 \sinh x + C$

35 $2\cos\sqrt{x} + 2\sqrt{x}\sin\sqrt{x} + C$

37 $x \cos^{-1} x - \sqrt{1 - x^2} + C$

43 $e^x(x^5 - 5x^4 + 20x^3 - 60x^2 + 120x - 120) + C$

45 2π **47** $(\pi/2)(e^2 + 1)$ **49** $(62.5\pi)/4 = 125\pi/8$

Exercises 9.2, page 413

1 $\sin x - \frac{1}{3}\sin^3 x + C$ **3** $\frac{1}{8}x - \frac{1}{32}\sin 4x + C$

5 $\frac{1}{5}\cos^5 x - \frac{1}{3}\cos^3 x + C$

7 $\frac{1}{8}(\frac{5}{2}x - 2\sin 2x + \frac{3}{8}\sin 4x + \frac{1}{6}\sin^3 2x) + C$

9 $\frac{1}{4}\tan^4 x + \frac{1}{6}\tan^6 x + C$ **11** $\frac{1}{5}\sec^5 x - \frac{1}{3}\sec^3 x + C$

13 $\frac{1}{5}\tan^5 x - \frac{1}{3}\tan^3 x + \tan x - x + C$

15 $\frac{2}{3}\sin^{3/2} x - \frac{2}{7}\sin^{7/2} x + C$

17 $\tan x - \cot x + C$ **19** $\frac{2}{3} - (5/6\sqrt{2})$

21 $\frac{1}{4}\sin 2x - \frac{1}{16}\sin 8x + C$ **23** $\frac{3}{5}$

25 $-\frac{1}{5}\cot^5 x - \frac{1}{7}\cot^7 x + C$

27 $-\ln|2 - \sin x| + C$ **29** $-1/(1 + \tan x) + C$

31 $3\pi^2/4$ **33** $\frac{5}{2}$

35 (b) $\int \sin mx \cos nx \, dx$

$$= \begin{cases} -\dfrac{\cos (m+n)x}{2(m+n)} - \dfrac{\cos (m-n)x}{2(m-n)} + C & \text{if } m \neq n \\ -\dfrac{\cos 2mx}{4m} + C & \text{if } m = n \end{cases}$$

$\int \cos mx \cos nx \, dx$

$$= \begin{cases} \dfrac{\sin (m+n)x}{2(m+n)} + \dfrac{\sin (m-n)x}{2(m-n)} + C & \text{if } m \neq n \\ \dfrac{x}{2} + \dfrac{\sin 2mx}{4m} & \text{if } m = n \end{cases}$$

Exercises 9.3, page 418

1 $2 \sin^{-1}(x/2) - \frac{1}{2}x\sqrt{4 - x^2} + C$

3 $\dfrac{1}{3} \ln \left| \dfrac{\sqrt{9 + x^2} - 3}{x} \right| + C$ **5** $\sqrt{x^2 - 25}/25x + C$

7 $-\sqrt{4 - x^2} + C$ **9** $-x/\sqrt{x^2 - 1} + C$

11 $\dfrac{1}{432}\left(\tan^{-1}\dfrac{x}{6} + \dfrac{6x}{36 + x^2}\right) + C$

13 $\frac{9}{4} \sin^{-1}(2x/3) + (x/2)\sqrt{9 - 4x^2} + C$

15 $1/(2(16 - x^2)) + C$

17 $\frac{1}{243}(9x^2 + 49)^{3/2} - \frac{49}{81}(9x^2 + 49)^{1/2} + C$

19 $(3 + 2x^2)\sqrt{x^2 - 3}/27x^3 + C$

21 $\frac{1}{2}x^2 + 8 \ln |x| - 8x^{-2} + C$

29 $25\pi(\sqrt{2} - \ln(\sqrt{2} + 1)) \approx 41.849$

31 $\sqrt{5} + \frac{1}{2} \ln(2 + \sqrt{5}) \approx 2.96$

33 $\frac{3}{2}\sqrt{13} + 2 \ln \frac{1}{2}(3 + \sqrt{13})$

35 $f(x) = \sqrt{x^2 - 16} - 4 \tan^{-1}(\sqrt{x^2 - 16}/4)$

37 $-\sqrt{25 + x^2}/25x + C$ **39** $-\sqrt{1 - x^2}/x + C$

Exercises 9.4, page 424

1 $3 \ln |x| + 2 \ln |x - 4| + C$, or $\ln |x|^3(x - 4)^2 + C$

3 $4 \ln |x + 1| - 5 \ln |x - 2| + \ln |x - 3| + C$, or

$\ln \dfrac{(x + 1)^4|x - 3|}{|x - 2|^5} + C$

5 $6 \ln |x - 1| + 5/(x - 1) + C$

7 $3 \ln |x - 2| - 2 \ln |x + 4| + C$, or $\ln \dfrac{|x - 2|^3}{(x + 4)^2} + C$

9 $2 \ln |x| - \ln |x - 2| + 4 \ln |x + 2| + C$, or

$\ln \dfrac{x^2(x + 2)^4}{|x - 2|} + C$

11 $5 \ln |x + 1| - \dfrac{1}{x + 1} - 3 \ln |x - 5| + C$, or

$\ln \dfrac{|x + 1|^5}{|x - 5|^3} - \dfrac{1}{x + 1} + C$

13 $5 \ln |x| - \dfrac{2}{x} + \dfrac{3}{2x^3} - \dfrac{1}{3x^2} + 4 \ln |x + 3| + C$

15 $\dfrac{1}{6} \ln |x - 3| - \dfrac{7}{2(x - 3)} + \dfrac{5}{6} \ln |x + 3| - \dfrac{3}{2(x + 3)} + C$

17 $3 \ln |x + 5| + \ln (x^2 + 4) + \frac{1}{2} \tan^{-1}(x/2) + C$, or
$\ln (x^2 + 4)|x + 5|^3 + \frac{1}{2} \tan^{-1}(x/2) + C$

19 $\ln \sqrt{\dfrac{x^2 + 1}{x^2 + 4}} + \dfrac{1}{2} \tan^{-1}(x/2) + C$

21 $\ln (x^2 + 1) - 4/(x^2 + 1) + C$

23 $(x^2/2) + x + 2 \ln |x| + 2 \ln |x - 1| + C$, or
$(x^2 + 2x)/2 + \ln (x^2 - x)^2 + C$

25 $(x^3/3) - 9x - (1/9x) - \frac{1}{2} \ln (x^2 + 9) + \frac{728}{27} \tan^{-1}(x/3) + C$

27 $2 \ln |x + 4| + 6(x + 4)^{-1} - 5(x - 3)^{-1} + C$

29 $2 \ln |x - 4| + 2 \ln |x + 1| - \frac{3}{2}(x + 1)^{-2} + C$

31 $\frac{3}{2} \ln (x^2 + 1) + \ln |x - 1| + x^2 + C$

37 $\frac{1}{2} \ln 3$ **39** $(\pi/27)(4 \ln 2 + 3) \approx 0.672$

Exercises 9.5, page 428

1 $\frac{1}{2} \tan^{-1}[(x - 2)/2] + C$ **3** $\sin^{-1}[(x - 2)/2] + C$

5 $-2\sqrt{9 - 8x - x^2} - 5 \sin^{-1}[(x + 4)/5] + C$

7 $\frac{1}{3} \ln |x - 1| - \frac{1}{6} \ln (x^2 + x + 1) -$
$\qquad (1/\sqrt{3}) \tan^{-1}[(2x + 1)/\sqrt{3}] + C$

9 $\frac{1}{2}[\tan^{-1}(x + 2) + (x + 2)/(x^2 + 4x + 5)] + C$

11 $(x + 3)/4\sqrt{x^2 + 6x + 13} + C$

13 $(2/3\sqrt{7}) \tan^{-1}[(4x - 3)/3\sqrt{7}] + C$

15 $1 + (\pi/4)$ **17** $\ln [(1 + e^x)/(2 + e^x)] + C$

19 $\frac{1}{3} \ln 2 + (2\pi/6\sqrt{3}) \approx 0.8356$

21 $\pi \ln (1.8) + (2\pi/3)(\tan^{-1}\frac{1}{3} - \pi/4) \approx 0.8755$

Exercises 9.6, page 432

1 $\frac{3}{7}(x + 9)^{7/3} - \frac{27}{4}(x + 9)^{4/3} + C$

3 $\frac{5}{81}(3x + 2)^{9/5} - \frac{5}{18}(3x + 2)^{4/5} + C$ **5** $2 + 8 \ln \frac{6}{7}$

7 $\frac{6}{7}x^{7/6} - \frac{6}{5}x^{5/6} + 2x^{1/2} - 6x^{1/6} + 6 \arctan x^{1/6} + C$

9 $(2/\sqrt{3}) \tan^{-1}\sqrt{(x - 2)/3} + C$

11 $\frac{3}{10}(x + 4)^{2/3}(2x - 7) + C$

13 $\frac{2}{7}(1 + e^x)^{7/2} - \frac{4}{5}(1 + e^x)^{5/2} + \frac{2}{3}(1 + e^x)^{3/2} + C$

15 $e^x - 4 \ln (e^x + 4) + C$

17 $2 \sin \sqrt{x + 4} - 2\sqrt{x + 4} \cos \sqrt{x + 4} + C$ **19** $\frac{137}{320}$

21 $(2/\sqrt{3}) \tan^{-1}[(2 \tan \frac{1}{2}x + 1)/\sqrt{3}] + C$

23 $\ln |1 + \tan (x/2)| + C$

25 $\dfrac{1}{5} \ln \left| \dfrac{\tan (x/2) + 2}{2 \tan (x/2) - 1} \right| + C$

27 $\frac{4}{3} \ln |\sin x - 4| + \frac{2}{3} \ln |\sin x + 2| + C$

Exercises 9.7, page 435

1 $\sqrt{4 + 9x^2} - 2 \ln \left| \dfrac{2 + \sqrt{4 + 9x^2}}{3x} \right| + C$

3 $-(x/8)(2x^2 - 80)\sqrt{16 - x^2} + 96 \sin^{-1}(x/4) + C$

5 $-\frac{2}{135}(9x + 4)(2 - 3x)^{3/2} + C$

7 $-\frac{1}{18} \sin^5 3x \cos 3x - \frac{5}{72} \sin^3 3x \cos 3x$
$\qquad\qquad - \frac{5}{48} \sin 3x \cos 3x + \frac{5}{16}x + C$

9 $-\frac{1}{3} \cot x \csc^2 x - \frac{2}{3} \cot x + C$

11 $\frac{1}{2}x^2 \sin^{-1} x + \frac{1}{4}x\sqrt{1 - x^2} - \frac{1}{4}\sin^{-1} x + C$

13 $\frac{1}{13}e^{-3x}(-3 \sin 2x - 2 \cos 2x) + C$

15 $\sqrt{5x - 9x^2} + \frac{5}{6}\cos^{-1}(5 - 18x)/5 + C$

17 $\dfrac{1}{4\sqrt{15}} \ln \left| \dfrac{\sqrt{5x^2} - \sqrt{3}}{\sqrt{5x^2} + \sqrt{3}} \right| + C$

19 $\frac{1}{4}(2e^{2x} - 1)\cos^{-1} e^x - \frac{1}{4}e^x\sqrt{1 - e^{2x}} + C$

21 $\frac{2}{315}(35x^3 - 60x^2 + 96x - 128)(2 + x)^{3/2} + C$

23 $\frac{2}{81}(4 + 9 \sin x - 4 \ln |4 + 9 \sin x|) + C$

25 $2\sqrt{9 + 2x} + 3 \ln |(\sqrt{9 + 2x} - 3)/(\sqrt{9 + 2x} + 3)| + C$

27 $\frac{3}{4} \ln |\sqrt[3]{x}/(4 + \sqrt[3]{x})| + C$

29 $\sqrt{16 - \sec^2 x} - 4 \ln |(4 + \sqrt{16 - \sec^2 x})/\sec x| + C$

Exercises 9.8, page 442

1 $-27, -46, \bar{x} = -\frac{23}{7}, \bar{y} = -\frac{27}{14}$

3 $\bar{x} = \frac{4}{5}, \bar{y} = \frac{2}{7}$ **5** $\bar{x} = \pi/2,$ **7** $\bar{x} = -\frac{1}{2}, \bar{y} = -\frac{3}{5}$
                                       $\bar{y} = \pi/8$

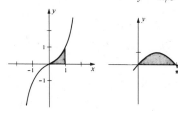

9 $\bar{x} = 1/\ln 2,$ **11** $\bar{x} = (3e^{-2} - 1)/2(1 - e^{-2}),$
    $\bar{y} = (\tan^{-1} \frac{3}{4})/(8 \ln 2)$ $\bar{y} = (1 - e^{-4})/4(1 - e^{-2})$

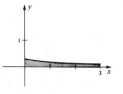

15 $\bar{x} = 0, \bar{y} = 4a/3\pi$ if the region is bounded by the x-axis and the upper half of the circle centered at $(0, 0)$.

17 $\bar{x} = 0, \bar{y} = -20a/3(8 + \pi)$. The figure is positioned vertically with the origin at the center of the circle.

19 $\bar{x} = \bar{y} = 256a/315\pi$

Exercises 9.9, page 448

1 $\bar{x} = 2 \ln 2$ **3** $\bar{x} = (e^2 + 1)/2(e^2 - 1)$

5 $\bar{x} = \frac{25}{48}$ **7** $\bar{y} = \frac{27}{8}$ **9** $\bar{y} = (7e^8 + 1)/8(3e^4 + 1)$

11 $\bar{y} = 1/(100 \ln 2)$ **13** $\bar{y} = (\pi + 2)/16$

15 If the base of the hemisphere is on the xz-plane, then
    $\bar{y} = 3(a^2 - 2h^2)/4(2a + 3h)$.

17 (a) 576π (b) 288π **19** $(4a/3\pi, 4a/3\pi)$

21 $512\pi/3$ **23** $\pi a^2 h/3$

Exercises 9.10, page 449

1 $\frac{1}{2}x^2 \sin^{-1} x - \frac{1}{4}\sin^{-1} x + \frac{1}{4}x\sqrt{1 - x^2} + C$

3 $2 \ln 2 - 1$ **5** $\frac{1}{6}\sin^3 2x - \frac{1}{10}\sin^5 2x + C$

7 $\frac{1}{5}\sec^5 x + C$ **9** $x/25\sqrt{x^2 + 25} + C$

11 $2 \ln |(2 - \sqrt{4 - x^2})/x| + \sqrt{4 - x^2} + C$

13 $2 \ln |x - 1| - \ln |x| - x/(x - 1)^2 + C$

15 $\ln \dfrac{(x + 3)^2(x^2 + 9)^2}{|x - 3|^5} + \dfrac{1}{3}\tan^{-1}\dfrac{x}{3} + C$

17 $-\sqrt{4 + 4x - x^2} + 2 \sin^{-1} \dfrac{x - 2}{\sqrt{8}} + C$

19 $3(x + 8)^{1/3} + \ln [(x + 8)^{1/3} - 2]^2$
     $- \ln [(x + 8)^{2/3} + 2(x + 8)^{1/3} + 4]$
     $- \dfrac{6}{\sqrt{3}}\tan^{-1}\dfrac{(x + 8)^{1/3} + 1}{\sqrt{3}} + C$

21 $\frac{1}{13}e^{2x}(2 \sin 3x - 3 \cos 3x) + C$

23 $\frac{1}{4}\sin^4 x - \frac{1}{6}\sin^6 x + C$ **25** $-\sqrt{4 - x^2} + C$

27 $\dfrac{1}{3}x^3 - x^2 + 3x - \dfrac{1}{2x} - \dfrac{1}{4}\ln |x| - \dfrac{23}{4}\ln |x + 2| + C$

29 $2 \tan^{-1}(x^{1/2}) + C$ **31** $\ln |\sec e^x + \tan e^x| + C$

33 $\frac{1}{125}[10x \sin 5x - (25x^2 - 2)\cos 5x] + C$

35 $\frac{2}{7}\cos^{7/2} x - \frac{2}{3}\cos^{3/2} x + C$ **37** $\frac{2}{3}(1 + e^x)^{3/2} + C$

39 $\frac{1}{16}[2x\sqrt{4x^2 + 25} - 25 \ln (2x + \sqrt{4x^2 + 25})] + C$

41 $\frac{1}{3}\tan^3 x + C$ **43** $-x \csc x + \ln |\csc x - \cot x| + C$

45 $-\frac{1}{4}(8 - x^3)^{4/3} + C$

47 $-2x \cos \sqrt{x} + 4 \cos \sqrt{x} + 4\sqrt{x} \sin \sqrt{x} + C$

49 $\frac{1}{2}e^{2x} - e^x + \ln (1 + e^x) + C$

51 $\frac{2}{5}x^{5/2} - \frac{8}{3}x^{3/2} + 6x^{1/2} + C$

53 $\frac{1}{3}(16 - x^2)^{3/2} - 16(16 - x^2)^{1/2} + C$

55 $\frac{11}{2}\ln |x + 5| - \frac{15}{2}\ln |x + 7| + C$

57 $x \tan^{-1} 5x - \frac{1}{10}\ln |1 + 25x^2| + C$ **59** $e^{\tan x} + C$

61 $(1/\sqrt{5})\ln |\sqrt{5}x + \sqrt{7 + 5x^2}| + C$

63 $-\frac{1}{5}\cot^5 x - \frac{1}{3}\cot^3 x - \cot x - x + C$

65 $\frac{1}{5}(x^2 - 25)^{5/2} + \frac{25}{3}(x^2 - 25)^{3/2} + C$

67 $\frac{1}{3}x^3 - \frac{1}{4}\tanh 4x + C$

69 $-\frac{1}{4}x^2e^{-4x} - \frac{1}{8}xe^{-4x} - \frac{1}{32}e^{-4x} + C$

71 $3 \sin^{-1}(x + 5)/6 + C$ **73** $-\frac{1}{7}\cos 7x + C$

75 $18 \ln |x - 2| - 9 \ln |x - 1| - 5 \ln |x - 3| + C$

77 $x^3 \sin x + 3x^2 \cos x - 6x \sin x - 6 \cos x + \sin x + C$

79 $(-1/x)\sqrt{9 - 4x^2} - 2 \sin^{-1}(2x/3) + C$

81 $24x - \frac{10}{3}\ln |\sin 3x| - \frac{1}{3}\cot 3x + C$

83 $-\ln x - 4/\sqrt[4]{x} + 4 \ln (1 + \sqrt[4]{x}) + C$

85 $-2\sqrt{1 + \cos x} + C$

87 $-x/2(25 + x^2) + \frac{1}{10}\tan^{-1}(x/5) + C$

89 $\frac{1}{3}\sec^3 x - \sec x + C$

91 $\ln (x^2 + 4) - \frac{3}{2}\tan^{-1}(x/2) + (7/\sqrt{5})\tan^{-1}(x/\sqrt{5}) + C$

93 $\frac{1}{4}x^4 - 2x^2 + 4 \ln |x| + C$

95 $\frac{2}{5}x^{5/2} \ln x - \frac{4}{25}x^{5/2} + C$

97 $\frac{3}{64}(2x + 3)^{8/3} - \frac{9}{20}(2x + 3)^{5/3} + \frac{27}{16}(2x + 3)^{2/3} + C$

99 $\frac{1}{2}e^{x^2}(x^2 - 1) + C$ **101** $2\pi^2$ **103** $\ln (2 + \sqrt{3})$

105 $\bar{x} = \frac{3}{5}, \bar{y} = \frac{12}{35}$ **107** $\bar{x} = \frac{8}{3}$

109 $\bar{y} = (1 - 7e^{-6})/8(1 - 4e^{-3})$

CHAPTER 10

Exercises 10.1, page 459

1 $\frac{1}{2}$ **3** $\frac{1}{40}$ **5** $\frac{3}{13}$ **7** 0   **9** $-\frac{1}{2}$
11 $-\frac{1}{2}$ **13** $\frac{1}{6}$ **15** ∞ **17** $\frac{1}{3}$ **19** ∞
21 1   **23** 0   **25** Does not exist   **27** $\frac{2}{5}$
29 ∞ **31** 0   **33** ∞ **35** 2
37 Does not exist   **39** $\frac{3}{5}$ **41** -3 **43** 0
45 ∞ **47** ∞ **49** 2   **51** 1   **53** 1
55 (a) $(E/L)t$ (b) E/R (c) E/R **57** 0

Exercises 10.2, page 463

1 0   **3** 0   **5** 0   **7** 0   **9** 1   **11** 0
13 e^5 **15** 1   **17** 1   **19** Does not exist
21 e^2 **23** 2   **25** 0   **27** 1
29 Does not exist   **31** $\frac{1}{2}$ **33** Does not exist
35 e **37** Does not exist   **39** $e^{1/3}$ **41** ∞
43 $f(e) = e^{1/e}$ is a local max; $\lim_{x \to 0^+} f(x) = 0$;
horizontal asymptote $y = 1$.

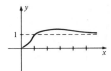

Exercises 10.3, page 467

1 3   **3** Diverges   **5** Diverges   **7** $\frac{1}{2}$ **9** $-\frac{1}{2}$
11 Diverges   **13** Diverges   **15** 0   **17** Diverges
19 Diverges   **21** π **23** $\ln 2$
25 (a) Does not exist (b) π **27** (a) 1 (b) $\pi/32$
29 π **31** (a) $n < -1$ (b) $n \geq -1$
33 4×10^5 mi-lb   **35** Converges   **37** Diverges
39 There are many possible answers; e.g., $f(x) = x$,
$f(x) = \sin x$, $f(x) = x^7 + x$.
43 $1/s, s > 0$ **45** $s/(s^2 + 1), s > 0$
47 $1/(s - a), s > a$ **49** (a) 1, 1, 2

Exercises 10.4, page 474

1 6   **3** Diverges   **5** Diverges   **7** Diverges
9 $3\sqrt[3]{4}$ **11** Diverges   **13** $\pi/2$ **15** Diverges
17 $-\frac{1}{4}$ **19** Diverges   **21** Diverges   **23** Diverges
25 0   **27** Diverges   **29** Diverges
31 (a) 2 (b) Value cannot be assigned.
33 Values cannot be assigned to either the area or the volume.
35 $n > -1$ **37** Converges   **39** Diverges

Exercises 10.5, page 483

1 $\sin x = 1 - \frac{1}{2}\left(x - \frac{\pi}{2}\right)^2 + \frac{\sin z}{4!}\left(x - \frac{\pi}{2}\right)^4$,
where z is between x and $\pi/2$.
3 $\sqrt{x} = 2 + \frac{1}{4}(x - 4) - \frac{1}{64}(x - 4)^2 + \frac{1}{512}(x - 4)^3$
$- \frac{5}{128}z^{-7/2}(x - 4)^4$, where z is between x and 4.
5 $\tan x = 1 + 2\left(x - \frac{\pi}{4}\right) + 2\left(x - \frac{\pi}{4}\right)^2 + \frac{8}{3}\left(x - \frac{\pi}{4}\right)^3$
$+ \frac{10}{3}\left(x - \frac{\pi}{4}\right)^4 + \frac{g(z)}{5!}\left(x - \frac{\pi}{4}\right)^5$,
where z is between x and $\pi/4$ and
$g(z) = 16 \sec^6 z + 88 \sec^4 z \tan^2 z + 16 \sec^2 z \tan^4 z$
7 $1/x = -\frac{1}{2} - \frac{1}{4}(x + 2) - \frac{1}{8}(x + 2)^2 - \frac{1}{16}(x + 2)^3$
$- \frac{1}{32}(x + 2)^4 - \frac{1}{64}(x + 2)^5 + z^{-7}(x + 2)^6$,
where z is between x and -2
9 $\tan^{-1} x = \frac{\pi}{4} + \frac{1}{2}(x - 1) - \frac{1}{4}(x - 1)^2$
$+ \frac{3z^2 - 1}{3(1 + z^2)^3}(x - 1)^3$,
where z is between 1 and x
11 $xe^x = -\frac{1}{e} + \frac{1}{2e}(x - 1)^2 + \frac{1}{3e}(x + 1)^3 + \frac{1}{8e}(x + 1)^4$
$+ \frac{ze^z + 5e^z}{120}(x + 1)^5$,
where z is between x and -1
13 $\ln(x + 1) = x - \frac{1}{2}x^2 + \frac{1}{3}x^3 - \frac{1}{4}x^4 + \frac{x^5}{5(z + 1)^5}$,
where z is between 0 and x
15 $\cos x = 1 - \frac{x^2}{2!} + \frac{x^4}{4!} - \frac{x^6}{6!} + \frac{x^8}{8!} - \frac{\sin z}{9!}x^9$,
where z is between x and 0
17 $e^{2x} = 1 + 2x + 2x^2 + \frac{4}{3}x^3 + \frac{2}{3}x^4 + \frac{4}{15}x^5 + \frac{4}{45}e^{2z}x^6$,
where z is between x and 0
19 $1/(x - 1)^2 = 1 + 2x + 3x^2 + 4x^3 + 5x^4 + 6x^5$
$+ 7x^6/(z - 1)^8$, where z is between 0 and x
21 $\arcsin x = x + \frac{1 + 2z^2}{6(1 - z^2)^{5/2}}x^3$,
where z is between 0 and x
23 $f(x) = 7 - 3x + x^2 - 5x^3 + 2x^4$ **25** 0.9998
27 2.0075   **29** 0.454545, error $\leq$ 0.0000005
31 0.223, error $\leq$ 0.0002
33 0.8660254, error $\leq (8.1)(10^{-9})$
35 Five decimal places since
$$|R_3(x)| \leq 4.2 \times 10^{-6} < 0.5 \times 10^{-5}$$
37 Three decimal places   **39** Four decimal places

Exercises 10.6, page 484

1 $\frac{1}{2}\ln 2$ **3** ∞ **5** $\frac{8}{3}$ **7** 0   **9** $-\infty$
11 e^8 **13** e **15** Diverges   **17** Diverges
19 $-\frac{9}{2}$ **21** Diverges   **23** $\pi/2$ **25** Diverges

27 (a) $\ln \cos x = \ln \dfrac{\sqrt{3}}{2} - \dfrac{1}{\sqrt{3}}\left(x - \dfrac{\pi}{6}\right) - \dfrac{2}{3}\left(x - \dfrac{\pi}{6}\right)^2$

$\qquad - \dfrac{4}{9\sqrt{3}}\left(x - \dfrac{\pi}{6}\right)^3 - \dfrac{1}{12}(\sec^4 z +$

$\qquad 2\sec^2 z \tan^2 z)\left(x - \dfrac{\pi}{6}\right)^4,$

where z is between x and $\pi/6$

(b) $\sqrt{x-1} = 1 + \frac{1}{2}(x-2) - \frac{1}{8}(x-2)^2 + \frac{1}{16}(x-2)^3$
$\qquad - \frac{5}{128}(x-2)^4 + \frac{7}{256}(z-1)^{-9/2}(x-2)^5,$
where z is between x and 2

29 0.4651 (with $n = 3$, $|R_3(x)| \leq 1.6 \times 10^{-6}$)

CHAPTER II

Exercises 11.1, page 496

1 $\frac{1}{5}, \frac{1}{4}, \frac{3}{11}, \frac{2}{7}; \frac{1}{3}$ **3** $\frac{3}{5}, -\frac{9}{11} -\frac{29}{21}, -\frac{57}{35}; -2$
5 $-5, -5, -5, -5; -5$ **7** $2, \frac{7}{3}, \frac{25}{14}, \frac{7}{5}; 0$
9 $2/\sqrt{10}, 2/\sqrt{13}, 2/\sqrt{18}, \frac{2}{5}; 0$
11 $\frac{3}{10}, -\frac{6}{17}, \frac{9}{26}, -\frac{12}{37}; 0$
13 $1.1, 1.01, 1.001, 1.0001; 1$
15 $2, 0, 2, 0$; the limit does not exist.   **17** 0 **19** $\pi/2$
21 Does not exist   **23** 0 **25** Does not exist
27 Does not exist   **29** e **31** 0 **33** $\frac{1}{2}$
35 Does not exist   **37** 1 **39** 0 **41** 0
43 Two of many possible answers are:

$a_n = \begin{cases} 2n & \text{if } 1 \leq n \leq 4 \\ (n-4)a & \text{if } n \geq 5 \end{cases}$

$a_n = 2n + (n-1)(n-2)(n-3)(n-4)(a-10)/24.$

Exercises 11.2, page 505

1 Converges, 4 **3** Converges, $\sqrt{5}/(\sqrt{5}+1)$
5 Converges, $\frac{37}{99}$ **7** Diverges   **9** Diverges
11 Converges, $\frac{1}{4}$ **13** Converges, 5 **15** Diverges
17 Diverges   **19** Converges, $\frac{8}{7}$ **21** Diverges
23 Converges   **25** Diverges   **27** Diverges
29 $\frac{1}{2}$ **35** $\frac{23}{99}$ **37** $\frac{16181}{4995}$ **39** 30 m
41 (b) $Q/(1 - e^{-ct})$ (c) $-(1/c)\ln[(M - Q)/M]$

Exercises 11.3, page 514

1 Converges   **3** Diverges   **5** Diverges
7 Diverges   **9** Converges   **11** Converges
13 Converges   **15** Diverges   **17** Converges
19 Converges   **21** Diverges   **23** Diverges
25 Converges   **27** Converges   **29** Converges
31 Converges   **33** Converges   **35** Diverges
37 Converges   **39** Converges   **41** Converges
43 Converges   **45** Converges
47 Converges if $k > 1$, diverges if $k \leq 1$ **53** 8

Exercises 11.4, page 518

1 Converges   **3** Converges   **5** Diverges
7 Converges   **9** Diverges   **11** Converges
13 Converges   **15** 0.368 **17** 0.901 **19** 0.306
21 141 **23** 5 **25** No. Consider $a_n = b_n = (-1)^n/\sqrt{n}$.

Exercises 11.5, page 525

1 Conditionally convergent   **3** Absolutely convergent
5 Absolutely convergent   **7** Absolutely convergent
9 Divergent   **11** Absolutely convergent
13 Absolutely convergent   **15** Conditionally convergent
17 Absolutely convergent   **19** Absolutely convergent
21 Absolutely convergent   **23** Divergent
25 Divergent   **27** Absolutely convergent
29 Absolutely convergent   **31** Divergent
33 Convergent   **35** Absolutely convergent
37 Divergent

Exercises 11.6, page 531

1 $[-1, 1)$ **3** $(-2, 2)$ **5** $(-1, 1]$ **7** $[-1, 1)$
9 $[-1, 1]$ **11** $(-6, 14)$
13 Converges only for $x = 0$ **15** $(-2, 2)$
17 $(-\infty, \infty)$ **19** $\left[\frac{17}{9}, \frac{19}{9}\right)$ **21** $(-12, 4)$
23 $(0, 2e)$ **25** $\left(-\frac{5}{2}, \frac{7}{2}\right]$ **27** $\frac{3}{2}$ **29** $1/e$ **31** ∞

Exercises 11.7, page 536

1 $\sum_{n=0}^{\infty} x^n; -1 < x < 1$ **3** $\sum_{n=1}^{\infty} nx^{n-1}; -1 < x < 1$
5 $\sum_{n=0}^{\infty} x^{2(n+1)}; -1 < x < 1$
7 $\displaystyle\sum_{n=0}^{\infty} \dfrac{3^n}{2^{n+1}} x^{n+1}; -\dfrac{2}{3} < x < \dfrac{2}{3}$
9 $-1 - x - 2\sum_{n=2}^{\infty} x^n; -1 < x < 1$
11 0.182 **13** $\displaystyle\sum_{n=0}^{\infty} \dfrac{(-1)^n}{n!} x^n$ **15** $\displaystyle\sum_{n=0}^{\infty} \dfrac{1}{(2n)!} x^{2n}$
17 $\displaystyle\sum_{n=0}^{\infty} \dfrac{3^n}{n!} x^{n+1}$ **19** 0.3333 **21** 0.0992
23 0.9677 **25** $\sum_{n=1}^{\infty} 2nx^{2n-1}$ **27** $\displaystyle\sum_{n=1}^{\infty} \dfrac{(-1)^n}{2n(n!)} x^{2n}$
29 $-\displaystyle\sum_{n=1}^{\infty} \dfrac{1}{n^2} x^n$

Exercises 11.8, page 545

1 $\displaystyle\sum_{n=0}^{\infty} \dfrac{(-1)^n}{(2n)!} x^{2n}$ **3** $\displaystyle\sum_{n=0}^{\infty} \dfrac{2^n}{n!} x^n$ **5** $\displaystyle\sum_{n=0}^{\infty} \dfrac{1}{n!} x^{n+2}; \infty$
7 $\displaystyle\sum_{n=0}^{\infty} \dfrac{1}{(2n+1)!} x^{2n+1}; \infty$
9 $\displaystyle\sum_{n=0}^{\infty} \dfrac{(-1)^n 3^{2n+1}}{(2n+1)!} x^{2n+2}; \infty$
11 $1 + \displaystyle\sum_{n=1}^{\infty} \dfrac{(-1)^n 2^{2n-1}}{2n!} x^{2n}; \infty$

13 $\displaystyle\sum_{n=0}^{\infty} \frac{(-1)^n}{\sqrt{2}(2n+1)!}\left(x-\frac{\pi}{4}\right)^{2n+1} + \sum_{n=0}^{\infty} \frac{(-1)^n}{\sqrt{2}(2n)!}\left(x-\frac{\pi}{4}\right)^{2n}$

15 $\displaystyle\sum_{n=0}^{\infty} \frac{(-1)^n}{2^{n+1}}(x-2)^n$ **17** $\displaystyle\sum_{n=0}^{\infty} \frac{(\ln 10)^n}{n!}x^n$

19 $\displaystyle\sum_{n=0}^{\infty} \frac{e^{-2}2^n}{n!}(x+1)^n$

21 $2 + 2\sqrt{3}\left(x-\frac{\pi}{3}\right) + 7\left(x-\frac{\pi}{3}\right)^2 + \frac{23\sqrt{3}}{3}\left(x-\frac{\pi}{3}\right)^3 + \cdots$

23 $\dfrac{\pi}{6} + \dfrac{2}{\sqrt{3}}\left(x-\dfrac{1}{2}\right) + \dfrac{2}{3\sqrt{3}}\left(x-\dfrac{1}{2}\right)^2 + \dfrac{8}{9\sqrt{3}}\left(x-\dfrac{1}{2}\right)^3$

25 $-(1/e) + (1/2e)(x+1)^2 + (1/3e)(x+1)^3 + (1/8e)(x+1)^4$

27 1.6487 **29** 0.9986 **31** 0.0997 **33** 0.5211
35 0.7468 **37** 0.4969 **39** 0.0115 **41** 0.4864
43 0.4484 **45** 0.1689

47 $2\left(x + \dfrac{x^3}{3} + \dfrac{x^5}{5} + \cdots + \dfrac{x^{2n-1}}{2n-1} + \cdots\right)$ if $|x| < 1$

49 $\pi = 4\left(1 - \dfrac{1}{3} + \dfrac{1}{5} - \dfrac{1}{7} + \cdots + (-1)^n \dfrac{1}{2n+1} + \cdots\right)$;
using five terms, $\pi \approx 3.34$; at least 40,000 terms are required.

Exercises 11.9, page 549

1 (a) $1 + \dfrac{1}{2}x + \displaystyle\sum_{n=2}^{\infty} (-1)^{n-1} \dfrac{1\cdot3\cdot5\cdots(2n-3)}{2^n n!} x^n$; 1

(b) $1 - \dfrac{1}{2}x^3 - \displaystyle\sum_{n=2}^{\infty} \dfrac{1\cdot3\cdot5\cdots(2n-3)}{2^n n!} x^{3n}$; 1

3 $1 + \displaystyle\sum_{n=1}^{\infty} (-1)^n \tfrac{1}{2}(n+1)(n+2)x^n$; 1

5 $2 + \dfrac{1}{12}x + 2\displaystyle\sum_{n=2}^{\infty} (-1)^{n-1} \dfrac{2\cdot5\cdot8\cdots(3n-4)}{3^n 8^n n!} x^n$; 8

7 $x + \displaystyle\sum_{n=1}^{\infty} \dfrac{1\cdot3\cdot5\cdots(2n-1)}{2^n n!(2n+1)} x^{2n+1}$; 1

9 0.508

Exercises 11.10, page 549

1 0 **3** Does not exist **5** 5 **7** Divergent
9 Convergent **11** Divergent **13** Divergent
15 Convergent **17** Divergent **19** Divergent
21 Absolutely convergent **23** Conditionally convergent
25 Absolutely convergent **27** Convergent
29 Convergent **31** Conditionally convergent
33 Convergent **35** Convergent **37** Divergent
39 0.159 **41** $(-3, 3)$ **43** $[-12, -8)$ **45** $\frac{1}{4}$

47 $\displaystyle\sum_{n=1}^{\infty} \dfrac{(-1)^{n+1}}{(2n)!} x^{2n-1}$; ∞ **49** $\displaystyle\sum_{n=0}^{\infty} \dfrac{(-1)^n 2^{2n}}{(2n+1)!} x^{2n+1}$; ∞

51 $1 + \dfrac{2}{3}x + 2\displaystyle\sum_{n=2}^{\infty} (-1)^{n-1} \dfrac{1\cdot4\cdot7\cdots(3n-5)}{3^n n!} x^n$; 1

53 $e^2 \displaystyle\sum_{n=0}^{\infty} \dfrac{(-1)^n}{n!}(x+2)^n$

55 $2 + \dfrac{x-4}{4} + \displaystyle\sum_{n=2}^{\infty} (-1)^{n-1} \dfrac{1\cdot3\cdot5\cdots(2n-3)}{2^{3n-1}n!}(x-4)^n$

57 0.189 **59** 1.002

CHAPTER 12

Exercises 12.2, page 560

1 $V(0, 0)$; $F(0, -3)$; $y = 3$ **3** $V(0, 0)$; $F(-\frac{3}{8}, 0)$; $x = \frac{3}{8}$

5 $V(0, 0)$; $F(0, \frac{1}{32})$; $y = -\frac{1}{32}$ **7** $V(-1, 0)$; $F(2, 0)$; $x = -4$

9 $V(2, -2)$; $F(2, -\frac{7}{4})$; $y = -\frac{9}{4}$ **11** $V(-4, 2)$; $F(-\frac{7}{2}, 2)$; $x = -\frac{9}{2}$

13 $V(-5, -6)$; $F(-5, -\frac{97}{16})$; $y = -\frac{95}{16}$ **15** $V(0, \frac{1}{2})$; $F(0, -\frac{9}{2})$; $y = \frac{11}{2}$

17 Let $y' = 2ax + b = 0$ to obtain the x-coordinate $-b/2a$ of the vertex. Given $x = ay^2 + by + c$ let $2ay + b = 0$ to obtain the y-coordinate $-b/2a$ of the vertex.
19 $y^2 = 8x$ **21** $(x-6)^2 = 12(y-1)$
23 $3x^2 = -4y$ **25** $\frac{9}{16}$ ft from the vertex
27 $y = 2x^2 - 3x + 1$
31 (a) $\frac{8}{3}$ (b) 2π (c) $16\pi/5$ **33** $\frac{200}{3}$ lb

Exercises 12.3, page 566

1 $V(\pm 3, 0); F(\pm\sqrt{5}, 0)$ **3** $V(0, \pm 4); F(0, \pm 2\sqrt{3})$

5 $V(0, \pm\sqrt{5}); F(0, \pm\sqrt{3})$ **7** $V(\pm\frac{1}{2}, 0); F(\pm\sqrt{21}/10, 0)$

9   Center (4, 2), vertices (1, 2) and (7, 2), $F(4 \pm \sqrt{5}, 2)$; endpoints of minor axis (4, 4) and (4, 0)

11   Center $(-3, 1)$, vertices $(-7, 1)$ and (1, 1), $F(-3 \pm \sqrt{7}, 1)$; endpoints of minor axis $(-3, 4)$ and $(-3, -2)$

13   Center (5, 2), vertices (5, 7) and $(5, -3)$, $F(5, 2 \pm \sqrt{21})$; endpoints of minor axis (3, 2) and (7, 2)

9   **11**   **13**

  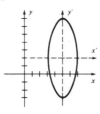

15 $x^2/64 + y^2/39 = 1$ **17** $4x^2/9 + y^2/25 = 1$
19 $8x^2/81 + y^2/36 = 1$ **21** $\{(2, 2), (4, 1)\}$

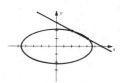

23 $2\sqrt{21}$ feet   **25** $y - 3 = \frac{5}{6}(x + 2)$ **27** $(\pm\sqrt{3}, \frac{3}{2})$
29 $\frac{4}{3}\pi ab^2$ **31**   (a) 864   (b) $216\sqrt{3}$
33 $2a/\sqrt{2}$ and $2b/\sqrt{2}$ **37** $\bar{x} = 0, \bar{y} = 4b/3\pi$

Exercises 12.4, page 572

1 $V(\pm 3, 0), F(\pm\sqrt{13}, 0)$,   **3** $V(0, \pm 3), F(0, \pm\sqrt{13})$,
    $y = \pm 2x/3$ $y = \pm 3x/2$

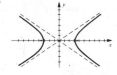

5 $V(0, \pm 4), F(0, \pm 2\sqrt{5})$,   **7** $V(\pm 1, 0), F(\pm\sqrt{2}, 0)$,
    $y = \pm 2x$ $y = \pm x$

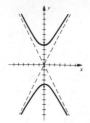 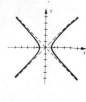

9 $V(\pm 5, 0), F(\pm\sqrt{30}, 0)$,   **11** $V(0, \pm\sqrt{3}), F(0, \pm 2)$,
    $y = \pm(\sqrt{5}/5)x$ $y = \pm\sqrt{3}x$

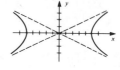 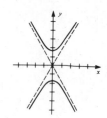

13   Center $(-5, 1)$, vertices $(-5 \pm 2\sqrt{5}, 1)$, $F(-5 \pm \sqrt{205/2}, 1)$, $y - 1 = \pm\frac{5}{4}(x + 5)$

15   Center $(-2, -5)$, vertices $(-2, -2)$ and $(-2, -8)$, $F(-2, -5 \pm 3\sqrt{5})$, $y + 5 = \pm\frac{1}{2}(x + 2)$

17   Center (6, 2), vertices (6, 4) and (6, 0), $F(6, 2 \pm 2\sqrt{10})$, $y - 2 = \pm\frac{1}{3}(x - 6)$

13   **15**   **17**

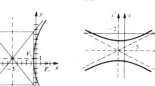

19 $15y^2 - x^2 = 15$ **21** $x^2/9 - y^2/16 = 1$
23 $y^2/21 - x^2/4 = 1$ **25** $x^2/9 - y^2/36 = 1$
27 $\{(0, 4), (\frac{8}{3}, \frac{20}{3})\}$
29   Conjugate hyperbolas have the same asymptotes.

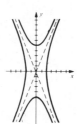

31 $y - 1 = -\frac{4}{5}(x + 2)$ **33** $(\pm 2\sqrt{2}, -6)$
35 $5x - 6y - 16 = 0$

37 $\pi b^2[\sqrt{a^2 + b^2}(b^2 - 2a^2) + 2a^3]/3a^2$

39 $(b/a)[bc - a^2 \ln(b + c) + a^2 \ln a]$

Exercises 12.5, page 576

The following answers contain equations in x' and y' resulting from a rotation of axis.

1 Ellipse, $(x')^2 + 16(y')^2 = 16$

3 Hyperbola, $4(x')^2 - (y')^2 = 1$

5 Ellipse, $(x')^2 + 9(y')^2 = 9$

7 Parabola, $(y')^2 = 4(x' - 1)$

9 Hyperbola, $2(x')^2 - (y')^2 - 4y' - 3 = 0$

11 Parabola, $(x')^2 - 6x' - 6y' + 9 = 0$

13 Ellipse, $(x')^2 + 4(y')^2 - 4x' = 0$

1   **3**

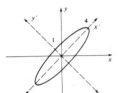

5   **7**

                        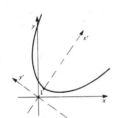

9   **11**

13

15 Sketch of proof: It can be shown (see page 575) that $B^2 - 4AC = B'^2 - 4A'C'$. For a suitable rotation of axes we obtain $B' = 0$ and the transformed equation has the form

$A'x'^2 + C'y'^2 + D'x' + E'y' + F = 0$. Except for degenerate cases, the graph of the last equation is an ellipse, hyperbola, or parabola if $A'C' > 0$, $A'C' < 0$, or $A'C' = 0$, respectively. However, if $B' = 0$, then $B^2 - 4AC = -4A'C'$ and hence the graph is an ellipse, hyperbola, or parabola if $B^2 - 4AC < 0, B^2 - 4AC > 0$, or $B^2 - 4AC = 0$, respectively.

Exercises 12.6, page 577

1 Parabola; $F(16, 0)$; $V(0, 0)$

3 Ellipse; $F(0, \pm\sqrt{7})$; $V(0, \pm 4)$

5 Hyperbola; $F(\pm 2\sqrt{2}, 0)$; $V(\pm 2, 0)$

1   **3**   **5**

7 Parabola; $F(0, -\frac{9}{4})$; $V(0, 4)$

9 Hyperbola: vertices $(-5, 5)$ and $(-3, 5)$;
foci $(-4 \pm \sqrt{10}/3, 5)$

7   **9**

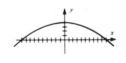

11 $9y^2 - 49x^2 = 441$ **13** $x^2 = -40y$

15 $4x^2 + 3y^2 = 300$ **17** $y^2 - 81x^2 = 36$

19 Ellipse; center $(-3, 2)$, vertices $(-6, 2)$ and $(0, 2)$, endpoints of minor axis $(-3, 0)$ and $(-3, 4)$

21 Parabola; vertex $(2, -4)$, focus $(4, -4)$

23 Hyperbola; center $(-4, 0)$, vertices $(-7, 0)$ and $(-1, 0)$, asymptotes $y = \pm(x + 4)/3$

25 $2 \pm 2\sqrt{2}$ **27** $64\pi/3$

29 Parabola; $(y')^2 - 3x' = 0$ is an equation after a rotation of axes.

CHAPTER 13

Exercises 13.1, page 585

1 $y = 2x + 7$ **3** $y = x - 2$ **5** $x = y^2 - 6y + 4$

   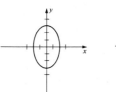   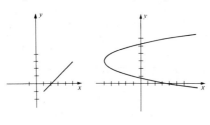

7 $y = 1/x^2$ **9** $x^2/4 + y^2/9 = 1$ **11** $x^2 - y^2 = 1$

      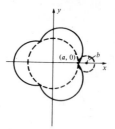

13 $y = \ln x$ **15** $y = 1/x$ **17** $x^2 - y^2 = 1$

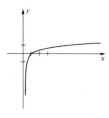   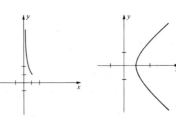

19 $y = \sqrt{x^2 - 1}$ **21**   **23**

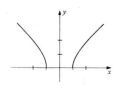   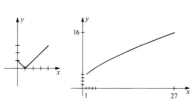

25

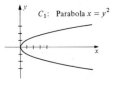 C_1: Parabola $x = y^2$

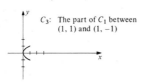 C_3: The part of C_1 between $(1, 1)$ and $(1, -1)$

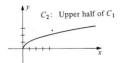 C_2: Upper half of C_1

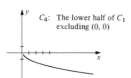 C_4: The lower half of C_1 excluding $(0, 0)$

27 (a) P moves counterclockwise along the circle $x^2 + y^2 = 1$ from $(1, 0)$ to $(-1, 0)$.
(b) P moves clockwise along the circle $x^2 + y^2 = 1$ from $(0, 1)$ to $(0, -1)$.
(c) P moves clockwise along the circle $x^2 + y^2 = 1$ from $(-1, 0)$ to $(1, 0)$.

29 $x = (x_2 - x_1)t^n + x_1$, $y = (y_2 - y_1)t^n + y_1$ are parametric equations for l if n is any odd positive integer.

35 $x = 4b \cos t - b \cos 4t$, $y = 4b \sin t - b \sin 4t$

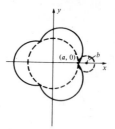

Exercises 13.2, page 590

1 2   **3** 1   **5** $\frac{1}{4}$ **7** $-2e^{-3}$ **9** $-\frac{3}{2} \tan 1$

11 $4x - y = 49$

13 $y - 1 = m(x - 4)$ where $m = 1/(12 \pm 8\sqrt{2})$

15 Horizontal at $(16, -16)$ and $(16, 16)$; vertical at $(0, 0)$; $d^2y/dx^2 = (3t^2 + 12)/64t^3$

17 Vertical at $(0, 0)$ and $(-3, 1)$; no horizontal; $d^2y/dx^2 = (1 - 3t)/144t^{3/2}(t - 1)^3$

19 $\sin t/(1 - \cos t)$; horizontal when $t = (2n + 1)\pi$; vertical when $t = 2n\pi$; slope 1 when $t = \pi/2 + 2n\pi$, where n is an integer; $d^2y/dx^2 = -1/a(1 - \cos t)^2$; CD if $0 < t < 2\pi$

Exercises 13.3, page 597

1   **3**   **5**

7   **9**   **11**

      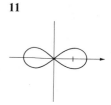

13   **15**   **17**

      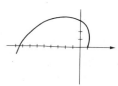

19   **21**   **23**

  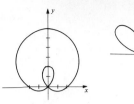 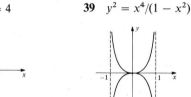

25 $r = -3 \sec \theta$ **27** $r = 4$ **29** $r = 6 \csc \theta$

31 $r^2 = 16 \sec 2\theta$

33 $x = 5$ **35** $x^2 + y^2 - 6y = 0$

     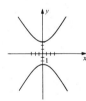

37 $x^2 + y^2 = 4$ **39** $y^2 = x^4/(1 - x^2)$

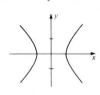   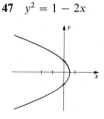

41 $y^2/9 - x^2/4 = 1$ **43** $x^2 - y^2 = 1$

45 $y - 2x = 6$ **47** $y^2 = 1 - 2x$

51 $\sqrt{3}/3$ **53** -1 **55** 2 **57** 0

59 $1/\ln 2 \approx 1.44$

Exercises 13.4, page 603

1 Ellipse; vertices $(\frac{3}{2}, \pi/2)$ and $(3, 3\pi/2)$, foci $(0, 0)$ and $(\frac{3}{2}, 3\pi/2)$

3 Hyperbola; vertices $(-3, 0)$ and $(\frac{3}{2}, \pi)$, foci $(0, 0)$ and $(-\frac{9}{2}, 0)$

5 Parabola; $V(\frac{3}{4}, 0)$, $F(0, 0)$

7 Ellipse; vertices $(-4, 0)$ and $(-\frac{4}{3}, \pi)$, foci $(0, 0)$ and $(-\frac{8}{3}, 0)$

9 Hyperbola (except for the points $(\pm 3, 0)$); vertices $(\frac{6}{5}, \pi/2)$ and $(-6, 3\pi/2)$, foci $(0, 0)$ and $(-\frac{36}{5}, 3\pi/2)$

11 $9x^2 + 8y^2 + 12y - 36 = 0$

13 $y^2 - 8x^2 - 36x - 36 = 0$ **15** $4y^2 = 9 - 12x$

17 $3x^2 + 4y^2 + 8x - 16 = 0$

19 $4x^2 - 5y^2 + 36y - 36 = 0$, $y \neq 0$

21 $r = 2/(3 + \cos \theta)$ **23** $r = 12/(1 - 4 \sin \theta)$

25 $r = 5/(1 + \cos \theta)$ **27** $r = 8/(1 + \sin \theta)$

31 $r = 2/(1 + \cos \theta)$, $e = 1$, $r = 2 \sec \theta$

33 $r = 4/(1 + 2 \sin \theta)$, $e = 2$, $r = 2 \csc \theta$

35 $r = 2/(3 + \cos \theta)$, $e = \frac{1}{3}$, $r = 2 \sec \theta$ **37** $-3\sqrt{3}/5$

39 3

Exercises 13.5, page 607

1 π **3** $3\pi/2$ **5** $\pi/2$ **7** $33\pi/2$ **9** $(e^\pi - 1)/4$

11 2 **13** $9\pi/20$ **15** $2\pi + 9\sqrt{3}/2$

17 $4\sqrt{3} - 4\pi/3$ **19** $5\pi/24 - \sqrt{3}/4$

21 $11 \sin^{-1}(\frac{1}{4}) + 3\pi/4 - \sqrt{15}/4$ **23** $4/\sqrt{3}$

25 $64\sqrt{2}/3$ **27** $a^2 \arccos(b/a) - b\sqrt{a^2 - b^2}$

Exercises 13.6, page 610

1 $\frac{2}{27}[34^{3/2} - 125]$ **3** $\sqrt{2}(e^{\pi/2} - 1)$

5 $\ln |2 + \sqrt{3}|$ **7** $\frac{483}{32}$ **9** $\sqrt{2}(1 - e^{-2\pi})$ **11** 2

13 $10\sqrt{26} + 2 \ln(5 + \sqrt{26})$ **15** $\pi^2/8$

19 $\sqrt{2} - \frac{1}{2}\sqrt{5} + \ln(2 + \sqrt{5}) - \ln(1 + \sqrt{2})$

Exercises 13.7, page 614

1 $(8\pi/3)(17^{3/2} - 1)$ **3** $11\pi/9$

5 $(\pi/27)(145^{3/2} - 10^{3/2})$ **7** $64\pi a^3/3$ **9** $536\pi/5$

11 $2\sqrt{2}\pi(2e^\pi + 1)/5$ **13** $128\pi(2\sqrt{2} - 1)/3$

15 $\pi[2\sqrt{5} + \ln(2 + \sqrt{5}) - \sqrt{2} - \ln(1 + \sqrt{2})]$

17 $128\pi/5$ **19** $4\pi^2 a^2$ **21** 10π **25** $4\pi^2 ab$

27 $12\pi a^2/5$

Exercises 13.8, page 615

1 $y = 2(x - 1) - 1/(x - 1)$; 3 **3** $y = e^{-x^2}$; $-2e^{-1}$

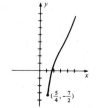   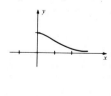

5

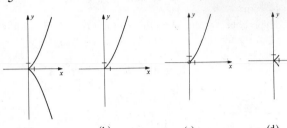

(a)   (b)   (c)   (d)

7 $(x^2 + y^2)^{3/2} = 6(x^2 - y^2)$ **9** $(x^2 + y^2)^2 + 8xy = 0$

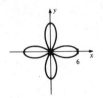

11 $3x - 2y = 6$ **13** $x^2 - y^2 = 1$

15 $x^2 + y^2 = 2(\sqrt{x^2 + y^2} + x)$
17 $8x^2 + 9y^2 + 16x - 64 = 0$
15   **17**

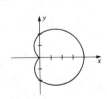   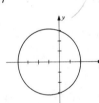

19 -1 **21** $r = 2\cos\theta\sec 2\theta$ **23** 2
25 $\sqrt{2} + \ln(1 + \sqrt{2})$ **27** $(\pi/2)(\sinh 2 + 2)$
29 $\frac{2}{5}\sqrt{2}\pi[e^2(2\cos 1 + \sin 1) - 2]$

CHAPTER 14

Exercises 14.1, page 626

11 $\langle 3, 1\rangle, \langle 1, -7\rangle, \langle 13, 8\rangle, \langle 3, -32\rangle$
13 $\langle -15, 6\rangle, \langle 1, -2\rangle, \langle -68, 28\rangle, \langle 12, -12\rangle$
15 $4\mathbf{i} - 3\mathbf{j}, -2\mathbf{i} + 7\mathbf{j}, 19\mathbf{i} - 17\mathbf{j}, -11\mathbf{i} + 33\mathbf{j}$
17 $-2\mathbf{i} - 5\mathbf{j}, -6\mathbf{i} + 7\mathbf{j}, -6\mathbf{i} - 26\mathbf{j}, -26\mathbf{i} + 34\mathbf{j}$
19 $-3\mathbf{i} + 2\mathbf{j}, 3\mathbf{i} + 2\mathbf{j}, -15\mathbf{i} + 8\mathbf{j}, 15\mathbf{i} + 8\mathbf{j}$ **21** $\langle 4, 7\rangle$
23 $\langle -6, 0\rangle$ **25** $\langle 9, -3\rangle$

27 $\mathbf{a} + \mathbf{b} = 2\mathbf{i} + 7\mathbf{j}, \mathbf{a} - \mathbf{b} = 4\mathbf{i} - 3\mathbf{j}; 2\mathbf{a} = 6\mathbf{i} + 4\mathbf{j};$
    $-3\mathbf{b} = 3\mathbf{i} - 15\mathbf{j}.$

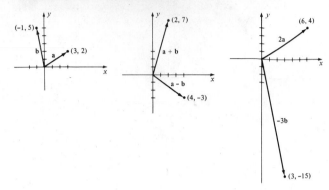

29 $\mathbf{a} + \mathbf{b} = \langle -6, 9\rangle; \mathbf{a} - \mathbf{b} = \langle -2, 3\rangle; 2\mathbf{a} = \langle -8, 12\rangle;$
    $-3\mathbf{b} = \langle 6, -9\rangle.$

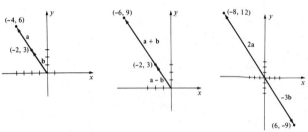

31 $3\sqrt{2}$ **33** 5   **35** $\sqrt{41}$ **37** 18
39 (a) $-\frac{8}{17}\mathbf{i} + \frac{15}{17}\mathbf{j}$ (b) $\frac{8}{17}\mathbf{i} - \frac{15}{17}\mathbf{j}$
41 (a) $\langle 2/\sqrt{29}, -5/\sqrt{29}\rangle$ (b) $\langle -2/\sqrt{29}, 5/\sqrt{29}\rangle$
43 (a) $\langle -12, 6\rangle$ (b) $\langle -3, 1.5\rangle$
45 $(24/\sqrt{65})\mathbf{i} - (42/\sqrt{65})\mathbf{j}$ **49** $p = 3, q = \frac{5}{2}$
53 A circle of radius c and center (x_0, y_0)

Exercises 14.2, page 631

1 (a) $\sqrt{104}$ (b) $(3, 1, -1)$
3 (a) $\sqrt{53}$ (b) $(-\frac{1}{2}, -1, 1)$
5 (a) $\sqrt{3}$ (b) $(\frac{1}{2}, \frac{1}{2}, \frac{1}{2})$
7 $(2, 5, 1), (-4, 2, -3), (-4, 5, 1), (2, 2, -3), (-4, 5, -3),$
    $(2, 2, 1)$
9 *Hint*: Show that $d(A, B)^2 + d(A, C)^2 = d(B, C)^2; 3\sqrt{2}/2$
11 $x^2 + y^2 + z^2 - 6x + 2y - 4z + 5 = 0$
13 $4x^2 + 4y^2 + 4z^2 + 40x - 8z + 103 = 0$
15 (a) $x^2 + y^2 + z^2 + 4x - 8y + 12z + 52 = 0$
    (b) $x^2 + y^2 + z^2 + 4x - 8y + 12z + 40 = 0$
    (c) $x^2 + y^2 + z^2 + 4x - 8y + 12z + 20 = 0$
17 $C(-2, 1, -1); 2$ **19** $C(4, 0, -4); 4$
21 $C(0, -2, 0); 2$ **23** $6x - 12y + 4z + 13 = 0$; a plane
25 (a) A plane parallel to the xy-plane
    (b) A plane parallel to the xz-plane
    (c) The yz-plane
27 All points within and on a sphere of radius 1 with center
    at the origin.

29 All points within and on a rectangular parallelepiped with center at the origin and having edges of lengths 2, 4, and 6.

31 All points inside and on a right circular cylinder with radius 1 and vertical axis the z-axis, excluding the points on the z-axis.

33 All points inside the sphere with center $(2, 1, 0)$ and radius 2.

Exercises 14.3, page 639

3 $\langle 1, 3, 0 \rangle$; $\langle -22, 42, 9 \rangle$; -25; $\sqrt{41}$; $3\sqrt{41}$; $3\sqrt{41}$; $\langle -5, 9, 2 \rangle$; $\sqrt{110}$

5 $4\mathbf{i} - 2\mathbf{j} - 3\mathbf{k}$; $11\mathbf{i} - 28\mathbf{j} + 30\mathbf{k}$; -15; $\sqrt{29}$; $3\sqrt{29}$; $3\sqrt{29}$; $2\mathbf{i} - 6\mathbf{j} + 7\mathbf{k}$; $\sqrt{89}$

7 $\mathbf{i} + \mathbf{k}$; $5\mathbf{i} + 9\mathbf{j} - 4\mathbf{k}$; -1; $\sqrt{2}$; $3\sqrt{2}$; $3\sqrt{2}$; $\mathbf{i} + 2\mathbf{j} - \mathbf{k}$; $\sqrt{6}$

11 3 **13** -12 **15** -99 **17** $-3/\sqrt{30}$

19 0 **21** $-3/\sqrt{534}$ **23** $\frac{6}{13}$ **25** 74

27 $\cos^{-1}(37/\sqrt{3081}) \approx 48.20°$ **29** $-82/\sqrt{126}$

31 $-2(x - 8) + 4(y + 3) - 12(z - 5) = 0$; a plane through P.

33 1 **35** $-12/\sqrt{3}$ ft-lb.   **37** $\frac{3}{10}$

39 $(1/\sqrt{1589})\langle 32, 23, 6 \rangle$

41 (a) $2\mathbf{a} = 28\mathbf{i} - 30\mathbf{j} + 12\mathbf{k}$ (b) $-\frac{1}{3}\mathbf{a} = -\frac{14}{3}\mathbf{i} + 5\mathbf{j} - 2\mathbf{k}$

43 (a) If $\mathbf{a}$ and $\mathbf{b}$ have the same or opposite directions
(b) If $\mathbf{a}$ and $\mathbf{b}$ have the same direction

Exercises 14.4, page 647

1 $\langle 5, 10, 5 \rangle$ **3** $-6\mathbf{i} - 8\mathbf{j} + 18\mathbf{k}$ **5** $\langle -4, 2, -1 \rangle$

7 $-40\mathbf{i} + 15\mathbf{k}$ **9** 0

11 $\langle 12, -14, 24 \rangle$, $\langle 16, -2, -5 \rangle$

13 $(\mathbf{a} \times \mathbf{b}) \cdot \mathbf{b} = \mathbf{a} \cdot (\mathbf{b} \times \mathbf{b}) = \mathbf{a} \cdot \mathbf{0} = 0$

17 (a) $13\mathbf{i} + 7\mathbf{j} + 5\mathbf{k}$ (b) $9\sqrt{3}/2$

19 (a) $c(5\mathbf{i} + 4\mathbf{j} + 10\mathbf{k})$ where $c \neq 0$ (b) $\sqrt{141}$ **23** 16

Exercises 14.5, page 649

1 $x = 5 - 3t$, $y = -2 + 8t$, $z = 4 - 3t$; $(1, \frac{26}{3}, 0)$, $(\frac{17}{4}, 0, \frac{13}{4})$, $(0, \frac{34}{3}, -1)$

3 $x = 2 - 8t$, $y = 0$, $z = 5 - 2t$; $(-18, 0, 0)$, $(0, 0, \frac{9}{2})$, (line lies on the xz-plane)

5 $x = 4 + \frac{1}{3}t$, $y = 2 + 2t$, $z = -3 + \frac{1}{2}t$

7 $x = 0$, $y = t$, $z = 0$

9 $x = -6 - 3s$, $y = 4 + s$, $z = -3 + 9s$

11 $\theta = \cos^{-1}(15/2\sqrt{171}) \approx 55°$ and $\pi - \theta$ **13** $(5, -7, 3)$

15 Do not intersect   **17** Intersect at $(2, -4, -1)$

19 $\sqrt{5411}/3\sqrt{10}$ **21** $\sqrt{474/17}$

23 $x = 3 + 2t$, $y = 1 + t$, $z = -2 - 3t$
(There are other forms for the equations.)

Exercises 14.6, page 656

9 (a) $z = 4$ (b) $x = 6$ (c) $y = -7$

11 $6x - 5y - z + 84 = 0$

13 $3x - y + 2z + 11 = 0$

15 $20x - 5y + 2z = 0$ **17** $11x - 2y - 6z - 69 = 0$

19 $(x - 5)/3 = (y + 2)/(-8) = (z - 4)/3$

21 $(x - 4)/(-7) = (z + 3)/8$; $y = 2$

23 (a) No   (b) Yes   (c) No

25 (a) No   (b) No   (c) Yes, at $(8, 0, 1)$ **27** $7/\sqrt{59}$

29 3 **31** $17/6\sqrt{14}$ **33** $6x + 11y + 4z = 38$

35 $71x - 15y + 17z = -10$ **37** $89/\sqrt{521}$

Exercises 14.7, page 661

1   **3**   **5**

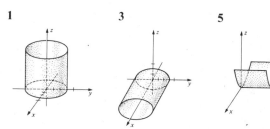

7   **9**   **11**

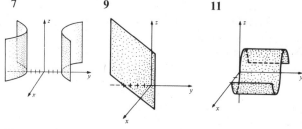

13   **15**

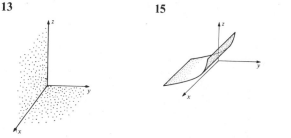

17 $x^2 + z^2 + 4y^2 = 16$ **19** $z = 4 - x^2 - y^2$

21 $y^2 + z^2 - x^2 = 1$

23   **25**

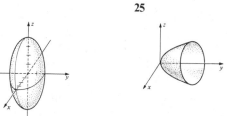

Exercises 14.8, page 665

1 Paraboloid; axis along the z-axis

3 Hyperboloid of one sheet; axis along the z-axis

5 Hyperboloid of two sheets; axis along the x-axis

7 Cone; axis along the x-axis   **9** Ellipsoid
11 Hyperboloid of one sheet; axis along the x-axis
13 Hyperboloid of two sheets; axis along the y-axis
15 Paraboloid; axis along the x-axis
17 Cone; axis along the y-axis
19 Hyperbolic paraboloid
21 $x^2 + y^2 = -4z$; paraboloid of revolution
23 $x^2 + y^2 = 3z^2$; cone

Exercises 14.9, page 669

1 (a) $(0, 5, 3)$ (b) $(3, 3\sqrt{3}, -5)$
3 (a) $(\sqrt{10}, \pi/4, \cos^{-1}(-2/\sqrt{5}))$ (b) $(2, \pi/3, \pi/2)$
5 (a) $(2\sqrt{3}, \pi/3, 2)$ (b) $(1, \pi/4, -\sqrt{3})$
7 $x^2 + y^2 = 16$; circular cylinder
9 $z = \sqrt{3x^2 + 3y^2}$; half cone
11 $x^2 + y^2 = 4x$; circular cylinder
13 $x^2 + y^2 + z^2 = 4z$; sphere   **15** The z-axis
17 A cylinder with a lemniscate as its directrix and rulings parallel to the z-axis
19 Spheres of radii 1 and 2 with centers at the origin
21 $r^2 + z^2 = 4, \rho = 2$
23 $3r \cos \theta + r \sin \theta - 4z = 12$,
$\rho(3 \sin \phi \cos \theta + \sin \phi \sin \theta - 4 \cos \phi) = 12$
25 $r^2 \sin^2 \theta + z^2 = 9, \rho^2(\sin^2 \phi \sin^2 \theta + \cos^2 \phi) = 9$
27 $r = 6 \cos \theta, \rho \sin \phi = 6 \cos \theta$
29 $z = \tan \theta, \rho \cos \phi = \tan \theta$

Exercises 14.10, page 670

1 $12\mathbf{i} + 19\mathbf{j}$ **3** -8 **5** $\sqrt{29} - \sqrt{17}$
7 $\tan^{-1}(\frac{5}{2})$ **9** $(1/\sqrt{29})(5\mathbf{i} - 2\mathbf{j})$
11 (a) $\sqrt{38}$ (b) $(2, -\frac{7}{2}, \frac{5}{2})$
(c) $x^2 + y^2 + z^2 + 2x + 8y - 6z + 10 = 0$
(d) $y = -4$
(e) $x = 5 + 6t, y = -3 + t, z = 2 - t$
(f) $6x + y - z - 25 = 0$
13 $6x - 15y + 5z = 30$ **15** $(x^2/64) + (y^2/9) + z^2 = 1$
17 Sphere with center $(7, -3, 4)$ and radius 8
19 Plane; x-, y-, and z-intercepts $\frac{10}{3}$, -2, and 5, respectively
21 Elliptic cylinder with rulings parallel to the y-axis
23 Hyperboloid of one sheet; axis along the x-axis
25 Hyperboloid of two sheets; axis along the z-axis
27 Hyperbolic paraboloid   **29** 36   **31** $\sqrt{33} + \sqrt{37}$
33 $3/\sqrt{26}, -1/\sqrt{26}, -4/\sqrt{26}$ **35** $22\mathbf{i} - 2\mathbf{j} + 17\mathbf{k}$
37 $9/\sqrt{33}$ **39** 26   **41** 0   **43** $-4\mathbf{i} - 10\mathbf{j} + 4\mathbf{k}$
45 (a) $(1/\sqrt{66})\langle 1, 4, 7 \rangle$ (b) $x + 4y + 7z - 5 = 0$
(c) $x = 2 + 7t, y = -1 - 7t, z = 1 + 3t$
(d) 59 (e) $\cos^{-1}(59/\sqrt{3745}) \approx 15.4°$ (f) $\sqrt{66}$
47 $x = 3 + 2t, y = -1 - 4t, z = 5 + 8t$;
$x = -1 + 7t, y = 6 - 2t, z = -\frac{7}{2} - 2t$
49 $\cos^{-1}(-25/\sqrt{2295}) \approx 121.46°$

51 $(2\sqrt{2}, -\pi/4, 1), (3, \cos^{-1}\frac{1}{3}, -\pi/4)$
53 $z = -\sqrt{(x^2 + y^2 + z^2)/2}$; bottom half of a cone
55 $x = 1$; plane   **57** $r = 1, \rho^2 \sin^2 \phi = 1$
59 $r^2 + z^2 - 2z = 0, \rho - 2 \cos \phi = 0$

CHAPTER 15

Exercises 15.1, page 675

1 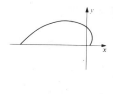   **3**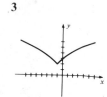

5   **7** Elliptic helix   **9** Twisted cubic

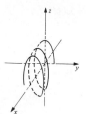   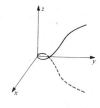

11 Parabola on the   **13**
plane $z = 3$

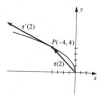   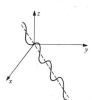

15 $5[4\sqrt{17} + \ln (4 + \sqrt{17})]/4 \approx 23.23$
17 $\sqrt{3}(e^{2\pi} - 1)$ **19** 7

Exercises 15.2, page 682

1 (a) Continuous throughout its domain $[1, 2]$
(b) $\mathbf{r}'(t) = \frac{1}{2}(t - 1)^{-1/2}\mathbf{i} - \frac{1}{2}(2 - t)^{-1/2}\mathbf{j}$,
$\mathbf{r}''(t) = -\frac{1}{4}(t - 1)^{-3/2}\mathbf{i} - \frac{1}{4}(2 - t)^{-3/2}\mathbf{j}$
3 (a) Continuous throughout its domain $\{t: t \neq \pi/2 + n\pi\}$
(b) $\mathbf{r}'(t) = \sec^2 t\mathbf{i} + (2t + 8)\mathbf{j}$,
$\mathbf{r}''(t) = 2 \sec^2 t \tan t\mathbf{i} + 2\mathbf{j}$
5 (b) $\mathbf{r}'(t) = -t^3\mathbf{i} + 2t\mathbf{j}$,
$\mathbf{r}(2) = -4\mathbf{i} + 4\mathbf{j}$,
$\mathbf{r}'(2) = -8\mathbf{i} + 4\mathbf{j}$

7 (b) $\mathbf{r}'(t) = -4\sin t\mathbf{i} + 2\cos t\mathbf{j}$,
$\mathbf{r}(3\pi/4) = -2\sqrt{2}\mathbf{i} + \sqrt{2}\mathbf{j}$,
$\mathbf{r}'(3\pi/4) = -2\sqrt{2}\mathbf{i} - \sqrt{2}\mathbf{j}$

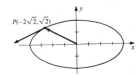

9 (b) $\mathbf{r}'(t) = 3t^2\mathbf{i} - 3t^{-4}\mathbf{j}$; $\mathbf{r}(1) = \mathbf{i} + \mathbf{j}$, $\mathbf{r}'(1) = 3\mathbf{i} - 3\mathbf{j}$
11 (b) $\mathbf{r}'(t) = 2\mathbf{i} - \mathbf{j}$; $\mathbf{r}(3) = 5\mathbf{i} + \mathbf{j}$
9   **11**

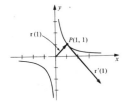 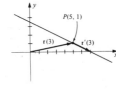

13 (a) Domain $= \{t: t \neq \pi/2 + k\pi, k$ any integer$\}$,
$\mathbf{r}$ is continuous on D.
(b) $\mathbf{r}'(t) = 2t\mathbf{i} + \sec^2 t\mathbf{j}$; $\mathbf{r}''(t) = 2\mathbf{i} + 2\sec^2 t \tan t\mathbf{j}$
15 (a) $t \geq 0, t \geq 0$
(b) $(1/2\sqrt{t})\mathbf{i} + 2e^{2t}\mathbf{j} + \mathbf{k}$, $(-1/4t\sqrt{t})\mathbf{i} + 4e^{2t}\mathbf{j}$
17 $x = 1 + 6t, y = -2 - 10t, z = 10 + 8t$
19 $x = 1 + s, y = s, z = 4$ **21** $\pm(1/\sqrt{5})\langle 2, -1, 0\rangle$
23 $16\mathbf{i} - 8\mathbf{j} + 6\mathbf{k}$
25 $(1 - 1/\sqrt{2})\mathbf{i} - (1/\sqrt{2})\mathbf{j} + \frac{1}{2}\ln 2\mathbf{k}$
27 $(\frac{1}{3}t^3 + 2)\mathbf{i} + (3t^2 + t - 3)\mathbf{j} + (2t^4 + 1)\mathbf{k}$
29 $(t^3 + t + 7)\mathbf{i} - (t^4 - 2t)\mathbf{j} + (\frac{1}{2}t^2 - 3t + 1)\mathbf{k}$
31 $x + y - 1 = 0$
45 $(1 + 5t^2)\sin t + (3t + 2t^3)\cos t$;
$[(t^3 + 4t)\sin t - t^2 \cos t]\mathbf{i}$
$+ [(3t^2 - 2)\sin t + (t^3 - 2t)\cos t]\mathbf{j}$
$+ [-3t \sin t + (1 - t^2)\cos t]\mathbf{k}$

Exercises 15.3, page 688

Answers for Exercises 1–7 are in the order, $\mathbf{r}'(t)$, $\mathbf{r}''(t)$, $|\mathbf{r}'(t)|$,
$\mathbf{r}'(a)$, $\mathbf{r}''(a)$ where a is the indicated time.
1 $2\mathbf{i} + 8t\mathbf{j}$, $8\mathbf{j}$, $2\sqrt{1 + 16t^2}$, $2\mathbf{i} + 8\mathbf{j}$, $8\mathbf{j}$
3 $\dfrac{-2}{t^2}\mathbf{i} - \dfrac{3}{(t + 1)^2}\mathbf{j}$, $\dfrac{4}{t^3}\mathbf{i} + \dfrac{6}{(t + 1)^3}\mathbf{j}$, $\sqrt{\dfrac{4}{t^4} + \dfrac{9}{(t + 1)^4}}$,
$-\frac{1}{2}\mathbf{i} - \frac{1}{3}\mathbf{j}$, $\frac{1}{2}\mathbf{i} + \frac{2}{9}\mathbf{j}$
5 $\cos t\mathbf{i} - 8\sin 2t\mathbf{j}$, $-\sin t\mathbf{i} - 16\cos 2t\mathbf{j}$,
$\sqrt{\cos^2 t + 64\sin^2 2t}$, $(\sqrt{3}/2)\mathbf{i} - 4\sqrt{3}\mathbf{j}$, $-\frac{1}{2}\mathbf{i} - 8\mathbf{j}$
7 $2e^{2t}\mathbf{i} - e^{-t}\mathbf{j}$, $4e^{2t}\mathbf{i} + e^{-t}\mathbf{j}$, $\sqrt{4e^{4t} + e^{-2t}}$, $2\mathbf{i} - \mathbf{j}$, $4\mathbf{i} + \mathbf{j}$
9 $-\sin t\mathbf{i} + \cos t\mathbf{j} + \mathbf{k}$, $-\cos t\mathbf{i} - \sin t\mathbf{j}$, $\sqrt{2}$
11 $2t\mathbf{i} + (1/\sqrt{t})\mathbf{j} + 6\sqrt{t}\mathbf{k}$, $2\mathbf{i} - (1/2\sqrt{t^3})\mathbf{j} + (3/\sqrt{t})\mathbf{k}$,
$\sqrt{4t^2 + (1/t) + 36t}$
13 $e^t(\cos t - \sin t)\mathbf{i} + e^t(\cos t + \sin t)\mathbf{j} + e^t\mathbf{k}$,
$-2e^t \sin t\mathbf{i} + 2e^t \cos t\mathbf{j} + e^t\mathbf{k}$, $\sqrt{3}e^t$

15 $\mathbf{i} + 2\mathbf{j} + 3\mathbf{k}, 0, \sqrt{14}$
19 (a) $750\sqrt{3}\mathbf{i} + (-gt + 750)\mathbf{j}$ (b) $(1500)^2/8g \approx 8{,}789$ ft
(c) $(1500)^2\sqrt{3}/2g \approx 60{,}892$ ft   (d) 1500 ft/sec
21 $40\sqrt{5} \approx 89.4$ ft/sec   **23** (a) $18{,}054$ mi/hr   (b) 86.7 min

Exercises 15.4, page 695

1 $6/10^{3/2}$ **3** 2 **5** 4 **7** $2/17^{3/2}$ **9** 0
11 $48/21^{3/2}$ **13** (a) 1 (b) $(\pi/2, 0)$
15 (a) $2\sqrt{2}$ (b) $(-2, 3)$ **17** $(\ln \sqrt{2}, 1/\sqrt{2})$
19 $(0, \pm 3)$ **21** $(\sqrt{2}/2, -\frac{1}{2}\ln 2)$ **27** $(\pm\sqrt{2}, -20)$
29 $(0, 0)$ **33** $|8 - 3\sin^2 2\theta|/(1 + 3\cos^2 2\theta)^{3/2}$
37 $(-\frac{14}{3}, \frac{29}{12})$ **39** $(4, -2)$ **41** $(0, -4)$

Exercises 15.5, page 700

1 $4t/(4t^2 + 9)^{1/2}, 6/(4t^2 + 9)^{1/2}$
3 $6t(t^2 + 2)/(t^4 + 4t^2 + 1)^{1/2}$,
$6(t^4 + t^2 + 1)^{1/2}/(t^4 + 4t^2 + 1)^{1/2}$
5 $t/(1 + t^2)^{1/2}, (2 + t^2)/(1 + t^2)^{1/2}$
7 $-65 \sin t \cos t/(16 \sin^2 t + 81 \cos^2 t + 1)^{1/2}$,
$\dfrac{(81 \sin^2 t + 16 \cos^2 t + 1296)^{1/2}}{(16 \sin^2 t + 81 \cos^2 t + 1)^{1/2}}$
9 $6/(4t + 9)^{3/2}$ **11** $2(t^4 + t^2 + 1)^{1/2}/3(t^4 + 4t^2 + 1)^{3/2}$
13 $(2 + t^2)/(1 + t^2)^{3/2}$
15 $\dfrac{(81 \sin^2 t + 16 \cos^2 t + 1296)^{1/2}}{(16 \sin^2 t + 81 \cos^2 t + 1)^{3/2}}$ **17** $36/\sqrt{5}, 18/\sqrt{5}$

Exercises 15.7, page 707

1 (b) $\mathbf{r}'(t) = 2t\mathbf{i} + (8t - 4t^3)\mathbf{j}$, $\mathbf{r}''(t) = 2\mathbf{i} + (8 - 12t^2)\mathbf{j}$
3 (a) $(1/\sqrt{3})(\mathbf{i} + \mathbf{j} + \mathbf{k})$ (b) $\sqrt{3}(e - 1) \approx 2.98$
5 $x = 4 + 6t, y = 4 + 4t, z = 1 + t$
7 $(72t^2 + 10t)\mathbf{i} + (2t - 16t^3)\mathbf{j} - (15t^2 + 12t)\mathbf{k}$; $15t^2 - 60t$
9 $2\mathbf{i} + \frac{1}{4}\mathbf{j} - \mathbf{k}$ **13** $\sqrt{2}/2$
15 $(4t^6 + 9t^4 + 1)^{1/2}/4(t^6 + t^2 + 1)^{3/2}$
17 (a) $x^2 + y^2 - 4y + 3 = 0$
19 $(\sin t \cos t - 8 \sin 2t \cos 2t)/(4 \cos^2 2t + \sin^2 t)^{1/2}$;
$(2|\cos 2t \cos t + 2 \sin 2t \sin t|)/(4 \cos^2 2t + \sin^2 t)^{1/2}$

CHAPTER 16

Exercises 16.1, page 716

1 $\mathbb{R} \times \mathbb{R}$; $-29, 6, -4$ **3** $\{(u, v): u \neq 2v\}$; $-\frac{3}{2}, \frac{4}{9}, 0$
5 $\{(x, y, z): x^2 + y^2 + z^2 \leq 25\}$; $4, 2\sqrt{3}$
7 $\{(r, s, v, p): v \neq \pi/2 + n\pi, p > 0\}$; $3 - \pi$
9 $2x + h, 2y + h$ **11** $y^2 + 3, 2xy + xh$
13 The top half of the sphere $x^2 + y^2 + z^2 = 1$
15 The plane with x-, y-, and z-intercepts 3, 2, and 6,
respectively
17 The plane $z = 5$

19 The upper half of a hyperboloid of one sheet having its axis along the x-axis

21 A hyperbolic paraboloid

23   **25**

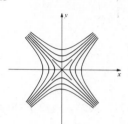 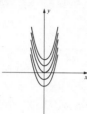

27   **29**

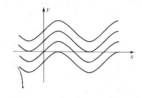

31 $y \arctan x = \pi$

33 The origin and all spheres with center at the origin

35 All planes with normal vector $\langle 1, 2, 3 \rangle$

37 The z-axis and all circular cylinders having axis along the z-axis.

39 (a) Circles with centers at the origin   (b) $x^2 + y^2 = 100$

41 Five; spheres with centers at the origin. The force F is constant if (x, y, z) moves along a level surface.

Exercises 16.2, page 722

1 $-\frac{2}{3}$ **3** Does not exist   **5** Does not exist

7 Does not exist   **9** 0

11 Continuous on $\{(x, y) : x + y > 1\}$

13 Continuous at (x, y, z) if $x^2 + y^2 \neq z^2$

15 Continuous on $\{(x, y) : x > 0, -1 < y < 1\}$

17 $(x^4 - 2x^2y^2 + y^4 - 4)/(x^2 - y^2)$; $\{(x, y) : y \neq \pm x\}$

19 $x^2 + 2x \tan y + \tan^2 y + 1$;

     $\{(x, y) : y \neq \pi/2 + k\pi, k \text{ any integer}\}$

21 $e^{x^2 + 2y}$, $(x^2 + 2y)(x^2 + 2y - 3)$, $e^{2t} + 2t^2 - 6t$

23 $(x - 2y)(2x + y) - 3(x - 2y) + (2x + y)$

27 $\lim_{(x, y, z, w) \to (a, b, c, d)} f(x, y, z, w) = L$ means that for every $\varepsilon > 0$ there corresponds a $\delta > 0$ such that if

     $0 < \sqrt{(x - a)^2 + (y - b)^2 + (z - c)^2 + (w - d)^2} < \delta$,

     then $|f(x, y, z, w) - L| < \varepsilon$.

Exercises 16.3, page 728

1 $f_x(x, y) = 8x^3y^3 - y^2$, $f_y(x, y) = 6x^4y^2 - 2xy + 3$

3 $f_r(r, s) = r/\sqrt{r^2 + s^2}$, $f_s(r, s) = s/\sqrt{r^2 + s^2}$

5 $f_x(x, y) = e^y + y \cos x$, $f_y(x, y) = xe^y + \sin x$

7 $f_t(t, v) = -v/(t^2 - v^2)$, $f_v(t, v) = t/(t^2 - v^2)$

9 $f_x(x, y) = \cos(x/y) - (x/y) \sin(x/y)$,

     $f_y(x, y) = (x/y)^2 \sin(x/y)$

11 $f_r(r, s, t) = 2re^{2s} \cos t$, $f_s(r, s, t) = 2r^2e^{2s} \cos t$,

     $f_t(r, s, t) = -r^2e^{2s} \sin t$

13 $f_x(x, y, z) = (y^2 + z^2)^x \ln(y^2 + z^2)$,

     $f_y(x, y, z) = 2xy(y^2 + z^2)^{x-1}$,

     $f_z(x, y, z) = 2xz(y^2 + z^2)^{x-1}$

15 $f_x(x, y, z) = e^z - ye^x$, $f_y(x, y, z) = -e^x - ze^{-y}$,

     $f_z(x, y, z) = xe^z + e^{-y}$

17 $f_q(q, v, w) = v/2\sqrt{qv}\sqrt{1 - qv}$,

     $f_v(q, v, w) = (q/2\sqrt{qv}\sqrt{1 - qv}) + w \cos vw$,

     $f_w(q, v, w) = v \cos vw$

25 $18xy^2 + 16y^3z$ **27** $t^2 \sec rt(\sec^2 rt + \tan^2 rt)$

29 $(1 - x^2y^2z^2) \cos xyz - 3xyz \sin xyz$

47 $w_{xx}, w_{xy}, w_{xz}, w_{yx}, w_{yy}, w_{yz}, w_{zx}, w_{zy}, w_{zz}$

49 ten   **51** In degrees per cm: (a) 200   (b) 400

53 In volts per in.: (a) $-\frac{100}{9}$ (b) $\frac{50}{9}$ (c) $-\frac{50}{9}$

55 $x = 1$, $y = t$, $z = -4t + 12$

Exercises 16.4, page 737

There are other correct answers to 1 and 3 (see Example 2).

1 $\varepsilon_1 = -3\Delta y$, $\varepsilon_2 = 4\Delta y$

3 $\varepsilon_1 = 3x\Delta x + (\Delta x)^2$, $\varepsilon_2 = 3y\Delta y + (\Delta y)^2$

5 $dw = (3x^2 - 2xy) \, dx + (6y - x^2) \, dy$

7 $dw = 2x \sin y \, dx + (x^2 \cos y + 3y^{1/2}) \, dy$

9 $dw = (x^2y + 2x)e^{xy} \, dx + (x^3e^{xy} - 2y^{-3}) \, dy$

11 $dw = 2x \ln(y^2 + z^2) \, dx + (2x^2y/(y^2 + z^2)) \, dy$

     $+ (2x^2z/(y^2 + z^2)) \, dz$

13 $dw = \dfrac{yz(y + z)}{(x + y + z)^2} \, dx + \dfrac{xz(x + z)}{(x + y + z)^2} \, dy$

     $+ \dfrac{xy(x + y)}{(x + y + z)^2} \, dz$

15 $dw = (2xz - z^2t) \, dx + 4t^3 \, dy + (x^2 - 2xzt) \, dz$

     $+ (12yt^2 - xz^2) \, dt$

17 Approximately 7.38   **19** Approximately 1.87

21 Approximately 18.006   **23** (a) $\frac{1}{4}$ ft²   (b) $\frac{47}{192}$ ft³

25 $57\pi(0.015)/4 \approx 0.67$ in.³   **27** 0.0185

29 $6W/(A - W)\%$ **31** 7%   **33** 2.96

35 1.7π in.²

Exercises 16.5, page 744

1 $\partial w/\partial x = 2x \sin(xy) + y(x^2 + y^2) \cos(xy)$;

     $\partial w/\partial y = 2y \sin(xy) + x(x^2 + y^2) \cos(xy)$

3 $\partial w/\partial r = 2r(\ln s)^2 + 8r \ln s + 2s \ln s$;

     $\partial w/\partial s = (2r^2 \ln s)/s + (4r^2/s) + 2r + 2r \ln s$

5 $\partial z/\partial x = 3x^2e^{3y} + ye^x + 4x^3y^2$;

     $\partial z/\partial y = 3x^3e^{3y} + e^x + 2x^4y$

7 $\partial r/\partial u = 3 \ln(uvt) + 3 + (vt/u)$;

     $\partial r/\partial v = t \ln(uvt) + (3u/v) + t$;

     $\partial r/\partial t = v \ln(uvt) + (3u/t) + v$

9 $-34y + 6r - 24s$ **11** $-3(t^2 + 1)/(t + 1)^4$

13 $4 \sin^3 t \cos t + \sin t \tan 4t - 4 \cos t \sec^2 4t$

15 $-(6x^2 + 2xy)/(x^2 + 3y^2)$

17 $-\left(6 + \dfrac{\sqrt{y}}{2\sqrt{x}}\right)\bigg/\left(\dfrac{\sqrt{x}}{2\sqrt{y}} - 3\right) = \dfrac{12\sqrt{xy} + y}{6\sqrt{xy} - x}$

19 $\partial z/\partial x = -(2z^3 + 2xy^2)/(6xz^2 - 6yz + 4)$;
$\partial z/\partial y = -(2x^2y - 3z^2)/(6xz^2 - 6yz + 4)$

21 $\partial z/\partial x = -(e^{yz} - 2yze^{xz} + 3yze^{xy})/(xye^{yz} - 2xye^{xz} + 3e^{xy})$;
$\partial z/\partial y = -(xze^{yz} - 2e^{xz} + 3xze^{xy})/(xye^{yz} - 2xye^{xz} + 3e^{xy})$

33 $0.88\pi \approx 2.76$ in.³/min; $0.46\pi \approx 1.44$ in.²/min

35 $dT/dt = (v/c)(dp/dt) + (p/c)(dv/dt)$

37 -6.4 in.³/min

Exercises 16.6, page 753

1 $-\frac{4}{5}\mathbf{i} + \frac{3}{5}\mathbf{j}$ **3** $3\mathbf{i} + 2\mathbf{j}$ **5** $-8\mathbf{i} + \mathbf{j} - 9\mathbf{k}$
7 $-10/\sqrt{2}$ **9** $-1/8\sqrt{13}$ **11** $67/8\sqrt{26}$
13 $1/2\sqrt{26}$ **15** $16\sqrt{14}$ **17** $15e^{-2}/\sqrt{35}$
19 $-12/\sqrt{10}$
21 (a) $-28/\sqrt{26}$ (b) $(1/\sqrt{5})(\mathbf{i} - 2\mathbf{j})$, $\sqrt{80}$
    (c) $(1/\sqrt{5})(-\mathbf{i} + 2\mathbf{j})$, $-\sqrt{80}$
23 (a) $-25/\sqrt{14}\sqrt{77}$
    (b) $(1/\sqrt{14})(-2\mathbf{i} + 3\mathbf{j} + \mathbf{k})$, 1
    (c) $(1/\sqrt{14})(2\mathbf{i} - 3\mathbf{j} - \mathbf{k})$, -1
25 $-28/\sqrt{2}$, $-12\mathbf{i} - 16\mathbf{j}$, $12\mathbf{i} + 16\mathbf{j}$, $c(4\mathbf{i} - 3\mathbf{j})$ for any $c \neq 0$
27 $-178/\sqrt{14}$, $4\mathbf{i} - 8\mathbf{j} + 54\mathbf{k}$, $\sqrt{2996} \approx 54.8$
37 (b) $5 + \sqrt{3}$

Exercises 16.7, page 759

1 $16(x - 2) + 6(y + 3) + 6(z - 1) = 0$;
$x = 2 + 16t, y = -3 + 6t, z = 1 + 6t$
3 $16(x + 2) + 18(y + 1) + (z - 25) = 0$;
$x = -2 + 16t, y = -1 + 18t, z = 25 + t$.
5 $4(x + 5) - 3(y - 5) + 20(z - 1) = 0$;
$x = -5 + 4t, y = 5 - 3t, z = 1 + 20t$
7 $x + \sqrt{3}(y - \pi/3) + (z - 1) = 0$;
$x = t, y = (\pi/3) + \sqrt{3}t, z = 1 + t$
9 $-x + \frac{1}{2}(y - 2) - (z - 1) = 0$;
$x = -t, y = 2 + \frac{1}{2}t, z = 1 - t$
15 $(8\sqrt{2}/\sqrt{5}, 2\sqrt{2}/\sqrt{5}, -2\sqrt{2}/\sqrt{5})$,
$(-8\sqrt{2}/\sqrt{5}, -2\sqrt{2}/\sqrt{5}, 2\sqrt{2}/\sqrt{5})$

21
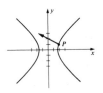

23
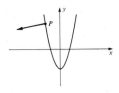

25

27
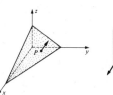

29

Exercises 16.8, page 764

Answers for Exercises 1–15 refer to local maxima and minima.
1 Min: $f(0, 0) = 0$ **3** Min: $f(1, -1) = -1$
5 Min: $f(\frac{1}{2}, -\frac{1}{4}) = -\frac{1}{2}$
7 Min: $f(-2, \sqrt{3}) = -48 - 6\sqrt{3}$ **9** No extrema
11 Min: $f(\sqrt[3]{2}, 2\sqrt[3]{2}) = 12/\sqrt[3]{2}$
13 Max: $f(\pi/3, \pi/3) = 3\sqrt{3}/2$; min: $f(5\pi/3, 5\pi/3) = -3\sqrt{3}/2$
15 At the point $(0, 0)$ f attains neither a maximum nor a minimum.
17 Max: $f(4, 3) = 67$; min: $f(0, 0) = 0$
19 Max: $f(-1, -2) = 13$; min: $f(1, -1) = -1 = f(1, 2)$
21 Max: $f(-\sqrt{2}/2, \sqrt{2}/4) = 1 + \sqrt{2}$;
    min: $f(\frac{1}{2}, -\frac{1}{4}) = -\frac{1}{2}$
23 $1/\sqrt{26}$ **25** $(\pm 2/\sqrt[4]{12}, \pm\sqrt[4]{12}, \pm 2\sqrt{2}/\sqrt[4]{12})$
27 Square base; altitude $\frac{1}{2}$ the length of the side of the base
29 $6/\sqrt{3}, 12/\sqrt{3}, 8/\sqrt{3}$ **31** $1, \frac{4}{3}, 4$
33 Square base of side $\sqrt[3]{4}$ ft, height $2\sqrt[3]{4}$ ft
35 $(14$ in.$) \times (14$ in.$) \times (28$ in.$)$ **39** $y = \frac{1}{2}x + \frac{8}{3}$
41 $y = mx + b$, where $m \approx 1.23, b \approx -18.13$; grade of 68

Exercises 16.9, page 773

1 Min: $f(1/\sqrt{5}, 2/\sqrt{5}) = 0 = f(-1/\sqrt{5}, -2/\sqrt{5})$;
    max: $f(2/\sqrt{5}, -1/\sqrt{5}) = 5 = f(-2/\sqrt{5}, 1/\sqrt{5})$
3 Min: $f(-5/\sqrt{3}, -5/\sqrt{3}, -5/\sqrt{3}) = -5/\sqrt{3}$;
    max: $f(5/\sqrt{3}, 5/\sqrt{3}, 5/\sqrt{3}) = 5/\sqrt{3}$
5 Min: $f(\frac{1}{3}, -\frac{1}{3}, \frac{1}{3}) = \frac{1}{3}$
7 Min: $f(0, -1, 0) = 1$ and $f(2, 1, 0) = 5$
9 Min: $f(1, 2/\sqrt{3}, -1, \frac{8}{3}) = -16/3\sqrt{3}$;
    max: $f(1, -2/\sqrt{3}, -1, \frac{8}{3}) = 16/3\sqrt{3}$
11 $(6/\sqrt{29}, 9/\sqrt{29}, 12/\sqrt{29})$
13 Square base $2/\sqrt[3]{7}$ by $2/\sqrt[3]{7}$, height $7/2\sqrt[3]{7}$
15 Height = twice the radius
29 Width $= 8\sqrt{3}$ in., depth $= \frac{16}{3}\sqrt{6}$ in.

Exercises 16.10, page 774

1 $\{(x, y): 4x^2 - 9y^2 \leq 36\}$ **3** $\{(x, y): z^2 > x^2 + y^2\}$
5 $f_x(x, y) = 3x^2 \cos y + 4$; $f_y(x, y) = -x^3 \sin y - 2y$
7 $f_x(x, y, z) = 2x/(y^2 + z^2)$;
    $f_y(x, y, z) = 2y(z^2 - x^2)/(y^2 + z^2)^2$;
    $f_z(x, y, z) = -(x^2 + y^2)2z/(y^2 + z^2)^2$

9 $f_x(x, y, z, t) = 2xz\sqrt{2y + t}$; $f_y(x, y, z, t) = x^2z/\sqrt{2y + t}$;
$f_z(x, y, z, t) = x^2\sqrt{2y + t}$; $f_t(x, y, z, t) = x^2z/2\sqrt{2y + t}$

11 $f_{xx}(x, y) = 6xy^2 + 12x^2$, $f_{yy}(x, y) = 2x^3 - 18xy$,
$f_{xy}(x, y) = 6x^2y - 9y^2$

15 $\Delta w = (2x + 3y)\Delta x + (3x - 2y)\Delta y$
$\quad + (\Delta x)^2 + 3\Delta x\Delta y - (\Delta y)^2$;
$dw = (2x + 3y)\,dx + (3x - 2y)\,dy$;
$\Delta w = -1.13$, $dw = -1.1$

17 $\partial s/\partial x = 12x + 18y$; $\partial s/\partial y = 18x - 22y$

19 $3e^{-t}(9t^2\cos 3t^3 - \sin 3t^3)$ **21** $-14\sqrt{41}$

23 $-16(x + 2) + 4(y + 1) - 7(z - 2) = 0$;
$x = -2 - 16t$, $y = -1 + 4t$, $z = 2 - 7t$

25 $(3x^2 - 4y^3 + 1)/(12xy^2 + 3)$

27 Min: $f(0, -1) = -2$ **29**

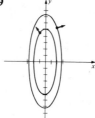

31 Min: $f(-\sqrt{8/3}, -\sqrt{2/3}, -\sqrt{4/3})$
$\quad = f(-\sqrt{8/3}, \sqrt{2/3}, \sqrt{4/3}) = f(\sqrt{8/3}, -\sqrt{2/3}, \sqrt{4/3})$
$\quad = f(\sqrt{8/3}, \sqrt{2/3}, -\sqrt{4/3}) = -8/3\sqrt{3}$;
$\quad$ max: $f(\sqrt{8/3}, \sqrt{2/3}, \sqrt{4/3})$
$\quad = f(-\sqrt{8/3}, -\sqrt{2/3}, \sqrt{4/3})$
$\quad = f(-\sqrt{8/3}, \sqrt{2/3}, -\sqrt{4/3})$
$\quad = f(\sqrt{8/3}, -\sqrt{2/3}, -\sqrt{4/3}) = 8/3\sqrt{3}$

33 If $a = 1 + 2^{2/3} + 3^{2/3}$, the point is $(a, 2^{1/3}a, 3^{1/3}a)$.

CHAPTER 17

Exercises 17.1, page 780

1 (a) 39 (b) 81 **3** 1769.13 **5** 240

Exercises 17.2, page 788

1 Both are -36 **3** $\frac{163}{120}$ **5** $\frac{36}{5}$ **7** $(4e - e^4)/2$

9 $(e^2 + 1)/4$

11 $\pi/24 + \ln[3/\sqrt{2}(\sqrt{2} + 1)] + 1/\sqrt{2} - \frac{1}{2} \approx 0.2087$

13 (a) $\int_0^4 \int_0^{\sqrt{x}} f(x, y)\,dy\,dx$

(b) $\int_0^2 \int_{y^2}^4 f(x, y)\,dx\,dy$

15 (a) $\int_0^2 \int_{x^3}^8 f(x, y)\,dy\,dx$

(b) $\int_0^8 \int_0^{y^{1/3}} f(x, y)\,dx\,dy$

17 (a) $\int_0^1 \int_{x^3}^{x^{1/2}} f(x, y)\,dy\,dx$

(b) $\int_0^1 \int_{y^2}^{y^{1/3}} f(x, y)\,dx\,dy$

19 $\int_{-1}^4 \int_{-1}^2 (y + 2x)\,dx\,dy = \int_{-1}^2 \int_{-1}^4 (y + 2x)\,dy\,dx = \frac{75}{2}$

21 $\int_0^1 \int_{-2y}^{3y} xy^2\,dx\,dy = \frac{1}{2}$

23 $\int_0^2 \int_0^{x^2} x^3 \cos xy\,dy\,dx = (1 - \cos 8)/3 \approx 0.38$

25 (a) $\int_1^2 \int_{9-4x}^{4+x} f(x, y)\,dy\,dx + \int_2^4 \int_{x^3/8}^{4+x} f(x, y)\,dy\,dx$

(b) $\int_1^5 \int_{(9-y)/4}^{2y^{1/3}} f(x, y)\,dx\,dy + \int_5^8 \int_{y-4}^{2y^{1/3}} f(x, y)\,dx\,dy$

27 (a) $\int_{-1}^1 \int_{-x-3}^{2x} f(x, y)\,dy\,dx + \int_1^3 \int_{-x-3}^{3-x^2} f(x, y)\,dy\,dx$

(b) $\int_{-6}^{-2} \int_{-y-3}^{\sqrt{3-y}} f(x, y)\,dx\,dy + \int_{-2}^2 \int_{y/2}^{\sqrt{3-y}} f(x, y)\,dx\,dy$

25 **27**

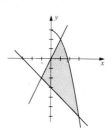

29 (a) $\int_0^1 \int_{1-x}^{e^x} f(x, y)\,dy\,dx + \int_1^e \int_{\ln x}^{1+e-x} f(x, y)\,dy\,dx$

(b) $\int_0^1 \int_{1-y}^{e^y} f(x, y)\,dx\,dy + \int_1^e \int_{\ln y}^{1+e-y} f(x, y)\,dx\,dy$

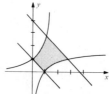

31

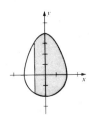

33

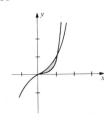

35

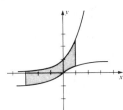

37

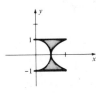

39 $\displaystyle\int_0^2 \int_0^{y/2} e^{y^2}\, dx\, dy = (e^4 - 1)/4$

41 $\displaystyle\int_0^4 \int_0^{\sqrt{x}} y \cos x^2\, dy\, dx = \tfrac14 \sin 16$

43 $\displaystyle\int_0^2 \int_0^{x^3} \frac{y}{\sqrt{16 + x^7}}\, dy\, dx = \frac{8}{7}$

Exercises 17.3, page 793

1 $\frac{17}{6}$　**3** $\frac{33}{2}$　**5** 2　**7** $e^\pi - e^{-\pi}$
9 $2(\tan^{-1} a^{1/2} - a^{3/2}/3)$ where $a = (-1 + \sqrt{5})/2$
11 $\frac{34}{3}$　**13** 18　**15** $\frac{16}{3}$　**17** $\frac{7488}{315} \approx 23.77$
19 $\frac{128}{9} \approx 14.2$　**21** $\frac{423}{20}$　**23** $16a^3/3$
25 The solid lies under the paraboloid $z = x^2 + y^2$ and over the region in the xy-plane bounded by the graphs of $y = x - 1$, $y = 1 - x^2$, $x = -2$, and $x = 1$.
27 The solid lies under the plane $z = x + y$ and over the region in the xy-plane bounded by the graphs of $x = y/4$, $x = \sqrt{y}$, $y = 0$, and $y = 4$.
29 The rectangular parallelepiped bounded by the planes $z = 0$, $z = 3$, $x = 0$, $x = 4$, $y = -1$, $y = 2$

Exercises 17.4, page 799

1 $M = \frac{2349}{20}$; $\bar{x} = \frac{1290}{203}$, $\bar{y} = \frac{38}{29}$
3 $M = 8k$ (k a proportionality constant); $\bar{x} = 0$, $\bar{y} = \frac{8}{3}$
5 $M = (1 - e^{-2})/4$;
　　$\bar{x} = 0$, $\bar{y} = 4(1 - e^{-3})/9(1 - e^{-2}) \approx 0.46$
7 $M = \frac{115}{6}$; $\bar{x} = \frac{20}{23}$, $\bar{y} = \frac12$
9 $M = 4\ln(\sqrt{2} + 1) - 4\ln(\sqrt{2} - 1) - \pi \approx 3.9$;
　　$\bar{x} = 0$, $\bar{y} = (16 - \pi)/4M \approx 0.8$
11 $M = \frac43$; $\bar{x} = \frac{21}{16}$, $\bar{y} = \frac{1}{32}(3 + 12\ln 2) \approx 0.35$

13 $M = (1 - e^{-3})/9 \approx 0.106$;
　　$\bar{x} = (1 - 4e^{-3})/3(1 - e^{-3}) \approx 0.28$,
　　$\bar{y} = 9(1 - e^{-4})/16(1 - e^{-3}) \approx 0.58$
15 $I_x = 3^5(\frac{31}{28})$, $I_y = 3^7(\frac{19}{8})$, $I_0 = 3^5(\frac{1259}{56})$
17 $I_x = 64k$, $I_y = \frac{32}{3}k$, $I_0 = \frac{224}{3}k$
19 (a) $\rho a^4/3$　(b) $\rho a^4/12$　(c) $\rho a^4/6$ (ρ = density)
21 $a/\sqrt{3}$　**23** $\rho a^4\pi/4$; $a/2$　**25** $\rho a^3 b/12$; $a/\sqrt{6}$
27 $\rho(b/96)(4a^2 - b^2)^{3/2}$; $\sqrt{4a^2 - b^2}/2\sqrt{6}$

Exercises 17.5, page 804

1 $\frac92$　**3** $\frac92\sqrt{3} - \pi$
5 $M = \frac{32}{9}k$; $\bar{x} = \frac65$, $\bar{y} = 0$
7 $M = 2k(\sqrt{3} - \pi/3)$;
　　$\bar{x} = 0$, $\bar{y} = (27\sqrt{3} - 4\pi)/(24\sqrt{3} - 8\pi) \approx 2.1$
9 $\rho a^2\pi/8$　**11** $\rho a^4\pi/4$; $a/2$　**13** $64\pi/5$
15 $(b^2 - a^2)\pi/2$　**17** π　**19** $(\pi/2)(1 - e^{-a^2})$
21 $\ln(\sqrt{2} + 1)$　**23** $(\pi/4)\sin 4$　**25** $256\pi/3$
27 $\frac{64}{9}$　**29** $\frac{128}{9}(3\pi - 4)$　**31** (a) π

Exercises 17.6, page 811

3 $-\frac{1}{12}$　**5** $\frac{513}{8}$

7 $\displaystyle\int_0^6 \int_0^{(6-x)/2} \int_0^{(6-x-2y)/3} f(x, y, z)\, dz\, dy\, dx$;

$\displaystyle\int_0^3 \int_0^{6-2y} \int_0^{(6-x-2y)/3} f(x, y, z)\, dz\, dx\, dy$;

$\displaystyle\int_0^6 \int_0^{(6-x)/3} \int_0^{(6-x-3z)/2} f(x, y, z)\, dy\, dz\, dx$;

$\displaystyle\int_0^2 \int_0^{6-3z} \int_0^{(6-x-3z)/2} f(x, y, z)\, dy\, dx\, dz$;

$\displaystyle\int_0^2 \int_0^{(6-3z)/2} \int_0^{6-2y-3z} f(x, y, z)\, dx\, dy\, dz$;

$\displaystyle\int_0^3 \int_0^{(6-2y)/3} \int_0^{6-2y-3z} f(x, y, z)\, dx\, dz\, dy$

9 $\displaystyle\int_{-3/2}^{3/2} \int_{-\sqrt{9-4x^2}}^{\sqrt{9-4x^2}} \int_0^{9-4x^2-y^2} f(x, y, z)\, dz\, dy\, dx$;

$\displaystyle\int_{-3}^3 \int_{-\sqrt{9-y^2}/2}^{\sqrt{9-y^2}/2} \int_0^{9-4x^2-y^2} f(x, y, z)\, dz\, dx\, dy$;

$\displaystyle\int_{-3/2}^{3/2} \int_0^{9-4x^2} \int_{-\sqrt{9-z-4x^2}}^{\sqrt{9-z-4x^2}} f(x, y, z)\, dy\, dz\, dx$;

$\displaystyle\int_0^9 \int_{-\sqrt{9-z}/2}^{\sqrt{9-z}/2} \int_{-\sqrt{9-z-4x^2}}^{\sqrt{9-z-4x^2}} f(x, y, z)\, dy\, dx\, dz$;

$\displaystyle\int_{-3}^3 \int_0^{9-y^2} \int_{-\sqrt{9-z-y^2}/2}^{\sqrt{9-z-y^2}/2} f(x, y, z)\, dx, dz, dy$;

$\displaystyle\int_0^9 \int_{-\sqrt{9-z}}^{\sqrt{9-z}} \int_{-\sqrt{9-z-y^2}/2}^{\sqrt{9-z-y^2}/2} f(x, y, z)\, dx\, dy\, dz$

11 $\frac{128}{5}$　**13** $\frac{32}{3}$　**15** 2π　**17** 108　**19** $\frac{1}{70}$

21 The region to the right of the rectangle with vertices $(2, 0, 0)$, $(2, 0, 1)$, $(3, 0, 0)$, $(3, 0, 1)$ and between the graphs of $y = \sqrt{1 - z}$ and $y = \sqrt{4 - z}$.

23 The region under the plane $z = x + y$ and over the region in the xy-plane bounded by the graphs of $y = x^2$, $y = 2x$, $x = 0$, and $x = 2$.

25 The region bounded by the paraboloid $z = x^2 + y^2$ and the planes $z = 1$ and $z = 2$.

Exercises 17.7, page 815

1 Let $I = \int_0^3 \int_0^{6-2x} \int_0^{9-x^2} f(x, y, z) \, dz \, dy \, dx$. To find M, M_{xy}, M_{xz}, M_{yz}, let $f(x, y, z)$ equal 1, z, y, and x, respectively. $\bar{x} = \frac{21}{25}$, $\bar{y} = \frac{54}{25}$, $\bar{z} = \frac{99}{25}$

3 Let $I = \int_{-1}^1 \int_{x^2}^1 \int_0^{4x^2+9z^2} f(x, y, z) \, dy \, dz \, dx$ and use the method given in the answer to Exercise 1.

5 $\bar{x} = \bar{y} = \bar{z} = 7a/12$

7 $M = \int_0^4 \int_{-\sqrt{x}/2}^{\sqrt{x}/2} \int_{-\sqrt{x-4z^2}}^{\sqrt{x-4z^2}} (x^2 + z^2) \, dy \, dz \, dx$,

$\bar{x} = \int_0^4 \int_{-\sqrt{x}/2}^{\sqrt{x}/2} \int_{-\sqrt{x-4z^2}}^{\sqrt{x-4z^2}} x(x^2 + z^2) \, dy \, dz \, dx/M$.

The $\bar{y}$ and $\bar{z}$ integrals have the same limits but the integrands are $y(x^2 + z^2)$ and $z(x^2 + z^2)$, respectively.

9 $2\rho a^5/3$; $(\sqrt{2/3})a$

11 $I_z = \int_{-3}^3 \int_{-\sqrt{36-4x^2}/3}^{\sqrt{36-4x^2}/3} \int_0^{36-4x^2-9y^2} (x^2 - y^2)z \, dz \, dy \, dx$

13 $I_z = \int_{-a}^a \int_{-\sqrt{a^2-x^2}}^{\sqrt{a^2-x^2}} \int_{-\sqrt{a^2-x^2-y^2}}^{\sqrt{a^2-x^2-y^2}} (x^2 + y^2)(x^2 + y^2 + z^2) \, dz \, dy \, dx$

15 $I_z = \int_0^a \int_0^{b(1-x/a)} \int_0^{c(1-x/a-y/b)} (x^2 + y^2)\rho \, dz \, dy \, dx$

Exercises 17.8, page 819

1 8π; $\bar{x} = 0$, $\bar{y} = 0$, $\bar{z} = \frac{4}{3}$

3 (a) $\frac{1}{2}h\pi a^4 k$ (b) $h\pi a^2 k(\frac{1}{4}a^2 + \frac{1}{3}h^2)$ (k = density)

5 $k\pi^2 a^4/4$ **7** $k\pi^2 a^6/8$; $a/\sqrt{2}$

9 $M = 9\pi k$; $\bar{x} = \bar{y} = 0$, $\bar{z} = \frac{9}{4}$

11 $ka^4\pi/2$ (where k is the proportionality constant) center of mass is $2a/5$ from base along the axis of symmetry

13 $2\pi a^6 k/9$ **15** 8π **17** $124\pi k/5$

19 $7\pi/16 \approx 1.374$ **21** $2^8\pi(\sqrt{2} - 1)/5 \approx 66.63$

Exercises 17.9, page 823

1 $[\sqrt{3} + 2\ln(1 + \sqrt{3}) - \ln 2]/2$

3 $\pi cd^2\sqrt{1/a^2 + 1/b^2 + 1/c^2}$ **5** $2a^2(\pi - 2)$

7 $\pi(5^{3/2} - 1)/6$ **9** $2\pi(5^{3/2} - 1)/3$

Exercises 17.10, page 823

1 $-\frac{5}{84}$ **3** $\frac{63}{4}$ **5** $-\frac{107}{210}$

7 $\int_2^4 \int_{-\sqrt{x^2-4}}^{\sqrt{x^2-4}} f(x, y) \, dy \, dx$

9 $\int_{-2}^2 \int_{y^2-4}^{-y^2+4} f(x, y) \, dx \, dy$

11 R is bounded by the graphs of $x = e^y$, $x = y^3$, $y = -1$, and $y = 1$.

13 $\int_0^9 \int_0^{\sqrt{x}} ye^{-x^2} \, dy \, dx = (1 - e^{-81})/4$

15 $M = 9k$ (k the constant of proportionality); $\bar{x} = \frac{9}{4}$, $\bar{y} = \frac{27}{8}$

17 $2\pi k$ **19** $\pi a^3/6$

21 $kab^4/20$ (k a constant); $\sqrt{3/10}b$ **23** 13

25 $V = \frac{256}{15}$; $\bar{x} = 0$, $\bar{y} = \frac{8}{7}$, $\bar{z} = \frac{12}{7}$

27 $\int_0^4 \int_{-\sqrt{1+y^2}}^{\sqrt{1+y^2}} \int_{-\sqrt{1+y^2-x^2}}^{\sqrt{1+y^2-x^2}} k(x^2 + z^2)^{3/2} \, dz \, dx \, dy$

29 $M = k\pi a^4$

CHAPTER 18

Exercises 18.1, page 831

1

3

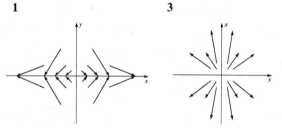

5

7

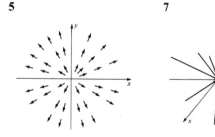

9

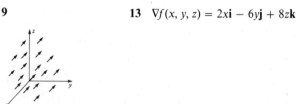

13 $\nabla f(x, y, z) = 2x\mathbf{i} - 6y\mathbf{j} + 8z\mathbf{k}$

15 $\mathbf{F}(x, y) = (1 + x^2y^2)^{-1}(y\mathbf{i} + x\mathbf{j})$
19 $\mathbf{i} + x^2\mathbf{j} + y^2\mathbf{k}$, $2xz + 2yx + 2$
21 $-y^2 \cos z\mathbf{i} + (6xyz - e^{2z})\mathbf{j} - 3xz^2\mathbf{k}$,
    $3yz^2 + 2y \sin z + 2xe^{2z}$

Exercises 18.2, page 841

1 $14(2\sqrt{2} - 1)$; 21; 14 **3** $\frac{34}{7}$ **5** $-\frac{16}{3}$
7 (a) $\frac{15}{2}$ (b) 6 (c) 7 (d) $\frac{29}{4}$
9 $(3e^4 + 6e^{-2} - 12e + 8e^3 - 5)/12$
11 (a) 19 (b) 35 (c) 27 **13** $3\sqrt{14}/2$
15 $\bar{x} = 0$, $\bar{y} = \pi a/4$ **19** $\frac{9}{2}$ (for all paths)
21 0 **23** $\frac{412}{15}$ **25** $I_x = 4a^3k/3$, $I_y = 2a^3k/3$
27 If the density at (x, y, z) is $f(x, y, z)$, then
    $I_x = \int_C (y^2 + z^2) f(x, y, z) \, ds$,
    $I_y = \int_C (x^2 + z^2) f(x, y, z) \, ds$,
    $I_z = \int_C (x^2 + y^2) f(x, y, z) \, ds$.

Exercises 18.3, page 849

1 $f(x, y) = x^3y + 2x + y^4 + c$
3 Not independent of path
5 Not independent of path
7 $f(x, y, z) = y \tan x - ze^x + c$
9 $f(x, y, z) = 4x^2z + y - 3y^2z^3 + c$ **11** 14
13 -31
15 If $\mathbf{F}(x, y, z) = (-c/|\mathbf{r}|^2)\mathbf{r}$, where $c > 0$, then
    $f(x, y, z) = -c \ln r + d = -(c/2) \ln (x^2 + y^2 + z^2) + d$
25 $W = c(d_2 - d_1)/d_1 d_2$

Exercises 18.4, page 857

1 $-\frac{7}{60}$ **3** $\frac{2}{3}$ **5** π **7** 3 **9** 0 **11** 0
13 -3π **15** $3\pi a^2/8$ **17** $\frac{128}{3}$
23 $\bar{x} = 0$, $\bar{y} = 4a/3\pi$

Exercises 18.5, page 864

1 $2\pi a^4/3$ **3** 5
5 (a) $\int_0^4 \int_0^{3-3y/4} (6 - 3y/2 - 2z)y^2z^3(\sqrt{29}/2) \, dz \, dy$

    (b) $\int_0^3 \int_0^{6-2z} x(4 - 2x/3 - 4z/3)^2(\sqrt{29}/3)z^3 \, dx \, dz$

7 (a) $\int_0^8 \int_0^6 (4 - 3y + y^2/16 + z)\sqrt{17}/4 \, dz \, dy$

    (b) $\int_0^2 \int_0^6 (x^2 + 8x - 16 + z)\sqrt{17} \, dz \, dx$

9 The value of the integral equals the volume of a cylinder
    of altitude c, rulings parallel to the z-axis, and whose
    base is the projection of S on the xy-plane.
11 $2\pi a^3$ **13** 3π **15** 18 **17** 8
21 $M = \frac{255}{2}\sqrt{2}\,\pi$, $\bar{x} = 0$, $\bar{y} = 0$, $\bar{z} = \frac{1364}{425} \approx 3.21$;
    $I_z = 1365\sqrt{2}\pi$

Exercises 18.6, page 870

1 24 **3** 20π **5** 0 **7** $136\pi/3$ **9** 24
11 Both integrals equal $4\pi a^3$.
13 Both integrals equal 4π.

Exercises 18.7, page 877

1 Both integrals equal -1.
3 Both integrals equal πa^2.   **5** 0 **7** -8π

Exercises 18.8, page 881

1 (a) Vertical lines; horizontal lines
    (b) $x = u/3$, $y = v/5$
3 (a) Straight lines $x - y = c$ and $3y + 2x = d$, where c
        and d are arbitrary constants
    (b) $x = \frac{3}{5}u + \frac{1}{5}v$, $y = -\frac{2}{5}u + \frac{1}{5}v$
5 (a) Straight lines $2x + y = c$ and hyperbolas $xy = d$
    (b) Not one-to-one
7 (a) Ellipses $x^2 + 4y^2 = c$ and hyperbolas $4x^2 - y^2 = d$
    (b) Not one-to-one
9 (a) Hyperbolas $x^2 - y^2 = c$ and $xy = d$
    (b) Not one-to-one
11 Rectangle with vertices $(0, 0)$, $(6, 0)$, $(6, 5)$, $(0, 5)$; the
     ellipse $u^2/9 + v^2/25 = 1$
15 Planes having equations of the form
     $2x + y = c$, $y + z = d$, $x + y + z = k$;
     $x = w - v$, $y = u + 2v - 2w$, $z = -u - v + 2w$

Exercises 18.9, page 886

1 $4u^2 + 4v^2$ **3** $2u(vu - 1)e^{-(2u+v)}$ **5** -6
7 $\int_0^2 \int_0^{2-u} (v - u)2 \, dv \, du$

9 $\int_{-1}^1 \int_{-\sqrt{1-v^2}}^{\sqrt{1-v^2}} (u^2 + v^2)6 \, du \, dv$

11 $\int_{-1}^1 \int_1^3 \frac{1}{2}u^2 \cos^2 v \, dv \, du = \frac{1}{3} + \frac{1}{12}\sin 6 - \frac{1}{12}\sin 2$

13 $\int_1^2 \int_{\sqrt{u}}^{\sqrt{2u}} (u^2v^{-3} + 2v) \, dv \, du = \frac{15}{8}$

15 $\int_2^4 \int_{-u/2}^{2u} \frac{v}{u}\frac{1}{5} \, dv \, du = \frac{9}{4}$

Exercises 18.10, page 887

1

3

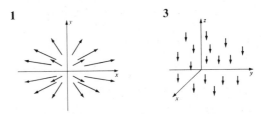

5 $\mathbf{F}(x, y) = y^2 \sec^2 x\mathbf{i} + 2y \tan x\mathbf{j}$ **7** $-\frac{56}{5}$ **9** $-\frac{40}{3}$

11 $(1025\sqrt{1025} - 17\sqrt{17})/144$ **13**   70   **15**   0
17 $\frac{25}{2}$ **19** $f(x, y, z) = x^2 e^{2y} + y^2 \cot z + c$ **21** $\frac{1}{6}$
23 $3x^2 z^4 + xz^2; (2x^2 y - 2xyz)\mathbf{i} + (4x^3 z^3 - 2xy^2)\mathbf{j} + yz^2\mathbf{k}$
25 $\sqrt{3}/5$ **27** $5\pi/2$ **29**   Both integrals equal 8π
31   (a) The lines $2x + 5y = c$ (b) The lines $3x - 4y = d$
     (c) $x = \frac{4}{23}u + \frac{5}{23}v, y = \frac{3}{23}u - \frac{2}{23}v$ (d) $-\frac{1}{23}$
     (e) $(4a + 3b)u + (5a - 2b)v + 23c = 0$
     (f) $25u^2 + 28uv + 29v^2 = 529a^2$

CHAPTER 19

Exercises 19.1, page 893

1 $\cos x + x \sin x - \ln|\sin y| = C$
3 $\tan^{-1}(\sin x) + \ln|\csc y - \cot y| = C$
5 $\ln|(x - 1)/x| + \arcsin y = C$
7 $y = (x + C)\csc x$ **9** $y = \frac{4}{3}x^3 \csc x + C \csc x$
11 $y = 2\sin x + C \cos x$ **13** $y = x \sin x + Cx$
15 $y = (-\frac{1}{2}x^3 + C/x^3)^{-1/3}$
17 $y = [-\frac{1}{2}(1 + x^2)\tan^{-1} x + \frac{1}{2}x + C(1 + x^2)]^{-1}$
19 $y = \frac{3}{2} + Ce^{-x^2}$ **21** $y = \frac{1}{2}\sin x + C/\sin x$
23 $4\tan x = 2\sin 2y + 4 - \sqrt{3}$
25 $\tan^{-1} y - \ln|\sec x| = \pi/4$ **27** $y = (x^2 + 3)\cos x$
29 $2yy' = 3x^2$ **31** $xy' = y \ln y$
33 $y'' - 2y' + y = 0$ **35** $xy' - y = 0$
37 $yy'' + (y')^2 + 1 = 0$ **39** $xy = k$
41 $2x^2 + y^2 = k$ **43** $2x^2 + 3y^2 = k$
45 $x^2 + y^2 = ky$
47 $s(t) = (1/c)\ln(e^{2\sqrt{gct}} + 1) - (\sqrt{g/c})t - (\ln 2)/c$,
49 $y = ab[e^{k(b-a)t} - 1]/[be^{k(b-a)t} - a]$

Exercises 19.2, page 898

1 $x^2 + xy + y^2 = C$ **3** $x - x^2 y + x^3 y^2 - y^3 = C$
5 $x^2 \sin y + 5x + 2y^2 = C$ **7** $xe^{xy} + y^2 = C$
9 $r \sin^2 \theta + r^2 \cos \theta = C$ **11** $e^{-x} \cos y + x^2 = C$
13 $\frac{1}{2}x^2 - y \ln x + x \ln y = C$
15 $x^2 y^3 + 4x^2 + 5y = 7$
17 $4y^2 e^{2x} - 8x + \ln(1 + 4y^2) = 1 + \ln 2$

Exercises 19.3, page 901

1 $x^4(1 + 4yx^{-1}) = C$ **3** $x \cos(y/x) = C$
5 $\ln x = (2/\sqrt{3})\tan^{-1}((2y - x)/\sqrt{3}x) + C$
7 $\ln((y/x) - 1) - 3\ln((3y/x) - 1) = \ln x^2 + C$
9 $\ln x = \sin^{-1}(y/x) + C$ **11** $y(1 + x^2 y^{-2}) = C$
13 $\ln y^2 + \frac{3}{2}\ln((2x/y) + 1) = C$
15 $\frac{1}{2}x^2 + 3xy - y^2 = C$

Exercises 19.4, page 908

1 $y = C_1 e^{2x} + C_2 e^{3x}$ **3** $y = C_1 + C_2 e^{3x}$
5 $y = C_1 e^{-2x} + C_2 xe^{-2x}$

7 $y = C_1 e^{(2+\sqrt{3})x} + C_2 e^{(2-\sqrt{3})x}$
9 $y = C_1 e^{-\sqrt{2}x} + C_2 xe^{-\sqrt{2}x}$
11 $y = C_1 e^{5x/4} + C_2 e^{-3x/2}$ **13** $y = C_1 e^{4x/3} + C_2 xe^{4x/3}$
15 $y = C_1 e^{(2+\sqrt{2})x/2} + C_2 e^{(2-\sqrt{2})x/2}$
17 $y = C_1 e^x \cos x + C_2 e^x \sin x$
19 $y = C_1 e^{2x} \sin 3x + C_2 e^{2x} \cos 3x$
21 $y = C_1 e^{(-3+\sqrt{7})x} + C_2 e^{(-3-\sqrt{7})x}$ **23** $y = 2e^{2x} - 2e^x$
25 $y = \cos x + 2\sin x$ **27** $y = (2 + 9x)e^{-4x}$
29 $y = \frac{1}{2}e^x \sin 2x$

Exercises 19.5, page 914

1 $y = C_1 \sin x + C_2 \cos x - \cos x \ln|\sec x + \tan x|$
3 $y = (C_1 + C_2 x + \frac{1}{12}x^4)e^{3x}$
5 $y = C_1 e^x + C_2 e^{-x} + \frac{2}{5}e^x \sin x - \frac{1}{5}e^x \cos x$
7 $y = (C_1 + \frac{1}{6}x)e^{3x} + C_2 e^{-3x}$
9 $y = C_1 e^{4x} + C_2 e^{-x} - \frac{1}{2}$
11 $y = C_1 e^x + C_2 e^{2x} + \frac{2}{3}e^{-x}$
13 $y = C_1 + C_2 e^{-2x} + \frac{1}{8}\sin 2x - \frac{1}{8}\cos 2x$
15 $y = C_1 e^x + C_2 e^{-x} + \frac{1}{9}(3x - 4)e^{2x}$
17 $y = e^{3x}(C_1 \cos 2x + C_2 \sin 2x) + \frac{1}{65}(7\cos x - 4\sin x)e^x$

Exercises 19.6, page 918

1 $y = -\frac{1}{3}\cos 8t$ **3** $y = (\sqrt{2}/8)e^{-8t}(e^{4\sqrt{2}t} - e^{-4\sqrt{2}t})$
5 $y = \frac{1}{3}e^{-8t}(\sin 8t + \cos 8t)$
7   If m is the mass of the weight, then the spring constant is $24m$ and the damping force is $-4m(dy/dt)$. The motion is begun by releasing the weight from 2 ft above the equilibrium position with an initial velocity of 1 ft/sec in the upward direction.
9 $-6\sqrt{2}(dy/dt)$

Exercises 19.7, page 921

1 $y = a_0 \sum_{n=0}^{\infty} \frac{(-1)^n}{(2n)!}x^{2n} + a_1 \sum_{n=0}^{\infty} \frac{(-1)^n}{(2n+1)!}x^{2n+1}$
     $(= a_0 \cos x + a_1 \sin x)$

3 $y = a_0\left[1 + \sum_{n=1}^{\infty} \frac{2^n(3n-2)(3n-5)\cdots 7 \cdot 4 \cdot 1}{(3n)!}x^{3n}\right]$
     $+ a_1\left[x + \sum_{n=1}^{\infty} \frac{2^n(3n-1)(3n-4)\cdots 8 \cdot 5 \cdot 2}{(3n+1)!}x^{3n+1}\right]$

5 $y = a_0(1 - x^2)$
     $+ a_1\left[x - \sum_{n=1}^{\infty} \frac{(2n-3)(2n-5)\cdots 5 \cdot 3 \cdot 1}{(2n+1)!}x^{2n+1}\right]$

7 $y = a_0(x + 1)^3$

9 $y = -5x + a_0 \sum_{n=0}^{\infty} \frac{x^n}{n!} + a_1 \sum_{n=0}^{\infty} \frac{(-x)^n}{n!}$
     $= -5x + a_0 e^x + a_1 e^{-x}$

11 $y = a_0 \sum_{n=0}^{\infty} (n+1)x^{2n} + a_1 \sum_{n=0}^{\infty} \left(\frac{2n+3}{3}\right)x^{2n+1}$
     $+ \sum_{n=1}^{\infty} (n+1)x^{2n}$

Exercises 19.8, page 922

1 $\sin x - x \cos x + e^{-y} = C$
3 $(1/\sqrt{3}) \tan^{-1} (y/\sqrt{3}x) + \frac{1}{2} \ln (3x^2 + y^2) = C$
5 $xy^2 + ye^{-x} + 3y = C$ **7** $y = 2 \sin x + C \cos x$
9 $\sqrt{1 - y^2} + \sin^{-1} x = C$ **11** $x^2 \sin (y/x) = C$
13 $y = (Cx^{-4} - \frac{2}{7}x^3)^{-1/2}$ **15** $y = (C_1 + C_2 x)e^{4x}$
17 $y = C_1 + C_2 e^{2x}$
19 $y = C_1 e^x + C_2 e^{-x} - \frac{1}{5}e^x(\sin x + 2 \cos x)$
21 $\sin^{-1} x - \sqrt{1 - x^2} - \frac{1}{2}y - \frac{1}{4} \sin 2y = C$
23 $y = Ce^{-x} + \frac{1}{5}e^{4x}$
25 $y = C_1 e^x + C_2 e^{2x} + \frac{1}{12}e^{5x}$
27 $y = (x - 2)^3/3x + C/x$
29 $y = C_1 e^{5x} + C_2 e^{-4x} + (\frac{1}{108} - \frac{1}{18}x)e^{-x}$
31 $y = e^{-(5/2)x}(C_1 \cos \sqrt{3}x/2 + C_2 \sin \sqrt{3}x/2)$
33 $e^x(\sin x - \cos x) + 2e^{-y} = C$
35 $y = \cos \frac{1}{2}x + C \sec x$ **37** $xy + e^{-2x} \sin y = C$
39 $y = (\ln |\sec x + \tan x| - x + C)/(\csc x - \cot x)$

APPENDICES

Exercises A.III, page A24

3 (a) II   (b) III   (c) IV

The order in Exercises 5 and 9 is sin, cos, tan, csc, sec, cot.
  5 (a) $1, 0, -, 1, -, 0$
     (b) $\sqrt{2}/2, -\sqrt{2}/2, -1, \sqrt{2}, -\sqrt{2}, -1$
     (c) $0, 1, 0, -, 1, -$
     (d) $-\frac{1}{2}, \sqrt{3}/2, -\sqrt{3}/3, -2, 2\sqrt{3}/3, -\sqrt{3}$
  7 $810°, -120°, 315°, 900°, 36°$
  9 (a) $-\frac{4}{5}, \frac{3}{5}, -\frac{4}{3}, -\frac{5}{4}, \frac{5}{3}, -\frac{3}{4}$
     (b) $2\sqrt{13}/13, -3\sqrt{13}/13, -\frac{2}{3}, \sqrt{13}/2, -\sqrt{13}/3, -\frac{3}{2}$
     (c) $-1, 0, -, -1, -, 0$
31 $\pi/2, 3\pi/2, \pi/4, 3\pi/4, 5\pi/5, 7\pi/4;$
    $90°, 270°, 45°, 135°, 225°, 315°$
33 $0, \pi; 0°, 180°$
35 $0, \pi, 2\pi/3, 4\pi/3; 0°, 180°, 120°, 240°$
37 $\pi/2, 7\pi/6, 11\pi/6; 90°, 210°, 330°$
39 $2\pi/3, \pi, 4\pi/3; 120°, 180°, 240°$
41 $\pi/3, 5\pi/3; 60°, 300°$ **43** $\frac{84}{85}$ **45** $-\frac{36}{77}$
47 $\frac{240}{289}$ **49** $\frac{24}{7}$ **51** $\frac{1}{3}$ **53** No; $|\sin t| \le 1$
55 (a) $\sqrt{2}/2$ (b) $-\frac{1}{4}$

INDEX

Trigonometric Forms

63 $\displaystyle\int \sin^2 u \, du = \tfrac{1}{2}u - \tfrac{1}{4}\sin 2u + C$

64 $\displaystyle\int \cos^2 u \, du = \tfrac{1}{2}u + \tfrac{1}{4}\sin 2u + C$

65 $\displaystyle\int \tan^2 u \, du = \tan u - u + C$

66 $\displaystyle\int \cot^2 u \, du = -\cot u - u + C$

67 $\displaystyle\int \sin^3 u \, du = -\tfrac{1}{3}(2 + \sin^2 u)\cos u + C$

68 $\displaystyle\int \cos^3 u \, du = \tfrac{1}{3}(2 + \cos^2 u)\sin u + C$

69 $\displaystyle\int \tan^3 u \, du = \tfrac{1}{2}\tan^2 u + \ln|\cos u| + C$

70 $\displaystyle\int \cot^3 u \, du = -\tfrac{1}{2}\cot^2 u - \ln|\sin u| + C$

71 $\displaystyle\int \sec^3 u \, du = \tfrac{1}{2}\sec u \tan u + \tfrac{1}{2}\ln|\sec u + \tan u| + C$

72 $\displaystyle\int \csc^3 u \, du = -\tfrac{1}{2}\csc u \cot u + \tfrac{1}{2}\ln|\csc u - \cot u| + C$

73 $\displaystyle\int \sin^n u \, du = -\frac{1}{n}\sin^{n-1} u \cos u + \frac{n-1}{n}\int \sin^{n-2} u \, du$

74 $\displaystyle\int \cos^n u \, du = \frac{1}{n}\cos^{n-1} u \sin u + \frac{n-1}{n}\int \cos^{n-2} u \, du$

75 $\displaystyle\int \tan^n u \, du = \frac{1}{n-1}\tan^{n-1} u - \int \tan^{n-2} u \, du$

76 $\displaystyle\int \cot^n u \, du = \frac{-1}{n-1}\cot^{n-1} u - \int \cot^{n-2} u \, du$

77 $\displaystyle\int \sec^n u \, du = \frac{1}{n-1}\tan u \sec^{n-2} u + \frac{n-2}{n-1}\int \sec^{n-2} u \, du$

78 $\displaystyle\int \csc^n u \, du = \frac{-1}{n-1}\cot u \csc^{n-2} u + \frac{n-2}{n-1}\int \csc^{n-2} u \, du$

79 $\displaystyle\int \sin au \sin bu \, du = \frac{\sin(a-b)u}{2(a-b)} - \frac{\sin(a+b)u}{2(a+b)} + C$

80 $\displaystyle\int \cos au \cos bu \, du = \frac{\sin(a-b)u}{2(a-b)} + \frac{\sin(a+b)u}{2(a+b)} + C$

81 $\displaystyle\int \sin au \cos bu \, du = -\frac{\cos(a-b)u}{2(a-b)} - \frac{\cos(a+b)u}{2(a+b)} + C$

82 $\displaystyle\int u \sin u \, du = \sin u - u \cos u + C$

83 $\displaystyle\int u \cos u \, du = \cos u + u \sin u + C$

84 $\displaystyle\int u^n \sin u \, du = -u^n \cos u + n\int u^{n-1}\cos u \, du$

85 $\displaystyle\int u^n \cos u \, du = u^n \sin u - n\int u^{n-1}\sin u \, du$

86 $\displaystyle\int \sin^n u \cos^m u \, du = -\frac{\sin^{n-1} u \cos^{m+1} u}{n+m} + \frac{n-1}{n+m}\int \sin^{n-2} u \cos^m u \, du$

$\displaystyle\qquad = \frac{\sin^{n+1} u \cos^{m-1} u}{n+m} + \frac{m-1}{n+m}\int \sin^n u \cos^{m-2} u \, du$

Inverse Trigonometric Forms

87 $\displaystyle\int \sin^{-1} u \, du = u \sin^{-1} u + \sqrt{1 - u^2} + C$

88 $\displaystyle\int \cos^{-1} u \, du = u \cos^{-1} u - \sqrt{1 - u^2} + C$

89 $\displaystyle\int \tan^{-1} u \, du = u \tan^{-1} u - \tfrac{1}{2}\ln(1 + u^2) + C$

90 $\displaystyle\int u \sin^{-1} u \, du = \frac{2u^2 - 1}{4}\sin^{-1} u + \frac{u\sqrt{1 - u^2}}{4} + C$

91 $\displaystyle\int u \cos^{-1} u \, du = \frac{2u^2 - 1}{4}\cos^{-1} u - \frac{u\sqrt{1 - u^2}}{4} + C$

92 $\displaystyle\int u \tan^{-1} u \, du = \frac{u^2 + 1}{2}\tan^{-1} u - \frac{u}{2} + C$

93 $\displaystyle\int u^n \sin^{-1} u \, du = \frac{1}{n+1}\left[u^{n+1}\sin^{-1} u - \int \frac{u^{n+1}\, du}{\sqrt{1 - u^2}} \right], \quad n \neq -1$

94 $\displaystyle\int u^n \cos^{-1} u \, du = \frac{1}{n+1}\left[u^{n+1}\cos^{-1} u + \int \frac{u^{n+1}\, du}{\sqrt{1 - u^2}} \right], \quad n \neq -1$

95 $\displaystyle\int u^n \tan^{-1} u \, du = \frac{1}{n+1}\left[u^{n+1}\tan^{-1} u - \int \frac{u^{n+1}\, du}{1 + u^2} \right], \quad n \neq -1$